# Mathematical Principles of the Internet

Volume 2
Mathematical Concepts

# Mathematical Principles
of the Internet

## Volume 2: Mathematical Concepts

Nirdosh Bhatnagar

**CRC Press**
Taylor & Francis Group
Boca Raton London New York

CRC Press is an imprint of the
Taylor & Francis Group, an **informa** business

CRC Press
Taylor & Francis Group
6000 Broken Sound Parkway NW, Suite 300
Boca Raton, FL 33487-2742

© 2019 by Taylor & Francis Group, LLC
CRC Press is an imprint of Taylor & Francis Group, an Informa business

No claim to original U.S. Government works

Printed on acid-free paper

International Standard Book Number Volume I: **978-1-1385-05483** (Hardback)
International Standard Book Number Volume II: **978-1-1385-05513** (Hardback)

**Visit the Taylor & Francis Web site at**
http://www.taylorandfrancis.com

**and the CRC Press Web site at**
http://www.crcpress.com

*...to the memory of my parents*:

*Smt. Shakuntla Bhatnagar & Shri Rai Chandulal Bhatnagar*

# Contents

# Preface

The author wrote this two-volume set of books to teach himself about the foundation and building blocks of the Internet. To both his delight and dismay, the author found that this is indeed a continual process. It is delightful, because there is so much to learn, and dismal because there is so little time. Nevertheless, this work is a partial report of the author's efforts to learn the *elements* (in a Euclidean sense) of Internet engineering.

As with any grand scientific endeavor, the Internet is a product of several persons. The true beauty of it is that this experiment, called the Internet is still evolving. It is the thesis of this work that mathematics provides a strong basis upon which the Internet can be studied, analyzed, and improved upon.

### Rationale for the work

The purpose of this work is to provide a mathematical basis for some aspects of Internet engineering. The Internet is a network of autonomous networks. It is also a young, unprecedented, unparalleled, and *continuously evolving* man-made commercial and social phenomenon. Thus, the Internet is in a state of *constant flux*. This italicized phrase appears to be an oxymoron, but it is indeed a recurring and prominent theme in the description of the Internet. However, the Internet has matured enough so that its structure deserves a careful study. In order to study the form and function of the Internet only an elementary knowledge of mathematics is required. Even though the form and functionality of any physical phenomenon are basically intertwined, the emphasis in this work is upon the form and its structure. In order to strive towards this goal, the universal language: *mathematics* is used. Moreover, as mathematician Richard Hamming opined, mathematics acts as a conscience of large projects (such as the Internet).

The goal of modeling any natural or man-made phenomenon is to advance our understanding of it. More specifically, the aim is to search for *invariants* that describe a phenomenon by using the *principle of parsimony* (or simplicity). The principle of parsimony states that the description of a phenomenon should be as simple as possible, be evocative, and use the least number of assumptions. There is certainly no other tool, language, or discipline besides mathematics, which is capable of meeting these requirements. The goal of this work is to offer the reader a glimpse of hidden laws buried deep inside the complex network called the Internet. As usual, this phenomenon of the Internet did not occur overnight, and so its success is indeed a testament to the human spirit.

Information theory, algebraic coding theory, cryptography, Internet traffic, dynamics and control of Internet congestion, and queueing theory are discussed in this work. In addition, stochastic networks, graph-theoretic algorithms, application of game theory to the Internet, Internet economics, data mining and knowledge discovery; and quantum computation, communication, and cryptography are also discussed. Some of the topics like information theory, algebraic coding theory, cryptography, queueing theory, and graph-theoretic algorithms are mature and relevant in both understanding and designing the tapestry of the Internet. On other hand, some subjects like, Internet traffic, dynamics and control of Internet congestion, stochastic structure of the Internet and the World Wide Web, application of game theory to the Internet, Internet economics, and data mining techniques are still evolving. However, the application of quantum computation, communication, and cryptography is still in its infancy.

Note that this work by no means attempts, or is ambitious enough, to provide a complete description of the mathematical foundations upon which the Internet is based. It simply offers a partial panorama and perspective of the principles upon which the Internet was envisioned. There are several outstanding treatises, books, and monographs on different topics covered in this work. However these are all targeted towards experts in the field. Furthermore, for those who are interested in simply the basic principles, these are much too detailed. In summary, no attempt was made in this work to be encyclopedic.

**Contents of the books**

As of today, the Internet is still in its infancy. Therefore, if it has to be regarded as a scientific discipline, then its edifice has to be built upon sound mathematical principles. This is essentially the *premise* and *raison d' être* of this work. This assumption is not new. In his exultation, Eugene P. Wigner a physicist by profession (physics Nobel Prize 1963) wrote an article: "The Unreasonable Effectiveness of Mathematics in the Natural Sciences," in the year 1959. He was sufficiently impressed and awed by the utility of mathematics in modeling physical phenomena.

The selection of topics in describing certain fundamental aspects of Internet engineering is entirely based upon the author's wisdom or lack thereof. Others could have selected entirely different themes to study. However, the author chose to study the subjects outlined below in more detail, simply because of their utility value, importance, and elegance; perhaps not in that order.

Different analytical tools, which are used in describing the Internet are also listed, discussed, and elaborated upon. These tools enable a more precise and deeper understanding of the concepts upon which the Internet is built. An effort has been made to make this work succinct yet *self-contained,* complete and readable. This work assumes that the reader is familiar with basic concepts in computer networking. Although a certain level of mathematical maturity and aptitude is certainly beneficial, all the relevant details have been summarized in this work.

In order to study the structure and function of the Internet only a basic knowledge of number theory, abstract algebra, matrices and determinants, graph theory, geometry, applied analysis, optimization, stability theory, and chaos theory; probability theory, and stochastic processes is required. These mathematical disciplines are defined and developed in this work to the extent that is needed to develop and justify their application to Internet engineering. The effort to do so is often substantial and not elementary. Occasionally, applications of the intermix of these different mathematical disciplines should also be evident in this work. A striking example is Wigner's semicircle law of random matrices, which uses elements of probability theory and matrices.

This work has been divided into two main parts. These are: certain fundamental aspects of Internet engineering, and the corresponding mathematical prerequisites. In general, the chapters on mathematical concepts complement those on the engineering fundamentals. An effort has been made to provide, justification and informal *proofs* of different assertions and propositions, as much as possible. The word "proof" has been italicized, because there appears to be no universally accepted definition of it. It is hoped that the reader does not accept the mathematical results on faith. He or she is encouraged to step through the proofs of the results. In order to lighten up the presentation, each chapter begins with a brief description of the life and works of a prominent mathematician or scientist or engineer.

The Internet engineering fundamentals are described in Volume 1. Each chapter in this volume essentially describes a single Internet engineering body of knowledge. The supporting mathematics is outlined in Volume 2. It is hoped that these later chapters provide *basic tools* to analyze different aspects of Internet engineering. These chapters on mathematics also go a long way towards making the chapters on Internet engineering self-contained. Occasionally, additional material has been included to improve the comprehensiveness of various definitions and theorems. Each volume has a list of commonly used symbols. A list of Greek symbols is also provided for ready reference.

The two-volume set is titled *Mathematical Principles of the Internet*. The Volume 1 is titled *Engineering Fundamentals* and Volume 2 *Mathematical Concepts*. The contents of the respective volumes are:

The chapter details, along with their dependencies on other chapters are listed below.

*Volume 1 - Engineering Fundamentals*

*Prologue*: The prologue is an introduction to modeling the Internet. It provides a brief history of communication, and an introduction to telecommunication networks, and the World Wide Web. Some empirical laws about the value of a network are also outlined. This is followed by a rationale for mathematical modeling of the Internet.

*Dependency*: This is independent of all other chapters.

*Chapter 1. Information theory*: The Internet is a vehicle for transportation of information. There-fore concepts developed in information theory are immediately applicable. Information theory as developed by Claude E. Shannon determines the ultimate transmission rate of a communication channel, which is its channel capacity. It also aids in determining the ultimate data compression rate. Concepts developed in information theory are also applicable to cryptography, and data min-ing.

*Dependency*: Applied analysis; optimization, stability, and chaos theory; probability theory, and stochastic processes.

*Chapter 2. Algebraic coding theory*: When information is transmitted on a noisy communication channel, packets become error prone. Consequently error correcting codes merit a careful study. This theory is based upon algebraic concepts developed by the likes of N. H. Abel and E. Galois.

*Dependency*: Number theory, abstract algebra, matrices and determinants, applied analysis, and basics of information theory.

*Chapter 3. Cryptography*: The study of cryptographic techniques is essential in order to have secure communication. The science and art of cryptographic techniques is as old as the history of mankind itself. Since time immemorial cryptographic techniques have been practiced by both good and bad persons (and organizations).

*Dependency*: Number theory, abstract algebra, matrices and determinants, applied analysis, probability theory, and basics of information theory.

*Chapter 4. Internet traffic*: The predecessor of the Internet in the United States of America was the telephone network. In the pre-Internet days it was designed to essentially carry voice traffic. Voice traffic was generally described by Poisson processes and exponential distributions. In contrast Internet traffic is self-similar and characterized by long-range dependence. The mathematical details required to model this type of traffic are significantly more complicated than those of the voice traffic. We also describe a model for Internet packet generation via nonlinear chaotic maps.

*Dependency*: Applied analysis; optimization, stability, and chaos theory; probability theory, and stochastic processes.

*Chapter 5. Dynamics, control, and management of Internet congestion*: Congestion control in the network addresses the problem of fair allocation of resources, so that the network can oper-ate at an acceptable performance level when the traffic demand is very near the capacity of the network. This is accomplished via a feedback mechanism. If Internet is modeled as a dynamical system, then the stability of the feedback schemes should be studied carefully. In the midst of chaos and congestion of the Internet, it is still possible to successfully manage traffic flow with different characteristics and requirements. This is done via intelligent scheduling of packets. An idealized scheme of packet scheduling called the generalized processor sharing (GPS) is discussed. Elements of network calculus are also outlined. Wireless networks are studied by using stochastic geometry.

*Dependency*: Applied analysis; optimization, stability, and chaos theory; probability theory, and stochastic processes.

*Chapter 6. Queueing theory*: Internet can be viewed as an extremely large queueing network. Given a specific level of traffic, queueing theory evaluates the utilization of physical resources. It also helps in modeling the response time characteristics of different types of Internet traffic. This discipline requires a sound knowledge of stochastic processes, including Markov processes, some knowledge of classical transformation techniques, and large deviation theory.

*Dependency*: Matrices and determinants, applied analysis; optimization, stability, and chaos theory; probability theory, stochastic processes, and basics of Internet traffic.

*Chapter 7. Stochastic structure of the Internet and the World Wide Web*: In order to study the structure of the Internet and the Web graph, it is assumed that these networks are stochastic. Use of stochastic and nondeterministic models might perhaps be an expression of our ignorance. Nevertheless, these models provide some useful insight into the structure of the Internet and the Web. The spread of computer virus, and its vulnerability to external attacks are also studied.

*Dependency*: Matrices and determinants, graph theory, geometry, applied analysis, probability theory, stochastic processes; and optimization, stability, and chaos theory.

*Chapter 8. Graph-theoretic algorithms*: A network can be modeled abstractly as a graph, where the vertices of the graph are the hosts and/or routers and its arcs are the transmission links. Using this model, routing algorithms for transporting packets can be developed. Furthermore, the reliability of the network layout can also be evaluated. The network coding technique is also described.

*Dependency*: Matrices and determinants, graph theory, geometry, applied analysis, and probability theory.

*Chapter 9. Game theory and the Internet*: Game theory is the study of nonaltruistic interaction between different entities (agents or users). This field of knowledge has found application in political science, sociology, moral philosophy, economics, evolution, and many other disciplines of human endeavor. Due to the distributed nature of the Internet, game theory appears to be eminently suitable to explore the selfish behavior of the Internet entities.

*Dependency*: Matrices and determinants, applied analysis; optimization, stability, and chaos theory; and probability theory.

*Chapter 10. Internet economics*: Economics has to be one of the important drivers of the Internet evolution. Interaction between network resources, congestion, and pricing are studied in this chapter. In addition, service differentiation techniques; applications of auction theory, coalitional game theory, and bargaining games are also elaborated upon.

*Dependency*: Matrices and determinants, applied analysis; optimization, stability, and chaos theory; and probability theory.

*Chapter 11. Data mining and knowledge discovery*: Internet makes the accessibility of information so much more convenient. It also generates and stores enormous amounts of information. Data mining and knowledge discovery processes help in extracting useful nuggets of information from this immense amount of data.

*Dependency*: Matrices and determinants, graph theory, applied analysis; optimization, stability, and chaos theory; probability theory, stochastic processes, and basics of information theory.

*Chapter 12. Quantum computation, communication, and cryptography*: The author is by no means suggesting that quantum computation, communication, and cryptography are the wave of the future. However, the study of this discipline and its application to the Internet is interesting in its own right. This paradigm offers immense parallelism. Furthermore, the algorithms developed under this discipline are fascinating. Peter Shor's factorization algorithm with possible application to cryptography and Lov K. Grover's search algorithm offer enough food for thought.

*Dependency*: Number theory, matrices and determinants, applied analysis, probability theory, and basics of information theory.

*Volume 2 - Mathematical Concepts*

*Chapter 1. Number theory*: Number theory is presumably the foundation of all branches of mathematics. Some mathematicians might also opine that it is the purest of all mathematical disciplines. However this subject has found several important applications, most notably in cryptography, signal processing, and error correcting codes. It has also made foray into quantum computation, communication, and cryptography. Therefore to the dismay of the purists and to the delight of the pragmatists the subject of number theory has indeed found practical applications.

*Dependency*: Abstract algebra, and matrices and determinants.

*Chapter 2. Abstract algebra*: This subject is used in developing error detecting and correcting codes, cryptography, and quantum computation and communication.

*Dependency*: Number theory, and applied analysis.

*Chapter 3. Matrices and determinants*: Elementary as it might appear, applications of matrices and determinants have produced several elegant results. The applications of matrices and determinants can be found in error detection and correction codes, cryptography, graph-theoretic algorithms, study of stochastic networks, queueing theory, game theory and the Internet, data mining and knowledge discovery, and quantum computation and communication.

*Dependency*: Applied analysis, and probability theory.

*Chapter 4. Graph theory*: The subject of graph theory lays down the foundation for graph-theoretic algorithms. It also includes study of routing algorithms, random graphs, power-law graphs, and network flow algorithms.

*Dependency*: Matrices and determinants.

*Chapter 5. Geometry*: Geometry is one of the most ancient branch of mathematics. This subject, more specifically hyperbolic geometry has found application in routing algorithms, and the study of structure of the Internet.

*Dependency*: Matrices and determinants, applied analysis; and optimization, stability, and chaos theory.

*Chapter 6. Applied analysis*: Analysis is a deep mathematical discipline. Its applications pervade all the topics discussed in this work including information theory, Internet traffic, queueing theory, and dynamics and control of Internet congestion. Other prominent applications of this subject are the application of game theory to the Internet, Internet economics, and data mining and knowledge discovery.

*Dependency*: Number theory, and matrices and determinants.

*Chapter 7. Optimization, stability, and chaos theory*: Concepts related to optimization, stability, and chaos theory can be found in information theory, application of game theory to the routing of packets, modeling packet dynamics, control of Internet congestion, Internet economics, and data mining and knowledge discovery.

*Dependency*: Matrices and determinants, and applied analysis.

*Chapter 8. Probability theory*: Probability theory, like any other theory has three parts. These are formal axiomatics and definitions, applications, and intuitive background. This subject provides the foundation for stochastic processes.

*Dependency*: Applied analysis; and optimization, stability, and chaos theory.

*Chapter 9. Stochastic processes*: This subject is used in describing the laws and results of information theory, Internet traffic, structure of the Internet and the Web graphs, queueing theory, and data mining and knowledge discovery.

*Dependency*: Applied analysis, and probability theory.

**Why read this work?**

The main features of this work are listed below.

(a) *Uniqueness of the work*: What is so special about this work? It is the author's humble belief, that the material presented in this work is the first of its type to describe the Internet. To the author's knowledge, several topics are discussed for the very first time in a student-friendly textbook form.

(b) *Precise presentation*: All the axioms, definitions, lemmas, theorems, corollaries, observations, and examples are clearly delineated in the chapters on mathematics. Proofs of important lemmas, theorems, and corollaries are also provided. These results are used to precisely state, prove, and develop significant engineering results. Important facts are presented as observations.

(c) *Self-contained*: The author has strived to make this work self-contained as much as possible. That is, sufficient mathematical background is provided to make the discussion of different topics complete. The author also does not assume that the reader has any prior knowledge of the subject. Sufficient mathematical details are also provided, so that the results of Internet related mathematics can be stated with plausible proofs.

(d) *Independence of chapters*: All chapters on Internet engineering are largely independent of each other. Each chapter is organized into sections and subsections. Concepts are made precise via the use of definitions. Notation is generally introduced in the definitions. Relatively easy consequences of the definitions are listed as observations, and important results are stated as theorems.

(e) *Historical introduction*: Prologue and each chapter begin with a brief biography of a mathematician or a scientist or an engineer. The purpose of providing biographies is two-fold. The first purpose is to provide inspiration to the student, and the second is to inform him or her about the lives of persons who developed important mathematical and engineering concepts. It is left to the individual student to evaluate the impact of these ideas upon the description and evolution of the Internet.

(f) *List of observations*: Several significant results are listed precisely as observations. Proofs of some of these observations are outlined in the problem section provided at the end of each chapter.

(g) *Examples and figures*: Each chapter is interspersed with several examples. These examples serve to clarify, enhance, and sometimes motivate different results. It is the author's belief that examples play a crucial role in getting a firm grasp of the subject. If and where necessary, figures are provided to improve the clarity of the exposition.

(h) *Algorithms*: Algorithms are provided in a pseudocode format for ready implementation in any high-level language.

(i) *Reference notes*: Each chapter is provided with a reference-notes section. This section lists the primary sources of the technical literature used in writing the chapter. These sources generally constitute the main citations. Sources for further study are also provided.

(j) *Problems*: Each chapter is provided with a problem section. Besides, enhancing the material presented in the main chapter, each problem states a significant result. A majority of the problems are provided with sufficient hints. In order to keep the continuity and not clutter with too many details, proofs of important observations made in the chapter are relegated to the problem section. It is strongly suggested that the reader peruse the problem section.

(k) *List of references*: A list of references is provided at the end of each chapter.

(l) *List of symbols*: A list of commonly used symbols and Greek letters is provided.

(m) *User-friendly index*: A comprehensive and user-friendly index of topics is provided at the end of each volume.

### Target audience of the work

This work evolved as the author taught courses on network modeling and related topics. These books can be used either in an upper-level undergraduate or first-year graduate class in electrical engineering or computer science. The choice and depth of topics actually depends upon the preparation of the students.

These books should also serve as a useful reference for networking researchers, professionals and practitioners of the art of Internet engineering, that will provide a better understanding and appreciation for the deeper engineering principles which are central to Internet engineering. This, in turn would provide the necessary guidelines for designing the next generation of the Internet.

### Commissions and omissions

The pseudocode for the different algorithms has been written with especial care, but is not guaranteed for any specific purpose. It is also quite possible, that the author has not provided complete credit to the different contributors of the subject. To them, the author offers a sincere apology for any such inadvertent omission. Receiving information about errors, and suggestions for improvement of the books will be greatly appreciated.

### Acknowledgments

Invention of the Internet was perhaps, one of the greatest technical achievements of the last century. Therefore, first and foremost I offer my sincere thanks to the inventors and builders of the Internet for their heroic work and dedication. I am also grateful to all the authors whose names appear in the list of references provided at the end of each chapter.

Further, these books were written in LaTeX, which is an extension of TeX. TeXis a trademark of American Mathematical Society. So a very special thanks to their respective creators: Leslie Lamport and Donald E. Knuth.

The impetus for the publication of this work was provided by Sartaj Sahni. I am forever grateful to him for his initial interest and consideration. His own work has certainly been a source of inspiration. My first contact at Taylor and Francis Publications was with Randi Cohen. She had been most courteous and extremely patient. The other staff members at Taylor and Francis Publications, including Robin Lloyd-Starkes, Alice Mulhern, and Veronica Rodriguez have been most cooperative. Expert typesetting was provided by Shashi Kumar. The constructive comments and suggestions provided by anonymous reviewers are also appreciated. Encouragement provided by Shyam P. Parekh is certainly treasured.

This work would not have been possible without the inspiration of my students and esteemed teachers. A very special thanks is extended to family, friends, and colleagues for their unwavering support. Occasional help in diction and artwork was provided by Rishi who makes all the work worthwhile.

NB
San Jose, California
Email address: nbhatnagar@alumni.stanford.edu

# List of Symbols

Different types of commonly used symbols are categorized as:

1. Logical operators
2. Set operators
3. Sets of numbers
4. Basic arithmetic operators
5. More arithmetic operators
6. Arithmetical relationships
7. Analysis
8. Complex numbers
9. Vectors
10. Matrices
11. Mappings
12. Combinatorial functions
13. Probability theory
14. Mathematical constants
15. Notation used in describing the algorithms

## Logical Operators

| | |
|---|---|
| $\wedge$ | logical and operator |
| $\vee$ | logical or operator |
| $\leftarrow$ | assignment operator |
| $\rightarrow$, $\Rightarrow$ | logical implication |
| $\Leftrightarrow$, iff | if and only if |

## Set Operators

| | |
|---|---|
| $\in$ | belongs to |
| $\notin$ | does not belong to, negation of $\in$ |
| $\ni$ | such that |
| $\forall$ | universal quantifier, for all |
| $\exists$ | existential quantifier, there exists |
| $\nexists$ | there does not exist |
| $\cap$ | set intersection operator |
| $\cup$ | set union operator |
| $\setminus$ | set difference operator |
| $\subset$ | proper subset containment operator |
| $\subseteq$ | subset operator |
| $\varnothing$ | empty set |
| $\oplus$ | set addition operator |
| $\square$ | end of: proof, definition, example, or observation |
| $\{\cdot, \cdots, \cdot\}$ | set list |
| $\sim$ | equivalence between sets |
| $A^c, \overline{A}$ | complement of the set $A$ |
| $|A|$ | cardinality of the set $A$ |
| $A^\perp$ | set orthogonal to the set $A$ |
| $A \times B$ | Cartesian product of sets $A$ and $B$ |
| $A^{(n)}, A^n$ | Cartesian product of the set $A$ with itself, $n$ times over |
| $\{x \mid R(x)\}$, $\{x : R(x)\}$ | set of all $x$ for which the relationship $R(x)$ is true |

## Sets of Numbers

| | |
|---|---|
| $\mathbb{B}$ | binary field, also $\mathbb{F}_2, GF_2$ |
| $\mathbb{C}$ | set of complex numbers |
| $\widehat{\mathbb{C}}$ | set of nonzero complex numbers $\mathbb{C} \setminus \{0\}$ |
| $\mathbb{C}^*$ | set of extended complex numbers $\mathbb{C} \cup \{+\infty\}$ |
| $\mathbb{F}_q$ | finite field of size $q$, also $GF_q$, $q = 2, 3, \ldots$ |
| $GF_q$ | Galois field of size $q$, also $\mathbb{F}_q$, $q = 2, 3, \ldots$ |
| $\mathbb{P}$ | set of positive numbers $\{1, 2, 3, \ldots\}$ |
| $\mathbb{N}$ | set of natural numbers $\{0, 1, 2, 3, \ldots\}$ |
| $\mathbb{Q}$ | set of rational numbers |
| $\mathbb{R}$ | set of real numbers |
| $\widehat{\mathbb{R}}$ | set of nonzero real numbers $\mathbb{R} \setminus \{0\}$ |
| $\mathbb{R}_0^+$ | set of nonnegative real numbers $\mathbb{R}^+ \cup \{0\}$ |
| $\mathbb{R}^+$ | set of positive real numbers |
| $\mathbb{R}^*$ | set of extended real numbers $\mathbb{R} \cup \{+\infty\}$ |
| $\mathbb{R}^n$ | $n$-dimensional real vector space, where $n \in \mathbb{P}$ |
| $\mathbb{Z}$ | set of integers $\{\ldots, -2, -1, 0, 1, 2, \ldots\}$ |
| $\mathbb{Z}_n$ | set of integers modulo $n$, the set $\{0, 1, 2, \ldots, n-1\}$ |
| $\mathbb{Z}_n^*$ | set of nonzero integers in the set $\mathbb{Z}_n$, which are relatively prime to $n$ |

## Basic Arithmetic Operators

| | |
|---|---|
| $+$ | addition operator |
| $-$ | subtraction operator |
| $\times, \cdot$ | multiplication operator |
| $\div, /$ | division operator |
| $\pm$ | plus or minus operator |
| $\sqrt{\cdot}$ | square root operator |
| $\lceil \cdot \rceil$ | ceiling operator; for $x \in \mathbb{R}$, $\lceil x \rceil$ = least integer greater than or equal to $x$ |
| $\lfloor \cdot \rfloor$ | floor operator; for $x \in \mathbb{R}$, $\lfloor x \rfloor$ = greatest integer less than or equal to $x$ |
| $[\cdot]$ | round-off operator; for $x \in \mathbb{R}$, $[x]$ = integer closest to $x$ |
| $\cdot \mid \cdot$ | divisibility operator; $a \mid m$ means nonzero integer $a$ can divide integer $m$ |
| $\cdot \nmid \cdot$ | nondivisibility operator; $a \nmid m$ means nonzero integer $a$ cannot divide integer $m$ |

## More Arithmetic Operators

| | |
|---|---|
| $\lvert a \rvert$ | absolute value (magnitude) of $a \in \mathbb{R}$ |
| $\langle n \rangle_p$ | modulus operator $n \pmod{p}$, $p \in \mathbb{P}$ |
| $\sum$ | discrete summation operator |
| $\prod$ | product operator |
| $*$ | convolution operator |
| $\gcd(a, b)$ | greatest common divisor of $a$ and $b$; $a, b \in \mathbb{P}$ |
| $\max\{\ldots\}, \max(\ldots)$ | maximum operator |
| $\min\{\ldots\}, \min(\ldots)$ | minimum operator |
| $\max(a, b)$ | maximum of $a$ and $b$; $a, b \in \mathbb{R}$ |
| $\min(a, b)$ | minimum of $a$ and $b$; $a, b \in \mathbb{R}$ |
| $\mod$ | modulo operator |
| $a^+$ | $\max(0, a)$, $a \in \mathbb{R}$ |
| $a^-$ | $\max(0, -a)$, $a \in \mathbb{R}$ |
| $\exp(\cdot)$ | exponential function with base $e$ |
| $\ln(\cdot)$ | natural logarithm |
| $\log_a(\cdot)$ | logarithm to the base $a$, where $a \in \mathbb{R}^+$ |
| $sgn(\cdot)$ | signum function |

## Arithmetical Relationships

| | |
|---|---|
| $=$ | equality operator |
| $\neq$ | not equal to |
| $\sim$ | asymptotically equal |
| $\simeq$ | approximate relationship between functions |
| $\approx$ | approximate relationship between numbers |
| $\asymp$ | approximate relationship between functions within a constant |
| $\geq$ | greater than or equal to |
| $\leq$ | less than or equal to |
| $\gg$ | much greater than |
| $\ll$ | much less than |
| $\gtrless$ | greater or less than |
| $\rightarrow$ | approaches, tends towards |
| $\propto$ | proportional to |
| $\equiv$ | congruent to |
| $\not\equiv$ | not congruent to |
| $\cong$ | isomorphism |

## Analysis

| | |
|---|---|
| $\infty$ | infinity |
| $\lim$ | limit |
| $\frac{d}{dt}$ | differentiation operator |
| $f'(t), \dot{f}(t)$ | $\frac{d}{dt}f(t), t \in \mathbb{R}$ |
| $\frac{\partial}{\partial t}$ | partial differentiation operator |
| $\int$ | integration operator |
| $\|\cdot\|$ | norm of a vector |
| $l^2$ | square summable sequences |
| $L^2(\mathbb{R})$ | set of square-integrable functions |
| $\leftrightarrow$ | Fourier transform pair |
| $\arg\max_x f(x)$ | $\{x \mid f(y) \le f(x) \ \forall y\}$ |
| $\arg\min_x f(x)$ | $\{x \mid f(y) \ge f(x) \ \forall y\}$ |
| $\delta_{ij}; \ i,j \in \mathbb{Z}$ | Kronecker's delta function. |
| $f \circ g(\cdot)$ | $f(g(\cdot))$ function composition |
| $\circ, \ \langle\cdot,\cdot\rangle, \ \langle\cdot\mid\cdot\rangle$ | inner (dot) product operators |
| $\times$ | cross product operator |

## Complex Numbers

| | |
|---|---|
| $i$ | $\sqrt{-1}$ |
| $\overline{z}$ | complex conjugate of $z \in \mathbb{C}$ |
| $\|z\|$ | magnitude of $z \in \mathbb{C}$ |
| $\mathrm{Re}(z)$ | real part of $z \in \mathbb{C}$ |
| $\mathrm{Im}(z)$ | imaginary part of $z \in \mathbb{C}$ |
| $\arg(z)$ | argument of $z \in \mathbb{C}$ |

## Vectors

| | |
|---|---|
| $\boxplus$ | vector addition |
| $\otimes$ | vector multiplication |
| $u^{\perp}$ | a vector orthogonal to vector $u$ |
| $x \perp y$ | vectors $x$ and $y$ are orthogonal |

## Matrices

| | |
|---|---|
| $A^T$ | transpose of matrix $A$ |
| $A^\dagger$ | Hermitian transpose of matrix $A$ |
| $A^{-1}$ | inverse of square matrix $A$ |
| $I$ | identity matrix |
| $[a_{ij}]$ | matrix with entries $a_{ij}$ |
| $tr\,(A)$ | trace of the square matrix $A$ |
| $\det A, |A|$ | determinant of the square matrix $A$ |

## Mappings

| | |
|---|---|
| $f : A \to B$ | $f$ is a mapping from the set $A$ to the set $B$ |
| $f\,(x)$ | image of $x \in A$ under the mapping $f$ |
| $f\,(X)$ | $\{f\,(x) \mid x \in X\}$ for $f : A \to B$ and $X \subset A$ |
| $\triangleq$ | definition, or alternate notation |

## Combinatorial Functions

| | |
|---|---|
| $n!$ | $n \in \mathbb{N}$, factorial of $n$ |
| $\binom{n}{k}$ | $k, n \in \mathbb{N}, 0 \le k \le n$, binomial coefficient |
| $(n)_k$ | $n \in \mathbb{N}, 0 \le k \le n$, falling factorial |

## Probability Theory

| | |
|---|---|
| $\stackrel{d}{=}$ | equality in distribution |
| $\stackrel{d}{\to}$ | convergence in distribution |
| $P\,(\cdot)$ | probability function |
| $\sim$ | distribution of a random variable |
| $\mathcal{E}\,(X), \mathcal{E}\,[X]$ | expectation of random variable $X$ |
| $Var\,(X)$ | variance of random variable $X$ |
| $Cov\,(X, Y)$ | covariance between random variables $X$ and $Y$ |

**Mathematical
Constants**

| | |
|---|---|
| $\pi$ | $3.141592653\ldots$ |
| $e$ | $2.718281828\ldots$, Euler's number |
| $\phi_g$ | $\left(\sqrt{5}+1\right)/2$, golden ratio |
| $\gamma$ | $0.577215664\ldots$, Euler's constant |

**Notation Used in Describing the Algorithms**

The algorithms are given in a pseudo-C language. Comments inside the algorithm are italicized. The language elements are: **all, begin, do, end, find, for, for all, go to label, if** ... **then, else, label, procedure, stop, such that, while, with, Case, Subcase, Step.**

# Greek Symbols

Greek letters are often used as symbols in mathematics and engineering. For example the English equivalent of the symbol $\Sigma$ is $S$. Therefore the symbol $\Sigma$ is used for summation. Similarly, the English equivalent of the symbol $\Pi$ is $P$. Therefore the symbol $\Pi$ is used for product. A list of lower- and upper-case Greek letters and their spelling in English language follows.

| Lower Case | Upper Case | Name | Lower Case | Upper Case | Name |
|---|---|---|---|---|---|
| $\alpha$ | A | alpha | $\nu$ | $N$ | nu |
| $\beta$ | B | beta | $\xi$ | $\Xi$ | xi |
| $\gamma$ | $\Gamma$ | gamma | $o$ | O | omicron |
| $\delta$ | $\Delta$ | delta | $\pi$ | $\Pi$ | pi |
| $\epsilon, \varepsilon$ | E | epsilon | $\rho$ | P | rho |
| $\zeta$ | Z | zeta | $\sigma, \varsigma$ | $\Sigma$ | sigma |
| $\eta$ | H | eta | $\tau$ | T | tau |
| $\theta, \vartheta$ | $\Theta$ | theta | $\upsilon$ | $\Upsilon$ | upsilon |
| $\iota$ | I | iota | $\phi, \varphi$ | $\Phi$ | phi |
| $\kappa$ | K | kappa | $\chi$ | X | chi |
| $\lambda$ | $\Lambda$ | lambda | $\psi$ | $\Psi$ | psi |
| $\mu$ | M | mu | $\omega$ | $\Omega$ | omega |

# Number Theory

$$\cfrac{1}{1+\cfrac{e^{-2\pi}}{1+\cfrac{e^{-4\pi}}{1+\cdots}}} = \left\{ \sqrt{2+\phi_g} - \phi_g \right\} e^{2\pi/5}$$

A Ramanujan's result relating
numbers $e$, $\pi$, and $\phi_g$

**Srinivasa Iyengar Ramanujan.** Ramanujan was born into a poor family on 22 December, 1887 in Erode, Tamilnadu, India. Ramanujan was unable to pass his college examinations, consequently he was able to obtain only a clerk's position in the city of Madras (now Chennai). However, he was fascinated with mathematics, and produced impressive results with his remarkable intuition. From 1914 through 1919, Ramanujan worked with G. H. Hardy at Cambridge University, England. Ramanujan's contributions were mainly in number theory and modular function theory.

Hardy a very modest and gifted mathematician himself, devised an informal scale of mathematical ability. On this scale he assigned himself a mere 25, J. Littlewood (Hardy's collaborator) a 30, David Hilbert a prominent mathematician of that time an 80, and Ramanujan a 100. Ramanujan fell ill in England and died soon after returning to India on 26 April, 1920 in Kumbakonam, Tamilnadu, India.

## 1.1 Introduction

Elementary number theory is discussed in this chapter. Sets, functions, and basic number-theoretic topics like countability, divisibility, prime numbers, greatest common divisor, and continued fractions are defined and discussed in this chapter. Basics of congruence arithmetic, Chinese remainder theorem, Moebius function, Euler's phi-function, modular arithmetic, quadratic residues, and Legendre and Jacobi symbols are also examined. Properties of cyclotomic polynomials are also stated. Finally, certain useful combinatorial results are outlined. Some of these topics are explained in terms of matrices, polynomials, and basic group theory. The reader who is not familiar with these terms and concepts is urged to refer to the appropriate chapters.

Number theory is perhaps one of the most beautiful and challenging branches of mathematics. In number theory we study properties of the set of numbers $\mathbb{Z} = \{\ldots, -3, -2, -1, 0, 1, 2, 3, \ldots\}$. It can safely be conjectured that number theory, which is essentially the art of counting numbers, fascinated human beings since times immemorial. Numbers are also the fountain of all mathematics. It was the German mathematician Leopold Kronecker (1823-1891) who said: "God made the integers, all else is the work of man."

## 1.2 Sets

Concepts and notation relating to sets are summarized in this section. Initially sets are defined. These are then illuminated via examples. Different set operations are subsequently described. Set theory was primarily developed by the mathematician Georg Cantor (1845-1918).

**Definitions 1.1.** *Definitions related to sets.*

1. *A set is a well-defined list or collection of objects. A set can be specified by listing all the objects in it. A set $S$ with elements $x$, and property $\alpha(x)$ is written as either $S = \{x \mid \alpha(x)\}$ or $S = \{x : \alpha(x)\}$.*

2. *An element or member of a set is an object which belongs to the list of objects of the set. If $S$ is a set, and $b$ is an element of this set, then it is customary to denote it by $b \in S$. The Greek symbol $\in$ is called the membership symbol. If an element $b$ does not belong to a set, then it is denoted by $b \notin S$.*

3. *An empty or null set is a set with no objects. It is denoted by $\varnothing$.*

4. *A set with a single object is called a singleton set.*

5. *Let $A$ and $B$ be two sets such that every element of $A$ is also an element of $B$, then the set $A$ is said to be a subset of the set $B$. This is denoted symbolically by $A \subseteq B$. It is also possible for these two sets $A$ and $B$ to be equal.*
   *Equivalently, $B$ is a superset of (includes) $A$. It is denoted by $B \supseteq A$.*

6. *Let $A$ and $B$ be two sets such that $A$ is a subset of the set $B$. Furthermore, there exists an element in the set $B$ that is not in the set $A$. Then the set $A$ is said to be a proper subset of the set $B$. This is denoted by $A \subset B$. Conversely, $B$ is a proper superset of $A$. This is denoted by $B \supset A$.*

7. *Two sets $U$ and $V$ are equal, if they contain identical elements. It is written as $U = V$.*

8. *The cardinality $|A|$ of a set $A$ is the number of elements in $A$. Sets can either have finite or infinite number of objects. Thus $|A|$ can be either finite or infinite.*

9. *The set of all subsets of set $A$ is called the power set of $A$. The power set of set $A$ is denoted by either $\wp(A)$ or $2^A$.*    □

For example, a set of some famous symbols in mathematics is:

$$\{0, 1, \infty, e, \sqrt{-1}, \pi, \aleph_0\}$$

In the above set, the symbol $\infty$ is read as 'infinity.' It is an entity which is larger than any real number. Infinity is often treated as a number. The symbol $e = \sum_{j \geq 0} 1/j!$ is called Euler's number. The symbol $\pi$ is the ratio of circumference and diameter of any circle. The symbol $\aleph_0$, read as 'aleph-zero' or 'aleph-naught' represents the cardinality of the set of natural numbers.

The set of positive even numbers $x$ less than 13 is written as:

$$\{x \mid x \text{ is a positive even number}, x < 13\}$$

This set is indeed equal to $\{2, 4, 6, 8, 10, 12\}$. This set has 6 elements. Therefore, its cardinality is 6. If the set $A$ has $n$ elements, then the power set has $2^n$ elements (sets) in it. That is $|\wp(A)| = 2^n$. Note that $|\varnothing| = 0$ and $|\wp(\varnothing)| = 1$. That is the only member of the power set $\wp(\varnothing)$ is the empty set $\varnothing$. Two sets $U$ and $V$ are equal if and only if $U \subseteq V$ and $V \subseteq U$.

**Examples 1.1.** Following is a list of some sets of infinite size.

1. The set of positive natural numbers $\mathbb{P} = \{1, 2, 3, \ldots\}$.
2. The set of natural numbers $\mathbb{N} = \{0, 1, 2, 3, \ldots\}$.
3. The set of integers $\mathbb{Z} = \{\ldots, -2, -1, 0, 1, 2, \ldots\}$.

4. Integers divisible by 2 are called even numbers, and integers not divisible by 2 are called odd numbers. The set of positive even numbers is $\{2, 4, 6, \ldots\}$. The set of positive odd numbers is $\{1, 3, 5, \ldots\}$.

5. The set of rational numbers $\mathbb{Q}$ is the set of all fractions $m/n$, where $m$ is any integer, and $n$ is any integer except $0$.

6. The set of all real numbers is denoted by $\mathbb{R}$. These numbers can be written either as terminating or as nonterminating decimal numbers.

7. The set of irrational numbers is the set of real numbers which are not rational. Some examples are: $e, \pi, \sqrt{2}$, and $6^{1/5}$.

8. The set of positive real numbers is $\mathbb{R}^+$. That is,

$$\mathbb{R}^+ = \{r \mid r > 0 \text{ and } r \in \mathbb{R}\}$$

9. The set of nonnegative real numbers is denoted by $\mathbb{R}_0^+$. Thus $\mathbb{R}_0^+ = \mathbb{R}^+ \cup \{0\}$.

10. The set of extended real numbers is $\mathbb{R}^*$. Thus $\mathbb{R}^* = \mathbb{R} \cup \{+\infty\}$.

11. The set of all complex numbers is denoted by $\mathbb{C}$. Complex numbers are of the form $(p + iq)$, where $p, q \in \mathbb{R}$ and $i = \sqrt{-1}$. Complex numbers of the form $(p + iq)$ where $p, q \in \mathbb{Z}$ are called Gaussian integers.

12. The set of extended complex numbers is $\mathbb{C}^*$. Thus $\mathbb{C}^* = \mathbb{C} \cup \{+\infty\}$. □

Note that
$$\mathbb{P} \subset \mathbb{N} \subset \mathbb{Z} \subset \mathbb{Q} \subset \mathbb{R} \subset \mathbb{C}$$

### 1.2.1 Set Operations

Set operations such as union, intersection, complement and Cartesian product are defined.

**Definitions 1.2.** *Definitions related to set operations.*

1. *The union of two sets $A$ and $B$ is written as $A \cup B$. It is the set of elements which belong to either $A$ or $B$.*
$$A \cup B = \{x \mid x \in A \ or \ x \in B\} \tag{1.1}$$

2. *The intersection of two sets $A$ and $B$ is written as $A \cap B$. It is the set of elements which belong to both $A$ and $B$.*
$$A \cap B = \{x \mid x \in A \ and \ x \in B\} \tag{1.2}$$

*If $A \cap B = \varnothing$, then the sets $A$ and $B$ are said to be disjoint. This intersection operation is sometimes simply denoted by $AB$.*

3. *All studied sets are usually subsets of some large fixed set $U$. This set is generally called a universal set, or universe of discourse, or space.*

4. *Let $A$ be a subset of some universal set $U$. Then the complement of the set $A$ is the set of all elements which do not belong to $A$. The complement of the set $A$ is denoted by $A^c$.*
$$A^c = \{x \mid x \in U \ and \ x \notin A\} \tag{1.3}$$

*Therefore $A \cup A^c = U$, and $A \cap A^c = \varnothing$. Alternate ways of denoting, the set $A^c$ are $(U - A)$, $\overline{A}$, and $A'$.*

5. *The difference of sets $A$ and $B$ is denoted by $A \backslash B$. It is the relative complement of set $B$ with respect to $A$. That is $A \backslash B$ is the set of elements which belong to set $A$, but not to set $B$.*

$$A \backslash B = \{x \mid x \in A \ and \ x \notin B\} \tag{1.4}$$

*This set is sometimes denoted by $(A - B)$. Actually $(A - B) = AB^c$.*

6. *Let $A$ and $B$ be any two sets. The Cartesian product of sets $A$ and $B$, denoted by $A \times B$ is the set of all ordered pairs $(a, b)$ where $a \in A$ and $b \in B$.*

$$A \times B = \{(a, b) \mid a \in A, b \in B\} \tag{1.5}$$

*The product of a set with itself, $A \times A$ is denoted by $A^{(2)}$ or $A^2$. Similarly*

$$A^{(n)} \triangleq A^n = \underbrace{A \times A \times \cdots \times A}_{n \text{ times}} \tag{1.6a}$$

*The Cartesian product of the sets $A_1, A_2, \ldots, A_n$ is denoted by*

$$\times_{i=1}^{n} A_i \tag{1.6b}$$

7. *Let $A$ and $B$ be any two sets. A relation (or binary relation or binary operation) $R$ from $A$ to $B$ is a subset of $A \times B$.*
   *The set $R$ is a set of ordered pairs, that is: $R = \{(a, b) \mid a \in A \ and \ b \in B\} \subseteq A \times B$.*
   *Therefore if $(a, b) \in R$, then $a$ is $R$-related to $b$, written $aRb$. However if $(a, b) \notin R$, then $a$ is not $R$-related to $b$, written $a\bar{R}b$.*

8. *Let $S$ be any set. A relation on a set $S$, is a binary relation from $S$ to $S$, which is a subset of $S \times S$.*

9. *Let $R$ be a binary relation on a nonempty set $S$ satisfying the following three properties:*
   (a) *Reflexive property: If $a \in S$, then $aRa$.*
   (b) *Symmetric property: If $aRb$ then $bRa$.*
   (c) *Transitive property: If $aRb$ and $bRc$, then $aRc$.*
   
   *The relationship $R$ is called an equivalence relation.*
   *If $a \in S$, then the subset of $S$ containing elements related to $a$ is called an equivalence class. The relationship $R$ is said to induce a partition of the set $S$ into equivalence classes.*
   *The equivalence class containing $a \in S$ is denoted by $[a]$. The set of equivalence classes induced by the relation $R$ is called the quotient set of $S$ by $R$. It is denoted by $S/R$. The equivalence class $[a]$ is an element of the quotient set $S/R$.*

10. *Let $R$ be a binary relation on a nonempty set $S$ satisfying the following three properties:*
    (a) *Reflexive property: If $a \in S$, then $aRa$.*
    (b) *Antisymmetric property: If $aRb$ and $bRa$, then $a = b$.*
    (c) *Transitive property: If $aRb$ and $bRc$, then $aRc$.*
    
    *The relationship $R$ is called a partial order and the set $S$ is called a partially ordered set. Alternately, the relationship $R$ is called an order relation and the set $S$ is called an ordered set.* □

An example of an order is the "less than or equal to" relation $\leq$, defined on any subset of real numbers $\mathbb{R}$.

## 1.2.2  Bounded Sets

Terms such as lower and upper bounds, greatest lower bound, and least upper bound are defined below.

>**Definitions. 1.3.** *Let $X \subseteq \mathbb{R}$ be a subset of real numbers.*
>
>1. *A number $l$ is said to be a lower bound of the set $X$, if for all $x \in X$, we have $l \leq x$.*
>2. *A number $l_0$ is said to be a infimum or greatest lower bound (g.l.b.), if for all lower bounds $l$, we have $l \leq l_0$.*
>   *If $l_0 \in X$, then $l_0$ is the minimum value of the points in the set $X$. In this case we also denote $l_0$ as $\min X$.*
>3. *A number $u$ is said to be an upper bound of the set $X$, if for all $x \in X$, we have $x \leq u$.*
>4. *A number $u_0$ is said to be a supremum or least upper bound (l.u.b.), if for all upper bounds $u$, we have $u_0 \leq u$.*
>   *If $u_0 \in X$, then $u_0$ is the maximum value of the points in the set $X$. In this case we also denote $u_0$ as $\max X$.*
>5. *If the set $X$ has both a lower and an upper bound, then it is said to be a bounded set.* □

The words infimum and supremum are sometimes abbreviated as "inf" and "sup" respectively. Limits superior and inferior of a sequence of real numbers are also defined in the chapter on applied analysis. We make the following *assumptions* in this chapter. If a nonempty set of real numbers has an upper bound, then it has a least upper bound. Similarly if a nonempty set of real numbers has a lower bound, then it has a greatest lower bound. It is also possible to define in general upper and lower bounds of a subset of a partially ordered set.

## 1.2.3  Interval Notation

Intervals on the real line are defined below. Recall that the points on the real line are denoted by $\mathbb{R}$.

>**Definitions 1.4.** *Let $a, b \in \mathbb{R}$, where $a < b$.*
>
>1. *Open interval $(a, b) = \{x \mid a < x < b\}$.*
>2. *Closed interval $[a, b] = \{x \mid a \leq x \leq b\}$, where $a$ and $b$ are called the end-points of the interval.*
>3. *Open-closed interval $(a, b] = \{x \mid a < x \leq b\}$, where $b$ is the end-point of the interval.*
>4. *Closed-open interval $[a, b) = \{x \mid a \leq x < b\}$, where $a$ is the end-point of the interval.*
>5. *The intervals $(a, b]$ or $[a, b)$ are half-open (or half-closed) intervals in $\mathbb{R}$.*
>6. *A single point in $\mathbb{R}$ is defined as a closed interval.* □

An infinite interval is best defined via examples. Let $a \in \mathbb{R}$. Some examples of infinite intervals are:

$$(a, \infty) = \{x \mid a < x, x \in \mathbb{R}\}$$
$$[a, \infty) = \{x \mid a \leq x, x \in \mathbb{R}\}$$
$$(-\infty, \infty) = \mathbb{R}$$

## 1.3  Functions

In this section, basic definitions of functions, sequences, permutation mappings, permutation matrices, unary and binary operations, and logical operations are given.

**Definitions 1.5.** *We define functions and permutations.*

1. *Let $A$ and $B$ be any two sets. Assign to each element $a$ of the set $A$, a unique element $b$ of the set $B$. The set of such assignments is called a function or mapping from $A$ into $B$. It is indicated as $f : A \to B$. The function $f$ is sometimes denoted by $f(\cdot)$.*
   *The specific element $b \in B$ assigned to $a \in A$ is denoted by $f(a)$. It is written as $f(a) = b$, or simply $a \mapsto b$. Furthermore, $f(a)$ is sometimes called the image of $a$ or the value of $f$ at $a$. Also $a$ is called the preimage of $b$. The set $A$ is called the domain of $f$ and the set $B$ is called the codomain of $f$. The range of $f$ is denoted by $f(A)$. It is the set of images $f(A) = \{f(a) \mid a \in A \text{ and } f(a) \in B\}$. Sometimes "codomain" and "range" are used synonymously. Note that $\{(a, b) \mid a \in A \text{ and } f(a) = b\} \subseteq A \times B$.*

2. *Types of functions:*
   (a) *A function $f : A \to B$ is surjective or onto if every element $b \in B$ is the image of at least one element $a \in A$. That is, $f(A) = B$.*
   (b) *A function $f : A \to B$ is injective or one-to-one if different elements of the domain $A$ are mapped to different elements of the codomain $B$. Therefore, if $a_1, a_2 \in A$, then $f(a_1) = f(a_2) \Rightarrow a_1 = a_2$.*
   (c) *A function $f : A \to B$ is bijective if it is both surjective and injective. If the sets $A$ and $B$ are finite, then $|A| = |B|$.*

3. *Inverse function: Let $f : A \to B$ be a bijective function. Its inverse is a function $f^{-1} : B \to A$ such that $f^{-1}(b)$ is equal to a unique $a \in A$ for each $b \in B$, and $f(a) = b$. Therefore a bijective function is said to be invertible.*

4. *Homogeneous function: A function $f(x_1, x_2, \ldots, x_n)$ is homogeneous of degree $d$, if for all values of the parameter $\lambda$ and some constant $d$, we have $f(\lambda x_1, \lambda x_2, \ldots, \lambda x_n) = \lambda^d f(x_1, x_2, \ldots, x_n)$.*

5. *Permutation: Let $S = \{a_1, a_2, \ldots, a_n\}$ be a set of $n$ real numbers. A one-to-one mapping of $S$ onto $S$ is called a permutation.*

6. *Support of a function: Let $f : A \to \mathbb{R}$ be a real-valued function, defined on an arbitrary set $A$. The support of function $f$ is $supp(f) = \{x \in A \mid f(x) \neq 0\}$.*   $\square$

### 1.3.1  Sequences

A sequence of objects from a set $S$ is a list of objects from it, where repetitions are permitted.

**Definitions 1.6.** *Definitions concerning sequences.*

1. *An infinite sequence from a set $S$ is a function $f : A \to S$, where $A$ is generally the set of positive integers $\mathbb{P}$, or the set of natural numbers $\mathbb{N}$. If $A = \mathbb{P}$, the sequence is generally represented as $s_1, s_2, s_3, \ldots$, such that each $s_j \in S$. If $A = \mathbb{N}$, the infinite sequence is represented as $s_0, s_1, s_2, \ldots$, such that each $s_j \in S$.*

2. *A finite sequence from a set $S$ is a function $f : A \rightarrow S$, where for example $A = \{1, 2, \ldots, n\}$. It is generally represented as $\{s_1, s_2, \ldots, s_n\}$, or $(s_1, s_2, \ldots, s_n)$, or $\langle s_1, s_2, \ldots, s_n \rangle$, or simply $s_1, s_2, \ldots, s_n$, where each $s_j \in S$. If the order of the elements in the sequence has to be emphasized, then the second and third representations are used.*

   *The finite sequence is also sometimes called an $n$-tuple. The value $n \in \mathbb{P}$ is said to be the length of the sequence.*

3. *Consider a sequence, $S = \{s_1, s_2, s_3, \ldots\}$. If $\{i_1, i_2, i_3, \ldots\}$ is a sequence of positive integers such that $i_1 < i_2 < i_3 < \cdots$, then $\{s_{i_1}, s_{i_2}, s_{i_3}, \ldots\}$ is a subsequence of the sequence $S$.* □

If there is no ambiguity, a sequence is sometimes denoted as $\{s_i \mid i \in A\}$ or simply $\{s_i\}$. An example of the set $S$ would be the set of real numbers $\mathbb{R}$.

### 1.3.2 Permutation Mappings

Consider the set of numbers $S = \{1, 2, \ldots, n\}$. The permutation mapping $\sigma$ of the numbers which belong to the set $S$ is an arrangement of these numbers (in some specified order) into a list. The language of matrix theory can be used to explain permutation mappings. A matrix is simply a rectangular array of numbers. A tabular representation of this permutation $\sigma$ is a matrix with two rows. The first row consists of the elements $1, 2, \ldots, n$ and the bottom row has elements which are the respective images. The image of the number $i$ is $\sigma_i \in S$, where $\sigma(i) \triangleq \sigma_i$, $1 \leq i \leq n$. Therefore

$$\sigma - \begin{pmatrix} 1 & 2 & \ldots & n \\ \sigma_1 & \sigma_2 & \ldots & \sigma_n \end{pmatrix} \tag{1.7}$$

In this map $\sigma_i \neq \sigma_j$ for all $i \neq j$, and $i, j \in S$. Therefore $\sigma$ is a bijective map. In this definition, the cardinality of the set $S$ is equal to $|S| = n$. The total number of permutations of this set is equal to $n!$.

In this discussion of permutation mappings, permutations were defined on a set of numbers $S$. Permutation mappings can as well be defined on a set of distinct objects. For example, the set $S$ can be equal to $\{s_1, s_2, \ldots, s_n\}$ where $s_i \neq s_j$ for $i \neq j$ for $1 \leq i, j \leq n$.

Permutation mappings can also be used to provide a definition of a determinant of a square matrix. The interested reader can see the chapter on matrices and determinants. These concepts are best illustrated via the following examples.

**Examples 1.2.** Some elementary examples.

1. Let $A = \{1, 2, 3, 4, 5, 6\}$. A permutation $\sigma : A \rightarrow A$ is defined as follows:

$$\sigma(1) - 4, \ \sigma(2) - 6, \ \sigma(3) - 2, \ \sigma(4) - 5, \ \sigma(5) - 1, \text{ and } \sigma(6) - 3$$

The permutation function can be conveniently displayed as an array.

$$\sigma = \begin{pmatrix} 1 & 2 & 3 & 4 & 5 & 6 \\ 4 & 6 & 2 & 5 & 1 & 3 \end{pmatrix}$$

The top row in the array represents the domain elements, and the bottom row is the image under the mapping $\sigma$. Since permutations are bijective mappings, they have inverses. Denote the inverse mapping of $\sigma$ by $\sigma^{-1}$. Then

$$\sigma^{-1} = \begin{pmatrix} 1 & 2 & 3 & 4 & 5 & 6 \\ 5 & 3 & 6 & 1 & 4 & 2 \end{pmatrix}$$

2. Let $A = \{a_1, a_2, a_3, a_4, a_5, a_6\}$. A permutation $\sigma : A \to A$ is also represented as follows:

$$\sigma(a_i) \triangleq a_{\sigma_i} \in A, \quad \text{where } a_{\sigma_i} \neq a_{\sigma_j}, \ i \neq j, \ \forall i, j, \ 1 \leq i, j \leq 6$$

The permutation function can be displayed as an array.

$$\sigma = \begin{pmatrix} a_1 & a_2 & a_3 & a_4 & a_5 & a_6 \\ a_{\sigma_1} & a_{\sigma_2} & a_{\sigma_3} & a_{\sigma_4} & a_{\sigma_5} & a_{\sigma_6} \end{pmatrix}$$

The sequence $a_{\sigma_1}, a_{\sigma_2}, \ldots, a_{\sigma_6}$ is also called the permutation-list of the list $a_1, a_2, \ldots, a_6$. □

Closely related to a permutation mapping is a permutation matrix.

### 1.3.3  Permutation Matrices

Special types of matrices called permutation matrices are discussed in this subsection. Given a permutation $\sigma$ of the set $A = \{1, 2, \ldots, m\}$, a corresponding $m \times m$ permutation matrix $\widetilde{M}_\sigma$ is defined below. Note that $\widetilde{M}_\sigma = [m_{ij}]$ where $1 \leq i, j \leq m$, and

$$m_{ij} = \begin{cases} 1, & i = \sigma(j) \\ 0, & \text{otherwise} \end{cases} \tag{1.8}$$

In a permutation matrix, each row and column contain exactly a single "1", and all other entries are "0" each. A permutation matrix can also be obtained by permuting either rows or columns of an identity matrix.

**Example 1.3.** Let $A = \{1, 2, 3, 4, 5, 6\}$. A permutation $\sigma : A \to A$ is defined as follows:

$$\sigma(1) = 4, \ \sigma(2) = 6, \ \sigma(3) = 2, \ \sigma(4) = 5, \ \sigma(5) = 1, \ \text{and } \sigma(6) = 3$$

The permutation matrix of the mapping $\sigma$ is $\widetilde{M}_\sigma$.

$$\widetilde{M}_\sigma = \begin{bmatrix} 0 & 0 & 0 & 0 & 1 & 0 \\ 0 & 0 & 1 & 0 & 0 & 0 \\ 0 & 0 & 0 & 0 & 0 & 1 \\ 1 & 0 & 0 & 0 & 0 & 0 \\ 0 & 0 & 0 & 1 & 0 & 0 \\ 0 & 1 & 0 & 0 & 0 & 0 \end{bmatrix}$$

The permutation matrix of the mapping $\sigma^{-1}$ is $\widetilde{M}_{\sigma^{-1}}$.

$$\widetilde{M}_{\sigma^{-1}} = \begin{bmatrix} 0 & 0 & 0 & 1 & 0 & 0 \\ 0 & 0 & 0 & 0 & 0 & 1 \\ 0 & 1 & 0 & 0 & 0 & 0 \\ 0 & 0 & 0 & 0 & 1 & 0 \\ 1 & 0 & 0 & 0 & 0 & 0 \\ 0 & 0 & 1 & 0 & 0 & 0 \end{bmatrix}$$

It can be verified that the permutation matrix $\widetilde{M}_{\sigma^{-1}}$ is equal to the inverse of the matrix $\widetilde{M}_\sigma$. That is

$$\widetilde{M}_{\sigma^{-1}} = \widetilde{M}_\sigma^{-1}$$

In addition, the value of determinant of each of these two matrices is equal to unity. That is

$$\det\left(\widetilde{M}_\sigma\right) = \det\left(\widetilde{M}_{\sigma^{-1}}\right) = 1$$

$\square$

### 1.3.4  Unary and Binary Operations

Unary operation is used in defining the complementation operation of a Boolean algebra, which is discussed in the chapter on abstract algebra. Binary operations are used in defining algebraic systems like groups and rings.

**Definitions 1.7.** *Definitions concerning unary and binary operations.*

1. *A unary or monadic operation is a function from a set into itself.*
2. *A binary operation $*$ on a nonempty set $A$ is a mapping from $A \times A$ to $A$. The operation $*$ is a rule which assigns each ordered pair $(a, b) \in A \times A$ to an element $c \in A$.* $\square$

### 1.3.5  Logical Operations

Certain logical operations of interest are defined. These operations help in properly inferring a conclusion from a specified set of facts.

**Definitions 1.8.** *The set of logical (truth) values is denoted by $\mathfrak{B}$, where $\mathfrak{B} = \{F, T\}$. The symbols $F$ and $T$ are abbreviations of "false" and "true" respectively.*

1. *Logical "and" operation: This is a mapping $\wedge : \mathfrak{B}^{(2)} \to \mathfrak{B}$. Let $x_1, x_2 \in \mathfrak{B}$, then $\wedge(x_1, x_2) = T$ if and only if $x_1 = x_2 = T$, otherwise it is equal to $F$. Sometimes it is convenient to denote this logical "and" operation by $x_1 \wedge x_2$.*
   *This operation is sometimes called conjunction operation; and $x_1$ and $x_2$ are called logical conjuncts.*
2. *Logical "or" operation: This is a mapping $\vee : \mathfrak{B}^{(2)} \to \mathfrak{B}$. Let $x_1, x_2 \in \mathfrak{B}$, then $\vee(x_1, x_2) = F$ if and only if $x_1 = x_2 = F$, otherwise it is equal to $T$. Sometimes it is convenient to denote this logical "or" operation by $x_1 \vee x_2$.*
   *This operation is sometimes called disjunction operation; and $x_1$ and $x_2$ are called logical disjuncts.*
3. *Logical "negation" operation: This is a mapping from the set $\mathfrak{B}$ to the set $\mathfrak{B}$. If $x \in \mathfrak{B}$, then the negation of $x$ is denoted by $x'$. In this map $x' = T$ if and only if $x = F$, and $x' = F$ if and only if $x = T$.*
4. *Logical "conditional" operation: This is a mapping $\to : \mathfrak{B}^{(2)} \to \mathfrak{B}$. Let $x_1, x_2 \in \mathfrak{B}$, then $\to(x_1, x_2) = F$ if and only if $x_1 = T$, and $x_2 = F$, otherwise it is equal to $T$. Sometimes it is convenient to denote this logical "conditional" operation by $x_1 \to x_2$.* $\square$

The statement $x_1 \to x_2$ is often read as "$x_1$ implies $x_2$." Sometimes the symbol "$\Rightarrow$" is also used instead of "$\to$."

## 1.4 Basic Number-Theoretic Concepts

Basic concepts of number theory such as countability, divisibility, prime numbers, greatest common divisor, and continued fractions are defined and discussed.

### 1.4.1 Countability

**Definitions 1.9.** *Definitions related to countability.*

1. *Consider two sets $A$ and $B$. These two sets are said to be equivalent if there is a one to one correspondence between $A$ and $B$. Equivalent sets $A$ and $B$ are denoted by $A \sim B$.*
2. *A set $A$ which is equivalent to the set of numbers $\{1, 2, \ldots, n\}$ for some $n \in \mathbb{P}$ is a finite set, otherwise it is called an infinite set.*
3. *Let $A$ be an infinite set, such that $A \sim \mathbb{P}$, then the set $A$ is denumerable, otherwise it is nondenumerable.*
4. *Empty, finite, or denumerable sets are called countable sets. A set which is not countable is called noncountable.* □

   **Examples 1.4.** Some illustrative examples.

1. The set of real numbers between $-1$ and $1$ is nondenumerable and therefore noncountable.
2. If $A \sim B$ and $A \sim C$, then $B \sim C$.
3. The set of rational numbers $\mathbb{Q}$ is denumerable and therefore countable.
4. The set of real numbers $\mathbb{R}$ is nondenumerable and therefore noncountable. □

### 1.4.2 Divisibility

Let $m \in \mathbb{Z}$, and $a \in \mathbb{Z} \backslash \{0\}$. Then $a$ is said to *divide* $m$ if $m = ab$, where $b$ is an integer. Furthermore, if $a$ divides $m$, then $a$ is said to be a *divisor* of $m$, and $m$ is called a *multiple* of $a$. This is denoted by $a \mid m$. If $m$ is not divisible by $a$, then this is denoted by $a \nmid m$.

Also if $a$ and $b$ are positive integers, such that $b \leq a$, then $a = bq + r$, where $0 \leq r < b$. The positive integer $a$ is called the *dividend*, $b$ the *divisor*, $q$ the *quotient*, and $r$ the *remainder*. It is customary to denote $q$ by $\lfloor a/b \rfloor$, where $\lfloor \cdot \rfloor$ is called the *floor function* (or floor operator).

More formally, if $r \in \mathbb{R}$ then its floor $\lfloor r \rfloor$ is defined as the largest integer less than or equal to $r$. For example

$$\lfloor 7.65 \rfloor = 7, \text{ and } \lfloor -7.65 \rfloor = -8$$

Similarly

$$\lfloor 7 \rfloor = 7, \ \lfloor -7 \rfloor = -7, \text{ and } \lfloor 0 \rfloor = 0$$

### 1.4.3 Prime Numbers

A positive number $p \in \mathbb{P}$ is said to be a *prime* number, if it is divisible by only two distinct positive integers. The two integers are 1 and itself. Note that by convention, the number 1 is not considered to be a prime number. Some examples of prime numbers are: $2, 3, 5, 7, 11, \ldots$.

A *composite* number is a positive integer, that has at least one factor besides the number one and itself. That is, a positive integer which is not one and a prime number is a composite number. Some examples are $4, 6, 8, 9, 10, \ldots$. Thus, any number which can be factored into prime numbers is called a composite number. In this case any number, greater than 1 but less than $n$, which divides $n$ is called its proper factor. The next theorem is called the *fundamental theorem of arithmetic*.

**Theorem 1.1.** *Every integer $n \in \mathbb{P} \setminus \{1\}$ can be represented as a product of prime factors. This representation is unique up to the order of the factors.*
*Proof.* See the problem section.                                                    □

**Theorem 1.2.** *There are infinitely many prime numbers.*
*Proof.* See the problem section.                                                    □

The last theorem asserts that there are infinitely many prime numbers. A more useful result is provided by the celebrated *prime number theorem*.

**Theorem 1.3.** *Let $\pi(x)$ be the number of primes $p$ such that $p \leq x$, then*

$$\lim_{x \to \infty} \frac{\pi(x)}{\dfrac{x}{\ln x}} = 1 \tag{1.9}$$

*Proof.* See the problem section.                                                    □

This theorem was first independently proved by the mathematicians Jacques Salomon Hadamard and de la Vallée Poussin in 1896. The prime number theorem states that if $x$ is a very large integer, and a number is selected at random between 1 and $x$, then this number is prime with probability $1/\ln x$.

### 1.4.4 Greatest Common Divisor

The greatest common divisor of two positive integers and the associated well-known Euclidean algorithm are next elucidated. This algorithm is named after the great ancient geometer, Euclid of Alexandria (325 BC-265 BC). The extended Euclidean algorithm is also outlined

**Definitions 1.10.** *Common divisor, greatest common divisor, and relatively prime integers.*

1. *Let $d$ divide two positive integers $a$ and $b$, then $d$ is called a common divisor of $a$ and $b$.*
2. *Let $a$ and $b$ be two positive integers. The largest positive integer $d$, that divides both $a$ and $b$ is called the greatest common divisor (gcd) of $a$ and $b$. It is written as $d = \gcd(a, b)$.*
3. *Let $a$ and $b$ be positive integers such that $\gcd(a, b) = 1$. This implies that the integers $a$ and $b$ have no factors in common, except $1$. Then $a$ and $b$ are said to be relatively prime (or coprime) to each other.*                                                    □

**Example 1.5.** The integer 6 is a common divisor of 36 and 48. Also $\gcd(36, 48) = 12$. The integers 6 and 19 are relatively prime to each other, because $\gcd(6, 19) = 1$.                □

**Observations 1.1.** Let $a, b, c \in \mathbb{P}$, and $\gcd(a, b) = d$.

1. $d \mid a$ and $d \mid b$.
2. $c \mid a$ and $c \mid b \Rightarrow c \mid d$.
3. There exist integers $\alpha, \beta \in \mathbb{Z}$, such that $\alpha a + \beta b = d$.  $\square$

The proof of the last observation follows immediately from the next theorem. The integers $\alpha$ and $\beta$ are determined via the extended Euclidean algorithm. We can also have integers $\alpha', \beta' \in \mathbb{Z}$, such that $\alpha' a + \beta' b = d$; where $\alpha' = (\alpha + kb/d)$, $\beta' = (\beta - ka/d)$, and $k \in \mathbb{Z}$.

### Euclidean and Extended Euclidean Algorithms

The Euclidean algorithm finds the greatest common divisor of two positive integers. The *extended Euclidean algorithm* finds the greatest common divisor of two positive integers $a$ and $b$, and expresses it in the form $\gcd(a, b) = (\alpha a + \beta b)$, where $\alpha$ and $\beta$ are some integers.

Let $a, b \in \mathbb{P}$, and $b < a$. The greatest common divisor, $d$ of the integers $a$ and $b$ is computed via an iterative procedure called the Euclidean algorithm. The procedure is as follows.

$$a_0 = a, \; b_0 = b$$

$$a_0 = b_0 q_1 + b_1, \; q_1 = \left\lfloor \frac{a_0}{b_0} \right\rfloor, \; 0 < b_1 < b_0, \; a_1 = b_0$$

$$a_1 = b_1 q_2 + b_2, \; q_2 = \left\lfloor \frac{a_1}{b_1} \right\rfloor, \; 0 < b_2 < b_1, \; a_2 = b_1$$

$$\cdots$$

$$\cdots$$

$$a_{n-1} = b_{n-1} q_n + b_n, \; q_n = \left\lfloor \frac{a_{n-1}}{b_{n-1}} \right\rfloor, \; 0 < b_n < b_{n-1}, \; a_n = b_{n-1}$$

$$a_n = b_n q_{n+1} + b_{n+1}, \; q_{n+1} = \left\lfloor \frac{a_n}{b_n} \right\rfloor, \; 0 = b_{n+1} < b_n, \; a_{n+1} = b_n$$

Note that the procedure terminates when the remainder $b_{n+1}$, is equal to zero. The last nonzero remainder, $b_n$ is the greatest common divisor of the integers $a$ and $b$. That is, $d = \gcd(a, b) = b_n$. This procedure terminates in a finite number of steps, because $0 = b_{n+1} < b_n < b_{n-1} < \ldots < b_2 < b_1 < b_0 = b$ and $b$ is finite in value.

Since $a = bq_1 + b_1$ it can be inferred that $d \mid b_1$. The relationship $a_1 = b_1 q_2 + b_2$ shows that $d \mid b_2$. It can be similarly shown that $d \mid b_3$. By induction $d$ divides each $b_i$, so $d \mid b_n$. Therefore $d \leq b_n$.

Since $b_{n+1} = 0$, we have $b_n \mid a_n$ which is equal to $b_n \mid b_{n-1}$. Therefore $b_n \mid a_{n-1}$, that is $b_n \mid b_{n-2}$. It follows by induction that $b_n$ divides each $b_i$ and $a_i$. Thus $b_n \mid b_0$ and $b_n \mid a_0$, that is $b_n \mid b$ and $b_n \mid a$. Therefore $b_n$ divides both $a$ and $b$. This implies $b_n \leq d$.

That is, $b_n$ is the gcd of $a$ and $b$. This algorithm is encapsulated in matrix notation via the following theorem.

**Theorem 1.4.** *Let $a, b \in \mathbb{P}$, and $b < a$. The greatest common divisor, $d$ of the numbers $a$ and $b$ is computed recursively as follows. The initializing step is*: $a_0 = a, b_0 = b$. *Define*

$$\Xi_i = \begin{bmatrix} a_i & b_i \end{bmatrix}^T, \quad 0 \leq i \leq (n+1) \tag{1.10a}$$

$$q_i = \left\lfloor \frac{a_{i-1}}{b_{i-1}} \right\rfloor, \qquad 1 \leq i \leq (n+1) \tag{1.10b}$$

$$\Psi_i = \begin{bmatrix} 0 & 1 \\ 1 & -q_i \end{bmatrix}, \quad 1 \le i \le (n+1) \tag{1.10c}$$

*The iterative step is*

$$\Xi_i = \Psi_i \Xi_{i-1}, \quad 1 \le i \le (n+1) \tag{1.10d}$$

*where $n$ is the smallest integer such that $b_{n+1} = 0$. Then $d = \gcd(a,b) = b_n = a_{n+1}$.*

*Proof.* Observe that $b$ is finite in value. Therefore iterations terminate in a finite number of steps, because at iterative step $i$, the value of the remainder $b_i$ is smaller than the remainder $b_{i-1}$ in the earlier step. That is, $0 = b_{n+1} < b_n < b_{n-1} < \ldots < b_2 < b_1 < b_0 = b$. Note that the iteration yields

$$\begin{bmatrix} a_{n+1} \\ 0 \end{bmatrix} = \Psi_{n+1} \Psi_n \ldots \Psi_1 \begin{bmatrix} a \\ b \end{bmatrix}$$

Therefore any divisor of both $a$ and $b$ divides $a_{n+1}$. Thus $d \mid a_{n+1}$. This implies $d \le a_{n+1} = b_n$. Since

$$\Psi_i^{-1} = \begin{bmatrix} q_i & 1 \\ 1 & 0 \end{bmatrix}, \quad 1 \le i \le (n+1)$$

it can be inferred that

$$\begin{bmatrix} a \\ b \end{bmatrix} = \Psi_1^{-1} \Psi_2^{-1} \ldots \Psi_{n+1}^{-1} \begin{bmatrix} a_{n+1} \\ 0 \end{bmatrix}$$

Therefore $a_{n+1}$ divides both $a$ and $b$. Consequently $a_{n+1} \mid d$. That is, $a_{n+1} \le d$. Thus $d = a_{n+1}$. $\square$

The Euclidean and extended Euclidean algorithms are next outlined. The extended Euclidean algorithm implicitly uses the Euclidean algorithm. If two positive integers $a$ and $b$ are given such that $\gcd(a,b) = d$, the extended Euclidean algorithm expresses the greatest common divisor as

$$d = (\alpha a + \beta b)$$

where $\alpha, \beta \in \mathbb{Z}$. The above result is called Bézout's theorem. There is an analogous Bézout's theorem for polynomials.

---

**Algorithm 1.1.** *Euclidean Algorithm.*

**Input:** $a, b \in \mathbb{P}$, and $b \le a$.
**Output:** $d = \gcd(a,b)$
**begin**
  **while** $b \ne 0$ **do**
  **begin**
    $q \leftarrow \lfloor \frac{a}{b} \rfloor$
    $r \leftarrow a - qb$
    $a \leftarrow b, b \leftarrow r$
  **end** (*end of while-loop*)
  $d \leftarrow a$
**end** (*end of Euclidean algorithm*)

---

**Algorithm 1.2.** *Extended Euclidean Algorithm.*

**Input:** $a, b \in \mathbb{P}$, and $b \leq a$.
**Output:** $d = \gcd(a, b)$, and $(\alpha a + \beta b) = d$, where $\alpha, \beta \in \mathbb{Z}$
**begin**
    (*initialization*)
    $\alpha_2 \leftarrow 1, \alpha_1 \leftarrow 0, \beta_2 \leftarrow 0, \beta_1 \leftarrow 1$
    **while** $b > 0$ **do**
    **begin**
        $q \leftarrow \left\lfloor \dfrac{a}{b} \right\rfloor, r \leftarrow a - qb$
        $u \leftarrow (\alpha_2 - q\alpha_1), v \leftarrow (\beta_2 - q\beta_1)$
        $a \leftarrow b, b \leftarrow r$
        $\alpha_2 \leftarrow \alpha_1, \alpha_1 \leftarrow u, \beta_2 \leftarrow \beta_1, \beta_1 \leftarrow v$
    **end** (*end of while-loop*)
    $d \leftarrow a, \alpha \leftarrow \alpha_2, \beta \leftarrow \beta_2$
**end** (*end of extended Euclidean algorithm*)

**Example 1.6.** Using the Euclidean algorithm we determine the greatest common divisor of $a$ and $b$, where

$$a = 160, \text{ and } b = 28$$

The steps in the Euclidean algorithm are:
*Iteration 1:*
$q \leftarrow \lfloor a/b \rfloor = \lfloor 160/28 \rfloor = 5, r \leftarrow 20, a \leftarrow 28$, and $b \leftarrow 20$
*Iteration 2:*
$q_2 \leftarrow \lfloor a/b \rfloor = \lfloor 28/20 \rfloor = 1, r \leftarrow 8, a \leftarrow 20$, and $b \leftarrow 8$
*Iteration 3:*
$q_3 \leftarrow \lfloor a/b \rfloor = \lfloor 20/8 \rfloor = 2, r \leftarrow 4, a \leftarrow 8$, and $b \leftarrow 4$
*Iteration 4:*
$q_4 \leftarrow \lfloor a/b \rfloor = \lfloor 8/4 \rfloor = 2, r \leftarrow 0, a \leftarrow 4$, and $b \leftarrow 0$
*End of iterations.*
The iterations terminate in four steps.
*Output.* The greatest common divisor is equal to $d \leftarrow a = 4$.

Using the extended Euclidean algorithm, the greatest common divisor $d$ is expressed as

$$(\alpha a + \beta b) = d$$

where $\alpha, \beta \in \mathbb{Z}$.
*Initialization:* $\alpha_2 \leftarrow 1, \alpha_1 \leftarrow 0, \beta_2 \leftarrow 0, \beta_1 \leftarrow 1$
*Iteration 1:*
$q \leftarrow \lfloor a/b \rfloor = \lfloor 160/28 \rfloor = 5, r \leftarrow 20$
$u \leftarrow 1, v \leftarrow -5$

$a \leftarrow 28$, and $b \leftarrow 20$
$\alpha_2 \leftarrow 0, \alpha_1 \leftarrow 1, \beta_2 \leftarrow 1, \beta_1 \leftarrow -5$
*Iteration* 2:
$q \leftarrow \lfloor a/b \rfloor = \lfloor 28/20 \rfloor = 1, r \leftarrow 8$
$u \leftarrow -1, v \leftarrow 6$
$a \leftarrow 20$, and $b \leftarrow 8$
$\alpha_2 \leftarrow 1, \alpha_1 \leftarrow -1, \beta_2 \leftarrow -5, \beta_1 \leftarrow 6$
*Iteration* 3:
$q \leftarrow \lfloor a/b \rfloor = \lfloor 20/8 \rfloor = 2, r \leftarrow 4$
$u \leftarrow 3, v \leftarrow -17$
$a \leftarrow 8$, and $b \leftarrow 4$
$\alpha_2 \leftarrow -1, \alpha_1 \leftarrow 3, \beta_2 \leftarrow 6, \beta_1 \leftarrow -17$
*Iteration* 4:
$q \leftarrow \lfloor a/b \rfloor = \lfloor 8/4 \rfloor = 2, r \leftarrow 0$
$u \leftarrow -7, v \leftarrow 40$
$a \leftarrow 4$, and $b \leftarrow 0$
$\alpha_2 \leftarrow 3, \alpha_1 \leftarrow -7, \beta_2 \leftarrow -17, \beta_1 \leftarrow 40$
*End of iterations.*
The iterations terminate in four steps.
*Output.* The greatest common divisor is equal to $d \leftarrow a = 4$
$\alpha \leftarrow \alpha_2 = 3, \beta \leftarrow \beta_2 = -17$
Thus $d = 160\alpha + 28\beta$, where $\alpha = 3$ and $\beta = -17$.                    $\square$

### 1.4.5  Continued Fractions

It can be said that continued fractions form a bridge between real numbers and integers. Continued fractions are best introduced via examples. The continued fraction representation of $16/7$ is

$$\frac{16}{7} = 2 + \cfrac{1}{3 + \cfrac{1}{2}}$$

A compact notation to represent this expansion is $[2, 3, 2]$. Similarly, the rational number $355/113$ has an expansion

$$\frac{355}{113} = 3 + \cfrac{1}{7 + \cfrac{1}{16}} = [3, 7, 16]$$

It is also possible to represent irrational numbers via continued fractions. In this case, the continued fractions are infinite in length. The following continuous fraction is infinite.

$$1 + \cfrac{1}{1 + \cfrac{1}{1 + \cdots}} = [1, 1, 1, \ldots]$$

Note that if this continued fraction is equal to $t$, then $t = 1 + 1/t$. That is, $t = \left(\sqrt{5} + 1\right)/2$. This number is often called the golden ratio. It is usually denoted by $\phi_g$.

**Definitions 1.11.** *Terminology of continued fractions is introduced via the following definitions.*

1. *A continued fraction is a representation of the form*

$$a_0 + \cfrac{1}{a_1 + \cfrac{1}{a_2 + \cfrac{1}{\ddots}}}$$

   *where* $a_0, a_1, a_2, \ldots$ *are positive integers. Its compact representation is* $[a_0, a_1, a_2, \ldots]$. *An alternate representation is*

$$a_0 + \cfrac{1}{a_1+} \, \cfrac{1}{a_2 + \cdots}$$

   *Sometimes* $a_0 = 0$ *is allowed.*
2. *The terms* $a_0, a_1, a_2, \ldots$ *are called partial quotients.*
3. *The continued fraction is said to be simple, if all its partial quotients are integers, and* $a_i \in \mathbb{P}$ *for* $i \geq 1$.
4. *A continued fraction is finite, if it is of the form* $[a_0, a_1, a_2, \ldots, a_N]$ *where* $N \in \mathbb{N}$.   $\square$

   Use of the above definitions yields

$$[a_0] = \frac{a_0}{1} \tag{1.11a}$$

$$[a_0, a_1] = \frac{a_1 a_0 + 1}{a_1} \tag{1.11b}$$

$$[a_0, a_1, a_2] = \frac{a_2 a_1 a_0 + a_2 + a_0}{a_2 a_1 + 1} \tag{1.11c}$$

**Lemma 1.1.** *Let* $x$ *be a rational number greater than or equal to one, then* $x$ *has a finite continued fraction representation.*

   *Proof.* See the problem section.   $\square$

   Note that if $a_0$ is allowed to be equal to zero, then the restriction $x \geq 1$ can be relaxed in the above lemma. In this case, the lemma can be extended to include all nonnegative rational numbers.

**Examples 1.7.** Continued fraction expansion of some well-known mathematical constants are listed below.

1. $22/7 = [3, 7]$
2. $355/113 = [3, 7, 16]$
3. $\pi = [3, 7, 15, 1, 292, 1, 1, \ldots]$
4. $e = [2, 1, 2, 1, 1, 4, 1, \ldots]$
5. $\gamma = [0, 1, 1, 2, 1, 2, 1, \ldots]$
6. $\sqrt{2} = [1, 2, 2, 2, 2, 2, 2, \ldots]$
7. $\sqrt{3} = [1, 1, 2, 1, 2, 1, 2, \ldots]$
8. $\sqrt{5} = [2, 4, 4, 4, 4, 4, 4, \ldots]$
9. $\ln 2 = [0, 1, 2, 3, 1, 6, 3, \ldots]$
10. $\phi_g = [1, 1, 1, 1, \ldots]$
11. $\phi_g^{-1} = [0, 1, 1, 1, \ldots]$
12. $e^{1/n} = [1, n - 1, 1, 1, 3n - 1, 1, 1, 5n - 1, 1, 1, 7n - 1, \ldots]$

13. $\tanh \frac{1}{n} = [0, n, 3n, 5n, \ldots]$
14. $\tan \frac{1}{n} = [0, n-1, 1, 3n-2, 1, 5n-2, 1, 7n-2, 1, 9n-2, \ldots]$

   Note that $\gamma$ is the celebrated Euler's constant.                          □

### Convergents

The theory of continued fractions can be further developed by introducing the concept of convergents. A convergent of a continued fraction is a number obtained by truncating its representation.

**Definition 1.12.** *Let* $x = [a_0, a_1, a_2, \ldots, a_N]$. *The nth convergent of this continued fraction representation is* $[a_0, a_1, a_2, \ldots, a_n]$, *where* $0 \le n \le N$.                          □

**Observations 1.2.** Some elementary properties of convergents are listed below.

1. Let $a_0, a_1, a_2, \ldots$ be a sequence of positive integers. Also let

$$[a_0, a_1, a_2, \ldots, a_n] = \frac{p_n}{q_n} \tag{1.12a}$$

   where $p_0 = a_0, q_0 = 1$ and $p_1 = (a_1 a_0 + 1), q_1 = a_1$. Then

$$p_n = a_n p_{n-1} + p_{n-2} \tag{1.12b}$$
$$q_n = a_n q_{n-1} + q_{n-2} \tag{1.12c}$$

   for $n \ge 2$.

2. For all $n \ge 1$, the convergents satisfy:

$$p_n q_{n-1} - p_{n-1} q_n = (-1)^{n-1} \tag{1.12d}$$
$$\frac{p_n}{q_n} - \frac{p_{n-1}}{q_{n-1}} = \frac{(-1)^{n-1}}{q_n q_{n-1}} \tag{1.12e}$$
$$\gcd(p_n, q_n) = 1 \tag{1.12f}$$

3. For all $n \ge 2$, the convergents satisfy:

$$p_n q_{n-2} - p_{n-2} q_n = (-1)^n a_n \tag{1.12g}$$

□

The above observations can generally be proved by induction.

**Example 1.8.** The convergents of

$$\pi = [3, 7, 15, 1, 292, 1, 1, \ldots]$$
$$= 3.141592653589793 \ldots$$

are

$$\frac{3}{1}, \frac{22}{7}, \frac{333}{106}, \frac{355}{113}, \ldots$$

Note that

$$\frac{355}{113} = 3.1415929 \ldots$$

What is so remarkable about the last rational approximation of $\pi$?                                              □

It is also known that for almost all real numbers,

$$\lim_{n \to \infty} (a_1 a_2 \ldots a_n)^{1/n} = K_k \approx 2.68545,$$

where $K_k$ is called Khintichine's constant.

## 1.5  Congruence Arithmetic

This section establishes basic results in congruence arithmetic. These are: Chinese remainder theorem, Moebius function, Euler's phi-function, modular arithmetic, quadratic residues, and the Legendre and Jacobi symbols.

**Definition 1.13.** *Let* $a, b \in \mathbb{Z}$, *and* $m \in \mathbb{Z} \backslash \{0\}$. *The integer* $a$ *is congruent to* $b$ *modulo* $m$, *if* $m$ *divides the difference* $(a - b)$. *Equivalently* $a \pmod{m} \equiv b \pmod{m}$. *The integer* $m$ *is called the modulus. The modulo operation is denoted by* $a \equiv b \pmod{m}$.

*However, if* $m$ *does not divide* $(a - b)$, *then* $a$ *and* $b$ *are incongruent modulo* $m$. *This relationship is denoted by* $a \not\equiv b \pmod{m}$.

*Typically* $m$ *is a positive integer.*                                              □

**Example 1.9.** $7 \equiv 2 \pmod{5}$, $15 \equiv 4 \pmod{11}$, and $23 \equiv 3 \pmod{5}$.                                              □

Some observations about congruences are listed below.

**Observations 1.3.** Let $a, b, c, a_1, a_2, b_1, b_2 \in \mathbb{Z}$, and $n \in \mathbb{P}$.

1. $a \equiv b \pmod{n}$, if the remainder obtained by dividing $a$ by $n$ is the same as the remainder obtained by dividing $b$ by $n$.
2. Reflexive property: $a \equiv a \pmod{n}$.
3. Symmetry property: If $a \equiv b \pmod{n}$, then $b \equiv a \pmod{n}$.
4. Transitive property: If $a \equiv b \pmod{n}$ and $b \equiv c \pmod{n}$, then $a \equiv c \pmod{n}$.
5. Let $a_1 \equiv a_2 \pmod{n}$, and $b_1 \equiv b_2 \pmod{n}$. Then

$$(a_1 + b_1) \equiv (a_2 + b_2) \pmod{n}$$
$$a_1 b_1 \equiv (a_2 b_2) \pmod{n}$$

6. If $0 < b < a$, and $a \not\equiv 0 \pmod{b}$, then $\gcd(a, b) = \gcd(b, a \pmod{b})$.                                              □

**Definition 1.14.** $\mathbb{Z}_m$ *is the set of integers* $\{0, 1, 2, \ldots, (m - 1)\}$, $m \in \mathbb{P} \backslash \{1\}$.                                              □

Modular arithmetical operations are well-defined on the set of integers $\mathbb{Z}_m$. This arithmetic is done by performing the usual real arithmetical operations, followed by the modulo operation.

**Examples 1.10.** The above ideas are illustrated via the following examples.

1. Addition and multiplication of integers 19 and 12 modulo 5 is performed. Note that

$$19 \,(\mathrm{mod}\,5) \equiv 4 \,(\mathrm{mod}\,5)\,, \text{ and } 12 \,(\mathrm{mod}\,5) \equiv 2 \,(\mathrm{mod}\,5)$$

Addition operation:

$$(19 + 12) \,(\mathrm{mod}\,5) \equiv 31 \,(\mathrm{mod}\,5) \equiv 1 \,(\mathrm{mod}\,5)$$

or

$$\begin{aligned}
(19 + 12) \,(\mathrm{mod}\,5) &\equiv (19 \,(\mathrm{mod}\,5)) + (12 \,(\mathrm{mod}\,5)) \\
&\equiv (4 \,(\mathrm{mod}\,5)) + (2 \,(\mathrm{mod}\,5)) \\
&\equiv 6 \,(\mathrm{mod}\,5) \equiv 1 \,(\mathrm{mod}\,5)
\end{aligned}$$

Multiplication operation:

$$(19 \times 12) \,(\mathrm{mod}\,5) \equiv 228 \,(\mathrm{mod}\,5) \equiv 3 \,(\mathrm{mod}\,5)$$

or

$$\begin{aligned}
(19 \times 12) \,(\mathrm{mod}\,5) &\equiv (19 \,(\mathrm{mod}\,5)) \times (12 \,(\mathrm{mod}\,5)) \\
&\equiv (4 \,(\mathrm{mod}\,5)) \times (2 \,(\mathrm{mod}\,5)) \\
&\equiv 8 \,(\mathrm{mod}\,5) \equiv 3 \,(\mathrm{mod}\,5)
\end{aligned}$$

2. In this example, the addition and multiplication tables for arithmetic in $\mathbb{Z}_2$ and $\mathbb{Z}_5$ are shown. See Tables 1.1 and 1.2.

| + | 0 | 1 |
|---|---|---|
| 0 | 0 | 1 |
| 1 | 1 | 0 |

| × | 0 | 1 |
|---|---|---|
| 0 | 0 | 0 |
| 1 | 0 | 1 |

Table 1.1. Addition and multiplication rules for arithmetic in $\mathbb{Z}_2$.

| + | 0 | 1 | 2 | 3 | 4 |
|---|---|---|---|---|---|
| 0 | 0 | 1 | 2 | 3 | 4 |
| 1 | 1 | 2 | 3 | 4 | 0 |
| 2 | 2 | 3 | 4 | 0 | 1 |
| 3 | 3 | 4 | 0 | 1 | 2 |
| 4 | 4 | 0 | 1 | 2 | 3 |

| × | 0 | 1 | 2 | 3 | 4 |
|---|---|---|---|---|---|
| 0 | 0 | 0 | 0 | 0 | 0 |
| 1 | 0 | 1 | 2 | 3 | 4 |
| 2 | 0 | 2 | 4 | 1 | 3 |
| 3 | 0 | 3 | 1 | 4 | 2 |
| 4 | 0 | 4 | 3 | 2 | 1 |

Table 1.2. Addition and multiplication rules for arithmetic in $\mathbb{Z}_5$.

3. The modulo 5 operation partitions the set of integers $\mathbb{Z}$ in to 5 classes (or sets). These are:

$$\begin{aligned}
&\{\ldots, -10, -5, 0, 5, 10, \ldots\}\,, \\
&\{\ldots, -9, -4, 1, 6, 11, \ldots\}\,, \\
&\{\ldots, -8, -3, 2, 7, 12, \ldots\}\,, \\
&\{\ldots, -7, -2, 3, 8, 13, \ldots\}\,, \\
&\text{and } \{\ldots, -6, -1, 4, 9, 14, \ldots\}
\end{aligned}$$

$\square$

The elements $a, b \in \mathbb{Z}_m \setminus \{0\}$, are said to be multiplicative inverses of each other if $ab \equiv 1 \pmod{m}$. If multiplicative inverse of an element exists, then it is unique. However, it is possible for the multiplicative inverse of $a \in \mathbb{Z}_m$ to not exist.

**Definition 1.15.** *Let $m$ be a positive integer greater than* $1$, $a \in \mathbb{Z}_m \setminus \{0\}$, *and* $\gcd(a, m) = 1$. *Then $b \in \mathbb{Z}_m$ is an inverse of $a$ modulo $m$ if $ab \equiv 1 \pmod{m}$. The element $b$ is sometimes denoted by $a^{-1}$.*  □

**Example 1.11.** The multiplicative inverse of $7 \in \mathbb{Z}_{10}$ is $3$, but the multiplicative inverse of $2 \in \mathbb{Z}_{10}$ does not exist.  □

Some useful observations are listed below. See the problem section for proofs of some of these observations.

**Observations 1.4.** Some useful observations about modulo arithmetic.

1. Let $x$ be an unknown integer. If $a, m \in \mathbb{P}$, and $\gcd(a, m) = 1$ then $ax \equiv 1 \pmod{m}$ has a unique solution modulo $m$.
2. Let $x$ be an unknown integer. If $a, m \in \mathbb{P}$, and $\gcd(a, m) = 1$ then $ax \equiv b \pmod{m}$ has a unique solution modulo $m$.
3. Let $a, b \in \mathbb{P}$, $\gcd(a, b) = 1$, and $n \in \mathbb{Z}$. If $a$ and $b$ each divide $n$, then $ab$ divides $n$.
4. Let $a, b, c \in \mathbb{P}$. Then $\gcd(a, bc) = 1$ if and only if $\gcd(a, b) = 1$ and $\gcd(a, c) = 1$.
5. Consider the set of $m$ integers $\{0, 1, 2, \ldots, (m-1)\}$. Multiply each element $n_j$ of this set by an integer $a \in \mathbb{P}$, where $\gcd(a, m) = 1$. Then the set of elements $b_j \equiv an_j \pmod{m}$, $0 \leq j \leq (m-1)$ spans the same values as $n_j$, $0 \leq j \leq (m-1)$, although possibly in some different permuted order.  □

**Example 1.12.** $\mathbb{Z}_7 = \{0, 1, 2, 3, 4, 5, 6\}$ is the set of integers modulo 7. Let $a \in \mathbb{Z}_7$, and $b \equiv (5a) \pmod{7}$. If $a$ takes values $0, 1, 2, 3, 4, 5$, and $6$, then the corresponding values of $b$ are $0, 5, 3, 1, 6, 4$, and $2$ respectively. Thus $b \in \mathbb{Z}_7$.  □

The following result is called Fermat's little theorem.

**Theorem 1.5.** *Let $p$ be a prime number.*

(a) *If $a \in \mathbb{P}$ and $\gcd(a, p) = 1$, then $a^{p-1} \equiv 1 \pmod{p}$.*
(b) *For every $a \in \mathbb{N}$, $a^p \equiv a \pmod{p}$.*

*Proof.* There are two parts in the statement of the theorem.
    Part (a): If $\gcd(a, p) = 1$, then the numbers $a, 2a, \ldots, (p-1)a \pmod{p}$ are a permutation of the numbers $1, 2, \ldots, (p-1)$. Then

$$\prod_{j=1}^{(p-1)} (aj) \equiv \prod_{j=1}^{(p-1)} j \pmod{p}$$

or

$$a^{p-1}(p-1)! \equiv (p-1)! \pmod{p}$$

Since $p$ and $(p-1)!$ are relatively prime, it follows that $a^{p-1} \equiv 1 \pmod{p}$.

Part (b): If $a = 0$, then the statement is true. Next assume that $a \neq 0$. If $p \nmid a$ then $a^p \equiv a$ $(\mathrm{mod}\, p)$ by part (a). However, if $p \mid a$ then $a^p \equiv a \equiv 0\,(\mathrm{mod}\, p)$.                                  □

**Definition 1.16.** *Let $b, n \in \mathbb{P}$ and $\gcd(b, n) = 1$. The least positive integer $m$ such that $b^m \equiv 1$ $(\mathrm{mod}\, n)$ is called the order of $b$ modulo $n$.*                                  □

**Theorem 1.6.** *If $b, n \in \mathbb{P}$, $\gcd(b, n) = 1$, $b^k \equiv 1\,(\mathrm{mod}\, n)$, and $m$ is the order of $b$ then $m \mid k$.* *Proof.* Assume that $m \nmid k$, that is $k = mq + r$, where $0 < r < m$. Then

$$1 \equiv b^k \equiv b^{mq+r}\ (\mathrm{mod}\, n)$$

The right-hand side of the above equation is equal to $b^r\,(\mathrm{mod}\, n)$. Thus $b^r \equiv 1\,(\mathrm{mod}\, n)$. That is, $r < m$ is the order of $b$. This is a contradiction.                                  □

**Note.** If there is no ambiguity, the addition operator in modulus 2 arithmetic is sometimes denoted by $\oplus$.

In the next subsection, the so-called Chinese remainder theorem is discussed. It is generally regarded as one of the numerous pearls in number theory. It has found widespread application in diverse fields such as signal processing, coding theory, and cryptography. There is also a version of Chinese remainder theorem for polynomials. This is discussed in the chapter on algebra.

### 1.5.1 Chinese Remainder Theorem

Following is the statement of the Chinese remainder theorem.

**Theorem 1.7.** *Let $m_1, m_2, \ldots, m_n \in \mathbb{P}$, be $n \in \mathbb{P} \setminus \{1\}$ positive integers, which are coprime in pairs, that is $\gcd(m_k, m_j) = 1$, $k \neq j, 1 \leq k, j \leq n$. Also let $m = \prod_{k=1}^{n} m_k$, and $x \in \mathbb{P}$. The $n$ integers $a_1, a_2, \ldots, a_n \in \mathbb{Z}$, with the congruences*

$$x \equiv a_k\ (\mathrm{mod}\, m_k), \quad 1 \leq k \leq n \tag{1.13a}$$

*are also given. These congruences have a single common solution*

$$x \equiv \sum_{k=1}^{n} a_k M_k N_k\,(\mathrm{mod}\, m),\ M_k = \frac{m}{m_k},\ (M_k N_k) \equiv 1\,(\mathrm{mod}\, m_k),\ 1 \leq k \leq n$$

$$\tag{1.13b}$$

*Proof.* Note that $M_k$ is mutually prime with $m_k$, that is $\gcd(m_k, M_k) = 1$, for $1 \leq k \leq n$. Consequently there exist integers $N_1, N_2, \ldots, N_n \in \mathbb{P}$ such that

$$(M_k N_k) \equiv 1\ (\mathrm{mod}\, m_k), \quad 1 \leq k \leq n$$

That is, each $M_k$ has a unique reciprocal $N_k$ modulo $m_k$. Define

$$x = a_1 M_1 N_1 + a_2 M_2 N_2 + \ldots + a_n M_n N_n$$

Since $M_k \equiv 0\,(\mathrm{mod}\, m_j)$, if $k \neq j, 1 \leq k, j \leq n$, we have

$$x\,(\mathrm{mod}\, m_k) \equiv (a_k M_k N_k)\,(\mathrm{mod}\, m_k) \equiv a_k\,(\mathrm{mod}\, m_k), \quad 1 \leq k \leq n$$

Therefore $x$ satisfies all congruences in the hypothesis of the theorem. If $x$ and $y$ are two solutions which satisfy the set of congruence equations, then $x \pmod{m_k} \equiv y \pmod{m_k}$ for $1 \leq k \leq n$. Also since the $m_k$'s are relatively prime in pairs, $x \pmod{m} \equiv y \pmod{m}$. Therefore the given system of congruences have a single solution.                                                                    $\square$

**Example 1.13.** The solution to the simultaneous congruences

$$x \equiv 2 \pmod{3}, \ x \equiv 3 \pmod{4}, \ x \equiv 4 \pmod{5}, \text{ and } x \equiv 6 \pmod{7}$$

is determined. Let $m_1 = 3, m_2 = 4, m_3 = 5$, and $m_4 = 7$. Then $m = 3 \cdot 4 \cdot 5 \cdot 7 = 420$, $M_1 = m/m_1 = 140, M_2 = m/m_2 = 105, M_3 = m/m_3 = 84$, and $M_4 = m/m_4 = 60$. Also $(M_1 N_1) \equiv 1 \pmod{3}$ implies $(140 N_1) \equiv 1 \pmod{3}$, that is $(2 N_1) \equiv 1 \pmod{3}$ gives $N_1 = 2$. Similarly $N_2 = 1, N_3 = 4$, and $N_4 = 2$. Therefore

$$x \equiv \{(2 \cdot 140 \cdot 2) + (3 \cdot 105 \cdot 1) + (4 \cdot 84 \cdot 4) + (6 \cdot 60 \cdot 2)\} \pmod{420}$$
$$\equiv 2939 \pmod{420} \equiv 419 \pmod{420}$$

The solution to the given congruences is $x \equiv 419 \pmod{420}$.                                          $\square$

The Chinese remainder theorem is used in the implementation of a fast version of the discrete Fourier transform. It uses a mapping of a positive integer $x$ modulo $m$, into $(a_1, a_2, \ldots, a_n)$ where $n$ is the number of relatively prime factors of $m \in \mathbb{P}$. The number $x$ and its representation are related as follows. Let $m_1, m_2, \ldots, m_n \in \mathbb{P}$ be coprime factors of $m$, where $m = \prod_{k=1}^{n} m_k$. Then $x \equiv a_k \pmod{m_k}$, for $1 \leq k \leq n$.

Another simpler decomposition of $x$ can be obtained. Define $M_k = m/m_k$, for $1 \leq k \leq n$, and also find integers $N_1, N_2, \ldots, N_n \in \mathbb{P}$ such that $(M_k N_k) \equiv 1 \pmod{m_k}$, for $1 \leq k \leq n$. Note that $a_k$'s span the set $\{0, 1, 2, \ldots, (m_k - 1)\}$, for $1 \leq k \leq n$. Let $b_k \equiv (a_k N_k) \pmod{m_k}$, then $b_k$'s also span the set $\{0, 1, 2, \ldots, (m_k - 1)\}$, because $\gcd(m_k, N_k) = 1$. It might help reiterating that the numbers $a_k, b_k, M_k$, and $N_k$ are all computed modulo $m_k$. Thus another representation of $x$ modulo $m$, is $(b_1, b_2, \ldots, b_n)$. These observations are summarized in the following lemma.

**Lemma 1.2.** *Let $m_1, m_2, \ldots, m_n \in \mathbb{P}$, be $n$ positive integers, which are coprime in pairs, that is $\gcd(m_k, m_j) = 1, k \neq j, 1 \leq k, j \leq n$. Furthermore, let $m = \prod_{k=1}^{n} m_k$. Define $M_k = m/m_k$, for $1 \leq k \leq n$, and also let $N_1, N_2, \ldots, N_n \in \mathbb{P}$ be such that $(M_k N_k) = 1 \pmod{m_k}$, for $1 \leq k \leq n$. Let $x \in \mathbb{P}$.*

*If $x \equiv a_k \pmod{m_k}, 1 \leq k \leq n$, then*

$$x \pmod{m} \mapsto (a_1, a_2, \ldots, a_n) \tag{1.14a}$$

*An alternate map is obtained by defining $b_k \equiv (a_k N_k) \pmod{m_k}, 1 \leq k \leq n$. Then*

$$x \pmod{m} \mapsto (b_1, b_2, \ldots, b_n) \tag{1.14b}$$

$\square$

### 1.5.2  Moebius Function

The Moebius and Euler's totient functions are examples of arithmetical functions that are used extensively in number theory. The Moebius function is named after Augustus Ferdinand Moebius (1790-1860). The Euler's totient function is discussed in the next subsection.

**Definition 1.17.** *Let $n \in \mathbb{P}$. If $n > 1$ then let*

$$n = p_1^{\alpha_1} p_2^{\alpha_2} \cdots p_m^{\alpha_m} \tag{1.15a}$$

*where $p_i$ is a prime number and $\alpha_i \in \mathbb{P}$ for $1 \leq i \leq m$. The Moebius function mapping $\mu(\cdot)$ from the set of positive integers $\mathbb{P}$ to the set of integers $\mathbb{Z}$ is defined as*

$$\mu(n) = \begin{cases} 1, & n = 1 \\ (-1)^m, & \alpha_1 = \alpha_2 = \ldots = \alpha_m = 1 \\ 0, & \text{otherwise} \end{cases} \tag{1.15b}$$

$\square$

Therefore, $\mu(n)$ is equal to zero, if $n$ has a square factor greater than one.

**Example 1.14.** Some basic values.

$$n: \quad 1 \quad 2 \quad 3 \quad 4 \quad 5 \quad 6 \quad 7 \quad 8 \quad 9 \quad 10$$
$$\mu(n): \quad 1 \;\; -1 \;\; -1 \quad 0 \;\; -1 \quad 1 \;\; -1 \quad 0 \quad 0 \quad 1$$

$\square$

**Observations 1.5.** Basic results about Moebius function are listed below.

1. Let $m, n \in \mathbb{P}$ and $\gcd(m, n) = 1$, then $\mu(mn) = \mu(m)\mu(n)$.
2. For $n \in \mathbb{P}$, we have

$$\sum_{d|n} \mu(d) = \begin{cases} 1, & n = 1 \\ 0, & n > 1 \end{cases} \tag{1.16}$$

where the summation is performed over all the positive divisors of $n$. $\square$

The Moebius inversion formulae are next established. There are two types of Moebius inversion formulae. These are the additive and multiplicative inversion formulae.

**Theorem 1.8.** *The additive Moebius inversion formula states: For all $n \in \mathbb{P}$*

$$G(n) = \sum_{d|n} g(d) \;\; \text{iff} \;\; g(n) = \sum_{d|n} \mu\left(\frac{n}{d}\right) G(d) = \sum_{d|n} \mu(d) G\left(\frac{n}{d}\right) \tag{1.17}$$

*Proof.* The proof consists of two parts. In the first part assume that $G(n) = \sum_{d|n} g(d)$. Then

$$\sum_{d|n} \mu(d) G\left(\frac{n}{d}\right) = \sum_{d|n} \mu(d) \sum_{c|\frac{n}{d}} g(c) = \sum_{cd|n} \mu(d) g(c) = \sum_{c|n} g(c) \sum_{d|\frac{n}{c}} \mu(d)$$

Note that $\sum_{d|\frac{n}{c}} \mu(d)$ is equal to unity if $n = c$, and 0 otherwise via observation number 2 of this subsection. Therefore the double summation reduces to $g(n)$, which is the desired result. In the second part assume that the right side of the statement is true.

$$\sum_{d|n} g(d) = \sum_{d|n} g\left(\frac{n}{d}\right) = \sum_{d|n} \sum_{c|\frac{n}{d}} \mu\left(\frac{n}{cd}\right) G(c) = \sum_{cd|n} \mu\left(\frac{n}{cd}\right) G(c)$$

$$= \sum_{c|n} G(c) \sum_{d|\frac{n}{c}} \mu\left(\frac{n}{cd}\right) = G(n)$$

$\square$

**Theorem 1.9.** *The multiplicative Moebius inversion formula states*: *For all* $n \in \mathbb{P}$

$$G(n) = \prod_{d|n} g(d) \;\; \text{iff} \; g(n) = \prod_{d|n} G(d)^{\mu(n/d)} = \prod_{d|n} G\left(\frac{n}{d}\right)^{\mu(d)} \tag{1.18}$$

*Proof.* The proof of this theorem is similar to the proof of the additive Moebius inversion formula. In that proof replace sums by products and multiples by exponents. $\square$

### 1.5.3 Euler's Phi-Function

The *Euler's phi-function* occurs extensively in number theory. This function is named in honor of Leonhard Euler. It also leads to a generalization of Fermat's little theorem.

**Definition 1.18.** *Let* $n$ *be a positive integer. The Euler's totient* $\varphi(n)$ *is the number of positive integers which do not exceed* $n$ *and are relatively prime to* $n$.
   *The Euler's totient function is also called Euler's phi-function.* $\square$

Some values of the $\varphi(\cdot)$ function are: $\varphi(1) = 1, \varphi(2) = 1, \varphi(3) = 2, \varphi(4) = 2, \varphi(5) = 4$, and $\varphi(6) = 2$. Following are some elementary properties of the totient function.

**Observations 1.6.** Some basic results related to the totient function.

1. Let $p$ be a prime number, then $\varphi(p) = (p-1)$.
2. Let $n = p^\alpha$, where $p$ is a prime number and $\alpha$ is a positive integer, then

$$\varphi(p^\alpha) = \left(p^\alpha - p^{\alpha-1}\right)$$

3. Let $n$ be a positive integer greater than 1, with its prime-power factorization

$$n = p_1^{\alpha_1} p_2^{\alpha_2} \dots p_m^{\alpha_m},$$

where $p_i$ is a prime number and $\alpha_i \in \mathbb{P}$ for $1 \le i \le m$. Then

$$\varphi(n) = \left(p_1^{\alpha_1} - p_1^{\alpha_1-1}\right) \left(p_2^{\alpha_2} - p_2^{\alpha_2-1}\right) \dots \left(p_m^{\alpha_m} - p_m^{\alpha_m-1}\right)$$

That is

$$\varphi(n) = n\left(1 - \frac{1}{p_1}\right)\left(1 - \frac{1}{p_2}\right)\dots\left(1 - \frac{1}{p_m}\right) \tag{1.19}$$

4. If $m$ and $n$ are positive integers, such that $\gcd(m, n) = 1$, then

$$\varphi(mn) = \varphi(m)\varphi(n)$$

5. Let $n \in \mathbb{P}$, then $\sum_{d|n} \varphi(d) = n$.
6. The following relationship can be obtained by applying additive Moebius inversion formula to the equation in the last observation. Let $n \in \mathbb{P}$, then

$$\frac{\varphi(n)}{n} = \sum_{d|n} \frac{\mu(d)}{d} \qquad (1.20)$$

$\square$

Euler-Fermat theorem is next established. Euler-Fermat theorem is a generalization of Fermat's little theorem. However, before this theorem is proved, the following lemma is established.

**Lemma 1.3.** *Let* $n > 1$. *The set of numbers*

$$\{a_1, a_2, \ldots, a_{\varphi(n)}\}$$

*have the following properties*:

(a) $a_j \in \mathbb{P}$, $a_j \neq a_k$, $j \neq k$ *for* $1 \leq j, k \leq \varphi(n)$. *That is, these numbers are all positive and distinct.*
(b) $a_j < n$, *for* $1 \leq j \leq \varphi(n)$. *These numbers are all less than* $n$.
(c) $\gcd(a_j, n) = 1$, *for* $1 \leq j \leq \varphi(n)$. *That is, these numbers are all relatively prime to* $n$.

*If* $a \in \mathbb{P}$, *such that* $\gcd(a, n) = 1$, *then the numbers* $aa_1, aa_2, \ldots, aa_{\varphi(n)}$ *are congruent modulo* $n$ *to* $a_1, a_2, \ldots, a_{\varphi(n)}$ *in some permuted order.*

*Proof.* Any two numbers in the sequence $aa_1, aa_2, \ldots, aa_{\varphi(n)}$ modulo $n$ are not identical. For if $aa_j \equiv aa_k \pmod{n}$, $j \neq k$, $1 \leq j, k \leq \varphi(n)$, then $a(a_j - a_k) \equiv 0 \pmod{n}$. This implies that $a_j = a_k$, which is a contradiction.

In the next step it is proved that all numbers $aa_j$, $1 \leq j \leq \varphi(n)$ are relatively prime to $n$. Note that $\gcd(a_j, n) = 1$ and $\gcd(a, n) = 1$, therefore $\gcd(aa_j, n) = 1$.

Let $aa_j = b \pmod{n}$, then $\gcd(b, n) = 1$. Therefore $b$ should belong to the set of integers $\{a_1, a_2, \ldots, a_{\varphi(n)}\}$. $\square$

**Theorem 1.10.** *Let* $n$ *and* $a$ *be positive integers such that* $\gcd(a, n) = 1$, *then*

$$a^{\varphi(n)} \equiv \begin{cases} 0 \pmod{n}, & n = 1 \\ 1 \pmod{n}, & n \in \mathbb{P} \setminus \{1\} \end{cases} \qquad (1.21)$$

*Proof.* Observe that $a^{\varphi(1)} \equiv 0 \pmod{1}$. If $n \in \mathbb{P} \setminus \{1\}$, the proof is similar to Fermat's little theorem. By hypothesis $\gcd(a, n) = 1$, then the numbers $aa_1, aa_2, \ldots, aa_{\varphi(n)}$ are a permutation of numbers $a_1, a_2, \ldots, a_{\varphi(n)}$. Thus

$$\prod_{j=1}^{\varphi(n)} (aa_j) \equiv \prod_{j=1}^{\varphi(n)} a_j \pmod{n}$$

or

$$a^{\varphi(n)} \prod_{j=1}^{\varphi(n)} a_j \equiv \prod_{j=1}^{\varphi(n)} a_j \pmod{n}$$

Since $\gcd(a_j, n) = 1$, for $1 \leq j \leq \varphi(n)$, then each side of the above equation can be divided by $a_1 a_2 \ldots a_{\varphi(n)}$. This yields $a^{\varphi(n)} \equiv 1 \pmod{n}$, for $n \in \mathbb{P} \backslash \{1\}$. $\qquad\qquad\qquad\square$

The above result is known as the Euler-Fermat theorem. Note that Fermat's little theorem follows from Euler-Fermat theorem by letting $n = p$, a prime number, and observing that $\varphi(p) = (p-1)$.

### 1.5.4 Modular Arithmetic

Implementation of arithmetical operations in modular arithmetic is explored in this subsection. This is actually modular arithmetic of numbers which belong to the set $\mathbb{Z}_m$, where $m > 1$. Operations of interest are addition, subtraction, multiplication, division, multiplicative inverse, and exponentiation. A scheme to implement multiplicative inverse is also outlined.

*Addition.* Let $a, b \in \mathbb{Z}_m$, then addition operation can be implemented as

$$(a+b) \pmod{m} \equiv \begin{cases} a+b, & (a+b) < m \\ a+b-m, & (a+b) \geq m \end{cases} \qquad (1.22)$$

*Subtraction.* Let $a, b \in \mathbb{Z}_m$, then subtraction operation can be implemented as

$$(a-b) \pmod{m} \equiv \begin{cases} a-b, & b \leq a \\ a-b+m, & b > a \end{cases} \qquad (1.23)$$

*Multiplication*: If $a, b \in \mathbb{Z}_m$, then the product of two numbers modulo $m$ can be obtained as

$$(ab) \pmod{m} \equiv ab - \left\lfloor \frac{ab}{m} \right\rfloor m \qquad (1.24)$$

This is actually obtained by multiplying the two numbers, dividing the product by $m$, and using the remainder.

*Division*: Let $a, b \in \mathbb{Z}_m$. The division operation $a/b$ in $\mathbb{Z}_m$ is actually the computation of $ab^{-1}$ in $\mathbb{Z}_m$, provided $b^{-1}$ exists. That is the division is performed in two stages. Initially multiplicative inverse of $b$ in $\mathbb{Z}_m$ is computed (if it exists). In the next step multiplication of $a$ and $b^{-1}$ is performed in $\mathbb{Z}_m$.

*Multiplicative Inverse.* Let $a \in \mathbb{Z}_m$. Using extended Euclidean algorithm, compute

$$\alpha a + \beta m = d; \quad \alpha, \beta \in \mathbb{Z}, \ \gcd(a, m) = d \qquad (1.25)$$

If $d > 1$, then $a^{-1}$ does not exist in $\mathbb{Z}_m$. However, if $d = 1$, then $a^{-1}$ in $\mathbb{Z}_m$ is given by $\alpha \pmod{m}$.

*Exponentiation*: Let $a \in \mathbb{Z}_m$. We compute $a^{16} \pmod{m}$ efficiently as follows. Compute sequentially $a^2 \pmod{m}$, $a^4 \pmod{m}$, $a^8 \pmod{m}$, and finally $a^{16} \pmod{m}$. In this sequence $a^2 \pmod{m}$ is obtained by squaring $a \pmod{m}$, $a^4 \pmod{m}$ is obtained by squaring $a^2 \pmod{m}$, and so on. This is the principle of repeated squaring. Thus, this technique can be used to compute efficiently $a^n \pmod{m}$, where $n$ is any positive power of 2.

More generally, $a^n \pmod{m}$ for any $n \in \mathbb{P}$ can be computed as follows. Given, $a \in \mathbb{Z}_m$, and $n \in \mathbb{P}$, the goal is to compute $a^n \pmod{m}$. Exponentiation can be implemented by first obtaining a binary expansion of $n$. That is, let

$$n = \sum_{i=0}^{k} u_i 2^i,$$

where each $u_i \in \{0, 1\} = \mathbb{Z}_2$. Denote it by

$$n = (u_k u_{k-1} \ldots u_0)_2$$

Then $a^n \pmod m$ is obtained by repeated square and multiply operations. This algorithm was originally discovered in ancient India.

---

**Algorithm 1.3.** *Exponentiation.*

---

**Input:** $a \in \mathbb{Z}_m$, $n \in \mathbb{P}$, and binary expansion of $n$ is $\sum_{i=0}^{k} u_i 2^i$
**Output:** $b \equiv a^n \pmod m$
**begin**
    (*initialization*)
    **if** $(u_0 = 1)$ **then** $b \leftarrow a$ **else** $b \leftarrow 1$
    **if** $k = 0$ **then stop**
    $c \leftarrow a$
    **for** $i = 1$ **to** $k$ **do**
    **begin**
        $c \leftarrow c^2 \pmod m$
        **if** $u_i = 1$ **then** $b \leftarrow (bc) \pmod m$
    **end** (*end of i for-loop*)
**end** (*end of exponentiation algorithm*)

---

The above algorithm is illustrated via an example.

**Example 1.15.** We compute $3^{58} \pmod{197}$. The binary expansion of $58 = (111010)_2$.
*Initialization.* $b \leftarrow 1, c \leftarrow 3$.
*Iteration* 1: $i = 1$
$3^2 \pmod{197} \equiv 9 \pmod{197}$. Therefore $c \leftarrow 9, u_1 = 1, b \leftarrow 9$.
*Iteration* 2: $i = 2$
$9^2 \pmod{197} \equiv 81 \pmod{197}$. Therefore $c \leftarrow 81, u_2 = 0, b$ remains unchanged.
*Iteration* 3: $i = 3$
$81^2 \pmod{197} \equiv 60 \pmod{197}$. Therefore $c \leftarrow 60, u_3 = 1, b \leftarrow (9 \times 60) \pmod{197}$. That is, $b \leftarrow 146$.
*Iteration* 4: $i = 4$
$60^2 \pmod{197} \equiv 54 \pmod{197}$. Therefore $c \leftarrow 54, u_4 = 1, b \leftarrow (146 \times 54) \pmod{197}$. That is, $b \leftarrow 4$.
*Iteration* 5: $i = 5$
$54^2 \pmod{197} \equiv 158 \pmod{197}$. Therefore $c \leftarrow 158, u_5 = 1, b \leftarrow (4 \times 158) \pmod{197}$. That is, $b \leftarrow 41$.
*End of iterations.*
*Output.* $b \equiv 3^{58} \pmod{197} \equiv 41 \pmod{197}$.                              □

### 1.5.5  Quadratic Residues

Consider the solution of the equation $x^2 \equiv \theta \pmod{2}$. If $\theta = 0$, then $x \equiv 0 \pmod{2}$. However, if $\theta = 1$, then $x \equiv 1 \pmod{2}$. Next consider the solution of the equation $x^2 \equiv \theta \pmod{p}$, where $p$ is an odd prime number. In this case, this congruence relationship has solution for some values of $\theta$, and no solution for other values of $\theta$. Let us discuss an example, where $p = 13$.

$$x: \ 0 \ 1 \ 2 \ 3 \ 4 \ \ 5 \ \ 6 \ \ 7 \ \ 8 \ 9 \ 10 \ 11 \ 12$$
$$x^2 \ (\mathrm{mod}\,13): \ 0 \ 1 \ 4 \ 9 \ 3 \ 12 \ 10 \ 10 \ 12 \ 3 \ \ 9 \ \ 4 \ \ 1$$

Note that the numbers, $0, 1, 3, 4, 9, 10$, and $12$ are squares, and the numbers $2, 5, 6, 7, 8$, and $11$ are not squares modulo 13. The number 0 has only one square root, but the other numbers which are squares each have two square roots. For example, the root of number 9 modulo 13, is both 3 and 10.

**Definition 1.19.** *The set* $\mathbb{Z}_n^* = \{a \mid a \in \mathbb{Z}_n \setminus \{0\}, \gcd(a, n) = 1\}$, *where* $n \in \mathbb{P} \setminus \{1\}$.              □

Note that $\mathbb{Z}_n^*$ is the set of nonzero integers which belong to $\mathbb{Z}_n$ and are relatively prime to $n$. Consequently, the integers which belong to the set $\mathbb{Z}_n^*$, have their multiplicative inverse in the set $\mathbb{Z}_n^*$. Also $|\mathbb{Z}_n^*| = \varphi(n)$.

**Example 1.16.** Consider the set

$$\mathbb{Z}_{10} = \{0, 1, 2, 3, 4, 5, 6, 7, 8, 9\}$$

Then $\mathbb{Z}_{10}^* = \{1, 3, 7, 9\}$. Note that $\varphi(10) = \varphi(5)\varphi(2) = 4 = |\mathbb{Z}_{10}^*|$. It can also be verified that the multiplicative inverses of the integers $1, 3, 7$, and $9$ are $1, 7, 3$, and $9$ respectively in modulo 10 arithmetic.              □

**Definitions 1.20.** *We define quadratic residue and nonresidue modulo a positive integer.*

1. *Let $n$ be a positive integer. An integer $\theta \in \mathbb{Z}_n^*$, is said to be a square modulo $n$, or quadratic residue modulo $n$, if there exists an $x \in \mathbb{Z}_n^*$, such that $x^2 \equiv \theta \pmod{n}$. If no such $x$ exists, then $\theta$ is called a quadratic nonresidue modulo $n$.*
   *Denote the set of all quadratic residues modulo $n$ by $Q_n$, and the set of all quadratic non-residues modulo $n$ by $\overline{Q}_n$.*
2. *If $x \in \mathbb{Z}_n^*$ and $\theta \in Q_n$, such that $x^2 \equiv \theta \pmod{n}$, then $x$ is called the square root of $\theta$ modulo $n$.*              □

In the solution of the congruence $x^2 \equiv \theta \pmod{13}$,

$$\mathbb{Z}_{13}^* = \{1, 2, 3, 4, 5, 6, 7, 8, 9, 10, 11, 12\}$$
$$Q_{13} = \{1, 3, 4, 9, 10, 12\}$$
$$\overline{Q}_{13} = \{2, 5, 6, 7, 8, 11\}$$

Note that $0 \notin Q_{13}$ and $0 \notin \overline{Q}_{13}$.

**Observations 1.7.** Following observations are immediate from the above definitions.

1. Let $p$ be a prime number, then $x^2 \equiv 1 \pmod{p}$ iff $x \equiv \pm 1 \pmod{p}$.
   Note that $x^2 \equiv 1 \pmod{p}$ implies $(x^2 - 1) \equiv 0 \pmod{p}$. This in turn implies

   $$(x+1)(x-1) \equiv 0 \pmod{p}$$

   This is $x \equiv 1 \pmod{p}$ or $x \equiv -1 \pmod{p} \equiv (p-1) \pmod{p}$.

2. Observation about the number of square roots:
   (a) Let $p$ be an odd prime. If $\theta \in Q_p$, then $\theta$ has exactly two square roots modulo $p$.
   (b) Let $n = p_1^{\alpha_1} p_2^{\alpha_2} \ldots p_m^{\alpha_m}$, where $p_i$'s are distinct odd prime numbers and $\alpha_i \in \mathbb{P}$ for $1 \leq i \leq m$. If $\theta \in Q_n$, then $\theta$ has $2^m$ distinct square roots modulo $n$.

3. Let $p$ be an odd prime, and $\beta$ be a generator of the group $(\mathbb{Z}_p^*, \times)$. It can be inferred that $\theta \in \mathbb{Z}_p^*$ is a quadratic residue modulo $p$ iff $\theta \equiv \beta^k \pmod{p}$, where $k$ is an even integer. This implies that $|Q_p| = |\overline{Q}_p| = (p-1)/2$. Also the numbers, which are evaluated modulo $p$:

   $$1^2, 2^2, \ldots, \left\{ (p-1)^2/2 \right\}$$

   form a complete set of quadratic residues.                               □

Note that $|Q_{13}| = |\overline{Q}_{13}| = 6$. The Legendre symbol is next defined. This symbol is named after the mathematician Adrien-Marie Legendre (1752-1833). It is generally used in noting that an integer $\theta \in \mathbb{Z}$ is a quadratic residue modulo an odd prime $p$.

**Definition 1.21.** *Let $p$ be an odd prime number, and $\theta \in \mathbb{Z}$. The Legendre symbol $\left( \frac{\theta}{p} \right)$ is defined as*

$$\left( \frac{\theta}{p} \right) = \begin{cases} 0, & \textit{if } p \mid \theta \\ 1, & \textit{if } \theta \in Q_p \\ -1, & \textit{if } \theta \in \overline{Q}_p \end{cases} \tag{1.26}$$

□

Thus $\left( \frac{\theta}{p} \right)$ is 1 if $\theta$ is a quadratic residue of $p$ and $-1$ if $\theta$ is a nonquadratic residue of $p$. The next two theorems are about Euler's criteria of quadratic residue.

**Theorem 1.11.** *Let $p$ be an odd prime. Then $\theta \in \mathbb{Z}$ is a quadratic residue modulo $p$ iff $\theta^{(p-1)/2} \equiv 1 \pmod{p}$.*

*Proof.* See the problem section.                                          □

This theorem provides a technique for exponentiation modulo $p$ via square and multiplication operations. Euler's criterion in terms of Legendre's symbol is given in the next theorem.

**Theorem 1.12.** *Let $p$ be an odd prime number. Then for any $\theta \in \mathbb{Z}$*

$$\left( \frac{\theta}{p} \right) \equiv \theta^{(p-1)/2} \pmod{p} \tag{1.27}$$

*Proof.* See the problem section.                                          □

This theorem gives us an efficient algorithm to compute Legendre's symbol. Some relevant properties of Legendre's symbol are listed below.

**Observations 1.8.** Let $p$ be an odd prime number, and $\alpha, \beta \in \mathbb{Z}$. The integers $\alpha$ and $\beta$ are not divisible by $p$.

1. The number of solutions of $x^2 \equiv \alpha \pmod{p}$ is equal to $\left(1 + \left(\frac{\alpha}{p}\right)\right)$.
2. $\left(\frac{\alpha\beta}{p}\right) = \left(\frac{\alpha}{p}\right)\left(\frac{\beta}{p}\right)$. Therefore, if $\alpha \in \mathbb{Z}_p^*$, then $\left(\frac{\alpha^2}{p}\right) = 1$.
3. If $\alpha \equiv \beta \pmod{p}$, then $\left(\frac{\alpha}{p}\right) = \left(\frac{\beta}{p}\right)$.                        $\square$

### 1.5.6 Jacobi Symbol

A Jacobi symbol is a generalization of the Legendre symbol to positive-odd integers which are not necessarily prime. These symbols are named after Carl Gustav Jacob Jacobi (1804-1851).

**Definition 1.22.** *Let $n$ be an odd positive integer with prime-power factorization*

$$n = p_1^{\alpha_1} p_2^{\alpha_2} \ldots p_m^{\alpha_m} \tag{1.28a}$$

*where $p_i$ is a prime number and $\alpha_i \in \mathbb{P}$ for $1 \le i \le m$; and $\theta \in \mathbb{Z}$. The Jacobi symbol $\left(\frac{\theta}{n}\right)$ for $n \ge 3$ is*

$$\left(\frac{\theta}{n}\right) = \left(\frac{\theta}{p_1}\right)^{\alpha_1} \left(\frac{\theta}{p_2}\right)^{\alpha_2} \cdots \left(\frac{\theta}{p_m}\right)^{\alpha_m} \tag{1.28b}$$

*Also*

$$\left(\frac{\theta}{1}\right) = 1, \quad \text{for each} \quad \theta \in \mathbb{Z} \tag{1.28c}$$

$\square$

Observe that, if $n$ is an odd prime number, then the Jacobi symbol is simply the Legendre symbol. Some useful properties of the Jacobi symbol are listed below.

**Observations 1.9.** Let $\theta, \alpha, \beta \in \mathbb{Z}$; and $m \ge 3, n \ge 3$ be odd integers.

1. $\left(\frac{\theta}{n}\right) \in \{-1, 0, 1\}$. Also for $\theta \ge 1$; $\left(\frac{\theta}{n}\right) - 0$ if and only if $\gcd(\theta, n) \neq 1$.
2. $\left(\frac{1}{n}\right) = 1$.
3. $\left(\frac{\theta}{mn}\right) = \left(\frac{\theta}{m}\right)\left(\frac{\theta}{n}\right)$.
4. $\left(\frac{\alpha\beta}{n}\right) = \left(\frac{\alpha}{n}\right)\left(\frac{\beta}{n}\right)$. Therefore, if $\alpha \in \mathbb{Z}_n^*$, then $\left(\frac{\alpha^2}{n}\right) = 1$.
5. If $\alpha \equiv \beta \pmod{n}$, then $\left(\frac{\alpha}{n}\right) = \left(\frac{\beta}{n}\right)$.                        $\square$

**Definition 1.23.** *Pseudosquares. Let $n \ge 3$ be an odd integer. An integer $\theta \in \mathbb{Z}_n^*$, is a square modulo $n$, or quadratic residue modulo $n$, if there exists an $x \in \mathbb{Z}_n^*$, such that $x^2 \equiv \theta \pmod{n}$. The set of all quadratic residues modulo $n$ is denoted by $Q_n$. Also let*

$$J_n = \left\{ \theta \in \mathbb{Z}_n^* \mid \left(\frac{\theta}{n}\right) = 1 \right\} \tag{1.29}$$

*The set of pseudosquares modulo $n$ is $(J_n - Q_n) \triangleq \widetilde{Q}_n$.*                        $\square$

Observe that, if $n = pq$, where $p$ and $q$ are two distinct odd prime numbers, then

$$|Q_n| = \left|\widetilde{Q}_n\right| = \frac{(p-1)(q-1)}{4}$$

**Example 1.17.** Let $n = 15$. Then

$$\mathbb{Z}_{15}^* = \{1, 2, 4, 7, 8, 11, 13, 14\}$$

$$Q_{15} = \{1, 4\}, \quad J_{15} = \{1, 2, 4, 8\}, \quad \text{and} \quad \widetilde{Q}_{15} = \{2, 8\}$$

$\square$

## 1.6 Cyclotomic Polynomials

Elements of cyclotomic polynomials are summarized in this section. It is shown in the chapter on abstract algebra that this theory of cyclotomic polynomials has a natural extension to finite fields.

**Definition 1.24.** *Let $n \in \mathbb{P}, i = \sqrt{-1}$, and $\omega = e^{2\pi i/n}$. The zeros of the polynomial equation $(z^n - 1) = 0$ are called the $n$th roots of unity. These are $\omega^k, 0 \le k \le (n-1)$. Note that*

$$\omega^k = \cos\left(\frac{2\pi k}{n}\right) + i\sin\left(\frac{2\pi k}{n}\right), \quad 0 \le k \le (n-1) \tag{1.30}$$

$\square$

Primitive $n$th root of unity is next defined.

**Definition 1.25.** *Let $n \in \mathbb{P}, \omega = e^{2\pi i/n}$, and $\omega^k$ is an $n$th root of unity where $0 \le k \le (n-1)$. Consider a root $\omega^k = \zeta$ such that $\zeta^0, \zeta^1, \ldots, \zeta^{n-1}$ are all distinct $n$ roots of unity. Then the root $\zeta$ is said to be a primitive $n$th root of unity.* $\square$

**Observations 1.10.** Let $\zeta$ be a primitive $n$th root of unity.

1. $\omega$ is a primitive $n$th root of unity.
2. Note that $\omega^k$ is a primitive $n$th root of unity iff $\gcd(k, n) = 1$. That is the numbers $k$ and $n$ are relatively prime to each other.
3. There are $\varphi(n)$ primitive $n$th roots of unity.
4. If $\omega^k$ is a primitive $n$th root of unity, then $\omega^{-k}$ is also a primitive $n$th root of unity.
5. If the primitive $n$th roots are

$$\left\{\zeta_1, \zeta_2, \ldots, \zeta_{\varphi(n)}\right\}$$

then

$$\left\{\zeta_1, \zeta_2, \ldots, \zeta_{\varphi(n)}\right\} = \left\{\zeta_1^{-1}, \zeta_2^{-1}, \ldots, \zeta_{\varphi(n)}^{-1}\right\} \tag{1.31}$$

$\square$

**Definition 1.26.** *For $n \in \mathbb{P}$, let $\zeta_1, \zeta_2, \ldots, \zeta_{\varphi(n)}$ be the primitive nth roots of unity. The polynomial $\Phi_n(x)$ defined over the complex field $\mathbb{C}$*

$$\Phi_n(x) = \prod_{k=1}^{\varphi(n)} (x - \zeta_k) \tag{1.32}$$

*is called the cyclotomic polynomial of index n. Note that the degree of this polynomial is $\varphi(n)$. The subscript n indicates that its zeros are the nth roots of unity.* $\square$

Despite this definition, the next two theorems demonstrate that the coefficients of the cyclotomic polynomial $\Phi_n(x)$ are indeed integers.

**Theorem 1.13.** *If $n \in \mathbb{P}$, then*

$$x^n - 1 = \prod_{d|n} \Phi_d(x) \tag{1.33}$$

*Proof.* The result follows by noting that

$$x^n - 1 = \prod_{k=0}^{n-1} (x - \omega^k)$$

$$\Phi_d(x) = \prod_{\substack{1 \le k \le (n-1), \\ \gcd(k,n) = n/d}} (x - \omega^k)$$

$\square$

**Theorem 1.14.** *For all $n \in \mathbb{P}$*

$$\Phi_n(x) = \prod_{d|n} (x^d - 1)^{\mu(n/d)} \tag{1.34}$$

*Proof.* This result can be established by applying the multiplicative Moebius inversion formula to the relationship $(x^n - 1) = \prod_{d|n} \Phi_d(x)$. $\square$

**Examples 1.18.** Some elementary results are listed below.

1. $\Phi_1(x) = (x - 1)$
2. If $p$ is prime number, then

$$\Phi_p(x) = \frac{(x^p - 1)}{(x - 1)} = x^{p-1} + x^{p-2} + \ldots + x + 1 \tag{1.35}$$

Specifically,

$$\Phi_2(x) = (x + 1)$$
$$\Phi_3(x) = (x^2 + x + 1)$$
$$\Phi_5(x) = (x^4 + x^3 + x^2 + x + 1)$$
$$\Phi_7(x) = (x^6 + x^5 + x^4 + x^3 + x^2 + x + 1)$$

3. Some other examples are

$$\Phi_4(x) = (x^2 + 1)$$
$$\Phi_6(x) = (x^2 - x + 1)$$
$$\Phi_8(x) = (x^4 + 1)$$
$$\Phi_9(x) = (x^6 + x^3 + 1)$$
$$\Phi_{10}(x) = (x^4 - x^3 + x^2 - x + 1)$$

4. The divisors of 3 are 1 and 3. Therefore

$$(x^3 - 1) = \Phi_1(x)\,\Phi_3(x)$$

5. The divisors of 6 are $1, 2, 3$, and 6. Therefore

$$(x^6 - 1) = \Phi_1(x)\,\Phi_2(x)\,\Phi_3(x)\,\Phi_6(x)$$

6. Let $n = 6$, and $\omega = e^{2\pi i/6}$. Since $d = 6$ is a divisor or 6, and $\varphi(6) = 2$

$$\Phi_6(x) = (x - \omega)\,(x - \omega^5) = (x^2 - x + 1)$$

$\square$

---

## 1.7 Some Combinatorics

The principle of inclusion and exclusion, and Stirling numbers are outlined in this section.

### 1.7.1 Principle of Inclusion and Exclusion

The principle of inclusion and exclusion counts the number of distinct elements in the union of sets of finite size. For example, if $A$ and $B$ are sets of finite size, then $|A \cup B| = |A| + |B| - |A \cap B|$. Similarly, if $A$, $B$ and $C$ are sets of finite size, then

$$|A \cup B \cup C| = |A| + |B| + |C| - |A \cap B| - |A \cap C| - |B \cap C| + |A \cap B \cap C|$$

A generalization of the above result is the principle of inclusion and exclusion.

**Theorem 1.15.** *Let $\{A_1, A_2, \ldots, A_n\}$ be a collection of sets, in which each set is of finite size, and $n \in \mathbb{P} \setminus \{1\}$. Then*

$$
\left| \bigcup_{i=1}^{n} A_i \right| = \sum_{1 \le i \le n} |A_i| - \sum_{1 \le i_1 < i_2 \le n} |A_{i_1} \cap A_{i_2}| + \sum_{1 \le i_1 < i_2 < i_3 \le n} |A_{i_1} \cap A_{i_2} \cap A_{i_3}|
$$
$$
- \sum_{1 \le i_1 < i_2 < i_3 < i_4 \le n} |A_{i_1} \cap A_{i_2} \cap A_{i_3} \cap A_{i_4}| + \ldots
$$
$$
+ (-1)^{n+1} |A_1 \cap A_2 \cap \ldots \cap A_n| \tag{1.36}
$$

*Proof.* Select an element $\alpha \in \bigcup_{i=1}^{n} A_i$. As the element $\alpha$ is counted only once in the left-hand side of the above expression, it has to be demonstrated that it is also counted only once in the right-hand side. Without loss of generality, assume that the element $\alpha$ is in the sets $A_1, A_2, \ldots, A_k$, where $1 \leq k \leq n$. It is known from the binomial theorem that $\sum_{j=0}^{k} (-1)^j \binom{k}{j} = 0$. This implies

$$
\begin{aligned}
1 &= \binom{k}{1} - \binom{k}{2} + \cdots + (-1)^{k+1} \binom{k}{k} \\
&= |\{A_i \mid 1 \leq i \leq k\}| - |\{A_{i_1} \cap A_{i_2} \mid 1 \leq i_1 < i_2 \leq k\}| + \cdots + \\
&\quad (-1)^{k+1} |A_1 \cap A_2 \cap \ldots \cap A_k|
\end{aligned}
$$

Therefore, the element $\alpha$ is also counted only once in the right-hand side of the stated equation. Note that the expression for $|\bigcup_{i=1}^{n} A_i|$ has $(2^n - 1)$ terms. □

**Observation 1.11.** Let $T(n, k)$ be equal to the number of onto (surjective) functions from a set of $n$ elements onto a set with $k$ elements, where $n, k \in \mathbb{P}$, and $1 \leq k \leq n$. Then

$$
T(n, k) = \sum_{j=0}^{k} (-1)^j \binom{k}{j} (k - j)^n
$$

□

See the problem section for a proof of the above observation. Observe that the number of partitions of a set of size $n$ elements into $k$ nonempty ordered subsets, where $n, k \in \mathbb{P}$, and $1 \leq k \leq n$ is also equal to $T(n, k)$.

### 1.7.2  Stirling Numbers

Stirling numbers are named after James Stirling (1692-1770). He developed Stirling numbers of the first and second kind. These are used extensively in combinatorics.

**Definitions 1.27.** *Stirling numbers of the first and second kind. Let $n, k \in \mathbb{N}$, and $0 \leq k \leq n$.*

1. *The Stirling number of the first kind $s(n, k)$ is the coefficient of $x^k$ in the polynomial $\prod_{i=0}^{(n-1)} (x - i)$. That is*

$$
\sum_{k=0}^{n} s(n, k) x^k = x (x - 1) (x - 2) \ldots (x - n + 1) \tag{1.37a}
$$

2. *The Stirling number of the second kind $S(n, k)$ can be obtained from the polynomial*

$$
x^n = \sum_{k=0}^{n} S(n, k) x (x - 1) (x - 2) \ldots (x - k + 1) \tag{1.37b}
$$

□

**Observations 1.12.** Some elementary properties of Stirling numbers.

1.  Stirling numbers of the first kind.

(a) $s(n, 0) = 0, \ \forall n \in \mathbb{P}$.

(b) $s(n, n) = 1, \ \forall n \in \mathbb{N}$.

(c) $s(n + 1, k) = -ns(n, k) + s(n, k - 1), \ \forall n \in \mathbb{P}$ and $1 \leq k \leq n$.

(d) $\sum_{k=0}^{n} (-1)^{n-k} s(n, k) = n!, \ \forall n \in \mathbb{N}$.

2. Stirling numbers of the second kind.

(a) $S(n, 0) = 0, \ \forall n \in \mathbb{P}$.

(b) $S(n, n) = 1, \ \forall n \in \mathbb{N}$.

(c) $S(n, 1) = 1, \ \forall n \in \mathbb{P}$.

(d) $S(n + 1, k) = kS(n, k) + S(n, k - 1), \ \forall n \in \mathbb{P}$ and $1 \leq k \leq n$.

(e) $S(n, 2) = (2^{n-1} - 1), \ n \in \mathbb{P} \backslash \{1\}$.

(f) $S(n, n - 1) = \binom{n}{2}, \ n \in \mathbb{P} \backslash \{1\}$.

(g) An explicit representation for Stirling number of second kind is

$$S(n, k) = \frac{1}{k!} \sum_{i=0}^{k} (-1)^{k-i} \binom{k}{i} i^n, \quad n, k \in \mathbb{N}, \text{ and } 0 \leq k \leq n$$

(h) Let $T(n, k)$ be equal to the number of partitions of a set of size $n$ elements into $k$ nonempty ordered subsets, where $n, k \in \mathbb{P}$, and $1 \leq k \leq n$. Then $T(n, k) = k!S(n, k)$. That is, Stirling number of the second kind $S(n, k)$ is equal to the number of partitions of $\{1, 2, \ldots, n\}$ into $k$ nonempty parts, where the $k$ parts are unordered. Sometimes, this result is taken as the definition of Stirling number of the second kind. $\square$

Some of the above results are established in the problem section

## Reference Notes

Number theory is perhaps the most ancient branch of mathematics. Seemingly elementary results are sometimes extremely hard to prove. Before the advent of computers and the Internet, this branch of mathematics was regarded as "pure." However, since the last few decades, this branch of mathematics has provided numerous applications in diverse fields such as cryptography, coding theory, digital signal processing, and graph theory.

The notes in this chapter on number theory are based upon the classical textbooks by Dudley (1969), Spiegel (1969), Apostol (1976), Burton (1976), Hardy, and Wright (1979), Hua (1982), and Allenby, and Redfern (1989). A readable exposition of cyclotomic polynomials is given in McEliece (1987), and Rivlin (1990).

## Problems

1. Prove the following statements about prime numbers.

(a) Every number $n \in \mathbb{P} \backslash \{1\}$ is either a prime number or a product of prime numbers.

(b) Let $p$ be a prime number, and $a \in \mathbb{P}$. If $p \nmid a$ then $\gcd(p, a) = 1$.

(c) Let $p$ be a prime number, and $a, b \in \mathbb{P}$. If $p \mid ab$ then $p \mid a$ and/or $p \mid b$. If the prime number $p$ divides $a_1 a_2 \ldots a_n$ where $a_i \in \mathbb{P}$ for $1 \leq i \leq n$, then $p$ divides at least one $a_i$.

Hint: See Apostol (1976).

2. Prove that every composite number $n$ has a proper factor less than or equal to $\sqrt{n}$.

Hint: As $n$ is composite, let $n = ab$, where $1 < a, b < n$. Assume that $a, b > \sqrt{n}$. Then $n = \sqrt{n}\sqrt{n} < ab = n$. This is a contradiction. Therefore at least either $a$ or $b$ is less than or equal to $\sqrt{n}$.

3. Prove the fundamental theorem of arithmetic. It asserts that every integer $n \in \mathbb{P}\backslash\{1\}$ can be represented as a product of prime factors. This representation is unique up to the order of its factors.

Hint: See Apostol (1976). This result is established by using induction on $n$. The theorem is true for $n = 2$. In the induction hypothesis, assume that the theorem is true for all integers greater than 1 but less than $n$. Our goal is to establish the correctness of the theorem for $n$. If $n$ is a prime integer, then the theorem is true. However, if $n$ is not a prime integer, then it is a composite number. Assume that it has two representations in factored form. Let these be

$$n = p_1 p_2 \cdots p_i \cdots p_s = q_1 q_2 \cdots q_j \cdots q_t$$

It is next shown that $s = t$ and each $p_i$ is equal to some $q_j$. Observe that $p_1$ must divide the product $q_1 q_2 \cdots q_t$. Consequently, it must divide at least one factor. Relabel $q_1, q_2, \cdots, q_t$ such that $p_1 \mid q_1$. Therefore $p_1 = q_1$ as the integers $p_1$ and $q_1$ are both prime. In the next step, we write

$$n/p_1 = p_2 \cdots p_i \cdots p_s = q_2 \cdots q_j \cdots q_t$$

If $s > 1$ or $t > 1$, then $1 < n/p_1 < n$. Invocation of the induction hypothesis implies that the two factorizations of $n/p_1$ must be identical, except for the order of the factors. Thus $s = t$, and the factorizations $n = p_1 p_2 \cdots p_i \cdots p_s = q_1 q_2 \cdots q_j \cdots q_t$ are identical, except for the order.

4. Prove that there are infinitely many prime numbers.

Hint: See Apostol (1976), and Baldoni, Ciliberto, and Cattaneo (2009). Assume that there are only a finite number of prime numbers $p_1 < p_2 < \cdots < p_n$. Let $N = p_1 p_2 \cdots p_n + 1$. Observe that $N$ is either a prime or a product of prime numbers. The number $N$ is not a prime number as it exceeds each $p_i$, where $1 \leq i \leq n$. However, if $p_i$ divides $N$, then $p_i$ also divides $(N - p_1 p_2 \cdots p_n) = 1$. This is not possible since $p_i > 1$.

5. Prove the prime number theorem.

Hint: This is one of the most well-known results in number theory. There are several proofs of this result. For example, see Apostol (1976), and Newman (1998).

6. Write the first six convergents of the infinite continued fraction

$$[1, 1, 1, 1, \ldots]$$

What do you notice about the denominators of these convergents?

Hint: See Allenby, and Redfern (1989). The convergents are

$$1, \frac{2}{1}, \frac{3}{2}, \frac{5}{3}, \frac{8}{5}, \frac{13}{8}$$

A Fibonacci sequence is recursively defined as: $F_0 = 0, F_1 = 1$, and $F_n = F_{n-1} + F_{n-2}, n \geq 2$. The first few terms in this sequence are: $1, 1, 2, 3, 5, 8, \ldots$. These are indeed the denominators in the above sequence of convergents. Similarly, the numerators of the convergents are: $F_1, F_2, F_3, \ldots$.

7. Show that if $a, m \in \mathbb{P}$, and $\gcd(a, m) = 1$ then $ax \equiv 1 \pmod{m}$ has a solution.

   Hint: As $\gcd(a, m) = 1$, there exist integers $\alpha, \beta \in \mathbb{Z}$, such that

$$\alpha a + \beta m = 1$$

   The inverse of $a$ is obtained by performing the modulo $m$ operation on both sides of the above equation. The required result follows.

8. Show that if $a, m \in \mathbb{P}$, and $\gcd(a, m) = 1$ then $ax \equiv b \pmod{m}$ has a unique solution.

   Hint: Let $x_1$ and $x_2$ be two different solutions of the congruence $ax \equiv b \pmod{m}$ such that $x_1 > x_2$. Then $ax_1 \equiv b \pmod{m}$ and $ax_2 \equiv b \pmod{m}$. This implies $a(x_1 - x_2) \equiv 0 \pmod{m}$, that is $a(x_1 - x_2) = mz, z \in \mathbb{P}$. Since $\gcd(a, m) = 1$, $(x_1 - x_2)$ is divisible by $m$. That is, $x_1 \pmod{m} \equiv x_2 \pmod{m}$ which is a contradiction.

9. Let $a, b \in \mathbb{P}$, $\gcd(a, b) = 1$, and $n \in \mathbb{Z}$. Show that if $a$ and $b$ each divide $n$, then $ab$ divides $n$.

   Hint: $\gcd(a, b) = 1$ implies $au + bv = 1$, where $u, v \in \mathbb{Z}$. Let $n = ra = sb$, where $r, s \in \mathbb{P}$. Then

$$n = nua + nvb$$
$$= suab + rvab$$
$$= (su + rv)ab$$

   This implies that $ab$ divides $n$.

10. Let $a, b, c \in \mathbb{P}$. Prove that $\gcd(a, bc) = 1$ if and only if $\gcd(a, b) = 1$ and $\gcd(a, c) = 1$.

    Hint: Assume that $\gcd(a, bc) = 1$, and let $d = \gcd(a, b)$. Then $d$ divides both $a$ and $b$, consequently $d$ divides both $a$ and $bc$. This in turn implies that $\gcd(a, bc) \geq d$. Therefore $d = 1$. It can be similarly established that $\gcd(a, c) = 1$.

    For the proof in the other direction, assume that $\gcd(a, b) = \gcd(a, c) = 1$, and $\gcd(a, bc) = \delta > 1$. Let $p$ be a prime divisor of $\delta$. Then $p$ divides $bc$, which in turn implies that $p$ divides either $b$ or $c$. If $p$ divides $b$, and it is also known that $p$ divides $a$. This in turn implies that $\gcd(a, b) \geq p$, which is a contradiction. Similar conclusion is reached, if it is assumed that $p$ divides $c$. Therefore $\delta = 1$.

11. Consider the set of $m$ integers $\{0, 1, 2, \ldots, (m-1)\}$. Multiply each element $n_j$ of this set by an integer $a \in \mathbb{P}$, where $\gcd(a, m) = 1$. Then the set of elements $b_j \equiv an_j \pmod{m}$, $0 \leq j \leq (m-1)$ spans the same values as $n_j$, although in some different permuted order. Prove this statement.

    Hint: Without loss of generality, assume that $n_0 = 0$. This implies that $b_0 = 0$. Then we need to prove that the set of elements $b_j \equiv an_j \pmod{m}$, $1 \leq j \leq (m-1)$ spans the set of integers $\{1, 2, \ldots, (m-1)\}$ in some permuted order. It is required to show that no two $b_j$'s are identical integers $\pmod{m}$.

    Assume that $b_j$ and $b_k$ are identical integers $\pmod{m}$ for $j \neq k$. Therefore $b_j \equiv b_k \pmod{m}$ implies $an_j \equiv an_k \pmod{m}$. That is, $a(n_j - n_k) \equiv 0 \pmod{m}$. As $\gcd(a, m) = 1$, we have $(n_j - n_k) \equiv 0 \pmod{m}$, which is a contradiction.

12. Let $\mu(\cdot)$ be the Moebius function. Establish the following results:

    (a) Let $m, n \in \mathbb{P}$ and $\gcd(m, n) = 1$, then $\mu(mn) = \mu(m)\mu(n)$.

(b) Let $n \in \mathbb{P}$, then

$$\sum_{d|n} \mu(d) = \begin{cases} 1, & n = 1 \\ 0, & n > 1 \end{cases}$$

where the summation is performed over all the positive divisors of $n$.

13. Let $\gcd(a, m) = 1$ and $b \equiv c \pmod{\varphi(m)}$, where $b, c \in \mathbb{P}$, then prove $a^b \equiv a^c \pmod{m}$.
    Hint: The result is obtained by using the Euler-Fermat theorem, and noting that $b = k\varphi(m) + c$, where $k \in \mathbb{P}$.

14. Let $p$ be an odd prime. Prove that $\theta$ is a quadratic residue modulo $p$ iff $\theta^{(p-1)/2} \equiv 1 \pmod{p}$.
    Hint: See Stinson (2006). The notion of primitive element is used in this problem. It is explained in the chapter on abstract algebra. Assume that $\theta \equiv x^2 \pmod{p}$. Also $\theta^{p-1} \equiv 1 \pmod{p}$ for any $\theta \not\equiv 0 \pmod{p}$. Consequently

    $$\theta^{(p-1)/2} \equiv \left(x^2\right)^{(p-1)/2} \pmod{p} \equiv x^{(p-1)} \pmod{p} \equiv 1 \pmod{p}$$

    Conversely, assume that $\theta^{(p-1)/2} \equiv 1 \pmod{p}$. Let $\beta \in \mathbb{Z}_p^*$ be a primitive element modulo $p$. Next let $\theta \equiv \beta^k \pmod{p}$ for some integer $k \in \mathbb{P}$. Then

    $$1 \pmod{p} \equiv \theta^{(p-1)/2} \pmod{p} \equiv \left(\beta^k\right)^{(p-1)/2} \pmod{p}$$
    $$\equiv \beta^{k(p-1)/2} \pmod{p}$$

    As the order of $\beta$ is $(p-1)$, then for the above relationship to be true, $k$ has to be an even integer, say $2j$. This yields $\theta \equiv \beta^{2j} \pmod{p}$. Therefore the square roots of integer $\theta$ are $\pm\beta^j \pmod{p}$.

15. Let $p$ be an odd prime number. Prove that for any $\theta \in \mathbb{Z}$

    $$\left(\frac{\theta}{p}\right) \equiv \theta^{(p-1)/2} \pmod{p}$$

    Hint: See Stinson (2006). If $\theta$ is a multiple of $p$, then $\theta^{(p-1)/2} \equiv 0 \pmod{p}$. Also, from the last problem, $\theta^{(p-1)/2} \equiv 1 \pmod{p}$ iff $\theta$ is a quadratic residue modulo $p$. Next consider the case when $\theta$ is a nonquadratic residue modulo $p$. We need to check if $\theta^{(p-1)/2} \equiv -1 \pmod{p}$. This is true because

    $$\left(\theta^{(p-1)/2}\right)^2 \equiv \theta^{p-1} \equiv 1 \pmod{p}$$

    and $\theta^{(p-1)/2} \not\equiv 1 \pmod{p}$. The proof is complete.

16. Let $T(n, k)$ be equal to the number of onto (surjective) functions from a set of $n$ elements onto a set with $k$ elements, where $n, k \in \mathbb{P}$, and $1 \leq k \leq n$. Prove that $T(n, k) = \sum_{j=0}^{k} (-1)^j \binom{k}{j} (k-j)^n$.
    Hint: See Rosen (2018). The number $T(n, k)$ is identical to the total number of ways that $n$ different objects can be placed among $k$ different boxes so that no box is empty. Let $B_i$ be the subset of distributions with box $i$ empty, where $1 \leq i \leq k$. Therefore:

    (a) $|B_i| = (k-1)^n$, with $\binom{k}{1}$ possible values for $i$.

    (b) $|B_{i_1} \cap B_{i_2}| = (k-2)^n$, with $\binom{k}{2}$ possible values for $i_1$ and $i_2$.

    (c) $|B_{i_1} \cap B_{i_2} \cap \cdots \cap B_{i_k}| = (k-k)^n$, with $\binom{k}{k}$ possible values for $i_1, i_2, \ldots, i_k$.

Use of the principle of inclusion and exclusion, yields the total number of distributions with at least one box empty to be $\sum_{j=1}^{k} (-1)^{j-1} \binom{k}{j} (k-j)^n$. As the total number of functions from a set of size $n$ elements to a set of size $k$ elements is $k^n$, the total number of onto functions $T(n,k)$ is

$$T(n,k) = k^n - \sum_{j=1}^{k} (-1)^{j-1} \binom{k}{j} (k-j)^n$$

$$= \sum_{j=0}^{k} (-1)^j \binom{k}{j} (k-j)^n$$

Stirling number of the second kind $S(n,k)$ is related to $T(n,k)$ by $T(n,k) = k! S(n,k)$.

17. Prove that Stirling number of the second kind $S(n,k)$ for $n, k \in \mathbb{N}$, and $0 \leq k \leq n$ is

$$S(n,k) = \frac{1}{k!} \sum_{i=0}^{k} (-1)^{k-i} \binom{k}{i} i^n$$

Hint: See Lovász (1979). The above expression yields $S(n,0) = 0, \forall \, n \in \mathbb{P}$. Assume that $1 \leq k \leq n$.

$$\frac{1}{k!} \sum_{i=0}^{k} (-1)^{k-i} \binom{k}{i} i^n$$

$$= \frac{1}{k!} \sum_{i=0}^{k} (-1)^{k-i} \binom{k}{i} \sum_{r=0}^{n} S(n,r) i (i-1) (i-2) \ldots (i-r+1)$$

$$= \sum_{r=0}^{n} S(n,r) \sum_{i=0}^{k} (-1)^{k-i} \frac{i (i-1) (i-2) \ldots (i-r+1)}{(k-i)! i!}$$

$$= \sum_{r=0}^{n} S(n,r) \sum_{i=r}^{k} (-1)^{k-i} \frac{i (i-1) (i-2) \ldots (i-r+1)}{(k-i)! i!}$$

$$= \sum_{r=0}^{n} S(n,r) \sum_{i=r}^{k} (-1)^{k-i} \frac{1}{(k-i)! (i-r)!}$$

$$= \sum_{r=0}^{n} \frac{S(n,r)}{(k-r)!} \sum_{j=0}^{k-r} (-1)^{k-r-j} \binom{k-r}{j}$$

$$= \sum_{r=0}^{n} \frac{S(n,r)}{(k-r)!} (1-1)^{k-r} = S(n,k)$$

18. Let Stirling number of the second kind be $S(n,k)$, where $n \in \mathbb{P}$ and $1 \leq k \leq n$. Prove that $S(n+1,k) = kS(n,k) + S(n,k-1)$.

Hint: See Cameron (1994). Consider the partition of $\{1, 2, \ldots, n+1\}$ into $k$ parts the following way. The number $(n+1)$ is a singleton part, and $\{1, 2, \ldots, n\}$ is partitioned into $(k-1)$ parts; or $(n+1)$ is appended to one of the $k$ partitions of the set $\{1, 2, \ldots, n\}$.

# References

1.  Allenby, R. B. J. T., and Redfern, E. J., 1989. *Introduction to Number Theory with Computing*, Edward Arnold, London, United Kingdom.
2.  Apostol, T. M., 1976. *Introduction to Analytic Number Theory,* Springer-Verlag, Berlin, Germany.
3.  Baldoni, M. W., Ciliberto, C., and Cattaneo, G. M. P., 2009. *Elementary Number Theory Cryptography and Codes*, Springer-Verlag, Berlin, Germany.
4.  Burton, D. M., 1976. *Elementary Number Theory*, Allyn and Bacon, Inc. Boston, Massachusetts.
5.  Cameron, P. J., 1994. *Combinatorics*: *Topics, Techniques, Algorithms,* Cambridge University Press, Cambridge, Great Britain.
6.  Clawson, C. C., 1996. *Mathematical Mysteries. The Beauty and Magic of Numbers.* Plenum Press, New York, New York.
7.  De Koninck, J. M., Mecier, A., 2007. 1001 *Problems in Classical Number Theory*, American Mathematical Society, Providence, Rhode Island.
8.  Dudley, U., 1969. *Elementary Number Theory*, W. H. Freeman and Company, San Francisco, California.
9.  Erickson, M., and Vazzana, A., 2008. *Introduction to Number Theory*, Chapman and Hall/CRC Press, New York, New York.
10. Graham, R. L., Knuth, D. E., and Patashnik, O., 1994. *Concrete Mathematics*: *A Foundation for Computer Science*, Second Edition, Addison-Wesley Publishing Company, New York, New York.
11. Hardy, G. H., and Wright, E. M. 1979. *An Introduction to the Theory of Numbers,* Fifth Edition, Oxford University Press, Oxford, Great Britain.
12. Hasse, H., 1978. *Number Theory*, Springer-Verlag, Berlin, Germany.
13. Hua, L. K., 1982. *Introduction to Number Theory*, Springer-Verlag, Berlin, Germany.
14. Kanigel, R., 1991. *The Man Who Knew Infinity, A Life of the Genius Ramanujan*, Charles Scribner's Sons, Maxwell Macmillan International, New York, New York.
15. Landau, E., 1927. *Elementary Number Theory,* Chelsea Publishing Company, New York, New York.
16. LeVeque, W. J., 1977. *Fundamentals of Number Theory*, Addison-Wesley Publishing Company, New York, New York.
17. Lipschutz, S., 1998. *Set Theory and Related Topics*, Schaum's Outline Series, McGraw-Hill Book Company, New York, New York.
18. Lovász, L., 1979. *Combinatorial Problems and Exercises*, Second Edition, North-Holland, New York, New York.
19. McEliece, R. J., 1987. *Finite Fields for Computer Scientists and Engineers*, Kluwer Academic Publishers, Norwell, Massachusetts.
20. Miller, S. J., and Takloo-Bighash, R., 2006. *An Invitation to Modern Number Theory*, Princeton University Press, Princeton, New Jersey.
21. Mollin, R. A., 1998. *Fundamental Number Theory with Applications,* Chapman and Hall/CRC Press: New York, New York.
22. Mollin, R. A., 1999. *Algebraic Number Theory*, Chapman and Hall/CRC Press: New York, New York.
23. Newman, D. J., 1998. *Analytic Number Theory,* Springer-Verlag, Berlin, Germany.
24. Niven, I., and Zuckerman, H. S., 1972. *An Introduction to the Theory of Numbers*, Third Edition, John Wiley & Sons, Inc., New York, New York.

25. Rivlin, T. J., 1990. *Chebyshev Polynomials, From Approximation Theory to Algebra and Number Theory,* Second Edition, John Wiley & Sons, Inc., New York, New York.

26. Rosen, K. H., Editor-in-Chief, 2018. *Handbook of Discrete and Combinatorial Mathematics,* Second Edition, CRC Press: Boca Raton, Florida.

27. Spiegel, M. R., 1969. *Real Variables*, Schaum's Outline Series, McGraw-Hill Book Company, New York, New York.

28. Stinson, D. R., 2006. *Cryptography, Theory and Practice,* Third Edition, CRC Press: New York, New York.

# Abstract Algebra

$$\int_u^v \frac{dx}{\sqrt{x^3 + ax^2 + bx + c}}$$

Elliptic integral

**Niels Henrik Abel.** Niels H. Abel was born on 5 August, 1802 in Frindoe (near Stavanger), Norway. His brief life was spent in abject poverty. Abel was a self-taught mathematician. At the young age of nineteen, he proved that there is no closed form solution for a general fifth degree polynomial equation in terms of radicals. His other contributions were in the area of infinite series and elliptic integrals. The phrase, "Abelian group" was coined in his honor. When Abel was asked how he developed his mathematical skills so quickly, he answered: *By studying the masters, not their pupils.*
He died of tuberculosis on 6 April, 1829 in Froland, Norway.

## 2.1 Introduction

Abstract algebra is the study of *structures* in mathematics. The following algebraic structures are studied in this chapter: groups, rings, subrings and ideals, fields, polynomial rings, and Boolean algebra. Vector spaces over fields, linear mappings, structure of finite fields, roots of unity in finite field, elliptic curves, and hyperelliptic curves are also studied. Some number-theoretic concepts are also generalized and extended in this chapter.

The goal of this chapter is to provide a concrete foundation for applications of abstract algebra. These concepts can then be applied to the study of cryptography, coding theory, and quantum computation and communication.

Several mathematical systems such as integers, rational numbers, real numbers, and complex numbers have a deep algebraic structure associated with them. In abstract algebra, familiar operations like addition and multiplication are replaced by general operations. These general operations are performed upon elements of abstract sets.

Many mathematicians see mathematics as a hierarchy of structures. Furthermore, abstract algebra synthesizes concepts from apparently different branches of mathematics like number theory, topology, analysis, geometry, and applied mathematics. Therefore, algebra appears to be a unifying force among almost all branches of mathematics. The use of the term "abstract" is of course subjective. Algebra that is applied is no longer abstract. Joseph Louis Lagrange (1736-1813), Carl Friedrich Gauss (1777-1855), Augustin-Louis Cauchy (1789-1857), Niels Henrik Abel (1802-1829), Evariste Galois (1811-1832), and several other mathematicians did early fundamental work in the field of algebra.

The symbol $p$ is generally used to denote a prime number in this chapter.

## 2.2 Algebraic Structures

Definitions and basic properties of algebraic structures such as groups, rings, subrings and ideals, fields, polynomial rings, and Boolean algebra are stated in this section.

### 2.2.1 Groups

Groups are the building blocks of more complex algebraic structures. We initially define a group in this subsection. This is followed by a list of some important observations, and examples.

**Definition 2.1.** *A group $\mathcal{G} = (G, *)$, is a nonempty set $G$ and a binary operation $*$ defined on $G$, such that the following properties are satisfied.*

(a) *Associative property: The group is associative. That is, $a*(b*c) = (a*b)*c$ for all $a, b, c \in G$.*
(b) *Identity property: The set $G$ has an element $e$, called the identity of $\mathcal{G}$, such that $e*a = a*e = a$ for all $a \in G$.*
(c) *Inverse property: For each element $a \in G$ there exists an element $a^{-1} \in G$ such that $a*a^{-1} = a^{-1}*a = e$. The element $a^{-1}$ is called the inverse of element $a$.*                     □

If the context is clear and there is no ambiguity, the symbols $\mathcal{G}$ and $G$ are sometimes used interchangeably. Thus the group $\mathcal{G} = (G, *)$, is sometimes simply referred to as the group $G$. If the binary operation $*$ is the addition operator $+$, then we have an *additive group*. In this group, the identity element is 0, and the inverse of $a \in G$ is $-a \in G$. However, if the binary operation $*$ is the multiplication operator $\times$, then we have a *multiplicative group*. In this group, the identity element is 1, and the inverse of $a \in G$ is $a^{-1} \in G$. To complete a preliminary study of groups, a few more concepts are introduced via the following definitions.

**Definitions 2.2.** *Let $\mathcal{G} = (G, *)$ be a group.*

1. *A group $\mathcal{G} = (G, *)$ is commutative or Abelian (after the mathematician Niels H. Abel), if it also satisfies the property: $a*b = b*a$ for all $a, b \in G$.*
2. *A group is finite, if $|G|$ is finite.*
3. *The number of elements in finite group is called the order of the group. Therefore the order of a finite group is equal to $|G|$.*
4. *A subgroup of a group $\mathcal{G} = (G, *)$, is a group $\mathcal{H} = (H, *)$ such that $H \subseteq G$. If $H \subset G$, then $\mathcal{H}$ is called a proper subgroup of $\mathcal{G}$.*
5. *Denote $a*a*\ldots*a \triangleq a^n$, where the binary operation $*$ is performed $(n-1)$ times.*
6. *A group is said to be cyclic, if there exists an element $a \in G$ such that for each $b \in G$ there exists an integer $n \in \mathbb{Z}$ such that $b = a^n$, where $a^0 = e$ and $a^{-n} = \left(a^{-1}\right)^n$. The element $a$ is called the generator of $\mathcal{G}$. Also the discrete logarithm of $b$ to the base $a$ is said to be $n$. This is formally denoted as $n = \log_a b$.*
7. *Let $a \in G$, then the order of element $a$ is the smallest positive integer $m$ such that $a^m = e$, provided such an integer exists. If such a positive number $m$ does not exists, then the order of $a$ is said to be infinite.*
8. *Two groups $\mathcal{G}_1 = (G_1, *)$ and $\mathcal{G}_2 = (G_2, \circ)$ are isomorphic to each other, if there exists a one-to-one and onto mapping $\theta$ of elements of $G_1$ to the elements of $G_2$. Also for all elements $a, b \in G_1$*

$$\theta(a*b) = \theta(a) \circ \theta(b) \tag{2.1}$$

*Isomorphism is usually denoted by $\mathcal{G}_1 \cong \mathcal{G}_2$.*
9. *Let $\mathcal{G} = (G, *)$ be an Abelian group, and $\mathcal{H} = (H, *)$ be its subgroup. For any $g \in G$ a set $J$ is called a coset of $H$, where*

$$J = g*H = \{g*h \mid h \in H\} \tag{2.2}$$

10. *The direct product of groups $\mathcal{G}_1 = (G_1, *_1)$ and $\mathcal{G}_2 = (G_2, *_2)$ is a group $\mathcal{G} = (G, *)$, where $G = G_1 \times G_2 = \{(\alpha_1, \alpha_2) \mid \alpha_1 \in G_1, \alpha_2 \in G_2\}$, and the binary operation $*$ is specified by $(\alpha_1, \alpha_2) * (\beta_1, \beta_2) = (\alpha_1 *_1 \beta_1, \alpha_2 *_2 \beta_2)$. The direct product is also sometimes called direct sum. It is possible to extend this definition to a direct product of $n$ groups.* □

The above definition of a coset can be generalized to a left coset and a right coset for an arbitrary group. However, this distinction is not of interest to us in this chapter. The following observations can be made about all groups.

**Observations 2.1.** Let $\mathcal{G} = (G, *)$ be a group, such that $a, b, c \in G$. The identity element of this group is $e$.

1. Every group has a single identity element.
2. Each element of the group has a single inverse.
3. The *cancellation laws* are:

    (a) If $a * b = a * c$ then $b = c$. This is the left cancellation law.
    (b) If $b * a = c * a$ then $b = c$. This is the right cancellation law.
4. $\left(a^{-1}\right)^{-1} = a$
5. $(a * b)^{-1} = \left(b^{-1}\right) * a^{-1}$
6. The groups $\mathcal{G}, (\{a\}, *)$, and $(\{e\}, *)$ are all subgroups of $\mathcal{G}$.
7. If $\mathcal{H} = (H, *)$ is a subgroup of $\mathcal{G}$, then the identity element of these two groups are same. Also if $a^{-1} \in II$ is the inverse element of $a \in II$, then $a^{-1}$ is also the inverse element of $a \in G$.
8. The set of all powers of $a$ forms a cyclic subgroup of $\mathcal{G}$. This subgroup is called a subgroup *generated* by $a$. It is denoted by $(\langle a \rangle, *)$. Let the order of the element $a$ be $m$, then $|\langle a \rangle| = m$.
9. Lagrange's theorem: Let $\mathcal{G} = (G, *)$ be a finite group, and $\mathcal{H} = (H, *)$ be its subgroup. Then $|II|$ divides $|G|$. This theorem is named after the mathematician, Joseph-Louis Lagrange (1736-1813).
10. If $\mathcal{G}$ is a finite group, and $p$ is a prime number which divides $|G|$, then $\mathcal{G}$ has a subgroup of order $p$.
11. The order of any element of the group divides the order of the group. This follows from Lagrange's theorem.
12. If the order of $a \in G$ is $m$, then the order of $a^j$ is $m / \gcd(m, j)$ where $1 \leq j \leq m$.
13. Let $\mathcal{G}$ be a cyclic group of order $n$. Denote its identity element by $\xi$.

    (a) If the group is generated by the element $a$, then

    $$G = \{a, a^2, \ldots, a^n\}$$

    such that $a^n = \xi$. Therefore $a^k$ is also a generator of the group iff $\gcd(k, n) = 1$.
    (b) The generators of the group are the elements $a^i$ where $\gcd(i, n) = 1$. Thus the group $\mathcal{G}$ has $\varphi(n)$ generators, where $\varphi(\cdot)$ is the Euler's phi-function.
    (c) Every subgroup of a cyclic group is also cyclic.
    (d) For each positive divisor $d$ of $n$:

        (i) There is exactly one subgroup of order $d$.
        (ii) The group $\mathcal{G}$ has $\varphi(d)$ elements of order $d$.

    (e) If the group is of prime order, then it is cyclic.
    (f) All cyclic groups are Abelian, but not vice-versa.
14. All subgroups of an Abelian group are also Abelian.

15. Elements $a$ and $b$ are in an Abelian group. The orders of elements $a$ and $b$ are $p$ and $q$ respectively, where $\gcd(p, q) = 1$. Then the order of $a * b$ is $pq$.

16. Two isomorphic groups have the same structure and the same number of elements.

17. Let $\mathcal{G} = (G, *)$ be an Abelian group, and $\mathcal{H} = (H, *)$ be its subgroup.

    (a) Every member of the group $\mathcal{G}$, say $g$ is in some coset (in $g * H$ for example).

    (b) The elements $a$ and $b$ are in the same coset iff $a * b^{-1} \in H$.

    (c) Two cosets of $H$ are either disjoint or coincide. See the problem section for its proof.

    (d) All cosets of the group $H$ have $|H|$ number of elements.

    (e) Consequently, if $m = |G| / |H|$ and the $g_j * H$'s are disjoint for $g_j \in G, 1 \leq j \leq m$, then

$$G = (g_1 * H) \cup (g_2 * H) \cup \ldots \cup (g_m * H) \qquad (2.3)$$

    That is, the union of all cosets of $H$ in $G$ is $G$ itself. $\qquad\qquad\qquad\qquad\qquad$ □

**Examples 2.1.** Some illustrative examples.

1. Some well-known examples of additive infinite groups:

$$(\mathbb{Z}, +), (\mathbb{Q}, +), (\mathbb{R}, +), \text{ and } (\mathbb{C}, +)$$

The identity element in these groups is 0, and the inverse of an element $a$ is $-a$.

2. Some well-known examples of multiplicative infinite groups:

$$(\mathbb{Q} - \{0\}, \times), (\mathbb{R} - \{0\}, \times), \text{ and } (\mathbb{C} - \{0\}, \times)$$

The identity element in these groups is 1, and the inverse of an element $a$ is $a^{-1}$.

3. Let $n \in \mathbb{P}$.

    (a) $\mathcal{Z}_n = (\mathbb{Z}_n, +)$ is a finite group, where the binary operation $+$ is addition modulo $n$. The order of this group is $n$.

    Note that $\mathbb{Z}_n$ with the operation of multiplication modulo $n$ is not a group. This is because all the elements of $\mathbb{Z}_n$ do not possess multiplicative inverses.

    Therefore, if the context is clear, the binary operator $+$ is generally dropped, and this group is simply denoted by $\mathbb{Z}_n$.

    (b) Let $p$ be a prime number. Then the group $(\mathbb{Z}_p, +)$ is cyclic. It is generated by each of the elements $1, 2, \ldots, (p - 1)$. Furthermore, if $a \in \mathbb{Z}_p$ and $a \neq 0$, then the order of $a$ is $p$.

    (c) The group $(\mathbb{Z}_n, +)$ is cyclic. It is generated by each element of $\mathbb{Z}_n$ that is relatively prime to $n$. If $a \in \mathbb{Z}_n \setminus \{0\}$, then the order of element $a$ is $n / \gcd(n, a)$.

    (d) Let $\mathcal{G}$ be a finite cyclic group of order $n$, then $\mathcal{G} \cong (\mathbb{Z}_n, +)$.

4. The group $(\mathbb{Z}, +)$ is an Abelian group.

5. Consider the set of all $m \times n$ matrices where the matrix elements are real numbers. The group operation is matrix addition. The zero matrix of order $m \times n$ is the identity element of the group. The inverse of an element $\widetilde{M}$ is the matrix $-\widetilde{M}$.

6. Consider the set of all $n \times n$ invertible matrices where the matrix elements are real numbers. The group operation is matrix multiplication. The identity matrix is the identity element of the group. The inverse of a matrix $\widetilde{M}$ is the matrix $\widetilde{M}^{-1}$. This group is not Abelian.

7. Let $G$ be the set

$$\{\omega^k \mid 0 \leq k \leq (n - 1)\},$$

where $\omega = e^{2\pi i/n}, n \in \mathbb{P}$, and $i = \sqrt{-1}$. Then the set $G$ and complex multiplication operation form a group. The identity element is 1, and the inverse of $\omega^k$ is $\omega^{-k}$. This is a cyclic group.

Let $n = 6$, then the set $H_1 = \{1, \omega^3\}$ and complex multiplication operation form a proper subgroup of $G$. Similarly, $H_2 = \{1, \omega^2, \omega^4\}$ and complex multiplication operation form another proper subgroup of $G$.

8. $(\mathbb{Z}_n^*, \times)$ is a group, where

$$\mathbb{Z}_n^* = \{k \mid k \in \mathbb{Z}_n \setminus \{0\}, \ \gcd(k, n) = 1\}, \ n > 1 \tag{2.4a}$$

The binary operation $\times$ is multiplication modulo $n$. The order of this group is $\varphi(n)$, and the identity element is 1. Also if

$$\mathbb{Z}_n^* = \{\alpha^k \ (\bmod \, n) \mid 0 \leq k \leq (\varphi(n) - 1)\} \tag{2.4b}$$

then the element $\alpha$ is the generator of the group. It is also called the *primitive element* of the group. If $a \in \mathbb{Z}_n^*$, then $a^{\varphi(n)} \equiv 1 \ (\bmod \, n)$. This group is generally denoted by $\mathbb{Z}_n^*$ by dropping the binary operator $\times$.

9. Let $p$ be a prime number, then $(\mathbb{Z}_p^*, \times)$ is a cyclic group, where the binary operation $\times$ is multiplication modulo $p$, and $\mathbb{Z}_p^* = \{k \mid 1 \leq k \leq (p-1)\}$. The order of this group is $\varphi(p) = (p-1)$. Let the generator of this group be $\alpha$. Then

$$\mathbb{Z}_p^* = \{\alpha^k \ (\bmod \, p) \mid 0 \leq k \leq (p-2)\} \tag{2.5}$$

If $\beta = \alpha^k \ (\bmod \, p)$, and $k \neq 0$, then its order is $\{(p-1) / \gcd((p-1), k)\}$.
It can be checked that the group $(\mathbb{Z}_{13}^*, \times)$ has $\varphi(12) = 4$ primitive elements. These are $2, 6, 7$, and 11.

10. Consider the cyclic group $(\mathbb{Z}_{11}^*, \times)$ where $\mathbb{Z}_{11}^* = \{1, 2, \ldots, 10\}$. The order of this group is 10. This group has $\varphi(10) = 4$ number of generators. These are $2, 6, 7$, and 8. The Table 2.1 has a list of its subgroups, and their respective generators.

| Subgroup | Generators | Order |
|----------|------------|-------|
| $\{1\}$ | 1 | 1 |
| $\{1, 10\}$ | 10 | 2 |
| $\{1, 3, 4, 5, 9\}$ | $3, 4, 5, 9$ | 5 |
| $\{1, 2, \ldots, 10\}$ | $2, 6, 7, 8$ | 10 |

Table 2.1. Subgroups and generators of group $(\mathbb{Z}_{11}^*, \times)$.

11. Let $S = \{1, 2, \ldots, n\}$ be a set of numbers. The set of all permutations of the set $S$, and the composition mapping form a symmetric group $S_n$. ◻

The above set of examples inspire the following observations about Abelian groups and their isomorphic representations.

**Observations 2.2.** Some useful facts about groups.

1. If $p$ is a prime number, then all groups of order $p$ are isomorphic to $\mathcal{Z}_p = (\mathbb{Z}_p, +)$.
2. Let $\mathcal{G}_i = (G_i, *)$ be an Abelian group, for $1 \leq i \leq n$. Then the direct product of groups $\mathcal{G} = (G, *)$ is an Abelian group, where $G$ is specified by

$$G = G_1 \times G_2 \times \ldots \times G_n \tag{2.6}$$

(a) Also $|G| = |G_1| |G_2| \ldots |G_n|$.

(b) If $e_i \in G_i$ is an identity element for $1 \leq i \leq n$, then the identity element of the group $\mathcal{G}$ is $(e_1, e_2, \ldots, e_n)$.

(c) If $b_i \in G_i$ and its inverse is $b_i^{-1} \in G_i$ for $1 \leq i \leq n$, then the inverse of $(b_1, b_2, \ldots, b_n) \in G$ is $\left(b_1^{-1}, b_2^{-1}, \ldots, b_n^{-1}\right) \in G$.

3. Denote the group $(\mathbb{Z}_n, +)$ by $\mathcal{Z}_n$. Then:

   (a) The group $(\mathbb{Z}_m \times \mathbb{Z}_n, +)$ is isomorphic to the group $\mathcal{Z}_{mn}$ iff $m$ and $n$ are relatively prime.

   (b) Let $n = n_1 n_2 \ldots n_m$, where $n_i = p_i^{\alpha_i}$, and $p_i$ is a prime number, $\alpha_i \in \mathbb{P}$, for $1 \leq i \leq m$. Furthermore, $p_i \neq p_j$ for $i \neq j$ and $1 \leq i, j \leq m$. Then the group

$$(\mathbb{Z}_{n_1} \times \mathbb{Z}_{n_2} \times \ldots \times \mathbb{Z}_{n_m}, +)$$

   is isomorphic to the group $\mathcal{Z}_n$.

4. Consider a finite Abelian group $\mathcal{G}$ of order greater than 2. This Abelian group is isomorphic to the direct product of cyclic groups where each of these cyclic groups has an order which is a prime power. Furthermore, any two such decompositions have identical number of factors of each order. This observation is called *the fundamental theorem of finite Abelian groups*.   □

## 2.2.2 Rings

A ring is initially defined in this subsection. Some important properties of rings are subsequently listed. This is followed by examples.

**Definition 2.3.** *A ring $\mathcal{R} = (R, +, \times)$, is a nonempty set $R$ and two binary operations $+$ (addition) and $\times$ (multiplication) defined on $R$, where*:

(a) *The algebraic structure $(R, +)$ is an Abelian group. It satisfies the following properties.*

   (i) *Associative property*: $a + (b + c) = (a + b) + c$ *for all* $a, b, c \in R$.

   (ii) *Identity property*: *The set $R$ has an element $0$, such that* $0 + a = a + 0 = a$ *for all* $a \in R$.

   (iii) *Inverse property*: *For each element $a \in R$ there exists an element $-a \in R$ such that* $-a + a = a + (-a) = 0$.

   (iv) *Commutative property*: $a + b = b + a$ *for all* $a, b \in R$.

(b) *The binary operation $\times$ is associative. That is*, $a \times (b \times c) = (a \times b) \times c$ *for all* $a, b, c \in R$.

(c) *The multiplication operation $\times$ is left and right distributive over addition operation $+$. That is*:

$$a \times (b + c) = (a \times b) + (a \times c) \tag{2.7a}$$

$$(b + c) \times a = (b \times a) + (c \times a) \tag{2.7b}$$

*for all* $a, b, c \in R$.   □

Sometimes the ring $\mathcal{R} = (R, +, \times)$, is simply referred to as the ring $R$.

**Definitions 2.4.** *Let $\mathcal{R} = (R, +, \times)$ be a ring.*

1. *The ring $\mathcal{R}$ is a ring with unity, if there is a multiplicative identity denoted by $1$, where $1 \times a = a \times 1 = a$ for all $a \in R$, and $1 \neq 0$.*

2. *A ring $\mathcal{R} = (R, +, \times)$ is said to be commutative, if the multiplication operation is commutative. That is*, $a \times b = b \times a$ *for all* $a, b \in R$.

3. *Consider a ring with unity. An element $x \in R$ is a unit if $x$ has a multiplicative inverse. That is, there is a $x^{-1} \in R$ such that $x \times x^{-1} = x^{-1} \times x = 1$.*
4. *Subtraction in a ring is defined as $(a - b) = \{a + (-b)\}$.*                        □

The multiplication operation $a \times b$ is often simply written as $a \cdot b$ or $ab$.

**Observations 2.3.** Some elementary properties of rings are listed below.

1. In all rings, $a \times 0 = 0 \times a = 0$ for all $a \in R$.
2. The set of all units of a ring form a group under the multiplication operation. It is called the *group of units* of $R$.
3. The precedence of operations in a ring is the same as that of real numbers. That is, multiplication operation precedes addition operation.
4. Subtraction properties. Assume that $a, b, c \in R$. Then
   (a) $-(-a) = a$
   (b) $(-1) \times a = -a$, assuming that the ring has the unit element.
   (c) $a \times (-b) = (-a) \times b = -(a \times b)$
   (d) $(-a) \times (-b) = a \times b$
   (e) $a \times (b - c) = a \times b - a \times c$ and $(b - c) \times a = b \times a - c \times a$              □

**Examples 2.2.** Some illustrative examples.

1. Some well-known examples of rings:

$$(\mathbb{Z}, +, \times), (\mathbb{Q}, +, \times), (\mathbb{R}, +, \times), \text{ and } (\mathbb{C}, +, \times)$$

   In these rings, $+$ and $\times$ are the usual addition and multiplication operations. The additive and multiplicative identities are 0 and 1 respectively.
2. $(\mathbb{Z}_n, +, \times)$ is a commutative ring, where $n \in \mathbb{P}$, $a + b \equiv (a + b) \pmod{n}$, and $a \times b \equiv (a \times b) \pmod{n}$. The group of units of this ring is $(\mathbb{Z}_n^*, \times)$, where $\mathbb{Z}_n^*$ is the set of integers in $\mathbb{Z}_n$ which are relatively prime to $n$.
3. The set of Gaussian integers $\{a + ib \mid a, b \in \mathbb{Z} \text{ and } i = \sqrt{-1}\}$. The additive and multiplicative identities are $(0 + i0)$ and $(1 + i0)$ respectively.
4. The set of all $n \times n$ real matrices with operations: matrix addition and matrix multiplication form a ring with unity. The zero and the identity matrices are the additive identity and multiplicative identity respectively. It is not a commutative ring.                    □

### 2.2.3  Subrings and Ideals

Subrings and ideals are defined in this subsection. This is followed by some observations and examples.

**Definitions 2.5.** *Let $\mathcal{R} = (R, +, \times)$ be a ring.*

1. *$S$ is a subset of $R$, and $\mathcal{S} = (S, +, \times)$ is a ring. Then $\mathcal{S}$ is a subring of the ring $\mathcal{R}$.*
2. *The structure $\mathcal{I} = (I, +, \times)$ is an ideal of the ring $\mathcal{R}$ iff:*
   (a) *$\mathcal{I}$ is a subring of $\mathcal{R}$.*
   (b) *The set $I$ is closed under left and right multiplication by elements of $R$. That is, if $r \in R$ and $a \in I$, then $a \times r, r \times a \in I$.*

3. *Assume that $R$ is a commutative ring, then an ideal $\mathcal{I}$ is principal, if there is $r \in R$ such that $I = Rr = \{a \times r \mid a \in R\}$. The ideal $\mathcal{I}$ is said to be generated by $r$, and simply denoted by $\langle r \rangle$.* □

**Observations 2.4.** Let $\mathcal{R} = (R, +, \times)$ be a ring, and $\mathcal{I} = (I, +, \times)$ be its ideal.

1. $\mathcal{R}$ and $(\{0\}, +, \times)$ are called trivial ideals.
2. The intersection of several ideals in a ring is an ideal.
3. $(I, +)$ is a subgroup of the group $(R, +)$, but the converse is not true in general.
4. Let $S$ be a nonempty subset of $R$, then $\mathcal{S} = (S, +, \times)$ is a subring of ring $\mathcal{R}$ if and only if the set $S$ is closed under the addition and multiplication operations. □

**Examples 2.3.** Some illustrative examples.

1. $\mathcal{Z} = (\mathbb{Z}, +, \times)$ is a ring. Then $(n\mathbb{Z}, +, \times)$ is a principal ideal in the ring $\mathcal{Z}$, where $n$ is an integer.
2. Let $i = \sqrt{-1}$, then the set of Gaussian integers $\{a + bi \mid a, b \in \mathbb{Z}\}$, where the addition and multiplication operations correspond to those of complex numbers, is a subring of the complex numbers. In this subring the additive identity is $(0 + i0)$ and multiplicative identity is $(1 + i0)$. □

**Definition 2.6.** *Let $\mathcal{I} = (I, +, \times)$ be an ideal in a commutative ring $\mathcal{R} = (R, +, \times)$. Assume that an ideal $\mathcal{J} = (J, +, \times)$ exists in the commutative ring $\mathcal{R} = (R, +, \times)$, so that $J$ contains $I$ as a subset. The ideal $\mathcal{I}$ is maximal, if either $J = I$ or $J = R$.* □

Thus $\mathcal{I}$ is a maximal ideal of a commutative ring $\mathcal{R}$, if there are no other ideals contained between $\mathcal{I}$ and $\mathcal{R}$.

**Definition 2.7.** *Quotient ring. Let $\mathcal{I} = (I, +, \times)$ be an ideal in a ring $\mathcal{R} = (R, +, \times)$. If $a \in R$, then the set $a + I = \{a + x \mid x \in I\}$ is a coset of $\mathcal{I}$ in $\mathcal{R}$.*

*The set of all cosets, $R/I = \{a + I \mid a \in R\}$, form a ring $\mathcal{R}/\mathcal{I}$. This ring is called the quotient ring. In this ring the addition and multiplication rules are respectively:*

(a) $(a + I) + (b + I) = (a + b) + I$.
(b) $(a + I) \times (b + I) - (a \times b) + I$. □

**Observations 2.5.** Some related basic observations.

1. If the ring $\mathcal{R}$ is commutative, then $\mathcal{R}/\mathcal{I}$ is also commutative.
2. If the ring $\mathcal{R}$ has the unity element 1, then $\mathcal{R}/\mathcal{I}$ has the coset $(1 + I)$ as unity.
3. Let $\mathcal{R}$ be a commutative ring with unity, and $\mathcal{I}$ be an ideal in $\mathcal{R}$, then $\mathcal{I}$ is a maximal ideal if and only if $\mathcal{R}/\mathcal{I}$ is a field. □

**Definitions 2.8.** *Divisor of zero, and integral domain. Let $\mathcal{R} = (R, +, \times)$ be a ring.*

1. *Let $a, b \in R$; where $a \neq 0$, and $b \neq 0$. If $ab = 0$, then $a$ is called a left divisor of zero, and $b$ is called a right divisor of $b$.*
2. *Let $\mathcal{R}$ be a commutative ring with unity, and with no zero divisors. Then $\mathcal{R}$ is an integral domain.* □

**Observations 2.6.** Some related basic observations.

1. If the ring $\mathcal{R}$ is an integral domain, then the polynomial ring $\mathcal{R}[x]$ is also an integral domain.
2. If the ring $\mathcal{R}$ is not an integral domain, then the polynomial ring $\mathcal{R}[x]$ is not an integral domain.

□

### 2.2.4  Fields

The concept of a field is introduced in this subsection. This is followed by a list of some useful observations related to fields. Some examples of fields are also given.

**Definition 2.9.** *A field $\mathcal{F} = (F, +, \times)$ consists of a nonempty set $F$ and two binary operations, $+$ and $\times$ defined on $F$. It satisfies the following properties.*

(a) *$(F, +, \times)$ is a ring.*
(b) *$(F - \{0\}, \times)$ is an Abelian group.*

*The binary operations, $+$ and $\times$ are called addition and multiplication operations respectively. Also $0 \in F$ is the additive identity, and $1 \in F$ is the multiplicative identity.*                □

The field $\mathcal{F} = (F, +, \times)$ is occasionally referred to as $F$, and vice-versa. If $a, b \in F$, then $a \times b$ is sometimes simply written as $a \cdot b$ or $ab$. Members of the set $F$ are called *scalars*.

**Definitions 2.10.** *Let $\mathcal{F} = (F, +, \times)$ and $\mathcal{M} = (M, +, \times)$ be fields.*

1. *$\mathcal{F}$ is a subfield of field $\mathcal{M}$ if $F$ is a subset of $M$.*
2. *If $\mathcal{F}$ is a subfield of field $\mathcal{M}$, then $\mathcal{M}$ is an extension field of $\mathcal{F}$. This is indicated by $\mathcal{M}/\mathcal{F}$.*
3. *The characteristic of a field $\mathcal{F}$ is the smallest positive integer $n$ such that*

$$\underbrace{1 + 1 + \ldots + 1}_{n \text{ times}} = 0 \tag{2.8a}$$

*If there is no such integer, the field $\mathcal{F}$ has a characteristic $0$ (sometimes called characteristic $\infty$)*
4. *The order of a field is the number of elements in $F$, which is $|F|$.*
5. *Let $\mathcal{F}$ be a subfield of field $\mathcal{M}$, that is $F \subseteq M$. Also let $G$ be a subset of $M$ such that $F \cap G = \varnothing$. Then $\mathcal{F}(G)$ is a field defined as the intersection of all subfields of $\mathcal{M}$ containing both $F$ and $G$. Therefore $\mathcal{F}(G)$ is the smallest subfield of $\mathcal{M}$ containing the sets $F$ and $G$. Actually $\mathcal{F}(G)$ is called the extension field of $\mathcal{F}$ obtained by adjoining the elements in $G$. If*

$$G = \{\alpha_1, \alpha_2, \ldots, \alpha_n\} \tag{2.8b}$$

*then*

$$\mathcal{F}(G) = \mathcal{F}(\alpha_1, \alpha_2, \ldots, \alpha_n) \tag{2,8c}$$

*However, if $G = \{\alpha\}$ then $\mathcal{F}(\alpha)$ is said to be a simple extension of $\mathcal{F}$ and $\alpha$ is called the defining element of $\mathcal{F}(\alpha)$.*                □

Thus the number of elements in the smallest subfield of a given field is its characteristic. Characteristic of the field $\mathbb{R}$ is 0.

If two fields are structurally similar, then these fields are said to be *isomorphic* to each other. The definition of field isomorphism is similar to the definition of group isomorphism.

**Definition 2.11.** *Two fields $\mathcal{F}_1 = (F_1, +, \times)$ and $\mathcal{F}_2 = (F_2, \oplus, \otimes)$ are isomorphic to each other, if there exists a one-to-one and onto mapping $\theta$ of elements of $F_1$ to elements of $F_2$. Then for all elements $a$ and $b$ in the set $F_1$ :*

$$\theta(a + b) = \theta(a) \oplus \theta(b) \tag{2.9a}$$
$$\theta(a \times b) = \theta(a) \otimes \theta(b) \tag{2.9b}$$

*Isomorphism is usually denoted by $\mathcal{F}_1 \cong \mathcal{F}_2$.*                                    $\square$

**Observations 2.7.** Consider a field $\mathcal{F} = (F, +, \times)$.

1. Every field $\mathcal{F}$ is a commutative ring with unity. In addition, every element $a \in F$ which is not zero has a multiplicative inverse $a^{-1}$.
   This fact can also be stated alternately. Consider a commutative ring $\mathcal{R} = (R, +, \times)$ with unity. The set $R$ satisfies all the rules of a field, except that nonzero elements of $R$ do not necessarily have multiplicative inverses.
2. Every ideal in $F[x]$ is a principal ideal.
3. For any $a, b \in F$, we have:
   (a) $(-1)a = -a$.
   (b) $ab = 0$ implies $a = 0$ or $b = 0$.
4. The characteristic of a field is either equal to 0 or a prime number.
5. If the characteristic of a field is a prime number $p$, then $p\alpha = 0$ for all $\alpha \in F$.
6. Let the field $\mathcal{F}$ have a characteristic $p \in \mathbb{P}$. Also let $a$ and $b$ be any elements in the field. This field has the following properties.
   (a) $(a \pm b)^p = a^p \pm b^p$.
   (b) $(a \pm b)^{p^n} = a^{p^n} \pm b^{p^n}$, where $n \in \mathbb{P}$.
7. Two isomorphic fields have the same structure, and consequently the same number of elements.
                                                                                              $\square$

The above important concepts and observations are further clarified via the following examples.

**Examples 2.4.** Some illustrative examples.

1. Some well-known examples of fields:

$$(\mathbb{Q}, +, \times), (\mathbb{R}, +, \times), \text{ and } (\mathbb{C}, +, \times)$$

In these fields, $+$ and $\times$ are the usual addition and multiplication operations. The order of these fields is infinite. The additive and multiplicative inverses of an element $a$ are, $-a$ and $a^{-1}$ respectively. Note that $a^{-1}$ is defined, only if $a \neq 0$. The characteristic of each of these fields is equal to 0.

2. $(\mathbb{Z}_m, +, \times)$ is a field, iff $m$ is a prime number. In this case, its characteristic is equal to $m$. The addition and multiplication is modulo $m$ in this field.
   Furthermore, if the number $m$ is prime and there is no ambiguity, then this field of numbers is simply denoted by $\mathbb{Z}_p$. Because of the importance of this observation, this result is stated as a theorem in the section on finite fields.
   Also note that, if $a \in \mathbb{Z}_p$, then $a^p = a$.
                                                                                              $\square$

### 2.2.5 Polynomial Rings

The concept of rings can be extended to polynomials. A polynomial and related terminology are first defined precisely.

**Definitions 2.12.** *Let $\mathcal{R} = (R, +, \times)$ be a ring.*

1. *A polynomial in the variable (or indeterminate) $x$ over a ring $\mathcal{R}$ is an expression of type*

$$f(x) = a_n x^n + a_{n-1} x^{n-1} + \ldots + a_1 x + a_0 \tag{2.10}$$

   *where $n \in \mathbb{N}$, $a_m \in R$ for $0 \le m \le n$.*
2. *The element $a_m$ is called the coefficient of $x^m$ in $f(x)$, for $0 \le m \le n$.*
3. *The largest integer $m$ for which $a_m \ne 0$ is called the degree of the polynomial $f(x)$. It is usually written as $\deg f(x)$, or as simply $\deg f$.*
4. *If $\deg f(x) = m$, and $a_m = 1$, then the polynomial $f(x)$ is a monic polynomial.*
5. *If $f(x) = a_0$, and $a_0 \ne 0$, then the polynomial is said to be a constant polynomial. Its degree is equal to $0$.*
6. *If all the coefficients of a polynomial are equal to $0$, then the polynomial $f(x)$ is said to be a zero polynomial. Its degree is said to be equal to $-\infty$.*
7. *The value of a polynomial at $b \in R$ is equal to $f(b) \in R$.*
8. *The element $b \in R$ is a root of the the equation $f(x) = 0$, if $f(b) = 0$.*
9. *The element $b \in R$ is a zero of the polynomial $f(x)$, if $f(b) = 0$.* $\square$

Sometimes, the terms root and zero are used interchangeably. A polynomial ring in a single variable $x$ is next defined.

**Definition 2.13.** *Let $\mathcal{R} = (R, +, \times)$ be a ring. The polynomial ring $\mathcal{R}[x]$ is the set of polynomials in $x$. The coefficients of these polynomials belong to the set $R$. The addition and multiplication of these polynomials are the standard polynomial addition and multiplication operations. The coefficient arithmetic of these operations is performed in the ring $\mathcal{R}$.* $\square$

The polynomial ring $\mathcal{R}[x]$ is sometimes denoted by $R[x]$.

Consider the polynomials $f(x) = \sum_{i=0}^{m} a_i x^i$, and $g(x) = \sum_{i=0}^{n} b_i x^i$. Assume that $m \ge n$. Then

$$f(x) + g(x) = \sum_{i=n+1}^{m} a_i x^i + \sum_{i=0}^{n} (a_i + b_i) x^i \tag{2.11a}$$

Evidently the first summand is equal to zero, if $m = n$. Also denote the product of the polynomials $f(x)$ and $g(x)$ by $f(x) g(x)$. Therefore

$$f(x) g(x) = \sum_{i=0}^{m+n} c_i x^i \tag{2.11b}$$

where

$$c_i = \sum_{\substack{0 \le j \le m,\, 0 \le k \le n, \\ j+k=i}} a_j b_k, \quad 0 \le i \le (m+n) \tag{2.11c}$$

**Observations 2.8.** Some basic related observations.

1. If $\mathcal{R} = (R, +, \times)$ is a commutative ring, then $R[x]$ is a commutative ring polynomial.
2. Subtraction notation is introduced as

$$\left\{a_j x^j - a_k x^k\right\} = \left\{a_j x^j + \left(-a_k x^k\right)\right\}$$

$\square$

Polynomial division and related concepts are next explored.

**Definitions 2.14.** *Let the ring of polynomials $F[x]$ be defined over the field $\mathcal{F} = (F, +, \times)$.*

1. *Let $f(x), d(x) \in F[x]$, where $d(x) \neq 0$. Then ordinary polynomial long division of $f(x)$ by $d(x)$ yields the unique polynomials $\psi(x)$ and $r(x)$. Also $\deg r(x) < \deg d(x)$, and $f(x) = \psi(x) d(x) + r(x)$. The polynomials $\psi(x)$ and $r(x)$ are called the quotient and remainder polynomials respectively.*
   *The polynomial $r(x)$ is denoted by $f(x) \pmod{d(x)}$, or $f(x) \equiv r(x) \pmod{d(x)}$. Also $\psi(x)$ is denoted by $f(x) \operatorname{div} d(x)$.*
2. *If $f(x) \in F[x]$, then $\alpha \in F$ is a zero of $f(x)$ iff $f(x) = (x - \alpha) g(x)$.*
3. *Let $f(x), d(x) \in F[x], d(x) \neq 0$, and $f(x) \equiv 0 \pmod{d(x)}$. Then $d(x)$ is said to divide $f(x)$. This is written as $d(x) \mid f(x)$.*
4. *Let $g_1(x), g_2(x), f(x) \in F[x]$, and $f(x) \neq 0$. Then $g_1(x)$ is said to be congruent to $g_2(x)$ modulo $f(x)$, iff $f(x)$ divides $(g_1(x) - g_2(x))$. This is indicated by $g_1(x) \equiv g_2(x) \pmod{f(x)}$.*
5. *Let $a(x), b(x) \in F[x], a(x) \neq 0$, and $b(x) \neq 0$, then the greatest common divisor of $a(x)$ and $b(x)$ is a monic polynomial $g(x)$ that divides both $a(x)$ and $b(x)$. The polynomial $g(x)$ is such that, if any polynomial $d(x)$ divides $a(x)$ and $b(x)$, then $d(x)$ divides $g(x)$. The greatest common divisor $g(x)$ is denoted by $\gcd(a(x), b(x))$.*
   *If $a(x) \in F[x], a(x) \neq 0$, then $\gcd(a(x), 0) = a(x)$.*
6. *Let $a(x), b(x) \in F[x]$ be monic polynomials. The polynomials $a(x)$ and $b(x)$ are said to be relatively prime, if $\gcd(a(x), b(x)) = 1$.*
7. *The least common multiple of a set of nonzero polynomials is a polynomial of smallest degree that is divisible by all polynomials which belong to this set. Let $a(x), b(x) \in F[x], a(x) \neq 0$, and $b(x) \neq 0$. The least common multiple of polynomials $a(x)$ and $b(x)$ is denoted by $\operatorname{lcm}(a(x), b(x))$.*
8. *Let $f(x) = \sum_{i=0}^{n} a_i x^i \in F[x]$. A formal derivative of $f(x)$ with respect to $x$ is denoted as $f'(x) = \sum_{i=1}^{n} i a_i x^{i-1}$.* $\square$

It can be shown that polynomial congruences observe: reflexivity, symmetry, and transitivity. This is analogous to congruences of integers modulo some positive integer. Some properties of polynomial congruences are next listed.

**Observations 2.9.** Let

$$f(x), a(x), a_1(x), a_2(x), b(x), b_1(x), b_2(x), c(x) \in F[x]$$

The polynomial $f(x)$ is nonzero.

1. $F[x]$ is not a field. For example the polynomials $x$ and $(x + 1)$ do not have multiplicative inverses in $F[x]$.

2. $a\left(x\right) \equiv b\left(x\right) \left(\text{mod} f\left(x\right)\right)$ if the remainder obtained by dividing $a\left(x\right)$ by $f\left(x\right)$ is the same as the remainder obtained by dividing $b\left(x\right)$ by $f\left(x\right)$.

3. Reflexive property: $a\left(x\right) \equiv a\left(x\right) \left(\text{mod} f\left(x\right)\right)$.

4. Symmetry property: If $a\left(x\right) \equiv b\left(x\right) \left(\text{mod} f\left(x\right)\right)$, then $b\left(x\right) \equiv a\left(x\right) \left(\text{mod} f\left(x\right)\right)$.

5. Transitive property: If $a\left(x\right) \equiv b\left(x\right) \left(\text{mod} f\left(x\right)\right)$ and $b\left(x\right) \equiv c\left(x\right) \left(\text{mod} f\left(x\right)\right)$, then $a\left(x\right) \equiv c\left(x\right) \left(\text{mod} f\left(x\right)\right)$.

6. Let $a_1\left(x\right) \equiv a_2\left(x\right) \left(\text{mod} f\left(x\right)\right)$, and $b_1\left(x\right) \equiv b_2\left(x\right) \left(\text{mod} f\left(x\right)\right)$. Then

$$\left(a_1\left(x\right) + b_1\left(x\right)\right) \equiv \left(a_2\left(x\right) + b_2\left(x\right)\right) \ \left(\text{mod} f\left(x\right)\right) \tag{2.12a}$$

$$a_1\left(x\right) b_1\left(x\right) \equiv \left(a_2\left(x\right) b_2\left(x\right)\right) \ \left(\text{mod} f\left(x\right)\right) \tag{2.12b}$$

7. The congruence modulo $f\left(x\right)$ relationship, partitions $F\left[x\right]$ into equivalence classes.

8. A Euclidean-type of algorithm can be used to determine the greatest common divisor of two nonzero polynomials defined over a finite field. Finite fields are discussed later in this chapter. Let nonzero polynomials $a\left(x\right), b\left(x\right) \in F\left[x\right]$ be given. These polynomials are defined over a finite field $\mathcal{F}$. An algorithm analogous to the extended Euclidean algorithm for positive numbers, determines polynomials $c\left(x\right), d\left(x\right) \in F\left[x\right]$ and $\gcd\left(a\left(x\right), b\left(x\right)\right)$ such that

$$a\left(x\right) c\left(x\right) + b\left(x\right) d\left(x\right) = \gcd\left(a\left(x\right), b\left(x\right)\right) \tag{2.13}$$

The extended Euclidean algorithm for polynomials is not outlined in this chapter.  $\square$

The concept of a polynomial ring modulo a given polynomial, is next developed. A fixed polynomial $f\left(x\right) \in F\left[x\right]$ of degree at least one is given. The equivalence class of a polynomial $g\left(x\right) \in F\left[x\right]$ is the set of all polynomials in $F\left[x\right]$ which are congruent to $g\left(x\right)$ modulo $f\left(x\right)$. Denote these equivalence classes of polynomials in $F\left[x\right]$ by $F\left[x\right] / \left(f\left(x\right)\right)$. The degree of these polynomials is less than $\deg f\left(x\right)$. It can be checked that the polynomials in $F\left[x\right] / \left(f\left(x\right)\right)$ form a ring. These ideas are formally stated below.

**Definition 2.15.** *Let $F\left[x\right]$ be a ring of polynomials defined over a field $\mathcal{F} = \left(F, +, \times\right)$. Also let $f\left(x\right) \in F\left[x\right]$ be a polynomial of degree at least one. Equivalence classes of polynomials are obtained by performing modulo $f\left(x\right)$ operation on any polynomial in $F\left[x\right]$. The set of equivalence classes of polynomials is denoted by $F\left[x\right] / \left(f\left(x\right)\right)$.*  $\square$

The following results are immediate from the above definition.

**Observations 2.10.** Some observations about polynomial-modulo arithmetic.

1. Let $f\left(x\right), g\left(x\right) \in F\left[x\right]$, and $f\left(x\right) \neq 0$. Then $g\left(x\right)$ can be written as $g\left(x\right) = \psi\left(x\right) f\left(x\right) + r\left(x\right)$, where $\deg r\left(x\right) < \deg f\left(x\right)$. The polynomial $r\left(x\right)$ can be used as a representative of the equivalence class of polynomials modulo $f\left(x\right)$ which contain $g\left(x\right)$.

2. Addition and multiplication of polynomials in $F\left[x\right] / \left(f\left(x\right)\right)$ can be performed modulo $f\left(x\right)$.

3. $F\left[x\right] / \left(f\left(x\right)\right)$ is the ring of polynomials modulo $f\left(x\right)$.  $\square$

**Example 2.5.** Let

$$f\left(x\right) = \left(x^4 + x^3 + x^2 + x + 1\right)$$
$$g\left(x\right) = \left(x^3 + x^2 + 1\right)$$

be polynomials which belong to the polynomial ring $\mathbb{Z}_2\left[x\right]$. Then

$$f\left(x\right) + g\left(x\right) \equiv \left(x^4 + x\right) \ \left(\mathrm{mod}\, 2\right)$$
$$f\left(x\right) g\left(x\right) \equiv \left(x^7 + x^4 + x^3 + x + 1\right) \ \left(\mathrm{mod}\, 2\right)$$
$$f\left(x\right) \ \mathrm{div}\ g\left(x\right) \equiv x \ \left(\mathrm{mod}\, 2\right)$$
$$f\left(x\right) \equiv \left(x^2 + 1\right) \ \left(\mathrm{mod}\, g\left(x\right)\right)$$

Note that the arithmetic is modulo 2, as the coefficient field is $\mathbb{Z}_2$.                        □

### Fundamental Theorem of Algebra

The *fundamental theorem of algebra* is next stated. It was first proved by Carl Friedrich Gauss in his doctoral dissertation.

**Theorem 2.1.** *Let $f\left(x\right) \in \mathbb{C}\left[x\right]$ be a polynomial of $\deg f\left(x\right) = d \in \mathbb{P}$. Then $f\left(x\right)$ has $n \in \mathbb{P}$ distinct roots in $\mathbb{C}$, where $1 \leq n \leq d$. However their total number, including multiplicities is equal to d. Thus*

$$f\left(x\right) = c \prod_{j=1}^{d} \left(x - c_j\right),  \tag{2.14}$$

*where $c \in \mathbb{C}\backslash\left\{0\right\}, c_j \in \mathbb{C}$ for $1 \leq j \leq d$.*
*Proof.* The proof is left to the reader.                        □

### Irreducible and Reducible Polynomials

Irreducible and reducible polynomials are next defined.

**Definition 2.16.** *A polynomial $f\left(x\right) \in F\left[x\right]$ of $\deg f\left(x\right) \geq 1$ is said to be irreducible over field F, if it cannot be written as a product of two polynomials in $F\left[x\right]$. These two polynomials are each required to have positive degree but less than $\deg f\left(x\right)$. Otherwise the polynomial $f\left(x\right)$ is said to be reducible.*
*A monic irreducible polynomial of degree at least one is called a prime polynomial.*                        □

Therefore, a polynomial is irreducible, if it has no divisors except itself and scalar quantities. The concept of irreducibility of monic polynomials is similar to the concept of primality of positive integers.

**Theorem 2.2.** *A monic polynomial defined over the field $\mathcal{F}$, of degree at least one is uniquely factorable into irreducible monic (prime) polynomials over the field $\mathcal{F}$.*
*Proof.* The proof is left to the reader.                        □

The above result is reminiscent of the prime-power factorization of a positive integer. It is shown in the section on finite fields that, if $F\left[x\right]$ is a ring of polynomials defined over a finite field $\mathcal{F} = \left(F, +, \times\right)$, and $f\left(x\right) \in F\left[x\right]$ is an irreducible polynomial, then the quotient ring $F\left[x\right]/\left(f\left(x\right)\right)$ is a field. The elements of this field are cosets of polynomials in $F\left[x\right]$ modulo $f\left(x\right)$. Further, $\left(f\left(x\right)\right)$ is the principal ideal generated by $f\left(x\right)$. Actually, the polynomials $f_1\left(x\right)$ and $f_2\left(x\right)$ lie in the same coset if and only if $f(x)$ divides $(f_1(x) - f_2(x))$. Further, $F[x] / (f(x))$ can be considered to be a field of all polynomials in $F[x]$ which are of degree less than $n$.

### Chinese Remainder Theorem for Polynomials

Analogous to the Chinese remainder theorem for integers, there exists a Chinese remainder theorem for polynomials.

**Theorem 2.3.** *Let $F$ be a field, and $m_1(x), m_2(x), \ldots, m_n(x) \in F[x]$, where $n \in \mathbb{P} \setminus \{1\}$ are nonzero polynomials and $\gcd(m_k(x), m_j(x))$ is a scalar for $k \neq j, 1 \leq k, j \leq n$. Also let $m(x) = \prod_{k=1}^{n} m_k(x)$, $f(x) \in F[x]$; and*

$$f(x) \equiv a_k(x) \pmod{m_k(x)}, \quad 1 \leq k \leq n \tag{2.15a}$$

*where $a_1(x), a_2(x), \ldots, a_n(x) \in F[x]$ are also given. Further*

$$M_k(x) = \frac{m(x)}{m_k(x)}, \text{ and } (M_k(x) N_k(x)) \equiv 1 \pmod{m_k(x)}, 1 \leq k \leq n \tag{2.15b}$$

*These congruences have a single common solution*

$$f(x) \equiv \sum_{k=1}^{n} a_k(x) M_k(x) N_k(x) \pmod{m(x)} \tag{2.15c}$$

*Proof.* The proof is left to the reader.                                                      $\square$

Observe in the above theorem that, if the polynomials

$$m_1(x), m_2(x), \ldots, m_n(x) \in F[x]$$

are all monic, and $\gcd(m_k(x), m_j(x)) = 1$ for $k \neq j, 1 \leq k, j \leq n$; then all these polynomials are relatively prime in pairs. Further, the polynomials $N_k(x)$ for $1 \leq k \leq n$ can be determined by using the extended Euclidean algorithm for polynomials. A more convenient technique uses the partial fraction expansion of $1/m(x)$. Let

$$\frac{1}{m(x)} = \sum_{k=1}^{n} \frac{N_k(x)}{m_k(x)}$$

This implies $\sum_{k=1}^{n} M_k(x) N_k(x) = 1$, from which the solution of $f(x)$ can be obtained as $f(x) \equiv \sum_{k=1}^{n} a_k(x) M_k(x) N_k(x) \pmod{m(x)}$.

**Example 2.6.** Let the field be $\mathbb{Z}_{11}$, $m(x) = x(x-1)(x-2)^2 \in \mathbb{Z}_{11}[x]$, and

$$f(x) \equiv 3 \pmod{x}$$
$$f(x) \equiv 6 \pmod{(x-1)}$$
$$f(x) \equiv (5x - 3) \left(\bmod (x-2)^2\right)$$

The polynomial $f(x) \pmod{m(x)}$ is obtained via the Chinese remainder theorem for polynomials. The partial fraction decomposition of $1/m(x)$ is

$$\frac{1}{x\,(x-1)\,(x-2)^2} = -\frac{1}{4x} + \frac{1}{(x-1)} + \frac{(-3x+8)}{4\,(x-2)^2}$$

$$\frac{1}{x\,(x-1)\,(x-2)^2} \equiv \left\{ \frac{8}{x} + \frac{1}{(x-1)} + \frac{2\,(x+1)}{(x-2)^2} \right\} (\mathrm{mod}\,11)$$

The use of Chinese remainder theorem for polynomials leads to

$$f\,(x) \equiv \Big( 3 \cdot 8\,(x-1)\,(x-2)^2$$

$$+6 \cdot 1x\,(x-2)^2 + (5x-3) \cdot 2\,(x+1)\,x\,(x-1) \Big)\,(\mathrm{mod}\,m\,(x))$$

$$\equiv \left( 10x^4 + 2x^3 + 2x + 3 \right)\,(\mathrm{mod}\,m\,(x))$$

$$\equiv \left( 8x^3 + 8x^2 + 9x + 3 \right)\,(\mathrm{mod}\,m\,(x))$$

where the coefficients of the polynomial are modulo 11.                                                    □

### 2.2.6 Boolean Algebra

Boolean algebra generalizes algebra of logical propositions and algebra of sets. It is so named after the logician George Boole (1813-1864). It has found applications in probability theory and abstract design of electrical circuits. Not surprisingly, these electrical circuits are sometimes called logical circuits.

**Definition 2.17.** *A Boolean algebra is* $\left( \widetilde{B}, +, \cdot, ', 0, 1 \right)$, *where* $\widetilde{B}$ *is a nonempty set with two binary operations* $+$ *(addition) and* $\cdot$ *(multiplication), a unary operation* $'$ *(complementation), and two distinct elements* $0, 1 \in \widetilde{B}$. *The element* $0$ *is called the zero element and* $1$ *is called the unit element. The following axioms also hold, where* $a, b, c \in \widetilde{B}$ :

1. *Commutative laws*:
   (a) $a + b = b + a$
   (b) $a \cdot b = b \cdot a$
2. *Distributive laws*:
   (a) $a + (b \cdot c) = (a + b) \cdot (a + c)$
   (b) $a \cdot (b + c) - a \cdot b + a \cdot c$
3. *Identity laws: The set* $\widetilde{B}$ *has two elements* $0$ *and* $1$, *such that* $0$ *is the identity element with respect to the* $+$ *operation, and* $1$ *is the identity element with respect to the* $\cdot$ *operation.*
   (a) $a + 0 = a$
   (b) $a \cdot 1 = a$
4. *Complement laws: For each* $a \in \widetilde{B}$, *there exists* $a' \in \widetilde{B}$ *such that*
   (a) $a + a' = 1$
   (b) $a \cdot a' = 0$
   *The element* $a' \in \widetilde{B}$ *is called the complement of* $a \in \widetilde{B}$.                    □

Generally, the multiplication operator $\cdot$ is omitted. Consequently, $a \cdot b$ is sometimes written as $ab$ without any ambiguity. The complementation operation $a'$ is also denoted by $\overline{a}$. Some useful laws which are applicable to Boolean algebras are listed below.

**Theorem 2.4.** *Consider a Boolean algebra* $\left( \widetilde{B}, +, \cdot, ', 0, 1 \right)$. *Let* $a, b, c \in \widetilde{B}$.

1. *Domination laws*:
   (a) $a + 1 = 1$
   (b) $a \cdot 0 = 0$

2. *Idempotent laws*:
   (a) $a + a = a$
   (b) $a \cdot a = a$

3. *Absorption laws*:
   (a) $a + a \cdot b = a$
   (b) $a \cdot (a + b) = a$

4. *Associative laws*:
   (a) $(a + b) + c = a + (b + c)$
   (b) $(a \cdot b) \cdot c = a \cdot (b \cdot c)$

5. *Involution law*: $(a')' = a$

6. *Uniqueness of complement*: If $(a + b) = 1$ and $a \cdot b = 0$ then $b = a'$.

7. *Complementation of* 0 *and* 1: $0' = 1$ *and* $1' = 0$.

8. *DeMorgan's laws*:
   (a) $(a + b)' = a' \cdot b'$
   (b) $(a \cdot b)' = a' + b'$

*Proof.* The proofs of these laws are left to the reader.                                   □

**Example 2.7.** Let $A$ be any set, and its power set be $\wp(A)$. This power set is a collection of sets which are closed under the operations of union, intersection, and complementation. The power set $\wp(A)$ is a Boolean algebra.

In this algebra, the addition operation is $\cup$, the multiplication operation is $\cap$, and complementation operation is $S^c$ where $S \subseteq \wp(A)$. In addition, the zero element is the null set $\varnothing$, and the unit element is the power set $\wp(A)$.                                   □

## 2.3 More Group Theory

In this section, we define and discuss:

(a) Group generation
(b) Normal subgroup, quotient group, centralizer, and center of a group
(c) Group homomorphism, isomorphism, and automorphism

Several useful examples of groups from linear algebra are also specified. Finally transformation group and group action are discussed.

### Group Generation

**Definition 2.18.** *Let* $\mathcal{G} = (G, *)$ *be a group. Also let* $S \subset G$. *If every element of* $G$ *can be written as* $a_1 * a_2 * \cdots * a_m$ *where* $a_i$ *or* $a_i^{-1} \in S$, *then* $S$ *generates* $G$, *or* $S$ *is the generating set of* $G$.

*If $G$ is generated by a set with a single element, that is $S = \{a\}$, then the group $G$ is cyclic. If $|S|$ is finite, then the group $G$ is said to be finitely generated.*    □

### Normal Subgroup, Quotient Group, Centralizer, and Center

**Definitions 2.19.** *Let $G = (G, *)$ be a group.*

1. *A normal subgroup $\mathcal{H} = (H, *)$ of group $G$ is a subgroup $\mathcal{H}$ of $G$ such that $a * H = H * a$ for all $a \in G$.*
2. *Let $\mathcal{H} = (H, *)$ be a normal subgroup of group $G$. The quotient group or factor group of $G$ modulo $\mathcal{H}$, is group $G/\mathcal{H}$ with elements in the set $\{a * H \mid a \in G\}$, where $(a * H) * (b * H) = (a * b) * H$.*
3. *Let $a \in G$. The set $\{x \mid x \in G, ax = xa\}$ is the centralizer (or normalizer) of $a$.*
4. *The set $\{x \mid x \in G, gx = xg, \ \forall \ g \in G\}$ is the center of group $G$.*    □

**Observations 2.11.** Let $G = (G, *)$ be a group. Let the identity element of the group $G$ be $e$.

1. The groups $G$ and $(\{e\}, *)$ are normal subgroups of $G$.
2. In the group $G/\mathcal{H}$, the identity is $e * H = H$ and the inverse of $a * H$ is $a^{-1} * H$.
3. For all $a \in G$, the centralizer of $a$, with the operator $*$ forms a subgroup of $G$.
4. The center of $G$, with the operator $*$ forms a subgroup of $G$.    □

### Group Homomorphism, Isomorphism, and Automorphism

**Definition 2.20.** *Homomorphism, isomorphism, and kernel of a mapping.*
   *Consider two groups $G = (G, *)$ and $\mathcal{H} = (H, \circ)$, and a function $\theta : G \to H$ such that $\theta(a * b) = \theta(a) \circ \theta(b)$ for all elements $a, b \in G$. The mapping $\theta(\cdot)$ is called a homomorphism.*

(a) *If the mapping $\theta(\cdot)$ is surjective, then it is a surjective group homomorphism.*
(b) *If the mapping $\theta(\cdot)$ is both one-to-one and onto, then $\theta(\cdot)$ is called an isomorphism. The group $G$ is said to be isomorphic to group $\mathcal{H}$, and it is written as $G \cong \mathcal{H}$.*
(c) *The isomorphic mapping $\theta : G \to G$ is called an automorphism.*
(d) *The kernel of the mapping $\theta(\cdot)$ is the set $\{a \in G \mid \theta(a) = e \in H\}$, where $e \in H$ is the identity element of the group $\mathcal{H}$.*
(e) *For some $a \in G$, let $i_a : G \to G$ be an automorphism, so that $i_a(x) = a * x * a^{-1}$ for all $x \in G$, then $i_a$ is called an inner automorphism of $G$.*
(f) *The set of all automorphisms of $G$ is denoted by $Aut(G)$, and the set of all inner automorphisms of $G$ is denoted by $Inn(G)$.*    □

**Observations 2.12.** Let $G = (G, *)$ be a group.

1. Consider a group $\mathcal{H} = (H, \circ)$. Also let a function $\theta : G \to H$ be a homomorphism.
   (a) The elements of the set $\theta(G) \subseteq H$ form a subgroup of $\mathcal{H}$.
   (b) The elements of the kernel of the mapping $\theta(\cdot)$ form a subgroup of $G$.
   (c) If $\theta(\cdot)$ is an isomorphism, then $\theta^{-1}(\cdot)$ is also an isomorphism.
2. $Aut(G)$ is a group under composition of functions.

3. Inner automorphisms.
   (a) Every inner automorphism of $\mathcal{G}$ is an automorphism of $\mathcal{G}$.
   (b) $Inn\,(\mathcal{G})$ is a normal subgroup of $Aut\,(\mathcal{G})$.                                                      □

Some of the above observations are proved in the problem section.

**Examples of Groups from Linear Algebra**

Special groups used in linear algebra are specified. In these examples, $\mathbb{F}$ is a field. Some examples of a field are the fields of real and complex numbers, which are denoted by $\mathbb{R}$ and $\mathbb{C}$ respectively. In these examples, the group operation is matrix multiplication.

1. *General linear group.* The general linear group $GL\,(n,\mathbb{F})$ consists of all $n \times n$ invertible (nonsingular) matrices with entries in the field $\mathbb{F}$, where $n \in \mathbb{P}$. Note that the determinants of these matrices are nonzero. Possible candidates for the field $\mathbb{F}$ are the set of real numbers $\mathbb{R}$, and the set of complex numbers $\mathbb{C}$. The corresponding groups are denoted by $GL\,(n,\mathbb{R})$ and $GL\,(n,\mathbb{C})$ respectively.
2. *Unitary group.* The set of all unitary $n \times n$ matrices along with matrix multiplication forms the group $U\,(n)$.
3. *Orthogonal group.* The set of all orthogonal real $n \times n$ matrices along with matrix multiplication forms the group $O\,(n)$. Therefore, $O\,(n)$ is a subgroup of the unitary group $U\,(n)$.
4. The matrices in the following groups have their determinants equal to unity.
   (a) *Special linear group.* The special linear group $SL\,(n,\mathbb{F})$ consists of all $n \times n$ matrices with entries in the field $\mathbb{F}$, where $n \in \mathbb{P}$, and whose determinant is equal to unity. Therefore $SL\,(n,\mathbb{F})$ is a subgroup of $GL\,(n,\mathbb{F})$. Some examples of special linear groups are $SL\,(n,\mathbb{R})$ and $SL\,(n,\mathbb{C})$.
   (b) *Special unitary group.* Let $A$ be a unitary matrix. Therefore the product of this matrix and its Hermitian transpose is equal to an identity matrix, and $|\det A| = 1$. The group of all unitary matrices with $\det A = 1$ is called the special unitary group $SU\,(n)$ of $n \times n$ complex matrices. Therefore $SU\,(n)$ is a subgroup of the unitary group $U\,(n)$.
   (c) *Special orthogonal group.* Let $A$ be an orthogonal matrix. Therefore the product of this matrix and its transpose is equal to an identity matrix, and $|\det A| = 1$. The group of all orthogonal matrices with $\det A = 1$ is called the special orthogonal group $SO\,(n)$ of $n \times n$ real matrices. Therefore $SO\,(n)$ is a subgroup of the unitary group $O\,(n)$.

Some examples of quotient groups are listed below. In these examples, $I$ is an identity matrix of size $n$. Also $\widehat{\mathbb{R}} \triangleq \mathbb{R}\backslash\{0\}$, and $\widehat{\mathbb{C}} \triangleq \mathbb{C}\backslash\{0\}$.

1. *Projective general linear group* $PGL\,(n,\mathbb{C}) = GL\,(n,\mathbb{C})\,/N$, where $N$ is the normal subgroup $\left\{kI : k \in \widehat{\mathbb{C}}\right\}$.
2. *Projective general linear group* $PGL\,(n,\mathbb{R}) = GL\,(n,\mathbb{R})\,/N$, where $N$ is the normal subgroup $\left\{kI : k \in \widehat{\mathbb{R}}\right\}$.
3. *Projective special linear group.* $PSL\,(n,\mathbb{C}) = SL\,(n,\mathbb{C})\,/N$, where $N$ is the normal subgroup $\{\omega I : \omega \text{ is an } n\text{th root of unity}\}$.
4. *Projective special linear group* $PSL\,(2n,\mathbb{R}) = SL\,(2n,\mathbb{R})\,/N$, where $N$ is the normal subgroup $\{\pm I\}$.

5. *Projective special unitary group.* $PSU(n) = SU(n)/N$, where $N$ is the normal subgroup $\{\omega I : \omega$ is an $n$th root of unity$\}$.

Also note that, $PSL(2,\mathbb{Z}) = SL(2,\mathbb{Z})/\{\pm I\}$ is called the *modular group,* where $I$ is an identity matrix of size 2. As per the general convention, we use the terms group, and the corresponding set upon which the group operator works on, synonymously. For example, $PGL(n,\mathbb{C})$ denotes both the group, and the corresponding set of matrices it operates upon.

### Transformation Group and Group Action

Transformation group is a family of bijections defined upon a set $X$. Group action allows us to consider the group elements as a collection of symmetries of the set $X$. Thus

**Definition 2.21.** *Transformation group. A transformation group on a set $X$ is $\mathcal{T} = (T, *)$, where each $\tau \in T$ is a bijection $\tau : X \to X$, such that*

(a) *If $\tau, \tau' \in T$ then $\tau * \tau' \in T$.*
(b) *The set $T$ contains the identity map.*
(c) *If $\tau \in T$ then its inverse $\tau^{-1} \in T$.*                                                         □

**Definition 2.22.** *Group action, and transitive group action. Let $\mathcal{G} = (G, *)$ be a group on the set $X$.*

1. *Consider the map $\varphi : G \times X \to X$. Also let $g \in G$ and $x \in X$. In this map, $(g, x)$ associates with $g(x) \in X$. The map $\varphi(\cdot, \cdot)$ is called the group action of G over $X$, provided for each $x \in X$:*

    (a) *$e(x) = x$, where $e \in G$ is the unit element of $G$.*
    (b) *$g(h(x)) = (g * h)(x)$ for all $g, h \in G$.*
2. *An action of a group $\mathcal{G}$ on a set $X$ is transitive, if and only if for each $x, y \in X$ there exists a $g \in G$ such that $g(x) = y$.*                                                         □

**Examples 2.8.** Some illustrative examples.

1. Let $\mathcal{G}$ be a group of permutations of the set $X$. The group $\mathcal{G}$ determines a group action of $\mathcal{G}$ on $X$.
2. Consider the metric space $(\mathcal{V}, d)$. Let the group of isometries of this set be $\mathcal{T} = (T, *)$. This fact can be stated alternately by saying that the *group of isometries* $\mathcal{T}$ acts on $(\mathcal{V}, d)$, or we have an *isometric action* of $\mathcal{T}$ on $(\mathcal{V}, d)$.
3. Let $O(n)$ be the set of all orthogonal matrices of size $n$. The orthogonal group of these matrices, with matrix multiplication operator $\times$ is denoted as $\mathcal{O}(n) = (O(n), \times)$. For each $m \leq n$, the action of $\mathcal{O}(n)$ on the $m$-dimensional vector subspaces of $\mathbb{R}^n$ is transitive.                                   □

## 2.4 Vector Spaces over Fields

A set of vectors form a vector space if: addition of any two vectors yields another vector, and multiplication of a vector by a scalar yields another vector. The reader should also consult the chapters on applied analysis, and matrices and determinants for further discussion on vector spaces. A formal definition of vector space over fields is given below.

**Definition 2.23.** *Let $\mathcal{F} = (F, +, \times)$ be a field. A vector space is $\mathcal{V} = (V, \mathcal{F}, \boxplus, \boxtimes)$, where $V$ is a nonempty set of vector elements, and $\boxplus$ and $\boxtimes$ are binary operations.*

(a) *The operation $\boxplus$ is called vector addition, where $\boxplus : V \times V \to V$. For any $u, v \in V$, the sum $u \boxplus v \in V$.*
(b) *The operation $\boxtimes$ is called vector multiplication by a scalar, where $\boxtimes : F \times V \to V$. For any $k \in F$ and $u \in V$, the product $k \boxtimes u \in V$.*

*The algebraic structure, $\mathcal{V}$ is called a vector space over $\mathcal{F}$ if the following axioms hold.*

[*Axiom A1*]  $\forall\, u, v, w \in V$, $(u \boxplus v) \boxplus w = u \boxplus (v \boxplus w)$.
[*Axiom A2*]  *There is a vector $0 \in V$, called the zero vector, such that $u \boxplus 0 = u$ for each $u \in V$.*
[*Axiom A3*]  *For each $u \in V$, there is a vector in $V$, denoted by $-u$, such that $u \boxplus (-u) = 0$. The vector $-u$ is called the inverse vector of $u$.*
[*Axiom A4*]  *Vector addition is commutative.* $\forall\, u, v \in V$, $u \boxplus v = v \boxplus u$.
[*Axiom M1*]  *For any $k \in F$, and any vectors $u, v \in V$, $k \boxtimes (u \boxplus v) = (k \boxtimes u) \boxplus (k \boxtimes v)$.*
[*Axiom M2*]  *For any $a, b \in F$, and any vector $u \in V$, $(a + b) \boxtimes u = (a \boxtimes u) \boxplus (b \boxtimes u)$.*
[*Axiom M3*]  *For any $a, b \in F$, and any vector $u \in V$, $(a \times b) \boxtimes u = a \boxtimes (b \boxtimes u)$.*
[*Axiom M4*]  $\forall\, u \in V$, *and for the unit element $1 \in F$, $1 \boxtimes u = u$.*                     $\square$

The first four axioms describe the additive structure of $\mathcal{V}$. The next four axioms describe the action of the field $\mathcal{F}$ on $V$.

The vector addition $\boxplus$ and the field addition $+$ are quite different, but they are both typically denoted by $+$. Similarly, if $a \in F$, and $u \in V$; $(a \boxtimes u)$ is denoted by $au$. The symbol $0$ is used to denote the additive identities of both $\mathcal{V}$ and $\mathcal{F}$.

A vector space is sometimes called a linear vector space or simply a linear space. The reader should be aware that occasionally it is convenient to specify (sometimes) unambiguously the vector space $\mathcal{V}$ and the field $\mathcal{F}$ by the symbols $V$ and $F$ respectively.

**Observations 2.13.** Let $\mathcal{V}$ be a vector space over a field $\mathcal{F}$.

1. For all $a \in F$ and $0 \in V$, $a0 = 0$.
2. For $0 \in F$ and any vector $u \in V$, $0u = 0$.
3. If $a \in F$, $u \in V$, and $au = 0$, then either $a = 0$ or $u = 0$, or both are equal to $0$.
4. For all $u \in V$, $(-1)\,u = -u$.
5. The difference of two vectors $u$ and $v$ is $u \boxplus (-v) \triangleq (u - v)$, where $-v$ is the negative of $v$.
6. For all $a \in F$ and $u, v \in V$, $a(u - v) = au - av$.
7. For all $u, v, w \in V$, if $u + w = v + w$ then $u = v$.
8. The algebraic structure $(V, \boxplus)$ is an Abelian group.                     $\square$

**Examples 2.9.** Some well-known examples of vector spaces are listed below.

1. The set of all $n$-tuples of real numbers

$$\mathbb{R}^n = \{(x_1, x_2, \ldots, x_n) \mid x_j \in \mathbb{R}, 1 \le j \le n\}$$

Note that $\mathbb{R}^1 = \mathbb{R}$. The zero vector in $\mathbb{R}^n$ is simply

$$(0, 0, \ldots, 0) \triangleq 0$$

2. The set of $n$-tuples of complex numbers

$$\mathbb{C}^n = \{(x_1, x_2, \ldots, x_n) \mid x_j \in \mathbb{C}, 1 \leq j \leq n\}$$

Note that $\mathbb{C}^1 = \mathbb{C}$. The zero vector in $\mathbb{C}^n$ is simply

$$(0, 0, \ldots, 0) \triangleq 0$$

3. The set of polynomials of degree less than $n$ with real coefficients. In this case the scalars belong to the set $\mathbb{R}$. Addition is ordinary polynomial addition and scalar multiplication is the usual scalar-by-polynomial multiplication.                                                   □

A vector can also be written as a single-column matrix or a single-row matrix without any ambiguity. In such cases it is called a column vector or a row vector respectively. With a little misuse of notation, the same symbol for the vector and the corresponding row or column vector is used. Also by convention, if a vector is specified as $u \geq 0$ then the vector $u$ is allowed to take a $0$ value.

**Definitions 2.24.** *Let* $\mathcal{U} = (U, \mathcal{F}, \boxplus, \boxtimes)$, $\mathcal{V} = (V, \mathcal{F}, \boxplus, \boxtimes)$, *and* $\mathcal{W} = (W, \mathcal{F}, \boxplus, \boxtimes)$ *be vector spaces defined over the same field. These spaces also have same addition and multiplication operations.*

1. *Let* $U \neq \varnothing$ *and* $U \subseteq V$, *then* $\mathcal{U}$ *is said to be a vector subspace of* $\mathcal{V}$.
2. *Let* $U$ *and* $W$ *be subsets of the set* $V$. *Their sum is the set* $U + W = \{u + w \mid u \in U, w \in W\}$. *The corresponding vector space is denoted by* $\mathcal{U} + \mathcal{W}$.
3. *If* $U \cap W = \{0\}$, *then* $U + W$ *is denoted by* $U \oplus W$. *The corresponding vector space is denoted by* $\mathcal{U} \oplus \mathcal{W}$. *This sum is called the direct sum.*                    □

**Observations 2.14.** Following observations can be made about the vector space $\mathcal{V}$ and the field $\mathcal{F}$. Also let $\mathcal{U}$ and $\mathcal{W}$ be vector subspaces of $\mathcal{V}$.

1. $\mathcal{U}$ is a vector subspace of $\mathcal{V}$ if and only if $U \neq \varnothing$ and for all $a, b \in F$ and $u, v \in U$, $(au + bv) \in U$.
   Equivalently, $\mathcal{U}$ is a vector subspace of $\mathcal{V}$ if and only if $U \neq \varnothing$ and for all $a \in F$ and $u, v \in U$, $(u + v) \in U$ and $au \in U$.
2. The vector space $\mathcal{V}$ is a vector subspace of itself.
3. The vector space $(\{0\}, \mathcal{F}, \boxplus, \boxtimes)$ is a vector subspace of $\mathcal{V}$.
4. All vector subspaces of $\mathcal{V}$ contain the zero vector $0$.
5. The sum of a collection of subspaces is a subspace.
6. The intersection of a collection of subspaces is a subspace.
7. If $U \subseteq V$, and $W \subseteq V$ then $(U + W) \subseteq V$.
8. Each element of the set $U \oplus W$ can be expressed as $u + w$, where $u \in U$ is unique and $w \in W$ is unique.                                                   □

The concept of linear combination of vectors, basis, independence, and dimension is introduced below.

**Definitions 2.25.** *Let* $\mathcal{V}$ *be a vector space over a field* $\mathcal{F}$.

1. *If* $u_1, u_2, \ldots, u_n \in V$, *then a vector* $u \in V$ *is a linear combination of* $u_1, u_2, \ldots, u_n$ *if* $u = \sum_{j=1}^{n} b_j u_j$, *where* $b_j \in F$ *for* $1 \leq j \leq n$.

2. *Let $S$ be a subset of $V$. The set of all finite linear combinations of vectors in $S$ is the span of the set $S$. Denote it by $L(S)$. Note that $L(S) \subseteq V$.*
   *$L(S)$ is called the space spanned or generated by the set $S$. Observe that $L(\varnothing) = \{0\}$.*

3. *If $S \subseteq V$, and $L(S) = V$ then the set $S$ is called the spanning set of $V$.*

4. *Vectors of a subset $S \subseteq V$ are said to be linearly independent, if for every finite subset $\{u_1, u_2, \ldots, u_n\}$ of $S$, $\sum_{j=1}^{n} b_j u_j = 0$ where $b_j \in F$ implies $b_j = 0$, for all $j = 1, 2, \ldots, n$. In other words, the vectors in the set $\{u_1, u_2, \ldots, u_n\}$ are linearly independent if and only if the vector $u_j$ cannot be represented as a linear combination of the other vectors of the set, where $j = 1, 2, \ldots, n$.*

5. *A subset $S \subseteq V$ is said to be linearly dependent, if it is not linearly independent. In other words, $S$ is linearly dependent if there exists a finite number of distinct vectors $\{u_1, u_2, \ldots, u_n\}$ in $S$ such that $\sum_{j=1}^{n} b_j u_j = 0$ for some combination of $b_j \in F$, $1 \leq j \leq n$, not all zero.*

6. *An independent spanning set of $V$ is called the basis of $V$.*

7. *The cardinality of any basis set of $V$ is called the dimension of $V$, or $\dim(V)$. The dimension of the vector space $V$ is finite, if it has a finite basis; or else $V$ is infinite-dimensional. The dimension of $V$ is sometimes denoted by $\dim(V)$.*

8. *Let $B = (u_1, u_2, \ldots, u_n)$ be an ordered basis set of the vector space $V$, then the coordinates of $u \in V$ with respect to $B$ are $b_1, b_2, \ldots, b_n$, where $u = \sum_{j=1}^{n} b_j u_j$ and $b_j \in F$, $1 \leq j \leq n$. The coordinate-vector $[u]_B$ of $u$ with respect to the ordered basis set $B$ is $(b_1, b_2, \ldots, b_n)$. Note that a basis is an ordered basis, if it is specified as an ordered set.*

9. *Let $x = (x_1, x_2, \ldots, x_n)$ and $y = (y_1, y_2, \ldots, y_n)$ be vectors defined over the field $\mathcal{F}$. The inner product of the vectors $x$ and $y$ is $x \circ y = \sum_{j=1}^{n} x_j y_j \in F$. The vectors $x$ and $y$ are said to be orthogonal, if $x \circ y = 0$. A convenient notation to indicate the orthogonality of two vectors $x$ and $y$ is $x \perp y$.* $\qquad\qquad\qquad\qquad$ $\square$

A general definition of inner product is given in the chapter on applied analysis.

**Observations 2.15.** Let the algebraic structures $\mathcal{T} = (T, \mathcal{F}, \boxplus, \boxtimes)$, $\mathcal{U} = (U, \mathcal{F}, \boxplus, \boxtimes)$, $\mathcal{V} = (V, \mathcal{F}, \boxplus, \boxtimes)$, and $\mathcal{W} = (W, \mathcal{F}, \boxplus, \boxtimes)$ be vector spaces.

1. Let $T$ be a nonempty subset of $V$. If $T \subseteq W \subseteq V$, then $L(T) \subseteq W$.

2. Let $U \subseteq V$, and $W \subseteq V$ then $(U + W) = L(U \cup W)$.

3. Let $U \subseteq V$, and $W \subseteq V$. If $\{u_j\}$ generates $U$, and $\{w_j\}$ generates $W$; then $\{u_j\} \cup \{w_j\} = \{u_j, w_j\}$ generates $U + W$.

4. Let $V$ be a finite-dimensional vector space, such that $\dim(V) = n$.

   (a) Every basis set of $V$ has $n$ elements.

   (b) Any linearly independent set of vectors with $n$ elements is a basis.

   (c) Any set of $m \geq (n + 1)$ vectors is linearly dependent.

   (d) Any set of $m < n$ linearly independent vectors, can be a part of a basis, and can be extended to form a basis of the vector space.

   (e) This vector space is sometimes denoted by $V^{(n)}$ or $V^n$.

5. Let $V$ be a finite-dimensional vector space, where $\dim(V) = n$. If $W \subseteq V$, then $\dim(W) \leq n$.

6. Let $V$ be a vector space, and $U \subseteq V$, and $W \subseteq V$. If the subspaces $\mathcal{U}$ and $\mathcal{W}$ are finite-dimensional, then $\mathcal{U} + \mathcal{W}$ has finite-dimension. Also

$$\dim(\mathcal{U} + \mathcal{W}) = \dim(\mathcal{U}) + \dim(\mathcal{W}) - \dim(\mathcal{U} \cap \mathcal{W})$$

where $\mathcal{U} \cap \mathcal{W} \triangleq (U \cap W, \mathcal{F}, \boxplus, \boxtimes)$. If $\mathcal{V} = \mathcal{U} \oplus \mathcal{W}$ then

$$\dim\left(\mathcal{V}\right) = \dim\left(\mathcal{U}\right) + \dim\left(\mathcal{W}\right)$$

$$\square$$

Some well-known examples of vector spaces are given below.

**Examples 2.10.** Some illustrative examples.

1. The set of complex numbers $\mathbb{C}$ is a two-dimensional vector space over $\mathbb{R}$. It has the ordered basis $\left(1, \sqrt{-1}\right)$. Any pair of complex numbers which are not a real multiple of the other form a basis.

2. The set of complex numbers $\mathbb{C}$ and real numbers $\mathbb{R}$ are infinite-dimensional vector spaces over the field of rational numbers $\mathbb{Q}$.

3. Consider the set of polynomials in $x$. Also assume that these polynomials have degree less than or equal to $n$. The dimension of space of such polynomials is $(n+1)$. Its ordered basis set is $\left(1, x, x^2, \ldots, x^n\right)$.

4. The vector space $F\left[x\right]$ defined over the field $\mathcal{F} = \left(F, +, \times\right)$ has infinite dimension. Its ordered basis set is $\left(1, x, x^2, x^3, \ldots\right)$.

5. In the space $\mathbb{R}^n$, the set of vectors $\{e_1, e_2, e_3, \ldots, e_n\}$ form a basis. These vectors are $e_1 = \left(1, 0, 0, \ldots, 0, 0\right)^T, e_2 = \left(0, 1, 0, \ldots, 0, 0\right)^T, \ldots$, and $e_n = \left(0, 0, 0, \ldots, 0, 1\right)^T$. This set is called the standard basis of $\mathbb{R}^n$ and the vectors are called unit vectors. Note that each of these unit vectors has $n$ elements, and $\dim\left(\mathbb{R}^n\right) = n$.

    The vector of all ones is called an all-1 vector. It is $e = \left(1, 1, 1, \ldots, 1, 1\right)^T$. Also $e = \sum_{i=1}^{n} e_i$.

6. The subspace of $\mathbb{R}^3$ with all 3-tuples of the form $(a, b, a+b)$ has dimension 2. A possible basis of this vector subspace is $\{(1, 0, 1), (0, 1, 1)\}$. Note that the vector $v = (2, -4, -2)$ is in this subspace. This is true, because $v = 2\left(1, 0, 1\right) - 4\left(0, 1, 1\right)$. Thus the coordinate-vector of $v$ with respect to this basis is $(2, -4)$.                                                    $\square$

## 2.5 Linear Mappings

Linear mappings or transformations or operators are functions that map one vector space to another.

**Definitions 2.26.** *Let* $\mathcal{U} = \left(U, \mathcal{F}, \boxplus, \boxtimes\right)$, *and* $\mathcal{V} = \left(V, \mathcal{F}, \boxplus, \boxtimes\right)$ *be vector spaces over the same field* $\mathcal{F} = \left(F, +, \times\right)$.

1. *Linear mapping: A mapping* $f : V \to U$ *is called a linear mapping (or linear transformation or vector space homomorphism) provided the following two conditions are true:*

    (a) $f\left(x \boxplus y\right) = f\left(x\right) \boxplus f\left(y\right)$ *for all* $x, y \in V$.
    (b) $f\left(k \boxtimes x\right) = k \boxtimes f\left(x\right)$ *for all* $x \in V$ *and all* $k \in F$.

    *Thus the mapping* $f$ *is linear if it preserves the two basic operations of vector addition and scalar multiplication.*

2. *Image (or range) and kernel of linear mapping: Let* $f : V \to U$ *be a linear mapping.*

    (a) *Image: The image of* $f$ *denoted by* im $f$, *is the set of image points in* $U$. *Thus*

$$\text{im } f = \{u \in U \mid f(v) = u \text{ for some } v \in V\} \tag{2.16a}$$

(b) *Kernel*: *The kernel of* $f$, *denoted by* $\ker f$ *is the set of points in* $V$ *which map into* $0 \in U$. *Thus*

$$\ker f = \{v \in V \mid f(v) = 0\} \tag{2.16b}$$

3. *Rank and nullity of a linear mapping*: *Let the vector space* $V$ *be of finite dimension and* $f$ : $V \to U$ *be a linear mapping.*

(a) *Rank of a linear mapping* $f$ *is equal to the dimension of its image. Thus*

$$\text{rank } (f) = \dim (\text{im } f) \tag{2.17a}$$

(b) *Nullity of a linear mapping* $f$ *is equal to the dimension of its kernel. Thus*

$$\text{nullity } (f) = \dim (\ker f) \tag{2.17b}$$

□

**Observations 2.16.** Let $\mathcal{U} = (U, \mathcal{F}, \boxplus, \boxtimes)$, and $\mathcal{V} = (V, \mathcal{F}, \boxplus, \boxtimes)$ be vector spaces over the same field $\mathcal{F} = (F, +, \times)$.

1. Let $f : V \to U$ be a linear mapping.
   (a) Let $\dim (\mathcal{V})$ be finite, then the relationships between different dimensions are

$$\dim (\mathcal{V}) = \dim (\text{im } f) + \dim (\ker f) = \text{rank } (f) + \text{nullity } (f) \tag{2.18}$$

   (b) The image of $f$ is a subset of $U$, and the kernel of $f$ is a subset of $V$.
   (c) Let $\{v_1, v_2, \ldots, v_n\}$ be a set of basis vectors of $\mathcal{V}$. Then the vectors $f(v_i) = u_i \in U$, for $1 \le i \le n$ generate im $f$.
   (d) $f(0) = 0$.
2. Let $\{v_1, v_2, \ldots, v_n\}$ be a set of basis vectors of $\mathcal{V}$. Also let $\{u_1, u_2, \ldots, u_n\}$ be any vectors in $U$. Then there exists a unique linear mapping $f : V \to U$ such that $f(v_i) = u_i \in U$, for $1 \le i \le n$. □

Linear mappings are also discussed in the chapter on matrices and determinants. Use of matrices to describe linear mappings makes the above abstract description more concrete.

The structure of finite fields is described in the next section. Finite fields have wide-ranging applications in diverse areas such as combinatorics, algebraic coding theory, cryptography, cryptanalysis, and the mathematical study of switching circuits.

## 2.6  Structure of Finite Fields

Definition of a field has been given in an earlier section. Recall that a field is an algebraic structure $(F, +, \times)$, where the nonempty set $F$ is closed under the two binary operations: *addition* operation denoted by $+$, and *multiplication* operation denoted by $\times$. The algebraic structure $(F, +)$ is an Abelian group, with additive identity called *zero*, which is denoted by $0$. Also the algebraic structure $(F - \{0\}, \times)$ is an Abelian group, with multiplicative identity called *one*, and denoted by $1$. In addition, the distributive law $a \times (b + c) = (a \times b + a \times c)$ holds for all $a, b, c \in F$. A field is finite, if the number of elements in the set $F$ is finite. Finite fields have several applications.

Some of these are: generating pseudorandom numbers, coding theory, switching circuit theory, and cryptography. Our goal in this section is to provide a constructive theory of finite fields. A formal definition of a finite field is given below.

**Definition 2.27.** *A finite field* $\mathbb{F}_q = (F, +, \times)$ *is a field with a finite number of elements in* $F$, *where* $|F| = q \in \mathbb{P}$ *is the order of the field.*                                                                   $\square$

If the context is clear, and there is no ambiguity the symbols $\mathbb{F}_q$ and $F$ are sometimes used interchangeably. It can easily be checked that $\mathbb{Z}_2$ is indeed a field of two elements. However, it can be observed from the addition and multiplication Table 2.2, that $\mathbb{Z}_4$ is not a field.

| + | 0 | 1 | 2 | 3 |  | $\times$ | 0 | 1 | 2 | 3 |
|---|---|---|---|---|---|---|---|---|---|---|
| 0 | 0 | 1 | 2 | 3 |  | 0 | 0 | 0 | 0 | 0 |
| 1 | 1 | 2 | 3 | 0 |  | 1 | 0 | 1 | 2 | 3 |
| 2 | 2 | 3 | 0 | 1 |  | 2 | 0 | 2 | 0 | 2 |
| 3 | 3 | 0 | 1 | 2 |  | 3 | 0 | 3 | 2 | 1 |

Table 2.2. Addition and multiplication arithmetic in $\mathbb{Z}_4$.

Observe in the multiplication table that the number 2 does not have a multiplicative inverse. A brief verification of the arithmetic can establish that $\mathbb{Z}_m$ is a field, if $m$ is a prime number. This verification can be done by completing the addition and multiplication tables. This important observation is stated below as a theorem.

**Theorem 2.5.** $\mathbb{Z}_m$ *is a finite field if and only if* $m$ *is a prime number.*
*Proof.* See the problem section.                                                                                                   $\square$

Next consider a finite field $\mathbb{F}_q$ with $q$ elements with characteristic $p$, a prime number. Therefore $q \geq p$. If $q = p$, then $\mathbb{F}_q$ is isomorphic to $\mathbb{Z}_p$. However, if $q > p$, the field $\mathbb{F}_q$ has $p$ elements which form a subfield isomorphic to $\mathbb{Z}_p$. Denote this subfield by $\mathbb{F}_p$ and call it the *prime subfield* of $\mathbb{F}_q$.

**Theorem 2.6.** *Let the characteristic of the finite field* $\mathbb{F}_q = (F, +, \times)$ *be the prime number* $p$. *The order of this field* $q$ *is a power of* $p$. *That is,* $q = p^n$, *where* $n \in \mathbb{P}$.
*Proof.* See the problem section.                                                                                                   $\square$

A useful representation of a finite field with $q$ elements can be obtained by using polynomial arithmetic. It is called the *polynomial representation* of the field.

Note that $\mathbb{Z}_p[x]$ is a ring of polynomials defined over the field $\mathbb{Z}_p$. If $f(x) \in \mathbb{Z}_p[x]$ is a polynomial, then $\mathbb{Z}_p[x]/(f(x))$ is the set of polynomials in $\mathbb{Z}_p[x]$ obtained by performing modulo $f(x)$ operation on polynomials in $\mathbb{Z}_p[x]$. Assume that the degree of the polynomial $f(x)$ is $n \in \mathbb{P}$. Then a representative of the equivalence class of polynomials in $\mathbb{Z}_p[x]/(f(x))$ has degree less than $n$.

If in addition, it is also assumed that the polynomial $f(x)$ is irreducible, then it is demonstrated in this section that $\mathbb{Z}_p[x]/(f(x))$ is a representation of a finite field of order $p^n$. It is demonstrated later in the section that, for given values of prime number $p$ and $n \in \mathbb{P}$, there always exists at least a single irreducible polynomial of degree $n$. This also guarantees the existence of a finite field of order $p^n$.

In the polynomial arithmetic $\mathbb{Z}_p[x]$ modulo $f(x)$, addition of two polynomials is performed by adding the two given polynomials and then performing the modulo $f(x)$ operation. Similarly, the multiplication of two polynomials is performed by multiplying the two given polynomials and then performing the modulo $f(x)$ operation. The multiplicative inverse of a polynomial can be obtained by the extended Euclidean algorithm applied to the polynomial ring $\mathbb{Z}_p[x]$. This procedure is similar to the extended Euclidean algorithm for integers. Rest of this section deals with discovering different techniques to generate finite fields with $p^n$ elements, where $p$ is a prime number and $n \in \mathbb{P} \setminus \{1\}$.

A technique to construct finite fields is next demonstrated. This is followed by a description of some constructs used in finite field theory like, primitive elements, primitive polynomials, minimal polynomials, and irreducible polynomials. Techniques to factorize polynomials in finite fields which use the theory of cyclotomic polynomials is also discussed. Finally, these ideas are elucidated via detailed examples.

### 2.6.1 Construction

The structure of finite fields can be studied via polynomial-modulo arithmetic performed over a ring of polynomials, where the coefficients of the polynomials belong to a finite field. This approach permits the construction of finite fields with a larger number of elements. For example, if the polynomial coefficients belong to a finite field $\mathbb{Z}_p$, and the degree of the irreducible polynomial used in modulo operation is $n$, then the new field has $p^n$ elements.

Let $F[x]$ be a polynomial ring. The coefficients of the polynomials in this ring belong to the field $\mathcal{F} = (F, +, \times)$. Also let $f(x) \in F[x]$ be a polynomial of degree $n \in \mathbb{P}$, where the polynomial $f(x)$ is typically monic. It has been stated in an earlier section that $F[x] / (f(x))$ is a ring of polynomials modulo $f(x)$. If the field $\mathcal{F}$ is equal to $\mathbb{Z}_p$ and the polynomial $f(x)$ is also irreducible, then the polynomials in $F[x] / (f(x))$ form a field. Denote this field by $\mathbb{F}_q$, where the number of elements in the finite field is equal to $q = p^n$, and the degree of the polynomial $f(x)$ is $n$. This important fact is stated in the following theorem.

**Theorem 2.7.** *Let $f(x) \in \mathbb{Z}_p[x]$ be an irreducible polynomial of degree $n \in \mathbb{P}$, over $\mathbb{Z}_p$, where $p$ is a prime number. Then $\mathbb{F}_q = \mathbb{Z}_p[x] / (f(x))$ is a field with $q = p^n$ elements.*

*Sketch of the proof.* The proof can be obtained by considering the polynomials as $n$-tuples. And each coefficient of a polynomial can take $p$ different values. Also note that $\mathbb{Z}_p[x] / (f(x))$ is a ring of polynomials. Therefore we only need to note that the nonzero elements of this set of polynomials form an Abelian group, under multiplication operation. Since the polynomial $f(x)$ is irreducible, existence of multiplicative inverses can be demonstrated by the use of an extended Euclidean algorithm for polynomials.

The elements of $\mathbb{F}_q$ are all polynomials of degree less than $n$, and the coefficients of these polynomials are in $\mathbb{Z}_p$. Therefore there are $q = p^n$ such polynomials. Alternately, these polynomials can be considered as $n$-tuples over $\mathbb{Z}_p$. These $n$-tuples form a vector space of dimension $n$. In either representation, addition and multiplication operations are performed modulo $f(x)$.                                                                                                       □

The above theorem implicitly demonstrates that the nonzero elements of the field form a multiplicative group.

**Definition 2.28.** *Let $\mathbb{F}_q = (F, +, \times)$ be a finite field. The multiplicative group $(F - \{0\}, \times)$ is denoted by $\mathbb{F}_q^*$.*                                                                                                       □

The order of this multiplicative group $\mathbb{F}_q^*$ is $(q-1)$. The polynomials $\mathbb{Z}_p[x]/(f(x))$ can also be considered to be polynomials in $\alpha$, where $\alpha$ is a zero of $f(x)$. As $\alpha$ is a zero of $f(x)$, we have $f(\alpha) = 0$. Consider a polynomial $g(x) \in \mathbb{Z}_p[x]$. Then $g(x) = \psi(x)f(x) + r(x)$, where $\deg r(x) < \deg f(x)$. Therefore $g(\alpha) = r(\alpha)$.

This field is denoted by either $\mathbb{F}_{p^n}$, or $GF_{p^n}$. Thus

$$GF_{p^n} = \left\{ \sum_{j=0}^{n-1} b_j \alpha^j \mid b_j \in \mathbb{Z}_p, \ 0 \leq j \leq (n-1) \right\} \tag{2.19}$$

This field is also called the Galois field, in honor of the mathematician Evariste Galois. Note that it is also convenient to consider $GF_{p^n}$ as an $n$-dimensional vector space over $\mathbb{Z}_p$. The components of the vector are the coefficients of the polynomial in $\mathbb{Z}_p[x]/(f(x))$. Observe trivially that the finite field $\mathbb{Z}_p \cong GF_p$. This discussion can be summarized into the following set of observations.

**Observations 2.17.** Let $\mathbb{F}_q$ (or $GF_q$) denote a finite field with $q$ elements, and characteristic $p$. The order of this field is $q$ and $p$ is a prime number.

1. Each and every Galois field has a unique smallest subfield, which has a prime number of elements. This subfield is denoted by $\mathbb{F}_p$. It is the prime subfield of the field $\mathbb{F}_q$. Therefore $\mathbb{F}_q$ is an extension field of $\mathbb{F}_p$.
2. Existence and uniqueness of finite fields:
   (a) $q = p^n$ for some positive integer $n$.
   (b) The finite field $\mathbb{F}_q$ is unique up to isomorphism.
3. Two finite fields with the same number of elements are isomorphic to each other. Therefore, every finite field is isomorphic to a Galois field.
4. $\mathbb{F}_q$ can be expressed as a vector space over $\mathbb{Z}_p$. The dimension of this vector space is $n$.    □

**Primitive Elements**

The multiplicative structure of a field is next examined. Let $a$ be an element of the field which is not equal to 0. Then the field must contain $a^2, a^3, a^4, \ldots$, and also the reciprocals $a^{-1}, a^{-2}, a^{-3}, \ldots$. If the field is finite then there exist two integers $k$ and $l$ such that $k \geq l$ and $a^k = a^l$. That is, $a^{k-l} = 1$. Therefore, if $a \subset F - \{0\}$, there exists a sequence of distinct elements $1, a, \ldots, a^{j-1}$, but $a^j = 1$. The integer $j$ is called the order of the element $a$.

**Definitions 2.29.** *Let $a \in GF_q \backslash \{0\}, q = p^n, n \in \mathbb{P}$. The characteristic of the field $p$ is a prime number.*

1. *The smallest possible positive integer $j$ for which $a^j = 1$ is called the order of the element $a$.*
2. *The element $a$ is a primitive element, iff its order is $(q-1)$. Thus, every nonzero element in the field is a power of the primitive element $a$.*
3. *An irreducible polynomial which has the primitive element as its root is called a primitive polynomial.*    □

Thus the primitive element has the highest possible order. Also, the primitive element of $GF_q$ is a generator of the cyclic group of all the nonzero elements of $GF_q$ under the multiplication operation. In general, all irreducible polynomials are not primitive. The existence of primitive elements of the finite field $\mathbb{F}_q$ is next demonstrated.

**Observations 2.18.** Some useful facts about order of nonzero field elements are noted.

1. The order of the element $b$ is $j$ if and only if $j$ divides $(q-1)$. This result follows by the application of Lagrange's theorem to the group $\mathbb{F}_q^*$, and its subgroup $\left(\left\{1, b, \dots, b^{j-1}\right\}, \times\right)$. Moreover, every divisor of $(q-1)$ is an order of some element in $\mathbb{F}_q^*$.
2. Let the order of the field element $a$ be $j$. Then $a^k = 1$ if and only if $j \mid k$.
3. Let $a$ and $b$ be two different field elements with orders $j$ and $k$ respectively, such that $\gcd(j, k) = 1$, then the order of the element $ab$ is $jk$.
4. Let $b$ be an element of the field. If its order is $j$, then the order of the element $b^k$ is $j / \gcd(j, k)$, where $k \in \mathbb{P}$. □

The next result is about the number of elements of a given order.

**Theorem 2.8.** *Let* $\varphi(\cdot)$ *denote the Euler's phi-function. Also let* $\mathbb{F}_q$ *be a finite field, and* $j$ *be any positive integer. If* $j \mid (q-1)$ *then there are* $\varphi(j)$ *elements of the field of order* $j$. *However, if* $j \nmid (q-1)$ *then there are* $0$ *elements of order* $j$ *in the field.*
*Proof.* See the problem section. □

Further observations can be made about the field $\mathbb{F}_q$.

**Observations 2.19.** Some basic observations.

1. As per the above theorem, there are $\varphi(q-1)$ field elements, each of order $(q-1)$. These are indeed the primitive field elements. Consequently, every finite field has a primitive element. Therefore the multiplicative group $\mathbb{F}_q^*$ is cyclic.
   Alternately, it can be stated that: a finite field of order $q$ has $\varphi(q-1)$ primitive elements.
2. Some of the following observations follow immediately from the definition of primitive element.
   (a) $\mathbb{F}_q^*$ is a cyclic group of order $(q-1)$. The binary operator of this group is the multiplication operation of $\mathbb{F}_q$. The cyclic group elements are $F - \{0\} = \left\{a, a^2, \dots, a^{q-1}\right\}$, where $a$ is a primitive element (generator) of the cyclic group. Note that $a^{q-1} = 1$.
   (b) The result $a^{q-1} = 1$, where $a \in F - \{0\}$ is in fact Fermat's little theorem applied to the finite field. It can be checked that any $a \in \mathbb{F}_q$ indeed satisfies the equation $x^q - x = 0$. That is, for all elements $a$ in the field $\mathbb{F}_q$, $a^q = a$. This equation is satisfied by $a = 0$, and also $a \neq 0$.
   Summarizing, if $a$ is a primitive element of $GF_q$, then every nonzero element of $GF_q$ can be expressed as a power of $a$. Furthermore, $a^{q-1} = 1$, where $q = p^n$. □

Some of the above statements are evidently tautologous. Nevertheless, these were stated explicitly for clarity.

**Observation 2.20.** Consider a polynomial $f(x) \in \mathbb{Z}_p[x]$, where $p$ is a prime number. Let $a$ be a root of the polynomial $f(x)$. Then

$$a^p, a^{p^2}, a^{p^3}, \dots$$

are also roots of $f(x)$. This result follows by successive application of Fermat's little theorem. See also the chapter on number theory. □

The above set of observations can be stated alternately as follows.

**Theorem 2.9.** *Let the order of a finite field $\mathbb{F}_q$ be $q$, where $q = p^n$, and $p$ a prime number is its characteristic.*

(a) *Every element of this field satisfies the equation $x^q = x$. That is, all elements of $\mathbb{F}_q$ are roots of the polynomial $(x^q - x)$. Also every nonzero element of this field satisfies the equation $x^{q-1} = 1$.*

(b) *Every element of this field satisfies the equation $x^{q^m} = x$ for every $m \in \mathbb{N}$.*        □

### Subfields

Results about subfields of the field $GF_{p^n}$ are next stated. Some of these results depend upon the following lemma.

**Lemma 2.1.** *Let $m, n \in \mathbb{P}$. Then*

(a) $(x^m - 1) \mid (x^n - 1)$ *iff $m \mid n$, where $x$ is an indeterminate.*

(b) $(s^m - 1) \mid (s^n - 1)$ *iff $m \mid n$, where $s \in \mathbb{P} \setminus \{1\}$.*

(c) $\gcd((x^m - 1), (x^n - 1)) = (x^d - 1)$, *where $\gcd(m, n) = d$.*

*Proof.* The proof is left to the reader.        □

**Observations 2.21.** Let $m, n \in \mathbb{P}$.

1. $GF_{p^m}$ is a subfield of $GF_{p^n}$ iff $\left(x^{p^m - 1} - 1\right) \mid \left(x^{p^n - 1} - 1\right)$.

   This observation follows from the last lemma applied to each field. All nonzero elements of $GF_{p^m}$ satisfy the equation $x^{p^m - 1} = 1$, and all nonzero elements of $GF_{p^n}$ satisfy the equation $x^{p^n - 1} = 1$.

2. $GF_{p^m}$ is a subfield of $GF_{p^n}$ iff $m \mid n$.

   This statement is true because $GF_{p^m}$ is a subfield of $GF_{p^n}$ iff

   $$\left(x^{p^m - 1} - 1\right) \mid \left(x^{p^n - 1} - 1\right) \Leftrightarrow (p^m - 1) \mid (p^n - 1) \Leftrightarrow m \mid n$$

3. Let $GF_{p^m}$ be a subfield of $GF_{p^n}$. The element $a \in GF_{p^n}$ is also in $GF_{p^m}$, iff $a^{p^m} = a$.

   This statement is true because the elements of $GF_{p^m}$ satisfy $x^{p^m} = x$.

4. For each $m \mid n$ there is only one subfield $GF_{p^m}$ of $GF_{p^n}$.        □

### 2.6.2 Minimal Polynomials

The structure of a finite field can be further examined by studying minimal polynomials. Assume that $f(x)$ is an irreducible polynomial of degree $n$, over $\mathbb{Z}_p$, where $p$ is a prime number. Then $\mathcal{F} = \mathbb{Z}_p[x] / (f(x))$ is a field with $q = p^n$ elements. Recall that every element $a$ in the field satisfies the equation $(x^q - x) = 0$. However, it is quite possible for the element $a$ to satisfy a smaller degree equation. This leads to the concept of the minimal polynomial of field element $a$.

A minimal polynomial of a field element $a$ is a monic polynomial of the smallest degree that has $a$ as a root. Existence of the minimal polynomial of element $a$ is guaranteed because the field element $a$ satisfies the equation $(x^q - x) = 0$. A formal definition of the minimal polynomial of the field element $a$ is given below.

**Definition 2.30.** *Consider the field $\mathbb{Z}_p[x]/(f(x))$. A minimal polynomial $\eta_a(x)$ of an element $a$ in this field is the smallest-degree, nonzero monic polynomial over $\mathbb{Z}_p$ such that $\eta_a(a) = 0$.* $\square$

Some properties of minimal and primitive polynomials are listed below.

**Observations 2.22.** Let $f(x) \in \mathbb{Z}_p[x]$ be an irreducible polynomial of $\deg f(x) = n$. The field

$$\mathcal{F} = \mathbb{Z}_p[x]/(f(x)) = GF_{p^n}$$

has $q = p^n$ elements.

1. Let $a \in \mathcal{F}$, and $\eta_a(x) \in \mathbb{Z}_p[x]$ be its minimal polynomial.
    (a) The minimal polynomial $\eta_a(x)$ is both unique and irreducible.
    (b) The degree of the minimal polynomial is a divisor of $n$. This implies $\deg \eta_a(x) \le n$. The equality occurs if $a$ is a primitive element of the finite field.
    (c) If $a$ is a root of the polynomial $g(x) \in \mathbb{Z}_p[x]$, then $\eta_a(x) \mid g(x)$.
    (d) Specifically, the polynomial $\eta_a(x)$ divides the polynomial $(x^q - x)$.
2. Basic properties of primitive polynomials.
    (a) If $a$ is a primitive element of the finite field, then the minimal polynomial $\eta_a(x)$ is also its primitive polynomial, and its degree is equal to $n$.
    (b) The roots of the $n$th degree primitive polynomial are the primitive elements of the finite field.
    (c) In a finite field $GF_{p^n}$, there are exactly $\varphi(q-1)/n$ primitive polynomials of degree $n$, where $\varphi(\cdot)$ is Euler's phi-function. This follows because there are $\varphi(q-1)$ primitive elements in the finite field.
    (d) As a primitive polynomial is irreducible by definition, it can be used to determine polynomial representation of the elements of $GF_{p^n}$. $\square$

See the problem-section for proofs of some of the above observations. A factorized representation of the minimal polynomials is next obtained. In addition to providing a constructive representation of a minimal polynomial, it also proves that the degree of the minimal polynomial is a divisor of $n$.

**Lemma 2.2.** *$g(x)$ is a polynomial with coefficients in the field $\mathbb{Z}_p$, where $p$ is prime number. Then $g\left(x^{p^r}\right) = \{g(x)\}^{p^r}$, $r \in \mathbb{P}$.*
*Proof.* See the problem section. $\square$

**Definition 2.31.** *Let $p$ be a prime number, and $m \in \mathbb{P}$, be relatively prime. The smallest possible positive integer $k$ which satisfies $p^k \equiv 1 \pmod{m}$ is called the multiplicative order of $p$ modulo $m$.* $\square$

The multiplicative order of $p$ modulo $m$ exists because of Euler-Fermat theorem. Let $a \in \mathbb{F}_q$, then the number of distinct elements in the sequence $a, a^p, a^{p^2}, \ldots$ is finite. This number depends upon the order of the element $a$.

**Theorem 2.10.** *Let $a \in \mathbb{F}_q^* \setminus \{1\}$, where $q = p^n$, $p$ is a prime number, and $n \in \mathbb{P}$. Also let the order of the element $a$ be $m \in \mathbb{P}$. Then $a^{p^k} = a$, where $k$ is the multiplicative order of $p$ modulo $m$. Moreover the $k$ elements*

$$a, a^p, a^{p^2}, \ldots, a^{p^{k-1}} \in \mathbb{F}_q^*$$

*are all different.*

*Proof.* By hypothesis $a^m = 1$, and $p^k \equiv 1 \pmod{m}$. This implies $a^{p^k} = a$.

Observe that

$$a^{p^i} = a^{p^j} \quad \text{iff} \quad a^{p^i - p^j} = 1$$

This is true iff $\left(p^i - p^j\right)$ is a multiple of $m$. This occurs iff $p^i \equiv p^j \pmod{m}$, which is true iff $p^{i-j} \equiv 1 \pmod{m}$. This in turn is true iff $(i - j)$ is a multiple of $k$. Therefore the $k$ elements

$$a, a^p, a^{p^2}, \ldots, a^{p^{k-1}} \in \mathbb{F}_q$$

are all different.                                                                         □

The following observation follows from the last lemma and theorem.

**Observation 2.23.** Let $p$ be a prime number, $n \in \mathbb{P}$, $q = p^n$, $a \in \mathbb{F}_q^* \setminus \{1\}$, $m \in \mathbb{P}$, $a^m = 1$, and $p^k \equiv 1 \pmod{m}$.

(a) As $a^{p^k} = a$, and also $a^{p^n} = a$, we have $k \mid n$.

(b) If $a$ is a root of the polynomial $g(x)$ over $\mathbb{Z}_p$. Then

$$a, a^p, a^{p^2}, \ldots, a^{p^{k-1}} \in \mathbb{F}_q$$

are all different roots of the polynomial $g(x)$.

Therefore the minimal polynomial $\eta_a(x)$ of $a$ must have at least $k$ roots.          □

Based upon the above observation, an important theorem about minimal polynomials can be stated.

**Theorem 2.11.** *Let $a \in \mathbb{F}_q^*$, $q = p^n$, where $p$ is a prime number, and $k$ is the smallest positive integer such that $a^{p^k} = a$. Then*

$$\eta_a(x) = \prod_{j=0}^{(k-1)} \left(x - a^{p^j}\right) \tag{2.20}$$

*where $\eta_a(x)$ is the minimal polynomial associated with the field element $a$.*          □

Note that a $k$, as stated in the theorem exists because $a^{p^n} = a$. This theorem is referred to as the factorization theorem of minimal polynomials.

Specifically, if $a \in \mathbb{F}_q^* \setminus \{1\}$, the order of the element $a$ is $m \in \mathbb{P}$, and $k$ is the smallest integer such that $p^k \equiv 1 \pmod{m}$. Then the minimal polynomial of this element $a$ is given by the above expression.

**Definition 2.32.** *Let* $a \in \mathbb{F}_q$, $q = p^n$, *where* $p$ *is a prime number, and* $n \in \mathbb{P}$. *Also* $k$ *is the smallest positive integer greater than one, such that* $a^{p^k} = a$. *The elements*

$$a^p, a^{p^2}, \ldots, a^{p^{k-1}} \in \mathbb{F}_q$$

*are called the conjugates of* $a$. *That is, the elements of a field with the same minimal polynomial are called conjugates. Note that the element* $a$ *and the above sequence of elements form its conjugacy class.*                                                                                                            □

This is analogous to the complex conjugate numbers $i = \sqrt{-1}$ and $-i$. These are the roots of the irreducible real polynomial, $(x^2 + 1)$. Note that if the conjugacy class of an element $a \in \mathbb{F}_q$ has $k$ elements, then $k \mid n$, where $q = p^n$, and $p$ is a prime number. This is true, because members of the conjugacy class form a subfield $GF_{p^k}$ of the field $GF_{p^n}$. This in turn implies $k \mid n$.

### 2.6.3 Irreducible Polynomials

Recall that irreducible polynomials are not factorable over the field $\mathbb{Z}_p$. Some useful facts about these polynomials are next listed.

**Theorem 2.12.** *The polynomial* $\left(x^{p^n} - x\right)$ *is equal to the product of all prime polynomials over* $\mathbb{Z}_p$ *whose degrees divide* $n$.
*Proof.* See the problem section.                                                                            ⊔

**Corollary 2.1.** *The polynomial* $\left(x^{q-1} - 1\right)$, *where* $q = p^n$ *is always divisible by a primitive polynomial of degree* $n$, *over* $\mathbb{Z}_p$. *However, the primitive polynomial does not divide* $(x^m - 1)$ *for any* $m < (q - 1)$.                                                                                                  □

**Observations 2.24.** Some basic observations about irreducible polynomials.

1. Every irreducible monic polynomial of degree $d$ with coefficients in the field $\mathbb{Z}_p$ divides $\left(x^{p^n} - x\right)$ if $d \mid n$.
2. Let $I_d$ be the number of distinct prime polynomials of degree $d$ in $\mathbb{Z}_p[x]$. Then

$$p^n = \sum_{d \mid n} d I_d \tag{2.21}$$

3. Use of the additive version of the Moebius inversion formula yields:

$$I_n = \frac{1}{n} \sum_{d \mid n} \mu(d) \, p^{n/d} \tag{2.22}$$

where $\mu(\cdot)$ is the Moebius function. See the chapter on number theory for its definition. For example,

$$I_1 = p, \ I_2 = \frac{1}{2}\left(p^2 - p\right), \ I_3 = \frac{1}{3}\left(p^3 - p\right), \ I_4 = \frac{1}{4}\left(p^4 - p^2\right), \ldots$$

For large $n$, the value of $I_n$ can be approximated by $p^n / n$.

4. Note that $nI_n$ is a sum of signed and distinct powers of $p$. This sum can never be equal to zero. Therefore $I_n > 0$ for all $p$ and $n \in \mathbb{P}$. That is, there exists at least a single irreducible polynomial of degree $n \in \mathbb{P}$ with coefficients in the field $\mathbb{Z}_p$.

5. The previous observation implies the existence of the field $\mathbb{Z}_p[x] / (f(x))$ with $q = p^n$ elements, where $f(x)$ is an irreducible polynomial of degree $n \in \mathbb{P}$ with coefficients in the field $\mathbb{Z}_p$.

6. All the roots of an irreducible polynomial have the same order.                                         □

### 2.6.4 Factoring Polynomials

In the previous subsection, it has been observed that $\left(x^{p^n} - x\right)$ can be represented as a product of prime polynomials whose degrees divide $n$. Let $W_d(x)$ represent the product of all prime polynomials of degree $d$, and with coefficients in $\mathbb{Z}_p$. Thus

$$x^{p^n} - x = \prod_{d|n} W_d(x) \tag{2.23a}$$

Use of the multiplicative Moebius inversion formula yields:

$$W_n(x) = \prod_{d|n} \left(x^{p^d} - x\right)^{\mu(n/d)} \tag{2.23b}$$

The astute reader will notice a strong similarity between these expressions, and the factorization of $(z^n - 1)$ via cyclotomic polynomials defined over the complex field. Cyclotomic polynomials have been discussed in the chapter on number theory.

**Examples 2.11.** The above results are elucidated via the following examples.

1. Let $p = 2$, and $n = 4$. This yields $q = 2^4 = 16$. The expression $\left(x^{16} - x\right)$ is factorized as follows. The divisors of 4 are $d = 1, 2$, and 4. Note that the number $-1$ in $GF_2$ is 1.

$$W_1(x) = \left(x^2 - x\right) = x(x - 1)$$

$$W_2(x) = \frac{\left(x^4 - x\right)}{\left(x^2 - x\right)} = \left(x^2 + x + 1\right)$$

$$W_4(x) = \frac{\left(x^{16} - x\right)}{\left(x^4 - x\right)} = \left(x^{12} + x^9 + x^6 + x^3 + 1\right)$$

Therefore

$$\left(x^{16} - x\right) = W_1(x) W_2(x) W_4(x)$$
$$= x(x - 1)\left(x^2 + x + 1\right)\left(x^{12} + x^9 + x^6 + x^3 + 1\right)$$

2. Let $p = 2$, $n = 6$. This yields $q = 64$. The expression $W_6(x)$ is factorized as follows. Note that the divisors of $n$ are $d = 1, 2, 3$, and 6. Then

$$W_6(x) = \frac{\left(x^{64} - x\right)\left(x^2 - x\right)}{\left(x^8 - x\right)\left(x^4 - x\right)} = \frac{\left(x^{63} - 1\right)(x - 1)}{\left(x^7 - 1\right)\left(x^3 - 1\right)}$$

The degree of the polynomial $W_6(x)$ is 54. Also notice that the terms in the expression for $W_6(x)$ can be factorized by using the theory of cyclotomic polynomials. This theory was developed for polynomials with coefficients in the complex field $\mathbb{C}$. A similar theory of cyclotomic polynomials $\Phi(\cdot)$ in any field can be developed. Assume that the coefficients in the cyclotomic polynomials are in the field $\mathbb{Z}_p$. Note that

$$(x-1) = \Phi_1(x), \quad (x^3 - 1) = \Phi_1(x)\Phi_3(x), \quad (x^7 - 1) - \Phi_1(x)\Phi_7(x),$$
$$(x^{63} - 1) = \Phi_1(x)\Phi_3(x)\Phi_7(x)\Phi_9(x)\Phi_{21}(x)\Phi_{63}(x)$$

This yields
$$W_6(x) = \Phi_9(x)\Phi_{21}(x)\Phi_{63}(x)$$

Space does not permit further discussion of factorization of the polynomials $W_n(x)$'s into cyclotomic polynomials. $\qquad\square$

A procedure to enumerate the elements of the field $GF_q$ is outlined below.

**Guideline to Find the Elements of the Field $GF_q$.**

(a) Let $p$ be a prime number, $n \in \mathbb{P}$, and $q = p^n$.
(b) Let $\alpha \in GF_q$ be a primitive element of this finite field. That is, all the nonzero elements of $GF_q$ are powers of this element.
(c) Let $\eta_\alpha(x)$ be a primitive polynomial corresponding to the primitive element $\alpha$. Furthermore, $\deg \eta_\alpha(x) = n$, and $\eta_\alpha(x) \mid (x^{q-1} - 1)$.
(d) Let $\xi^j \triangleq x^j \pmod{\eta_\alpha(x)}$ for $0 \le j \le (q-2)$. Then $GF_q \setminus \{0\} = \{\xi^j \mid 0 \le j \le (q-2)\}$.
(e) Also, since $\alpha$ is a primitive element of $GF_q$, and its every nonzero element can be expressed as a power of $\alpha$, then $\xi$ is a primitive element. $\qquad\square$

### 2.6.5  Examples

The different definitions and observations of this section are clarified via several examples.

**Examples 2.12.** Some properties of the field $\mathbb{F}_4 = GF_4$ are studied below.

1. Note that the elements of the field $\mathbb{F}_2 = \mathbb{Z}_2 = GF_2$ belong to the set $\{0, 1\}$. Members of this set are the coefficients of the polynomials which represent the elements of the field $GF_4$.
2. The irreducible monic polynomials of degree 1 are $x$ and $(x+1)$. The irreducible monic polynomial of degree 2 is $(x^2 + x + 1)$. The polynomial $(x^2 + x + 1)$ is irreducible, because its value at either $x = 0$ or 1 is equal to $1 \ne 0$. It can also be verified that it is the only irreducible monic polynomial of degree 2. Then

$$x(x+1)(x^2 + x + 1) \equiv (x^4 + x) \pmod{2}$$

Therefore let
$$f(x) = (x^2 + x + 1)$$

Note that the divisors of $n = 2$ are $d = 1$ and 2. Therefore, the number of irreducible polynomials of degree 1 that divide $(x^4 + x)$ is $I_1 = 2$. Similarly the number of irreducible polynomials of degree 2 that divide $(x^4 + x)$ is $I_2 = 1$. It can be checked that $(I_1 + 2I_2) = 2^2 = 4$.
3. Different types of representations of elements of the field $\mathbb{F}_4 = GF_4$ are described below.

(a) Elements of the field $\mathbb{F}_4 = GF_4$ belong to the set

$$\{b_1 x + b_0 \mid b_i \in \{0, 1\}, \ 0 \le i \le 1\}$$

This set is $\{0, 1, x, (x + 1)\}$. The number of elements in this set is 4.

(b) For notational convenience, the members of the above set can also be represented by a vector $(b_1 b_0)$ of length 2. Therefore, this set is equivalent to

$$\{(b_1 b_0) \mid b_i \in \{0, 1\}, \ 0 \le i \le 1\}$$

(c) Another way to represent these elements is to convert the binary representation $(b_1 b_0)$ into its decimal value.

(d) One more convenient rule to remember is to substitute $x^2 + x + 1 = 0$ in the arithmetic of these polynomials. Let $\alpha$ be a root of this polynomial, then $\alpha^2 = \alpha + 1$, where the coefficient arithmetic is in $GF_2$. Note that $\alpha^3 = \alpha(\alpha + 1) = \alpha^2 + \alpha = 1 = \alpha^0$.

Different representations of the elements of $GF_4$ are shown in the Table 2.3.

| Polynomial representation | Power representation | Binary representation | Decimal representation |
|---|---|---|---|
| 0 | 0 | 00 | 0 |
| 1 | $\alpha^0$ | 01 | 1 |
| $x$ | $\alpha^1$ | 10 | 2 |
| $x + 1$ | $\alpha^2$ | 11 | 3 |

Table 2.3. Representation of elements of $GF_4$.

4. The element $\alpha$ is called the generator because all the nonzero elements of $GF_4$ are obtained as a power of $\alpha$. The order of the element $\alpha$ is 3. Therefore the nonzero elements of $GF_4$ can be obtained from the set

$$\{\alpha^i \ (\mathrm{mod}\ (\alpha^2 + \alpha + 1)) \mid 0 \le i \le 2\}$$

Since $\alpha$ is a generator of the multiplicative group of $GF_4$, then $(x^2 + x + 1)$ is the corresponding primitive polynomial. Note that the degree of the primitive polynomial is 2.

5. The field $GF_4$ is completely specified by the addition and multiplication Tables 2.4 and 2.5. For example, consider the two elements: $(x + 1)$ and $x$. The binary representations of these elements are 11 and 10 respectively. The corresponding decimal representations are 3 and 2 respectively. The addition and multiplication of these two elements modulo $(x^2 + x + 1)$ is:

$$\{(x + 1) + x\} \equiv 1 \ (\mathrm{mod}\ (x^2 + x + 1))$$
$$= 01 \ \text{in binary notation}$$
$$= 1 \ \text{in decimal notation}$$
$$\{(x + 1) \times x\} \equiv 1 \ (\mathrm{mod}\ (x^2 + x + 1))$$
$$= 01 \ \text{in binary notation}$$
$$= 1 \ \text{in decimal notation}$$

Therefore in the addition table, the entry in the fourth row and third column in decimal notation is 1. Similarly, the corresponding entry in the multiplication table in decimal notation is 1. It can be observed that the polynomial or binary representation of elements is more convenient to generate the addition table. Each addition requires only vector addition modulo 2. Also the multiplication table can be generated by using the power representation of the elements of $GF_4$. The above addition and multiplication rules complete the specification of the field $GF_4$.

| + | 00 | 01 | 10 | 11 |
|----|----|----|----|----|
| 00 | 00 | 01 | 10 | 11 |
| 01 | 01 | 00 | 11 | 10 |
| 10 | 10 | 11 | 00 | 01 |
| 11 | 11 | 10 | 01 | 00 |

| × | 00 | 01 | 10 | 11 |
|----|----|----|----|----|
| 00 | 00 | 00 | 00 | 00 |
| 01 | 00 | 01 | 10 | 11 |
| 10 | 00 | 10 | 11 | 01 |
| 11 | 00 | 11 | 01 | 10 |

Table 2.4. Addition and multiplication rules for arithmetic in $GF_4$
in binary notation.

| + | 0 | 1 | 2 | 3 |
|---|---|---|---|---|
| 0 | 0 | 1 | 2 | 3 |
| 1 | 1 | 0 | 3 | 2 |
| 2 | 2 | 3 | 0 | 1 |
| 3 | 3 | 2 | 1 | 0 |

| × | 0 | 1 | 2 | 3 |
|---|---|---|---|---|
| 0 | 0 | 0 | 0 | 0 |
| 1 | 0 | 1 | 2 | 3 |
| 2 | 0 | 2 | 3 | 1 |
| 3 | 0 | 3 | 1 | 2 |

Table 2.5. Addition and multiplication rules for arithmetic in $GF_4$
in decimal notation.

6. The minimal polynomials of the elements of $GF_4$ are listed in Table 2.6.

| Polynomial representation | Power representation | Minimal polynomial |
|---|---|---|
| $0$ | $0$ | $x$ |
| $1$ | $\alpha^0$ | $x + 1$ |
| $x$ | $\alpha^1$ | $x^2 + x + 1$ |
| $x + 1$ | $\alpha^2$ | $x^2 + x + 1$ |

Table 2.6. Minimal polynomials of elements of $GF_4$.

Use of the factorization theorem of minimal polynomials yields

$$\eta_0 (x) = x$$
$$\eta_1 (x) = (x - \alpha^0) = (x + 1)$$
$$\eta_\alpha (x) = \eta_{\alpha^2} (x) = (x - \alpha)(x - \alpha^2) = (x^2 + x + 1)$$

These results follow by noting that $(\alpha^2 + \alpha) \equiv 1 \pmod 2$, and $\alpha^3 \equiv 1 \pmod 2$. The degrees of the minimal polynomial of the element $\alpha$ and its conjugate $\alpha^2$ are each 2.  □

**Examples 2.13.** Some properties of the field $\mathbb{F}_8 = GF_8$ are next studied.

1. Members of the set $\mathbb{Z}_2 = \{0, 1\}$ are the coefficients of the polynomials which represent the elements of the field $GF_8$.
2. The irreducible monic polynomials of degree 1 are $x$, and $(x + 1)$. The irreducible monic polynomials of degree 3 are

$$(x^3 + x + 1), \quad \text{and} \quad (x^3 + x^2 + 1)$$

These are the only two irreducible monic polynomials of degree 3. Then

$$x (x + 1) (x^3 + x + 1) (x^3 + x^2 + 1) \equiv (x^8 + x) \pmod 2$$

Select

$$f (x) = (x^3 + x + 1)$$

3. Different types of representations of elements of the field $\mathbb{F}_8 = GF_8$ are described below. Different representations of the elements of $GF_8$ are shown in the Table 2.7.

| Polynomial representation | Power representation | Binary representation | Decimal representation |
|:---:|:---:|:---:|:---:|
| 0 | 0 | 000 | 0 |
| 1 | $\alpha^0$ | 001 | 1 |
| $x$ | $\alpha^1$ | 010 | 2 |
| $x + 1$ | $\alpha^3$ | 011 | 3 |
| $x^2$ | $\alpha^2$ | 100 | 4 |
| $x^2 + 1$ | $\alpha^6$ | 101 | 5 |
| $x^2 + x$ | $\alpha^4$ | 110 | 6 |
| $x^2 + x + 1$ | $\alpha^5$ | 111 | 7 |

Table 2.7. Representation of elements of $GF_8$.

(a) Elements of the field $\mathbb{F}_8 = GF_8$ belong to the set

$$\{b_2 x^2 + b_1 x + b_0 \mid b_i \in \{0, 1\}, \ 0 \le i \le 2\}$$

This set is

$$\{0, 1, x, (x + 1), x^2, (x^2 + 1), (x^2 + x), (x^2 + x + 1)\}$$

The number of elements in this set is 8.

(b) For notational convenience, the members of the above set can also be represented by a vector $(b_2 b_1 b_0)$ of length 3. Therefore, this set is equivalent to

$$\{(b_2 b_1 b_0) \mid b_i \in \{0, 1\}, \ 0 \le i \le 2\}$$

(c) Another way to represent these elements is to convert the binary representation $(b_2 b_1 b_0)$ into its decimal value.

(d) One more convenient rule to remember is to substitute $x^3 + x + 1 = 0$ in the arithmetic of these polynomials. Let $\alpha$ be a root of this polynomial, then $\alpha^3 = \alpha + 1$, as the coefficient arithmetic is in $GF_2$. Note that

$$\alpha^4 = \alpha (\alpha + 1) = \alpha^2 + \alpha$$
$$\alpha^5 = \alpha (\alpha^2 + \alpha) = \alpha^3 + \alpha^2 = \alpha^2 + \alpha + 1$$
$$\alpha^6 = \alpha (\alpha^2 + \alpha + 1) = \alpha^3 + \alpha^2 + \alpha = \alpha^2 + 1$$
$$\alpha^7 = \alpha (\alpha^2 + 1) = \alpha^3 + \alpha = 1$$

4. The element $\alpha$ is the generator because, all the nonzero elements of $GF_8$ can be obtained from the set

$$\{\alpha^i \ (\mathrm{mod} \ (\alpha^3 + \alpha + 1)) \mid 0 \le i \le 6\}$$

Since $\alpha$ is a generator of the multiplicative group of $GF_8$, then $(x^3 + x + 1)$ is the corresponding primitive polynomial. Note that the degree of the primitive polynomial is 3.

5. The field $GF_8$ is completely specified by the addition and multiplication Table 2.8.

| + | 0 | 1 | 2 | 3 | 4 | 5 | 6 | 7 |
|---|---|---|---|---|---|---|---|---|
| 0 | 0 | 1 | 2 | 3 | 4 | 5 | 6 | 7 |
| 1 | 1 | 0 | 3 | 2 | 5 | 4 | 7 | 6 |
| 2 | 2 | 3 | 0 | 1 | 6 | 7 | 4 | 5 |
| 3 | 3 | 2 | 1 | 0 | 7 | 6 | 5 | 4 |
| 4 | 4 | 5 | 6 | 7 | 0 | 1 | 2 | 3 |
| 5 | 5 | 4 | 7 | 6 | 1 | 0 | 3 | 2 |
| 6 | 6 | 7 | 4 | 5 | 2 | 3 | 0 | 1 |
| 7 | 7 | 6 | 5 | 4 | 3 | 2 | 1 | 0 |

| × | 0 | 1 | 2 | 3 | 4 | 5 | 6 | 7 |
|---|---|---|---|---|---|---|---|---|
| 0 | 0 | 0 | 0 | 0 | 0 | 0 | 0 | 0 |
| 1 | 0 | 1 | 2 | 3 | 4 | 5 | 6 | 7 |
| 2 | 0 | 2 | 4 | 6 | 3 | 1 | 7 | 5 |
| 3 | 0 | 3 | 6 | 5 | 7 | 4 | 1 | 2 |
| 4 | 0 | 4 | 3 | 7 | 6 | 2 | 5 | 1 |
| 5 | 0 | 5 | 1 | 4 | 2 | 7 | 3 | 6 |
| 6 | 0 | 6 | 7 | 1 | 5 | 3 | 2 | 4 |
| 7 | 0 | 7 | 5 | 2 | 1 | 6 | 4 | 3 |

Table 2.8. Addition and multiplication rules for arithmetic in $GF_8$
in decimal notation.

6. The minimal polynomials of the elements of $GF_8$ are listed in Table 2.9.

| Polynomial representation | Power representation | Minimal polynomial |
|---|---|---|
| 0 | 0 | $x$ |
| 1 | $\alpha^0$ | $x + 1$ |
| $x$ | $\alpha^1$ | $x^3 + x + 1$ |
| $x + 1$ | $\alpha^3$ | $x^3 + x^2 + 1$ |
| $x^2$ | $\alpha^2$ | $x^3 + x + 1$ |
| $x^2 + 1$ | $\alpha^6$ | $x^3 + x^2 + 1$ |
| $x^2 + x$ | $\alpha^4$ | $x^3 + x + 1$ |
| $x^2 + x + 1$ | $\alpha^5$ | $x^3 + x^2 + 1$ |

Table 2.9. Minimal polynomials of elements of $GF_8$.

Use of the factorization theorem of minimal polynomials yields

$$\eta_0(x) = x$$
$$\eta_1(x) = \left(x - \alpha^0\right) = (x + 1)$$
$$\eta_\alpha(x) = \eta_{\alpha^2}(x) = \eta_{\alpha^4}(x)$$
$$= (x - \alpha)\left(x - \alpha^2\right)\left(x - \alpha^4\right)$$
$$= \left(x^3 + x + 1\right)$$
$$\eta_{\alpha^3}(x) = \eta_{\alpha^5}(x) = \eta_{\alpha^6}(x)$$
$$= \left(x - \alpha^3\right)\left(x - \alpha^5\right)\left(x - \alpha^6\right)$$
$$= \left(x^3 + x^2 + 1\right)$$

These results follow by noting that

$$\left(\alpha^3 + \alpha\right) \equiv 1\,(\mathrm{mod}\,2), \quad \alpha^7 \equiv 1\,(\mathrm{mod}\,2)$$

The degree of the minimal polynomial of the element $\alpha$ and its conjugates $\alpha^2$ and $\alpha^4$, is 3. Also, the degree of the minimal polynomial of the element $\alpha^3$ and its conjugates $\alpha^5$ and $\alpha^6$, is 3. $\qquad\square$

## 2.7 Roots of Unity in Finite Field

In this section, the $n$th root of unity and the primitive root of unity are defined. Roots of unity in the complex number field $\mathbb{C}$, is discussed in the chapter on applied analysis.

**Definition 2.33.** *Let $\mathbb{F}$ be a field, then an $n$th root of unity is a root of the polynomial $(x^n - 1) \in \mathbb{F}[x]$, where $n \in \mathbb{P}$.* $\qquad\square$

Evidently, $1$ is an $n$th root of unity. However, the other $(n-1)$ roots of unity will not in general belong to the field $\mathbb{F}$. If $\mathbb{F}$ is the Galois field $GF_p$, then we show in this section that there exists an integer $m \in \mathbb{P}$ such that the $n$ roots of unity belong to an extension field $GF_{p^m}$. The relationship between the integers $m, n$, and $p$ is next explored.

Assume that $n$ and $p$ are relatively prime. Therefore, by Euler-Fermat theorem, there exists a smallest integer $m$ such that $p^m \equiv 1 \pmod{n}$. This implies $n \mid (p^m - 1)$. The integer $m$ is called the multiplicative order of $p$ modulo $n$. Thus $(x^n - 1)$ divides $\left(x^{(p^m-1)} - 1\right)$, but does not divide $\left(x^{(p^s-1)} - 1\right)$ for $0 < s < m$. Therefore the roots of the polynomial $(x^n - 1)$ lie in the extension field $GF_{p^m}$ and in no smaller field.

Also observe that the first derivative of $(x^n - 1)$ with respect to $x$ is $n x^{n-1}$. This derivative is relatively prime to $(x^n - 1)$, since $n$ and $p$ are relatively prime. Thus the roots of the polynomial $(x^n - 1)$ are all distinct. This discussion is summarized in the following set of observations.

**Observations 2.25.** Some basic observations about roots of unity in a finite field.

1. The unit element $1$ of the field $\mathbb{F}$, is an $n$th root of unity.
2. The roots of unity, do not always belong to the field $\mathbb{F}$, but to an extension field of $\mathbb{F}$.
3. Let $p$ be a prime number. The $n$th roots of unity of $(x^n - 1) \in \mathbb{Z}_p[x]$ belong to the field $GF_{p^m} \triangleq \mathbb{F}_q$, where $m$ is the smallest positive integer such that $n \mid (p^m - 1)$. Thus there are $n$ distinct elements

$$\alpha_i \in GF_{p^m}, \ \ 0 \le i \le (n-1)$$

    such that

$$(x^n - 1) = \prod_{i=0}^{n-1} (x - \alpha_i)$$

where $\alpha_0 = 1$. Therefore $GF_{p^m}$ is called the *splitting field* of $(x^n - 1)$. $\qquad\square$

**Definition 2.34.** *Let the field $\mathbb{F}$ be the Galois field $GF_p$ where $p$ is a prime number. Also let $q = p^m$, where $m \in \mathbb{P}$, and $\alpha \in GF_q$ be a root of the polynomial $(x^n - 1) \in \mathbb{Z}_p[x]$. If $\alpha, \alpha^2, \ldots, \alpha^n$ are all distinct roots of this polynomial, then $\alpha$ is called a primitive $n$th root of unity.* $\qquad\square$

**Example 2.14.** The fifth roots of unity in $GF_2$ are determined. In this example $n = 5$, and $p = 2$. As $(2^4 - 1) = 15, m = 4$. Therefore the roots belong to the extension field $GF_q$ where $q = 2^4 = 16$. Observe that

$$\left(x^5 - 1\right) = (x + 1)\left(x^4 + x^3 + x^2 + x + 1\right)$$

The minimal polynomial of 1 is $(x+1)$. The minimal polynomial of the remaining four fifth roots of unity is $(x^4 + x^3 + x^2 + x + 1)$. This polynomial is irreducible over $GF_2$. That is, the four roots of $(x^4 + x^3 + x^2 + x + 1)$ in $GF_{16}$ are the primitive fifth roots of unity in $GF_2$.  □

## 2.8 Elliptic Curves

Elliptic curves have found wide spread applications in cryptography, primality testing of integers, and factorization of positive integers. Recall from high school geometry that an ellipse is obtained by a cutting plane which does not pass through the vertex of a cone, and is not parallel to the generators of the cone. Ellipses are described by quadratic equations. However, elliptic curves are not ellipses. Actually elliptic curves are cubic equations. Elliptic curves are so named because, these arose in the computation of arc-lengths of ellipses. The arc-length of an ellipse in turn is specified by an integral of type:

$$\int_u^v \frac{dx}{\sqrt{x^3 + ax^2 + bx + c}}$$

This type of integral is called an elliptic integral. Elliptic curves over real and finite fields are studied in this section. It turns out that for arithmetic in these fields, the points on the elliptic curve form an Abelian group (H. Poincaré). Elliptic curves have a rich theory, however only sufficient theoretical details are provided in this section to make the application of elliptic curves to cryptography comprehensible. Elliptic curves are related to cubic equations. A general form of a cubic equation in two variables $x$ and $y$ is

$$ax^3 + bx^2 y + cxy^2 + dy^3 + ex^2 + fxy + gy^2 + hx + iy + j = 0 \qquad (2.24)$$

where $a, b, c, d, e, f, g, h, i, j \in \mathbb{R}$ are constants; $a, b, c,$ and $d$ are not simultaneously equal to zero; and $x, y \in \mathbb{R}$. A more specialized form of the cubic curve is

$$y^2 + axy + by = (x^3 + cx^2 + dx + e) \,;\; a, b, c, d, e \in \mathbb{R},\; x, y \in \mathbb{R} \qquad (2.25a)$$

This form of equation is called the Weierstrass form of the cubic curve. It is named after the mathematician Karl Weierstrass (1815-1897), who made major contributions to the theory of elliptic functions. Completing the square on the left side and making a linear change of variables in the above equation, results in a curve of form

$$y^2 = (x^3 + ax^2 + bx + c) \,;\; a, b, c \in \mathbb{R},\; x, y \in \mathbb{R} \qquad (2.25b)$$

The above equation represents the standard form of an elliptic curve. A group law on the elliptic curve can be defined by drawing chords and tangents to an elliptic curve. For simplicity, consider the elliptic curve $y^2 = f(x)$, where $f(x)$ is a cubic polynomial. Denote the derivative of $y$ and $f(x)$ with respect to $x$ by $dy/dx$ and $f'(x)$ respectively. Therefore $f'(x) = 2y\, dy/dx$. Observe that, if at a point $(x_0, y_0)$

$$y_0 = f'(x_0) = 0$$

the first derivative $dy/dx$ is indeterminate at the point $(x_0, y_0)$. Therefore, the tangent line to the curve at a point $(x_0, y_0)$ is undefined if $y_0 = f'(x_0) = 0$. This occurs when $f(x_0) = f'(x_0) = 0$, which in turn implies that $f(x)$ has at least a double root at $x = x_0$. For example, if $f(x)$ is a cubic, say $x^3$, then a tangent line at $(0, 0)$ cannot be defined. See the Figure 2.1.

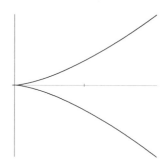

Figure 2.1. $y^2 = x^3$.

This example suggests that the behavior of a curve can be evaluated by studying its roots. The roots of a curve in turn can be studied by evaluating its discriminant. This approach also helps in discovering the remarkable existence of the group law for the elliptic curves.

### Study of a Cubic Polynomial via its Discriminant

Special types of Weierstrass form of the cubic curves are studied in this subsection. As mentioned earlier in this subsection, a polynomial can be conveniently studied by examining its roots. The roots of a polynomial in turn can be studied via its discriminant. This method can also be used to study the multiplicity of roots of a polynomial. Discriminant of a polynomial has been defined in the chapter on applied analysis.

Let the degree of a polynomial $f(x) \in \mathbb{R}[x]$ be $d \geq 2$, and

$$f(x) = \alpha \prod_{k=1}^{d} (x - x_k)$$

where $\alpha \in \mathbb{R} \setminus \{0\}$. The roots of the polynomial $f(x)$ are $x_1, x_2, \ldots, x_d \in \mathbb{C}$. These roots are not necessarily distinct. The discriminant of the polynomial $f(x)$ is

$$\Delta(f) = \alpha^{2d-2} \prod_{1 \leq j < k \leq d} (x_j - x_k)^2$$

**Lemma 2.3.** *Let $f(x) \in \mathbb{R}[x]$, and $r \in \mathbb{C}$ be a root of the polynomial $f(x)$, where the imaginary part of $r$ is nonzero. Then the complex conjugate of $r$ is also a root of the polynomial $f(x)$.*

*Proof.* The proof is left to the reader.                                                    $\square$

**Theorem 2.13.** *Let $f(x) \in \mathbb{R}[x]$ be a monic polynomial of degree $d \geq 2$, and*

$$f(x) = \prod_{k=1}^{d} (x - x_k) \tag{2.26}$$

*The roots of this polynomial $x_1, x_2, \ldots, x_d \in \mathbb{C}$, are not necessarily distinct. Let the discriminant of this polynomial be $\Delta(f)$.*

*Then $\Delta(f) = 0$ iff $f(x)$ has a root of multiplicity at least 2.*

*If all the roots of $f(x)$ are real, then $\Delta(f) \geq 0$. If $d = 2, 3$, then the converse is true.*
*Proof.* The proof is left to the reader.                                                    □

**Examples 2.15.** Some illustrative examples.

1. Consider the quadratic polynomial, $f(x) = ax^2 + bx + c$, where $a, b, c \in \mathbb{R}$ and $a \neq 0$. Let $f(x) = a(x - x_1)(x - x_2)$. Then

$$(x_1 + x_2) = -\frac{b}{a}$$

$$x_1 x_2 = \frac{c}{a}$$

$$\Delta(f) = a^2 (x_1 - x_2)^2 = (b^2 - 4ac)$$

Several cases occur:

(a) If $\Delta(f) > 0$, that is if $(b^2 - 4ac) > 0$, then the polynomial $f(x)$, has distinct real roots.
(b) If $\Delta(f) < 0$, that is if $(b^2 - 4ac) < 0$, then the polynomial $f(x)$, has complex conjugate roots.
(c) If $\Delta(f) = 0$, that is if $(b^2 - 4ac) = 0$, then the polynomial $f(x)$, has equal real roots.

2. Consider a cubic polynomial $f(x) = x^3 + ax + b$, where $a, b \in \mathbb{R}$. Let

$$f(x) = (x - x_1)(x - x_2)(x - x_3)$$

Then

$$(x_1 + x_2 + x_3) = 0$$

$$(x_1 x_2 + x_2 x_3 + x_3 x_1) = a$$

$$x_1 x_2 x_3 = -b$$

It can be shown that

$$(x_1^2 + x_2^2 + x_3^2) = -2a$$

$$(x_1^2 x_2^2 + x_2^2 x_3^2 + x_3^2 x_1^2) = a^2$$

Note that

$$\Delta(f) = (x_1 - x_2)^2 (x_1 - x_3)^2 (x_2 - x_3)^2$$

Also the derivative of $f(x)$ with respect to $x$, evaluated at $x_i$ is

$$(x_i - x_j)(x_i - x_k) = (3x_i^2 + a); \quad i \neq j \neq k, \quad 1 \leq i, j, k \leq 3$$

Thus

$$\Delta(f) = (-1)^3 (3x_1^2 + a)(3x_2^2 + a)(3x_3^2 + a)$$
$$= -(4a^3 + 27b^2)$$

Different cases for the range of values of $\Delta(f)$ are:

(a) If $\Delta(f) > 0$, that is if $(4a^3 + 27b^2) < 0$, then the polynomial $f(x)$, has three distinct real roots.
(b) If $\Delta(f) < 0$, that is if $(4a^3 + 27b^2) > 0$, then the polynomial $f(x)$, has only one real root, and two complex roots. The complex roots are conjugate of each other.
(c) If $\Delta(f) = 0$, that is if $(4a^3 + 27b^2) = 0$, then the polynomial $f(x)$, has all real and at least two equal roots.                                                                           □

Elliptic curves of the form $y^2 = f(x)$ where $\Delta(f) \neq 0$, are called *nonsingular*.

### 2.8.1 Elliptic Curves over Real Fields

In this subsection, the study of elliptic curves is focused on the equation:

$$y^2 = \left(x^3 + ax + b\right); \; a, b \in \mathbb{R}, \; x, y \in \mathbb{R} \tag{2.27}$$

A plot of this curve requires the computation of $y = \pm \left(x^3 + ax + b\right)^{1/2}$. It can be observed that this curve is symmetric about the $x$-axis, which is $y = 0$. See Figure 2.2.

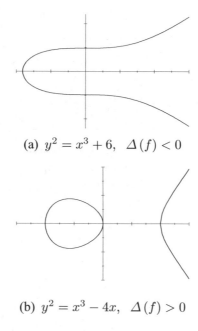

(a) $y^2 = x^3 + 6, \;\; \Delta\left(f\right) < 0$

(b) $y^2 = x^3 - 4x, \;\; \Delta\left(f\right) > 0$

Figure 2.2. Elliptic curves.

Our aim is to define a group structure on this curve. To succeed in this attempt, define a *point at infinity* or the *zero-point* as $O$. This point turns out to be the identity element of the group. Also, this point at infinity $O$ should be imagined to be lying infinitely far up the $y$-axis. A precise meaning of this zero-point will become clear shortly. In order to define a group structure, it is also required that $\left(x^3 + ax + b\right)$ has no repeated factors. This implies t hat $\left(4a^3 + 27b^2\right) \neq 0$. Therefore the elliptic curve is nonsingular.

Denote all the points on this elliptic curve and the point at infinity by $\mathbb{E}\left(a, b\right)$. The group structure is defined on these points. Our next goal is to define the group operation. Denote it by $\uplus$. A geometric, and then an analytic description of this group operation is given.

#### Geometric Description of the Group Operation

The principle of group operation can be stated as follows: *Let $P, Q$, and $R$ be any three points which belong to the set $\mathbb{E}\left(a, b\right)$. If these three points are on the same straight line $L$, then $\left(P \uplus Q \uplus R\right) = O$. That is the group sum of any three collinear points is equal to the zero-point.* This rule is next elaborated.

(a) Assume that $\left(x^3 + ax + b\right)$ has no repeated factors. That is, the elliptic curve is nonsingular.

(b) $O$ is the additive identity. Also $O = -O$. In addition, if $P \in \mathbb{E}(a,b)$, then $(P \uplus O) = (O \uplus P) = P$.

(c) Two equivalent methods for finding the negative of a point which belong to the set $\mathbb{E}(a,b)$ are:

(i) Let $P = (x,y) \neq O$, then $-P = (x,-y)$. That is the negative of a point is the mirror image of the point $P$, across the $x$-axis. Observe from the definition of the elliptic curve that if the point $(x,y)$ is on the curve, then the point $(x,-y)$ is also on the curve. Note that $P \uplus (-P) = O$, and these two points can be joined to produce a vertical line $L$. In this context, the third point on the line joining points $P$ and $-P$, can be considered to be the zero-point $O$.

(ii) Assume that $P, Q \in \mathbb{E}(a,b)$, and the points $P$ and $Q$ have the *same x-coordinates but* $P \neq Q$, then define the addition operation as $(P \uplus Q) = O$.

(d) Assume that $P, Q \in \mathbb{E}(a,b)$ and the points $P$ and $Q$ have *different x-coordinates*, then the line $PQ$ intersects the curve at one more point $R$. Let this third point $R$ be $(x,y)$. Define the addition operation as $(P \uplus Q) = (x,-y)$.

(e) If $P = (x,0) \in \mathbb{E}(a,b)$ then define $(P \uplus P) = O$.

(f) If $P = Q = (x_0, y_0)$ with $y_0 \neq 0$, then let the line $L$ be tangent to the curve at $P$. Also let the third point of intersection of this line with the elliptic curve be at $(x,y)$. Then define $(P \uplus P) = (x,-y)$.

The above set of rules can be used to prove that the algebraic structure $(\mathbb{E}(a,b), \uplus)$ is indeed an Abelian group.

### Algebraic Description of the Group Operation

An algebraic description of the group operation $\uplus$ is specified below.

**Definition 2.35.** *Let* $\mathbb{E}(a,b)$ *include the zero-point* $O$, *and the set of points on the elliptic curve* $E$:

$$y^2 = \left(x^3 + ax + b\right) \tag{2.28}$$

*The variables* $x,y \in \mathbb{R}$, *constants* $a,b \in \mathbb{R}$, *and* $\left(4a^3 + 27b^2\right) \neq 0$. *Let* $P = (x_1, y_1)$ *and* $Q = (x_2, y_2)$ *be two points in* $\mathbb{E}(a,b)$. *The operation* $P \uplus Q$ *is:*

(a) *If* $P = Q = O$, *then* $P \uplus Q = O$.

(b) *If* $Q = O$, *then* $P \uplus Q = P$. *Similarly, if* $P = O$, *then* $P \uplus Q = Q$.

(c) *If* $P \neq O, Q \neq O$; *and* $x_1 = x_2$:

(i) *If* $y_1 = y_2$, *that is* $P = Q$.

(A) *If* $y_1 = 0$, *then* $P \uplus Q = O$.

(B) *If* $y_1 \neq 0$, *then* $P \uplus Q = (x_3, y_3)$, *where*

$$x_3 = m^2 - (x_1 + x_2), \quad y_3 = -m(x_3 - x_1) - y_1$$

*and* $m = \left(3x_1^2 + a\right) / (2y_1)$.

(ii) *If* $y_1 \neq y_2$, *then* $P \uplus Q = O$. *In this case* $y_1 = -y_2$, *and* $P \neq Q$.

(d) *If* $P \neq O, Q \neq O$; *and* $x_1 \neq x_2$, *then* $P \uplus Q = (x_3, y_3)$, *where*

$$x_3 = m^2 - (x_1 + x_2), \quad y_3 = -m(x_3 - x_1) - y_1$$

*and* $m = (y_2 - y_1) / (x_2 - x_1)$.

*Notation.* $(P \uplus P) \triangleq [2] P, (P \uplus P \uplus P) \triangleq [3] P, \ldots$ *and so on.* □

A justification of the above definition is next provided. Only nontrivial cases are considered. The coordinates $(x_3, y_3)$ of the point $P \uplus Q$, are next evaluated. Assume that $P \neq Q \neq O$ and $x_1 \neq x_2$. Then the equation of the line $L$ connecting points $P$ and $Q$ is: $y = (mx + y_1 - mx_1)$, where

$$m = \frac{(y_2 - y_1)}{(x_2 - x_1)}$$

Substituting this expression for $y$ in the equation $y^2 = (x^3 + ax + b)$ yields a cubic equation in $x$. The zeros of this cubic equation are $x_1, x_2$, and $x_3$. The coefficient of $x^2$ in this cubic equation is $-m^2$. Therefore the sum of the three $x$-coordinates is

$$(x_1 + x_2 + x_3) = m^2$$

Consequently

$$x_3 = m^2 - (x_1 + x_2)$$

The corresponding $y_3$ value is obtained by substituting $x_3$ for $x$ in the relationship: $y = (mx + y_1 - mx_1)$ and then multiplying by $-1$. Check the corresponding geometric description. This yields

$$y_3 = -m(x_3 - x_1) - y_1$$

If $P = Q = (x_1, y_1)$, and $y_1 \neq 0$, then slope can be obtained by differentiating

$$y^2 = (x^3 + ax + b)$$

and evaluating the derivative

$$\left. \frac{dy}{dx} \right|_{(x_1, y_1)} = m$$

This yields $m = (3x_1^2 + a) / (2y_1)$. The $y_3$ coordinate is computed as in the last case. See Figure 2.3 for a geometric interpretation of the group operation on an elliptic curve. Before the group law for elliptic curve is stated, some notation is introduced. In addition, certain notions about polynomials in two variables are also developed.

**Definitions 2.36.** *Let $P$ and $Q$ be points on an elliptic curve $E$.*

1. *If $P \uplus Q = O$, then $Q = -P$ is the inverse of the point $P$.*
2. *A line drawn through the two points $P$ and $Q$ intersects the elliptic curve at another point on the curve. Denote it by $P \circ Q$. That is*

$$P \circ Q = -(P \uplus Q), \quad -(P \circ Q) = (P \uplus Q), \quad (P \circ Q) \uplus (P \uplus Q) = O \qquad (2.29)$$

□

For example: in Figure 2.3 (a), $P \circ Q = R$; and in Figure 2.3 (b), $P \circ P = R$. Before the group law of elliptic curves is stated, certain properties of curves in two variables have to be discussed. Note that $f(x, y)$ is a polynomial in $x$ and $y$ if and only if it can be represented as a finite sum of terms of the form $ax^i y^j$, where $a$ is called the coefficient of $x^i y^j$ and $i, j \in \mathbb{N}$. The number $(i + j)$ is called the degree of the term, and the maximum value of the degree of all terms in the polynomial is called the degree of $f(x, y)$. A polynomial in which all terms have the same degree is called an homogeneous polynomial.

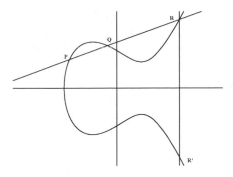

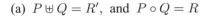

(a) $P \uplus Q = R'$, and $P \circ Q = R$

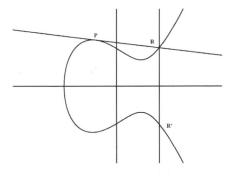

(b) $[2] P = R'$, and $P \circ P = R$

Figure 2.3. Group operation $\uplus$ on elliptic curve.

The proof of the group law of elliptic curves depends upon a theorem due to Étienne Bézout (1730-1783). A weaker form of the Bézout's theorem has been stated and proved in the chapter on applied analysis. This theorem, and some of the related useful results, are listed below for immediate reference. Of interest to us are application of this theorem to cubic curves. Recall that a cubic curve is of degree three.

**Observations 2.26.** Some basic observations.

1. Two curves of degree $m$ and $n$, without any common components meet (intersect) in $mn$ points. This is Bézout's theorem stripped of all the technical details.
2. Two cubic curves $C_1$ and $C_2$ have at most nine points of intersection. This follows directly from the Bézout's theorem. If the points at infinity, multiplicities of points of intersection, and the use of complex coordinates are also considered, then it can be shown that the total number points of intersections are exactly nine.
3. Let $C_1$ and $C_2$ be two cubic curves which consist of nine distinct points of intersection $\{p_i \mid 1 \leq i \leq 9\}$. Then any cubic passing through the eight points in the set $\{p_i \mid 1 \leq i \leq 8\}$ also passes though the ninth point $p_9$. $\square$

See the problem section for a proof of the last observation.

**Theorem 2.14.** *Points on an elliptic curve $E$ and the zero-point $O$, form an Abelian group. The group operation is $\uplus$, the identity element is $O$, and the inverse of a point $P$ on the elliptic curve is $-P$.*

*Sketch of the proof.* Note that the binary operation $\uplus$, the identity element $O$, and the inverse of a point on the elliptic curve are well-defined. Furthermore, the binary operation $\uplus$ is commutative. That is, if $P$ and $Q$ are two points on an elliptic curve, $(P \uplus Q) = (Q \uplus P)$.

In order to complete the sketch of the proof of the theorem, the associative law has to be established. That is, if $P, Q$, and $R$ are points on an elliptic curve $E$, it has to be shown that $(P \uplus Q) \uplus R = P \uplus (Q \uplus R)$. This can be directly checked by algebraic calculations using the addition laws defined earlier in this subsection. A brief outline of a direct proof of the associative law is given below. In order to prove the equality $(P \uplus Q) \uplus R = P \uplus (Q \uplus R)$, it is sufficient to prove the equality $(P \uplus Q) \circ R = P \circ (Q \uplus R)$. Consider the following lines:

$L_1$ is a line through points $P$ and $Q$. It intersects $E$ at the point $-(P \uplus Q)$.

$L_2$ is a line through points $(P \uplus Q)$ and $R$. It intersects $E$ at the point $(P \uplus Q) \circ R$.

$L_3$ is a line through points $(Q \uplus R)$ and $O$. It intersects $E$ at the point $-(Q \uplus R)$.

$M_1$ is a line through points $Q$ and $R$. It intersects $E$ at the point $-(Q \uplus R)$.

$M_2$ is a line through points $P$ and $(Q \uplus R)$. It intersects $E$ at the point $P \circ (Q \uplus R)$.

$M_3$ is a line through points $(P \uplus Q)$ and $O$. It intersects $E$ at the point $-(P \uplus Q)$.

See Figure 2.4. In this figure $A = -(P \uplus Q)$, $B = P \uplus Q$, $C = -(Q \uplus R)$, $D = Q \uplus R$, $F = (P \uplus Q) \circ R$, and $F' = P \circ (Q \uplus R)$.

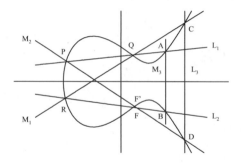

Figure 2.4. Associative group law on elliptic curve.

Next observe that a line is specified by a linear equation. Therefore, if three linear equations are multiplied, a cubic equation is obtained. The set of solutions to this cubic equation is indeed the union of these three lines. Define informally two cubic curves $C_1$ and $C_2$ as

$$C_1 \triangleq L_1 \cup L_2 \cup L_3 \text{ and } C_2 \triangleq M_1 \cup M_2 \cup M_3$$

The given elliptic curve $E$, and the cubic curve $C_1$ have no common components, because the later cubic curve is formed as a union of three lines. As per Bézout's theorem, the elliptic curve $E$ and the cubic curve $C_1$ intersect at nine points. These are

$$O, P, Q, R, (P \uplus Q), -(P \uplus Q), (Q \uplus R), -(Q \uplus R), \text{ and } (P \uplus Q) \circ R$$

The curve $C_2$ also intersects $E$ at the first 8 of the above 9 points. Therefore, $C_2$ also intersects at the ninth common point. However, the curve $C_2$ intersects the elliptic curve $E$ at the following nine points

$$O, P, Q, R, (P \uplus Q), -(P \uplus Q), (Q \uplus R), -(Q \uplus R), \text{ and } P \circ (Q \uplus R)$$

Therefore $(P \uplus Q) \circ R = P \circ (Q \uplus R)$. That is, the points $F$ and $F'$ coincide. This completes the sketch of the proof.                                                                                               □

There are several other rigorous and insightful proofs of the associative law in mathematical literature. But these rigorous proofs are highly technical, and are therefore left to the cognoscente.

**Examples 2.16.** The above results are clarified via the following examples.

1.  Consider the elliptic curve, $y^2 = (x^3 - 10x)$. In this curve $a = -10$, and $b = 0$. Therefore

    $$(4a^3 + 27b^2) = -4000 \neq 0$$

    We add the points $P = (-1, 3)$ and $Q = (0, 0)$.
    The slope of the line joining points $P$ and $Q$ is $m = -3$. The equation of this line is $y = -3x$.
    The point $(P \circ Q) = (10, -30)$ is on this line and the given cubic curve. Therefore $(P \uplus Q) = (10, 30)$. It can be checked that the points $(10, 30)$ and $(10, -30)$ lie on the cubic curve.
2.  Consider the elliptic curve, $y^2 = (x^3 + 3)$. In this curve $a = 0$, and $b = 3$. Therefore

    $$(4a^3 + 27b^2) = 243 \neq 0$$

    Let $P = Q = (1, 2)$. We compute $[2]\,P = P \uplus P$.
    The slope of the tangent at the point $P$ is $m = 0.75$. The equation of this line is $y = (0.75x + 1.25)$. The point $(P \circ Q) = (-23/16, 11/64)$ is on this line and the given cubic curve. Therefore $[2]\,P = (-23/16, -11/64)$.
    It can be checked that the points $(-23/16, 11/64)$ and $(-23/16, -11/64)$ lie on the cubic curve.                                                                                               □

Note that, if $a, b \in \mathbb{Q}$, and the coordinates of points $P$ and $Q$ on the elliptic curve are rational numbers, then the coordinates of $(P \uplus Q)$ are rational numbers, unless $(P \uplus Q) = O$. The above results can be conveniently extended to elliptic curves over finite field. This is actually the primary goal of the section.

## 2.8.2  Elliptic Curves over Finite Fields

The concepts developed for elliptic curves over real fields can generally be extended to such curves over finite fields. However, some concepts are not easily extensible. For example, the concept of a tangent line to the curve is not extensible to the finite field. Furthermore, elliptic curves over finite fields are a set of noncontiguous points.

The choice of an elliptic curve to obtain a group from the set of points, really depends upon the type of finite field over which it is defined. The finite fields can be either $\mathbb{Z}_p$, where $p$ is a prime number or $\mathbb{F}_q = GF_q$, where $q$ is not necessarily a prime number. However, different types of cubic equations have to be used for different fields. Once again denote the zero-point, or the point at infinity by $O$. Some examples of elliptic curves which can be used with different fields are listed below.

(a) If the finite field is $\mathbb{Z}_p$, where $p$ is a prime number greater than 3, a useful elliptic curve is:

$$y^2 \equiv (x^3 + ax + b) \pmod{p}; \; a, b \in \mathbb{Z}_p, \; x, y \in \mathbb{Z}_p \qquad (2.30)$$

It is also required that the equation $(x^3 + ax + b) \equiv 0 \pmod{p}$ have no repeated zeros. This can be checked by computing the greatest common divisor of $(x^3 + ax + b)$ and its "formal" derivative $(3x^2 + a)$. The condition of no repeated zeros translates to $(4a^3 + 27b^2) \not\equiv 0 \pmod{p}$.

(b) If the finite field is $GF_{2^n}$, the elliptic curve used is:

$$y^2 + (c + x)y = (x^3 + ax^2 + b); \quad a, b, c \in GF_{2^n}, \; x, y \in GF_{2^n} \tag{2.31}$$

Note that this field has a characteristic 2, and $n$ is a positive integer.

(c) If the finite field is $GF_{3^n}$, the elliptic curve used is:

$$y^2 = (x^3 + ax^2 + bx + c); \quad a, b, c \in GF_{3^n}, \; x, y \in GF_{3^n} \tag{2.32}$$

Note that this field has a characteristic 3, and $n$ is a positive integer.

### 2.8.3  Elliptic Curves over $\mathbb{Z}_p$, $p > 3$

The definition of group operation on elliptic curves over $\mathbb{Z}_p$ for prime numbers $p > 3$ is similar to the definition of group operation over real fields. However, the arithmetic operations should be performed modulo $p$. Also the division operation should be converted to the corresponding multiplicative inverse in $\mathbb{Z}_p$. Define $\mathbb{E}_p(a, b)$ as the union of the zero-point $O$, and the set of points on the elliptic curve $y^2 \equiv (x^3 + ax + b) \pmod{p}$, where $a, b \in \mathbb{Z}_p$ and $x, y \in \mathbb{Z}_p$.

A simple expression for the number of points on an elliptic curve modulo a prime number is next established.

**Theorem 2.15.** *The number of points $N$ on an elliptic curve*

$$y^2 \equiv (x^3 + ax + b) \pmod{p}, \quad a, b, x, y \in \mathbb{Z}_p, \quad p > 3 \tag{2.33a}$$

*is*

$$N = (p + 1) + \sum_{x=0}^{(p-1)} \left( \frac{x^3 + ax + b}{p} \right) \tag{2.33b}$$

*where $\left( \frac{\cdot}{p} \right)$ is the Legendre symbol.*

*Proof.* Each $x \in [0, p-1]$ results in a single value of $(x^3 + ax + b) \pmod{p}$. Therefore the number of $y \in \mathbb{Z}_p$, such that $y^2 \equiv (x^3 + ax + b) \pmod{p}$ is either $0, 1$, or $2$. The number of such $y$'s for each value of $x$ are:

(a) $0$, if $(x^3 + ax + b) \pmod{p}$ is a quadratic nonresidue.
(b) $1$, if $(x^3 + ax + b) \equiv 0 \pmod{p}$.
(c) $2$, if $(x^3 + ax + b) \pmod{p}$ is a quadratic residue.

Therefore use of an observation about Legendre's symbol yields

$$N = 1 + \sum_{x=0}^{(p-1)} \left\{ 1 + \left( \frac{x^3 + ax + b}{p} \right) \right\}$$

The number 1 on the right-hand side of the above equation, which is outside the summation is the contribution to $N$, due to the zero-point $O$. The final result follows.                  □

A theorem by H. Hasse gives an estimate of the value of $N$. According to his estimate, the number of points in a finite Abelian group defined over the elliptic curve, is given by

$$p + 1 - 2\sqrt{p} \leq N \leq p + 1 + 2\sqrt{p}$$

This estimate implies that the number of points in $\mathbb{E}_p(a, b)$ is approximately equal to the number of elements in $\mathbb{Z}_p$. A proof of this estimate is beyond the scope of this textbook. If the finite field is $GF_q$, where $q = p^n$, $p$ is a prime number, and $n \in \mathbb{P}$; the number $p$ in the above result should be replaced by $q$.

**Example 2.17.** Find group-points on the elliptic curve over $\mathbb{Z}_7$. The congruence is

$$y^2 \equiv \left(x^3 + 2x + 6\right) \pmod 7, \quad x, y \in \mathbb{Z}_7$$

In this congruence relationship, $a = 2$, and $b = 6$. Then $\left(4a^3 + 27b^2\right) \equiv 1004 \pmod 7 \equiv 3 \pmod 7 \not\equiv 0 \pmod 7$. All points in $\mathbb{E}_7(2, 6)$ except the zero-point are listed in Table 2.10.

| $x$ | $\left(x^3 + 2x + 6\right) \pmod 7$ | $\left(\frac{x^3 + 2x + 6}{7}\right)$ | $y$ |
|---|---|---|---|
| 0 | 6 | $-1$ | no |
| 1 | 2 | 1 | 3, 4 |
| 2 | 4 | 1 | 2, 5 |
| 3 | 4 | 1 | 2, 5 |
| 4 | 1 | 1 | 1, 6 |
| 5 | 1 | 1 | 1, 6 |
| 6 | 3 | $-1$ | no |

Table 2.10. Group points on the elliptic curve over $\mathbb{Z}_7$.

The number of points in the set $\mathbb{E}_7(2, 6)$, which includes the zero-point $O$, is equal to $N = 11$. This can also be checked from the above theorem. Therefore this Abelian group has a prime order. Consequently it is a cyclic group isomorphic to $(\mathbb{Z}_{11}, +)$. Assume that the generator of this group is $\beta = (1, 3)$. Using arithmetic modulo 7 yields:

$$\beta = (1, 3), \quad [2]\beta = (2, 2), [3]\beta = (5, 1), [4]\beta = (3, 5), [5]\beta = (4, 1),$$
$$[6]\beta = (4, 6), [7]\beta = (3, 2), [8]\beta = (5, 6), [9]\beta = (2, 5), [10]\beta = (1, 4)$$

Note that $[11]\beta = \beta \uplus [10]\beta = (1, 3) \uplus (1, 4) = O$. It can directly be checked that $(2, 2) \uplus (4, 1) = (3, 2)$. This also follows from the observation that $(2, 2) = [2]\beta$, and $(4, 1) = [5]\beta$. Thus

$$(2, 2) \uplus (4, 1) = [2]\beta + [5]\beta = [7]\beta = (3, 2)$$

$\square$

**Example 2.18.** Find group-points on the elliptic curve over $\mathbb{Z}_{13}$. The congruence is

$$y^2 \equiv \left(x^3 + x + 1\right) \pmod{13}, \quad x, y \in \mathbb{Z}_{13}$$

In this congruence relationship, $a = b = 1$. Then

$$\left(4a^3 + 27b^2\right) \equiv 31 \,(\mathrm{mod}\,13) \equiv 5 \,(\mathrm{mod}\,13) \not\equiv 0 \,(\mathrm{mod}\,13)$$

All points in $\mathbb{E}_{13}\,(1,1)$ except the zero-point are listed in Table 2.11.

The number of points in the set $\mathbb{E}_{13}\,(1,1)$, which includes the zero-point $O$, is equal to $N = 18$. This can be verified from the above theorem and the table. By Lagrange's theorem, the possible orders of elements of this group are $1, 2, 3, 6, 9$, and $18$. The only element of order $1$ is $O$. The element with the $y$-coordinate equal to $0$ is $(7,0)$. It is the only such point on this elliptic curve which is of order $2$. This is because $(7,0) \uplus (7,0) = O$. Let $\alpha = (1,4)$, then:

$$\begin{array}{llll}
\alpha = (1,4), & [2]\,\alpha = (8,12), & [3]\,\alpha = (0,12), & [4]\,\alpha = (11,11), \\
[5]\,\alpha = (5,1), & [6]\,\alpha = (10,6), & [7]\,\alpha = (12,8), & [8]\,\alpha = (4,2), \\
[9]\,\alpha = (7,0), & [10]\,\alpha = (4,11), & [11]\,\alpha = (12,5), & [12]\,\alpha = (10,7), \\
[13]\,\alpha = (5,12), & [14]\,\alpha = (11,2), & [15]\,\alpha = (0,1), & [16]\,\alpha = (8,1), \\
[17]\,\alpha = (1,9)
\end{array}$$

Note that $[18]\,\alpha = \alpha \uplus [17]\,\alpha = (1,4) \uplus (1,9) = O$. It can directly be checked that $(5,1) \uplus (0,12) = (4,2)$. This also follows from the observation that $(5,1) = [5]\,\alpha$, and $(0,12) = [3]\,\alpha$. Thus

$$(5,1) \uplus (0,12) = [5]\,\alpha + [3]\,\alpha = [8]\,\alpha = (4,2)$$

| $x$ | $\left(x^3 + x + 1\right)\,(\mathrm{mod}\,13)$ | $\left(\frac{x^3+x+1}{13}\right)$ | $y$ |
|-----|-----|-----|-----|
| 0 | 1 | 1 | 1, 12 |
| 1 | 3 | 1 | 4, 9 |
| 2 | 11 | −1 | no |
| 3 | 5 | −1 | no |
| 4 | 4 | 1 | 2, 11 |
| 5 | 1 | 1 | 1, 12 |
| 6 | 2 | −1 | no |
| 7 | 0 | 0 | 0 |
| 8 | 1 | 1 | 1, 12 |
| 9 | 11 | −1 | no |
| 10 | 10 | 1 | 6, 7 |
| 11 | 4 | 1 | 2, 11 |
| 12 | 12 | 1 | 5, 8 |

Table 2.11. Group points on the elliptic curve over $\mathbb{Z}_{13}$.

□

We next state a crucial theorem without proof to describe the group structure of points in $\mathbb{E}_p\,(a,b)$. According to this theorem, elliptic curve groups modulo a prime number, are either a cyclic group or the product of two cyclic groups.

**Theorem 2.16.** *Let $\mathbb{E}_p\,(a,b)$ be the union of the zero-point $O$, and the set of points on the elliptic curve $y^2 \equiv \left(x^3 + ax + b\right)\,(\mathrm{mod}\,p)$, where $a,b,x,y \in \mathbb{Z}_p$, $\left(4a^3 + 27b^2\right) \not\equiv 0 \,(\mathrm{mod}\,p)$, and $p > 3$. Let $\mathcal{G}_{\mathbb{E}_p} = \left(\mathbb{E}_p\,(a,b),\uplus\right)$ be the group formed by the points in $\mathbb{E}_p\,(a,b)$, and $\uplus$ be the group operation defined over these points. Then there exist $m, n \in \mathbb{P}$ such that $\mathcal{G}_{\mathbb{E}_p}$ is isomorphic to*

*the product of two cyclic groups of order $m$ and $n$, where $m \mid n$ and $m \mid (p-1)$. Furthermore, $m$
is allowed to take a value of unity.*                                                                    □

Observe that $m = 1$ iff the group $\mathcal{G}_{\mathbb{E}_p}$ is cyclic. Furthermore, if the number of elements in the
group is prime or the product of distinct primes, then the group is cyclic.

### 2.8.4  Elliptic Curves over $GF_{2^n}$

Let $n$ be a large positive integer. Elliptic curves over finite field $GF_{2^n}$ use arithmetic in $GF_{2^n}$. A
possible polynomial equation used in elliptic cryptosystem is

$$y^2 + xy = \left(x^3 + ax^2 + b\right) ; \; a, b \in GF_{2^n}, \; x, y \in GF_{2^n} \tag{2.34}$$

The set of points $\mathbb{E}_{2^n}\left(a, b\right)$, except the zero-point $O$, satisfy the above equation, and the addition operation defined below. Furthermore, these set of points $\mathbb{E}_{2^n}\left(a, b\right)$ form an Abelian group.

**Definition 2.37.** *Let $n$ be a large positive integer, and $\mathbb{E}_{2^n}\left(a, b\right)$ be the set of points on the
elliptic curve $\left(y^2 + xy\right) = \left(x^3 + ax^2 + b\right)$ over $GF_{2^n}$, and the zero-point $O$. The variables $x, y \in
GF_{2^n}$; constants $a, b \in GF_{2^n}$; and $b \neq 0$. Let*

$$P = (x_1, y_1), \; and \; Q = (x_2, y_2)$$

*be two points in $\mathbb{E}_{2^n}\left(a, b\right)$. We let*

(a) *Addition of point $P$ and the zero-point.*

$$P \uplus O = P \tag{2.35a}$$

(b) *Negative of point $P$.*

$$P \uplus Q = O, \; where \; x_2 = x_1, \; \; y_2 = (x_1 + y_1) \tag{2.35b}$$

(c) *If $x_1 \neq 0$, and $[2]\,P = (x_3, y_3)$ then*

$$x_3 = \left(\lambda^2 + \lambda + a\right) \tag{2.35c}$$
$$y_3 = x_1^2 + (\lambda + 1)\,x_3 \tag{2.35d}$$

   *where*

$$\lambda = x_1 + \frac{y_1}{x_1} \tag{2.35e}$$

(d) *If $P \neq Q$, $P \neq -Q$, and $P \uplus Q = (x_3, y_3)$ then*

$$x_3 = \left(\lambda^2 + \lambda + a + x_1 + x_2\right) \tag{2.35f}$$
$$y_3 = \lambda\,(x_1 + x_3) + x_3 + y_1 \tag{2.35g}$$

   *where*

$$\lambda = \frac{(y_1 + y_2)}{(x_1 + x_2)} \tag{2.35h}$$

                                                                                                        □

## 2.9 Hyperelliptic Curves

Basic elements of hyperelliptic curves are considered in this section. The aim is to explore the applications of this curve in cryptography. Hyperelliptic curves are a generalization of elliptic curves. Consider the Weierstrass form of the cubic curve over a field $\mathbb{F}$.

$$y^2 + (ax + b)\, y = \left(x^3 + cx^2 + dx + e\right);\ a, b, c, d, e \in \mathbb{F}$$

Let

$$h\left(x\right) = \left(ax + b\right),\ \text{ and }\ f\left(x\right) = \left(x^3 + cx^2 + dx + e\right)$$

where $\deg h\left(x\right) = 1$, and $\deg f\left(x\right) = 3$. Then

$$y^2 + h\left(x\right) y = f\left(x\right)$$

This cubic is a special case of hyperelliptic curve for proper values of the parameters $a, b, c, d,\,, e \in \mathbb{F}$.

In this section, basics of hyperelliptic curves are initially outlined. In describing hyperelliptic curves: polynomials, rational functions, zeros, and poles; divisors, Mumford representation of divisors; and Jacobian group, and its order are considered.

### 2.9.1  Basics of Hyperelliptic Curves

Definitions, and basic properties of hyperelliptic curves are covered in this subsection.

**Definition 2.38.** *Algebraic closure. A field $\mathbb{F}$ is algebraically closed if each and every nonconstant polynomial $g\left(x\right) \in \mathbb{F}\left[x\right]$ has a root in the field $\mathbb{F}$. Equivalently, a field is closed if and only if every polynomial in $\mathbb{F}\left[x\right]$ factorizes completely into linear polynomials (which are polynomials of degree one).*

*The algebraic closure of the field $\mathbb{F}$ is the smallest algebraically closed extension of the field. It is denoted as $\overline{\mathbb{F}}$.*                                                                   □

The field of complex numbers $\mathbb{C}$ is an algebraically closed field. Finite fields are not algebraically closed. However, there exist algebraically closed infinite fields with characteristic $p$, where $p$ is a prime number. The field $\overline{\mathbb{F}}_p$ is the algebraic closure of each finite field $\mathbb{F}_{p^n}$, where $n \in \mathbb{P}$, and

$$\mathbb{F}_p \subset \mathbb{F}_{p^2} \subset \mathbb{F}_{p^3} \subset \cdots \subset \overline{\mathbb{F}}_p$$

**Definitions 2.39.** *Hyperelliptic curve, and singular point.*

1. *Hyperelliptic curve. Consider a field $\mathbb{F}$, and let $\overline{\mathbb{F}}$ be its algebraic closure. A hyperelliptic curve $C$ of genus $g \in \mathbb{P}$ over the field $\mathbb{F}$ is:*

$$C : y^2 + h\left(x\right) y = f\left(x\right)\quad in\quad \mathbb{F}\left[x, y\right] \tag{2.36a}$$

   *where:*

(a) $h\left(x\right) \in \mathbb{F}\left[x\right]$ is a polynomial of $\deg h\left(x\right) \le g$.
(b) $f\left(x\right) \in \mathbb{F}\left[x\right]$ is a monic polynomial of $\deg f\left(x\right) = \left(2g + 1\right)$.
(c) There are no solutions $\left(x, y\right) \in \overline{\mathbb{F}} \times \overline{\mathbb{F}}$ that satisfy the equation $\left(y^2 + h\left(x\right)y\right) = f\left(x\right)$, and the partial derivative equations

$$2y + h\left(x\right) = 0, \quad and \quad h'\left(x\right)y = f'\left(x\right) \tag{2.36b}$$

simultaneously, where $h'\left(x\right)$ and $f'\left(x\right)$ are the first derivatives of $h\left(x\right)$ and $f\left(x\right)$ respectively, with respect to $x$.

2. Singular point. A singular point on a curve $C$ is a solution $\left(x, y\right) \in \overline{\mathbb{F}} \times \overline{\mathbb{F}}$ that simultaneously satisfies the equations $\left(y^2 + h\left(x\right)y\right) = f\left(x\right)$, and the partial derivative equations: $\left(2y + h\left(x\right)\right) = 0$, and $h'\left(x\right)y = f'\left(x\right)$. □

It can be inferred from the above definition that a hyperelliptic curve does not have any singular points. Also note that a hyperelliptic curve of genus $g = 1$ is an elliptic curve.

**Observations 2.27.** Let $C$ be a hyperelliptic curve over the field $\mathbb{F}$.

1. If $h\left(x\right) = 0$, then the characteristic of the field $\mathbb{F}$ is not equal to 2.
2. Let the characteristic of the field $\mathbb{F}$ be not equal to 2. Then a change of variables $x \to x$, and $y \to \left(y - h\left(x\right)/2\right)$ transforms $C$ to the form $y^2 = k\left(x\right)$, where $\deg k\left(x\right) = \left(2g + 1\right)$.
3. Let $h\left(x\right) = 0$, and the characteristic of the field $\mathbb{F}$ be not equal to 2. Then $C$ is a hyperelliptic curve if and only if $f\left(x\right)$ has no repeated roots in the field $\overline{\mathbb{F}}$. □

Proofs of the above observations are provided in the problem section.

**Definition 2.40.** Rational points, point at infinity, and finite points. Consider a hyperelliptic curve $C \in \mathbb{F}\left[x, y\right]$. Let $\mathbb{K}$ be an extension field of $\mathbb{F}$, where $\mathbb{F} \subseteq \mathbb{K} \subseteq \overline{\mathbb{F}}$. The set of $\mathbb{K}$-rational points on the curve $C$ is the set

$$C\left(\mathbb{K}\right) = \left\{\left(x, y\right) \mid \left(x, y\right) \in \mathbb{K} \times \mathbb{K}, \left(y^2 + h\left(x\right)y\right) = f\left(x\right)\right\} \cup \left\{O\right\} \tag{2.37}$$

where $O$ is a special point at infinity. All points other than $O$ in the set $C\left(\mathbb{K}\right)$ are called finite points. Sometimes, the point at infinity $O$ is also denoted as $P_\infty$. Also we denote $\left(P - P_\infty\right)$ by $\left[P\right]$.
    The set of points $C\left(\overline{\mathbb{F}}\right)$ is simply denoted by $C$. □

If $g \ge 2$, then $P_\infty$ is a singular point. This is permitted, as $P_\infty \notin \overline{\mathbb{F}} \times \overline{\mathbb{F}}$.

**Definitions 2.41.** Opposite, special, and ordinary points.

1. Let $P = \left(x, y\right)$ be a finite point on the curve $C$.
   (a) The opposite of point $P$ is the point $\widetilde{P} = \left(x, -y - h\left(x\right)\right)$. (Observe that $\widetilde{P}$ lies on the curve $C$).
   (b) If $P = \widetilde{P}$, then the point $P$ is special, otherwise the point $P$ is said to be ordinary.
2. The opposite of point at infinity $O$, is $O$ itself. □

It can be checked that the opposite of $\widetilde{P}$ is $P$.

**Example 2.19.** Consider the hyperelliptic curve $C$, over the field $\mathbb{Z}_7$, where

$$y^2 = x\,(x - 1)\,(x + 1)\,(x - 2)\,(x + 2)$$

In this curve, the genus $g = 2$, $h\,(x) = 0$, and $f\,(x) = \left(x^5 - 5x^3 + 4x\right)$. The points that satisfy $y^2 = f\,(x)$, over the field $\mathbb{Z}_7$ are:

$$(0, 0)\,,(\pm 1, 0)\,,(\pm 2, 0)\,,(3, \pm 1)$$

The partial derivative equations are

$$2y = 0, \quad \text{and} \quad \left(5x^4 - 15x^2 + 4\right) = 0$$

The above equations in the field $\mathbb{Z}_7$ are

$$2y \equiv 0\,(\mathrm{mod}\,7)\,, \quad \text{and} \quad \left(5x^4 + 6x^2 + 4\right) \equiv 0\,(\mathrm{mod}\,7)$$

It can be checked that the curve $C$ has no singular points (other than $O$). Consequently, $C$ is a hyperelliptic curve. The set of $\mathbb{Z}_7$-rational points on the curve $C$ is

$$C\,(\mathbb{Z}_7) = \{(0, 0)\,,(1, 0)\,,(-1, 0)\,,(2, 0)\,,(-2, 0)\,,(3, 1)\,,(3, -1)\} \cup \{O\}$$

Note that in the curve $C$: $(0, 0)\,,(1, 0)\,,(-1, 0)\,,(2, 0)\,$, and $(-2, 0)$ are special points; and $(3, 1)$ and $(3, -1)$ are ordinary points.                                                                        $\square$

### 2.9.2 Polynomials, Rational Functions, Zeros, and Poles

Certain special properties of polynomials and rational functions are required in the further description of hyperelliptic curves. We initially introduce some terminology, and related observations. These terms are: coordinate ring, and polynomial function; conjugate and norm of a polynomial function; function field, and rational functions; and degree of a polynomial function. In addition, zeros and poles of a rational function are also studied.

#### Coordinate Ring and Polynomial Function

**Definition 2.42.** *Coordinate ring, polynomial function. The coordinate ring of the hyperelliptic curve $C$ over the field $\mathbb{F}$ is denoted by $\mathbb{F}\,[C]$. It is the quotient ring*

$$\mathbb{F}\,[C] = \mathbb{F}\,[x, y]\,/\,(r\,(x, y)) \tag{2.38a}$$

*where $r\,(x, y) = y^2 + h\,(x)\,y - f\,(x)\,$, and $(r\,(x, y))$ is the ideal in $\mathbb{F}\,[x, y]$ generated by the polynomial $r\,(x, y)$.*

*Analogously, the coordinate ring of the hyperelliptic curve $C$ over the field $\overline{\mathbb{F}}$ is denoted by $\overline{\mathbb{F}}\,[C]$. It is the quotient ring*

$$\overline{\mathbb{F}}\,[C] = \overline{\mathbb{F}}\,[x, y]\,/\,(r\,(x, y)) \tag{2.38b}$$

*An element of $\overline{\mathbb{F}}\,[C]$ is called a polynomial function on $C$.*                                    $\square$

**Observation 2.28.** The polynomial $r\,(x, y) = \left(y^2 + h\,(x)\,y - f\,(x)\right)$ is irreducible over the field $\overline{\mathbb{F}}$. Consequently, $\overline{\mathbb{F}}\,[C]$ is an integral domain.                                    $\square$

### Conjugate and Norm of a Polynomial Function

From the definition of the quotient ring $\overline{\mathbb{F}}[C]$, it can be observed that for each polynomial $G(x,y) \in \overline{\mathbb{F}}[C]$, it is possible to repeatedly replace any occurrence of $y^2$ by $(f(x) - h(x)y)$. This would eventually lead to a polynomial $G(x,y)$ to have the form

$$G(x,y) = (a(x) - b(x)y), \quad \text{where} \quad a(x), b(x) \in \overline{\mathbb{F}}[x]$$

Moreover, this representation of $G(x,y)$ is unique.

**Definitions 2.43.** *Let* $G(x,y) = (a(x) - b(x)y)$ *be a polynomial function in* $\overline{\mathbb{F}}[C]$.

1. *The conjugate of* $G(x,y)$ *is the polynomial function*

$$\overline{G}(x,y) = (a(x) + b(x)(h(x) + y)) \tag{2.39}$$

2. *The norm of* $G(x,y)$ *is the polynomial function* $N(G) = G \cdot \overline{G}$.   □

**Observation 2.29.** Properties of the norm. Let $G$, and $H$ be polynomial functions in $\overline{\mathbb{F}}[C]$.

(a) $N(G)$ is a polynomial in $\overline{\mathbb{F}}[x]$.
(b) $N(\overline{G}) = N(G)$.
(c) $N(GH) = N(G)N(H)$.   □

### Function Field and Rational Functions

**Definitions 2.44.** *Function field, and rational functions.*

1. *The function field* $\mathbb{F}(C)$ *of* $C$ *over* $\mathbb{F}$ *is the field of fractions of* $\mathbb{F}[C]$. *Analogously, the function field* $\overline{\mathbb{F}}(C)$ *of* $C$ *over* $\overline{\mathbb{F}}$ *is the field of fractions of* $\overline{\mathbb{F}}[C]$.
2. *The elements of* $\overline{\mathbb{F}}(C)$ *are called rational functions on* $C$.   □

Observe that $\overline{\mathbb{F}}[C]$ is a subring of $\overline{\mathbb{F}}(C)$. This implies that each polynomial function is also a rational function.

**Definition 2.45.** *Value of a rational function at a finite point. Let* $R \in \overline{\mathbb{F}}(C)$, *and* $P \in C$, *where* $P \neq P_\infty$. *The rational function* $R$ *is defined at* $P$, *if there exist polynomial functions* $G, H \in \overline{\mathbb{F}}[C]$ *so that* $R = G/H$ *and* $H(P) \neq 0$. *However, if no such* $G, H \in \overline{\mathbb{F}}[C]$ *exist, then* $R$ *is not defined at* $P$. *If* $R$ *is defined at the point* $P$, *the value of* $R$ *at* $P$ *is* $R(P) = G(P)/H(P)$.   □

Note that the value of $R(P)$ is properly defined. That is, it is independent of choice of $G$ and $H$, as it should be.

### Degree of a Polynomial Function

**Definition 2.46.** *Degree of a polynomial function. Let*

$$G(x,y) = (a(x) - b(x)y) \tag{2.40a}$$

*be a nonzero polynomial function in* $\overline{\mathbb{F}}[C]$. *The degree of* $G$ *is*

$$\deg G = \max\{2\deg a, 2g + 1 + 2\deg b\} \tag{2.40b}$$

□

Motivation for the above definition of the degree of a polynomial function $G$ is provided in the following observation.

**Observation 2.30.** Let $G, H \in \overline{\mathbb{F}}[C]$

(a) $\deg G = \deg_x N(G)$, *where* $\deg_x \cdot$ *is the degree with respect to the variable* $x$.
(b) $\deg GH = \deg G + \deg H$.
(c) $\deg G = \deg \overline{G}$.                                                □

**Definition 2.47.** *Let* $R = G/H \in \overline{\mathbb{F}}(C)$ *be a rational function, and* $O$ *be the point at infinity.*

(a) $\deg G < \deg H \Rightarrow R(O) = 0$.
(b) $\deg G > \deg H \Rightarrow R(O)$ *is not defined.*
(c) $\deg G = \deg H \Rightarrow R(O) = $ *ratio of the leading coefficients* (*determined with respect to the* $\deg$ *function*) *of* $G$ *and* $H$.                                □

**Example 2.20.** Consider the rational function $R = G/H$, where

$$G(x, y) = (a(x) - b(x)y) = 3x^4 - 5x^2y$$
$$H(x, y) = (c(x) - d(x)y) = 4x^5 - (6x^3 + 2x + 1)y$$

Further, the hyperelliptic curve has genus $g = 3$. Therefore the degree of $G$ is

$$\deg G = \max\{2\deg a, 2g + 1 + 2\deg b\}$$
$$= \max\{2 \cdot 4, 2 \cdot 3 + 1 + 2 \cdot 2\}$$
$$= \max\{8, 11\} = 11$$

Similarly, the degree of $H$ is

$$\deg H = \max\{2\deg c, 2g + 1 + 2\deg d\}$$
$$= \max\{2 \cdot 5, 2 \cdot 3 + 1 + 2 \cdot 3\}$$
$$= \max\{10, 13\} = 13$$

As $\deg G < \deg H$, $R(O) = 0$.                                                □

### Zeros and Poles of a Rational Function

Ideas related to zeros and poles of a rational function are formalized.

**Definition 2.48.** *Zero and pole of a rational function. Let* $R \in \overline{\mathbb{F}}(C)$ *be a nonzero rational function. Also let* $P \in C$.

(a) $R(P) = 0 \Rightarrow R$ *has a zero at* $P$.
(b) *If* $R(P)$ *is not defined at* $P$ *then* $R$ *is said to have a pole at* $P$. *This is stated as* $R(P) = \infty$. □

**Observation 2.31.** Let $G \in \overline{\mathbb{F}}[C]$ be a nonzero polynomial function. Also let $P \in C$. If $G(P) = 0$, then $\overline{G}\left(\widetilde{P}\right) = 0$.                                □

### 2.9.3 Divisors

In contrast to points on elliptic curves, there is no obvious way to provide a group structure for points on $C(\mathbb{K})$ for genus $g \geq 2$. An alternate scheme would be to introduce a different entity which is related to the curve $C$, and then define the group operation on such entities. This entity, and its associated group operation is called the Jacobian of $C$. Jacobians of hyperelliptic curves in turn are studied by introducing an entity called divisor. Definition and elementary properties of divisors are initially introduced.

**Definitions 2.49.** *Divisor related terminology.*

1. *A divisor $D$ is a formal sum of points on $C$, where*

$$D = \sum_{P \in C} m_P P, \quad m_P \in \mathbb{Z} \tag{2.41}$$

   *and only a finite number of $m_P$'s are nonzero.*
   (a) *The degree of divisor $D$ is denoted by $\deg D$. It is equal to the integer sum $\sum_{P \in C} m_P$.*
   (b) *The order of $D$ at the point $P$ is the integer $m_P$. It is denoted as $\mathrm{ord}_P(D) = m_P$.*
   (c) *The support of the divisor $D$ is the set $\mathrm{supp}(D) = \{P \in C \mid m_P \neq 0\}$.*
2. *The set of all divisors is denoted by $\mathbb{D}(C)$.*
3. *The set of all divisors of degree $0$ is denoted by $\mathbb{D}^0(C)$. That is,*

$$\mathbb{D}^0(C) = \{D \in \mathbb{D}(C) \mid \deg D = 0\} \tag{2.42}$$

$\square$

**Observations 2.32.** Some basic observations about divisors.

1. As the opposite of a point $P$ is $\widetilde{P}$, let $\widetilde{D} \triangleq \sum_{\widetilde{P} \in C} m_{\widetilde{P}} \widetilde{P}$. Therefore, if $D = \sum_{P \in C} m_P P$, then $\widetilde{D} = \sum_{P \in C} m_P \widetilde{P}$.
2. A representation of $\mathbb{D}^0(C)$ is

$$\mathbb{D}^0(C) = \left\{ \sum_{P \in C} m_P[P] \mid m_P \in \mathbb{Z}, \text{ where } m_P = 0 \text{ for almost all } P \right\}$$

   where $[P] = P - P_\infty$.
3. The set of all divisors $\mathbb{D}(C)$ form an additive Abelian group under the divisor sum operation. Let $D$ and $D'$ be two divisors $D = \sum_{P \in C} m_P P$, and $D' = \sum_{P \in C} m'_P P$ respectively. The sum $(D + D')$ is

$$D + D' = \sum_{P \in C} m_P P + \sum_{P \in C} m'_P P = \sum_{P \in C} (m_P + m'_P) P$$

4. $\mathbb{D}^0(C)$ is a subgroup of $\mathbb{D}(C)$. $\square$

**Definition 2.50.** *Divisor of a nonzero rational function. Let $R \in \overline{\mathbb{F}}(C)$ be a nonzero rational function. If*

$$R(x, y) = \frac{G(x, y)}{H(x, y)} \tag{2.43a}$$

*then*

$$\mathrm{div}\left(R\left(x,y\right)\right) = \sum_{i=1}^{k_z} m_i G_i - \sum_{i=1}^{k_p} n_i H_i \tag{2.43b}$$

*where $G_i$ is a zero of $R\left(x,y\right)$ with multiplicity $m_i$, for $1 \leq i \leq k_z$; and $H_i$ is a pole of $R\left(x,y\right)$ with multiplicity $n_i$, for $1 \leq i \leq k_p$. Thus $\mathrm{div}\left(R\left(x,y\right)\right)$ is a finite formal sum.*    □

**Observations 2.33.** Some observations related to rational and polynomial functions.

1. The divisor of a nonzero rational function $R \in \overline{\mathbb{F}}\left(C\right)$ can also be specified as $\mathrm{div}\left(R\right) = \sum_{P \in C} \left(ord_P\left(R\right)\right) P$.
2. The degree of $\mathrm{div}\left(R\right)$ is always zero. This is true because of the definition of the value of a rational function at $O$ (the point at infinity).
3. If $R = G/H$ then $\mathrm{div}\left(R\right) = \mathrm{div}\left(G\right) - \mathrm{div}\left(H\right)$.
4. Let $G \in \overline{\mathbb{F}}\left[C\right]$ be a nonzero polynomial function, where $\mathrm{div}\left(G\right) = \sum_{P \in C} m_P P$.
   Then $\mathrm{div}\left(\overline{G}\right) = \sum_{P \in C} m_P \widetilde{P}$.    □

**Example 2.21.** Let $P = \left(x_0, y_0\right)$ be a point on the curve $C$. We find $\mathrm{div}\left(x - x_0\right)$. Observe that the straight line $\left(x - x_0\right) = 0$ is parallel to the $y$-axis, and passes through the point $\left(x_0, 0\right)$. This straight line intersects the curve $C$ at either one or two finite points. If the point $P$ is ordinary, then it intersects the curve $C$ at two finite points. However, if the point $P$ is special, then it intersects the curve $C$ at a single finite point. Therefore

$$\mathrm{div}\left(x - x_0\right) = \begin{cases} P + \widetilde{P} - 2P_\infty, & P \text{ is ordinary} \\ 2P - 2P_\infty, & P \text{ is special} \end{cases}$$

□

The product of two rational functions that are not necessarily distinct is also a rational function. Consequently, the set of divisors corresponding to rational functions form an Abelian group under the divisor sum operation. The divisor of a constant function is the identity element of this group. Thus the set of divisors of rational functions is a subgroup of $\mathbb{D}^0\left(C\right)$. These ideas are next stated formally.

**Definitions 2.51.** *Principal divisor.*

1. *A divisor $D \in \mathbb{D}^0\left(C\right)$ is a principal divisor, if $D = \mathrm{div}\left(R\right)$ for some nonzero rational function $R \in \overline{\mathbb{F}}\left(C\right)$.*
2. *The set of all principal divisors is denoted by $\mathbb{P}_{pd}\left(C\right)$.*    □

**Observation 2.34.** Some observations related to principal divisor.

1. An explicit representation of $\mathbb{P}_{pd}\left(C\right)$ is

$$\mathbb{P}_{pd}\left(C\right) = \left\{ \sum_{P \in C} m_P\left(R\right)\left[P\right] \mid R \in \overline{\mathbb{F}}\left(C\right) \right\}$$

where $\left[P\right] = P - P_\infty$, and $\overline{\mathbb{F}}\left(C\right)$ is the function field of $C$. Further, for any rational function $R \in \mathbb{F}\left(C\right), m_P\left(R\right)$ is the multiplicity of the point $P \in C\left(\overline{\mathbb{F}}\right)$ at $R$. That is:

   (a) If $P$ is neither a zero nor a pole of $R$, then $m_P(R) = 0$.

   (b) If $P$ is a zero of $R$, then $m_P(R)$ is a multiplicity of this zero.

   (c) If $P$ is a pole of $R$, then $-m_P(R)$ is a multiplicity of this pole.

2. The set of all principal divisors $\mathbb{P}_{pd}(C)$ is a subgroup of $\mathbb{D}^0(C)$         $\square$

**Definitions 2.52.** *Jacobian of the curve $C$, and equivalent divisors.*

1. *The Jacobian of the curve $C$ is the quotient group $\mathbb{J}(C) = \mathbb{D}^0(C)/\mathbb{P}_{pd}(C)$.*

2. *If $D_1, D_2 \in \mathbb{D}^0(C)$, then $D_1$ and $D_2$ are said to be equivalent divisors if $(D_1 - D_2) \in \mathbb{P}_{pd}(C)$. This is indicated as $D_1 \sim D_2$.*      $\square$

The set of points on the hyperelliptic curve $C$ embeds into the Jacobian of $C$. This occurs by assigning each point $P \in C(\overline{\mathbb{F}})$ the coset of $[P]$ in $\mathbb{J}(C)$. The identity element of $\mathbb{J}(C)$ is the coset of $[P_\infty]$. For genus $g = 1$, this embedding is surjective. However, for genus $g \geq 2$, this embedding is not surjective.

Recall that, in the case of elliptic curves, three collinear points on the elliptic curve sum to zero. The corresponding generalization of this rule for hyperelliptic curves is as follows. "All the points on any function that occur on the curve $C$ sum to zero." More precisely, let $P_1, P_2, \ldots, P_r$ be the complete collection of points of intersection of the curve $C$ with some rational function $\varrho \in \mathbb{F}(C)$. Also let the multiplicities of these points be $m_{P_i}(\varrho)$, for $1 \leq i \leq r$. Then the divisor

$$D = \sum_{i=1}^{r} m_{P_i}(\varrho)[P_i]$$

is indeed a principal divisor. Further, for any point $P = (x_0, y_0) \in C(\overline{\mathbb{F}})$, the straight line $x = x_0$ intersects the curve $C$ at only $P, \tilde{P}$, and $P_\infty$; and the inverse of the class of $[P]$ is simply the class of $-[P] = \left[\tilde{P}\right]$. Therefore, the inverse of the class of a divisor $D = \sum m_P[P]$ in the Jacobian $\mathbb{J}(C)$ is the class of $\tilde{D} = \sum m_P\left[\tilde{P}\right]$. These facts are summarized in the following observations.

**Observations 2.35.** Some observations about Jacobian of the curve $C$.

1. The identity element in the Jacobian $\mathbb{J}(C)$ is the coset of $[P_\infty]$.

2. The inverse of the class of $[P]$ in the Jacobian $\mathbb{J}(C)$ is the class of $\left[\tilde{P}\right]$.

3. Let $P \in C$.

   (a) If $P$ is an ordinary point, that is $P \neq \tilde{P}$; then $[P] + \left[\tilde{P}\right] = [P_\infty]$.

   (b) If $P$ is a special point, that is $P = \tilde{P}$; then $2[P] = [P_\infty]$.

4. The inverse of the class of a divisor $D = \sum m_P[P]$ in the Jacobian $\mathbb{J}(C)$ is the class of $\tilde{D} = \sum m_P\left[\tilde{P}\right]$.      $\square$

**Definition 2.53.** *A semi-reduced divisor $D$ is a divisor*

$$D = \sum_{P_i \in C} m_i[P_i], \quad \text{where} \quad m_i \in \mathbb{N} \tag{2.44}$$

*with the following properties.*

(a) *The $P_i$'s are finite points*
(b) *If $P_i$ is an ordinary point, that is $P_i \neq \tilde{P}_i$; and $m_i > 0$, then $m_{\tilde{P}_i} = 0$. That is, $\tilde{P}_i \notin supp(D)$.*
   *Alternately, if $P_i \neq \tilde{P}_i$, then only either one can occur in the summation.*
(c) *If $P_i$ is a special point, that is $P_i = \tilde{P}_i$; and $m_i > 0$, then $m_i = 1$.*                       □

**Observation 2.36.** For each divisor $D \in \mathbb{D}^0(C)$ there exists a semi-reduced divisor $D_1 \in \mathbb{D}^0(C)$, where $D \sim D_1$.                       □

Observe that semi-reduced divisors are not unique. It is possible to show that, each coset in the quotient group $\mathbb{J}(C) = \mathbb{D}^0(C)/\mathbb{P}_{pd}(C)$ has exactly a single reduced divisor. Consequently every element of the Jacobian of the curve $C$ can be identified by a reduced divisor.

**Definition 2.54.** *Let $g$ be the genus of the hyperelliptic curve $C$, and $D = \sum_{P_i \in C} m_i[P_i]$ be a semi-reduced divisor. If $\sum_{P_i \in C} m_i \leq g$, then $D$ is called a reduced divisor.*                       □

Note that the support of a reduced divisor contains at most $g$ number of points.

**Theorem 2.17.** *Each divisor $D \in \mathbb{D}^0(C)$ is associated with a unique reduced divisor $D_1$ so that $D \sim D_1$.*
*Proof.* See the problem section.                       □

We next address the problem of evaluating the sum of two reduced divisors $D_1$ and $D_2$. The sum of these two reduced divisors is $D_1 + D_2$. This is initially illustrated via an example.

**Example 2.22.** Let $C$ be a hyperelliptic curve of genus $g = 2$, defined over the field $\mathbb{Q}$. It is

$$C : y^2 = f(x) = x(x^2 - 1)(x^2 - 4) + 1$$

Observe that $f(x)$ has no repeated roots in the field $\overline{\mathbb{Q}}$, where $\overline{\mathbb{Q}}$ is the algebraic closure of the field $\mathbb{Q}$. Let $P_1, P_2, Q_1$, and $Q_2$ be four points on the curve $C$. These are:

$$P_1 = (-1, 1), \quad P_2 = (0, 1), \quad Q_1 = (1, -1), \quad \text{and} \quad Q_2 = (2, -1)$$

Consider the two reduced divisors $D_1$ and $D_2$, where $D_1 = [P_1] + [P_2]$, and $D_2 = [Q_1] + [Q_2]$. Our goal is to determine the sum of the two reduced divisors $D_1$ and $D_2$, which is $D_1 + D_2 \triangleq D$.

Observe that the four points $P_1, P_2, Q_1$, and $Q_2$ all lie on a unique polynomial $y = v(x) \in \overline{\mathbb{Q}}(C)$, where

$$v(x) = \frac{1}{3}(2x^3 - 3x^2 - 5x + 3)$$

The degree of this polynomial is 3. Also, the curve $y = v(x)$ intersects the hyperelliptic curve $C$. Substitute $y = v(x)$ in the equation for the hyperelliptic curve, which is $y^2 = f(x)$. This leads to

$$0 = f(x) - y^2 = f(x) - (v(x))^2$$

The right hand side of the above equation is a polynomial of degree 6. Therefore, the curve $y = v(x)$ intersects the hyperelliptic curve $C$, in addition to points $P_1, P_2, Q_1$, and $Q_2$; in two more points. Let these two points be $R_1 = (x_1, y_1)$ and $R_2 = (x_2, y_2)$. Thus the $x$-coordinates of the six points $P_1, P_2, Q_1, Q_2, R_1$, and $R_2$ satisfy the above equation. We have

$$0 = f(x) - y^2 = f(x) - (v(x))^2$$
$$= Ax(x-1)(x+1)(x-2)(x-x_1)(x-x_2)$$

where $A$ is a constant. It turns out that $A = -1/9$ and

$$a(x) = (x-x_1)(x-x_2) = 4x^2 - 13x - 33$$

The roots of the above equation are

$$x_1 = \frac{1}{8}\left(13 + \sqrt{697}\right), \quad \text{and} \quad x_2 = \frac{1}{8}\left(13 - \sqrt{697}\right)$$

The corresponding $y$-coordinates are obtained from the equation $y = v(x)$. These are

$$y_1 = v(x_1) = \frac{1}{64}\left(1473 + 61\sqrt{697}\right)$$
$$y_2 = v(x_2) = \frac{1}{64}\left(1473 - 61\sqrt{697}\right)$$

Observe that the six points $P_1, P_2, Q_1, Q_2, R_1$, and $R_2$ form a complete intersection of the hyperelliptic curve $C$ with $y = v(x)$, and sum to zero. See Figure 2.5.

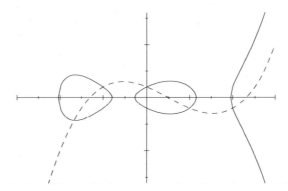

Figure 2.5. The curves $y^2 = f(x)$ (solid curve), and $y = v(x)$ (dotted curve) on the $xy$-plane. Only the points of intersection of the two curves: $P_1, P_2, Q_1, Q_2$, and $R_2$ are shown. The point $R_1$ is not shown.

Therefore

$$[P_1] + [P_2] + [Q_1] + [Q_2] + [R_1] + [R_2] = [P_\infty]$$

This leads to

$$[P_1] + [P_2] + [Q_1] + [Q_2] = -[R_1] - [R_2]$$
$$= \left[\widetilde{R_1}\right] + \left[\widetilde{R_2}\right]$$

and

$$D = D_1 + D_2 = \left[\widetilde{R_1}\right] + \left[\widetilde{R_2}\right]$$

In the above expression $\widetilde{R}_1 = (x_1, -y_1)$, and $\widetilde{R}_2 = (x_2, -y_2)$. Note that the support of $D$ has two distinct points: $\widetilde{R}_1$ and $\widetilde{R}_2$. As the genus $g$ of the hyperelliptic curve is also two, $D$ is a reduced divisor.                                                                                □

**Notation.** Let $D_1$ and $D_2$ be two reduced divisors. The reduced representation of the class of $D_1 + D_2$ in the Jacobian $\mathbb{J}(C)$ is denoted by $D_1 \oplus D_2$. Thus $\oplus$ is the group operation of the Jacobian $\mathbb{J}(C)$. $\qquad\qquad\qquad\qquad\qquad\qquad\qquad\qquad\qquad\qquad\qquad\qquad\qquad\qquad\qquad\qquad\qquad\qquad\square$

### A General Technique to Find the Reduced Sum

Let $D_1$ and $D_2$ be two reduced divisors on the hyperelliptic curve $C$. The goal is to determine the reduced sum $D_1 \oplus D_2$. This is accomplished in three steps.

*Step* 0: The semi-reduced sum $D$, of $D_1$ and $D_2$ is determined in this step. First compute $D_1 + D_2$. It is quite possible that the sum $D_1 + D_2$ is semi-reduced, and therefore equal to $D$. If it is not, then there exists a $\widetilde{P} \in supp(D_2)$ for some $P \in supp(D_1)$. In this later case, for each such $P$, replace $-[P]$ by $\left[\widetilde{P}\right]$, in the sum $D_1 + D_2$ and do:

*Step* 0.1: If $P$ is an ordinary point, that is $P \neq \widetilde{P}$; then set $[P] + \left[\widetilde{P}\right] = [P_\infty]$ in the sum $D_1 + D_2$. This ensures that in this case, $P$ and $\widetilde{P}$ do not occur simultaneously in the sum $D_1 + D_2$.

*Step* 0.2: If $P$ is a special point, that is $P = \widetilde{P}$; then set $2[P] = [P_\infty]$ in the sum $D_1 + D_2$. This ensures that in this case, double of $P$ does not occur in the sum $D_1 + D_2$.

*Step* 1: The support of the semi-reduced sum $D$ is $|supp(D)| = r$. If $r \leq g$, then $D$ is indeed the reduced sum of the divisors $D_1$ and $D_2$. However, the value of $r$ can be as large as $2g$. If $r \geq (g + 1)$ then go to the next step.

*Step* 2: Perform the following iterations, until $r \leq g$.

*Step* 2.1: The points $P_i$ for $1 \leq i \leq r$, all lie on an interpolating curve $y = v(x)$, where $\deg v(x) \leq (r - 1)$. Using the equations for the curves $C$, and $y = v(x)$ obtain the polynomial

$$w(x) = f(x) - (v(x))^2 - h(x)v(x)$$

where $\deg w(x) \triangleq d$. Of the $d$ zeros of the polynomial $w(x)$ (which are counted with multiplicities), $r$ are actually the $x$-coordinates of the points in $supp(D)$. Determine the remaining $(d - r)$ zeros of the polynomial $w(x)$. Let these be $x_i$, where $1 \leq i \leq (d - r)$. Also obtain the $(d - r)$ new points on the curve $C$, which are $(x_i, v(x_i))$, for $1 \leq i \leq (d - r)$.

*Step* 2.2: Replace the $r$ points in $supp(D)$ by the opposites

$$(x_i, -v(x_i) - h(x_i)), \quad \text{for} \quad 1 \leq i \leq (d - r)$$

These $(d - r)$ new points form a new semi-reduced divisor in the divisor class of $D_1 + D_2$. This new semi-reduced divisor is called $D$. Note that after this step $|supp(D)| = (d - r)$. Next consider two cases.

*Step* 2.2.1: If $d \geq (2g + 2)$, then $\deg w(x)$ is determined by the degree of $-(v(x))^2$. Therefore $(d - r) = 2\deg v(x) - r \leq \{2(r - 1) - r\} = (r - 2)$. The last inequality is true because $\deg v(x) \leq (r - 1)$.

*Step* 2.2.2: If $d \leq (2g + 1)$, then $\deg w(x)$ is determined by the degree of $f(x)$. Therefore

$$(d - r) = \{(2g + 1) - r\} \leq \{(2g + 1) - (g + 1)\} = g$$

The last inequality is true because $r \geq (g+1)$ in Step 2.

Thus the number of points on the curve $C$ in the new divisor decreases by 2 in each iteration, except possibly in the last iteration, where it decreases by at least unity.

As $r \leq 2g$ at the beginning of this iterative process, $D_1 \oplus D_2$ is obtained in at most $\lceil g/2 \rceil$ steps.

In the last example, the genus $g$ was equal to 2. Therefore only a single step was required in the reduction process. The above outlined technique is not convenient for actual implementation. However, this can be facilitated via Mumford representation of divisors.

### 2.9.4 Mumford Representation of Divisors

It is both convenient and efficient to implement operations in the Jacobian of the hyperelliptic curve $C$ via Mumford representation of a semi-reduced divisor. This representation is named after the mathematician David Mumford. An algorithmic technique to define the group operation $\oplus$, of the Jacobian group $\mathbb{J}(C)$ is described in this subsection.

**Definition 2.55.** *Let $D = \sum_{i=1}^{r} m_i [P_i]$ be a nonzero semi-reduced divisor on a hyperelliptic curve $C$. Also let $P_i = (x_i, y_i)$, where $x_i, y_i \in \overline{\mathbb{F}}$ for $1 \leq i \leq r$. The Mumford representation of the divisor $D$ is a pair of polynomials $u(x), v(x) \in \overline{\mathbb{F}}[x]$. These are*

$$u(x) = \prod_{i=1}^{r} (x - x_i)^{m_i} \tag{2.45a}$$

$$\left( \frac{d}{dx} \right)^{j} \left( f(x) - (v(x))^2 - v(x) h(x) \right) \Bigg|_{x=x_i} = 0,$$

$$\text{where } 0 \leq j \leq (m_i - 1), \ 1 \leq i \leq r \tag{2.45b}$$

*The Mumford representation of the divisor $D$ is $[u, v]$.*                                        □

In the above definition, the polynomial $u(x)$ completely specifies the $x$-coordinates along with their multiplicities (the $m_i$'s). The zeros of $u(x)$ are actually the $x$-coordinates of the points in the support of the divisor $D$. Further, $y = v(x)$ is the interpolating polynomial through all the $P_i$'s. More specifically, $y_i = v(x_i)$ for $1 \leq i \leq r$.

Recall that in the original definition of a divisor, $D$ is specified as a formal sum of points on the curve $C$; and its Mumford representation is $[u, v]$. In a misuse of notation, we use the specification $D \triangleq [u, v]$.

**Observations 2.37.** Some facts about Mumford's polynomial representation of a divisor.

1. $u(x_i) = 0$, and $v(x_i) = y_i$ with multiplicity $m_i$, where $1 \leq i \leq r$.
2. The polynomial $u(x)$ is monic, and

$$u(x) \mid \left( f(x) - (v(x))^2 - v(x) h(x) \right)$$

3. The divisor $D$ uniquely determines the polynomials

$$u(x), \text{ and } v(x) \, (\mathrm{mod} \, u(x))$$

4. The polynomials $u(x), v(x) \in \overline{\mathbb{F}}[x]$ specify a semi-reduced divisor $D = \sum_{i=1}^{r} m_i [P_i]$, where $P_i = (x_i, v(x_i)) \in C$ for $1 \le i \le r$.

5. As the polynomial $v(x)$ is determined modulo $u(x)$, its uniqueness is ensured by selecting it so that it has least nonnegative degree in its congruence class modulo $u(x)$. This implies that if $D = [u, v]$ is a reduced divisor, then $\deg v < \deg u \le g$.                                                                              □

**Examples 2.23.** Let $C$ be a hyperelliptic curve.

1. Let $P = (x_0, y_0)$ be a point on $C$, and $D = [P]$. The Mumford representation of $D$ is $[u, v]$, where $u(x) = (x - x_0)$, and $v(x) = y_0$.

2. If $D = [u(x), v(x)]$, then $\widetilde{D} = [u(x), -(v(x) + h(x)) \pmod{u(x)}]$.

3. Let $D = [P_\infty]$. Note that $[P_\infty] = P_\infty - P_\infty$. Therefore, the reduced divisor consists of empty sum of points. The Mumford representation of $D$ is $[1, 0]$.

4. Let $P_1 = (x_1, y_1)$, and $P_2 = (x_2, y_2)$ be two different points on $C$; and $D = [P_1] + [P_2] = [u(x), v(x)]$. Then $u(x) = (x - x_1)(x - x_2)$, and $y = v(x)$ is the straight line through points $P_1$ and $P_2$.

5. Let $C$ be a hyperelliptic curve, and $P = (x_0, y_0)$ be an ordinary point on it. That is, $P \ne \widetilde{P}$. Mumford representation of the divisor $D = 2[P] = [u(x), v(x)]$ is determined. Note that $u(x) = (x - x_0)^2$. Let

$$v(x) = y_0 + a(x - x_0)$$

where $a \in \overline{\mathbb{F}}$. Also, $v(x)$ should satisfy

$$f(x) - (v(x))^2 - v(x) h(x) \Big|_{x=x_0} = 0$$

$$\left(\frac{d}{dx}\right) \left(f(x) - (v(x))^2 - v(x) h(x)\right) \Big|_{x=x_0} = 0$$

Of the above two equations, the first one should be satisfied by $(x_0, y_0)$. The second equation yields

$$f' - 2vv' - v'h - vh' = 0 \quad \text{at} \quad x = x_0$$

where $f', h'$, and $v'$ are derivatives of $f, h$, and $v$ respectively, with respect to $x$. Substituting $v(x_0) = y_0$ and $v'(x) = a$ in the above equation leads to

$$a = \frac{f'(x_0) - y_0 h'(x_0)}{2y_0 + h(x_0)}$$

□

**Example 2.24.** This is a continuation of the example covered in the last subsection, where $C$ is a hyperelliptic curve of genus $g = 2$, defined over the field $\mathbb{Q}$. It is

$$C : y^2 = f(x) = x(x^2 - 1)(x^2 - 4) + 1$$

Observe that $f(x)$ has no repeated roots in the field $\overline{\mathbb{Q}}$. Let $P_1, P_2, Q_1$, and $Q_2$ be four points on the curve $C$. These are:

$$P_1 = (-1, 1), \quad P_2 = (0, 1), \quad Q_1 = (1, -1), \quad \text{and} \quad Q_2 = (2, -1)$$

Consider the two reduced divisors $D_1$ and $D_2$, where $D_1 = [P_1] + [P_2]$, and $D_2 = [Q_1] + [Q_2]$. Also let $D = D_1 + D_2$. We determine Mumford representation of divisors $D_1, D_2$, and $D$.

(a) Let the Mumford representation of the divisor $D_1$ be $[u_1(x), v_1(x)]$. We have

$$u_1(x) = (x+1)(x-0) = x(x+1)$$

The polynomial $v_1(x)$ is the equation of the straight line joining points $P_1$ and $P_2$. It is $v_1(x) = 1$. Therefore

$$D_1 = [x(x+1), 1]$$

(b) Let the Mumford representation of the divisor $D_2$ be $[u_2(x), v_2(x)]$. We have

$$u_2(x) = (x-1)(x-2)$$

The polynomial $v_2(x)$ is the equation of the straight line joining points $Q_1$ and $Q_2$. It is $v_2(x) = -1$. Therefore

$$D_2 = [(x-1)(x-2), -1]$$

(c) Let the Mumford representation of the divisor $D = D_1 + D_2$ be $[u(x), v(x)]$. We have

$$u(x) = (x+1)(x-0)(x-1)(x-2)$$
$$= x(x^2-1)(x-2)$$

The polynomial $v(x)$ is the equation of the curve which passes through the four points $P_1, P_2, Q_1$, and $Q_2$. This has been determined in an example stated in the last subsection. It is

$$v(x) = \frac{1}{3}(2x^3 - 3x^2 - 5x + 3)$$

Therefore

$$D = \left[x(x^2-1)(x-2), \frac{1}{3}(2x^3 - 3x^2 - 5x + 3)\right]$$

The above representation of $D$ is not reduced, as $\deg u = 4 > g = 2$.                                      □

An algorithm is next introduced to add two semi-reduced divisors so that the result is a reduced divisor. Let $D_1$ and $D_2$ be two semi-reduced divisors. We determine $D \sim D_1 + D_2$, so that $D$ is a reduced divisor. In this algorithm, the Mumford representation of divisors is used for convenience. The algorithm in the stated form is due to Koblitz. It is actually a generalization of an algorithm developed earlier by Cantor.

A motivation for the algorithm is initially developed. In the first step, a semi-reduced divisor $D \sim D_1 + D_2$ is obtained. In the second step, $D$ is reduced. Let the Mumford representation of semi-reduced divisors $D_1$ and $D_2$ be $[u_1, v_1]$ and $[u_2, v_2]$ respectively.

*Step* 1: Consider the following two cases. Let $P$ be a point on the curve $C$.

*Case* 1: Let $[P]$ occur in the divisor $D_1$ and $\left[\tilde{P}\right]$ does not occur in the divisor $D_2$, and vice-versa. In this case $D_1 + D_2$ is a semi-reduced divisor. Let its Mumford representation be $[u, v]$. Then

$$u = u_1 u_2$$

$$v \equiv v_1 \pmod{u_1}, \quad \text{and} \quad v \equiv v_2 \pmod{u_2}$$

In the next step $v \pmod{u}$ is determined. Note that the polynomials $u_1$ and $u_2$ are monic, and relatively prime to each other. Use of the extended Euclidean algorithm for polynomials leads to

$$1 = s_1 u_1 + s_2 u_2$$

where $s_1(x), s_2(x) \in \overline{\mathbb{F}}[x]$. This implies $s_1 u_1 \equiv 1 \pmod{u_2}$ and $s_2 u_2 \equiv 1 \pmod{u_1}$. Use of Chinese remainder theorem for polynomials results in

$$v \equiv s_1 u_1 v_2 + s_2 u_2 v_1 \pmod{u}$$

*Case* 2: Let $[P]$ occur in the divisor $D_1$, and $\left[\widetilde{P}\right]$ occur in the divisor $D_2$. In this case $D_1 + D_2$ is not a semi-reduced divisor. More precisely, the sum $D_1 + D_2$ is not semi-reduced if $\widetilde{P} \in supp(D_2)$ for some $P \in supp(D_1)$. Note that for each $(x_0, y_0) \in supp(D_1) \cap supp\left(\widetilde{D_2}\right)$ we have

$$u_1(x_0) = u_2(x_0) = 0, \quad \text{and} \quad y_0 = v_1(x_0) = -v_2(x_0) - h(x_0)$$

Therefore, $(x - x_0)$ occurs in $u_1(x), u_2(x)$, and $(v_1(x) + v_2(x) + h(x))$ as a factor. Let the Mumford representation of the sum of divisors $D_1$ and $D_2$ be $D$. As the sum $D = [u, v]$ is not semi-reduced in general, it is possible to obtain an expression for $u$ via the use of extended Euclidean algorithm for polynomials. We have

$$d = \gcd(u_1, u_2, v_1 + v_2 + h) = s_1 u_1 + s_2 u_2 + s_3(v_1 + v_2 + h)$$

where $d(x), s_1(x), s_2(x), s_3(x) \in \overline{\mathbb{F}}[x]$, and

$$u = \frac{u_1 u_2}{d^2}$$

$$v \equiv \left\{ \frac{s_1 u_1 v_2 + s_2 u_2 v_1 + s_3(v_1 v_2 + f)}{d} \right\} \pmod{u}$$

If the original sum $D$ is semi-reduced, then $d = 1$, and $s_3 = 0$. A justification of expressions for $u$ and $v$ is provided in the problem section. At the end of Step 1, the divisor $D$ is semi-reduced. However, it may be reduced or not.

*Step* 2: If the divisor $D = [u, v]$ is not reduced then $\deg u > g$. In this case compute

$$u' = \frac{(f - vh - v^2)}{u}, \quad \text{and} \quad v' \equiv -(h + v) \pmod{u'}$$

Therefore the polynomial $u'$ does not have roots of the polynomial $(f - vh - v^2)$ that are the $x$-coordinates of the points in $supp(D)$. The polynomial $u'$ has roots which are the $x$-coordinates of the remaining intersection points of $C$ with $y = v(x)$. The expression for $v'$ implies that these remaining intersection points are replaced by their opposites. Let

$$u \leftarrow u', \quad \text{and} \quad v \leftarrow v'$$

Iterate the steps in Step 2 finitely many times, till $\deg u \leq g$.

At this stage $\deg v < \deg u \leq g$. Finally, make the polynomial $u$ monic by dividing it by its leading coefficient. The unique reduced divisor is $[u, v]$.

An algorithm is next outlined to add two semi-reduced divisors so that the result is a reduced divisor.

---

**Algorithm 2.1.** *Cantor-Koblitz Algorithm.*

---

**Input:** Two semi-reduced divisors $D_1$ and $D_2$, where
$D_1 = [u_1, v_1]$, and $D_2 = [u_2, v_2]$
**Output:** Determine $D \sim D_1 + D_2$, where $D = [u, v]$ is a reduced divisor.
**begin**
    **Step 1:**
    (*Use extended Euclidean algorithm of polynomials twice to*)
    (*determine d*)
    $d = \gcd(u_1, u_2, v_1 + v_2 + h) = s_1 u_1 + s_2 u_2 + s_3 (v_1 + v_2 + h)$
    $u = u_1 u_2 / d^2$
    $v \equiv \{(s_1 u_1 v_2 + s_2 u_2 v_1 + s_3 (v_1 v_2 + f)) / d\} \pmod{u}$
    **Step 2**:
    **while** $\deg u > g$ **do**
    **begin**
        $u' \leftarrow (f - vh - v^2) / u$
        $v' \leftarrow -(h + v) \pmod{u'}$
        $u \leftarrow u'$, and $v \leftarrow v'$
    **end** (*end of while loop*)
    make the polynomial $u$ monic by dividing it by its leading coefficient
    $D \leftarrow [u, v]$
**end** (*end of Cantor-Koblitz algorithm*)

---

It is possible to optimize the above algorithm for special values of the genus $g$. The above algorithm is illustrated via the following example.

**Example 2.25.** This example is a continuation of the example covered earlier in the subsection, where $C$ is a hyperelliptic curve of genus $g = 2$, defined over the field $\mathbb{Q}$. It is

$$C : y^2 = f(x) = x(x^2 - 1)(x^2 - 4) + 1$$

Observe that $h(x) = 0$ for all values of $x$ in the field $\mathbb{Q}$. Let the Mumford representation of the divisor $D_1$ be

$$[u_1(x), v_1(x)] = [x(x + 1), 1]$$

As $\deg u_1 = 2 = g$, $D_1$ is a reduced divisor. Similarly, let the Mumford representation of the divisor $D_2$ be

$$[u_2(x), v_2(x)] = [(x - 1)(x - 2), -1]$$

As $\deg u_2 = 2 = g$, $D_2$ is a reduced divisor. Our goal is to determine $D$; which is the reduced sum of divisors $D_1$ and $D_2$. This is accomplished by using the Cantor-Koblitz algorithm. That is, $D = D_1 \oplus D_2$ is determined.

*Step* 1 *Computations*:

Observe that $(v_1 + v_2 + h) = 0$. By using extended Euclidean algorithm for polynomials, we determine

$$d = \gcd(u_1, u_2, v_1 + v_2 + h) = s_1 u_1 + s_2 u_2 + s_3 (v_1 + v_2 + h)$$
$$= (s_1 u_1 + s_2 u_2)$$

where

$$d(x) = \frac{3}{4}, \quad s_1(x) = \frac{1}{8}(-2x + 5), \quad s_2(x) = \frac{1}{8}(2x + 3)$$

Therefore

$$u(x) = \frac{u_1(x) u_2(x)}{(d(x))^2} = \frac{16}{9} x (x^2 - 1)(x - 2)$$

and

$$v(x) \equiv \left\{ \frac{(s_1(x) u_1(x) v_2(x) + s_2(x) u_2(x) v_1(x))}{d(x)} \right\} \pmod{u(x)}$$
$$\equiv \frac{1}{3}(2x^3 - 3x^2 - 5x + 3) \pmod{u(x)}$$

*Step 2 Computations*:
*Iteration* 1: $\deg u = 4 > g = 2$. We have

$$u'(x) = \frac{\left( f(x) - v(x) h(x) - (v(x))^2 \right)}{u(x)} = -\frac{1}{16}(4x^2 - 13x - 33)$$

Also

$$v'(x) \equiv -(h(x) + v(x)) \pmod{u'(x)}$$
$$\equiv -\frac{1}{8}(61x + 85) \pmod{u'(x)}$$

We let

$$u \leftarrow u', \quad \text{and} \quad v \leftarrow v'$$

Observe that $\deg u = 2 = g$. Therefore computations of Step 2 are finished after the first iteration. In the final stage, the polynomial $u$ is made monic. This leads to

$$u(x) = x^2 - \frac{13}{4} x - \frac{33}{4}$$
$$v(x) = -\frac{1}{8}(61x + 85)$$

This is the end of the Cantor-Koblitz algorithm. It terminates after only a single iteration of the while-loop, as $g = 2$. As a final check, we verify if the curve $C$ and the polynomial $v(x)$ intersect at the roots of the polynomial $u(x)$. Let the roots of the polynomial $u(x)$ be $x_1$ and $x_2$. These are

$$x_1, x_2 = \frac{1}{8}\left(13 \pm \sqrt{697}\right)$$

Therefore

$$v(x_1) = -\frac{1}{64}\left(1473 + 61\sqrt{697}\right)$$

$$v(x_2) = -\frac{1}{64}\left(1473 - 61\sqrt{697}\right)$$

It can be numerically verified that $(v(x_i))^2 = f(x_i)$ for $i = 1, 2$. Finally, the reduced sum $D$ of divisors $D_1$ and $D_2$ is

$$D = [u(x), v(x)] = \left[\left(x^2 - \frac{13}{4}x - \frac{33}{4}\right), \left(-\frac{1}{8}\{61x + 85\}\right)\right]$$

$\square$

### 2.9.5 Order of the Jacobian

Order of the Jacobian of a hyperelliptic curve plays a key role in its application to cryptography. Consider a hyperelliptic curve $C$ of genus $g$. The Jacobian of the curve $C$ is an Abelian group $\mathbb{J}(C)$, which is the group of divisors of degree zero modulo the set of principal divisors. The group operation $\oplus$ of $\mathbb{J}(C)$ is best described via the Cantor-Koblitz algorithm. Let $J$ be the set of points on which this group operation acts. In this notation, the group $\mathbb{J}(C) = (J, \oplus)$. Denote the cardinality of the set $J$ by $|J| = N$. Thus the order of the Jacobian is $N$.

Assume that the hyperelliptic curve is defined over a finite field $\mathbb{F}_q$; where $q = p^n$, $p$ is a prime number, and $n \in \mathbb{P}$. An estimate of the cardinality of the set $J$ is given by the Hasse-Weil theorem. It is named after the mathematicians H. Hasse and André Weil. It is

$$(\sqrt{q} - 1)^{2g} \leq N \leq (\sqrt{q} + 1)^{2g}$$

For large values of $q$, $N \simeq q^g$. This estimate of $N$ is useful in determining appropriate hyperelliptic curves for cryptographic applications.

## Reference Notes

The mathematics discussed in this chapter provides ideas and techniques for reliable data communication and storage. In addition, modern cryptographic methods and techniques rely heavily upon algebra. Therefore issues of both data integrity and security depend upon basic concepts of abstract algebra.

The description of elementary algebraic structures is based upon books by Menezes, Oorschot, and Vanstone (1996), and Rosen (2018). The section on further results on group theory is based upon Beachy, and Blair (1996), Ratcliffe (1994), and Rosen (2018). Details pertaining to the discussion on vector algebra are from Lipschutz (1968). The discussion on finite fields follows: Berlekamp (1968), Lidl, and Niederreiter (1986), MacWilliams, and Sloane (1988), Pless (1998), and Hardy, and Walker (2003). The theorem describing the group structure of points on an elliptic curve over finite field is proved in Kumanduri, and Romero (1998), Schmitt, and Zimmer (2003), and Washington (2003). Interested reader can find the proof of the Bézout's theorem in Silverman, and Tate (1992). A weaker version of this theorem is proved in the chapter on applied analysis. A proof of an estimate of the number of points on an elliptic curve defined over a finite field can be found in Washington (2003).

The discussion on hyperelliptic curves is essentially based upon the comprehensive and elegant development of this topic by Menezes, Wu, and Zuccherato in Koblitz (1999). A readable introduction to this subject is also provided by Gruenewald (2009), and Scheidler (2016). Another useful reference on this topic is the book by Galbraith (2012). A comprehensive description of elliptic and hyperelliptic curves can also be found in Cohen and Frey (2006).

## Problems

1. Let $\mathcal{G} = (G, *)$ be an Abelian group, and $\mathcal{H} = (H, *)$ be its subgroup. Prove that two cosets of $H$ are either disjoint or coincide.

   Hint: Let $a, b \in G$. If the cosets $a * H$ and $b * H$ overlap, let $u \in (a * H) \cap (b * H)$. Assume $u = a * x = b * y$, where $x, y \in H$. Consequently $b = a * x * y^{-1} = a * x'$, where $x' \in H$. Therefore

   $$b * H = \{b * h \mid h \in H\} = \{(a * x') * h \mid h \in H\} \subseteq a * H$$

   since $(a * x') * h = a * x''$, and $x'' \in H$. It can be similarly shown that $a * H \subseteq b * H$. These results imply $b * H = a * H$.

2. Let $\mathcal{G} = (G, *)$ be a group, and $Aut(\mathcal{G})$ be the set of all automorphism of $\mathcal{G}$. Prove that $Aut(\mathcal{G})$ is a group under composition of functions.

   Hint: See Beachy, and Blair (1996). It is known that composition of a function is always associative. Furthermore, composition of two isomorphisms is also an isomorphism, and inverse of an isomorphism is also an isomorphism. Therefore, $Aut(\mathcal{G})$ is a group under composition of functions.

3. Let $\mathcal{G} = (G, *)$ be a group. Establish the following assertions.

   (a) Every inner automorphism of $\mathcal{G}$ is an automorphism of $\mathcal{G}$.

   (b) $Inn(\mathcal{G})$ is a normal subgroup of $Aut(\mathcal{G})$.

   Hint: See Beachy, and Blair (1996).

   (a) Let $x, y \in G$, and $i_a(\cdot)$ be an inner automorphism of $\mathcal{G}$ for $a \in G$. Then

   $$i_a(x * y) = a * (x * y) * a^{-1} = \left(a * x * a^{-1}\right) * \left(a * y * a^{-1}\right)$$
   $$= i_a(x) * i_a(y)$$

   Therefore $i_a(\cdot)$ is a homomorphism. If $i_a(x) = e$, the identity element of the group $\mathcal{G}$, then $a * x * a^{-1} = e$. This implies $x = e$ and $i_a(\cdot)$ is a one-to-one mapping as its kernel is trivial. Also for $y \in G$, observe that $y = i_a\left(a^{-1} * y * a\right)$. Therefore, $i_a(\cdot)$ is an onto mapping.

   (b) For any $a, b, x \in G$ we have

   $$i_a(i_b(x)) = a * \left(b * x * b^{-1}\right) * a^{-1} = (a * b) * x * (a * b)^{-1} = i_{a*b}(x)$$

   Therefore $i_a(i_b(\cdot)) = i_{a*b}(\cdot)$. Note that $i_e(\cdot)$ is the identity mapping, and $i_a^{-1}(\cdot) = i_{a^{-1}}(\cdot)$. Consequently, $Inn(\mathcal{G})$ is a subgroup of $Aut(\mathcal{G})$.

   We next demonstrate that $Inn(\mathcal{G})$ is a normal subgroup of $Aut(\mathcal{G})$. Let $\alpha(\cdot) \in Aut(\mathcal{G})$, and $i_a(\cdot) \in Inn(\mathcal{G})$. For $x \in G$ we have

$$\alpha \left( i_a \left( \alpha^{-1} \left( x \right) \right) \right) = \alpha \left( a * \left( \alpha^{-1} \left( x \right) \right) * a^{-1} \right)$$
$$= \left( \alpha \left( a \right) \right) * \left( \alpha \left( \alpha^{-1} \left( x \right) \right) \right) * \left( \alpha \left( a^{-1} \right) \right)$$
$$= \left( \alpha \left( a \right) \right) * \left( x \right) * \left( \alpha \left( a^{-1} \right) \right) = b * x * b^{-1} = i_b \left( x \right)$$

where $b = \alpha \left( a \right)$. Thus $\alpha \left( i_a \left( \alpha^{-1} \left( \cdot \right) \right) \right) \in Inn \left( \mathcal{G} \right)$. Consequently, $Inn \left( \mathcal{G} \right)$ is a normal subgroup of $Aut \left( \mathcal{G} \right)$.

4. Show that for any field $\mathbb{F}$, the center of $SL \left( n, \mathbb{F} \right)$ is the set nonzero matrices with determinant equal to unity.

Hint: See Beachy, and Blair (1996). Let $I$ be an identity matrix of size $n$. Also let $I_{ij}$ be a matrix of size $n$, whose $ij$th element is equal to unity, and all other elements are equal to zero. Observe that $\left( I + I_{ij} \right) \in SL \left( n, \mathbb{F} \right)$. Also let $A = \left[ a_{ij} \right]$ be the center of $SL \left( n, \mathbb{F} \right)$. As $A$ is a center we should have

$$\left( I + I_{ij} \right) A \left( I + I_{ij} \right)^{-1} = A$$

Comparing the $ii$th element in this matrix equation yields $\left( a_{ii} + a_{ji} \right) = a_{ii}$. This implies that $A$ is a diagonal matrix. Using this result, and again comparing the $ij$th element in this matrix yields $\left( a_{jj} - a_{ii} \right) = 0$. That is, all the diagonal elements of matrix $A$ are equal. Its determinant $\det \left( A \right) = 1$, as it is the center of $SL \left( n, \mathbb{F} \right)$.

5. If $p$ is a prime number, then prove that the binomial coefficient $\binom{p}{j}$ is divisible by $p$ for $1 \leq j \leq \left( p - 1 \right)$.

Hint: Note that
$$\binom{p}{j} = \frac{p \left( p - 1 \right) \ldots \left( p - j + 1 \right)}{1 \cdot 2 \cdot \ldots \cdot j}, \quad 1 \leq j \leq \left( p - 1 \right)$$

In the above representation of the binomial coefficient, the denominator does not have $p$ as a factor, however the numerator does. The result follows by noting that the binomial coefficient is an integer.

6. Prove that the characteristic of a field is either 0 or a prime number.

Hint: See Moon (2005). If the field is infinite, then its characteristic is 0. Assume that the field is finite, and its characteristic is a composite number, say $rs$. Then either $r - 0$, or $s = 0$. This is a contradiction. Therefore, the characteristic of a finite field is a prime number.

7. The characteristic of a field $p$ is positive; and $a$ and $b$ are any elements (or variables) in a field. Prove that $\left( a \pm b \right)^p = \left( a^p \pm b^p \right)$.

Hint: This can be proved by using the binomial theorem, and noting that each binomial coefficient $\binom{p}{r}$ for $1 \leq r \leq \left( p - 1 \right)$ are multiples of $p$ (which is a prime number).

8. The characteristic of a field $p$ is positive; $a$ and $b$ are any elements (or variables) in a field; and $n \in \mathbb{P}$. Prove that

$$\left( a \pm b \right)^{p^n} = a^{p^n} \pm b^{p^n}$$

Hint: If $n = 1$, then this is the previous problem. Let $n = 2$, then $\left( a \pm b \right)^{p^2} = \left\{ \left( a \pm b \right)^p \right\}^p = \left( a^p \pm b^p \right)^p = a^{p^2} \pm b^{p^2}$. The result follows by repeating this procedure $\left( n - 2 \right)$ more times.

9. Justify the definition of a zero of the polynomial $f \left( x \right)$.

Hint: We need to justify that: If $f \left( x \right)$ is a polynomial over a field $\mathcal{F}$, then $\alpha$ is a zero of $f \left( x \right)$ iff $f \left( x \right) = \left( x - \alpha \right) g \left( x \right)$.

Note that $f \left( x \right)$ can be written as $f \left( x \right) = \left( x - \alpha \right) g \left( x \right) + r \left( x \right)$, where $r \left( x \right)$ is a polynomial of degree less than 1. This implies that $r \left( x \right)$ is a constant. Also since $r \left( \alpha \right) = 0$, then $r \left( x \right) = 0$. Therefore $f \left( x \right) = \left( x - \alpha \right) g \left( x \right)$.

Also, if $f \left( x \right) = \left( x - \alpha \right) g \left( x \right)$, then $f \left( \alpha \right) = 0$.

10. Let $\mathcal{U} = (U, \mathcal{F}, \boxplus, \boxtimes)$, and $\mathcal{V} = (V, \mathcal{F}, \boxplus, \boxtimes)$ be vector spaces over the same field $\mathcal{F} = (F, +, \times)$. Also let $f : V \to U$ be a linear mapping, and $\dim(\mathcal{V})$ finite. Prove that $\dim(\mathcal{V}) = \dim(\operatorname{im} f) + \dim(\ker f)$.

Hint: See Lipschutz (1968). As per the hypothesis $\dim(\mathcal{V}) = n$. Let the image of the mapping be $\widetilde{I}$ and the kernel be $\widetilde{K}$. That is, $\widetilde{I} \subseteq U$ and $\widetilde{K} \subseteq V$. Since $\dim(\mathcal{V}) = n$ is finite, assume that $\dim(\ker f) = s \leq n$. Therefore we need to establish that $\dim(\operatorname{im} f) = (n - s)$. Let $\{k_1, k_2, \ldots, k_s\}$ be a basis of the kernel space $\widetilde{K}$. This basis set can be extended to form a basis for $V$. Let this basis set of $V$ be

$$\widetilde{C} = \{k_1, k_2, \ldots, k_s, v_1, v_2, \ldots, v_{n-s}\}$$

Next define $\widetilde{D} = \{f(v_1), f(v_2), \ldots, f(v_{n-s})\}$. It remains to prove that the set $\widetilde{D}$ is a basis of the image $\widetilde{I}$. This is done in two steps. In the first step it is proved that the set $\widetilde{D}$ generates the image and in the second step it is established that the members of the set $\widetilde{D}$ are linearly independent.

*Step* 1: Assume that $u \in \widetilde{I}$. Thus there exists $v \in V$ such that $u = f(v)$. As the set $\widetilde{C}$ generates $V$, we have

$$v = \sum_{i=1}^{s} a_i k_i + \sum_{i=1}^{n-s} b_i v_i, \ \forall \, a_i, b_i \in F$$

$$u = f(v) = \sum_{i=1}^{s} a_i f(k_i) + \sum_{i=1}^{n-s} b_i f(v_i)$$

Note that the $f(k_i)$'s are each equal to zero, since $k_i \in \widetilde{K}$. Consequently $u = \sum_{i=1}^{n-s} b_i f(v_i)$. Thus the set $\widetilde{D}$ generates the image of $f$.

*Step* 2: Define $d = \sum_{i=1}^{n-s} d_i v_i$ where all $d_i \in F$, and assume that $\sum_{i=1}^{n-s} d_i f(v_i) = 0$, then $f(d) = 0$, thus $d \in \widetilde{K}$. Therefore $d = \sum_{i=1}^{s} c_i k_i$, where all $c_i \in F$. This yields

$$\sum_{i=1}^{n-s} d_i v_i - \sum_{i=1}^{s} c_i k_i = 0$$

Since the set $\widetilde{C}$ generates $V$, the members of the set $\widetilde{C}$ are linearly independent. Hence the coefficients $d_i$'s and $c_i$'s are all equal to zero. That is, $d_i = 0$ for $1 \leq i \leq (n - s)$. Thus the $f(v_i)$'s are linearly independent. Consequently the set $\widetilde{D}$ is a basis of the image $\widetilde{I}$. Therefore $\dim(\operatorname{im} f) = (n - s)$.

11. Prove that $\mathbb{Z}_m$ is a finite field if and only if $m$ is a prime number.

Hint: See Ling, and Xing (2004). Assume that $m = ab$, where $a, b \in \mathbb{P}$, and $1 < a, b < m$. Observe that $0 = m = ab$ is in $\mathbb{Z}_m$. This implies that $a = 0$ or $b = 0$ or $a = b = 0$ in modulo $m$ arithmetic. This is a contradiction. Therefore $m$ is not a composite number.

Next let $m$ be a prime number. Select $a \in \mathbb{Z}_m$ such that $0 < a < m$. Therefore $\gcd(a, m) = 1$. This in turn implies that there exist two integers $u$ and $v$ such that $(ua + vm) = 1$. Therefore $ua \equiv 1 \pmod{m}$. Thus there exists a multiplicative inverse of element $a$. Therefore $\mathbb{Z}_m$ is a field.

12. Let the characteristic of the finite field $\mathbb{F}_q = (F, +, \times)$ be the prime number $p$. Prove that the order of this field $q$ is a power of $p$. That is $q = p^n$, where $n \in \mathbb{P}$.

Hint: See MacWilliams, and Sloane (1988), and Moon (2005). Assume that $q \geq p$. The field $\mathbb{F}_q$ has a prime subfield $\mathbb{F}_p$ of order $p$. It is demonstrated that $\mathbb{F}_q$ can be represented as a vector space over its subfield $\mathbb{F}_p$.

Let $\lambda_1 \in F \setminus \{0\}$. Form the elements $a_1 \lambda_1$, where $a_1$ varies over the elements of subfield $\mathbb{F}_p$. Note that $a_1 \lambda_1$ takes on $p$ distinct values. This is true because $x\lambda_1 \neq y\lambda_1$ if $x \neq y$. If the $p$ distinct products are equal to the $q$ elements of the field $\mathbb{F}_q$, then the elements of the field form a vector space over $\mathbb{F}_p$, and $q = p$. The dimension of this vector space is unity.

If the elements of the field $\mathbb{F}_q$ are not covered, let $\lambda_2$ be an element which has not yet been covered. Form the elements $(a_1\lambda_1 + a_2\lambda_2)$, where $a_1$ and $a_2$ vary over the elements of subfield $\mathbb{F}_p$. This results in $p^2$ distinct elements of the field $\mathbb{F}_q$. At this stage, if all the elements of the field $\mathbb{F}_q$ are not yet covered, then keep trying linear combinations of the form $\sum_{i=1}^n a_i \lambda_i$ till all the elements of $\mathbb{F}_q$ are covered. As each $a_i$ takes $p$ distinct values in the subfield $\mathbb{F}_p$, we have $|F| = q = p^n$.

13. Let $\mathbb{F}_q$ be a finite field. Prove the following statements.
   (a) Let the order of the field element $a$ be $j$. Then $a^k = 1$ if and only if $j \mid k$.
   (b) Let $a$ and $b$ be two different field elements of orders $j$ and $k$ respectively, such that $\gcd(j, k) = 1$, then the order of the element $ab$ is $jk$.
   (c) If the order of $a \in \mathbb{F}_q$ is $m$, then the order of $a^j$ is $m/\gcd(m, j)$ where $1 \leq j \leq m$.

   Hint: See Moon (2005).

   (a) Left to the reader.
   (b) As per the hypothesis of the problem, $a^j = 1$, and $b^k = 1$. Therefore

   $$(ab)^{jk} = \left(a^j\right)^k \left(b^k\right)^j = 1$$

   It is quite possible that the order of $ab$ can be smaller than $jk$. Let it be $u$. Therefore from part (a) of the problem, $u \mid jk$. As $\gcd(j, k) = 1$, this implies $u = cj$, and $u = dk$ for some positive integers $c$ and $d$. Thus $u = ejk$, for some positive integer $e$. The smallest such value of $e$ is 1.
   (c) As per the hypothesis of the problem, $a^m = 1$. Let $a^j \triangleq b$, and its order be $u$. Therefore $b^u = 1$. It needs to established that indeed $u = m/\gcd(m, j)$. Thus

   $$b^{m/\gcd(m,j)} = \left(a^j\right)^{m/\gcd(m,j)} = (a^m)^{j/\gcd(m,j)} = 1$$

   This implies $u \mid m/\gcd(m, j)$ as per part (a) of the problem. Also

   $$1 = b^u = \left(a^j\right)^u$$

   Therefore $m \mid ju$. This in turn implies $m/\gcd(m, j) \mid u$. Finally $u \mid m/\gcd(m, j)$ together with $m/\gcd(m, j) \mid u$ yields $u = m/\gcd(m, j)$.

14. Let $\varphi(\cdot)$ denote the Euler's phi-function. Also let $\mathbb{F}_q$ be a finite field, and $j$ be any positive integer. If $j \mid (q - 1)$ then there are $\varphi(j)$ elements of the field of order $j$. However, if $j \nmid (q - 1)$ then there are 0 elements of order $j$ in the field. Prove this assertion.

   Hint: See Moon (2005). Observe that if $j \nmid (q - 1)$ then there are 0 elements of order $j$ in the field, since the order of an element has to divide $(q - 1)$. Therefore assume that $j \mid (q - 1)$. The task now is to determine the number of elements of order $j$.

   Assume that the field element $a$ has order $j$. Therefore, if $b = a^i$ and $\gcd(i, j) = 1$, then the order of $b$ is also $j$. The number of such $i$ is $\varphi(j)$.

   It might also be possible for an element $b$ to be not of the form $a^i$, and yet have order $j$. However, any element of order $j$ is a root of the polynomial $\left(x^j - 1\right)$. The $j$ roots of this polynomial are $a, a^2, \ldots, a^j$, and this polynomial of degree $j$ has evidently no more than $j$ roots. Therefore there are no elements of order $j$ that are not in the set $\left\{a, a^2, \ldots, a^j\right\}$.

   This result also implies that the number of primitive elements in $\mathbb{F}_q$ is equal to $\varphi(q - 1)$.

15. $g(x)$ is a polynomial with coefficients in the field $\mathbb{Z}_p$, where $p$ is prime number. Prove that $g(x^{p^r}) = \{g(x)\}^{p^r}, r \in \mathbb{P}$.

    Hint: Let

$$g(x) = (b_m x^m + b_{m-1} x^{m-1} + \ldots + b_1 x + b_0); \; b_i \in \mathbb{Z}_p, \; 0 \le i \le m$$

   then

$$\begin{aligned}
\{g(x)\}^{p^r} &= (b_m x^m + b_{m-1} x^{m-1} + \ldots + b_1 x + b_0)^{p^r} \\
&= b_m^{p^r}\left(x^{p^r}\right)^m + b_{m-1}^{p^r}\left(x^{p^r}\right)^{m-1} + \ldots + b_1^{p^r}\left(x^{p^r}\right) + b_0^{p^r} \\
&= b_m\left(x^{p^r}\right)^m + b_{m-1}\left(x^{p^r}\right)^{m-1} + \ldots + b_1\left(x^{p^r}\right) + b_0 \\
&= g\left(x^{p^r}\right)
\end{aligned}$$

   where the relationship $(x+y)^{p^r} = \left(x^{p^r} + y^{p^r}\right)$, and Fermat's little theorem are used.

16. Let $f(x) \in F[x]$ be a polynomial defined over a field $\mathcal{F} = (F, +, \times)$. The formal derivative of $f(x)$ is $f'(x)$. Prove that $f(x)$ has no repeated irreducible factors if and only if the polynomials $f(x)$ and $f'(x)$ are relatively prime.

17. The minimal polynomial associated with an element $a \in GF_{p^n}$, is $\eta_a(x)$. Prove that $\eta_a(x)$ is irreducible.

    Hint: See Pless (1998). Assume that $\eta_a(x)$ is reducible. Let $\eta_a(x) = u(x)v(x)$. Since $\eta_a(a) = 0$, then $u(a) = 0$ or $v(a) = 0$, or $u(a) = v(a) = 0$. This contradicts the hypothesis that $\eta_a(x)$ is a polynomial of the smallest degree such that $a$ is its root.

18. Establish the following results.

    (a) If $a$ is a root of the polynomial $g(x)$ with coefficients in $\mathbb{Z}_p$, prove that the minimal polynomial $\eta_a(x)$ is a divisor of polynomial $g(x)$.

    (b) The minimal polynomial associated with an element $a \in GF_{p^n}$, is $\eta_a(x)$. Prove that $\eta_a(x)$ divides the polynomial $(x^q - x)$, where $q = p^n$, and $n$ is the degree of the irreducible polynomial $f(x)$ which is used in generating the field $\mathbb{Z}_p[x]/(f(x))$.

    Hint: See Pless (1998).

    (a) Let $g(x) = \psi(x)\eta_a(x) + r(x)$ where degree of $r(x)$ is less than the degree of $\eta_a(x)$. Substituting $x = a$, in this equation yields $r(a) = 0$, since $g(a) = \eta_a(a) = 0$. Therefore $\deg r(x) < \deg \eta_a(x)$ and $r(a) = 0$. This is a contradiction unless $r(x) = 0$. Therefore $g(x)$ is divisible by $\eta_a(x)$.

    (b) This part is actually a special instance of part (a) of the problem. Write $(x^q - x) = \eta_a(x)\psi(x) + r(x)$, where the degree of the polynomial $r(x)$ is smaller than the degree of the polynomial $\eta_a(x)$. Substitute $x = a$ in this equation, where $a$ is an element of the field $\mathbb{Z}_p[x]/(f(x))$. This yields $r(a) = 0$. But $\eta_a(x)$ is a polynomial with the smallest degree such that $\eta_a(a) = 0$. Therefore $r(x) = 0$.

19. Establish the following results.

    (a) The degree of the minimal polynomial $\eta_a(x)$ associated with the element $a \in \mathbb{F}_q$, is less than or equal to $n$.

    (b) If $a$ is a primitive element of the finite field, then prove that the minimal polynomial $\eta_a(x)$ is also its primitive polynomial, and its degree is equal to $n$.

    Hint: See MacWilliams, and Sloane (1988), and Pless (1998).

(a) Let $a$ be any element in $\mathbb{F}_q = GF_{p^n}$.

Consider the elements $1, a, a^2, \ldots, a^n$ which belong to the field $GF_{p^n}$. As $GF_{p^n}$ is an $n$-dimensional vector space over the field $\mathbb{Z}_p$, it follows that these $(n+1)$ elements are linearly dependent. Therefore, there exist $\beta_0, \beta_1, \ldots, \beta_n \in \mathbb{Z}_p$, not all zeros such that $\sum_{j=0}^{n} \beta_j a^j = 0$. That is, if $\beta(x) = \sum_{j=0}^{n} \beta_j x^j$, then $a$ satisfies the equation $\beta(x) = 0$. Therefore, the degree of the minimal polynomial is less than or equal to $n$.

(b) If $a$ is a primitive element of the finite field, then the minimal polynomial $\eta_a(x)$ of degree $d$ can be used to generate a finite field $\mathbb{F}'$ of order $p^d$. However, $a \in \mathbb{F}'$ and therefore $\mathbb{F}'$ also contains $\mathbb{F}_q$, that is $d \geq n$. However from part (a) of the problem $\deg \eta_a(x) \leq n$. Therefore $d = n$.

20. The polynomial $\left(x^{p^n} - x\right)$ is equal to the product of all prime polynomials over $\mathbb{Z}_p$ whose degree divide $n$. Prove this statement.

Hint: See MacWilliams, and Sloane (1988). Recall that a prime polynomial is a monic irreducible polynomial.

*Step* 1: Assume that $\varrho(x)$ is an irreducible polynomial over $\mathbb{Z}_p$, where $\deg \varrho(x) = d$, and $d \mid n$. The case $\varrho(x) = x$ is trivial. Therefore assume that $\varrho(x) \neq x$. If the polynomial $\varrho(x)$ is used to construct a field $GF_{p^d}$, then $\varrho(x)$ is indeed a minimal polynomial of one of the elements.

Therefore from a property of minimal polynomials $\varrho(x) \mid \left(x^{p^d-1} - 1\right)$.

Also since $d \mid n \Rightarrow (p^d - 1) \mid (p^n - 1) \Rightarrow \left(x^{p^d-1} - 1\right) \mid \left(x^{p^n-1} - 1\right)$, we have

$$\varrho(x) \mid \left(x^{p^n-1} - 1\right)$$

*Step* 2: Conversely, assume that $\varrho(x)$ is an irreducible polynomial over $\mathbb{Z}_p$, where $\deg \varrho(x) = d$, and $\varrho(x) \mid \left(x^{p^n-1} - 1\right)$. Our goal is to demonstrate that $d \mid n$. As in Step 1 assume that $\varrho(x) \neq x$, such that $\varrho(x) \mid \left(x^{p^n-1} - 1\right)$. Since $\varrho(x)$ is an irreducible polynomial, use $\varrho(x)$ to construct the field $GF_{p^d}$. Let $\alpha \in GF_{p^d}$ be a root of the polynomial $\varrho(x)$ and let $\beta$ be a primitive element of the field $GF_{p^d}$. That is

$$\beta = \sum_{i=0}^{d-1} a_i \alpha^i; \quad a_i \in \mathbb{Z}_p, \ 0 \leq i \leq (d-1)$$

Also $\varrho(\alpha) = 0$ implies $\alpha^{p^n} = \alpha$. Applying this result to the above equation yields $\beta^{p^n} = \beta$. Therefore $\beta^{p^n-1} = 1 \Rightarrow$ the order of the element $\beta$, which is $(p^d - 1)$ must divide $(p^n - 1)$. This in turn implies $d \mid n$.

This completes the proof.

21. Prove that all the roots of an irreducible polynomial have the same order.

Hint: See Wicker (1995), and Moon (2005). Let $g(x) \in \mathbb{F}_q[x]$ be an irreducible polynomial, where $q$ is a power of some prime number. Let $GF_{q^m}$ be the smallest field containing all the roots of the polynomial $g(x)$. Let $\beta \in GF_{q^m}$ be a root of $g(x)$. Let the order of the element $\beta$ be $j$. Therefore $j \mid (q^m - 1)$ by Lagrange's theorem. Furthermore, the roots of $g(x)$ are the conjugates of $\beta$, and are of the form $\left\{\beta, \beta^q, \beta^{q^2}, \ldots\right\}$. Let $q = p^r$, where $p$ is a prime number and $r \in \mathbb{P}$. Then $\gcd(q, q^m - 1) = 1$. Furthermore, if $t \mid (q^m - 1)$, then $\gcd(q, t) = 1$. Therefore,

$$\text{order of } \beta^{q^k} = \frac{j}{\gcd(q^k, j)} = j$$

for any $k$. Therefore, each root of the irreducible polynomial $g(x)$ has the same order.

22. Find different representations of the field $GF_{16}$. Also determine the minimal polynomial of each field element.

Hint: See Hardy, and Walker (2003). Let $GF_{16}$ be an extension field of $GF_2$. Note that

$$(x^{16} + x) \equiv x\,(x + 1)\,(x^2 + x + 1)\,(x^4 + x^3 + 1)\,(x^4 + x + 1)$$
$$\cdot\,(x^4 + x^3 + x^2 + x + 1) \pmod{2}$$

Denote the set of elements $GF_{16} \setminus \{0\} = \{\alpha^j \mid 0 \le j \le 14\}$. The primitive polynomial of the element $\alpha$ is selected to be

$$\eta_\alpha(x) = (x^4 + x + 1)$$

Therefore $(\alpha^4 + \alpha + 1) = 0$. The polynomial representation of $\alpha^j$ is obtained by computing $x^j \pmod{(x^4 + x + 1)}$ for $0 \le j \le 14$. The polynomial, power, binary, and decimal representations of the elements which belong to the field $GF_{16}$ are listed in Table 2.12.

| Polynomial representation | Power representation | Binary representation | Decimal representation |
|:---:|:---:|:---:|:---:|
| 0 | 0 | 0000 | 0 |
| 1 | $\alpha^0$ | 0001 | 1 |
| $x$ | $\alpha^1$ | 0010 | 2 |
| $x^2$ | $\alpha^2$ | 0100 | 4 |
| $x^3$ | $\alpha^3$ | 1000 | 8 |
| $x + 1$ | $\alpha^4$ | 0011 | 3 |
| $x^2 + x$ | $\alpha^5$ | 0110 | 6 |
| $x^3 + x^2$ | $\alpha^6$ | 1100 | 12 |
| $x^3 + x + 1$ | $\alpha^7$ | 1011 | 11 |
| $x^2 + 1$ | $\alpha^8$ | 0101 | 5 |
| $x^3 + x$ | $\alpha^9$ | 1010 | 10 |
| $x^2 + x + 1$ | $\alpha^{10}$ | 0111 | 7 |
| $x^3 + x^2 + x$ | $\alpha^{11}$ | 1110 | 14 |
| $x^3 + x^2 + x + 1$ | $\alpha^{12}$ | 1111 | 15 |
| $x^3 + x^2 + 1$ | $\alpha^{13}$ | 1101 | 13 |
| $x^3 + 1$ | $\alpha^{14}$ | 1001 | 9 |

Table 2.12. Representation of elements of $GF_{16}$.

Use of the factorization theorem of minimal polynomials yields

$$\eta_0(x) = x$$
$$\eta_1(x) = (x - \alpha^0) = (x + 1)$$
$$\eta_\alpha(x) = \eta_{\alpha^2}(x) = \eta_{\alpha^4}(x) = \eta_{\alpha^8}(x)$$
$$= (x - \alpha)(x - \alpha^2)(x - \alpha^4)(x - \alpha^8)$$
$$= (x^4 + x + 1)$$
$$\eta_{\alpha^3}(x) = \eta_{\alpha^6}(x) = \eta_{\alpha^9}(x) = \eta_{\alpha^{12}}(x)$$
$$= (x - \alpha^3)(x - \alpha^6)(x - \alpha^9)(x - \alpha^{12})$$
$$= (x^4 + x^3 + x^2 + x + 1)$$

$$\eta_{\alpha^5}(x) = \eta_{\alpha^{10}}(x)$$
$$= (x - \alpha^5)(x - \alpha^{10})$$
$$= (x^2 + x + 1)$$
$$\eta_{\alpha^7}(x) = \eta_{\alpha^{11}}(x) = \eta_{\alpha^{13}}(x) = \eta_{\alpha^{14}}(x)$$
$$= (x - \alpha^7)(x - \alpha^{11})(x - \alpha^{13})(x - \alpha^{14})$$
$$= (x^4 + x^3 + 1)$$

23. This problem has two parts.

(a) Explain why equations of the type $y^2 = (x^3 + a_2 x^2 + a_1 x + a_0)$ are called elliptic curves.

(b) Find the perimeter of an ellipse.

Hint: See Abramowitz, and Stegun (1965), and Brown (2000). Let the ellipse centered at the origin with semimajor axis $a$, and semiminor axis $b$ be

$$\frac{x^2}{a^2} + \frac{y^2}{b^2} = 1, \quad 0 < b < a$$

The value $\epsilon = (1 - b^2/a^2)^{1/2}$ is called the eccentricity of the ellipse. A parametric representation of the ellipse is $x = x(\theta) = a\cos\theta$, and $y = y(\theta) = b\sin\theta$, where $\theta$ varies from 0 to $2\pi$. Let $\dot{x}(\theta) \triangleq dx(\theta)/d\theta$, and $\dot{y}(\theta) \triangleq dy(\theta)/d\theta$. The length of the curve for $\theta \in [\theta_0, \theta_1]$ is

$$L(\theta_0, \theta_1) = \int_{\theta_0}^{\theta_1} \sqrt{\left(\dot{x}(\theta)\right)^2 + \left(\dot{y}(\theta)\right)^2} \, d\theta$$

(a) We show that cubic polynomials of the type

$$f(x) = (x^3 + a_2 x^2 + a_1 x + a_0)$$

occur in the computation of arc-lengths of an ellipse. The arc-length $L(\theta_0, \theta_1)$ is

$$L(\theta_0, \theta_1) = a \int_{\theta_0}^{\theta_1} \sqrt{1 - \epsilon^2 \cos^2\theta} \, d\theta$$

Substituting $\cos\theta = u$ in the above integral yields

$$L(\theta_0, \theta_1) = a \int_{u_1}^{u_0} \frac{(1 - \epsilon^2 u^2)}{\sqrt{(1 - u^2)(1 - \epsilon^2 u^2)}} \, du$$

where $u_0 = \cos\theta_0$, and $u_1 = \cos\theta_1$. In order to evaluate $L(\theta_0, \theta_1)$, we need to evaluate the integral

$$I(u) = \int \frac{du}{\sqrt{(1 - u^2)(1 - \epsilon^2 u^2)}}$$

This integral is called an elliptic integral. Let

$$v^2 = h(u) = (1 - u^2)(1 - \epsilon^2 u^2)$$

Observe that $h(u)$ is a quartic polynomial. That is, $h(u)$ is a polynomial of degree 4. If $\epsilon \neq \pm 1$, then $h(u)$ has four distinct roots. This facilitates the conversion of $v^2 = h(u)$

to $y^2 = f(x)$, where $f(x)$ is a cubic polynomial. The degree of a cubic polynomial is 3. This conversion can be seen as follows.

Let the four distinct roots of the polynomial $h(u)$ be $\alpha_i$, where $0 \le i \le 3$. Therefore

$$v^2 = h(u) = \prod_{i=0}^{3} (u - \alpha_i)$$

Divide both sides of the above equation by $(u - \alpha_0)^4$. This yields

$$\left\{ \frac{v}{(u - \alpha_0)^2} \right\}^2 = \prod_{i=1}^{3} \left\{ 1 - \frac{(\alpha_i - \alpha_0)}{(u - \alpha_0)} \right\}$$

In the above equation, substitute

$$x = \frac{1}{(u - \alpha_0)}, \quad \text{and} \quad y = \frac{v}{(u - \alpha_0)^2}$$

This results in

$$y^2 = \prod_{i=1}^{3} \left\{ 1 - (\alpha_i - \alpha_0) x \right\}$$

The right-hand side of the above equation is a cubic polynomial in $x$. Therefore, in order to evaluate the length of an arc of an ellipse, we need to evaluate integrals of the type

$$J(x) = \int \frac{dx}{\sqrt{x^3 + a_2 x^2 + a_1 x + a_0}}$$

Thus

$$y^2 = \left( x^3 + a_2 x^2 + a_1 x + a_0 \right)$$

is called an elliptic curve because of its relationship to the computation of arc-length of an ellipse.

(b) The perimeter $L_{perim}$ of the ellipse is equal to $L(0, 2\pi)$. Therefore

$$L_{perim} = 4a \int_0^{\pi/2} \sqrt{1 - \epsilon^2 \cos^2 \theta} \, d\theta$$

It is known that

$$\int_0^{\pi/2} \cos^{2n} \theta \, d\theta = \frac{1}{2^{2n}} \binom{2n}{n} \frac{\pi}{2}, \quad n \in \mathbb{P}$$

$$(1 + z)^\alpha = 1 + \frac{\alpha}{1!} z + \frac{\alpha(\alpha - 1)}{2!} z^2 + \frac{\alpha(\alpha - 1)(\alpha - 2)}{3!} z^3 + \cdots, \quad |z| < 1$$

From the last expansion, we have

$$\sqrt{1 - z} = 1 - \sum_{n \in \mathbb{P}} \frac{1}{(2n - 1)} \binom{2n}{n} \frac{z^n}{2^{2n}}, \quad |z| < 1$$

Thus

$$L_{perim} = 2\pi a \left\{ 1 - \sum_{n \in \mathbb{P}} \frac{1}{(2n - 1)} \left\{ \binom{2n}{n} \right\}^2 \frac{\epsilon^{2n}}{2^{4n}} \right\}$$

The first few terms in the above expression can be obtained explicitly as

$$L_{perim} = 2\pi a \left\{ 1 - \left(\frac{1}{2}\right)^2 \frac{\epsilon^2}{1} - \left(\frac{1 \cdot 3}{2 \cdot 4}\right)^2 \frac{\epsilon^4}{3} - \left(\frac{1 \cdot 3 \cdot 5}{2 \cdot 4 \cdot 6}\right)^2 \frac{\epsilon^6}{5} - \cdots \right\}$$

Note that if $a = b$, ellipse is a circle. In this case $\epsilon = 0$, and $L_{perim} = 2\pi a$.

24. Let $C_1$ and $C_2$ be two cubic curves which consist of nine distinct points of intersection $\{p_i \mid 1 \leq i \leq 9\}$.

Prove that any cubic passing through the eight points in the set $\{p_i \mid 1 \leq i \leq 8\}$ also passes though the ninth point $p_9$.

Hint: See Kumanduri, and Romero (1998). Consider a cubic polynomial which is made up of the following ten monomials:

$$1, \; x, \; y, \; x^2, \; xy, \; y^2, \; x^3, \; x^2y, \; xy^2, \; \text{and } y^3$$

Therefore an equation of a cubic polynomial $f(x, y) = 0$ determines a point in 9-dimensional space $\Psi$. Let $f(p) = 0$ be the equation of a cubic curve $C$ which corresponds to a point $f(\cdot, \cdot)$ in space $\Psi$. Let the points in the set $\{p_i = (x_i, y_i) \mid 1 \leq i \leq 8\}$ be on the cubic curve $C$. The solution consists of a set of points of dimension $(9 - 8) = 1$. Therefore, the cubic curve which passes through the set of eight points corresponds to a straight line in space $\Psi$.

Let the equations which correspond to the curve $C_1$ and $C_2$ be $f_1(p) = 0$ and $f_2(p) = 0$ respectively.

Therefore, a line joining these two points in space $\Psi$ has an equation $(\psi_1 f_1 + \psi_2 f_2) = 0$ for some constants $\psi_1$ and $\psi_2$. Therefore $f = (\psi_1 f_1 + \psi_2 f_2)$. However by hypothesis, the point $p_9$ also lies on the curves $C_1$ and $C_2$. That is, $f_1(p_9) = f_2(p_9) = 0$. This implies $f(p_9) = 0$. Thus the point $p_9$ also lies on the curve $C$.

25. Prove the following observations about hyperelliptic curve $C$.

   (a) If $h(x) = 0$, then the characteristic of the field $\mathbb{F}$ is not equal to 2.
   (b) Let the characteristic of the field $\mathbb{F}$ be not equal to 2. Then a change of variables $x \to x$, and $y \to (y - h(x)/2)$ transforms $C$ to the form $y^2 = k(x)$, where $\deg k(x) = (2g + 1)$.
   (c) Let $h(x) = 0$, and the characteristic of the field $\mathbb{F}$ be not equal to 2. Then $C$ is a hyperelliptic curve if and only if $f(x)$ has no repeated roots in the field $\overline{\mathbb{F}}$.

   Hint: See the Appendix by Menezes, Wu, and Zuccherato in Koblitz (1999).

   (a) This result is proved by contradiction. As $h(x) = 0$, the hyperelliptic curve $C$ equation and its partial derivative equations are

$$y^2 = f(x), \, 2y = 0, \text{ and } f'(x) = 0$$

   Let $x_0 \in \overline{\mathbb{F}}$ satisfy $f'(x) = 0$, and let $y_0 \in \overline{\mathbb{F}}$ satisfy $y^2 = f(x_0)$. Further assume that the characteristic of field $\mathbb{F}$ is equal to 2. Thus $(x_0, y_0) \in \overline{\mathbb{F}} \times \overline{\mathbb{F}}$ simultaneously satisfy the above three equations.

   That is, $(x_0, y_0)$ is a singular point on the curve $C$. This is a contradiction, because $C$ is a hyperelliptic curve, and it does not have any singular points.

   (b) In the equation $(y^2 + h(x)y) = f(x)$, substitute $(y - h(x)/2)$ for $y$. This results in

$$(y - h(x)/2)^2 + h(x)(y - h(x)/2) = f(x)$$

   That is

$$y^2 = f(x) + \frac{(h(x))^2}{4} \triangleq k(x)$$

Observe that $\deg k(x) = (2g+1)$.

(c) As $h(x) = 0$, the hyperelliptic curve $C$ equation and its partial derivative equations are

$$y^2 = f(x), \, 2y = 0, \text{ and } f'(x) = 0$$

If $(x_0, y_0)$ is a singular point on $C$, then it must satisfy the three equations $y^2 = f(x)$, $2y = 0$, and $f'(x) = 0$ simultaneously. These are satisfied at $y_0 = 0$, and by the repeated root of $f(x)$. Therefore, $C$ is a hyperelliptic curve if and only if $f(x)$ has no repeated roots in $\overline{\mathbb{F}}$.

26. This problem is on hyperelliptic curves. Prove the following statement about a polynomial used in studying hyperelliptic curves. The polynomial $r(x, y) = (y^2 + h(x)y - f(x))$ is irreducible over the field $\mathbb{F}$. Consequently, $\mathbb{F}[C]$ is an integral domain.

    Hint: See the Appendix by Menezes, Wu, and Zuccherato in Koblitz (1999). This result is proved by contradiction. Assume that the polynomial $r(x, y)$ is reducible. Let $r(x, y) = (y - a(x))(y - b(x))$, where $a(x), b(x) \in \overline{\mathbb{F}}[x]$. Therefore

$$\deg(a(x)b(x)) = \deg f(x) = (2g+1)$$
$$\deg(a(x) + b(x)) = \deg h(x) \leq g$$

    The later two conclusions are not simultaneously true. Consequently, the polynomial $r(x, y)$ is irreducible.

27. This problem is on hyperelliptic curves. Establish the following properties of the norm. Let $G$, and $H$ be polynomial functions in $\overline{\mathbb{F}}[C]$.

    (a) $N(G)$ is a polynomial in $\overline{\mathbb{F}}[x]$.
    (b) $N(\overline{G}) = N(G)$.
    (c) $N(GH) = N(G)N(H)$.

    Hint: See the Appendix by Menezes, Wu, and Zuccherato in Koblitz (1999). For brevity, dependence of a polynomial on the independent variables is not stated explicitly. For example, the polynomial $h(x)$ is simply stated as $h$; and $G(x, y) = (a(x) - b(x)y)$ is stated as $G = (a - by)$.

    (a) As $\overline{G} = (a + b(h + y))$, we have

$$N(G) = G \cdot \overline{G} = (a - by)\{a + b(h + y)\}$$
$$= a(a + bh) - b^2(hy + y^2) = \{a(a + bh) - b^2 f\} \in \overline{\mathbb{F}}[x]$$

    (b) Conjugate of $\overline{G} = (a + bh + by)$ is

$$\overline{\overline{G}} = (a + bh) - b(h + y) = a - by = G$$

    Therefore

$$N(\overline{G}) = \overline{G} \cdot \overline{\overline{G}} = \overline{G} \cdot G = N(G)$$

    (c) Let $G = (a - by)$, and $H = (c - dy)$. We have

$$GH = ac - (ad + bc)y + bdy^2$$

    Use of the relationship $(y^2 + hy - f) = 0$ leads to

$$GH = (ac + bdf) - (ad + bc + bdh)\, y$$

Conjugate of $GH$ is

$$\overline{GH} = (ac + bdf) + (ad + bc + bdh)\,(h + y)$$
$$= ac + bdf + adh + bch + bdh^2 + ady + bcy + bdhy$$
$$= ac + bc\,(h + y) + ad\,(h + y) + bd\,\left(h^2 + hy + f\right)$$
$$= ac + bc\,(h + y) + ad\,(h + y) + bd\,\left(h^2 + 2hy + y^2\right)$$

$$= ac + bc\,(h + y) + ad\,(h + y) + bd\,(h + y)^2$$
$$= \{a + b\,(h + y)\}\,\{c + d\,(h + y)\} = \overline{G} \cdot \overline{H}$$

Therefore

$$N\,(GH) = GH \cdot \overline{GH} = G \cdot \overline{G} \cdot H \cdot \overline{H} = N\,(G)\,N\,(II)$$

28. This problem is on hyperelliptic curves. Let $G, H \in \mathbb{F}\,[C]$. Establish the following observations.

   (a) $\deg G = \deg_x N\,(G)$, where $\deg_x \cdot$ is the degree with respect to the variable $x$.
   (b) $\deg GH = \deg G + \deg H$.
   (c) $\deg G = \deg \overline{G}$.

   Hint: See the Appendix by Menezes, Wu, and Zuccherato in Koblitz (1999).

   (a) Let $G\,(x, y) = (a\,(x) - b\,(x)\,y)$. The norm of this polynomial is

$$N\,(G) = \left(a^2 + abh - b^2 f\right)$$

   For brevity in notation, let $d_1 = \deg a$, and $d_2 = \deg b$. We also have $\deg h \leq g$, and $\deg f = (2g + 1)$. Consider two cases.
   *Case* 1: Assume that $2d_1 > (2g + 1 + 2d_2)$. This implies $2d_1 \geq (2g + 2 + 2d_2)$. This in turn implies $d_1 \geq (g + 1 + d_2)$. Thus

$$\deg a^2 = 2d_1 \geq d_1 + (g + 1 + d_2) > (d_1 + d_2 + g) \geq \deg abh$$

   *Case* 2: Assume that $2d_1 < (2g + 1 + 2d_2)$. This implies $2d_1 \leq (2g + 2d_2)$. This in turn implies $d_1 \leq (g + d_2)$. Thus

$$\deg abh \leq (d_1 + d_2 + g) \leq (2g + 2d_2) < (2g + 2d_2 + 1) = \deg b^2 f$$

   Therefore
$$\deg_x N\,(G) = \max\,[2\deg a, 2g + 1 + 2\deg b] = \deg G$$

   (b) Use of part (a) of the problem gives

$$\deg GH = \deg_x N\,(GH)$$

   From the last problem, $N\,(GH) = N\,(G)\,N\,(H)$. Therefore

$$\deg GH = \deg_x N\,(G)\,N\,(H)$$
$$= \deg_x N\,(G) + \deg_x N\,(H)$$
$$= \deg G + \deg H$$

(c) As $N\left(\overline{G}\right) = N\left(G\right)$, we have $\deg G = \deg_x N\left(G\right) = \deg_x N\left(\overline{G}\right) = \deg \overline{G}$.

29. This problem is on hyperelliptic curves. Let $G \in \overline{\mathbb{F}}\left[C\right]$ be a nonzero polynomial function. Also let $P \in C$. If $G\left(P\right) = 0$, then prove that $\overline{G}\left(\widetilde{P}\right) = 0$.

   Hint: See the Appendix by Menezes, Wu, and Zuccherato in Koblitz (1999).
   Let $G\left(x, y\right) = \left(a\left(x\right) - b\left(x\right)y\right)$, $P = \left(x, y\right)$. Then $\overline{G}\left(x, y\right) = \left(a\left(x\right) + b\left(x\right)\left(h\left(x\right) + y\right)\right)$, and $\widetilde{P} = \left(x, -y - h\left(x\right)\right)$. We have

$$\overline{G}\left(\widetilde{P}\right) = \overline{G}\left(x, -y - h\left(x\right)\right) = a\left(x\right) + b\left(x\right)\left(h\left(x\right) - y - h\left(x\right)\right)$$
$$= a\left(x\right) - b\left(x\right)y = G\left(P\right) = 0$$

30. This problem is on hyperelliptic curves. For each divisor $D \in \mathbb{D}^0\left(C\right)$ there exists a semi-reduced divisor $D_1 \in \mathbb{D}^0\left(C\right)$, where $D \sim D_1$. Establish this assertion.

   Hint: See the Appendix by Menezes, Wu, and Zuccherato in Koblitz (1999). Let $D = \sum_{P \in C} m_P P$. The set $C$ is made up of ordinary points, special points, and the point at infinity. Denote the set of ordinary points by $C_o$, and the set of special points be $C_s$. The set of points in $C_o$ is partitioned into sets $C_{o1}$ and $C_{o2}$, so that: if $P \in C_{o1}$ then $\widetilde{P} \in C_{o2}$; and if $P \in C_{o1}$ then $m_P \geq m_{\widetilde{P}}$. We have

$$D = \sum_{P \in C_{o1}} m_P P + \sum_{P \in C_{o2}} m_P P + \sum_{P \in C_s} m_P P - m P_\infty$$

   Also let

$$D_1 = D - \sum_{P = \left(x_0, y_0\right) \in C_{o2}} m_P \operatorname{div}\left(x - x_0\right) - \sum_{P = \left(x_0, y_0\right) \in C_s} \left\lfloor \frac{m_P}{2} \right\rfloor \operatorname{div}\left(x - x_0\right)$$

   This implies $D_1 \sim D$. Next, let $P = \left(x_0, y_0\right)$. We know from the example in the section on divisors that: if $P$ is an ordinary point, then

$$\operatorname{div}\left(x - x_0\right) = \left(P + \widetilde{P} - 2P_\infty\right)$$

   and if $P$ is a special point, then $\operatorname{div}\left(x - x_0\right) = \left(2P - 2P_\infty\right)$. Therefore

$$D_1 = \sum_{P \in C_{o1}} \left(m_P - m_{\widetilde{P}}\right) P + \sum_{P \in C_s} \left(m_P - 2\left\lfloor \frac{m_P}{2} \right\rfloor\right) P - m_1 P_\infty$$

   for some $m_1 \in \mathbb{N}$. Consequently $D_1$ is a semi-reduced divisor.

31. This problem is on hyperelliptic curves. Prove that each divisor $D \in \mathbb{D}^0\left(C\right)$ is associated with a unique reduced divisor $D_1$ so that $D \sim D_1$.

   Hint: See the Appendix by Menezes, Wu, and Zuccherato in Koblitz (1999).

32. This problem is on hyperelliptic curves. Let $u_1\left(x\right), v_1\left(x\right), u_2\left(x\right), v_2\left(x\right) \in \overline{\mathbb{F}}\left[x\right]$. Using the extended Euclidean algorithm for polynomials, show that

$$d = \gcd\left(u_1, u_2, v_1 + v_2 + h\right)$$
$$= s_1 u_1 + s_2 u_2 + s_3\left(v_1 + v_2 + h\right)$$

   where $d\left(x\right), s_1\left(x\right), s_2\left(x\right), s_3\left(x\right) \in \overline{\mathbb{F}}\left[x\right]$.

   Hint: See the Appendix by Menezes, Wu, and Zuccherato in Koblitz (1999). Using the extended Euclidean algorithm for polynomials obtain

$$d_1 = \gcd(u_1, u_2) = e_1 u_1 + e_2 u_2$$

where $d_1(x), e_1(x), e_2(x) \in \mathbb{F}[x]$. Again using the extended Euclidean algorithm for polynomials obtain

$$d = \gcd(d_1, v_1 + v_2 + h)$$
$$= c_1 d_1 + c_2 (v_1 + v_2 + h)$$

where $d(x), c_1(x), c_2(x) \in \mathbb{F}[x]$. Combining the above two results, we obtain the stated result, where $s_1 = c_1 e_1$, $s_2 = c_1 e_2$, and $s_3 = c_2$.

33. This problem is on hyperelliptic curves. Let the Mumford representation of semi-reduced divisors $D_1$ and $D_2$ be $[u_1, v_1]$ and $[u_2, v_2]$ respectively. The sum of these two divisors is $D$. Its Mumford representation is $D = [u, v]$. Establish if

$$d = \gcd(u_1, u_2, v_1 + v_2 + h)$$
$$= s_1 u_1 + s_2 u_2 + s_3 (v_1 + v_2 + h)$$

and $d(x), s_1(x), s_2(x), s_3(x) \in \mathbb{F}[x]$, then

$$u = \frac{u_1 u_2}{d^2}$$

$$v = \left\{ \frac{s_1 u_1 v_2 + s_2 u_2 v_1 + s_3 (v_1 v_2 + f)}{d} \right\} \pmod{u}$$

Further $D \sim D_1 + D_2$.

Hint: See the Appendix by Menezes, Wu, and Zuccherato in Koblitz (1999). The stated result is established in three steps.

*Step* 1: In this step, we establish that $v$ is indeed a polynomial. Using the expression for $d$, we have

$$s_1 u_1 = d - s_2 u_2 - s_3 (v_1 + v_2 + h)$$

$$\frac{s_1 u_1 v_2 + s_2 u_2 v_1 + s_3 (v_1 v_2 + f)}{d}$$
$$= \frac{(d - s_2 u_2 - s_3 (v_1 + v_2 + h)) v_2 + s_2 u_2 v_1 + s_3 (v_1 v_2 + f)}{d}$$
$$= v_2 + \frac{s_2 u_2 (v_1 - v_2) - s_3 (v_2^2 + v_2 h - f)}{d}$$

Note that

$$d \mid u_2, \text{ and } u_2 \mid (v_2^2 + v_2 h - f)$$

Therefore $v$ is a polynomial.

*Step* 2: In this step, we establish that $u \mid (v^2 + vh - f)$. Let

$$v = \frac{s_1 u_1 v_2 + s_2 u_2 v_1 + s_3 (v_1 v_2 + f)}{d} + su$$

where $s(x) \in \mathbb{F}[x]$. We also have

$$v - y = \frac{s_1 u_1 v_2 + s_2 u_2 v_1 + s_3 (v_1 v_2 + f) - dy}{d} + su$$

$$= \frac{s_1 u_1 v_2 + s_2 u_2 v_1 + s_3 \left(v_1 v_2 + f\right) - s_1 u_1 y - s_2 u_2 y - s_3 \left(v_1 + v_2 + h\right) y}{d} + su$$

$$= \frac{s_1 u_1 \left(v_2 - y\right) + s_2 u_2 \left(v_1 - y\right) + s_3 \left(v_1 - y\right) \left(v_2 - y\right)}{d} + su$$

Observe that $(v - y)$ is a polynomial, and its conjugate is $(v + y + h)$. Multiplication of these two polynomials results in $\left(v^2 + vh - f\right)$. After some tedious algebra, it can be checked that the product of

$$\left(s_1 u_1 \left(v_2 - y\right) + s_2 u_2 \left(v_1 - y\right) + s_3 \left(v_1 - y\right) \left(v_2 - y\right)\right)$$

and its conjugate is divisible by $u_1 u_2$. In this conclusion, we used the following facts

$$\left(v_1 - y\right)\left(v_1 + y + h\right) = \left(v_1^2 + v_1 h - f\right), \text{ and } u_1 \mid \left(v_1^2 + v_1 h - f\right)$$
$$\left(v_2 - y\right)\left(v_2 + y + h\right) = \left(v_2^2 + v_2 h - f\right), \text{ and } u_2 \mid \left(v_2^2 + v_2 h - f\right)$$

Therefore, it can be inferred that $u \mid \left(v^2 + vh - f\right)$.

*Step* 3: In this step, $D \sim D_1 + D_2$ has to be established. For a proof of this result, the reader can also see Galbraith (2012), in addition to the Appendix by Menezes, Wu, and Zuccherato in Koblitz (1999).

# References

1. Abramowitz, M., and Stegun, I. A., 1965. *Handbook of Mathematical Functions*, Dover Publications, Inc., New York, New York.
2. Beachy, J. A., and Blair, W. D., 1996. *Abstract Algebra*, Second Edition, Waveland Press, Inc., Prospect Heights, Illinois.
3. Berlekamp, E. R., 1968. *Algebraic Coding Theory*, McGraw-Hill Book Company, New York, New York.
4. Brown, E., 2000. "Three Fermat Trails to Elliptic Curves," The College Mathematics Journal, Vol. 31, No. 3, pp. 162-172.
5. Cantor, D., 1987. "Computing the Jacobian of a Hyperelliptic Curve," Mathematics of Computation, Vol. 48, No. 177, pp. 95-101.
6. Cohen, H., and Frey, G., 2006. *Handbook of Elliptic and Hyperelliptic Curve Cryptography*, CRC Press, New York.
7. Galbraith, S. D., 2012. *Mathematics of Public Key Cryptography*, Cambridge University Press, Cambridge, Great Britain.
8. Gruenewald, D., 2009. "An Introduction to Hyperelliptic Curves," Crypto'Puces, île de Porquerolles.
9. Hardy, D. W., and Walker, C. L., 2003. *Applied Algebra, Codes, Ciphers, and Discrete Algorithms*, Prentice-Hall, Upper Saddle River, New Jersey.
10. Howie, J. M., 2006. *Fields and Galois Theory*, Springer-Verlag, Berlin, Germany.
11. Koblitz, N., 1989. "Hyperelliptic Cryptosystems," Journal of Cryptology, Vol. 1, No. 3, pp. 139-150.
12. Koblitz, N., 1999. *Algebraic Aspects of Cryptography*, Springer-Verlag, Berlin, Germany.
13. Kumanduri, R., and Romero, C., 1998. *Number Theory with Computer Applications*, Prentice-Hall, Englewood Cliffs, New Jersey.

14. Lidl, R., and Niederreiter, H., 1986. *Introduction to Finite Fields and Their Applications*, Cambridge University Press, Cambridge, Great Britain.

15. Ling, S., and Xing, C., 2004. *Coding Theory A First Course*, Cambridge University Press, Cambridge, Great Britain.

16. Lipschutz, S., 1968. *Linear Algebra*, Schaum's Outline Series, McGraw-Hill Book Company, New York, New York.

17. Lipschutz, S., 1998. *Set Theory and Related Topics*, Schaum's Outline Series, McGraw-Hill Book Company, New York, New York.

18. MacWilliams, F. J., and Sloane, N. J. A., 1988. *The Theory of Error-Correcting Codes*, North-Holland, New York, New York.

19. Menezes, A., Oorschot, P. von, and Vanstone, S., 1996. *Handbook of Applied Cryptography*, CRC Press, New York.

20. Moon, T. K., 2005. *Error Correction Coding Mathematical Methods and Algorithms*, John Wiley & Sons, Inc., New York, New York.

21. Pless, V., 1998. *Introduction to the Theory of Error-Correcting Codes*, Third Edition, John Wiley & Sons, Inc., New York, New York.

22. Ratcliffe, J. G., 1994. *Foundations of Hyperbolic Manifolds*, Springer-Verlag, Berlin, Germany.

23. Rosen, K. H., Editor-in-Chief, 2018. *Handbook of Discrete and Combinatorial Mathematics*, Second Edition, CRC Press: Boca Raton, Florida.

24. Scheidler, R., 2016. "An Introduction to Hyperelliptic Curve Arithmetic," in *Contemporary Developments in Finite Fields and Applications*, World Scientific Publishers, Singapore, pp. 321-340.

25. Schmitt, S., Zimmer, H. G., 2003. *Elliptic Curves, A Computational Approach*, Walter De Gruyter, New York, New York.

26. Silverman, J. H., and Tate, J., 1992. *Rational Points on Elliptic Curves*, Springer-Verlag, Berlin, Germany.

27. van der Waerden, B. L., 1970. *Algebra Vols. 1 & 2*, Frederick Ungar Publishing Co., New York, New York.

28. Washington, L. C., 2003. *Elliptic Curves, Number Theory and Cryptography*, Chapman and Hall/CRC Press, New York, New York.

29. Wicker, S. B., 1995. *Error Control Systems for Digital Communication and Storage*, Prentice-Hall, Englewood Cliffs, New Jersey.

# Matrices and Determinants

$$W_4 = \frac{1}{2} \begin{bmatrix} 1 & 1 & 1 & 1 \\ 1 & -1 & 1 & -1 \\ 1 & 1 & -1 & -1 \\ 1 & -1 & -1 & 1 \end{bmatrix}$$

A $4 \times 4$ Hadamard matrix

**Jacques Salomon Hadamard.** Hadamard was born on 8 December, 1865 in Versailles, France. He graduated from the École Normale Supérieure on 30 October, 1888. He subsequently earned his doctorate in 1892.

Hadamard proved the prime number theorem (PNT) independently of de la Vallée Poussin in 1896. Besides number theory, he contributed to calculus of variations, and developed Hadamard matrices (named after him). Hadamard matrices are important in coding theory, and picture compression. Among his nontechnical works, *The Psychology of Invention in the Mathematical Field* (1945), is generally regarded as outstanding. Hadamard, died on 17 October, 1963 in Paris, France.

## 3.1 Introduction

The subject of matrix theory can be regarded as a bridge between the abstract structures of mathematics, and their engineering applications. Definitions, and elementary properties of matrices and determinants are briefly discussed in this chapter. Matrices as examples of linear mappings or transformations (operators) are also explored. Spectral analysis of matrices, Hermitian matrices and their eigenstructures, Perron-Frobenius theory of positive and nonnegative matrices, singular value decomposition, matrix calculus, and random matrices are also studied.

Matrices find their application in diverse areas, ranging from error correcting codes, graph theory, routing algorithms to stochastic description of the Internet and the World Wide Web topology, queueing theory, and quantum mechanics. The phrase *matrix* was coined by the mathematician James Joseph Sylvester (1814-1897) in the year 1850 to denote a rectangular array of numbers. This subject was developed by several mathematicians, including William Rowan Hamilton (1805-1865), Hermann Grassmann (1809-1877), James Joseph Sylvester, Arthur Cayley (1821-1895), Ferdinand Georg Frobenius (1849-1917), Oskar Perron (1880-1975), and John von Neumann (1903-1957).

Certain parts of this chapter depend upon abstract algebra, applied analysis, and probability theory. These topics are covered in different chapters.

## 3.2 Basic Matrix Theory

Basics of matrices are discussed in this section. Matrix notation is initially introduced. This is followed by a description of different matrix operations. Different types of matrices are next defined. The concept of a matrix norm is also discussed. Basic matrix definitions are next given.

**Definitions 3.1.** *In the following definitions, $m, n \in \mathbb{P}$.*

1. *Matrix*: *An $m \times n$ matrix $A$ is a rectangular array of $mn$ real or complex numbers arranged into $m$ rows and $n$ columns. The array elements are called its elements. A matrix of $m$ rows and $n$ columns is of order (size) $m \times n$ (read as $m$ by $n$).*
   *The matrix element in the i-th row and j-th column is $a_{ij}$, where $1 \leq i \leq m$ and $1 \leq j \leq n$. The matrix $A$ is also written as $[a_{ij}]$ .*

$$A = \begin{bmatrix} a_{11} & a_{12} & \cdots & a_{1j} & \cdots & a_{1n} \\ a_{21} & a_{22} & \cdots & a_{2j} & \cdots & a_{2n} \\ \vdots & \vdots & \ddots & \vdots & \ddots & \vdots \\ a_{i1} & a_{i2} & \cdots & a_{ij} & \cdots & a_{in} \\ \vdots & \vdots & \ddots & \vdots & \ddots & \vdots \\ a_{m1} & a_{m2} & \cdots & a_{mj} & \cdots & a_{mn} \end{bmatrix} \tag{3.1}$$

2. *Column vector*: *A column vector is an $m \times 1$ matrix. It is a matrix with $m$ rows and a single column. The size or length of this vector is equal to $m$.*
   *The $m \times n$ matrix $A$ is said to be an array of $n$ column vectors, where the length of each column vector is $m$.*
3. *Row vector*: *A row vector is a $1 \times n$ matrix. It is a matrix with a single row and $n$ columns. The size or length of this vector is equal to $n$.*
   *The $m \times n$ matrix $A$ is said to be an array of $m$ row vectors, where the length of each row vector is $n$.*
4. *Square matrix*: *An $n \times n$ matrix with the same number of rows and columns is called a square matrix. It is sometimes simply said to be of order $n$, or of size $n$.*
5. *Diagonal elements of a matrix*: *If a matrix $A$ is of size $m \times n$, then the matrix elements $a_{ii}, 1 \leq i \leq \min(m, n)$ are called its diagonal elements. Therefore, if $A$ is a square matrix of order $n$, then the matrix elements $a_{ii}, 1 \leq i \leq n$ are called its diagonal elements. The diagonal is sometimes called the main diagonal. The elements $a_{ij}$, with $i \neq j$ and $1 \leq i, j \leq n$ are called its off-diagonal elements.*
6. *Diagonal matrix*: *An $n \times n$ matrix $D$ is called a diagonal matrix, if all its off-diagonal elements are equal to zero. If the diagonal matrix $D$ has diagonal entries $d_1, d_2, \ldots, d_n$ then the matrix $D$ is represented as $diag(d_1, d_2, \ldots, d_n)$ .*
7. *Identity matrix*: *An $n \times n$ matrix $A$ is called an identity matrix, if all its diagonal elements $a_{ii}$, $1 \leq i \leq n$ are each equal to unity, and all other elements are each equal to zero. It is usually denoted by either $I$ or $I_n$.*
8. *Trace of a square matrix*: *Trace of a square matrix $A$ is the sum of its diagonal elements. The trace of an $n \times n$ matrix $A = [a_{ij}]$, denoted by $tr(A)$ , is equal to $\sum_{i=1}^{n} a_{ii}$.*
9. *Zero or null matrix*: *If all the elements of a matrix are equal to zero, then it is called a zero or a null matrix. If there is no ambiguity and the context is clear, then it is simply represented as $0$ (not to be confused with the real number $0$).*
10. *Equal matrices*: *Let $A = [a_{ij}]$ and $B = [b_{ij}]$ be two $m \times n$ matrices. The matrix $A$ is equal to matrix $B$, iff $a_{ij} = b_{ij}$, for all values of $i$ and $j$, where $1 \leq i \leq m$, and $1 \leq j \leq n$. This equality of matrices is simply represented (denoted) as $A = B$.*
11. *Upper triangular matrix*: *It is a square matrix in which all the elements below the main diagonal are all zeros.*
12. *Lower triangular matrix*: *It is a square matrix in which all the elements above the main diagonal are all zeros.*

13. *Submatrix*: *The submatrix of a matrix $A$ is a matrix obtained by deleting from it a specified set of rows and columns.*

14. *Principal submatrix*: *The principal submatrix of a matrix $A$ is a submatrix obtained from $A$ by deleting rows and columns, such that the remaining rows and columns have identical indices.*

15. *Block-diagonal matrix*: *A block-diagonal matrix is a square matrix with square matrices on its diagonal, and the rest of the elements are all zeros.*                                                $\square$

In the above definition of a matrix, the matrix elements were surrounded by the square brackets $[\,\cdot\,]$. Alternately, the matrix elements can also be surrounded by the curved brackets $(\cdot)$. Occasionally, a row vector $\begin{bmatrix} x_1 & x_2 & \cdots & x_n \end{bmatrix}$ is represented as $(x_1, x_2, \ldots, x_n)$. This is in conformance with the vector notation described in the chapter on abstract algebra. In general, if a vector is specified as $x \geq 0$, then the vector is allowed to take a $0$ value. Also, the zero vector $0$ is simply $\begin{bmatrix} 0 & 0 & \cdots & 0 \end{bmatrix}$.

### 3.2.1 Basic Matrix Operations

Following are the basic operations of matrix algebra

*Addition and subtraction of matrices*: Let the matrices $A = [a_{ij}]$ and $B = [b_{ij}]$ be each of order $m \times n$. The matrices $A$ and $B$ of the same order are said to be conformable (compatible) for addition and subtraction.

The sum of matrices $A$ and $B$ is a matrix $C = [c_{ij}]$, where $c_{ij} = (a_{ij} + b_{ij})$, $1 \leq i \leq m$ and $1 \leq j \leq n$. The matrix $C$ is also of order $m \times n$. This addition operation is denoted by $C = (A + B)$.

Similarly the subtraction of matrices $A$ and $B$ is a matrix $C = [c_{ij}]$, where $c_{ij} = (a_{ij} - b_{ij})$, $1 \leq i \leq m$ and $1 \leq j \leq n$. The matrix $C$ is also of order $m \times n$. This subtraction operation is denoted by $C = (A - B)$.

*Matrix multiplication by a constant*: Let $\alpha$ be a complex number and $A = [a_{ij}]$ be a matrix of order $m \times n$. Then $\alpha A = C = [c_{ij}]$, where $c_{ij} = \alpha a_{ij}$, $1 \leq i \leq m$ and $1 \leq j \leq n$. The matrix $C$ is also of order $m \times n$.

*Scalar product of row vectors*: Let $x$ and $y$ be row vectors, each with $n$ columns,

$$x = \begin{bmatrix} x_1 & x_2 & \cdots & x_n \end{bmatrix}, \text{ and } y = \begin{bmatrix} y_1 & y_2 & \cdots & y_n \end{bmatrix}$$

The scalar product of the two row vectors is $x \circ y = \sum_{i=1}^{n} x_i y_i$. If the elements of these two row vectors are real numbers, then this definition is identical to the *dot* or *inner product* of the two vectors.

*Multiplication of matrices*: Let $A = [a_{ij}]$ be a matrix of order $m \times k$, and $B = [b_{ij}]$ be a matrix of order $k \times n$. Then the product of matrices $A$ and $B$ is a matrix $C = [c_{ij}]$ of order $m \times n$, where $c_{ij} = \sum_{l=1}^{k} a_{il} b_{lj}$, $1 \leq i \leq m$ and $1 \leq j \leq n$. In other words $c_{ij}$ is the scalar product of row $i$ of the matrix $A$ and column $j$ of matrix $B$. The matrix $A$ is said to be conformable (compatible) to matrix $B$ for multiplication when the number of columns of $A$ is equal to the number of rows of $B$. The matrix $C$ is also denoted by $AB$.

*Inverse of a matrix*: If $A$ and $B$ are square matrices such that $AB = BA = I$, then the matrix $B$ is called the inverse matrix of $A$. Generally $B$ is denoted by $A^{-1}$. Conversely, the inverse of matrix $A^{-1}$ is $A$. If the inverse of a matrix $A$ exists, then the matrix $A$ is called a *nonsingular matrix*. If the inverse does not exist, then $A$ is called a *singular matrix*.

*Conjugate of a matrix*: If $A = [a_{ij}]$, where $a_{ij} \in \mathbb{C}$, then the conjugate of matrix $A$ is $\overline{A} = [\overline{a}_{ij}]$.

*Transpose of a matrix*: If $A = [a_{ij}]$ is a matrix of order $m \times n$, then a matrix obtained by interchanging the rows and columns of the matrix $A$ is called the transpose of $A$. It is of order $n \times m$. It is generally denoted by $A^T$. Note that $A^T = [a_{ji}]$.

*Hermitian transpose of a matrix*: If $A = [a_{ij}]$ is a complex matrix of order $m \times n$, then a matrix obtained by interchanging the rows and columns of the matrix $A$ and taking the complex conjugate of the elements is called the Hermitian transpose of $A$. It is of order $n \times m$, and denoted by $A^\dagger$ ($\dagger$ is the dagger symbol). Note that $A^\dagger = [\bar{a}_{ji}] = \overline{A}^T$. The Hermitian transpose of a matrix is named after the mathematician Charles Hermite (1822-1901).

*Elementary row operations (or transformations)*: The elementary row operations on a matrix are:

(a) Multiplication of any row by a field element which is not equal to zero.
(b) Interchange any two rows.
(c) Substitute any row by the sum of itself and a multiple of any other row.

*Elementary column operations (or transformations)*: The elementary column operations are defined analogous to the elementary row operations.

### 3.2.2  Different Types of Matrices

The power of matrix algebra is further illustrated in this subsection.

**Definitions 3.2.** *Different types of matrices are defined below.*

1. *Similar matrices*: Let $A$ and $B$ be square matrices of order $n$. Let $P$ be an invertible matrix of order $n$ such that $A = P^{-1}BP$. Then the matrices $A$ and $B$ are termed similar matrices. The operation $P^{-1}BP$ is the similarity transformation of the matrix $B$.

2. *Normalized vector*: $A$ is a row vector of size $n$. Let $A = \begin{bmatrix} a_1 & a_2 & \cdots & a_n \end{bmatrix}$, then this row vector is normalized, if $\sum_{j=1}^{n} |a_j|^2 = 1$. A similar definition can be extended to a normalized column vector.

3. *Symmetric matrix*: $A$ is a symmetric matrix if $A = A^T$.

4. *Orthogonal matrix*: A real square matrix $A$ is orthogonal, if $A^T A = AA^T = I$, that is if $A^T = A^{-1}$.

5. *Hermitian matrix*: A complex square matrix $A$ is Hermitian if $A^\dagger = A$.

6. *Unitary matrix*: A complex square matrix $A$ is unitary, if $A^\dagger A = AA^\dagger = I$, that is if $A^\dagger = A^{-1}$.

7. *Orthogonal vectors*: Two complex row vectors $A$ and $B$ of the same size are orthogonal to each other, if $AB^\dagger = 0$

8. *Orthonormal set of vectors*: The complex row vectors $x_1, x_2, \ldots, x_n$ are an orthonormal set, if the vectors $x_j, 1 \leq j \leq n$ are normalized to unity, and $x_i x_j^\dagger = 0$, for all $i \neq j$ and $1 \leq i, j \leq n$. A similar definition can be extended to a set of complex column vectors.

9. *Normal matrix*: A complex square matrix $A$ is normal if and only if $A^\dagger A = AA^\dagger$.

10. *Quadratic forms and definiteness*: Let $A$ be a Hermitian matrix of order $n$, and $x$ is a complex column vector of size $n$. Let $f(x) = x^\dagger Ax$. The Hermitian matrix $A$ and the quadratic form $f(x)$ associated with matrix $A$ are said to be:

    (a) *Negative definite if $f(x) < 0$, for all $x \neq 0$.*
    (b) *Negative semidefinite if $f(x) \leq 0$, for all $x$; and $f(x) = 0$, for some $x \neq 0$.*

(c) *Positive definite if $f(x) > 0$, for all $x \neq 0$.*
(d) *Positive semidefinite if $f(x) \geq 0$, for all $x$; and $f(x) = 0$, for some $x \neq 0$.*
(e) *Indefinite if $f(x) > 0$, for some $x$; and $f(x) < 0$, for some $x$.*

11. *Diagonalizable matrix: A square matrix A is diagonalizable, if there exists an invertible matrix $P$ such that $PAP^{-1} = \Lambda$, where $\Lambda$ is a diagonal matrix.*
12. *Square root of a matrix: A matrix A is a square root of matrix B if $A^2 = B$, where A and B are square matrices. Sometimes the notation $A \triangleq \sqrt{B}$ is also used.*                               □

**Observations 3.1.** Few properties of matrix operations are summarized below.

1. Transposition properties.
   (a) $(\alpha A)^T = \alpha A^T, \alpha \in \mathbb{C}$
   (b) $(A^T)^T = A$
   (c) $(\alpha A + \beta B)^T = \alpha A^T + \beta B^T; \alpha, \beta \in \mathbb{C}$
   (d) $(AB)^T = B^T A^T$
   (e) $A^T A$ and $AA^T$ are symmetric matrices.

2. Let the matrices $A$, $B$ and $C$ be conformable, and $\alpha \in \mathbb{C}$. Then
   (a) $A + B = B + A$
   (b) $A + (B + C) = (A + B) + C$
   (c) $\alpha(A + B) = \alpha A + \alpha B$
   (d) $A(B + C) = AB + AC$, and $(B + C)A = BA + CA$

3. Matrices of different dimensions cannot be added or subtracted.

4. Matrix multiplication is not commutative in general. That is, if $A$ and $B$ are compatible matrices, then $AB$ is not equal to $BA$ in general.

5. Properties of the trace operator. Let $A$ and $B$ be square matrices of the same order.
   (a) Cyclic property of trace: $tr(AB) = tr(BA)$
   (b) Linearity of trace: $tr(A + B) = tr(A) + tr(B)$
   (c) $tr(zA) = z\, tr(A), z \in \mathbb{C}$
   (d) $tr(B^{-1}AB) = tr(A)$

6. Properties of Hermitian operators and matrices.
   (a) $(A^\dagger)^\dagger = A$
   (b) $(AB)^\dagger = B^\dagger A^\dagger$
   (c) Let $A$ be a Hermitian matrix, and $R$ be another matrix of the same order, then $R^\dagger A R$ is a Hermitian matrix.

7. Orthogonal expansions and linearly independent vectors. The concept of independence of a set of vectors has been discussed in the chapter on abstract algebra.
   (a) The set of vectors which are orthogonal to each other are linearly independent.
   (b) A set of $n$ orthogonal column vectors $x_i \neq 0, 1 \leq i \leq n$ is given. Let $u$ be a column vector of size $n$. Then the column vector $u$ can be expressed uniquely as a linear combination of the given orthogonal set of vectors.

$$u = \sum_{i=1}^{n} \beta_i x_i; \quad \beta_i \in \mathbb{C}, \quad 1 \leq i \leq n$$

$$\overline{\beta}_i = \frac{u^\dagger x_i}{x_i^\dagger x_i}, \quad 1 \leq i \leq n$$

If the column vectors $x_i, 1 \leq i \leq n$ are normalized to unity, then $\overline{\beta}_i = u^\dagger x_i, 1 \leq i \leq n$.

8. Properties of inverse matrices.

(a) $A^{-1}A = AA^{-1} = I$, where $A$ is a nonsingular matrix.

(b) $\left(A^{-1}\right)^{-1} = A$, where $A$ is a nonsingular matrix.

(c) The inverse of a nonsingular matrix is unique.

(d) If $A$ and $B$ are nonsingular matrices, then $(AB)^{-1} = B^{-1}A^{-1}$.

(e) If $A$ is a nonsingular matrix, then $A^T$ is also a nonsingular matrix. Also $\left(A^T\right)^{-1} = \left(A^{-1}\right)^T$.

(f) The inverse of a matrix $A$ exists, if its rows (or columns) form a linearly independent set of vectors.

(g) The inverse of a matrix $A$ exists, if there is no nonzero $x$ such that $Ax = 0$.

9. Properties of square root matrices.

(a) Every diagonalizable matrix has a square root. Thus $A = P^{-1}\Lambda P$, implies $\sqrt{A} = P^{-1}\sqrt{\Lambda}P$.

(b) If $D = [d_{ij}]$ is a diagonal matrix, then $\sqrt{D} = \left[\sqrt{d_{ij}}\right]$.

(c) If $B$ is a positive definite Hermitian matrix, then there exists a unique positive definite Hermitian matrix $A$ such that $B = A^2$. However, if $B$ is a semidefinite matrix, then $A$ is also semidefinite. ☐

### 3.2.3  Matrix Norm

The concept of matrix norm is analogous to the concept of vector norm. Vector norms are discussed in the chapter on applied analysis. Since matrices and vectors generally occur together, it is desirable that the matrix and vector norms be in consonance with each other. For example, if $\|\cdot\|$ is the norm operator, then we should have

$$\|Ax\| \leq \|A\| \, \|x\|$$

where $A$ and $x$ are compatible matrix and vector respectively. Similarly, we should have

$$\|AB\| \leq \|A\| \, \|B\|$$

where $A$ and $B$ are compatible matrices.

**Definition 3.3.** *Norm of a matrix*: *The norm function* $\|\cdot\|$ *assigns a nonnegative real number, to each complex matrix A, subject to the following axioms.*

(a) $\|A\| = 0$ *if and only if* $A = 0$.

(b) $\|A\| > 0$ *for* $A \neq 0$.

(c) $\|aA\| = |a| \, \|A\|$, *where* $|a|$ *is the magnitude of* $a \in \mathbb{C}$.

(d) $\|A + B\| \leq \|A\| + \|B\|$, *where the matrices A and B are of the same size. This is the triangle inequality.*

(e) $\|AB\| \leq \|A\| \, \|B\|$, *where the matrices A and B are compatible.* ☐

The most commonly used norms in matrix analysis are the Frobenius norm and the $p$-norm.

**Definitions 3.4.** *Let* $A = [a_{ij}]$ *be an* $m \times n$ *complex matrix.*

1. *Frobenius norm of a matrix. The Frobenius norm, also called the F-norm is*

$$\|A\|_F = \left\{ \sum_{i=1}^{m} \sum_{j=1}^{n} |a_{ij}|^2 \right\}^{1/2} = \sqrt{tr\left(AA^\dagger\right)} \tag{3.2}$$

*Alternate names are: $\ell_2$, Euclidean, Hilbert-Schmidt, or Schur norm.*
2. *p-norm of a matrix. Note that vector p-norm has been discussed in the chapter on applied analysis. A vector p-norm induces a matrix p-norm by setting*

$$\|A\|_p = \sup_{\|x\|_p=1} \|Ax\|_p \tag{3.3}$$

*where $x$ is a complex vector of size $n$, and $p \geq 1$.*
3. *Trace norm of a matrix. The trace norm of a complex matrix $A$ is*

$$\|A\|_{tr} = tr\left(\sqrt{AA^\dagger}\right) \tag{3.4}$$

*Notice that the square root of the matrix $AA^\dagger$ is well-defined.*  □

It can be shown that the Frobenius norm, the $p$-norm, and the trace norm do indeed satisfy the requirements for the norm of a matrix.

**Observations 3.2.** Let $A = [a_{ij}]$ be an $m \times n$ complex matrix. Some properties of well-known norms are listed below.

1. It is evident that,

$$\|Ax\|_p \leq \|A\|_p \|x\|_p$$

   for all vectors $x$ of size $n$.
2. If the matrix $A$ is nonsingular, then

$$\min_{\|x\|_p=1} \|Ax\|_p = \frac{1}{\|A^{-1}\|_p}$$

3. We have

$$\|A\|_1 = \max_{1 \leq j \leq n} \sum_{i=1}^{m} |a_{ij}| = \text{the largest absolute column sum}$$

$$\|A\|_\infty = \max_{1 \leq i \leq m} \sum_{j=1}^{n} |a_{ij}| = \text{the largest absolute row sum}$$

$$\|A\|_2 \leq \|A\|_F \leq \sqrt{n} \|A\|_2$$

$$\|A\|_2 \leq \sqrt{\|A\|_1 \|A\|_\infty}$$

4. The Frobenius norm and the 2-norm are invariant with respect to orthogonal transformations.
5. Let the size of complex vectors $x$ and $y$ be $n$ and $m$ respectively. Then

$$\|A\|_2 = \max_{\substack{\|x\|_2=1 \\ \|y\|_2=1}} |y^\dagger Ax|$$

□

As in the case of the vector norms, if the type of the norm is not indicated, then it is assumed to be a 2-norm.

## 3.3 Determinants

A square matrix has a very special number associated with it. It is called its determinant. These are introduced in this section. A very special type of determinant called the Vandermonde determinant is also discussed. The well-known Binet-Cauchy theorem is also stated in this section. This theorem has application in graph theory.

### 3.3.1 Definitions

The notion of the determinant of a square matrix is initially introduced. This is followed by a summary of some basic properties of determinants.

**Definitions 3.5.** *Let $A = [a_{ij}]$ be an $n \times n$ square matrix of either real or complex numbers.*

1. *Determinant of a matrix: The determinant $\det A$ of the matrix $A$ is defined recursively as follows:*
   - (a) *If $n = 1$, $A = [a]$, then $\det A = a$.*
   - (b) *Let $n > 1$, and $A_{ij}$ be an $(n-1) \times (n-1)$ matrix obtained by deleting row $i$ and column $j$ of matrix $A$. Then $\det A = \sum_{j=1}^{n} (-1)^{j+1} a_{1j} \det A_{1j}$.*

   *The value $n$ is called the order of the determinant. This definition is due to Laplace.*
2. *Minor, and cofactor: $A_{ij}$ is the submatrix obtained from $A$ by deleting the $i$th row and the $j$th column. The minor of the element $a_{ij}$ is the determinant of the matrix $A_{ij}$. It is denoted by $M_{ij}$. Therefore $M_{ij} = \det A_{ij}$.*

   *The order of this minor is $(n-1)$. The cofactor of $a_{ij}$ is defined by $(-1)^{i+j} M_{ij}$. Denote this cofactor by $\beta_{ij}$.*
3. *Notation. It is customary to denote the determinant of the matrix $A$ as*

$$\det A \triangleq |A| \triangleq \begin{vmatrix} a_{11} & a_{12} & \cdots & a_{1n} \\ a_{21} & a_{22} & \cdots & a_{2n} \\ \vdots & \vdots & \ddots & \vdots \\ a_{n1} & a_{n2} & \cdots & a_{nn} \end{vmatrix} \tag{3.5}$$

   *The vertical lines in the above definition are not related to the absolute value or the modulus of a complex number.*
4. *Principal minor: The principal minor is the determinant of a principal submatrix of a matrix $A$. The principal submatrix is obtained by deleting a certain column, and the same numbered row of $A$.*
5. *An alternate and useful notation for determinants and its minors is as follows. Let $A = [a_{ij}]$ be an $m \times n$ matrix. From this matrix, strike out all but rows $i_1, i_2, \ldots, i_p$ and columns $k_1, k_2, \ldots, k_p$. The determinant of the resulting matrix is a minor of order $p$ and it is written as*

$$A \begin{pmatrix} i_1 & i_2 & \cdots & i_p \\ k_1 & k_2 & \cdots & k_p \end{pmatrix} \triangleq \det \begin{bmatrix} a_{i_1 k_1} & a_{i_1 k_2} & \cdots & a_{i_1 k_p} \\ a_{i_2 k_1} & a_{i_2 k_2} & \cdots & a_{i_2 k_p} \\ \vdots & \vdots & \ddots & \vdots \\ a_{i_p k_1} & a_{i_p k_2} & \cdots & a_{i_p k_p} \end{bmatrix}$$

*If $A$ is a square matrix of order $n$, then in this notation*

$$\det A = A \begin{pmatrix} 1 & 2 & \cdots & n \\ 1 & 2 & \cdots & n \end{pmatrix}$$

$\square$

See the problem section for an alternate definition of the determinant of a square matrix.

**Observations 3.3.** Some basic observations on determinants.

1. The determinant of the identity matrix $I$ is equal to $\det I = 1$.
2. The determinant of matrix $A = [a_{ij}]$ in terms of its cofactors is

$$\det A = \sum_{k=1}^{n} a_{ik}\beta_{ik} = \sum_{k=1}^{n} a_{kj}\beta_{kj}; \quad \text{for each } i, j, \text{ where } 1 \le i, j \le n$$

   The above representation of a determinant is called the *Laplace expansion* of the determinant, after the mathematician Pierre-Simon Laplace.
3. $\det A = \det A^T$, where $A$ is any square matrix.
4. $\det AB = \det A \det B = \det BA$, where matrices $A$ and $B$ are any $n \times n$ matrices.
5. $\det \alpha A = \alpha^n \det A$, where $A$ is any $n \times n$ matrix and $\alpha \in \mathbb{C}$.
6. Let $A$ be an invertible matrix, then $\det A^{-1} = (\det A)^{-1}$
7. Let $D = [d_{ij}]$ be an $n \times n$ diagonal matrix. Then $\det D = \prod_{i=1}^{n} d_{ii}$.
8. If $A = [a_{ij}]$ is an $n \times n$ lower or upper triangular matrix, then $\det A = \prod_{i=1}^{n} a_{ii}$.
9. Let $A$ be a matrix with at least two identical rows (or columns), then $\det A = 0$.
10. If two columns (or two rows) of a matrix are interchanged, then the sign of the determinant changes
11. If a column (or row) of a matrix is multiplied by $\alpha \in \mathbb{C}$, then the determinant is multiplied by $\alpha$.
12. If a multiple of a column (row) is added to another column (row), then the value of the determinant remains unchanged.
13. If the determinant of a matrix is equal to zero, then it is a singular matrix; otherwise it is a nonsingular matrix.
14. Let $A = [a_{ij}]$ be a $2 \times 2$ matrix, then $\det A = a_{11}a_{22} - a_{12}a_{21}$.
15. Let $A = [a_{ij}]$ be a $3 \times 3$ matrix, then

$$\det A = a_{11}a_{22}a_{33} + a_{12}a_{23}a_{31} + a_{13}a_{21}a_{32}$$
$$-a_{11}a_{23}a_{32} - a_{12}a_{21}a_{33} - a_{13}a_{22}a_{31}$$

16. The determinant of an orthogonal matrix is equal to either $1$ or $-1$.
17. If the sum of entries of every row and every column of a square matrix $M$ is equal to $0$, then all the cofactors of the matrix $M$ are equal.

$\square$

**Example 3.1.** Let $M$ be a $3 \times 3$ matrix.

$$M = \begin{bmatrix} a & b & -(a+b) \\ c & d & -(c+d) \\ -(a+c) & -(b+d) & (a+b+c+d) \end{bmatrix}$$

It can be verified that the sum of each and every row and column of the matrix $M$ is equal to $0$. Furthermore, the 9 cofactors are each equal to $(ad - bc)$. A more general version of this example is established in the problem section.

$\square$

### 3.3.2  Vandermonde Determinant

An interesting determinant which occurs repeatedly in classical analysis is the Vandermonde determinant.

**Definition 3.6.** *An* $n \times n$ *determinant* $V_n$ *is Vandermonde, if it is given by*

$$
V_n = \begin{vmatrix}
1 & 1 & \cdots & 1 \\
\alpha_1 & \alpha_2 & \cdots & \alpha_n \\
\alpha_1^2 & \alpha_2^2 & \cdots & \alpha_n^2 \\
\vdots & \vdots & \ddots & \vdots \\
\alpha_1^{n-1} & \alpha_2^{n-1} & \cdots & \alpha_n^{n-1}
\end{vmatrix}
\tag{3.6}
$$

*In the above determinant,* $n \geq 2$*, and* $\alpha_1, \alpha_2, \ldots, \alpha_n \in \mathbb{C}$*. The corresponding matrix is called the Vandermonde matrix.*  □

**Theorem 3.1.** *Let* $n \geq 2$*, then* $V_n$ *is the product of all* $(\alpha_i - \alpha_j)$ *for* $i > j$*. That is*

$$
V_n = \prod_{i>j} (\alpha_i - \alpha_j) = \prod_{j=1}^{(n-1)} \prod_{i=j+1}^{n} (\alpha_i - \alpha_j)
\tag{3.7}
$$

*Proof.* Induction is used in the proof. Notice that $V_2 = (\alpha_2 - \alpha_1)$. If $n = 3$, then

$$
V_3 = \begin{vmatrix}
1 & 1 & 1 \\
\alpha_1 & \alpha_2 & \alpha_3 \\
\alpha_1^2 & \alpha_2^2 & \alpha_3^2
\end{vmatrix} = \begin{vmatrix}
1 & 1 & 1 \\
0 & \alpha_2 - \alpha_1 & \alpha_3 - \alpha_1 \\
0 & \alpha_2^2 - \alpha_1\alpha_2 & \alpha_3^2 - \alpha_1\alpha_3
\end{vmatrix}
$$

where each third row element was obtained by subtracting from it, the corresponding $\alpha_1$ multiple of the row two element. Similarly each second row element was obtained by subtracting from it the corresponding $\alpha_1$ multiple of the row one element. Therefore

$$
V_3 = \begin{vmatrix}
\alpha_2 - \alpha_1 & \alpha_3 - \alpha_1 \\
\alpha_2^2 - \alpha_1\alpha_2 & \alpha_3^2 - \alpha_1\alpha_3
\end{vmatrix} = (\alpha_2 - \alpha_1)(\alpha_3 - \alpha_1) \begin{vmatrix} 1 & 1 \\ \alpha_2 & \alpha_3 \end{vmatrix}
$$
$$
= (\alpha_2 - \alpha_1)(\alpha_3 - \alpha_1)(\alpha_3 - \alpha_2)
$$

It can also be shown that

$$
V_n = (\alpha_n - \alpha_1)(\alpha_n - \alpha_2)\ldots(\alpha_n - \alpha_{n-1}) V_{n-1}, \quad n \geq 2
$$

where $V_1 = 1$.  □

**Corollary 3.1.** $V_n = 0$ *iff* $\alpha_i = \alpha_j$ *for some* $i \neq j$.  □

### 3.3.3  Binet-Cauchy Theorem

Binet-Cauchy theorem is used in computing the determinant of the product of two matrices, which are not necessarily square. The computation is performed in terms of the determinants of their submatrices. This theorem is used in determining the number of spanning trees of a graph. It was independently discovered by the French mathematicians Jacques P. M. Binet (1786-1856) and Augustin-Louis Cauchy (1789-1857) in the year 1812.

**Theorem 3.2.** *Let $A$ and $B$ be $m \times n$ and $n \times m$ matrices respectively, where $m \leq n$. Major determinants of matrices $A$ and $B$ are determinants of their submatrices of size $m \times m$. An $m \times m$ submatrix $A_i$ of $A$ corresponds to the $m \times m$ submatrix $B_i$ of $B$, if the column numbers of $A$ that determine $A_i$ are also the row numbers of $B$ that determine $B_i$. That is, the matrix $A_i$ is formed by selecting the $m$ columns $i_1, i_2, \ldots, i_m$ of matrix $A$, where $i_p \in \{1, 2, \ldots, n\}$ for $1 \leq p \leq m$; and $B_i$ is formed by selecting the $i_1, i_2, \ldots, i_m$ corresponding rows of the matrix $B$. It can be observed that the number of such submatrices is $N = \binom{n}{m}$. Then*

$$\det(AB) = \sum_{i=1}^{N} \det(A_i B_i) \tag{3.8a}$$

*An alternate and useful statement of this result is*

$$\det(AB) = \sum_{1 \leq j_1 < j_2 < \cdots < j_m \leq n} A\begin{pmatrix} 1 & 2 & \cdots & m \\ j_1 & j_2 & \cdots & j_m \end{pmatrix} B\begin{pmatrix} j_1 & j_2 & \cdots & j_m \\ 1 & 2 & \cdots & m \end{pmatrix}$$

$$\tag{3.8b}$$

*where the summation is over products of all possible minors of order $m$ of the matrix $A$ with corresponding minors of matrix $B$ of order $m$.*

*Proof.* See the problem section. $\qquad\qquad\square$

This theorem is also illustrated via an example.

**Example 3.2.** Let

$$A = \begin{bmatrix} 5 & -3 & 2 \\ 4 & 1 & -6 \end{bmatrix}, \quad \text{and } B = \begin{bmatrix} 4 & -1 \\ 3 & 2 \\ -5 & 7 \end{bmatrix}$$

Then $n = 3$ and

$$A_1 = \begin{bmatrix} 5 & -3 \\ 4 & 1 \end{bmatrix}, \quad A_2 = \begin{bmatrix} 5 & 2 \\ 4 & -6 \end{bmatrix}, \quad \text{and } A_3 = \begin{bmatrix} -3 & 2 \\ 1 & -6 \end{bmatrix}$$

$$B_1 = \begin{bmatrix} 4 & -1 \\ 3 & 2 \end{bmatrix}, \quad B_2 = \begin{bmatrix} 4 & -1 \\ -5 & 7 \end{bmatrix}, \quad \text{and } B_3 = \begin{bmatrix} 3 & 2 \\ -5 & 7 \end{bmatrix}$$

and

$$\det A_1 = 17, \quad \det A_2 = -38, \text{ and } \det A_3 = 16$$
$$\det B_1 = 11, \quad \det B_2 = 23, \text{ and } \det B_3 = 31$$

Use of the Binet-Cauchy theorem yields

$$\det(AB) = \sum_{i=1}^{3} \det A_i \det B_i$$
$$= 17(11) + (-38)23 + 16(31) = -191$$

A direct computation gives

$$AB = \begin{bmatrix} 1 & 3 \\ 49 & -44 \end{bmatrix},$$

$$\det(AB) = \det \begin{bmatrix} 1 & 3 \\ 49 & -44 \end{bmatrix} = -191$$

$\square$

## 3.4 More Matrix Theory

Some more concepts from matrix theory are defined and discussed in this section. These are the rank of a matrix, adjoint of a square matrix, and nullity of a matrix. Some elementary facts about the theory of linear equations are also summarized. Use of the matrix inversion lemma occurs in control theory. The tensor product of matrices is also defined.

### 3.4.1 Rank of a Matrix

The concept of the rank of a matrix is introduced.

**Definitions 3.7.** *Let $A$ be an $m \times n$ matrix. Let the elements of the matrix belong to a field $\mathcal{F}$.*

1. *The rank of the matrix $A$ is the size of the largest square nonsingular (invertible) submatrix of $A$. It is denoted by $rank\ A$, or $r_A$.*
2. *Let $R_1, R_2, \ldots, R_m$ be the rows of matrix $A$, and the set of rows be $R$. That is, $R = \{R_1, R_2, \ldots, R_m\}$. Then the space $L(R)$ spanned by the vectors which belong to the set $R$, is called the row space of the matrix $A$. The dimension of the space $L(R)$ is denoted by $\dim(L(R))$.*
3. *Let $C_1, C_2, \ldots, C_n$ be the columns of matrix $A$, and the set of columns be $C$. That is, $C = \{C_1, C_2, \ldots, C_n\}$. Then the space $L(C)$ spanned by the vectors which belong to the set $C$, is called the column space of the matrix $A$. The dimension of the space $L(C)$ is denoted by $\dim(L(C))$.* $\square$

**Observations 3.4.** Some basic observations on ranks.

1. The rank of a matrix $A$ is equal to its maximum number of linearly independent rows (or columns).
2. Let $A$ be an $n \times n$ matrix. Then $r_A = n$ if and only if the matrix $A$ is nonsingular. That is, the inverse of a matrix exists if and only if $r_A = n$.
3. $r_A = r_{A^T}$.
4. $r_{(A+B)} \leq r_A + r_B$.
5. $A$ and $B$ are matrices such that their product is well-defined, then $r_{AB} \leq \min\{r_A, r_B\}$. If $B$ is a square and nonsingular matrix, then $r_{AB} = r_A$. Similarly, if $A$ is a square and nonsingular matrix, then $r_{AB} = r_B$.
6. Let $A$ be $m \times n$ and $B$ be $n \times q$ matrices, then $(r_A + r_B - n) \leq r_{AB}$.
7. $r_A = \dim(L(R)) = \dim(L(C))$. $\square$

### 3.4.2  Adjoint of a Square Matrix

Adjoint of a square matrix is next defined.

**Definition 3.8.** *Let $A = [a_{ij}]$ be a square matrix of order $n$. Also let $\beta_{ij}$ be the cofactor of $a_{ij}$. Then the adjoint of matrix $A$ is a matrix of order $n$. It is:*

$$adj\ A = \left[\beta_{ji}\right] \tag{3.9}$$

□

**Observations 3.5.** Let $A = [a_{ij}], B = [b_{ij}]$, and $I$ be an identity matrix. Each matrix is of order $n$.

1. The adjoint of a diagonal matrix is a diagonal matrix.
2. If $A$ is a Hermitian matrix, then its adjoint is also a Hermitian matrix.
3. $A\,(adj\ A) = |A|\,I = (adj\ A)\,A$.
4. $|A|\,|(adj\ A)| = |A|^{n}$.
5. If the matrix $A$ is nonsingular, then $|(adj\ A)| = |A|^{n-1}$.
6. $(adj\ AB) = (adj\ B)\,(adj\ A)$.
7. $adj\,(adj\ A) = |A|^{n-2}\,A$, if $\det A \neq 0$.                                         □

**Theorem 3.3.** *Let $A$ be a square nonsingular matrix of order $n$. Then*

$$A^{-1} = \frac{adj\ A}{|A|} \tag{3.10}$$

*Proof.* The proof is left to the reader.                                                 □

### 3.4.3  Nullity of a Matrix

Consider a system of equations $AX = 0$, where $A$ is a square matrix of order $n$, and $X$ is a column vector, where $X^{T} = \left[\,x_1\ x_2\ \cdots\ x_n\,\right]$. These are called homogeneous equations. It can be established that:

(a) If $rank\ A = r_A = n$, then $\det A \neq 0$ and $X = 0$. This solution is called a trivial solution.
(b) If $r_A < n$, then:
   (i) $\det A = 0$.
   (ii) The system of homogeneous equations has precisely $(n - r_A)$ linearly independent solutions. Every other solution is a linear combination of these solutions.

The above facts motivate the following definition.

**Definitions 3.9.** *Let $A$ be a matrix.*

1. *Null space of $A$: Let $AX = 0$ be a system of homogeneous equations. The null space of matrix $A$ is the vector space formed by the solution vectors $X$.*
2. *The dimension $\eta_A$, of this vector space is called the nullity of matrix $A$.*                □

**Observations 3.6.** $A$ is an $m \times n$ matrix of rank $r_A$, and nullity $\eta_A$.

1. The set of equations $AX = 0$ has $\eta_A$ linearly independent solutions. Every other solution is a linear combination of these solutions.
2. The set of basis vectors of the null space of the matrix $A$ is any set of $\eta_A$ linearly independent solutions of the equation $AX = 0$.
3. $(r_A + \eta_A) = n$.
4. Let $A$ and $B$ be matrices of order $n$. Then
   (a) $\eta_A \leq \eta_{AB}$ and $\eta_B \leq \eta_{AB}$.
   (b) $\eta_{AB} \leq \eta_A + \eta_B$.
   The above inequalities are called the Sylvester's laws of nullity. These are named after the mathematician James J. Sylvester.
5. Let $A$ be a matrix of order $n$. The inverse of the matrix $A$ exists if and only if $\eta_A = 0$.     □

### 3.4.4  System of Linear Equations

In this subsection, we briefly outline the necessary facts to solve a set of linear equations. These results can be stated succinctly in terms of the rank and nullity of a matrix. A given set of linear equations can be either consistent or inconsistent. Furthermore, if the equations are consistent, then these can either have a unique solution or an infinite number of solutions.

**Definitions 3.10.** *Coefficient and augmented matrices, homogeneous and nonhomogeneous equations are defined.*

1. *A set of $m$ linear equations in $n$ unknowns $x_1, x_2, \ldots, x_n$ is $\sum_{j=1}^{n} a_{ij} x_j = b_i$, where $1 \leq i \leq m$; and $a_{ij}, b_i \in \mathbb{C}$ for $1 \leq i \leq m$, and $1 \leq j \leq n$. These linear equations can be represented more compactly as $Ax = b$, where $A = [a_{ij}]$ is a matrix of size $m \times n$, $b^T = \begin{bmatrix} b_1 & b_2 & \cdots & b_m \end{bmatrix}$, and $x^T = \begin{bmatrix} x_1 & x_2 & \cdots & x_n \end{bmatrix}$. The matrix $A$ is called the coefficient matrix, and the $m \times (n+1)$ matrix $A_b = [A : b]$ is called the augmented matrix.*
2. *The $m$ set of equations are homogeneous if the column vector $b = 0$, otherwise it is nonhomogeneous.*     □

A set of equations specified by $Ax = b$ is consistent if there is a solution for $x$, or inconsistent if there is no solution. A consistent set of equations can either have a single (unique) solution or an infinite number of solutions. Some useful observations are listed below.

**Observations 3.7.** On solution(s) of a system of linear equations.

1. Let $A$ be a nonsingular square matrix of size $n$, and $b$ be a column vector of size $n$. A unique solution of $Ax = b$ is $x = A^{-1}b$. This is possible if and only if $\det A \neq 0$.
2. A linear set of equations are specified by $Ax = b$, where $A$ is a matrix of size $m \times n$. Let $A_b$ be the corresponding augmented matrix. Let the rank of the matrices $A$ and $A_b$, be $r_A$ and $r_{A_b}$ respectively. The nullity of matrix $A$ is $\eta_A = (n - r_A)$.
   (a) If $r_A < r_{A_b}$, the set of equations are inconsistent. Thus the set of equations has no solution.
   (b) If $r_A = r_{A_b}$, the set of equations is consistent. Thus the set of equations is solvable.
      (i) Let $r_A = n$, the number of unknowns. The set of equations has a unique solution.
      (ii) Let $r_A < n$, the number of unknowns. The set of equations has infinitely many solutions. The solution set forms a space of dimension $\eta_A$.     □

### 3.4.5  Matrix Inversion Lemma

The following result is called the *matrix inversion lemma.* It finds use in the derivation of the Kalman filtering algorithm.

**Lemma 3.1.** *Let $A$ be an $n \times n$ invertible matrix, $B$ is an $n \times m$ matrix, $C$ is an $m \times m$ invertible matrix, and $D$ is an $m \times n$ matrix. Then*

$$(A + BCD)^{-1} = A^{-1} - A^{-1}B\left(C^{-1} + DA^{-1}B\right)^{-1}DA^{-1} \tag{3.11}$$

□

This result is established in the problem section.

### 3.4.6  Tensor Product of Matrices

The tensor product of two matrices is also called the *Kronecker product.* This product is named after the number theorist L. Kronecker (1823-1891).

**Definitions 3.11.** *Tensor product of matrices.*

1. *Let $A = [a_{ij}]$ be an $m \times n$ matrix, and $B$ be a $p \times q$ matrix, then the Kronecker product $A \otimes B$ of these two matrices is defined as*

$$A \otimes B = \begin{bmatrix} a_{11}B & a_{12}B & \cdots & a_{1n}B \\ a_{21}B & a_{22}B & \cdots & a_{2n}B \\ \vdots & \vdots & \ddots & \vdots \\ a_{m1}B & a_{m2}B & \cdots & a_{mn}B \end{bmatrix} \tag{3.12a}$$

*where the matrix $A \otimes B$ is $mp \times nq$.*
2. *Denote the $m$-fold tensor product of a matrix $A$ by itself as*

$$A^{\otimes m} = \underbrace{A \otimes A \otimes \ldots \otimes A}_{m \text{ times}} \tag{3.12b}$$

*Sometimes, $A^{\otimes m}$ is also denoted by $A^{(m)}$.*   □

**Examples 3.3.** Some examples on tensor product of matrices.

1. Let

$$A = \begin{bmatrix} 2 & -3 \\ -8 & 5 \end{bmatrix}, \quad B = \begin{bmatrix} 4 & 0 & -9 \\ 1 & -6 & 7 \end{bmatrix}$$

Then

$$A \otimes B = \begin{bmatrix} 8 & 0 & -18 & -12 & 0 & 27 \\ 2 & -12 & 14 & -3 & 18 & -21 \\ -32 & 0 & 72 & 20 & 0 & -45 \\ -8 & 48 & -56 & 5 & -30 & 35 \end{bmatrix}$$

2. Let

$$C = \begin{bmatrix} 1 & 2 \\ 2 & -1 \end{bmatrix}$$

Then

$$C^{\otimes 2} = C^{(2)} = C \otimes C = \begin{bmatrix} 1 & 2 & 2 & 4 \\ 2 & -1 & 4 & -2 \\ 2 & 4 & -1 & -2 \\ 4 & -2 & -2 & 1 \end{bmatrix}$$

3. Let

$$H = \begin{bmatrix} 1 & 1 \\ 1 & -1 \end{bmatrix}$$

Then

$$H^{\otimes 2} = H^{(2)} = H \otimes H = \begin{bmatrix} 1 & 1 & 1 & 1 \\ 1 & -1 & 1 & -1 \\ 1 & 1 & -1 & -1 \\ 1 & -1 & -1 & 1 \end{bmatrix}$$

$\square$

Tensor products provide a convenient framework for representing Hadamard matrices. These matrices are discussed in the chapter on applied analysis.

## 3.5 Matrices as Linear Transformations

Matrices can be viewed as examples of linear mappings. Matrix transformation is initially defined in this section. This is followed by a description of a technique for coordinates and change of basis. Linear mappings or transformations have been defined in the chapter on abstract algebra.

**Definition 3.12.** *A matrix transformation is a function* $T : \mathbb{C}^n \to \mathbb{C}^m$ *for which there exists a complex* $m \times n$ *matrix* $A$ *such that* $T(x) = Ax$ *where* $x \in \mathbb{C}^n$ *and* $T(x) \in \mathbb{C}^m$. $\square$

**Lemma 3.2.** *Each and every matrix transformation is a linear transformation.*
*Proof.* The proof is left to the reader. $\square$

In a minor misuse of notation, the matrix transformation $T$ is occasionally denoted by the matrix $A$.

**Definitions 3.13.** *Types of matrix transformations.*

1. *Let* $A$ *and* $B$ *two matrix transformations from the space* $\mathbb{C}^n$ *to* $\mathbb{C}^m$.
   (a) *Equality of transformations*: *The transformations* $A$ *and* $B$ *are equal if* $Ax = Bx$ *for every* $x \in \mathbb{C}^n$. *This equality of transformation is denoted by* $A = B$.
   (b) *Sum of transformations*: *The sum transformation* $(A + B)$ *is*:

$$(A + B)x = Ax + Bx, \quad \forall x \in \mathbb{C}^n \tag{3.13a}$$

2. *Let $A$ be a matrix transformation from the space $\mathbb{C}^n$ to $\mathbb{C}^m$, and $\alpha \in \mathbb{C}$. The transformation $\alpha A$ is defined by*

$$(\alpha A)\, x = \alpha\,(Ax)\,, \quad \forall\, x \in \mathbb{C}^n \tag{3.13b}$$

3. *Product transformation: Let $A$ be a matrix transformation from the space $\mathbb{C}^n$ to $\mathbb{C}^m$, and let $B$ be a matrix transformation from the space $\mathbb{C}^m$ to $\mathbb{C}^q$. The product transformation $BA$ is:*

$$(BA)\, x = B\,(Ax)\,, \quad \forall\, x \in \mathbb{C}^n \tag{3.13c}$$

*Note that the matrix $B$ is of size $q \times m$ and $A$ is of size $m \times n$. That is, the matrix $B$ is conformable to matrix $A$.*

4. *Zero transformation: Let $A$ be a matrix transformation from the space $\mathbb{C}^n$ to $\mathbb{C}^m$. The matrix $A$ is a zero transformation if*

$$Ax = 0, \quad 0 \in \mathbb{C}^m, \quad \forall\, x \in \mathbb{C}^n \tag{3.13d}$$

*This transformation is generally denoted by $0$.*

5. *Identity transformation: Let $A$ be a matrix transformation from the space $\mathbb{C}^n$ to $\mathbb{C}^n$. The matrix $A$ is an identity transformation if*

$$Ax = x, \quad \forall\, x \in \mathbb{C}^n \tag{3.13e}$$

*This transformation is generally denoted by $I$.*

6. *Unitary transformation: If the linear transformation matrix is unitary, then it is called a unitary transformation.*

7. *Self-adjoint transformation: A linear transformation specified by a complex matrix $A$ is a self-adjoint transformation if the matrix $A$ is Hermitian.* ◻

### Coordinates and Change of Basis

Let a vector $x$ represent a point in the $n$-dimensional space $\mathbb{R}^n$, where

$$x = \begin{bmatrix} x_1 & x_2 & \cdots & x_n \end{bmatrix}^T$$

and $x_i$ is the $i$-th coordinate (element) of this column vector. Recall that the set of vectors $\{e_1, e_2, e_3, \ldots, e_n\}$ is the standard basis of the vector space $\mathbb{R}^n$, where $e_1 = (1, 0, 0, \ldots, 0, 0)^T$, $e_2 = (0, 1, 0, \ldots, 0, 0)^T, \ldots$, and $e_n = (0, 0, 0, \ldots, 0, 1)^T$. In this case $x = \sum_{i=1}^{n} x_i e_i$. The coordinates of the vector $x$ with respect to the standard basis are specified by the $n$-tuple $(x_1, x_2, \ldots, x_n)$.

Let $\{u_1, u_2, \ldots, u_n\}$ be another basis set. Assume that the basis vectors $u_1, u_2, \ldots, u_n$ are ordered. Therefore this basis is called an *ordered basis*. It is actually specified as $(u_1, u_2, \ldots, u_n)$. If $x = \sum_{i=1}^{n} \alpha_i u_i$, then $\alpha_1, \alpha_2, \ldots, \alpha_n$ can be considered to be the coordinates of the vector $x$ with respect to this ordered basis. Thus, $(\alpha_1, \alpha_2, \ldots, \alpha_n)$ is the coordinate-vector of the vector $x$ with respect to this ordered basis. Let $U = \begin{bmatrix} u_1 & u_2 & \cdots & u_n \end{bmatrix}$, and $a_u^T = \begin{bmatrix} \alpha_1 & \alpha_2 & \cdots & \alpha_n \end{bmatrix}$. Then

$$x = U a_u$$

Similarly assume that $(v_1, v_2, \ldots, v_n)$ is another ordered basis. Also let $x = \sum_{i=1}^{n} \beta_i v_i$. The coordinate-vector of the vector $x$ with respect to this ordered basis is $(\beta_1, \beta_2, \ldots, \beta_n)$. Let $V = \begin{bmatrix} v_1 & v_2 & \cdots & v_n \end{bmatrix}$, and $b_v^T = \begin{bmatrix} \beta_1 & \beta_2 & \cdots & \beta_n \end{bmatrix}$. Then

$$x = V b_v$$

Note that, since the columns of the matrix $U$ are independent, the matrix $U$ is invertible. The matrix $V$ is similarly invertible. Thus

$$a_u = U^{-1} V b_v$$

Another interpretation of the above representation is as follows. Since $(u_1, u_2, \ldots, u_n)$ is an ordered basis, the vector $v_j$ is expanded as

$$v_j = \sum_{i=1}^{n} d_{ij} u_i, \quad 1 \leq j \leq n$$

If $D = [d_{ij}]$, then $V = UD$. This yields

$$a_u = D b_v$$

Thus $D = U^{-1} V$. The above discussion is summarized in the following theorem.

**Theorem 3.4.** *Let $x$ be a vector in the vector space $\mathbb{R}^n$. Also let the coordinates of the vector $x$ with respect to the ordered bases $(u_1, u_2, \ldots, u_n)$ and $(v_1, v_2, \ldots, v_n)$ be $(\alpha_1, \alpha_2, \ldots, \alpha_n)$ and $(\beta_1, \beta_2, \ldots, \beta_n)$ respectively. If*

$$U = \begin{bmatrix} u_1 \ u_2 \ \cdots \ u_n \end{bmatrix}, \quad V = \begin{bmatrix} v_1 \ v_2 \ \cdots \ v_n \end{bmatrix} \tag{3.14a}$$

$$a_u = \begin{bmatrix} \alpha_1 \ \alpha_2 \ \cdots \ \alpha_n \end{bmatrix}^T, \quad b_v = \begin{bmatrix} \beta_1 \ \beta_2 \ \cdots \ \beta_n \end{bmatrix}^T \tag{3.14b}$$

*then*

$$x = U a_u = V b_v, \quad \text{and} \quad a_u = D b_v, \quad \text{where} \quad D = U^{-1} V$$

$\square$

The matrix transformation, in which an ordered basis is changed to another, is next examined.

**Theorem 3.5.** *Let $A$ be a matrix of transformation from the vector space $\mathbb{R}^n$ to itself, where the coordinate-vectors are specified with respect to the ordered basis $\widetilde{U} = (u_1, u_2, \ldots, u_n)$. Let $\widetilde{V} = (v_1, v_2, \ldots, v_n)$ be another ordered basis. Also let $a_u$ and $b_v$ be the coordinate-vectors of a point in space $\mathbb{R}^n$ with respect to the ordered bases $\widetilde{U}$ and $\widetilde{V}$ respectively. If $a_u = D b_v$, then the matrix of transformation with respect to the ordered basis $\widetilde{V}$ is $B = D^{-1} A D$.*
      *Proof.* Let

$$y_u = A a_u$$

Also, let $y_v$ be the transformed vector with respect to the ordered basis $\widetilde{V}$. Substituting $a_u = D b_v$ and $y_u = D y_v$ in the above equation yields

$$D y_v = A D b_v$$

That is

$$y_v = D^{-1} A D b_v = B b_v$$

The result follows.                                                                                  $\square$

## 3.6 Spectral Analysis of Matrices

Properties of a square matrix can be studied effectively via its *eigenvalues* and *eigenvectors*. These words are derivatives of the German word *eigen*, meaning *peculiar to*. Eigenvalue is also sometimes referred to as *characteristic value*, or *proper value*, or *latent value*. Similarly, eigenvector is also referred to as *characteristic vector*, or *proper vector*, or *latent vector*. This body of knowledge associated with square matrices is called its *spectral analysis*.

**Definitions 3.14.** *Let $A$ be a square matrix of size $n$. Its elements can possibly be complex numbers.*

1. *Eigenvalue and eigenvector: A scalar $\lambda$ is an eigenvalue of matrix $A$, if $Ax = \lambda x$, where $x$ is a nonzero column vector of size $n$. The vector $x$ is called an eigenvector of the matrix $A$. The vector $x$ is unique to within a constant. Also $(\lambda, x)$ is called an eigenpair of $A$.*
2. *Simple eigenvalue: An eigenvalue which occurs only once is called a simple eigenvalue.*
3. *Multiple eigenvalue: An eigenvalue which is not simple is a multiple eigenvalue.*
4. *Left-hand eigenvector: Nonzero row vectors $y^\dagger$ of length $n$, such that $y^\dagger A = \lambda y^\dagger$ are called left-hand eigenvectors. That is, $A^\dagger y = \overline{\lambda} y$.*
   *These are different than the usual eigenvectors $x$, where $Ax = \lambda x$. The vector $x$ can be called right-hand eigenvector.*
5. *Dominant eigenvalue: An eigenvalue of a matrix of the largest modulus.*
6. *Dominant eigenvector: The eigenvector corresponding to the dominant eigenvalue.*
7. *Eigenspace: It is the set of all column vectors $\{x \in \mathbb{C}^n \mid Ax - \lambda x\}$ associated with $\lambda$. The vectors which belong to an eigenspace constitute a vector space.*
8. *Spectrum: The set of distinct eigenvalues of $A$ is called its spectrum.*
9. *Spectral radius of a matrix: It is the maximum value of the modulus of the eigenvalue of a matrix. The spectral radius of matrix $A$ is denoted by $\varrho(A)$.*
10. *Characteristic polynomial: The characteristic polynomial $p_A(\lambda)$ of matrix $A$ is equal to $\det(\lambda I - A)$, where $I$ is an $n \times n$ identity matrix.*
11. *Characteristic equation: The characteristic equation of matrix $A$ is specified by $p_A(\lambda) = 0$.* □

Unless stated explicitly, all eigenvectors in this chapter are right-hand eigenvectors. Based upon the above definitions, the following elementary observations are listed.

**Observations 3.8.** Let $A$ be a square matrix of size $n$. Its elements are permitted to be complex numbers.

1. The characteristic polynomial $p_A(\lambda)$ of matrix $A$ is a monic polynomial of degree $n$ in $\lambda$.
2. The zeros of the characteristic equation are the eigenvalues of the matrix $A$. Consequently, the eigenvalues are also called the characteristic roots of $A$. Therefore the number of eigenvalues of $A$ is equal to $n$. Even if the elements of the matrix $A$ are real numbers, the $\lambda$'s can be imaginary. Furthermore, the $n$ eigenvalues of the matrix $A$ are not necessarily all distinct.
3. The coefficient of $\lambda^{n-1}$ in the polynomial $p_A(\lambda)$ is equal to negative of the trace of matrix $A$.
4. The trace of matrix $A$ is equal to the sum of all the eigenvalues.
5. The constant term in the polynomial $p_A(\lambda)$ is equal to $(-1)^n \det A$.
6. The determinant of a matrix $A$ is equal to the product of all eigenvalues.

7. The $\det A = 0$ if and only if $0$ is an eigenvalue of the matrix $A$.
8. If $p_A(\lambda)$ is a characteristic polynomial of a matrix $A$, then $p_A(A) = 0$. This statement is the so-called Cayley-Hamilton theorem.
9. The eigenvalues of the matrices $A$ and $A^T$ are identical.
10. The eigenspace associated with an eigenvalue $\lambda$ is the null space of the matrix $(A - \lambda I)$. In addition, it is a subspace of $\mathbb{C}^n$. The multiplicity of the eigenvalue $\lambda$ is the dimension of this vector space.
11. Let $S$ be a nonsingular matrix, then the eigenvalues of the matrix $SAS^{-1}$ and the matrix $A$ are identical. That is, similar matrices have identical spectra.
12. The eigenvalues of the matrix $A^k, k \in \mathbb{P}$ are $k$th powers of the eigenvalues of the matrix $A$.
13. Assume that the matrix $A$ is invertible. The eigenvalues of the matrix $A^{-k}, k \in \mathbb{P}$ are the negative $k$th powers of the eigenvalues of the matrix $A$.
14. Let $g(x)$ be a polynomial in a single indeterminate $x$ with real coefficients. Also $A$ is an $n \times n$ matrix with real elements. Let the eigenvalues of this matrix be $\lambda_1, \lambda_2, \ldots, \lambda_n$. Then the eigenvalues of the matrix $g(A)$ are $g(\lambda_1), g(\lambda_2), \ldots, g(\lambda_n)$.
15. Let $A$ be a square matrix of order $n$. Let its distinct eigenvalues be $\lambda_1, \lambda_2, \ldots, \lambda_k$, and the corresponding eigenvectors be $x_1, x_2, \ldots, x_k$ respectively, where $k \leq n$. Then the set of vectors $x_1, x_2, \ldots, x_k$ are linearly independent. If $k = n$,

$$R = \begin{bmatrix} x_1 & x_2 & \cdots & x_n \end{bmatrix}, \quad \text{and} \quad \Lambda = diag(\lambda_1, \lambda_2, \ldots, \lambda_n)$$

then $R^{-1}AR = \Lambda$, and $A = R\Lambda R^{-1}$. Therefore the matrix $A$ is diagonalizable, if it has $n$ linearly independent eigenvectors. If in addition $R$ is unitary, then

$$R^\dagger AR = \Lambda, \quad \text{and} \quad A = R\Lambda R^\dagger$$

Note that the matrices $A$ and $\Lambda$ are similar.                                                    □

**Example 3.4.** Consider the matrix

$$A = \begin{bmatrix} 4 & 6 \\ -1 & -3 \end{bmatrix}$$

Its characteristic polynomial is

$$p_A(\lambda) = \begin{vmatrix} \lambda - 4 & -6 \\ 1 & \lambda + 3 \end{vmatrix}$$

Therefore $p_A(\lambda) = \lambda^2 - \lambda - 6 = (\lambda - 3)(\lambda + 2)$. Thus the eigenvalues are $\lambda = 3$ and $\lambda = -2$. It can be verified that the sum of the eigenvalues is equal to the trace of the matrix $A$, which is equal to $1$. Furthermore, the product of the eigenvalues is equal to $\det A$, which is equal to $-6$.

An eigenvector associated with the eigenvalue $3$ is $\begin{bmatrix} 6 & -1 \end{bmatrix}^T$. Similarly, an eigenvector associated with the eigenvalue $-2$ is $\begin{bmatrix} 1 & -1 \end{bmatrix}^T$. Also check that

$$\begin{aligned} p_A(A) &= A^2 - A - 6I \\ &= \begin{bmatrix} 10 & 6 \\ -1 & 3 \end{bmatrix} - \begin{bmatrix} 4 & 6 \\ -1 & -3 \end{bmatrix} - \begin{bmatrix} 6 & 0 \\ 0 & 6 \end{bmatrix} \\ &= \begin{bmatrix} 0 & 0 \\ 0 & 0 \end{bmatrix} \end{aligned}$$

Thus $p_A(A) = 0$. $\qquad\qquad\qquad\qquad\qquad\qquad\qquad\qquad\qquad\qquad\qquad\qquad\qquad$ $\square$

**Observations 3.9.** Some properties of real symmetric matrices are:

1. The eigenvalues of a real symmetric matrix are all real numbers.
2. The eigenvectors corresponding to distinct eigenvalues of a real symmetric matrix are mutually orthogonal.
3. If $A$ is a real symmetric matrix, then there exists a real orthogonal matrix $P$ such that $P^T AP$ is a diagonal matrix, with eigenvalues on the diagonal. $\qquad\qquad\qquad\qquad\qquad\qquad$ $\square$

**Observations 3.10.** Some properties of unitary matrices are:

1. If a matrix $R$ is unitary, then it is nonsingular, and $R^{-1} = R^\dagger$.
2. The rows of a unitary matrix form an orthonormal set of vectors. Similarly, the columns of a unitary matrix form an orthonormal set of vectors.
3. The product of two unitary matrices is a unitary matrix.
4. If a matrix $R$ is unitary, then $|\det(R)| = 1$.
5. All eigenvalues of a unitary matrix have a unit modulus (magnitude).
6. Let $A$ be a square matrix. Using a unitary transformation, it can be transformed to an upper triangular matrix with eigenvalues of the matrix $A$ on the diagonal.
7. Let $R$ be a unitary matrix. If matrices $A$ and $B$ are related to each other via a unitary transformation, that is if $A = R^\dagger BR$, then the matrices $A$ and $B$ have the same eigenvalues. $\qquad$ $\square$

**Observations 3.11.** Some properties of normal matrices are:

1. Diagonal, real symmetric, orthogonal, Hermitian, and unitary matrices are some examples of normal matrices.
2. Let $R$ be a unitary matrix, and the matrix $A$ be normal, then the matrix $R^\dagger AR$ is normal.
3. If $A$ is a square matrix, then it can be reduced to a diagonal matrix via a unitary transformation if and only if $A$ is a normal matrix.
4. Let $A$ be a normal matrix, then the eigenvectors corresponding to different eigenvalues are orthogonal.
5. Let $A$ be a complex square matrix of size $n$. If the matrix $A$ is normal, then

$$\|A\|_{tr} = \sum_{i=1}^{n} |\lambda_i|$$

where $\lambda_i, 1 \leq i \leq n$ are the eigenvalues of the matrix $A$. $\qquad\qquad\qquad\qquad$ $\square$

**Bounds for the Eigenvalues of Matrices**

Computation of eigenvalues can be numerically complicated and expensive. Therefore, it is necessary to evaluate bounds on the eigenvalues of matrices. Some well-known bounds for the eigenvalues of a square matrix are listed below. These set of observations are generally attributed to S. A. Gerschgorin (1901-1933).

**Observations 3.12.** The matrix $A = [a_{ij}]$ is a complex square matrix of order $n$, with spectral radius $\varrho(A)$. Define

$$P_i = \sum_{\substack{j=1 \\ i \neq j}}^{n} |a_{ij}|, \quad 1 \leq i \leq n, \quad \text{and} \quad Q_j = \sum_{\substack{i=1 \\ i \neq j}}^{n} |a_{ij}|, \quad 1 \leq j \leq n$$

1. Every eigenvalue $\lambda$ of the matrix $A$:
   (a) Satisfies at least one of the following inequalities: $|\lambda - a_{ii}| \leq P_i$, for $1 \leq i \leq n$.
   (b) All the eigenvalues $\lambda$ of the matrix $A$ lie within the union of discs $|\lambda - a_{ii}| \leq P_i, 1 \leq i \leq n$, in the complex plane.
2. Every eigenvalue $\lambda$ of the matrix $A$:
   (a) Satisfies at least one of the following inequalities: $|\lambda - a_{jj}| \leq Q_j$, for $1 \leq j \leq n$.
   (b) All the eigenvalues $\lambda$ of the matrix $A$ lie within the union of discs $|\lambda - a_{jj}| \leq Q_j, 1 \leq j \leq n$, in the complex plane.
3. $\varrho(A) \leq \min \left\{ \max_{1 \leq i \leq n} \sum_{j=1}^{n} |a_{ij}|, \max_{1 \leq j \leq n} \sum_{i=1}^{n} |a_{ij}| \right\}$.
   The discs which are mentioned in these observations are called the *Gerschgorin's discs*. $\square$

## 3.7 Hermitian Matrices and Their Eigenstructures

The eigenstructures of Hermitian matrices are next examined. Recall that a Hermitian matrix is a generalization of a symmetric matrix. A square matrix $A$ is Hermitian if $A = A^\dagger$. Notice that all the diagonal elements of a Hermitian matrix are real numbers.

**Observations 3.13.** Let $A$ be a Hermitian matrix of order $n$.

1. Eigenstructure of a Hermitian matrix.
   (a) There exists a unitary matrix $R$ such that $R^\dagger A R = \Lambda$, where $\Lambda$ is a diagonal matrix whose elements are the eigenvalues of the matrix $A$, that is $A = R\Lambda R^\dagger$. This result is known as the *spectral theorem* of Hermitian matrices.
   (b) The diagonal matrix $\Lambda$ is real. That is, all the eigenvalues of a Hermitian matrix are real numbers.
   (c) The eigenvectors corresponding to distinct eigenvalues of a Hermitian matrix are orthogonal.
   (d) If an eigenvalue repeats itself $m$ times, then associated with this eigenvalue are exactly $m$ linearly independent eigenvectors, which can be orthogonal.
   (e) The matrix has $n$ linearly independent eigenvectors which form an orthonormal set.
2. Let the Hermitian matrix $A$ have distinct eigenvalues. Also, let the eigenvalue and a corresponding eigenvector be $\lambda_i$ and $x_i$ respectively for $1 \leq i \leq n$. The length of each eigenvector is also assumed to be unity. Define $P_i = x_i x_i^\dagger$ for $1 \leq i \leq n$. Then
   (a) The matrix $P_i$ is Hermitian, for $1 \leq i \leq n$.
   (b) $P_i P_j = \delta_{ij} P_i$, where $\delta_{ij} = 1$ if $i = j$, and $\delta_{ij} = 0$ if $i \neq j$, for $1 \leq i, j \leq n$.
   (c) $A = \sum_{i=1}^{n} \lambda_i P_i$.
   This result is another form of spectral decomposition of a Hermitian matrix.
3. Definite, semidefinite, and indefinite characterization of a Hermitian matrix $A$.

(a) Matrix $A$ is positive (negative) definite if and only if every eigenvalue of $A$ is positive (negative).
(b) Matrix $A$ is positive (negative) semidefinite if and only if every eigenvalue of $A$ is nonnegative (nonpositive).
(c) Matrix $A$ is indefinite if and only if the matrix $A$ has both positive and negative eigenvalues.

4. If the Hermitian matrix $A$ is semidefinite, then $\|A\|_{tr} = tr\,(A)$. $\qquad\qquad$ □

Bounds on the eigenvalues of Hermitian matrices are next obtained.

**Lemma 3.3.** *Let $A$ be a Hermitian matrix, $\lambda$ a scalar, and $x$ a vector. The expression*

$$E\,(x) = (Ax - \lambda x)^{\dagger}\,(Ax - \lambda x) \qquad\qquad (3.15\text{a})$$

*is minimized if*

$$\lambda = \frac{x^{\dagger} A x}{x^{\dagger} x}, \qquad x \neq 0 \qquad\qquad (3.15\text{b})$$

*The minimized value of $E\,(x)$ is equal to $\left\{(Ax)^{\dagger}\,(Ax) - \lambda\,(x^{\dagger} A x)\right\}$. If in addition $x^{\dagger} x$ is constrained to be equal to 1, then the minimizing value of $\lambda$ is equal to $x^{\dagger} A x$.*
*Proof.* The proof is left to the reader. $\qquad\qquad$ □

In the above lemma, if $A$ is any matrix (not necessarily Hermitian), then it can be shown that $E\,(x)$ is minimized if

$$\lambda = \frac{1}{2}\frac{x^{\dagger}\,\left(A + A^{\dagger}\right) x}{x^{\dagger} x}, \qquad x \neq 0$$

**Theorem 3.6.** *Let $x$ be a column vector of size $n$, and $A$ be a Hermitian matrix of order $n$. The eigenvalues of matrix $A$ are $\lambda_1 \leq \lambda_2 \leq \ldots \leq \lambda_n$. Define*

$$\vartheta\,(x) = \frac{x^{\dagger} A x}{x^{\dagger} x}, \qquad x \neq 0 \qquad\qquad (3.16\text{a})$$

*Then*

$$\lambda_1 = \min_{x \neq 0} \vartheta\,(x), \quad \lambda_n = \max_{x \neq 0} \vartheta\,(x) \qquad\qquad (3.16\text{b})$$

*Proof.* Let the orthonormal set of eigenvectors of the Hermitian matrix $A$ be $x_i,\ 1 \leq i \leq n$ such that the eigenvalue $\lambda_i$ corresponds to the eigenvector $x_i$. A representation of a nonzero vector $x$ in terms of these eigenvectors is next obtained. It is

$$x = \sum_{j=1}^{n} b_j x_j$$

Therefore

$$Ax = \sum_{j=1}^{n} b_j \lambda_j x_j,$$

$$\vartheta\,(x) = \frac{\sum_{j=1}^{n} \bar{b}_j b_j \lambda_j}{\sum_{j=1}^{n} \bar{b}_j b_j}$$

Thus

$$\lambda_n - \vartheta(x) = \frac{\sum_{j=1}^{n-1} \bar{b}_j b_j (\lambda_n - \lambda_j)}{\sum_{j=1}^{n} \bar{b}_j b_j}$$

Since $(\lambda_n - \lambda_j) \geq 0$ for all $1 \leq j \leq n$, it follows that $\lambda_n \geq \vartheta(x)$. Furthermore, if $x = x_n$ then $\vartheta(x_n) = \lambda_n$. The minimum estimate of $\vartheta(x)$ is obtained similarly by considering $(\vartheta(x) - \lambda_1)$.
□

The above theorem is generally attributed to John William Strutt Rayleigh (1842-1919) and Walther Ritz (1878-1909).

**Corollary 3.2.** *In the above theorem,* $\lambda_1 = \min_{x^\dagger x = 1} \vartheta(x)$ *and* $\lambda_n = \max_{x^\dagger x = 1} \vartheta x$      □

The next theorem gives a *min-max characterization* of the eigenvalues of a Hermitian matrix. This result is due to Richard Courant (1888-1972) and Ernst Fischer (1875-1954).

**Theorem 3.7.** *Let $x$ be a column vector of size $n$, and $A$ be a Hermitian matrix of order $n$. The eigenvalues of matrix $A$ are $\lambda_1 \leq \lambda_2 \leq \ldots \leq \lambda_n$. If $\vartheta(x) = x^\dagger A x / (x^\dagger x)$, where $x \neq 0$ then*

$$\lambda_k = \max_{\substack{a_1, a_2, \ldots, a_{k-1} \in \mathbb{C}^n}} \min_{\substack{x \neq 0, x \in \mathbb{C}^n \\ x \perp a_1, a_2, \ldots, a_{k-1}}} \vartheta(x) \tag{3.17a}$$

*That is, the minimization is taken over all vectors $x \neq 0$, such that $a_i^\dagger x = 0$, $i = 1, 2, \ldots, (k-1)$, where the $a_i$'s are considered to be fixed vectors of size $n$, and $k \neq 1$. The maximization is then taken over all $a_i$'s, where $i = 1, 2, \ldots, (k-1)$. If $k = 1$, then the outer maximization should be omitted, and the result reduces to the theorem due to Rayleigh and Ritz. Similarly,*

$$\lambda_k = \min_{\substack{a_{k+1}, a_{k+2}, \ldots, a_n \in \mathbb{C}^n}} \max_{\substack{x \neq 0, x \in \mathbb{C}^n \\ x \perp a_{k+1}, a_{k+2}, \ldots, a_n}} \vartheta(x) \tag{3.17b}$$

*The maximization is taken over all vectors $x \neq 0$, such that $a_i^\dagger x = 0$, $i = (k+1), \ldots, n$, where the $a_i$'s are considered to be fixed vectors, and $k \neq n$. The minimization is then taken over all $a_i$'s, where $i = (k+1), \ldots, n$. If $n = k$, then the outer minimization should be omitted and the result reduces to the theorem due to Rayleigh and Ritz.*

*Proof.* See the problem section.      □

**Theorem 3.8.** *Let $A$ and $B$ be Hermitian matrices of order $n$ each. Let their eigenvalues be $\lambda_i(A)$ and $\lambda_i(B)$ for $1 \leq i \leq n$ respectively, such that $\lambda_1(A) \leq \lambda_2(A) \leq \ldots \leq \lambda_n(A)$ and $\lambda_1(B) \leq \lambda_2(B) \leq \ldots \leq \lambda_n(B)$. These matrices are also constrained by the inequality $x^\dagger A x \leq x^\dagger B x$, for all vectors $x$ of size $n$. Then $\lambda_i(A) \leq \lambda_i(B)$ for $1 \leq i \leq n$.*

*Proof.* See the problem section.      □

The following theorem states the relationship between the eigenvalues of the Hermitian matrices $A, B$, and $(A + B)$.

**Theorem 3.9.** *Let $A$ and $B$ be Hermitian matrices of order $n$ each. Let their eigenvalues be $\lambda_i(A)$ and $\lambda_i(B)$ for $1 \leq i \leq n$ respectively, such that $\lambda_1(A) \leq \lambda_2(A) \leq \ldots \leq \lambda_n(A)$ and*

$\lambda_1(B) \leq \lambda_2(B) \leq \ldots \leq \lambda_n(B)$. *Let the eigenvalues of the matrix* $C = (A + B)$ *be* $\lambda_i(C)$ *for* $1 \leq i \leq n$, *such that* $\lambda_1(C) \leq \lambda_2(C) \leq \ldots \leq \lambda_n(C)$. *Then*

$$\lambda_k(A) + \lambda_1(B) \leq \lambda_k(C) \leq \lambda_k(A) + \lambda_n(B), \quad 1 \leq k \leq n \qquad (3.18)$$

*Proof.* For any $x \neq 0$ and $x \in \mathbb{C}^n$, define

$$\vartheta_A(x) = \frac{x^\dagger A x}{(x^\dagger x)}, \quad \vartheta_B(x) = \frac{x^\dagger B x}{(x^\dagger x)}, \quad \text{and} \quad \vartheta_C(x) = \frac{x^\dagger C x}{(x^\dagger x)}$$

Then

$$\lambda_1(B) \leq \vartheta_B(x) \leq \lambda_n(B)$$

Furthermore, for $1 \leq k \leq n$ we have

$$
\begin{aligned}
\lambda_k(C) &= \min_{a_{k+1}, a_{k+2}, \ldots, a_n \in \mathbb{C}^n} \quad \max_{\substack{x \neq 0, x \in \mathbb{C}^n \\ x \perp a_{k+1}, a_{k+2}, \ldots, a_n}} \vartheta_C(x) \\
&= \min_{a_{k+1}, a_{k+2}, \ldots, a_n \in \mathbb{C}^n} \quad \max_{\substack{x \neq 0, x \in \mathbb{C}^n \\ x \perp a_{k+1}, a_{k+2}, \ldots, a_n}} (\vartheta_A(x) + \vartheta_B(x)) \\
&\geq \min_{a_{k+1}, a_{k+2}, \ldots, a_n \in \mathbb{C}^n} \quad \max_{\substack{x \neq 0, x \in \mathbb{C}^n \\ x \perp a_{k+1}, a_{k+2}, \ldots, a_n}} (\vartheta_A(x) + \lambda_1(B)) = \lambda_k(A) + \lambda_1(B)
\end{aligned}
$$

The upper bound for $\lambda_k(C)$ can be similarly proved.                                        □

The above result is sometimes referred to as the *interlacing eigenvalue theorem*. We have not explicitly explored methods to compute eigenvalues of a matrix. In principle, the eigenvalues of a matrix can be determined by finding the zeros of its characteristic equation. However, this technique is not always practical. Eigenvalues of a matrix are computed in general by iterative techniques. These techniques can be found in any standard textbook on numerical analysis.

## 3.8 Perron-Frobenius Theory

The Perron-Frobenius theory of positive and nonnegative matrices was developed by mathematicians Oskar Perron and Ferdinand Georg Frobenius. For simplicity, we assume that the matrix elements of the matrices considered in this section are all real numbers. Properties of positive and negative matrices are studied in this section. These results form a basis for the search techniques used in the Internet.

**Definitions 3.15.** *Some basic definitions.*

1. *Nonnegative matrix: A matrix A is nonnegative if all of its elements are nonnegative, however* $A \neq 0$. *It is denoted by* $A \geq 0$, *where* $A \neq 0$.
2. *Positive matrix: A matrix A is positive, if all of its elements are positive. It is denoted by* $A > 0$.
3. *Let the matrices A and B be of the same order. Then* $(A - B) \geq 0$ *can be written as* $A \geq B$. *Similarly, if* $(A - B) > 0$ *then write* $A > B$.

4. *Permutation matrix*: *A permutation matrix is a square matrix whose elements are either $0$'s or $1$'s, in which the entry $1$ occurs exactly once in each column, and exactly once in each row. Each remaining entry is equal to $0$.*

5. *Reducible matrix*: *An $n \times n$ matrix $A$ is reducible if it is the $1 \times 1$ zero matrix, or if there exists a permutation matrix $P$ such that*

$$P^T A P = \begin{bmatrix} X & Y \\ 0 & Z \end{bmatrix} \tag{3.19}$$

*where $X$ and $Z$ are square matrices not necessarily of the same order.*

6. *Irreducible matrix*: *A square matrix, which is not reducible is an irreducible matrix.*

7. *Primitive matrix*: *Let $A$ be a nonnegative square matrix. The matrix $A$ is primitive if $A^k > 0$ for some positive integer $k$.*

8. *Modulus of a matrix*: *Let $A = [a_{ij}]$, then the matrix $[|a_{ij}|]$ is the modulus of matrix $A$. It is denoted by $m(A)$.*   □

Some authors define a nonnegative matrix $A$ as a matrix in which all of its elements are nonnegative. They do not impose the additional condition that $A$ be not equal to an all-zero matrix. However, they call a matrix in which $A \geq 0$ and $A \neq 0$ as a semipositive matrix.

Note that the relationship between permutation mapping and the corresponding permutation matrix has also been discussed in the chapter on number theory.

**Observations 3.14.** Following facts are immediate from the above set of definitions.

1. Let $A$ be a positive square matrix of order $n$, and $x$ is a nonnegative column vector of size $n$. That is, if $A > 0$ and $x \geq 0$ (but $x \neq 0$) then $Ax > 0$.

2. Let $A$ be a $l \times n$ matrix, and $x$ and $y$ be column vectors of size $n$. Then:
   (a) If $A \geq 0$, and $x \geq y$, then $Ax \geq Ay$.
   (b) If $A > 0$, and $x > y$, then $Ax > Ay$.

3. A primitive matrix is irreducible.   □

Positive matrices are studied in the next subsection. A generalization of a positive matrix is a nonnegative matrix. A well-known example of a nonnegative matrix is a stochastic matrix. These type of matrices are examined in subsequent subsections.

### 3.8.1  Positive Matrices

Some inequalities about positive matrices are stated in this subsection. Some of these results were originally discovered by Perron, and later extended by Frobenius. Therefore, this group of results is sometimes termed the Perron-Frobenius theorem for positive matrices.

**Observations 3.15.** Let $A = [a_{ij}]$ be an $n \times n$ square and positive matrix, with spectral radius $\varrho(A)$.

1. Let $x = \begin{bmatrix} x_1 & x_2 & \cdots & x_n \end{bmatrix}^T$ be a nonnegative column vector of length $n$. Define

$$\phi(x) = \min_{\substack{1 \leq j \leq n \\ x_j \neq 0}} \frac{\sum_{k=1}^{n} a_{jk} x_k}{x_j}$$

$$\Phi = \left\{ x \mid x \geq 0, x^T x = 1 \right\}$$
$$\Psi = \left\{ y \mid y = Ax, x \in \Phi \right\}$$

then

(a) $\eta = \max_{x \in \Phi} \left\{ \phi(x) \right\} = \max_{y \in \Psi} \left\{ \phi(y) \right\}$ is an eigenvalue of the matrix $A$.

(b) Let the value of $y$ corresponding to $\eta$ be equal to $y_0$. Then $y_0 > 0$ is an eigenvector associated with the eigenvalue $\eta$.

2. The eigenvalue $\eta > 0$.

3. The eigenvalue $\eta$ is equal to the spectral radius $\varrho(A)$ of the matrix $A$.

4. Let $\lambda$ be any other eigenvalue of matrix $A$, besides $\eta$. Then $\eta > |\lambda|$.

5. Let $B$ be a positive square matrix of order $n$, such that $A \geq B > 0$. Let $\eta = \varrho(A)$ and $\lambda$ be an eigenvalue of $B$. Then

(a) $\varrho(A) \geq |\lambda|$.

(b) If $\varrho(A) = |\lambda|$, then $A = B$.

6. The multiplicity of the eigenvalue $\eta$ is equal to one.

7. If any element of the matrix $B > 0$ increases, then its spectral radius also increases.

8. The following facts also hold true.

(a) $\min_{1 \leq i \leq n} \sum_{j=1}^{n} a_{ij} \leq \varrho(A) \leq \max_{1 \leq i \leq n} \sum_{j=1}^{n} a_{ij}$

(b) If any row sum of matrix $A$ is equal to $\varrho(A)$, then all row sums are equal.

(c) Similar statement applies to maximal and minimal column sums.              $\square$

Proofs of the above observations can be found in the problem section.

### 3.8.2  Nonnegative Matrices

The Perron-Frobenius theory of nonnegative matrices is used to study graphs. Graph theory in turn is used in studying reducible and irreducible matrices. Let $A = [a_{ij}]$ be any $n \times n$ matrix, then a digraph $G$ associated with matrix $A$ can be defined on $n$ vertices. Label these vertices as $v_1, v_2, \ldots, v_n$. In this graph, there is a directed edge (arc) leading from $v_i$ to $v_j$ if and only if $a_{ij} \neq 0$.

The sequence of vertices $(v_1, v_2, \ldots, v_k)$ such that $(v_1, v_2), (v_2, v_3), \ldots, (v_{k-1}, v_k)$ are directed arcs, is called a directed walk. If all the vertices in this walk are different from each other, the walk is called a directed path. The number of arcs in a directed path or walk is called the length of the path or walk respectively. For more observations about the properties of digraphs see the chapter on graph theory. Based upon these concepts, the following observations are immediate.

**Observations 3.16.** Let $A$ be a matrix of order $n$, and the associated digraph be $G$.

1. Let $P$ be a permutation matrix, and the digraph associated with the matrix $P^T A P$ be $G'$. Then the digraph $G'$ is simply a relabeling of the digraph $G$.

2. The matrix $A$ is a irreducible if and only if the digraph $G$ is strongly connected. That is, in the digraph $G$ for each vertex-pair $(v_i, v_j)$ there is a sequence of directed edges leading from vertex $v_i$ to vertex $v_j$.

3. Let $k$ be a positive integer. Then the $ij$th entry of the matrix $A^k$ is nonzero if and only if there is a walk of length $k$ from the vertex $v_i$ to vertex $v_j$ in the digraph $G$.

4. The matrix $A$ is irreducible, if and only if there exists an integer $k > 0$ such that, for each $(i, j)$ pair the $ij$th entry of the matrix $A^k$ is positive.              $\square$

The next set of observations are useful in proving certain results about nonnegative matrices.

**Observations 3.17.** Let $A$ be a nonnegative square matrix of order $n$, with eigenvalues $\lambda_i, 1 \leq i \leq n$.

1. The eigenvalues of the matrix $(I + A)^j$ are $(1 + \lambda_i)^j, 1 \leq i \leq n$. The converse of this statement is also true for each real eigenvalue $\lambda_i$.
2. If the multiplicity of the eigenvalue $\lambda_i$ in $A$ is $k$, then the multiplicity of the eigenvalue $(1 + \lambda_i)^j$ in $(I + A)^j$ is at least $k$.
3. Let an eigenvector corresponding to the eigenvalue $(1 + \lambda_i)$ of the matrix $(I + A)$ be $x$, then $Ax = \lambda_i x$.
4. Let $A$ and $B$ be matrices of the same order, such that $B \geq m(A)$, then $\varrho(B) \geq \varrho(A)$, where $m(A)$ is the modulus of the matrix $A$. More specifically $\varrho(m(A)) \geq \varrho(A)$.                   $\square$

The following observations are associated with nonnegative matrices.

**Observations 3.18.** Let $A$ be a square, nonnegative, and irreducible matrix of order $n$.

1. If $k_i > 0$, for $0 \leq i \leq (n - 1)$, then
   (a) $\sum_{i=0}^{n-1} k_i A^i > 0$
   (b) $(I + A)^{n-1} > 0$
2. If $u$ is a nonnegative column vector of length $n$, and $\beta$ is a real number such that $Au > \beta u$, then $\varrho(A) > \beta$.                   $\square$

See the problem section for a proof of some of these observations. In the next step we explore, whether it is possible for nonnegative irreducible matrices to have a maximum value of $\phi(x)$, as defined in the subsection on positive matrices.

**Lemma 3.4.** *The matrix $A = [a_{ij}]$ is a nonnegative irreducible matrix of order $n$. Let $x$ be a nonnegative column vector of length $n$, where $x^T = \begin{bmatrix} x_1 & x_2 & \cdots & x_n \end{bmatrix}$. Define*

$$\phi(x) = \min_{\substack{1 \leq j \leq n \\ x_j \neq 0}} \frac{\sum_{k=1}^{n} a_{jk} x_k}{x_j} \tag{3.20a}$$

$$\Phi = \{x \mid x \geq 0, x^T x = 1\} \tag{3.20b}$$

$$\Psi = \{y \mid y = Ax, x \in \Phi\} \tag{3.20c}$$

$$\Upsilon = \left\{z \mid z = (I + A)^{n-1} y, y \in \Psi\right\} \tag{3.20d}$$

*Then*

$$\eta = \max_{z \in \Upsilon} \{\phi(z)\} \tag{3.20e}$$

*is an eigenvalue of the matrix $A$. If the corresponding value of $z$ is $z_0$ then $Az_0 = \eta z_0$. That is, $z_0$ is an eigenvector of the matrix $A$ associated with the eigenvalue $\eta$.*

*Proof.* See the problem section.                   $\square$

The Perron-Frobenius theorem for nonnegative matrices is next stated.

**Theorem 3.10.** *The matrix $A = [a_{ij}]$ is a nonnegative irreducible matrix of order $n$. Then:*

(a) *The matrix $A$ has a positive eigenvalue $\eta$, which is equal to the spectral radius $\varrho(A)$.*
(b) *The eigenvector $z_0$ corresponding to the eigenvalue $\eta$ is positive.*
(c) *The eigenvalue $\eta$ is simple.*

   *Proof.* See the problem section.                                                                    □

### 3.8.3  Stochastic Matrices

The theory of nonnegative matrices developed in the last subsection is applied to stochastic and doubly stochastic matrices. Stochastic matrices find useful application in the study of Markov chains and several other applications in engineering, economics, and operations research. One of its recent application is in the interpretation of a certain popular search algorithm.

   **Definitions 3.16.** *Stochastic vector, stochastic matrix, and doubly stochastic matrix are defined.*

1. *Let $z$ be a vector of length $n$, such that all its elements are real and nonnegative numbers. In addition, the sum of its components is equal to 1, then this vector is called a stochastic vector.*
2. *Let $A = [a_{ij}]$ be an $n \times n$ matrix of real elements, $e = \begin{bmatrix} 1 & 1 & \cdots & 1 \end{bmatrix}^T$ is an all-1 vector of length $n$. The matrix $A$ is stochastic provided, $a_{ij} \geq 0$ for $1 \leq i, j \leq n$, and $Ae = e$. If in addition, $A^T$ is also stochastic, then the matrix $A$ is doubly stochastic.*                                                            □

   As per the above definition, a stochastic matrix is nonnegative, and the sum of matrix elements of each row is equal to unity. A doubly stochastic matrix is stochastic, and in addition the sum of elements of each column is unity.

   **Observations 3.19.** Let $A$ be an $n \times n$ stochastic matrix, and $e = \begin{bmatrix} 1 & 1 & \cdots & 1 \end{bmatrix}^T$ be an all-1 vector of length $n$.

1. The spectral radius $\varrho(A)$ of the stochastic matrix $A$ is equal to 1, and a corresponding eigenvector is equal to $e$.
2. A nonnegative matrix $A$ is stochastic if and only if an eigenvector corresponding to the eigenvalue 1 is $e$.
3. If the matrix $A$ is stochastic, then so are the matrices $A^k$, $k \in \mathbb{P}$.
4. Let $A$ be an irreducible primitive stochastic matrix, and $z$ be an eigenvector of $A^T$, such that the corresponding eigenvalue is equal to unity. Also let the eigenvector $z$ be stochastic, that is $z^T e = 1$. Then

$$\lim_{k \to \infty} A^k = e z^T$$

   This matrix has rank one.                                                                            □

## 3.9  Singular Value Decomposition

This section is devoted to the singular value decomposition of a matrix which is not square. The singular value decomposition of a matrix is useful in search techniques. Consider an $m \times n$ matrix $\widehat{A} = [a_{ij}]$. If $m = n$, then the eigenvalues of the matrix $\widehat{A}$ are its spectral representation. However, if $m \neq n$ the singular value decomposition of a rectangular matrix is a means for its spectral representation.

**Definition 3.17.** *Singular value decomposition (SVD): Let $\widehat{A}$ be an $m \times n$ matrix of real elements, and $\widehat{p} = \min(m, n)$. Also let $\widehat{U}$ and $\widehat{V}$ be orthogonal matrices of order $m$ and $n$ respectively. Then $\widehat{A} = \widehat{U}\Sigma\widehat{V}^T$ is a SVD of matrix $\widehat{A}$, where the matrix $\Sigma = [\sigma_{ij}]$ is of size $m \times n$, $\sigma_{ij} = 0$ if $i \neq j$ (nondiagonal elements), and the diagonal elements $\sigma_{ii} \triangleq \sigma_i$ for $1 \leq i \leq \widehat{p}$. The nonzero $\sigma_i$'s are called the singular values of the matrix $\widehat{A}$.* $\qquad\qquad\qquad\qquad\qquad\qquad\qquad\square$

**Observations 3.20.** Let $\widehat{A}$ be an $m \times n$ matrix of real elements, with a SVD equal to $\widehat{U}\Sigma\widehat{V}^T$, and $\widehat{p} = \min(m, n)$. Also, let the rank of the matrix $\widehat{A}$ be $r_{\widehat{A}}$.

1. A SVD of the matrix $\widehat{A}$ exists, where

$$\Sigma = \begin{bmatrix} \widehat{S} & 0 \\ 0 & 0 \end{bmatrix}$$

In the matrix $\Sigma$, the submatrix $\widehat{S}$ is a diagonal matrix with all positive diagonal entries. The size of this matrix is $r_{\widehat{A}}$, where $r_{\widehat{A}} \leq \widehat{p}$. Also $\sigma_i \in \mathbb{R}^+$ for $1 \leq i \leq r_{\widehat{A}}$, and $\sigma_i = 0$ for $(r_{\widehat{A}} + 1) \leq i \leq \widehat{p}$. In the matrix $\widehat{S}$, the null submatrices are included to obtain the $m \times n$ matrix $\Sigma$. For a given matrix $\widehat{A}$, the matrix $\widehat{S}$ is unique except for the order of its diagonal elements.

2. *A Fourier interpretation of SVD*: Let $\widehat{U} = \begin{bmatrix} u_1 & u_2 & \cdots & u_m \end{bmatrix}$, where the $u_i$'s are orthonormal column vectors of length $m$; and $\widehat{V} = \begin{bmatrix} v_1 & v_2 & \cdots & v_n \end{bmatrix}$, where the $v_i$'s are orthonormal column vectors of length $n$. Define $\widehat{Z}_i = u_i v_i^T$, for $1 \leq i \leq r_{\widehat{A}}$. Note that $\widehat{Z}_i$ is an $m \times n$ real matrix. Then:

   (a) The set of matrices $\left\{ \widehat{Z}_i \mid 1 \leq i \leq r_{\widehat{A}} \right\}$ constitute an orthonormal set because

$$tr\left(\widehat{Z}_i^T \widehat{Z}_j\right) = \begin{cases} 0, & i \neq j \\ 1, & i = j \end{cases} \quad \text{for } 1 \leq i \leq r_{\widehat{A}}.$$

   (b) $\widehat{A} = \sum_{i=1}^{r_{\widehat{A}}} \sigma_i \widehat{Z}_i$, where $\sigma_i = tr\left(\widehat{Z}_i^T \widehat{A}\right)$ for $1 \leq i \leq r_{\widehat{A}}$.

   Thus the SVD can be regarded as an abstract Fourier expansion of the matrix $\widehat{A}$; where $\sigma_i$ can be considered as the proportion of $\widehat{A}$ lying in the "direction" of $\widehat{Z}_i$.

3. A consequence of the Fourier interpretation is

$$\sum_{1 \leq i \leq m} \sum_{1 \leq j \leq n} a_{ij}^2 = \sum_{i=1}^{r_{\widehat{A}}} \sigma_i^2$$

4. $\widehat{A}\widehat{A}^T = \widehat{U}\left(\Sigma\Sigma^T\right)\widehat{U}^T$, then the $\sigma_i^2$'s are the eigenvalues of the matrix $\widehat{A}\widehat{A}^T$. These are the diagonal elements of the matrix $\Sigma\Sigma^T$ and the associated eigenvectors are given by the columns of the matrix $\widehat{U}$.
   Similarly, $\widehat{A}^T\widehat{A} = \widehat{V}\left(\Sigma^T\Sigma\right)\widehat{V}^T$, then the $\sigma_i^2$'s are also the eigenvalues of the matrix $\widehat{A}^T\widehat{A}$. These are the diagonal elements of the matrix $\Sigma^T\Sigma$ and the associated eigenvectors are given by the columns of the matrix $\widehat{V}$.

5. Norm properties:

(a) The Frobenius norm is

$$\left\| \widehat{A} \right\|_F^2 = \sum_{i=1}^{r_{\widehat{A}}} \sigma_i^2$$

(b) The trace norm of the matrix is

$$\left\| \widehat{A} \right\|_{tr} = tr\left( \sqrt{\widehat{A}\widehat{A}^T} \right) = \sum_{i=1}^{r_{\widehat{A}}} \sigma_i$$

□

See the problem section for proofs of some of these observations. Consider an example below.

**Example 3.5.** The SVD of the matrix $\widehat{A}$ is determined. Let

$$\widehat{A} = \frac{1}{2}\begin{bmatrix} 1 & \sqrt{2} & 1 \\ 1 & \sqrt{2} & 1 \end{bmatrix}$$

The SVD is of the form $\widehat{A} = \widehat{U}\Sigma\widehat{V}^T$, where $\widehat{U}$ is an orthogonal matrix of order 2, and $\widehat{V}$ is an orthogonal matrix of order 3. That is, $\widehat{U}\widehat{U}^T = I_2$, and $\widehat{V}\widehat{V}^T = I_3$.

Let $\widehat{U} = \begin{bmatrix} u_1 & u_2 \end{bmatrix}$, where $u_1$ and $u_2$ are column vectors of size 2 each. Also let $\widehat{V} = \begin{bmatrix} v_1 & v_2 & v_3 \end{bmatrix}$, where $v_1, v_2,$ and $v_3$ are column vectors of size 3 each. First compute $\widehat{A}\widehat{A}^T$.

$$\widehat{A}\widehat{A}^T = \begin{bmatrix} 1 & 1 \\ 1 & 1 \end{bmatrix}$$

The eigenvalues of $\widehat{A}\widehat{A}^T$ are 2 and 0. That is, $\sigma_1 = \sqrt{2}$. The respective eigenvalue and eigenvector are:

$$\text{eigenvalue } 2, \quad \text{corresponding eigenvector} = u_1 = \begin{bmatrix} 1/\sqrt{2} & 1/\sqrt{2} \end{bmatrix}^T$$
$$\text{eigenvalue } 0, \quad \text{corresponding eigenvector} = u_2 = \begin{bmatrix} 1/\sqrt{2} & -1/\sqrt{2} \end{bmatrix}^T$$

Next compute $\widehat{A}^T\widehat{A}$.

$$\widehat{A}^T\widehat{A} = \begin{bmatrix} 1/2 & 1/\sqrt{2} & 1/2 \\ 1/\sqrt{2} & 1 & 1/\sqrt{2} \\ 1/2 & 1/\sqrt{2} & 1/2 \end{bmatrix}$$

The eigenvalues of $\widehat{A}^T\widehat{A}$ are $2, 0,$ and $0$. That is, $\sigma_1 = \sqrt{2}$. The respective eigenvalue and eigenvector are:

$$\text{eigenvalue } 2, \quad \text{eigenvector} = v_1 = \begin{bmatrix} 1/2 & 1/\sqrt{2} & 1/2 \end{bmatrix}^T$$
$$\text{eigenvalue } 0, \quad \text{eigenvector} = v_2 = \begin{bmatrix} 1/\sqrt{2} & 0 & -1/\sqrt{2} \end{bmatrix}^T$$
$$\text{eigenvalue } 0, \quad \text{eigenvector} = v_3 = \begin{bmatrix} 1/2 & -1/\sqrt{2} & 1/2 \end{bmatrix}^T$$

Summarizing, the SVD of the matrix $\widehat{A}$ is $\widehat{U}\Sigma\widehat{V}^T$ where

$$\widehat{U} = \frac{1}{\sqrt{2}}\begin{bmatrix} 1 & 1 \\ 1 & -1 \end{bmatrix}$$

$$\Sigma = \begin{bmatrix} \sqrt{2} & 0 & 0 \\ 0 & 0 & 0 \end{bmatrix}$$

$$\widehat{V} = \begin{bmatrix} 1/2 & 1/\sqrt{2} & 1/2 \\ 1/\sqrt{2} & 0 & -1/\sqrt{2} \\ 1/2 & -1/\sqrt{2} & 1/2 \end{bmatrix}$$

It can be checked that the matrix $\widehat{A}$ is indeed equal to $\widehat{U}\Sigma\widehat{V}^T$.                               $\square$

## 3.10  Matrix Calculus

Matrix calculus is essentially a scheme to describe certain special notation used in multivariable calculus. Methods of matrix calculus are sometimes convenient to use in certain optimization problems. Vector and matrix derivatives are briefly described in this section. Further, all functions in this section are assumed to be differentiable.

The reader should note that there are several possible definitions of these types of derivative-specifications. Alternate definitions can be found in the literature.

### Vector Derivatives

**Definition 3.18.** *Derivative of a vector with respect to another vector. Let $x \in \mathbb{R}^n$, and $y \in \mathbb{R}^m$; where $n \in \mathbb{P}$, and $m \in \mathbb{P}$. That is*

$$x = \begin{bmatrix} x_1 \ x_2 \ \cdots \ x_n \end{bmatrix}^T, \quad and \quad y = \begin{bmatrix} y_1 \ y_2 \ \cdots \ y_m \end{bmatrix}^T \tag{3.21a}$$

*Observe that the vectors $x$ and $y$ are represented as column matrices. Derivative of vector $y$ with respect to vector $x$ is an $m \times n$ matrix*

$$\frac{\partial y}{\partial x} = [c_{ij}], \quad and \quad c_{ij} = \frac{\partial y_i}{\partial x_j}, \qquad 1 \le i \le m, \ and \ 1 \le j \le n \tag{3.21b}$$

*where $y_i$ is a function of all the elements of vector $x$, for $1 \le i \le m$.*                               $\square$

Noting that a scalar is only a degenerate form of a vector, we have the following observations.

**Observations 3.21.** Let $x \in \mathbb{R}^n$, and $z \in \mathbb{R}$.

1. Let the vector $x$ be a function of the scalar $z$. Derivative of the vector $x$ with respect to a scalar $z$ is the column vector

$$\frac{\partial x}{\partial z} = \begin{bmatrix} \dfrac{\partial x_1}{\partial z} \ \dfrac{\partial x_2}{\partial z} \ \cdots \ \dfrac{\partial x_n}{\partial z} \end{bmatrix}^T$$

2. Let the scalar $z$ be a function of all the elements of the vector $x$. Derivative of the scalar $z$ with respect to a vector $x$ is the row vector

$$\frac{\partial z}{\partial x} = \begin{bmatrix} \dfrac{\partial z}{\partial x_1} \ \dfrac{\partial z}{\partial x_2} \ \cdots \ \dfrac{\partial z}{\partial x_n} \end{bmatrix}$$

This is actually the gradient of $z$. That is

$$\frac{\partial z}{\partial x} = \nabla z \, (x)$$

$\square$

**Observations 3.22.** Let $x \in \mathbb{R}^n$, and $y \in \mathbb{R}^m$. Thus, $x$ and $y$ are real-valued vectors of size $n \in \mathbb{P}$, and $m \in \mathbb{P}$ respectively. These are represented as column matrices. Also $W = [w_{ij}]$ is an $m \times n$ matrix, where $w_{ij} \in \mathbb{R}$, for $1 \leq i \leq m$, and $1 \leq j \leq n$.

1. The matrix $W$ is independent of the vector $x$. Let

$$y = Wx$$

then

$$\frac{\partial y}{\partial x} = W$$

2. The matrix $W$ is independent of the vector $x$. Also, the vector $x$ is a function of the vector $z$, and $W$ is independent of $z$. If

$$y = Wx$$

then

$$\frac{\partial y}{\partial z} = W \frac{\partial x}{\partial z}$$

3. The matrix $W$ is independent of vectors $x$ and $y$. Further, vectors $x$ and $y$ are independent of each other. Let

$$\beta = y^T W x$$

then

$$\frac{\partial \beta}{\partial x} = y^T W, \quad \text{and} \quad \frac{\partial \beta}{\partial y} = x^T W^T$$

4. The matrix $W$ is independent of the vector $x$. Also let $m = n$. That is, $W$ is a square matrix. Further

$$\beta = x^T W x$$

Then

$$\frac{\partial \beta}{\partial x} = x^T \left( W + W^T \right)$$

If the matrix $W$ is symmetric, then

$$\frac{\partial \beta}{\partial x} = 2x^T W$$

5. Let $m = n$. That is, both vectors $x$ and $y$ have $n$ elements each. Further, both vectors $x$ and $y$ are functions of vector $z$. Also if

$$\beta = y^T x$$

then

$$\frac{\partial \beta}{\partial z} = x^T \frac{\partial y}{\partial z} + y^T \frac{\partial x}{\partial z}$$

If $x = y$, then

$$\frac{\partial \beta}{\partial z} = 2x^T \frac{\partial x}{\partial z}$$

6. The Jacobian of a multivariable transformation can also be specified in the notation of matrix differentiation. Let $x$ and $y$ be vectors of the same size. The Jacobian matrix is $\partial y/\partial x$, and the Jacobian determinant of the square matrix $\partial y/\partial x$ is

$$J = \left| \frac{\partial y}{\partial x} \right|$$

In the above expression, $J$ is called the Jacobian determinant of the transformation specified by $y = y(x)$. □

### Matrix Derivatives

Elementary cases of derivatives with respect to matrices are next defined. These are derivative of a matrix with respect to a scalar, and derivative of a scalar with respect to a matrix.

**Definitions 3.19.** Let $W = [w_{ij}]$ be an $m \times n$ matrix, where $w_{ij} \in \mathbb{R}$, for $1 \leq i \leq m$, and $1 \leq j \leq n$. Also, $z \in \mathbb{R}$.

1. *Derivative of the matrix $W$ with respect to the scalar $z$ is the matrix*

$$\frac{\partial W}{\partial z} = \left[ \frac{\partial w_{ij}}{\partial z} \right] \tag{3.22}$$

   *where $w_{ij}$ is a function of the scalar $z$, for $1 \leq i \leq m$, and $1 \leq j \leq n$.*
2. *Derivative of the scalar $z$ with respect to the matrix $W$ is the matrix*

$$\frac{\partial z}{\partial W} = \left[ \frac{\partial z}{\partial w_{ij}} \right] \tag{3.23}$$

   *where, $z$ is a function of all the elements of matrix $W$, and $n \geq 2$.* □

Note that the definition of derivative of matrix $W$ with respect to a scalar $z$ is compatible with the representation of derivative of a vector (column matrix) with respect to a scalar $z$, if $W$ is a column matrix (which is a vector).

If in the expression $\partial z/\partial W$ we allow $n = 1$, then this will be a derivative of a scalar, with respect to a column matrix. As per this definition, $\partial z/\partial W$ will be a column matrix. However, as per the definition of the derivative of a scalar, with respect to a vector (column matrix), then this ought to be a row vector. Therefore we allow $\partial z/\partial W$ to be properly defined only for $n \geq 2$.

Based upon the above definitions, some useful results are summarized.

**Observations 3.23.** Some useful matrix differentiation formulae involving trace of a matrix are considered. Matrix $A$ is independent of matrix $X$.

1. Let the size of matrices $X$ and $A$ be $n \times m$ and $m \times n$ respectively. Then

$$\frac{\partial\, tr\,(XA)}{\partial X} = A^T$$

2. Let the size of matrices $X$ and $A$ be $n \times m$ and $m \times m$ respectively. Then

$$\frac{\partial\, tr\,\left(XAX^T\right)}{\partial X} = X\left(A + A^T\right)$$

If the matrix $A$ is symmetric, then

$$\frac{\partial \, tr\left(XAX^T\right)}{\partial X} = 2XA$$

$\square$

## 3.11 Random Matrices

Random matrices are useful in studying large networks, of which the Internet is a prime example. There is a rich literature on random matrices in diverse subjects such as physics and number theory. A celebrated result in the theory of random matrices, is the semicircle law originally due to Eugene P. Wigner. This law asserts that the spectral density (probability distribution of eigenvalues) of a very large and appropriately normalized random matrix, whose elements are independent, and identically distributed Gaussian random variables, converges to a semicircular probability density function.

Assume that the random matrices are real-valued, symmetric, and the matrix elements are independently distributed zero-mean Gaussian random variables. The spectral density of such Gaussian ensemble of matrices of very large size exhibits the semicircular probability density function. Such class of matrices are termed Gaussian ensemble. A Gaussian ensemble of such matrices whose probability density function is invariant under orthogonal transformation is called a Gaussian orthogonal ensemble (GOE). Gaussian orthogonal ensembles and Wigner's semicircle law are discussed in this section.

### 3.11.1  Gaussian Orthogonal Ensemble

The *Gaussian orthogonal ensemble* is a set of random matrices whose elements have a Gaussian distribution. Furthermore, the joint probability density function of the matrix elements is invariant under orthogonal transformation.

Consider an ensemble of real-valued symmetric random matrices in which the matrix elements are independently distributed, and have a Gaussian distribution with zero mean. If $A = [A_{ij}]$ is such a random matrix, and $O$ is an orthogonal and deterministic matrix such that $OO^T = I$, and $A' = O^T AO$, then the ensemble of matrices $A$ is a GOE if the joint probability density function of the elements of matrix $A$ is equal to the joint probability density function of the elements of matrix $A'$. This is possible only if the joint probability density function of the elements of matrix $A$ is proportional to

$$\exp\left\{-\theta_0 tr\left(a\right) - \theta_1 tr\left(a^2\right)\right\}$$

where $a$ is an instance of $A$, and $\theta_0$ and $\theta_1$ are constants. We are only interested in a specific case of GOE of random matrices. It is defined below. In this definition, $\mathcal{N}\left(\mu, \sigma^2\right)$ is used to a denote a normally (Gaussian) distributed random variable mean $\mu$ and variance $\sigma^2$.

**Definition 3.20.** *The Gaussian orthogonal ensemble is a set of* $n \times n$ *symmetric random matrices* $A = [A_{ij}]$, *such that*:

(a) *Symmetric condition*: $A_{ij} = A_{ji}, 1 \leq j < i \leq n$.

(b) *Independence*: The matrix elements $A_{ij}, 1 \leq j \leq i, 1 \leq i \leq n$ are independent of each other.

(c) *Distribution*: $A_{ii} \sim \mathcal{N}(0,1), 1 \leq i \leq n,$ and $A_{ij} \sim \mathcal{N}\left(0, \frac{1}{2}\right), 1 \leq j < i \leq n.$ $\qquad\square$

**Observations 3.24.** The following observations about the matrix $A$ are immediate from the above definition.

1. The joint probability density function of the matrix elements is given by

$$\frac{1}{2^{n/2}\pi^{n(n+1)/4}} \exp\left\{-\frac{1}{2}\sum_{1 \leq i,j \leq n} a_{ij}^2\right\}, \quad a_{ij} \in \mathbb{R}, \; 1 \leq i,j \leq n$$

   where $a_{ij}$ is an instance of the random variable $A_{ij}$.

   Let $a$ be an instance of the random matrix $A$. Observe from the above expression, that the joint probability density function of the elements of the matrix $A$ depends upon $tr\left(a^2\right)$. This is invariant under an orthogonal transformation of $a$. Therefore the joint distribution is invariant under a deterministic orthogonal transformation of $A$. Consequently, the ensemble is a GOE of random matrices.

2. Since the matrix $A$ is real and symmetric, its eigenvalues $\lambda_i, 1 \leq i \leq n$ are real. Furthermore, the matrix $A$ can be diagonalized to matrix $\Lambda$ via an orthogonal matrix $R$, such that $R^T A R = \Lambda$, and $RR^T = I$. The diagonal entries of the matrix $\Lambda$ are the random eigenvalues $\lambda_i, 1 \leq i \leq n$. Also, the elements of the matrix $R$ are random variables.

3. We have

$$\sum_{1 \leq i,j \leq n} A_{ij}^2 = tr\left(A^2\right) = \sum_{1 \leq i \leq n} \lambda_i^2$$

$\qquad\qquad\qquad\qquad\qquad\qquad\qquad\qquad\qquad\qquad\qquad\qquad\qquad\qquad\qquad\square$

The joint probability density function of the eigenvalues is next determined. It is based upon the above set of observations. Therefore the Jacobian $J$ of the map $A \to (R, \Lambda)$ has to be determined. Note that the matrix $A$ has $n(n+1)/2$ independent random variables, the diagonal matrix $\Lambda$ has $n$ random variables, and the orthogonal transformation $R$ has $n(n-1)/2$ random variables. It turns out that the magnitude of the Jacobian $J$ is equal to the magnitude of the determinant of a Vandermonde matrix. See the problem section for its derivation. Thus

$$|J| = \prod_{i>j} |\lambda_i - \lambda_j|$$

Therefore, the joint probability density function of the eigenvalues $\lambda_i, 1 \leq i \leq n$ is:

$$f_{\lambda_1,\lambda_2,\ldots,\lambda_n}(\lambda_1, \lambda_2, \ldots, \lambda_n) = N_n \exp\left\{-\frac{1}{2}\sum_{1 \leq i \leq n}\lambda_i^2\right\}\prod_{i>j}|\lambda_i - \lambda_j|$$

In the above expression for the joint probability density function of the random variables $\lambda_1, \lambda_2, \ldots, \lambda_n$, the same notation is used for their instances. This is simply to facilitate the notation. Finally, the constant $N_n$ is determined from the relationship

$$\int_{-\infty}^{\infty}\int_{-\infty}^{\infty}\cdots\int_{-\infty}^{\infty} f_{\lambda_1,\lambda_2,\ldots,\lambda_n}(\lambda_1, \lambda_2, \ldots, \lambda_n)\, d\lambda_1 d\lambda_2 \ldots d\lambda_n = 1$$

Spectral density of GOE of random matrices exhibit a semicircular probability density function. This is studied in the next subsection.

### 3.11.2  Wigner's Semicircle Law

Let $A = [A_{ij}]$ be a real-valued random symmetric matrix of size $n$. Assume that each matrix element $A_{ij}$ has zero mean, unit variance, and finite higher moments. Also assume that the random variables $A_{ij}$, for $1 \le i \le j \le n$ are stochastically independent of each other. It is well-known that the eigenvalues $\lambda_1, \lambda_2, \ldots, \lambda_n$ of the matrix $A$ are real numbers. Define the cumulative distribution function $F(x) = n^{-1} \widetilde{N}(x)$, where $\widetilde{N}(x)$ is the number of $\lambda_j$'s less than or equal to $x$. We evaluate the behavior of $F(x)$ as $n \to \infty$. Wigner showed that the spectral density function $f(x) \triangleq dF(x)/dx$ follows a semicircle law as $n \to \infty$. A derivation of Wigner's semicircle law for matrix elements with Gaussian elements is given in this subsection.

Instead of directly studying eigenvalues of the matrix $A$, we study the eigenvalues of the normalized matrix $B = A/(2\sqrt{n})$. A motivation for this normalization is provided as follows. Observe that

$$\sum_{i=1}^{n} \lambda_i^2 = tr\left(AA^T\right)$$

$$= \sum_{i=1}^{n} \sum_{j=1}^{n} A_{ij}^2 \sim n^2 \cdot 1$$

The last approximation follows because $A_{ij}^2$ is approximately equal to unity and there are $n^2$ such terms. Therefore approximate average value of a squared eigenvalue is equal to $n$. With another big leap of faith, we assume that the absolute value of the average eigenvalue is approximately equal to $\sqrt{n}$. Therefore, we study a matrix, whose eigenvalues are $\lambda_i/(c\sqrt{n}), 1 \le i \le n$. The value of $c$ is selected to be 2 to get mathematically elegant results.

**Definitions 3.21.** *Consider an $n \times n$ symmetric matrix $A = [A_{ij}]$ whose entries $A_{ij}, 1 \le i \le j \le n$ are independent, and identically distributed random variables. The distribution of each of these elements is Gaussian, with zero mean and unit variance. Let the random eigenvalues of this matrix be $\lambda_j, 1 \le j \le n$. Let a matrix $B = A/(2\sqrt{n})$.*

1. *The spectral density of the matrix $B$ is*

$$\xi_n(x) = \mathcal{E}\left\{ \frac{1}{n} \sum_{j=1}^{n} \delta\left(x - \frac{\lambda_j}{2\sqrt{n}}\right) \right\}, \quad x \in \mathbb{R} \tag{3.24a}$$

*where $\mathcal{E}(\cdot)$ is the expectation operator, and $\delta(\cdot)$ is the delta function.*
2. *The $m$th moment of the eigenvalues of the matrix $B$ is $\varphi_n(m)$.*

$$\varphi_n(m) = \mathcal{E}\left\{ \frac{1}{n} \sum_{j=1}^{n} \left(\frac{\lambda_j}{2\sqrt{n}}\right)^m \right\}, \quad \forall\, m \in \mathbb{P} \tag{3.24b}$$

*Note that $\varphi_n(0)$ can be defined to be equal to 1.*                                                  □

It is known that $\sum_{j=1}^{n} \lambda_j^m = tr(A^m), \forall\, m \in \mathbb{N}$. Thus

$$\varphi_n(m) = \mathcal{E}\left\{ \frac{1}{2^m n^{1+m/2}} tr(A^m) \right\}, \quad \forall\, m \in \mathbb{N}$$

The following variables are useful in the asymptotic analysis of the spectral density. Let

$$\lim_{n \to \infty} \xi_n (x) \triangleq f(x), \quad x \in \mathbb{R} \tag{3.25a}$$

$$\lim_{n \to \infty} \varphi_n (m) \triangleq \beta_m, \quad \forall\, m \in \mathbb{N} \tag{3.25b}$$

It is established in the remaining part of this subsection that $f(\cdot)$ is the semicircle probability density function, where

$$f(x) = \begin{cases} \dfrac{2}{\pi} \sqrt{1 - x^2}, & |x| \le 1 \\[2mm] 0, & |x| \ge 1 \end{cases} \tag{3.26}$$

A random variable specified by this density function has zero mean, and variance equal to $1/4$. We next compute the moments.

**Moment Computation**

It can be inferred that $\beta_0 = 1$. Observe, that the average value of the traces of the matrices $A^{2k+1}$ is equal to zero. Thus $\varphi_n (2k + 1) = \beta_{2k+1} = 0$, for all $k \in \mathbb{N}$. In the next step $\varphi_n (2)$ is determined.

$$\varphi_n (2) = \mathcal{E} \left\{ \frac{1}{2^2 n^2} tr\left(A^2\right) \right\} = \frac{1}{4n^2} \mathcal{E} \left\{ \sum_{1 \le i, j \le n} A_{ij}^2 \right\} = \frac{n^2}{4n^2} = \frac{1}{4}$$

Thus $\beta_2 = 1/4$. Computation of higher even-numbered moments is delicate. This is facilitated by the following observation. Let $\widetilde{X}$ be a centered (zero mean) Gaussian random variable. Also let $h(x)$ be a real-valued nonrandom function, such that it is differentiable. Denote the first derivative of $h(x)$ with respect to $x$ by $h'(x)$. Then it can be shown that

$$\mathcal{E}\left( \widetilde{X} h\left( \widetilde{X} \right) \right) = \mathcal{E}\left( \widetilde{X}^2 \right) \mathcal{E}\left( h'\left( \widetilde{X} \right) \right)$$

A generalization of this observation is the following lemma.

**Lemma 3.5**. *The random variables $\widetilde{X}_1, \widetilde{X}_2, \ldots, \widetilde{X}_n$ each have zero mean, and have a joint Gaussian distribution. Let $h(\cdot, \cdot, \ldots, \cdot)$ be a real-valued nonrandom function defined on $\mathbb{R}^n$, such that $\partial h(\cdot, \cdot, \ldots, \cdot)/\partial x_i$ exists for $1 \le i \le n$. Then*

$$\mathcal{E}\left( \widetilde{X}_j h\left( \widetilde{X}_1, \widetilde{X}_2, \ldots, \widetilde{X}_n \right) \right) = \sum_{i=1}^{n} \mathcal{E}\left( \widetilde{X}_j \widetilde{X}_i \right) \mathcal{E}\left( \frac{\partial h\left( \widetilde{X}_1, \widetilde{X}_2, \ldots, \widetilde{X}_n \right)}{\partial \widetilde{X}_i} \right) \tag{3.27}$$

*for $1 \le j \le n$. It is assumed in the above result that all integrals in it exist.*
    *Proof.* The result can be established by integration by parts. Details are left to the reader.    $\square$

For convenience in notation define $M_m = \mathcal{E}\left\{ tr\left(A^{2m}\right) \right\}, \forall\, m \in \mathbb{N}$. Therefore

$$\varphi_n (2m) = \frac{M_m}{2^{2m} n^{m+1}}, \quad \forall\, m \in \mathbb{N}$$

Note that $M_0 = n$, which is equal to the size of the symmetric matrix $A$. Using the above lemma, it has been shown in the problem section that as $n \to \infty$

$$M_m = \sum_{r=0}^{m-1} M_r M_{m-r-1}, \ \forall \, m \in \mathbb{P}$$

Thus

$$\beta_0 = 1$$

$$\beta_{2m} = \frac{1}{4} \sum_{r=0}^{m-1} \beta_{2r} \beta_{2(m-r-1)}, \ \forall \, m \in \mathbb{P}$$

For convenience in notation define $\eta_m = \beta_{2m}$, for each $m \in \mathbb{N}$. Finally

$$\eta_0 = 1$$

$$\eta_m = \frac{1}{4} \sum_{r=0}^{m-1} \eta_r \eta_{m-r-1}, \ \forall \, m \in \mathbb{P}$$

Using these recursive equations, the expression for the spectral density function of the random matrix $B$ as $n \to \infty$, is derived from first principles.

### Spectral Density Function

Based upon the above computation of moments, an informal derivation of Wigner's semicircle law is next presented. Recall that

$$\delta(x) = \lim_{\epsilon \to 0} \frac{\epsilon}{\pi \left( x^2 + \epsilon^2 \right)}$$

$$= -\lim_{\epsilon \to 0} \mathrm{Im} \left\{ \frac{1}{\pi \left( x + i\epsilon \right)} \right\}$$

Therefore

$$\xi_n(x) = -\mathcal{E} \left\{ \lim_{\epsilon \to 0} \mathrm{Im} \, \frac{1}{\pi n} \sum_{j=1}^{n} \frac{1}{\left( x - \frac{\lambda_j}{2\sqrt{n}} + i\epsilon \right)} \right\}$$

$$= -\mathcal{E} \left\{ \lim_{\epsilon \to 0} \mathrm{Im} \, \frac{2}{\pi \sqrt{n}} \sum_{j=1}^{n} \frac{1}{\left( 2\sqrt{n}x - \lambda_j + i\epsilon 2\sqrt{n} \right)} \right\}$$

Note that

$$\sum_{j=1}^{n} \frac{1}{\left( 2\sqrt{n}x - \lambda_j \right)} = tr \left\{ \frac{1}{\left( 2\sqrt{n}xI - A \right)} \right\}$$

Thus

$$\xi_n(x) = -\mathcal{E} \left[ \mathrm{Im} \, \frac{2}{\pi \sqrt{n}} tr \left\{ \frac{1}{\left( 2\sqrt{n}xI - A \right)} \right\} \right]$$

Observe that the eigenvalues do not appear explicitly in this equation. Next assume that a Taylor-series type of expansion is permissible with matrices. This yields

$$tr\left\{\frac{1}{(2\sqrt{n}xI - A)}\right\} = \sum_{m \in \mathbb{N}} \frac{1}{(2\sqrt{n}x)^{m+1}} tr\left(A^m\right)$$

Therefore the series for $\xi_n(x)$ is convergent, provided $|2\sqrt{n}x|$ is larger than the magnitude of the largest eigenvalue of $A$. Therefore

$$\xi_n(x) = -\frac{1}{\pi} \text{Im}\left\{\sum_{m \in \mathbb{N}} \frac{\varphi_n(m)}{x^{m+1}}\right\}$$

Recall that $\lim_{n \to \infty} \xi_n(x) = f(x)$, $x \in \mathbb{R}$ and $\lim_{n \to \infty} \varphi_n(m) = \beta_m$, $\forall\, m \in \mathbb{N}$. Therefore

$$f(x) = -\frac{1}{\pi} \text{Im}\left\{\sum_{m \in \mathbb{N}} \frac{\beta_m}{x^{m+1}}\right\}$$

Let

$$f(x) = -\frac{1}{\pi} \text{Im}\left\{F(x)\right\}, \quad \text{where} \quad F(x) = \sum_{m \in \mathbb{N}} \frac{\beta_m}{x^{m+1}}$$

Note that $\beta_m$ is equal to zero for odd values of $m$. Thus

$$F(x) = \sum_{m \in \mathbb{N}} \frac{\eta_m}{x^{2m+1}}$$

Substituting the recursive relationship for $\eta_m$'s yields

$$xF(x) = 1 + \sum_{m \in \mathbb{P}} \frac{\eta_m}{x^{2m}}$$

$$= 1 + \frac{1}{4} \sum_{m \in \mathbb{P}} \frac{1}{x^{2m}} \sum_{r=0}^{m-1} \eta_r \eta_{m-1-r}$$

$$= 1 + \frac{1}{4} \sum_{r \in \mathbb{N}} \sum_{m \geq r+1} \frac{1}{x^{2m}} \eta_r \eta_{m-1-r}$$

$$= 1 + \frac{1}{4} \sum_{r \in \mathbb{N}} \frac{1}{x^{2r+1}} \eta_r \sum_{m \in \mathbb{N}} \frac{1}{x^{2m+1}} \eta_m$$

$$= \left(1 + \frac{1}{4}\left\{F(x)\right\}^2\right)$$

Thus a quadratic equation in $F(x)$ is obtained. Solving for $F(x)$ yields

$$F(x) = 2\left\{x \pm \sqrt{x^2 - 1}\right\}$$

It is required that $F(x) \to 0$ as $x \to \infty$. Therefore the negative sign of the square root is selected. That is

$$F(x) = 2\left\{x - \sqrt{x^2 - 1}\right\}$$

Since $f(x) = -\text{Im}\left\{F(x)/\pi\right\}$, we finally obtain

$$f(x) = \begin{cases} \dfrac{2}{\pi}\sqrt{1 - x^2}, & |x| \leq 1 \\[2mm] 0, & |x| \geq 1 \end{cases}$$

The above discussion is summarized in the following theorem.

**Theorem 3.11.** *Consider a real-valued symmetric random matrix $A = [A_{ij}]$ of size $n$. The matrix elements $A_{ij}, 1 \leq i \leq j \leq n$ are stochastically independent Gaussian random variables, where $A_{ij} \sim \mathcal{N}(0,1), 1 \leq i \leq j \leq n$. Define $B = A/(2\sqrt{n})$. As $n \to \infty$ the spectral density function $f(\cdot)$ of the matrix $B$ is given by $f(x) = (2/\pi)\sqrt{1-x^2}$ for $|x| \leq 1$, and $f(x) = 0$ elsewhere.* $\square$

It should also be noted that the semicircle law is also found to hold true for other types matrix ensembles, where matrix elements are chosen from a specific probability density function with zero mean, and finite higher moments. Consider a real-valued symmetric random matrix $A = [A_{ij}]$ of size $n$. The matrix elements $A_{ij}, 1 \leq i \leq j \leq n$ are stochastically independent Gaussian random variables, where $A_{ii} \sim \mathcal{N}(0, 2\sigma^2), 1 \leq i \leq n$, and $A_{ij} \sim \mathcal{N}(0, \sigma^2), 1 \leq j < i \leq n$. This is a GOE of random matrices. Define $B = A/\sqrt{n}$. As $n \to \infty$ the spectral density function $f(\cdot)$ of the matrix $B$ is given by

$$f(x) = \begin{cases} \dfrac{1}{2\pi\sigma^2}\sqrt{4\sigma^2 - x^2}, & |x| \leq 2\sigma \\ 0, & |x| \geq 2\sigma \end{cases} \tag{3.28}$$

See the problem section for a proof of this assertion.

## Reference Notes

Summary of results on matrix theory and determinants is from Ayers (1962), Lancaster (1969), Lipschutz (1968), Noble (1969), and Rosen (2018). The results on matrix spectral analysis essentially follow Horn, and Johnson (1987), and Noble (1969). The discussion on Perron-Frobenius theory and stochastic matrices is from Graham (1987). An encyclopedic and comprehensive reference on linear algebra is the handbook by Hogben (2007).

A lively discussion of spectral analysis of matrices, stochastic matrices, SVD and related topics can be found in Golub, and van Loan (1983), Fiedler (1986), and Meyer (2000). Books by Graybill (1969), and Lütkepohl (1996) are useful references on matrix calculus. The section on random matrices is based upon Mehta (1991), Stöckmann (1999), Boutet de Monvel, and Khorunzhy (1999), Lévêque (2015), and Miller, and Takloo-Bighash (2006).

## Problems

1. Let $A$ be an $m \times n$ matrix. Establish the following statements about matrix norms.

    (a) $\|A\|_1 = \max_{1 \leq j \leq n} \sum_{i=1}^{m} |a_{ij}| = $ the largest absolute column sum.

    (b) $\|A\|_\infty = \max_{1 \leq i \leq m} \sum_{j=1}^{n} |a_{ij}| = $ the largest absolute row sum.

    Hint: See Meyer (2000). Let $x = \begin{bmatrix} x_1 & x_2 & \cdots & x_n \end{bmatrix}^T$ be a complex vector.

(a) Let $\|x\|_1 = 1$. Then

$$\|Ax\|_1 = \sum_{i=1}^{m} \left| \sum_{j=1}^{n} a_{ij} x_j \right| \leq \sum_{i=1}^{m} \sum_{j=1}^{n} |a_{ij}| \, |x_j| = \sum_{j=1}^{n} |x_j| \sum_{i=1}^{m} |a_{ij}|$$

$$\leq \left\{ \sum_{j=1}^{n} |x_j| \right\} \left\{ \max_{1 \leq j \leq n} \sum_{i=1}^{m} |a_{ij}| \right\}$$

$$= \max_{1 \leq j \leq n} \sum_{i=1}^{m} |a_{ij}|$$

Equality is obtained because of the following reason. Let the $k$th column be the column with the largest absolute sum. Select $x = e_k$, be the $k$th unit vector, then $\|Ae_k\|_1 = \sum_{i=1}^{m} |a_{ik}| = \max_{1 \leq j \leq n} \sum_{i=1}^{m} |a_{ij}|$.

(b) Let $\|x\|_\infty = 1$. Then

$$\|Ax\|_\infty = \max_{1 \leq i \leq m} \left| \sum_{j=1}^{n} a_{ij} x_j \right| \leq \max_{1 \leq i \leq m} \sum_{j=1}^{n} |a_{ij}| \, |x_j| \leq \max_{1 \leq i \leq m} \sum_{j=1}^{n} |a_{ij}|$$

Equality can occur as follows. Let the $k$th row be the row with the largest absolute sum. Select the vector $x$ such that $x_j = 1$ if $a_{kj} \geq 0$, otherwise it is equal to $-1$. Denote the rows of the matrix $A$ by $A_i$ for $1 \leq i \leq m$. Then

$$|A_i x| = \sum_{j=1}^{n} |a_{ij} x_j| \leq \sum_{j=1}^{n} |a_{ij}|, \quad 1 \leq i \leq m$$

$$|A_k x| = \sum_{j=1}^{n} |a_{kj}| = \max_{1 \leq i \leq m} \sum_{j=1}^{n} |a_{ij}|$$

Since $\|x\|_\infty = 1$, and

$$\|Ax\|_\infty = \max_{1 \leq i \leq m} |A_i x| = \max_{1 \leq i \leq m} \sum_{j=1}^{n} |a_{ij}|$$

2. Let $A$ be a square matrix. Also let $\lambda_{\max}$ and $\lambda_{\min}$ be the largest and smallest eigenvalues of the matrix product $A^\dagger A$. Prove that

$$\|A\|_2 = \sqrt{\lambda_{\max}} \quad \text{and} \quad \|A^{-1}\|_2 = \frac{1}{\sqrt{\lambda_{\min}}}$$

Hint: See Meyer (2000). Use the method of Lagrange multipliers. It is discussed in the chapter on optimization, stability, and chaos theory. Maximize $\|Ax\|_2^2 = x^\dagger A^\dagger A x$ subject to $x^\dagger x = 1$.

3. Let $A$ and $B$ be $n \times n$ square matrices. Establish the following results.
   (a) $\det A^T = \det A$.
   (b) If the matrix $A$ has either two identical columns (or rows) then $\det A = 0$.
   (c) If $A = [a_{ij}]$ is a lower or upper triangular matrix, then $\det A = \prod_{i=1}^{n} a_{ii}$.
   (d) $\det AB = \det A \det B = \det BA$.

4. Let $A = [a_{ij}]$ be a square matrix in which the sum of elements of each row is equal to zero. Similarly, the sum of elements of each column is equal to zero. An example of this type of matrix is the Laplacian matrix of an undirected and simple graph. Prove that the cofactors of the different elements of the matrix $A$ are identical.

Hint: We use the elegant proof provided by Bailey and Cameron (2009). Let the size of the square matrix $A$ be $n$. Also let $Q$ be an all-1 square matrix of size $n$. It is proved that the cofactor of any element of the matrix $A$ is equal to $n^{-2} \det (A + Q)$. This result is useful in computing the total number of spanning trees of an undirected and simple graph.

The cofactor of the matrix element $a_{ij}$ is determined, via the following row and column operations on the matrix $(A + Q)$.

Initially add all rows of the matrix $(A + Q)$ to the $i$th row. After this operation, each element of row $i$ is equal to $n$. Next add all other columns to the $j$th column. After this operation, the $(i, j)$th entry is equal to $n^2$, the other entries of the $i$th row and the $j$th column are each equal to $n$, and the remaining entries are unchanged.

In the next step, take out a factor $n$ from the $i$th row. Subtract the $i$th row from all other rows. After this operation, the $(i, j)$th entry is equal to $n$, the other entries of the $i$th row are each equal to $1$, the other entries of the $j$th column are each equal to $0$, and all other entries are each equal to that in the original matrix $A$.

Then $\det (A + Q)$ is equal to $n^2$ times $(-1)^{i+j} \det A_{ij}$, where $A_{ij}$ is the submatrix obtained from $A$ by deleting the $i$th row and the $j$th column. Note that, $(-1)^{i+j} \det A_{ij}$ is equal to the cofactor of the matrix element $a_{ij}$. Thus the cofactor of $a_{ij}$ is equal to

$$\frac{1}{n^2} \det (A + Q)$$

which is independent of $i$ and $j$.

5. $A$ is a $2 \times 2$ matrix

$$A = \begin{bmatrix} a & b \\ c & d \end{bmatrix}$$

such that $\det A = (ad - bc) \neq 0$. Prove that

$$A^{-1} = \frac{1}{(ad - bc)} \begin{bmatrix} d & -b \\ -c & a \end{bmatrix}$$

6. Let $A = [a_{ij}]$ be a square matrix of size $n$. Also $S_n$ is the set of all permutations on the set of integers $\{1, 2, \ldots, n\}$. Note that $|S_n| = n!$. Let $\sigma(1), \sigma(2), \ldots, \sigma(n)$ be a permutation of these $n$ integers. Define $sgn(\sigma)$ to be $+1$ if the integers $\sigma(1), \sigma(2), \ldots, \sigma(n)$ can be transformed into $1, 2, \ldots, n$ by an *even* number of interchanges; otherwise $sgn(\sigma) = -1$. In the later case, there are an *odd* number of interchanges. Let the number of such interchanges or transpositions be $t(\sigma)$, then $sgn(\sigma) = (-1)^{t(\sigma)}$. It can be demonstrated that $sgn(\cdot)$ is independent of the sequence in which the interchanges are performed. Using the definition of the determinant, prove that

$$\det A = \sum_{\sigma \in S_n} sgn(\sigma) a_{1,\sigma(1)} a_{2,\sigma(2)} \cdots a_{n,\sigma(n)}$$

7. A function $f(\cdot)$ is *additive* iff $f(a + b) = f(a) + f(b)$, for all $a, b$ in the domain of $f(\cdot)$. A function $f(\cdot)$ is *homogeneous* over the set of complex numbers $\mathbb{C}$, if $f(\lambda x) = \lambda f(x)$, $\forall$

$\lambda \in \mathbb{C}$. Let $c_1, c_2, \ldots, c_n$ be the $n$ columns of a square matrix $C$ of size $n$. Also $a_1$ and $b_1$ are columns of length $n$. Let

$$A = \begin{bmatrix} a_1 \; c_2 \; \cdots \; c_n \end{bmatrix}$$
$$B = \begin{bmatrix} b_1 \; c_2 \; \cdots \; c_n \end{bmatrix}$$

$$C = \begin{bmatrix} c_1 \; c_2 \; \cdots \; c_n \end{bmatrix}, \quad c_1 = (\alpha a_1 + \beta b_1)$$

Prove that

$$\det C = \alpha \det A + \beta \det B$$

Hint: Expand the determinant of the matrix $C$ by its first column.

This result shows that the determinant is an additive and homogeneous function of the first column. It can also be shown similarly that the determinant is an additive and homogeneous function of any column $j$, where $1 \leq j \leq n$. A similar statement is true for the rows of the matrix $A$.

8. Prove Binet-Cauchy theorem.

Hint: See Lancaster (1969). We use results and notation from the last two problems. The matrix product $AB = [\sum_{k=1}^{n} a_{ik} b_{kj}]$, is an $m \times m$ matrix, where

$$AB = \begin{bmatrix} \sum_{\beta_1=1}^{n} a_{1\beta_1} b_{\beta_1 1} & \cdots & \sum_{\beta_m=1}^{n} a_{1\beta_m} b_{\beta_m m} \\ \vdots & \cdots & \vdots \\ \sum_{\beta_1=1}^{n} a_{m\beta_1} b_{\beta_1 1} & \cdots & \sum_{\beta_m=1}^{n} a_{m\beta_m} b_{\beta_m m} \end{bmatrix}$$

In the last problem, it has been shown that a determinant is an additive and homogeneous function of each and every column. Thus $\det (AB)$ is

$$\det (AB) = \sum_{\beta_1=1}^{n} \cdots \sum_{\beta_m=1}^{n} \det \begin{bmatrix} a_{1\beta_1} b_{\beta_1 1} & \cdots & a_{1\beta_m} b_{\beta_m m} \\ \vdots & \cdots & \vdots \\ a_{m\beta_1} b_{\beta_1 1} & \cdots & a_{m\beta_m} b_{\beta_m m} \end{bmatrix}$$

$$= \sum_{\beta_1=1}^{n} \cdots \sum_{\beta_m=1}^{n} A \begin{pmatrix} 1 & 2 & \cdots & m \\ \beta_1 & \beta_2 & \cdots & \beta_m \end{pmatrix} b_{\beta_1 1} b_{\beta_2 2} \ldots b_{\beta_m m}$$

Observe that there are $n^m$ determinants in this multiple summation. However, if there is a repetition of two or more subscripts in the sequence $\beta_1, \beta_2, \ldots, \beta_m$ in the determinant, then the corresponding determinant is equal to zero. In such scenarios, the determinant has two or more identical columns. Therefore there will be actually $n!/(n-m)!$ terms in which the $\beta_j$'s are all different. These are sorted into $N = \binom{n}{m}$ groups, each of $m!$ terms. In each such group, the terms differ only in the sequence in which the subscripts $\beta_1, \beta_2, \ldots, \beta_m$ occur.

If $j_1 < j_2 < \cdots < j_m$ and $\beta_1, \beta_2, \ldots, \beta_m$ is the permutation $\sigma(\cdot)$ of $j_1, j_2, \ldots, j_m$ then

$$A \begin{pmatrix} 1 & 2 & \cdots & m \\ \beta_1 & \beta_2 & \cdots & \beta_m \end{pmatrix} = (-1)^{t(\sigma)} A \begin{pmatrix} 1 & 2 & \cdots & m \\ j_1 & j_2 & \cdots & j_m \end{pmatrix}$$

where $t(\sigma)$ is the number of transpositions of the permutation. Collecting the $m!$ terms in which $\beta_1, \beta_2, \ldots, \beta_m$ is a permutation of $j_1, j_2, \ldots, j_m$ results in

$$A \begin{pmatrix} 1 & 2 & \cdots & m \\ j_1 & j_2 & \cdots & j_m \end{pmatrix} \sum_{\sigma} (-1)^{t(\sigma)} b_{\beta_1 1} b_{\beta_2 2} \cdots b_{\beta_m m}$$

The terms in the above summation are next arranged so that the first subscripts occur in ascending order. This results in

$$A \begin{pmatrix} 1 & 2 & \cdots & m \\ j_1 & j_2 & \cdots & j_m \end{pmatrix} \sum_{\sigma'} (-1)^{t(\sigma')} b_{j_1 k_1} b_{j_2 k_2} \cdots b_{j_m k_m}$$

where $\sigma'(\cdot)$ is a permutation $k_1, k_2, \ldots, k_m$ of $1, 2, \ldots, m$. It is evident that $t(\sigma) = t(\sigma')$. Therefore, the above expression is simply equal to

$$A \begin{pmatrix} 1 & 2 & \cdots & m \\ j_1 & j_2 & \cdots & j_m \end{pmatrix} B \begin{pmatrix} j_1 & j_2 & \cdots & j_m \\ 1 & 2 & \cdots & m \end{pmatrix}$$

The stated result follows.

9. $A$ and $B$ are matrices such that their product is well-defined. Prove that $r_{AB} \leq \min\{r_A, r_B\}$. If $B$ is a square and nonsingular matrix, then $r_{AB} = r_A$. Similarly, if $A$ is a square and nonsingular matrix, then $r_{AB} = r_B$.

   Hint: See Fiedler (1986). The columns of $AB$ are linear combinations of the columns of $A$, which yields $r_{AB} \leq r_A$. Similarly, by using rows it can be inferred that $r_{AB} \leq r_B$. Therefore $r_{AB} \leq \min\{r_A, r_B\}$.

   If $B$ is nonsingular, then $r_A = r_{ABB^{-1}} \leq r_{AB}$, thus $r_A = r_{AB}$. If $A$ is nonsingular, then $r_B = r_{A^{-1}AB} \leq r_{AB}$, thus $r_B = r_{AB}$.

10. Establish Sylvester's laws of nullity.

11. Prove that a nonzero polynomial of degree $n$ has at most $n$ distinct roots.

    Hint: See Arora, and Barak (2009). Assume that the degree $n$ polynomial $p(x) = \sum_{i=0}^{n} c_i x^i$ has $(n+1)$ distinct roots $\alpha_j$, $1 \leq j \leq (n+1)$ in some field $\mathcal{F}$. Then

    $$p(\alpha_j) = \sum_{i=0}^{n} c_i \alpha_j^i = 0, \quad 1 \leq j \leq (n+1)$$

    The above set of equations can be written in matrix form as $cA = 0$, where

    $$A = \begin{bmatrix} 1 & 1 & \cdots & 1 \\ \alpha_1 & \alpha_2 & \cdots & \alpha_n \\ \alpha_1^2 & \alpha_2^2 & \cdots & \alpha_n^2 \\ \vdots & \vdots & \ddots & \vdots \\ \alpha_1^n & \alpha_2^n & \cdots & \alpha_n^n \end{bmatrix}$$

    and $c = \begin{bmatrix} c_0 & c_1 & \cdots & c_n \end{bmatrix}$. Note that $\det A$ is a Vandermonde determinant, where

    $$\det A = \prod_{i>j} (\alpha_i - \alpha_j)$$

    Further, $\det A$ is nonzero because the roots $\alpha_j$'s are all distinct. Thus the rank of the matrix $A$ is $r_A = (n+1)$. This implies that $cA = 0$ is true if $c = 0$. This is a contradiction, because $c_n \neq 0$.

12. The Binet-Cauchy theorem considers the product of two matrices $A$ and $B$, which is $C = AB$. The matrices $A$ and $B$ are $m \times n$ and $n \times m$ respectively. In this theorem, it was assumed that $m \leq n$.

However, if $m > n$, prove that $\det C = 0$.

Hint: See Lancaster (1969). Note that the matrix $C$ is $m \times m$. Let the ranks of the matrices $A$ and $C$ be $r_A$ and $r_C$ respectively. Each and every column of the matrix $C$ is a linear combination of the columns of the matrix $A$. Therefore $r_C \leq r_A$. However $r_A \leq \min(m,n)$. Thus $r_C \leq n$. However, the matrix $C$ is $m \times m$ and $m > n$. Therefore $\det C = 0$.

13. The *matrix inversion lemma* is established in this problem. Let $A$ be an $n \times n$ invertible matrix, $B$ is an $n \times m$ matrix, $C$ is an $m \times m$ invertible matrix, and $D$ is an $m \times n$ matrix. Then

$$(A + BCD)^{-1} = A^{-1} - A^{-1}B \left( C^{-1} + DA^{-1}B \right)^{-1} DA^{-1}$$

There is a practical reason for the utility of this lemma. For example, if $m$ is much less than $n$, then it might be relatively easier to invert $\left( C^{-1} + DA^{-1}B \right)$, than invert $(A + BCD)$ directly. Moreover, if $A$ and $C$ are also easy to invert, then the right hand side of the above expression can also be easier to compute, than a direct inversion of $(A + BCD)$.

Hint: See Stengel (1986); and Kailath, Sayed, and Hassibi, (2000). This problem is addressed in several steps.

*Step* 1: Consider a partitioned, square, and invertible matrix $E$ of size $(m + n)$ It is

$$E = \begin{bmatrix} P & Q \\ R & S \end{bmatrix}$$

where $P$ is an $m \times m$ invertible matrix, $Q$ is an $m \times n$ matrix, $R$ is an $n \times m$ matrix, and $S$ is an $n \times n$ invertible matrix. Let the inverse of matrix $E$ be $F$, where

$$F = \begin{bmatrix} T & U \\ V & W \end{bmatrix}$$

where $T$ is an $m \times m$ matrix, $U$ is an $m \times n$ matrix, $V$ is an $n \times m$ matrix, and $W$ is an $n \times n$ matrix.

*Step* 2: Let $I_m$, $I_n$, and $I_{m+n}$ be identity matrices of size $m$, $n$, and $(m + n)$ respectively. We have

$$EF = I_{m+n}$$

That is

$$\begin{bmatrix} P & Q \\ R & S \end{bmatrix} \begin{bmatrix} T & U \\ V & W \end{bmatrix} = \begin{bmatrix} I_m & 0 \\ 0 & I_n \end{bmatrix}$$

Thus

$$\begin{bmatrix} (PT + QV) & (PU + QW) \\ (RT + SV) & (RU + SW) \end{bmatrix} = \begin{bmatrix} I_m & 0 \\ 0 & I_n \end{bmatrix}$$

Comparing both sides of the above equation leads to

$$(RT + SV) = 0 \Rightarrow V = -S^{-1}RT$$
$$(PU + QW) = 0 \Rightarrow U = -P^{-1}QW$$

$$(PT + QV) = I_m \Rightarrow \left( P - QS^{-1}R \right) T = I_m; \quad T = \left( P - QS^{-1}R \right)^{-1}$$
$$(RU + SW) = I_n \Rightarrow \left( -RP^{-1}Q + S \right) W = I_n; \quad W = \left( -RP^{-1}Q + S \right)^{-1}$$

Therefore

$$V = -S^{-1}R\left(P - QS^{-1}R\right)^{-1}$$
$$U = -P^{-1}Q\left(-RP^{-1}Q + S\right)^{-1}$$

Thus

$$F = \begin{bmatrix} T & U \\ V & W \end{bmatrix}$$

$$= \begin{bmatrix} \left(P - QS^{-1}R\right)^{-1} & -P^{-1}Q\left(-RP^{-1}Q + S\right)^{-1} \\ -S^{-1}R\left(P - QS^{-1}R\right)^{-1} & \left(-RP^{-1}Q + S\right)^{-1} \end{bmatrix}$$

*Step* 3: We also have

$$FE = I_{m+n}$$

That is

$$\begin{bmatrix} T & U \\ V & W \end{bmatrix} \begin{bmatrix} P & Q \\ R & S \end{bmatrix} = \begin{bmatrix} I_m & 0 \\ 0 & I_n \end{bmatrix}$$

Thus

$$\begin{bmatrix} (TP + UR) & (TQ + US) \\ (VP + WR) & (VQ + WS) \end{bmatrix} = \begin{bmatrix} I_m & 0 \\ 0 & I_n \end{bmatrix}$$

*Step* 4: From Step 3, we have

$$(VQ + WS) = I_n \Rightarrow W = (I_n - VQ)\,S^{-1}$$

In the above result, substitute expressions for $V$ and $W$ obtained in Step 2. This leads to

$$\left(-RP^{-1}Q + S\right)^{-1} = S^{-1} + S^{-1}R\left(P - QS^{-1}R\right)^{-1}QS^{-1}$$

*Step* 5: Substitute

$$S = A, \quad R = -B, \quad P^{-1} = C, \quad \text{and} \quad Q = D$$

in the last equation of Step 4. It leads to the stated result. There are several other derivations of this result in the literature. This result is useful in the derivation of the discrete Kalman filter algorithm.

14. Prove that, if $B$ is a positive definite Hermitian matrix, then there exists a unique positive definite Hermitian matrix $A$ such that $B = A^2$. However, if $B$ is semidefinite, then $A$ is semidefinite. (The matrix $A$ is the square root of the matrix $B$.)
    Hint: See Noble (1969). Let $B = P^\dagger \Lambda P$, and $A = P^\dagger \sqrt{\Lambda} P$ where $PP^\dagger = I$. To establish uniqueness assume that the matrix $A$ is Hermitian and $B = A^2$. This implies that $A = S^\dagger \Omega S$ and $B = S^\dagger \Omega^2 S$ where $\Omega$ is a diagonal matrix, and the diagonal elements of $\Omega^2$ must be eigenvalues of matrix $B$ in some permuted order.

15. Let $B$ be a positive definite matrix. Prove that $B$ is nonsingular
    Hint: See Watkins (2002). This result is proved by contradiction. Assume that $B$ is a singular matrix. Then there is a nonzero $x$ such that $Bx = 0$. This implies $x^T Bx = 0$, where $x \neq 0$. That is, $B$ is not positive definite. This is a contradiction. Therefore $B$ is nonsingular.

16. Prove that the diagonal elements of a positive definite matrix are all positive.
    Hint: See Noble (1969). Follows from the last problem.

17. Let $A$ be a positive definite matrix, then there is a unique lower triangular matrix $L$ with positive elements on the main diagonal, so that $A = LL^T$.
    Hint: See Stewart (1973). Use recursion.

18. Let $A$ be a square matrix of order $n$. If $p_A(\lambda)$ is a characteristic polynomial of a matrix $A$, then prove that $p_A(A) = 0$.
    Hint: See Lipschutz (1968). The characteristic polynomial

$$p_A(\lambda) = \det(\lambda I - A)$$
$$= \sum_{k=0}^{n} a_k \lambda^{n-k}$$

where $a_0 = 1$. Note that

$$(\lambda I - A) \ adj \ (\lambda I - A) = |\lambda I - A| \ I$$

The matrix $D(\lambda) = adj \ (\lambda I - A)$ is a matrix of cofactors of the matrix $(\lambda I - A)$. It can be noted that each cofactor is a polynomial in $\lambda$ of degree less than or equal to $(n-1)$. Therefore $D(\lambda)$ is a matrix whose elements are polynomials in $\lambda$ of degree less than or equal to $(n-1)$. Let

$$D(\lambda) = \sum_{k=0}^{n-1} D_k \lambda^k$$

Substituting this in the above equation results in

$$(\lambda I - A) \sum_{k=0}^{n-1} D_k \lambda^k = \sum_{k=0}^{n} a_k \lambda^{n-k} I$$

where $a_0 = 1$. Equating coefficients of different powers of $\lambda$ yields

$$D_{n-1} = I$$
$$D_{n-2} - AD_{n-1} = a_1 I$$
$$D_{n-3} - AD_{n-2} = a_2 I$$
$$\vdots$$
$$D_0 - AD_1 = a_{n-1} I$$
$$-AD_0 = a_n I$$

Multiply the above equations by $A^n, A^{n-1}, \ldots, A, I$ respectively. This leads to

$$A^n D_{n-1} = A^n$$
$$A^{n-1} D_{n-2} - A^n D_{n-1} = a_1 A^{n-1}$$
$$A^{n-2} D_{n-3} - A^{n-1} D_{n-2} = a_2 A^{n-2}$$
$$\vdots$$
$$AD_0 - A^2 D_1 = a_{n-1} A$$
$$-AD_0 = a_n I$$

Adding the above equations results in $0 = \sum_{k=0}^{n} a_k A^{n-k}$. This proves the theorem.

19. The Fibonacci sequence of numbers is defined recursively as:

$$F_0 = 0, \quad F_1 = 1$$
$$F_n = F_{n-1} + F_{n-2}, \quad n \geq 2$$

Find an explicit expression for $F_n$ in terms of the golden ratio $\phi_g = \left(\sqrt{5} + 1\right)/2$ and $\widehat{\phi}_g = -\phi_g^{-1} = \left(1 - \sqrt{5}\right)/2$.

Hint: Observe that

$$\begin{bmatrix} F_{n+1} \\ F_n \end{bmatrix} = A \begin{bmatrix} F_n \\ F_{n-1} \end{bmatrix}; \quad n \geq 1, \quad A = \begin{bmatrix} 1 & 1 \\ 1 & 0 \end{bmatrix}$$

Thus

$$\begin{bmatrix} F_n \\ F_{n-1} \end{bmatrix} = A^{n-1} \begin{bmatrix} 1 \\ 1 \end{bmatrix}, \quad n \geq 1$$

The eigenvalues and corresponding eigenvectors of the matrix $A$ are first determined. The characteristic polynomial of the matrix $A$ is $p_A\left(\lambda\right) = \det\left(\lambda I - A\right) = \left(\lambda^2 - \lambda - 1\right)$, and the roots of this polynomial are $\phi_g$ and $\widehat{\phi}_g$. Corresponding eigenvectors are $\nu_1$ and $\nu_2$, where

$$\nu_1 = \begin{bmatrix} \phi_g & 1 \end{bmatrix}^T, \quad \text{and} \quad \nu_2 = \begin{bmatrix} \widehat{\phi}_g & 1 \end{bmatrix}^T$$

respectively. Define $V = \begin{bmatrix} \nu_1 & \nu_2 \end{bmatrix}$, and

$$\Lambda = \begin{bmatrix} \phi_g & 0 \\ 0 & \widehat{\phi}_g \end{bmatrix}$$

Then $A = V\Lambda V^{-1}$, $A^{n-1} = V\Lambda^{n-1}V^{-1}$, and

$$V^{-1} = \frac{1}{\sqrt{5}} \begin{bmatrix} 1 & -\widehat{\phi}_g \\ -1 & \phi_g \end{bmatrix}$$

Using the relationships

$$\left(\phi_g^2 - \phi_g - 1\right) = 0, \quad \left(\widehat{\phi}_g^2 - \widehat{\phi}_g - 1\right) = 0, \quad \text{and} \quad \phi_g\widehat{\phi}_g = -1$$

finally results in

$$F_n = \frac{1}{\sqrt{5}} \left(\phi_g^n - \widehat{\phi}_g^n\right), \quad \forall n \in \mathbb{N}$$

20. Let $A$ be a square matrix of order $n$. Let its distinct eigenvalues be $\lambda_1, \lambda_2, \ldots, \lambda_k$, and the corresponding eigenvectors be $x_1, x_2, \ldots, x_k$ respectively, where $k \leq n$. Prove that the set of vectors $x_1, x_2, \ldots, x_k$ are linearly independent.

Hint: Let $a_1, a_2, \ldots, a_k$ be numbers such that

$$\sum_{j=1}^{k} a_j x_j = 0$$

Multiplication of the above equation by $A$, and use of the relationship $Ax_j = \lambda_j x_j, 1 \leq j \leq k$ yields

$$\sum_{j=1}^{k} a_j \lambda_j x_j = 0$$

Multiplying the above equation by $A$ results in

$$\sum_{j=1}^{k} a_j \lambda_j^2 x_j = 0$$

Repeating the process yields

$$\sum_{j=1}^{k} a_j \lambda_j^m x_j = 0, \quad 0 \le m \le (k-1)$$

These equations can be rewritten as

$$\begin{bmatrix} a_1 x_1 & a_2 x_2 & \cdots & a_k x_k \end{bmatrix} \begin{bmatrix} 1 & \lambda_1 & \lambda_1^2 & \dots & \lambda_1^{k-1} \\ 1 & \lambda_2 & \lambda_2^2 & \dots & \lambda_2^{k-1} \\ \vdots & \vdots & \vdots & \ddots & \vdots \\ 1 & \lambda_k & \lambda_k^2 & \dots & \lambda_k^{k-1} \end{bmatrix} = 0$$

The second factor in the above equation is a nonsingular Vandermonde matrix, because all the $k$ eigenvalues are distinct. Thus

$$\begin{bmatrix} a_1 x_1 & a_2 x_2 & \cdots & a_k x_k \end{bmatrix} = 0$$

Since $x_j \ne 0, 1 \le j \le k$ it can be concluded that $a_j = 0, 1 \le j \le k$. Thus the vectors $x_j, 1 \le j \le k$ are linearly independent.

21. Prove that the eigenvalues of a real symmetric matrix are all real.

    Hint: See Ayres (1962). Let $\lambda = a + ib$ be a complex eigenvalue of a real symmetric matrix $A$. Consider a matrix

$$C = (\lambda I - A) (\overline{\lambda} I - A)$$
$$= (aI - A)^2 + b^2 I$$

Note that the matrix $C$ is real and singular since $(\lambda I - A)$ is singular. Therefore there exists a real vector $Y$, such that $Y \ne 0$ and $CY = 0$. Thus

$$Y^T C Y = Y^T (aI - A)^2 Y + b^2 Y^T Y = 0$$

Observe that the vector $(aI - A) Y$ is real, therefore $Y^T (aI - A)^2 Y \ge 0$. Since $Y^T Y > 0$, it follows that $b$ is equal to 0. Thus there are no complex eigenvalues of a real symmetric matrix.

22. Prove that the eigenvalues of a Hermitian matrix are all real numbers.

    Hint: See Lancaster (1969). This result is a generalization of the last problem. Let an eigenpair of a Hermitian matrix $A$ be $(\lambda, x)$ where $x \ne 0$. Therefore

$$Ax = \lambda x$$

Take the Hermitian transpose of both sides of the above equation. This yields

$$x^\dagger A = \overline{\lambda} x^\dagger$$

Pre-multiplication of the first equation by $x^\dagger$ and post-multiplication of the second equation by $x$ results in $\lambda x^\dagger x = \overline{\lambda} x^\dagger x$. However $x \ne 0$, consequently $x^\dagger x > 0$. The result follows.

23. Prove that eigenvectors corresponding to distinct eigenvalues of a real symmetric matrix are mutually orthogonal.

Hint: Let $(\lambda_1, Y_1)$ and $(\lambda_2, Y_2)$ be two eigenpairs of a real symmetric matrix $A$, where $\lambda_1 \neq \lambda_2$. Thus

$$AY_1 = \lambda_1 Y_1$$
$$AY_2 = \lambda_2 Y_2$$

Therefore

$$Y_2^T A Y_1 = \lambda_1 Y_2^T Y_1$$
$$Y_1^T A Y_2 = \lambda_2 Y_1^T Y_2$$

Transposing both sides of the first relationship yields

$$Y_1^T A Y_2 = \lambda_1 Y_1^T Y_2$$

Thus $\lambda_1 Y_1^T Y_2 = \lambda_2 Y_1^T Y_2$. As $\lambda_1 \neq \lambda_2$, it can be inferred that $Y_1^T Y_2 = 0$. The result follows.

24. Prove that all eigenvalues of a unitary matrix have a unit modulus (magnitude).

Hint: Let $A$ be a unitary matrix, that is $AA^\dagger = I$. Let $\lambda$ be an eigenvalue and a corresponding eigenvector be $x$, that is $Ax = \lambda x$. Applying the Hermitian operation yields $x^\dagger A^\dagger = x^\dagger \overline{\lambda}$. Thus

$$\left(x^\dagger \overline{\lambda}\right)\left(\lambda x\right) = \left(x^\dagger A^\dagger\right)\left(Ax\right)$$

That is, $|\lambda|^2 x^\dagger x = x^\dagger x$. As $x \neq 0$, this relationship implies that the eigenvalue $\lambda$ has unit magnitude.

25. Prove the list of observations on the Gerschgorin's discs (bounds) on the eigenvalues of a square matrix $A$.

Hint: See Noble (1969).

Observation 1: Let an eigenvalue and eigenvector pair of the matrix $A$ be $\lambda$, and $x$ respectively, where $x^T = \begin{bmatrix} x_1 & x_2 & \cdots & x_n \end{bmatrix}$. An eigenvector $x$ is chosen such that $|x_k| = \max_{1 \leq i \leq n} |x_i| = 1$. Since $(\lambda I - A) x = 0$

$$(\lambda - a_{kk}) x_k + \sum_{\substack{j=1 \\ j \neq k}}^{n} (-a_{kj}) x_j = 0$$

Thus

$$|\lambda - a_{kk}| \leq \sum_{\substack{j=1 \\ j \neq k}}^{n} |a_{kj}|\, |x_j| \leq \sum_{\substack{j=1 \\ j \neq k}}^{n} |a_{kj}| = P_k$$

The result follows.

Observation 2: For $A$ substitute $A^\dagger$ in Observation 1.

Observation 3: Follows from Observations 1 and 2.

26. Establish the min-max characterization of the eigenvalues of the Hermitian matrix.

Hint: The proof essentially follows Noble (1969), and Horn, and Johnson (1985). It is given that $A$ is a Hermitian matrix. Then there exists a unitary matrix $R$ such that $R^\dagger A R = \Lambda$, where $\Lambda$ is a diagonal matrix, with eigenvalues of the matrix $A$ on its diagonal. Let $x$ and $y$ be column vectors of size $n$, such that $x = Ry$. Then

$$x^\dagger A x = y^\dagger \Lambda y$$

$$x^\dagger x = y^\dagger y$$

Also define $b_i = R^\dagger a_i$, then $a_i^\dagger x = b_i^\dagger y$. Define $y^T = \begin{bmatrix} y_1 & y_2 & \cdots & y_n \end{bmatrix}$, and $M$ to be:

$$M = \max \left\{ \min \frac{\sum_{i=1}^n \lambda_i \left| y_i \right|^2}{\sum_{i=1}^n \left| y_i \right|^2} \right\}$$

In the above expression for $M$, the minimization over $y$ is constrained by the relationships $b_i^\dagger y = 0, i = 1, 2, \ldots, (k-1), k \neq 1$, where the $b_i$'s are fixed vectors. The maximization is then taken over all the $b_i$'s where $i = 1, 2, \ldots, (k-1)$. The $(k-1)$ set of equations $b_i^\dagger y = 0, i = 1, 2, \ldots, (k-1)$ are in the $n$ unknowns, $y_1, y_2, \ldots, y_n$. A nonzero solution can still be obtained by letting $y_{k+1} = \ldots = y_n = 0$. This is so, because there are $k$ unknowns $y_1, y_2, \ldots, y_k$ and there are only exactly $(k-1)$ equations. Thus there is a nonzero solution. Then

$$M = \max \left\{ \min \frac{\sum_{i=1}^k \lambda_i \left| y_i \right|^2}{\sum_{i=1}^k \left| y_i \right|^2} \right\}$$

$$= \max \left\{ \min \left[ \lambda_k - \frac{\sum_{i=1}^{k-1} (\lambda_k - \lambda_i) \left| y_i \right|^2}{\sum_{i=1}^k \left| y_i \right|^2} \right] \right\}$$

But $\lambda_1 \leq \lambda_2 \leq \ldots \leq \lambda_k$. Thus $M \leq \lambda_k$ for any vector $x$, and consequently for any vector $y$. In the next step consider maximization over the entire domain of $x$. Select $a_i = Re_i$, where $e_i$ is a unit column vector, then $b_i = e_i$ and $b_i^\dagger y = 0$ imply $y_i = 0$ for $i = 1, 2, \ldots, (k-1)$. Consequently $M = \lambda_k$. The case of $k = 1$ corresponds to the result due to Rayleigh and Ritz. This proves the first part of the theorem. The second part of the theorem can be proved by using the same line of reasoning.

27. Let $A$ and $B$ be Hermitian matrices of order $n$ each. Let their eigenvalues be $\lambda_i (A)$ and $\lambda_i (B)$ for $1 \leq i \leq n$ respectively, such that $\lambda_1 (A) \leq \lambda_2 (A) \leq \ldots \leq \lambda_n (A)$ and $\lambda_1 (B) \leq \lambda_2 (B) \leq \ldots \leq \lambda_n (B)$. These matrices are also constrained by the inequality $x^\dagger A x \leq x^\dagger B x$, for all vectors $x$ of size $n$. Prove that $\lambda_i (A) \leq \lambda_i (B)$ for $1 \leq i \leq n$.

    Hint: For any $x \neq 0$ and $x \in \mathbb{C}^n$, define $\vartheta_A (x) = x^\dagger A x / (x^\dagger x)$, and $\vartheta_B (x) = x^\dagger B x / (x^\dagger x)$, then by hypothesis $\vartheta_A (x) \leq \vartheta_B (x)$. Using the result about the min-max (or max-min) characterization of eigenvalues of Hermitian matrices, it can be inferred that $\lambda_i (A) \leq \lambda_i (B)$, for $1 \leq i \leq n$.

28. Let $A$ be a positive square matrix of order $n$, and $x$ is a nonnegative column vector of size $n$. That is, if $A > 0$ and $x \geq 0$ (but $x \neq 0$) then prove that $Ax > 0$.

    Hint: See Graham (1987). Let $A = [a_{ij}], y = Ax$, and

$$x = \begin{bmatrix} x_1 & x_2 & \cdots & x_n \end{bmatrix}^T \quad \text{and} \quad y = \begin{bmatrix} y_1 & y_2 & \cdots & y_n \end{bmatrix}^T$$

Then

$$y_i = \sum_{j=1}^n a_{ij} x_j \geq a_{ik} \sum_{j=1}^n x_j$$

where $a_{ik} = \min \{ a_{i1}, a_{i2}, \ldots, a_{in} \}$. Since $A > 0$, it follows that $a_{ik} > 0$. Also since $x \geq 0$ (but $x \neq 0$), it follows that $\sum_{j=1}^n x_j > 0$. Therefore $y_i > 0$. This result is true for any $i$, where $1 \leq i \leq n$.

29. Prove the observations about positive matrices (Perron-Frobenius theorem for positive matrices).

Hints: Graham (1987).

Observation 1: Note that $\phi(kx) = \phi(x)$ for any scalar $k > 0$. Therefore while finding $\phi(x)$, only consider values of $x$ for which $x^T x = 1$. Furthermore, this condition implies that $x \neq 0$. Thus

$$Ax \geq \phi(x)\,x$$

Define $y = Ax$. From the last problem $y > 0$. Multiplying both sides of the above inequality by $A$ results in

$$Ay \geq \phi(x)\,y$$

Also

$$Ay \geq \phi(y)\,y, \;\; \text{and} \;\; y > 0$$

where $\phi(y)$ is the *largest* number for which the inequality holds. Therefore $\phi(y)\,y \geq \phi(x)\,y$. Since $y > 0$, it can be concluded that

$$\phi(y) \geq \phi(x)$$

The set $\Psi$ is a continuous mapping $y = Ax$ of the set $\Phi$. Therefore the set $\Psi$ is closed and bounded. Consequently, $\phi(x)$ achieves its maximum value $\eta$. Thus

$$\eta - \max_{x \in \Phi}\{\phi(x)\} - \max_{y \in \Psi}\{\phi(y)\}$$

Let the value of $y$ corresponding to $\eta$ be $y_0$. That is, $\eta = \phi(y_0)$ and $Ay_0 \geq \eta y_0$. We next prove that $\eta$ is the eigenvalue of the matrix $A$.

Assume that $Ay_0 > \eta y_0$, then $Au > \eta u$, where $u = Ay_0$. But as per the definition $\phi(u)$ is the largest value for which the inequality $Au \geq \phi(u)\,u$ holds. Thus $\phi(u) > \eta$. This is a contradiction of the definition of $\eta$. Thus

$$Ay_0 = \eta y_0$$

This equation demonstrates that $\eta$ is an eigenvalue of the matrix $A$ and $y_0$ is a corresponding eigenvector. Observe that this eigenvector is positive as per the definition of the set $\Psi$.

Observation 2: Let $x = \begin{bmatrix} 1/\sqrt{n} & 1/\sqrt{n} & \cdots & 1/\sqrt{n} \end{bmatrix}^T$. Since $A > 0$, it can be inferred that

$$\phi(x) = \min_{1 \leq j \leq n} \sum_{k=1}^{n} a_{jk} > 0$$

As $\eta \geq \phi(x)$, the result follows.

Observation 3: Let $\lambda$ and $x$ be the eigenvalue and an associated eigenvector of matrix $A$ respectively. Then $Ax = \lambda x$. Thus

$$\lambda x_j = \sum_{k=1}^{n} a_{jk} x_k, \quad 1 \leq j \leq n$$

Therefore

$$|\lambda|\,|x_j| = \left| \sum_{k=1}^{n} a_{jk} x_k \right| \leq \sum_{k=1}^{n} a_{jk}\,|x_k|, \quad 1 \leq j \leq n$$

Define
$$m(x) = \left[\, |x_1|\ |x_2|\ \cdots\ |x_n|\,\right]^T$$

That is, $m(x)$ is a vector obtained by taking the magnitude of each of the element of the vector $x$. Thus $|\lambda|\, m(x) \leq Am(x)$, but $\phi(m(x))$ is the largest value such that $Am(x) \geq \phi(m(x))\, m(x)$. Therefore $\phi(m(x)) \geq |\lambda|$. Finally, $|\lambda| \leq \phi(m(x)) \leq \eta$, which implies $\varrho(A) = \eta$.

Observation 4: It is given that the spectral radius of the matrix $A > 0$, is equal to $\varrho(A) = \eta$. Assume that $\lambda$ is another eigenvalue of the matrix $A$ such that $|\lambda| = \eta$. Since $\lambda x = Ax$

$$|\lambda|\, |x_j| = \left| \sum_{k=1}^{n} a_{jk} x_k \right| \leq \sum_{k=1}^{n} a_{jk} |x_k|, \quad 1 \leq j \leq n$$

In observation number 3, the inequality $|\lambda| \leq \phi(m(x)) \leq \eta$ was established. Substitute $|\lambda| = \eta$, in it. Thus $|\lambda| = \phi(m(x)) = \eta$. Using the definition of $\phi(x)$ yields $\eta m(x) = Am(x)$. This implies

$$\eta\, |x_j| = \sum_{k=1}^{n} a_{jk} |x_k|, \quad 1 \leq j \leq n$$

Therefore, when $|\lambda| = \eta$,

$$\left| \sum_{k=1}^{n} a_{jk} x_k \right| = \sum_{k=1}^{n} a_{jk} |x_k|, \quad 1 \leq j \leq n$$

Since $A > 0$, and using elementary theory of complex variables, it can be concluded that $x = m(x)\, e^{i\theta}$. Thus the vector $x$ is a constant multiple of the vector $m(x)$. This implies that $\lambda = \eta$. That is, there is no other eigenvalue of $A$ whose magnitude is equal to $\eta$, except $\eta$ itself. Therefore $\eta > |\lambda|$ implies that $\lambda$ is any other eigenvalue of the matrix $A$.

Observation 5: Part (a): The eigenvalues of the matrix $A$ and its transpose $A^T$ are equal. Thus $A^T x = \eta x$, where $x > 0$. Also let $By = \lambda y$, then

$$|\lambda|\, m(y) \leq Bm(y)$$

where
$$m(y) = \left[\, |y_1|\ |y_2|\ \cdots\ |y_n|\,\right]^T$$

Consequently
$$|\lambda|\, m(y) \leq Bm(y) \leq Am(y)$$

That is
$$|\lambda|\, x^T m(y) \leq x^T Bm(y) \leq x^T Am(y) = \eta x^T m(y)$$

However $x^T m(y) > 0$, because $x > 0$ and $m(y) \geq 0$. Thus $|\lambda| < \eta$.

Part (b): As per the hypothesis $|\lambda| = \eta$. Thus from Part (a), $\eta m(y) \leq Am(y)$. That is

$$u = Am(y) - \eta m(y) \geq 0$$

Then either $u = 0$ or $u \neq 0$. Assume that $u \neq 0$, then $Au > 0$. Define $m(y) = t$, then $At > \eta t$. If $t = \left[\, t_1\ t_2\ \cdots\ t_n\,\right]^T$ then

$$\frac{\sum_{k=1}^{n} a_{jk} t_k}{t_j} > \eta, \quad 1 \le j \le n$$

The above relationship is in contradiction with the definition of $\eta$. Thus $u = 0$. This leads us to

$$\eta m(y) = Bm(y) = Am(y)$$

Since $m(y) > 0$ and $A \ge B$, the result follows.

Observation 6: The characteristic polynomial of matrix $A$ is $p(\lambda) = \det(\lambda I - A)$. Let $p'(\lambda)$ be the first derivative of $p(\lambda)$ with respect to $\lambda$. Then $p'(\lambda)$ is a linear combination of the principal minors of the matrix $(\lambda I - A)$. Note that if $p'(\lambda) \ne 0$, then the eigenvalue of the matrix $A$ is simple.

Denote a typical principal minor of the matrix $(\lambda I - A)$ by $\det(\lambda I - A_{ps})$, where $A_{ps}$ is a principal submatrix of $A$. Let

$$\det(\lambda I - A_{ps}) = \prod_i (\lambda - \lambda_i)$$

then by observation number 5, $\eta = \varrho(A) > |\lambda_i|$.

Consequently $\det(\lambda I - A_{ps}) \ne 0$ if $\lambda \ge \eta$. That is, $\det(\eta I - A_{ps}) > 0$. Thus $p'(\eta) > 0$. The result follows by noting that, if $p'(\eta) \ne 0$, then the eigenvalue $\eta$ is simple.

Observation 7: Assume that a matrix $A$ is obtained by increasing any element of the matrix $B$. Therefore $A \ge B > 0$. The result follows by using observation number 5.

Observation 8: Part (a). Using the hint for observation number 1, for any $x \ge 0$ results in

$$\eta \ge \phi(x) = \min_{\substack{1 \le j \le n \\ x_j \ne 0}} \frac{\sum_{k=1}^{n} a_{jk} x_k}{x_j}$$

Substitute $x = e = \begin{bmatrix} 1 & 1 & \cdots & 1 \end{bmatrix}^T$ in the above inequality. This results in

$$\eta \ge \phi(e) = \min_{1 \le j \le n} \sum_{k=1}^{n} a_{jk}$$

It is also known that for $x \ge 0$, $\phi(x) x \le Ax$. Thus $e^T \phi(x) x \le e^T Ax$. However $e^T x > 0$, which yields

$$\phi(x) \le \frac{e^T Ax}{e^T x}$$

Define $\psi_j = \sum_{i=1}^{n} a_{ij}$ = sum of the elements of the $j$th column for $1 \le j \le n$, and $\psi = \max_{1 \le j \le n} \psi_j$. This yields

$$e^T A = \begin{bmatrix} \psi_1 & \psi_2 & \cdots & \psi_n \end{bmatrix} \le \psi e^T$$

Therefore

$$\phi(x) \le \frac{e^T Ax}{e^T x} \le \frac{\psi e^T x}{e^T x}$$
$$= \psi = \max_{1 \le j \le n} \sum_{i=1}^{n} a_{ij}$$

Finally, since $\eta = \max_{x \in \Phi} \{\phi(x)\}$

$$\min_{1 \le j \le n} \sum_{k=1}^{n} a_{jk} \le \eta \le \max_{1 \le j \le n} \sum_{i=1}^{n} a_{ij}$$

Noting that $A$ and its transpose $A^T$ have same eigenvalues, using the above inequalities yields

$$\min_{1 \le j \le n} \sum_{k=1}^{n} a_{kj} \le \eta \le \max_{1 \le j \le n} \sum_{i=1}^{n} a_{ji}$$

Combining the above two pairs of inequalities yields

$$\min_{1 \le j \le n} \sum_{k=1}^{n} a_{jk} \le \eta \le \max_{1 \le j \le n} \sum_{i=1}^{n} a_{ji}$$

Part (b). This observation is proved by contradiction. Assume that the elements of each row of matrix $A$ sum to $\eta$ except a single row, which sums to less than $\eta$. Increase the sum of a single element of this row, such that the elements of this row also sum to $\eta$, then $\eta = \max_{1 \le j \le n} \sum_{i=1}^{n} a_{ji}$. However, by observation number 5, the spectral radius of the new matrix is greater than $\eta$. Thus there is a contradiction. This in turn implies that the initial assumption of the sum of elements of a single row is smaller than $\eta$ is incorrect. Therefore all row sums are equal to $\eta$. This line of reasoning can be extended to cases when more than a single row-sum is smaller than $\eta$, and that the row-sums are $\gneq$ than $\eta$.

Proof of Part (c) of the observation is similar to Part (b).

30. The matrix $A$ is square, irreducible, nonnegative and of order $n$. Also let $k_i > 0$, for $0 \le i \le (n-1)$. Prove that $\sum_{i=0}^{n-1} k_i A^i > 0$.

    Hint: See Fiedler (1986). Let $B = \sum_{i=0}^{n-1} k_i A^i \ge 0$. Note that $B \ge 0$, and the diagonal entries of this matrix are positive because the diagonal entries of the matrix $k_0 A^0 = k_0 I$ are positive. Next consider the nondiagonal entries of the matrix $B$. Let $G$ be the digraph associated with the matrix $A$. Note that there is a path from vertex $v_i$ to vertex $v_j$, because it is strongly connected (due to irreducibility of matrix $A$), where $i \ne j$. Furthermore, the length of the path is not more than $(n-1)$. If the length of this path is $s$ then the $ij$th entry of the matrix $A^s$ is positive.

31. Prove the lemma in the subsection on nonnegative matrices.

    Hint: See Graham (1987). Note that it is possible for an element of vector $y$, for example $y_k$ to be equal to 0 (for some $k$). Thus $\phi(y)$ is not necessarily continuous for values of $y \in \Psi$. Observe that $\Upsilon \subset \Psi$, and for $z \in \Upsilon$, we have $z > 0$. Consequently, $\phi(z)$ is continuous on the closed and bounded set $\Upsilon$. Therefore $\max_{z \in \Upsilon} \{\phi(z)\} = \eta$ exists. Denote the corresponding value of $z$ by $z_0$.

    Observe that for each $y \in \Psi$, $Ay \ge \phi(y) y$. Therefore $Az \ge \phi(y) z$, where $z = (I + A)^{n-1} y$. However, as per the definition of the set $\Upsilon$, $\phi(z)$ is the *largest* value for which the inequality $Az > \phi(z) z$ is valid. This leads us to

    $$\phi(z) \ge \phi(y) \ge \phi(x)$$

    $$\eta = \phi(z_0) = \max_{x \in \Phi} \{\phi(x)\}$$

    It is next established that $\eta$ is indeed an eigenvalue of the matrix $A$. From the above discussion, it can be inferred that $Az_0 \ge \phi(z_0) z_0$.

Consider the case when $Az_0 \neq \phi(z_0) z_0$, that is when $(A - \phi(z_0) I) z_0 \geq 0$. This implies that at least one element of this vector is not equal to zero. Since $(I + A)^{n-1} > 0$, it can be noted that $(I + A)^{n-1} (A - \phi(z_0) I) z_0 > 0$. Observe that the matrices $(I + A)^{n-1}$ and $A$ commute.

Thus $(A - \phi(z_0) I) u > 0$ where $u = (I + A)^{n-1} z_0$. Therefore $Au > \phi(z_0) u$, however $\phi(u)$ is the largest number such that $Au \geq \phi(u) u$. Consequently $\phi(u) \geq \phi(z_0)$. This conclusion is in contradiction with the definition of $\phi(z_0)$. Therefore $Az_0 = \phi(z_0) z_0$, that is $Az_0 = \eta z_0$.

32. Establish the Perron-Frobenius theorem about nonnegative matrices.

Hint: See Graham (1987). This theorem has three parts.

Part (a): Let $\lambda$ be an eigenvalue of the matrix $A$, and a corresponding eigenvector be $w$, where $w^T = \begin{bmatrix} w_1 & w_2 & \cdots & w_n \end{bmatrix}$. This eigenvector is assumed to be normalized. Thus $Aw = \lambda w$. That is

$$\lambda w_i = \sum_{j=1}^{n} a_{ij} w_j, \quad 1 \leq i \leq n$$

$$|\lambda| |w_i| \leq \sum_{j=1}^{n} a_{ij} |w_j|, \quad 1 \leq i \leq n$$

The above inequality is valid for each component of $w$. This implies that

$$|\lambda| \leq \frac{\sum_{j=1}^{n} a_{ij} |w_j|}{|w_i|}$$

where $|w_i| > 0$. Define $x = m(w)$, which is the modulus of the vector $w$. Thus

$$|\lambda| \leq \min_{\substack{1 \leq i \leq n \\ x_i > 0}} \frac{\sum_{j=1}^{n} a_{ij} x_j}{x_i} = \phi(x)$$

Consequently $|\lambda| \leq \phi(x) \leq \max_{x>0} \{\phi(x)\} = \eta$. Therefore $\eta = \varrho(A)$. Also note that for an irreducible matrix $A$, it is possible that there exist other eigenvalues $\tilde{\lambda}_i$ such $\left| \tilde{\lambda}_i \right| = \eta$.

Part (b): An eigenvector associated with the eigenvalue $\eta$ is $z_0$. It has been proved in the lemma (last problem) that the eigenvector $z_0 \in \Upsilon$. Consequently $z_0 > 0$.

Part (c): It is known that $(I + A)^{n-1} > 0$, therefore the spectral radius of this positive matrix is

$$\varrho\left\{ (I + A)^{n-1} \right\} = (1 + \eta)^{n-1}$$

This is a simple eigenvalue of the matrix $(I + A)^{n-1}$. This follows from the observation about positive matrices. Using Part (b) of this problem, it can be inferred that the multiplicity of the eigenvalue $\eta$ is also equal to unity.

33. Let the matrix $A = [a_{ij}]$ be an irreducible square matrix of order $n$, such that $a_{ii} > 0$, for $i = 1, 2, \ldots, n$. Prove that the matrix $A$ is primitive.

Hint: See Graham (1987). Let the smallest value of the $a_{ii}$'s be $a_{kk}$. Define a matrix $C = [c_{ij}]$ such that

$$c_{ij} = \frac{a_{ij}}{a_{kk}}, \quad 1 \leq i, j \leq n$$

Therefore $c_{ii} \geq 1$, for $1 \leq i \leq n$. Define $C = (I + B)$, then $C \geq \alpha(I + B)$ for $0 < \alpha < 1$. That is

$$C^{n-1} \geq \alpha^{n-1} (I + B)^{n-1}$$

The right-hand side expression in the above inequality is positive from a property of nonnegative matrices. That is, $C^{n-1} > 0$. This in turn implies that $A^{n-1} = (a_{kk})^{n-1} C^{n-1} > 0$.

34. Prove the observations about stochastic matrices.

Hint: See Graham (1987), and Fiedler (1986). The following four observations are established

Observation 1: Assume that $\lambda$ is an eigenvalue of the matrix $A$. Use of Gerschgorin's observation about the bound on the eigenvalue of the matrix $A$, results in the inequality $|\lambda| \leq 1$. However, since $A$ is a stochastic matrix $Ae = 1e$. Thus $\varrho(A) = 1$.

Observation 2: If $A$ is a stochastic matrix, then $Ae = 1e$. Thus 1 is the eigenvalue of the matrix $A$ and a corresponding eigenvector is $e$.

However if 1 and $e$ are the eigenvalue and eigenvector pair of the matrix $A$, then $Ae = 1e$, which implies that the matrix $A$ is stochastic.

Observation 3: As per the hypothesis $Ae = e$. Thus $A^2e = A(Ae) = Ae = e$. The statement is established using induction.

Observation 4: Let $A$ and $B$ be similar matrices such that $QAQ^{-1} = B$, and

$$B = \begin{bmatrix} 1 & Z_1 \\ Z_2 & C \end{bmatrix}$$

where $Z_1$ is a $1 \times (n-1)$ matrix of all zeros, $Z_2$ is a $(n-1) \times 1$ matrix of all zeros, and $C$ is a $(n-1) \times (n-1)$ matrix such that $\varrho(C) < 1$. Thus for $k \in \mathbb{P}$

$$B^k = \begin{bmatrix} 1 & Z_1 \\ Z_2 & C^k \end{bmatrix}$$

Consequently

$$E \triangleq \lim_{k \to \infty} B^k = \begin{bmatrix} 1 & Z_1 \\ Z_2 & Z_3 \end{bmatrix} = e_1 e_1^T$$

where $Z_3$ is a $(n-1) \times (n-1)$ matrix of all zeros, and $e_1 = \begin{bmatrix} 1 & 0 & \cdots & 0 \end{bmatrix}^T$ is a column vector of length $n$. Define

$$\widetilde{E} \triangleq \lim_{k \to \infty} A^k = \lim_{k \to \infty} Q^{-1} B^k Q = Q^{-1} E Q = Q^{-1} e_1 e_1^T Q$$

As per the hypothesis $Ae = e$, $A^T z = z$, and $z^T e = 1$. It can be established that

$$(QAQ^{-1}) Qe = Qe$$
$$(QAQ^{-1})^T (Q^T)^{-1} z = (Q^T)^{-1} z$$

That is

$$BQe = Qe$$
$$B^T (Q^T)^{-1} z = (Q^T)^{-1} z$$

Notice that $Be_1 = e_1$, and $B^T e_1 = e_1$, and the eigenvectors corresponding to unit eigenvalue are determined uniquely up to a multiplicative factor. This yields

$$Qe = \alpha_1 e_1$$
$$(Q^T)^{-1} z = \alpha_2 e_1$$

Using the relationship $z^T e = 1$ results in $1 = z^T e = \left(z^T Q^{-1}\right)(Qe) = \alpha_2 e_1^T \alpha_1 e_1 = \alpha_2 \alpha_1$. Therefore

$$\widetilde{E} = Q^{-1} e_1 e_1^T Q = \alpha_1^{-1} e \alpha_2^{-1} z^T = e z^T$$

35. Let $\widehat{A}$ be an $m \times n$ matrix of real elements and rank $r_{\widehat{A}}$. Its SVD is equal to $\widehat{U} \Sigma \widehat{V}^T$. In this decomposition, the matrices $\widehat{U}$ and $\widehat{V}$ are orthogonal, and

$$\Sigma = \begin{bmatrix} \widehat{S} & 0 \\ 0 & 0 \end{bmatrix}$$

In the matrix $\Sigma$, the matrix $\widehat{S}$ is a diagonal matrix with all positive diagonal entries. Therefore the order of the matrix $\widehat{S}$ is equal to its rank $r_{\widehat{S}}$. Note that the matrix $\Sigma$ is an $m \times n$ matrix. It is determined uniquely except for the order of its diagonal elements (singular values). Prove that this SVD of the matrix $\widehat{A}$ exists.

Hint: See Fiedler (1986). It should be immediately evident that the ranks of the matrices $\Sigma$ and $\widehat{S}$ are equal, that is $r_\Sigma = r_{\widehat{S}}$. Without any loss of generality assume that $m \le n$. Then $\widehat{A}^T \widehat{A}$ is a square matrix of order $n$, and it can be represented as $\widehat{V} \widehat{D} \widehat{V}^T$, where $\widehat{V}$ is an orthogonal matrix and $\widehat{D}$ is a diagonal matrix each of size $n$. Thus

$$\left(\widehat{A}\widehat{V}\right)^T \left(\widehat{A}\widehat{V}\right) = \widehat{D}$$

The matrix $\widehat{A}\widehat{V}$ is of order $m \times n$. Each diagonal entry of matrix $\widehat{D}$ is a sum of squares. Assume that the first $r$ diagonal entries of the matrix $\widehat{D}$ are positive, and the remaining entries are zero. Let these positive diagonal elements be $c_i, 1 \le i \le r$. That is, if $\widehat{D} = [d_{ij}]$, then $d_{ii} \triangleq c_i$ for $1 \le i \le r$, and $d_{ii} = 0$ for $(r+1) \le i \le n$.

Define

$$w_i = c_i^{-1/2} \left(\widehat{A}\widehat{V}\right)_i, \quad 1 \le i \le r$$

where $\left(\widehat{A}\widehat{V}\right)_i$ is the $i$th column of the matrix $\widehat{A}\widehat{V}$ for $1 \le i \le r$. Note that the $w_i$'s are column vectors of length $m$. Thus the vectors $w_i$'s are orthonormal as per their construction. Using the Gram-Schmidt orthogonalization procedure, generate vectors $w_{r+1}, \ldots, w_m$ such that $\widehat{U} = \begin{bmatrix} w_1 & w_2 & \cdots & w_m \end{bmatrix}$ is an orthogonal matrix of order $m$. It is next demonstrated that

$$\widehat{A}\widehat{V} = \widehat{U}\Sigma$$

where $\Sigma$ is an $m \times n$ matrix, with the elements $\sqrt{c_i}, 1 \le i \le r$ on its main diagonal, and the remaining elements are all equal to zero. The above equation holds true for the first $r$ columns of $\widehat{A}\widehat{V}$ by the construction of the vectors $w_i, 1 \le i \le r$. However, the other columns of $\widehat{A}\widehat{V}$ are null for $i > r$. This follows from the observation that $d_{ii} = 0$ for $(r+1) \le i \le n$ and $\left(\widehat{A}\widehat{V}\right)^T \left(\widehat{A}\widehat{V}\right) = \widehat{D}$.

Finally, use of the relationship $\widehat{A}\widehat{V} = \widehat{U}\Sigma$ results in $\widehat{A} = \widehat{U}\Sigma\widehat{V}^T$. Note that the rank of the matrix $\widehat{A}$ is equal to $r_{\widehat{A}} = r_\Sigma = r_{\widehat{S}} = r$. Also observe that the nonzero diagonal entries of the matrix $\Sigma$ are unique because these are equal to the nonzero square roots of the eigenvalues of the matrix $\widehat{A}^T \widehat{A}$.

36. Let $\widehat{A}$ be an $m \times n$ matrix of real elements, with SVD equal to $\widehat{U}\Sigma\widehat{V}^T$, and $b$ is a column vector of length $m$. Find a vector $x$ of length $n$ such that $\|s\|_2$ is minimized, where $s = \left(\widehat{A}x - b\right)$.

This is the principle of least squares. For simplicity assume that $\widehat{p} = \min\left(m, n\right) = n = \text{rank}$ of the matrix $\widehat{A}$.

Hint: $\|s\|_2^2 = \left(\widehat{A}x - b\right)^T \left(\widehat{A}x - b\right) = x^T \widehat{A}^T \widehat{A} x - 2x^T \widehat{A}^T b + b^T b$. Note that $\widehat{A}^T \widehat{A}$ is a symmetric matrix, and $\widehat{A}^T \widehat{A} = \widehat{V}\left(\Sigma^T \Sigma\right)\widehat{V}^T$, where $\Sigma^T \Sigma$ is a diagonal matrix with positive values $\sigma_i^2$ for $1 \le i \le n$ on its diagonal. Define $z = \widehat{V}^T x$. Thus $x = \widehat{V}z$, and

$$\|s\|_2^2 = x^T \widehat{V}\left(\Sigma^T \Sigma\right)\widehat{V}^T x - 2x^T \widehat{A}^T b + b^T b$$
$$= z^T \left(\Sigma^T \Sigma\right) z - 2z^T \widehat{V}^T \widehat{A}^T b + b^T b$$

Let $d = \widehat{V}^T \widehat{A}^T b$, then

$$\|s\|_2^2 = z^T \left(\Sigma^T \Sigma\right) z - 2z^T d + b^T b$$
$$= \sum_{i=1}^{n} \left\{\sigma_i^2 z_i^2 - 2z_i d_i\right\} + \sum_{i=1}^{m} b_i^2$$

where $z_i$, $1 \le i \le n$ are elements of the vector $z$, and $b_i, 1 \le i \le m$ are elements of the vector $b$.

The minimum of $\|s\|_2^2$ is achieved by finding the minimum of each quadratic term. This yields $z_i = d_i/\sigma_i^2$ for $1 \le i \le n$. Finally, $x$ can be evaluated via the expression $x = \widehat{V}z$, which is $x = \widehat{V}\left(\Sigma^T \Sigma\right)^{-1}\Sigma^T \widehat{U}^T b$. Note that the matrix $\left(\Sigma^T \Sigma\right)$ is invertible.

In summary:

(a)  The eigenvalues of the matrix $\widehat{A}^T \widehat{A}$ are equal to $\sigma_i^2$, for $1 \le i \le n$.

(b)  Define $d = \widehat{V}^T \widehat{A}^T b$, a column vector of length $n$, where

$$d = \begin{bmatrix} d_1 & d_2 & \dots & d_n \end{bmatrix}^T$$

Then $x = \widehat{V}z$, where $z = \begin{bmatrix} z_1 & z_2 & \dots & z_n \end{bmatrix}^T$, and $z_i = d_i/\left(\sigma_i^2\right)$ for $1 \le i \le n$. An alternate expression for $x$ is

$$\widehat{V}\left(\Sigma^T \Sigma\right)^{-1}\Sigma^T \widehat{U}^T b$$

37.  Let $x \in \mathbb{R}^n$, and $y \in \mathbb{R}^m$. Thus, $x$ and $y$ are vectors of size $n \in \mathbb{P}$, and $m \in \mathbb{P}$ respectively. These are represented as column matrices. Also $W = [w_{ij}]$ is an $m \times n$ matrix, where $w_{ij} \in \mathbb{R}$, for $1 \le i \le m$, and $1 \le j \le n$. Prove the following results.

(a)  The matrix $W$ is independent of the vector $x$. If $y = Wx$, then

$$\frac{\partial y}{\partial x} = W$$

(b)  The matrix $W$ is independent of the vector $x$. Also, the vector $x$ is a function of the vector $z$, and $W$ is independent of $z$. If $y = Wx$, then

$$\frac{\partial y}{\partial z} = W\frac{\partial x}{\partial z}$$

(c)  The matrix $W$ is independent of vectors $x$ and $y$. Further, vectors $x$ and $y$ are independent of each other. If $\beta = y^T W x$, then

$$\frac{\partial \beta}{\partial x} = y^T W, \quad \text{and} \quad \frac{\partial \beta}{\partial y} = x^T W^T$$

(d) The matrix $W$ is independent of the vector $x$. Also let $m = n$. That is, $W$ is a square matrix. If $\beta = x^T W x$, then

$$\frac{\partial \beta}{\partial x} = x^T \left( W + W^T \right)$$

(e) Let $m = n$. That is, both vectors $x$ and $y$ have $n$ elements each. Further, both vectors $x$ and $y$ are functions of vector $z$. Also if $\beta = y^T x$, then

$$\frac{\partial \beta}{\partial z} = x^T \frac{\partial y}{\partial z} + y^T \frac{\partial x}{\partial z}$$

Hint:

(a) Observe that

$$y_i = \sum_{k=1}^{n} w_{ik} x_k, \quad 1 \le i \le m$$

Therefore

$$\frac{\partial y_i}{\partial x_j} = w_{ij}, \quad \text{for } 1 \le j \le n, \text{ and } 1 \le i \le m$$

The result follows.

(b) Observe that

$$y_i = \sum_{k=1}^{n} w_{ik} x_k, \quad 1 \le i \le m$$

Let $z_j$ be the $j$th element of the vector $z$. Therefore

$$\frac{\partial y_i}{\partial z_j} = \sum_{k=1}^{n} w_{ik} \frac{\partial x_k}{\partial z_j}, \quad \text{for } 1 \le i \le m$$

The right hand side expression in the above equation is simply the $(i, j)$th element of the matrix $W \partial x / \partial z$. The result follows.

(c) Using part (a) of the problem, it can be inferred that

$$\frac{\partial \beta}{\partial x} = y^T W$$

Note that

$$\beta = \beta^T = \left( y^T W x \right)^T = x^T W^T y$$

Using part (a) of the problem again, it can be inferred that

$$\frac{\partial \beta}{\partial y} = x^T W^T$$

(d) The result follows by using part (c) of the problem.

(e) Observe that

$$\beta = y^T x = \sum_{i=1}^{n} x_i y_i$$

Let $z_j$ be the $j$th element of the vector $z$. Therefore

$$\frac{\partial \beta}{\partial z_j} = \sum_{i=1}^{n} \left\{ x_i \frac{\partial y_i}{\partial z_j} + \frac{\partial x_i}{\partial z_j} y_i \right\}$$

The result follows. This result can also be proved by using part (b) of the problem.

38. Matrix $A$ is independent of matrix $X$. Prove the following results.

   (a) Let the size of matrices $X$ and $A$ be $n \times m$ and $m \times n$ respectively. Then

$$\frac{\partial \, tr\,(XA)}{\partial X} = A^T$$

   (b) Let the size of matrices $X$ and $A$ be $n \times m$ and $m \times m$ respectively. Then

$$\frac{\partial \, tr\,\left(XAX^T\right)}{\partial X} = X\left(A + A^T\right)$$

   Hint: Let $X = [x_{ij}]$, and $A = [a_{ij}]$.

   (a) We have

$$tr\,(XA) = \sum_{i=1}^{n} \left\{ \sum_{k=1}^{m} x_{ik} a_{ki} \right\}$$

   Therefore

$$\frac{\partial \, tr\,(XA)}{\partial X} = \left[ \frac{\partial}{\partial x_{ij}} \sum_{i=1}^{n} \left\{ \sum_{k=1}^{n} x_{ik} a_{ki} \right\} \right] = [a_{ji}] = A^T$$

   (b) We have

$$\frac{\partial \, tr\,\left(XAX^T\right)}{\partial X} = \left. \frac{\partial \, tr\,\left(XAB^T\right)}{\partial X} \right|_{B=X} + \left. \frac{\partial \, tr\,\left(CAX^T\right)}{\partial X} \right|_{C=X}$$

   As trace of a matrix is a scalar, $tr\,\left(CAX^T\right) = tr\,\left(XA^TC^T\right)$. This leads to

$$\begin{aligned}
\frac{\partial \, tr\,\left(XAX^T\right)}{\partial X} &= \left. \frac{\partial \, tr\,\left(XAB^T\right)}{\partial X} \right|_{B=X} + \left. \frac{\partial \, tr\,\left(XA^TC^T\right)}{\partial X} \right|_{C=X} \\
&= \left. \left(AB^T\right)^T \right|_{B=X} + \left. \left(A^TC^T\right)^T \right|_{C=X} \\
&= \left. \left(BA^T\right) \right|_{B=X} + \left. (CA) \right|_{C=X} \\
&= XA^T + XA
\end{aligned}$$

39. Let $\Xi$ be a $2 \times 2$ random symmetric matrix, whose elements take real values. That is

$$\Xi = \begin{bmatrix} A & B \\ B & C \end{bmatrix}$$

where $A, B,$ and $C$ are independent normally distributed random variables. Furthermore, $A, C \sim \mathcal{N}\,(0,1)$, and $B \sim \mathcal{N}\,(0,1/2)$. Since $\Xi$ is a real symmetric matrix, its eigenvalues $\lambda$ and $\mu$ are real random variables. Determine the joint probability density function of the random eigenvalues $\lambda$ and $\mu$.

Hint: See Lévêque (2015). The matrix $\Xi$ can be diagonalized to matrix $\Lambda$ via an orthogonal matrix $R$, such that $R^T \Xi R = \Lambda, RR^T = I$, and

$$\Lambda = \begin{bmatrix} \lambda & 0 \\ 0 & \mu \end{bmatrix}, \quad R = \begin{bmatrix} \cos\theta & \sin\theta \\ -\sin\theta & \cos\theta \end{bmatrix}, \quad \theta \in \left[0, \frac{\pi}{2}\right]$$

Use of the above relationships results in

$$A = \lambda \cos^2 \theta + \mu \sin^2 \theta$$
$$B = (\mu - \lambda) \sin \theta \cos \theta$$
$$C = \lambda \sin^2 \theta + \mu \cos^2 \theta$$

Assume that the joint probability density function of the random variables $A, B$, and $C$ is $f_{A,B,C}(a, b, c)$, where $a, b, c \in \mathbb{R}$. Then

$$f_{A,B,C}(a, b, c) = \frac{1}{2\pi^{3/2}} \exp\left\{-\frac{1}{2}\left(a^2 + 2b^2 + c^2\right)\right\}$$

Note that $(A^2 + 2B^2 + C^2) = tr(\Xi^2) = (\lambda^2 + \mu^2)$. In the rest of the discussion, the same notation for the instances of the random variables $\lambda, \mu$, and $\theta$, as the random variables themselves is used. In order to find the joint probability density function of the random variables $\lambda, \mu$, and $\theta$, the Jacobian of the transformation of $(a, b, c)$ to $(\lambda, \mu, \theta)$ has to be computed. It is

$$J = \frac{\partial(a, b, c)}{\partial(\lambda, \mu, \theta)} = \begin{vmatrix} \partial a/\partial \lambda & \partial a/\partial \mu & \partial a/\partial \theta \\ \partial b/\partial \lambda & \partial b/\partial \mu & \partial b/\partial \theta \\ \partial c/\partial \lambda & \partial c/\partial \mu & \partial c/\partial \theta \end{vmatrix}$$

Use of the above relationships, results in $J = (\lambda - \mu)$. Thus

$$f_{\lambda,\mu,\theta}(\lambda, \mu, \theta) = \frac{|J|}{2\pi^{3/2}} \exp\left\{-\frac{1}{2}\left(\lambda^2 + \mu^2\right)\right\}$$
$$= \frac{|\lambda - \mu|}{2\pi^{3/2}} \exp\left\{-\frac{1}{2}\left(\lambda^2 + \mu^2\right)\right\}$$

In the expression for the joint probability density function $f_{\lambda,\mu,\theta}(\lambda, \mu, \theta)$, $\lambda, \mu \in \mathbb{R}$. Note that $f_{\lambda,\mu,\theta}(\lambda, \mu, \theta)$ does not depend upon $\theta$. Therefore, the random variable $\theta$ is uniformly distributed between $0$ and $\pi/2$, with probability density function $2/\pi$. Thus

$$f_{\lambda,\mu}(\lambda, \mu) = \frac{|\lambda - \mu|}{4\pi^{1/2}} \exp\left\{-\frac{1}{2}\left(\lambda^2 + \mu^2\right)\right\}, \quad \lambda, \mu \in \mathbb{R}$$

Also observe that $(\lambda + \mu) = (A + C)$, and $\lambda\mu = (AC - B^2)$. Thus $\mathcal{E}(\lambda + \mu) = 0$, and $\mathcal{E}(\lambda\mu) = -1/2$, where $\mathcal{E}(\cdot)$ is the expectation operator.

40. Determine the magnitude of the Jacobian of the transformation $A \to (R, \Lambda)$, where $A$ is a symmetric random matrix which belongs to a GOE. The matrix $\Lambda$ is a diagonal matrix, $R^T A R = \Lambda$, and $R^T R = I$.

Hint: See Lévêque (2015). Let $dA, d\Lambda$, and $dR$ be infinitesimal variations in the matrices $A, \Lambda$, and $R$ respectively. The joint probability density function of the eigenvalues is of immediate interest. Therefore due to invariance under orthogonal transformation, we consider infinitesimal variations near $R = I$, that is $A = \Lambda$. Thus

$$A + dA = (I + dR)(\Lambda + d\Lambda)(I + dR)^T$$

Note that $(R + dR) = (I + dR)$ is orthogonal. Therefore

$$(I + dR)(I + dR)^T = I$$

Ignoring second order term yields $dR + dR^T = 0$. Substituting this relationship in the above equation, and ignoring second and third order terms yields

$$dA = (dR)\,\Lambda + d\Lambda - \Lambda\,(dR)$$

Thus

$$dA_{ii} = (dR_{ii})\,\lambda_i + d\lambda_i - \lambda_i\,(dR_{ii}) = d\lambda_i, \quad 1 \le i \le n$$
$$dA_{ij} = (dR_{ij})\,\lambda_j - \lambda_i\,(dR_{ij}) = (\lambda_j - \lambda_i)\,(dR_{ij}), \quad 1 \le i < j \le n$$

Finally

$$\prod_{1 \le i \le n} dA_{ii} \prod_{1 \le i < j \le n} dA_{ij} = \prod_{1 \le i \le n} d\lambda_i \prod_{i < j} (\lambda_j - \lambda_i)\, dR_{ij}$$

Then the magnitude of the Jacobian is given by $|J| = \prod_{i>j} |\lambda_i - \lambda_j|$, which is the absolute value of a Vandermonde determinant. It can also be proved that the probability of equal eigenvalues is zero.

41. Let $A = [A_{ij}]$ be a real-valued symmetric random matrix of size $n$. The matrix elements $A_{ij}, 1 \le i \le j \le n$ are stochastically independent Gaussian random variables, where $A_{ij} \sim \mathcal{N}(0,1), 1 \le i \le j \le n$. Let $M_m = \mathcal{E}\{tr(A^{2m})\}$ for each $m \in \mathbb{N}$. Prove that $M_0 = n$, and

$$M_m = \sum_{r=0}^{m-1} M_r M_{m-r-1}, \text{ as } n \to \infty, \ \forall\, m \in \mathbb{P}$$

Hint: It is evident that $M_0 = \mathcal{E}\{tr(A^0)\} = n$. For any $m \in \mathbb{P}$ write $M_m$ as

$$M_m = \mathcal{E}\{tr(A^{2m})\} = \mathcal{E}\left\{\sum_i (A^{2m})_{ii}\right\} = \sum_{i,j} \mathcal{E}\left\{A_{ij}(A^{2m-1})_{ji}\right\}$$

Use of Lemma 3.5 yields

$$\mathcal{E}\left\{A_{ij}(A^{2m-1})_{ji}\right\} = \sum_{k,l} \mathcal{E}(A_{ij}A_{kl})\,\mathcal{E}\left\{\frac{\partial(A^{2m-1})_{ji}}{\partial A_{kl}}\right\}$$

Also

$$\frac{\partial(A^{2m-1})_{ji}}{\partial A_{kl}} = \sum_{r=0}^{2m-2} (A^{2m-2-r})_{jk}(A^r)_{li}$$

Therefore

$$\mathcal{E}\left\{A_{ij}(A^{2m-1})_{ji}\right\} = \sum_{k,l} \mathcal{E}(A_{ij}A_{kl}) \sum_{r=0}^{2m-2} \mathcal{E}\left\{(A^{2m-2-r})_{jk}(A^r)_{li}\right\}$$

Thus

$$M_m = \sum_{i,j}\sum_{k,l} \mathcal{E}(A_{ij}A_{kl}) \sum_{r=0}^{2m-2} \mathcal{E}\left\{(A^{2m-2-r})_{jk}(A^r)_{li}\right\}$$

Observe that $\mathcal{E}(A_{ij}A_{kl}) = 1$ if: $i = j = k = l$, or $i \ne j, k = i, l = j$, or $i \ne j, k = j, l = i$. In all other cases $\mathcal{E}(A_{ij}A_{kl}) = 0$. As $n \to \infty$, significant contribution to $M_m$ occurs if $k = j$ and $l = i$. This results in

$$M_m \simeq \sum_{i,j} \sum_{r=0}^{2m-2} \mathcal{E}\left\{ \left(A^{2m-2-r}\right)_{jj} \left(A^r\right)_{ii} \right\}$$

$$= \sum_{r=0}^{2m-2} \mathcal{E}\left\{ tr\left(A^{2m-2-r}\right) tr\left(A^r\right) \right\}$$

In the above equation,

$$\mathcal{E}\left\{ tr\left(A^{2m-2-r}\right) tr\left(A^r\right) \right\}$$
$$\simeq \mathcal{E}\left\{ tr\left(A^{2m-2-r}\right) \right\} \mathcal{E}\left\{ tr\left(A^r\right) \right\}, \quad \text{for } 0 \le r \le (2m-2)$$

Note that if $r$ is an odd integer and $n \to \infty$, then $\mathcal{E}\left\{ tr\left(A^r\right) \right\} \to 0$. Therefore, $M_m$ can finally be expressed as $M_m = \sum_{r=0}^{m-1} M_r M_{m-r-1}$, $m \in \mathbb{P}$ as $n \to \infty$.

42. Consider a real-valued symmetric random matrix $A = [A_{ij}]$ of size $n$. The matrix elements $A_{ij}, 1 \le i \le j \le n$ are stochastically independent Gaussian random variables, where $A_{ii} \sim \mathcal{N}\left(0, 2\sigma^2\right), 1 \le i \le n$, and $A_{ij} \sim \mathcal{N}\left(0, \sigma^2\right), 1 \le j < i \le n$. Define $B = A/\sqrt{n}$. Prove that, as $n \to \infty$ the spectral density $f(x)$ of the matrix $B$ is given by

$$f(x) = \begin{cases} \dfrac{1}{2\pi\sigma^2}\sqrt{4\sigma^2 - x^2}, & |x| \le 2\sigma \\ 0, & |x| \ge 2\sigma \end{cases}$$

Hint: See A. Boutet de Monvel, and A. Khorunzhy (1999).

# References

1. Arora, S., and Barak, B., 2009. *Computational Complexity: A Modern Approach*, Cambridge University Press, Cambridge, Great Britain.

2. Ayres Jr., F., 1962. *Matrices*, Schaum's Outline Series, McGraw-Hill Book Company, New York, New York.

3. Bellman, R., 1995. *Introduction to Matrix Analysis*, Society of Industrial and Applied Mathematics, Philadelphia, Pennsylvania.

4. Boutet De Monvel, Λ., and Khorunzhy, Λ., 1999. "On the Norm and Eigenvalue Distribution of Large Random Matrices," Ann. Probability, Vol. 27, No. 2, pp. 913-944.

5. Brualdi, R. A., and Ryser, H., 1991. *Combinatorial Matrix Theory*, Cambridge University Press, Cambridge, Great Britain.

6. Davis, P. J., 1984. *The Mathematics of Matrices, A First Book on Matrix Theory and Linear Algebra*, Robert E. Krieger Publishing Company, Malabar, Florida.

7. Eves, H., 1966. *Elementary Matrix Theory*, Dover Publications, Inc., New York, New York.

8. Fiedler, M., 1986. *Special Matrices and their Applications in Numerical Mathematics*, Martinus Nijhoff Publishers, Dordrecht, The Netherlands.

9. Gelfand, I. M., Kapranov, M. M., and Zelevinsky, 1994. *Determinants, Resultants, and Multidimensional Determinants*, Birkhauser, Boston, Massachusetts.

10. Golub, G. H., and Loan, C. F. V., 1983. *Matrix Computations*, The John Hopkins University Press, Baltimore, Maryland.

11. Graham, A., 1987. *Nonnegative Matrices and Applicable Topics in Linear Algebra*, John Wiley & Sons, Inc., New York, New York.

12. Graybill, F. A., 1969. *Introduction to Matrices with Applications in Statistics*, Wadsworth Publishing Company, Inc., Belmont, California.

13. Hogben, L., Editor-in-Chief, 2007. *Handbook of Linear Algebra,* CRC Press: New York, New York.

14. Horn, R. A., and Johnson, C. R., 1985. *Matrix Analysis,* Cambridge University Press, Cambridge, Great Britain.

15. Korn, G. A., and Korn, T. M., 1968. *Mathematical Handbook for Scientists and Engineers*, Second Edition, McGraw-Hill Book Company, New York, New York.

16. Kreyszig, E., 2011. *Advanced Engineering Mathematics*, Tenth Edition, John Wiley & Sons, Inc., New York, New York.

17. Lancaster, P., 1969. *Theory of Matrices*, Academic Press, New York, New York.

18. Lévêque, O., 2015. "Random Matrices and Communication Systems: An Unexpected Journey: There and Back Again," available at http://ipg.epfl.ch/~leveque/Matrix/. Retrieved February 20, 2015.

19. Lipschutz, S., 1968. *Linear Algebra*, Schaum's Outline Series, McGraw-Hill Book Company, New York, New York.

20. Lütkepohl, H., 1996. *Handbook of Matrices*, John Wiley & Sons, Inc., New York, New York.

21. Marcus, M., and Minc, H., 1964. *A Survey of Matrix Theory and Matrix Inequalities*, Dover Publications, Inc., New York, New York.

22. Mehta, M. L., 1991. *Random Matrices*, Second Edition, Academic Press, New York, New York.

23. Meyer, C., 2000. *Matrix Analysis and Applied Linear Algebra,* Society of Industrial and Applied Mathematics, Philadelphia, Pennsylvania.

24. Miller, S. J., and Takloo-Bighash, R., 2006. *An Invitation to Modern Number Theory*, Princeton University Press, Princeton, New Jersey.

25. Minc, H., 1988. *Nonnegative Matrices*, John Wiley & Sons, Inc., New York, New York.

26. Noble, B., 1969, *Applied Linear Algebra*, Prentice-Hall, Englewood Cliffs, New Jersey.

27. Rosen, K. H., Editor-in-Chief, 2018. *Handbook of Discrete and Combinatorial Mathematics,* Second Edition, CRC Press: Boca Raton, Florida.

28. Seneta, E., 1981. *Nonnegative Matrices and Markov Chains*, Second Edition, Springer-Verlag, Berlin, Germany.

29. Stewart, G. W., 1973. *Introduction to Matrix Computations*, Academic Press, New York, New York.

30. Stöckmann, H.-J., 1999. *Quantum Chaos, An Introduction*, Cambridge University Press, Cambridge, Great Britain.

31. Varga, R. S., 1962. *Matrix Iterative Analysis*, Prentice-Hall, Englewood Cliffs, New Jersey.

32. Watkins, D. S., 2002. *Fundamentals of Matrix Computations*, Second Edition, John Wiley & Sons, Inc., New York, New York.

# Graph Theory

$$|V| - |E| + |F| = 2$$

In a connected planar graph,
$V$ = vertex set, $E$ = edge set,
$F$ = set of faces.
Euler's formula

**Leonhard Euler.** Euler was born on 15 April, 1707 in Basel, Switzerland. He was tutored by Johann Bernoulli (1667-1748). Euler contributed to several branches of mathematics including number theory, real analysis, complex analysis, calculus of variations, graph theory, and differential equations. He also worked in the general area of mechanics, optics, electricity, and magnetism.

Euler lost sight in his right eye when he was twenty eight years old, and in the left eye before he was sixty years old. In spite of his blindness, he continued to publish his mathematical results prodigiously by dictating them. Euler is generally regarded as the most prolific mathematics researcher of all time. He published over 500 papers in his lifetime, and another 350 papers posthumously. Besides, he published several outstanding textbooks on algebra, trigonometry, calculus, mechanics, dynamics, calculus of variations, astronomy, optics, and several other subjects.

Pierre Simon De Laplace (1749-1827) told his students: *Read Euler, read Euler, he is our master in everything.* During his lifetime, Euler worked at the St. Petersburg (now Leningrad) Academy and the Berlin Academy of Sciences. He won the Paris Academy Prize twelve times. Euler died on 18 September, 1783 in St. Petersburg, Russia.

## 4.1 Introduction

A graph is a collection of dots which are connected (and sometimes not connected) by lines in some pattern. These lines can be either straight or curved. Graph theory is the study of such spatial structures. Undirected and directed graphs are discussed in this chapter. Several special graphs are also defined. In addition, different graph operations, representations, and transformations are also outlined. Planar graphs, and some useful observations about graphs, and spanning trees are also discussed. The $\mathcal{K}$-core, $\mathcal{K}$-crust, and $\mathcal{K}$-shell of a graph are introduced. Basics of matroid theory, and graph spectrum are also studied. Some results in this chapter are based upon the theory of matrices and determinants. This later subject is discussed in a different chapter.

This theoretical discipline is also a rich branch of applied mathematics. It has recently found several applications, which occur wherever connectedness of a system is important. Some of the applications of graph theory are: design of computer networks; routing of packets in a network; optimal chip layout, that is circuit design; minimum cost layout of telephone cables; and routing of tasks in a computer system. It is also useful in the study of reliability of networks; biology, information theory, operations research, stochastic processes, set theory, and sociology.

## 4.2 Undirected and Directed Graphs

A graph is generally used to model the connectivity of a networking system. A graph is made up of dots, called vertices and the lines connecting them. These lines are called edges or links or arcs. If these lines are not assigned a direction, then the corresponding graph is called an *undirected graph*. However, if these lines are assigned a direction, the associated graph is called a *directed graph* or simply a *digraph*. An undirected graph with multiple edges between a pair of vertices is called a *multigraph*. Similarly a directed graph with multiple edges between a pair of vertices is called a *multidigraph*. Undirected and directed graphs are studied in this section.

### 4.2.1  Undirected Graphs

Some concepts related to undirected graphs are described in the following definitions.

**Definitions 4.1.** *Undirected graphs.*

1. *An undirected graph is $G = (V, E)$, where $V$ is the set of vertices (vertex set) and $E$ is the set of edges (edge set). A member of the set $V$ is called a vertex. It is also referred to as a node or a point. A vertex or a node is typically represented pictorially as a point or a small circle, while an edge (link or arc) is represented as a line. An edge joins either a pair of different vertices or joins a vertex to itself.*

2. *A proper edge joins one vertex to another, while a self-loop or simply a loop joins a vertex to itself. A proper edge has two different endpoints, while a self-loop has a single endpoint.*

3. *The cardinality of sets $V$ and $E$ is denoted by $|V|$ and $|E|$ respectively, where $|V|$ is called the order of the graph $G$, and $|E|$ is called the size of the graph $G$.*

4. *The graph $G$ is finite, if both $|V|$ and $|E|$ are finite; otherwise it is infinite.*

5. *An edge $e \in E$ is defined by a vertex pair $(u, v)$, that is $e = (u, v)$, where $u, v \in V$. The edge joins the two vertices $u$ and $v$. Observe that no direction has been assigned to the edge. That is, the edge $(u, v)$ of an undirected graph can also be represented by $(v, u)$.*

6. *Vertices $u$ and $v$ are adjacent to each other if there is an edge $e = (u, v) \in E$. The edge $e$ is said to be incident upon vertices $u$ and $v$. Adjacent vertices are also called neighbors.*

7. *The degree of a vertex $v$ is the number of edges incident upon $v$. That is, the degree of a vertex $v$ is the number of proper edges incident upon it plus two times the number of self-loops. It is denoted by $d_v \triangleq \deg(v)$.*

8. *An isolated point of a graph $G$ is a vertex which is not joined by an edge to any other vertex of the graph.*

9. *A pendant vertex or a leaf is a vertex with degree one.*

10. *In a weighted graph, a number is assigned to each edge. This number is called the weight of the edge. Sometimes, the weight of an edge is also referred to as its cost or length.*

11. *Walks, paths, and circuits:*

    (a) *A walk in the graph $G$ is a sequence of nodes $u_1, u_2, \ldots, u_m$ where $m \geq 2$ and $(u_j, u_{j+1}) \in E$ for $1 \leq j \leq (m - 1)$.*

       (i) *If the graph is not weighted, then the length of this walk is $(m - 1)$.*

       (ii) *If the graph is weighted, and the weight (length) of each edge is a nonnegative number, then the length of this walk is the sum of the weights of all the edges in the walk.*

       *The walk is considered to be open, if its end vertices are distinct, otherwise it is closed.*

    (b) *If the walk has no repeated nodes, then it is called an elementary path, or simply a path.*

    (c) *A circuit or a cycle is a walk (sequence of vertices) with identical first and last vertices, and no repeated intermediate vertices. Some graph theorists require the definition of a cycle to have a minimum of three different vertices that has no repeated vertices except the first and the last.*

12. *Hamiltonian paths and cycles*:
    (a) *A Hamiltonian path is a path which contains all the vertices of the graph.*
    (b) *A Hamiltonian cycle is a cycle which contains all the vertices of the graph. The corresponding graph is called Hamiltonian.*
13. *A multi-edge in an undirected graph $G$ is a set of more than one edge between the same pair of vertices. If an undirected graph has multiple edges between the same pair of nodes, then the graph is called a multigraph.*
14. *A graph without self-loops is called a loopless graph.*
15. *A graph without multi-edges and self-loops is called a simple graph.*                    ☐

The Hamiltonian paths and cycles are named after the Irish mathematician William Rowan Hamilton (1805-1865).

**Example 4.1.** Consider an undirected graph $G = (V, E)$, where $V = \{v_1, v_2, v_3, v_4\}$ is the vertex set, and $E = \{a, b, c, d, e\}$ is the edge set. See Figure 4.1.

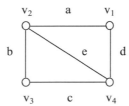

Figure 4.1. Undirected graph.

In this graph $|V| = 4$, and $|E| = 5$. Also,

$$a = (v_1, v_2), b = (v_2, v_3), c = (v_3, v_4), d = (v_1, v_4), \text{ and } e = (v_2, v_4)$$

The degree of nodes $v_1, v_2, v_3$, and $v_4$ are $2, 3, 2$, and $3$ respectively. Note that the sum of degrees of all the nodes in the graph is equal to 10, which is equal to twice number of edges in the graph.    ☐

A generalization of an undirected graph is a directed graph.

### 4.2.2  Directed Graphs

Directions are assigned to an edge in a directed graph. In this case the corresponding graph is called a directed graph, or simply a digraph. Actually, an undirected graph can be transformed to a directed graph by replacing each edge by two edges in opposite directions. Several concepts about directed graphs are similar to those of undirected graphs.

**Definitions 4.2.** *Directed graphs.*

1. *A digraph is $G = (V, E)$, where $V$ is a set of vertices (vertex set), and $E$ is a set of ordered pairs of vertices called arcs (edge set). An arc is simply a directed edge. If the context is clear, an arc is sometimes simply referred to as an edge or a link.*

2. *The tail of an arc is the vertex (or node) at which the arc originates and the head of an arc is the vertex at which the arc terminates. Let an arc be $e = (u, v)$ where $u, v \in V$. The vertices $u$ and $v$ are the tail and head of the arc $e$ respectively. The vertex $v$ is called the successor of vertex $u$. The vertex $u$ is called the predecessor of vertex $v$. An edge in a directed graph is represented by a line with a tip of an arrow at the head of the arc.*

3. *The set of arcs which terminate on a vertex $v$ are called its in-edges or in-arcs. Similarly, the set of arcs which emanate from a vertex $v$ are called its out-edges or out-arcs.*

4. *If $e = (u, u)$ then the edge $e$ is called a self-loop or simply a loop.*

5. *The out-degree of a vertex $v$ is the number of arcs in the set $E$ which emanate from the vertex $v$. Similarly, the in-degree of a vertex $v$ is the number of arcs in the set $E$ that are incident on the vertex $v$.*

   *Denote the out-degree of a vertex $v$, by $d_v^+$. It is equal to the number of arcs with tail at $v$. Also denote the in-degree of a vertex $v$, by $d_v^-$. It is equal to the number of arcs with head incident at $v$. The degree of a vertex $v$, is denoted by $d_v \triangleq \deg(v)$, where*

$$d_v = d_v^+ + d_v^- \tag{4.1}$$

6. *Directed walks, directed paths, and directed circuits:*

   (a) *A directed walk in the graph $G$ is a sequence of nodes $u_1, u_2, \ldots, u_m$ where $m \geq 2$ and $(u_j, u_{j+1}) \in E$ for $1 \leq j \leq (m-1)$.*

      (i) *If the graph is not weighted, then the length of this walk is $(m-1)$.*

      (ii) *If the graph is weighted, and the weight (length) of each edge is a nonnegative number, then the length of this walk is the sum of the weights of all the edges in the walk.*

      *The directed walk is considered to be open, if its end vertices are distinct, otherwise it is closed.*

   (b) *If the directed walk has no repeated nodes, then it is called an elementary directed path or simply a directed path.*

   (c) *A directed circuit or cycle is a directed walk (sequence of vertices) with identical first and last vertices, and no repeated intermediate vertices.*

7. *A multi-edge in a directed graph $G$ is a set of more than one arc between the same pair of vertices. These arcs have the same head and tail. The corresponding graph is called a multidigraph.*

8. *A digraph without self-loops is called a loopless digraph.*

9. *A directed graph without multi-edges and self-loops is called either a strict or simple digraph.*

$\square$

**Example 4.2.** Consider a directed graph $G = (V, E)$, where $V = \{v_1, v_2, v_3, v_4\}$ is the vertex set, and $E = \{a, b, c, d, e\}$ is the edge set. See Figure 4.2.

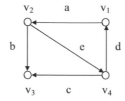

Figure 4.2. Directed graph.

In this graph $|V| = 4$, and $|E| = 5$. Also,

$$a = (v_1, v_2), b = (v_2, v_3), c = (v_4, v_3), d = (v_4, v_1), \quad \text{and} \quad e = (v_2, v_4)$$

The in-degrees of nodes $v_1, v_2, v_3$, and $v_4$ are $1, 1, 2$, and $1$ respectively. Similarly the out-degrees of nodes $v_1, v_2, v_3$, and $v_4$ are $1, 2, 0$, and $2$ respectively. Note that the sum of in-degrees and out-degrees of all the nodes in the graph is equal to 10, which is equal to twice number of edges in the graph. □

## 4.3 Special Graphs

We sometimes study graphs with special structure, because these might be convenient to analyse. It is hoped that these special graphs would provide more insight into the properties of graphs without any obvious structure. Some special graphs are listed below.

**Definitions 4.3.** *Some special graphs.*

1. *Null graph and empty graph*:
    (a) *A graph with both empty vertex and edge sets is called a null graph.*
    (b) *A graph with no edges is sometimes called an empty graph. Other times, a null graph is also called an empty graph.*
2. *Connected, disconnected, and biconnected graph*:
    (a) *A graph is connected if every pair of distinct nodes $v_i$ and $v_j$ is connected by a path. If the graph is a digraph and connected, then it is also sometimes referred to as strongly connected. In the later case, the path has to exist from both $v_i$ to $v_j$, and $v_j$ to $v_i$.*
    (b) *A graph is disconnected, if there is at least one pair of nodes which cannot be linked. A disconnected graph can be split into several components in which each component graph is connected.*
    (c) *A biconnected undirected graph is a connected graph, which is not disconnected by deleting any single vertex, and its incident edges.*
    *A digraph is biconnected if every pair of distinct nodes $v_i$ and $v_j$ is connected by at least a pair of paths which have no vertices in common, except $v_i$ and $v_j$.*
3. *The diameter of a connected and weighted graph, with nonnegative weights, is defined as the largest distance between any two vertices of the graph. In this context, the distance between any two vertices $u$ and $v$ is the smallest length of a path from $u$ to $v$. A disconnected graph is said to have an infinite diameter.*
4. *A graph is said to be labeled, if each vertex in the graph is assigned a unique label. The graph is unlabeled, if the vertices of the graph are not assigned labels.*
5. *A tree is a connected graph without cycles. Let $G = (V, T)$ be a tree graph, then $G$ is said to span its set of nodes $V$.*
    (a) *A tree in which one vertex is specifically identified as a root node is called the rooted tree. All other nodes in the tree are called the descendents of the root node.*
    (b) *Each edge of this tree divides the set of vertices which are attached to either side of this edge into two groups. The set of nodes away from the root node are called the descendent nodes with respect to this edge. Similarly, the set of nodes towards the root node are called the ancestral nodes with respect to this edge.*

(c) *In a rooted tree, child of a vertex $v$ is a vertex, whose immediate ancestor is $v$. Vertex with children is called an internal vertex.*

(d) *The depth of a vertex in a rooted tree is the number of edges in the unique path from the root of the tree to the vertex.*

(e) *An ordered tree is a rooted tree in which the children (or branches) of every internal vertex are linearly ordered. That is, if the vertex has $k$ children, then it has a first child, second child, ..., and a $k$th child.*

(f) *A binary tree is an ordered rooted tree in which each internal vertex has at most two children (branches).*

(g) *An $m$-ary tree is an ordered rooted tree in which each internal vertex has at most $m$ children (branches), where $m = 2, 3, 4, \ldots$. Parameter $m$, is called the branching factor of the tree.*

(h) *The subtree rooted at a specific node of a tree, is a tree, which consists of all its descendent nodes, and the corresponding connecting edges.*

(i) *The height of a rooted tree is equal to the maximum number of edges in the unique path from its root to any vertex.*

(j) *A balanced tree of height $b$ is an $m$-ary tree in which all leaves are of height either $b$ or $(b-1)$, where $b = 2, 3, \ldots$.*

(k) *In a $d$-regular tree (typically infinite) each internal vertex has exactly $d$ children (branches).*

(l) *In a complete $d$-regular tree, all leaves are at the same depth.*

6. *If $G$ is any undirected and connected graph, then a spanning tree is a tree that includes (spans) all the nodes in the vertex set $V$.*

7. *A forest is an ensemble of tree graphs. It contains no cycles.*

8. *An undirected graph or a digraph which has no cycles is called acyclic.*

9. *The following graphs are undirected and simple.*

(a) *$K_n$ is a complete graph with $n$ vertices, where all pairs of vertices are adjacent.*

(b) *A star graph $S_n$ with $n$ vertices is a graph in which a single vertex is connected to all other vertices. There are no other edges in the graph. Thus this graph is a tree with $(n-1)$ edges.*

(c) *A ring graph $R_n$ with $n$ vertices is a graph in which the vertices are laid out on a circle. This graph has $n$ edges, and the degree of each vertex is equal to 2.*

(d) *A wheel graph $W_{n+1}$ is a ring graph $R_n$ and an additional vertex. This additional vertex is connected to all the vertices of the ring graph, where $n \geq 3$.*

(e) *An $n$-path graph $P_n$ is a graph with $n$ vertices, and $(n-1)$ edges, such that the graph appears like a path or a line.*

(f) *A ring-lattice $R_{n,2k}$ is a graph with $n$ vertices in which each vertex is connected to its first $2k$ neighbors ($k$ on either side).*

(g) *A bipartite graph is a graph in which the vertices are split into two mutually exclusive subsets, so that the vertices at the ends of any edge are in the two different subsets. That is, any pair of vertices within the same set are not adjacent to each other.*

(h) *A complete bipartite graph $K_{r,s}$ is a simple bipartite graph, in which the set of vertices is divided into two mutually exclusive subsets of size $r$ and $s$ vertices. Furthermore, each vertex in one subset is adjacent to every vertex in the other subset of vertices.*

(i) *A $t$-dimensional hypercube graph $H_t$ is a graph with $2^t$ vertices. In this graph each vertex is labeled by a bit string of length $t$. Further, two vertices are adjacent if and only if their bit-labels differ exactly in a single bit.*

    (j) *Hyperplane, closed half-space, and polyhedron or polyhedral set.*

      (i) *A hyperplane in the space $\mathbb{R}^t$ is the set of all points on a $(t-1)$-dimensional plane.*

     (ii) *A closed half-space in $\mathbb{R}^t$ is the union of the set of points on one side of a $(t-1)$-dimensional plane (hyperplane), and the set of points on this hyperplane.*

    (iii) *A polyhedron or polyhedral set is the intersection of a finite number of closed half-spaces in $\mathbb{R}^t$.*

10. *An arborescence is a directed rooted tree in which there is a unique node called the root, and all edges point away from it.*     □

Sometimes a digraph is defined to be connected if the path exists, either from $v_i$ to $v_j$, or from $v_j$ to $v_i$, or both, for $i \neq j$ and $1 \leq i, j \leq n$, where $|V| = n$. Note that it is also possible to define a wheel graph with either two or three vertices.

## 4.4 Graph Operations, Representations, and Transformations

In order for the graph theory to be an effective tool in understanding networks, it is necessary to study its efficient representation. Furthermore, a study of relatively simple graph transformations can be used to increase the repertoire of graph-theoretic algorithms. Thus in order to make graph theory a practical tool, a knowledge of graph operations, representations, and transformations is necessary.

### 4.4.1  Graph Operations

Before we describe graph operations, the notion of a subgraph of a graph is introduced.

    **Definitions 4.4.** *Subgraphs.*

1. *A subgraph of a graph $G = (V, E)$ is a graph $G_s = (V_s, E_s)$ where $V_s \subseteq V$ and $E_s \subseteq E$. Note that if $e = (u, v) \in E_s$, then $u, v \in V_s$.*
2. *A spanning subgraph of $G$ is a subgraph where $V_s = V$.*
3. *A clique in a graph $G$ is a complete subgraph of $G$ which is not contained in a larger and complete subgraph of $G$.*
4. *A subtree is a subgraph of a tree which is also a tree.*     □

Based upon the definition of a subgraph of a graph, certain graph operations are defined. These operations have many applications.

    **Definitions 4.5.** *Let $G$ be a graph.*

1. *Deletion of an edge $e$ in a graph $G$ results in a subgraph denoted by $G - e$. This subgraph contains all the vertices of the graph $G$, and all edges of $G$ except the edge $e$.*
2. *Deletion of a vertex $v$ in a graph $G$ results in a subgraph denoted by $G - v$. This subgraph contains all the vertices of graph $G$ except $v$, and all the edges of $G$ except those incident upon the vertex $v$.*

3. *Contraction of an edge e in a graph G results in a graph in which the edge e is shrunk to a point. That is, the endpoints of the edge e are merged to form a single vertex. The rest of the graph remains unchanged. This graph is denoted by $G/e$.*                                        □

Next subsection describes different types of graph representations.

### 4.4.2 Graph Representations

Different representations of a graph are initially studied. These are: endpoint table, incident-edge table, list-of neighbors representation, adjacency set, adjacency matrix, and incidence matrix. Some graph representations like the adjacency matrix, and the incidence matrix are used in studying algebraic properties of graphs. Some other representations can be used appropriately for the computer representation of graphs.

(a) *Endpoint Table.* This is a representation of the graph in a table. It lists the endpoints of all edges.
(b) *Incident-Edge Table.* This is a tabular representation of the graph, where for each vertex $v$, a list of the edges incident upon this vertex is given. For a digraph, there are two sublists of edges for each vertex. The lists correspond to whether $v$ is a tail or a head.
(c) *List-of-Neighbors Representation.* For each vertex $v$, the immediate neighbors of the vertex $v$ are listed.
(d) *Adjacency Set.* The adjacency set $A_v$, of a vertex $v$ is the set of all arcs leaving this vertex. $A_v = \{(v, v_j) \mid (v, v_j) \in E\}$.
(e) *Adjacency Matrix.* Let $G$ be an undirected graph $(V, E)$. Let $V = \{v_1, v_2, \ldots, v_n\}$ and $E = \{e_1, e_2, \ldots, e_m\}$. The adjacency matrix $A$ is an $n \times n$ matrix. Let the elements of the matrix $A$ be $a_{ij}, 1 \leq i, j \leq n$. Then

$$a_{ij} = \begin{cases} \text{number of edges between } v_i \text{ and } v_j, & \text{if } i \neq j \\ \text{number of self-loops}, & \text{if } i = j \end{cases}$$

A row-sum or column-sum in the adjacency matrix of a *loopless* graph is equal to the degree of the corresponding vertex. If the graph is directed, then the elements of the corresponding adjacency matrix $A$ are

$$a_{ij} = \text{ number of arcs from } v_i \text{ to } v_j$$

A row-sum in the adjacency matrix of a digraph is equal to the out-degree of the corresponding vertex. And the column-sum is equal to the in-degree.

(f) *Incidence Matrix.* Let $G$ be an undirected graph $(V, E)$, where $V = \{v_1, v_2, \ldots, v_n\}$ and $E = \{e_1, e_2, \ldots, e_m\}$. The incidence matrix $I_G$ is an $n \times m$ matrix. Let the elements of this matrix be $b_{ij}, 1 \leq i \leq n$, and $1 \leq j \leq m$. Then

$$b_{ij} = \begin{cases} 0, & \text{if } v_i \text{ is not an endpoint of } e_j \\ 1, & \text{if } v_i \text{ is an endpoint of a proper edge } e_j \\ 2, & \text{if } e_j \text{ is a self-loop at } v_i \end{cases}$$

In this matrix every row-sum is equal to the degree of the corresponding vertex, and each column-sum is equal to 2. For a digraph without self-loops, the elements of the corresponding incidence matrix $I_G$ are

$$b_{ij} = \begin{cases} +1, & \text{if } v_i \text{ is the tail of } e_j \\ -1, & \text{if } v_i \text{ is the head of } e_j \\ 0, & \text{otherwise} \end{cases}$$

Some authors define the incidence matrix of a digraph by interchanging the $+1$ and $-1$ entries in the above specification.

The above representations are next illustrated via an example.

**Example 4.3.** Consider an undirected graph $G = (V, E)$, where $V = \{v_1, v_2, v_3, v_4\}$, and $E = \{e_1, e_2, e_3, e_4, e_5, e_6\}$, and $e_1 = (v_1, v_3)$, $e_2 = (v_2, v_4)$, $e_3 = (v_2, v_3)$, $e_4 = (v_1, v_4)$, $e_5 = (v_1, v_4)$, and $e_6 = (v_1, v_1)$. See Figure 4.3.

We obtain the endpoint and the incident-edge tables of the graph. The list-of-neighbors representation, adjacency matrix, and the incidence matrix of the graph are also specified.

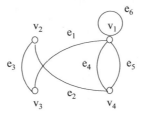

Figure 4.3. Graph $G = (V, E)$.

The endpoint table of this graph is

| $e_1$ | $e_2$ | $e_3$ | $e_4$ | $e_5$ | $e_6$ |
|-------|-------|-------|-------|-------|-------|
| $v_1$ | $v_2$ | $v_2$ | $v_1$ | $v_1$ | $v_1$ |
| $v_3$ | $v_4$ | $v_3$ | $v_4$ | $v_4$ |       |

The incident-edge table is

$$
\begin{aligned}
v_1 &: \quad e_1 \quad e_4 \quad e_5 \quad e_6 \\
v_2 &: \quad e_2 \quad e_3 \\
v_3 &: \quad e_1 \quad e_3 \\
v_4 &: \quad e_2 \quad e_4 \quad e_5
\end{aligned}
$$

The list-of-neighbors representation is

$$
\begin{aligned}
v_1 &: \quad v_1 \quad v_3 \quad v_4 \quad v_4 \\
v_2 &: \quad v_3 \quad v_4 \\
v_3 &: \quad v_1 \quad v_2 \\
v_4 &: \quad v_1 \quad v_1 \quad v_2
\end{aligned}
$$

The adjacency set of each vertex should be easily observable from the list-of-neighbors representation. The adjacency matrix of this graph is

$$
\begin{array}{c c}
& \begin{array}{cccc} v_1 & v_2 & v_3 & v_4 \end{array} \\
A = \begin{array}{c} v_1 \\ v_2 \\ v_3 \\ v_4 \end{array} &
\begin{bmatrix}
1 & 0 & 1 & 2 \\
0 & 0 & 1 & 1 \\
1 & 1 & 0 & 0 \\
2 & 1 & 0 & 0
\end{bmatrix}
\end{array}
$$

Note that the adjacency matrix $A$ of the undirected graph is symmetric. Furthermore, the sum of the elements in the first row (or column) of the matrix $A$ is equal to 4. This is not equal to the degree of the vertex $v_1$, which is actually 5. The incidence matrix of the graph is $I_G$.

Observe that the row-sums are equal to the degree of the corresponding vertex, and each column-sum is equal to 2.

$$
\begin{array}{c}
\phantom{I_G = v_1} e_1 \ \ e_2 \ \ e_3 \ \ e_4 \ \ e_5 \ \ e_6 \\[4pt]
I_G = \begin{array}{c} v_1 \\ v_2 \\ v_3 \\ v_4 \end{array}
\begin{bmatrix}
1 & 0 & 0 & 1 & 1 & 2 \\
0 & 1 & 1 & 0 & 0 & 0 \\
1 & 0 & 1 & 0 & 0 & 0 \\
0 & 1 & 0 & 1 & 1 & 0
\end{bmatrix}
\end{array}
$$

$\square$

### 4.4.3  Graph Transformations

Occasionally graph transformations can be used to translate the problem to another problem for which the solution is known. We discuss:

(a)  The node-splitting transformation
(b)  Transformation of an undirected graph to a directed graph

These transformations have both practical and theoretical implications.

**Node-Splitting Transformation**

Assume that some metrics like cost and capacity are assigned to each arc in the original graph. The node-splitting-transformation splits each node $v_i$ in a digraph to two nodes $v_{i'}$ and $v_{i''}$. One of these nodes corresponds to the node's *input* stream, and the other the *output* stream.

In this transformation, each arc $(v_i, v_j)$ is replaced by an arc $(v_{i'}, v_{j''})$ of the same cost and capacity. The nodes $v_{i''}$ and $v_{i'}$ are connected by an arc $(v_{i''}, v_{i'})$ of zero cost and infinite capacity. All the arcs which emanate (output) from the node $v_i$, emanate from the node $v_{i'}$, and all the arcs which are incident (input) upon the node $v_i$, are incident upon the node $v_{i''}$. Further, the arc $(v_{i''}, v_{i'})$ carries flow (stream) from the input side to the output side. See Figure 4.4 (a) and (b). Node-splitting transformation can possibly find use in computing disjoint paths on a digraph.

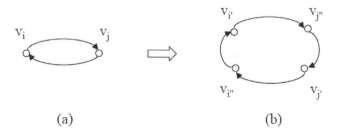

(a)                                                          (b)

Figure 4.4. Node-splitting transformation.

### Transformation of an Undirected Graph to a Directed Graph

Sometimes it is necessary to transform an undirected graph to a directed graph. This is accomplished by converting an edge in the undirected graph between the vertices $v_i$ and $v_j$ to two arcs $(v_i, v_j)$ and $(v_j, v_i)$ in the directed graph. If a cost is also assigned to the edge in the undirected graph, then the same values can be assigned to each of the two arcs in the directed graph.

In addition, assume that a capacity of $\tilde{u}$ units is assigned to the link between the two vertices $v_i$ and $v_j$ in the undirected graph. Capacity of a link is the maximum flow it can carry. If the traffic flow between the vertices $v_i$ and $v_j$, and $v_j$ and $v_i$, are $x_{ij}$ and $x_{ji}$ respectively, then $(x_{ij} + x_{ji}) \leq \tilde{u}$.

## 4.5 Plane and Planar Graphs

An undirected graph is a plane graph, if it can be drawn on a plane without its edges crossing each other. An undirected graph isomorphic to a plane graph is called a planar graph. An undirected graph which is not planar is called a nonplanar graph. Note that all graphs in this subsection are undirected. Some immediately useful properties of plane graphs are determined in this subsection.

**Definitions 4.6.** *Embedded, plane, planar, and nonplanar graphs.*

1. *Embedded graph. An undirected graph $G = (V, E)$ is embedded in a plane, if it can be drawn on the plane with no two intersecting edges, except possibly at the vertices.*
2. *Plane graph. Let $G = (V, E)$ be an undirected graph, where $V$ is the vertex set, and $E$ is the edge set. Th graph $G$ is plane if it can be drawn on the plane $\mathbb{R}^2$, where the vertex $v \in V$ is represented as a point $f(v) \in \mathbb{R}^2$, and an edge $(u, v) \in E$ is drawn as a continuous curve between points $f(u)$ and $f(v)$, such that no two edges intersect (except possibly at the vertices). More simply, a graph is plane, if it can be drawn on the plane without edge crossing. Further, the plane graph is said to be embedded in the plane.*
3. *Isomorphic graphs. Let $G_1 = (V_1, E_1)$, and $G_2 = (V_2, E_2)$, be two undirected graphs, where $V_1$ and $V_2$ are the sets of vertices of graphs $G_1$ and $G_2$ respectively. Similarly, $E_1$ and $E_2$ are the sets of edges of graphs $G_1$ and $G_2$ respectively. Further, there is a bijective map of their vertices: $h : V_1 \to V_2$ such that*

$$\forall\, u, v \in V_1, \ (u, v) \in E_1 \Rightarrow (h(u), h(v)) \in E_2 \tag{4.2a}$$
$$\forall\, u, v \in V_1, \ (u, v) \notin E_1 \Rightarrow (h(u), h(v)) \notin E_2 \tag{4.2b}$$

4. *An undirected graph is planar, if it is isomorphic to a plane graph.*
5. *An undirected graph is nonplanar, if it has no embedding in the plane.*
6. *Region, boundary of a region, and faces of a plane graph.*
    (a) *A region is the maximal subsurface of the plane which contains neither a vertex nor an edge of the graph representation.*
    (b) *A boundary of the region $R$ is a subgraph which contains all the vertices and edges incident upon $R$.*
    (c) *A face consists of a region and its boundary.* □

Thus a plane representation of a planar graph divides the plane into a number of connected regions, where each region is bounded by some edges and vertices of the graph. Properties of planar graphs can be studied from their corresponding plane embeddings. An important result in the theory of planar graphs is the Euler's formula which relates the number of vertices, edges, and faces of its plane embedding.

**Theorem 4.1.** *Let $G = (V, E)$ be a planar graph with $n$ vertices, $m$ edges, $q$ number of faces of its plane embedding, and $k$ connected components. Then*

$$n - m + q = k + 1$$

*Proof.* The result is proved by induction on the number of edges of the graph. The graph $G$ is built by the addition of edges successively. Initially we have an empty graph with $n$ vertices (which is a graph with zero edges). In this graph, $m = 0, q = 1$, and $k = n$. These values satisfy the stated result.

Next consider the planar graph $G_{j-1}$ with $n$ vertices, $(j - 1)$ edges, where $j \in \mathbb{P}$. The number of faces in its plane embedding is $q_{j-1}$, and the number of disconnected components is $k_{j-1}$. The corresponding relationship is

$$n - (j - 1) + q_{j-1} = k_{j-1} + 1$$

We next create a new graph $G_j$ by adding a new edge $e \in E$ to the graph $G_{j-1}$. The graph $G_j$ has $j$ edges. Two possibilities can occur.

*Case* 1: Let $e$ connect two disconnected components of $G_{j-1}$. In the graph $G_j$, we have $q_j = q_{j-1}$ and $k_j = (k_{j-1} - 1)$. Therefore

$$n - j + q_j = n - j + q_{j-1} = k_{j-1} = k_j + 1$$

Thus the stated result is satisfied.

*Case* 2: Let $e$ connect two vertices of the same component of $G_{j-1}$. In the graph $G_j$, we have $q_j = (q_{j-1} + 1)$ and $k_j = k_{j-1}$. Therefore

$$n - j + q_j = n - j + (q_{j-1} + 1) = k_{j-1} + 1 = k_j + 1$$

Thus the stated result is satisfied.                                                                    □

For a connected graph $(k = 1)$ the above result yields the celebrated Euler's formula (1752).

**Corollary 4.1.** *Euler's formula. Let $G = (V, E)$ be a connected planar graph with $n$ vertices, $m$ edges, and $q$ number of faces in its plane embedding. Then $(n - m + q) = 2$.*                            □

**Example 4.4.** We verify Euler's formula for a tree graph. A tree graph with $n$ vertices is planar. Note that, in this graph, the number edges $m = (n - 1)$, and $q = 1$. Thus

$$(n - m + q) = \{n - (n - 1) + 1\} = 2$$

□

**Observations 4.1.** Let $G$ be a simple, connected, and planar graph with $n$ vertices, $m$ edges, and $q$ number of faces in its plane embedding. Assume that $n \geq 3$.

1. $2m \geq 3q$. This result is also called the edge-face inequality.
2. $m \leq (3n - 6)$.                                                                          □

The above observations are established in the problem section.

**Example 4.5.** Consider the complete graph $K_5$. This graph has 5 vertices. That is, in this graph $n = 5$, and $m = \binom{5}{2} = 10$. These values do not satisfy the inequality $m \leq (3n - 6)$. Therefore $K_5$ is not a planar graph.                                                                □

### Dual Graphs

A close relative of a plane graph $G$ is its dual. The dual graph $G_D$ of the graph $G$ is obtained by making each face of $G$ a vertex. An edge in the graph $G_D$ is formed by linking two faces (regions) with a common edge of $G$.

**Definition 4.7.** *A bridge (edge) or a cut-edge. An edge $e$ of a graph $G$ is a cut-edge if $G - e$ has more components than $G$.*                                                        □

**Definition 4.8.***Dual graph. Let $G = (V, E)$ be a plane graph. Let the set of faces of the graph $G$ be $F$. The dual graph of $G$ is $G_D = (V_D, E_D)$, where $V_D$ and $E_D$ are the set of vertices and edges respectively of the graph $G_D$.*

*Each face $f \in F$ of the graph $G$ corresponds to a vertex $f_D \in V_D$ of $G_D$. For each edge $e \in F$ there is a corresponding edge $e_D \in E_D$.*

*If the edge $e \in E$ occurs on the boundary of faces $f$ and $g$, where $f, g \in F$, then an edge $e_D \in E_D$ joins the faces $f$ and $g$. However, if the edge $e \in E$ is a cut-edge, that is the edge lies completely in one face $f \in F$, then the corresponding edge $e_D \in E_D$ is a self-loop incident on the vertex $f_D \in V_D$.*                                                        □

The term "dual" is used in the above definition, because if $G_D$ is a dual of the graph $G$, then $G$ itself is a dual of the graph $G_D$ if $G$ is connected.

**Observations 4.2.** Let $G = (V, E)$ be a plane graph, and its corresponding dual be $G_D = (V_D, E_D)$. Let the set of faces of the graph $G$ be $F$.

1. Let the graph $G$ be connected. Then, the graphs $G$ and $G_D$ are duals of each other.
2. The dual graph $G_D$ is a multigraph in general.
3. If $G$ is a connected graph, then it is isomorphic to its dual $G_D$.
4. The graph $G_D$ is plane. This essentially follows by the construction of the graph $G_D$.
5. The total number of edges in the two graphs are equal. That is, $|E| = |E_D|$.
6. Let $G$ be a simple, connected, and plane graph. Also, let $d_v$ denote the degree of vertex $v \in V$. Further, let $f_D \in V_D$ be the vertex in the dual graph $G_D$ which corresponds to the face $f \in F$. Then

$$2|E| = \sum_{v \in V} d_v, \quad 2|E_D| = \sum_{f_D \in V_D} d_{f_D}$$

Thus

$$2|E| = \sum_{v \in V} d_v = \sum_{f_D \in V_D} d_{f_D}$$

□

## 4.6 Some Useful Observations

Relatively simple yet important observations about undirected and directed graphs are summarized in this section. The reader should verify the validity of these statements.

**Observations 4.3.** Some observations about undirected graphs are listed below.

1. *Euler's theorem.* In a graph $G$, the sum of the degrees of vertices is equal to twice the number of edges.
2. In a graph $G$, the number of vertices of odd degree is even.
3. Let $A$ be the adjacency matrix of a simple graph $G$ with $n$ vertices.
   (a) The $ij$th element of the matrix $A^k, 1 \leq k \leq (n-1)$ is the number of different edge sequences (walks) of $k$ edges between the vertices $v_i$ and $v_j$.
   (b) Let $A^2 = \left[ a_{ij}^{(2)} \right]$, then

      (i) $a_{ij}^{(2)}, i \neq j$ is the number of walks with two 2 edges between vertices $v_i$ and $v_j$.
      (ii) $a_{ii}^{(2)} = $ degree of vertex $v_i$.
4. Let $A$ be the adjacency matrix of a simple graph $G$ with $n$ vertices. Define

$$Y = \sum_{k=1}^{(n-1)} A^k$$

The graph $G$ is a disconnected graph if and only if there exists at least one nondiagonal element of the matrix $Y$ that is zero. Use of this equation to test the connectedness of a graph is computationally inefficient.
5. A graph is said to be disconnected if and only if its vertices can be ordered such that its adjacency matrix $A$ can be represented as

$$A = \begin{bmatrix} A_1 & 0 \\ 0 & A_2 \end{bmatrix}$$

where $A_1$ and $A_2$ are adjacency matrices of the disconnected subgraphs $G_1$ and $G_2$ respectively, and $0$'s in the above expression for matrix $A$ are matrices (of appropriate size) of all zeros. That is, there is no edge joining any pair of vertices in the two subgraphs.
6. Some properties of a tree are summarized below
   (a) The number of edges in a tree with $n$ vertices is equal to $(n-1)$.
   (b) The tree has no cycles. That is, it is an acyclic graph.
   (c) There is no more than a single path between a pair of vertices of a tree.
   (d) The tree graph is connected. Furthermore, if an edge is removed from the graph, then the remaining graph is disconnected.
   (e) A tree with $n \geq 2$ number of vertices has at least two leaves.
   (f) The total number of spanning trees in a labeled $K_n$ graph (with $n$ vertices) is $n^{n-2}$, for $n \geq 2$. This result is derived later in the chapter. $\qquad\square$

**Observations 4.4.** Some observations about digraphs are listed below.

1. In any digraph $G = (V, E)$, the sum of in-degrees, and the sum of out-degrees are each equal to $|E|$, the total number of arcs in the graph. That is

$$\sum_{v \in V} d_v^- = \sum_{v \in V} d_v^+ = |E|$$

2. All trees are acyclic graphs, but the converse is not true.
3. A digraph is acyclic if and only if its vertices can be sequenced such that its adjacency matrix is an upper triangular matrix.
   Equivalently, a digraph is acyclic if and only if the vertices can be numbered such that there exists an arc from $v_i$ to $v_j$ only if $i < j$.
4. An arborescence is a directed acyclic graph. However every directed acyclic graph is not an arborescence. $\qquad\square$

Two more special matrices are of especial interest to us. These are the degree and the Laplacian matrices of an undirected graph. We demonstrate later in the chapter that the eigenvalues of the Laplacian matrix of an undirected and simple graph characterize its connectivity.

**Definitions 4.9.** *Let $G = (V, E)$ be an undirected and simple graph, where $|V| = n$. Denote its adjacency matrix by $A$.*

1. *Let $D = [d_{ij}], 1 \leq i, j \leq n$ be a matrix with degrees of vertices on its diagonal, and the remaining elements are each equal to $0$. Let the degree of a vertex $v_i$ be $\deg(v_i), 1 \leq i \leq n$. Therefore*

$$d_{ij} = \begin{cases} \deg(v_i), & \text{if } i = j \\ 0, & \text{if } i \neq j \end{cases} \quad 1 \leq i, j \leq n \qquad (4.3)$$

   *$D$ is called the degree matrix of the graph $G$.*
2. *The Laplacian matrix of the graph $G$ is $L = (D - A)$.* $\qquad\square$

**Observations 4.5.** Let $A$ and $L$ be the adjacency and Laplacian matrices respectively of an undirected and simple graph $G$ with $n$ vertices.

1. The Laplacian matrix is a square matrix in which each row and each column sum to zero. Thus $\det L = 0$.
2. All the cofactors of the matrix $L$ are identical. $\qquad\square$

The second of the above observations is true for any square matrix with each row and each column sum zero. The eigenvalues and eigenvectors of the Laplacian matrix are studied in the section on graph spectral analysis. These further provide insight into the connectivity aspects of the graph. Lemmas which are useful in counting the spanning trees of a graph are next established.

**Lemma 4.1.** *The determinant of any square submatrix of the incidence matrix $I_G$ of a digraph without self-loops is equal to either $1, -1,$ or $0$.*
*Proof.* See the problem section. $\qquad\square$

**Lemma 4.2.** *Let $G$ be a connected digraph with $n$ nodes, but without self-loops. The rank of its incidence matrix $I_G$ is $(n - 1)$.*
*Proof.* See the problem section. $\qquad\square$

**Definitions 4.10.** *Unimodular matrix, and reduced incidence matrix of a connected digraph without self-loops.*

1. *If the determinant of all the square submatrices of a matrix are equal to $1, -1,$ or $0,$ then the matrix is said to be unimodular.*
2. *The reduced incidence matrix $I_\alpha$ of a connected digraph without self-loops, is the matrix obtained from the incidence matrix $I_G$ by deleting some row $k,$ where $1 \leq k \leq n.$*   □

Using this definition of unimodularity, Lemmas 4.1 and 4.2 can be restated. That is, the incident matrix of a digraph without self-loops is unimodular. Also, the ranks of the matrices $I_G$ and $I_\alpha$ are identical. The following lemmas about the reduced incidence matrix readily follow.

**Lemma 4.3.** *The determinant of any nonsingular square submatrix $Q$ of $I_G,$ (or of the reduced incidence matrix $I_\alpha$) is equal to either $1$ or $-1.$*
   *Proof.* See the problem section.   □

**Lemma 4.4.** *Let $Q$ be a square submatrix of the reduced incidence matrix $I_\alpha.$ Let its order be equal to $(n-1).$ Then the matrix $Q$ is nonsingular if and only if the edges corresponding to its columns form a spanning tree of $G.$*
   *Proof.* The $(n-1)$ edges corresponding to the columns of the matrix $Q$ form a subgraph of $G.$ Denote this subgraph by $H.$ Then the matrix $Q$ is the reduced incidence matrix of the subgraph $H.$ Therefore $Q$ is nonsingular iff rank of matrix $Q$ is equal to $(n-1)$ iff $H$ is connected iff $H$ is a tree by Lemmas 4.2 and 4.3.   □

## 4.7 Spanning Trees

Recall that a tree is a connected graph without cycles. Assume that this graph is undirected. Removal of a single edge makes this graph disconnected. The number of edges in a tree with $n$ vertices is equal to $(n-1),$ where we assume that $n \geq 2.$ Furthermore, there is no more than a single path between a pair of vertices of a tree. The total number of spanning trees of a connected graph can be computed via the *matrix-tree theorem*. This matrix-tree theorem was originally discovered by Gustav Robert Kirchoff (1824-1887). The proof of this theorem utilizes Binet-Cauchy theorem. As a consequence of the matrix-tree theorem, the formula for the number of labeled trees with $n$ vertices is also established.

An alternate technique to compute the number of spanning trees of a graph is also outlined in this section. The spanning trees of a graph can be explicitly generated via the so-called *variable matrix-tree theorem*. The utility of this theorem is also demonstrated via an example. We shall assume in this section that the number of vertices $n$ in the graph is greater than one.

### 4.7.1 Matrix-Tree Theorem

Techniques to determine the number of spanning trees of an undirected graph are discussed in this subsection. Let $G = (V, E),$ $|V| = n,$ $|E| = m,$ and $n \geq 2,$ be an undirected, connected, simple and labeled graph. Also let $D$ and $L$ be its degree and Laplacian matrices respectively. The total number of spanning trees of the graph $G$ is generally determined by the matrix-tree theorem.

Denote the total number of spanning trees of the graph $G$ by $\eta(G)$. The following lemmas are initially established.

**Lemma 4.5.** *Let $G = (V, E)$ be an undirected, simple, and connected graph, with $|V| \geq 2$. This graph is specified by its adjacency matrix $A$. Let the incidence and Laplacian matrices of the graph $G$ be $I_G$ and $L$ respectively. The elements of each column of this matrix $I_G$ are all $0$'s except two entries which are $1$'s. Form a matrix $C$ by replacing anyone of the two $1$'s by $-1$ in all the columns of matrix $I_G$. Then*

$$CC^T = L \tag{4.4}$$

*Proof.* Let $|V| = n$, and $|E| = m$. The matrix $C$ is of size $n \times m$. Let $C = [c_{ij}]$, then the $(i, j)$th element of the matrix $CC^T$ is equal to $\sum_{k=1}^{m} c_{ik}c_{kj}$. It can be observed by direct computation that this $(i, j)$th element is equal to $\deg(v_i)$ if $i = j$; and equal to $-1$ if $i \neq j$, $(v_i, v_j) \in E$; and $0$ otherwise. These are indeed the elements of the Laplacian matrix $L$, which is equal to $(D - A)$. $\qquad\square$

**Lemma 4.6.** *Let $G = (V, E)$ be an undirected, simple, and connected graph with $|V| \geq 2$. Its adjacency and incidence matrices are $A$ and $I_G$ respectively. The elements of each column of this matrix $I_G$ are all $0$'s except two entries which are $1$'s. Form a matrix $C$ by replacing any one of the two $1$'s by $-1$ in each of the columns. The matrix $C$ is an incidence matrix of some directed graph. Form a reduced incidence matrix of this graph by deleting any single row of this matrix $C$. Denote this matrix by $B$. The total number of spanning trees of the graph $G$ is given by $\det\left(BB^T\right)$.*

*Proof.* Let $|V| = n$, and $|E| = m$. The matrix $B$ is $(n-1) \times m$. Note that $(n-1) \leq m$. Then use of Binet-Cauchy theorem yields

$$\det\left(BB^T\right) = \sum_{i=1}^{J} \det\left(B_i B_i^T\right), \quad J = \binom{m}{n-1}$$

where $B_i$ is a major submatrix of matrix $B$ of size $(n-1) \times (n-1)$ for $1 \leq i \leq J$. Then

$$\det\left(BB^T\right) = \sum_{i=1}^{J} \det(B_i)\det\left(B_i^T\right) = \sum_{i=1}^{J} \{\det(B_i)\}^2$$

where $\{\det(B_i)\}^2$ contributes a value of $1$ for each nonsingular matrix $B_i$, which in turn corresponds to a single spanning tree of the graph $G$ as per Lemma 4.4. $\qquad\square$

The well-known matrix-tree theorem is next stated. It follows by combining Lemmas 4.5 and 4.6.

**Theorem 4.2.** *Let $G = (V, E)$ be an undirected, simple, and connected graph with $|V| \geq 2$, and adjacency matrix $A$. Its degree and Laplacian matrices are $D$ and $L$ respectively. Then the total number of spanning trees of the graph $G$ is equal to cofactor of any element of the Laplacian matrix $L$. Let $L_{rs}$ be a matrix obtained by deleting the row $r$ and column $s$ of matrix $L$. Then the total number of spanning trees of the graph $G$ is*

$$\eta(G) = (-1)^{r+s} \det L_{rs} \tag{4.5}$$

*Proof.* Let $|V| = n$, and $|E| = m$. Observe that $\eta(G)$ is equal to $0$ if $m < (n-1)$. Therefore assume $m \geq (n-1)$.

The sum of each row and column of the Laplacian matrix $L$ is equal to $0$. Therefore all the cofactors of this matrix are equal. The stated result follows by using Lemmas 4.5 and 4.6.                                   $\square$

### 4.7.2  Numerical Algorithm

An algorithm for computing $\eta(G)$, the number of spanning trees of an undirected, simple, and connected graph is outlined in this subsection. It uses the result of the matrix-tree theorem, which requires the evaluation of a determinant of order $(n-1)$. Therefore this technique can be used if the order of this determinant is large. This is accomplished via a technique called Gaussian method in numerical analysis. Initially, the determinant is converted to an upper triangular form. The value of the determinant is next obtained by multiplying its diagonal elements.

Let $L = [l_{ij}]$, and the cofactor of element $l_{kt}$ be $(-1)^{k+t} \det M$, where $M$ is a $(n-1) \times (n-1)$ square matrix. The matrix $M$ is obtained from the Laplacian matrix $L$ by deleting its row $k$ and column $t$. Let $M = [m_{ij}]$, $1 \leq i, j \leq (n-1)$. An algorithm to compute the number of spanning trees is next outlined. It essentially computes the determinant of a matrix.

---

**Algorithm 4.1.** *Computation of Number of Spanning Trees of an Undirected Graph.*

---

**Input:** Undirected, simple, and connected graph $G = (V, E)$,
where $|V| = n$, and $n \geq 2$. Its Laplacian matrix is $L = [l_{ij}]$.
**Output:** $\eta(G)$, the number of spanning trees of graph $G$.
**begin**
    (*initialization*)
    (*find the cofactor of element $l_{kt}$, and let it be* $(-1)^{k+t} \det M$)
    $\eta(G) \leftarrow (-1)^{k+t}$
    **for** $r = 2$ **to** $(n-1)$ **do**
    **begin**
        $s \leftarrow (r-1)$
        **for** $i = r$ **to** $(n-1)$ **do**
        **begin**
            **for** $j = s$ **to** $(n-1)$ **do**
            **begin**
$$m_{ij} \leftarrow m_{ij} - \left( \frac{m_{is}}{m_{ss}} \right) m_{sj}$$
            **end** (*end of $j$ for-loop*)
        **end** (*end of $i$ for-loop*)
    **end** (*end of $r$ for-loop*)
    **for** $w = 1$ **to** $(n-1)$ **do**
    **begin**
        $\eta(G) \leftarrow \eta(G) \, m_{ww}$
    **end** (*end of $w$ for-loop*)
**end** (*end of algorithm to compute the number of spanning trees*)

---

The computational complexity of this algorithm is $O\left(n^{3}\right)$. The above facts are illuminated via the following example.

**Example 4.6.** Consider the undirected graph $G = (V, E)$, $|V| = 5$, and $|E| = 6$. The graph is completely specified by its adjacency matrix $A$.

$$
A = \begin{bmatrix}
0 & 1 & 1 & 1 & 0 \\
1 & 0 & 1 & 0 & 0 \\
1 & 1 & 0 & 1 & 0 \\
1 & 0 & 1 & 0 & 1 \\
0 & 0 & 0 & 1 & 0
\end{bmatrix}
$$

The total number of spanning trees $\eta\left(G\right)$, of this graph is computed. The degree matrix $D$ of this graph has $3, 2, 3, 3$, and $1$ on its diagonal. Therefore

$$
L = [l_{ij}] = \begin{bmatrix}
3 & -1 & -1 & -1 & 0 \\
-1 & 2 & -1 & 0 & 0 \\
-1 & -1 & 3 & -1 & 0 \\
-1 & 0 & -1 & 3 & -1 \\
0 & 0 & 0 & -1 & 1
\end{bmatrix}
$$

The cofactor of the element $l_{54}$ is

$$
\eta\left(G\right) = - \begin{vmatrix}
3 & -1 & -1 & 0 \\
-1 & 2 & -1 & 0 \\
-1 & -1 & 3 & 0 \\
-1 & 0 & -1 & -1
\end{vmatrix}
$$

$$
= \begin{vmatrix}
3 & -1 & -1 \\
-1 & 2 & -1 \\
-1 & -1 & 3
\end{vmatrix} = 8
$$

Therefore the total number spanning trees of this graph is equal to $8$.

The computation of $\eta\left(G\right)$ is verified via Gaussian elimination. The matrix obtained by deleting the first row and first column of matrix $L$ is

$$
M = \begin{bmatrix}
2 & -1 & 0 & 0 \\
-1 & 3 & -1 & 0 \\
0 & -1 & 3 & -1 \\
0 & 0 & -1 & 1
\end{bmatrix}
$$

After the iteration $r = 2$ in the algorithm, the matrix $M$ is

$$
M = \begin{bmatrix}
2 & -1 & 0 & 0 \\
0 & 2.5 & -1 & 0 \\
0 & -1 & 3 & -1 \\
0 & 0 & -1 & 1
\end{bmatrix}
$$

After the iteration $r = 3$ in the algorithm

$$
M = \begin{bmatrix}
2 & -1 & 0 & 0 \\
0 & 2.5 & -1 & 0 \\
0 & 0 & 2.6 & -1 \\
0 & 0 & -1 & 1
\end{bmatrix}
$$

Finally, after the iteration $r = 4$

$$M = \begin{bmatrix} 2 & -1 & 0 & 0 \\ 0 & 2.5 & -1 & 0 \\ 0 & 0 & 2.6 & -1 \\ 0 & 0 & 0 & \frac{8}{13} \end{bmatrix}$$

The product of the diagonal terms of the last matrix is $\eta(G) = 8$ as expected.                                          □

The celebrated result which counts the number of labeled trees on $n$ vertices is outlined in the next subsection.

### 4.7.3  Number of Labeled Trees

The number of labeled trees on $n$ vertices is determined via the matrix-tree theorem.

**Theorem 4.3.** *The total number of labeled trees with $n$ vertices is $n^{n-2}$, for $n \geq 2$.*

*Proof.* This expression is proved by determining the total number of spanning trees of a complete graph with $n$ vertices. The total number of spanning trees is obtained by evaluating the $(n-1) \times (n-1)$ determinant.

$$\begin{vmatrix} (n-1) & -1 & -1 & \cdots & -1 & -1 \\ -1 & (n-1) & -1 & \cdots & -1 & -1 \\ \vdots & \vdots & \vdots & \ddots & \vdots & \vdots \\ -1 & -1 & -1 & \cdots & -1 & (n-1) \end{vmatrix}$$

Adding all rows to the first row yields

$$\begin{vmatrix} 1 & 1 & 1 & \cdots & 1 & 1 \\ -1 & (n-1) & -1 & \cdots & -1 & -1 \\ \vdots & \vdots & \vdots & \ddots & \vdots & \vdots \\ -1 & -1 & -1 & \cdots & -1 & (n-1) \end{vmatrix}$$

Adding the first row to the other rows yields

$$\begin{vmatrix} 1 & 1 & 1 & \cdots & 1 & 1 \\ 0 & n & 0 & \cdots & 0 & 0 \\ 0 & 0 & n & \cdots & 0 & 0 \\ \vdots & \vdots & \vdots & \ddots & \vdots & \vdots \\ 0 & 0 & 0 & \cdots & n & 0 \\ 0 & 0 & 0 & \cdots & 0 & n \end{vmatrix}$$

which is equal to $n^{n-2}$.                                                                                             □

The above result was discovered by C. W. Borchardt in 1860. A. Cayley provided an independent derivation in 1889.

### 4.7.4 Computation of Number of Spanning Trees

Note that $\eta\,(G)$ is the number of spanning trees of a graph. Next, recall that the deletion of an edge $e$ in a graph $G$ results in a subgraph denoted by $(G - e)$. This subgraph contains all the vertices and edges of the graph $G$, except the edge $e$. Also the contraction of an edge $e$ in a graph $G$ results in a graph in which the edge $e$ is shrunk to a point. That is, the endpoints of this edge are merged to form a single vertex. The rest of the graph remains unchanged. This graph is denoted by $G/e$.

In summary "$-e$" denotes *edge deletion*, and "$/e$" denotes *edge contraction*. This yields the recursive formula

$$\eta\,(G) = \eta\,(G - e) + \eta\,(G/e)$$

The two graphs $(G - e)$ and $G/e$ are smaller in size than $G$. Therefore repeated application of this recursive relationship yields $\eta\,(G)$.

### 4.7.5 Generation of Spanning Trees of a Graph

A stronger version of the matrix-tree theorem called the variable matrix-tree theorem is outlined. The use of this theorem helps in listing the spanning trees of an undirected, simple, and connected graph $G = (V, E)$, where $|V| = n, |E| = m, (n - 1) \leq m$, and $n \geq 2$. Let the edges of this graph be $E = \{y_i \mid 1 \leq i \leq m\}$. Then a spanning tree can be specified as $\tau_k$.

$$\tau_k = y_{k_1} y_{k_2} \cdots y_{k_{n-1}} \triangleq \prod_{i=1}^{n-1} y_{k_i}, \quad 1 \leq k \leq \eta\,(G)$$

where the $y_{k_i}$'s are edges of the tree. All the spanning trees of the graph $G$ can be represented compactly as $\wp\,(G) = \sum_{k=1}^{\eta(G)} \tau_k$. Note that $\wp\,(G)$ is called the *spanning-tree polynomial* of the graph $G$.

**Lemma 4.7.** *Let $G = (V, E)$ be an undirected, simple, and connected graph. Let $|V| = n, |E| = m, (n - 1) \leq m$, and $n \geq 2$. Let the edges of the graph be $E = \{y_i \mid 1 \leq i \leq m\}$. This graph is specified by its adjacency matrix $A$. Let the incidence matrix of this graph be $I_G$. Then the elements of each column of this matrix $I_G$ are all $0$'s except two entries which are $1$'s. Form a matrix $C$ by replacing any one of the two $1$'s by $-1$ in each of the columns. Let $W$ be a diagonal matrix of size $m \times m$, with $w_{ii} = y_i, 1 \leq i \leq m$. Then*

$$N = CWC^T \tag{4.6}$$

*where $N = [n_{ij}]$ is an $n \times n$ matrix. The element $n_{ij} = -y_p$ if the edge $y_p$ joins the vertices $v_i$ and $v_j$, and $i \neq j$; $n_{ii} = y_{p_1} + y_{p_2} + \ldots + y_{p_h}$, where $\deg\,(v_i) \triangleq h$, and the edges $y_{p_t}$ for $1 \leq t \leq h$ are incident on the vertex $v_i$; and $n_{ij} = 0$ otherwise.*

*Proof.* The proof is left to the reader.                                                      $\square$

The next theorem is called the variable matrix-tree theorem.

**Theorem 4.4.** *The cofactor of any element of the matrix $N$ is the spanning-tree polynomial of the graph $G$.*

*Proof.* The proof is left to the reader. This proof is similar to the matrix-tree theorem, albeit a little more complicated.                                                      $\square$

**Corollary 4.2.** *Substitute* $y_i = 1$, *for* $1 \leq i \leq m$ *in the expression for the spanning-tree polynomial* $\wp(G)$. *The resulting value of the summation is equal to the number of spanning trees* $\eta(G)$, *of the graph* $G$. □

The use of this theorem is demonstrated explicitly by generating the spanning trees of a graph.

**Example 4.7.** Generate the spanning trees of the graph of Example 4.6. The vertex set of this graph is $V = \{v_1, v_2, v_3, v_4, v_5\}$, and the edge set $E = \{a, b, c, d, e, f\}$ where

$$a = (v_1, v_2), \quad b = (v_2, v_3), \quad c = (v_3, v_4),$$
$$d = (v_4, v_5), \quad e = (v_1, v_3), \text{ and } f = (v_1, v_4)$$

Thus

$$N = \begin{bmatrix} (a+e+f) & -a & -e & -f & 0 \\ -a & (a+b) & -b & 0 & 0 \\ -e & -b & (b+c+e) & -c & 0 \\ -f & 0 & -c & (c+d+f) & -d \\ 0 & 0 & 0 & -d & d \end{bmatrix}$$

The cofactor of the $(5,4)$th element of the matrix $N$ is the spanning tree polynomial $\wp(G)$, of the graph $G$. Therefore

$$\wp(G) = - \begin{vmatrix} (a+e+f) & -a & -e & 0 \\ -a & (a+b) & -b & 0 \\ -e & -b & (b+c+e) & 0 \\ -f & 0 & -c & -d \end{vmatrix}$$

Expanding the above determinant via the last column yields the spanning-tree polynomial of the graph. Thus

$$\wp(G) = (abcd + abdf + acde + acdf + adef + bcde + bcdf + bdef)$$

The number of terms $\eta(G)$, in this polynomial is equal to 8. This value can also be obtained by substituting

$$a = b = c = d = e = f = 1$$

in the expression for $\wp(G)$. The spanning trees of the graph $G$ are:

$$abcd, abdf, acde, acdf, adef, bcde, bcdf, \text{ and } def$$

□

## 4.8 The $\mathcal{K}$-core, $\mathcal{K}$-crust, and $\mathcal{K}$-shell of a Graph

Graphs of large order and size can be studied via their $\mathcal{K}$-shell. Study of $\mathcal{K}$-shell of a graph helps in analysing and visualizing large graphs. The $\mathcal{K}$-shell of the graph can be explained in terms of another parameter of the graph, called the $\mathcal{K}$-core. The $\mathcal{K}$-core of a graph is obtained as follows.

Remove all vertices (along with their incident edges) from a graph with vertices of degree less than $\mathcal{K}$. After this process is accomplished, it is possible to have some vertices with degree less than $\mathcal{K}$. Then remove these vertices, and continue this process recursively till all the vertices in the remaining graph have at least degree $\mathcal{K}$. At the end of this process, it is possible to obtain a graph in which all of its vertices have a degree at least $\mathcal{K}$, or have a null graph. The nonnull graph, if it exists is called a $\mathcal{K}$-core. Thus the $\mathcal{K}$-core technique to study a graph consists of simplifying the graph by annihilation of vertices (and the associated edges). The concept of $\mathcal{K}$-core is initially formalized.

**Definition 4.11.** *Let $G = (V, E)$ be an undirected finite graph. A subgraph $H_{\mathcal{K}} = (V_{\mathcal{K}}, E_{\mathcal{K}})$ of the graph $G$ is a $\mathcal{K}$-core or a core of order $\mathcal{K}$ if and only if the degree of all vertices $v \in V_{\mathcal{K}}$ is greater than or equal to $\mathcal{K}$, and $H_{\mathcal{K}}$ is the maximum subgraph (largest-order subgraph) with this property.* □

The next set of examples clarify the above definition.

**Examples 4.8.** Some basic examples.

1. Any tree graph is a 1-core.
2. A ring graph is a 2-core.
3. A completely connected graph with $n$ vertices is an $(n-1)$-core.
4. A graph is $\mathcal{K}$-connected if there are $\mathcal{K}$ disjoint paths between any pair of vertices. Thus a $\mathcal{K}$-connected graph is a $\mathcal{K}$-core. The converse of this statement is not necessarily true. □

The $\mathcal{K}$-shell and the $\mathcal{K}$-crust of a graph can be defined in terms of the $\mathcal{K}$-core of the graph. The $\mathcal{K}$-shell of a graph is the set of vertices, which belong to the $\mathcal{K}$-core, but not to the $(\mathcal{K}+1)$-core. Similarly, the $\mathcal{K}$-crust is the set of vertices in the graph which are not in the set $(\mathcal{K}+1)$-core. Thus the set of vertices in the $\mathcal{K}$-crust is $V \backslash V_{\mathcal{K}+1}$, where $V_{\mathcal{K}+1}$ is the set of vertices in the $(\mathcal{K}+1)$-core.

**Definitions 4.12.** *Let $G = (V, E)$ be an undirected finite graph, and the subgraph $H_{\mathcal{K}} = (V_{\mathcal{K}}, E_{\mathcal{K}})$ be the $\mathcal{K}$-core of the graph $G$.*

1. *Concept of shells:*
   (a) *A vertex $v \in V_{\mathcal{K}}$ has a shell index $\mathcal{K}$, if it belongs to the $\mathcal{K}$-core but not to the $(\mathcal{K}+1)$-core.*
   (b) *A $\mathcal{K}$-shell, $\Psi_{\mathcal{K}}$ is the set of all vertices whose shell index is $\mathcal{K}$.*
   (c) *The maximum value of $\mathcal{K}$ such that $\Psi_{\mathcal{K}} \neq \varnothing$ is denoted $\mathcal{K}_{\max}$.*
   (d) *Each connected set of vertices in the shell $\Psi_{\mathcal{K}}$ form a cluster. The $i$th cluster in the shell $\Psi_{\mathcal{K}}$ is denoted by $C_{\mathcal{K},i}$, where $1 \leq i \leq m_{\mathcal{K}}$, and $m_{\mathcal{K}}$ is the number of clusters.*
2. *Concept of crusts: The $\mathcal{K}$-crust is the union of all $c$-shells, where $c \leq \mathcal{K}$. That is, the $\mathcal{K}$-crust is the set of vertices*

$$\bigcup_{1 \leq c \leq \mathcal{K}} \Psi_c$$

3. *The set of vertices in the $\mathcal{K}_{\max}$-shell is called the nucleus.* □

The concept of shell hierarchy, and core and crust is illustrated in Figure 4.5 (a) and (b). Based upon the above set of definitions, the following set of elementary observations about the $\mathcal{K}$-core decomposition of a graph can be made.

**Observations 4.6.** Let $G = (V, E)$ be an undirected graph, $V_{\mathcal{K}}$ be the set of vertices in the $\mathcal{K}$-core, and $\Psi_{\mathcal{K}}$ be the set of all vertices whose shell index is $\mathcal{K}$. Each connected set of vertices in a shell form a cluster. There can be more than a single cluster of vertices with the same shell index.

1. The cores have the following properties.
    (a) The cores are nested. That is, $i < j \Rightarrow V_j \subseteq V_i$.
    (b) A core can be either connected or not connected.
    (c) The vertices of the $\mathcal{K}$-core graph are equal to the union of all vertices in the shells: $\Psi_c$, $\mathcal{K}_{\max} \ge c \ge \mathcal{K}$. Thus

$$V_{\mathcal{K}} = \bigcup_{\mathcal{K} \le c \le \mathcal{K}_{\max}} \Psi_c$$

2. The shells have the following properties.
    (a) Vertices in a shell are not necessarily connected.
    (b) Each shell $\Psi_{\mathcal{K}}$ is the union of all clusters $C_{\mathcal{K},i}$ where $1 \le i \le m_{\mathcal{K}}$, and $m_{\mathcal{K}}$ is the number of clusters in the shell. Thus

$$\Psi_{\mathcal{K}} = \bigcup_{1 \le i \le m_{\mathcal{K}}} C_{\mathcal{K},i}$$

3. The union of the vertices in the sets $\mathcal{K}$-core and $(\mathcal{K} - 1)$-crust is the vertex set $V$ of the graph $G$. $\qquad\square$

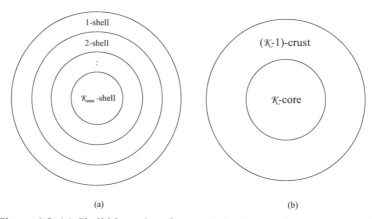

(a)                                                  (b)

Figure 4.5. (a) Shell hierarchy of a graph. (b) Core and crust of a graph.

## 4.9 Matroids

It can safely be said that matroid theory was founded by Hassler Whitney (1907-1989) in 1935. Matroids are a generalization of concepts used in matrix algebra, and graph theory. The strength of matroid theory depends upon the number of its applications. Matroid theory is developed in this chapter to apply it to lexicographic ordering, which is simply dictionary ordering. This is useful in computing the reliability of a communication network. Algorithms to compute the reliability of a communication network are developed in the chapter on graph-theoretic algorithms. Use of concepts from matroid theory reduce the computational complexity of the algorithm. To develop a motivation for the development of matroids consider the following examples.

Let $A$ be a real $m \times n$ matrix, where the set of its columns is $S = \{C_i \mid 1 \leq i \leq n\}$. It is known from linear algebra that any subset of the columns of the matrix are either linearly dependent or linearly independent. These subsets possess some structure. For example, let $F \subseteq S$ be an independent set of columns. Then $F'$ is also a set of independent columns of the matrix $A$ if $F' \subset F$. Furthermore, if $I_k$ and $I_{k+1}$ are independent-column subsets of $S$, such that $|I_k| = k$ and $|I_{k+1}| = (k+1)$, then the columns of $I_k$ and some column of $I_{k+1}$ form an independent set of $(k+1)$ columns.

There exist several algebraic systems, besides matrix algebra, which satisfy similar interesting properties. Matroid theory is a generalization of such systems. There are multiple equivalent axiom systems which characterize a matroid. Following is one such definition of a matroid.

**Definition 4.13.** *A matroid $\mathcal{M}$ is a two-tuple $(S, \mathcal{I})$, where $S$ is a finite set, and $\mathcal{I}$ is a collection of subsets of $S$, such that the following axioms are observed.*

$[I1] : \varnothing \in \mathcal{I}$.
$[I2] : $ *If $\mathcal{U} \in \mathcal{I}$ and $\mathcal{V} \subseteq \mathcal{U}$ then $\mathcal{V} \in \mathcal{I}$.*
$[I3] : $ *If $\mathcal{U}, \mathcal{V} \in \mathcal{I}$, such that $|\mathcal{U}| = |\mathcal{V}| + 1$, then there exists an element $u \in (\mathcal{U} - \mathcal{V})$ such that $(\mathcal{V} \cup u) \in \mathcal{I}$.*                                                                                                   ☐

These axioms are called the *independence axioms*. The following terminology is reminiscent of matrix algebra and graph theory. The set $S$ is called the *ground set* of $\mathcal{M}$. The elements of $\mathcal{I}$ are called *independent sets*. Subsets of $S$, which are not members of $\mathcal{I}$ are called *dependent sets*. A *basis* or *base* of a matroid $\mathcal{M}$ is a maximal independent subset of $S$. A collection of bases is denoted by $\mathfrak{B}$. A *circuit* of a matroid is a minimal dependent set. It is denoted by $\mathcal{C}$. A *loop* of a matroid is an element $s \in S$ such that $\{s\}$ is dependent.

If $\mathcal{A} \subseteq S$, then the *rank* of $\mathcal{A}$ is a nonnegative integer $\beta(\mathcal{A})$ defined as

$$\beta(\mathcal{A}) = \max\{|X| \mid X \subseteq \mathcal{A}, X \in \mathcal{I}\}$$

The rank of the matroid $\mathcal{M}$ is the rank of the set $S$. It is denoted by $\beta(\mathcal{M})$, where $\beta(\mathcal{M}) = \beta(S)$. The *spanning set* of a matroid $\mathcal{M}$ is a subset of the ground set $S$ of rank $\beta(S)$. These definitions are clarified via the following examples.

**Examples 4.9.** Some examples on matroids.

1. A *matric matroid* $\mathcal{M}_A$ is associated with a real $m \times n$ matrix $A$. In this matroid, the ground set is the set of its columns $S = \{C_i \mid 1 \leq i \leq n\}$. It is known from linear algebra that any subset of the columns of the matrix are either linearly dependent or linearly independent. Let $\mathcal{I}$ be the set of linearly independent subsets of $S$. It can be verified that all the three axioms are satisfied. Furthermore, the rank of the matrix $A$ is given by the rank of the matric matroid.

2. Let $G = (V, E)$ be an undirected graph, with the edge set $E$. Let $\mathcal{I}$ be the collection of all subsets of $E$ which do not contain any circuits. It can be verified that the two-tuple $\mathcal{M}_G = (E, \mathcal{I})$ is a matroid. In this matroid, a base of $\mathcal{M}_G$ is an edge-set of spanning tree of $G$. Therefore, $\mathcal{M}_G$ is called the *graphic* or *forest matroid* of the graph $G$.

3. Let $S$ be any finite subset of a vector space. Each and every subset of vectors of $S$ is either linearly independent or linearly dependent. The collection of all the subsets of linearly independent vectors in $S$ satisfy the three matroid axioms. Therefore, this collection of subsets of $S$

which are linearly independent form the set $\mathcal{I}$. Furthermore, the rank of a subset $Y$ of $S$, which is linearly independent, is equal to the dimension of the vector space spanned by $Y$.

4. Let $G = (V, E)$ be an undirected graph, with

$$V = \{v_1, v_2, v_3, v_4, v_5, v_6\}$$

The graph is completely specified by its adjacency matrix $A$.

$$A = \begin{bmatrix} 0 & 1 & 1 & 0 & 0 & 0 \\ 1 & 0 & 1 & 1 & 1 & 0 \\ 1 & 1 & 0 & 0 & 1 & 0 \\ 0 & 1 & 0 & 0 & 1 & 1 \\ 0 & 1 & 1 & 1 & 0 & 1 \\ 0 & 0 & 0 & 1 & 1 & 0 \end{bmatrix}$$

In this graph, the arcs $(v_1, v_2)$, $(v_1, v_3)$, $(v_4, v_6)$, and $(v_5, v_6)$ form an independent set. These arcs plus any one of the arcs $(v_2, v_4)$, $(v_2, v_5)$, and $(v_3, v_5)$ form a base. The arcs

$$(v_1, v_2), (v_1, v_3), (v_4, v_6), (v_5, v_6), (v_2, v_4), \text{ and } (v_2, v_5)$$

form a dependent set, as it contains a cycle:

$$(v_2, v_4), (v_4, v_6), (v_6, v_5), (v_5, v_2)$$

This cycle is a circuit of the matroid.                                               □

Observe that if $\mathcal{M}$ is a matroid on $S$, and $\mathcal{A}$ is an independent set, then there exists a base $\mathcal{B}$ such that $\mathcal{A} \subseteq \mathcal{B}$. However, a much stronger result can be established. It is called the *augmentation theorem*.

**Theorem 4.5.** *Let $\chi$ and $\psi$ be any two independent sets in a matroid $\mathcal{M}$, such that $|\chi| < |\psi|$. There exists $\xi \subseteq (\psi - \chi)$ such that $|\chi \cup \xi| = |\psi|$ and $\chi \cup \xi$ is independent in $\mathcal{M}$.*
*Proof.* See the problem section.                                                     □

**Corollary 4.3.** *The cardinality of all bases of a matroid $\mathcal{M}$ on a set $S$ is equal to the rank of $\mathcal{M}$, which is $\beta(\mathcal{M})$.*
*Proof.* Assume $\mathcal{B}_1, \mathcal{B}_2$ to be a set of bases such that $|\mathcal{B}_1| < |\mathcal{B}_2|$. Then augmentation theorem implies that $\exists\, Y \subseteq (\mathcal{B}_2 - \mathcal{B}_1)$ such that the set $(\mathcal{B}_1 \cup Y)$ is independent and belongs to $\mathcal{I}$. This contradicts the maximality of the basis set $\mathcal{B}_1 \in \mathcal{I}$.                     □

**Corollary 4.4.** *If $\mathcal{B}_1$ and $\mathcal{B}_2$ are the bases of a matroid $\mathcal{M}$ and $x \in (\mathcal{B}_1 - \mathcal{B}_2)$, then there exists an element $y \in (\mathcal{B}_2 - \mathcal{B}_1)$ such that $(\mathcal{B}_1 - \{x\}) \cup \{y\}$ is a base of $\mathcal{M}$.*
*Proof.* The proof is left to the reader. It is similar to the last corollary.             □

The application of this corollary to graphic matroid yields the next result.

**Corollary 4.5.** *Let $T_1$ and $T_2$ be the edge sets of two spanning trees of a connected graph. If $x$ is an edge in $(T_1 - T_2)$, then there exists an edge $y \in (T_2 - T_1)$ such that $(T_1 - \{x\}) \cup \{y\}$ is also the edge set of a spanning tree.*                                                             □

### Lexicographic Ordering and Matroids

Use of lexicographic ordering of sequences sometimes improves the efficiency of matroid-theoretic algorithms. *Lexicographic order* means dictionary-order. This is formally defined below.

**Definition 4.14.** *Let $\widetilde{H}$ be an ordered set. That is, the elements of this set are sorted in some order $<_r$. Let*

$$
\begin{aligned}
I_1 &= \{a_1, a_2, \ldots, a_p\}, \quad a_i \in \widetilde{H}, \quad 1 \le i \le p \\
I_2 &= \{b_1, b_2, \ldots, b_q\}, \quad b_i \in \widetilde{H}, \quad 1 \le i \le q
\end{aligned}
\tag{4.7}
$$

*where $a_1 <_r a_2 <_r \ldots <_r a_p$, and $b_1 <_r b_2 <_r \ldots <_r b_q$. The ordered set $I_1$ precedes the ordered set $I_2$ if any of the following conditions occur:*

(a) $a_1 <_r b_1$.
(b) *There is a positive integer $k$ such that $a_1 = b_1, a_2 = b_2, \ldots, a_k = b_k$ and $a_{k+1} <_r b_{k+1}$.*
(c) $p < q$ *and* $a_1 = b_1, a_2 = b_2, \ldots, a_p = b_p$.                         □

**Examples 4.10.** Illustrative examples.

1. The lexicographic order of permutation of the elements in the set $\{1, 2, 3\}$ is

$$
123, \ 132, \ 213, \ 231, \ 312, \ \text{and } 321
$$

2. The lexicographic order of the elements of the set $\{4, 2, 1, 3\}$ is $1, 2, 3$, and $4$.
3. The lexicographic order of six 2-tuples of the set $\{1, 2, 3, 4\}$ is

$$
12, \ 13, \ 14, \ 23, \ 24, \ \text{and } 34
$$

Note that $\binom{4}{2} = 6$.                                                       □

The *lexico-exchange property* is useful in studying the reliability of networks. It is next defined.

**Definition 4.15.** *Let $\widetilde{H}$ be an ordered set. That is, the elements of this set are sorted in some order $<_r$. Let*

$$
E_k = \left\{a_{k_1}, a_{k_2}, \ldots, a_{k_p}\right\}, \ a_{k_i} \in \widetilde{H}; \ k_i \in \mathbb{P}, \ 1 \le i \le p;
$$

$$
\text{where } k_1 < k_2 < \ldots < k_p
\tag{4.8}
$$

*Also $E_1 <_r E_2 <_r \ldots <_r E_w$, that is the sets $E_k$'s are lexicographically ordered. Note that these sets have the same cardinality $p$.*

*Then for any $1 \le i < k < w$ and $j^* = \min\{j \mid a_j \in (E_i - E_k)\}$ there exists an $i^* < k$ such that $\{a_{j^*}\} = (E_{i^*} - E_k)$.*

*The system is said to possess lexico-exchange property if the above condition holds true.*     □

This property is next applied to the spanning trees of a graph. Let $G = (V, E)$ be an undirected and connected graph, where $|V| = n$, $|E| = m$, $n \ge 2$, and the edge set $E = \{e_1, e_2, \ldots, e_m\}$. The total number of spanning trees of this graph is $\eta(G)$. Furthermore, any spanning tree of this graph has $(n - 1) \triangleq p$ edges. Let the edge-set of the $i$-th tree be $E_i = \left\{e_{i_1}, e_{i_2}, \ldots, e_{i_p}\right\}$, for $1 \le i \le \eta(G)$. The edges $e_{i_1}, e_{i_2}, \ldots, e_{i_p}$ of this tree are listed in order, that is $i_1 < i_2 < \ldots <$

$i_p$. Furthermore, let $\mathcal{M} = (E, \mathcal{I})$, where the set $\mathcal{I}$ is the set of independent sets which satisfies the axioms of a matroid. Therefore $\mathcal{M}$ is a matroid. The bases set $\mathfrak{B}$ of this matroid is given by $\{E_1, E_2, \ldots, E_{\eta(G)}\}$. The list of the trees $E_1, E_2, \ldots, E_{\eta(G)}$ is lexicographically ordered. That is, $E_i <_r E_{i+1}$ for $1 \leq i \leq (\eta(G) - 1)$.

**Lemma 4.8.** *In the graph $G$, the ordered edges of the lexicographically ordered spanning trees $E_1, E_2, \ldots, E_{\eta(G)}$ satisfy the lexico-exchange property.*

*Proof.* For any $1 \leq i < k < \eta(G)$ let $j^* = \min\{j \mid e_j \in (E_i - E_k)\}$. It is known from a corollary to the augmentation theorem that there exists a $e_{j'} \in (E_k - E_i)$ such that $E^* = (E_k \cup \{e_{j^*}\} - \{e_{j'}\}) \in \mathfrak{B}$. As $E_i <_r E_k$ we should have $j^* < j'$. Thus $E^* <_r E_k$. Consequently $E^*$ can be used as $E_{i^*}$ in order to satisfy the lexico-exchange property.    $\square$

## 4.10 Spectral Analysis of Graphs

A brief review of spectral theory of graphs is provided in this section. The spectrum of a graph is an algebraically defined concept. However, it does not characterize a graph uniquely. This theory relies upon spectral theory of matrices.

The spectrum of a graph is studied via its adjacency and Laplacian matrices. The spectral analysis of a graph via its adjacency matrix is simply called spectral analysis, and the spectral analysis of a graph via its Laplacian matrix is called Laplacian spectral analysis. Spectrum of a graph is one of several techniques to characterize it.

Graphs with only finite numbers of vertices and edges are studied in this section. Spectral analysis of asymptotically large graphs is studied via stochastic techniques in the chapter on stochastic structure of the Internet and the World Wide Web. These techniques in turn rely upon the theory of spectral analysis of large random matrices. Spectral analysis of large random matrices is studied in the chapter on matrices and determinants.

### 4.10.1 Spectral Analysis via Adjacency Matrix

We study the characteristics of a graph via the eigenvalues and eigenvectors of its adjacency matrix.

**Definitions 4.16.** *Let $G = (V, E)$ be an undirected and simple graph. It is specified by its adjacency matrix $A$.*

1. *The eigenvalue and eigenvector pair of a graph $G$ is the eigenvalue and eigenvector pair of its adjacency matrix $A$ respectively.*
2. *The characteristic polynomial of a graph $G$ is the characteristic polynomial of the adjacency matrix $A$.*
3. *The spectrum of a graph is specified by the eigenvalues of the adjacency matrix $A$ along with their multiplicities.*
4. *The maximum eigenvalue of a graph $G$ is called its spectral index.*    $\square$

**Observations 4.7.** Let $G = (V, E)$ be an undirected and simple graph with $n$ vertices. This graph is specified by its adjacency matrix $A$.

1. The eigenvalues of a graph are all real numbers. This follows from the observation that the adjacency matrix is symmetric.
2. The eigenvalues of a graph do not depend upon a specific labeling of the vertices.
3. The set of eigenvalues of a graph is the union of the eigenvalues of all its connected components.
4. The graph with zero number of edges has a single eigenvalue, which is equal to zero. The multiplicity of this eigenvalue is equal to $n$.
5. There are no self-loops in the graph $G$, therefore the sum of the eigenvalues of the matrix $A$ is zero. This follows from the observation that the trace of the matrix $A$ is equal to zero.
6. Let $\lambda_1, \lambda_2, \ldots, \lambda_n$ be the eigenvalues of the graph $G$. Then $\sum_{i=1}^{n} \lambda_i^2 = 2\,|E|$.
7. The eigenvalues of the following special graphs are listed.
    (a) The eigenvalues of a completely connected graph $K_n$ are: $(n-1)$ and $-1$. The multiplicities of these eigenvalues are $1$ and $(n-1)$ respectively.
    (b) The eigenvalues of a star graph $S_n$, for $n \geq 3$ are: $0$ and $\pm\sqrt{n-1}$. The multiplicity of eigenvalue $0$ is $(n-2)$.
    (c) The eigenvalues of a ring graph $R_n$ are :

$$\{2 \cos\left(2k\pi/n\right) \mid 1 \leq k \leq n\}$$

    (d) The eigenvalues of a path graph $P_n$ are:

$$\{2 \cos\left(k\pi/\left(n+1\right)\right) \mid 1 \leq k \leq n\}$$

8. Let $G_{\widetilde{A}} = \left(V, E_{\widetilde{A}}\right)$ and $G_{\widetilde{B}} = \left(V, E_{\widetilde{B}}\right)$ be two undirected and simple graphs with the same vertex set, but $E_{\widetilde{A}} \cap E_{\widetilde{B}} = \varnothing$. Also let $|V| = n$, and the adjacency matrices of these graphs be $\widetilde{A}$ and $\widetilde{B}$ respectively.
   The union of these two graphs is $G_{\widetilde{C}} = \left(V, E_{\widetilde{C}}\right)$, where $E_{\widetilde{C}} = E_{\widetilde{A}} \cup E_{\widetilde{B}}$. The adjacency matrix of this graph is

$$\widetilde{C} - \left(\widetilde{A} + \widetilde{B}\right)$$

Note that the matrices $\widetilde{A}, \widetilde{B}$ and $\widetilde{C}$ are symmetric. Let their eigenvalues be

$$\lambda_1\left(\widetilde{A}\right) \leq \lambda_2\left(\widetilde{A}\right) \leq \ldots \leq \lambda_n\left(\widetilde{A}\right),$$
$$\lambda_1\left(\widetilde{B}\right) \leq \lambda_2\left(\widetilde{B}\right) \leq \ldots \leq \lambda_n\left(\widetilde{B}\right),$$
$$\lambda_1\left(\widetilde{C}\right) \leq \lambda_2\left(\widetilde{C}\right) \leq \ldots \leq \lambda_n\left(\widetilde{C}\right)$$

respectively. Use of the interlacing eigenvalue theorem, yields

$$\lambda_k\left(\widetilde{A}\right) + \lambda_1\left(\widetilde{B}\right) \leq \lambda_k\left(\widetilde{C}\right) \leq \lambda_k\left(\widetilde{A}\right) + \lambda_n\left(\widetilde{B}\right), \quad 1 \leq k \leq n$$

$\square$

Bounds on the spectral values of the graph $G$ are next established.

**Observations 4.8.** Let $G = (V, E)$ be an undirected and simple graph, with $n$ vertices and $m$ edges. This graph is specified by its adjacency matrix $A$. Let the maximum eigenvalue of the graph $G$ be $\lambda_{\max}$.

1. Using Perron-Frobenius theory of nonnegative matrices, some statements about the maximum eigenvalues of a graph follow immediately. These are:
   (a) The maximum eigenvalue of a graph $\lambda_{\max} \geq 0$. An eigenvector associated with this eigenvalue is nonnegative.
   (b) If the graph $G$ is connected, then the eigenvalue $\lambda_{\max}$ is simple. That is, the multiplicity of this eigenvalue is unity. An eigenvector associated with this eigenvalue is strictly positive. Moreover, it is the only such eigenvector.
   (c) The eigenvalue $\lambda_{\max}$, is greater than or equal to the absolute value of any other eigenvalue $\mu$. That is, $\lambda_{\max} \geq |\mu|$.
2. Bounds on the maximum eigenvalue are

$$\frac{2m}{n} \leq \lambda_{\max} \leq \left\{ \frac{2m\,(n-1)}{n} \right\}^{1/2}$$

   This result also demonstrates that the maximum value of the eigenvalue is nonnegative.
3. Let $G'$ be a subgraph of the graph $G$. Also let the maximum eigenvalues of these graphs be $\lambda'_{\max}$ and $\lambda_{\max}$ respectively, then $\lambda_{\max} \geq \lambda'_{\max}$.
4. Let the minimum and maximum degrees of the vertices of the graph $G$ be $d_{\min}$ and $d_{\max}$ respectively. Then

$$\max\left(d_{\min}, \sqrt{d_{\max}}\right) \leq \lambda_{\max} \leq d_{\max}$$

$\square$

The concept of spectral density of a graph is useful when the number of vertices in a graph is very large.

**Definitions 4.17.** *Let $G = (V, E)$ be an undirected and simple graph with $n$ vertices. Its adjacency matrix is $A$. The eigenvalues of this matrix are $\lambda_1, \lambda_2, \ldots, \lambda_n$.*

1. *The spectral density $\rho(\lambda)$ of the graph $G$ is*

$$\rho(\lambda) = \frac{1}{n} \sum_{i=1}^{n} \delta(\lambda - \lambda_i) \tag{4.9a}$$

   *where $\delta(\cdot)$ is the Dirac's delta function.*
2. *The kth moment, $M_k$ of $\rho(\lambda)$ is*

$$M_k = \frac{1}{n} \sum_{i=1}^{n} \lambda_i^k, \quad k \in \mathbb{P} \tag{4.9b}$$

$\square$

Also, the spectral density $\rho(\cdot)$ converges to a continuous function as $n \to \infty$. More specifically

$$M_2 = \frac{tr\left(A^2\right)}{n} = \frac{2\,|E|}{n} \tag{4.10a}$$

and

$$M_k = \frac{tr\left(A^k\right)}{n} \tag{4.10b}$$

That is, $nM_k$ is equal to the number of paths that return to their starting node after $k$ steps.

### 4.10.2 Laplacian Spectral Analysis

We study the characteristics of a graph via the eigenvalues and eigenvectors of its Laplacian matrix. The Laplacian matrix of an undirected and simple graph, was defined in an earlier section in terms of its adjacency and degree matrices. Let $A, D$, and $L$ be the adjacency, degree, and Laplacian matrices respectively of an undirected and simple graph $G$. In this graph $L = (D - A)$.

**Definition 4.18.** *Let $G = (V, E)$ be an undirected and simple graph with $|V| = n$ vertices. Denote its Laplacian matrix by $L$. Let the eigenvalues of the matrix $L$ be $\varrho_1, \varrho_2, \ldots, \varrho_n$. These eigenvalues form the Laplacian spectrum of the graph $G$.* $\square$

Some basic observations about the eigenvalues of a Laplacian matrix are listed below.

**Observations 4.9.** Let $A$ and $L$ be the adjacency and Laplacian matrices respectively of an undirected and simple graph $G$ with $n$ vertices. Also let the eigenvalues of the matrix $L$ be $\varrho_1 \leq \varrho_2 \leq \cdots \leq \varrho_n$.

1. The vector $e = \begin{bmatrix} 1 & 1 & \cdots & 1 \end{bmatrix}^T \in \mathbb{R}^n$ is called an all-1 vector. The sum of elements of each row in the matrix $L$ is equal to zero. This implies that $Le = 0 = 0e$, which in turn implies that $e$ is an eigenvector of the matrix $L$, with the corresponding eigenvalue equal to 0.

2. Let $A = [a_{ij}]$. For any vector $g = \begin{bmatrix} g_1 & g_2 & \cdots & g_n \end{bmatrix}^T \in \mathbb{R}^n$

$$g^T L g = \frac{1}{2} \sum_{i=1}^{n} \sum_{j=1}^{n} a_{ij} (g_i - g_j)^2$$

3. $L$ is a real symmetric and positive semidefinite matrix.
   (a) The eigenvalues of the matrix $L$ are all real and nonnegative numbers.
   (b) The smallest eigenvalue of the matrix $L$ is equal to $\varrho_1 = 0$, and a corresponding eigenvector is an all-1 vector.
   (c) The number of connected components of the graph $G$ is equal to the multiplicity of the eigenvalue 0 of matrix $L$. Thus the graph is connected if $\varrho_2 > 0$.
   (d) The cofactor of any element of the matrix $L$ is equal to $n^{-1} \prod_{i=2}^{n} \varrho_i$.

      If the graph $G$ is connected then the matrix-tree theorem yields the above expression for $\eta(G)$, which is the number of spanning trees of the graph $G$. $\square$

The above observations suggest an approach to evaluate the connectivity of an undirected and simple graph. That is, if the eigenvalue $\varrho_2$ of the Laplacian matrix is positive, then the graph is connected. Furthermore, the number of connected components of a graph is equal to the multiplicity of the zero eigenvalue of the corresponding Laplacian matrix. The chapter on graph-theoretic algorithms lists an alternate algorithm to determine the connectivity of a graph.

## Reference Notes

The basic definitions and properties of graphs are excerpted from Deo (1974), Gondran, and Minoux (1984), Lawler (1976), Papadimitriou, and Steiglitz (1982), and Rosen (2018). The section on planar graphs is based upon Diestel (2000), Tucker (2007), and West (2001). Discussion about spanning trees and matrix-tree theorem is after Moon (1970), and Gibbons (1985). Ahuja, Magnanti, and Orlin (1993) provide an excellent survey on graph transformations.

The $\mathcal{K}$-cores in graph theory were first studied by Seidman (1983), and Bollobás (1984). Matroid concepts and definitions are borrowed from Lawler (1976), Rosen (2018), Swamy, and Thulasiraman (1981), and Welsh (1976). The augmentation theorem has been proved in Swamy, and Thulasiraman. Introduction to the lexico-exchange property is from Ball, and Provan (1988).

Discussion on spectral graph theory is based upon the textbooks by Lovász (1979), Behzad, Chartrand, and Lesniak-Foster (1979), Biggs (1993), and Merris (2001). The textbooks by Cvetković, Doob, and Sachs (1980), Brualdi, and Ryser (1991), and Godsil, and Royle (2001) are standard references on algebraic graph theory.

## Problems

1. Let $G$ be a simple, connected, and planar graph with $n$ vertices, $m$ edges, and $q$ number of faces in its plane embedding. It is also specified that $n \geq 3$. Prove the following results.

   (a) $2m \geq 3q$. This result is also called the edge-face inequality.

   (b) $m \leq (3n - 6)$.

   Hint: See West (2001).

   (a) If $n = 3$, the result is easily verified. Each face boundary in a simple graph with $n > 3$ has at least three edges. In addition, every edge bounds at most 2 faces in its embedding. Therefore, $2m \geq 3q$.

   (b) If $n = 3$, then the number of edges $m$ is equal to at most 3. Therefore the inequality is satisfied.

   Use of part (a) and Euler's formula $q = (m - n + 2)$ yields the stated result.

2. Prove that the determinant of any square submatrix of the incidence matrix $I_G$ of a digraph without self-loops is equal to either $1, -1$, or $0$.

   Hint: See Swamy, and Thulsiraman (1981). This result is proved by induction on the order of a square submatrix of $I_G$.

   Evidently, the determinant of every square submatrix of order 1 is either $1, -1$, or $0$. As per the induction hypothesis assume that the determinant of any square submatrix of order less than $n$ is equal to $1, -1$, or $0$.

   Select any square submatrix of order $n$. Each column of this matrix contains at most two nonzero entries, either $1$ or $-1$. If any column of this submatrix is all $0$'s then the determinant of this matrix is $0$. Also if this submatrix has two nonzero ($1$ and $-1$) entries in all the columns, then its determinant is also equal to $0$. This is because, if all the rows are added, a row with all $0$'s is obtained. This implies that the rows are dependent.

   However if this square matrix is nonsingular, then there is not a single column in the incidence matrix with all-zero entries in it. There will be some (but not all) columns of this matrix which contain two nonzero entries which are $1$ and $-1$. Also there will be some columns with a single nonzero entry, which can be either $1$ or $-1$.

   Therefore, a nonsingular square submatrix has at least one column with exactly one nonzero entry. Expand the determinant of this submatrix by this column, then the determinant of this submatrix is either $1$ or $-1$ when the induction hypothesis is used. Hence the desired determinant is equal to $\pm 1$.

3. Let $G$ be a connected digraph without self-loops, and with $n$ nodes. Prove that the rank of its incidence matrix $I_G$ is $(n-1)$.

   Hint: See Moon (1970). Sum any $j$ rows of the matrix, where $j < n$. This sum should have at least a single nonzero entry for the graph to be connected. This implies that no $j$ rows are linearly dependent if $j < n$. The result follows from the observation that the sum of all the $n$ rows of the incidence matrix $I_G$ has all-zero elements.

   Actually, if the graph $G$ has $k$ connected components, then the rank of the matrix $I_G$ is $(n-k)$.

4. Prove that the determinant of any nonsingular square submatrix $Q$ of $I_G$, and the reduced incidence matrix $I_\alpha$ is equal to either $1$ or $-1$.

   Hint: See Moon (1970). Moon attributes this result to Henri Poincaré. The proof follows from Lemma 4.1 and 4.2. By hypothesis, $Q$ is a nonsingular square submatrix. Consequently, each column of this matrix must contain at least one nonzero entry. However not all columns can contain two nonzero entries. Thus some columns of the matrix $Q$ must contain just one nonzero entry. The stated result follows by induction if the determinant of $Q$ is expanded along this column.

5. Verify Lemma 4.7 for the graph in Example 4.7.

6. The ring graph $R_n$ is an undirected graph with $n$ vertices and $n$ edges. In this graph, the degree of each vertex is equal to 2. Let the vertices in this graph be $v_1, v_2, \ldots, v_n$. This graph has the edges $(v_i, v_{i+1})$, $1 \leq i \leq (n-1)$ and the edge $(v_n, v_1)$. Prove that the number of spanning trees of a ring graph is equal to $n$.

7. A wheel graph $W_{n+1}, n \geq 3$ is a ring graph $R_n$ with an additional vertex $v_{n+1}$ which is connected to all other vertices. Therefore the wheel graph $W_{n+1}$ has $(n+1)$ vertices, and $2n$ edges. It has the following edges: $(v_i, v_{i+1})$, $1 \leq i \leq (n-1)$; $(v_n, v_1)$, and $(v_i, v_{n+1})$, $1 \leq i \leq n$. Prove that:

$$\eta(W_4) - 16$$
$$\eta(W_5) = 45$$
$$\eta(W_{n+1}) = \left(\frac{3+\sqrt{5}}{2}\right)^n + \left(\frac{3-\sqrt{5}}{2}\right)^n - 2, \; n \geq 3$$

   Also express $\eta(W_{n+1})$ in terms of the golden ratio $\phi_g = \left(\sqrt{5} + 1\right)/2$ and its reciprocal.
   Hint: See Moon (1970). Use matrix-tree theorem and generating functions.

8. The number of labeled trees on $n$ vertices was derived to be equal to $n^{n-2}$. This value was derived by using the matrix-tree theorem. Derive this result by using a different technique.
   Hint: See Aigner, and Ziegler (1998). This textbook has three other elegant proofs.

9. Let $\chi$ and $\psi$ be any two independent sets in a matroid $\mathcal{M}$, such that $|\chi| < |\psi|$. Prove that there exists $\xi \subseteq (\psi - \chi)$ such that $|\chi \cup \xi| = |\psi|$ and $\chi \cup \xi$ is independent in $\mathcal{M}$.
   Hint: See Swamy, and Thulasiraman (1981). Select $\xi_0$ such that, it has the following properties.

   (a) $\xi_0 \subseteq (\psi - \chi)$.
   (b) $\chi \cup \xi_0$ is independent in the matroid.
   (c) If $\chi \cup \xi$ is independent in $\mathcal{M}$ for any $\xi \subseteq (\psi - \chi)$, then $|\chi \cup \xi_0| \geq |\chi \cup \xi|$.

   Assume that $|\chi \cup \xi_0| < |\psi|$, then there exists $\psi_0 \subseteq \psi$, where $|\psi_0| = |\chi \cup \xi_0| + 1$. Furthermore, since $\psi_0$ is independent in $\mathcal{M}$, we have by the independent axiom $[I3]$ : there exists $\varphi \in$

$\psi_0 - (\chi \cup \xi_0)$ such that $\chi \cup \xi_0 \cup \varphi$ is independent in $\mathcal{M}$. Consequently, the existence of the set $\xi_0 \cup \varphi$ contradicts the choice of $\xi_0$. Therefore $|\chi \cup \xi_0| \geq |\psi|$, and the theorem follows.

10. Prove that an edgeless graph with $n$ vertices has only one eigenvalue, and that it is zero with multiplicity $n$.

11. Let $\lambda_1, \lambda_2, \ldots, \lambda_n$ be the eigenvalues of an undirected and simple graph $G = (V, E)$. Prove that

$$\sum_{i=1}^{n} \lambda_i^2 = 2\,|E|$$

    Hint: Let $A$ be the adjacency matrix of the graph $G$. Then the eigenvalues of the matrix $A^2$ are $\lambda_1^2, \lambda_2^2, \ldots, \lambda_n^2$. Furthermore, the trace of the matrix $A^2$ is equal to $2\,|E|$. The result follows.

12. Find the eigenvalues of the graphs $K_n, S_n, R_n$, and $P_n$.

    Hint: In each case, use induction on $n$.

13. Let $G = (V, E)$ be an undirected and simple graph, with $n$ vertices, and $m$ edges. This graph is specified by its adjacency matrix $A$. Let $\lambda_{\max}$ be the maximum eigenvalue of the graph $G$. Prove that

$$\frac{2m}{n} \leq \lambda_{\max} \leq \left\{ \frac{2m\,(n-1)}{n} \right\}^{1/2}$$

    Hint: See Lovász (1979). Define $\vartheta\,(x) = \left( x^T A x \right) / \left( x^T x \right)$, and $x \neq 0$ is a vector of size $n$, then $\lambda_{\max} = \max_{x^T x = 1} \vartheta\,(x)$. Select $x^T = n^{-1/2}\left[\, 1\ 1\ \cdots\ 1 \,\right]$, then $x^T x = 1$. Also recall that the adjacency matrix $A$ of the graph $G$ is symmetric. Furthermore, the sum of all the elements of the adjacency matrix $A$ is equal to $2m$. Thus

$$\lambda_{\max} \geq x^T A x = \frac{1}{n} \sum_{i=1}^{n} \sum_{j=1}^{n} a_{ij} = \frac{2m}{n}$$

    The upper bound is established as follows. Let $\lambda_1, \lambda_2, \ldots, \lambda_n$ be the eigenvalues of the graph $G$, such that $\lambda_1 \leq \lambda_2 \leq \ldots \leq \lambda_n = \lambda_{\max}$. It is known that

$$\sum_{i=1}^{n} \lambda_i = 0, \quad \text{and} \quad \sum_{i=1}^{n} \lambda_i^2 = 2m$$

    Therefore

$$\lambda_{\max}^2 = \left\{ \sum_{i=1}^{n-1} \lambda_i \right\}^2$$

$$\leq (n-1) \sum_{i=1}^{n-1} \lambda_i^2$$

$$= (n-1)\left(2m - \lambda_{\max}^2\right)$$

    where we have used the relationship

$$\left\{ \sum_{i=1}^{n-1} \frac{\lambda_i}{\sqrt{n-1}} \right\} \leq \left\{ \sum_{i=1}^{n-1} \lambda_i^2 \right\}^{1/2} \left\{ \sum_{i=1}^{n-1} \frac{1}{n-1} \right\}^{1/2}$$

    This result follows from the Bunyakovsky-Cauchy-Schwartz inequality. The required result is immediate.

14. Let $G'$ be a subgraph of an undirected graph $G$. Also let the maximum eigenvalues of these graphs be $\lambda'_{\max}$ and $\lambda_{\max}$ respectively. Prove that $\lambda_{\max} \geq \lambda'_{\max}$.

    Hint: See Lovász (1979). If the graph $G'$ is not a spanning subgraph of $G$, add the remaining vertices of $G$ to $G'$ to make it a spanning graph. These vertices are added as isolated points. Thus these vertices contribute only 0's to the spectrum of $G'$, and do not modify the maximum eigenvalue of the graph $G'$.

    Let the adjacency matrix of the graph $G$ of order $n$, be $A$. From this matrix obtain a matrix $A'$ by setting elements to 0 which are absent in the graph $G'$. Then $A \geq A'$, and for all vectors $x$, $x^T A x \geq x^T A' x$. The final result follows by using results from spectral theory of nonnegative matrices.

15. Let the minimum and maximum degrees of the vertices of an undirected, and simple graph $G = (V, E)$ be $d_{\min}$ and $d_{\max}$ respectively. The maximum eigenvalue of the graph $G$ is $\lambda_{\max}$. Prove that

$$\max\left(d_{\min}, \sqrt{d_{\max}}\right) \leq \lambda_{\max} \leq d_{\max}$$

    Hint: See Lovász (1979). Let $A$ be the adjacency matrix of the graph $G = (V, E)$, where $|V| = n$. Let $e$ be a vector of size $n$. The elements of this vector are all 1's. Therefore the elements of the column vector $Ae$ represent the degrees of the vertices of the graph $G$. Thus $Ae \geq d_{\min}e$, therefore $e^T Ae \geq nd_{\min}$. Note that

$$\lambda_{\max} = \max_{x \neq 0} \frac{x^T A x}{x^T x} \geq \frac{e^T Ae}{e^T e} \geq d_{\min}$$

    Next consider a subgraph $G'$ of $G$ which is a star graph of a vertex of the maximum degree. Then $\lambda_{\max} \geq \sqrt{d_{\max}}$. This establishes the lower bound on $\lambda_{\max}$.

    The upper bound is established as follows. Let $x = \begin{bmatrix} x_1 & x_2 & \cdots & x_n \end{bmatrix}^T$ be an eigenvector corresponding to the eigenvalue $\lambda_{\max}$. Also let $|x_p| = \max_{1 \leq i \leq n} |x_i|$. Then

$$\lambda_{\max} |x_p| \leq |x_p| \sum_{i=1}^{n} a_{pi} \leq |x_p| d_{\max}$$

    As $|x_p| > 0$, the result follows.

16. Let $A$ and $L$ be the adjacency and Laplacian matrices respectively of an undirected and simple graph $G$ with $n$ vertices. Prove the following assertions.

    (a) Let $A = [a_{ij}]$. For any vector $g = \begin{bmatrix} g_1 & g_2 & \cdots & g_n \end{bmatrix}^T \in \mathbb{R}^n$

$$g^T Lg = \frac{1}{2} \sum_{i=1}^{n} \sum_{j=1}^{n} a_{ij} \left(g_i - g_j\right)^2$$

    (b) $L$ is a real symmetric and positive semidefinite matrix.

    (c) The number of connected components of the graph $G$ is equal to the multiplicity $k$ of the eigenvalue 0 of matrix $L$.

    (d) The eigenvalues of the Laplacian matrix $L$ are $\varrho_1, \varrho_2, \ldots, \varrho_n$, where $\varrho_1 \leq \varrho_2 \leq \cdots \leq \varrho_n$, and $\varrho_1 = 0$. Show that the cofactor of any element of the matrix $L$ is equal to $n^{-1} \prod_{i=2}^{n} \varrho_i$

    Hint: See Luxburg (2007).

(a) Let the degree matrix of the graph $G$ be $D = [d_{ij}]$ where $d_{ii} \triangleq d_i$, $1 \leq i \leq n$. Thus

$$
\begin{aligned}
g^T L g &= g \left( D - A \right) g^T \\
&= \sum_{i=1}^{n} d_i g_i^2 - \sum_{i=1}^{n} \sum_{j=1}^{n} g_i g_j a_{ij} \\
&= \frac{1}{2} \left\{ \sum_{i=1}^{n} d_i g_i^2 - 2 \sum_{i=1}^{n} \sum_{j=1}^{n} g_i g_j a_{ij} + \sum_{j=1}^{n} d_j g_j^2 \right\} \\
&= \frac{1}{2} \sum_{i=1}^{n} \sum_{j=1}^{n} a_{ij} \left( g_i - g_j \right)^2
\end{aligned}
$$

(b) The matrix $L$ is real and symmetric by its definition. Furthermore, from part (a) of the problem, we have $g^T L g \geq 0$. Further, $g^T L g = 0$ if $g$ is an all-1 vector. This implies that the matrix $L$ is positive semidefinite.

(c) Initially assume that $k = 1$. This implies that the graph $G$ is connected. If $g$ is an eigenvector of the matrix $L$ with 0 eigenvalue, then

$$
0 = g^T L g = \frac{1}{2} \sum_{i=1}^{n} \sum_{j=1}^{n} a_{ij} \left( g_i - g_j \right)^2
$$

The summation in the above equation is equal to zero, if each term $a_{ij} \left( g_i - g_j \right)^2$ is equal to zero. Note that $a_{ij}$ is equal to either 0 or 1. Thus if $a_{ij} = 0$, we have $a_{ij} \left( g_i - g_j \right)^2 = 0$, else if $a_{ij} = 1$, we should have $g_i = g_j$. Again note that $a_{ij} = 1$ implies that the vertices $v_i$ and $v_j$ of the graph are directly connected to each other. Thus the $g_i$ values of all the vertices which are connected by a path in the graph should have a constant value. Therefore if the graph $G$ is connected, then all the $g_i$ values should be identical. Select $g_i = 1$ for $1 \leq i \leq n$. Thus the all-1 vector

$$
g = \begin{bmatrix} 1 \ 1 \ \cdots \ 1 \end{bmatrix}^T \in \mathbb{R}^n
$$

is an eigenvector of the Laplacian matrix $L$ of a connected graph. The corresponding eigenvalue is equal to 0.

Next assume that the graph $G$ has $k > 1$ disconnected components. That is, the graph $G$ has $k$ proper subgraphs. Further assume that the vertices are labeled such that the Laplacian matrix $L$ of the graph $G$ can be written as

$$
L = \begin{bmatrix}
L_1 & 0 & \cdots & 0 \\
0 & L_2 & \cdots & 0 \\
\vdots & \vdots & \ddots & \vdots \\
0 & 0 & \cdots & L_k
\end{bmatrix}
$$

where $L_i$ is the Laplacian matrix of the $i$th connected component of the graph $G$, and $1 \leq i \leq k$. The matrix $L$ is said to be of block-diagonal type because of square matrices on its diagonal, and other elements are all null matrices. Each Laplacian matrix $L_i$ has a 0 eigenvalue with multiplicity one. A corresponding eigenvector is an all-1 vector of appropriate size.

The set of eigenvalues of the matrix $L$ is equal to the union of the eigenvalues of the Laplacian matrices $L_i$'s. Thus the Laplacian matrix $L$ has 0 eigenvalue with multiplicity

$k$. An eigenvector of $L$ which corresponds to the 0 eigenvalue of $L_i$ is determined as follows. This eigenvector of $L$ has all 1's in positions corresponding to the $L_i$'s block and 0's in all other positions.

Thus the graph $G$ with $k$ number of connected components has a 0 eigenvalue with multiplicity $k$.

(d) Consider the characteristic polynomial of the matrix $L = [l_{ij}]$. It is equal to

$$\det(\varrho I - L) = \prod_{i=1}^{n}(\varrho - \varrho_i)$$

The linear term in $\varrho$ on the left-hand side of the above equation is equal to

$$(-1)^{n-1}\sum_{i=1}^{n}\det L_{ii},$$

where $\det L_{ii}$ is the minor of the element $l_{ii}$. Since $\varrho_1 = 0$, the linear term in $\varrho$ on the right-hand side of the above equation is equal to the product of the eigenvalues $\varrho_2, \varrho_3, \cdots, \varrho_n$ and $(-1)^{n-1}$. The result follows by noting that all the cofactors of this matrix are equal because all row and column sums of matrix $L$ are zero.

# References

1. Ahuja, R. K., Magnanti, T. L., and Orlin, J. B., 1993. *Network Flows, Theory, Algorithms, and Applications,* Prentice-Hall, Englewood Cliffs, New Jersey.
2. Aigner, M., and Ziegler, G. M., 1998. *Proofs from THE BOOK*, Springer-Verlag, Berlin, Germany.
3. Bailey, R. A., and Cameron, P. J., 2009. "Combinatorics of Optimal Designs," Surveys in Combinatorics 2009. Editors: S. Huczynska, J.D. Mitchell, and C.M. Roney-Dougal, London Mathematical Society Lecture Notes, London, Great Britain, pp. 19-73.
4. Ball, M. O., and Provan, J. S., 1988. "Disjoint Products and Efficient Computation of Reliability," Operations Research, Vol. 36, No. 5, pp. 703-715.
5. Behzad, M., Chartrand, G., and Lesniak-Foster, L., 1979. *Graphs and Digraphs*, Wadsworth International Group, Belmont, California.
6. Biggs, N., 1993. *Algebraic Graph Theory*, Second Edition, Cambridge University Press, Cambridge, Great Britain.
7. Bollobás, B., 1984. "The Evolution of Sparse Graphs," in Graph Theory and Combinatorics, Proceedings of Cambridge Combinatorial Conference in Honor of Paul Erdös, Academic Press, London, Great Britain, pp. 35-37.
8. Brualdi, R. A., and Ryser, H., 1991. *Combinatorial Matrix Theory*, Cambridge University Press, Cambridge, Great Britain.
9. Christofides, N., 1975. *Graph Theory - An Algorithmic Approach*, Academic Press, New York, New York.
10. Cvetković, D., Doob, M., and Sachs, H., 1980. *Spectra of Graphs*, Second Edition, Academic Press, New York, New York.
11. Deo, N., 1974. *Graph Theory with Applications to Engineering and Computer Science*, Prentice-Hall, Englewood Cliffs, New Jersey.

12. Diestel, R., 2000. *Graph Theory*, Second Edition, Springer-Verlag, Berlin, Germany.

13. Even, S., 1979. *Graph Algorithms*, Computer Science Press, Rockville, Maryland.

14. Gibbons, A., 1985. *Algorithmic Graph Theory*, Cambridge University Press, Cambridge, Great Britain.

15. Godsil, C., and Royle, G., 2001. *Algebraic Graph Theory*, Springer-Verlag, Berlin, Germany.

16. Gondran, M., and Minoux, M., 1984. *Graphs and Algorithms*, John Wiley & Sons, Inc., New York, New York.

17. Gross, J. L., and Yellen, J., Editors, 2003. *Handbook of Graph Theory*, CRC Press: New York, New York.

18. Kocay, W., and Kreher, D. L., 2005. *Graphs, Algorithms, and Optimization*, Chapman and Hall/CRC Press, New York, New York.

19. Lawler, E. L., 1976. *Combinatorial Optimization: Networks and Matroids*, Holt, Rinehart and Winston, New York, New York.

20. Lovász, L., 1979. *Combinatorial Problems and Exercises*, Second Edition, North-Holland, New York, New York.

21. Luxburg, U. V., 2007. "A Tutorial on Spectral Clustering," Statistics and Computing, Vol. 17, No. 4, pp. 395-416.

22. Merris, R., 2001. *Graph Theory*, John Wiley & Sons, Inc., New York, New York.

23. Moon, J. W., 1970, *Counting Labeled Trees*, Canadian Mathematical Monographs, William Clowes and Sons, Limited, London, U. K.

24. Papadimitriou, C. H., and Steiglitz, K., 1982. *Combinatorial Optimization - Algorithms and Complexity*, Prentice-Hall,, Englewood Cliffs, New Jersey.

25. Rosen, K. H., Editor-in-Chief, 2018. *Handbook of Discrete and Combinatorial Mathematics,* Second Edition, CRC Press: Boca Raton, Florida.

26. Seidman, S. B., 1983. "Network Structure and Minimum Degree," Social Networks Vol. 5, No. 3, pp. 269-287.

27. Swamy, M. N. S., and Thulasiraman, K., 1981. *Graphs, Networks, and Algorithms*, John Wiley & Sons, Inc., New York, New York.

28. Tucker, A., 2007. *Applied Combinatorics*, Fifth Edition, John Wiley & Sons, Inc., New York, New York.

29. Welsh, D., 1976. *Matroid Theory*, Academic Press, New York, New York.

30. West, D. B., 2001. *Introduction to Graph Theory*, Second Edition, Prentice-Hall, Upper Saddle River, New Jersey.

# Geometry

$$d_{\mathbb{H}}(a, b) = 2\tanh^{-1}\left|\frac{b - a}{b - \bar{a}}\right|$$

Hyperbolic distance
between two points $a, b \in \mathbb{H}$

**János Bolyai**. János Bolyai was born on 15 December 1802 in Cluj, Romania (formerly Kolozsvár, Hungary). János' father, Farkas Bolyai was a mathematician and a friend of Gauss. János studied at Royal Engineering College in Vienna (1818-1822) and graduated with the highest honors. Around the year 1820 the young János began to develop an axiomatic system for hyperbolic geometry, eventually replacing Euclid's parallel postulate. He called his work *Absolute Science of Space*. On 3 November 1823 he wrote to his father that he had "... created a new, another world out of nothing...." Farkas Bolyai published his son's ideas in an Appendix in his treatise titled *Tentamen* in the year 1832.

Farkas Bolyai, mailed his son's work to Gauss. Gauss' reply to János' work was that he has also developed independently similar ideas about replacing Euclid's parallel postulate. Nevertheless, he had high praise for the young János. János Bolyai died on 27 January 1860 in Târgu-Mure, Romania (formerly Marosvásárhely, Hungary).

$$\frac{\sin\alpha}{\sinh a_\ell} = \frac{\sin\beta}{\sinh b_\ell} = \frac{\sin\gamma}{\sinh c_\ell}$$

Lengths of sides of a hyperbolic
triangle in disc $\mathbb{D}$ are: $a_\ell, b_\ell$, and $c_\ell$;
and angles $\alpha, \beta$, and $\gamma$.
The hyperbolic sine rule.

**Nikolai Ivanovich Lobachevsky**. Nikolai Ivanovich Lobachevsky was born on 1 December 1792 in Nizhny Novgorod, Russia. Lobachevsky's father died when he was only seven years old. He was raised by his mother Praskovia Alexandrovna Lobachevskaya, along with his two brothers in the city of Kazan, in Russia under financially trying circumstances. Lobachevsky received a Master's Degree in physics and mathematics from the Kazan University in 1811.

In the year 1829, Lobachevsky published an article *On the Principles of Geometry* in the Kazan Messenger. In the years between 1835 and 1855 he published his work on non-Euclidean geometry in *New Foundations of Geometry* in Russian (1835-1838), *Geometrical Investigations on the Theory of Parallels* in German (1840), and *Pangeometry* in French and Russian (1855). Gauss became acquainted with Lobachevsky's work, and became duly impressed. Based upon his recommendation, Lobachevsky was elected to the Göttingen Scientific Society in 1842. Lobachevsky died on 24 February 1856 in Kazan, Russia.

In summary, both János Bolyai and Nikolai Ivanovich Lobachevsky can be regarded as the "Copernicus of geometry." They demonstrated that Euclidean geometry is not the absolute truth. More importantly, their work showed that any mathematical system is neither the absolute truth nor the last word.

## 5.1  Introduction

Mathematics is an important branch of human thought and endeavor, and geometry is certainly one of its most influential and older components. Components of the word *geometry*: *geo* and *metry,* mean earth and measurement respectively. Therefore geometry is the science of measurements of figures in space. In brief, Euclidean geometry, elementary differential geometry, basics of surface geometry, and hyperbolic geometry are studied in this chapter.

Geometry was originally an applied science. One can safely say that in earlier times applications preceded the sound development of geometry as a mathematical discipline. The mathematical development of geometry closely follows its applications. Geometry became a branch of mathematics, after its axiomatic foundations were laid. There are several branches of geometry. Some examples of it are: Euclidean, hyperbolic, and elliptic geometries. A primary goal of this chapter is to study hyperbolic geometry. Before a formal study of hyperbolic geometry is undertaken, we set up it motivation.

We initially study Euclidean geometry. It is important to summarize important notions about Euclidean geometry, before we venture into the fascinating world of non-Euclidean geometry. This is followed by the study of certain basic geometric constructions. These constructions are useful in circle inversions. Circle inversions are useful in the study of hyperbolic geometry.

We next present basic concepts from differential geometry of curves and surfaces. Differential geometry is a convenient vehicle to analytically study different types of geometries. The characteristics of a surface are studied via its first and second fundamental forms. A surface can also be characterized by studying properties of curves on it. Local isometries between a pair of surfaces can also be characterized in terms of the coefficients of the first fundamental form. Geodesics on a surface are versions of straight lines on a plane. These are curves with shortest lengths between a pair of points on a surface. Geodesics can also be used to specify coordinates on a surface. These coordinates are therefore called geodesic coordinates. The next step in our understanding of the evolution of non-Euclidean geometries, is the study of surfaces with constant Gaussian curvature. Finally the upper half-plane and Poincaré disc models of hyperbolic geometry are studied.

**Notation.** The symbols $\circ$ and $\times$ are used in vector algebra to denote the inner and cross product operators respectively. Let $a, b, c \in \mathbb{R}^3$, then $a \circ (b \times c) \triangleq [a\ b\ c]$. The Euclidean norm is denoted by $\|\cdot\|$. Prime ($'$) denotes derivative. The symbol $\circ$ is also used to denote the function composition operator.                                                                                    □

## 5.2 Euclidean Geometry

Euclidean geometry is the discipline that we studied in high school. It is a forerunner of several other types of geometries. Euclidean geometry consists of some terminology, certain commonly accepted notions, and postulates. Useful propositions are then derived from these concepts. However the main feature of Euclidean geometry is its emphasis on the use of axioms, from which all other results are derived. In the interest of completeness, certain immediately useful basic geometric constructions are also outlined.

### 5.2.1  Requirements for an Axiomatic System

All results in Euclidean geometry follow from a set of axioms. Axioms, also called postulates, are a set of self-evident facts. All the propositions (theorems) of the Euclidean geometry are logically deduced from a set of axioms. Therefore, it is imperative that we clarify, the essential ingredients of the axiomatic technique of constructing a mathematical structure. The axiomatic technique consists of:

- A set of undefined technical terms. The meanings of these technical terms are left to the user (interpreter) of the system.
- The other technical terms are defined in terms of the aforementioned undefined terms. These terms are said to be the definitions of the system.
- The axioms of the system are statements made by using the undefined and defined terms. These statements are deemed unproven.
- It is required that all other statements about the system are logical consequences of the axioms. Such statements (deductions) are called the theorems of the system.

If the undefined technical terms are assigned specific meanings, then we obtain an *interpretation* of the system. If the assigned meanings are sensible statements about a system, then the interpretation is called a *model* of the system. There are two types of models. These are the *concrete* and *abstract* models.

- Concrete models are related to the real world.
- Models which are not immediately related to real world are abstract. Some examples of abstract model systems are groups, rings, fields, and complex number systems.

An axiomatic system has to say something which is interesting. In addition, it needs to possess certain necessary properties. These are consistency, independence and completeness. In addition, the axiomatic system has to be interesting.

- *Something Interesting*: Axioms have to be about something which is interesting. If it is not interesting then what follows will certainly be of no consequence. This is certainly not a mathematical criterion.
- *Consistency*: The axioms of a system are consistent, if it is not possible to infer contradictory results from the axioms. That is, an axiomatic system is consistent, if it is not possible to prove that an assertion is both true and false.
- *Independence*: An axiom is independent of other axioms, if it is not derivable from other axioms. An axiomatic system is independent, if all of its axioms are independent of each other. Requirement of independence disallows duplication.
- *Completeness*: An axiomatic system is said to be complete, if it is not possible to add another consistent and independent axiom to the system without the addition of undefined terms.
  It is usually difficult to establish that an axiomatic system is complete. However, it is easier to establish that the axioms of the system are *categorical*. An axiomatic system is said to be categorical, if there is a one-to-one correspondence (isomorphism) between each of its models. Thus, a categorical axiomatic system describes a unique mathematical object.

### 5.2.2  Axiomatic Foundation of Euclidean Geometry

Euclidean geometry was first formulated axiomatically by Euclid of Alexandria in the *The Elements* in the year 300 BC. It is one of the most important books ever written in the history of mankind. Not much is known about Euclid's life. He was born around 325 BC, and died about 265 BC in Alexandria, Egypt. The fame and glory of Euclid rests upon his deductions of the principles of geometry from a set of axioms.

A precise description of the Euclidean geometry consists of specification of certain defined terms and the axioms of the system. More precisely, certain terms have to be introduced in order to specify axioms and some of its consequences.

In order to complete the specification, certain assumed common notions also have to be stated precisely. These fall under the category of some commonly accepted notions.

## Terminology

The axioms are defined in certain terminology. Some of these terms are undefined and some are defined. For plane Euclidean geometry, the five undefined terms are: point (that which has no part), line (as in "breadthless length"), lie on (as in "two points lie on a specific line"), between (for example, point C is between points A and B), and congruent (as in "similar").

For Euclidean geometry in three-dimensional space, the undefined term, plane, has to be introduced. Also the meaning of the term "lie on" has to be extended to also mean, "points and lines lie on planes."

Defined terms in Euclidean geometry are: extremities of a line, straight line, surface, extremities of a surface, plane surface, plane angle, rectilineal angle, right angle, perpendicular, obtuse angle, acute angle, boundary, figure, circle, centre of a circle, diameter, semicircle, rectilineal figure (trilateral, quadrilateral, and multilateral figures), equilateral triangle, isosceles triangle, scalene triangle, right-angled triangle, obtuse-angled triangle, acute-angled triangle, square, oblong, rhombus, rhomboid, trapezia, and parallel straight lines. Precise meaning of these terms can be found in standard textbooks on geometry.

## Common Notions

In addition to the axioms, Euclid also specified certain common notions. Common notions are assumptions which are acceptable by all reasonable human beings. On the other hand, Euclidean axioms or postulates are specific to the science of geometry. The five common notions are:

1. Objects which are equal to the same object are themselves equal to one another.
2. If equals are added to equals, then the wholes are also equals.
3. If equals are subtracted from equals, then the remainders are also equals.
4. Objects which coincide with one another are also equal to one another.
5. The whole is always greater than the part.

### Postulates of Euclidean Geometry

We first state the postulates due to Euclid, and then discuss their ramifications. There are five postulates in Euclidean geometry.

[$Axiom\ 1$]  For each pair of points $A \neq B$, there exists a unique line $\ell$ that passes through points $A$ and $B$.

[$Axiom\ 2$]  It is always possible to produce a finite straight line continuously in a straight line.

[$Axiom\ 3$]  For any given point $C$ and any other point $A \neq C$ there exists a circle with center $C$ and radius $CA$ (radius is half the diameter). The circle is the set of points $B$ such that $CB = CA$.

[$Axiom\ 4$]  All right angles are equal to each other.

[$Axiom\ 5$]  For every line $\ell$ and for every point $A$ not on the line $\ell$; there exists exactly a single line $m$ that passes through the point $A$, and is also parallel to line $\ell$.

Axiom number 5, as stated above is called Playfair's axiom, after John Playfair (1748-1819). Playfair's axiom is easier to state than Euclid's original and equivalent axiom number 5. As it turns out, Playfair's axiom has also been stated earlier by other mathematicians.

The mathematician David Hilbert (1862-1943) also stated the above postulates in a more rigorous manner. Euclid's postulate number 5 is called Euclid's *parallel postulate*. It is not immediately obvious that this postulate is independent of the other four postulates. Several mathematicians over the span of several centuries unsuccessfully tried to prove that Euclid's parallel postulate depends upon the first four postulates. Modification of the fifth postulate results in different types of geometries.

**Types of Two-Dimensional Geometries**

There exist three two-dimensional geometries. These are: Euclidean, hyperbolic, and elliptic geometries. In all these geometries, the first four postulates are identical. However, each of these geometries has a unique fifth postulate. These are:

- *Euclidean geometry*: For every line $\ell$ and for every point $A$ not on the line $\ell$; there exists exactly a single line $m$ that passes through the point $A$, and is also parallel to line $\ell$.
- *Hyperbolic geometry*: For every line $\ell$ and for every point $A$ not on the line $\ell$; there exist two or more lines that pass through the point $A$, and are also parallel to line $\ell$.
- *Elliptic geometry*: For every line $\ell$ and for every point $A$ not on the line $\ell$; there does not exist any line that passes through the point $A$ and is also parallel to line $\ell$.

The hyperbolic and elliptic geometries are quite counter-intuitive. For example, the sum of interior angles of a triangle in hyperbolic geometry is strictly less than $180°$. There are infinitely many parallel lines passing through a specific point. There are several models which describe hyperbolic space. We shall discuss some of these subsequently.

The sum of interior angles of a triangle in elliptic geometry is strictly greater than $180°$. Also parallel lines do not exist in elliptic geometry. A specific example of elliptic geometry is spherical geometry. In spherical geometry, points are represented as points on the sphere, and lines are Euclidean circles which pass through the points.

The Euclidean geometry is also sometimes called parabolic geometry. These three different geometries can be qualified based upon the curvature of surfaces.

### 5.2.3  Basic Definitions and Constructions

Basic definitions and certain geometric constructions in the Euclidean plane using straightedge and compass are outlined. This is a prelude to the study of circle inversions, which are studied in the next section. Some geometric objects are initially defined. These definitions are from Euclid's *Elements*.

1. A *point* is that which has no parts. Points in the Euclidean plane are denoted by $A, B, C, \ldots$.
2. A *line* is a breadthless length.
3. The *extremities of a line* are points.
4. A *straight line* is a line which lies evenly with the points on itself. A *straight line* segment joining two points $A$ and $B$ is denoted by $AB$. The *length* of the line segment $AB$ is also often denoted by simply $AB$.
5. A *surface* is that which has length and breadth only.
6. The *extremities of a surface* are lines.
7. A *plane surface* is a surface which lies evenly with the straight line on itself.

8. A *plane angle* is the inclination to one another of two lines in a plane which meet one another and do not lie in a straight line. An *angle* between two line segments $OA$ and $OB$ at the point $O$ is denoted as: angle $AOB$ or $\angle AOB$.

9. And when the lines containing the angle are straight, the angle is called *rectilinear*.

10. When a straight line standing on a straight line makes the adjacent angles equal to one another, each of the equal angles is *right*, and the straight line standing on the other is called a *perpendicular* to that on which it stands.

11. An *obtuse angle* is an angle greater than a right angle.

12. An *acute angle* is an angle less than a right angle.

13. A *boundary* is that which is an extremity of anything.

14. A *figure* is that which is contained by any boundary or boundaries.

15. The set of points which are equidistant from two fixed points $A$ and $B$ is the *perpendicular bisector* of the line segment $AB$.

16. Three points $A, B$, and $C$ are *collinear* if these three points all lie on the same straight line. Otherwise, the points are said to be *noncollinear*.

17. Let $O$ be a point in the Euclidean plane, and $r$ be a positive number. The set of all points which are at a distance of $r$ from the point $O$ is called a *circle* with *center* $O$ and *radius* $r$. If $A$ is a point on the circle, then the length of the line segment $OA$ is the radius of the circle. In our notation $OA$ is also equal to the length $r$.

18. Let $A$ and $B$ are any two points on the circle. The line segment $AB$ is called the *chord* of the circle. A chord which passes through the center $O$ is called the *diameter* of the circle. The length of each diameter of a circle of radius $r$ is equal to $2r$. An *arc* is a portion of a circle. An arc which is half a circle is called a *semicircle*.

19. Consider a circle and a line in a Euclidean plane. The line is a *tangent* to the circle if the circle and the line have exactly one point in common. If the line intersects the circle at more than one point, then it is called a *secant* line.

20. *Parallel* straight lines are straight lines which, being in the same plane and being produced indefinitely in both directions, do not meet one another in either direction.

21. *Rectilinear figures* are those which are contained by straight lines, *trilateral figures* being those contained by three, *quadrilateral* those contained by four, and *multilateral* those contained by more than four straight lines.

22. The *triangle* $\triangle ABC$ is a trilateral figure which consists of the union of three line segments $AB, BC$, and $CA$.

    A *right-angled triangle* is that which has a right angle. An *obtuse-angled triangle* is that which has an obtuse angle. An *acute-angled triangle* is that which has three acute angles.

    An *equilateral triangle* is that which has its three sides equal, an *isosceles triangle* that which has two of its sides alone are equal, and a *scalene triangle* that which has three sides unequal.

23. The triangles $\triangle ABC$ and $\triangle DEF$ are *similar* if $\angle ABC = \angle DEF, \angle BCA = \angle EFD$, and $\angle CAB = \angle FDE$.

    The triangles $\triangle ABC$ and $\triangle DEF$ are *congruent* if $AB = DE$, $BC = EF$, and $CA = FD$.

    It can be shown that congruence of two triangles implies their similarity.

24. Of quadrilateral figures, a *square* is that which is both equilateral and right-angled; an *oblong* that which is right-angled but not equilateral; a *rhombus* that which is equilateral but not right-angled; and a *rhomboid* that which has its opposite sides and angles equal to one another but is neither equilateral nor right-angled. And let quadrilaterals other than these be called *trapezia*.

We assume that the reader is familiar with the following constructions using straightedge and compass.

- Determine the midpoint of a line segment $AB$.
- Bisect an angle $AOB$.
- Given a line $AB$, and a point $P$, either *on* or *not on* the line $AB$, construct a line passing through $P$ and perpendicular to the line $AB$.

**Observations 5.1.** The following observations are useful in the geometric construction of the inverse of a point with respect to a circle.

1. Let $AB$ be a diameter of a circle. The perpendicular at any point $C$ in the line $AB$ intersects the circle at point $D$. Then $CD^2 = AC \cdot CB$.
2. Consider a circle $C$ and a point $P$ outside $C$. A tangent line from the point $P$ to the circle $C$ touches it at point $T$. A secant through $P$ meets circle at $S$ and $S'$. Then $PS \cdot PS' = PT^2$.
3. Sides of similar triangles are proportional.  □

## 5.3 Circle Inversion

Circle inversion from a geometric perspective is studied. Circle inversion is also briefly discussed in the chapter on applied analysis. Study of this subject is useful in the construction of geodesics in hyperbolic geometry.

**Definition 5.1.** *Circle inversion. Let $C_{a,r}$ be a circle with center $a \in \mathbb{C}$, and radius $r \in \mathbb{R}^+$. That is, $C_{a,r} = \{z \in \mathbb{C} \mid |z - a| = r\}$. The inversion transformation $I_{a,r}(\cdot)$ with respect to circle $C_{a,r}$ takes a point $z \in \mathbb{C} \setminus \{a\}$, and maps it to a point $z' \in \mathbb{C}$, where*

$$z' - I_{a,r}(z) = a + \frac{r^2}{(\bar{z} - \bar{a})}, \quad z \in \mathbb{C} \setminus \{a\} \tag{5.1}$$

*The point $I_{a,r}(z) = z'$ is said to be the inverse of point $z$ with respect to the circle $C_{a,r}$.*  □

From a geometric perspective the point $I_{a,r}(z) = z'$ is selected so that the product of the distance of $z$ and $z'$ from $a$ is equal to $r^2$. That is, $|z - a| |z' - a| = r^2$. Note that $|z - a| |z' - a| = r^2$ if and only if $(z' - a) = k(z - a)$ for some $k > 0$. That is, the inverse point $z'$ is on the radius of the circle $C_{a,r}$ passing through $z$. Thus $a$, $z$, and $z'$ are collinear points.

**Example 5.1.** Let $C_{0,r}$ be a circle with center $0 \in \mathbb{C}$, and radius $r > 0$. In the Euclidean plane, this circle is specified by $(x^2 + y^2) = r^2$. The inverse of a point $(x, y) \neq (0, 0)$ is $(x', y')$, where

$$x' = \frac{x}{(x^2 + y^2)} r^2, \quad y' = \frac{y}{(x^2 + y^2)} r^2$$

□

**Observations 5.2.** Consider a circle $C_{a,r}$ with center $a \in \mathbb{C}$, and radius $r \in \mathbb{R}^+$. Results related to inversion of points with respect to the circle $C_{a,r}$ are summarized. Let $C_{\beta,\rho}$ be another circle with center $\beta \in \mathbb{C}$, and radius $\rho \in \mathbb{R}^+$.

1. If $z \in C_{a,r}$, then $z' \in C_{a,r}$.
2. The circle $C_{\beta,\rho}$ does not pass through the center of the circle $C_{a,r}$. That is, $\rho \neq |a - \beta|$. Let $\delta \triangleq r^2 / \left( |\beta - a|^2 - \rho^2 \right)$. The inverse image of points $z \in C_{\beta,\rho}$ with respect to the circle $C_{a,r}$, form a circle with center at $\{a + (\beta - a)\,\delta\}$, and radius $|\delta|\,\rho$. The center of this later circle does not pass through the center of the circle $C_{a,r}$. That is, $\{a + (\beta - a)\,\delta\} \neq a$. If $\delta = 1$, the inverse image of points in the circle $C_{\beta,\rho}$ is $C_{\beta,\rho}$ itself. Also $\left( r^2 + \rho^2 \right) = |\beta - a|^2$. This condition implies that the circles $C_{a,r}$ and $C_{\beta,\rho}$ are orthogonal to each other.
3. If $\rho = |a - \beta|$, then the inverse image of points $z \in C_{\beta,\rho}$ is a straight line orthogonal to the line joining the centers $a$ and $\beta$. Distance of this straight line from center $a$ is $r^2 / (2\rho)$.
4. Converse of the last observation. The image of any straight line which does not pass through $a$, is a circle which contains the center $a$.
5. The inverse of points on a straight line which passes through the point $a$ are also on this straight line. However, a point does not map to itself. $\qquad\qquad\qquad\square$

See the problem section for proofs of some of the above observations.

**Examples 5.2.** In the following examples, let $C_{a,r}$ be a circle with center $a \in \mathbb{C}$, and radius $r \in \mathbb{R}^+$.

1. Let $z_1$ and $z_2$ be an inverse pair with respect to the circle $C_{a,r}$. We show that every circle that passes through the points $z_1$ and $z_2$ is orthogonal to $C_{a,r}$.
   Let $C'$ be a circle which passes through points $z_1$ and $z_2$. Also let $z'$ be a point of intersection of the circle $C'$ with circle $C_{a,r}$. Therefore $|z' - a| = r$.
   As $z_1$ and $z_2$ are a pair of inverse points with respect to the circle $C_{a,r}$, we have

$$|z_1 - a|\,|z_2 - a| = r^2$$

   Therefore
$$|z_1 - a|\,|z_2 - a| = |z' - a|^2$$

   Observe that $|z_1 - a|$ and $|z_2 - a|$ are lengths of segments of a secant to the circle $C'$. Furthermore, $|z' - a| = \sqrt{|z_1 - a|\,|z_2 - a|}$ is equal to the length of the tangent from the point $a$ to the circle $C'$. Therefore, the line segment joining points $a$ and $z'$ is tangent to the circle $C'$.
2. The last example can also be restated as follows. Let $C$ and $C'$ be circles. A point $P$ lies on the circle $C'$. The inverse of point $P$ with respect to circle $C$ is $P'$.
   (a) If the circles $C$ and $C'$ are orthogonal, then point $P'$ lies on $C'$.
   (b) The converse is also true. That is, if the point $P'$ lies on $C'$, then the circles $C$ and $C'$ are orthogonal.
3. Let $C$ be a circle with center at point $O$. Also let $P$ be another point not equal to $O$. Then the locus of centers of all circles which are orthogonal to $C$, and that pass through the point $P$ is a straight line.
   More specifically, let the inverse of point $P$ with respect to circle $C$ be $P'$.
   (a) Let the point $P$ lie in the interior of the circle $C$. The locus is the perpendicular bisector of the line joining points $P$ and $P'$. This locus is in the exterior of circle $C$.
   (b) Let the point $P$ lie in the exterior of the circle $C$. The locus is the perpendicular bisector of the line joining points $P$ and $P'$. This locus is in the exterior of circle $C$.
   (c) Let the point $P$ lie on the circle $C$, then the locus is the tangent line at $P$.

4. We show that inversion preserves angles.

   Let $z_1$ and $z_2$ be an inverse pair with respect to the circle $C_{a,r}$. Also let $C'$ and $C''$ be two circles, each of which passes through points $z_1$ and $z_2$. Let $\alpha_1$ be the angle between circles $C'$ and $C''$ at the point $z_1$. Note that $\alpha_1$ is the angle between the tangent lines to the circles $C'$ and $C''$ at the point $z_1$. Similarly, let $\alpha_2$ be the angle between circles $C'$ and $C''$ at the point $z_2$. As inversion maps circles into circles, the inverse of $C'$ in $C_{a,r}$ is a circle which passes through points $z_1$ and $z_2$, and intersects $C_{a,r}$ at the same point as $C'$. However, these points uniquely determine the circle $C'$. Consequently, the inverse of $C'$ with respect to $C_{a,r}$ is $C'$ itself. Analogously, the inverse of $C''$ with respect to $C_{a,r}$ is $C''$ itself. Therefore, inversion maps angle $\alpha_1$ to angle $\alpha_2$. Furthermore, by symmetry the magnitude of angles $\alpha_1$ and $\alpha_2$ must be same, as these are the angles between circles $C'$ and $C''$ at points $z_1$ and $z_2$. Thus $|\alpha_1| = |\alpha_2|$. Inversion does not preserve the sense of the angles.

5. The inverse of two orthogonal circles with respect to $C_{a,r}$, and not passing through the point $a$ is a pair of two orthogonal circles not intersecting the point $a$. This is an immediate consequence of the last example.                                                                   □

**Constructions** Geometric construction of the inverse of a point. Our goal is to construct the inverse of a point $P$ in the Euclidean plane, with respect to a circle $C$ with center $O$.

1. *Construction of the inverse of a point $P$ with respect to the circle $C$, where $P$ is inside the circle.*

   *Method*:
   - Draw line $OP$.
   - Construct a perpendicular on the line $OP$ at point $P$. The perpendicular line intersects the circle of inversion $C$ at two points. Let one of these points be labeled $A$.
   - Join points $O$ and $A$ by a straight line.
   - Construct a perpendicular on the line $OA$ at point $A$. The perpendicular line intersects the extension of the line $OP$ at the point $P'$. Inverse of the point $P$ is $P'$.

2. *Construction of the inverse of a point $P$ with respect to the circle $C$, where $P$ is outside the circle.*

   *Method*:
   - Bisect the line $OP$. Let the middle point of the line $OP$ be $D$.
   - Draw a circle with center $D$ and radius $DO$. This circle intersects the circle of inversion $C$ at two points $B$ and $B'$.
   - Join the points $B$ and $B'$ via a straight line. The lines $OP$ and $BB'$ intersect at $P'$. Inverse of the point $P$ is $P'$.

3. *Construction of the inverse of a point $P$ with respect to the circle $C$, where $P$ is on a circle $C'$, which is orthogonal to the circle $C$.*

   *Method*:
   - Let the line passing through points $P$ and the center $O$ of the circle $C$, intersect the circle $C'$ at $P'$. Inverse of the point $P$ is $P'$.

   *Justification*: The proof of correctness of these constructions via certain basic results of Euclidean geometry is left to the reader.

4. *Construction of orthogonal circles.*

   Let $C$ be a circle with center $O$. Also let $P$ be a point in the exterior of circle of $C$. Construct a circle $C'$ with center at $P$, and which is orthogonal to $C$.

*Method*:
Construct a circle $C''$ with the center at the middle point of the line segment $OP$, and passing through points $O$ and $P$. Let the circle $C''$ intersect the circle $C$ at points $Q$ and $R$. The required circle has center $P$, and passes through points $Q$ and $R$.
*Justification*: This is left to the reader.                                                $\square$

---

## 5.4 Elementary Differential Geometry

Differential geometry is the study of geometry via methods and techniques borrowed from differential and integral calculus. It also borrows techniques from mathematical disciplines like complex variables, linear algebra, and vector algebra. The goal in studying differential geometry is to characterize curves and surfaces in different types of spaces. Certain immediately useful mathematical facts are initially summarized in this section. This is followed by a study of lines and planes, and curves in plane and three-dimensional space.

**Notation.** Note that $\mathbb{R}^1 = \mathbb{R}$, and $\mathbb{R}^2$ is simply the $xy$-plane.                    $\square$

### 5.4.1 Mathematical Preliminaries

Euclidean $m$-dimensional space $\mathbb{R}^m$ is an $m$-dimensional vector space. It consists of all $m$-tuples:

$$\mathbb{R}^m = \{(x_1, x_2, \ldots, x_m) \mid x_j \in \mathbb{R}, 1 \leq j \leq m\}$$

Members of the set $\mathbb{R}^m$ are called vectors. This implies that vector addition of two elements in the set $\mathbb{R}^m$ is properly defined. Further, scalar multiplication is also defined. Therefore if

$$x = (x_1, x_2, \ldots, x_m), (y_1, y_2, \ldots, y_m) \in \mathbb{R}^m$$

then $(x + y) \in \mathbb{R}^m$, and

$$x + y = (x_1 + y_1, x_2 + y_2, \ldots, x_m + y_m)$$

Also if $\lambda \in \mathbb{R}$, and $x \in \mathbb{R}^m$, then $\lambda x \in \mathbb{R}^m$. That is,

$$\lambda x = (\lambda x_1, \lambda x_2, \ldots, \lambda x_m)$$

The length of the vector $x$ is $\|x\|$, where $\|\cdot\|$ is the Euclidean norm. The Euclidean norm function $\|\cdot\|$ defined on the space $\mathbb{R}^m$ is

$$\|x\| = \left(x_1^2 + x_2^2 + \ldots + x_m^2\right)^{1/2}, \quad \text{for} \quad x \in \mathbb{R}^m$$

Furthermore, the distance between vectors $x$ and $y$ is $\|x - y\|$.

In order to make these concepts precise, we initially introduce some notation and terminology. Some of these terms have also been described in the chapter on applied analysis. Nevertheless, some terms are repeated for accessibility.

**Definition 5.2.** *Open subset of $\mathbb{R}^m$, where $m \geq 1$. Let $U \subseteq \mathbb{R}^m$. The subset $U$ is open, if $u \in U$, there exists a real number $\epsilon > 0$ such that each $x \in \mathbb{R}^m$ is within a distance $\epsilon$ of $u$ is also in $U$. That is*

$$u \in U \quad and \quad \|x - u\| < \epsilon \Rightarrow x \in U \tag{5.2}$$

$\square$

An example of an open subset of $\mathbb{R}^m$ is $\mathbb{R}^m$ itself. Another example of an open set is an open ball with center $x_0 \in \mathbb{R}^m$ and radius $r \in \mathbb{R}^+$. An open ball is the set

$$B(x_0, r) = \{x \in \mathbb{R}^m \mid \|x - x_0\| < r\}$$

Note that, if $m = 1$, the open ball is an open interval. However, if $m = 2$, we have an open disc.

A continuous map is next described. Let $X \subseteq \mathbb{R}^m$, and $Y \subseteq \mathbb{R}^n$. A map $f : X \to Y$ is continuous at a point $x_0 \in X$ if points in the neighborhood of $x_0$ are mapped onto points in $Y$ near the neighborhood of points $f(x_0) \in Y$.

**Definition 5.3.** *Continuous map (function). Let $X \subseteq \mathbb{R}^m$, and $Y \subseteq \mathbb{R}^n$. A map $f : X \to Y$ is continuous at $x_0 \in X$ if, for every $\epsilon > 0$, there exists $\delta_{x_0, \epsilon} > 0$ such that for*

$$x \in X \quad and \quad \|x - x_0\| < \delta_{x_0, \epsilon} \Rightarrow \|f(x) - f(x_0)\| < \epsilon \tag{5.3}$$

*Furthermore, $f(\cdot)$ is continuous, if it is continuous at every point of $X$.* $\square$

Using the definition of an open set, an equivalent definition of continuous function can be obtained. A function $f(\cdot)$ is continuous if and only if for any open set $V$ of $\mathbb{R}^n$, there is an open set $U$ of $\mathbb{R}^m$ such that $U \cap X = \{x \in X \mid f(x) \in V\}$.

**Definition 5.4.** *Homeomorphism and homeomorphic sets. Let $f : X \to Y$ be a continuous and bijective function. If its inverse map $f^{-1} : Y \to X$ is also continuous, then the map $f(\cdot)$ is said to be a homeomorphism. Further, $X$ and $Y$ are said to be homeomorphic sets.* $\square$

**Definition 5.5.** *A real-valued function of real variables is differentiable or smooth, if it has derivatives of all orders.* $\square$

The above definition of a smooth function is different than that given in the chapter on applied analysis. In the chapter on applied analysis a smooth function belongs to class $C^1$. A smooth function in the above definition belongs to class $C^\infty$. We do this in order to keep the terminology in consonance with standard textbooks on differential geometry.

**Definition 5.6.** *Diffeomorphism. Let $f : X \to Y$ be a smooth and bijective function. If its inverse map $f^{-1} : Y \to X$ is also smooth, then the map $f(\cdot)$ is said to be a diffeomorphism.* $\square$

**Definition 5.7.** *Let $U$ be an open subset of $\mathbb{R}^2$. The map $g_i : U \to \mathbb{R}$ is smooth, if $g_i(u, v)$ has continuous partial derivatives of all orders, where $i = 1, 2, 3$. Consider the map $g : U \to \mathbb{R}^3$ where*

$$g(u, v) = (g_1(u, v), g_2(u, v), g_3(u, v)), \quad (u, v) \in U \tag{5.4}$$

*The map $g(\cdot, \cdot)$ is smooth as $g_1(\cdot, \cdot), g_2(\cdot, \cdot)$, and $g_3(\cdot, \cdot)$ are all smooth functions.* $\square$

**Notation.**

$$\frac{\partial g\,(u,v)}{\partial u} = \left( \frac{\partial g_1\,(u,v)}{\partial u}, \frac{\partial g_2\,(u,v)}{\partial u}, \frac{\partial g_3\,(u,v)}{\partial u} \right)$$

$$\frac{\partial g\,(u,v)}{\partial v} = \left( \frac{\partial g_1\,(u,v)}{\partial v}, \frac{\partial g_2\,(u,v)}{\partial v}, \frac{\partial g_3\,(u,v)}{\partial v} \right)$$

Higher derivatives are similarly defined. We adapt the following abbreviations. Let $g\,(u,v) = g$, and

$$\frac{\partial g}{\partial u} = g_u, \quad \frac{\partial g}{\partial v} = g_v$$

$$\frac{\partial^2 g}{\partial u^2} = g_{uu}, \quad \frac{\partial^2 g}{\partial u \partial v} = g_{uv}, \quad \frac{\partial^2 g}{\partial v \partial u} = g_{vu}, \quad \frac{\partial^2 g}{\partial v^2} = g_{vv}$$

□

**Notation.** Differentiation of a function with respect to the parameter (independent variable) $t$ is denoted by primes. That is,

$$\xi'\,(t) = \frac{d\xi\,(t)}{dt}, \quad \xi''\,(t) = \frac{d^2\xi\,(t)}{dt^2}, \quad \xi'''\,(t) = \frac{d^3\xi\,(t)}{dt^3}, \quad \text{and so on.}$$

□

### 5.4.2  Lines and Planes

Lines and planes are basic geometric entities.

**Definition 5.8.** *Line. Let $a, u \in \mathbb{R}^m$, where $u \neq 0$. A straight line that passes through $a$, and parallel to $u$ is the set of points $x \in \mathbb{R}^m$ such that*

$$x\,(t) = tu + a, \quad \text{where} \quad t \in \mathbb{R} \tag{5.5}$$

*This is a parametric equation of a line. The point $x$ generates the line as the parameter $t$ varies over all real numbers. If $m = 2$, the line is in the plane $\mathbb{R}^2$, and if $m = 3$, the line is in space $\mathbb{R}^3$.*

□

**Definition 5.9.** *Parallel lines. Two lines $x\,(t_1) = t_1 u + a$, $u \neq 0$, where $t_1 \in \mathbb{R}$; and $y\,(t_2) = t_2 v + b$, $v \neq 0$ where $t_2 \in \mathbb{R}$; are parallel to each other, if $u$ and $v$ are multiples of each other.*   □

**Example 5.3.** We find the equation of a line that passes through two different points $c$ and $d$. Thus $(c - d) \neq 0$ is a vector which is parallel to the line. Therefore, the equation of a line that passes through the points $c$ and $d$ is

$$x\,(t) = t\,(c - d) + c, \quad \text{where} \quad t \in \mathbb{R}$$

□

**Definitions 5.10.** *Plane, vectors parallel and normal to the plane. Let $u$ and $v$ be two linearly independent vectors in $\mathbb{R}^3$.*

1. *A plane parallel to the vectors $u$ and $v$, and passing through a point $a \in \mathbb{R}^3$ is the set of points $x \in \mathbb{R}^3$ such that*

$$x \triangleq x(t_1, t_2) = t_1 u + t_2 v + a, \quad \text{where} \quad t_1, t_2 \in \mathbb{R} \tag{5.6}$$

*This is a parametric equation of a plane, where $x(\cdot, \cdot)$ generates the plane as the parameters $t_1$ and $t_2$ vary independently over all real numbers.*
2. *A vector is parallel to the plane if it is linearly dependent upon $u$ and $v$.*
3. *A vector is normal to the plane if it is orthogonal to both $u$ and $v$.*                □

**Example 5.4.** *Let $u, v \in \mathbb{R}^3$, and $n \in \mathbb{R}^3$ be a vector normal to the plane $x(t_1, t_2) = t_1 u + t_2 v + a$, where $a \subset \mathbb{R}^3$, and $t_1, t_2 \in \mathbb{R}$. If $y \in \mathbb{R}^3$ lies in the plane, then $(y - a) \circ n = 0$.*                □


### 5.4.3   Curves in Plane and Space

A curve in a real $m$-dimensional space is a set of points with certain specific properties. We explore properties of a curve in plane and space like: length of a curve, tangents, curvature, and torsion. There are two types of curves in general. These are the level and parameterized curves.

**Definition 5.11.** *Level curves in two- and three-dimensional spaces.*

(a) *Level curve in two-dimensional space (plane). Let $f : \mathbb{R}^2 \to \mathbb{R}$. A level curve $C$ in the plane $\mathbb{R}^2$ is $C = \{(x, y) \in \mathbb{R}^2 \mid f(x, y) = c, \text{ where } c \in \mathbb{R}\}$.*
(b) *Level curve in three-dimensional space. Let $f_i : \mathbb{R}^3 \to \mathbb{R}; i = 1, 2$. A level curve $C$ in three-dimensional space $\mathbb{R}^3$ is*

$$C = \{(x, y, z) \in \mathbb{R}^3 \mid f_1(x, y, z) = c_1, f_2(x, y, z) = c_2, \text{ where } c_1, c_2 \in \mathbb{R}\} \tag{5.7}$$

□

**Example 5.5.** Some level curves in a plane are: $x + y = 1$, $y = x^2$, and $x^2 + y^2 = 1$.                □

Parameterization of a curve is sometimes mathematically very useful. Before we define a parameterized curve, we introduce certain notation. Consider a function $f : I \to \mathbb{R}$, where $I$ is a closed interval in $\mathbb{R}$. Denote the $r$-th derivative of $f(x)$ by $f^{(r)}(x)$, where $r \in \mathbb{P}$. The function $f(\cdot)$ is of class $C^r$ on $I$ if $f^{(r)}(x)$ exists and is continuous at all $x \in I$. The definition of a function of class $C^r$, where $r \in \mathbb{P}$, has also been given in the chapter on applied analysis.

**Definition 5.12.** *The vector-valued function $x(t) \in \mathbb{R}^m$, where $t \in I \subseteq \mathbb{R}$ is denoted by*

$$x(t) = (x_1(t), x_1(t), \ldots, x_m(t)) \tag{5.8a}$$

*Its derivatives with respect to $t$ are*

$$\frac{dx(t)}{dt} = \left(\frac{dx_1(t)}{dt}, \frac{dx_2(t)}{dt}, \ldots, \frac{dx_m(t)}{dt}\right) \triangleq x'(t) \tag{5.8b}$$

$$\frac{d^2x(t)}{dt^2} = \left(\frac{d^2x_1(t)}{dt^2}, \frac{d^2x_2(t)}{dt^2}, \ldots, \frac{d^2x_m(t)}{dt^2}\right) \triangleq x''(t) \tag{5.8c}$$

*and so on. For $r \in \mathbb{P}$, the vector $x(t)$ is said to be in class $C^r$, if $x_i(t) \in C^r$, $\forall i = 1, 2, \ldots m$.*  □

**Definition 5.13.** *Parameterized curve. Let $\alpha, \beta \in \mathbb{R}$, $\alpha < \beta$, and $I$ be an open interval. That is, let $I = (\alpha, \beta) = \{ t \in \mathbb{R} \mid \alpha < t < \beta \}$. A parameterized curve in $\mathbb{R}^m$ is a map $\xi : I \to \mathbb{R}^m$. The variable $t$ is called the parameter of the curve.*

*The curve $C$ is said to have a regular parametric representation, if the curve $C$ is of class $C^1$ in $I$ and $\xi'(t) \neq 0$, $\forall t \in I$.*  □

**Examples 5.6.** Some illustrative examples.

1. Parameterization of the level curve $v = u^3$. It is $\xi : \mathbb{R} \to \mathbb{R}^2$, where $\xi(t) = (t, t^3)$.
2. Parameterization of the circle $u^2 + v^2 = 1$. It is $\xi : (0, 2\pi) \to \mathbb{R}^2$, where $\xi(t) = (\cos t, \sin t)$.

□

**Definition 5.14.** *The first derivative of $\xi(t)$ with respect to $t$, $\xi'(t)$ is called the tangent vector of $\xi(\cdot)$ at the point $\xi(t)$. The tangent vector at the point $\xi(t)$ is $\widetilde{T}(t) = \xi'(t)$.*

*The value $\left\| \xi'(t) \right\|$ is called the speed of the parameterized curve $C$ at point $\xi(t)$. Also $C$ is called a unit-speed curve if $\left\| \xi'(t) \right\| = 1$ for all values of $t$ in the interval $I$.*  □

**Observation 5.3.** If $\xi(t)$, is a vector of constant length, then

$$\xi(t) \circ \xi'(t) = 0$$

That is, $\xi(t)$ is orthogonal to $\xi'(t)$.  □

See the problem section for a proof of the above observation. More specifically, if $\xi(t)$ is a unit vector function, then $\xi(t)$ is orthogonal to $\xi'(t)$.

**Length of a Curve and Tangents**

**Definition 5.15.** *The arc-length of a curve $C$ beginning at the point $\xi(t_0)$ and ending at the point $\xi(t)$ is $s(t)$. It is*

$$s(t) = \int_{t_0}^{t} \left\| \xi'(\zeta) \right\| d\zeta \tag{5.9}$$

*where $t \geq t_0$.*  □

Note that $s(t_0) = 0$; $s(t) > 0$, if $t > t_0$. The above definition also yields

$$\frac{ds(t)}{dt} = \left\| \xi'(t) \right\|$$

**Example 5.7.** Let $\xi(t) = (\cos t, \sin t)$. Observe that $\left\| \xi'(t) \right\| = 1$. The arc-length starting at $\xi(t_0)$ is

$$s(t) = \int_{t_0}^{t} d\zeta = (t - t_0)$$

If $t_0 = 0$, and $t = 2\pi$, then $s(2\pi) = 2\pi$.  □

If the parameter of the curve $C$ is $s$, then the representation of the curve is said to be *natural*. In this case, the *tangent vector* to the curve at $s$ is $\widetilde{T}(s) = d\xi(s)/ds$. Thus

$$\widetilde{T}(s) = \frac{d\xi(s)}{ds} = \frac{d\xi(s(t))}{dt}\frac{dt}{ds} = \frac{\xi'(t)}{\left\|\xi'(t)\right\|}$$

This implies that the length of the vector $\widetilde{T}(s)$ is unity. That is,

$$\left\|\widetilde{T}(s)\right\| = \left\|\frac{d\xi(s)}{ds}\right\| = 1$$

The *tangent line* to the regular curve $C$ at a point $\xi(t_0)$ is

$$\eta(u) = \xi(t_0) + u\xi'(t_0), \quad u \in \mathbb{R}$$

A plane through $\xi(t_0)$, which is orthogonal to the tangent line $\eta(u)$ is called the *normal plane* to the curve $C$ at $\xi(t_0)$. The equation of this normal plane is

$$(\zeta - \xi(t_0)) \circ \xi'(t_0) = 0$$

where $\zeta \in \mathbb{R}^m$.

**Example 5.8.** We use the notation: $(1,0,0) \triangleq i$, $(0,1,0) \triangleq j$, and $(0,0,1) \triangleq k$. The vectors $i, j$, and $k$ are along $x$-axis, $y$-axis, and $z$-axis respectively. Consider the curve $\xi(t) = \left(it^3 + jt^2 + kt\right)$, in $\mathbb{R}^3$, where $t \in \mathbb{R}$. The tangent vector at $t = 2$ is $\xi'(2) = (12i + 4j + k)$. The tangent line to the curve at $\xi(2) = (8i + 4j + 2k)$ is

$$\eta(u) = \xi(2) + u\xi'(2) = (8 + 12u)i + (4 + 4u)j + (2 + u)k, \quad u \in \mathbb{R}$$

The equation of the plane which is normal to the tangent line at $t = 2$ is

$$(\zeta - \xi(t_0)) \circ \xi'(t_0) = 0$$

which is

$$(\zeta_1 - 8)12 + (\zeta_2 - 4)4 + (\zeta_3 - 2) = 0$$

That is, the equation of the plane which is normal to the tangent line is $(12\zeta_1 + 4\zeta_2 + \zeta_3) = 114$. $\square$

### Curvature

We discuss results related to curvature of curves. Let $\alpha < \beta$, where $\alpha, \beta \in \mathbb{R}$. Consider a curve $\xi : (\alpha, \beta) \to \mathbb{R}^3$ parameterized by arc length $s$. In this case, the tangent vector $\widetilde{T}(s) = \xi'(s)$ has unit length, and $\left\|\xi''(s)\right\|$ is a measure of rate of change of the angle which nearby tangents make with the tangent at $s$.

**Definition 5.16.** *Let $(\alpha, \beta) = \{s \mid \alpha < s < \beta\}$ be an open interval, where $\alpha, \beta \in \mathbb{R}$ and $\alpha < \beta$.. A regular curve $C$ of class $C^r$, where $r \geq 2$ parameterized by its arc length $s \in (\alpha, \beta)$ is a map $\xi : (\alpha, \beta) \to \mathbb{R}^3$.*

*The tangent vector to the curve at $s$ is $\widetilde{T}(s) = \xi'(s)$. The curvature vector of $C$ at $s$ is $k(s) = \widetilde{T}'(s) = \xi''(s)$, the curvature of $C$ at $s$ is $\kappa(s) = \left\|\xi''(s)\right\|$. If $\kappa(s) \neq 0$, then the radius of curvature at $s$ is $\rho = 1/\kappa(s)$.* $\qquad\qquad\square$

If $\xi''(s) = 0$, the corresponding point on the curve is called a *point of inflection*. At this point, the curvature is zero, and the radius of curvature is infinite.

**Example 5.9.** We compute the curvature of the circular helix

$$\xi(t) = (a\cos t, a\sin t, bt), \quad a,b \in \mathbb{R}, \quad \text{and} \quad t \in \mathbb{R}$$

We have

$$\xi'(t) = (-a\sin t, a\cos t, b)$$
$$\left\|\xi'(t)\right\| = \left(a^2 + b^2\right)^{1/2}$$

The unit tangent vector

$$\widetilde{T}(t) = \frac{\xi'(t)}{\left\|\xi'(t)\right\|} = \frac{(-a\sin t, a\cos t, b)}{\left(a^2 + b^2\right)^{1/2}}$$

and

$$k(s) = \widetilde{T}'(s) = \frac{d\widetilde{T}}{ds} = \frac{d\widetilde{T}}{dt}\frac{dt}{ds} = \frac{d\widetilde{T}}{dt} \div \frac{ds}{dt} = \frac{d\widetilde{T}}{dt}\frac{1}{\left\|\xi'(t)\right\|}$$
$$= \frac{-a(\cos t, \sin t, 0)}{\left(a^2 + b^2\right)}$$

The curvature is

$$\kappa(s) = \|k(s)\| = \frac{|a|}{\left(a^2 + b^2\right)}$$

If $a \neq 0$, and $b = 0$, then the curvature $\kappa(s) = 1/|a|$; and $\xi(t)$ is simply a circle in the $xy$-plane with radius $|a|$. However, if $a = 0$, and $b \neq 0$, then the curvature $\kappa(s) = 0$, and $\xi(t) = (0, 0, bt)$ is simply the $z$-axis. $\qquad\qquad\square$

If the parameterization of a curve is not its arc length $s$, but some other parameter $t$, then its curvature at $t$ can be determined via the following observation.

**Observation 5.4.** Let $\alpha, \beta \in \mathbb{R}$, $\alpha < \beta$; and $(\alpha, \beta) = \{t \in \mathbb{R} \mid \alpha < t < \beta\}$ be an open interval. A curve $C$ of class greater than or equal to 2, parameterized by $t \in (\alpha, \beta)$ is a map $\xi : (\alpha, \beta) \to \mathbb{R}^3$. Assume that $\xi'(t) \neq 0$. The curvature of the curve at $t \in (\alpha, \beta)$ is given by

$$\kappa(t) = \frac{\left\|\xi''(t) \times \xi'(t)\right\|}{\left\|\xi'(t)\right\|^3}$$

$\qquad\qquad\square$

See the problem section for a proof of the above observation.

**Torsion**

In 2-dimensional space plane curves are essentially described by curvature. Another type of curvature is required to describe curves in 3-dimensional space. Consider a curve $C$ in 3-dimensional space, which is parameterized by its arc-length $s$, measured from some fixed point on the curve $C$. Let $\widetilde{T}(s) = \xi'(s)$ be its unit tangent vector. The derivative of $\widetilde{T}(s)$ with respect to $s$ is a measure of its curvature. This is the curvature vector $\widetilde{T}'(s) = \xi''(s)$. If the curvature $\kappa(s) = \left\|\xi''(s)\right\| \neq 0$, then the direction of $\widetilde{T}'(s)$ is specified by the unit vector $\widetilde{N}(s)$. That is,

$$\widetilde{N}(s) = \frac{\widetilde{T}'(s)}{\left\|\widetilde{T}'(s)\right\|} = \frac{\widetilde{T}'(s)}{\left\|\xi''(s)\right\|} = \frac{\widetilde{T}'(s)}{\kappa(s)}$$

As $\widetilde{T}(s)$ is a unit vector, we also have $\widetilde{T}(s) \circ \widetilde{T}'(s) = 0$. Therefore, the vectors $\widetilde{T}(s)$ and $\widetilde{N}(s)$ are orthogonal to each other. This implies that the direction of $\widetilde{N}(s)$ is normal to the curve $C$ at $s$. Therefore, the unit vector $\widetilde{N}(s)$ is called the *principal normal* to the curve $C$ at the point $s$. The vector

$$\widetilde{B}(s) = \widetilde{T}(s) \times \widetilde{N}(s)$$

is perpendicular to the plane of $\widetilde{T}(s)$ and $\widetilde{N}(s)$. As the vectors $\widetilde{T}(s)$ and $\widetilde{N}(s)$ have unit length, the magnitude of the vector $\widetilde{B}(s)$ is unity. The vector $\widetilde{B}(s)$ is called *binormal vector* of the curve $C$ at the point $s$. Therefore, $\left\{\widetilde{T}(s), \widetilde{N}(s), \widetilde{B}(s)\right\}$ form an orthonormal basis of $\mathbb{R}^3$, and is right-handed. That is,

$$\widetilde{B}(s) = \widetilde{T}(s) \times \widetilde{N}(s)$$
$$\widetilde{N}(s) = \widetilde{B}(s) \times \widetilde{T}(s)$$
$$\widetilde{T}(s) = \widetilde{N}(s) \times \widetilde{B}(s)$$

The plane containing the tangent and the principal normal vectors at a point $P$ of the curve $C$ is called the *osculating plane*. It is perpendicular to the binormal vector. The *normal plane* passes through the point $P$ and is perpendicular to the tangent vector. The *rectifying plane* passes through the point $P$ and is perpendicular to the principal normal vector.

Consider a curve $C$ which is parameterized by the arc-length $s$. At a point $\xi(s_0)$ on curve $C$, the equations of the tangent, principal normal, and binormal lines are:

- Tangent line: $(\zeta - \xi(s_0)) \times \widetilde{T}(s_0) = 0$.
- Principal normal line: $(\zeta - \xi(s_0)) \times \widetilde{N}(s_0) = 0$, $\kappa(s_0) > 0$.
- Binormal line: $(\zeta - \xi(s_0)) \times \widetilde{B}(s_0) = 0$, $\kappa(s_0) > 0$.

The corresponding equations of the normal, rectifying, and osculating planes are:

- Normal plane: $(\zeta - \xi(s_0)) \circ \widetilde{T}(s_0) = 0$.
- Rectifying plane: $(\zeta - \xi(s_0)) \circ \widetilde{N}(s_0) = 0$, $\kappa(s_0) > 0$.
- Osculating plane: $(\zeta - \xi(s_0)) \circ \widetilde{B}(s_0) = 0$, $\kappa(s_0) > 0$.

The relationship between the unit vectors $\widetilde{T}(s), \widetilde{N}(s), \widetilde{B}(s)$ and their derivatives with respect to the parameter $s$ is described by the Frenet-Serret equations. These equations were discovered independently by Frenchmen Jean Frédéric Frenet in 1847 and by Joseph Alfred Serret in 1851. These are

$$\widetilde{T}'(s) = \kappa(s)\,\widetilde{N}(s)$$
$$\widetilde{B}'(s) = -\tau(s)\,\widetilde{N}(s)$$
$$\widetilde{N}'(s) = \tau(s)\,\widetilde{B}(s) - \kappa(s)\,\widetilde{T}(s)$$

where $\tau(s)$ is called the *torsion* at the point $s$. The *radius of torsion* is equal to $\sigma(s) = 1/\tau(s)$. See the problem section for a derivation of the Frenet-Serret equations.

**Observation 5.5.** Let $\alpha < \beta$, where $\alpha, \beta \in \mathbb{R}$, and

$$(\alpha, \beta) = \{t \in \mathbb{R} \mid \alpha < t < \beta\}$$

be an open interval. A curve $C$ of class greater than or equal to 3, parameterized by $t \in (\alpha, \beta)$ is a map $\xi : (\alpha, \beta) \to \mathbb{R}^3$. Assume that $\kappa(t) \neq 0$. The torsion of the curve $C$ at $t \in (\alpha, \beta)$ is given by

$$\tau(t) = \frac{\xi'(t) \circ \xi''(t) \times \xi'''(t)}{\left\| \xi'(t) \times \xi''(t) \right\|^2}$$

In the above expression, the primes denote derivatives with respect to the independent variable $t$. If the parameter $t$ is the arc-length $s$, then

$$\tau(s) = \frac{\xi'(s) \circ \xi''(s) \times \xi'''(s)}{\left\| \xi'(s) \times \xi''(s) \right\|^2} = \tau(t)$$

In the above expression, the primes denote derivatives with respect to the independent variable $s$. $\square$

See the problem section for a proof of the above observation.

**Example 5.10.** We compute the torsion of the circular helix

$$\xi(t) = (a \cos t, a \sin t, bt), \quad a, b \in \mathbb{R}, \quad \text{and} \quad t \in \mathbb{R}$$

We have

$$\xi'(t) = (-a \sin t, a \cos t, b)$$
$$\xi''(t) = (-a \cos t, -a \sin t, 0)$$
$$\xi'''(t) = (a \sin t, -a \cos t, 0)$$

Therefore

$$\xi''(t) \times \xi'''(t) = (0, 0, a^2)$$
$$\xi'(t) \circ \xi''(t) \times \xi'''(t) = a^2 b$$
$$\left\| \xi'(t) \times \xi''(t) \right\|^2 = a^2(a^2 + b^2)$$

This results in the torsion

$$\tau(t) = \frac{\xi'(t) \circ \xi''(t) \times \xi'''(t)}{\left\| \xi'(t) \times \xi''(t) \right\|^2}$$

$$= \frac{b}{(a^2 + b^2)}$$

If $a \neq 0$, and $b = 0$, then the torsion $\tau(t) = 0$. In this case $\xi(t)$ is simply a circle in the $xy$-plane with radius $|a|$. However, if $a = 0$, and $b \neq 0$, then the torsion $\tau(t) = 1/b$, and $\xi(t) = (0, 0, bt)$ is simply the $z$-axis. $\square$

## 5.5 Basics of Surface Geometry

In the last section, we studied basics of lines and curves in space. In this section, we introduce the subject of surface geometry. Basic concepts related to surface geometry are initially described. This is followed by techniques to describe surfaces via first fundamental form, conformal mapping of surfaces, and second fundamental form. These enable us to determine arc-lengths, areas, angle between curves, and "curved" properties of surfaces.

### 5.5.1 Preliminaries

A surface in three-dimensional space $\mathbb{R}^3$ appears locally like a subset of space $\mathbb{R}^2$ in the neighborhood of a point. This is also a generalization of a curve in two-dimensional space. A simple 2-dimensional generalization of a curve in $\mathbb{R}^m$ is called a *surface patch* or *local surface*. We state the definition of a *surface* in the space $\mathbb{R}^3$.

**Definition 5.17.** *Surface in* $\mathbb{R}^3$. *A subset* $S \subseteq \mathbb{R}^3$ *is a surface if, for every point* $P \in S$, *there exists an open set* $U \subseteq \mathbb{R}^2$ *and an open set* $V \subseteq \mathbb{R}^3$ *containing* $P$ *such that* $S \cap V$ *is homeomorphic to* $U$. *This type of homeomorphism* $x : U \to S \cap V$ *is called a coordinate chart or surface patch or parameterization of the open subset* $S \cap V$ *of* $S$. $\qquad\square$

Unless stated otherwise, all surfaces in this chapter are smooth. As per the above definition, a surface has a collection of homeomorphisms. Collection of all surface patches whose images cover the entire surface $S$ is called an *atlas* of $S$.

We next introduce the idea of a *regular surface* $S$ in the 3-dimensional space $\mathbb{R}^3$. Intuitively speaking, regular surface is a subset of points in $\mathbb{R}^3$, where at each point in $S$ we can define a plane, which is tangential to the surface $S$. Moreover, functions which are defined on a regular surface are differentiable.

**Definition 5.18.** *Regular surface patch and smooth surface.*

1. *Let* $U$ *be an open subset of* $\mathbb{R}^2$. *A surface patch* $x : U \to \mathbb{R}^3$ *is regular, if it is smooth, and the vectors* $x_u$ *and* $x_v$ *are linearly independent of each other at all points* $(u, v) \in U$.
2. *If all the surface patches in the atlas* $\mathcal{A}$ *of a surface are regular, such that every point of* $S$ *lies in at least in a single image of one patch in* $\mathcal{A}$, *then the surface* $S$ *is smooth.* $\qquad\square$

Let $x(u,v) = (x_1(u,v), x_2(u,v), x_3(u,v))$, where $x(u,v) \in U$. Independence of the first partial derivatives $x_u$ and $x_v$ imply that these derivative vectors are nonzero, and in the $3 \times 2$ matrix $dx_{(u,v)}$, where

$$dx_{(u,v)} = \begin{bmatrix} \partial x_1(u,v)/\partial u & \partial x_1(u,v)/\partial v \\ \partial x_2(u,v)/\partial u & \partial x_2(u,v)/\partial v \\ \partial x_3(u,v)/\partial u & \partial x_3(u,v)/\partial v \end{bmatrix}$$

the two columns are linearly independent. That is, the two columns are not multiples of each other. This implies that the vector cross product of $x_u$ and $x_v$ is not equal to zero. That is, $x_u \times x_v \neq 0$. Linear independence of the two columns also means that the rank of this matrix is equal to 2. The rank of the matrix is equal to 2 if and only if the determinant of at least one of the three $2 \times 2$ minors of the matrix is nonzero. That is, at least one of the following three Jacobian determinants is nonzero.

$$\frac{\partial (x_1, x_2)}{\partial (u, v)} = \begin{vmatrix} \partial x_1/\partial u & \partial x_1/\partial v \\ \partial x_2/\partial u & \partial x_2/\partial v \end{vmatrix}, \quad \frac{\partial (x_2, x_3)}{\partial (u, v)}, \quad \frac{\partial (x_1, x_3)}{\partial (u, v)}$$

Also observe that $x_u (u_0, v_0)$ is the derivative of $x$ at $(u_0, v_0)$ in the direction of the $u$-axis. That is, $x_u (u_0, v_0)$ is tangent to the $u$-parameter curve $x (u, v_0)$ at $x (u_0, v_0)$ in the direction of increasing $u$. Similarly, $x_v (u_0, v_0)$ is tangent to the $v$-parameter curve $x (u_0, v)$ at $x (u_0, v_0)$ in the direction of increasing $v$. This discussion is summarized in the following observation.

**Observation 5.6.** Let the surface patch $x : U \to S$ be regular. Then $x_u \times x_v \neq 0$ at every point of $(u, v) \in U$.                                                                                        □

**Definition 5.19.** *Tangent plane, and unit normal vector to the tangent plane.*

1.  *The tangent plane at a point* $x (u, v) \triangleq x \in S$ *is*

$$y = x + \lambda x_u + \mu x_v \tag{5.10}$$

    *where* $\lambda, \mu \in \mathbb{R}$ *are called the parameters of the tangent plane.*
2.  *As* $x_u \times x_v \neq 0$, *the direction of the unit normal vector to the tangent plane at* $x (u, v)$ *is given by the vector* $\mathcal{N}$, *where*

$$\mathcal{N} = \frac{x_u \times x_v}{\|x_u \times x_v\|} \tag{5.11}$$

□

**Example 5.11.** Consider a sphere of radius $r \in \mathbb{R}^+$. It is specified by the set of points

$$\left\{ (x_1, x_2, x_3) \in \mathbb{R}^3 \mid x_1^2 + x_2^2 + x_3^2 = r^2 \right\}$$

A possible parameterization is

$$\varsigma (\theta, \phi) = (r \cos \theta \cos \phi, r \cos \theta \sin \phi, r \sin \theta)$$

where $\theta \in [-\pi/2, \pi/2]$, and $\phi \in [0, 2\pi)$. The $\theta\phi$-coordinate system is used on the globe to locate a point on it. The parameters $\theta$ and $\phi$ specify the *latitude* and *longitude* respectively of a point. The curve $\theta =$ constant, specifies a *parallel*, and the curve $\phi =$ constant specifies a *meridian*. The curve $\theta = 0$ specifies the *equator,* and $\theta = \pm\pi/2$ the *poles.*

Observe that the set of points $\{(\theta, \phi) \mid \theta \in [-\pi/2, \pi/2], \text{ and } \phi \in [0, 2\pi)\}$ do not form an open subset of $\mathbb{R}^2$. Therefore these set of points are not admissible as the domain of a surface patch. Let us try the open set

$$U = \{(\theta, \phi) \mid \theta \in (-\pi/2, \pi/2), \text{ and } \phi \in (0, 2\pi)\}$$

In this case the image of $U$ under the mapping $\varsigma : U \to \mathbb{R}^3$ clearly does not cover the entire sphere. That is, the set $\varsigma (U)$ covers only a part of the sphere. Consider another mapping $\widetilde{\varsigma} : U \to \mathbb{R}^3$, where $U$ is the open set as described above. It is obtained from the mapping $\varsigma$ by initially rotating it by $\pi$ about the $z$-axis, and subsequently by $\pi/2$ about the $x$-axis. Thus

$$\widetilde{\varsigma}\left(\theta, \phi\right) = \left(-r \cos\theta \cos\phi, -r \sin\theta, -r \cos\theta \sin\phi\right)$$

The union of the sets $\varsigma\left(U\right)$ and $\widetilde{\varsigma}\left(U\right)$ covers the entire sphere. Furthermore, the intersection of these two sets is nonempty.                                                                            □

In constructing atlas of a surface $S$, it is possible to describe the atlas via several different sets of surface patches. In several cases, it is advisable to use the *maximal atlas*. A maximal atlas of a surface $S$ consists of *all* regular surface patches $x : U \to V \cap S$, where $U$ and $V$ are open subsets of $\mathbb{R}^2$ and $\mathbb{R}^3$ respectively. Surface patches of this type are called allowable surface patches for $S$. Unless stated otherwise, all surface patches are allowable in this chapter.

### 5.5.2  First Fundamental Form

The first fundamental form of a surface patch enables us to compute lengths, angles, and areas on a surface. Let $x\left(u, v\right) \in \mathbb{R}^3$ be a patch on a surface of class $C^\infty$, where $u$ and $v$ are real parameters. The differential of the mapping of $x\left(u, v\right)$ at $\left(u, v\right) \in \mathbb{R}^2$ is one-to-one and linear. It is

$$dx\left(u, v\right) = x_u\left(u, v\right) du + x_v\left(u, v\right) dv$$

where

$$x\left(u + du, v + dv\right) = x\left(u, v\right) + dx\left(u, v\right)$$

Compute

$$
\begin{aligned}
I &= dx \circ dx \\
&= \left(x_u du + x_v dv\right) \circ \left(x_u du + x_v dv\right) \\
&= \left(x_u \circ x_u\right) du^2 + 2\left(x_u \circ x_v\right) dudv + \left(x_v \circ x_v\right) dv^2
\end{aligned}
$$

Let

$$E \triangleq x_u \circ x_u, \quad F \triangleq x_u \circ x_v, \quad G \triangleq x_v \circ x_v$$

Therefore

$$I = E du^2 + 2F dudv + G dv^2$$

The function $I$ is called the *first fundamental form* of $x\left(\cdot, \cdot\right)$; and $E, F$, and $G$ are called the *first fundamental coefficients*. The first fundamental form and the first fundamental coefficients are all functions of $u$ and $v$.

**Definition 5.20.** *First fundamental form. Let* $x : U \to S$ *be a surface patch, where $U$ is an open subset of $\mathbb{R}^2$, and $S \subseteq \mathbb{R}^3$ is a regular surface. Let $E, F, G : U \to \mathbb{R}$, where for $\left(u, v\right) \in U$*

$$E\left(u, v\right) = \left\|x_u\left(u, v\right)\right\|^2 \tag{5.12a}$$

$$F\left(u, v\right) = x_u\left(u, v\right) \circ x_v\left(u, v\right) \tag{5.12b}$$

$$G\left(u, v\right) = \left\|x_v\left(u, v\right)\right\|^2 \tag{5.12c}$$

*and*

$$
\begin{aligned}
I\left(u, v\right) &= dx\left(u, v\right) \circ dx\left(u, v\right) \\
&= E\left(u, v\right) du^2 + 2F\left(u, v\right) dudv + G\left(u, v\right) dv^2
\end{aligned}
\tag{5.12d}
$$

*is called the Riemann metric or first fundamental form.of the surface patch* $x\left(\cdot,\cdot\right)$ *induced from* $\mathbb{R}^3$ *at* $(u,v)$. *Also* $E\left(u,v\right)$, $F\left(u,v\right)$, *and* $G\left(u,v\right)$ *are called the first fundamental coefficients.*   $\square$

Certain characteristics of the first fundamental form of a surface patch are summarized.

**Observation 5.7.** Characteristics of the first fundamental form $I\left(\cdot,\cdot\right)$ of a surface patch $x\left(\cdot,\cdot\right)$ and its coefficients. Let $(u,v)\in U$.

1. $I\left(\cdot,\cdot\right)=dx\left(\cdot,\cdot\right)\circ dx\left(\cdot,\cdot\right)$.
2. The first fundamental form $I\left(\cdot,\cdot\right)$ is a homogeneous function of second degree in $du$ and $dv$.
3. The function $I\left(\cdot,\cdot\right)$ can be written as a quadratic defined on the the vectors $(du,dv)$ in the $uv$-plane. That is

$$I = \begin{bmatrix} du & dv \end{bmatrix} \begin{bmatrix} E & F \\ F & G \end{bmatrix} \begin{bmatrix} du & dv \end{bmatrix}^T$$

Also, $I\left(\cdot,\cdot\right)$ is positive definite.
4. $EG - F^2 = \|x_u \times x_v\|^2$.
5. $E > 0, G > 0$, and $\left(EG - F^2\right) > 0$.
6. The function $I\left(\cdot,\cdot\right)$ depends only on the surface and not on its parameterization.
7. The function $I\left(\cdot,\cdot\right)$ occurs in the calculation of arc-length, angle, and surface area.

   (a) Let $\alpha\left(t\right) = x\left(u\left(t\right),v\left(t\right)\right)$ where $t\in\mathbb{R}$. The arc-length starting at $x\left(u\left(t_0\right),v\left(t_0\right)\right)$ is

$$s\left(t\right) = \int_{t_0}^{t}\left\|\alpha'\left(\zeta\right)\right\|d\zeta$$

$$= \int_{t_0}^{t}\left\{E\left(\frac{du}{d\zeta}\right)^2 + 2F\left(\frac{du}{d\zeta}\right)\left(\frac{dv}{d\zeta}\right) + G\left(\frac{dv}{d\zeta}\right)^2\right\}^{1/2}d\zeta$$

   Thus

$$\frac{ds\left(t\right)}{dt} = \left\{E\left(\frac{du}{dt}\right)^2 + 2F\left(\frac{du}{dt}\right)\left(\frac{dv}{dt}\right) + G\left(\frac{dv}{dt}\right)^2\right\}^{1/2}$$

   (b) If the angle between the vectors $x_u$ and $x_v$ is $\beta$ then

$$\cos\beta = \frac{x_u \circ x_v}{\|x_u\|\,\|x_v\|} = \frac{F}{\sqrt{EG}}$$

   Therefore, the $u$- and $v$-parameter curves are orthogonal at a point, if and only if $F = 0$.
   (c) The area of a region on the surface which corresponds to a set of points $W \subseteq U$ is given by

$$A = \int\int_{W}\left\{EG - F^2\right\}^{1/2}dudv$$

   The above expression is also independent of the parameterization.   $\square$

Some of the above observations are established in the problem section.

**Examples 5.12.** Computation of first fundamental form and its coefficients.

1. Plane. Let $a$ be a point on the plane. Also let $p$ and $q$ be two unit vectors parallel to the plane, and orthogonal to each other. The plane is described by

$$x(u, v) = up + vq + a$$

Therefore

$$x_u(u, v) = p, \quad x_v(u, v) = q$$

Consequently

$$E = \|x_u(u, v)\|^2 = 1, \quad F = x_u(u, v) \circ x_v(u, v) = 0, \quad G = \|x_v(u, v)\|^2 = 1$$

$$I(u, v) = du^2 + dv^2$$

2. Sphere. A sphere is described by

$$\left\{ (x_1, x_2, x_3) \in \mathbb{R}^3 \mid x_1^2 + x_2^2 + x_3^2 = r^2 \right\}$$

A parameterization is

$$x(\theta, \phi) = (r\cos\theta\cos\phi, r\cos\theta\sin\phi, r\sin\theta)$$

Therefore

$$x_\theta(\theta, \phi) = (-r\sin\theta\cos\phi, -r\sin\theta\sin\phi, r\cos\theta)$$
$$x_\phi(\theta, \phi) = (-r\cos\theta\sin\phi, r\cos\theta\cos\phi, 0)$$

The first fundamental coefficients are

$$E = \|x_\theta(\theta, \phi)\|^2 = r^2$$
$$F = x_\theta(\theta, \phi) \circ x_\phi(\theta, \phi) = 0$$
$$G = \|x_\phi(\theta, \phi)\|^2 = r^2\cos^2\theta$$

The first fundamental form is

$$I(\theta, \phi) = r^2 \left( d\theta^2 + \cos^2\theta \, d\phi^2 \right)$$

$\square$

### 5.5.3  Conformal Mapping of Surfaces

The first fundamental form provides us with methods to measure arc-lengths, angles, and areas on surfaces. We next try to understand properties of angles between intersecting curves on surfaces.

**Definition 5.21.** *Let $S$ be a surface, and $\xi_1$ and $\xi_2$ be two curves on a surface patch $x(\cdot, \cdot)$ which intersect at a point $P$ on the surface. Let $\xi_1(t) = x(u_1(t), v_1(t))$, and $\xi_2(t) = x(u_2(t), v_2(t))$, where $u_1, v_1, u_2,$ and $v_2$ are smooth functions. For some parameters $t_1$ and $t_2$, let $P = \xi_1(t_1) = \xi_2(t_2)$.*

*The angle $\theta$ between the two intersecting curves $\xi_1$ and $\xi_2$ at the point $P$ is the angle between the tangent vectors $\xi_1'(t_1)$ and $\xi_2'(t_2)$.*  $\square$

Use of the above definition yields

$$\xi'_1 = x_{u_1} u'_1 + x_{v_1} v'_1, \quad \text{and} \quad \xi'_2 = x_{u_2} u'_2 + x_{v_2} v'_2$$

$$\cos \theta = \frac{\xi'_1 \circ \xi'_2}{\|\xi'_1\| \, \|\xi'_2\|}$$

where a prime denotes derivative with respect to $t$. Let the coefficients of the first fundamental form of a surface patch $x\,(\cdot, \cdot)$ be $E$, $F$, and $G$. Simplification of the expression for $\cos \theta$ yields

$$\cos \theta = \frac{E u'_1 u'_2 + F\,(u'_1 v'_2 + u'_2 v'_1) + G v'_1 v'_2}{\left(E u'^2_1 + 2F u'_1 v'_1 + G v'^2_1\right)^{1/2} \left(E u'^2_2 + 2F u'_2 v'_2 + G v'^2_2\right)^{1/2}}$$

**Definition 5.22.** *Conformal mapping of surfaces. Let $S$ and $\widetilde{S}$ be two surfaces, and $f : S \to \widetilde{S}$ be a diffeomorphism. Let $\xi_1$ and $\xi_2$ be two intersecting curves on the surface $S$. These two curves intersect at a point $P$ on the surface $S$.*

*The two intersecting curves $\xi_1$ and $\xi_2$ are mapped by the function $f$ into two intersecting curves $\widetilde{\xi}_1$ and $\widetilde{\xi}_2$ on the surface $\widetilde{S}$, respectively. The two curves $\widetilde{\xi}_1$ and $\widetilde{\xi}_2$ intersect at a point $f\,(P) = \widetilde{P}$ on $\widetilde{S}$.*

*If the angle of intersection of the two curves $\xi_1$ and $\xi_2$ at point $P$ is equal to the angle of intersection of the two curves $\widetilde{\xi}_1$ and $\widetilde{\xi}_2$ at the point $\widetilde{P}$, then the mapping $f$ is said to be conformal.*
□

Thus, a conformal mapping $f$ preserves angles.

**Theorem 5.1.** *Conformal mapping and the coefficients of the first fundamental forms. Let $S$ and $\widetilde{S}$ be two surfaces, and $f : S \to \widetilde{S}$ be a diffeomorphism. Let $\xi_1$ and $\xi_2$ be two intersecting curves on the surface patch $x\,(\cdot, \cdot)$ on $S$. Also let $f\,(x\,(\cdot, \cdot)) = \widetilde{x}\,(\cdot, \cdot)$. The corresponding mapped intersecting curves on the surface patch $\widetilde{x}\,(\cdot, \cdot)$ on the surface $\widetilde{S}$ are $\widetilde{\xi}_1$ and $\widetilde{\xi}_2$ respectively.*

*The mapping $f$ is conformal if and only if the coefficients of the first fundamental forms of $x\,(\cdot, \cdot)$ and $\widetilde{x}\,(\cdot, \cdot)$ are proportional.*

*Proof.* See the problem section.
□

### 5.5.4  Second Fundamental Form

The second fundamental form of a surface patch enables us to determine "curved" properties of surfaces. Let $x\,(\cdot, \cdot)$ be a surface patch in $\mathbb{R}^3$. The surface patch is parameterized by $(u, v) \in \mathbb{R}^2$. There exists a unit normal vector $\mathcal{N}$ at each point on the surface patch. It is

$$\mathcal{N} = \frac{x_u \times x_v}{\|x_u \times x_v\|} = \frac{x_u \times x_v}{\left(EG - F^2\right)^{1/2}}$$

The vector $\mathcal{N}$ is a function of $u$ and $v$. Let $P$ and $Q$ be points on a surface patch, and $P$ be $x\,(u, v)$, $Q$ be $x\,(u + du, v + dv)$, and $\mathcal{N}$ be the unit normal vector at the point $P$. The projection of the vector $PQ$ on to the normal $\mathcal{N}$ is $d = PQ \circ \mathcal{N}$. Note that it is possible for $d$ to be either positive or negative. Actually, $|d|$ is equal to the perpendicular distance from the point $Q$ to the tangent plane at the point $P$. Use of Taylor's series in two variables results in

$$x\,(u + du, v + dv) = x\,(u, v) + dx\,(u, v) + \frac{1}{2} d^2 x\,(u, v) + o\,(du^2 + dv^2)$$

where $dx(u,v)$ and $d^2x(u,v)$ are respectively the first and second order derivatives of $x(u,v)$. Further $\left(du^2 + dv^2\right)$ tends to zero. Therefore

$$d = PQ \circ \mathcal{N}$$
$$= \left(x(u+du, v+dv) - x(u,v)\right) \circ \mathcal{N}$$

Note that $dx \circ \mathcal{N} = 0$, as the vector $dx$ is tangent to the surface at $P$, and hence perpendicular to the normal vector $\mathcal{N}$. Therefore

$$d = \frac{1}{2}\left(x_{uu}du^2 + 2x_{uv}dudv + x_{vv}dv^2\right) \circ \mathcal{N} + o\left(du^2 + dv^2\right)$$
$$= \frac{1}{2}II + o\left(du^2 + dv^2\right)$$

In the above expression

$$II = Ldu^2 + 2Mdudv + Ndv^2$$
$$L = x_{uu} \circ \mathcal{N}$$
$$M = x_{uv} \circ \mathcal{N}$$
$$N = x_{vv} \circ \mathcal{N}$$

The function $II$ is called the *second fundamental form* of $x(u,v)$; and $L, M$, and $N$ are called the *second fundamental coefficients*. The second fundamental form and the second fundamental coefficients are all functions of $u$ and $v$.

**Definition 5.23.** *Second fundamental form. Let $x : U \to S$ be a surface patch, where $U$ is an open subset of $\mathbb{R}^2$, and $S \subseteq \mathbb{R}^3$ is a regular surface. Let $\mathcal{N}$ be the unit normal vector at a point $x(u,v)$, where $(u,v) \in U$. Let $L, M, N : U \to \mathbb{R}$, where for $(u,v) \in U$*

$$L(u,v) = x_{uu}(u,v) \circ \mathcal{N} \tag{5.13a}$$
$$M(u,v) = x_{uv}(u,v) \circ \mathcal{N} \tag{5.13b}$$
$$N(u,v) = x_{vv}(u,v) \circ \mathcal{N} \tag{5.13c}$$

*and*

$$II(u,v) = L(u,v)\,du^2 + 2M(u,v)\,dudv + N(u,v)\,dv^2 \tag{5.13d}$$

*is called the second fundamental form.of the surface patch $x(\cdot, \cdot)$ induced from $\mathbb{R}^3$ at $(u,v)$. Also $L(u,v), M(u,v)$, and $N(u,v)$ are called the second fundamental coefficients.* $\square$

In contrast to the first fundamental form, the second fundamental form is not necessarily positive or definite. Certain characteristics of the second fundamental form of a surface patch are summarized.

**Observation 5.8.** Characteristics of the second fundamental form $II(\cdot, \cdot)$ of a surface patch $x(\cdot, \cdot)$ and its coefficients. Let $(u,v) \in U$, and $\mathcal{N}$ be the unit normal vector.

1. $II(\cdot, \cdot) = d^2x(\cdot, \cdot) \circ \mathcal{N}$.
2. The second fundamental form $II(\cdot, \cdot)$, is a homogeneous function of second degree in $du$ and $dv$.

3. The function $II\left(\cdot,\cdot\right)$ can be written as a quadratic defined on the the vectors $(du, dv)$ in the $uv$-plane. That is

$$II = \begin{bmatrix} du & dv \end{bmatrix} \begin{bmatrix} L & M \\ M & N \end{bmatrix} \begin{bmatrix} du & dv \end{bmatrix}^{T}$$

4. The function $II\left(\cdot,\cdot\right)$ depends only on the surface and not upon its parameterization, if the orientation is preserved.                                                                                    □

Some of the above observations are established in the problem section. The function

$$\delta = \frac{1}{2}II = \frac{1}{2}\left\{Ldu^2 + 2Mdudv + Ndv^2\right\}$$

is called the *osculating paraboloid* at the surface point $x\left(u,v\right)$. It is a polynomial of degree two in $du$ and $dv$. This function can be interpreted qualitatively in the language of conic sections, which is: circles, ellipses, hyperbolas, and parabolas. Let $\delta = 0$, then

$$Ldu^2 + 2Mdudv + Ndv^2 = 0$$

This yields

$$du = \frac{-M \pm \left(M^2 - LN\right)^{1/2}}{L}dv$$

We consider several cases which depend upon the value of the discriminant $\left(M^2 - LN\right)$, and examine the osculating paraboloid at the surface point $x\left(u,v\right) = P$.

1. *Elliptic case*: The surface point $P$ is called elliptic if $\left(LN - M^2\right) > 0$. The function $\delta$ has the same sign for all $(du, dv)$ pairs. In the neighborhood of the point $P$, the surface lies on only one side of the tangent plane at point $P$. As $\delta$ is a function of $du$ and $dv$, it forms an elliptic paraboloid.

2. *Hyperbolic case*: The surface point $P$ is called hyperbolic if $\left(LN - M^2\right) < 0$. The surface near point $P$ intersects the tangent plane at point $P$ with two distinct straight lines. The two straight lines intersect in the tangent plane at point $P$. The value of $\delta$ is equal to zero along these two lines. The surface $\delta$ lies on both sides of the tangent plane in the neighborhood of $P$. As $\delta$ is a function of $du$ and $dv$, it forms a hyperbolic paraboloid.

3. *Parabolic case*: The surface point $P$ is called parabolic if $\left(LN - M^2\right) = 0$; and $L, M$, and $N$ are not all simultaneously equal to zero. There is a single line in the tangent plane passing through point $P$. Along this line the value of $\delta = 0$, otherwise $\delta$ maintains the same sign. However, it is possible for the surface to lie on both sides of the tangent plane. As $\delta$ is a function of $du$ and $dv$, it forms a parabolic cylinder.

4. *Planar case*: The surface point $P$ is called planar if $L = M = N = 0$. In this case $\delta = 0$ for all values of $(du, dv)$.

**Example 5.13.** Computation of second fundamental form and its coefficients.

1. Plane. Let $a$ be a point on the plane. Also let $p$ and $q$ be two unit vectors parallel to the plane, and orthogonal to each other. The plane is described by

$$x\left(u,v\right) = up + vq + a$$

Therefore

$$x_u(u, v) = p, \quad x_v(u, v) = q$$
$$x_{uu}(u, v) = x_{uv}(u, v) = x_{vv}(u, v) = 0$$

It implies

$$L(u, v) = M(u, v) = N(u, v) = 0$$

and consequently $II(u, v) = 0$.

2. Sphere. A sphere is described by

$$\left\{(x_1, x_2, x_3) \in \mathbb{R}^3 \mid x_1^2 + x_2^2 + x_3^2 = r^2\right\}$$

A parameterization is

$$x(\theta, \phi) = (r \cos\theta \cos\phi, r \cos\theta \sin\phi, r \sin\theta)$$

Therefore

$$x_\theta(\theta, \phi) = (-r \sin\theta \cos\phi, -r \sin\theta \sin\phi, r \cos\theta)$$
$$x_\phi(\theta, \phi) = (-r \cos\theta \sin\phi, r \cos\theta \cos\phi, 0)$$
$$x_\theta(\theta, \phi) \times x_\phi(\theta, \phi) = \left(-r^2 \cos^2\theta \cos\phi, -r^2 \cos^2\theta \sin\phi, -r^2 \cos\theta \sin\theta\right)$$

$$\mathcal{N} = \frac{x_\theta(\theta, \phi) \times x_\phi(\theta, \phi)}{\|x_\theta(\theta, \phi) \times x_\phi(\theta, \phi)\|} = (-\cos\theta \cos\phi, -\cos\theta \sin\phi, -\sin\theta)$$

Also

$$x_{\theta\theta}(\theta, \phi) = (-r \cos\theta \cos\phi, -r \cos\theta \sin\phi, -r \sin\theta)$$
$$x_{\theta\phi}(\theta, \phi) = (r \sin\theta \sin\phi, -r \sin\theta \cos\phi, 0)$$
$$x_{\phi\phi}(\theta, \phi) = (-r \cos\theta \cos\phi, -r \cos\theta \sin\phi, 0)$$

The second fundamental coefficients are

$$L(\theta, \phi) = x_{\theta\theta}(\theta, \phi) \circ \mathcal{N} = r$$
$$M(\theta, \phi) = x_{\theta\phi}(\theta, \phi) \circ \mathcal{N} = 0$$
$$N(\theta, \phi) = x_{\phi\phi}(\theta, \phi) \circ \mathcal{N} = r \cos^2\theta$$

The second fundamental form is

$$II(\theta, \phi) = r \left(d\theta^2 + \cos^2\theta d\phi^2\right)$$

$\square$

## 5.6 Properties of Surfaces

Relevant properties of surfaces, like: curves on a surface, local isometry of surfaces, and geodesics on a surface are studied in this section.

### 5.6.1  Curves on a Surface

A surface can also be quantified by examining curves on it. For example, curvature of curves on a surface at different points on it can also be evaluated.

Let $x(\cdot, \cdot)$ be a regular surface patch, and $C$ be a curve on it. Also, let the parameter of this curve be $s$. The parameter $s$ measures the length of the curve from some starting point on it. A point on this curve is specified by $\xi(s) = x(u(s), v(s))$. In this parameterization $\|\xi'(s)\| = 1$.

The tangent vector of $\xi(s)$ is $\widetilde{T}(s) = \xi'(s)$. The length of this vector is unity. The unit normal vector on the surface at the point $s$ is $\mathcal{N}(s)$. The vector $\mathcal{N}(s)$ is perpendicular to the tangent vector $\widetilde{T}(s)$. Therefore the vector $\mathcal{N}(s) \times \widetilde{T}(s)$ is of unit length, and it is perpendicular to both the vectors $\mathcal{N}(s)$ and $\widetilde{T}(s)$. That is, the vectors $\widetilde{T}(s)$, $\mathcal{N}(s)$, and $\mathcal{N}(s) \times \widetilde{T}(s)$ are mutually perpendicular.

Also the vector $\widetilde{T}'(s) = \xi''(s)$ is perpendicular to $\widetilde{T}(s)$. Therefore $\widetilde{T}'(s)$ lies in the plane formed by the orthogonal vectors $\mathcal{N}(s)$ and $\mathcal{N}(s) \times \widetilde{T}(s)$. Consequently, $\widetilde{T}'(s)$ can be written as a linear combination of vectors $\mathcal{N}(s)$ and $\mathcal{N}(s) \times \widetilde{T}(s)$. That is,

$$\widetilde{T}'(s) = \xi''(s) = \kappa_n(s)\,\mathcal{N}(s) + \kappa_g(s)\,\mathcal{N}(s) \times \widetilde{T}(s)$$

where $\kappa_n(s) \in \mathbb{R}$ and $\kappa_g(s) \in \mathbb{R}$ are called the *normal curvature* and *geodesic curvature* of the curve $C$ at the point $s$, respectively. As $\mathcal{N}(s)$ and $\mathcal{N}(s) \times \widetilde{T}(s)$ are unit orthogonal vectors, we have

$$\kappa_n(s) = \xi''(s) \circ \mathcal{N}(s)$$
$$\kappa_g(s) = \xi''(s) \circ \left(\mathcal{N}(s) \times \widetilde{T}(s)\right)$$
$$\|\xi''(s)\|^2 = \kappa_n(s)^2 + \kappa_g(s)^2$$

The geodesic curvature $\kappa_g(s)$ can also be expressed as

$$\kappa_g(s) = \left[\xi''(s)\ \mathcal{N}(s)\ \widetilde{T}(s)\right] = \left[\xi'(s)\ \xi''(s)\ \mathcal{N}(s)\right]$$

The curvature $\kappa(s) = \|\xi''(s)\|$ of the curve $C$ at $s$ can be computed from

$$\kappa(s)^2 = \kappa_n(s)^2 + \kappa_g(s)^2$$

The principal normal of the curve $C$ at $s$ is given by

$$\widetilde{N}(s) = \frac{\widetilde{T}'(s)}{\kappa(s)} = \frac{\xi''(s)}{\kappa(s)}$$

Therefore

$$\kappa_n(s) = \xi''(s) \circ \mathcal{N}(s) = \kappa(s)\,\widetilde{N}(s) \circ \mathcal{N}(s) = \kappa(s)\cos\theta$$

where $\theta$ is the angle between the unit vectors $\widetilde{N}(s)$ and $\mathcal{N}(s)$. This implies

$$\kappa_g(s) = \pm\kappa(s)\sin\theta$$

The above discussion is condensed in the following definition.

**Definition 5.24.** *Normal and geodesic curvature. Let $x(\cdot, \cdot)$ be a regular surface patch. Also let $C$ be a curve on the surface, and the parameter of this curve be $s$. The parameter $s$ measures the* length of the curve from some starting point on it. A point on the curve $C$ is specified by $\xi(s) = x(u(s), v(s))$. Let $\widetilde{N}(s)$ and $\mathcal{N}(s)$ be the principal normal of the curve and normal to the surface at $s$, respectively. Also let the curvature of the curve $C$ at $s$ be $\kappa(s)$.

The normal curvature and geodesic curvature of the curve $C$ at $s$ are $\kappa_n(s)$ and $\kappa_g(s)$ respectively. Then

$$\kappa_n(s) = \kappa(s)\,\widetilde{N}(s) \circ \mathcal{N}(s), \quad \text{and} \quad \kappa_g(s) = \pm \left(\kappa(s)^2 - \kappa_n(s)^2\right)^{1/2} \tag{5.14}$$

In general, only magnitudes of $\kappa_n(s)$ and $\kappa_g(s)$ are well defined.   □

**Theorem 5.2.** Let $\xi(s) = x(u(s), v(s))$ be a parameterization of a curve $C$ on a regular surface, where $s$ is the arc-length of the curve from some starting point. The normal curvature $\kappa_n(s)$ of the curve $C$ at the point $s$ is

$$\kappa_n(s) = L\,(u')^2 + 2Mu'v' + N\,(v')^2 \tag{5.15}$$

In the above expression, a prime denotes derivative with respect to $s$.

*Proof.* A point on the curve $C$ is specified by $\xi(s) = x(u(s), v(s))$. Let $\mathcal{N}(s)$ be the unit normal to the surface at $s$. In the following steps we do not show the explicit dependence of functions upon the parameters $u, v$, and $s$. Therefore

$$\kappa_n = \mathcal{N} \circ \xi'' = \mathcal{N} \circ \frac{d\xi'}{ds}$$
$$= \mathcal{N} \circ \frac{d}{ds}(x_u u' + x_v v')$$
$$= \mathcal{N} \circ \{x_u u'' + x_v v'' + (x_{uu}u' + x_{uv}v')\,u' + (x_{uv}u' + x_{vv}v')\,v'\}$$
$$= L\,(u')^2 + 2Mu'v' + N\,(v')^2$$

In the above steps, we used the fact that $L, M$, and $N$ are the coefficients of the second fundamental form, and $\mathcal{N}$ is perpendicular to $x_u$ and $x_v$.   □

**Corollary 5.1.** Let $\xi(t) = x(u(t), v(t))$ be a parameterization of a curve $C$ on a regular surface, where $t$ is a parameterization which is not necessarily equal to the arc-length of the curve $C$. The normal curvature $\kappa_n(t)$ of the curve $C$ at the point $t$ is equal to $II(u(t), v(t))/I(u(t), v(t))$ where $I(\cdot, \cdot)$ and $II(\cdot, \cdot)$ are the first and second fundamental forms of the surface at $t$.   □

See the problem section for a proof of the above corollary.

### Principal Curvatures and Principal Vectors

Our next goal is to further study the *principal curvatures* at a point on a regular surface.

**Definitions 5.25.** *Consider a point $P$ on a surface patch $x(\cdot, \cdot)$. At the point $P$:*

1. *The principal curvatures are the maximum and minimum values of the normal curvature $\kappa_n$ among all curves on the surface that pass through $P$.*
2. *The principal vectors are the tangent vectors of the curves that yield the maximum and minimum values of the normal curvature $\kappa_n$.*   □

The normal curvature $\kappa_n$ is given by

$$\kappa_n = \frac{II}{I}$$
$$= \frac{L\,du^2 + 2M\,du\,dv + N\,dv^2}{E\,du^2 + 2F\,du\,dv + G\,dv^2}$$

The above expression can be written as

$$\kappa_n = \frac{L + 2M\lambda + N\lambda^2}{E + 2F\lambda + G\lambda^2}$$

where $\lambda \triangleq dv/du$. The extreme values of $\kappa_n$ can be determined by letting $d\kappa_n/d\lambda = 0$. Some crucial observations related to the principal curvatures and the corresponding principal vectors are summarized.

**Observations 5.9.** Let $x\,(\cdot, \cdot)$ be a surface patch. At a point $P$ on a surface patch, the coefficients of the first and second fundamental forms are $E, F, G$; and $L, M, N$ respectively. The first and second fundamental forms are $I$ and $II$ respectively. Let $\mathcal{F}_I$ and $\mathcal{F}_{II}$ be two square matrices, each of size 2, where

$$\mathcal{F}_I = \begin{bmatrix} E & F \\ F & G \end{bmatrix}, \quad \text{and} \quad \mathcal{F}_{II} = \begin{bmatrix} L & M \\ M & N \end{bmatrix}$$

The normal curvature $\kappa_n = II/I$.

1. Substitute $\lambda = dv/du$ in the expression for $\kappa_n$ and let $d\kappa_n/d\lambda = 0$. This yields a quadratic equation in $\kappa_n$

$$\det\left(\mathcal{F}_{II} - \kappa_n \mathcal{F}_I\right) = 0$$

The zeros of this equation give the principal curvatures at the point $P$. Furthermore, let $\begin{bmatrix} \psi & \varrho \end{bmatrix}^T \triangleq \mathcal{T}$ be the column vector which satisfies

$$\left(\mathcal{F}_{II} - \kappa_n \mathcal{F}_I\right)\mathcal{T} = 0$$

The corresponding *tangent vector* $T = (\psi x_u + \varrho x_v)$ at point $P$ is the principal vector.

2. The eigenvalues of the matrix $\mathcal{F}_I^{-1}\mathcal{F}_{II}$ are the principal curvatures.
3. Let the quadratic equation in $\kappa_n$ be

$$\kappa_n^2 - 2H\kappa_n + K = 0$$

Its zeros (principal curvatures) are $\kappa_1$ and $\kappa_2$, where

$$\kappa_1, \kappa_2 = H \pm \left(H^2 - K\right)^{1/2}$$
$$K = \kappa_1 \kappa_2 = \frac{\left(LN - M^2\right)}{\left(EG - F^2\right)} = \frac{\det \mathcal{F}_{II}}{\det \mathcal{F}_I}$$
$$H = \frac{1}{2}\left(\kappa_1 + \kappa_2\right) = \frac{EN + GL - 2FM}{2\left(EG - F^2\right)}$$

4. The principal curvatures $\kappa_1$ and $\kappa_2$ are real numbers.

5. If $\kappa_1 = \kappa_2 = \kappa$ then $\mathcal{F}_{II} = \kappa\mathcal{F}_I$. Any vector $\begin{bmatrix} \psi & \varrho \end{bmatrix}^T \triangleq \mathcal{T}$ satisfies the equation $(\mathcal{F}_{II} - \kappa\mathcal{F}_I)\mathcal{T} = 0$. Therefore any tangent vector $T$ at point $P$ is a principal vector. Also

$$\kappa = \frac{L}{E} = \frac{M}{F} = \frac{N}{G}$$

6. If $\kappa_1 \neq \kappa_2$ then any two nonzero principal vectors which correspond to $\kappa_1$ and $\kappa_2$ are orthogonal to each other.

7. The principal vectors at the point $P$ can be determined from $\det A = 0$, where

$$A = \begin{bmatrix} \lambda^2 & -\lambda & 1 \\ E & F & G \\ L & M & N \end{bmatrix}$$

8. At a point $P$ on a surface $S$, the nonzero principal vectors associated with the principal curvatures $\kappa_1$ and $\kappa_2$, are $T_1$ and $T_2$ respectively. Also consider a curve $C$ on the surface $S$, which passes through the point $P$. The curve $C$ is specified by $\xi(s) = x(u(s), v(s))$, where $s$ is its distance parameter. Let $\theta$ be the angle between $T_1$ and $\xi'(s)$. Then the normal curvature of $C$ at $s$ is

$$\kappa_n = \kappa_1 \cos^2 \theta + \kappa_2 \sin^2 \theta$$

This result is generally called Euler's theorem.

9. As $(EG - F^2) > 0$, the sign of $K$ depends only upon the sign of $(LN - M^2)$. Furthermore, from earlier discussion, the sign of $(LN - M^2)$ determines whether the point $P$ is either elliptic, or hyperbolic, or parabolic. As $K = \kappa_1\kappa_2$, and $\kappa_1, \kappa_2 = H \pm (H^2 - K)^{1/2}$, we have the following classification of the point $P$.

   (a) $K > 0$. This implies $(LN - M^2) > 0$. This in turn implies that the point $P$ is elliptic. Also, $\kappa_1$ and $\kappa_2$ are both simultaneously positive, or both simultaneously negative.

   (b) $K < 0$. This implies $(LN - M^2) < 0$. This in turn implies that the point $P$ is hyperbolic. Also, $\kappa_1$ and $\kappa_2$ are both nonzero, but have opposite signs.

   (c) $K = 0$. This implies $(LN - M^2) = 0$.

      (i) Let $L, M$, and $N$ all be not simultaneously equal to zero. The point $P$ is parabolic. Also, one of $\kappa_1$ and $\kappa_2$ is equal to zero, and the other is nonzero.

      (ii) Let $L = M = N = 0$. The point $P$ is planar. Also, both $\kappa_1$ and $\kappa_2$ are equal to zero.

10. Dilation. A dilation is applied to the surface $S$. If the coordinates in $\mathbb{R}^3$ are scaled by a real constant $a \neq 0$, then $E, F, G$ are multiplied by $a^2$, and $L, M, N$ are multiplied by $a$. Therefore $K$ is scaled by $a^{-2}$, and $H$ by $a^{-1}$.                                    $\square$

Some of the above observations are proved in the problem section. The mean and Gaussian curvatures of a curve on a surface are next defined. Gaussian curvature is named after Carl Friedrich Gauss (1777-1855).

**Definition 5.26.** *Let $\kappa_1$ and $\kappa_2$ be the principal curvatures at point $P$ on a surface patch. Let*

$$H = \frac{1}{2}(\kappa_1 + \kappa_2) \tag{5.16a}$$

$$K = \kappa_1\kappa_2 \tag{5.16b}$$

*Then $H$ and $K$ are called the mean curvature and the Gaussian curvature respectively at the point $P$.*                                    $\square$

**Examples 5.14**. Computation of Gaussian curvatures.

1. Plane. Let $a$ be a point on the plane. Also let $p$ and $q$ be two unit vectors parallel to the plane, and orthogonal to each other. The plane is described by

$$x(u,v) = up + vq + a$$

The coefficients of the first and second fundamental forms computed in earlier examples are: $E = 1, F = 0, G = 1$; and $L = M = N = 0$. Therefore $K = 0$.

2. A sphere is described by

$$\left\{ (x_1, x_2, x_3) \in \mathbb{R}^3 \mid x_1^2 + x_2^2 + x_3^2 = r^2 \right\}$$

The coefficients of the first and second fundamental forms computed in earlier examples are: $E = r^2, F = 0, G = r^2 \cos^2 \theta$; and $L = r, M = 0, N = r \cos^2 \theta$. Therefore $K = 1/r^2$.   $\square$

**Example 5.15**. We compute the principal curvatures and principal vectors of a cylinder. A parameterization of the cylinder is

$$x(u,v) = (r \cos u, r \sin u, v)$$

where $r \in \mathbb{R}^+$. Therefore

$$x_u(u,v) = (-r \sin u, r \cos u, 0)$$
$$x_v(u,v) = (0,0,1)$$

The first fundamental coefficients are

$$E = \|x_u(u,v)\|^2 = r^2$$
$$F = x_u(u,v) \circ x_v(u,v) = 0$$
$$G = \|x_v(u,v)\|^2 = 1$$

$$x_u(u,v) \times x_v(u,v) = (r \cos u, r \sin u, 0)$$

$$\mathcal{N} = \frac{x_u(u,v) \times x_v(u,v)}{\|x_u(u,v) \times x_v(u,v)\|} = (\cos u, \sin u, 0)$$

Also

$$x_{uu}(u,v) = (-r \cos u, -r \sin u, 0)$$
$$x_{uv}(u,v) = (0,0,0)$$
$$x_{vv}(u,v) = (0,0,0)$$

The second fundamental coefficients are

$$L(u,v) = x_{uu}(u,v) \circ \mathcal{N} = -r$$
$$M(u,v) = x_{uv}(u,v) \circ \mathcal{N} = 0$$
$$N(u,v) = x_{vv}(u,v) \circ \mathcal{N} = 0$$

Therefore

$$\mathcal{F}_I = \begin{bmatrix} r^2 & 0 \\ 0 & 1 \end{bmatrix}, \quad \text{and} \quad \mathcal{F}_{II} = \begin{bmatrix} -r & 0 \\ 0 & 0 \end{bmatrix}$$

The Gaussian and the mean curvature are

$$K = \frac{(LN - M^2)}{(EG - F^2)}$$

$$= \frac{\det \mathcal{F}_{II}}{\det \mathcal{F}_I} = \frac{0}{r^2} = 0$$

$$H = \frac{EN + GL - 2FM}{2(EG - F^2)}$$

$$= -\frac{1}{2r}$$

Zeros of the equation $\det(\mathcal{F}_{II} - \kappa \mathcal{F}_I) = 0$ give the principal curvatures. This is obtained from

$$\begin{vmatrix} -r - \kappa r^2 & 0 \\ 0 & -\kappa \end{vmatrix} = 0$$

Therefore $\kappa(1 + r\kappa) = 0$. This implies $\kappa_1 = -1/r$ and $\kappa_2 = 0$. The vectors $\mathcal{T}_i$, for $i = 1, 2$ are found from

$$(\mathcal{F}_{II} - \kappa_i \mathcal{F}_I)\mathcal{T}_i = 0, \quad i = 1, 2$$

For $\kappa_1 = -1/r$, we have

$$\begin{bmatrix} 0 & 0 \\ 0 & 1/r \end{bmatrix} \mathcal{T}_1 = 0$$

Thus $\mathcal{T}_1$ is a multiple of $\begin{bmatrix} 1 & 0 \end{bmatrix}^T$, and the corresponding principal vector $T_1$ is a multiple of

$$(x_u 1 + x_v 0) = x_u = (-r \sin u, r \cos u, 0)$$

For $\kappa_2 = 0$, we have

$$\begin{bmatrix} -r & 0 \\ 0 & 0 \end{bmatrix} \mathcal{T}_2 = 0$$

Thus $\mathcal{T}_2$ is a multiple of $\begin{bmatrix} 0 & 1 \end{bmatrix}^T$, and the corresponding principal vector $T_2$ is a multiple of $(x_u 0 + x_v 1) = x_v = (0, 0, 1)$.                                                                          □

The following theorem due to Gauss provides an explicit representation of the Gaussian curvature in terms of the coefficients of first fundamental form and its derivatives only. It does not depend explicitly upon the coefficients of the second fundamental form. Gauss called his result *Theorema egrigium*. The phrase "egrigium" means remarkable in Latin language. Before this theorem is stated, we introduce the necessary notation.

**Notation.** Let $H(u, v)$ represent: $E(u, v)$, $F(u, v)$, and $G(u, v)$. The first partial derivative of $H$ with respect to $u$ and $v$ is denoted by $H_u$, and $H_v$ respectively. The second partial derivative of $H$ with respect to $u$ and $v$ is denoted by $H_{uu}$, and $H_{vv}$ respectively. The partial derivative of $H$ with respect to $u$ and $v$ is denoted by $H_{uv}$.                                                □

**Theorem 5.3.** *Theorema egregium. The Gaussian curvature $K$ of a surface depends only upon the coefficients of first fundamental form $E$, $F$, and $G$ and their partial derivatives. It is*

$$K = \frac{\det A - \det B}{(EG - F^2)^2} \tag{5.17a}$$

*where*

$$A = \begin{bmatrix} \left(-\frac{1}{2}E_{vv} + F_{uv} - \frac{1}{2}G_{uu}\right) & \left(F_v - \frac{1}{2}G_u\right) & \frac{1}{2}G_v \\ \frac{1}{2}E_u & E & F \\ \left(F_u - \frac{1}{2}E_v\right) & F & G \end{bmatrix} \tag{5.17b}$$

$$B = \begin{bmatrix} 0 & \frac{1}{2}E_v & \frac{1}{2}G_u \\ \frac{1}{2}E_v & E & F \\ \frac{1}{2}G_u & F & G \end{bmatrix} \tag{5.17c}$$

*Proof.* See the problem section.  □

**Observations 5.10.** Special and useful cases for values of the Gaussian curvature immediately follow from Gauss' theorem.

1. If $F = 0$ we have

$$K = -\frac{1}{2\sqrt{EG}} \left\{ \frac{\partial}{\partial u}\left(\frac{G_u}{\sqrt{EG}}\right) + \frac{\partial}{\partial v}\left(\frac{E_v}{\sqrt{EG}}\right) \right\}$$

2. If $E = 1$, and $F = 0$ we have

$$K = -\frac{1}{\sqrt{G}} \frac{\partial^2 \sqrt{G}}{\partial u^2}$$

□

### 5.6.2  Local Isometry of Surfaces

Two different surfaces can be compared and evaluated, by examining the existence of local isometry between them.

**Definition 5.27.** *Let $S$ and $\widetilde{S}$ be two surfaces. A bijective smooth mapping $f : S \to \widetilde{S}$ is a local isometric mapping or simply a local isometry, if the length of any arbitrary regular arc $\xi = \xi(t)$ on $S$ is equal to the length of its image $\widetilde{\xi} = \widetilde{\xi}(t) = f(\xi(t))$ on $\widetilde{S}$. If a local isometry $f$ exists, then the surfaces $S$ and $\widetilde{S}$ are said to be locally isometric.*  □

**Theorem 5.4.** *Let $S$ and $\widetilde{S}$ be two surfaces, and $f : S \to \widetilde{S}$ be a smooth bijective mapping. Let the coefficients of the first fundamental form of the two surfaces be $E, F, G$ and $\widetilde{E}, \widetilde{F}, \widetilde{G}$ respectively. The mapping $f$ is a local isometry if and only if $E = \widetilde{E}$, $F = \widetilde{F}$, and $G = \widetilde{G}$.*
*Proof.* See the problem section.  □

Combination of the last theorem and Gauss' remarkable theorem yields the following result.

**Theorem 5.5.** *Let $S$ and $\widetilde{S}$ be two surfaces, and $f : S \to \widetilde{S}$ be a local isometric mapping. Corresponding points on the two surfaces have the same Gaussian curvature.*  □

### 5.6.3 Geodesics on a Surface

The shortest distance between two points on a plane is a straight line. Similarly, the shortest distance between two points on a surface is always a geodesic. A curve $\xi(\cdot)$ with parameter $t$ is a straight line if $\xi''(t) = 0$ for all values of $t$. This in turn implies that its geodesic curvature $\kappa_g(t) = 0$ for all values of its parameter $t$.

A geodesic, which passes through any point on the surface, in any specific direction, can be shown to exist by using the theory of ordinary differential equations. Furthermore, geodesics on a surface also allow us to construct an atlas for the surface

**Definition 5.28.** *A curve $\xi(\cdot)$ on a surface $S$ with parameter $t$ is a geodesic, if $\xi''(t) = 0$ or $\xi''(t)$ is perpendicular to the tangent plane of the surface at the point $\xi(t)$. That is, $\xi''(t)$ is equal to zero or parallel to the unit normal to the surface $\mathcal{N}(t)$ at the point $\xi(t)$ for all values of its parameter $t$.*                                                                                                    □

**Observations 5.11.** Let $\xi(\cdot)$ be a geodesic on a surface $S$ with parameter $s$.

1. $\xi'(s)$ is constant for all values of parameter $s$.
2. The geodesic curvature $\kappa_g(s) = 0$ for all values of parameter $s$. This result is also often used as a definition of a geodesic.
3. Each and every straight line on a surface is a geodesic.
4. Let $\widetilde{N}(s)$ be the principal normal at a point on the geodesic $C$, parameterized by $s$. The direction of the surface normal $\mathcal{N}(s)$ is equal to that of $\pm\widetilde{N}(s)$ for each value of $s$ on the geodesic.
5. Furthermore, if geodesic $C$ is not a straight line, the osculating plane of $C$ is perpendicular to the tangent plane to the surface at each point $s$.

    Alternately, the intersection $C$ of the surface $S$ with a plane $P_{plane}$ which is perpendicular to the tangent plane of the surface $S$ at every point of $C$ is a geodesic. This intersection is also called the *normal section* of the surface. This result is true because $\kappa_g = 0$ for a normal section.                                                                                                      □

**Example 5.16.** Geodesics on a sphere. All great circles of a sphere are geodesics. A great circle is a curve formed by the intersection of the sphere and a plane that passes through the center of the sphere. These planes are also the osculating planes of the geodesics.                             □

The following result states the well-known geodesic equations.

**Theorem 5.6.** *Geodesic differential equations. Let $\xi(t) = x(u(t), v(t))$ be a parameterization of a curve $C$ on a regular surface $S$ with parameter $t \in \mathbb{R}$. The coefficients of first fundamental form are $E(u(t), v(t))$, $F(u(t), v(t))$, and $G(u(t), v(t))$. The curve $C$ is a geodesic if:*

$$\frac{d}{dt}(Eu' + Fv') = \frac{1}{2}(E_u u'^2 + 2F_u u'v' + G_u v'^2) \tag{5.18a}$$

$$\frac{d}{dt}(Fu' + Gv') = \frac{1}{2}(E_v u'^2 + 2F_v u'v' + G_v v'^2) \tag{5.18b}$$

*where $E_u, F_u, G_u$ are the first partial derivatives of $E, F, G$ with respect to $u$; $E_v, F_v, G_v$ are the first partial derivatives of $E, F, G$ with respect to $v$; and $u'$ and $v'$ are the derivatives of $u$ and $v$ with respect to $t$.*

*Proof.* See the problem section.                                                                                                     □

The above two equations are referred to as the first and second *geodesic equations* for future reference.

**Observations 5.12.** Some observations about geodesics.

1. The parameter $t$ that satisfies the geodesic equations is directly proportional to the arc-length.
2. The geodesic curvature $\kappa_g$ of a curve $C$ on a surface $S$ depends only upon the coefficients of the first fundamental form, and its partial derivatives.
3. A local isometric mapping between two surfaces, maps the geodesic of one surface to a geodesic of the other surface.
4. Assume that the parametric curves which specify the surface $S$ form an orthogonal system. The $u$-curves are geodesics if and only if $E$ is a function of only $u$. Similarly, the $v$-curves are geodesics if and only if $G$ is a function of only $v$. Note that on the $u$-curve, the parameter $v$ is constant; and on the $v$-curve, the parameter $u$ is constant.                                                            $\square$

The above observations are established in the problem section. Geodesics on surfaces can also be determined via variational techniques.

### Geodesics via Variational Techniques

It is known that the geodesic between two points on a Euclidean plane is a straight line. A straight line also has the smallest length of all curves between the two points on a Euclidean plane. Similarly, segments of great circles are also the shortest paths between two points on a sphere. Therefore it should not come as a surprise that the geodesic between two points on a surface can be obtained by finding a path on the surface with the smallest length. We next establish necessary conditions for a curve $C$ between two points on a surface to be an arc of minimum length.

This problem can be stated as a variational problem. Let $C$ be a curve on the surface $x\,(\cdot,\cdot)$ specified by $\xi\,(t) = x\,(u\,(t)\,,v\,(t))$, where $t \in [t_1,t_2]$. The length $\mathcal{L}\,(C)$ of this curve $C$ is specified by

$$\mathcal{H}\,(u,v,u'v') = \left\{ Eu'^2 + 2Fu'v' + Gv'^2 \right\}^{1/2}$$

$$\mathcal{L}\,(C) = \int_{t_1}^{t_2} \mathcal{H}\,(u,v,u'v')\,dt$$

where primes denote derivatives with respect to the parameter $t$. The length of the curve is an extremum, if the following Euler-Lagrange equations are satisfied. These are

$$\frac{d}{dt}\left(\frac{\partial \mathcal{H}}{\partial u'}\right) - \frac{\partial \mathcal{H}}{\partial u} = 0, \text{ and } \frac{d}{dt}\left(\frac{\partial \mathcal{H}}{\partial v'}\right) - \frac{\partial \mathcal{H}}{\partial v} = 0$$

Note that

$$\frac{\partial \mathcal{H}}{\partial u'} = \frac{(Eu' + Fv')}{\mathcal{H}}, \text{ and } \frac{\partial \mathcal{H}}{\partial u} = \frac{(E_u u'^2 + 2F_u u'v' + G_u v'^2)}{2\mathcal{H}}$$

In order to proceed further, we let the parameter $t$ be equal to the arc-length $s$. In this case we have $\mathcal{H} = 1$. Letting primes denote derivatives with respect to the parameter $s$, the Euler-Lagrange equations imply

$$\frac{d}{ds}\left(Eu' + Fv'\right) = \frac{1}{2}\left(E_u u'^2 + 2F_u u'v' + G_u v'^2\right)$$

The other Euler-Lagrange equation similarly yields

$$\frac{d}{ds}\left(Fu' + Gv'\right) = \frac{1}{2}\left(E_v u'^2 + 2F_v u'v' + G_v v'^2\right)$$

The last two equations are the necessary conditions for the curve $C$ to be a geodesic as per the theorem on geodesic differential equations. This discussion is summarized in the following theorem.

**Theorem 5.7.** *Let $\xi(s) = x(u(s), v(s))$ be a parameterization of a curve $C$ on a regular surface $S$, with arc-length parameter $s \in \mathbb{R}$. The coefficients of first fundamental are $E(u(s), v(s))$, $F(u(s), v(s))$, and $G(u(s), v(s))$. Let*

$$\mathcal{H}(u, v, u'v') = \left\{Eu'^2 + 2Fu'v' + Gv'^2\right\}^{1/2} \tag{5.19a}$$

*where primes denote derivatives with respect to the parameter $s$. The following Euler-Lagrange equations are necessary for the curve $C$ to be a geodesic.*

$$\frac{d}{ds}\left(\frac{\partial \mathcal{H}}{\partial u'}\right) - \frac{\partial \mathcal{H}}{\partial u} = 0 \tag{5.19b}$$

$$\frac{d}{ds}\left(\frac{\partial \mathcal{H}}{\partial v'}\right) - \frac{\partial \mathcal{H}}{\partial v} = 0 \tag{5.19c}$$

$\square$

### Geodesic Coordinates

Coordinate systems on a surface $S$ can be used to specify the relative location of a point on it. This can be achieved by using geodesics to specify coordinates on a surface. We study two types of geodesic coordinate systems. These are geodesic parallel and polar coordinate systems. The geodesic parallel coordinate system is analogous to the rectangular coordinate system on a plane. Similarly, the geodesic polar coordinate system is analogous to the polar coordinate system on a plane. A geodesic coordinate system in turn can be used to study the Gaussian curvature of surfaces.

### Geodesic Parallel Coordinates

The generalization of orthogonal parallel coordinates in a plane is the generalized parallel coordinates on a surface $S$. The concept of field of geodesics is initially introduced.

**Definition 5.29.** *Field of geodesics. Consider a curve $C$ on a surface $S$. In a sufficiently small region $S'$ of the surface $S$ which contains $C$, the family of one-parameter geodesics is said to form a field of geodesics if exactly a single geodesic of the family passes through each point of $S'$.* $\square$

A family of parallel straight lines on a plane is a field of geodesics. A geodesic parallel coordinate system can be constructed as follows. Consider a curve $C$ on a surface $S$. At each point $P$ on the curve $C$, exactly a single geodesic passes through it and is orthogonal to $C$. These geodesics constitute a field of geodesics in a sufficiently small patch $S'$ of the surface $S$, where the patch $S'$ contains the curve $C$. Therefore, these geodesics can be assumed to be geodesic coordinate curves

$v = $ constant. The orthogonal trajectories of these geodesics can be assumed to be geodesic coordinate curves $u = $ constant. The coordinates thus created are called geodesic parallel coordinates.

As the two family of curves $u = $ constant, and $v = $ constant are orthogonal to each other, we have $F = 0$. Furthermore, for the geodesic curve $v = $ constant, the second geodesic equation yields $E_v = 0$. It is shown in the problem section that on the surface $S$

$$\kappa_g \big|_{v= \text{ constant}} = -\frac{E_v}{2E\sqrt{G}}$$

$$= -\frac{1}{\sqrt{G}} \frac{\partial \ln \sqrt{E}}{\partial v}$$

This implies that on the curve $v = $ constant, $\kappa_g = 0$. Also $E$ is independent of $v$. That is, $E$ depends only upon the parameter $u$. Thus

$$ds^2 = E(u, v) \, du^2 + G(u, v) \, dv^2$$

We define a new set of coordinates $(\widetilde{u}, \widetilde{v})$ as

$$\widetilde{u} = \int_0^u \sqrt{E(u, v)} du, \quad \widetilde{v} = v$$

These newly introduced coordinates have the same coordinate curves as the original coordinates. The first fundamental form in terms of the new coordinates is

$$\widetilde{ds}^2 = d\widetilde{u}^2 + G(\widetilde{u}, \widetilde{v}) \, d\widetilde{v}^2$$

Next consider a curve $C$ on a surface $S$, and the field of geodesics which are orthogonal to $C$. On this orthogonal family of curves, we mark off equal distances (measured along the geodesics) on the same side of the curve $C$. The curve thus created by these marked points is *geodesically parallel* to the curve $C$.

On each of the geodesic parallels, let $v = $ constant. The length of the curve on this geodesic between $u = u_1$ and $u = u_2$ is

$$\int_{u_1}^{u_2} du = (u_2 - u_1)$$

is constant on each geodesic parallel. Thus, we have the following result due to Gauss.

**Observation 5.13.** Let $C$ be a curve on a surface $S$. Also, let the family of the geodesic parallels of the curve $C$ be $C_{\parallel}$, and the field of geodesics orthogonal to the curve $C$ be $C_{\perp}$. Then the curves in the set $C_{\parallel}$ are orthogonal trajectories of the geodesics in $C_{\perp}$. $\qquad\square$

We next state an important result about geodesics. It specifies sufficient conditions for a curve between two points on a surface to be a geodesic. It is stated in terms of the arc-length.

**Theorem 5.8.** *Geodesic has minimum arc-length. Let $C$ be a geodesic between two points $P_1$ and $P_2$ on a surface $S$. If there exists a geodesic field $\mathfrak{F}$ on $S$ containing the geodesic $C$, and is indeed the only geodesic joining the points $P_1$ and $P_2$ in the field $\mathfrak{F}$, then the arc-length of the geodesic $C$ joining the points $P_1$ and $P_2$ is smallest of all arcs in the field $\mathfrak{F}$ which join points $P_1$ and $P_2$.*

*Proof.* See the problem section. $\qquad\square$

**Example 5.17.** Let $P$ be a point on the equator of a sphere of radius one. The equator is parameterized by the longitude $\phi$, where $\phi \in [0, 2\pi)$. Let $\alpha_\phi(\cdot)$ be the meridian which passes through the longitude $\phi$. This meridian is parameterized by the latitude $\theta$, where $\theta \in [-\pi/2, \pi/2]$. The corresponding geodesic coordinate pair on the sphere is $(\theta, \phi)$. The first fundamental form is $\left(d\theta^2 + \cos^2\theta \, d\phi^2\right)$. $\qquad\square$

### Geodesic Polar Coordinates

The generalization of polar coordinates in a plane to a surface $S$ is the geodesic polar coordinates. We first provide a motivation for the existence of geodesic polar coordinates. Consider an $x_1x_2$-plane. This plane surface, excluding the origin, is specified by $x = x(r, \theta) = (r\cos\theta, r\sin\theta)$, where $r \in \mathbb{R}^+$, and $\theta \in \mathbb{R}$. We have

$$x_r = (\cos\theta, \sin\theta), \quad x_\theta = (-r\sin\theta, r\cos\theta)$$

The corresponding coefficients of the first fundamental form are

$$E = x_r \circ x_r = 1, \quad F = x_r \circ x_\theta = 0, \quad G = x_\theta \circ x_\theta = r^2$$

Its first fundamental form is

$$I = dr^2 + r^2 d\theta^2$$

We generalize this idea to any surface $S$. Let $P$ be a point on the surface $S$. Also let $g_1$ and $g_2$ be a pair of orthonormal vectors parallel to the tangent plane at the point $P$. The origin of the $g_1g_2$-plane is at the point $P$. It can be demonstrated that, at each angle $\theta_0$ in the $g_1g_2$-plane, there exists a unique geodesic $x(r, \theta_0)$ passing through the origin in the direction of the tangent vector $((\cos\theta_0)g_1 + (\sin\theta_0)g_2)$. Using a careful argument, it can be shown that there exists an $\epsilon > 0$, such that the surface in the neighborhood of point $P$ but excluding it, can be parameterized as $x = x(r, \theta)$, where $0 < r < \epsilon$, $\theta \in [0, 2\pi)$, and $x_r \circ x_\theta = 0$. The $\theta$-parameter curves with $r$ equal to a constant are called *geodesic circles*, and $r$ is called the *radius* of the geodesic circle.

The first fundamental form of $x = x(r, \theta)$ where $r \in \mathbb{R}^+$, is similar to the first fundamental form of $x$ with geodesic parallel coordinates. It is

$$I = dr^2 + G(r, \theta) d\theta^2$$

See the problem section for a justification of the above form of the first fundamental form of $x$. For small values of $r$ it can be shown that

$$I = dr^2 + r^2 d\theta^2$$

Therefore we surmise that for small values of $r$, a surface $S$ behaves as a plane. Some observations about geodesic polar coordinates, and Gaussian curvature are listed.

**Observations 5.14.** Let $x = x(r, \theta)$ be a set of geodesic polar coordinates at a point $P$ on a surface $S$, where $r \in \mathbb{R}^+$, and $\theta \in [0, 2\pi)$. The Gaussian curvature at point $P$ on the surface $S$ is $K(P)$. Also, $G(r, \theta) = x_\theta \circ x_\theta$ is a coefficient of the first fundamental form of the surface $S$.

1. Then

$$\sqrt{G(r, \theta)} = r - \frac{1}{6}K(P)r^3 + R(r, \theta)$$

where $\lim_{r \to 0}\left(R(r, \theta)/r^3\right) = 0$.

2. Let $r, C\left(r\right)$, and $A\left(r\right)$ be the radius, circumference, and surface area enclosed by a geodesic circle about the point $P$ on surface $S$. Then

$$K\left(P\right) = \lim_{r \to 0} \frac{3}{\pi}\left(\frac{2\pi r - C\left(r\right)}{r^3}\right)$$

$$K\left(P\right) = \lim_{r \to 0} \frac{12}{\pi}\left(\frac{\pi r^2 - A\left(r\right)}{r^4}\right)$$

$\square$

The above observations are established in the problem section.

## 5.7 Prelude to Hyperbolic Geometry

Euclid's first four axioms about plane geometry appear to be self-explanatory. However, his fifth axiom invited some skepticism. Mathematicians over several centuries wondered whether, it could be derived from the other four axioms. Near the end of the eighteenth century, the celebrated mathematician Carl Friedrich Gauss surmised that Euclid's fifth axiom was really an axiom. He began investigating other types of geometries by replacing Euclid's fifth postulate, but never published them.

The Russian mathematician Nicolai Lobachevsky (1792-1856) was the first mathematician to publish (1829) a geometry which was independent of Euclid's fifth postulate. The Hungarian mathematician János Bolyai (1802-1860) independently published (1829) similar results. This subject was later developed by mathematicians Eugenio Beltrami (1835-1899), Felix Klein (1849-1925), and Henri Poincaré (1854-1912).

Lobachevsky and Bolyai investigated. the so-called *hyperbolic non-Euclidean geometry* (or simply hyperbolic geometry). Another type of geometry called *elliptic non-Euclidean geometry* (or simply elliptic geometry) was developed by Bernhard Riemann (1826-1866). Lines in these geometries are actually geodesics on surfaces.

Hyperbolic geometry is closely related to the local geometry of surfaces of constant *negative* Gaussian curvature ($K < 0$). In this geometry, there exist at least two lines which pass through a point, and are parallel to a given line. Furthermore, the sum of interior angles of a triangle is less than $\pi$.

Elliptic geometry is also closely related to the local geometry of surfaces. However, these surfaces have constant *positive* Gaussian curvature ($K > 0$). In this geometry, there does not exist a line which passes through a point and parallel to a given line. Furthermore, the sum of interior angles of a triangle is greater than $\pi$. Plane geometry is related to surfaces of constant Gaussian curvature, which is identically equal to zero ($K = 0$).

Surfaces with constant Gaussian curvature are studied in this section from several different perspectives. More specifically, surfaces of revolution are considered. These in turn are related to different types of non-Euclidean geometry. Consequence of constant Gaussian curvature on local isometry are also examined. We also study surfaces of constant Gaussian curvature from the perspective of conformal mappings. It is established that, if a surface $S$ is mapped conformally to a plane $\widetilde{S}$ so that geodesics on $S$ are mapped to circles on the plane $\widetilde{S}$, then the surface $S$ has constant Gaussian curvature. The ramifications of this result to hyperbolic geometry are also investigated.

### 5.7.1  Surfaces of Revolution

A surface of revolution is generated by revolving a plane curve, around a straight line. The plane curve is often called the *profile curve*, and the corresponding straight line is called the *axis of revolution*. Thus a single point on the profile curve generates a single circle. Such circles are called the *parallels* of the surface. A *meridian* of a surface of revolution is a curve formed by the intersection of the surface with a plane which passes through the axis of revolution. A surface of revolution $S$ can be parameterized as

$$x\,(u,v) = (f\,(u)\cos v, f\,(u)\sin v, g\,(u))$$

where $f > 0$, $\left(f'^2 + g'^2\right) = 1$, and prime denotes derivative with respect to $u$. The coefficients of the first and second fundamental form of the surface are

$$E = 1, \quad F = 0, \quad G = f^2$$

$$L = (f'g'' - f''g'), \quad M = 0, \quad N = fg'$$

The Gaussian curvature $K$ is

$$K = \frac{g'\,(f'g'' - f''g')}{f}$$

Differentiating both sides of the equation $\left(f'^2 + g'^2\right) = 1$ with respect to $u$ yields $f'f'' + g'g'' = 0$. Therefore

$$K = -\frac{f''}{f}$$

We consider three cases. These are: $K = 0$, $K > 0$, and $K < 0$.

   *Case* 1: The Gaussian curvature is equal to zero. That is, $K = 0$. This implies $f'' = 0$, which in turn implies $f\,(u) = (\alpha u + \beta)$, where $\alpha$ and $\beta$ are constants. Further, $\left(f'^2 + g'^2\right) = 1$ implies $g' = \pm\sqrt{1 - \alpha^2}$. Thus $g\,(u) = \pm\sqrt{1 - \alpha^2}u + \gamma$, where $|\alpha| \leq 1$. If we let $\gamma = 0$, and select positive sign in the expression for $g\,(u)$, the surface $x\,(\cdot,\cdot)$ is

$$x\,(u,v) = (\beta\cos v, \beta\sin v, 0) + u\left(\alpha\cos v, \alpha\sin v, \sqrt{1 - \alpha^2}\right)$$

   If $|\alpha| = 1$, this surface is a plane; if $\alpha = 0$, this surface is a circular cylinder; and if $|\alpha| \in (0,1)$ this surface is a circular cone.

   *Case* 2: The Gaussian curvature is positive. That is, $K > 0$. An example of such surface is a sphere. Therefore a surface of revolution with constant positive curvature is called a *spherical surface*. Let $K = 1/c^2$, where $c > 0$ is a constant. Therefore

$$f'' + \frac{f}{c^2} = 0$$

The solution of the above differential equation is

$$f = \alpha\cos\left(\frac{u}{c} + \beta\right)$$

where $\alpha$ and $\beta$ are constants. For simplicity, let $\beta = 0$. In this case

$$g\,(u) = \pm\int\left\{1 - \frac{\alpha^2}{c^2}\sin^2\frac{u}{c}\right\}^{1/2} du$$

Again select the positive sign before the integral. The above integral with the positive sign before it can be evaluated easily, if $\alpha = 0$, or $\alpha = \pm c$. As $f > 0$, $\alpha = 0$ is not admissible. However, if $\alpha = c$, we have $f(u) = c\cos(u/c)$, and $g(u) = c\sin(u/c)$. In this case, the surface of revolution is a sphere, as

$$x(u, v) = \left( c\cos\frac{u}{c}\cos v, c\cos\frac{u}{c}\sin v, c\sin\frac{u}{c} \right)$$

The corresponding first fundamental form is

$$I = du^2 + c^2\cos^2\left(\frac{u}{c}\right)dv^2 = du^2 + \frac{1}{K}\cos^2\left(u\sqrt{K}\right)dv^2$$

*Case* 3: The Gaussian curvature is negative. That is, $K < 0$. A surface of revolution with constant negative curvature is called a *pseudospherical surface*. For simplicity, assume that $K = -1$. Therefore, this surface may be regarded as a "sphere" with imaginary radius $\sqrt{-1}$. Hence the name pseudosphere. The equation $f'' = f$ has a solution

$$f(u) = \alpha e^u + \beta e^{-u}$$

where $\alpha$ and $\beta$ are constants. Therefore

$$g(u) = \pm \int \left\{ 1 - \left(\alpha e^u - \beta e^{-u}\right)^2 \right\}^{1/2} du$$

Again, for simplicity assume the positive sign before the integral, and let $\alpha = 1$, and $\beta = 0$. In this case

$$f(u) = e^u, \quad g(u) = \int \left(1 - e^{2u}\right)^{1/2} du$$

The integral $\int \left(1 - e^{2u}\right)^{1/2} du$ can be evaluated for nonpositive values of $u$. This integral is evaluated in two steps. In the first step substitute $e^u = \cos\theta$, and evaluate the integral. In the next step substitute back $\theta = \cos^{-1}e^u$. This results in

$$g(u) = \left(1 - e^{2u}\right)^{1/2} - \ln\left\{ e^{-u} + \left(e^{-2u} - 1\right)^{1/2} \right\}$$

$$= \left(1 - e^{2u}\right)^{1/2} - \cosh^{-1}\left(e^{-u}\right), \quad \text{where } u \le 0$$

where the constant of integration has been assumed to be equal to zero. In the above expression, we used the relationship $\cosh^{-1}y = \ln\left\{ y + \left(y^2 - 1\right)^{1/2} \right\}$. Let $f(u) = e^u = r$, where $r \in (0, 1]$, then

$$g(u) = \left(1 - r^2\right)^{1/2} - \cosh^{-1}\left(\frac{1}{r}\right) \triangleq z(r), \quad \text{where } e^u = r, \ u \le 0$$

Therefore a surface of revolution $S$ with constant curvature $K = -1$ can be parameterized as

$$x(u, v) = \left(e^u\cos v, e^u\sin v, g(u)\right), \quad u < 0$$

The corresponding first fundamental form is

$$I = du^2 + e^{2u}dv^2$$

The curve

$$z(r) = \left(1 - r^2\right)^{1/2} - \cosh^{-1}\left(\frac{1}{r}\right), \quad r \in (0, 1]$$

is called a *tractrix*.

**Observations 5.15.** Properties of the tractrix curve.

1. The range of values of $z(r)$, for $r \in (0, 1]$ is $(-\infty, 0]$.
2. Let $A$ be a point on the curve $z(\cdot)$ in the $rz$-plane. The tangent to the curve $z$ at point $A$ intersects $z$-axis at point $B$. Then the distance between points $A$ and $B$ is a constant, and equal to unity. $\qquad\square$

The above observations are established in the problem section.

### 5.7.2 Constant Gaussian Curvature Surfaces

We characterize two small neighborhoods of surfaces with constant Gaussian curvature. It is concluded that these surfaces are locally isometric. Let $x = x(r, \theta)$ be a set of geodesic polar coordinates at a point $P$ on a surface of constant geodesic curvature. In this case, the coefficients of first fundamental form of the surface are $E = 1$, $F = 0$, and $G = x_\theta \circ x_\theta = G(r, \theta)$. The corresponding first fundamental form is

$$I = dr^2 + G(r, \theta)\, d\theta^2$$

Using an earlier result, it can be shown that the Gaussian curvature $K$ satisfies the differential equation and initial conditions

$$\frac{\partial^2 \sqrt{G}}{\partial r^2} + K\sqrt{G} = 0, \quad \lim_{r \to 0} \sqrt{G} = 0, \quad \text{and} \quad \lim_{r \to 0} \left(\partial \sqrt{G}/\partial r\right) = 1$$

Depending upon the values of $K$, we get different solutions of the above differential equation. We consider three cases. These are: $K = 0$, $K > 0$, and $K < 0$.

*Case* 1: The Gaussian curvature is equal to zero. That is, $K = 0$. In this case $G(r, \theta) = r^2$. The corresponding first fundamental form is

$$I - dr^2 + r^2 d\theta^2$$

This is the first fundamental form of a plane in polar coordinates.

*Case* 2: The Gaussian curvature is positive. That is, $K > 0$. In this case

$$G(r, \theta) = (1/K)\sin^2\left(r\sqrt{K}\right)$$

The corresponding first fundamental form is

$$I = dr^2 + \frac{1}{K}\sin^2\left(r\sqrt{K}\right) d\theta^2$$

*Case* 3: The Gaussian curvature is negative. That is, $K < 0$. In this case

$$G(r, \theta) = (-1/K)\sinh^2\left(r\sqrt{-K}\right)$$

The corresponding first fundamental form is

$$I = dr^2 - \frac{1}{K}\sinh^2\left(r\sqrt{-K}\right) d\theta^2$$

Thus we observe that the coefficients of first fundamental form of a surface described in polar coordinates, and with constant curvature uniquely depend upon on $K$. Therefore, if $\widetilde{P}$ is another point on the surface with constant Gaussian curvature, then the neighborhoods of points $P$ and $\widetilde{P}$ are locally isometric. The above results are summarized in the next theorem.

**Theorem 5.9.** *Consider any two small neighborhoods of sufficiently smooth surfaces with the same Gaussian curvatures. Then these neighborhoods are locally isometric.* $\qquad\square$

### 5.7.3  Isotropic Curves

Sometimes, it is mathematically convenient to study curves and surfaces via methods borrowed from the theory of complex variables. For example, the parameter $t$ of a curve $\xi(\cdot)$ can be complex of the form $t = (t_1 + it_2)$, where $t_1, t_2 \in \mathbb{R}$ and $i = \sqrt{-1}$. In this case it is possible for the length of the curve between two points on it to be equal to zero. Such curves are called isotropic or minimal curves.

**Definition 5.30.** *Isotropic or minimal curve. A curve $\xi(\cdot)$ with parameter $t \in \mathbb{C}$ is isotropic if the length of the arc between any two points on the curve is zero.* □

**Observation 5.16.** A curve $\xi(t)$, $t \in \mathbb{C}$ is isotropic if and only if $ds^2 = 0$, where $\xi'(t) \neq 0$. □

The above observation is true if

$$ds^2 = Edu^2 + 2Fdudv + Gdv^2 = 0$$

A real solution of the above equation is not possible. However, a complex solution is possible. The first fundamental form of this curve can be factored as

$$ds^2 = \left\{ \sqrt{E}du + \frac{\left(F + i\sqrt{D}\right)}{\sqrt{E}}dv \right\} \left\{ \sqrt{E}du + \frac{\left(F - i\sqrt{D}\right)}{\sqrt{E}}dv \right\}$$

where $D = \left(EG - F^2\right)$. The above two factors are complex conjugates of each other. Let

$$a(u,v) = \left\{ \sqrt{E}du + \frac{\left(F + i\sqrt{D}\right)}{\sqrt{E}}dv \right\}$$

$$\overline{a}(u,v) = \left\{ \sqrt{E}du + \frac{\left(F - i\sqrt{D}\right)}{\sqrt{E}}dv \right\}$$

In the next step we multiply the $a(u,v)$ by $(\sigma_1(u,v) + i\sigma_2(u,v))$, where $\sigma_1(\cdot,\cdot)$ and $\sigma_2(\cdot,\cdot)$ are real-valued functions of $u$ and $v$. It is hoped that it will be easier to integrate the product

$$a(u,v)(\sigma_1(u,v) + i\sigma_2(u,v))$$

Therefore, $(\sigma_1(u,v) + i\sigma_2(u,v))$ is called integrating factor in calculus. Similarly,

$$(\sigma_1(u,v) - i\sigma_2(u,v))$$

is the integrating factor of $\overline{a}(u,v)$. Let

$$d\mu = a(u,v)(\sigma_1(u,v) + i\sigma_2(u,v))$$
$$d\nu = \overline{a}(u,v)(\sigma_1(u,v) - i\sigma_2(u,v))$$

The curves $\mu = $ constant, and $\nu = $ constant are called isotropic curves on the surface $S$. Thus, $\mu = \mu(u,v)$ and $\nu = \nu(u,v)$ is a coordinate transformation that introduces isotropic curves as parametric curves. Also, the first fundamental form with respect to the isotropic coordinates $(\widetilde{u}, \widetilde{v})$ is

$$ds^2 = \lambda \, d\mu \, d\nu$$

where $\lambda = 1/\left(\sigma_1^2 + \sigma_2^2\right)$. As $\mu$ and $\nu$ are complex conjugates of each other, we have

$$\mu = u_1(u, v) + i v_1(u, v)$$
$$\nu = u_1(u, v) - i v_1(u, v)$$

where $u_1(\cdot, \cdot)$ and $u_2(\cdot, \cdot)$ are real-valued functions of $u$ and $v$. The corresponding first fundamental form is

$$ds^2 = \lambda \left(du_1^2 + dv_1^2\right)$$

The coordinate curves $u_1 = $ constant, and $u_2 = $ constant are real-valued curves.

### 5.7.4  A Conformal Mapping Perspective

A study of surfaces with constant Gaussian curvature provide a suitable backdrop to examine non-Euclidean geometries. Surfaces of constant Gaussian curvature were studied earlier in the section. These surfaces were formed by revolution of plane curve about a straight line. Surfaces of constant Gaussian curvature are characterized from yet another perspective in this subsection.

**Notation.** For ease in notation, the first, second, and third derivatives of $v$ with respect to $u$ are denoted by $v_u, v_{uu}$, and $v_{uuu}$ respectively                                                                         □

**Theorem 5.10.** *Consider a surface $S$ which is mapped into a plane $\widetilde{S}$. Further, this mapping is conformal, and the geodesics in $S$ correspond to circles in $\widetilde{S}$. Then the Gaussian curvature $K$ of the surface $S$ is a constant.*

*Proof.* This result is established in several steps.

*Step* 1: Assume that the surfaces $S$ and $\widetilde{S}$ use the same coordinates. As the mapping is conformal, the coefficients of the first fundamental form of the two surfaces at corresponding points are proportional. Therefore isotropic curves on the surface $S$ are mapped into isotropic curves on the surface $\widetilde{S}$. We use isotropic curves on $S$ as coordinate curves. Let the corresponding coordinates be $(u, v)$. In this coordinate system, the coefficients $E = G = 0$. The differential equations of the geodesics are

$$u'' + \frac{F_u}{F} u'^2 = 0, \quad v'' + \frac{F_v}{F} v'^2 = 0$$

where the primes denote derivatives with respect to the parameter $t$. The above differential equations are converted into differential equations, with independent variable as $u$. The second of the above equations transforms to

$$\frac{d^2 v}{du^2} u'^2 + \frac{dv}{du} u'' + \frac{F_v}{F}\left(\frac{dv}{du}\right)^2 u'^2 = 0$$

Substituting $u'' = -u'^2 F_u/F$ in the above equation and removing the common factor $u'^2$ results in

$$\frac{d^2 v}{du^2} + A\left(\frac{dv}{du}\right)^2 + B\frac{dv}{du} = 0$$

where

$$A = \frac{F_v}{F}, \quad B = -\frac{F_u}{F}$$

The above differential equation can be rewritten as

$$v_{uu} + A\{v_u\}^2 + Bv_u = 0$$

Use of Gauss' expression for the curvature of the surface $S$ yields

$$K = -\frac{1}{F^3}\left(FF_{uv} - F_uF_v\right)$$
$$= -\frac{1}{F}\frac{\partial^2 \ln F}{\partial u \partial v}$$

*Step* 2: Let $(x, y)$ be the rectangular coordinates in the plane $\widetilde{S}$. The isotropic coordinates $(u, v)$ in this plane are

$$u = (x + iy), \quad v = (x - iy), \quad \text{where } i = \sqrt{-1}$$

Consequently, we have identical coordinates in $S$ and $\widetilde{S}$. The equation of a circle in the plane $\widetilde{S}$ is

$$x^2 + y^2 + a_1 x + a_2 y + a_3 = 0$$

In terms of the isotropic coordinates, the equation of the circle is

$$uv + au + bv + k = 0$$

The differential equation in $u$ and $v$ in Step 1, and the above equation of a circle specified in terms of $u$ and $v$ should match. Therefore, in the later equation, we take derivative with respect to $u$ on both sides. This results in

$$v_u = -\frac{(v + a)}{(u + b)}$$

Successive derivatives of the above expression with respect to $u$ yield

$$v_{uu} = 2\frac{(v + a)}{(u + b)^2}, \quad v_{uuu} = -6\frac{(v + a)}{(u + b)^3}$$

Therefore, circles on the surface $\widetilde{S}$ can be specified by

$$v_{uuu} = \frac{3}{2}\{v_{uu}\}^2\{v_u\}^{-1}$$

Observe that the above equation is independent of $a, b$, and $k$.

*Step* 3: The differential equation in Step 1 is differentiated with respect to $u$. This results in

$$v_{uuu} + \left\{\frac{\partial A}{\partial u} + \frac{\partial A}{\partial v}v_u\right\}\{v_u\}^2 + 2Av_uv_{uu} + \left\{\frac{\partial B}{\partial u} + \frac{\partial B}{\partial v}v_u\right\}v_u + Bv_{uu} = 0$$

In the above equation, substitute the expression for $v_{uu}$ from Step 1. This yields

$$v_{uuu} + \left\{\frac{\partial A}{\partial v} - 2A^2\right\}\{v_u\}^3 + \left\{\frac{\partial A}{\partial u} - 3AB + \frac{\partial B}{\partial v}\right\}\{v_u\}^2 + \left\{\frac{\partial B}{\partial u} - B^2\right\}v_u = 0$$

The coordinates systems used on the surfaces $S$ and $\widetilde{S}$ are same. Furthermore, the geodesics on the surface $S$ are mapped to circles on the surface $\widetilde{S}$. Therefore $v_{uuu}$ can be eliminated from the last equation, and also from the expression for $v_{uuu}$ in Step 2. Furthermore, the expression for $v_{uu}$ can also be used from Step 1. This results in

$$\left\{ \frac{\partial A}{\partial v} - \frac{A^2}{2} \right\} \{v_u\}^3 + \left\{ \frac{\partial A}{\partial u} + \frac{\partial B}{\partial v} \right\} \{v_u\}^2 + \left\{ \frac{\partial B}{\partial u} + \frac{B^2}{2} \right\} v_u = 0$$

The coefficient of $\{v_u\}^2$ in the above equation is equal to zero. The coefficients of $\{v_u\}^3$ and $v_u$ should each be equal to zero. Therefore

$$\frac{\partial A}{\partial v} = \frac{A^2}{2}, \quad \frac{\partial B}{\partial u} = -\frac{B^2}{2}$$

*Step* 4: Integration yields

$$\frac{1}{A} = -\frac{1}{2} \left( v + k_1 \left( u \right) \right), \quad \frac{1}{B} = \frac{1}{2} \left( u + k_2 \left( v \right) \right)$$

where $k_1 \left( u \right)$ and $k_2 \left( v \right)$ are constants of integration. As $A = F_v / F$ and $B = -F_u / F$, we have

$$\frac{\partial \ln F}{\partial v} = -\frac{2}{\left( v + k_1 \left( u \right) \right)}, \quad \frac{\partial \ln F}{\partial u} = -\frac{2}{\left( u + k_2 \left( v \right) \right)}$$

Integration of the above equations results in

$$\ln F = -2 \ln \left\{ k_3 \left( u \right) \left( v + k_1 \left( u \right) \right) \right\}, \quad \ln F = -2 \ln \left\{ k_4 \left( v \right) \left( u + k_2 \left( v \right) \right) \right\}$$

where $k_3 \left( u \right)$ and $k_4 \left( v \right)$ are constants of integration. The above two expression for $\ln F$ are satisfied, if it is of the form

$$\ln F = -2 \ln \left( \alpha_{12} uv + \alpha_1 u + \alpha_2 v + \alpha_0 \right)$$

where $\alpha_{12}, \alpha_1, \alpha_2,$ and $\alpha_0$ are constants. The Gaussian curvature is finally computed as

$$K = -\frac{1}{F} \frac{\partial^2 \ln F}{\partial u \partial v} = 2 \left( \alpha_0 \alpha_{12} - \alpha_1 \alpha_2 \right)$$

That is, the Gaussian curvature $K$ is a constant.                                                   □

The consequences of the above theorem are next investigated. In the last theorem, select the constants as

$$\alpha_{12} = 0, \quad \alpha_1 = ic, \quad \alpha_2 = -ic, \quad \alpha_0 = 0$$

This results in the first fundamental form of the surface $S$ as

$$2F dudv = -\frac{2}{c^2 \left( u - v \right)^2} dudv$$

The first fundamental form of the surface $\widetilde{S}$ is

$$ds^2 = \frac{\left( dx^2 + dy^2 \right)}{2c^2 y^2}$$

The Gaussian curvature is

$$K = -2c^2$$

Observe that $K < 0$. The differential equation of the geodesic on the surface $S$ which is obtained in Step 1 (of the proof of the last theorem) is

$$v_{uu} + \frac{2}{(u-v)} \left\{ (v_u)^2 + v_u \right\} = 0$$

Substitution of

$$v_u = -\frac{(v+a)}{(u+b)}$$

$$v_{uu} = 2\frac{(v+a)}{(u+b)^2}$$

from Step 2 (of the proof of the last theorem), in the previous equation yields $a = b$. Therefore, the equation of the circle on the surface $\widetilde{S}$ is

$$uv + a(u+v) + k = 0$$

or equivalently

$$x^2 + y^2 + 2ax + k = 0$$

The center of this circle lies on the $x$-axis. The case of a geodesic on $S$ which maps to $x = \text{constant}$ on the surface $\widetilde{S}$ is permissible. This is a limiting case of the circle on the surface $\widetilde{S}$. If the image of the conformal mapping is restricted to the upper half $xy$-plane ($y > 0$), we obtain a mapping due to Poincaré. It is also called the *upper half-plane of Poincaré*. It uses the metric (the first fundamental form) $ds^2 = (dx^2 + dy^2) / (2c^2y^2)$. This is one of the models of hyperbolic geometry. Therefore, if $P$ and $Q$ are two points on a geodesic on the surface $S$, then their images $\widetilde{P}$ and $\widetilde{Q}$ on the surface $\widetilde{S}$ lie on a semicircle. Furthermore, the geodesic on the surface $S$ which passes through the points $P$ and $Q$ on it, maps to the semicircle which passes through points $\widetilde{P}$ and $\widetilde{Q}$ on the surface $\widetilde{S}$.

---

## 5.8 Hyperbolic Geometry

Discovery of hyperbolic geometry in the nineteenth century is one of the greatest intellectual achievements in mathematics. The pseudosphere discussed in an earlier section can be used to describe hyperbolic geometry. Geodesics on the pseudosphere mimic the role of straight lines in Euclidean geometry. Well-known models of hyperbolic geometry are: the upper half-plane model, Poincaré disc model, hyperboloid model, and Beltrami-Klein model. Of these four models, we only discuss the first two. The upper half-plane model is also sometimes referred to as the Poincaré upper half-plane model.

Hyperbolic geometry can also be described via a set of axioms. In hyperbolic and Euclidean geometry, the first four axioms are identical. However, the "parallel" (fifth) axiom is different. In Euclidean geometry, for every line $\ell$ and for every point $A$ not on the line $\ell$; there exists exactly a single line $m$ that passes through the point $A$, and is also parallel to line $\ell$. In hyperbolic geometry, for every line $\ell$ and for every point $A$ not on the line $\ell$; there exist two or more lines that pass through the point $A$, and are also parallel to line $\ell$. The use of the adjective "hyperbolic" in hyperbolic geometry will become clear subsequently. Isometries, surfaces of different constant curvature, tessellations, and certain geometric constructions of hyperbolic geometry are also described in this section.

### 5.8.1  Upper Half-Plane Model

The upper half-plane model of hyperbolic geometry is described in this subsection. Motivation for this model was developed earlier by examining distance metrics on surfaces of revolutions. Consider a surface of revolution $S$ which is parameterized as

$$x\,(u,v) = (f\,(u)\cos v, f\,(u)\sin v, g\,(u))$$

where $f > 0$, $\left(f'^2 + g'^2\right) = 1$, and prime denotes derivative with respect to $u$. Its Gaussian curvature is given by

$$K = -\frac{f''}{f}$$

Let $K = -1$, then a possible solution is $f\,(u) = e^u = r$, where $u < 0$; and if $g\,(u) = z\,(r)$, then

$$z\,(r) = \left(1 - r^2\right)^{1/2} - \cosh^{-1}\left(\frac{1}{r}\right), \quad r \in (0,1]$$

Therefore a surface of revolution $S$ with constant curvature $K = -1$ can be parameterized as

$$x\,(u,v) = (e^u \cos v, e^u \sin v, g\,(u)), \quad u < 0$$

The corresponding first fundamental form is

$$I = du^2 + e^{2u}dv^2$$

The above results were obtained earlier in the chapter. Substitute $w = e^{-u}, u < 0$ in the parametric representation of the surface $S$. This yields an alternate representation of the surface $S$. It is

$$\widetilde{x}\,(v,w) = \left(\frac{1}{w}\cos v, \frac{1}{w}\sin v, \left(1 - \frac{1}{w^2}\right)^{1/2} - \cosh^{-1} w\right), \quad w > 1$$

**Observation 5.17.** The first fundamental form of the above representation is

$$\frac{dv^2 + dw^2}{w^2}, \quad w > 1$$

$\square$

The above result is proved in the problem section. It can also be verified that the above first fundamental form results in a negative curvature of the surface. The curvature of this surface is equal to $K = -1$.

The surface $S$ is described via the parametric representation $\widetilde{x}\,(v,w)$ for $w > 1$. However, its first fundamental form is well-defined for $w > 0$. Nevertheless, the characterizations of surfaces which depend only upon the above first fundamental form can be studied for the upper half-plane

$$\mathbb{H} = \{(v,w) \in \mathbb{R}^2 \mid w > 0\}$$

The upper half-plane can also be specified via the set of complex numbers

$$\mathbb{H} = \{z \in \mathbb{C} \mid \operatorname{Im} z > 0\}$$

where $z = (v + iw)\,;\,i = \sqrt{-1},\, v,w \in \mathbb{R}, w > 0$. Thus, $\mathbb{H}$ is the set of points with positive imaginary part.

**Definitions 5.31.** *Upper half-plane and its boundary.*

1. *The upper half-plane* $\mathbb{H}$ *is the set of complex numbers*

$$\mathbb{H} = \{z \in \mathbb{C} \mid \operatorname{Im} z > 0\} \tag{5.20}$$

2. *The boundary of* $\mathbb{H}$ *is the set* $\partial\mathbb{H} = \{z \in \mathbb{C} \mid \operatorname{Im}(z) = 0\} \cup \{\infty\}$. $\qquad\square$

Note that $\partial\mathbb{H}$ is the real axis and the point at infinity. Also, $\partial\mathbb{H}$ is called the circle at infinity, because the points in $\partial\mathbb{H}$ are at an infinite "hyperbolic distance" from each and every point in $\mathbb{H}$. Members of the set $\partial\mathbb{H}$ are also called *ideal points*.

**Observations 5.18.** Observations about the upper half-plane model of the hyperbolic geometry.

1. In the upper half-plane model of hyperbolic geometry, the first fundamental form is:

$$ds^2 = \frac{dv^2 + dw^2}{w^2}, \quad (v,w) \in \mathbb{R}^2, w > 0$$

Also
$$\mathbb{H} = \{(v,w) \in \mathbb{R}^2 \mid w > 0\}$$

2. Hyperbolic angles in the space $\mathbb{H}$ and Euclidean angles in a plane are identical.
3. The geodesics in $\mathbb{H}$ are called *hyperbolic lines*. The geodesics in $\mathbb{H}$ are: semicircles with centers on the real axis, and half-lines parallel to the imaginary axis. A half-line is a straight line which extends indefinitely in one direction from a point.
4. Let $p, q \in \mathbb{H}$, where $p \neq q$. A unique hyperbolic line (geodesic) passes through $p$ and $q$.
5. The parallel axiom of the Euclidean geometry is not valid in hyperbolic geometry. Actually, if $a$ is a point which is not on a hyperbolic line $l$, then there are infinitely many hyperbolic lines which pass through $a$ that do not intersect $l$.
6. Let $l$ and $m$ be two different geodesics in $\mathbb{H}$ that do not intersect each other at any point in $\mathbb{H}$.
   (a) If $l$ and $m$ are both half-lines, then they are said to be *parallel* lines.
   (b) Either of the two lines is a half-line, and the other is a semicircle:
      (i) If these intersect at a point on the real axis, then $l$ and $m$ are said to be *parallel*.
      (ii) If these do not intersect at any point on the real axis, then $l$ and $m$ are said to be *ultra-parallel*.
   (c) Both $l$ and $m$ are semicircles with centers on the real axis:
      (i) If these intersect at a point on the real axis, then $l$ and $m$ are said to be *parallel*.
      (ii) If these do not intersect at any point on the real axis, then $l$ and $m$ are said to be *ultra-parallel*.

   In summary, if the geodesics $l$ and $m$ do not intersect in $\mathbb{H}$, but at the plane's boundary of infinity, then the two geodesics are said to be *parallel* to each other. However, if the two geodesics neither intersect in $\mathbb{H}$ nor at the boundary at infinity, then the geodesics are said to be *ultra-parallel*.
7. The *hyperbolic distance* between two points $a, b \in \mathbb{H}$ is

$$d_\mathbb{H}(a,b) = 2\tanh^{-1}\left|\frac{b-a}{b-\bar{a}}\right|$$

The presence of $\tanh^{-1}(\cdot)$ on the right-hand side of the above expression is one of the reasons, the geometry in space $\mathbb{H}$ is called "hyperbolic geometry."

8. A *hyperbolic polygon* is a polygon whose sides are hyperbolic lines. Let $\mathcal{P}$ be a hyperbolic polygon in $\mathbb{H}$ with internal angles $\alpha_1, \alpha_2, \ldots, \alpha_n$. The hyperbolic area of this polygon is

$$A\left(\mathcal{P}\right) = (n-2)\,\pi - \sum_{i=1}^{n} \alpha_i$$

Note that the area of a hyperbolic polygon depends only upon its angles. If $n = 3$, the hyperbolic polygon is a hyperbolic triangle. As $A\left(\mathcal{P}\right) > 0$, we have

$$\left(\alpha_1 + \alpha_2 + \alpha_3\right) < \pi$$

That is, the sum of internal angles of a hyperbolic triangle is upper-bounded by $\pi$.

9. The *hyperbolic circle* $C_{a,R}$ with center $a \in \mathbb{H}$ and radius $R > 0$ is the set of points in $\mathbb{H}$ which are at a hyperbolic distance $R$ from $a$. That is,

$$C_{a,R} = \{z \in \mathbb{H} \mid d_{\mathbb{H}}\left(a, z\right) = R\}$$

Then $C_{a,R}$ is also a Euclidean circle.

10. The Euclidean center of the hyperbolic circle $C_{ic,R}$, where $c > 0$, is $ib$, and its Euclidean radius is $r$, where

$$c = \sqrt{b^2 - r^2}, \quad R = \frac{1}{2}\ln\frac{(b+r)}{(b-r)}$$

The hyperbolic length of the circumference of $C_{ic,R}$ is equal to $2\pi \sinh R$, and the hyperbolic area of this circle is equal to $2\pi\left(\cosh R - 1\right)$. Observe that these values of circumference and area are independent of $c$.                                                                    □

Some of the above observations are proved in the problem section. Isometries of the upper half-plane model are developed in the next subsection. Using these isometries, it can be shown that the circumference and area of the hyperbolic circle $C_{a,R}$ are independent of the center $a$, and depend only upon the radius $R$.

### 5.8.2  Isometries of Upper Half-Plane Model

Recall that isometries are special length preserving transformations. Isometries of the upper half-plane model are explored in this subsection. Some isometries of $\mathbb{H}$ are: translations parallel to the real axis, reflections in lines parallel to the imaginary axis, dilations by a positive real number, and inversions in circles with centers on the real axis.

1. *Translation*: Translation parallel to the real axis is:

$$T_a\left(z\right) = z + a, \quad a \in \mathbb{R}$$

That is, $z = (x + iy)$ is mapped to $(x + a + iy)$.

2. *Reflection*: Reflection in line parallel to the imaginary axis is:

$$R_a\left(z\right) = 2a - \bar{z}, \quad a \in \mathbb{R}$$

Thus $R_a\left(z\right)$ is the "reflection" of $z$ in the line $\mathrm{Re}\left(z\right) = a$. That is, $z = (x + iy)$ is mapped to $(2a - x + iy)$.

3. *Dilation*: Dilation by a positive real number is:

$$D_a(z) = az, \quad a \in \mathbb{R}^+$$

That is, $z = (x + iy)$ is mapped to $(ax + iay)$.

4. *Inversion in circles*: Inversion in circle with center on the real axis. Consider a circle with center $a \in \mathbb{R}$, and radius $r \in \mathbb{R}^+$. The inversion in this circle is:

$$I_{a,r}(z) = a + \frac{r^2}{\overline{z} - a}$$

In the first three mappings, each point $z \in \mathbb{H}$ is mapped to a point in $\mathbb{H}$. In addition, the first fundamental form is also preserved. It is proved in the problem section that the fourth mapping: inversion in circles also preserves isometry. The maps: translation, reflection, dilation, and inversion in circles are called *elementary isometries* of $\mathbb{H}$.

**Observation 5.19.** A composition of a finite number of elementary isometries is an isometry of $\mathbb{H}$. □

It has been stated in an earlier section, that isometries map geodesics to geodesics. Therefore, elementary isometries map geodesics to geodesics. It is known that the geodesics in $\mathbb{H}$ are: semicircles with centers on the real axis, and half-lines parallel to the imaginary axis. It should be evident that translations, dilations, and reflections map half-lines and semicircles (with centers on the real axis), to other half-lines and semicircles (with centers on the real axis). However, the role of inversion map has to be clarified.

It has been established in the chapter on applied analysis that the inversion map $I_{a,r}(\cdot)$ is a conjugate-Moebius transformation. Therefore, it maps "circles" to "circles." Note that $I_{a,r}(a) = \infty$. Consequently $I_{a,r}(\cdot)$ maps "circles" that pass through the point $a$ to lines, and all other "circles" to circles.

**Observation 5.20.** Consider the inversion map $I_{a,r}(\cdot)$ with $a \in \mathbb{R}$ and $r > 0$. This map takes hyperbolic lines that intersect the real axis perpendicular at $a$ to half-lines, and all other hyperbolic lines to semicircles. □

The next set of observations establish that, the hyperbolic distance between two points in $\mathbb{H}$ is the length of the geodesic that passes through these two points.

**Observations 5.21.** Some useful observations.

1. Let $l_1$ and $l_2$ be hyperbolic lines in the upper half-plane $\mathbb{H}$. Also, $z_1$ and $z_2$ are points on $l_1$ and $l_2$ respectively. Then there exists an isometry, which takes $l_1$ to $l_2$ and $z_1$ to $z_2$.
2. Let $a, b \in \mathbb{H}$. The hyperbolic distance between these two points $d_{\mathbb{H}}(a, b)$ is the length of the shortest curve in $\mathbb{H}$ which joins $a$ and $b$. □

The above observations are established in the problem section. Some more properties of hyperbolic geometry, which utilize isometries of the upper half-plane $\mathbb{H}$ are summarized.

**Observations 5.22.** Some useful observations.

1. Let $l$ be a half-line in $\mathbb{H}$ and $p$ be a point not on $l$. There are an infinite number of hyperbolic lines that pass through the point $p$ that do not intersect $l$.
2. Let $l$ be a hyperbolic line in $\mathbb{H}$ and $p$ be a point not on $l$. There are an infinite number of hyperbolic lines that pass through the point $p$ that do not intersect $l$. $\qquad\square$

The above observations are established in the problem section.

**Moebius Transformations and Isometries of $\mathbb{H}$**

We relate Moebius transformations to isometries of the upper half-plane $\mathbb{H}$. Moebius transformations have been discussed in the chapter on applied analysis. A Moebius transformation $M(z)$, $z \in \mathbb{C}$ is *real*, if it is of the form

$$M(z) = \frac{az + b}{cz + d}, \quad \text{where} \quad a, b, c, d \in \mathbb{R}, \text{ and } (ad - bc) \neq 0$$

Several observations about real Moebius transformations are listed. The proofs of some of these observations are provided in the problem section.

**Observations 5.23.** Some useful facts about Moebius transformations and upper half-plane $\mathbb{H}$.

1. Inverse of any real Moebius transformation is also real. Composition of real Moebius transformations is also real.
2. The Moebius transformations that map $\mathbb{H}$ to $\mathbb{H}$ are precisely the real Moebius transformations, where $(ad - bc) > 0$. The coefficients $a, b, c$, and $d$ can be normalized so that $(ad - bc) = 1$.
3. Moebius isometry.
   (a) Each real Moebius transformation can be expressed as a composition of elementary isometries of $\mathbb{H}$. Consequently, a real Moebius transformation is an isometry of $\mathbb{H}$.
   (b) Let $M(\cdot)$ be a real Moebius transformation, and define a function $J(\cdot)$ as $J(z) = -\overline{z}$, where $z \in \mathbb{C}$. Then $M \circ J(\cdot)$ is an isometry of $\mathbb{H}$, where $\circ$ is a function composition operator.
   (c) The isometries described in observations (a) and (b) are called *Moebius isometries*. The composition of Moebius isometries is a Moebius isometry.
   (d) Each isometry of the upper half-plane $\mathbb{H}$ is a Moebius isometry.
4. The set of all Moebius transformations of $\mathbb{H}$ is denoted by $\mathcal{M}(\mathbb{H})$. The set $\mathcal{M}(\mathbb{H})$ forms a group under composition of Moebius transformations.
   If the coefficients $a, b, c$, and $d$ are normalized so that $(ad - bc) = 1$, then the group $\mathcal{M}(\mathbb{H})$ can be identified with $PSL(2, \mathbb{R})$, the projective special linear group. $\qquad\square$

### 5.8.3  Poincaré Disc Model

The Poincaré disc model of hyperbolic geometry is considered in this subsection. Let $P(\cdot)$ be a function, where

$$P(z) = \frac{z - i}{z + i}$$

where $z \in \mathbb{C} \backslash \{-i\}$. This map is a bijective mapping between the complex plane without the point $-i$ and the complex plane without the point $1$. The corresponding inverse mapping is

$$P^{-1}(z) = \frac{z+1}{i(z-1)}$$

where $z \in \mathbb{C} \backslash \{1\}$. Observe that the function $P(\cdot)$ is properly defined for all points in the upper half-plane $\mathbb{H}$ and also for all points on the real axis. For $(v, w) \in \mathbb{H}$ we have

$$P(v + iw) = \frac{v + i(w-1)}{v + i(w+1)}$$

Thus

$$|P(v + iw)| = \left\{ \frac{v^2 + (w-1)^2}{v^2 + (w+1)^2} \right\}^{1/2}$$

Consider the following cases:

(a) For $w < 0$, $|P(v + iw)| > 1$.
(b) For $w = 0$, $|P(v + iw)| = 1$.
(c) For $w > 0$, $|P(v + iw)| < 1$.

As $w > 0$ for all points in $\mathbb{H}$, the function $P(\cdot)$ maps all points in the space $\mathbb{H}$ to the unit disc

$$\mathbb{D} = \{z \in \mathbb{C} \mid |z| < 1\}$$

Also the points on the real axis are mapped to points on a circle $\mathcal{C}$ of unit radius, and center $(0,0)$. It is defined by $|z| = 1$. The circle $\mathcal{C}$ is the boundary of $\mathbb{D}$.

**Definitions 5.32.** *The Poincaré disc and its boundary.*

1. *The Poincaré disc $\mathbb{D}$ is the set of complex numbers $\mathbb{D} = \{z \in \mathbb{C} \mid |z| < 1\}$.*
2. *The boundary of $\mathbb{D}$ is the set $\partial\mathbb{D} = \{z \in \mathbb{C} \mid |z| = 1\}$.*                    □

Also, $\partial\mathbb{D}$ is called the circle at infinity, because the points in $\partial\mathbb{D}$ are at an infinite "hyperbolic distance" from each and every point in $\mathbb{D}$. Members of the set $\partial\mathbb{D}$ are also called *ideal points*.

**Definition 5.33.** *The Poincaré disc model $\mathbb{D}_P$ of hyperbolic geometry. Let $\mathbb{H}$ be the upper half of the complex $z$-plane, and the disc $\mathbb{D} = \{z \in \mathbb{C} \mid |z| < 1\}$. Also, let $P(z) = (z - i)/(z + i)$. The mapping $P(\cdot)$ maps points in $\mathbb{H}$ to points in $\mathbb{D}$. Also it maps points on the real line to points on the unit circle $\partial\mathbb{D}$. The first fundamental form in this mapping is an isometry.*                    □

In the above discussion and definition, $P(z)$ can also be replaced by $Q(z) = -P(z)$. Thus

$$Q(z) = -\frac{(z-i)}{(z+i)} = -\frac{(iz+1)}{(iz-1)}, \quad z \in \mathbb{C} \backslash \{-i\}$$

**Observations 5.24.** Some observations about Poincaré disc model of hyperbolic geometry $\mathbb{D}_P$.

1. The first fundamental form of $\mathbb{D}_P$ is

$$ds^2 = \frac{4(dv^2 + dw^2)}{(1 - v^2 - w^2)^2}, \quad (v, w) \in \mathbb{R}^2, (v^2 + w^2) < 1$$

That is, $(v, w) \in \mathbb{D}$. In addition, $\mathbb{D}_P$ is a conformal model.

2. As $P(\cdot)$ is an isometry, for each geodesic (hyperbolic line) in $\mathbb{D}_P$ there is a corresponding geodesic in $\mathbb{H}$, and vice-versa.

3. The distance between two points $a, b \in \mathbb{D}$ is $d_{\mathbb{D}}(a, b)$, where

$$d_{\mathbb{D}}(a, b) = d_{\mathbb{H}}\left(P^{-1}(a), P^{-1}(b)\right)$$

$$= 2\tanh^{-1}\frac{|b - a|}{|1 - \bar{a}b|} = \cosh^{-1}\left\{1 + 2\frac{|b - a|^2}{\left(1 - |a|^2\right)\left(1 - |b|^2\right)}\right\}$$

4. Let $\mathcal{C}$ be a circle of unit radius defined by $|z| = 1$. The lines (diameters) and arcs of circles that intersect the circle $\mathcal{C}$ perpendicularly are the hyperbolic lines (geodesics) of $\mathbb{D}_P$.

5. As mentioned earlier, the parallel axiom of the Euclidean geometry is not valid in hyperbolic geometry. In Euclidean geometry two parallel lines are a constant distance apart. In contrast, two different hyperbolic lines are never a constant distance apart.

6. Let $l$ and $m$ be two different lines in $\mathbb{D}$ that do not intersect each other at any point in $\mathbb{D}$. If these intersect at a point on $\partial\mathbb{D}$ (boundary of $\mathbb{D}$), then $l$ and $m$ are said to be *parallel*; otherwise, they are said to be *ultra-parallel*.

Consider a point $a$ in $\mathbb{D}$, which does not lie on a given hyperbolic line $l$. There are an infinite number of hyperbolic lines that pass through the point $a$. However, there are exactly two hyperbolic lines which pass through $a$ that are parallel to the hyperbolic line $l$.

7. The isometries of $\mathbb{D}_P$ are the maps $P \circ G \circ P^{-1}(\cdot)$ where $G(\cdot)$ is any isometry of $\mathbb{H}$.

8. The three sides of a hyperbolic triangle are geodesics. Let the vertices of the hyperbolic triangle be $a, b, c \in \mathbb{D}$; where the angles at these vertices are $\alpha, \beta$, and $\gamma$ respectively. Also, let the hyperbolic length of sides opposite angles $\alpha, \beta$, and $\gamma$ be $a_\ell, b_\ell$, and $c_\ell$ respectively.

(a) The "hyperbolic cosine rule" is:

$$\cosh c_\ell = \cosh a_\ell \cosh b_\ell - \sinh a_\ell \sinh b_\ell \cos\gamma$$

Expressions for $\cosh a_\ell$, and $\cosh b_\ell$ can be obtained similarly. This result is often called the *first hyperbolic cosine rule*.

(i) For small values of $a_\ell$, we have $\cosh a_\ell \simeq \left(1 + a_\ell^2/2\right)$, and $\sinh a_\ell \simeq a_\ell$. Therefore for small values of $a_\ell, b_\ell$, and $c_\ell$ the above expression simplifies to

$$c_\ell^2 = a_\ell^2 + b_\ell^2 - 2a_\ell b_\ell \cos\gamma$$

(ii) If the angle $\gamma$ is equal to a right angle, then

$$\cosh c_\ell = \cosh a_\ell \cosh b_\ell$$

This result is the equivalent of Pythogoras' theorem of Euclidean geometry, in hyperbolic geometry.

(iii) The triangle inequality $c_\ell \leq (a_\ell + b_\ell)$ holds true, with equality if and only if $c$ is on the hyperbolic line segment joining $a$ and $b$.

(b) The *hyperbolic sine rule*:

$$\frac{\sin\alpha}{\sinh a_\ell} = \frac{\sin\beta}{\sinh b_\ell} = \frac{\sin\gamma}{\sinh c_\ell}$$

(c) If $\gamma = \pi/2$, then

(i)  $\cos \alpha = \sinh b_\ell \cosh a_\ell / \sinh c_\ell$

(ii)  $\cosh a_\ell = \cos \alpha / \sin \beta$

(iii)  $\sinh a_\ell = \tanh b_\ell / \tan \beta$

(d) The length of a side of a hyperbolic triangle in terms of its angles:

$$\cosh a_\ell = \frac{\cos \alpha + \cos \beta \cos \gamma}{\sin \beta \sin \gamma}$$

This result is often called the *second hyperbolic cosine rule.*

(e) Unlike in Euclidean geometry, similarity of two hyperbolic triangles implies their congruence. Observe in the above expression for $\cosh a_\ell$, that the hyperbolic length $a_\ell$ is uniquely determined, once the angles $\alpha, \beta$, and $\gamma$ are specified.

(f) Consider a special case of the right-angled hyperbolic triangle, where $\gamma = \pi/2$. In this hyperbolic triangle, we further let $\alpha \to 0$. This occurs if the sides with lengths $b_\ell$ and $c_\ell$ become parallel. Furthermore, $b_\ell \to \infty$, and $\tanh b_\ell \to 1$.

In this hyperbolic triangle, let $a_\ell \triangleq m$, and $\beta \triangleq \Pi$. The angle $\Pi$ is called the *angle of parallelism.* Using results from part (c), we have

$$\sin \Pi = \operatorname{sech} m, \quad \tan \Pi = \operatorname{csch} m, \quad \cos \Pi = \tanh m$$

The above expressions yield the well-known Bolyai-Lobachevsky formula

$$\tan \left( \frac{\Pi}{2} \right) = e^{-m}$$

9.  As the map $P(\cdot)$ from the upper half-plane $\mathbb{H}$ to the unit disc $\mathbb{D}$ is an isometry, the circumference and area of hyperbolic circle $C_{a,R}$ with center $a \in \mathbb{D}$ and radius $R > 0$ are the same as in the case of the upper half-plane model of hyperbolic geometry. Therefore the corresponding circumference and area of the hyperbolic circle in unit disc $\mathbb{D}$ are $2\pi \sinh R$, and $2\pi (\cosh R - 1)$ respectively.                                                                                                           □

### Moebius Transformations and Isometries of $\mathbb{D}_P$

We relate Moebius transformations to isometries of the hyperbolic disc model $\mathbb{D}_P$. Let $\mathbb{C}^* = \mathbb{C} \cup \{+\infty\}$. A Moebius transformation $M(z), z \in \mathbb{C}^*$ is *hyperbolic Moebius transformation*, if it is of the form

$$M(z) = \frac{az + b}{\bar{b}z + \bar{a}}, \quad \text{where} \quad a, b \in \mathbb{C}, \text{ and } |a| > |b|$$

Several observations about hyperbolic Moebius transformations are listed. The proofs of these observations are provided in the problem section.

**Observations 5.25.** Some observations about hyperbolic Moebius transformations.

1.  The Moebius transformations that map unit disc to itself are the hyperbolic Moebius transformations.

2.  The isometries of $\mathbb{D}_P$ are the hyperbolic Moebius, and conjugate hyperbolic Moebius transformations.

3.  The set of all Moebius transformations of $\mathbb{D}$ is denoted by $\mathcal{M}(\mathbb{D})$. The set $\mathcal{M}(\mathbb{D})$ forms a group under composition of the Moebius transformation.

    If the coefficients $a$ and $b$ are normalized so that $\left( |a|^2 - |b|^2 \right) = 1$, then the group $\mathcal{M}(\mathbb{D})$ can be identified with $PSL(2, \mathbb{C})$, the projective special linear group.                                               □

### 5.8.4  Surface of Different Constant Curvature

Consider a hyperbolic surface $S$ of constant Gaussian curvature equal to $-1$. Let $\xi$ be a positive constant. Dilate $(x, y, z) \in \mathbb{R}^3$ by $\xi$ to $(\xi x, \xi y, \xi z) \in \mathbb{R}^3$. We examine the effect of this dilation upon Gaussian curvature of the surface $S$.

It can be inferred that the coefficients of the first fundamental form $E$, $F$, and $G$ are each multiplied by $\xi^2$. Similarly, the coefficients of the second fundamental form $L$, $M$, and $N$ are each multiplied by $\xi$. As Gaussian curvature $K$ is

$$K = \frac{\left(LN - M^2\right)}{\left(EG - F^2\right)}$$

dilation modifies $K$ by a factor $\xi^{-2}$. That is, Gaussian curvature of this modified surface is equal to $-\xi^{-2}$. It can also be checked that if the first fundamental form of the modified surface is

$$\frac{\xi^2 \left(dv^2 + dw^2\right)}{w^2},$$

we obtain a surface of constant Gaussian curvature equal to $-\xi^{-2}$. Therefore the circumference and area of the hyperbolic circle of radius $R > 0$ are $2\pi\xi \sinh\left(R/\xi\right)$, and $2\pi\xi^2 \left(\cosh\left(R/\xi\right) - 1\right)$ respectively. As expected, the circumference and area of the circle for small values of radius $R$ are approximately $2\pi R$ and $\pi R^2$ respectively. The corresponding first hyperbolic cosine rule is:

$$\cosh\frac{c_\ell}{\xi} = \cosh\frac{a_\ell}{\xi} \cosh\frac{b_\ell}{\xi} - \sinh\frac{a_\ell}{\xi} \sinh\frac{b_\ell}{\xi} \cos\gamma$$

where $a_\ell$, $b_\ell$, and $c_\ell$ are lengths of sides of an hyperbolic triangle on the original undilated hyperbolic surface. Also, $\alpha$, $\beta$, and $\gamma$ are angles opposite the sides $a_\ell$, $b_\ell$, and $c_\ell$ respectively.

Notion of tessellation is developed in the next subsection.

### 5.8.5  Tessellations

Tessellation means tiling of a surface. For example, the surface can either be Euclidean or hyperbolic. The goal is to tile a surface via regular $n$-gon. A regular $n$-gon is a polygon with $n \in \mathbb{P} \backslash \{1, 2\}$ sides, each of equal length. In this regular polygon, all of its internal angles are also equal. In tiling the surface, we also specify that $k \in \mathbb{P} \backslash \{1, 2\}$ polygons meet at each vertex. This type of tiling is called an $(n, k)$ or $\{n, k\}$ tessellation. The notation $\{n, k\}$ is also called the Schläfli symbol, after the mathematician Ludwig Schläfli (1814-1895).

The dual of the $\{n, k\}$ tessellation is the $\{k, n\}$ tessellation. The dual tessellation is generated by taking the center of each $n$-polygon (in the $\{n, k\}$ tessellation) as a vertex and joining centers of adjacent polygons via geodesics to produce new polygons in order to tile the plane.

It can be shown that a Euclidean surface $\mathbb{R}^2$ has only three such types of tessellations. These are: equilateral triangles, squares, and regular hexagons. These types of tilings are $(3, 6)$, $(4, 4)$, and $(6, 3)$ tessellations respectively.

It can be shown that an infinite number of tessellations are possible with regular hyperbolic polygons on hyperbolic surfaces. In hyperbolic polygons, the sides are geodesics between its vertices. In each such tessellation, the polygons can be generated recursively. In the following observation, a hyperbolic plane is either a upper half-plane or a Poincaré disc.

**Observation 5.26.** If the tessellation of a hyperbolic plane, via regular hyperbolic $n$-gon with $k$ polygons meeting at each vertex exists, then

$$\frac{1}{n} + \frac{1}{k} < \frac{1}{2}$$

where $n, k \in \mathbb{P} \backslash \{1, 2\}$. It is clear that infinite number of $(n, k)$ pairs exist which satisfy the above inequality. The above inequality is often expressed as

$$(n - 2)(k - 2) > 4$$

$\square$

The validity of the above observation is established in the problem section. It can also be shown that the converse of the above observation is also true. If $n$ tends to infinity, then the hyperbolic plane is tessellated via an infinite regular tree of degree $k$. This is indeed the $\{\infty, k\}$ tessellation. Its dual is the $\{k, \infty\}$ tessellation. This dual tessellates the hyperbolic plane with an infinite number of $k$-gons. This later tiling splits the hyperbolic plane into distinct spaces.

The infinite hyperbolic tree created with the $\{\infty, k\}$ tessellation can be used for embedding the spanning tree of a graph which represents a communication network.

### 5.8.6 Geometric Constructions

Certain geometric constructions used in hyperbolic geometry are outlined in this subsection.

**Definitions 5.34.** *Pole and polar.*

1. *Let $C$ be a circle with center at $O$. Also let $P$ be a point which is not equal to $O$. The inverse of the point $P$ with respect to the circle $C$ is $P'$. The line $\ell$ is perpendicular to the line $OP$ at the point $P'$.*

   (a) *$\ell$ is called polar of point $P$ with respect to the circle $C$.*
   (b) *$P$ is called pole of line $\ell$ with respect to the circle $C$.*

   *The line $\ell$ is also denoted by $P^{\perp}$, and the point $P$ by $\ell^{\perp}$.*

2. *In the hyperbolic disc $\mathbb{D}_P$, let $\gamma$ be a geodesic. This geodesic is the arc of a circle with center $A$. Pole of the arc $\gamma$ is the point $A$. The point $A$ is denoted by $\gamma^{\perp}$. The arc $\gamma$ is also denoted by $A^{\perp}$.* $\square$

Some methods of constructing hyperbolic geodesics in the Poincaré disc are next stated.

**Constructions** Certain useful constructions in the Poincaré disc. Let $\mathcal{C}$ be the circle with unit radius, the corresponding Poincaré disc be $\mathbb{D}$, and its boundary $\partial \mathbb{D}$.

1. *Construction of geodesic through points $A, B \in \partial \mathbb{D}$. Let $A$ and $B$ be points on the boundary of the Poincaré disc (ideal points). Hyperbolic geodesic passing through points $A$ and $B$ has to be constructed.*

   *Method*:
   - Let the mid-point of the Euclidean line $AB$ be $M$.
   - Let the inverse of point $M$ with respect to the circle $\mathcal{C}$ be $M'$.
   - Draw a circle $\mathcal{C}'$ with center $M'$ and radius $M'A$.

The circle $C'$ is the required geodesic passing through points $A$ and $B$.

2. *Construction of geodesic through points* $A, B \in \mathbb{D}$. Let $A$ and $B$ be points in the Poincaré disc. Hyperbolic geodesic passing through points $A$ and $B$ has to be constructed.

   This is equivalent to constructing a unique circle passing through points $A$ and $B$ that is orthogonal to circle $C$.

   *Method*:
   - Let $A'$ be the inverse of the point $A$ with respect to the circle $C$.
   - Let $\gamma$ be the circle passing through points $A, B$, and $A'$. The desired geodesic is an arc of the circle $\gamma$.

     *Justification*: It can be shown that the circle $\gamma$ is orthogonal to the circle $C$. The method is justified via an observation about inversion with respect to a circle.

3. *The polars of the points* $A, B \in \mathbb{D}$ *are* $A^{\perp}$ *and* $B^{\perp}$ *respectively. Geodesic through* $A$ *and* $B$ *has to be constructed.* Given $A, B \in \mathbb{D}$, $A^{\perp}$, and $B^{\perp}$, a hyperbolic geodesic passing through points $A$ and $B$ has to be constructed.

   *Method*:
   - Let $P$ be the point of intersection of the polars $A^{\perp}$ and $B^{\perp}$. The desired geodesic is the circle with center $P$ which passes through points $A$ and $B$.

     *Justification*: Points within the Poincaré disc are in precise correspondence with Euclidean lines outside of the disc, and the hyperbolic geodesics are in precise correspondence with points outside of the disc.

     Note that the perpendicular bisector of the Euclidean line segment $AB$ also passes through the point $P$.

4. *Construction of hyperbolic circle, with center at $A$ and which passes through point $B$, where* $A, B \in \mathbb{D}$. Let $A$ and $B$ be points in the Poincaré disc. Let the center of this disc be the point $O$. We construct the hyperbolic circle $C$ with hyperbolic center $A$, which also passes through point $B$.

   *Method*:
   - Let the geodesic passing through points $A$ and $B$ be an arc of the circle $\gamma$.
   - Let the tangent to the circle $\gamma$ at point $B$ intersect the line segment $OA$ at the point $E$.
   - The required circle $C$ is the unique Euclidean circle with center $E$ and radius $EB$.

     *Justification*: The required circle $C$ is the unique Euclidean circle that is orthogonal to every geodesic through the point $A$. More specifically, the circle $C$ must be orthogonal to the Euclidean line segment $OA$. Therefore, the center of the circle $C$ should be on this line. Furthermore, the circle $C$ must be orthogonal to the geodesic $\gamma$ passing through points $A$ and $B$. Consequently, the center of circle $C$ should lie on the tangent to $\gamma$ at the point $B$. Denote the intersection of this tangent line, and the Euclidean line segment $OA$ as $E$. Therefore, $C$ is the Euclidean circle with center at point $E$, and radius $EB$.

5. *Construction of perpendicular bisector of the geodesic, which passes through points $A$ and $B$, where* $A, B \in \mathbb{D}$. Let $A$ and $B$ be points in the Poincaré disc. Construct the perpendicular bisector of the geodesic passing through points $A$ and $B$. Also determine the midpoint of this geodesic segment.

   *Method*:
   - This method is similar to the construction of perpendicular bisector of a line segment in Euclidean geometry. Let $C_1$ be a hyperbolic circle centered at $A$, which passes through point $B$. Similarly, let $C_2$ be a hyperbolic circle centered at $B$, which passes through point $A$. Let $E$ and $F$ be points of intersection of these two circles.

- The hyperbolic line segment $EF$ is the perpendicular bisector of hyperbolic line segment $AB$. The point of intersection of these hyperbolic segments is the mid point of the hyperbolic line segment $AB$.

    *Justification:* The justification of this construction is left to the reader.

6. *Construction of reflection point of* $A \in \mathbb{D}$ *across the geodesic* $\gamma$. Let $A$ be a point in the Poincaré disc, and $\gamma$ be a geodesic. Construct the reflection of point $A$ across $\gamma$.

    *Method:*
    - The point $A$ is reflected across $\gamma$ by inverting $A$ through $\gamma$.

    *Justification:* The justification of this construction is left to the reader.  $\square$

## Reference Notes

The section on Euclidean geometry and its foundations is based upon the books by Wolfe (1966), Greenberg (1974), Henle (1997), and Venema (2011). A good source of Euclid's *Elements* is the book by Heath (1956). It is also available on the World Wide Web. Neither Heath nor the corresponding material on the Web is copyrighted.

Accessible accounts of differential geometry are given in Graustein (1966), Kreyszig (1968), Lipschutz (1969), do Cormo (1976), Struik (1988), Kühnel (2002), Reid, and Szendrői (2005), and Pressley (2001, 2010). The author owes an immense debt of gratitude to Andrew Pressley. His eminently readable and instructive textbook has been extremely useful. Its influence upon the material in this chapter should be clearly evident to the perceptive reader. The geometric constructions are based upon the elegant paper by Goodman-Strauss (2001).

A readable account of hyperbolic geometry is given in Stillwell (1992), Ramsay, and Richtmyer (1995), and Anderson (2005).

## Problems

1. Consider a circle $C_{a,r}$ with center $a \in \mathbb{C}$, and radius $r \in \mathbb{R}^+$. We determine inversion of points with respect to the circle $C_{a,r}$. Let $C_{\beta,\rho}$ be another circle with center $\beta \in \mathbb{C}$, and radius $\rho \in \mathbb{R}^+$. Establish the following results.

    (a) The circle $C_{\beta,\rho}$ does not pass through the center of the circle $C_{a,r}$. That is, $\rho \neq |a - \beta|$. Let $\delta \triangleq r^2 / \left( |\beta - a|^2 - \rho^2 \right)$. The inverse image of points $z \in C_{\beta,\rho}$ with respect to the circle $C_{a,r}$, form a circle with center at $\{a + (\beta - a)\,\delta\}$, and radius $|\delta|\,\rho$. The center of this later circle does not pass through the center of the circle $C_{a,r}$. That is, $\{a + (\beta - a)\,\delta\} \neq a$.

    If $\delta = 1$, the inverse image of points in the circle $C_{\beta,\rho}$ is $C_{\beta,\rho}$ itself. Also $\left( r^2 + \rho^2 \right) = |\beta - a|^2$. This condition implies that the circles $C_{a,r}$ and $C_{\beta,\rho}$ are orthogonal to each other.

(b) If $\rho = |a - \beta|$, then the inverse image of points $z \in C_{\beta,\rho}$ is a straight line orthogonal to the line joining the centers $a$ and $\beta$. Distance of this straight line from center $a$ is $r^2 / (2\rho)$.

(c) Converse of part (b). The image of any straight line which does not pass through $a$, is a circle which contains the center $a$.

(d) The inverse of points on a straight line which passes through the point $a$ are also on this straight line. However, a point does not map to itself.

Hint: The equation of the circle $C_{\beta,\rho}$ is

$$(z - \beta)(\bar{z} - \bar{\beta}) = \rho^2$$

That is

$$z\bar{z} - (\bar{\beta}z + \beta\bar{z}) + |\beta|^2 = \rho^2$$

Inverse of $z \neq a$ with respect to $C_{a,r}$ is $z'$, where

$$z' = a + \frac{r^2}{(\bar{z} - \bar{a})}$$

Therefore

$$z = a + \frac{r^2}{(\bar{z}' - \bar{a})}$$

Substitute the above expression for $z$ in $\left\{ z\bar{z} - (\bar{\beta}z + \beta\bar{z}) + |\beta|^2 \right\} = \rho^2$. Consider two cases.

(a) Let $\rho \neq |a - \beta|$. Substitution yields

$$\left\{ (z' - a) - (\beta - a)\delta \right\} \left\{ (\bar{z}' - \bar{a}) - (\bar{\beta} - \bar{a})\delta \right\} = \delta^2 \rho^2$$

This is the equation of a circle with center at $\{a + (\beta - a)\delta\}$, and radius $|\delta|\,\rho$.

(b) Let $\rho = |a - \beta|$. Substitution yields equation of a straight line

$$z'(\bar{\beta} - \bar{a}) + \bar{z}'(\beta - a) - a(\bar{\beta} - \bar{a}) - \bar{a}(\beta - a) - r^2 = 0$$

The equation of straight line joining points $a$ and $\beta$ is

$$z(\bar{\beta} - \bar{a}) - \bar{z}(\beta - a) + (\beta\bar{a} - \bar{\beta}a) = 0$$

Therefore the straight lines specifying the set of inversion points (inverse image), and the join of points $a$ and $\beta$ are perpendicular to each other. It can be shown that the distance between the straight line specifying the inverse image and point $a$ is $r^2 / (2\rho)$.

(c) Reverse the steps in part (b)

(d) Let

$$\bar{b}z + b\bar{z} + c = 0, \quad \text{where} \quad c \in \mathbb{R},\ b \in \mathbb{C} \backslash \{0\}$$

be the equation of a straight line. If this straight line passes through the center $a$, then we have

$$\bar{b}a + b\bar{a} + c = 0$$

The last two equations yield

$$\bar{b}(z - a) + b(\bar{z} - \bar{a}) = 0$$

The inverse of a point $z$ with respect to the circle $C_{a,r}$ is $z'$. Substituting $z = a + r^2 / \left( \overline{z'} - \overline{a} \right)$ in the last equation results in

$$\overline{b} \left( z' - a \right) + b \left( \overline{z'} - \overline{a} \right) = 0$$

This is the original equation of the straight line.

2. If $\xi(t)$, is a vector of constant length, then prove that

$$\xi(t) \circ \xi'(t) = 0$$

Hint: See Lipschutz (1969). If $\|\xi(t)\|$ is constant, then $\xi(t) \circ \xi(t)$ is constant. Differentiating $\xi(t) \circ \xi(t)$ with respect to $t$, yields

$$\xi(t) \circ \xi'(t) + \xi'(t) \circ \xi(t) = 0$$

The result follows.

3. Prove that a regular curve of class $C^r$, where $r \geq 2$, defined over an interval $I$ is a straight line if and only if its curvature $\kappa(s) = 0$ over the entire interval $I$.

   Hint: See Lipschutz (1969). We first prove that, if the curvature $\kappa(s) = 0$ over the entire interval $I$, then the curve is a straight line. Note that $\kappa(s) = 0$ implies that $\xi''(s) = 0$. This in turn implies that $\xi(s) = (as + b)$, where $a \neq 0$ and $b$ are constant vectors. This is the equation of a straight line which passes through $b$, and parallel to $a$.

   In the next step, we prove the converse. That is, if the curve $C$ is a straight line, then its curvature $\kappa(s) = 0$. The curve $C$ is

   $$\xi(t) = at + b, \quad a \neq 0$$

   Therefore

   $$\widetilde{T}(s) = \frac{\xi'(t)}{\|\xi'(t)\|} = \frac{a}{|a|}$$

   $$\kappa(s) = \left| \widetilde{T}'(s) \right| = 0$$

4. Let $\alpha < \beta$, where $\alpha, \beta \in \mathbb{R}$. Also $(\alpha, \beta) = \{t \in \mathbb{R} \mid \alpha < t < \beta\}$ is an open interval. A curve $C$ of class greater than or equal to 2, parameterized by $t \in (\alpha, \beta)$ is a map $\xi : (\alpha, \beta) \to \mathbb{R}^3$. Assume that $\xi'(t) \neq 0$. Prove that the curvature of the curve at $t \in (\alpha, \beta)$ is given by $\kappa(t) = \|\xi''(t) \times \xi'(t)\| / \|\xi'(t)\|^3$.

   Hint: See Lipschutz (1969).

   $$\xi'(t) = \frac{d\xi(t)}{dt} = \frac{d\xi}{ds} \frac{ds(t)}{dt} = \frac{d\xi}{ds} s'(t)$$

   $$\xi''(t) = \frac{d^2 \xi(t)}{dt^2} = \frac{d}{dt}\left( \frac{d\xi}{ds} s'(t) \right) = \frac{d}{dt}\left( \frac{d\xi}{ds} \right) s'(t) + \frac{d\xi}{ds} s''(t)$$

   $$= \frac{d^2 \xi}{ds^2} (s'(t))^2 + \frac{d\xi}{ds} s''(t)$$

   Therefore

   $$\xi''(t) \times \xi'(t) = \left\{ \frac{d^2 \xi}{ds^2} (s'(t))^2 + \frac{d\xi}{ds} s''(t) \right\} \times \frac{d\xi}{ds} s'(t)$$

$$= \left\{ \frac{d^2\xi}{ds^2} \times \frac{d\xi}{ds} \right\} (s'(t))^3$$

The result follows by observing that $s'(t) = \|\xi'(t)\|$, $\|d^2\xi/ds^2\| = \kappa(s)$, $\|d\xi/ds\| = 1$; and $d^2\xi/ds^2$ and $d\xi/ds$ are orthogonal to each other.

5. Consider the plane curve $y = f(x)$, and $z = 0$. This curve lies in the $xy$-plane. Let $y'$ and $y''$ denote the first and second derivatives of $y$ with respect to $x$ respectively. Prove that the radius of the curvature $\rho$ is given by

$$\rho = \frac{\left\{ 1 + (y')^2 \right\}^{3/2}}{|y''|}$$

Hint: See Spiegel (1963).

6. Prove the Frenet-Serret equations.

(a) $\widetilde{T}'(s) = \kappa(s)\widetilde{N}(s)$

(b) $\widetilde{B}'(s) = -\tau(s)\widetilde{N}(s)$

(c) $\widetilde{N}'(s) = \tau(s)\widetilde{B}(s) - \kappa(s)\widetilde{T}(s)$

Hint: See Spiegel (1959).

(a) As $\widetilde{T}(s) \circ \widetilde{T}(s) = 1$, we have $\widetilde{T}(s) \circ \widetilde{T}'(s) = 0$. That is, $\widetilde{T}'(s)$ is orthogonal to $\widetilde{T}(s)$. If $\widetilde{N}(s)$ is a unit vector in the direction of $\widetilde{T}'(s)$, we have $\widetilde{T}'(s) = \kappa(s)\widetilde{N}(s)$, where $\kappa(s)$ is the curvature.

(b) Assume $\widetilde{B}(s) = \widetilde{T}(s) \times \widetilde{N}(s)$. Differentiating it yields

$$\begin{aligned}
\widetilde{B}'(s) &= \widetilde{T}(s) \times \widetilde{N}'(s) + \widetilde{T}'(s) \times \widetilde{N}(s) \\
&= \widetilde{T}(s) \times \widetilde{N}'(s) + \kappa(s)\widetilde{N}(s) \times \widetilde{N}(s) \\
&= \widetilde{T}(s) \times \widetilde{N}'(s)
\end{aligned}$$

The above result implies that $\widetilde{B}'(s)$ is orthogonal to $\widetilde{T}(s)$.
As $\widetilde{B}(s) \circ \widetilde{B}(s) = 1$, we have $\widetilde{B}(s) \circ \widetilde{B}'(s) = 0$. That is, $\widetilde{B}'(s)$ is orthogonal to $\widetilde{B}(s)$. Thus $\widetilde{B}'(s)$ is orthogonal to both $\widetilde{T}(s)$ and $\widetilde{B}(s)$. Consequently $\widetilde{B}'(s)$ is parallel to $\widetilde{N}(s)$. Therefore let $\widetilde{B}'(s) = -\tau(s)\widetilde{N}(s)$, where $\tau(s)$ is the torsion, and the negative sign is chosen by convention.

(c) Differentiating $\widetilde{N}(s) = \widetilde{B}(s) \times \widetilde{T}(s)$ yields

$$\begin{aligned}
\widetilde{N}'(s) &= \widetilde{B}(s) \times \widetilde{T}'(s) + \widetilde{B}'(s) \times \widetilde{T}(s) \\
&= \widetilde{B}(s) \times \kappa(s)\widetilde{N}(s) - \tau(s)\widetilde{N}(s) \times \widetilde{T}(s) \\
&= -\kappa(s)\widetilde{T}(s) + \tau(s)\widetilde{B}(s)
\end{aligned}$$

7. Let $\alpha < \beta$, where $\alpha, \beta \in \mathbb{R}$; and $(\alpha, \beta) = \{t \in \mathbb{R} \mid \alpha < t < \beta\}$ is an open interval. A curve $C$ of class greater than or equal to 3, parameterized by $t \in (\alpha, \beta)$ is a map $\xi : (\alpha, \beta) \to \mathbb{R}^3$. Assume that $\kappa(t) \neq 0$. Prove that the torsion of the curve $C$ at $t \in (\alpha, \beta)$ is given by

$$\tau(t) = \frac{\xi'(t) \circ \xi''(t) \times \xi'''(t)}{\|\xi'(t) \times \xi''(t)\|^2}$$

If the parameter $t$ is the arc-length $s$, then

$$\tau(s) = \frac{\xi'(s) \circ \xi''(s) \times \xi'''(s)}{\|\xi'(s) \times \xi''(s)\|^2} = \tau(t)$$

Hint: See Spiegel (1959), and Pressley (2010). We first establish the expression for $\tau(s)$. We have

$$\xi'(s) = \widetilde{T}(s)$$
$$\xi''(s) = \widetilde{T}'(s) = \kappa(s)\,\widetilde{N}(s)$$
$$\xi'''(s) = \kappa(s)\,\widetilde{N}'(s) + \frac{d\kappa(s)}{ds}\widetilde{N}(s)$$
$$= \kappa(s)\left\{-\kappa(s)\,\widetilde{T}(s) + \tau(s)\,\widetilde{B}(s)\right\} + \frac{d\kappa(s)}{ds}\widetilde{N}(s)$$

Therefore

$$\xi'(s) \circ \xi''(s) \times \xi'''(s)$$
$$= \widetilde{T}(s) \circ \kappa(s)\,\widetilde{N}(s) \times \left\{\kappa(s)\left\{-\kappa(s)\,\widetilde{T}(s) + \tau(s)\,\widetilde{B}(s)\right\} + \frac{d\kappa(s)}{ds}\widetilde{N}(s)\right\}$$
$$= \widetilde{T}(s) \circ \left\{\kappa(s)^3\,\widetilde{B}(s) + \kappa(s)^2\,\tau(s)\,\widetilde{T}(s)\right\} = \kappa(s)^2\,\tau(s)$$

Also

$$\xi'(s) \times \xi''(s) = \widetilde{T}(s) \times \kappa(s)\,\widetilde{N}(s) = \kappa(s)\,\widetilde{B}(s)$$

Therefore

$$\frac{\xi'(s) \circ \xi''(s) \times \xi'''(s)}{\left\|\xi'(s) \times \xi''(s)\right\|^2} = \frac{\kappa(s)^2\,\tau(s)}{\kappa(s)^2} = \tau(s)$$

We next find expression for $\tau(t)$.

$$\xi'(t) = \frac{d\xi(t)}{dt} = \frac{d\xi}{ds}\frac{ds}{dt} = \xi'(s)\frac{ds}{dt}$$
$$\xi''(t) = \xi''(s)\left\{\frac{ds}{dt}\right\}^2 + \xi'(s)\frac{d^2s}{dt^2}$$
$$\xi'''(t) = \xi'''(s)\left\{\frac{ds}{dt}\right\}^3 + 3\xi''(s)\frac{ds}{dt}\frac{d^2s}{dt^2} + \xi'(s)\frac{d^3s}{dt^3}$$

This yields

$$\xi'(t) \circ \xi''(t) \times \xi'''(t)$$
$$= \xi'(s)\frac{ds}{dt} \circ \left\{\xi''(s)\left\{\frac{ds}{dt}\right\}^2 + \xi'(s)\frac{d^2s}{dt^2}\right\}$$
$$\times \left\{\xi'''(s)\left\{\frac{ds}{dt}\right\}^3 + 3\xi''(s)\frac{ds}{dt}\frac{d^2s}{dt^2} + \xi'(s)\frac{d^3s}{dt^3}\right\}$$
$$= \xi'(s) \circ \xi''(s) \times \xi'''(s)\left\{\frac{ds}{dt}\right\}^6$$

and

$$\xi'(t) \times \xi''(t) = \xi'(s)\frac{ds}{dt} \times \left\{\xi''(s)\left\{\frac{ds}{dt}\right\}^2 + \xi'(s)\frac{d^2s}{dt^2}\right\}$$
$$= \xi'(s) \times \xi''(s)\left\{\frac{ds}{dt}\right\}^3$$

Hence

$$\frac{\xi'(t) \circ \xi''(t) \times \xi'''(t)}{\left\| \xi'(t) \times \xi''(t) \right\|^2} = \frac{\xi'(s) \circ \xi''(s) \times \xi'''(s)}{\left\| \xi'(s) \times \xi''(s) \right\|^2}$$

8. Prove that the Frenet-Serret equations can be written as

$$\widetilde{T}'(s) = \omega(s) \times \widetilde{T}(s), \quad \widetilde{N}'(s) = \omega(s) \times \widetilde{N}(s), \quad \widetilde{B}'(s) = \omega(s) \times \widetilde{B}(s)$$

where $\omega(s) = \tau(s)\widetilde{T}(s) + \kappa(s)\widetilde{B}(s)$.

Hint: See Spiegel (1959). This is straightforward verification.

9. This problem is about the first fundamental form $I(\cdot, \cdot)$. Establish the following results.

   (a) $I(\cdot, \cdot)$ is positive definite.
   (b) $EG - F^2 = \left\| x_u \times x_v \right\|^2$.
   (c) $E > 0, G > 0$, and $\left( EG - F^2 \right) > 0$
   (d) $I(\cdot, \cdot)$ depends only on the surface and not on its parameterization.

   Hint: See Lipschutz (1969).

   (a) It is required to be shown that: $I(u, v) \geq 0$; and $I(u, v) = 0 \Leftrightarrow du = dv = 0$. By its definition $I(u, v) = \left\| dx \right\|^2 \geq 0$. Furthermore, as $x_u$ and $x_v$ are linearly independent (because of regularity) $I(u, v) = \left\| x_u du + x_v dv \right\|^2 = 0$ is true if and only if $du = dv = 0$.

   (b) We have

   $$EG - F^2 = \left\| x_u \right\|^2 \left\| x_v \right\|^2 - \left\| x_u \circ x_v \right\|^2 = \left\| x_u \times x_v \right\|^2$$

   (c) As $x_u$ and $x_v$ are linearly independent, $x_u \neq 0$ and $x_v \neq 0$, we have $E = x_u \circ x_u > 0$. Similarly, $G = x_v \circ x_v > 0$. Using part (b), we have $EG - F^2 > 0$, as $x_u \times x_v \neq 0$ at every point.

   (d) A surface patch has a parameterization $x(u, v)$. Consider another parameterization of a surface patch $\widetilde{x}(\widetilde{u}, \widetilde{v})$, which is in the neighborhood of $x(u, v)$. Let $\widetilde{u} = \widetilde{u}(u, v)$, and $\widetilde{v} = \widetilde{v}(u, v)$. Therefore

   $$d\widetilde{u} = \widetilde{u}_u du + \widetilde{u}_v dv, \quad \text{and} \quad d\widetilde{v} = \widetilde{v}_u du + \widetilde{v}_v dv$$

   and

   $$\begin{aligned}
   \widetilde{I}(\widetilde{u}, \widetilde{v}) &= d\widetilde{x} \circ d\widetilde{x} = \left\| d\widetilde{x} \right\|^2 = \left\| \widetilde{x}_{\widetilde{u}} d\widetilde{u} + \widetilde{x}_{\widetilde{v}} d\widetilde{v} \right\|^2 \\
   &= \left\| \widetilde{x}_{\widetilde{u}} \left( \widetilde{u}_u du + \widetilde{u}_v dv \right) + \widetilde{x}_{\widetilde{v}} \left( \widetilde{v}_u du + \widetilde{v}_v dv \right) \right\|^2 \\
   &= \left\| \left( \widetilde{x}_{\widetilde{u}} \widetilde{u}_u + \widetilde{x}_{\widetilde{v}} \widetilde{v}_u \right) du + \left( \widetilde{x}_{\widetilde{u}} \widetilde{u}_v + \widetilde{x}_{\widetilde{v}} \widetilde{v}_v \right) dv \right\|^2 \\
   &= \left\| x_u du + x_v dv \right\|^2 = I(u, v)
   \end{aligned}$$

10. A surface $x(u, v)$ is reparameterized as $\widetilde{x}(\widetilde{u}, \widetilde{v})$. The corresponding first fundamental forms are

$$I = E du^2 + 2F du dv + G dv^2, \quad \text{and} \quad \widetilde{I} = \widetilde{E} d\widetilde{u}^2 + 2\widetilde{F} d\widetilde{u} d\widetilde{v} + \widetilde{G} d\widetilde{v}^2$$

respectively. Let $J$ be the Jacobian matrix which specifies the transformation from the $\widetilde{u}\widetilde{v}$-plane to $uv$-plane. It is

$$J = \begin{bmatrix} \partial u / \partial \widetilde{u} & \partial u / \partial \widetilde{v} \\ \partial v / \partial \widetilde{u} & \partial v / \partial \widetilde{v} \end{bmatrix} = \begin{bmatrix} u_{\widetilde{u}} & u_{\widetilde{v}} \\ v_{\widetilde{u}} & v_{\widetilde{v}} \end{bmatrix}$$

Show that

(a)

$$\begin{bmatrix} \widetilde{E} & \widetilde{F} \\ \widetilde{F} & \widetilde{G} \end{bmatrix} = J^T \begin{bmatrix} E & F \\ F & G \end{bmatrix} J$$

(b) $\left\{ \widetilde{E}\widetilde{G} - \widetilde{F}^2 \right\}^{1/2} = \left\{ EG - F^2 \right\}^{1/2} |\det J|$

(c) $\widetilde{x}_{\widetilde{u}}(\widetilde{u}, \widetilde{v}) \times \widetilde{x}_{\widetilde{v}}(\widetilde{u}, \widetilde{v}) = (\det J)\, x_u(u, v) \times x_v(u, v)$

(d) The standard unit normal of the surface patch with parameterizations $x(u, v)$ and $\widetilde{x}(\widetilde{u}, \widetilde{v})$ at a point $P$ on the surface is

$$\mathcal{N}(u, v) = \frac{x_u(u, v) \times x_v(u, v)}{\|x_u(u, v) \times x_v(u, v)\|}$$

$$\widetilde{\mathcal{N}}(\widetilde{u}, \widetilde{v}) = \frac{\widetilde{x}_{\widetilde{u}}(\widetilde{u}, \widetilde{v}) \times \widetilde{x}_{\widetilde{v}}(\widetilde{u}, \widetilde{v})}{\|\widetilde{x}_{\widetilde{u}}(\widetilde{u}, \widetilde{v}) \times \widetilde{x}_{\widetilde{v}}(\widetilde{u}, \widetilde{v})\|}$$

respectively. Show that

$$\widetilde{\mathcal{N}}(\widetilde{u}, \widetilde{v}) = sgn\,(\det J)\, \mathcal{N}(u, v)$$

where $sgn\,(\cdot)$ is the sign function.

Hint: See Pressley (2010).

(a) We have $\widetilde{x}_{\widetilde{u}} = x_u u_{\widetilde{u}} + x_v v_{\widetilde{u}}$, and $\widetilde{x}_{\widetilde{v}} = x_u u_{\widetilde{v}} + x_v v_{\widetilde{v}}$. Therefore

$$\widetilde{E} = \|\widetilde{x}_{\widetilde{u}}\|^2 = E u_{\widetilde{u}}^2 + 2F u_{\widetilde{u}} v_{\widetilde{u}} + G v_{\widetilde{u}}^2$$
$$\widetilde{F} = \widetilde{x}_{\widetilde{u}} \circ \widetilde{x}_{\widetilde{v}} = E u_{\widetilde{u}} u_{\widetilde{v}} + F(u_{\widetilde{u}} v_{\widetilde{v}} + v_{\widetilde{u}} u_{\widetilde{v}}) + G v_{\widetilde{u}} v_{\widetilde{v}}$$
$$\widetilde{G} = \|\widetilde{x}_{\widetilde{v}}\|^2 = E u_{\widetilde{v}}^2 + 2F u_{\widetilde{v}} v_{\widetilde{v}} + G v_{\widetilde{v}}^2$$

These results can be checked by multiplying the three specified matrices on the right-hand side.

An alternate proof of the result $I(u, v) = \widetilde{I}(\widetilde{u}, \widetilde{v})$ is also possible by using the result of this problem. Premultiplication and post multiplication of the left-hand side matrix by $\begin{bmatrix} d\widetilde{u} & d\widetilde{v} \end{bmatrix}$ and $\begin{bmatrix} d\widetilde{u} & d\widetilde{v} \end{bmatrix}^T$ respectively yields $\widetilde{I}$.

Noting that $J \begin{bmatrix} d\widetilde{u} & d\widetilde{v} \end{bmatrix}^T = \begin{bmatrix} du & dv \end{bmatrix}^T$, and premultiplication and post multiplication of the right-hand side matrices by $\begin{bmatrix} d\widetilde{u} & d\widetilde{v} \end{bmatrix}$ and $\begin{bmatrix} d\widetilde{u} & d\widetilde{v} \end{bmatrix}^T$ respectively yields $I$. Therefore $\widetilde{I}(\widetilde{u}, \widetilde{v}) = I(u, v)$.

(b) Take determinants of both sides in the result of part (a), and then take the square roots.

(c) Use of the relationships $\widetilde{x}_{\widetilde{u}} = x_u u_{\widetilde{u}} + x_v v_{\widetilde{u}}$, and $\widetilde{x}_{\widetilde{v}} = x_u u_{\widetilde{v}} + x_v v_{\widetilde{v}}$ results in

$$\widetilde{x}_{\widetilde{u}}(\widetilde{u}, \widetilde{v}) \times \widetilde{x}_{\widetilde{v}}(\widetilde{u}, \widetilde{v}) = (\det J)\, x_u(u, v) \times x_v(u, v)$$

(d) Follows from part (c) of the problem.

11. The first fundamental coefficients are $E$, $F$, and $G$. Show that the area of a region on the surface which corresponds to a set of points $W \subseteq U$ is given by

$$A = \int\!\!\int_W \left\{ EG - F^2 \right\}^{1/2} du\, dv$$

The above expression is also independent of the parameterization.

Hint: See Lipschutz (1969), and Pressley (2010). Let $x : U \to S$ be a surface patch. Consider a small region $\Delta R$ on the surface patch. Also let $(u_0, v_0) \in U$, and $\Delta u$ and $\Delta v$ be small changes in $u$ and $v$ respectively. The corresponding changes in $x(u, v)$ are $x_u \Delta u$ and $x_v \Delta v$ respectively. The derivatives are evaluated at $(u_0, v_0)$.

The surface area bordered by the parameter curves $u = u_0, u = u_0 + \Delta u, v = v_0$, and $v = v_0 + \Delta v$ can be approximated by the area of a parallelogram specified by the vectors $x_u \Delta u$ and $x_v \Delta v$. If we assume, $\Delta u > 0$ and $\Delta v > 0$, the area of this parallelogram is

$$\|x_u \Delta u \times x_v \Delta v\| = \|x_u \times x_v\| \Delta u \Delta v = \left\{EG - F^2\right\}^{1/2} \Delta u \Delta v$$

Therefore, the area of the surface corresponding to the point in $W \subseteq U$ is $A$.

We next show that the area $A$ is independent of parameterization. Let the surface $x(u, v)$ be reparameterized as $\widetilde{x}(\widetilde{u}, \widetilde{v})$. The relationship between the corresponding first fundamental coefficients of the two parameterizations is given in the last problem. It is

$$\left\{\widetilde{E}\widetilde{G} - \widetilde{F}^2\right\}^{1/2} = \left\{EG - F^2\right\}^{1/2} |\det J|$$

where $\det J = \partial(u, v) / \partial(\widetilde{u}, \widetilde{v})$ is the Jacobian determinant. Let $\widetilde{W}$ be the image of $W$ in the $\widetilde{u}\widetilde{v}$-plane. Thus

$$\begin{aligned}
\widetilde{A} &= \iint_{\widetilde{W}} \left\{\widetilde{E}\widetilde{G} - \widetilde{F}^2\right\}^{1/2} d\widetilde{u}d\widetilde{v} \\
&= \iint_{W} \left\{EG - F^2\right\}^{1/2} |\det J| \left|\frac{\partial(\widetilde{u}, \widetilde{v})}{\partial(u, v)}\right| dudv \\
&= \iint_{W} \left\{EG - F^2\right\}^{1/2} dudv = A
\end{aligned}$$

12. Let $S$ and $\widetilde{S}$ be two surfaces, and $f : S \to \widetilde{S}$ be a diffeomorphism. Let $\xi_1$ and $\xi_2$ be two intersecting curves on the surface patch $x(\cdot, \cdot)$ on $S$. Also let $f(x(\cdot, \cdot)) = \widetilde{x}(\cdot, \cdot)$. The corresponding mapped intersecting curves on the surface patch $\widetilde{x}(\cdot, \cdot)$ on the surface $\widetilde{S}$ are $\widetilde{\xi}_1$ and $\widetilde{\xi}_2$ respectively.

Prove that the mapping $f$ is conformal if and only if the coefficients of the first fundamental forms of $x(\cdot, \cdot)$ and $\widetilde{x}(\cdot, \cdot)$ are proportional.

Hint: See Lipschutz (1969), and Pressley (2001). Without loss of generality, assume that the surfaces $S$ and $\widetilde{S}$ are covered by single patches $x(\cdot, \cdot)$ and $\widetilde{x}(\cdot, \cdot) = f(x(\cdot, \cdot))$ respectively. The two intersecting curves on the surface $S$ are: $\xi_1(t) = x(u_1(t), v_1(t))$, and $\xi_2(t) = x(u_2(t), v_2(t))$. The corresponding mapped intersecting curves on the surface $\widetilde{S}$ are: $\widetilde{\xi}_1(t) = \widetilde{x}(u_1(t), v_1(t))$, and $\widetilde{\xi}_2(t) = \widetilde{x}(u_2(t), v_2(t))$ respectively. Let the coefficients of the first fundamental forms of the two surfaces be $E, F, G$ and $\widetilde{E}, \widetilde{F}, \widetilde{G}$ respectively.

Initially assume that the coefficients of the first fundamental forms are proportional That is,

$$E = \lambda \widetilde{E}, \quad F = \lambda \widetilde{F}, \quad E = \lambda \widetilde{G}$$

Note that, as $E$ and $\widetilde{E}$ are both positive, $\lambda > 0$. Let the curves $\xi_1$ and $\xi_2$ intersect at point $P$ on the surface $S$ at an angle $\theta$. Similarly, let the curves $\widetilde{\xi}_1$ and $\widetilde{\xi}_2$ intersect at point $\widetilde{P}$ on the surface $\widetilde{S}$ at an angle $\widetilde{\theta}$. It is readily observed by substituting $E = \lambda \widetilde{E}$, $F = \lambda \widetilde{F}$, and $E = \lambda \widetilde{G}$

in the expression for $\cos\theta$ we obtain $\cos\theta = \cos\widetilde{\theta}$. This implies that $\theta = \widetilde{\theta}$. Consequently the mapping $f$ is conformal.

Conversely, assume that the mapping $f$ is conformal. That is, $\cos\theta = \cos\widetilde{\theta}$ for *all* pairs of intersecting curves $\xi_1(t) = x(u_1(t), v_1(t))$, and $\xi_2(t) = x(u_2(t), v_2(t))$ in $S$. Let $x: U \to \mathbb{R}^3$, where $U$ is an open subset of $\mathbb{R}^2$. For fixed values of $(a, b) \in U$, let

$$\xi_1(t) = x(a + t, b), \quad \text{and} \quad \xi_2(t) = x(a + t\cos\phi, b + t\sin\phi)$$

where $\phi$ is a constant. Observe that

$$u_1' = 1, \quad v_1' = 0, \quad u_2' = \cos\phi, \quad v_2' = \sin\phi$$

Substituting this in the equation $\cos\theta = \cos\widetilde{\theta}$, which is

$$\frac{Eu_1'u_2' + F(u_1'v_2' + u_2'v_1') + Gv_1'v_2'}{\left(Eu_1'^2 + 2Fu_1'v_1' + Gv_1'^2\right)^{1/2}\left(Eu_2'^2 + 2Fu_2'v_2' + Gv_2'^2\right)^{1/2}}$$
$$= \frac{\widetilde{E}u_1'u_2' + \widetilde{F}(u_1'v_2' + u_2'v_1') + \widetilde{G}v_1'v_2'}{\left(\widetilde{E}u_1'^2 + 2\widetilde{F}u_1'v_1' + \widetilde{G}v_1'^2\right)^{1/2}\left(\widetilde{E}u_2'^2 + 2\widetilde{F}u_2'v_2' + \widetilde{G}v_2'^2\right)^{1/2}}$$

It results in

$$\frac{E\cos\phi + F\sin\phi}{E^{1/2}\left(E\cos^2\phi + 2F\cos\phi\sin\phi + G\sin^2\phi\right)^{1/2}}$$
$$= \frac{\widetilde{E}\cos\phi + \widetilde{F}\sin\phi}{\widetilde{E}^{1/2}\left(\widetilde{E}\cos^2\phi + 2\widetilde{F}\cos\phi\sin\phi + \widetilde{G}\sin^2\phi\right)^{1/2}}$$

Square both sides of the above equation, and use of the relationship.

$$(E\cos\phi + F\sin\phi)^2$$
$$= E\left(E\cos^2\phi + 2F\cos\phi\sin\phi + G\sin^2\phi\right) - \left(EG - F^2\right)\sin^2\phi$$

yields

$$\left(EG - F^2\right)\widetilde{E}\left(\widetilde{E}\cos^2\phi + 2\widetilde{F}\cos\phi\sin\phi + \widetilde{G}\sin^2\phi\right)$$
$$= \left(\widetilde{E}\widetilde{G} - \widetilde{F}^2\right)E\left(E\cos^2\phi + 2F\cos\phi\sin\phi + G\sin^2\phi\right)$$

Substitution of

$$\lambda = \frac{\left(\widetilde{E}\widetilde{G} - \widetilde{F}^2\right)E}{(EG - F^2)\widetilde{E}} \neq 0$$

in the above equation results in

$$\left(\widetilde{E} - \lambda E\right)\cos^2\phi + 2\left(\widetilde{F} - \lambda F\right)\cos\phi\sin\phi + \left(\widetilde{G} - \lambda G\right)\sin^2\phi = 0$$

Substitution of $\phi = 0$ in the above equation yields $\widetilde{E} = \lambda E$. Similarly, substitution of $\phi = \pi/2$ in the above equation yields $\widetilde{G} = \lambda G$. Finally substituting these results in the above equation yields $\widetilde{F} = \lambda F$.

13. A surface $x\,(u, v)$ is reparameterized as $\widetilde{x}\,(\widetilde{u}, \widetilde{v})$. The corresponding second fundamental forms are

$$II = L\,du^2 + 2M\,du\,dv + N\,dv^2, \quad \text{and} \quad \widetilde{II} = \widetilde{L}\,d\widetilde{u}^2 + 2\widetilde{M}\,d\widetilde{u}\,d\widetilde{v} + \widetilde{N}\,d\widetilde{v}^2$$

respectively. Let $J$ be the Jacobian matrix which specifies the transformation from the $\widetilde{u}\widetilde{v}$-plane to $uv$-plane. It is

$$J = \begin{bmatrix} \partial u/\partial \widetilde{u} & \partial u/\partial \widetilde{v} \\ \partial v/\partial \widetilde{u} & \partial v/\partial \widetilde{v} \end{bmatrix} = \begin{bmatrix} u_{\widetilde{u}} & u_{\widetilde{v}} \\ v_{\widetilde{u}} & v_{\widetilde{v}} \end{bmatrix}$$

Show that

(a)

$$\begin{bmatrix} \widetilde{L} & \widetilde{M} \\ \widetilde{M} & \widetilde{N} \end{bmatrix} = \pm J^T \begin{bmatrix} L & M \\ M & N \end{bmatrix} J$$

The positive sign in the right-hand side of the above expression is taken if $\det J > 0$, and negative sign if $\det J < 0$.

(b) $II\,(u, v) = \pm\widetilde{II}\,(\widetilde{u}, \widetilde{v})$.

(c) As $\det J \neq 0$, we have

$$\left(\widetilde{L}\widetilde{N} - \widetilde{M}^2\right) = \pm \left|\frac{\partial\,(u, v)}{\partial\,(\widetilde{u}, \widetilde{v})}\right|^2 \left(LN - M^2\right)$$

Hint: See Pressley (2010).

(a) We have $\widetilde{x}_{\widetilde{u}} = x_u u_{\widetilde{u}} + x_v v_{\widetilde{u}}$, and $\widetilde{x}_{\widetilde{v}} = x_u u_{\widetilde{v}} + x_v v_{\widetilde{v}}$. Therefore

$$\widetilde{x}_{\widetilde{u}\widetilde{u}} - x_u u_{\widetilde{u}\widetilde{u}} + x_v v_{\widetilde{u}\widetilde{u}} + x_{uu}\,(u_{\widetilde{u}})^2 + 2x_{uv} u_{\widetilde{u}} v_{\widetilde{u}} + x_{vv}\,(v_{\widetilde{u}})^2$$

As $\widetilde{\mathcal{N}}\,(\widetilde{u}, \widetilde{v}) = sgn\,(\det J)\,\mathcal{N}\,(u, v)$ from an earlier problem, and using the fact $x_u \circ \mathcal{N}\,(u, v) = x_v \circ \mathcal{N}\,(u, v) = 0$ yields

$$\widetilde{L} = \widetilde{x}_{\widetilde{u}\widetilde{u}} \circ \widetilde{\mathcal{N}} = \pm \left(x_{uu}\,(u_{\widetilde{u}})^2 + 2x_{uv} u_{\widetilde{u}} v_{\widetilde{u}} + x_{vv}\,(v_{\widetilde{u}})^2\right) \circ \mathcal{N}$$
$$= \pm \left(L u_{\widetilde{u}}^2 + 2M u_{\widetilde{u}} v_{\widetilde{u}} + N v_{\widetilde{u}}^2\right)$$

Similar expressions for $\widetilde{M}$, and $\widetilde{N}$ can be obtained. Equivalent expressions are stated in the problem in matrix notation.

(b) Premultiplication and post multiplication of the left-hand side matrix by $\begin{bmatrix} d\widetilde{u} & d\widetilde{v} \end{bmatrix}$ and $\begin{bmatrix} d\widetilde{u} & d\widetilde{v} \end{bmatrix}^T$ respectively yields $\widetilde{II}$.

Noting that $J \begin{bmatrix} d\widetilde{u} & d\widetilde{v} \end{bmatrix}^T = \begin{bmatrix} du & dv \end{bmatrix}^T$, and premultiplication and post multiplication of the right-hand side matrices by $\begin{bmatrix} d\widetilde{u} & d\widetilde{v} \end{bmatrix}$ and $\begin{bmatrix} d\widetilde{u} & d\widetilde{v} \end{bmatrix}^T$ respectively yields $II$. Therefore $II\,(u, v) = \pm\widetilde{II}\,(\widetilde{u}, \widetilde{v})$.

(c) This result follows by taking determinant on both sides of the matrix equation.

14. Let $\xi\,(t) = x\,(u\,(t), v\,(t))$ be a parameterization of a curve $C$ on a regular surface, where $t$ is a parameterization which is not necessarily equal to the arc-length of the curve $C$. Show that the normal curvature $\kappa_n\,(t)$ of the curve $C$ at the point $t$ is equal to $II\,(u\,(t), v\,(t))\,/I\,(u\,(t), v\,(t))$ where $I\,(\cdot, \cdot)$ and $II\,(\cdot, \cdot)$ are the first and second fundamental forms of the surface at $t$.

Hint: See Pressley (2010). Let $s$ be the arc length along the curve $C$. From the theorem which assumes that the parameter of the curve is $s$, we have

$$\kappa_n = L\left(u'\right)^2 + 2Mu'v' + N\left(v'\right)^2$$

where the primes denote derivatives with respect to $s$. Then

$$
\begin{aligned}
\kappa_n &= L\left(u'\right)^2 + 2Mu'v' + N\left(v'\right)^2 \\
&= \frac{L\left(du/dt\right)^2 + 2M\left(du/dt\right)\left(dv/dt\right) + N\left(dv/dt\right)^2}{\left(ds/dt\right)^2} \\
&= \frac{II\left(u\left(t\right), v\left(t\right)\right)}{I\left(u\left(t\right), v\left(t\right)\right)}
\end{aligned}
$$

15. Show that the zeros of the quadratic equation in $\kappa_n$: $\det\left(\mathcal{F}_{II} - \kappa_n\mathcal{F}_I\right) = 0$ give the principal curvatures at the point $P$.

    Hint: See Struik (1988), Lipschutz (1969), and Pressley (2010). Use of the relationship $d\kappa_n/d\lambda = 0$ yields

    $$\left(E + 2F\lambda + G\lambda^2\right)\left(N\lambda + M\right) - \left(L + 2M\lambda + N\lambda^2\right)\left(G\lambda + F\right) = 0$$

    Write

    $$
    \begin{aligned}
    I &= \left(E + 2F\lambda + G\lambda^2\right) = \left(E + F\lambda\right) + \lambda\left(F + G\lambda\right) \\
    II &= \left(L + 2M\lambda + N\lambda^2\right) = \left(L + M\lambda\right) + \lambda\left(M + N\lambda\right)
    \end{aligned}
    $$

    Therefore

    $$\left(E + F\lambda\right)\left(M + N\lambda\right) = \left(L + M\lambda\right)\left(F + G\lambda\right)$$

    This implies

    $$
    \begin{aligned}
    \kappa_n &= \frac{II}{I} \\
    &= \frac{\left(M + N\lambda\right)}{\left(F + G\lambda\right)} = \frac{\left(L + M\lambda\right)}{\left(E + F\lambda\right)}
    \end{aligned}
    $$

    Note that $\lambda \triangleq dv/du$, and $dv/du$ is not the derivative of $v$ with respect $u$. As $\lambda \triangleq dv/du$ we have from the above ratios

    $$
    \begin{aligned}
    \left(L - \kappa_n E\right)du + \left(M - \kappa_n F\right)dv &= 0 \\
    \left(M - \kappa_n F\right)du + \left(N - \kappa_n G\right)dv &= 0
    \end{aligned}
    $$

    Let $\begin{bmatrix} du & dv \end{bmatrix}^T \triangleq \begin{bmatrix} \psi & \varrho \end{bmatrix}^T \triangleq \mathcal{T}$. In matrix notation, the above set of equations can be written as

    $$\left(\mathcal{F}_{II} - \kappa_n\mathcal{F}_I\right)\mathcal{T} = 0$$

    These equations are satisfied provided $\det\left(\mathcal{F}_{II} - \kappa_n\mathcal{F}_I\right) = 0$. The zeros of this equation give the principal curvatures at the point $P$. Furthermore, the corresponding vectors $\mathcal{T}$ can be used in determining the principal vectors.

16. Prove that the eigenvalues of the matrix $\mathcal{F}_I^{-1}\mathcal{F}_{II}$ are the principal curvatures.

    Hint: Matrix $\mathcal{F}_I$ is invertible, as $\det\mathcal{F}_I = \left(EG - F^2\right) > 0$.

    Therefore, $\det\left(\mathcal{F}_{II} - \kappa_n\mathcal{F}_I\right) = 0$ can be written as

    $$\det\left(\mathcal{F}_I\left(\mathcal{F}_I^{-1}\mathcal{F}_{II} - \kappa_n I_2\right)\right) = 0 \Rightarrow \det\mathcal{F}_I\det\left(\mathcal{F}_I^{-1}\mathcal{F}_{II} - \kappa_n I_2\right) = 0$$

where $I_2$ is the identity matrix of size 2. As $\det \mathcal{F}_I > 0$, we have

$$\det \left( \mathcal{F}_I^{-1} \mathcal{F}_{II} - \kappa_n I_2 \right) = 0$$

That is, the principal curvatures are the eigenvalues of the matrix $\mathcal{F}_I^{-1} \mathcal{F}_{II}$. The corresponding eigenvector satisfies $\left( \mathcal{F}_I^{-1} \mathcal{F}_{II} - \kappa_n I_2 \right) \mathcal{T} = 0$, which in turn implies $\mathcal{F}_I^{-1} \left( \mathcal{F}_{II} - \kappa_n \mathcal{F}_I \right) \mathcal{T} = 0$. Consequently $\left( \mathcal{F}_{II} - \kappa_n \mathcal{F}_I \right) \mathcal{T} = 0$.

17. Prove that the quadratic equation for $\kappa_n$ can be written explicitly as

$$\kappa_n^2 - 2H\kappa_n + K = 0$$

where

$$\kappa_1, \kappa_2 = H \pm \left( H^2 - K \right)^{1/2}$$

$$K = \kappa_1 \kappa_2 = \frac{\left( LN - M^2 \right)}{\left( EG - F^2 \right)} = \frac{\det \mathcal{F}_{II}}{\det \mathcal{F}_I}$$

$$H = \frac{1}{2} \left( \kappa_1 + \kappa_2 \right) = \frac{EN + GL - 2FM}{2 \left( EG - F^2 \right)}$$

Hint: Use the result $\det \left( \mathcal{F}_{II} - \kappa_n \mathcal{F}_I \right) = 0$.

18. Prove that the principal curvatures $\kappa_1$ and $\kappa_2$ are real numbers.

Hint: Note that $\mathcal{F}_I$ and $\mathcal{F}_{II}$ are real-valued symmetric matrices. Therefore $\mathcal{F}_I^{-1}$ is a symmetric matrix. This in turn implies that $\mathcal{F}_I^{-1} \mathcal{F}_{II}$ is a real-valued symmetric matrix. It is known that the eigenvalues of a real-valued symmetric matrix are all real-valued.

19. Prove that if the principal curvatures $\kappa_1 = \kappa_2 = \kappa$ then $\mathcal{F}_{II} = \kappa \mathcal{F}_I$. Also show that any vector $\begin{bmatrix} \psi & \varrho \end{bmatrix}^T \triangleq \mathcal{T}$ satisfies the equation

$$\left( \mathcal{F}_{II} - \kappa \mathcal{F}_I \right) \mathcal{T} = 0$$

This in turn implies that any tangent vector $\mathcal{T}$ at the point $P$ is a principal vector.

Hint: See Pressley (2001). The principal curvatures are the eigenvalues of the real-valued-symmetric matrix $\mathcal{F}_I^{-1} \mathcal{F}_{II}$. Therefore $A^T \mathcal{F}_I^{-1} \mathcal{F}_{II} A = \Lambda$, where $A$ is an orthonormal matrix, and $\Lambda$ is a diagonal matrix of size 2, with the eigenvalues on its main diagonal. Let $I_2$ be an identity matrix of size 2. Therefore we have $\left( A^T \mathcal{F}_I^{-1} \mathcal{F}_{II} A - \kappa A^T A \right) = 0$. This implies $A^T \left( \mathcal{F}_I^{-1} \mathcal{F}_{II} - \kappa I_2 \right) A = 0$. Therefore $A^T \mathcal{F}_I^{-1} \left( \mathcal{F}_{II} - \kappa \mathcal{F}_I \right) A = 0$. This in turn implies $\left( A^T \mathcal{F}_I^{-1} A \right) \left\{ A^T \left( \mathcal{F}_{II} - \kappa \mathcal{F}_I \right) A \right\} = 0$. Note that $A^T \mathcal{F}_I^{-1} A \neq 0$, as $\det \mathcal{F}_I = \left( EG - F^2 \right) > 0$. Therefore $A^T \left( \mathcal{F}_{II} - \kappa \mathcal{F}_I \right) A = 0$. This is possible if and only if $\left( \mathcal{F}_{II} - \kappa \mathcal{F}_I \right) = 0$. As $\left( \mathcal{F}_{II} - \kappa \mathcal{F}_I \right) = 0$, we have for any vector $\mathcal{T}$

$$\left( \mathcal{F}_{II} - \kappa \mathcal{F}_I \right) \mathcal{T} = 0$$

Therefore any tangent vector $\mathcal{T}$ at point $P$ is a principal vector.

20. Prove that if $\kappa_1 \neq \kappa_2$ then any two nonzero principal vectors which correspond to $\kappa_1$ and $\kappa_2$ are orthogonal to each other.

Hint: See Pressley (2001). We have

$$\begin{bmatrix} \psi_i & \varrho_i \end{bmatrix}^T \triangleq \mathcal{T}_i$$
$$\left( \mathcal{F}_{II} - \kappa_i \mathcal{F}_I \right) \mathcal{T}_i = 0$$
$$\mathcal{T}_i = \psi_i x_u + \varrho_i x_v$$

for $i = 1, 2$. Thus

$$T_1 \circ T_2 = T_1^T \mathcal{F}_I T_2$$

Also

$$\mathcal{F}_{II} T_1 = \kappa_1 \mathcal{F}_I T_1, \quad \text{and} \quad \mathcal{F}_{II} T_2 = \kappa_2 \mathcal{F}_I T_2$$

Therefore

$$T_2^T \mathcal{F}_{II} T_1 = \kappa_1 T_2^T \mathcal{F}_I T_1 = \kappa_1 (T_1 \circ T_2)$$
$$T_1^T \mathcal{F}_{II} T_2 = \kappa_2 T_1^T \mathcal{F}_I T_2 = \kappa_2 (T_1 \circ T_2)$$

As the matrix $\mathcal{F}_{II}$ is symmetric, we have $T_2^T \mathcal{F}_{II} T_1 = T_1^T \mathcal{F}_{II} T_2$. Therefore

$$\kappa_1 (T_1 \circ T_2) = \kappa_2 (T_1 \circ T_2)$$

As $\kappa_1 \neq \kappa_2$, we have $T_1 \circ T_2 = 0$.

21. Prove that the principal vectors at a point $P$ can be determined from $\det A = 0$, where

$$A = \begin{bmatrix} \lambda^2 & -\lambda & 1 \\ E & F & G \\ L & M & N \end{bmatrix}$$

Hint: See Struik (1988), Lipschutz (1969), and Pressley (2010). It is known that

$$(L - \kappa_n E)\, du + (M - \kappa_n F)\, dv = 0$$
$$(M - \kappa_n F)\, du + (N - \kappa_n G)\, dv = 0$$

The above set of equations can be written as

$$\begin{bmatrix} L\,du + M\,dv & E\,du + F\,dv \\ M\,du + N\,dv & F\,du + G\,dv \end{bmatrix} \begin{bmatrix} 1 \\ -\kappa_n \end{bmatrix} = \begin{bmatrix} 0 \\ 0 \end{bmatrix}$$

The above set of equations are true if and only if the determinant of the $2 \times 2$ matrix on the left-hand side is equal to zero. Expansion of the determinant results in

$$(EM - LF)\, du^2 + (EN - LG)\, du\,dv + (FN - MG)\, dv^2 = 0$$

As $dv/du \triangleq \lambda$, the above result can be stated compactly as a determinant (as specified in the problem statement).

22. Establish Euler's theorem.

Hint: See Struik (1988), Lipschutz (1969), and Pressley (2001).

Let $T = \xi' = (x_u u' + x_v v')$ be a tangent vector of the curve $C$. The length of the vector $T$ is unity. Also let $\begin{bmatrix} u' & v' \end{bmatrix}^T = \mathcal{T}$. It follows that

$$\kappa_n = \mathcal{T}^T \mathcal{F}_{II} \mathcal{T}$$

The vector $\mathcal{T}$ also satisfies $(\mathcal{F}_{II} - \kappa \mathcal{F}_I)\, \mathcal{T} = 0$. Two tangent vectors $T_1$ and $T_2$ at a point on the surface $x(\cdot, \cdot)$ also satisfy $T_1 \circ T_2 = T_1^T \mathcal{F}_I T_2$. Also assume that the tangent vectors $T_1$ and $T_2$ are of unit length.

We consider two cases.

*Case* 1: Let $\kappa_1 = \kappa_2 = \kappa$, and $T_1 = T_2 = T$. Therefore

$$\kappa_n = T^T \mathcal{F}_{II} T = \kappa T^T \mathcal{F}_I T = \kappa T \circ T = \kappa$$

This result satisfies the problem statement.

*Case* 2: Let $\kappa_1 \neq \kappa_2$. In this case, the two principal vectors $T_1$ and $T_2$ are orthogonal to each other. Let

$$\begin{bmatrix} u_i' & v_i' \end{bmatrix}^T \triangleq T_i, \quad \text{and} \quad T_i = x_u u_i' + x_v v_i', \quad \text{for } i = 1, 2$$

Also

$$\xi' = \cos\theta T_1 + \sin\theta T_2$$

Furthermore, $T = \xi' = (x_u u' + x_v v')$ implies

$$(x_u u' + x_v v') = \cos\theta (x_u u_1' + x_v v_1') + \sin\theta (x_u u_2' + x_v v_2')$$

This implies

$$\begin{bmatrix} u' & v' \end{bmatrix}^T = \cos\theta \begin{bmatrix} u_1' & v_1' \end{bmatrix}^T + \sin\theta \begin{bmatrix} u_2' & v_2' \end{bmatrix}^T$$
$$T = \cos\theta T_1 + \sin\theta T_2$$

Therefore

$$\begin{aligned}
\kappa_n &= T^T \mathcal{F}_{II} T \\
&= \left(\cos\theta T_1^T + \sin\theta T_2^T\right) \mathcal{F}_{II} \left(\cos\theta T_1 + \sin\theta T_2\right) \\
&= \cos^2\theta T_1^T \mathcal{F}_{II} T_1 + \cos\theta \sin\theta \left(T_1^T \mathcal{F}_{II} T_2 + T_2^T \mathcal{F}_{II} T_1\right) + \sin^2\theta T_2^T \mathcal{F}_{II} T_2
\end{aligned}$$

However $\kappa_1$ and $\kappa_2$ satisfy $(\mathcal{F}_{II} - \kappa \mathcal{F}_I) T = 0$. Therefore

$$\begin{aligned}
\kappa_n &= \kappa_1 \cos^2\theta T_1^T \mathcal{F}_I T_1 \\
&\quad + \cos\theta \sin\theta \left(\kappa_2 T_1^T \mathcal{F}_I T_2 + \kappa_1 T_2^T \mathcal{F}_I T_1\right) + \kappa_2 \sin^2\theta T_2^T \mathcal{F}_I T_2
\end{aligned}$$

Use of the result $T_1 \circ T_2 = T_1^T \mathcal{F}_I T_2$ yields

$$\begin{aligned}
\kappa_n &= \kappa_1 \cos^2\theta T_1 \circ T_1 \\
&\quad + \cos\theta \sin\theta \left(\kappa_2 T_1 \circ T_2 + \kappa_1 T_2 \circ T_1\right) + \kappa_2 \sin^2\theta T_2 \circ T_2 \\
&= \kappa_1 \cos^2\theta + \kappa_2 \sin^2\theta
\end{aligned}$$

23. Show that the Gaussian curvature of a surface described by $z = f(x, y)$ is

$$K = \frac{z_{xx} z_{yy} - z_{xy}^2}{\left(z_x^2 + z_y^2 + 1\right)^2}$$

where $z_x$, and $z_y$ are the first partial derivatives of $z$ with respect $x$ and $y$ respectively. Also $z_{xx}$, and $z_{yy}$ are the second partial derivatives of $z$ with respect $x$ and $y$ respectively; and $z_{xy}$ is the partial derivative of $z$ with respect to $x$ and $y$.

Hint: We describe the surface by $X(x, y) = (x, y, z)$ where $z = f(x, y)$. We have

$$X_x = (1, 0, z_x), \quad X_y = (0, 1, z_y)$$

$$X_{xx} = (0, 0, z_{xx}), \quad X_{xy} = (0, 0, z_{xy}), \quad X_{yy} = (0, 0, z_{yy})$$

The coefficients of the first fundamental form are

$$E = \|X_x\|^2 = \left(1 + z_x^2\right), \quad F = X_x \circ X_y = z_x z_y, \quad G = \|X_y\|^2 = \left(1 + z_y^2\right)$$

The unit normal at the surface is

$$\mathcal{N} = \frac{X_x \times X_y}{\|X_x \times X_y\|} = \frac{(-z_x, -z_y, 1)}{\left(z_x^2 + z_y^2 + 1\right)^{1/2}}$$

The second fundamental coefficients are

$$L = X_{xx} \circ \mathcal{N} = \frac{z_{xx}}{\left(z_x^2 + z_y^2 + 1\right)^{1/2}}$$

$$M = X_{xy} \circ \mathcal{N} = \frac{z_{xy}}{\left(z_x^2 + z_y^2 + 1\right)^{1/2}}$$

$$N = X_{yy} \circ \mathcal{N} = \frac{z_{yy}}{\left(z_x^2 + z_y^2 + 1\right)^{1/2}}$$

Therefore the Gaussian curvature $K$ is

$$
\begin{aligned}
K &= \frac{\left(LN - M^2\right)}{\left(EG - F^2\right)} \\
&= \frac{z_{xx} z_{yy} - z_{xy}^2}{\left(z_x^2 + z_y^2 + 1\right)} \frac{1}{\left\{\left(z_x^2 + 1\right)\left(z_y^2 + 1\right) - z_x^2 z_y^2\right\}} \\
&= \frac{z_{xx} z_{yy} - z_{xy}^2}{\left(z_x^2 + z_y^2 + 1\right)^2}
\end{aligned}
$$

24. Prove Gauss' theorema egrigium.
    Hint: See Struik (1988), Lipschutz (1969), and Pogorelov (1987). We have

$$K = \frac{\left(LN - M^2\right)}{\left(EG - F^2\right)}$$

where $E = x_u \circ x_u$, $F = x_u \circ x_v$, $G = x_v \circ x_v$; $\mathcal{N} = x_u \times x_v / \left(EG - F^2\right)^{1/2}$; and $L = x_{uu} \circ \mathcal{N}$, $M = x_{uv} \circ \mathcal{N}$, $N = x_{vv} \circ \mathcal{N}$. Therefore

$$\left(EG - F^2\right)^{1/2} L = x_{uu} \circ x_u \times x_v = [x_{uu}\ x_u\ x_v]$$

$$\left(EG - F^2\right)^{1/2} M = x_{uv} \circ x_u \times x_v = [x_{uv}\ x_u\ x_v]$$

$$\left(EG - F^2\right)^{1/2} N = x_{vv} \circ x_u \times x_v = [x_{vv}\ x_u\ x_v]$$

The final result is established in several steps.
*Step* 1: We write

$$\left(EG - F^2\right) LN = [x_{uu}\ x_u\ x_v][x_{vv}\ x_u\ x_v]$$

Using the determinant representation of $[x_{uu}\ x_u\ x_v]$ and $[x_{vv}\ x_u\ x_v]$ we obtain

$$\left(EG - F^2\right) LN = \begin{vmatrix} x_{uu} \circ x_{vv} & x_u \circ x_{vv} & x_v \circ x_{vv} \\ x_{uu} \circ x_u & x_u \circ x_u & x_v \circ x_u \\ x_{uu} \circ x_v & x_u \circ x_v & x_v \circ x_v \end{vmatrix}$$

It can similarly be shown that

$$\left(EG - F^2\right) M^2 = \begin{vmatrix} x_{uv} \circ x_{uv} & x_u \circ x_{uv} & x_v \circ x_{uv} \\ x_{uv} \circ x_u & x_u \circ x_u & x_v \circ x_u \\ x_{uv} \circ x_v & x_u \circ x_v & x_v \circ x_v \end{vmatrix}$$

*Step* 2: Some of the terms in the above two determinants are evaluated in this step

$$x_{uu} \circ x_u = \frac{1}{2} \left(x_u \circ x_u\right)_u = \frac{1}{2} E_u, \quad x_{uv} \circ x_u = \frac{1}{2} \left(x_u \circ x_u\right)_v = \frac{1}{2} E_v$$

$$x_{vv} \circ x_v = \frac{1}{2} \left(x_v \circ x_v\right)_v = \frac{1}{2} G_v, \quad x_{uv} \circ x_v = \frac{1}{2} \left(x_v \circ x_v\right)_u = \frac{1}{2} G_u$$

Also

$$F_u = \left(x_u \circ x_v\right)_u = x_{uu} \circ x_v + x_u \circ x_{uv} = x_{uu} \circ x_v + \frac{1}{2} E_v$$

$$F_v = \left(x_u \circ x_v\right)_v = x_{uv} \circ x_v + x_u \circ x_{vv} = \frac{1}{2} G_u + x_u \circ x_{vv}$$

These imply

$$x_{uu} \circ x_v = F_u - \frac{1}{2} E_v, \quad x_u \circ x_{vv} = F_v - \frac{1}{2} G_u$$

*Step* 3: Using results from Steps 1 and 2, we obtain

$$\left(EG - F^2\right) \left(LN - M^2\right)$$

$$= \begin{vmatrix} x_{uu} \circ x_{vv} & F_v - \frac{1}{2} G_u & \frac{1}{2} G_v \\ \frac{1}{2} E_u & E & F \\ F_u - \frac{1}{2} E_v & F & G \end{vmatrix} - \begin{vmatrix} x_{uv} \circ x_{uv} & \frac{1}{2} E_v & \frac{1}{2} G_u \\ \frac{1}{2} E_v & E & F \\ \frac{1}{2} G_u & F & G \end{vmatrix}$$

The above result can also be written as

$$\left(EG - F^2\right) \left(LN - M^2\right)$$
$$= \left(x_{uu} \circ x_{vv} - x_{uv} \circ x_{uv}\right) \left(EG - F^2\right)$$

$$+ \begin{vmatrix} 0 & F_v - \frac{1}{2} G_u & \frac{1}{2} G_v \\ \frac{1}{2} E_u & E & F \\ F_u - \frac{1}{2} E_v & F & G \end{vmatrix} - \begin{vmatrix} 0 & \frac{1}{2} E_v & \frac{1}{2} G_u \\ \frac{1}{2} E_v & E & F \\ \frac{1}{2} G_u & F & G \end{vmatrix}$$

*Step* 4: The expression $\left(x_{uu} \circ x_{vv} - x_{uv} \circ x_{uv}\right)$ is evaluated. Using results from Step 2, we have

$$\left(F_u - \frac{1}{2} E_v\right)_v = \left(x_{uu} \circ x_v\right)_v = x_{uuv} \circ x_v + x_{uu} \circ x_{vv}$$

$$\left(\frac{1}{2} G_u\right)_u = \left(x_{uv} \circ x_v\right)_u = x_{uvu} \circ x_v + x_{uv} \circ x_{uv}$$

Therefore

$$(x_{uu} \circ x_{vv} - x_{uv} \circ x_{uv}) = \left(F_u - \frac{1}{2}E_v\right)_v - \left(\frac{1}{2}G_u\right)_u$$

$$= F_{uv} - \frac{1}{2}E_{vv} - \frac{1}{2}G_{uu}$$

25. Let $S$ and $\widetilde{S}$ be two surfaces, and $f : S \to \widetilde{S}$ be a smooth bijective mapping. Let the coefficients of the first fundamental form of the two surfaces be $E, F, G$ and $\widetilde{E}, \widetilde{F}, \widetilde{G}$ respectively. Prove that the mapping $f$ is local isometry if and only if $E = \widetilde{E}, F = \widetilde{F}$, and $G = \widetilde{G}$.

Hint: See Pressley (2001). The length of a curve can be computed as the sum of all the lengths of curves, each of which lies on a single surface patch. Therefore, without any loss of generality we can assume that $S$ and $\widetilde{S}$ are each single surface patches. As the mapping $f$ is smooth, the surface patches are of the form $x : U \to \mathbb{R}^3$ and $\widetilde{x} = f(x(\cdot, \cdot))$, where $U \subseteq \mathbb{R}^2$ is an open set. Assume that coefficients of the first fundamental form of the two surfaces $S$ and $\widetilde{S}$: $E, F, G$ and $\widetilde{E}, \widetilde{F}, \widetilde{G}$ are equal respectively. Let $\xi(t) = x(u(t), v(t))$ and $\widetilde{\xi}(t) = \widetilde{x}(u(t), v(t))$ be the corresponding curves on the surfaces $S$ and $\widetilde{S}$ respectively. Then we have

$$f(\xi(t)) = f(x(u(t), v(t))) = \widetilde{x}(u(t), v(t)) = \widetilde{\xi}(t)$$

Therefore $f$ maps the curve $\xi(\cdot)$ to $\widetilde{\xi}(\cdot)$. As the coefficients of the first fundamental form of the two surfaces are equal by assumption, lengths of corresponding curves in the two spaces are equal, as arc-lengths in each case are found by integrating $\left(Eu'^2 + 2Fu'v' + Gv'^2\right)^{1/2}$ and $\left(\widetilde{E}u'^2 + 2\widetilde{F}u'v' + \widetilde{G}v'^2\right)^{1/2}$ respectively with appropriate limits. The primes in these expressions denote derivatives with respect to $t$. These two integrands are equal by assumption. Therefore the arc-lengths are equal.

We next establish the theorem in the opposite direction. Assume that the mapping $f$ is a local isometry. This implies that the lengths of the curves $\xi(t) = x(u(t), v(t))$ and $\widetilde{\xi}(t) = \widetilde{x}(u(t), v(t))$ over some interval $t \in (\alpha, \beta)$ are equal. That is,

$$\int_{t_0}^{t_1} \left(Eu'^2 + 2Fu'v' + Gv'^2\right)^{1/2} dt = \int_{t_0}^{t_1} \left(\widetilde{E}u'^2 + 2\widetilde{F}u'v' + \widetilde{G}v'^2\right)^{1/2} dt$$

where $t_0 < t_1$, for all $t_0, t_1 \in (\alpha, \beta)$. This in turn implies that

$$\left(Eu'^2 + 2Fu'v' + Gv'^2\right) = \left(\widetilde{E}u'^2 + 2\widetilde{F}u'v' + \widetilde{G}v'^2\right)$$

For the parameters $(u(t), v(t))$ in $U$, where $t \in (\alpha, \beta)$, we have the following three cases.

(i) Keep $v = v_0$ fixed, and vary $u$. The above equation yields $E = \widetilde{E}$.
(ii) Keep $u = u_0$ fixed, and vary $v$. The above equation yields $G = \widetilde{G}$.
(iii) Using results (i) and (ii), the above equation yields $F = \widetilde{F}$.

26. A surface $S$ is described by

$$x(\theta, u) = (\cos\theta \cosh u, \sin\theta \cosh u, u), \quad \theta \in (0, 2\pi), \ u \in \mathbb{R}$$

Another surface $\widetilde{S}$ is described by

$$\widetilde{x}(\phi, v) = (\cos\phi \sinh v, \sin\phi \sinh v, \phi), \quad \phi \in (0, 2\pi), \ v \in \mathbb{R}$$

Find the local isometric mapping between the surfaces $S$ and $\widetilde{S}$.
Hint: See Lipschutz (1969). For the surface $S$ we have

$$x_\theta = (-\sin\theta\cosh u, \cos\theta\cosh u, 0)$$
$$x_u = (\cos\theta\sinh u, \sin\theta\sinh u, 1)$$

$$E = \cosh^2 u, \quad F = 0, \quad G = \cosh^2 u$$

Similarly, for surface $\widetilde{S}$ we have

$$\widetilde{x}_\phi = (-\sin\phi\sinh v, \cos\phi\sinh v, 1)$$
$$\widetilde{x}_v = (\cos\phi\cosh v, \sin\phi\cosh v, 0)$$

$$\widetilde{E} = \cosh^2 v, \quad \widetilde{F} = 0, \quad \widetilde{G} = \cosh^2 v$$

In order to find local isometric mapping, we should have $E = \widetilde{E}$, $F = \widetilde{F}$, and $G = \widetilde{G}$. This is possible, if and only if $u = v$ for all values of $u, v \in \mathbb{R}$.

27. Prove the theorem about geodesic differential equations.
    Hint: See Pressley (2010). In proving the theorem, the following results are used.

    $$E_u = (x_u \circ x_u)_u = 2x_u \circ x_{uu}$$
    $$G_u = (x_v \circ x_v)_u = 2x_v \circ x_{uv}$$
    $$F_u = (x_u \circ x_v)_u = x_{uu} \circ x_v + x_u \circ x_{uv}$$

Recall that $\{x_u, x_v\}$ form a basis of the tangent plane of $x(\cdot, \cdot)$. Furthermore, the curve $C$ specified by $\xi$ is a geodesic if and only if $\xi''$ is orthogonal to both $x_u$ and $x_v$. This implies

$$\xi'' \circ x_u = \xi'' \circ x_v = 0$$

As $\xi' = (u'x_u + v'x_v)$, we have

$$\left\{\frac{d}{dt}(u'x_u + v'x_v)\right\} \circ x_u = 0$$

$$\left\{\frac{d}{dt}(u'x_u + v'x_v)\right\} \circ x_v = 0$$

Also

$$0 = \left\{\frac{d}{dt}(u'x_u + v'x_v)\right\} \circ x_u$$
$$= \frac{d}{dt}\{(u'x_u + v'x_v) \circ x_u\} - (u'x_u + v'x_v) \circ \frac{dx_u}{dt}$$
$$= \frac{d}{dt}(Eu' + Fv') - (u'x_u + v'x_v) \circ (u'x_{uu} + v'x_{uv})$$
$$= \frac{d}{dt}(Eu' + Fv')$$
$$\quad - \{u'^2(x_u \circ x_{uu}) + u'v'(x_u \circ x_{uv} + x_v \circ x_{uu}) + v'^2(x_v \circ x_{uv})\}$$
$$= \frac{d}{dt}(Eu' + Fv') - \frac{1}{2}(E_u u'^2 + 2F_u u'v' + G_u v'^2)$$

This yields the first geodesic equation. The second geodesic equation is derived similarly.

28. Prove that the parameter that satisfies the geodesic equations is directly proportional to the arc-length.

    Hint: See Pressley (2001). Consider

$$\left(Eu'^2 + 2Fu'v' + Gv'^2\right)'$$
$$= (E_u u' + E_v v') u'^2 + 2Eu'u'' + 2(F_u u' + F_v v') u'v' + 2F (u''v' + u'v'')$$
$$+ (G_u u' + G_v v') v'^2 + 2Gv'v''$$
$$= \left(E_u u'^2 + 2F_u u'v' + G_u v'^2\right) u' + \left(E_v u'^2 + 2F_v u'v' + G_v v'^2\right) v'$$
$$+2 \left\{ Eu'u'' + F (u''v' + u'v'') + Gv'v'' \right\}$$

    Use of the geodesic equations and the above result yields

$$\left(Eu'^2 + 2Fu'v' + Gv'^2\right)'$$
$$= 2 (Eu' + Fv')' u' + 2(Fu' + Gv')' v'$$
$$+2 (Eu' + Fv') u'' + 2(Fu' + Gv') v''$$
$$= 2 \left\{ (Eu' + Fv') u' \right\}' + 2 \left\{ (Fu' + Gv') v' \right\}'$$
$$= 2 \left(Eu'^2 + 2Fu'v' + Gv'^2\right)'$$

    Therefore

$$\left(Eu'^2 + 2Fu'v' + Gv'^2\right)' = 0$$

    This implies that

$$\left\|\xi'\right\|^2 = \left(Eu'^2 + 2Fu'v' + Gv'^2\right)$$

    is a constant.

29. Prove that the geodesic curvature $\kappa_g$ of a curve $C$ on a surface $S$ depends only upon the coefficients of the first fundamental form, and its partial derivatives.

    Hint: See Pressley (2001). The geodesic curvature $\kappa_g(s)$

$$\kappa_g(s) = \xi''(s) \circ \left(\mathcal{N}(s) \times \widetilde{T}(s)\right)$$

    where

$$\mathcal{N}(s) = \frac{x_u(s) \times x_v(s)}{\|x_u(s) \times x_v(s)\|}, \quad \text{and} \quad \widetilde{T}(s) = \xi'(s)$$

    In rest of the steps, we drop the functional dependence upon the distance parameter $s$ for clarity. Note that $\xi' = (u'x_u + v'x_v)$. Therefore

$$\mathcal{N} \times x_u = \frac{(x_u \times x_v) \times x_u}{\|x_u \times x_v\|} = \frac{(x_u \circ x_u) x_v - (x_v \circ x_u) x_u}{(EG - F^2)^{1/2}} = \frac{Ex_v - Fx_u}{(EG - F^2)^{1/2}}$$

$$\mathcal{N} \times x_v = \frac{(x_u \times x_v) \times x_v}{\|x_u \times x_v\|} = \frac{(x_u \circ x_v) x_v - (x_v \circ x_v) x_u}{(EG - F^2)^{1/2}} = \frac{Fx_v - Gx_u}{(EG - F^2)^{1/2}}$$

$$\mathcal{N} \times \xi' = \frac{u' (Ex_v - Fx_u) + v' (Fx_v - Gx_u)}{(EG - F^2)^{1/2}}$$

$$\xi'' = u''x_u + v''x_v + u'^2 x_{uu} + 2u'v'x_{uv} + v'^2 x_{vv}$$

    Also

$$E_u = (x_u \circ x_u)_u = 2x_u \circ x_{uu}$$
$$G_u = (x_v \circ x_v)_u = 2x_v \circ x_{uv}$$
$$F_u = (x_u \circ x_v)_u = x_{uu} \circ x_v + x_u \circ x_{uv}$$
$$E_v = (x_u \circ x_u)_v = 2x_u \circ x_{uv}$$
$$G_v = (x_v \circ x_v)_v = 2x_v \circ x_{vv}$$
$$F_v = (x_u \circ x_v)_v = x_{uv} \circ x_v + x_u \circ x_{vv}$$

Substituting the above expressions in $\kappa_g = \xi'' \circ \mathcal{N} \times \xi'$ yields an expression for $\kappa_g$ in terms of the coefficients of the first fundamental form, and its partial derivatives.

30. Prove that a local isometric mapping between two surfaces, maps the geodesic of one surface to the geodesic surface of the other surface.

    Hint: See Pressley (2001). Consider two surfaces $S$ and $\widetilde{S}$, where $f : S \to \widetilde{S}$ is a local isometric mapping. Further, let $\xi$ be a geodesic on $S$. Also let $x(\cdot, \cdot)$ be a surface patch in $S$ such that $\xi(t) = x(u(t), v(t))$. The parameters $u$ and $v$ satisfy the geodesic equations, where $E$, $F$, and $G$ are coefficients of the first fundamental form of the surface patch $x(\cdot, \cdot)$. As the map $f$ is a local isometry, $f(x(\cdot, \cdot))$ is a surface patch in $\widetilde{S}$, and coefficients of the first fundamental form of the space $\widetilde{x}(\cdot, \cdot)$ are identical to those of the space $x(\cdot, \cdot)$. This implies that $\widetilde{\xi}(\cdot) = f(\xi(\cdot))$ is a geodesic on $\widetilde{S}$.

31. Assume that the parametric curves which specify the surface $S$ form an orthogonal system. The $u$-curves are geodesics if and only if $E$ is a function of only $u$. Similarly, the $v$-curves are geodesics if and only if $G$ is a function of only $v$. Note that on the $u$-curve, the parameter $v$ is constant; and on the $v$-curve, the parameter $u$ is constant. Establish this assertion.

    Hint: See Graustein (1966). The parametric curves are orthogonal. This implies $F = 0$. On the $u$-curves, the parameter $v$ is constant. Use of the second geodesic equation yields $E_v = 0$. This implies that $E$ is a function of the parameter $u$ only. It can similarly be shown by using the $v$-curves that $G$ is a function of $v$ only.

32. Prove that the geodesic curvature $\kappa_g$ on a surface $S$ along a curve $v$ — constant, and $F = 0$ is $-E_v / \{2E\sqrt{G}\}$.

    Hint: See Lipschutz (1969). The geodesic curvature is

$$\kappa_g = \left[ \xi'(s)\ \xi''(s)\ \mathcal{N}(s) \right]$$

As $v = $ constant, we have

$$\xi' = x_u \frac{du}{ds}, \quad \xi'' = x_{uu} \left( \frac{du}{ds} \right)^2 + x_u \frac{d^2u}{ds^2}$$

$$ds^2 = Edu^2 + 2Fdudv + Gdv^2 = Edu^2 \Rightarrow \frac{du}{ds} = E^{-1/2}$$

Therefore

$$\kappa_g = \left[ \xi'\ \xi''\ \mathcal{N} \right] = \left[ x_u\ x_{uu}\ \mathcal{N} \right] \left( \frac{du}{ds} \right)^3$$

As $x_u, x_v$, and $\mathcal{N}$ are linearly independent, we can write

$$x_{uu} = ax_u + bx_v + c\mathcal{N}$$

where $a, b$, and $c$ are coefficients. Therefore

$$\kappa_g = b \left[ x_u \ x_v \ \mathcal{N} \right] \left( \frac{du}{ds} \right)^3$$

Note that

$$[x_u \ x_v \ \mathcal{N}] = [\mathcal{N} \ x_u \ x_v] = \frac{(x_u \times x_v) \circ (x_u \times x_v)}{\|x_u \times x_v\|} = \|x_u \times x_v\|$$

$$= \left( EG - F^2 \right)^{1/2} = (EG)^{1/2}$$

Therefore

$$\kappa_g = b \, (EG)^{1/2} \, E^{-3/2} = b \frac{G^{1/2}}{E}$$

It remains to determine $b$. We have

$$E_v = 2 x_u \circ x_{uv}$$

$$F_u = (x_u \circ x_v)_u = x_{uu} \circ x_v + x_u \circ x_{uv}$$

$$= (a x_u + b x_v + c \mathcal{N}) \circ x_v + \frac{1}{2} E_v$$

$$= a F + b G + \frac{1}{2} E_v$$

As $F = 0$, we have $b = -E_v/(2G)$. Therefore

$$\kappa_g = - \frac{E_v}{2 E G^{1/2}}$$

33. Prove that a geodesic is an arc of minimum length.

   Hint: See Kreyszig (1968). Let $C$ be a geodesic on a surface $S$, which connects points $P_1$ and $P_2$ on it. Also, let $S$ be specified by $x \, (u, v)$; where $u, v$ are geodesic parallel coordinates in the neighborhood of $C$ such that $v = v_0 = $ constant. Next consider an arbitrary arc $x \, (u \, (t) , v \, (t))$, $t \in [t_1, t_2]$ which lies in the field $\mathfrak{F}$ and joins points $P_1$ and $P_2$. The length of this arc is

$$s \, (t_1, t_2) = \int_{t_1}^{t_2} \left\{ u'^2 + G \, (u, v) \, v'^2 \right\}^{1/2} dt \geq \int_{t_1}^{t_2} u' dt = u \, (t_2) - u \, (t_1)$$

   The equality sign in the above relationship holds if and only if $G \, (u, v) \, v'^2 = 0$. As $G \, (u, v) > 0$, this implies that we require $v'^2 = 0$. This is true, provided $v = $ constant.

34. Show that the first fundamental form of $x = x \, (r, \theta)$, where $r \in \mathbb{R}^+$ and $\theta \in [0, 2\pi)$, in terms of geodesic polar coordinates is

$$I = dr^2 + G \, (r, \theta) \, d\theta^2$$

   Hint: See Pressley (2010). Let $P$ be a point on the surface $x$. Also, let $\xi_\theta \, (r)$ be a geodesic with unit-speed passing through the point $P$. At the point $P$ we have $r = 0$. Also $\theta$ is the oriented angle between a unit tangent vector on the surface $S$ at the point $P$ and $\xi'_\theta \, (r)$ (where prime denotes derivative with respect to $r$). Furthermore, $x \, (r, \theta) = \xi_\theta \, (r)$, is smooth for all values of $r \in (-\epsilon, \epsilon)$, and $\theta \in [0, 2\pi)$. It is also an allowable surface patch for $0 < r < \epsilon$, and for all values of $\theta$ in any open interval of length less than or equal to $2\pi$. Let

$$\int_0^R \left\| \frac{d\xi_\theta \, (r)}{dr} \right\|^2 dr = R, \quad 0 < R < \epsilon$$

As the geodesic $\xi_\theta(r)$ has unit speed, we have $x_r \circ x_r = 1$. Consequently

$$\int_0^R \|x_r\|^2 \, dr = R, \quad 0 < R < \epsilon$$

Differentiating both sides of the above result with respect to $\theta$ yields

$$\int_0^R x_r \circ x_{r\theta} dr = 0$$

That is,

$$x_r \circ x_\theta \big|_{r=0}^{r=R} - \int_0^R x_{rr} \circ x_\theta dr = 0$$

Recall that $x(0,\theta) = P$ for all values of $\theta$. Therefore, $x_\theta|_{r=0} = 0$. Note that $x_{rr} = \xi_\theta''(r)$ is parallel to the unit normal $\mathcal{N}$ of surface $x$. As $x_\theta \circ \mathcal{N} = 0$, we have $x_{rr} \circ x_\theta = 0$. Therefore $x_r \circ x_\theta = 0$. This implies that the first fundamental form coefficients $E = 1$, and $F = 0$. Therefore the first fundamental form is $I = \{dr^2 + G(r,\theta) \, d\theta^2\}$. The construction of $x(r,\theta)$ is called the *geodesic polar patch* on $S$.

35. Prove the observations about geodesic polar coordinates, and Gaussian curvature.

(a) Hint: See Kreyszig (1968), Lipschutz (1969), and Pressley (2010). In the last problem substitute $u = r\cos\theta$, and $v = r\sin\theta$. This results in a reparameterization $\widetilde{x}(u,v)$ of $x(r,\theta)$. Therefore, $\widetilde{x}$ is an allowable surface patch for $S$ defined on the open set $(u^2 + v^2) < \epsilon^2$. We obtain the desired expressions via the following steps.

*Step* 1: Let the first fundamental form of $\widetilde{x}$ be $\widetilde{I} = \widetilde{E} du^2 + 2\widetilde{F} du dv + \widetilde{G} dv^2$. The coefficients of first fundamental form of $x$ are $E, F, G$; where $E = 1$, and $F = 0$. The coefficients $E, F, G$ and $\widetilde{E}, \widetilde{F}, \widetilde{G}$ are related via

$$\begin{bmatrix} \widetilde{E} & \widetilde{F} \\ \widetilde{F} & \widetilde{G} \end{bmatrix} = J^T \begin{bmatrix} E & F \\ F & G \end{bmatrix} J$$

and the Jacobian matrix $J$ is

$$J = \begin{bmatrix} \partial r/\partial u & \partial r/\partial v \\ \partial\theta/\partial u & \partial\theta/\partial v \end{bmatrix} = \begin{bmatrix} u/r & v/r \\ -v/r^2 & u/r^2 \end{bmatrix}$$

The last two equations yield

$$\widetilde{E} - 1 = \frac{v^2}{r^2}\left(\frac{G}{r^2} - 1\right), \quad \widetilde{F} = \left(1 - \frac{G}{r^2}\right)\frac{uv}{r^2}, \quad \widetilde{G} - 1 = \frac{u^2}{r^2}\left(\frac{G}{r^2} - 1\right)$$

*Step* 2: These equations in turn yield

$$u^2\left(\widetilde{E} - 1\right) = v^2\left(\widetilde{G} - 1\right)$$

As $\widetilde{E}$ and $\widetilde{G}$ are smooth functions of $u$ and $v$, their Taylor series expansion about $u = v = 0$ can be expressed as

$$\widetilde{E} - 1 = \sum_{\substack{i=0 \\ 1\leq i+j\leq 2}}^{2}\sum_{j=0}^{2} e_{ij}u^i v^j + o\left(r^2\right), \quad \widetilde{G} - 1 = \sum_{\substack{i=0 \\ 1\leq i+j\leq 2}}^{2}\sum_{j=0}^{2} g_{ij}u^i v^j + o\left(r^2\right)$$

Substitute these expansions in $u^2\left(\tilde{E}-1\right)=v^2\left(\tilde{G}-1\right)$, and compare coefficients on either sides. It can be concluded that $e_{02}=g_{20}\triangleq k$, and all other coefficients are equal to zero. Thus

$$\tilde{E}=1+kv^2+o\left(r^2\right)$$

This implies

$$G=r^2+kr^4+o\left(r^4\right)$$

and

$$\sqrt{G}=r+\frac{1}{2}kr^3+o\left(r^3\right)$$

The Gaussian curvature $K$ with $E=1$, and $F=0$ is

$$K=-\frac{1}{\sqrt{G}}\frac{\partial^2\sqrt{G}}{\partial r^2}\simeq-3k$$

Therefore

$$\sqrt{G\left(r,\theta\right)}=r-\frac{1}{6}K\left(P\right)r^3+o\left(r^3\right)$$

(b)  Hint: See Lipschutz (1969).On the geodesic circle $r=$ constant, $dr=0$. Therefore, the circumference of the geodesic circle is

$$C\left(r\right)=\int_0^{2\pi}\sqrt{G\left(r,\theta\right)}d\theta$$
$$=2\pi r-\frac{1}{3}K\left(P\right)\pi r^3+o\left(r^3\right)$$

Consequently,

$$K\left(P\right)=\frac{3}{\pi}\left(\frac{2\pi r-C\left(r\right)}{r^3}\right)+o\left(1\right)$$

This implies

$$K\left(P\right)=\lim_{r\to0}\frac{3}{\pi}\left(\frac{2\pi r-C\left(r\right)}{r^3}\right)$$

The surface area is given by

$$A=\iint_W\left\{EG-F^2\right\}^{1/2}dudv$$

For the area of the geodesic circle is

$$A\left(r\right)=\int_0^r\int_0^{2\pi}\sqrt{G\left(r,\theta\right)}d\theta dr$$
$$=\pi r^2-\frac{\pi}{12}K\left(P\right)r^4+o\left(r^4\right)$$

This implies

$$K\left(P\right)=\lim_{r\to0}\frac{12}{\pi}\left(\frac{\pi r^2-A\left(r\right)}{r^4}\right)$$

36. Using the notation of the last problem, show that for polar and planar geodesic coordinates

$$\lim_{r \to 0} \sqrt{G(r, \theta)} = 0, \quad \lim_{r \to 0} \frac{\partial \sqrt{G(r, \theta)}}{\partial r} = 1$$

$$\tilde{G}(0, v) = 1, \quad \lim_{u \to 0} \frac{\partial \sqrt{\tilde{G}(u, v)}}{\partial u} = 0$$

Hint: Use results from the last problem.

37. Establish the two observations about the tractrix curve

$$z(r) = \left(1 - r^2\right)^{1/2} - \cosh^{-1}\left(\frac{1}{r}\right), \quad r \in (0, 1]$$

Hint: See Pressley (2010).

(a) Note that $\lim_{r \to 0} z(r) \to -\infty$, $z(1) = 0$, and

$$\frac{dz}{dr} = \frac{\left(1 - r^2\right)^{1/2}}{r} \geq 0$$

(b) Let the point $A$ on the tractrix be $(r_0, z_0)$. The equation of the tangent line at the point $A$ is

$$z - z_0 = \frac{\left(1 - r_0^2\right)^{1/2}}{r_0}(r - r_0)$$

This tangent intersects the $z$-axis at point $B = (0, z_1)$ where

$$z_1 - z_0 = \frac{\left(1 - r_0^2\right)^{1/2}}{r_0}(0 - r_0) = -\left(1 - r_0^2\right)^{1/2}$$

Therefore the square of the length of the line segment $AB$ is equal to

$$(0 - r_0)^2 + (z_1 - z_0)^2 = r_0^2 + \left(1 - r_0^2\right) = 1$$

38. A surface is parameterized as

$$\tilde{x}(v, w) = \left(\frac{1}{w} \cos v, \frac{1}{w} \sin v, \left(1 - \frac{1}{w^2}\right)^{1/2} - \cosh^{-1} w\right), \quad w > 1$$

Show that its first fundamental form is

$$\frac{dv^2 + dw^2}{w^2}$$

Hint: See Pressley 2001. Consider a surface $S$ parameterized as

$$x(u, v) = (e^u \cos v, e^u \sin v, g(u)), \quad u \leq 0$$

$$g(u) = \left(1 - r^2\right)^{1/2} - \cosh^{-1}\left(\frac{1}{r}\right), r = e^u, \quad u \leq 0$$

The corresponding first fundamental form is

$$I = du^2 + e^{2u} dv^2$$

The coefficients of the first fundamental form are $E = 1, F = 0$, and $G = e^{2u}$.

In the representation $x\,(\cdot, \cdot)$, map the coordinates $(u, v)$ to $(v, w)$, where $w = e^{-u}$. This results in the reparameterization of the surface $S$ as $\widetilde{x}\,(\cdot, \cdot)$. The corresponding coefficients of the first fundamental form $\widetilde{E}, \widetilde{F}$, and $\widetilde{G}$ are to be determined. The coefficients $E, F, G$ and $\widetilde{E}, \widetilde{F}, \widetilde{G}$ are related via

$$\begin{bmatrix} \widetilde{E} & \widetilde{F} \\ \widetilde{F} & \widetilde{G} \end{bmatrix} = J^T \begin{bmatrix} E & F \\ F & G \end{bmatrix} J$$

and the Jacobian matrix $J$ is

$$J = \begin{bmatrix} \partial u/\partial v & \partial u/\partial w \\ \partial v/\partial v & \partial v/\partial w \end{bmatrix} = \begin{bmatrix} 0 & -1/w \\ 1 & 0 \end{bmatrix}$$

This yields $\widetilde{E} = 1/w^2, \widetilde{F} = 0$, and $\widetilde{G} = 1/w^2$.

39. Establish the following observations about the upper half-plane model of the hyperbolic geometry.

   (a) Hyperbolic angles in the space $\mathbb{H}$ and Euclidean angles in a plane are identical.
   (b) The geodesics in $\mathbb{H}$ are called *hyperbolic lines*. The geodesics in $\mathbb{H}$ are: semicircles with centers on the real axis, and half-lines parallel to the imaginary axis.
   (c) Let $p, q \in \mathbb{H}$, where $p \neq q$. A unique hyperbolic line (geodesic) passes through $p$ and $q$.
   (d) The parallel axiom of the Euclidean geometry is not valid in hyperbolic geometry.

   Hint: See Pressley 2010.

   (a) The first fundamental forms in the Euclidean and the hyperbolic planes are $(dv^2 + dw^2)$ and $(dv^2 + dw^2)/w^2$ respectively. Note that these fundamental forms are proportional to each other. This in turn implies that the mapping between the Euclidean and hyperbolic planes is conformal..
   (b) Let $\gamma\,(t) = \widetilde{x}\,(v\,(t), w\,(t))$ be a unit-speed geodesic. This gives

$$v'^2 + w'^2 = w^2$$

   The coefficients of the first fundamental form are $E = G = 1/w^2$, and $F = 0$. The first geodesic equation results in

$$\frac{d}{dt}\,(Ev') = \frac{1}{2}\,(E_v v'^2 + G_v w'^2) = 0$$

   That is

$$v' = kw^2$$

   where $k$ is a constant. If $k = 0$, we get $v = $ constant. Next assume that $k \neq 0$. Use of $(v'^2 + w'^2) = w^2$ and the last two equations results in

$$w' = \pm w\sqrt{1 - k^2 w^2}$$

   Use of $v' = kw^2$ and the last equation yields

$$\frac{dv}{dw} = \pm \frac{kw}{\sqrt{1 - k^2 w^2}}$$

   Therefore

$$v = v_0 \mp \frac{1}{k}\sqrt{1 - k^2 w^2}$$

$$(v - v_0)^2 + w^2 = \frac{1}{k^2}$$

where $v_0$ is a constant. The above result is the equation of a circle in the $vw$-plane. These circles have their centers on the $v$-axis. Therefore the geodesics are points on the surface $\widetilde{x}(\cdot, \cdot)$ which satisfy the above equation. Also $v = $ constant corresponds to straight lines which are perpendicular to the $v$-axis.

(c) Let $p, q \in \mathbb{H}$, where $p \neq q$. We consider two cases. If the Euclidean line passing through $p$ and $q$ is parallel to the imaginary axis, then the unique hyperbolic line passing through these points is the half-line containing them. However, if the Euclidean line passing through $p$ and $q$ is not parallel to the imaginary axis, then let its perpendicular bisector intersect the real axis at say some point $a$. The unique hyperbolic line passing through $p$ and $q$ is the semicircle with center $a$, and radius $|p - a| = |q - a|$.

(d) Denote the imaginary axis in $\mathbb{H}$ by $l$. Let $a \in \mathbb{H}$ be a point not on $l$, where $\text{Re}(a) > 0$. Let the perpendicular bisector of the line joining $a$ and the origin intersect the real axis at $b$. Note that $b > 0$. Let $c$ be any point on the real axis which is greater than $b$. Then the semicircle with center $c$, and passing through the point $a$ is indeed a hyperbolic line in $\mathbb{H}$. Furthermore, it does not intersect line $l$. Note that the half-line passing through $a$, and parallel to the imaginary axis is also another hyperbolic line passing through it. This half-line is also obtained by letting $c \to \infty$.

40. Prove that the hyperbolic distance between two points $a, b \in \mathbb{H}$ is

$$d_{\mathbb{H}}(a, b) = 2\tanh^{-1}\left|\frac{b - a}{b - \overline{a}}\right|.$$

Hint: See Pressley 2010. We consider two cases. In the first case the hyperbolic line joining the two points $a$ and $b$ is a semicircle. In the second case, the two points $a$ and $b$ lie on a half-line. Let $d_{\mathbb{H}}(a, b) \triangleq d$.

*Case* 1: The two points $a$ and $b$ lie on a semicircle with center $c$ on the $v$-axis (real axis). The points on the semicircle can be described by

$$v = c + r\cos\theta, \quad \text{and} \quad w = r\sin\theta$$

Let prime ($'$) denote derivative with respect to $\theta$. Also let $\varphi = \arg(a - c)$, and $\psi = \arg(b - c)$, where $\varphi, \psi \in [0, \pi]$. Thus

$$d = \int_\varphi^\psi \left\{\frac{v'^2 + w'^2}{w^2}\right\}^{1/2} d\theta = \int_\varphi^\psi \left\{\frac{r^2\sin^2\theta + r^2\cos^2\theta}{r^2\sin^2\theta}\right\}^{1/2} d\theta$$

$$= \int_\varphi^\psi \frac{d\theta}{\sin\theta} = \ln\frac{\tan(\psi/2)}{\tan(\varphi/2)}$$

Observe that $d$ is independent of $r$, the radius of the semicircle. Also

$$\tanh\frac{d}{2} = \frac{e^d - 1}{e^d + 1} = \frac{\tan(\psi/2) - \tan(\varphi/2)}{\tan(\psi/2) + \tan(\varphi/2)} = \frac{\sin\{(\psi - \varphi)/2\}}{\sin\{(\psi + \varphi)/2\}}$$

It can be shown that

$$|b-a|^2 = r^2 \left\{ (\cos\psi - \cos\varphi)^2 + (\sin\psi - \sin\varphi)^2 \right\}$$

$$= 4r^2 \sin^2 \left( \frac{\psi - \varphi}{2} \right)$$

and

$$|b - \bar{a}|^2 = 4r^2 \sin^2 \left( \frac{\psi + \varphi}{2} \right)$$

The result follows.

*Case* 2: The two points $a$ and $b$ lie on a half-line. Let $a = r + is$, and $b = r + it$, where $r, s, t \in \mathbb{R}$, $t > s$, and $i = \sqrt{-1}$. Thus

$$d = \int_s^t \frac{dw}{w} = \ln \left( \frac{t}{s} \right)$$

Therefore $e^d = t/s$. Check that

$$2 \tanh^{-1} \left| \frac{b - a}{b - \bar{a}} \right| = 2 \tanh^{-1} \left\{ \frac{t - s}{t + s} \right\} = 2 \tanh^{-1} \left\{ \frac{e^d - 1}{e^d + 1} \right\}$$

$$= 2 \tanh^{-1} \left( \tanh \frac{d}{2} \right) = d$$

41. Let $\Delta$ be a hyperbolic triangle in $\mathbb{H}$, with internal angles $\alpha, \beta$, and $\gamma$. Show that its area $A(\Delta)$ is

$$A(\Delta) = \pi - (\alpha + \beta + \gamma)$$

This result implies that the sum of internal angles of a hyperbolic triangle is less than $\pi$. Also observe that the area $A(\Delta)$ is independent of the length of geodesics joining the vertices of the hyperbolic triangle.

This result is in contrast to Euclidean geometry, where the sum of internal angles of a triangle is always equal to $\pi$. Furthermore, the area of a Euclidean triangle depends upon the lengths of the sides of the triangle.

Hint: See Walkden (2015). The area $A(\Delta)$ of the hyperbolic triangle is evaluated in three steps. *Step* 0: Label the vertices of the hyperbolic triangle by $P, Q$, and $R$. Let the internal angles at these vertices be $\alpha, \beta$, and $\gamma$ respectively. Denote this hyperbolic triangle by $\Delta_{PQR} \triangleq \Delta$. See Figure 5.1.

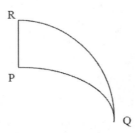

Figure 5.1. Hyperbolic triangle with vertices $P, Q$, and $R$.

*Step* 1: Initially assume that the vertex $R$ is at infinity. Label this vertex $R^\infty$. Therefore, the angle $\gamma$ is equal to zero. In this case the geodesics $PR^\infty$ and $QR^\infty$ are Euclidean straight lines parallel to the imaginary axis. We call this triangle $\Delta_{PQR\infty}$.

The geodesic joining vertices $P$ and $Q$ is the arc of a circle. By applying a Moebius transformation (translation) of $\mathbb{H}$ translate the center of this circle to the origin of the complex plane $\mathbb{C}$. Again applying a Moebius transformation (dilation) map the radius of the circle to unity.

In this construction, the angle between the tangent to the circle at $P$ and $PR^\infty$ is $\alpha$, and the angle between the tangent to the circle at $Q$ and $QR^\infty$ is $\beta$. Also the Cartesian coordinates of the vertices $P$ and $Q$ are $(\cos(\pi - \alpha), \sin(\pi - \alpha))$, and $(\cos\beta, \sin\beta)$ respectively. Observe that the internal angles of the hyperbolic triangle $\Delta_{PQR\infty}$ are $\alpha, \beta$, and $0$. Therefore the area of the triangle thus formed is

$$A(\Delta_{PQR\infty}) = \int\int_{\Delta_{PQR\infty}} \frac{dvdw}{w^2}$$

$$= \int_{\cos(\pi-\alpha)}^{\cos\beta} \int_{\sqrt{1-v^2}}^{\infty} \frac{dw}{w^2} dv = \int_{\cos(\pi-\alpha)}^{\cos\beta} \frac{dv}{\sqrt{1-v^2}}$$

Substitute $v = \cos\theta$ in the above expression. This results in

$$A(\Delta_{PQR\infty}) = -\int_{(\pi-\alpha)}^{\beta} d\theta = \pi - (\alpha + \beta)$$

*Step* 2: In this step we determine area of the hyperbolic triangle $\Delta_{PQR}$. Apply a Moebius transformation of $\mathbb{H}$ so that the geodesic between the vertices $P$ and $R$ is a Euclidean straight line parallel to the imaginary axis. Furthermore, the vertex $R$ is not at infinity. Observe that

$$A(\Delta) = A(\Delta_{PQR}) - A(\Delta_{PQR\infty}) - A(\Delta_{RQR\infty})$$

Let the internal angles of the triangle $\Delta_{RQR\infty}$ be $(\pi - \gamma), \delta$, and $0$. Similarly, let the internal angles of the triangle $\Delta_{PQR\infty}$ be $\alpha, (\beta + \delta)$, and $0$. Therefore, using result from Step 1, we have

$$A(\Delta_{PQR\infty}) = \pi - \{\alpha + (\beta + \delta)\}$$
$$A(\Delta_{RQR\infty}) = \pi - \{(\pi - \gamma) + \delta\}$$

Use of the above expressions yields

$$A(\Delta) = A(\Delta_{PQR}) = \pi - (\alpha + \beta + \gamma)$$

42. Let $\mathcal{P}$ be a hyperbolic polygon in $\mathbb{H}$ with internal angles $\alpha_1, \alpha_2, \ldots, \alpha_n$. Show that the hyperbolic area of this polygon is

$$A(\mathcal{P}) = (n-2)\pi - \sum_{i=1}^{n} \alpha_i$$

Hint: See Pressley 2010. This result can either be established by successive use of the last problem, or we proceed as follows.

*Step* 1: Consider points $a$ and $b$ on a hyperbolic line $l$ in $\mathbb{H}$. These two points lie on a semicircle with center $c$. Let the radius vectors $ca$ and $cb$ make angles $\varphi$ and $\psi$ respectively, with the positive real axis. Then it is established that

$$\int_l \frac{dv}{w} = \varphi - \psi$$

Observe that this result is independent of the radius of the semicircle. This result is also true, if the hyperbolic line is a half-line in $\mathbb{H}$.

Proof of the statement. Let $r$ be the radius of the semicircle. Also let $v = r\cos\theta$ and $w = r\sin\theta$. Then

$$\int_l \frac{dv}{w} = \int_\varphi^\psi \frac{-r\sin\theta}{r\sin\theta}d\theta = \varphi - \psi$$

*Step* 2: The main result is established in this step by using Green-Ostrogradski theorem form the theory of complex variables. Let the vertices of the hyperbolic polygon $\mathcal{P}$ be $b_1, b_2, \ldots, b_n$. Its boundary $\mathcal{C}$ consists of hyperbolic line segments $b_1b_2, b_2b_3, \ldots, b_nb_1$. The internal angle at vertex $b_i$ is $\alpha_i$ where $1 \le i \le n$. As the first fundamental form is $\left(dv^2 + dw^2\right)/w^2$, the area of the polygon $\mathcal{P}$ is

$$\int_\mathcal{P} \frac{dvdw}{w^2}$$

Next consider the following result obtained via Green-Ostrogradski theorem.

$$\oint_\mathcal{C} \{P(v,w)\,dv + Q(v,w)\,dw\} = \int_\mathcal{P} \left\{\frac{\partial Q(v,w)}{\partial v} - \frac{\partial P(v,w)}{\partial w}\right\} dvdw$$

where $P(\cdot,\cdot)$ and $Q(\cdot,\cdot)$ are continuous real-valued functions of $v$ and $w$. Substitute $P(v,w) = 1/w$ and $Q(v,w) = 0$. It yields

$$\int_\mathcal{P} \frac{dvdw}{w^2} = \int_\mathcal{C} \frac{dv}{w}$$

The result established in Step 1 is next used. Let $\varphi_i$ and $\psi_i$ be the angles defined in Step 1 with endpoints $b_i$ and $b_{i+1}$, where $1 \le i \le n$. Note that the vertex $b_{n+1}$ is $b_1$. Therefore, we have

$$\int_\mathcal{P} \frac{dvdw}{w^2} = \sum_{i=1}^n (\varphi_i - \psi_i)$$

The right hand side of the above equation can also be alternately evaluated. Examine the change in direction of a normal of the polygon $\mathcal{P}$ which points outwards as we walk around its boundary in an anticlock direction. Consider a walk from the vertex $b_i$ to vertex $b_{i+1}$. The normal which points outwards, rotates anticlockwise through an angle $(\psi_i - \varphi_i)$. In addition, the normal at vertex $b_i$ rotates by $(\pi - \alpha_i)$. Thus the net change in direction of the normal, as we walk around the edges of the polygon $\mathcal{P}$ is

$$n\pi + \sum_{i=1}^n (\psi_i - \varphi_i - \alpha_i)$$

As the polygon $\mathcal{P}$ is a nonintersecting closed curve, this net angle of rotation is equal to $2\pi$. Therefore

$$2\pi = n\pi + \sum_{i=1}^n (\psi_i - \varphi_i - \alpha_i)$$

Therefore

$$\int_\mathcal{P} \frac{dvdw}{w^2} = \sum_{i=1}^n (\varphi_i - \psi_i) = (n-2)\pi - \sum_{i=1}^n \alpha_i$$

43. The *hyperbolic circle* $C_{a,R}$ with center $a \in \mathbb{H}$ and radius $R > 0$ is the set of points in $\mathbb{H}$ which are at a hyperbolic distance $R$ from $a$. That is,

$$C_{a,R} = \{z \in \mathbb{H} \mid d_{\mathbb{H}}(a, z) = R\}$$

Show that $C_{a,R}$ is also a Euclidean circle.
Hint: See Pressley (2010). Let $z \in C_{a,R}$, then

$$R = 2\tanh^{-1}\left|\frac{z - a}{z - \bar{a}}\right|$$

Let $\beta = \tanh(R/2)$. As $R > 0$, we have $\beta < 1$. This implies

$$\left(1 - \beta^2\right)|z|^2 - \left(\bar{a} - \beta^2 a\right)z - \left(a - \beta^2\bar{a}\right)\bar{z} + \left(1 - \beta^2\right)|a|^2 = 0$$

This is the equation of a circle provided

$$\left(1 - \beta^2\right) > 0, \quad \text{and} \quad \left|a - \beta^2\bar{a}\right|^2 > \left(1 - \beta^2\right)^2|a|^2$$

This implies $2|a|^2 > \left(a^2 + \bar{a}^2\right)$. Let $a = |a|\,e^{i\theta}$. This implies $1 > \cos 2\theta$. This inequality is true because $a \in \mathbb{H}$ implies $0 < \theta < \pi$.

44. Establish the following results.

(a) The Euclidean center of the hyperbolic circle $C_{ic,R}$, where $c > 0$, is $ib$, and its Euclidean radius is $r$, where

$$c = \sqrt{b^2 - r^2}, \quad R = \frac{1}{2}\ln\frac{(b + r)}{(b - r)}$$

(b) For $c, R \in \mathbb{R}^+$, the hyperbolic length of the circumference of $C_{ic,R}$ is equal to $2\pi\sinh R$, and the hyperbolic area of this circle is equal to $2\pi(\cosh R - 1)$.

(c) The circumference and area of hyperbolic circle $C_{a,R}$ are independent of $a \in \mathbb{H}$, and depend only upon $R \in \mathbb{R}^+$.

Hint: See Pressley (2010).

(a) The Euclidean circle corresponding to the hyperbolic circle $C_{ic,R}$ has a center on the imaginary axis at $ib$. This circle intersects the imaginary axis at $i(b \pm r)$. Therefore the hyperbolic distance between these two points is $2R$. Furthermore,

$$2R = 2\tanh^{-1}\left|\frac{2ir}{2ib}\right|$$

This implies $r = b\tanh R$, which yields

$$R = \frac{1}{2}\ln\frac{(b + r)}{(b - r)}$$

Observe that the hyperbolic distance between the points $ic$ and $i(b + r)$ is equal to the hyperbolic distance between the points $ic$ and $i(b - r)$. This implies

$$\left|\frac{i(b + r) - ic}{i(b + r) + ic}\right| = \left|\frac{i(b - r) - ic}{i(b - r) + ic}\right|$$

Simplification of the above relationship results in

$$\left| b^2 - (c-r)^2 \right| = \left| b^2 - (c+r)^2 \right|$$

Therefore

$$(c-r)^2 - b^2 = b^2 - (c+r)^2$$

This implies

$$c^2 = b^2 - r^2$$

(b) *Step* 1: In this step it is shown that

$$\int_0^{2\pi} \frac{d\theta}{(1+t\sin\theta)} = \frac{2\pi}{\sqrt{1-t^2}}$$

The following results are used

$$\int_0^{\pi/2} \sin^{2m} x \, dx = \frac{1\cdot 3 \cdot \cdots \cdot (2m-1)}{2\cdot 4 \cdot \cdots \cdot (2m)} \frac{\pi}{2}, \quad m = 1,2,3,\ldots$$

$$\left(1-x^2\right)^{-1/2} = 1 + \frac{1}{2}x^2 + \frac{1\cdot 3}{2\cdot 4}x^4 + \frac{1\cdot 3\cdot 5}{2\cdot 4\cdot 6}x^6 + \ldots, \quad x\in(-1,1)$$

Let $f(t) = (1+t\sin\theta)^{-1}$, for $t \in (-1,1)$. Therefore

$$f(\theta) = \sum_{n\geq 0} (-1)^n (t\sin\theta)^n$$

Use of the above series expansions yields

$$\int_0^{2\pi} f(\theta)\,d\theta = \frac{2\pi}{\sqrt{1-t^2}}$$

*Step* 2: The hyperbolic circle $C_{ic,R}$ is parameterized as: $v = r\cos\theta$, and $w = (b+r\sin\theta)$. Denote the first derivatives of $v$ and $w$ with respect to $\theta$ by $v'$ and $w'$ respectively. Therefore the circumference is equal to

$$\int_0^{2\pi} \frac{\sqrt{v'^2 + w'^2}}{w}\,d\theta = \int_0^{2\pi} \frac{r}{(b+r\sin\theta)}\,d\theta$$
$$= \frac{2\pi r}{\sqrt{b^2 - r^2}} = 2\pi\sinh R$$

where results from Step 1, and part (a) of the problem were used. The area of the hyperbolic circle is

$$\int_{interior(C_{ic,R})} \frac{dv\,dw}{w^2} = \int_{C_{ic,R}} \frac{dv}{w}$$

The last equality is obtained via application of Green-Ostrogradski theorem. Therefore the area is

$$\int_0^{2\pi} \frac{-r\sin\theta}{(b+r\sin\theta)}\,d\theta = \frac{2\pi b}{\sqrt{b^2 - r^2}} - 2\pi = 2\pi(\cosh R - 1)$$

(c) Let $a, b \in \mathbb{H}$. Also let $F_I$ be an isometric mapping of $\mathbb{H}$ that maps $a$ to $b$. Consequently, $F_I$ maps the hyperbolic circle $C_{a,R}$ to $C_{b,R}$ for all positive values of $R$. Therefore, these circles related via this isometric mapping have identical circumference and area.

45. Prove that the mapping: inversion in circles with centers on the real axis in $\mathbb{H}$ is isometric.
    Hint: See Pressley (2010). The stated result is established in two steps. In the first step, we establish that $I_{0,1}\left(\cdot\right)$ is an isometry of $\mathbb{H}$. In the next step, the result is generalized.
    *Step* 1: Let $a = 0, r = 1$, and $I_{0,1}\left(\cdot\right) \triangleq I$. Let $z = \left(v + iw\right)$. Therefore

    $$I\left(v + iw\right) = \frac{\left(v + iw\right)}{\left(v^2 + w^2\right)}$$

    Thus $I\left(\cdot\right)$ maps any point in $\mathbb{H}$ to another point in $\mathbb{H}$, Furthermore, any nonzero point on the real axis is mapped to another point on the real axis. The first fundamental form is next computed. The circle is reparameterized under the mapping $I\left(\cdot\right)$ as

    $$\widetilde{v} = \frac{v}{\left(v^2 + w^2\right)}$$
    $$\widetilde{w} = \frac{w}{\left(v^2 + w^2\right)}$$

    This yields

    $$\left(\widetilde{v}^2 + \widetilde{w}^2\right) = \frac{1}{\left(v^2 + w^2\right)}$$

    and

    $$v = \frac{\widetilde{v}}{\left(\widetilde{v}^2 + \widetilde{w}^2\right)}, \quad w = \frac{\widetilde{w}}{\left(\widetilde{v}^2 + \widetilde{w}^2\right)}$$

    The transformation from the $\widetilde{v}\widetilde{w}$-plane to $vw$-plane is obtained by the Jacobian matrix

    $$J = \begin{bmatrix} \partial v / \partial \widetilde{v} & \partial v / \partial \widetilde{w} \\ \partial w / \partial \widetilde{v} & \partial w / \partial \widetilde{w} \end{bmatrix}$$

    where

    $$\frac{\partial v}{\partial \widetilde{v}} = \frac{\widetilde{w}^2 - \widetilde{v}^2}{\left(\widetilde{v}^2 + \widetilde{w}^2\right)^2} = -\frac{\partial w}{\partial \widetilde{w}}, \quad \frac{\partial v}{\partial \widetilde{w}} = \frac{\partial w}{\partial \widetilde{v}} = -\frac{2\widetilde{v}\widetilde{w}}{\left(\widetilde{v}^2 + \widetilde{w}^2\right)^2}$$

    The coefficients $\widetilde{E}, \widetilde{F}$, and $\widetilde{G}$ of the first fundamental form of the mapping are obtained via

    $$\begin{bmatrix} \widetilde{E} & \widetilde{F} \\ \widetilde{F} & \widetilde{G} \end{bmatrix} = J^T \begin{bmatrix} E & F \\ F & G \end{bmatrix} J$$

    where $E = G = 1/w^2$, and $F = 0$. This yields $\widetilde{E} = \widetilde{G} = 1/\widetilde{w}^2$, and $\widetilde{F} = 0$. It also follows from the expressions for $\widetilde{v}$ and $\widetilde{w}$ in terms of $v$ and $w$ that

    $$d\widetilde{v} = \frac{\left(w^2 - v^2\right)dv - 2vwdw}{\left(v^2 + w^2\right)^2}, \quad d\widetilde{w} = \frac{-2vwdv + \left(v^2 - w^2\right)dw}{\left(v^2 + w^2\right)^2}$$

    Simplification yields

    $$\frac{d\widetilde{v}^2 + d\widetilde{w}^2}{\widetilde{w}^2} = \frac{dv^2 + dw^2}{w^2}$$

    Therefore the first fundamental form with parameters $v$ and $w$ is equal to the first fundamental form with parameters $\widetilde{v}$ and $\widetilde{w}$.
    *Step* 2: The result of Step 1 is generalized. Let $a \in \mathbb{R}$, and $r \in \mathbb{R}^+$. We write $I_{a,r}\left(z\right)$ as

$$I_{a,r}(z) = T_a \left( \frac{r^2}{\bar{z} - a} \right) = T_a \circ D_{r^2} \left( \frac{1}{\bar{z} - a} \right)$$
$$= T_a \circ D_{r^2} \circ I(z - a) = T_a \circ D_{r^2} \circ I \circ T_{-a}(z)$$

where $\circ$ is the composition operator of functions. Thus $I_{a,r}(\cdot)$ can be represented as a composite function, where the functions $T_a(\cdot), D_{r^2}(\cdot)$, and $I(\cdot)$ are isometries of $\mathbb{H}$. Therefore, as composition of several isometries is an isometry, $I_{a,r}(\cdot)$ is indeed an isometry of $\mathbb{H}$.

46. Establish the following assertions.

    (a) Let $l_1$ and $l_2$ be hyperbolic lines in the upper half-plane $\mathbb{H}$. Also, $z_1$ and $z_2$ are points on $l_1$ and $l_2$ respectively. Then there exists an isometry, which takes $l_1$ to $l_2$ and $z_1$ to $z_2$.

    (b) Let $a, b \in \mathbb{H}$. The hyperbolic distance between these two points $d_{\mathbb{H}}(a, b)$ is the length of the shortest curve in $\mathbb{H}$ which joins $a$ and $b$.

    Hint: See Pressley (2010).

    (a) The stated result is established in two steps.

       *Step* 1: Assume that the line $l_2$ is a half-line $l$, which passes through the origin, and $z_2 = i$. We have to consider two cases: whether line $l_1$ is a half-line or a semicircle.

          *Case* (i): Let the line $l_1$ be the half-line $v = a$. Then the translation $T_{-a}(\cdot)$ maps $l_1$ to $l$, and the point $z_1$ to, say some point $ib$ on line $l$, where $b > 0$. Then the dilation operation $D_{b^{-1}}$ maps $l$ to itself, and the point $ib$ to $i$. Therefore, the required isometric transformation is $D_{b^{-1}} \circ T_{-a}(\cdot)$.

          *Case* (ii): Let the line $l_1$ be a semicircle. The semicircle intersects the real axis at two points. Let one of these two points be $a$. Then the inversion $I_{a,1}(\cdot)$ maps $l_1$ to a half-line geodesic $l'$, and the point $z_1$ to a point $z'$ on $l'$. Then by Case (i), there exists an isometric mapping $G$ which maps $l'$ to $l$ and $z'$ to $i$. Therefore, the the required isometric transformation is $G \circ I_{a,1}(\cdot)$.

       Combining the two cases, we infer that there exists an isometric mapping $G_1(\cdot)$ that takes $l_1$ to $l$ and $z_1$ to $i$.

       *Step* 2: Let the isometric transformation $G_2(\cdot)$ map $l_2$ to $l$, and $z_2$ to $i$. Then the isometric transformation $G_2^{-1} \circ G_1(\cdot)$ maps $l_1$ to $l_2$ and $z_1$ to $z_2$.

    (b) From part (a) of the problem, assume that there exists an isometric transformation that takes $a$ to $i$ and $b$ to $ir$, where $r > 0$. As distances, are invariant under isometric transformations, the stated result is established for $a = i$, and $b = ir$. Initially assume that $r > 1$. Then

$$d_{\mathbb{H}}(a, b) = 2 \tanh^{-1} \left| \frac{r - 1}{r + 1} \right| = \ln r$$

       Next assume that $\beta(t) = v(t) + iw(t)$ is any curve in the upper half-plane $\mathbb{H}$ with $\beta(t_0) = i$, and $\beta(t_1) = ir$. Let $v'(t) = dv(t)/dt$, and $w'(t) = dw(t)/dt$. Then the length of the curve between points $a$ and $b$ is

$$\int_{t_0}^{t_1} \frac{\sqrt{(v'(t))^2 + (w'(t))^2}}{w(t)} dt \geq \int_{t_0}^{t_1} \frac{w'(t)}{w(t)} dt = \int_1^r \frac{dw}{w} = \ln r$$

       The case for $r < 1$ is similar. However, in this case $d_{\mathbb{H}}(a, b) = -\ln r$.

47. Prove the following assertions about hyperbolic lines in $\mathbb{H}$.

    (a) Let $l$ be a half-line in $\mathbb{H}$ and $p$ be a point not on $l$. There are infinitely many hyperbolic lines that pass through the point $p$ that do not intersect $l$.

(b) Let $l$ be a hyperbolic line in $\mathbb{H}$ and $p$ be a point not on $l$. There are infinitely many hyperbolic lines that pass through the point $p$ that do not intersect $l$.

Hint: See Pressley (2010).

(a) Let the half-line $l$ meet the real axis at $q \in \mathbb{R}$. Consider two cases.

*Case* (i): Let $\mathrm{Re}\,(p) > q$. Consider a semicircle with center $c$ on the real axis, with radius is $|p - c|$. This semicircle does not meet $l$ provided $|p - c| < |q - c|$. This is true if $c > \left( |p|^2 - q^2 \right) / \{2\,(\mathrm{Re}\,(p) - q)\}$.

*Case* (ii): Let $\mathrm{Re}\,(p) < q$. The discussion is similar to Case (i).

(b) An isometry is used to transform the hyperbolic line $l$ to a half-line. The result follows by using part (a) of the problem.

48. Prove that the Moebius transformation $M\,(\cdot)$ that maps $\mathbb{H}$ to $\mathbb{H}$ is precisely the real Moebius transformation, where $(ad - bc) > 0$.

Hint: See Pressley (2010). This result is proved in two steps.

*Step* 1: Let $a, b, c, d \in \mathbb{R}$, then

$$\mathrm{Im}\,(M\,(z)) = \frac{(ad - bc)}{|cz + d|^2}\,\mathrm{Im}\,(z)$$

Therefore, if $(ad - bc) > 0$, and $\mathrm{Im}\,(z) > 0$, we have the specified mapping.

*Step* 2: We establish the converse of the result in Step 1. Assume that $M\,(\cdot)$ maps $\mathbb{H}$ to $\mathbb{H}$. Further assume that $c \neq 0$ and $d \neq 0$. Then the mapping $M\,(\cdot)$ takes real-axis to itself. That is, $M\,(z) \in \mathbb{R}$ if $z \in \mathbb{R}$. Let $\alpha = b/d, \beta = a/c$, and $\gamma = c/d$. Note that $\alpha, \beta, \gamma \in \mathbb{R}$. Therefore, $M\,(0) = \alpha$, and $M\,(z) \to \beta$ as $z \to \infty$. Furthermore, $M\,(z)$ can be written as

$$M\,(z) = \beta + \frac{(\alpha - \beta)}{(\gamma z + 1)}$$

Therefore, $M\,(z)$ takes real values if $z \in \mathbb{R}$. Consequently, $a, b, c, d$ are equal to real multiples of $\beta\gamma, \alpha, \gamma, 1$ respectively. The result $(ad - bc) > 0$ follows from earlier computations. The cases, when $c = 0$ or $d = 0$ follow similarly.

49. Establish the following results about Moebius isometry in the upper half plane $\mathbb{H}$.

(a) Each real Moebius transformation can be expressed as a composition of elementary isometries of $\mathbb{H}$. Consequently, a real Moebius transformation is an isometry of $\mathbb{H}$.

(b) Let $M\,(\cdot)$ be a real Moebius transformation, and define a function $J\,(\cdot)$ as $J\,(z) = -\overline{z}$, where $z \in \mathbb{C}$. Then $M \circ J\,(\cdot)$ is an isometry of $\mathbb{H}$.

(c) The isometries described in observations (a) and (b) are called Moebius isometries. The composition of Moebius isometries is a Moebius isometry.

(d) Each isometry of the upper half-plane $\mathbb{H}$ is a Moebius isometry.

Hint: See Pressley (2010). It has been established in the chapter on applied analysis that a Moebius transformation is a composition of translations, dilations, and inversion. These mappings for $z \in \mathbb{C}$ are:

Translation: $T_a\,(z) = (z + a)\,, a \in \mathbb{C}$.

Complex dilations: $D_a\,(z) = az, a \in \mathbb{C} \setminus \{0\}$.

Inversion: $K\,(z) = 1/z$.

We also use the mappings:

Reflection: $R_a\,(z) = 2a - \overline{z}, a \in \mathbb{R}$

Circle inversion: Let $a \in \mathbb{R}$, and $r \in \mathbb{R}^+$

$$I_{a,r}(z) = a + \frac{r^2}{\bar{z} - a}$$

(a) Consider two cases.

*Case* (i): Let $c = 0$. Then $d \neq 0$. We have

$$\begin{aligned} M(z) &= \frac{a}{d}z + \frac{b}{d} \\ &= T_{b/d} \circ D_{a/d}(z) \end{aligned}$$

*Case* (ii): Let $c \neq 0$. Then we have

$$\begin{aligned} M(z) &= \frac{a}{c} - \frac{(ad - bc)/c}{cz + d} \\ &= T_{a/c} \circ D_{(ad-bc)/c^2} \circ (-K) \circ T_{d/c}(z) \end{aligned}$$

It can be shown that $-K$ is the composite map $R_0 \circ I_{0,1}$, where $R_0$ and $I_{0,1}$ are elementary isometries of $\mathbb{H}$.

(b) The mapping $J(\cdot)$ is a reflection across the imaginary axis. Therefore it is an isometry of $\mathbb{H}$. Note that $J(z) = -\bar{z}$. As inverse of any real Moebius transformation is also real, and composition of real Moebius transformations is also real, the assertion is true.

(c) Observe that, if $M(\cdot)$ is a real Moebius transformation, then so is $J \circ M \circ J(\cdot)$. For instance, if $M_1$ and $M_2$ are real Moebius transformation, then $(M_1 \circ J) \circ (M_2 \circ J) = M_1 \circ (J \circ M_2 \circ J)$ is a composition of real Moebius transformations, which in turn is a real Moebius transformation. The result follows because the composition of real Moebius transformations is a real Moebius transformation.

(d) We assume without proof that every isometry of $\mathbb{H}$ is a composition of elementary isometries. If this assumption is true, then it is sufficient to prove that every isometry of $\mathbb{H}$ is a Moebius isometry. Let $a \in \mathbb{R}$, then $T_a(\cdot)$ is a real Moebius transformation. Similarly, for $a \in \mathbb{R}$ we have $R_a(\cdot) = T_{2a} \circ J(\cdot)$. Also $D_a(\cdot)$ is a real Moebius transformation if $a \in \mathbb{R}^+$. Next assume that $a \in \mathbb{R}$, and $r \in \mathbb{R}^+$, then $I_{a,r}(\cdot) = T_a \circ D_{r^2} \circ I_{0,1} \circ T_{-a}(\cdot)$. It remains to prove that $I_{0,1}(\cdot)$ is a Moebius. Note that $I_{0,1}(\cdot) = (-K) \circ J(\cdot)$, where $-K(z) = -1/z$ is a real Moebius transformation.

50. Prove that the first fundamental form of the Poincaré disc model of hyperbolic geometry $\mathbb{D}_P$ is

$$\frac{4\left(dv^2 + dw^2\right)}{\left(1 - v^2 - w^2\right)^2}$$

where $(v, w) \in \mathbb{D}$. Also show that $\mathbb{D}_P$ is a conformal model.
Hint: See Pressley (2010). Let $(\tilde{v}, \tilde{w}) \in \mathbb{H}$. As

$$(\tilde{v} + i\tilde{w}) = P^{-1}(v + iw)$$

we have

$$\tilde{v} = \frac{-2w}{(v - 1)^2 + w^2}$$

$$\tilde{w} = \frac{1 - v^2 - w^2}{(v - 1)^2 + w^2}$$

This results in

$$d\widetilde{v} = \frac{4\left(v-1\right)w\,dv - 2\left(\left(v-1\right)^2 - w^2\right)dw}{\left(\left(v-1\right)^2 + w^2\right)^2}$$

$$d\widetilde{w} = \frac{2\left(\left(v-1\right)^2 - w^2\right)dv + 4\left(v-1\right)w\,dw}{\left(\left(v-1\right)^2 + w^2\right)^2}$$

The above equations yield

$$\frac{d\widetilde{v}^2 + d\widetilde{w}^2}{\widetilde{w}^2} = \frac{4\left(dv^2 + dw^2\right)}{\left(1 - v^2 - w^2\right)^2}$$

Note that the first fundamental form of $\mathbb{D}_P$ is a multiple of $\left(dv^2 + dw^2\right)$. Therefore $\mathbb{D}_P$ is conformal.

51. Let the distance between two points $a, b \in \mathbb{D}$ be $d_{\mathbb{D}}\left(a,b\right)$. Establish the following results.

   (a)

   $$d_{\mathbb{D}}\left(a,b\right) = d_{\mathbb{H}}\left(P^{-1}\left(a\right), P^{-1}\left(b\right)\right) = 2\tanh^{-1}\frac{|b-a|}{|1-\overline{a}b|}$$

   (b)

   $$d_{\mathbb{D}}\left(a,b\right) = \cosh^{-1}\left\{1 + 2\frac{|b-a|^2}{\left(1-|a|^2\right)\left(1-|b|^2\right)}\right\}$$

   (a) Hint: See Pressley (2010). We have

   $$d_{\mathbb{D}}\left(a,b\right) = d_{\mathbb{H}}\left(P^{-1}\left(a\right), P^{-1}\left(b\right)\right)$$

   $$= 2\tanh^{-1}\left|\frac{P^{-1}\left(b\right) - P^{-1}\left(a\right)}{P^{-1}\left(b\right) - \overline{P^{-1}\left(a\right)}}\right|$$

   As $P^{-1}\left(z\right) = \left(z+1\right)/\left\{i\left(z-1\right)\right\}$, simplification of the above expression yields the stated result.

   (b) Hint: Use the result from part (a) and the relationship

   $$\cosh\theta = \frac{1 + \tanh^2\left(\theta/2\right)}{1 - \tanh^2\left(\theta/2\right)}, \quad \theta \in \mathbb{R}$$

52. Let $\mathcal{C}$ be a circle of unit radius defined by $|z| = 1$. In the Poincaré disc model show that the lines (diameters) and arcs of circles that intersect the circle $\mathcal{C}$ perpendicularly are the hyperbolic lines of $\mathbb{D}_P$.

   Hint: See Pressley (2010). It has been proved earlier that the geodesics in $\mathbb{H}$ are: semicircles with centers on the real axis, and half-lines parallel to the imaginary axis. Also the transformation $P\left(\cdot\right)$ is Moebius and maps the boundary of $\mathbb{H}$ to that of $\mathbb{D}$. Therefore, this transformation preserves angles (in Euclidean sense), and maps lines and circles to lines and circles.

53. Prove that the isometries of $\mathbb{D}_P$ are the maps $P \circ G \circ P^{-1}\left(\cdot\right)$ where $G\left(\cdot\right)$ is any isometry of $\mathbb{H}$. Hint: See Pressley (2010). Note that $P\left(\cdot\right)$ is an isometry. It is specified that $G\left(\cdot\right)$ is any isometry of $\mathbb{H}$. Therefore the isometries of $\mathbb{D}_P$ are the maps: $P \circ G \circ P^{-1}\left(\cdot\right)$. Conversely, if $F$ is any isometry of $\mathbb{D}_P$, then $G\left(\cdot\right) = P^{-1} \circ F \circ P\left(\cdot\right)$ is an isometry of $\mathbb{H}$, and $F = P \circ G \circ P^{-1}\left(\cdot\right)$.

54. The three sides of a hyperbolic triangle are geodesics. Let the angles of the hyperbolic triangle be $\alpha, \beta$, and $\gamma$; and the length of sides opposite these angles be $a_\ell, b_\ell$, and $c_\ell$ respectively. Prove that

$$\cosh c_\ell = \cosh a_\ell \cosh b_\ell - \sinh a_\ell \sinh b_\ell \cos \gamma$$

This result is called the  first hyperbolic cosine rule.
Hint: See Pressley (2010). In this analysis, we shall use the following results.

$$\cosh \theta = \cosh^2 (\theta/2) + \sinh^2 (\theta/2)$$
$$= \frac{1 + \tanh^2 (\theta/2)}{1 - \tanh^2 (\theta/2)}$$
$$\sinh \theta = 2 \sinh (\theta/2) \cosh (\theta/2)$$

Let the vertices of the hyperbolic triangle be $a, b, c \in \mathbb{D}$; where the angles at these vertices are $\alpha, \beta$, and $\gamma$ respectively. By appropriate isometric transformation of the hyperbolic triangle, we select $c = 0 \in \mathbb{D}$, and the vertex $a$ along the positive real axis, so that $a > 0$. Therefore using distance formula of hyperbolic geometry, we have

$$a = \tanh (b_\ell/2), \quad |b| = \tanh (a_\ell/2), \quad b = e^{i\gamma} \tanh (a_\ell/2)$$

$$\tanh (c_\ell/2) = \frac{|b - a|}{|1 - \bar{a}b|}$$

Therefore

$$\cosh a_\ell = \frac{1 + \tanh^2 (a_\ell/2)}{1 - \tanh^2 (a_\ell/2)} = \frac{1 + |b|^2}{1 - |b|^2}$$

and

$$\cosh c_\ell = \frac{1 + \tanh^2 (c_\ell/2)}{1 - \tanh^2 (c_\ell/2)} = \frac{|1 - \bar{a}b|^2 + |b - a|^2}{|1 - \bar{a}b|^2 - |b - a|^2}$$
$$= \frac{\left(1 + |a|^2\right) \left(1 + |b|^2\right) - 2 \left(\bar{a}b + a\bar{b}\right)}{\left(1 - |a|^2\right) \left(1 - |b|^2\right)}$$
$$= \cosh a_\ell \cosh b_\ell - 4 \cos \gamma \frac{\tanh (a_\ell/2) \tanh (b_\ell/2)}{\left(1 - \tanh^2 (a_\ell/2)\right) \left(1 - \tanh^2 (b_\ell/2)\right)}$$
$$= \cosh a_\ell \cosh b_\ell - \sinh a_\ell \sinh b_\ell \cos \gamma$$

55. Establish the triangle inequality $c_\ell \leq (a_\ell + b_\ell)$ in the last problem. Equality occurs if and only if $c$ is on the hyperbolic line segment joining $a$ and $b$.
Hint: See Reid, and Szendrői (2005). We need to prove that

$$\cosh c_\ell \leq \cosh (a_\ell + b_\ell)$$

It is known that

$$\cosh (a_\ell + b_\ell) = \cosh a_\ell \cosh b_\ell + \sinh a_\ell \sinh b_\ell$$

Using result from the last problem, we have

$$\cosh (a_\ell + b_\ell) = \cosh c_\ell + (1 + \cos \gamma) \sinh a_\ell \sinh b_\ell$$

Also $\sinh a_\ell > 0$, and $\sinh b_\ell > 0$, therefore $\cosh(a_\ell + b_\ell) \geq \cosh c_\ell$. Equality occurs if and only if $\gamma = \pi$. As $\cosh c_\ell$ is an increasing function of $c_\ell$, we have $(a_\ell + b_\ell) \geq c_\ell$, where equality occurs if and only if $c$ is on the hyperbolic line segment joining $a$ and $b$.

56. The three sides of a hyperbolic triangle are geodesics. Let the angles of the hyperbolic triangle be $\alpha, \beta$, and $\gamma$; and the lengths of sides opposite these angles be $a_\ell, b_\ell$, and $c_\ell$ respectively. Prove that

$$\frac{\sin \alpha}{\sinh a_\ell} = \frac{\sin \beta}{\sinh b_\ell} = \frac{\sin \gamma}{\sinh c_\ell}$$

This result is called the hyperbolic sine rule.

Hint: See Pressley (2010). This result is established in two steps.

*Step* 1: Initially assume that $\gamma = \pi/2$. Therefore

$$\cosh c_\ell = \cosh a_\ell \cosh b_\ell$$

We also have

$$\cosh a_\ell = \cosh b_\ell \cosh c_\ell - \sinh b_\ell \sinh c_\ell \cos \alpha$$

Elimination of $\cosh c_\ell$ in the above two equations yields

$$\cos \alpha = \frac{\cosh a_\ell \sinh b_\ell}{\sinh c_\ell}$$

Therefore

$$\begin{aligned}
\sin^2 \alpha \sinh^2 c_\ell &= \sinh^2 c_\ell - \cosh^2 a_\ell \sinh^2 b_\ell \\
&= \cosh^2 a_\ell \cosh^2 b_\ell - 1 - \cosh^2 a_\ell \sinh^2 b_\ell \\
&= \cosh^2 a_\ell - 1 = \sinh^2 a_\ell
\end{aligned}$$

The above result implies

$$\frac{\sin \alpha}{\sinh a_\ell} = \frac{1}{\sinh c_\ell}$$

Similarly, we can obtain

$$\frac{\sin \beta}{\sinh b_\ell} = \frac{1}{\sinh c_\ell}$$

*Step* 2: In this step we consider the case, when $\gamma$ is not necessarily equal to $\pi/2$. Let the vertices of the hyperbolic triangle be $a, b, c \in \mathbb{D}$; where the angles at these vertices are $\alpha, \beta$, and $\gamma$ respectively. Also, let the hyperbolic line through the vertex $a$ intersect the opposite side at a point, which divides the side $a_\ell$ into hyperbolic line segments of length $a'_\ell$ and $a''_\ell$. Denote this line segment by $m$. Note that as $\beta$ is the angle at vertex $b$, it is also equal to the angle between hyperbolic line segments of length $c_\ell$ and $a'_\ell$. Assume that the length of this hyperbolic line segment $m$ be equal to $d_\ell$.

With this construction, the original hyperbolic triangle is split into two hyperbolic triangles. One hyperbolic triangle has line segments, $d_\ell, a'_\ell$, and $c_\ell$; with angles $\beta, \alpha'$, and $\pi/2$. The other hyperbolic triangle has line segments, $d_\ell, a''_\ell$, and $b_\ell$; with angles $\gamma, \alpha''$, and $\pi/2$.

We next apply the results of Step 1 to each of these two hyperbolic triangles. This results in

$$\frac{\sin \beta}{\sinh d_\ell} = \frac{1}{\sinh c_\ell}, \quad \text{and} \quad \frac{\sin \gamma}{\sinh d_\ell} = \frac{1}{\sinh b_\ell}$$

Therefore, we have

$$\frac{\sin \beta}{\sinh b_\ell} = \frac{\sin \gamma}{\sinh c_\ell}$$

We similarly obtain

$$\frac{\sin \alpha}{\sinh a_\ell} = \frac{\sin \beta}{\sinh b_\ell}$$

The final result follows. The analysis of the result, when the hyperbolic line through vertex $a$ intersects the hyperbolic line segment of length $a_\ell$ outside of it is similar.

57. The three sides of a hyperbolic triangle are geodesics. Let the angles of the hyperbolic triangle be $\alpha, \beta$, and $\gamma = \pi/2$; and the lengths of sides opposite these angles be $a_\ell, b_\ell$, and $c_\ell$ respectively. Prove that

   (a) $\cos \alpha = \sinh b_\ell \cosh a_\ell / \sinh c_\ell$
   (b) $\cosh a_\ell = \cos \alpha / \sin \beta$
   (c) $\sinh a_\ell = \tanh b_\ell / \tan \beta$

   Hint: See Pressley (2010).

   (a) If $\gamma = \pi/2$, the first hyperbolic cosine and hyperbolic sine rules yield

$$\cosh c_\ell = \cosh a_\ell \cosh b_\ell$$

$$\frac{\sin \alpha}{\sinh a_\ell} = \frac{\sin \beta}{\sinh b_\ell} = \frac{1}{\sinh c_\ell}$$

   Therefore

$$\sinh^2 c_\ell \cos^2 \alpha = \sinh^2 c_\ell \left\{ 1 - \frac{\sinh^2 a_\ell}{\sinh^2 c_\ell} \right\} = \cosh^2 c_\ell - \cosh^2 a_\ell$$
$$= \cosh^2 a_\ell \left\{ \cosh^2 b_\ell - 1 \right\} = \cosh^2 a_\ell \sinh^2 b_\ell$$

   (b) We have

$$\sin \beta = \frac{\sinh b_\ell}{\sinh c_\ell}$$

$$\cosh a_\ell = \cosh b_\ell \cosh c_\ell - \sinh b_\ell \sinh c_\ell \cos \alpha$$

   Therefore

$$\frac{\cos \alpha}{\sin \beta} = \frac{\cosh b_\ell \cosh c_\ell - \cosh a_\ell}{\sinh^2 b_\ell} = \frac{\cosh a_\ell \cosh^2 b_\ell - \cosh a_\ell}{\sinh^2 b_\ell}$$
$$= \cosh a_\ell$$

   (c) We have

$$\sin \beta = \frac{\sinh b_\ell}{\sinh c_\ell}$$

$$\cosh b_\ell = \cosh a_\ell \cosh c_\ell - \sinh a_\ell \sinh c_\ell \cos \beta$$

   Therefore

$$\cot \beta = \frac{\cosh a_\ell \cosh c_\ell - \cosh b_\ell}{\sinh a_\ell \sinh b_\ell} = \frac{\cosh^2 a_\ell \cosh b_\ell - \cosh b_\ell}{\sinh a_\ell \sinh b_\ell}$$
$$= \frac{\sinh^2 a_\ell \cosh b_\ell}{\sinh a_\ell \sinh b_\ell} = \frac{\sinh a_\ell}{\tanh b_\ell}$$

58. The three sides of a hyperbolic triangle are geodesics. Let the angles of the hyperbolic triangle be $\alpha, \beta$, and $\gamma$; and the lengths of sides opposite these angles be $a_\ell, b_\ell$, and $c_\ell$ respectively. Establish the following result

$$\cosh a_\ell = \frac{\cos\alpha + \cos\beta\cos\gamma}{\sin\beta\sin\gamma}$$

Hint: See Pressley (2010). This result is established in two steps.
*Step* 1: Observe that, if $\gamma = \pi/2$, we can use the results from the last problem. In this case

$$\cosh a_\ell = \frac{\cos\alpha}{\sin\beta}, \quad \sinh a_\ell = \frac{\tanh b_\ell}{\tan\beta}$$

$$\cosh b_\ell = \frac{\cos\beta}{\sin\alpha}, \quad \sinh b_\ell = \frac{\tanh a_\ell}{\tan\alpha}.$$

*Step* 2: Also, let the hyperbolic line through the vertex $a$ intersect the opposite side at a point, which divides the side $a_\ell$ into hyperbolic line segments of length $a'_\ell$ and $a''_\ell$. Denote this line segment by $m$. Note that as $\beta$ is the angle at vertex $b$, it is also equal to the angle between hyperbolic line segments of length $c_\ell$ and $a'_\ell$. Assume that the length of this hyperbolic line segment $m$ be equal to $d_\ell$.

With this construction, the original hyperbolic triangle is split into two hyperbolic triangles. One hyperbolic triangle has line segments, $d_\ell, a'_\ell$, and $c_\ell$; with angles $\beta, \alpha'$, and $\pi/2$. The other hyperbolic triangle has line segments, $d_\ell, a''_\ell$, and $b_\ell$; with angles $\gamma, \alpha''$, and $\pi/2$.

We next apply the results of Step 1 to each of these two hyperbolic triangles. This results in

$$\cosh a'_\ell = \frac{\cos\alpha'}{\sin\beta}, \quad \sinh a'_\ell = \frac{\tanh d_\ell}{\tan\beta}$$

$$\cosh a''_\ell = \frac{\cos\alpha''}{\sin\gamma}, \quad \sinh a''_\ell = \frac{\tanh d_\ell}{\tan\gamma}$$

Therefore

$$\begin{aligned}
\cosh a_\ell &= \cosh\left(a'_\ell + a''_\ell\right) \\
&= \cosh a'_\ell \cosh a''_\ell + \sinh a'_\ell \sinh a''_\ell \\
&= \frac{\cos\alpha'\cos\alpha''}{\sin\beta\sin\gamma} + \frac{\tanh^2 d_\ell}{\tan\beta\tan\gamma}
\end{aligned}$$

As in Step 1

$$\cosh d_\ell = \frac{\cos\beta}{\sin\alpha'} = \frac{\cos\gamma}{\sin\alpha''}$$

Therefore

$$\begin{aligned}
\cosh a_\ell \sin\beta\sin\gamma &= \cos\alpha'\cos\alpha'' + \cos\beta\cos\gamma\tanh^2 d_\ell \\
&= \cos\alpha + \sin\alpha'\sin\alpha'' + \cos\beta\cos\gamma\tanh^2 d_\ell \\
&= \cos\alpha + \cos\beta\cos\gamma\,\text{sech}^2 d_\ell + \cos\beta\cos\gamma\tanh^2 d_\ell \\
&= \cos\alpha + \cos\beta\cos\gamma
\end{aligned}$$

59. Prove the following statements about hyperbolic Moebius transformations.

(a) The Moebius transformations that map unit disc to itself are the hyperbolic Moebius transformations.

(b) The isometries of $\mathbb{D}_P$ are the hyperbolic Moebius, and conjugate hyperbolic Moebius transformations.

Hint: See Pressley (2010).

(a) Let $M(\cdot)$ be a real Moebius transformation that maps $\mathbb{H}$ to itself. As $P(\cdot)$ is also a Moebius transformation, the Moebius transformations that maps unit disc to itself are of the type $P \circ M \circ P^{-1}(\cdot)$. Let

$$M(z) = \frac{az + b}{cz + d}, \quad \text{where} \quad a, b, c, d \in \mathbb{R}, \text{ and } (ad - bc) > 0$$

Then
$$P \circ M \circ P^{-1}(z) = \frac{\{(a + d) + i(b - c)\}z + \{(a - d) - i(b + c)\}}{\{(a - d) + i(b + c)\}z + \{(a + d) - i(b - c)\}}$$

Observe that

$$|(a + d) + i(b - c)|^2 - |(a - d) - i(b + c)|^2 = 4(ad - bc) > 0$$

Therefore, the map $P \circ M \circ P^{-1}(\cdot)$ is a hyperbolic Moebius transformation.

Proceed as follows, for the proof in the opposite direction. Assume that $M(\cdot)$ is a hyperbolic Moebius transformation, then we have to show that $P \circ M \circ P^{-1}(\cdot)$ is a real Moebius transformation. Similar details are left to the reader.

(b) It is known from an earlier problem that the isometric transformations of $\mathbb{H}$ are of the type $M(\cdot)$ or $M \circ J(\cdot)$, where $M(\cdot)$ is a real Moebius transformation and $J(z) = -\bar{z}$. Consequently, the isometries of $\mathbb{D}_P$ are $P \circ M \circ P^{-1}(\cdot)$, and $P \circ (M \circ J) \circ P^{-1}(\cdot) = \left(P \circ M \circ P^{-1}\right) \circ \left(P \circ J \circ P^{-1}\right)(\cdot)$. Note that $P \circ M \circ P^{-1}(\cdot)$ is a hyperbolic Moebius transformation, and

$$P \circ J \circ P^{-1}(z) = P \circ J\left(\frac{z + 1}{i(z - 1)}\right)$$
$$= P\left(\frac{\bar{z} + 1}{i(\bar{z} - 1)}\right) = P \circ P^{-1}(\bar{z}) = \bar{z}$$

60. Show that the only possible tessellations of a Euclidean plane using regular polygons are via: equilateral triangles, squares, and regular hexagons.

Hint: Let the regular polygon which tiles the surface have $n \in \mathbb{P} \backslash \{1, 2\}$ sides. Assume that $k \in \mathbb{P} \backslash \{1, 2\}$ polygons meet at each vertex. Let $\alpha$ be the internal angle of the regular $n$-gon. Therefore

$$\alpha = \frac{(\pi n - 2\pi)}{n}, \quad \text{and} \quad \alpha k = 2\pi$$

The above two equations yield $nk - 2(n + k) = 0$. It can be rewritten as

$$(n - 2)(k - 2) = 4$$

The only possible solutions of the above equation are: $(n, k) = (3, 6), (4, 4),$ and $(6, 3)$. These results correspond to polygons which are: the equilateral triangles, squares, and regular hexagons respectively.

61. Prove that if the tessellation of a hyperbolic plane via regular hyperbolic $n$-gon with $k$ polygons meeting at each vertex exists, then

$$\frac{1}{n} + \frac{1}{k} < \frac{1}{2}$$

where $n, k \in \mathbb{P}\backslash\{1, 2\}$.

Hint: A regular hyperbolic polygon is a polygon, whose sides are hyperbolic lines of equal length. Let $\mathcal{P}$ be a regular hyperbolic polygon with internal angle equal to $\alpha$. The hyperbolic area $A(\mathcal{P})$ of this polygon is

$$A(\mathcal{P}) = (n-2)\pi - n\alpha > 0$$

Use of the relationship $\alpha k = 2\pi$ and the above inequality yields the stated result.

62. The Poincaré disc can be recursively tiled by $\{n, k\}$ tessellations. Show that the number of vertices of the tessellating polygons grows exponentially as the number of levels of recursion. Hint: See Csernai, Gulyás, Körösi, Sonkoly, and Biczók (2013). The $\{n, k\}$ tessellation is possible if $(n-2)(k-2) > 4$. Let the number of polygons and vertices at the "perimeter" of the $m$th level be $p_m$, and $v_m$ respectively. Also let $n > 3$. Then for $m \in \mathbb{N}$ we have

$$v_{m+1} + v_m = (n-2)p_{m+1}$$
$$p_{m+1} + p_m = (k-2)v_m$$

These equations can be written in matrix form as

$$\begin{bmatrix} v_{m+1} \\ p_{m+1} \end{bmatrix} = \begin{bmatrix} (k-2)(n-2)-1 & -(n-2) \\ (k-2) & -1 \end{bmatrix} \begin{bmatrix} v_m \\ p_m \end{bmatrix}$$

Denote the $2 \times 2$ matrix in the above equation by $M$. This results in

$$\begin{bmatrix} v_m \\ p_m \end{bmatrix} = M^m \begin{bmatrix} v_0 \\ p_0 \end{bmatrix}, \quad m \in \mathbb{N}$$

where $v_0 = n$, and $p_0 = 0$. Also $M^m = Q\Lambda^m Q^{-1}$. The diagonal elements of the diagonal matrix $\Lambda$ are the eigenvalues of the matrix $M$. The matrix $Q$ has the corresponding eigenvectors in its columns. For brevity, we introduce the notation

$$\alpha = (k-2)(n-2)$$
$$\beta = \left\{(\alpha-2)^2 - 4\right\}^{1/2}$$

Thus

$$\Lambda = \begin{bmatrix} (\alpha-2-\beta)/2 & 0 \\ 0 & (\alpha-2+\beta)/2 \end{bmatrix}$$

$$Q = \begin{bmatrix} \dfrac{(\alpha-\beta)}{2(k-2)} & \dfrac{(\alpha+\beta)}{2(k-2)} \\ 1 & 1 \end{bmatrix}$$

As $\beta \simeq \alpha$, we have $v_m \simeq n(\alpha-1)^m$. This result implies that the number of vertices in the tessellated Poincaré disc grows exponentially as $m$.

# References

1. Abramowitz, M., and Stegun, I. A., 1965. *Handbook of Mathematical Functions*, Dover Publications, Inc., New York, New York.
2. Anderson, J. W., 2005. *Hyperbolic Geometry*, Second Edition, Springer-Verlag, Berlin, Germany.
3. Csernai, M., Gulyás, A., Korösi, A., Sonkoly, B., and Biczók, G., 2013. "Incrementally Upgradable Data Center Architecture Using Hyperbolic Tessellations," Computer Networks, Vol. 57, No. 6, pp. 1373-1393.
4. do Carmo, M. P., 1976. *Differential Geometry of Curves and Surfaces*, Prentice-Hall, Englewood Cliffs, New Jersey.
5. Goodman-Strauss, C., 2001. "Compass and Straightedge in the Poincaré Disk," The Mathematical Association of America, Monthly, Vol. 108, No. 1, pp. 38-49.
6. Graustein, W. C., 1966. *Differential Geometry*, Dover Publications, Inc., New York, New York.
7. Greenberg, M. J., 1974. *Euclidean, and Non-Euclidean Geometries*: *Development and History*, Second Edition, W. H. Freeman and Company, San Francisco, California.
8. Heath, T. L., 1956. *Euclid's Elements,* Dover Publications, Inc., New York, New York.
9. Henle, M., 1997. *Modern Geometries*: *The Analytical Approach*, Prentice-Hall Inc., Simon & Schuster/A Viacom Company, Upper Saddle River, New Jersey.
10. Iversen, B., 1992. *Hyperbolic Geometry*, Cambridge University Press, Cambridge, Great Britain.
11. Keen, L., and Lakic, N., 2007. *Hyperbolic Geometry from a Local Viewpoint,* Cambridge University Press, Cambridge, Great Britain.
12. Kreyszig, E., 1968. *Introduction to Differential Geometry, and Riemannian Geometry,* University of Toronto Press, Toronto, Canada.
13. Kreyszig, E., 2011. *Advanced Engineering Mathematics,* Tenth Edition, John Wiley & Sons, Inc., New York, New York.
14. Kühnel, W. 2002. *Differential Geometry, Curves - Surfaces - Manifolds,* American Mathematical Society, Providence, Rhode Island.
15. Lipschutz, M. M., 1969. *Differential Geometry*, Schaum's Outline Series, McGraw-Hill Book Company, New York, New York.
16. Pogorelov, A., 1987. *Geometry*, Mir Publishers, Moscow.
17. Pressley, A., 2001. *Elementary Differential Geometry,* Springer-Verlag, Berlin, Germany.
18. Pressley, A., 2010. *Elementary Differential Geometry*, Second Edition, Springer-Verlag, Berlin, Germany.
19. Ramsay, A., and Richtmyer, R. D., 1995. *Introduction to Hyperbolic Geometry*, Springer-Verlag, Berlin, Germany.
20. Ratcliffe, J. G., 1994. *Foundations of Hyperbolic Manifolds*, Springer-Verlag, Berlin, Germany.
21. Reid, M., and Szendrői, B., 2005. *Geometry and Topology*, Cambridge University Press, Cambridge, Great Britain.
22. Spiegel, M. R., 1959. *Vector Analysis*, Schaum's Outline Series, McGraw-Hill Book Company, New York, New York.
23. Spiegel, M. R., 1963. *Advanced Calculus*, Schaum's Outline Series, McGraw-Hill Book Company, New York, New York.
24. Spiegel, M. R., 1964. *Complex Variables*, Schaum's Outline Series, McGraw-Hill Book Company, New York, New York.
25. Stillwell, J., 1992. *Geometry of Surfaces*, Springer-Verlag, Berlin, Germany.

26.  Struik, D. J., 1988. *Lectures on Classical Differential Geometry*, Dover Publications, Inc., New York, New York.

27.  Venema, G. A., 2011. *The Foundations of Geometry*, Second Edition, Pearson Prentice-Hall, Upper Saddle River, New Jersey.

28.  Walkden, C., 2015. "Hyperbolic Geometry," available at http://www.maths.manchester.ac.uk/~cwalkden/hyperbolic-geometry/hyperbolic_geometry.pdf. Retrieved February 16, 2015.

29.  Weinstock, R., 1974. *Calculus of Variations, with Applications to Physics and Engineering,* Dover Publications, Inc., New York, New York.

30.  Wolfe, H. E., 1966. *Non-Euclidean Geometry*, Holt, Rinehart and Winston, New York, New York.

# Applied Analysis

$$f(t) = \sum_{n \in \mathbb{Z}} c_n e^{in\omega_0 t}, \quad t \in \mathbb{R}$$

$$T_0 \in \mathbb{R}^+, \quad \omega_0 = 2\pi/T_0$$

Complex Fourier series

**Jean Baptiste Joseph Fourier.** Fourier was born on 21 March, 1768 in Auxerre, Bourgogne, France. He was orphaned at the age of nine. Besides contemplating a career in priesthood at a very young age, Fourier demonstrated an aptitude for mathematics beginning at the age of thirteen. Young Fourier was nominated to study at the École Normale in Paris in the year 1794. In the year 1795, he was appointed to a position in École Polytechnique.

Fourier discovered that any periodic motion can be expressed as a superposition of sinusoidal and cosinusoidal waves. Based upon partial differential equations, Fourier developed a mathematical theory of heat and expounded it in *Théorie Analytique de la Chaleur*, (1822). It is in this treatise that he developed the concept of Fourier series. He also initiated the study of Fourier integrals.

Fourier series has found multiple applications in mathematics, mathematical physics, engineering, and several other scientific disciplines. William Thompson Kelvin, a British mathematician called Fourier's work a *mathematical poem*. Fourier died on 16 May, 1830 in Paris, France.

## 6.1 Introduction

Analysis is the fountain of several powerful techniques in applied mathematics. The mathematical concepts defined and developed in this chapter find a variety of applications. Basic concepts in analysis, complex analysis, basics of Cartesian geometry, and Moebius transformation, are discussed in some depth. Properties of polynomials, asymptotic behavior and algorithmic-complexity classes, vector algebra and, vector spaces are revisited from the chapter on abstract algebra. Generalized functions, Fourier series, continuous transform techniques, discrete transform techniques, Faà di Bruno's formula, and special mathematical functions are also elucidated. Basics of number theory, algebra, and matrices and determinants are also used in this chapter.

Like any other field of human endeavor, analysis was developed by several persons. This branch of mathematics essentially flourished and developed significantly in the nineteenth century AD. Some of the mathematicians who nurtured this field were: Abraham De Moivre (1667-1754), Leonhard Euler (1707-1783), Joseph-Louis Lagrange (1736-1813), Pierre-Simon Laplace (1749-1827), Joseph Fourier (1768-1830), and Augustin-Louis Cauchy (1789-1857).

## 6.2 Basic Analysis

Several basic concepts in analysis are outlined in this section. These are: point sets, limits, continuous functions, derivatives, monotonicity, partial derivatives, Jacobians and singularities, hyperbolic functions, and ordinary differential equations.

### 6.2.1  Point Sets

Concepts such as neighborhoods, interior points, interior of a set, exterior point, boundary points, limit points, open set, closure of a set, closed set, dense set, and compact set are defined. These concepts are defined on subsets of the real line $\mathbb{R}$. These in turn can be conveniently extended to other spaces.

**Definitions 6.1.** *All the defined points and sets are on the real line $\mathbb{R}$.*

1. *Absolute value of $a \in \mathbb{R}$ is denoted by $|a|$. It is equal to $a$ if $a \geq 0$ and $-a$ if $a < 0$.*
2. *$\delta$-neighborhood: Let $\delta$ be a positive number. A $\delta$-neighborhood of a point $x_0$ is the set of all points $x$ such that $|x - x_0| < \delta$.*
3. *Deleted $\delta$-neighborhood: A deleted $\delta$-neighborhood of a point $x_0$ is the set of all points $x$ such that $0 < |x - x_0| < \delta$. It excludes the point $x_0$ itself.*
4. *Interior point: A point $x_0 \in X$ is an interior point of the set $X$ if and only if there exists a $\delta$-neighborhood of the point $x_0$, such that all the points in this neighborhood belong to the set $X$.*
5. *Interior of a set $X$ : The interior of a set is the set of all its interior points.*
6. *Exterior point: A point $x_0 \in X$ is an exterior point of set $X$ if and only if all the $\delta$-neighborhoods of the point $x_0$, belong to the complement of the set $X$.*
7. *Boundary point: A point $x_0 \in X$ is a boundary point of set $X$ if and only if all the $\delta$-neighborhoods of the point $x_0$, belong to both the set $X$ and its complement.*
8. *Limit points: A point $x_0 \in X$ is a limit point of a set $X$ if and only if all deleted $\delta$-neighborhoods of $x_0$ contain points which belong to $X$.*
9. *Open set: A set is open, if every point in it is an interior point.*
10. *Closure of a set: The union of a set of points $X$ and all its limit points is called its closure.*
11. *Closed set: A set $X$ is closed, if it contains all its limit points.*
12. *Dense set: Let $\widetilde{X}$ be a subset of $X$. The subset $\widetilde{X}$ is dense if the closure of the set $\widetilde{X}$ is equal to $X$.*
13. *Compact set: A set of points is compact, if and only if it is closed and bounded.*
14. *Let $x \in \mathbb{R}$ and $\epsilon \in \mathbb{R}^+$. As $\epsilon \to 0$ then:*
    (a) *$(x + \epsilon)$ is denoted by $x_+$. Thus $x_+$ is the right limiting value of $x$.*
    (b) *$(x - \epsilon)$ is denoted by $x_-$. Thus $x_-$ is the left limiting value of $x$.*                     $\square$

Observe that a finite union of closed sets is also closed. However, an infinite union of closed sets is not necessarily closed. For example, let $I_n = [1/n, 1]$. Then $\bigcup_{n=1}^{\infty} I_n = (0, 1]$. This infinite union is not closed, since $0$ is a limit point of this union, which is not in this set. Note that an empty set is closed by definition. Intersection of closed sets yields a closed set.

Infinite unions of open intervals are open sets in $\mathbb{R}$. In contrast, infinite intersections of open intervals are not open sets. For example if $J_n = (-1/n, 1/n)$, Then $\bigcap_{n=1}^{\infty} J_n = \{0\}$ is closed.

An open set $X$ is dense in its closure. The set of rational numbers $\mathbb{Q}$ is dense in $\mathbb{R}$. The set of irrational numbers is also dense in $\mathbb{R}$.

### 6.2.2  Limits, Continuity, Derivatives, and Monotonicity

The concepts of limits, continuity, derivative are interlinked. Different types of monotonic functions are also described.

**Definitions 6.2.** *About limits.*

1. *Limit of a function*: *A function $f : \mathbb{R} \to \mathbb{R}$ has a limit $L$ at a point $\widetilde{x}$, if for every real number $\epsilon > 0$ there exists a real number $\delta > 0$ such that for all $x \in \mathbb{R}$ with*

$$0 < |x - \widetilde{x}| < \delta \;\; \Rightarrow \;\; |f(x) - L| < \epsilon \qquad (6.1a)$$

*The limit is denoted by $\lim_{x \to \widetilde{x}} f(x) = L$.*

2. *Right-hand limit of a function*: *A function $f : \mathbb{R} \to \mathbb{R}$ has a right-hand limit $L$ at a point $\widetilde{x}$, if for every real number $\epsilon > 0$ there exists a real number $\delta > 0$ such that for all $x \in \mathbb{R}$ with*

$$\widetilde{x} < x < \widetilde{x} + \delta \;\; \Rightarrow \;\; |f(x) - L| < \epsilon \qquad (6.1b)$$

*The limit is denoted by $\lim_{x \to \widetilde{x}_+} f(x) = L$.*

3. *Left-hand limit of a function*: *A function $f : \mathbb{R} \to \mathbb{R}$ has a left-hand limit $L$ at a point $\widetilde{x}$, if for every real number $\epsilon > 0$ there exists a real number $\delta > 0$ such that for all $x \in \mathbb{R}$ with*

$$\widetilde{x} - \delta < x < \widetilde{x} \;\; \Rightarrow \;\; |f(x) - L| < \epsilon \qquad (6.1c)$$

*The limit is denoted by $\lim_{x \to \widetilde{x}_-} f(x) = L$.*

4. *Limit superior and limit inferior of a sequence of real numbers*: *Consider a sequence of real numbers $\ldots, x_{-2}, x_{-1}, x_0, x_1, x_2, \ldots$. Let $\epsilon$ be any positive real number.*

   (a) *A real number $\overline{x}$ is called a limit superior, or greatest limit, or upper limit* (lim sup) *of the sequence, if infinite number of terms of the sequence are greater than $(\overline{x} - \epsilon)$ and only a finite number of terms are greater than $(\overline{x} + \epsilon)$.*

   (b) *A real number $\underline{x}$ is called a limit inferior, or least limit, or lower limit* (lim inf) *of the sequence, if infinite number of terms of the sequence are less than $(\underline{x} + \epsilon)$ and only a finite number of terms are less than $(\underline{x} - \epsilon)$.* □

In simple terms, $f(x)$ has a limit $L$ at $x = \widetilde{x}$, if for the numbers $x$ near $\widetilde{x}$, the value of $f(x)$ is close to $L$. The right-hand and left-hand limits are generally called the one-sided limits, and $\lim_{x \to \widetilde{x}} f(x)$ is called the two-sided limit. These three limits are related by the following lemma.

**Lemma 6.1.** *A function $f : \mathbb{R} \to \mathbb{R}$ has a limit $L$ at a point $\widetilde{x}$ if and only if the right-hand and left-hand limits at the point $\widetilde{x}$ exist and are equal. That is:*

$$\lim_{x \to \widetilde{x}} f(x) = L \;\; \Leftrightarrow \;\; \lim_{x \to \widetilde{x}_+} f(x) = L \;\; and \;\; \lim_{x \to \widetilde{x}_-} f(x) = L \qquad (6.2)$$

□

A sequence of real numbers converge if and only if its limit superior and limit inferior are equal and finite. A continuous function is next defined.

**Definitions 6.3.** *About continuity.*

1. *Right-hand continuity. A function $f : \mathbb{R} \to \mathbb{R}$ is continuous on the right at point $\widetilde{x}$ if: both $\lim_{x \to \widetilde{x}_+} f(x)$ and $f(\widetilde{x})$ exist, and $\lim_{x \to \widetilde{x}_+} f(x) = f(\widetilde{x})$.*

2. *Left-hand continuity. A function $f : \mathbb{R} \to \mathbb{R}$ is continuous on the left at point $\widetilde{x}$ if: both $\lim_{x \to \widetilde{x}_-} f(x)$ and $f(\widetilde{x})$ exist, and $\lim_{x \to \widetilde{x}_-} f(x) = f(\widetilde{x})$.*

3. *A function $f : \mathbb{R} \to \mathbb{R}$ is continuous at point $\widetilde{x}$ if: both $\lim_{x \to \widetilde{x}} f(x)$ and $f(\widetilde{x})$ exist, and $\lim_{x \to \widetilde{x}} f(x) = f(\widetilde{x})$.*
*Equivalently, a function $f : \mathbb{R} \to \mathbb{R}$ is continuous at point $\widetilde{x}$ if for every $\epsilon > 0$, there exists $\delta_{\widetilde{x},\epsilon} > 0$ such that $x \in \mathbb{R}$ and $|x - \widetilde{x}| < \delta_{\widetilde{x},\epsilon} \Rightarrow |f(x) - f(\widetilde{x})| < \epsilon$.*
*A function $f(\cdot)$ which is not continuous at $\widetilde{x}$ is said to be discontinuous at $\widetilde{x}$.*
4. *A function $f : \widetilde{S} \to \mathbb{R}$ is a continuous function on a set $\widetilde{S} \subseteq \mathbb{R}$, if $f(\cdot)$ is continuous at every point of $\widetilde{S}$.*
5. *Piecewise-continuous functions: A function $f : \mathbb{R} \to \mathbb{R}$ is piecewise-continuous in a finite interval $I \subseteq \mathbb{R}$, if:*

   (a) *The interval $I$ can be divided into a finite number of subintervals. Furthermore, in each such subinterval the function $f(\cdot)$ is continuous.*

   (b) *The limits of $f(x)$ as $x$ approaches the end-points of each subinterval are finite.*

   *Thus a piecewise-continuous function is one which has at most a finite number of finite discontinuities in every finite subinterval of $\mathbb{R}$.*
6. *A function $f : \mathbb{R} \to \mathbb{R}$ is uniformly continuous on a set $H \subseteq \mathbb{R}$ if for every $\epsilon > 0$ there exists a $\delta > 0$ such that $|x - y| < \delta \Rightarrow |f(x) - f(y)| < \epsilon$ where $x, y \in H$.* $\qquad\square$

It can be shown that, if a real-valued function $f(\cdot)$ is continuous on a closed bounded set $H$, then it is also uniformly continuous on the set $H$. The derivative of a function is defined as follows.

**Definitions 6.4.** *Let $f : \mathbb{R} \to \mathbb{R}$ be a function.*

1. *Let $a, b \in \mathbb{R}$ such that $a < b$, and $f(\cdot)$ is defined at any point $x_0 \in (a, b)$. The first derivative of $f(x)$ at $x = x_0$ is defined as*

$$f'(x_0) = \lim_{h \to 0} \frac{f(x_0 + h) - f(x_0)}{h} \tag{6.3a}$$

*if the limit exists. Other convenient notations for the first derivative of $f(x)$ at $x_0$ are*

$$\left. \frac{df(x)}{dx} \right|_{x=x_0} \quad and \quad \dot{f}(x_0) \tag{6.3b}$$

*If there is no ambiguity, the first derivative of $f(x)$ is simply referred to as the derivative of $f(x)$.*
2. *A function $f(\cdot)$ is differentiable at a point $x = x_0$ if $f'(x_0)$ exists.*
3. *If the first derivative of a function exists at all points of an interval, then its is said to be differentiable in the interval.*
4. *Second derivative: The second derivative of $f(x)$ at $x = x_0$, if it exists is the first derivative of $f'(x)$. This second derivative is denoted by either $f''(x_0)$ or $\ddot{f}(x_0)$.*
5. *Higher derivatives: Higher order derivatives can be defined recursively. The $n$th derivative of $f(x)$ at $x = x_0$, if it exists is the first derivative of the $(n-1)$th derivative of $f(x)$. It is denoted by $f^{(n)}(x_0), n \in \mathbb{P}$. The $n$th derivative of $f(x)$ at $x_0$ is also denoted by*

$$\left. \frac{d^n f(x)}{dx^n} \right|_{x=x_0} \tag{6.3c}$$

*Note that the notation $f^{(0)}(x) \triangleq f(x)$ is often used.* $\qquad\square$

It should be noted that if $f(x)$ is differentiable at $x = x_0$ then it is continuous at that point. Functions can also be classified based upon the existence of its derivatives.

**Definitions 6.5.** *Consider a function $f : I \to \mathbb{R}$, where $I$ is a closed interval in $\mathbb{R}$.*

1. *The function $f(\cdot)$ is of class $C^0$ on $I$ if $f(x)$ is continuous at all $x \in I$.*
2. *The function $f(\cdot)$ is of class $C^r$ on $I$ if $f^{(r)}(x)$ exists and is continuous at all $x \in I$, where $r$ is a positive integer.*
3. *The function $f(\cdot)$ is smooth (or continuously differentiable) on the closed interval $I$, if it belongs to class $C^1$.*
4. *The function $f(\cdot)$ is of class $C^\infty$ on the closed interval, if all its derivatives exist and are continuous.* □

One of the most important theorems in calculus is the *mean value theorem*. This theorem was proved by Lagrange in 1787.

**Theorem 6.1.** *Let $f(x)$ be continuous in $[a, b]$ and differentiable in $(a, b)$, then there exists at least a single number $\xi \in (a, b)$ such that*

$$\frac{f(b) - f(a)}{b - a} = f'(\xi) \tag{6.4}$$

*Proof.* The proof can be found in any standard textbook on calculus. □

A function $f(\cdot)$ which is infinitely differentiable (that is all its derivatives exist) has a Taylor's series expansion. It is named after the mathematician Brook Taylor (1685-1731).

**Theorem 6.2.** *Taylor's theorem of the mean. Let $f(x)$ and its first $n$ derivatives $f'(x)$, $f''(x)$, $\ldots, f^{(n)}(x)$ be continuous in $[a, b]$ and differentiable in $(a, b)$, then there exists a point $\xi \in (a, b)$ such that*

$$f(b) = \sum_{m=0}^{n} \frac{(b-a)^m}{m!} f^{(m)}(a) + R_n \tag{6.5a}$$

$$R_n = \frac{(b-a)^{n+1}}{(n+1)!} f^{(n+1)}(\xi), \quad a < \xi < b \tag{6.5b}$$

*where $R_n$ is called the remainder.*
*Proof.* The proof can be found in any standard textbook on calculus. □

An alternate and well-known representation of the above result is as follows. Let $x, (x + h) \in (a, b)$, then

$$f(x + h) = \sum_{m=0}^{n} \frac{h^m}{m!} f^{(m)}(x) + \frac{h^{n+1}}{(n+1)!} f^{(n+1)}(\xi), \quad a < \xi < b \tag{6.6a}$$

The above result is called the Taylor's series for $f(x)$ with a remainder. If $\lim_{n \to \infty} R_n \to 0$, an infinite series is obtained.

$$f(x + h) = \sum_{m \in \mathbb{N}} \frac{h^m}{m!} f^{(m)}(x) \tag{6.6b}$$

Taylor series is an example of *power series*. If the power series exists in some interval, then it is a *convergent series* in that interval. Furthermore, the corresponding interval is called the *interval of convergence*.

**Example 6.1.** A binomial series expansion.

$$(1+x)^\alpha = 1 + \alpha x + \frac{\alpha(\alpha - 1)}{2!}x^2 + \ldots + \frac{\alpha(\alpha - 1)\ldots(\alpha - n + 1)}{n!}x^n + \ldots$$

for $|x| < 1$, and any $\alpha \in \mathbb{R}$. □

Different types of monotonic functions are described below.

**Definition 6.6.** *Monotonic functions*: Let $f : \mathbb{R} \to \mathbb{R}$, $S \subseteq \mathbb{R}$, and $x_1, x_2 \in S$.

(a) *The function $f(\cdot)$ is monotonically increasing on the set $S$ iff for each pair of numbers $x_1, x_2$, $x_1 < x_2$ implies $f(x_1) < f(x_2)$.*
(b) *The function $f(\cdot)$ is monotonically nondecreasing on the set $S$ iff for each pair of numbers $x_1, x_2$, $x_1 < x_2$ implies $f(x_1) \leq f(x_2)$.*
(c) *The function $f(\cdot)$ is monotonically decreasing on the set $S$ iff for each pair of numbers $x_1, x_2$, $x_1 < x_2$ implies $f(x_1) > f(x_2)$.*
(d) *The function $f(\cdot)$ is monotonically nonincreasing on the set $S$ iff for each pair of numbers $x_1, x_2$, $x_1 < x_2$ implies $f(x_1) \geq f(x_2)$.* □

### 6.2.3 Partial Derivatives

Functions of two or more variables are defined and discussed in this subsection. Difference between dependent and independent variables is also stated. Neighborhoods, limits, continuity, and partial derivatives are defined.

**Definitions 6.7.** *All the defined points and sets are on the real line $\mathbb{R}$.*

1. *Real-valued function of two real-variables. Let $I, J, K \subseteq \mathbb{R}$. A function of two variables is $f : I \times J \to K$, where $(x, y) \in I \times J$ is assigned a unique element $z \subset K$. The assignment of the specific pair $(x, y)$ to $z$ is denoted as $f(x, y) = z$. This function is sometimes denoted by $f(\cdot, \cdot)$.*
2. *Dependent and independent variables. If $z = f(x, y)$, then $x$ and $y$ are called the independent variables, and $z$ the dependent variable.*
3. *Neighborhoods. Let $\delta$ be a positive real number. A rectangular $\delta$-neighborhood of a point $(x_0, y_0)$ is the set of all points $(x, y)$ such that $|x - x_0| < \delta$, and $|y - y_0| < \delta$. A circular $\delta$-neighborhood of a point $(x_0, y_0)$ is the set of all points $(x, y)$ such that $(x - x_0)^2 + (y - y_0)^2 < \delta^2$. Deleted $\delta$-neighborhood is the set of all points in the $\delta$-neighborhood, except the point $(x_0, y_0)$.*
4. *Limits. Consider a real-valued function $f : \mathbb{R}^2 \to \mathbb{R}$ defined in a deleted $\delta$-neighborhood of $(x_0, y_0)$. The limit of the function $f(x, y)$ as $(x, y)$ approaches $(x_0, y_0)$ is $L$, if for every real number $\epsilon > 0$ there exists a real number $\delta > 0$ such that for all $x, y \in \mathbb{R}$ with*

$$0 < |x - x_0| < \delta \quad and \quad 0 < |y - y_0| < \delta \quad \Rightarrow \quad |f(x, y) - L| < \epsilon \qquad (6.7a)$$

*In general, $\delta$ depends upon $\epsilon$ and $(x_0, y_0)$. The above condition can also be replaced by an alternate condition. It is called the deleted circular $\delta$-neighborhood of the point $(x_0, y_0)$. This is*

$$0 < (x - x_0)^2 + (y - y_0)^2 < \delta^2 \quad \Rightarrow \quad |f(x, y) - L| < \epsilon \tag{6.7b}$$

*The limit is denoted by $\lim_{(x,y)\to(x_0,y_0)} f(x, y) = L$.*

5. *Continuity. Let $f(\cdot, \cdot)$ be a real-valued function of two real variables. It is defined at $(x_0, y_0)$ and also in a $\delta$-neighborhood of $(x_0, y_0)$, where $\delta > 0$. The function $f(\cdot, \cdot)$ is continuous at $(x_0, y_0)$, if the following three conditions hold*

   (i) *$\lim_{(x,y)\to(x_0,y_0)} f(x, y) = L$. That is, the limit exists as $(x, y) \to (x_0, y_0)$*
   (ii) *$f(x_0, y_0)$ is defined at $(x_0, y_0)$.*
   (iii) *$L = f(x_0, y_0)$.*

   *If the function is not continuous at $f(x_0, y_0)$, then it is said to be discontinuous at $f(x_0, y_0)$. In this case, $(x_0, y_0)$ is called a point of discontinuity.*  □

Next consider the limits

$$\lim_{x\to x_0} \left\{ \lim_{y\to y_0} f(x, y) \right\} \triangleq L_1$$

$$\lim_{y\to y_0} \left\{ \lim_{x\to x_0} f(x, y) \right\} \triangleq L_2$$

Note that $L_1 \neq L_2$ in general. However, it is necessary that $L_1 = L_2$ for $L$ to exist. Furthermore, the equality $L_1 = L_2$ does not guarantee that $L$ exists.

Consider a function of two variables. These variables are assumed to be independent of each other. The ordinary derivative of the function with respect to a single variable, while keeping all other variables fixed is called the partial derivative of the function with respect to this variable.

**Definition 6.8.** *Partial derivatives. Let $f : \mathbb{R}^2 \to \mathbb{R}$ be a function.*

1. *Let $a, b, c, d \in \mathbb{R}$ such that $a < b$ and $c < d$, and $f(\cdot, \cdot)$ is defined at any point $x_0 \in (a, b)$, and $y_0 \in (c, d)$. The first partial derivative of $f(x, y)$ at $(x_0, y_0)$ with respect to $x$ is defined as*

$$\frac{\partial f(x_0, y_0)}{\partial x} = \lim_{\Delta x \to 0} \frac{f(x_0 + \Delta x, y_0) - f(x_0, y_0)}{\Delta x} \tag{6.8a}$$

*if the limit exists. Similarly, the first partial derivative of $f(x, y)$ at $(x_0, y_0)$ with respect to $y$ is defined as*

$$\frac{\partial f(x_0, y_0)}{\partial y} = \lim_{\Delta y \to 0} \frac{f(x_0, y_0 + \Delta y) - f(x_0, y_0)}{\Delta y} \tag{6.8b}$$

*if the limit exists. Other convenient notations for the first partial derivative of $f(x, y)$ with respect to $x$ at $(x_0, y_0)$ are*

$$\left. \frac{\partial f(x, y)}{\partial x} \right|_{(x=x_0, y=y_0)} , \text{ and } f_x(x_0, y_0) \tag{6.8c}$$

*Similarly, the other convenient notations for the first partial derivative of $f(x, y)$ with respect to $y$ at $(x_0, y_0)$ are*

$$\left. \frac{\partial f(x, y)}{\partial y} \right|_{(x=x_0, y=y_0)} , \text{ and } f_y(x_0, y_0) \tag{6.8d}$$

*If $f_x$ and $f_y$ are also continuous in a region $\mathcal{R} \subseteq \mathbb{R}^2$ then $f$ is continuously differentiable in region $\mathcal{R}$.*

2. *Higher order partial derivatives. If the partial derivatives $f_x(x,y)$ and $f_y(x,y)$ exist at all points in a region $\mathcal{R} \subseteq \mathbb{R}^2$ then these partial derivatives are also functions of $x$ and $y$. Therefore both $f_x(x,y)$ and $f_y(x,y)$ may have partial derivatives with respect to $x$ and $y$. If these exist, then they are called the second-order partial derivatives of $f(x,y)$. These are specified as*

$$\frac{\partial}{\partial x}\left(\frac{\partial f(x,y)}{\partial x}\right) = \frac{\partial^2 f(x,y)}{\partial x^2} = f_{xx}(x,y) \tag{6.9a}$$

$$\frac{\partial}{\partial y}\left(\frac{\partial f(x,y)}{\partial y}\right) = \frac{\partial^2 f(x,y)}{\partial y^2} = f_{yy}(x,y) \tag{6.9b}$$

$$\frac{\partial}{\partial y}\left(\frac{\partial f(x,y)}{\partial x}\right) = \frac{\partial^2 f(x,y)}{\partial y\partial x} = f_{xy}(x,y) \tag{6.9c}$$

$$\frac{\partial}{\partial x}\left(\frac{\partial f(x,y)}{\partial y}\right) = \frac{\partial^2 f(x,y)}{\partial x\partial y} = f_{yx}(x,y) \tag{6.9d}$$

*If $f_{xy}(\cdot,\cdot)$ and $f_{yx}(\cdot,\cdot)$ are continuous functions, then $f_{yx}(\cdot,\cdot) = f_{yx}(\cdot,\cdot)$. Third, fourth, and other high-ordered derivatives can similarly be defined.* □

### 6.2.4  Jacobians, Singularity, and Related Topics

Jacobians, bounded function, bounded variation of a function, and singularities of a function are next defined. Jacobians are useful in the transformation of multiple integrals.

**Definition 6.9.** *Let $f(u,v)$ and $g(u,v)$ be differentiable in a region. The Jacobian determinant of $f(u,v)$ and $g(u,v)$ with respect to $u$ and $v$ is the determinant*

$$\frac{\partial(f,g)}{\partial(u,v)} = \begin{vmatrix} \partial f/\partial u & \partial f/\partial v \\ \partial g/\partial u & \partial g/\partial v \end{vmatrix} \tag{6.10a}$$

*Similarly, the third order Jacobian determinant is defined as*

$$\frac{\partial(f,g,h)}{\partial(u,v,w)} = \begin{vmatrix} \partial f/\partial u & \partial f/\partial v & \partial f/\partial w \\ \partial g/\partial u & \partial g/\partial v & \partial g/\partial w \\ \partial h/\partial u & \partial h/\partial v & \partial h/\partial w \end{vmatrix} \tag{6.10b}$$

*The corresponding matrices are called Jacobian matrices.* □

These determinants are sometimes simply termed Jacobians.

**Example 6.2.** Let $x = r\cos\theta$, and $y = r\sin\theta$. Then $\partial(x,y)/\partial(r,\theta) = r$. □

A Jacobian is a convenient construct for transforming multiple integrals. Consider two types of planes: the $xy$-plane and the $uv$-plane. The points in these two planes are related as $x = f(u,v)$, and $y = g(u,v)$. That is, a point in the $uv$-plane is mapped into a point in the $xy$-plane by these two relationships.

If the mappings $f(\cdot, \cdot)$ and $g(\cdot, \cdot)$ are continuously differentiable and the Jacobian is not equal to zero in the region, then this mapping or transformation is one to one. Under this condition, a closed region $\mathcal{R}'$ in the $uv$-plane is mapped into a closed region $\mathcal{R}$ in the $xy$-plane. Let $\widetilde{B}(u, v) = \widetilde{A}(f(u, v), g(u, v))$, then

$$\int\int_{\mathcal{R}} \widetilde{A}(x, y)\, dx\, dy = \int\int_{\mathcal{R}'} \widetilde{B}(u, v) \left| \frac{\partial(x, y)}{\partial(u, v)} \right| du\, dv \tag{6.11}$$

This technique can be easily extended to triple integrals, and higher-ordered multiple integrals. Singularity of a function is next defined as follows.

**Definitions 6.10.** *Bounded function, function of bounded variation, and singularities of a function.*

1. *A real-valued function $f(\cdot)$ is bounded in an interval $(a, b)$ if there exists $M \in \mathbb{R}_0^+$ such that $|f(x)| < M$ for all $x \in (a, b)$.*
2. *A real-valued function $f(\cdot)$ is of bounded variation in an interval $(a, b)$ if and only if there exists $M \in \mathbb{R}_0^+$ such that $\sum_{i=1}^m |f(x_i) - f(x_{i-1})| < M$ for all partitions $a = x_0 < x_1 < x_2 < \cdots < x_m = b$.*
3. *If a function $f(\cdot)$ is unbounded at one or more points of the interval $a \le x \le b$, then such points are called the singularities of $f(\cdot)$.* $\qquad\square$

A function $f(\cdot)$ is of bounded variation in every finite open interval if and only if $f(x)$ is bounded and possesses a finite number of relative maximum and minimum values and discontinuities. That is, the function can be represented as a curve of finite length in any finite interval.

### 6.2.5 Hyperbolic Functions

Hyperbolic functions are specified in terms of exponential functions. The domain of these functions is the set of real numbers $\mathbb{R}$. Let $e^x = \sum_{j \in \mathbb{N}} x^j / j!$, where $x \in \mathbb{R}$. These functions are:

$$\sinh x = \frac{e^x - e^{-x}}{2}$$

$$\cosh x = \frac{e^x + e^{-x}}{2}$$

$$\tanh x = \frac{\sinh x}{\cosh x} = \frac{e^x - e^{-x}}{e^x + e^{-x}}$$

$$\operatorname{csch} x = \frac{1}{\sinh x} = \frac{2}{e^x - e^{-x}}$$

$$\operatorname{sech} x = \frac{1}{\cosh x} = \frac{2}{e^x + e^{-x}}$$

$$\coth x = \frac{\cosh x}{\sinh x} = \frac{e^x + e^{-x}}{e^x - e^{-x}}$$

Some useful identities concerning these functions are:

$$\sinh(-x) = -\sinh x, \quad \cosh(-x) = \cosh x, \quad \tanh(-x) = -\tanh x$$

$$\cosh^2 x - \sinh^2 x = 1, \quad 1 - \tanh^2 x = \operatorname{sech}^2 x, \quad \coth^2 x - 1 = \operatorname{csch}^2 x$$

The inverse hyperbolic functions are specified in terms of natural logarithms.

$$\sinh^{-1} x = \ln\left(x + \sqrt{x^2 + 1}\right), \quad \forall\, x \in \mathbb{R}$$

$$\cosh^{-1} x = \ln\left(x + \sqrt{x^2 - 1}\right), \quad x \geq 1$$

$$\tanh^{-1} x = \frac{1}{2} \ln\left\{\frac{1 + x}{1 - x}\right\}, \quad |x| < 1$$

The addition formulae:

$$\sinh(x \pm y) = \sinh x \cosh y \pm \cosh x \sinh y$$

$$\cosh(x \pm y) = \cosh x \cosh y \pm \sinh x \sinh y$$

$$\tanh(x \pm y) = \frac{\tanh x \pm \tanh y}{1 \pm \tanh x \tanh y}$$

The hyperbolic functions $\sinh(\cdot)$, $\cosh(\cdot)$, and $\tanh(\cdot)$; and trigonometric functions $\sin(\cdot)$, $\cos(\cdot)$, and $\tan(\cdot)$ are related as

$$\sinh(ix) = i \sin x,$$

$$\cosh(ix) = \cos x,$$

$$\tanh(ix) = i \tan x$$

where $i = \sqrt{-1}$.

### 6.2.6 Ordinary Differential Equations

A differential equation is an equation or a set of equations relating unknown functions (dependent variables) of one or more than one real-valued variable (independent variables); and the derivatives of unknown functions with respect to the independent variables. If the unknown functions depend only on a single independent variable, then we have *ordinary differential equation(s)*. However, if the unknown functions depend on more than one independent variable, then we have *partial differential equation(s)*. Differential equations generally occur in geometric and physical problems.

An ordinary differential equation of *order* $r \in \mathbb{P}$ in unknown function $y$, and independent variable $x \in \mathbb{R}$ is

$$F\left[x, y(x), y'(x), y''(x), \ldots, y^{(r)}(x)\right] = 0$$

where $y'(x)$ is the first derivative of $y$ with respect to $x$, $y''(x)$ is the second derivative of $y$ with respect to $x$, and $y^{(j)}(x)$, $j \in \mathbb{P}$ is the $j$th derivative of $y$ with respect to $x$. Note that the $r$th derivative is the highest derivative occurring in the above differential equation. Further, the above equation is called *implicit*. If the differential equation is linear in $y, y', y'', \ldots, y^{(r)}$, then it is called *linear*.

The function $y(\cdot)$ which satisfies the above differential equation is said to *solve* it for values of $x$ in the interval $I$, where $I \subseteq \mathbb{R}$. The function $y(x)$, $x \in I$ is called the *solution* of the differential equation. A *general solution* of the above differential equation of order $r$ is of the form

$$y = y(x, C_1, C_2, \ldots, C_r)$$

where $C_1, C_2, \ldots, C_r$ are *arbitrary constants*. Specific choice of the constants yields a *particular solution* of the differential equation.

The *initial value problem* determines a particular solution if it satisfies the $r$ *initial conditions*

$$y(x_0) = y_0,\ y'(x_0) = y_0',\ y''(x_0) = y_0'', \ldots, y^{(r-1)}(x_0) = y_0^{(r-1)}$$

for some value $x_0 \in I$. These $r$ initial conditions determine the arbitrary constants $C_1, C_2, \ldots, C_r$. The *boundary value problem* determines particular solution if it satisfies $r$ boundary conditions on $y$ and its derivatives at the end points $x = a$ or $x = b$ of the interval $a \leq x \leq b$. We next explore the existence and uniqueness of solution of initial value problem.

**Existence and Uniqueness of Solution of Initial Value Problem**

Consider an initial value problem of the form

$$y' = f(x, y), \quad y(x_0) = y_0$$

A solution to this problem exists, if $f$ is continuous and bounded in some region $R$ which contains the point $(x_0, y_0)$ in the $xy$-plane. Under this condition, the above problem has at least one solution. However if in addition, the partial derivative $\partial f / \partial y$ exists and is continuous in the region $R$, then the solution to the problem is unique. The solution is obtained by the so-called Picard's iteration method. This iteration method is named after the mathematician Émile Picard (1856-1941). The iteration generates the sequence $y_0, y_1, \ldots, y_n, \ldots$ where

$$y_n(x) = y_0 + \int_{x_0}^{x} f(t, y_{n-1}(t))\, dt, \quad n = 1, 2, 3, \ldots$$

This sequence converges to $y(x)$. Note that the stated conditions required for the existence and uniqueness of the stated problem are sufficient, but not necessary. More relaxed conditions can be obtained by the use of mean value theorem. The use of this later theorem results in the Lipschitz condition. This condition is named after the German mathematician Rudolf Lipschitz (1831-1903).

**Theorem 6.3.** *Consider the first-order differential equation*

$$y' = f(x, y), \quad y(x_0) = y_0 \tag{6.12a}$$

*It satisfies the following conditions:*

(a) *In the region $R$*

$$R = \{(x, y) \mid |x - x_0| < a, |y - y_0| < b\} \tag{6.12b}$$

   *the function $f(\cdot, \cdot)$ is continuous, and bounded. That is, $|f(x, y)| \leq A$ for all $(x, y) \in R$.*

(b) *Lipschitz condition: The inequality*

$$|f(x, y) - f(x, \widetilde{y})| \leq M |y - \widetilde{y}| \tag{6.12c}$$

   *is satisfied for all $(x, y), (x, \widetilde{y}) \in R$. The real constant $M$ is independent of points in $R$, and is called the Lipschitz's constant.*

*Under these condition, there exists exactly a single solution of the first-order differential equation for $|x - x_0| \leq \alpha$ where $\alpha = \min(a, b/A)$ in R.*

*Proof.* The proof can be found in any standard textbook on differential equations.  □

The above theorem can be generalized to a system of first-order differential equations.

**Example 6.3.** Solve a linear first-order differential equation of the form

$$y' + p(x) y = g(x)$$

Assume that $p(\cdot)$ and $g(\cdot)$ are continuous functions of $x$ in some interval $I$. Using the method of integrating factors, it can be shown that

$$y(x) = e^{-\mu(x)} \left\{ \int e^{\mu(x)} g(x) \, dx + c \right\}, \quad \mu(x) = \int p(x) \, dx$$

where $c$ is a constant of integration. The method of integrating factors is discussed in textbooks on ordinary differential equations.  □

## 6.3 Complex Analysis

A complex number is an ordered pair $(a, b)$ where $a, b \in \mathbb{R}$, and the operations $+$ (addition) and $\times$ (multiplication) are defined by

$$(a, b) + (c, d) = (a + c, b + d) \tag{6.13a}$$

$$(a, b) \times (c, d) = (ac - bd, ad + bc) \tag{6.13b}$$

$$m(a, b) = (ma, mb), \quad m \in \mathbb{R} \tag{6.13c}$$

In addition,

$$(a, b) = (c, d) \quad \Leftrightarrow \quad a = c \text{ and } b = d \tag{6.13d}$$

The set of all complex numbers is denoted by $\mathbb{C}$. It can easily be checked that this definition satisfies all the axioms of a field. The notion of field is developed in the chapter on abstract algebra. Furthermore, $(0, 0)$ and $(1, 0)$ are additive and multiplicative identities respectively. Also the additive inverse of $(a, b)$ is $(-a, -b)$, and its multiplicative inverse is

$$\left( \frac{a}{a^2 + b^2}, \frac{-b}{a^2 + b^2} \right)$$

The existence of multiplicative inverse assumes that $a$ and $b$ are not simultaneously equal to 0. Also if $(a, 0)$ is represented by $a$ and $(0, b)$ by $ib$, where $i = \sqrt{-1}$ and $i^2 = -1$, then

$$(a, b) = (a, 0) + (0, b) = a + ib$$

Therefore an alternate representation of the complex number $(a, b)$ is $(a + ib)$. It can be checked that the definitions of complex addition and multiplication operations are consistent in this representation. The existence of additive and multiplicative identities can be similarly verified.

**Definitions 6.11.** *Let $a, b \in \mathbb{R}$, $i = \sqrt{-1}$ and $(a, b) = (a + ib) = z \in \mathbb{C}$ be a complex number.*

1. $a$ and $b$ are the real and imaginary parts of $z$ respectively. The real component of $z$ is denoted by $\operatorname{Re}(z) = a$. Similarly, the imaginary component of $z$ is denoted by $\operatorname{Im}(z) = b$.
2. $|z| = \left(a^2 + b^2\right)^{1/2}$ is the absolute value or modulus of $z$.
3. $\bar{z} = (a - ib) \in \mathbb{C}$ is the complex conjugate or simply conjugate of $z$.                                    □

**Observations 6.1.** Some basic facts.

1. $a = (z + \bar{z})/2$, and $b = (z - \bar{z})/(2i)$.
2. $|z| = |\bar{z}|$.
3. If $z_1, z_2 \in \mathbb{C}$ then $|z_1 z_2| = |z_1| |z_2|$.
4. The triangle inequality: $|z_1 + z_2| \leq |z_1| + |z_2|$. Also $|z_1| - |z_2| \leq |z_1 - z_2|$.                        □

Topics such as Argand diagram, polar representation of a complex number, De Moivre's and Euler's identities, are discussed in this section. In addition, concepts related to limits, continuity, derivatives, analyticity, contours, conformal mapping, integration, infinite series are also stated. Lagrange's series expansion and the saddle point technique of integration are also discussed at length.

### 6.3.1 Coordinate Representation

Rectangular coordinates, Argand diagram, and polar representation of a complex number are discussed in this subsection.

#### Rectangular Coordinates

Consider two mutually perpendicular lines on a plane, which intersect at a point $O$. Call these lines $x$-axis and $y$-axis respectively. The corresponding plane is called the $xy$-plane, and the point $O$ is called the origin. Further, any point $P$ on the plane is specified by an ordered pair $(a, b)$, where $a \in \mathbb{R}$ is measured along the $x$-axis from the origin, and $b \in \mathbb{R}$ is measured along the $y$-axis from the origin. The ordered pair of numbers $(a, b)$ are called *rectangular* or *Cartesian coordinates*. The origin $O$ is specified as $(0, 0)$.

#### Argand Diagram

Since a complex number $z = (x, y) = (x + iy) \in \mathbb{C}$ is a two-tuple, it can be represented graphically in an $xy$-plane. This plane is also called the *complex plane* or the $z$-plane, and the representation of the point $(x, y)$ in this plane is called the *Argand diagram*. Since $(x, y)$ represents a point in the $xy$-plane, the complex number $z$ is referred to as the *point $z$*. It can be verified that the distance between two complex points

$$z_1 = (x_1, y_1), \text{ and } z_2 = (x_2, y_2)$$

is indeed given by

$$|z_1 - z_2| = \left\{(x_1 - x_2)^2 + (y_1 - y_2)^2\right\}^{1/2}$$

These diagrams are named after Jean Robert Argand (1768-1822) of Geneva. Consider the complex variable $w = f(z) \in \mathbb{C}$. Important information about the variable $w$ can be obtained by its Argand diagram in the $w$-plane.

### Polar Representation

A complex number $z = (x, y) = (x + iy) \in \mathbb{C}$ has a polar form representation, where

$$x = r \cos \theta, \quad y = r \sin \theta; \qquad r \in \mathbb{R}_0^+, \ 0 \leq \theta \leq 2\pi$$
$$r = \left( x^2 + y^2 \right)^{1/2}$$

In this representation $r$ is called the *absolute value* or *modulus* of $z$, and $\theta$ is called the *argument* of $z$. The argument is simply represented as $\arg(z) = \theta$. Also $r$ and $\theta$ are called the *polar coordinates* of the complex number $z$. For any $z \neq 0$ there exists only a single value of $\theta$ in the interval $[0, 2\pi]$. Note that any other interval of length $2\pi$, for example $[-\pi, \pi]$ is equally valid for representing $\theta$. A prespecified range of values of $\theta$ is called the *principal range,* and the value of $\theta$ in this range is called the *principal value.*

### 6.3.2  De Moivre's and Euler's Identities

In this subsection, certain celebrated results are discussed. These are: De Moivre's and Euler's identities.

### De Moivre's Identity

Observe that if $z_1, z_2 \in \mathbb{C}$, such that

$$z_1 = r_1 \left( \cos \theta_1 + i \sin \theta_1 \right), \quad \text{and} \quad z_2 = r_2 \left( \cos \theta_2 + i \sin \theta_2 \right)$$

then

$$z_1 z_2 = r_1 r_2 \left\{ \cos \left( \theta_1 + \theta_2 \right) + i \sin \left( \theta_1 + \theta_2 \right) \right\}$$
$$\frac{z_1}{z_2} = \frac{r_1}{r_2} \left\{ \cos \left( \theta_1 - \theta_2 \right) + i \sin \left( \theta_1 - \theta_2 \right) \right\}, \quad r_2 \neq 0$$

Extension of these results yields the De Moivre's theorem. It is named after the mathematician Abraham De Moivre.

**Theorem 6.4.** *De Moivre's theorem. Let $z = r \left( \cos \theta + i \sin \theta \right)$, then for any $n \in \mathbb{Z}$*

$$z^n = r^n \left( \cos n\theta + i \sin n\theta \right) \tag{6.14}$$

*Proof.* The result can be established by induction.  □

**Definition 6.12.** *Let $u, z \in \mathbb{C}$, and $n \in \mathbb{P}$. A number $u$ is called an $n$th root of $z$ if $z = u^n$.*  □

If $z = r \left( \cos \theta + i \sin \theta \right)$, the above definition for the $n$th root of a complex number and De Moivre's theorem yields

$$u = z^{1/n} = \left\{ r \left( \cos \theta + i \sin \theta \right) \right\}^{1/n} \tag{6.15a}$$
$$= r^{1/n} \left\{ \cos \left( \frac{\theta + 2\pi k}{n} \right) + i \sin \left( \frac{\theta + 2\pi k}{n} \right) \right\}, \ \forall \, k \in \mathbb{Z}_n \tag{6.15b}$$

Therefore it can be concluded that there are $n$ different values of the $n$th root of $z$ iff $z \neq 0$.

**Euler's Identity**

The following series expansions are well-known.

$$e^x = \sum_{j \in \mathbb{N}} \frac{x^j}{j!}; \quad e = 2.718281828\ldots, \quad x \in \mathbb{R} \tag{6.16a}$$

$$\sin x = \sum_{j \in \mathbb{N}} (-1)^j \frac{x^{2j+1}}{(2j+1)!}, \quad x \in \mathbb{R} \tag{6.16b}$$

$$\cos x = \sum_{j \in \mathbb{N}} (-1)^j \frac{x^{2j}}{(2j)!}, \quad x \in \mathbb{R} \tag{6.16c}$$

The number $e$ is called Euler's number. The series expansion of

$$e^x \triangleq \exp\left(x\right)$$

is also valid if $x$ is a complex number. Substituting $x = i\theta$ in the series expansion of $e^x$, yields the well-known Euler's identity.

$$e^{i\theta} = \cos\theta + i\sin\theta \tag{6.17}$$

**Observations 6.2.** Use of Euler's identity yields:

1. If $z = x + iy$, then $e^z = e^x \left(\cos y + i \sin y\right)$.
2. $\sin x = \left(e^{ix} - e^{-ix}\right) / (2i)$, and $\cos x = \left(e^{ix} + e^{-ix}\right) / 2$.
3. An alternate proof of De Moivre's theorem.

$$\left(\cos\theta + i\sin\theta\right)^n = \left(e^{i\theta}\right)^n = e^{in\theta} = \cos n\theta + i\sin n\theta$$

4. The *$n$th roots of unity*: Let $z^n = 1$, then the $n$ roots are

$$z = \cos\left(\frac{2\pi k}{n}\right) + i\sin\left(\frac{2\pi k}{n}\right) = e^{2\pi ik/n}, \quad k \in \mathbb{Z}_n$$

Let

$$\cos\left(\frac{2\pi}{n}\right) + i\sin\left(\frac{2\pi}{n}\right) = e^{2\pi i/n} \triangleq \omega$$

Thus the $n$ roots of unity are $1, \omega, \omega^2, \ldots, \omega^{n-1}$.                         □

## 6.3.3  Limits, Continuity, Derivatives, and Analyticity

The definitions of neighborhoods, limit points, closed sets, bounded sets, interior and exterior points, boundary points, and open sets in the complex plane are similar to those defined on the real line. The definitions of limits, and continuity in the complex domain are also similar to those in real number domain. Therefore, these are not repeated.

**Definitions 6.13.** *Let $z \in \mathbb{C}$.*

1. *Assume that $f(z)$ is single-valued in some region $\mathcal{R}$ of the $z$-plane. The derivative of $f(z)$ is defined as*

$$f'(z) = \lim_{\Delta z \to 0} \frac{f(z + \Delta z) - f(z)}{\Delta z} \qquad (6.18)$$

2. *A function $f(\cdot)$ is analytic at a point $z_0 \in \mathbb{C}$, if its first derivative $f'(z)$ exists at all points in the neighborhood of $z_0$. That is, $f'(z)$ exists at all points in the region $|z - z_0| < \delta$, where $\delta > 0$.*

3. *If the derivative $f'(z)$ exists at all points $z$ of a region $\mathcal{R}$, then the function $f(\cdot)$ is analytic in $\mathcal{R}$.*

4. *A function which is analytic over the entire complex plane (except at infinity) is called an entire function.* □

A necessary and sufficient condition for a function to be analytic in a region is specified by the Cauchy-Riemann theorem. As we shall see, existence of a complex derivative leads to a pair of partial differential equations.

**Theorem 6.5.** *Let $z = x + iy$, where $x, y \in \mathbb{R}$ and $w = f(z) = u(x, y) + iv(x, y)$. The necessary and sufficient conditions that the function $f(\cdot)$ be analytic in a region $\mathcal{R}$ is that functions $u(\cdot, \cdot)$ and $v(\cdot, \cdot)$ satisfy the Cauchy-Riemann equations*

$$\frac{\partial u}{\partial x} = \frac{\partial v}{\partial y}, \quad and \quad \frac{\partial u}{\partial y} = -\frac{\partial v}{\partial x} \qquad (6.19)$$

*and these partial derivatives be continuous in the region $\mathcal{R}$.*

*Proof.* See the problem section. □

**Observation 6.3.** Let $w = f(z) = u(x, y) + iv(x, y)$ be analytic in a region $\mathcal{R}$. Then

$$f'(z) \triangleq \frac{dw}{dz} = \frac{\partial u}{\partial x} + i\frac{\partial v}{\partial x} = \frac{\partial v}{\partial y} - i\frac{\partial u}{\partial y} \qquad (6.20)$$

□

### 6.3.4  Contours or Curves

A contour is a curve in the complex $z$-plane. It can be either smooth or piecewise-smooth.

**Definitions 6.14.** *Let $\alpha(\cdot)$ and $\beta(\cdot)$ be real functions of a real variable $t$, defined over the interval $t_1 \leq t \leq t_2$, such that $z(t) = \alpha(t) + i\beta(t)$, where $t_1 \leq t \leq t_2$*

1. *If the functions $\alpha(\cdot)$ and $\beta(\cdot)$ are continuous in the interval $[t_1, t_2]$, then the complex function $z(\cdot)$ is a continuous curve or arc in the complex plane, which starts at $a = z(t_1)$ and ends at $b = z(t_2)$. Therefore, an orientation can also be assigned to the curve, as it moves from $t = t_1$ to $t = t_2$.*

2. *If $t_1 \neq t_2$ but $a = b$, that is the end-points coincide, then the curve is closed.*

3. *A closed curve which does not intersect itself at any point in the complex plane is called a simple closed curve.*

4. *If $\alpha(t)$ and $\beta(t)$, and consequently $z(t)$, have continuous derivatives in the specified interval, then the curve is called a smooth curve or arc.*

5. *A curve which consists of a finite number of smooth arcs is called sectionally or piecewise-smooth curve or a contour.* □

### 6.3.5 Conformal Mapping

Consider the set of equations

$$u = u(x, y), \text{ and } v = v(x, y)$$

This mapping or transformation defines a relationship between points in the $xy$- and the $uv$-planes. Assume that in this mapping for each point in the $uv$-plane, there corresponds only a single point in the $xy$-plane, and vice-versa.

Let a point $(x_0, y_0)$ in the $xy$-plane be mapped into a point $(u_0, v_0)$ in the $uv$-plane. Further assume that the curves $C_1$ and $C_2$ in the $xy$-plane which intersect at the point $(x_0, y_0)$, map to the curves $C_1'$ and $C_2'$ respectively in the $uv$-plane and intersect at the point $(u_0, v_0)$.

The angle between two intersecting curves at their point of intersection, with existing tangents, is defined to be equal to the angle between their tangents at the point of intersection.

If the angle between two curves $C_1$ and $C_2$ in the $xy$-plane at their point of intersection $(x_0, y_0)$, and the angle between two curves $C_1'$ and $C_2'$ in the $uv$-plane at their corresponding point of intersection $(u_0, v_0)$, are equal in magnitude and sense (either clockwise or counter clockwise), then the mapping is said to be conformal at the point $(x_0, y_0)$.

**Definition 6.15.** *Let $\Omega \subseteq \mathbb{C}$. A mapping $f : \Omega \to \mathbb{C}$ is conformal at $z_0 \in \Omega$ if and only if the mapping preserves both magnitude and sense of the angle at $z_0$.*  $\square$

**Theorem 6.6.** *Let $\Omega \subseteq \mathbb{C}$, and the mapping $f : \Omega \to \mathbb{C}$ be analytic. If $z_0 \in \Omega$ and $f'(z_0) \neq 0$, then $f(\cdot)$ is conformal at $z_0$.*

*Proof.* See the problem section.  $\square$

### 6.3.6 Integration

It is possible to integrate a complex function $f(\cdot)$ along a curve $C$ in the complex plane. Denote this integral by $\int_C f(z)\, dz$. The integral can be defined as the limit of a sum. Let $f(\cdot)$ be a continuous function at all points on the curve $C$. The end points of the curve $C$ are $a$ and $b$. Divide $C$ arbitrarily into $n$ parts via points $z_1, z_2, \ldots, z_{n-1}$, and call $a = z_0$, and $b = z_n$. Let

$$\Delta z_k = (z_k - z_{k-1}), \quad 1 \le k \le n$$

If $\xi_k$ is a point on the curve $C$ between $z_{k-1}$ and $z_k$, then

$$\int_a^b f(z)\, dz = \lim_{\substack{n \to \infty \\ \max|\Delta z_k| \to 0}} \sum_{k=1}^n f(\xi_k)\, \Delta z_k$$

Therefore, if a function $f(\cdot)$ is analytic at all points in a region $\mathcal{R}$ of the complex plane, and $C$ is a curve lying in the region $\mathcal{R}$, then $f(\cdot)$ is integrable along the curve $C$. The integration around the boundary $C$ of a region $\mathcal{R}$ is denoted by

$$\oint_C f(z)\, dz$$

It is next demonstrated that, if $f(z)$ is analytic at all points in a region $\mathcal{R}$ and on its boundary $C$, then the value of the above contour integral is equal to zero. This conclusion is facilitated by Green-Ostrogradski theorem. This theorem is named after the mathematicians George Green (1793-1841) and Michel Ostrogradski (1801-1861).

**Theorem 6.7.** *Let* $P\left(\cdot,\cdot\right)$ *and* $Q\left(\cdot,\cdot\right)$ *be continuous real-valued functions and have continuous partial derivatives in a region* $\mathcal{R}$ *and on its boundary* $C$*. Then*

$$\oint_C \left\{ P\left(x,y\right)dx + Q\left(x,y\right)dy \right\} = \iint_{\mathcal{R}} \left\{ \frac{\partial Q\left(x,y\right)}{\partial x} - \frac{\partial P\left(x,y\right)}{\partial y} \right\} dxdy \qquad (6.21)$$

*Proof.* See the problem section.                                                                      □

The celebrated Cauchy-Goursat integral theorem is stated below. This theorem is named after Augustin-Louis Cauchy and Édourad Goursat (1858-1936). A special case of this result uses Green-Ostrogradski theorem

**Theorem 6.8.** *If* $f\left(z\right)$*,* $z \in \mathbb{C}$ *is analytic in a region* $\mathcal{R}$ *and on its boundary* $C$*. Then*

$$\oint_C f\left(z\right)dz = 0 \qquad (6.22)$$

*Proof.* See the problem section.                                                                      □

### 6.3.7 Infinite Series

Some terminology about infinite series is introduced via the following definitions.

**Definitions 6.16.** *Let* $f_1\left(\cdot\right), f_2\left(\cdot\right), \ldots, f_n\left(\cdot\right), \ldots$*, be a sequence of functions defined on some region* $\mathcal{R}$ *of the complex* $z$*-plane. Denote this sequence by* $\left\{ f_n\left(\cdot\right) \right\}$*.*

1. *Limit of sequence of functions: The function* $f\left(\cdot\right)$ *is the limit of* $f_n\left(\cdot\right)$ *as* $n \to \infty$*. That is,* $\lim_{n \to \infty} f_n\left(z\right) = f\left(z\right)$*, iff for any positive number* $\epsilon$ *there is a number* $N$ *such that*

$$\left| f_n\left(z\right) - f\left(z\right) \right| < \epsilon, \quad \forall\, n > N \qquad (6.23a)$$

   *If this condition is satisfied, the sequence is said to converge to* $f\left(z\right)$*. In general, the number* $N$ *can be a function of both* $\epsilon$ *and* $z$*. If this convergence occurs for all points in the region* $\mathcal{R}$*, then* $\mathcal{R}$ *is called the region of convergence.*
   *If the sequence does not converge at some point* $z$*, then it is called divergent at* $z$*.*

2. *Convergence of series of functions: Using the sequence of functions* $\left\{ f_n\left(\cdot\right) \right\}$*, generate a new sequence of functions* $\left\{ g_n\left(\cdot\right) \right\}$*, where* $g_n\left(z\right) = \sum_{i=1}^{n} f_i\left(z\right)$*,* $g_n\left(z\right)$ *is called the* $n$*th partial sum, and*

$$f_1\left(z\right) + f_2\left(z\right) + \ldots = \sum_{n \in \mathbb{P}} f_n\left(z\right) \qquad (6.23b)$$

   *is called an infinite series. If* $\lim_{n \to \infty} g_n\left(z\right) = g\left(z\right)$*, the infinite series is called convergent, otherwise the series is divergent. If this series converges for all points in the region* $\mathcal{R}$*, then* $\mathcal{R}$ *is called the region of convergence of the series.*
   *If the series does not converge at some point* $z$*, then it is called divergent at* $z$*.*

3. *Absolute convergence of series: The series* $\sum_{n \in \mathbb{P}} f_n\left(z\right)$ *is said to be absolutely convergent, if* $\sum_{n \in \mathbb{P}} \left| f_n\left(z\right) \right|$ *converges.*

4. *Conditional convergence of series: The series* $\sum_{n \in \mathbb{P}} f_n\left(z\right)$ *is said to be conditionally convergent, if* $\sum_{n \in \mathbb{P}} f_n\left(z\right)$ *converges, but* $\sum_{n \in \mathbb{P}} \left| f_n\left(z\right) \right|$ *does not converge.*

5. *Uniform convergence of sequence of functions*: *In the definition of the limit of sequence of functions, if the number $N$ depends only on $\epsilon$, and is independent of $z \in \mathcal{R}$, then $\{f_n(z)\}$ is said to converge uniformly to $f(z)$ for all points $z \in \mathcal{R}$.*

6. *Uniform convergence of series of functions*: *In the definition of the convergence of series of functions, if the sequence of partial sums $\{g_n(z)\}$ converges uniformly, for all points $z \in \mathcal{R}$, then the infinite series $\sum_{n \in \mathbb{P}} f_n(z)$ converges uniformly for all points $z \in \mathcal{R}$.*                    □

### 6.3.8  Lagrange's Series Expansion

If $\widetilde{\varphi}(\cdot)$ is an analytic function inside and on a circle $C$ containing a point $z = a$, and $z$ has a dependence of the form $z = a + \xi\widetilde{\varphi}(z)$, then an explicit series representation of $z$ is possible. This series representation of $z \in \mathbb{C}$ is called Lagrange's series expansion.

**Theorem 6.9.** *Let $z = a + \xi\widetilde{\varphi}(z)$, $z = a$ when $\xi = 0$. If $\widetilde{\varphi}(\cdot)$ is analytic inside and on a circle $C$ containing $z = a$, then*

$$z = a + \sum_{n \in \mathbb{P}} \frac{\xi^n}{n!} \frac{d^{n-1}}{da^{n-1}} \left[\widetilde{\varphi}(a)\right]^n \tag{6.24}$$

*Proof.* The proof can be found in any standard textbook on complex analysis.                    □

### 6.3.9  Saddle Point Technique of Integration

We next learn to approximate integrals of the form

$$I(M) = \int_C g(z) e^{Mf(z)} dz$$

where $C$ is some path (contour) in the complex $z$-plane, and $M$ is a large real number. The basic idea behind this technique is to evaluate $I(M)$ on a modified path which will contribute most to its value. This technique is also known as the *method of steepest descent*. We first analyse the technique in which the path $C$ is the real axis. In this instance, the method is also called *Laplace's method*. Let

$$I(M) = \int_{-\infty}^{\infty} g(x) e^{Mf(x)} dx$$

Assume that the first and second derivatives of $f(x)$ with respect to $x$ exist. These are denoted by $f'(x)$ and $f''(x)$ respectively. In addition, assume that $f(x)$ has only a single maximum at $x = x_0$ on the entire real axis, and that the function $g(\cdot)$ is continuous at least near $x = x_0$. Consider the Taylor's series expansion of $f(x)$ about the point $x_0$.

$$f(x) = f(x_0) + (x - x_0) f'(x_0) + \frac{1}{2}(x - x_0)^2 f''(x_0) + O(x^3)$$

where the $O(x^3)$ term gives an upper bound on the error in the above expansion to within a constant factor. As the global maximum of $f(x)$ occurs at $x = x_0$, we have $f'(x_0) = 0$, and $f''(x_0) < 0$. Points at which $f'(x) = 0$ are called *saddle points*, hence the name *method of saddle points*. For large values of $M$, contribution to the integral $I(M)$ comes from the region around $x_0$. Thus for large values of $M$

$$I(M) \sim g(x_0) e^{Mf(x_0)} \int_{-\infty}^{\infty} e^{Mf''(x_0)(x-x_0)^2/2} dx$$

where the symbol "$\sim$" means asymptotically equal to. The integrand in the above relationship is of the Gaussian (probability density function) type. Therefore

$$I(M) \sim g(x_0) e^{Mf(x_0)} \left( \frac{-2\pi}{Mf''(x_0)} \right)^{1/2}$$

The integral is next evaluated for a path in the complex plane. Assume that the first derivative of the complex function $f(z)$, $z \in \mathbb{C}$ has a simple zero (a zero of multiplicity one) at the point $z_0$, that is $f'(z_0) = 0$. Thus $z_0$ is a saddle point of $f(z)$. If $f(z)$ is analytic in a neighborhood of $z_0$, Taylor's series expansion yields

$$f(z) = f(z_0) + (z - z_0) f'(z_0) + \frac{1}{2} (z - z_0)^2 f''(z_0) + \ldots$$
$$= f(z_0) - v^2$$

Note that a Taylor's series expansion of $f(z)$, where $z \in \mathbb{C}$ is analogous to the Taylor's series expansion of a function of real variable. Deform the contour $C$ such that: it passes through $z_0$, $\mathrm{Re}(f(z))$ is largest at $z_0$, and $\mathrm{Im}(f(z))$ is equal to a constant $\mathrm{Im}(f(z_0))$ in the neighborhood of $z_0$. Under these assumptions, the variable $v$ in the above equation is real, and

$$I(M) \sim g(z_0) e^{Mf(z_0)} \int_{-\infty}^{\infty} e^{-Mv^2} \left( \frac{dz}{dv} \right) dv$$
$$\sim g(z_0) e^{Mf(z_0)} \left( \frac{2\pi}{M|f''(z_0)|} \right)^{1/2}$$

This technique can be extended if there are multiple saddle points. The above discussion is condensed in the following set of observations.

**Observations 6.4.** Let $M$ be a large real number.

1. Let $f(\cdot)$ and $g(\cdot)$ be real-valued functions. Let the first and second derivatives of $f(x)$ with respect to $x$ be $f'(x)$ and $f''(x)$ respectively. Assume that the global maximum of $f(x)$ occurs at $x = x_0$, and $g(x)$ is continuous at least near $x = x_0$. Then $f'(x_0) = 0$, and the integral

$$I(M) = \int_{-\infty}^{\infty} g(x) e^{Mf(x)} dx \qquad (6.25a)$$

is approximated by

$$g(x_0) e^{Mf(x_0)} \left( \frac{-2\pi}{Mf''(x_0)} \right)^{1/2} \qquad (6.25b)$$

Note that $f''(x_0)$ is negative.

2. Let $f(\cdot)$ and $g(\cdot)$ be complex valued functions. The function $f(\cdot)$ has a simple zero at $z = z_0$, and the function $g(\cdot)$ is analytic near $z = z_0$. If the second derivative of $f(z)$ at $z_0$ is denoted by $f''(z_0)$, and $C$ is some contour in the $z$-plane, then the integral

$$I(M) = \int_C g(z) e^{Mf(z)} dz \qquad (6.25c)$$

is approximated by

$$g(z_0) e^{Mf(z_0)} \left( \frac{2\pi}{M|f''(z_0)|} \right)^{1/2} \qquad (6.25d)$$

$\square$

See the problem section for examples of use of this technique. We shall use these techniques in determining large deviation bounds in probability theory. These are discussed in the chapter on probability theory.

## 6.4 Cartesian Geometry

Basics of Cartesian geometry: straight lines, circles, orthogonal circles, and conic sections are outlined in this section.

### 6.4.1 Straight Line and Circle

Some properties of straight lines and circles in the complex plane are summarized.

**Definitions 6.17.** *Straight line and a circle in the complex $z$-plane.*

1. *Straight line*:
$$\bar{b}z + b\bar{z} + c = 0, \quad where\ c \in \mathbb{R},\ b \in \mathbb{C}\backslash\{0\} \tag{6.26}$$

2. *Circle*:
$$az\bar{z} + \bar{b}z + b\bar{z} + c - 0, \quad where\ a, c \in \mathbb{R},\ a \neq 0,\ b \in \mathbb{C},\ |b|^2 > ac \tag{6.27}$$

□

The above definitions are justified in the problem section.

**Observations 6.5.** Some observations about straight lines.

1. Consider the straight lines

$$\bar{b}z + b\bar{z} + c = 0, \quad \text{where } c \in \mathbb{R},\ b \in \mathbb{C}\backslash\{0\}$$
$$\bar{e}z + e\bar{z} + f = 0, \quad \text{where } f \in \mathbb{R},\ e \in \mathbb{C}\backslash\{0\}$$

These straight lines are orthogonal (perpendicular) to each other if $e = ikb$, where $i = \sqrt{-1}$ and $k \in \mathbb{R}\backslash\{0\}$. An alternate form of orthogonal straight lines is

$$\bar{b}z + b\bar{z} + c = 0$$
$$\bar{b}z - b\bar{z} + \frac{if}{k} = 0$$

2. Shortest distance between point $\alpha \in \mathbb{C}$ and straight line $\bar{b}z + b\bar{z} + c = 0$, where $c \in \mathbb{R}$, $b \in \mathbb{C}\backslash\{0\}$ is
$$\frac{|\bar{b}\alpha + b\bar{\alpha} + c|}{2|b|}$$

3. Equation of a straight line passing through point $z_1 \in \mathbb{C}$ is

$$\bar{b}(z - z_1) + b(\bar{z} - \bar{z}_1) = 0, \quad \text{where } b \in \mathbb{C}\backslash\{0\}$$

4. Equation of a straight line passing through two distinct points $z_1, z_2 \in \mathbb{C}$ is

$$
\begin{vmatrix}
z & \overline{z} & 1 \\
z_1 & \overline{z}_1 & 1 \\
z_2 & \overline{z}_2 & 1
\end{vmatrix} = 0
$$

or

$$
(\overline{z}_1 - \overline{z}_2) z + (z_2 - z_1) \overline{z} + (z_1 \overline{z}_2 - \overline{z}_1 z_2) = 0
$$

Observe that $(z_1 \overline{z}_2 - \overline{z}_1 z_2)$ is imaginary. Therefore that above equation can also be expressed as

$$
i (\overline{z}_1 - \overline{z}_2) z + i (z_2 - z_1) \overline{z} + i (z_1 \overline{z}_2 - \overline{z}_1 z_2) = 0
$$

□

**Observation 6.6.** A circle $C_{a,r}$ with center $a \in \mathbb{C}$, and radius $r \in \mathbb{R}^+$ is

$$
C_{a,r} = \{ z \in \mathbb{C} \mid |z - a| = r \}
$$

□

The above observation can also be taken as the definition of a circle.

**Definition 6.18.** *Orthogonal circles. Two intersecting circles are orthogonal to each other, if the tangents to the two circles at the points of intersection are orthogonal to each other.*   □

**Observation 6.7.** Consider two intersecting circles $C_{a_1, r_1}$, and $C_{a_2, r_2}$. That is, the triangle inequality: $|a_1 - a_2| < (r_1 + r_2)$ holds true. These two circles are orthogonal to each other, if

$$
|a_1 - a_2|^2 = r_1^2 + r_2^2
$$

□

### 6.4.2  Conic Sections

We briefly summarize certain useful characteristics and terminology of a conic section. We introduce curves of type: hyperbola, ellipse, and parabola. These are precursors to the terms: hyperbolic, elliptic, and parabolic respectively in the corresponding geometries. A conic section is a second-degree polynomial equation in $x$ and $y$ of the form

$$
g (x, y) = ax^2 + bxy + cy^2 + dx + ey + f = 0
$$

where $a, b, c, d, e, f \in \mathbb{R}$; and $a, b$, and $c$ are simultaneously not equal to zero. The polynomial $g (x, y)$ can also be expressed as a quadratic form. It is

$$
g (x, y) = X A X^T, \quad \text{where } X = \begin{bmatrix} x & y & 1 \end{bmatrix}, \; A = \begin{bmatrix} a & b/2 & d/2 \\ b/2 & c & e/2 \\ d/2 & e/2 & f \end{bmatrix}
$$

The polynomial $g (x, y)$ can be studied based upon the value of the determinant of the matrix $A$. The conic section is degenerate if $\det A = 0$, and nondegenerate if $\det A \neq 0$. It can also be further analyzed by examining the sign of the expression $\Delta (g) = (b^2 - 4ac)$. If the conic section is nondegenerate, the graph of $g (x, y) = 0$ in the $xy$-plane is either an ellipse, or parabola, or a hyperbola, depending upon the sign of $\Delta (g)$. More precisely, this classification is:

1. Nondegenerate conic section: $\det A \neq 0$.
   (a) Ellipse if $\Delta(g) < 0$.
       (i) Ellipse is a circle if $b = 0$, and $a = c$. Example: $\left(x^2 + y^2\right) = 1$.
       (ii) Real ellipse if $c \det A < 0$. Example: $\left(x^2 + 4y^2\right) = 1$.
       (iii) No graph if $c \det A > 0$. Example: $\left(x^2 + y^2 + 1\right) = 0$.
   (b) Parabola if $\Delta(g) = 0$. Example: $x^2 = y$.
   (c) Hyperbola if $\Delta(g) > 0$. Example: $\left(x^2 - 4y^2\right) = 1$.
2. Degenerate conic section: $\det A = 0$.
   (a) If $\Delta(g) < 0$, we have a single point. Example: $\left(x^2 + y^2\right) = 0$.
   (b) If $\Delta(g) = 0$.
       (i) Two parallel lines. Example: $x^2 = 4$.
       (ii) Single line: Example: $y^2 = 0$.
       (iii) No graph: Example: $x^2 = -1$.
   (c) If $\Delta(g) > 0$, we have two intersecting lines. Example: $\left(x^2 + xy\right) = 0$.

## 6.5 Moebius Transformation

We study a special type of transformation called Moebius transformation in this subsection. These transformations are named after Augustus Ferdinand Moebius (1790-1860), a German-born astronomer and mathematician. This transformation is defined on the extended complex plane $\mathbb{C}^* = \mathbb{C} \cup \{+\infty\}$. Such transformations are also called bilinear transformations and linear fractional transformations.

The Moebius transformation is a generalization of transformations like: identity, translation, dilation, rotation, and inversion. These elementary transformations are initially defined.

**Definitions 6.19.** *Identity, translation, multiplication, dilation (or scaling), rotation, inversion, and linear (or affine).*

1. *The function $f(\cdot)$ is an identity if $f(z) = z$, where $z \in \mathbb{C}$.*
2. *For $a \in \mathbb{C}$, the function $f(\cdot)$ is a translation if $f(z) = z + a$, where $z \in \mathbb{C}$.*
3. *For $a \in \mathbb{C}$, the function $f(\cdot)$ is a multiplication if $f(z) = az$, where $z \in \mathbb{C}$.*
4. *For $a \in \mathbb{C} \setminus \{0\}$, the function $f(\cdot)$ is a dilation if $f(z) = az$, where $z \in \mathbb{C}$.*
5. *For $\theta \in \mathbb{R}$, the function $f(\cdot)$ is a rotation if $f(z) = e^{i\theta}z$, where $i = \sqrt{-1}$, and $z \in \mathbb{C}$.*
6. *The function $f(\cdot)$ is an inversion if $f(z) = 1/z$, where $z \in \mathbb{C} \setminus \{0\}$.*
7. *For $a, b \in \mathbb{C}$, the function $f(\cdot)$ is linear if $f(z) = az + b$, where $z \in \mathbb{C}$.*    □

We next define the celebrated Moebius transformation.

**Definition 6.20.** *A Moebius transformation is a map $f : \mathbb{C}^* \to \mathbb{C}^*$,*

$$f(z) = \frac{az + b}{cz + d}, \quad \text{where } a, b, c, d \in \mathbb{C}, \text{ and } (ad - bc) \neq 0 \tag{6.28}$$

*Furthermore*

(a) *If $c = 0$, then $f(\infty) = \infty$.*
(b) *If $c \neq 0$, then $f(\infty) = a/c$, and $f(-d/c) = \infty$.*
(c) *The Moebius transformation is said to be real, if $a, b, c, d \in \mathbb{R}$.*                                  □

The reader should infer that if $(ad - bc) = 0$, then $f(z)$ takes either a constant value, or is of the form $0/0$ (indeterminate). Such values are not interesting. Observe that $(ad - bc)$ is equal to the determinant of the matrix

$$\begin{bmatrix} a & b \\ c & d \end{bmatrix}$$

It is also possible to define composition of Moebius transformations.

**Definition 6.21.** *Let $f : \mathbb{C}^* \to \mathbb{C}^*$ and $g : \mathbb{C}^* \to \mathbb{C}^*$ be two Moebius transformations. The composition of functions $f(\cdot)$ and $g(\cdot)$ is denoted by $f \circ g(z) = f(g(z))$, where $z \in \mathbb{C}^*$.*                □

We shall also qualify the fixed points of a Moebius transformation.

**Definition 6.22.** *Let $f(\cdot)$ be a Moebius transformation. The fixed points of this transformation are the set of all points $z \in \mathbb{C}^*$ which satisfy $f(z) = z$.*                □

**Observations 6.8.** Let $f(\cdot)$ be a Moebius transformation

1. Let $\lambda \in \mathbb{C} \setminus \{0\}$, then

$$f(z) = \frac{az + b}{cz + d} = \frac{(\lambda a)z + (\lambda b)}{(\lambda c)z + (\lambda d)}$$

That is, the coefficients $a, b, c, d$ are not unique.

2. The inverse of $f(z)$ is

$$f^{-1}(z) = \frac{dz - b}{-cz + a}$$

It satisfies $f(f^{-1}(z)) = f^{-1}(f(z)) = z$. That is, $f^{-1}(\cdot)$ is the inverse mapping of $f(\cdot)$, and the inverse is also a Moebius transformation.

3. The mapping $f(\cdot)$ is bijective.

4. Let $f(\cdot)$ and $g(\cdot)$ be two Moebius transformations. Then the function $f \circ g(\cdot)$ is also a Moebius transformation. For if

$$f(z) = \frac{az + b}{cz + d}, \quad \text{and} \quad g(z) = \frac{\alpha z + \beta}{\gamma z + \delta}$$

then

$$f \circ g(z) = f(g(z)) = \frac{ag(z) + b}{cg(z) + d} = \frac{(a\alpha + b\gamma)z + (a\beta + b\delta)}{(c\alpha + d\gamma)z + (c\beta + d\delta)}$$

The coefficients of the composition of the transformations can be obtained from product of the corresponding coefficient matrices. That is,

$$\begin{bmatrix} a & b \\ c & d \end{bmatrix} \begin{bmatrix} \alpha & \beta \\ \gamma & \delta \end{bmatrix} = \begin{bmatrix} (a\alpha + b\gamma) & (a\beta + b\delta) \\ (c\alpha + d\gamma) & (c\beta + d\delta) \end{bmatrix}$$

Note that the determinant associated with this composition is

$$(ad - bc)(\alpha\delta - \beta\gamma) \neq 0$$

5. The Moebius transformation

$$f(z) = \frac{az + b}{cz + d}$$

is a composition of translations, dilations, and the inversion.

6. The fixed points of the Moebius transformation satisfy $f(z) = z$. This implies

$$cz^2 + (d - a)z - b = 0$$

Therefore the Moebius transformation has at most two fixed points, unless $f(z) = z$ (identity mapping). In this later case the transformation fixes all points.

7. The Moebius transformation is conformal, except at $z = -d/c, \infty$.

8. The Moebius transformation maps a "circle" in $\mathbb{C}^*$ into a "circle" in $\mathbb{C}^*$. In this observation, a "circle" means either a straight line or a circle.

9. Let $z_1, z_2$, and $z_3$ be three distinct points in $\mathbb{C}^*$. Also let $w_1, w_2$, and $w_3$ be three distinct points in $\mathbb{C}^*$. Then there exists a unique Moebius transformation $f(\cdot)$ such that $f(z_k) = w_k$ for $k = 1, 2$, and 3. This result is often called the fundamental theorem of Moebius transformation.

10. Let $z_1, z_2$, and $z_3$ be three distinct points in $\mathbb{C}^*$. Also let $w_1, w_2$, and $w_3$ be three distinct points in $\mathbb{C}^*$. Then the Moebius transformation $w = f(z)$ such that $f(z_k) = w_k$ for $k = 1, 2$, and 3 is determined by

$$\frac{(w_1 - w_3)(w - w_2)}{(w_1 - w_2)(w - w_3)} = \frac{(z_1 - z_3)(z - z_2)}{(z_1 - z_2)(z - z_3)}$$

That is, the Moebius transformation acts transitively on a triple of distinct points in $\mathbb{C}$.    □

See the problem section for proofs of some of the above observations. In the above set of observations, the type of expression in the last observation is important. It is called cross ratio.

**Definition 6.23.** *Let $z_0, z_1, z_2, z_3$ be distinct points in $\mathbb{C}^*$. Their cross ratio is*

$$(z_0, z_1, z_2, z_3) = \frac{(z_0 - z_2)(z_1 - z_3)}{(z_0 - z_3)(z_1 - z_2)} \tag{6.29}$$

□

Observe that the cross ratio $(z, z_1, z_2, z_3)$ maps: $z_1$ to $1$, $z_2$ to $0$, and $z_3$ to $\infty$.

**Observations 6.9.** Let $z_0, z_1, z_2, z_3$ be distinct points in $\mathbb{C}^*$.

1. The Moebius transformation $f(\cdot)$ preserves cross ratio. That is,

$$(z_0, z_1, z_2, z_3) = (f(z_0), f(z_1), f(z_2), f(z_3))$$

2. The cross ratio $(z_0, z_1, z_2, z_3)$ is real if and only if the points $z_0, z_1, z_2, z_3$ lie on a Euclidean circle or a straight line.    □

The above observations are proved in the problem section.

**Conjugate-Moebius Transformation**

Conjugate-Moebius transformations are used in proving certain important results in hyperbolic geometry.

**Definition 6.24.** *A conjugate Moebius transformation is a map* $f : \mathbb{C}^* \to \mathbb{C}^*$,

$$f(z) = \frac{a\bar{z} + b}{c\bar{z} + d}, \quad \text{where } a, b, c, d \in \mathbb{C}, \text{ and } (ad - bc) \neq 0 \tag{6.30}$$

$\square$

Let $g(\cdot)$ be a Moebius transformation, and $c_{conj}(\cdot)$ be the complex conjugation map $c_{conj}(z) = \bar{z}$, where $z \in \mathbb{C}$. The conjugate-Moebius transformation $f(\cdot)$ can be expressed as $f(\cdot) = g \circ c_{conj}(\cdot)$.

**Observations 6.10.** Some observations about conjugate-Moebius transformations.

1. The composition of Moebius and conjugate-Moebius transformations in either order yields a conjugate-Moebius transformation. Further, the composition of two conjugate-Moebius transformations is a Moebius transformation.
2. Conjugate-Moebius transformations map "circles" to "circles" and are also conformal. $\square$

See the problem section for proofs of the above observations. Some examples of conjugate-Moebius transformations are geometric transformations: reflections and inversions.

**Reflection and Inversion**

Reflection of a point in a straight line is a conjugate-Moebius transformation. The equation of a straight line is:

$$\bar{b}z + b\bar{z} + c = 0, \quad \text{where } c \in \mathbb{R}, \ b \in \mathbb{C} \setminus \{0\}$$

The reflection of a point $z \in \mathbb{C}$ is $R(z) \in \mathbb{C}$, where the perpendicular distance of the points $z$ and $R(z)$ from the straight line are equal. It can be shown that

$$R(z) = -\frac{1}{|b|^2} \left( b^2 \bar{z} + bc \right)$$

Inversion is analogous to the reflection transformation. Consider a circle $C_{a,r}$ with center $a \in \mathbb{C}$, and radius $r \in \mathbb{R}^+$. That is

$$C_{a,r} = \{ z \in \mathbb{C} \mid |z - a| = r \}$$

The inversion transformation $I_{a,r}(\cdot)$ with respect to circle $C_{a,r}$ takes a point $z \in \mathbb{C} \setminus \{a\}$, and maps it to a point $z' \in \mathbb{C}$. The point $z'$ is selected so that it is on the radius of the circle which passes through $z$, and the product of the distance of $z$ and $z'$ from $a$ is equal to $r^2$. That is,

$$|z - a| \, |z' - a| = r^2$$

Therefore $(z' - a) = \beta (z - a)$ for some $\beta > 0$. This implies $\beta = r^2 / |z - a|^2$, and

$$I_{a,r}(z) = a + \frac{r^2}{(\bar{z} - \bar{a})}$$

Note that:

- $I_{a,r}(\cdot)$ is a conjugate-Moebius transformation. Consequently, it maps "circles" to "circles."

- $I_{a,r}(a) = \infty$. Therefore $I_{a,r}(\cdot)$ maps "circles" that pass through the point $a$ to lines, and all other "circles" to circles.
- $I_{a,r}(z) = z$ if $z \in C_{a,r}$.
- $I_{a,r} \circ I_{a,r}(z) = z$ for all $z \neq a$ in $\mathbb{C}$.

### Unitary Moebius Transformation

Unitary Moebius transformations are useful in the study of hyperbolic geometry.

**Definition 6.25.** *A Moebius transformation,*

$$f(z) = \frac{az + b}{cz + d}, \quad \text{where } a, b, c, d \in \mathbb{C}, \text{ and } (ad - bc) \neq 0 \tag{6.31}$$

*is unitary if $d = \bar{a}$, and $c = -\bar{b}$.*                                    $\square$

**Observations 6.11.** Basic observations about unitary Moebius transformations.

1. Composition of two unitary Moebius transformations is also unitary.
2. Inverse of unitary Moebius transformation is also unitary.                       $\square$

These observations are established in the problem section.

### Group Structure of the Moebius Transformation

The Moebius transformations form an algebraic group. The group operation is composition of functions, which acts upon all the Moebius transformations. It can be observed that this operation is associative. The identity element is the identity mapping $f(z) = z$. Existence of the inverse of a transformation has also been established in an earlier observation. Thus the set of Moebius transformations $\mathcal{M}(\mathbb{C}^*)$ under the operation of composition of functions forms a group. This group can also be identified with the general linear group $GL(2, \mathbb{C})$, which consists of all invertible matrices of size 2 in the field $\mathbb{C}$.

**Theorem 6.10.** *Let the set of Moebius transformations $\mathcal{M}(\mathbb{C}^*)$ be*

$$\mathcal{M}(\mathbb{C}^*) = \left\{ f : \mathbb{C}^* \to \mathbb{C}^* \mid f(z) = \frac{az + b}{cz + d}, \ a, b, c, d \in \mathbb{C}, \ (ad - bc) \neq 0 \right\} \tag{6.32a}$$

*The set $\mathcal{M}(\mathbb{C}^*)$ forms a group under function composition. In addition $\theta : GL(2, \mathbb{C}) \to \mathcal{M}(\mathbb{C}^*)$, where*

$$\begin{bmatrix} a & b \\ c & d \end{bmatrix} \mapsto \frac{az + b}{cz + d} \tag{6.32b}$$

*and $\theta$ is a surjective group homomorphism. The kernel of the mapping is the set of diagonal matrices $\{kI \mid k \in \widehat{\mathbb{C}}\}$, where $I$ is an identity matrix of size 2, and $\widehat{\mathbb{C}} = \mathbb{C} - \{0\}$.*

    *Proof.* The proof is left to the reader.                                    $\square$

The mapping $\theta$ induces a group isomorphism $\theta : PGL\,(2,\mathbb{C}) \to \mathcal{M}\,(\mathbb{C}^*)$, where

$$PGL\,(2,\mathbb{C}) = GL\,(2,\mathbb{C}) \,/\, \left\{ kI : k \in \widehat{\mathbb{C}} \right\}$$

is the projective general linear group. If we further restrict the mapping $\theta\,(\cdot)$ so that $(ad - bc) = 1$, then the mapping $\theta : SL\,(2,\mathbb{C}) \to \mathcal{M}\,(\mathbb{C}^*)$ is onto. The corresponding kernel of the mapping is $\pm I$. The mapping $\theta : PSL\,(2,\mathbb{C}) \to \mathcal{M}\,(\mathbb{C}^*)$ is also an isomorphism.

## 6.6 Polynomial Properties

Elementary properties of polynomials like: roots of a polynomial, resultant of two polynomials, discriminant of a polynomial, and the Bézout's theorem are outlined in this section. The Bézout's theorem is named after the mathematician Etienne Bézout (1730-1783). This theorem is a statement about the total number of intersecting points of two curves which do not have a common component. Polynomials have also been introduced in the chapter on abstract algebra. An important application of these properties is the study of elliptic curves. Elliptic curves have interesting group-theoretic properties. Properties of elliptic curves are studied in the chapter on abstract algebra.

The set of polynomials in indeterminate $x$ with coefficients in the field $\mathbb{R}$ is denoted by $\mathbb{R}\,[x]$. Similarly, the set of polynomials in indeterminate $x$ with coefficients in the field $\mathbb{C}$ is denoted by $\mathbb{C}\,[x]$. Basics of polynomials are also discussed in the chapter on abstract algebra.

### 6.6.1  Roots of a Polynomial

Let $p\,(z) = \sum_{j=0}^{n} a_j z^j \in \mathbb{C}\,[z]$, be a polynomial where $a_n \neq 0$, and $a_j \in \mathbb{C}$ for $0 \leq j \leq n$. The symbol $z$ is called the *indeterminate*. The polynomial $p\,(z)$ is said to be of *degree* $n$. Denote the degree of this polynomial by $\deg p\,(z)$, or as simply $\deg p$. Therefore $\deg p\,(z) = n$. If $n \in \mathbb{P}$, then the polynomial is nonconstant. The $a_j$'s are called the *coefficients* of the polynomial.

The $z$'s which satisfy $p\,(z) = 0$ are called the *zeros* of the polynomial $p\,(z)$; and $z^*$ is a *root* of the equation $p\,(z) = 0$, if $p\,(z^*) = 0$. Sometimes, the terms root and zero are used interchangeably.

If a zero of the polynomial equation occurs only once, then it is called a *simple zero*. If a zero occurs $k \geq 1$ times, then it is said to have a *multiplicity* $k$. It can be demonstrated that a polynomial of degree $n$ has exactly $n$ roots, possibly not all distinct. This fact is often called *the fundamental theorem of algebra*. This theorem is stated in the chapter on algebra.

### 6.6.2  Resultant of Two Polynomials

The theory of resultants is useful in solving a system of polynomial equations. It is also used to study the relative properties of these polynomials. The terminology used in this subsection is next introduced.

**Definitions 6.26.** *Monomial, polynomial, homogeneous polynomial, and greatest common divisor of two nonzero polynomials are defined.*

1. *A monomial in indeterminates $x$ and $y$ is a product of the form $x^\lambda y^\mu$ where the exponents $\lambda$ and $\mu$ are nonnegative integers. The degree of this monomial is $(\lambda + \mu)$.*

2. *A polynomial $f(x,y)$ in indeterminates $x$ and $y$ with coefficients in the field of complex numbers is a finite linear combination of monomials. The polynomial $f(x,y)$ is specified by*

$$f(x,y) = \sum_{\lambda,\mu} a_{\lambda\mu} x^{\lambda} y^{\mu}, \quad a_{\lambda\mu} \in \mathbb{C} \tag{6.33}$$

   (a) *The coefficient of the monomial $x^{\lambda} y^{\mu}$ is $a_{\lambda\mu}$.*
   (b) *$a_{\lambda\mu} x^{\lambda} y^{\mu}$ is a term of $f(x,y)$ provided $a_{\lambda\mu} \neq 0$. The degree of this term is equal to $(\lambda + \mu)$.*
   (c) *The degree of $f(x,y)$ is the maximum value of $(\lambda + \mu)$ such that $a_{\lambda\mu} \neq 0$. It is denoted by $\deg f(x,y)$.*

3. *A polynomial in which all of its terms are of the same degree is called a homogeneous polynomial.*

4. *Let $f(x), g(x) \in \mathbb{C}[x]$; where $f(x) \neq 0$ and $g(x) \neq 0$. The greatest common divisor of $f(x)$ and $g(x)$ is a monic polynomial of greatest degree in $\mathbb{C}[x]$ which divides both $f(x)$ and $g(x)$. It is denoted by $\gcd(f(x), g(x))$.* $\qquad\square$

The resultant of two nonconstant polynomials is directly proportional to the product of the differences of their roots. Therefore if the polynomials have a common root, then the value of the resultant of the two polynomials is equal to zero.

**Definition 6.27.** *Let $m, n \in \mathbb{P}$, and $f(x), g(x) \in \mathbb{C}[x]$, where*

$$f(x) = \sum_{i=0}^{m} a_i x^i = a_m \prod_{i-1}^{m} (x - \alpha_i), \quad a_m \neq 0 \tag{6.34a}$$

$$g(x) = \sum_{j=0}^{n} b_j x^j = b_n \prod_{j=1}^{n} (x - \beta_j), \quad b_n \neq 0 \tag{6.34b}$$

*The coefficients of the polynomial $f(x)$ of degree $m$, are $a_i \in \mathbb{C}, 0 \leq i \leq m$. The roots of this polynomial are $\alpha_i \in \mathbb{C}, 1 \leq i \leq m$. Similarly, the coefficients of the polynomial $g(x)$ of degree $n$ are $b_j \in \mathbb{C}, 0 \leq j \leq n$. The roots of this polynomial are $\beta_j \in \mathbb{C}, 1 \leq j \leq n$. The resultant of the two polynomials $R(f,g)$ is*

$$R(f,g) = a_m^n b_n^m \prod_{i=1}^{m} \prod_{j=1}^{n} (\alpha_i - \beta_j) \tag{6.34c}$$

*The resultant is also sometimes denoted by*

$$\widetilde{R}(a_m, a_{m-1}, \ldots, a_1, a_0; b_n, b_{n-1}, \ldots, b_1, b_0) \tag{6.34d}$$

$\qquad\square$

Some properties of the resultant of two polynomials are immediate.

**Observations 6.12.** Let $f(\cdot)$ and $g(\cdot)$ be polynomials as defined above.

1. Vanishing of the resultant. $R(f,g) = 0$ if and only if the two polynomials have a common root.

2. Symmetry. $R(g, f) = (-1)^{mn} R(f, g)$
3. Representation via roots of polynomial, either $f(\cdot)$ or $g(\cdot)$

$$R(f, g) = a_m^n \prod_{i=1}^{m} g(\alpha_i) = (-1)^{mn} b_n^m \prod_{j=1}^{n} f(\beta_j)$$

4. Factorization. Let $f_1(x), f_2(x) \in \mathbb{C}[x]$ be nonconstant polynomials. If $f(x) = f_1(x) f_2(x)$, then $R(f, g) = R(f_1, g) R(f_2, g)$.
5. Homogeneity and quasi-homogeneity. The representation of the resultant

$$R(f, g) = \widetilde{R}(a_m, a_{m-1}, \ldots, a_1, a_0; b_n, b_{n-1}, \ldots, b_1, b_0)$$

is homogeneous of degree $n$ in the $a_i$'s and of degree $m$ in the $b_j$'s. It also exhibits the following quasi-homogeneity property. For any nonzero scalar $\lambda$

$$\widetilde{R}\left(\lambda^m a_m, \lambda^{m-1} a_{m-1}, \ldots, \lambda^1 a_1, \lambda^0 a_0; \lambda^n b_n, \lambda^{n-1} b_{n-1}, \ldots, \lambda^1 b_1, \lambda^0 b_0\right)$$
$$= \lambda^{mn} \widetilde{R}(a_m, a_{m-1}, \ldots, a_1, a_0; b_n, b_{n-1}, \ldots, b_1, b_0)$$

$\square$

It turns out that the resultant of the two polynomials $f(\cdot)$ and $g(\cdot)$ is equal to the determinant of its Sylvester matrix. This matrix is named after the mathematician James J. Sylvester (1814-1897). This remarkable result is next proved. In order to prove this result, the concept of greatest common divisor of two polynomials is necessary.

**Definitions 6.28.** Let $f(x), g(x) \in \mathbb{C}[x]$, $f(x) = \sum_{i=0}^{m} a_i x^i$, $g(x) = \sum_{j=0}^{n} b_j x^j$, and $m, n \in \mathbb{P}$. The degrees of the polynomials $f(x)$ and $g(x)$ are $m$ and $n$ respectively. The Sylvester matrix $S(f, g)$ of these two polynomials is a square matrix of size $(m + n)$. It has the first $n$ rows involving $a_i$'s and the next $m$ rows involving $b_j$'s.

$$S(f, g) = \begin{bmatrix} a_m & a_{m-1} & a_{m-2} & \cdots & a_0 & 0 & \cdots & \cdots & 0 \\ 0 & a_m & a_{m-1} & \cdots & a_1 & a_0 & 0 & \cdots & 0 \\ \vdots & \vdots & \vdots & \ddots & \vdots & \vdots & \vdots & \ddots & \vdots \\ 0 & 0 & \cdots & 0 & a_m & a_{m-1} & a_{m-2} & \cdots & a_0 \\ b_n & b_{n-1} & b_{n-2} & \cdots & b_1 & b_0 & 0 & \cdots & 0 \\ 0 & b_n & b_{n-1} & \cdots & \cdots & b_1 & b_0 & \cdots & 0 \\ \vdots & \vdots & \vdots & \ddots & \vdots & \vdots & \vdots & \ddots & \vdots \\ 0 & 0 & \cdots & \cdots & 0 & b_n & b_{n-1} & \cdots & b_0 \end{bmatrix}$$

(6.35)

$\square$

**Theorem 6.11.** Let $f(x), g(x) \in \mathbb{C}[x]$. If the Sylvester matrix of these two polynomials is $S(f, g)$, then

$$R(f, g) = \det S(f, g) \tag{6.36}$$

*Proof.* Let $f(x) = \sum_{i=0}^{m} a_i x^i$ and $g(x) = \sum_{j=0}^{n} b_j x^j$, and $m, n \in \mathbb{P}$. The degrees of the polynomials $f(x)$ and $g(x)$ are $m$ and $n$ respectively.

Observe that $R(f, g) = 0$ if and only if the polynomials $f(x)$ and $g(x)$ have a common root. This is true if and only if $\gcd(f(x), g(x))$ is a nonconstant polynomial. This in turn is true if and only if the two polynomials have a common divisor $h(x)$ of positive degree, such that $f(x) = h(x) f_1(x)$, $\deg f_1(x) \leq (m-1)$, $g(x) = h(x) g_1(x)$, $\deg g_1(x) \leq (n-1)$, and $f(x) g_1(x) = f_1(x) g(x)$. Let $f_1(x) = \sum_{i=0}^{m-1} c_i x^i$ and $g_1(x) = \sum_{j=0}^{n-1} d_j x^j$.

In the next step substitute the different explicit polynomial expressions in the relationship $f(x) g_1(x) = f_1(x) g(x)$, and compare the coefficients of $x^{m+n-1}, x^{m+n-2}, \ldots, x, 1$ on the left and right-hand side. This results in a homogeneous linear system of equations in the unknowns

$$d_{n-1}, d_{n-2}, \ldots, d_1, d_0, -c_{m-1}, -c_{m-2}, \ldots, -c_1, -c_0$$

There are $(m + n)$ homogeneous set of equations in $(m + n)$ unknowns. These unknowns have a nontrivial solution if and only if the determinant of the corresponding coefficient matrix vanishes. It can be observed that the coefficient matrix is indeed equal to the Sylvester matrix $S(f, g)$. Therefore the condition for the nontrivial solution is $\det S(f, g) = 0$. Also observe that the resultant $R(f, g)$ and $\det S(f, g)$ have terms of degree $n$ in the $a_i$'s and of degree $m$ in the $b_j$'s. Therefore $R(f, g)$ is equal to $\det S(f, g)$ up to a constant. Furthermore, the coefficient of $a_m^n b_0^m$ in both $R(f, g)$ and $\det S(f, g)$ is equal to unity.

Note that $-c_i$'s and not $c_i$'s were chosen as unknowns in order to avoid negative signs in the definition of the Sylvester matrix.                                                                    $\square$

The above discussion assumes that the polynomials are defined over the complex field. Similar results are true for polynomials defined over any field.

**Examples 6.4.** Let $f(x), g(x) \in \mathbb{C}[x]$. Assume that these polynomials are nonconstant.

1. Computation of the resultant.
   (a) Let $f(x) = \sum_{i=0}^{1} a_i x^i$ and $g(x) = \sum_{j=0}^{1} b_j x^j$. We compute the resultant of the two polynomials in several different ways.
      (i) Note that if $f(x) = a_1 (x - \alpha_1)$, then $\alpha_1 = -a_0/a_1$. If $g(x) = b_1 (x - \beta_1)$, then $\beta_1 = -b_0/b_1$. Therefore the expression $R(f, g) = a_1 b_1 (\alpha_1 - \beta_1) = (a_1 b_0 - a_0 b_1)$.
      (ii) It can be shown that $R(g, f) = -R(f, g)$.
      (iii) $R(f, g) = a_1 g(\alpha_1) = a_1 \{b_1 (-a_0/a_1) + b_0\} = (a_1 b_0 - a_0 b_1)$.
      (iv) $R(f, g) = -b_1 f(\beta_1) = -b_1 \{a_1 (-b_0/b_1) + a_0\} = (a_1 b_0 - a_0 b_1)$.
      (v) $R(f, g)$ is also computed via the determinant of the Sylvester matrix.

$$R(f, g) = \begin{vmatrix} a_1 & a_0 \\ b_1 & b_0 \end{vmatrix} = (a_1 b_0 - a_0 b_1)$$

   (b) Let $f(x) = \sum_{i=0}^{1} a_i x^i$ and $g(x) = \sum_{j=0}^{2} b_j x^j$. We compute the resultant of the two polynomials in two different ways.
      (i) Note that if $f(x) = a_1 (x - \alpha_1)$, then $\alpha_1 = -a_0/a_1$. Therefore

$$R(f, g) = a_1^2 g(\alpha_1) = \left(a_0^2 b_2 - a_1 a_0 b_1 + a_1^2 b_0\right)$$

      (ii) $R(f, g)$ is also computed via the determinant of the Sylvester matrix.

$$R(f, g) = \begin{vmatrix} a_1 & a_0 & 0 \\ 0 & a_1 & a_0 \\ b_2 & b_1 & b_0 \end{vmatrix} = \left(a_0^2 b_2 - a_1 a_0 b_1 + a_1^2 b_0\right)$$

2. Let $f(x) = \sum_{i=0}^{3} a_i x^i$ and $g(x) = \sum_{j=0}^{2} b_j x^j$. The Sylvester matrix $S(f,g)$ is of size 5.

$$S(f,g) = \begin{bmatrix} a_3 & a_2 & a_1 & a_0 & 0 \\ 0 & a_3 & a_2 & a_1 & a_0 \\ b_2 & b_1 & b_0 & 0 & 0 \\ 0 & b_2 & b_1 & b_0 & 0 \\ 0 & 0 & b_2 & b_1 & b_0 \end{bmatrix}$$

□

### 6.6.3  Discriminant of a Polynomial

The discriminant of a polynomial is directly proportional to the product of the differences of the roots of the polynomial. A standard definition of the discriminant of a polynomial multiplies this product by a normalization constant.

**Definition 6.29.** *Let $f(x) \in \mathbb{C}[x]$, and $\deg f(x) \geq 2$, where*

$$f(x) = \sum_{i=0}^{m} a_i x^i = a_m \prod_{i=1}^{m} (x - \alpha_i), \quad a_m \neq 0 \tag{6.37a}$$

*The discriminant of the polynomial $f(x)$ is*

$$\Delta(f) = (-1)^{m(m-1)/2} a_m^{2m-2} \prod_{\substack{1 \leq i,j \leq m \\ i \neq j}} (\alpha_i - \alpha_j) \tag{6.37b}$$

□

An alternate expression for $\Delta(f)$ is

$$\Delta(f) = a_m^{2m-2} \prod_{1 \leq i < j \leq m}^{m} (\alpha_i - \alpha_j)^2 \tag{6.38}$$

It is also possible to derive a relationship between the discriminant of a polynomial, and the resultant of the polynomial and its derivative.

**Observations 6.13.** Let $f(x) \in \mathbb{C}[x]$, $f(x) = \sum_{i=0}^{m} a_i x^i$, $\deg f(x) = m \geq 2$, and $f'(x)$ be the derivative of $f(x)$ with respect to $x$.

1. We have
$$R(f, f') = (-1)^{m(m-1)/2} a_m \Delta(f)$$

2. If the polynomial $f(x)$ has a repeated root, then $f(x)$ and $f'(x)$ have a common root. Therefore the resultant $R(f, f') = 0$. This implies that the discriminant $\Delta(f) = 0$ if and only if the polynomial $f(x)$ has a repeated root. □

The proof of the first observation is given in the problem section.

**Examples 6.5.** The discriminant of a quadratic and cubic polynomial are computed.

1. Let $a \neq 0$, and $f(x) = ax^2 + bx + c$ (quadratic) and $f'(x) = 2ax + b$. Therefore the resultant

$$R(f, f') = \begin{vmatrix} a & b & c \\ 2a & b & 0 \\ 0 & 2a & b \end{vmatrix} = -a\left(b^2 - 4ac\right)$$

Thus $\Delta(f) = \left(b^2 - 4ac\right)$.

2. Let $f(x) = x^3 + ax + b$ (cubic) and $f'(x) = 3x^2 + a$. Therefore the resultant

$$R(f, f') = \begin{vmatrix} 1 & 0 & a & b & 0 \\ 0 & 1 & 0 & a & b \\ 3 & 0 & a & 0 & 0 \\ 0 & 3 & 0 & a & 0 \\ 0 & 0 & 3 & 0 & a \end{vmatrix} = \left(4a^3 + 27b^2\right)$$

Thus $\Delta(f) = -\left(4a^3 + 27b^2\right)$.    □

Discriminants are used in the chapter on abstract algebra to study elliptic curves.

### 6.6.4 Bézout's Theorem

The Bézout's theorem provides us with a numerical value for the number of points of intersection of two algebraic curves. This theorem in turn lets us study the group law of elliptic curves. There is an analogous Bézout's theorem in number theory. Assume that the curves $f(x, y) = 0$ and $g(x, y) = 0$ have no common components, then according to this theorem the number of points of intersection of the two curves is equal to the product of their degrees. Counting of the number of points of intersection is done by including a proper interpretation of points at infinity, multiplicities of points of intersection, and complex solutions. Bézout's theorem is a fundamental result of algebraic geometry. We only concern ourselves with curves in two variables with real coefficients. A motivation for this result is provided by considering several examples.

**Examples 6.6.** Some elementary examples.

1. Let the two curves be two straight lines. The degree of each curve is one. According to the Bézout's theorem, the total number of points of intersection is one. Consider the following cases.
   (a) Let the two straight lines have different slopes. Then the two straight lines intersect at a single point. This is the value predicted by the Bézout's theorem.
   (b) Let the two straight lines have identical slopes. Then the two straight lines are parallel to each other. It is assumed that the two straight lines intersect each other "at infinity." Thus the two straight lines intersect at a single point, which is the value predicted by the Bézout's theorem.
   (c) If the two straight lines are identical, then there are an infinite number of common points. The Bézout's theorem is not applicable because the two curves have a common component.
2. Let the two curves be a straight line, and a circle. The degrees of these two curves are one and two respectively. According to the Bézout's theorem, the total number of points of intersection is two. Consider the following cases.

(a) Let the equation of the straight line and the circle be $y = x$ and $x^2 + y^2 = 2$ respectively. The points of intersection are $(1, 1)$ and $(-1, -1)$. Thus there are two points of intersection as predicted by the Bézout's theorem.

(b) Let the equation of the straight line and the circle be $x + y = 2$ and $x^2 + y^2 = 2$ respectively. The straight line is tangent to the circle at $(1, 1)$. Elimination of $x$ from the two equations yields $(y - 1)^2 = 0$. Thus the point $(1, 1)$ has a multiplicity two. Therefore counting multiplicities, there are two points of intersection as predicted by the Bézout's theorem.

(c) Let the equation of the straight line and the circle be $x = 4$ and $x^2 + y^2 = 2$ respectively. These two equations are satisfied by $(x, y) = \left(4, \pm i\sqrt{14}\right)$, where $i = \sqrt{-1}$. Thus the two curves have an empty intersection over $\mathbb{R}$. However, there are two points of intersection over $\mathbb{C}$.

(d) Let the equation of the straight line and the circle be $x + y = 4$ and $x^2 + y^2 = 2$ respectively. These two equations are satisfied by $(x, y) = \left(2 \pm i\sqrt{3}, 2 \mp i\sqrt{3}\right)$. Thus the curves have an empty intersection over $\mathbb{R}$. However, there are two points of intersection over $\mathbb{C}$.

3. Consider two concentric circles with center at the origin in the $xy$-plane, and different radii. There are no apparent points of intersection of the two curves in this plane. The degree of the polynomial representing a circle is two. Therefore, as per the Bézout's theorem the total number of points of intersection is 4. These points can be considered to be "at infinity."

4. Consider the parabola $y = x^2$ and a vertical straight line $x = 2$. According to the Bézout's theorem, the total number of points of intersection is two. However one point of intersection over $\mathbb{R}$ is $(2, 4)$. As multiplicities appear to require tangency, the point $(2, 4)$ has a multiplicity one. Therefore the second point of intersection is considered to be "at infinity."  $\qquad\Box$

As mentioned earlier, the goal of this subsection is to determine the number of common points of two curves which have only finitely many points in common. Let $f(x, y)$ and $g(x, y)$ be two nonconstant polynomials with real coefficients. Let the degrees of these polynomials be $m$ and $n$ respectively, where $m, n \in \mathbb{P}$. The corresponding curves are specified by $f(x, y) = 0$ and $g(x, y) = 0$. Write the polynomials $f(x, y)$ and $g(x, y)$ in powers of $y$ as follows.

$$f(x, y) = \sum_{i=0}^{m} f_i(x) y^{m-i}, \quad f_0(x) \neq 0$$

$$g(x, y) = \sum_{j=0}^{n} g_j(x) y^{n-j}, \quad g_0(x) \neq 0$$

Note that $f_0(x) \neq 0$ and $g_0(x) \neq 0$, however $f_0(x)$ and $g_0(x)$ each can possibly be equal to zero for particular values of $x$. The resultant of the polynomials $\sum_{i=0}^{m} f_i(x) y^{m-i}$ and $\sum_{j=0}^{n} g_j(x) y^{n-j}$ with respect to $y$ is

$$\widetilde{R}\left(f_0(x), f_1(x), \ldots, f_{m-1}(x), f_m(x); g_0(x), g_1(x), \ldots, g_{n-1}(x), g_n(x)\right)$$

Denote it by $R(f, g)(x)$. Let $x = a$ and $y = b$ be the solution of the equations $f(x, y) = 0$ and $g(x, y) = 0$. Thus the polynomials $\sum_{i=0}^{m} f_i(a) y^{m-i}$ and $\sum_{j=0}^{n} g_j(a) y^{n-j}$ have a common root $y = b$. Therefore if $f_0(a) \neq 0$ and $g_0(a) \neq 0$ then $R(f, g)(a) = 0$.

Conversely, if $f_0(a) \neq 0, g_0(a) \neq 0$, and $R(f, g)(a) = 0$ then the two polynomials $\sum_{i=0}^{m} f_i(a) y^{m-i}$ and $\sum_{j=0}^{n} g_j(a) y^{n-j}$ are of exactly degrees $m$ and $n$ respectively, and their resultant vanishes. This implies that there exists a $b$ such that $x = a$ and $y = b$ is a solution of the equations $f(x, y) = 0$ and $g(x, y) = 0$.

Observe from the determinant form of expression for $R(f, g)(x)$ that the above results are true if either $f_0(a) = 0$ or $g_0(a) = 0$. However if $f_0(a) = g_0(a) = 0$ then $R(f, g)(a) = 0$, but the resultant of the polynomials $f(x, y)$ and $g(x, y)$ need not be equal to zero. This discussion is condensed into the following theorem.

**Theorem 6.12.** *Let $f(x, y)$ and $g(x, y)$ be two polynomials with real coefficients. Also let the degrees of these polynomials be $m$ and $n$ respectively, where $m, n \in \mathbb{P}$. The corresponding curves are specified by $f(x, y) = 0$ and $g(x, y) = 0$. Express these polynomials in powers of $y$ as follows.*

$$f(x, y) = \sum_{i=0}^{m} f_i(x) y^{m-i}, \quad f_0(x) \neq 0 \tag{6.39a}$$

$$g(x, y) = \sum_{j=0}^{n} g_j(x) y^{n-j}, \quad g_0(x) \neq 0 \tag{6.39b}$$

*If $x = a$ and $y = b$ is a solution of the equations $f(x, y) = 0$ and $g(x, y) = 0$, and if either $f_0(a) \neq 0$ or $g_0(a) \neq 0$, then the resultant $R(f, g)(a) = 0$.*

*Conversely, if either $f_0(a) \neq 0$ or $g_0(a) \neq 0$, and $R(f, g)(a) = 0$, then there exists a number $b$ such that $x = a$ and $y = b$ satisfy the equations $f(x, y) = 0$ and $g(x, y) = 0$.* $\qquad \square$

This theorem is illustrated via examples.

**Examples 6.7.**

1. Let
$$f(x, y) = \left(-3x^2 + 2xy + y^2\right), \quad \text{and} \quad g(x, y) = \left(x^2 - y\right)$$

We find $(x, y)$ which satisfy $f(x, y) = 0$ and $g(x, y) = 0$. The polynomials $f(x, y)$ and $g(x, y)$ are written as

$$f(x, y) = y^2 + 2xy - 3x^2$$
$$g(x, y) = -y + x^2$$

Thus
$$R(f, g)(x) = \begin{vmatrix} 1 & 2x & -3x^2 \\ -1 & x^2 & 0 \\ 0 & -1 & x^2 \end{vmatrix} = x^2(x-1)(x+3)$$

Similarly write $f(x, y)$ and $g(x, y)$ as

$$f(x, y) = -3x^2 + 2yx + y^2$$
$$g(x, y) = x^2 - y$$

Thus
$$R(f, g)(y) = \begin{vmatrix} -3 & 2y & y^2 & 0 \\ 0 & -3 & 2y & y^2 \\ 1 & 0 & -y & 0 \\ 0 & 1 & 0 & -y \end{vmatrix} = y^2(y-9)(y-1)$$

The possible values of $x$ are $0, 1$, and $-3$. Similarly, the possible values of $y$ are $0, 1$, and $9$. Finally, the possible values of $(x, y)$ are

$$(0,0), (0,1), (0,9), (1,0), (1,1), (1,9), (-3,0), (-3,1), \text{ and } (-3,9)$$

Of these values, only $(0,0), (1,1),$ and $(-3,9)$ satisfy $f(x,y) = 0$ and $g(x,y) = 0$.

2. Let $f(x,y) = (x+y-1)$, and $g(x,y) = (x+y-2)$. It is readily evident that $f(x,y) = 0$ and $g(x,y) = 0$ are an inconsistent set of equations. It can also be determined that $R(f,g)(x) = R(f,g)(y) = -1$ is a constant. Therefore the conditions of the above theorem are not applicable.

3. Let $f(x,y) = (x^2y^2 - 1)$, and $g(x,y) = (x^2y^2 - 4)$. It is again evident that $f(x,y) = 0$ and $g(x,y) = 0$ are an inconsistent set of equations. It can also be determined that $R(f,g)(x) = 9x^4$ and $R(f,g)(y) = 9y^4$. Thus a possible value of $(x,y)$ is $(0,0)$. However, it does not satisfy the equations $f(x,y) = 0$ and $g(x,y) = 0$. Nevertheless, express $f(x,y)$ and $g(x,y)$ as

$$f(x,y) = f_0(x) y^2 + f_1(x) y + f_2(x)$$
$$g(x,y) = g_0(x) y^2 + g_1(x) y + g_2(x)$$

Observe that $f_0(x) = g_0(x) = x^2$, and therefore $f_0(0) = g_0(0) = 0$. This implies that the conditions of the above theorem are not applicable.    □

A weaker form of the Bézout's theorem is proved with the aid of the following two lemmas.

**Lemma 6.2.** *Let*

$$f(x,y;t) = \sum_{i=0}^{m} a_{m-i}(x,y) t^i, \quad and \quad g(x,y;t) = \sum_{j=0}^{n} b_{n-j}(x,y) t^j \tag{6.40}$$

*be polynomials with real coefficient, where $m, n \in \mathbb{P}$. Also $a_i(x,y)$ is either equal to zero or a homogeneous polynomial of degree $i$ in variables $x$ and $y$, where $0 \leq i \leq m$. Similarly, $b_j(x,y)$ is either equal to zero or a homogeneous polynomial of degree $j$ in variables $x$ and $y$, where $0 \leq j \leq n$. Let $a_r(x,y)$ and $b_s(x,y)$ be the first nonvanishing coefficients of the polynomials $f(x,y;t)$ and $g(x,y;t)$ respectively, where $0 \leq r \leq m$ and $0 \leq s \leq n$. Then the resultant $R(f,g)(x,y)$ with respect to $t$ is a homogeneous polynomial in $x$ and $y$ of degree $(mn - rs)$.*

*Proof.* See the problem section.    □

Consider the two nonconstant polynomials $f(x,y)$ and $g(x,y)$ with real coefficients. Let the degrees of these polynomials be $m$ and $n$ respectively, where $m, n \in \mathbb{P}$. The resultant of these two polynomials with respect to $y$ is $R(f,g)(x)$. We establish that the degree of this resultant is less than or equal to $mn$. In order to prove this result, the concept of *homogenization* of a polynomial is required.

Let $f(x,y) = \sum_{\lambda,\mu} a_{\lambda\mu} x^\lambda y^\mu$, where $a_{\lambda\mu} \in \mathbb{R}$, be a nonconstant polynomial of degree $d$. The homogenized version of the polynomial $f(x,y)$ is the polynomial

$$F(x,y,z) = \sum_{\lambda,\mu} a_{\lambda\mu} x^\lambda y^\mu z^{d-\lambda-\mu}$$

Thus the homogenized version of the polynomial $f(x,y)$ is obtained by multiplying each term of the polynomial by a power of $z$ required to produce a term of degree $d$. This polynomial $F(x,y,z)$ is homogeneous because $F(tx,ty,tz) = t^d F(x,y,z)$. Observe that $f(x,y) = F(x,y,1)$.

**Lemma 6.3.** *Let* $f(x, y)$ *and* $g(x, y)$ *be two polynomials with real coefficients. Also let the degrees of these polynomials be* $m$ *and* $n$ *respectively, where* $m, n \in \mathbb{P}$. *Write these polynomials in powers of* $y$ *as follows.*

$$f(x, y) = \sum_{i=0}^{m} f_i(x) y^{m-i}, \quad and \quad g(x, y) = \sum_{j=0}^{n} g_j(x) y^{n-j} \qquad (6.41)$$

*The resultant of the polynomials* $\sum_{i=0}^{m} f_i(x) y^{m-i}$ *and* $\sum_{j=0}^{n} g_j(x) y^{n-j}$ *with respect to* $y$ *is* $R(f, g)(x)$. *Then* $\deg R(f, g)(x) \leq mn$.

*Proof.* The polynomials $f(x, y)$ and $g(x, y)$ are homogenized as $F(x, y, z)$ and $G(x, y, z)$ respectively. Thus $F(x, y, 1) = f(x, y)$ and $G(x, y, 1) = g(x, y)$. The polynomials $F(x, y, z)$ and $G(x, y, z)$ can be expressed as

$$F(x, y, z) = \sum_{i=0}^{m} F_i(x, z) y^{m-i}, \quad and \quad G(x, y, z) = \sum_{j=0}^{n} G_j(x, z) y^{n-j}$$

where $F_i(x, z)$ is either a homogeneous polynomial of degree $i$ or equal to zero for $1 \leq i \leq m$. Similarly, $G_j(x, z)$ is either a homogeneous polynomial of degree $j$ or equal to zero for $1 \leq j \leq n$. The resultant of the polynomials $F(x, y, z)$ and $G(x, y, z)$ with respect to $y$ is a homogeneous polynomial of degree $\leq mn$ by the previous lemma. Substituting $z = 1$ we obtain $\deg R(f, g)(x) \leq mn$. $\qquad\square$

The above result is not sufficient to count the number of points of intersection of the two curves. For example a root $x_0$ of the resultant $R(f, g)(x)$ might correspond to several values of the $y$-coordinate. The following result is a weak form of the Bézout's theorem.

**Theorem 6.13.** *Let* $f(x, y)$ *and* $g(x, y)$ *be two polynomials with real coefficients. Also let the degrees of these polynomials be* $m$ *and* $n$ *respectively, where* $m, n \in \mathbb{P}$. *Further assume that the two polynomials do not have any common factor of positive degree. Then there are at most* $mn$ *common points of the curves* $f(x, y) = 0$ *and* $g(x, y) = 0$.

*Proof.* The proof is by contradiction. Assume that there are $(mn + 1)$ common points of the two curves. Let these be $(\alpha_u, \beta_u)$, where $1 \leq u \leq (mn + 1)$. Select a number $c$ such that

$$\alpha_u + c\beta_u \neq \alpha_k + c\beta_k, \quad 1 \leq u < k \leq (mn + 1)$$

Note that the number $c$ does exist, and substitute $x = (x' - cy')$, and $y = y'$ in the polynomials $f(x, y)$ and $g(x, y)$. This yields polynomials $f(x' - cy', y')$ and $g(x' - cy', y')$. Let $\widetilde{f}(x', y') \triangleq f(x' - cy', y')$ and $\widetilde{g}(x', y') \triangleq g(x' - cy', y')$. Since

$$f(x, y) = \sum_{i=0}^{m} f_i(x) y^{m-i}, \quad and \quad g(x, y) = \sum_{j=0}^{n} g_j(x) y^{n-j}$$

we have

$$\widetilde{f}(x', y') = \sum_{i=0}^{m} f_i(x' - cy') y'^{m-i}, \quad and \quad \widetilde{g}(x', y') = \sum_{j=0}^{n} g_j(x' - cy') y'^{n-j}$$

Furthermore, $\deg \widetilde{f}(x',y') = m$ and $\deg \widetilde{g}(x',y') = n$. Observe that the set of solutions $(\alpha'_u, \beta'_u) = (\alpha_u + c\beta_u, \beta_u)$, where $1 \le u \le (mn+1)$ satisfy

$$\widetilde{f}(\alpha'_u, \beta'_u) = \widetilde{g}(\alpha'_u, \beta'_u) = 0, \text{ for } 1 \le u \le (mn+1)$$
$$\text{where } \alpha'_u \ne \alpha'_k, \text{ for } 1 \le u < k \le (mn+1)$$

Therefore the numbers $\alpha'_u$, $1 \le u \le (mn+1)$ are the roots of the resultant $R(f', g')(x')$ of the polynomials $\widetilde{f}(x',y')$ and $\widetilde{g}(x',y')$ with respect to $y'$. However, $\deg R(f',g')(x') \le mn$ as per the last lemma. Therefore as $R(f',g')(x')$ has $(mn+1)$ roots it should be the zero polynomial. Thus the polynomials $\widetilde{f}(x',y')$ and $\widetilde{g}(x',y')$ have a common factor of positive degree. This in turn implies that the polynomials $f(x,y)$ and $g(x,y)$ have a common factor of positive degree, which is a contradiction.                                                                      □

We established in the last theorem that the two curves, $f(x,y) = 0$ where $\deg f(x,y) = m \in \mathbb{P}$, and $g(x,y) = 0$ where $\deg g(x,y) = n \in \mathbb{P}$, with no common components, have at most $mn$ common points. Using techniques from algebraic geometry, it can be proved that the total number of intersecting points is indeed equal to $mn$ points. This is done by counting: points at infinity, multiplicities of points of intersection, and complex solutions.

---

## 6.7 Asymptotic Behavior and Algorithmic-Complexity Classes

Asymptotic behavior of functions, and different algorithmic-complexity classes are studied in this section. Typically, an algorithm is a finite step-by-step procedure to execute a computational task on a computer. Such steps are known as an *algorithm*. The word, algorithm is named after Muhammad ibn Mūsā al-Khwārizmī. He was a Persian mathematician who lived in the ninth century. In the year 825 AD, he wrote a scholarly book: *On the Calculation with Hindu Numerals*. This book was largely responsible for the spread of Indian numerical system throughout Europe and Middle East. The Latin translation of al-Khwārizmī's book was *Algoritmi de numero Indorum*. The Latinized name of al-Khwārizmī was Algoritmi, which later transformed into *algorithm*.

Asymptotic behavior of functions is usually used to describe the computational complexity of algorithms and the amount of computer memory needed to execute them. Study of algorithmic-complexity classes helps in classifying the algorithms based upon their complexity.

### 6.7.1 Asymptotic Behavior

It is generally useful to specify the asymptotic behavior of continuous functions, convergence of series and sequences, or the computational complexity of algorithms. These provide a comprehensive insight into the behavior of functions. There are several different measures (and corresponding notations) to describe their asymptotic behavior.

**Definitions 6.30.** *The asymptotic behavior of a sequence of real numbers $a_n$ and $b_n$ as $n \to \infty$ is defined below. Let $b_n > 0$ for sufficiently large $n$.*

1. *O-Notation. It is also called the big-oh notation. For a specified sequence $b_n$, $O(b_n)$ is a set of sequences*

$$O\left(b_n\right) = \{a_n \mid \textit{there exist positive constants } K \textit{ and } n_0 \textit{ such that}$$
$$0 \leq |a_n| \leq K b_n \textit{ for all } n \geq n_0\} \tag{6.42a}$$

*The O-notation provides an asymptotic upper bound for a sequence to within a constant factor.*

2. *$\Omega$-Notation. For a specified sequence $b_n$, $\Omega\left(b_n\right)$ is a set of sequences*

$$\Omega\left(b_n\right) = \{a_n \mid \textit{there exist positive constants } k \textit{ and } n_0 \textit{ such that}$$
$$0 \leq k b_n \leq |a_n| \textit{ for all } n \geq n_0\} \tag{6.42b}$$

*Therefore $\Omega\left(b_n\right)$ is the set of sequences that grow at least as rapidly as a positive multiple of $b_n$. This notation provides an asymptotic lower bound for a sequence to within a constant factor.*

3. *$\Theta$-Notation. For a specified sequence $b_n$, $\Theta\left(b_n\right)$ is a set of sequences*

$$\Theta\left(b_n\right) = \{a_n \mid \textit{there exists positive constants } K, k, \textit{ and } n_0 \textit{ such that}$$
$$0 \leq k b_n \leq |a_n| \leq K b_n \textit{ for all } n \geq n_0\} \tag{6.42c}$$

*That is, $a_n \in O\left(b_n\right)$ and $a_n \in \Omega\left(b_n\right)$ iff $a_n \in \Theta\left(b_n\right)$. This notation implies that the sequences $a_n$ and $b_n$ have the same order of magnitude. Therefore $\Theta\left(b_n\right)$ is the set of sequences that grow at the same rate as a positive multiple of $b_n$.*

4. *o-Notation. It is also called the little-oh (or small-oh) notation. For a specified sequence $b_n$, $o\left(b_n\right)$ is a set of sequences*

$$o\left(b_n\right) = \{a_n \mid \textit{for any } k > 0, \textit{there exists a positive number } n_0 \textit{ such that}$$
$$0 \leq |a_n| < k b_n \textit{ for all } n \geq n_0\} \tag{6.42d}$$

*That is, $a_n \in o\left(b_n\right)$ if $\lim_{n \to \infty} |a_n/b_n| = 0$. Thus the sequence $a_n$ becomes insignificant relative to $b_n$ as $n$ gets larger.*                                                                        □

The $O$-notation gives an upper bound to within a constant factor. The set of functions that grow no more rapidly than a positive multiple of $b_n$ is called $O\left(b_n\right)$. This notation is often used in stating the running time of an algorithm. The big-oh symbol is sometimes called the Landau symbol after the mathematician Edmund Landau (1877-1938). However, this symbol was first introduced in the year 1892 by Paul Bachmann (1837-1920). Even though $O\left(b_n\right)$ is a set, and $a_n$ belongs to this set, it is customary to write $a_n = O\left(b_n\right)$. This convention is extended to all other notations: $\Omega\left(\cdot\right), \Theta\left(\cdot\right)$, and $o\left(\cdot\right)$.

Notice the difference in the definitions of the $O$-notation and the $o$-notation. In the $O$-notation, $a_n = O\left(b_n\right)$, the bound $0 \leq |a_n| \leq K b_n$ holds for some constant $K > 0$. However in the $o$-notation, $a_n = o\left(b_n\right)$, the bound $0 \leq |a_n| < k b_n$ holds for all constants $k > 0$. Thus asymptotic upper bound provided by the $O$-notation may or may not be asymptotically tight. In contrast, the $o$-notation is used to denote an upper bound which is not asymptotically tight.

The asymptotic bound $5n^2 = O\left(n^2\right)$ is asymptotically tight, but the bound $5n = O\left(n^2\right)$ is not asymptotically tight. Also $o\left(b_n\right)$ is the set of functions that grow less rapidly than a positive multiple of $b_n$. In summary, if a positive sequence $b_n$ is given, then:

(a) $O\left(b_n\right)$ is the set of all $a_n$ such that $|a_n/b_n|$ is bounded from above as $n \to \infty$. Therefore, this notation is a convenient way to express an upper bound of a sequence within a constant.

(b) $\Omega\left(b_n\right)$ is the set of all $a_n$ such that $|a_n/b_n|$ is bounded from below by a strictly positive number as $n \to \infty$. This notation is used in expressing a lower bound of a sequence within a constant.

(c) $\Theta\left(b_n\right)$ is the set of all $a_n$ such that $\left|a_n/b_n\right|$ is bounded from both above and below as $n \to \infty$. This notation is used to express matching upper and lower bounds.

(d) $o\left(b_n\right)$ is the set of all $a_n$ such that $\left|a_n/b_n\right| \to 0$ as $n \to \infty$. This notation is used to express bound which is not asymptotically tight.

These definitions about asymptotic sequences have been defined in terms of sequences. However, these can easily be extended to continuous functions. Alternate simplified notation is given below.

**Definitions 6.31.** *More notation.*

1. *As $n \to \infty$:*

   a) $a_n \asymp b_n$ *iff* $a_n = \Theta\left(b_n\right)$.

   b) $a_n \ll b_n$ *or* $b_n \gg a_n$ *iff* $a_n \geq 0$ *and* $a_n = o\left(b_n\right)$.

   c) *Asymptotically equivalent sequences:* $a_n \sim b_n$ *iff* $a_n/b_n \to 1$.

2. *Asymptotic equality (approximation) between two functions is denoted by $\simeq$ .*

3. *Approximation between numbers is denoted by $\approx$ .* $\qquad\square$

**Examples 6.8.** Let $n \in \mathbb{P}$.

1. $f\left(n\right) = \sum_{j=0}^{m} a_j n^j$, $a_m \neq 0$. Then $f\left(n\right) \in O\left(n^m\right)$.
2. $\cos\left(x\right) \in O\left(1\right)$.
3. $\ln\left(n!\right) \in O\left(n \ln n\right)$.
4. $\sum_{j=1}^{n} j \in O\left(n^2\right)$.
5. $n! \in o\left(n^n\right)$. This follows from the observation $n! \leq n^n$.
6. $x^n \in o\left(x^{n+1}\right)$, $x^{n+1} \notin o\left(x^n\right)$.
7. $x^n \in O\left(x^{n+1}\right)$, $x^{n+1} \notin O\left(x^n\right)$.
8. $x^n \notin o\left(x^n\right)$.
9. If $a_n \in O\left(1\right)$, then the sequence $a_n$ is bounded.
10. If $a_n \in o\left(1\right)$, then $a_n \to 0$ as $n \to \infty$.
11. $2n \in o\left(n^2\right)$, but $2n^2 \notin o\left(n^2\right)$.
12. Stirling's formula: $n! \sim \sqrt{2\pi n}\left(n/e\right)^n$ .
13. If $p_n$ is the $n$th prime number, then $p_n \sim n \ln n$.
14. $7x^4 \in \Theta\left(2x^4\right)$.
15. $x^5 \in \Omega\left(x^4\right)$.
16. $\pi \approx 3.14$. However, it is incorrect to state: $\pi \simeq 3.14$. $\qquad\square$

**Binomial Coefficients**

Some bounds on the binomial coefficients are obtained. These coefficients are defined in terms of factorials and falling factorials.

**Definitions 6.32.** *Let $n \in \mathbb{N}$.*

1. *Factorial of a nonnegative integer $n$ is denoted by $n!$. It is:*

$$0! = 1$$
$$n! = (n-1)!n, \quad n = 1, 2, 3, \ldots$$

2. *Binomial coefficients arise in the expansion of the series*

$$(x + y)^n = \sum_{k=0}^{n} \binom{n}{k} x^k y^{n-k}$$

$$\binom{n}{k} = \frac{n!}{k!\,(n-k)!}, \quad 0 \le k \le n$$

*The coefficients $\binom{n}{k}$, of the above polynomial in two variables $x$ and $y$ are called the binomial coefficients. The above series expansion is often referred to as the binomial theorem.*

3. *The binomial coefficient $\binom{n}{k}$ is assumed to be zero if $k > n$.*
4. *The falling factorial $(n)_k$ is*

$$(n)_0 = 1$$
$$(n)_k = n\,(n-1)\cdots(n-k+1), \quad 1 \le k \le n$$

$\square$

Some useful bounds on binomial coefficients and falling factorials are summarized in the following observations.

**Observations 6.14.** Let $n \in \mathbb{P}$, and $1 \le k \le n$.

1.
$$\binom{n}{k} \le \frac{n^k}{k!} \le \left(\frac{en}{k}\right)^k$$

2.
$$\frac{(n)_k}{n^k} \le \exp\left\{-(k-1) \sum_{m \in \mathbb{P}} \frac{1}{m\,(m+1)} \left(\frac{k-1}{n}\right)^m\right\}$$

3.
$$\frac{(n)_k}{n^k} \sim \exp\left\{-\frac{k^2}{2n} - \frac{k^3}{6n^2}\right\}, \quad k = o\left(n^{3/4}\right)$$

$$\frac{(n)_k}{n^k} = 1 + o\,(1), \quad k = o\left(n^{1/2}\right)$$

$$\frac{(n)_k}{n^k} = O\,(1) \exp\left\{-\frac{k^2}{2n} - \frac{k^3}{6n^2}\right\}, \quad \text{any } k$$

$\square$

**Multinomial Expansion**

Multinomial expansion (theorem) is a generalization of the binomial expansion (theorem). It is

$$(x_1 + x_2 + \ldots + x_r)^n = \sum_{\substack{n_i \in \mathbb{N}, 1 \le i \le r, \\ (n_1 + n_2 + \ldots + n_r) = n}} \binom{n}{n_1, n_2, \ldots, n_r} x_1^{n_1} x_2^{n_2} \ldots x_r^{n_r}$$

$$(6.43a)$$

where

$$\binom{n}{n_1, n_2, \ldots, n_r} = \frac{n!}{n_1! n_2! \ldots, n_r!}, \quad \text{and} \quad n = \sum_{k=1}^{r} n_k \tag{6.43b}$$

is called the multinomial coefficient. Observe that there are $\widetilde{n} = \binom{n+r-1}{r-1}$ number of distinct non-negative integer-valued vectors $(n_1, n_2, \ldots, n_r)$, such that $\sum_{i=1}^{r} n_i = n$. Therefore the multinomial expansion has $\widetilde{n}$ terms.

### 6.7.2 Algorithmic-Complexity Classes

This subsection is a brief tour of computational complexity of algorithms. This is a vast and deep subject in its own right. The computational complexity of an algorithm is a measure of the *computational resources* required to execute it. A practical measure of the computational complexity of an algorithm is its *running time* and the required *computational space* (memory) to execute it. Both these measures are computational resources. The running time of an algorithm depends upon the type of computer on which it is executed. Therefore, it is certainly advisable to obtain a complexity measure which is independent of the actual physical resources. This is possible if the complexity of the algorithm is specified in terms of the *number of computations* and/or the *computational space* required to execute it. These measures are termed *computational-complexity (or time-complexity)* and *space-complexity* respectively. More formally, the computational complexity of an algorithm is an asymptotic measure of the number of computational operations required to execute it. And, the space-complexity of an algorithm is an asymptotic measure of the amount of computational space required to execute it.

These complexity measures are generally specified as a big-oh expression, whose argument is a function of the *input size* of the algorithm. The input size of an algorithm is the number of bits required to specify this argument. In lieu of bits other alternate forms of encoding of the data can also be used. It should however be noted in general that different representations of the input data do not alter the complexity of the algorithm.

Depending upon its complexity, algorithms can generally be divided into two classes. These are *polynomial-time* and *exponential-time* algorithms. A polynomial-time algorithm is an algorithm whose worst-case computational complexity is of the form $O\left(n^k\right)$, where $n$ is the input size of the algorithm, and $k$ is an integer constant. An important type of algorithm whose computational complexity cannot be so bounded is called an exponential-time algorithm.

Polynomial-time algorithms are generally considered to be efficient, while algorithms with exponential-time complexity are considered to be inefficient. This is a fair assessment, provided the input size $n$ is large. However, if the value of $n$ is small, the degree of polynomial can possibly be significant in practical cases. For example, if the value of $n$ is small, the algorithm with computational-complexity of $O\left(n^{\ln \ln n}\right)$ might out perform an algorithm with a computational-complexity of $O\left(n^{100}\right)$.

Further, a *subexponential-time algorithm* is an algorithm whose worst-case computational-complexity function is of the form $e^{o(n)}$, where $n$ is the input size. The computational complexity of a subexponential-time algorithm generally lies between the computational complexity of a polynomial-time algorithm, and an exponential-time algorithm.

A modern-day, and highly relevant measure of algorithmic complexity is the energy used in the execution of an algorithm. The energy can be of any type. It can either be electrical, or wind, or solar, or any other form of energy. This aspect of algorithmic complexity is not discussed in this chapter.

## 6.8 Vector Algebra

Elementary notions from vector algebra are summarized in this section. Vector spaces have also been studied in the chapter on abstract algebra. A vector is a quantity which has both *magnitude* and *direction*. Vectors can themselves be added, and multiplied by scalars. We shall assume that the elements of a vector are real numbers.

A vector in 3-dimensional space $\mathbb{R}^3$ is represented as a point $u = (u_1, u_2, u_3)$, where $u_1, u_2, u_3 \in \mathbb{R}$. That is, $u_1$, $u_2$, and $u_3$ are the *coordinates* of a point specified by $u$ in three-dimensional space. These are also called the *components* of vector $u$. A vector is represented as a row in this section. We shall use the notation: $(1, 0, 0) \triangleq i$, $(0, 1, 0) \triangleq j$, and $(0, 0, 1) \triangleq k$. The vectors $i, j$, and $k$ are along $x$-axis, $y$-axis, and $z$-axis respectively. Also the set of vectors $\{i, j, k\}$ form a basis of the 3-dimensional vector space $\mathbb{R}^3$. The vector $u$ is also written as

$$u = u_1 i + u_2 j + u_3 k$$

The vector $u$ is also called the *position vector* or *radius vector* from the origin $(0, 0, 0)$ to the point $(u_1, u_2, u_3)$. The magnitude or length or Euclidean norm of this vector is

$$\|u\| = \sqrt{u_1^2 + u_2^2 + u_3^2}$$

Observe that $\|u\|$ is the distance from the point $u = (u_1, u_2, u_3)$, to the origin $0 = (0, 0, 0)$. The vector $(0, 0, 0)$ is often called the *null vector*. A *unit vector* is a vector of unit length. The *direction* of a nonnull vector $u$ is specified by $u/\|u\|$. Two vectors $u$ and $v$ are said to be *parallel* to each other, if their directions are identical. There are two kinds of vector products. These are the dot product and cross product.

### 6.8.1 Dot Product

The *dot (or inner) product* of two vectors $u = (u_1, u_2, u_3)$ and $v = (v_1, v_2, v_3)$ is defined as

$$u \circ v = u_1 v_1 + u_2 v_2 + u_3 v_3 \qquad (6.44)$$

It can readily be inferred that

$$|u \circ v| \leq \|u\| \cdot \|v\|$$

For the purpose of visual clarity, the symbol "$\cdot$" is used for denoting scalar multiplication. The above inequality yields

$$-1 \leq \frac{u \circ v}{\|u\| \cdot \|v\|} \leq 1$$

Therefore, we can also specify the dot product of the two vectors $u$ and $v$ as

$$u \circ v = \|u\| \cdot \|v\| \cdot \cos\theta, \quad \text{where} \quad \theta \in [0, \pi]$$

It can also be shown that $\theta$ is the angle between the line segments $0u$ and $0v$.

**Observations 6.15.** We list some useful results related to dot product of vectors. Let $u, v$, and $w$ be vectors; and $a \in \mathbb{R}$.

1. Commutative law for dot products: $u \circ v = v \circ u$
2. Distributive law for dot products: $u \circ (v + w) = u \circ v + u \circ w$
3. $a(u \circ v) = (au) \circ v = u \circ (av) = (u \circ v)a$
4. $i \circ i = j \circ j = k \circ k = 1$, and $i \circ j = j \circ k = k \circ i = 0$
5. Let $u = u_1 i + u_2 j + u_3 k$, then

$$u_1 = u \circ i, \; u_2 = u \circ j, \; u_3 = u \circ k$$

6. If $u \circ v = 0$, and $u$ and $v$ are not null vectors, then $u$ and $v$ are *orthogonal* or *perpendicular* or *normal* to each other. $\qquad\square$

In a three-dimensional coordinate system, the basis is ordered as $(i, j, k)$. These basis vectors form an orthonormal basis because:

$$i \circ i = j \circ j = k \circ k = 1, \quad \text{and} \quad i \circ j = j \circ k = k \circ i = 0$$

Let the basis vectors $i$, $j$, and $k$ be ordered as $(i, j, k)$. These basis vectors can be represented graphically in a rectangular coordinate system. The vectors $(i, j, k)$ are a *right-handed basis* if the respective vectors have the same direction in space as the thumb, index finger, and middle finger of the *right hand*. If this is not the case, $(i, j, k)$ form a *left-handed basis*.

### 6.8.2  Vector Product

We specify the *vector cross product*, or simply *vector product* in $\mathbb{R}^3$.

**Definition 6.33.** *Let* $u, v \in \mathbb{R}^3$, *where* $u = (u_1, u_2, u_3)$ *and* $v = (v_1, v_2, v_3)$. *The vector product of* $u$ *and* $v$ (*in that order*) *is the unique vector* $u \times v \in \mathbb{R}^3$ *which satisfies*

$$w \circ (u \times v) = \det \begin{bmatrix} w \\ u \\ v \end{bmatrix}$$

$$= \begin{vmatrix} w_1 & w_2 & w_3 \\ u_1 & u_2 & u_3 \\ v_1 & v_2 & v_3 \end{vmatrix}, \quad \forall \, w = (w_1, w_2, w_3) \in \mathbb{R}^3 \tag{6.45}$$

$\qquad\square$

It follows from the above definition that

$$u \times v = \begin{vmatrix} u_2 & u_3 \\ v_2 & v_3 \end{vmatrix} i - \begin{vmatrix} u_1 & u_3 \\ v_1 & v_3 \end{vmatrix} j + \begin{vmatrix} u_1 & u_2 \\ v_1 & v_2 \end{vmatrix} k$$

Thus $u \times v$ is a vector. This is in contrast to $u \circ v$, which is a scalar quantity. With a little misuse of notation, we write

$$u \times v = \begin{vmatrix} i & j & k \\ u_1 & u_2 & u_3 \\ v_1 & v_2 & v_3 \end{vmatrix}$$

**Observations 6.16.** Some useful results related to cross product of vectors are listed. Let $u, v, w, z \in \mathbb{R}^3$, and $a \in \mathbb{R}$.

1. $u \times u = 0$
2. If $u \times v = 0$, and $u$ and $v$ are not null vectors, then $u$ and $v$ are linearly dependent (scalar multiples of each other).
3. Anticommutative law for cross products: $u \times v = -v \times u$.
4. Distributive law for cross products: $u \times (v + w) = u \times v + u \times w$.
5. $a(u \times v) = (au) \times v = u \times (av) = (u \times v)a$
6. $i \times i = j \times j = k \times k = 0$, and $i \times j = k, j \times k = i, k \times i = j$.
7. $\|u\|^2 \cdot \|v\|^2 = \|u \circ v\|^2 + \|u \times v\|^2$. This result implies: $\|u \times v\| = \|u\| \cdot \|v\| \, |\sin \theta|$. where $\theta \in [0, \pi]$ is the angle between the line segments $0u$ and $0v$. Note that $\|u \times v\|$ is the area of the parallelogram determined by the line segments $0u$ and $0v$.
8. Triple products:

   (a) $u \circ (v \times w) = v \circ (w \times u) = w \circ (u \times v)$. The product $u \circ (v \times w)$ is referred to as scalar triple product. Note that $|u \circ (v \times w)|$ is the volume of the parallelepiped determined by the line segments $0u, 0v$ and $0w$.

   (b) The vector product is not associative: $(u \times v) \times w \neq u \times (v \times w)$

      (i) $(u \times v) \times w = (u \circ w)v - (v \circ w)u$

      (ii) $u \times (v \times w) = (u \circ w)v - (u \circ v)w$

      (iii) $(u \times v) \circ (w \times z) = (u \circ w)(v \circ z) - (u \circ z)(v \circ w)$

9. $(u \times v) \circ u = 0$, and $(u \times v) \circ v = 0$. This result implies that $u \times v$ is perpendicular (normal) to both $u$ and $v$. That is, $u \times v$ is perpendicular (normal) to a plane generated by vectors $u$ and $v$. $\qquad\qquad\qquad\qquad\qquad\qquad\qquad\qquad\qquad\qquad\qquad\qquad\qquad\qquad\qquad \square$

The result $\|u \times v\| = \|u\| \cdot \|v\| \, |\sin \theta|$ lets us provide a possible alternate definition of cross product of two vectors $u$ and $v$. Thus

$$u \times v = \|u\| \cdot \|v\| \sin \theta \, n, \quad \theta \in [0, \pi]$$

where $n$ is a normal vector of unit length. It is perpendicular to the plane generated by the vectors $u$ and $v$. It also specifies the direction of the vector $u \times v$. The direction of the unit normal vector $n$ is such that, $u, v$, and $n$ form a right-handed system.

**Notation.** Let $u, v, w \in \mathbb{R}^3$. Then $w \circ (u \times v) \triangleq [w \, u \, v]$. $\qquad\qquad\qquad\qquad\qquad\qquad \square$

---

## 6.9  Vector Spaces Revisited

A vector space can be defined informally as an ensemble of objects which can be added, and multiplied by scalar elements. These operations in turn yield another object of the ensemble. Vector spaces have been defined formally in the chapter on abstract algebra. This subject is revisited in this chapter. In this section the following topics are discussed: normed vector space, complete vector space, concept of compactness of a set, inner product space, orthogonality, Gram-Schmidt orthogonalization process, projections, and isometry.

### 6.9.1  Normed Vector Space

After defining a vector space, it is possible to introduce a metric related to the size (length) of the vector. This is done by introducing the concept of a norm. A representation of the distance between any two vectors can also be defined by the notion of norm.

> **Definitions 6.34.** *Let* $\mathcal{V} = (V, \mathcal{F}, +, \times)$ *be a vector space over a field* $\mathcal{F} = (F, +, \times)$.
>
> 1. *Norm of a vector: The norm of a vector is a function* $\|\cdot\| : V \to \mathbb{R}_0^+$. *The norm of the vector* $u \in V$ *is a nonnegative real number, denoted by* $\|u\|$. *It is subject to the following conditions.*
>    *[N1]* $\|u\| \geq 0$, *with equality if and only if* $u = 0$. *That is, the norm of a vector is a nonnegative number.*
>    *[N2]* $\|au\| = |a| \cdot \|u\|$, *where* $|a|$ *is the magnitude of* $a$, *and* $a \in F$. *This is the homogeneity property of the norm.*
>    *[N3]* $\|u + v\| \leq \|u\| + \|v\|$. *This is the triangle inequality.*
>    *This vector space* $\mathcal{V}$, *along with its norm* $\|\cdot\|$ *is called a normed vector space. That is, the two-tuple* $(\mathcal{V}, \|\cdot\|)$ *is called a normed space.*
> 2. *Distance function: For* $u, v \in V$, *the distance or metric function from* $u$ *to* $v$ *is* $d(u, v)$. *That is,* $d : V \times V \to \mathbb{R}$. *Let* $u, v, w \in V$, *then the distance function satisfies the following axioms:*
>    *[D1]* $d(u, v) \geq 0$, *with equality if and only if* $u = v$.
>    *[D2]* $d(u, v) = d(v, u)$. *This is the symmetry of the distance function.*
>    *[D3]* $d(u, w) \leq d(u, v) + d(v, w)$. *This is the triangle inequality.*
> 3. *Let* $\mathcal{V}$ *be a normed vector space, and* $d(\cdot, \cdot)$ *be a distance function. The two-tuple* $(\mathcal{V}, d)$ *is called a metric space.*
>    *Also let* $u, v \in V$. *The function* $d(\cdot, \cdot)$ *defined by* $d(u, v) = \|u - v\|$ *is called the induced metric on* $\mathcal{V}$.
> 4. *Let the vector space be* $\mathbb{C}^n$. *Also let* $x = (x_1, x_2, \ldots, x_n) \in \mathbb{C}^n$. *Then the Euclidean norm is given by*
>
> $$\|x\|_2 = \left( \sum_{j=1}^{n} |x_j|^2 \right)^{1/2} \tag{6.46a}$$
>
> *If the norm is Euclidean and the context is unambiguous, the subscript $2$ in the above equation is generally dropped.*
> 5. *Matrix norm: A matrix $A$ is a rectangular array of complex numbers. The matrix norm induced by the vector norm* $\|.\|_2$ *is defined by*
>
> $$\|A\|_2 = \sup_{\|x\|_2 = 1} \|Ax\|_2 \tag{6.46b}$$
>
> $\square$

In a misuse of notation, the vector space $\mathcal{V}$, and the field $\mathcal{F}$ are generally denoted by $V$ and $F$ respectively. Matrix norms are revisited in the chapter on matrices and determinants.

### 6.9.2  Complete Vector Space

The concepts of convergence and complete vector space are introduced in this subsection. This enables us to extend concepts from finite-dimensional spaces to infinite-dimensional spaces. Let

$\mathcal{V} = (V, \mathcal{F}, +, \times)$ be a normed vector space over a field $\mathcal{F} = (F, +, \times)$, where $\|\cdot\|$ is the norm function of a vector. Also let $\{g_n\}$ denote a sequence of vectors $g_1, g_2, \ldots, g_n, \ldots$, which belong to the vector space $\mathcal{V}$. This sequence converges to $g$ if $\|g_n - g\|$ tends to 0 for very large values of $n$.

A sequence $\{g_n\}$ is called a *Cauchy sequence* if $\|g_n - g_m\|$ tends to 0 for very large values of $m$ and $n$. More precisely, a sequence $\{g_n\}$ in a normed vector space $\mathcal{V}$ is said to be a Cauchy sequence if for each $\epsilon > 0$, there exists $n_0$ such that $\|g_n - g_m\| < \epsilon$ for all $m, n > n_0$.

**Definition 6.35.** *A normed vector space $\mathcal{V}$ is complete, if every Cauchy sequence in the vector space converges.*                                                                    □

Examples of normed vector spaces are presented below. All the vector spaces in these examples are complete.

**Examples 6.9.** Some illustrative examples.

1. Let the vector space be $\mathbb{R}^n$. Also let $x = (x_1, x_2, \ldots, x_n) \in \mathbb{R}^n$. Then the $p$-norm of vector $x$ is defined as

$$\|x\|_p = \left( \sum_{j=1}^{n} |x_j|^p \right)^{1/p}, \quad 1 \leq p < \infty \tag{6.47}$$

(a) $p = 1$: This is the Manhattan norm. $\|x\|_1 = \sum_{j=1}^{n} |x_j|$.
(b) $p = 2$: This is the Euclidean norm.

$$\|x\|_2 = \left( \sum_{j=1}^{n} x_j^2 \right)^{1/2}$$

(c) As $p$ tends towards $\infty$, $\|x\|_p$ tends towards

$$\|x\|_\infty = \max_{1 \leq j \leq n} \{|x_j|\}$$

The extension of the definition of the $p$-norm to $x \in \mathbb{C}^n$ should be immediately evident.

2. *Sequence space $l^p$, $1 \leq p < \infty$*: Let the sequence $x = (x_1, x_2, \ldots, x_n, \ldots)$, $x_j \in \mathbb{R}$, $j = 1, 2, 3, \ldots$ satisfy

$$\sum_{j \in \mathbb{P}} |x_j|^p < \infty \tag{6.48a}$$

Sequences which satisfy the above condition belong to the space $l^p$. The following norm can be defined, for $x = (x_1, x_2, \ldots, x_n, \ldots) \in l^p$.

$$\|x\|_p = \left( \sum_{j \in \mathbb{P}} |x_j|^p \right)^{1/p}, \quad 1 \leq p < \infty \tag{6.48b}$$

This norm is commonly referred to as the $l^p$ *norm*.
The sequence $(x_1, x_2, \ldots, x_n, \ldots)$ is bounded if there exists a finite $M \in \mathbb{R}$ such that $|x_j| < M$ for all values $j = 1, 2, 3, \ldots$. The space $l^\infty$ consists of bounded sequences.

3. *Function space $L^p, 1 \leq p < \infty$*: Let the function $f(t)$, $t \in \mathbb{R}$, satisfy

$$\int_{-\infty}^{\infty} |f(t)|^p \, dt < \infty \tag{6.49a}$$

This function $f(\cdot)$ is said to be $p$-integrable. Then $L^p$ is a function space in which the following norm can be defined

$$\|f\|_p = \left( \int_{-\infty}^{\infty} |f(t)|^p \, dt \right)^{1/p}, \ 1 \leq p < \infty. \tag{6.49b}$$

This norm is commonly referred to as $L^p$ or $L^p (\mathbb{R})$ *norm.*                      □

### 6.9.3 Compactness

The concept of compactness on a metric space is introduced in this subsection. It is established that a continuous mapping of a compact subset of a metric space into the field of real numbers attains a maximum and a minimum at some points of the compact subset.

**Definitions 6.36.** *Bounded metric space, and compact metric space.*

1. *A subset $A \subseteq V$ of the metric space is said to be totally bounded, if for every $\epsilon > 0$, $A$ contains a finite set $A_\epsilon$ such that for each $x \in V$, there is a $y \in A_\epsilon$ such that $d(x, y) < \epsilon$.*
2. *A metric space $(\mathcal{V}, d)$ is said to be compact, if it is complete and totally bounded.*            □

Continuous operators on compact sets are next defined.

**Definition 6.37.** *Let $\mathcal{V}_X = (V_X, \mathcal{F}_X, +, \times)$ be a vector space over a field $\mathcal{F}_X = (F_X, +, \times)$, and $(\mathcal{V}_X, d_X)$ be a metric space. Similarly, let $\mathcal{V}_Y = (V_Y, \mathcal{F}_Y, +, \times)$ be a vector space over a field $\mathcal{F}_Y = (F_Y, +, \times)$, and $(\mathcal{V}_Y, d_Y)$ be a metric space. A mapping $T : V_X \rightarrow V_Y$ is said to be continuous at $x_0 \in V_X$, if for every $\epsilon > 0$, there exists $\delta_{x_0,\epsilon} > 0$ such that $d_X(x, x_0) < \delta_{x_0,\epsilon} \Rightarrow d_Y(T(x), T(x_0)) < \epsilon$.*
*The mapping $T$ is said to be continuous, if it is continuous at every point $x_0 \in V_X$.*            □

The above definition is equivalent to the definition of continuity of a function of a single variable, if $\mathcal{V}_X = \mathcal{V}_Y = \mathbb{R}$. This definition also implies that the mapping $T(\cdot)$ is continuous at $x_0 \in V_X$: if for any sequence, $\{x_n \mid x_n \in V_X, n = 1, 2, 3, \ldots\} \rightarrow x_0$, then

$$\{T(x_n) \mid T(x_n) \in V_Y, n = 1, 2, 3, \ldots\} \rightarrow T(x_0) \in V_Y$$

**Observations 6.17.** Observations on continuous mappings.

1. If $T(\cdot)$ is a continuous mapping, then the image of a compact set is also compact.
2. If $T(\cdot)$ is a continuous mapping from a metric space to the real line $\mathbb{R}$, then the image of a compact set is closed and bounded. Consequently, $T(x)$ achieves its supremum and infimum.            □

### 6.9.4 Inner Product Space

The definition of inner product of two vectors within a general setting has been given in the chapter on abstract algebra. Using the same ideas, but different notation, this concept is revisited.

By placing additional structure on vector space $\mathcal{V} = (V, \mathcal{F}, +, \times)$ over a field $\mathcal{F} = (F, +, \times)$, inner product space can be defined. The concept of angle between two vectors is introduced via the concept of inner product. This is followed by the concept of orthogonality of vectors.

**Definition 6.38.** *Inner product space*: *A complex inner product on a vector space $\mathcal{V}$ over field $\mathbb{C}$ is a function $\langle \cdot, \cdot \rangle : V \times V \to \mathbb{C}$ such that for all $u, v, w \in V$ and $a, b \in \mathbb{C}$ the following axioms hold*:

[$I1$] $\langle u, u \rangle \geq 0$, *where $\langle u, u \rangle = 0$ if and only if $u = 0$.*
[$I2$] $\langle u, v \rangle = \overline{\langle v, u \rangle}$ *(the bar indicates complex conjugation).*
[$I3$] $\langle au + bv, w \rangle = a \langle u, w \rangle + b \langle v, w \rangle$. *That is, the inner product function $\langle \cdot, \cdot \rangle$ is linear in the first argument.*

*The inner product vector space is denoted by the two-tuple $(\mathcal{V}, \langle \cdot, \cdot \rangle)$.* □

The axioms [$I1$] and [$I2$] imply that $\langle u, u \rangle$ is real. Observe that the axiom [$I2$] implies that the inner product is commutative over its two components, except for the conjugacy. Furthermore, axiom [$I3$] implies that the inner product is linear in its first component. The vector space $\mathcal{V}$ along with the defined inner product is called an *inner product vector space*. It can easily be shown that the inner product as defined is also linear in its second component, except for conjugacy.

$$\langle u, av + bw \rangle = \overline{a} \langle u, v \rangle + \overline{b} \langle u, w \rangle$$

Some textbooks define the axiom [$I3$] alternately. Let us call this modified axiom [$I3'$]. It is $\langle u, av + bw \rangle = a \langle u, v \rangle + b \langle u, w \rangle$. That is, the inner product $\langle \cdot, \cdot \rangle$ is linear in the second argument. This definition is equally valid, if its usage is consistent with it. Also observe that this distinction is unnecessary, if the inner product space is defined over the set of real numbers $\mathbb{R}$. Unless stated otherwise, the axioms [$I1$], [$I2$], and [$I3$] are used in this chapter to define the inner product vector space. Some of its useful properties are summarized below.

**Observations 6.18.** Let $(\mathcal{V}, \langle \cdot, \cdot \rangle)$ be an inner product vector space.

1. Bunyakovsky-Cauchy-Schwartz inequality: If $u, v \in V$, then $|\langle u, v \rangle| \leq \|u\| \cdot \|v\|$. Equality is obtained if $u = 0$, or $v = 0$, or $u = \alpha v$, where $\alpha \in \mathbb{C}$. The last condition shows linear dependence.

2. The inner product $\langle \cdot, \cdot \rangle$ defined on the vector space $\mathcal{V}$ induces a norm $\|\cdot\|$ on $\mathcal{V}$. The norm is given by the relationship $\|u\| = \sqrt{\langle u, u \rangle}$ for all $u \in V$. Thus every inner product space is a normed space, but not vice-versa. The nonnegative number $\|u\|$ is called the *length* of the vector $u$.

3. The norm on an inner product space satisfies the parallelogram equality. Let $u, v \in V$, then

$$\|u + v\|^2 + \|u - v\|^2 = 2 \left( \|u\|^2 + \|v\|^2 \right)$$

This result derives its name from a similar result about parallelograms in plane geometry.

4. If a norm does not satisfy the above parallelogram equality, it cannot be obtained from an inner product by the use of relationship $\|u\| = \sqrt{\langle u, u \rangle}$. Therefore, it can be inferred that not all normed spaces are inner product spaces. Conversely, all inner product spaces are normed. For example, the Manhattan norm is not induced by an inner product.

5. Let $u, v \in V$. Then the *polar forms* of $\langle u, v \rangle$ are:

$$\langle u, v \rangle = \frac{1}{4} \|u + v\|^2 - \frac{1}{4} \|u - v\|^2, \quad \text{over field } \mathbb{R}$$

$$\langle u, v \rangle = \frac{1}{4} \|u + v\|^2 - \frac{1}{4} \|u - v\|^2 + \frac{i}{4} \|u + iv\|^2 - \frac{i}{4} \|u - iv\|^2, \quad \text{over field } \mathbb{C}$$

These results imply that, if the norm is known in an inner product space, then the inner product can be derived from it. $\qquad\square$

The above observations are further clarified by the following examples.

**Examples 6.10.** Some elementary examples.

1. Let $\mathcal{V}$ be a vector space such that $V = \mathbb{R}^n$. Also let $x, y \in V$, where $x = (x_1, x_2, \ldots, x_n)$ and $y = (y_1, y_2, \ldots, y_n)$. The function defined by $x \circ y \triangleq \langle x, y \rangle = \sum_{j=1}^{n} x_j y_j$ is an inner product on $\mathcal{V}$. This is actually the definition of the *dot product* in $\mathbb{R}^n$. It is also reminiscent of the dot product operation introduced in the section on vector algebra. This function is also called the *real* or *standard* or *Euclidean inner product* on $V = \mathbb{R}^n$.
Note that $\langle x, x \rangle = \sum_{j=1}^{n} x_j^2 = \|x\|_2^2$, where $\|\cdot\|_2$ is the Euclidean norm on $\mathbb{R}^n$. This space is called *real Euclidean space*. If $n = 2$, then the space is called Euclidean plane.

2. Let $\mathcal{V}$ be a vector space such that $V = \mathbb{C}^n$. Let $x, y \in V$, where $x = (x_1, x_2, \ldots, x_n)$ and $y = (y_1, y_2, \ldots, y_n)$. The function defined by $\langle x, y \rangle = \sum_{j=1}^{n} x_j \overline{y}_j$ is an inner product on $\mathcal{V}$. This representation of the inner product is also called the dot product. The function $\langle \cdot, \cdot \rangle$ is also called the *standard inner product* on $V = \mathbb{C}^n$. This space is called *complex Euclidean space*.

3. The inner product $\langle \cdot, \cdot \rangle$ defined on a vector space $\mathcal{V}$ is also an inner product on any vector subspace $\mathcal{U}$ of $\mathcal{V}$. $\qquad\square$

One of the most important inner-product spaces is the Hilbert space. The two most well-known examples of Hilbert spaces are the $l^2$ and $L^2$ spaces. Hilbert space is a natural extension of the Euclidean space. These spaces were extensively studied by the mathematician David Hilbert (1862-1943). This is why the mathematician von Neumann designated them Hilbert spaces.

**Definitions 6.39.** *Some useful definitions related to Hilbert space are:*

1. *An inner-product space $\mathcal{H}$ is called a Hilbert space, if it is complete with respect to the norm induced by the inner product.*
2. *A Hilbert space of finite dimension $n$, is denoted by $\mathcal{H}_n$.* $\qquad\square$

### 6.9.5 Orthogonality

The concept of orthogonality of two vectors is next introduced. In a real inner product space, the angle between two nonzero vectors $u, v \in V$ is the real number $\theta$, such that

$$\cos \theta = \frac{\langle u, v \rangle}{\|u\| \cdot \|v\|}, \quad 0 \leq \theta \leq \pi \qquad (6.50)$$

Therefore two vectors $u$ and $v$ are orthogonal to each other, if $\langle u, v \rangle = 0$. This concept of orthogonality can be extended to any inner product vector space.

**Definitions 6.40.** *Let* $(\mathcal{V}, \langle \cdot, \cdot \rangle)$ *be an inner product vector space.*

1. *Orthogonal vectors: Two vectors* $u, v \in V$ *are orthogonal if and only if* $\langle u, v \rangle = 0$. *This relationship is sometimes denoted by* $u \perp v$.
2. *Orthogonal set: A subset* $S \subseteq V$ *is an orthogonal set, if all the vectors in the set $S$ are mutually orthogonal. That is, for all vectors* $u, v \in S$, $\langle u, v \rangle = 0$, *where* $u \neq v$.
3. *Orthonormal set: A subset* $S \subseteq V$ *is an orthonormal set if $S$ is an orthogonal set and all vectors of the set $S$ have unit length.*
4. *Orthogonal complement: If* $U \subseteq V$, *then the orthogonal complement of the subset $U$ is the set of all vectors* $v \in V$, *where $v$ is orthogonal to all the vectors in $U$. Denote this set by* $U^\perp = \{v \in V \mid \langle u, v \rangle = 0, \ \forall \, u \in U\}$. *The sets $U$ and $U^\perp$ are said to be orthogonal to each other, and this relationship is indicated by* $U \perp U^\perp$. $\qquad\square$

**Observations 6.19.** Let $(\mathcal{V}, \langle \cdot, \cdot \rangle)$ be an inner product vector space.

1. Let $u, v \in V$, then $\langle u, v \rangle = 0$ if and only if $\|u + v\|^2 = \|u\|^2 + \|v\|^2$. This is the Pythogorean theorem.
2. If $u \neq v$, and $u, v \in V$, then $\|u\| = \|v\| \Leftrightarrow \langle u + v, u - v \rangle = 0$. This result is reminiscent of a property of a rhombus in plane geometry, which is: The diagonals of a rhombus are orthogonal to each other.
3. An orthogonal set $S$ of nonzero vectors, is independent. That is, all the vectors in the set $S$ are independent of each other. The concept of independence of vectors has been discussed in the chapter on abstract algebra.
4. If the inner product space is defined over real numbers, then two nonzero vectors are orthogonal if and only if the angle between them is equal to $\pi/2$.
5. Let the dimension of the vector space be $\dim(\mathcal{V}) = n$. Then:
   (a) Any orthonormal set contains at most $n$ vectors.
   (b) Any orthonormal set of $n$ vectors is a basis of $\mathcal{V}$.
6. Let $U \subseteq V, u \in U$, and $(u_1, u_2, \ldots, u_n)$ be an orthonormal basis of $U$. Consequently $u = \sum_{j=1}^{n} \langle u, u_j \rangle u_j$.
7. Generalized Pythogorean theorem: Let $(u_1, u_2, \ldots, u_n)$ be an orthogonal set of vectors in $V$, which is defined over the field of complex numbers $\mathbb{C}$, then for $b_j \in \mathbb{C}, 1 \leq j \leq n$

$$\left\| \sum_{j=1}^{n} b_j u_j \right\|^2 = \sum_{j=1}^{n} |b_j|^2 \|u_j\|^2 \qquad (6.51)$$

8. Let $(u_1, u_2, \ldots, u_r)$ be an orthonormal subset of $V$. Then for any $v \in V$,

$$\sum_{j=1}^{r} |\langle v, u_j \rangle|^2 \leq \|v\|^2 \qquad (6.52)$$

This is known as the Bessel's inequality.
9. If $U \subseteq V$, then the orthogonal complement $U^\perp$ is a subset of $V$. Also $V$ is a direct sum of the sets $U$ and $U^\perp$, that is $V = U \oplus U^\perp$. $\qquad\square$

### 6.9.6  Gram-Schmidt Orthogonalization Process

Gram-Schmidt orthogonalization process is a procedure for constructing an orthonormal set of vectors from an arbitrary linearly independent set of vectors. It is named after Jörgen Gram (1850-1916) and Erhardt Schmidt (1876-1959). This construction follows from the following observation. Let $(\mathcal{V}, \langle \cdot, \cdot \rangle)$ be an inner product vector space over the field $\mathbb{C}$. Also let $\{u_1, u_2, \ldots, u_r\}$ be an orthonormal set of vectors which belong to the set $V$. These vectors are linearly independent. Furthermore, for any $v \in V$, the vector $w$ given by

$$w = v - \langle v, u_1 \rangle u_1 - \langle v, u_2 \rangle u_2 - \cdots - \langle v, u_r \rangle u_r$$

is orthogonal to each $u_j, 1 \leq j \leq r$. This observation is used in the Gram-Schmidt orthonormalization process. It is summarized in the following theorem.

**Theorem 6.14.** *Let* $(\mathcal{V}, \langle \cdot, \cdot \rangle)$ *be an inner product vector space over* $\mathbb{C}$. *Also let*

$$\{u_1, u_2, \ldots, u_m\}$$

*be a linearly independent set of vectors which belong to* $V$. *Define*

$$w_1 = u_1 \tag{6.53a}$$

$$v_1 = \frac{w_1}{\|w_1\|} \tag{6.53b}$$

$$w_r = u_r - \sum_{j=1}^{r-1} \langle u_r, v_j \rangle v_j \tag{6.53c}$$

$$v_r = \frac{w_r}{\|w_r\|} \tag{6.53d}$$

*where* $2 \leq r \leq m$. *Then the set* $\{v_1, v_2, \ldots, v_m\}$ *is an orthonormal set in* $V$, *and the spaces spanned by the vector-sets* $\{u_1, u_2, \ldots, u_m\}$ *and* $\{v_1, v_2, \ldots, v_m\}$ *are identical for all values of* $m$.

*Proof.* The proof is left to the reader.                                                                     $\square$

### 6.9.7  Projections

Projection mappings are specially useful in studying quantum mechanics. Let $\oplus$ be the direct sum operator. These operators are discussed in the chapter on abstract algebra.

**Definition 6.41.** *Let* $\mathcal{V}$ *be a vector space such that*

$$\mathcal{V} = \mathcal{U}_1 \oplus \mathcal{U}_2 \oplus \ldots \oplus \mathcal{U}_n \tag{6.54a}$$

*That is,*

$$V = U_1 \oplus U_2 \oplus \ldots \oplus U_n \tag{6.54b}$$

*and the sets* $U_i$*'s are mutually disjoint. The projection of* $\mathcal{V}$ *into its subspace* $\mathcal{U}_j$ *is the mapping* $E :$ $V \to V$ *such that for a* $v \in V, E(v) = u_j$ *where* $v = \sum_{i=1}^n u_i$ *and* $u_i \in U_i$.                                   $\square$

**Observations 6.20.** Some useful observations.

1. The mapping $E$ is linear.
2. $E^2 = E$.                                                                                                  $\square$

See the problem section for a proof of these observations.

### 6.9.8  Isometry

Isometry is a map which preserves distances between two metric spaces.

**Definition 6.42.** *Let $(X, d_X)$ and $(Y, d_Y)$ be metric spaces. In these metric spaces, $d_X$ and $d_Y$ are distance functions defined upon the sets $X$ and $Y$ respectively. The bijective map $I : X \to Y$ is called an isometry or distance preserving, if*

$$d_Y\left(I\left(u\right), I\left(v\right)\right) = d_X\left(u, v\right) \quad \text{for each } u, v \in X \tag{6.55}$$

$\square$

**Observations 6.21.** Some useful observations about isometries.

1. A composition of isometries is an isometry.
2. Let $\|\cdot\|$ be a Euclidean norm. An isometry of $\mathbb{R}^n$ is a map $I : \mathbb{R}^n \to \mathbb{R}^n$ that preserves the distance between any two points in $\mathbb{R}^n$. That is,

$$\|I\left(u\right) - I\left(v\right)\| = \|u - v\|, \quad \forall\, u, v \in \mathbb{R}^n$$

$\square$

**Examples 6.11.** Some elementary examples.

1. Let $M$ be an orthogonal matrix of size $n$. Recall that a real square matrix $M$ is orthogonal if its inverse is equal to its transpose. That is $M^{-1} = M^T$. A vector $a \in \mathbb{R}^n$ is specified as a column vector of size $n$. We show that the map $I : \mathbb{R}^n \to \mathbb{R}^n$, where $I\left(u\right) = Mu + a, \forall\, u \in \mathbb{R}^n$ is an isometry of $\mathbb{R}^n$. For any $u, v \in \mathbb{R}^n$ we have

$$\begin{aligned}
\|I\left(u\right) - I\left(v\right)\|^2 &= \left(Mu - Mv\right)^T \left(Mu - Mv\right) \\
&= \left(u^T - v^T\right) M^T M \left(u - v\right) \\
&= \|u - v\|^2
\end{aligned}$$

Therefore the map $I$ is an isometry.
2. The isometries in the Euclidean plane are: translation, rotation around a point, reflections or mirror isometry, and glide reflection. Glide reflection is a reflection in a line, followed by a translation. $\square$

## 6.10  Fourier Series

Fourier series is an expansion of a periodic function in terms of trigonometric functions. Such expansions are generally termed Fourier series. The series are named in honor of the French mathematician Jean Baptiste Joseph Fourier.

Some examples of periodic functions are $\sin x$, $\cos x$, and $\tan x$, where $x \in \mathbb{R}$. A linear combination of these elementary trigonometric functions is also periodic. A generalization of this concept is the expansion of a periodic function in terms of elementary trigonometric functions like $\sin\left(\cdot\right)$

and $\cos{(\cdot)}$. Fourier series also provide a stepping stone to study transform techniques. Transform techniques are studied later in the chapter.

Generalized functions are a useful artifice to study Fourier series and continuous transform techniques. In addition to generalized functions, the following topics are also discussed in this section: conditions for the existence of the Fourier series, complex Fourier series, and trigonometric Fourier series.

### 6.10.1 Generalized Functions

A systematic study of generalized functions is useful in the study of continuous transform techniques, like Fourier and wavelet transforms.

**Definition 6.43.** *The unit impulse function* $\delta(t)$, $t \in \mathbb{R}$, *is also called the delta function ($\delta$-function), or Dirac's delta function. It is traditionally defined as:*

$$\delta(t) = \begin{cases} 0, & t \neq 0 \\ \infty, & t = 0 \end{cases} \tag{6.56a}$$

$$\int_{-\infty}^{\infty} \delta(t)\, dt = \int_{-\epsilon}^{\epsilon} \delta(t)\, dt = 1, \quad \epsilon > 0 \tag{6.56b}$$

□

This definition is mathematically not sound. This is because a function, which is zero everywhere except at a single point, must have the integral value equal to zero. The integration is assumed to be Riemannian. Therefore, it is alternately defined in terms of generalized functions.

**Definition 6.44.** $\delta(\cdot)$ *is a generalized (or symbolic) function, and* $\phi(\cdot)$ *is a test function, such that*

$$\int_{-\infty}^{\infty} \delta(t)\, \phi(t)\ dt = \phi(0) \tag{6.57}$$

□

Observe in the above definition that we can only interpret about the *values of integrals* involving $\delta(t)$, and *not* about *its* value.

**Observations 6.22.** Let $t \in \mathbb{R}$.

1. $\delta(t) = \delta(-t)$.
2. $f(t)\,\delta(t) = f(0)\,\delta(t)$, where it is assumed that $f(t)$ is continuous at $t = 0$.
3. We have

$$\int_{-\infty}^{\infty} \delta(t - t_0)\, \phi(t)\, dt = \phi(t_0)$$

4. $\delta(at) = \delta(t) / |a|$, where $a \in \mathbb{R} \backslash \{0\}$.
5. Denote the $n$th derivative of $\delta(t)$ and $\phi(t)$ with respect to $t$ by $\delta^{(n)}(t)$ and $\phi^{(n)}(t)$ respectively. Then

$$\int_{-\infty}^{\infty} \delta^{(n)}(t)\, \phi(t)\, dt = (-1)^n\, \phi^{(n)}(0), \quad \forall\, n \in \mathbb{P}$$

6. Let $f(x) = 0$ have real zeros $x_n$, where $1 \leq n \leq K$. Then

$$\delta(f(x)) = \sum_{n=1}^{K} \frac{\delta(x - x_n)}{|f'(x_n)|}$$

where it is assumed that $f'(x_n) \neq 0$ (that is, no repeated zeros) for $1 \leq n \leq K$.

7. Integral representation of $\delta(t)$

$$\delta(t) = \frac{1}{2\pi} \int_{-\infty}^{\infty} e^{i\omega t} d\omega = \frac{1}{\pi} \int_{0}^{\infty} \cos(\omega t) d\omega \qquad (6.58)$$

where $i = \sqrt{-1}$.                                                                                  □

### 6.10.2  Conditions for the Existence of Fourier Series

Before we state the conditions for the existence of Fourier series, a periodic function is defined.

**Definition 6.45.** *Let $f(t)$, $t \in \mathbb{R}$ be a real-valued function and $T_0 \in \mathbb{R}^+$. The function $f(\cdot)$ is periodic, if $f(t) = f(t + T_0)$, $\forall\, t \in \mathbb{R}$. The smallest such value of $T_0$ is called the period of the function $f(\cdot)$.*                                                                                  □

**Examples 6.12.** Elementary examples.

1. Some examples of periodic functions are $\sin(\omega_0 t), \cos(\omega_0 t)$, and $e^{i\omega_0 t}$; where $\omega_0 = 2\pi/T_0$, $i = \sqrt{-1}$, and $t \in \mathbb{R}$.

2. Let $f(t), t \in \mathbb{R}$ be a periodic function with period $T_0$. For any $a, b, c \in \mathbb{R}$

$$\int_{0}^{T_0} f(t) dt = \int_{a}^{T_0 + a} f(t) dt$$

$$\int_{b}^{c} f(t) dt = \int_{T_0 + b}^{T_0 + c} f(t) dt$$

3. The sum, difference, product, or quotient (if defined properly) of two functions each of period $T_0$, is also a function of period $T_0$.                                                                                  □

A periodic function $f(t), t \in \mathbb{R}$, of period $T_0 \in \mathbb{R}^+$ has a Fourier series representation if the so-called *Dirichlet conditions* are satisfied. These conditions are named after the mathematician Peter Gustav Lejeune Dirichlet (1805-1859). The Dirichlet conditions are sufficient but not necessary for the Fourier series representation (existence) of a periodic function.

**Dirichlet Conditions** The function $f(t)$ is defined over $t \in \mathbb{R}$.

1. The function $f(\cdot)$ is periodic with period $T_0$.
2. The function $f(\cdot)$ is well-defined and single-valued except possibly at a finite number of points in a single periodic interval $I = (a, a + T_0)$, where $a \in \mathbb{R}$.
3. Let $f'(t)$ be the first derivative of $f(t)$ with respect $t$. The functions $f(\cdot)$ and $f'(\cdot)$ are piecewise-continuous in the periodic interval $I$. This implies that the function $f(\cdot)$ is piecewise-smooth in the periodic interval $I$.

Then the Fourier series expansion of the function $f(\cdot)$ converges to:

(a) $f(t)$ if the function is continuous at point $t$.

(b) $\{f(t_+) + f(t_-)\}/2$ if the function is discontinuous at point $t$.                    □

The proof of correctness of these conditions can be found in any standard textbook on harmonic analysis.

### 6.10.3  Complex Fourier Series

There are two equivalent forms of Fourier series. These are the complex and trigonometric forms of Fourier series representation. Each series representation can be transformed to the other. Complex Fourier series is discussed in this subsection.

**Definition 6.46.** *Let $T_0 \in \mathbb{R}^+$ be the period of a real-valued periodic function $f(t), t \in \mathbb{R}$. Also, let $\omega_0 = 2\pi/T_0$, then for any $a \in \mathbb{R}$.*

$$f(t) = \sum_{n \in \mathbb{Z}} c_n e^{in\omega_0 t}, \quad c_n = \frac{1}{T_0} \int_a^{a+T_0} f(t) e^{-in\omega_0 t} dt, \quad \forall\, n \in \mathbb{Z} \tag{6.59}$$

*In this series representation, it is assumed that Dirichlet conditions hold, and that $f(t)$ is continuous at $t$. If $f(t)$ is discontinuous at $t$, then $f(t)$ in the above expansion should be replaced by $\{f(t_+) + f(t_-)\}/2$. The $c_n$'s are called coefficients of the complex Fourier series.*                    □

The above expansion is evident from the following relationship for any $m, n \in \mathbb{Z}$.

$$\frac{1}{T_0} \int_a^{a+T_0} e^{i\omega_0 t(m-n)} dt = \delta_{mn}$$

where

$$\delta_{mn} = \begin{cases} 1, & m = n \\ 0, & m \neq n \end{cases}$$

The function $\delta_{mn}$ is called the Kronecker's delta function. It is named after the mathematician Leopold Kronecker (1823-1891). The next set of observations follows immediately from the complex Fourier series representation of the periodic function $f(t), t \in \mathbb{R}$.

**Observations 6.23.** Some basic observations about complex Fourier series.

1. For any $a \in \mathbb{R}$, Parseval's relationship is

$$\frac{1}{T_0} \int_a^{a+T_0} |f(t)|^2 dt = \sum_{n \in \mathbb{Z}} |c_n|^2 \tag{6.60}$$

   This relationship is named after the French mathematician Marc-Antoine Parseval (1755-1836).

2. The coefficients of the complex Fourier series of the periodic function $f(\cdot)$ are related as $c_n = \bar{c}_{-n}, \forall\, n \in \mathbb{Z}$.

3. We have $f(t) = f(t + nT_0), \forall\, n \in \mathbb{Z}$,                    □

An alternative to a complex Fourier series expansion of a periodic function is its trigonometric Fourier series representation.

### 6.10.4  Trigonometric Fourier Series

Let $T_0 \in \mathbb{R}^+$, and $\omega_0 = 2\pi/T_0$. For $t \in \mathbb{R}$, and any $a \in \mathbb{R}$, the sequence of functions

$$\{\cos n\omega_0 t \mid n \in \mathbb{N}\}, \quad \text{and} \quad \{\sin n\omega_0 t \mid n \in \mathbb{P}\}$$

form an orthogonal set of functions in the interval $[a, a + T_0]$. Observe that

$$\frac{2}{T_0} \int_0^{T_0} \cos m\omega_0 t \, \cos n\omega_0 t \, dt = \delta_{mn}, \quad \forall \, m, n \in \mathbb{P}$$

$$\frac{2}{T_0} \int_0^{T_0} \sin m\omega_0 t \, \sin n\omega_0 t \, dt = \delta_{mn}, \quad \forall \, m, n \in \mathbb{P}$$

$$\int_0^{T_0} \sin m\omega_0 t \, \cos n\omega_0 t \, dt = 0, \quad \forall \, m, n \in \mathbb{N}$$

Similarly

$$\int_0^{T_0} \cos m\omega_0 t \, dt = 0, \quad \forall \, m \in \mathbb{P}$$

$$\int_0^{T_0} \sin m\omega_0 t \, dt = 0, \quad \forall \, m \in \mathbb{N}$$

**Definition 6.47.** *If $f(t)$, $t \in \mathbb{R}$, is a periodic real-valued function with period $T_0 \in \mathbb{R}^+$, and $\omega_0 = 2\pi/T_0$. Then for any $a \in \mathbb{R}$*

$$f(t) = \frac{a_0}{2} + \sum_{n \in \mathbb{P}} (a_n \cos n\omega_0 t + b_n \sin n\omega_0 t) \tag{6.61a}$$

$$a_n = \frac{2}{T_0} \int_a^{a+T_0} f(t) \cos n\omega_0 t \, dt, \quad \forall \, n \in \mathbb{N} \tag{6.61b}$$

$$b_n = \frac{2}{T_0} \int_a^{a+T_0} f(t) \sin n\omega_0 t \, dt, \quad \forall \, n \in \mathbb{P} \tag{6.61c}$$

*In this series representation, it is assumed that Dirichlet conditions hold, and that $f(t)$ is continuous at $t$. If $f(t)$ is discontinuous at $t$, then $f(t)$ in the above expansion should be replaced by $\{f(t_+) + f(t_-)\}/2$.* $\square$

Some textbooks use the period $T_0 = 2\pi$, in the Fourier series representation. In this case, the above Fourier series representation and the corresponding coefficients can be suitably modified.

**Observations 6.24.** Let $t \in \mathbb{R}$.

1. For any $a \in \mathbb{R}$, Parseval's relationship is

$$\frac{1}{T_0} \int_a^{a+T_0} |f(t)|^2 \, dt = \frac{a_0^2}{4} + \frac{1}{2} \sum_{n \in \mathbb{P}} (a_n^2 + b_n^2) \tag{6.62}$$

2. If $f(\cdot)$ is an even function, that is $f(t) = f(-t), \forall \, t \in \mathbb{R}$, then $b_n = 0$ for all values of $n \in \mathbb{P}$. However, if $f(\cdot)$ is an odd function, that is $f(t) = -f(-t), \forall \, t \in \mathbb{R}$, then $a_n = 0$ for all values of $n \in \mathbb{N}$. $\square$

The Fourier coefficients of the complex and trigonometric Fourier series are related as follows.

**Observation 6.25.** Note that

$$c_0 = \frac{1}{2}a_0, \ c_n = \frac{1}{2}\left(a_n - ib_n\right), \ c_{-n} = \frac{1}{2}\left(a_n + ib_n\right), \ \forall\, n \in \mathbb{P} \tag{6.63a}$$

$$a_0 = 2c_0, \ a_n = \left(c_n + c_{-n}\right), \ b_n = i\left(c_n - c_{-n}\right), \quad \forall\, n \in \mathbb{P} \tag{6.63b}$$

$\square$

The following examples are simultaneously instructive and elegant.

**Examples 6.13.** Let $t \in \mathbb{R}$.

1. Let $\delta_{T_0}\left(\cdot\right)$ be a periodic train of impulse functions with period $T_0$. Let $\omega_0 = 2\pi/T_0$ and

$$\delta_{T_0}\left(t\right) \triangleq \sum_{n \in \mathbb{Z}} \delta\left(t - nT_0\right)$$

This is an even function. Its Fourier series expansion is

$$\delta_{T_0}\left(t\right) = \frac{1}{T_0} \sum_{n \in \mathbb{Z}} e^{in\omega_0 t} = \frac{1}{T_0} + \frac{2}{T_0} \sum_{n \in \mathbb{P}} \cos n\omega_0 t$$

2. Let $f(\cdot)$ be a periodic function with period $T_0$.

$$f(t) = \begin{cases} -1, & -\dfrac{T_0}{2} < t < 0 \\[2mm] 1, & 0 < t < \dfrac{T_0}{2} \end{cases}$$

The function $f\left(\cdot\right)$ is odd. Its trigonometric Fourier expansion is

$$f(t) = \frac{4}{\pi} \sum_{n \in \mathbb{P}} \frac{1}{\left(2n - 1\right)} \sin\left(2n - 1\right)\omega_0 t$$

Letting $t = T_0/4$ in the above equation yields the following well-known identity.

$$1 - \frac{1}{3} + \frac{1}{5} - \frac{1}{7} + \cdots = \frac{\pi}{4}$$

The discovery of the above series for $\pi$ is generally attributed to Gottfried Wilhelm von Leibniz (1646-1716). However, this result was discovered three centuries earlier by Madhava of Sangamagramma (1350-1425). Use of Parseval's relationship results in

$$\frac{1}{1^2} + \frac{1}{3^2} + \frac{1}{5^2} + \cdots = \frac{\pi^2}{8}$$

3. Let $f(\cdot)$ be a periodic function with period $T_0$. It is triangular in shape within this interval.

$$f(t) = \begin{cases} 1 + \dfrac{4t}{T_0}, & -\dfrac{T_0}{2} \le t < 0 \\[2mm] 1 - \dfrac{4t}{T_0}, & 0 \le t < \dfrac{T_0}{2} \end{cases}$$

This function is an even function. Its trigonometric Fourier expansion is

$$f(t) = \frac{8}{\pi^2} \sum_{n \in \mathbb{P}} \frac{1}{(2n-1)^2} \cos(2n-1)\,\omega_0 t$$

Letting $t = 0$ in the above equation yields the following well-known identity.

$$\frac{1}{1^2} + \frac{1}{3^2} + \frac{1}{5^2} + \cdots = \frac{\pi^2}{8}$$

Use of Parseval's relationship results in

$$\frac{1}{1^4} + \frac{1}{3^4} + \frac{1}{5^4} + \cdots = \frac{\pi^4}{96}$$

$\square$

A complement of the Fourier series technique is the Fourier transform of a function. Fourier series expansion is useful in studying periodic functions. In contrast, Fourier transform technique is used to study nonperiodic functions. In addition to the Fourier transforms, other continuous and discrete transforms are also studied in the next section.

## 6.11 Transform Techniques

A transform is a mapping of a function from one space to another. Specially crafted transforms help us see patterns. A problem or a physical scenario can sometimes be addressed more easily in the transform domain than in the time domain. The purpose of transform analysis is to represent a function as a linear combination of some basis functions. This is in a sense a decomposition of a function into some "elementary" functions. These elementary functions are the building blocks of the transform. Transforms can be either continuous or discrete.

Some examples of continuous transforms are: the Fourier, Laplace, Wigner-Ville and wavelet transforms. All of these transforms are useful in studying functions in different domains. The different transform techniques also complement each other in their applications.

The discrete transform techniques are evidently related to their continuous counterpart. Furthermore, the discrete transforms are well-suited for computer implementation. The Hadamard transform and discrete Fourier transform are discussed in this section.

### 6.11.1 Fourier Transform

Recall that a Fourier series is used to study periodic functions. In contrast, Fourier transforms are used to study nonperiodic functions. An alternate view-point is to think that Fourier series is used to analyse functions which are defined only over a finite interval, while Fourier transform is used to examine functions defined over the entire real line $\mathbb{R}$. A Fourier transform converts a time-domain function into a function defined over frequency-domain.

One of the conditions for the existence of Fourier transform of $f(t), t \in \mathbb{R}$ is that $\int_{-\infty}^{\infty} |f(t)|\, dt$ be convergent. If this integral exists, then the function $f\,(\cdot)$ is said to be *absolutely integrable* on $\mathbb{R}$.

**Definition 6.48.** *The Fourier transform of* $f(t), t \in \mathbb{R}$ *is defined by*

$$F(\omega) = \Im\left[f(t)\right] = \int_{-\infty}^{\infty} f(t)e^{-i\omega t} dt \qquad (6.64)$$

*where* $i = \sqrt{-1}$, $\omega \in \mathbb{R}$, $f(\cdot)$ *is piecewise-continuous* $\forall\, t \in \mathbb{R}$, *and* $f(\cdot)$ *is absolutely integrable on* $\mathbb{R}$. *The functions* $f(\cdot)$ *and* $F(\cdot)$ *are called a Fourier transform pair. This relationship is denoted by* $f(t) \leftrightarrow F(\omega)$. $\qquad\qquad\square$

The transformed domain is often referred to as the frequency domain.

**Observations 6.26.** Let $f(t), t \in \mathbb{R}$, be a piecewise-continuous function and absolutely integrable on $\mathbb{R}$. Also $f(t) \leftrightarrow F(\omega)$.

1. The function $f(\cdot)$ is absolutely integrable if $\int_{-\infty}^{\infty} |f(t)|\, dt < \infty$, that is $f(t) \in L^1(\mathbb{R})$.
2. The condition of absolute integrability of a function is sufficient but not necessary for the existence of Fourier transform. For example, functions which are not absolutely integrable but have Fourier transforms are: $\sin \omega_0 t$, and $\cos \omega_0 t$.
3. The function $f(\cdot)$ is piecewise-continuous if $f(\cdot)$ is continuous, except for finite number of finite jumps in every finite subinterval of $\mathbb{R}$. Thus the function $f(\cdot)$ can possibly have an infinite number of discontinuities, but only a finite number in each finite subinterval.
4. It can be proved that:
    (a) $F(\omega)$ is properly defined (bounded) $\forall\, \omega \in \mathbb{R}$.
    (b) $F(\omega)$ is a continuous function defined on $\mathbb{R}$.
    (c) $F(\omega)$ tends to $0$ as $\omega \to \pm\infty$.
5. The *inverse Fourier transform* of $F(\omega)$ is given by

$$f(t) = \Im^{-1}\left[F(\omega)\right] = \frac{1}{2\pi} \int_{-\infty}^{\infty} F(\omega)\, e^{i\omega t} d\omega \qquad (6.65a)$$

If $f(t)$ is discontinuous at a point $t$, and the discontinuity is finite, then the Fourier inverse transformation yields

$$f(t) = \frac{(f(t_+) + f(t_-))}{2} \qquad (6.65b)$$

$\qquad\qquad\square$

The constants $1$ and $1/(2\pi)$ preceding the integration signs in the definition of the Fourier transform, and the expression for the inverse Fourier transform respectively, can be replaced by any two constants whose product is equal to $1/(2\pi)$. In the interest of symmetry some authors prefer to use the constant $1/\sqrt{2\pi}$ preceding the definition of the Fourier transform, and the expression for the inverse Fourier transform.

**Properties of the Fourier Transform**

Let $a, t, \omega \in \mathbb{R}$. Also let $\alpha_1, \alpha_2 \in \mathbb{C}$, and $f(t) \leftrightarrow F(\omega)$, $f_1(t) \leftrightarrow F_1(\omega)$, and $f_2(t) \leftrightarrow F_2(\omega)$.

1. Linearity: $\alpha_1 f_1(t) + \alpha_2 f_2(t) \leftrightarrow \alpha_1 F_1(\omega) + \alpha_2 F_2(\omega)$
2. Time reversal: $f(-t) \leftrightarrow F(-\omega)$

3. Conjugate function: $\overline{f(t)} \leftrightarrow \overline{F(-\omega)}$
4. Symmetry or duality: $F(t) \leftrightarrow 2\pi f(-\omega)$
5. Time shift: $f(t - t_0) \leftrightarrow e^{-i\omega t_0} F(\omega)$
6. Frequency shift: $e^{i\omega_0 t} f(t) \leftrightarrow F(\omega - \omega_0)$
7. Time scaling:

$$f(at) \leftrightarrow \frac{1}{|a|} F\left(\frac{\omega}{a}\right), \quad \text{where} \quad a \neq 0$$

8. Convolution:

$$\int_{-\infty}^{\infty} f_1(\tau) f_2(t - \tau) \, d\tau \triangleq f_1(t) * f_2(t) \leftrightarrow F_1(\omega) F_2(\omega)$$

where $*$ is called the convolution operator.
9. Multiplication:

$$f_1(t) f_2(t) \leftrightarrow \frac{1}{2\pi} F_1(\omega) * F_2(\omega)$$

where $*$ is called the convolution operator.
10. Time differentiation:

$$\frac{d^n}{dt^n} f(t) \leftrightarrow (i\omega)^n F(\omega), \quad \forall n \in \mathbb{P}$$

11. Frequency differentiation:

$$(-it)^n f(t) \leftrightarrow \frac{d^n}{d\omega^n} F(\omega), \quad \forall n \in \mathbb{P}$$

12. Integration:

$$\int_{-\infty}^{t} f(\tau) \, d\tau \leftrightarrow \frac{F(\omega)}{i\omega} + \pi F(0) \delta(\omega)$$

13. Modulation identities: Let $\omega_0 \in \mathbb{R}$, then

$$f(t) \cos \omega_0 t \leftrightarrow \frac{1}{2} \{F(\omega - \omega_0) + F(\omega + \omega_0)\}$$

$$f(t) \sin \omega_0 t \leftrightarrow \frac{1}{2i} \{F(\omega - \omega_0) - F(\omega + \omega_0)\}$$

14. Parseval's relationships:

$$\int_{-\infty}^{\infty} |f(t)|^2 \, dt = \frac{1}{2\pi} \int_{-\infty}^{\infty} |F(\omega)|^2 \, d\omega$$

$$\int_{-\infty}^{\infty} f_1(t) f_2(t) \, dt = \frac{1}{2\pi} \int_{-\infty}^{\infty} F_1(-\omega) F_2(\omega) \, d\omega$$

$$\int_{-\infty}^{\infty} \overline{f_1(t)} f_2(t) \, dt = \frac{1}{2\pi} \int_{-\infty}^{\infty} \overline{F_1(\omega)} F_2(\omega) \, d\omega$$

$$\int_{-\infty}^{\infty} f_1(t) F_2(t) \, dt = \int_{-\infty}^{\infty} F_1(\omega) f_2(\omega) \, d\omega$$

15. Fourier transform of a series: Let $f(\cdot)$ be a periodic function with period $T_0 \in \mathbb{R}^+$. For the periodic function $f(\cdot)$, $\int_{-\infty}^{\infty} |f(t) \, dt| \to \infty$. However, assume that the Fourier transform of $f(\cdot)$ exists in the sense of a generalized function. Let $\omega_0 = 2\pi/T_0$, and $f(t) \leftrightarrow F(\omega)$. If

$$f(t) = \sum_{n \in \mathbb{Z}} c_n e^{in\omega_0 t}$$

then

$$F(\omega) = 2\pi \sum_{n \in \mathbb{Z}} c_n \delta(\omega - n\omega_0)$$

$\square$

Before the Fourier transform pairs are listed, certain useful functions are first defined. In all these functions $\alpha, t \in \mathbb{R}$.

*Signum function sgn $(\cdot)$:*

$$sgn(t) = \begin{cases} 1, & t > 0 \\ -1, & t < 0 \end{cases}$$

$sgn(\cdot)$ is not defined at $t = 0$. It follows that

$$\frac{d}{dt} sgn(t) = 2\delta(t)$$

*Unit step function $u(\cdot)$:*

$$u(t) = \begin{cases} 1, & t > 0 \\ 0, & t < 0 \end{cases}$$

$u(t)$ is not defined at $t = 0$. It follows that

$$u(t) = \frac{1}{2} + \frac{1}{2} sgn(t)$$

*Gate function $g_\alpha(\cdot)$, $\alpha > 0$:*

$$g_\alpha(t) = \begin{cases} 1, & |t| < \alpha \\ 0, & |t| > \alpha \end{cases}$$

*Sinc function $sinc(\cdot)$:*

$$sinc(t) = \frac{\sin t}{t}$$

Some useful Fourier transform pairs are listed below.

1. $\delta(t) \leftrightarrow 1$
2. $\delta(t - t_0) \leftrightarrow e^{-i\omega t_0}$
3. Let $\delta_{T_0}(t) = \sum_{n \in \mathbb{Z}} \delta(t - nT_0)$ and $\delta_{\omega_0}(\omega) = \sum_{n \in \mathbb{Z}} \delta(\omega - n\omega_0)$, where $\omega_0 = 2\pi/T_0$, then $\delta_{T_0}(t) \leftrightarrow \omega_0 \delta_{\omega_0}(\omega)$.
4. $1 \leftrightarrow 2\pi\delta(\omega)$
5. $e^{i\omega_0 t} \leftrightarrow 2\pi\delta(\omega - \omega_0)$
6. $\sin \omega_0 t \leftrightarrow i\pi [\delta(\omega + \omega_0) - \delta(\omega - \omega_0)]$
7. $\cos \omega_0 t \leftrightarrow \pi [\delta(\omega + \omega_0) + \delta(\omega - \omega_0)]$
8. $sgn(t) \leftrightarrow \frac{2}{i\omega}$
9. $u(t) \leftrightarrow \pi\delta(\omega) + \frac{1}{i\omega}$
10. Let $\alpha > 0$, then:
    (a) $g_\alpha(t) \leftrightarrow 2\alpha\, sinc(\omega\alpha)$
    (b) $\frac{\alpha}{\pi} sinc(\alpha t) \leftrightarrow g_\alpha(\omega)$

11. $\frac{1}{\sqrt{2\pi}}e^{-t^2/2} \leftrightarrow e^{-\omega^2/2}$                                                                        □

Poisson's summation formulae are next derived.

**Theorem 6.15.** *Poisson's theorem. Let $T_0$ and $\tau$ be positive real numbers. Also let $\omega_0 = 2\pi/T_0, \Omega_0 = 2\pi/\tau, t \in \mathbb{R}$ and $f(t)$ be an arbitrary function such that $f(t) \leftrightarrow F(\omega)$. Define*

$$f_s(t) = \sum_{n \in \mathbb{Z}} f(t + nT_0), \quad t \in \mathbb{R} \tag{6.66a}$$

$$F_s(\omega) = \sum_{n \in \mathbb{Z}} F(\omega + n\Omega_0), \quad \omega \in \mathbb{R} \tag{6.66b}$$

*Then*

$$f_s(t) = \frac{1}{T_0} \sum_{n \in \mathbb{Z}} F(n\omega_0) e^{in\omega_0 t} \tag{6.66c}$$

$$F_s(\omega) = \tau \sum_{n \in \mathbb{Z}} f(n\tau) e^{-in\omega\tau} \tag{6.66d}$$

*Proof.* See the problem section.                                                                           □

Note that $f_s(t)$ and $F_s(\omega)$ are not a Fourier transform pair. The following formulae are immediate from the above theorem. These formulae are known as *Poisson's summation formulae* after the mathematician Siméon-Denis Poisson (1781-1840).

$$\sum_{n \in \mathbb{Z}} f(nT_0) = \frac{1}{T_0} \sum_{n \in \mathbb{Z}} F(n\omega_0) \tag{6.67a}$$

$$\sum_{n \in \mathbb{Z}} F(n\Omega_0) = \tau \sum_{n \in \mathbb{Z}} f(n\tau) \tag{6.67b}$$

### 6.11.2  Laplace Transform

Laplace transforms are useful in the study of queueing theory, and renewal theory. Renewal theory in turn is useful in understanding Internet traffic. Two types of Laplace transforms are studied. These are the single-sided (unilateral) and the two-sided (bilateral, or double-sided) Laplace transforms.

#### Single-Sided Laplace Transform

Before a formal definition of Laplace transform is given, recall that if $\epsilon > 0$, then $0_+$ is the limit of $(0 + \epsilon)$ as $\epsilon$ goes to 0. Similarly $0_-$ is the limit of $(0 - \epsilon)$ as $\epsilon$ goes to 0.

**Definition 6.49.** *Let $f(t), t \in \mathbb{R}_0^+$ be a real-valued piecewise-continuous function. Also let $s = (\sigma + i\omega) \in \mathbb{C}$, where $\sigma$ and $\omega$ are real numbers, and $i = \sqrt{-1}$. The single-sided Laplace transform of $f(t)$, is*

$$\widehat{f}(s) = \mathcal{L}[f(t)] = \int_0^\infty f(t)e^{-st}dt \tag{6.68}$$

*where*:

(a) *The function $f\left(\cdot\right)$ satisfies $\left|f\left(t\right)\right| \leq Me^{\alpha t}, \forall\, t > t_0$; where $M$ and $t_0$ are positive real numbers, and $\alpha$ is a real number. This function is said to have an exponential order.*

(b) $\operatorname{Re}\left(s\right) > \alpha$.

(c) *The lower limit in this integral is $0_-$.*

$f(t)$ and $\widehat{f}(s)$ *are called Laplace transform pair. This relationship is denoted by* $f(t) \Longleftrightarrow \widehat{f}(s)$.

$\square$

In the above definition, the Laplace transform $\widehat{f}(s)$ is analytic at least in the complex half-plane $\operatorname{Re}\left(s\right) > \alpha$. Note that the symbol $\Longleftrightarrow$ is different from the "if and only if" symbol $\Leftrightarrow$ . If $\widehat{f}\left(s\right)$ exists for some $s_0 = \left(\sigma_0 + i\omega_0\right)$, where $\sigma_0 > \alpha$, then it exists $\forall\, s \in \mathbb{C}$ such that $\operatorname{Re}\left(s\right) \geq \sigma_0$. The smallest such value of $\sigma_0$ is called the *abscissa of convergence*. Therefore the line $\operatorname{Re}\left(s\right) = \alpha$ is called the *axis of convergence*, and the region $\operatorname{Re}\left(s\right) > \alpha$ is called the *region of convergence* of the Laplace transform $\widehat{f}(s)$.

**Observation 6.27.** The inverse of Laplace transform $\widehat{f}\left(s\right), s \in \mathbb{C}$ is given by

$$f(t) = \mathcal{L}^{-1}\left[\widehat{f}\left(s\right)\right] = \frac{1}{2\pi i} \int_{\sigma-i\infty}^{\sigma+i\infty} \widehat{f}\left(s\right) e^{ts} ds, \quad t \in \mathbb{R}_0^+ \tag{6.69}$$

where the integration is within the region of convergence. The integral is actually a contour integral in the complex plane. It is a vertical line in the half-plane $\sigma > \alpha$. This result can be proved by using the Fourier inversion formula. $\qquad\square$

### Properties of the Single-Sided Laplace Transform

Some properties of single-sided Laplace transforms are listed below. In these properties the domain of functions is $\mathbb{R}_0^+$. Denote the $n$th derivative of $f\left(t\right)$ with respect to $t$, by $f^{(n)}\left(t\right)$. Also $f^{(0)}\left(t\right) \triangleq f\left(t\right)$, and $u\left(t\right)$ is the unit step function. Let $f_1(t) \Longleftrightarrow \widehat{f}_1(s)$ and $f_2(t) \Longleftrightarrow \widehat{f}_2(s)$.

1. Linearity: If $a_1, a_2 \in \mathbb{R}$, then $a_1 f_1(t) + a_2 f_2(t) \Longleftrightarrow a_1\widehat{f}_1(s) + a_2\widehat{f}_2(s)$

2. Scaling property:
$$f\left(at\right) \Longleftrightarrow \frac{1}{a}\widehat{f}\left(\frac{s}{a}\right), \quad a > 0$$

3. Time shifting: $f\left(t-a\right)u\left(t-a\right) \Longleftrightarrow e^{-as}\widehat{f}\left(s\right),\ a > 0$

4. Shifting in transform domain: $e^{-at}f\left(t\right) \Longleftrightarrow \widehat{f}\left(s+a\right)$

5. Derivative in transform domain:
$$t^n f\left(t\right) \Longleftrightarrow \left(-1\right)^n \frac{d^n \widehat{f}\left(s\right)}{ds^n}, \quad \forall\, n \in \mathbb{P}$$

6. Derivative in time domain:
$$f^{(n)}\left(t\right) \Longleftrightarrow s^n\widehat{f}\left(s\right) - \sum_{j=1}^{n} s^{n-j} f^{(j-1)}\left(0_-\right), \quad \forall\, n \in \mathbb{P}$$

7. Integral with zero initial condition:
$$\int_0^t f\left(u\right) du \Longleftrightarrow \frac{\widehat{f}\left(s\right)}{s}$$

8. Multiple integrals with zero initial condition:

$$\int_0^t \int_0^{t_1} \cdots \int_0^{t_{n-1}} f\left(t_n\right) dt_n \ldots dt_2 dt_1 \iff \frac{\widehat{f}\left(s\right)}{s^n}$$

9. Division by $t$ :

$$\frac{f\left(t\right)}{t} \iff \int_s^\infty \widehat{f}\left(s'\right) ds'$$

10. Convolution theorem:

$$\int_0^t f_1\left(t-u\right) f_2\left(u\right) du \iff \widehat{f_1}\left(s\right) \widehat{f_2}\left(s\right)$$

11. Initial value theorem: $\lim_{t \to 0} f\left(t\right) = \lim_{s \to \infty} s\widehat{f}\left(s\right)$
12. Final value theorem: $\lim_{t \to \infty} f\left(t\right) = \lim_{s \to 0} s\widehat{f}\left(s\right)$, if $s\widehat{f}\left(s\right)$ is analytic for $\operatorname{Re} s \geq 0$.     ☐

**Useful Laplace Transform Pairs**

Some useful unilateral Laplace transform pairs are listed below.

1. $c \iff \frac{c}{s}$, $c$
2. $t^n \iff \frac{n!}{s^{n+1}}$, $\forall n \in \mathbb{N}$
3. $t^{-1/2} \iff \sqrt{\frac{\pi}{s}}$
4. $e^{-at} \iff \frac{1}{\left(s+a\right)}$, $\operatorname{Re}\left(s\right) > a$
5. $\delta\left(t\right) \iff 1$
6. $t^{\varrho-1} \iff \frac{\Gamma\left(\varrho\right)}{s^\varrho}$, $\operatorname{Re}\left(\varrho\right) > 0$, and $\Gamma\left(\cdot\right)$ is the gamma function.     ☐

**Bilateral Laplace Transform**

The double-sided or bilateral Laplace transform is useful in queueing theory.

**Definition 6.50.** *The double-sided Laplace transform of $f(t)$, where $t \in \mathbb{R}$ is*

$$\widetilde{f}\left(s\right) = \int_{-\infty}^\infty f(t)e^{-st} dt, \quad s \in \mathbb{C} \tag{6.70}$$

*where $f\left(t\right)$ is piecewise-continuous and of exponential order.*     ☐

**Observation 6.28**. The inverse double-sided Laplace transform of $\widetilde{f}\left(s\right)$, $s \in \mathbb{C}$ is given by

$$f(t) = \frac{1}{2\pi i} \int_{\sigma-i\infty}^{\sigma+i\infty} \widetilde{f}\left(s\right) e^{ts} ds \tag{6.71}$$

where the integration is within the region of convergence in the complex plane. This region is a vertical strip $\sigma_1 < \operatorname{Re}\left(s\right) < \sigma_2$.     ☐

**Relationship Between Single-Sided and Bilateral Laplace Transforms**

It is useful to determine the relationship between the single-sided and bilateral Laplace transforms. Define

$$f_-(t) = \begin{cases} f(t), & t < 0 \\ 0, & t \geq 0 \end{cases}$$

$$f_+(t) = \begin{cases} 0, & t < 0 \\ f(t), & t \geq 0 \end{cases}$$

Therefore

$$f(t) = f_-(t) + f_+(t)$$

Notice that $f_-(-t)$ is nonzero only for positive values of $t$, and $f_+(t)$ is nonzero only for nonnegative values of $t$. Next define the one-sided Laplace transforms of $f_-(-t)$ and $f_+(t)$.

$$\mathcal{L}[f_-(-t)] = \widehat{f}_-(s), \quad \text{Re}(s) > \sigma_-$$
$$\mathcal{L}[f_+(t)] = \widehat{f}_+(s), \quad \text{Re}(s) > \sigma_+$$

However, the Laplace transform of $f_-(t)$ is required, which is equal to $\widehat{f}_-(-s)$, where $\text{Re}(s) < \sigma_-$. Therefore, the two sided Laplace transform $\widetilde{f}(s)$ is:

$$\widetilde{f}(s) = \widehat{f}_-(-s) + \widehat{f}_+(s), \quad \sigma_+ < \text{Re}(s) < \sigma_-$$

Therefore the two-sided Laplace transform exists if and only if $\sigma_+ < \sigma_-$.

### 6.11.3  Mellin Transform

The Mellin transform is a linear transformation of a real-valued function. For example the gamma function $\Gamma(s)$ where $\text{Re}(s) > 0$, is the Mellin transform of $e^{-t}$, where $t \geq 0$. This transform is named after the Finnish mathematician Robert Hjalmar Mellin (1854-1933). The Mellin transform and its inverse were however used earlier by the celebrated mathematican Bernhard Riemann (1826-1866) in his famous memoir on prime numbers. The Mellin transform is closely related to the Fourier and bilateral Laplace transforms.

**Definition 6.51.** *Let* $f(t), t \in \mathbb{R}_0^+$ *be a real-valued function. The Mellin transform of* $f(\cdot)$*, is*

$$F_M(s) = \mathcal{M}(f(t)) = \int_0^\infty t^{s-1} f(t) dt \tag{6.72}$$

*where* $s \in \Omega_f$*, and* $\Omega_f \subseteq \mathbb{C}$ *is the domain of convergence of the transform.*     $\square$

**Observations 6.29.** In this observation, we state a domain of convergence of the Mellin transform. An expression for the inverse of the transform is also provided.

1. If the Mellin transform $F_M(s)$ of $f(t), t \in \mathbb{R}_0^+$ exists at two points $s_1, s_2 \in \mathbb{C}$, where $\text{Re}\, s_1 < \text{Re}\, s_2$ then the Mellin transform exists in the vertical strip $\text{Re}\, s_1 < \text{Re}\, s < \text{Re}\, s_2$.
2. The inverse of Mellin transform $F_M(s)$ is given by

$$f(t) = \frac{1}{2\pi i} \int_{\sigma - i\infty}^{\sigma + i\infty} F_M(s) t^{-s} ds \tag{6.73}$$

where $i = \sqrt{-1}$, and $\sigma$ is within the vertical strip of analyticity $\sigma_1 < \text{Re}(s) < \sigma_2$.     $\square$

We next identify the relationship between the Mellin transform, bilateral Laplace and the Fourier transforms.

**Observations 6.30.** Let $F_M(s)$, $s \in \Omega_f$ be the Mellin transform of $f(t)$, $t \in \mathbb{R}_0^+$.

1. Let $g(t) = f(e^{-t})$, $t \in \mathbb{R}$, and $\tilde{g}(s)$, $s \in \Omega_f$ be its bilateral Laplace transform, then $F_M(s) = \tilde{g}(s)$.
2. Let $s = (\sigma + i\omega)$, where $\sigma, \omega \in \mathbb{R}$, and $i = \sqrt{-1}$. Also let $h(t) = e^{-\sigma t} f(e^{-t})$, $t \in \mathbb{R}$. If $H(\omega)$ is the Fourier transform of $h(t)$, then $F_M(s) = H(\omega)$. $\qquad\square$

Some elementary and useful properties of the Mellin transform are listed below.

**Properties of the Mellin Transform**

Let $a, \omega \in \mathbb{R}$, and $\alpha, \beta \in \mathbb{R}$. Also $F_M(s) = \mathcal{M}(f(t))$ and $G_M(s) = \mathcal{M}(g(t))$ are the Mellin transforms of $f(t)$ and $g(t)$ respectively, where $t \in \mathbb{R}_0^+$ and $s \in \Omega_f \subseteq \mathbb{C}$.

1. Linearity:
$$\mathcal{M}(\alpha f(t) + \beta g(t)) = \alpha F_M(s) + \beta G_M(s)$$

2. Scaling: Let $a \in \mathbb{R}^+$

$$\mathcal{M}\left(f\left(\frac{t}{a}\right)\right) = a^s F_M(s)$$

$$\mathcal{M}(f(t^a)) = \frac{1}{a} F_M\left(\frac{s}{a}\right)$$

3. Translation:
$$\mathcal{M}(t^a f(t)) = F_M(s + a), \quad \text{where} \quad (s + a) \in \Omega_f$$

4. Inverse of independent variable:
$$\mathcal{M}\left(t^{-1} f\left(t^{-1}\right)\right) = F_M(1 - s), \quad \text{where} \quad (1 - s) \in \Omega_f$$

5. Differentiation: Let $f(\cdot) \in L^1(0, \infty)$ be a differentiable function such that
$$\lim_{t \to 0_+} t^{s-1} f(t) \to 0, \quad \text{and} \quad \lim_{t \to \infty} t^{s-1} f(t) \to 0; \quad \alpha < \operatorname{Re}(s) < \beta$$

and if $f'(t)$ denotes the first derivative of $f(t)$ with respect to $t$, then
$$\mathcal{M}(f'(t)) = -(s - 1) F_M(s - 1), \quad (s - 1) \in \Omega_f$$

$\qquad\square$

An attractive property of the Mellin transform is the multiplicative convolution, or simply the Mellin convolution.

**Definition 6.52.** *Let $f(t), g(t), t \in \mathbb{R}_0^+$ be real-valued functions. The multiplicative convolution operation $*$, is defined as*

$$(f * g)(x) = \int_0^\infty f\left(\frac{x}{u}\right) g(u) \frac{du}{u} \tag{6.74}$$

*whenever the integral exists.*                                                                 □

The Mellin transform of $(f * g)(x)$ is given in the next observation.

**Observation 6.31.** Let $f(t)$ and $g(t)$, where $t \in \mathbb{R}_0^+$ be real-valued functions. Their Mellin transforms are $F_M(s)$ and $G_M(s)$ respectively. Then

$$\mathcal{M}((f * g)(x)) = F_M(s) G_M(s)$$

□

The above details are elucidated via the following examples.

**Examples 6.14.** In these examples, we compute $\mathcal{M}(f(t))$, where $f(t), t \in \mathbb{R}_0^+$.

1. If $a \in \mathbb{R}^+$, then
    (a) $\mathcal{M}(\delta(t-a)) = a^{s-1}$.
    (b) $\mathcal{M}(u(t-a)) = -a^s/s$, where $\mathrm{Re}(s) < 0$.
    (c) $\mathcal{M}(e^{-at}) = a^{-s}\Gamma(s)$, where $\mathrm{Re}(s) > 0$.
    (d) $\mathcal{M}\left(e^{-at^2}\right) = a^{-s/2}\Gamma(s/2)/2$, where $\mathrm{Re}(s) > 0$.

2. If $\theta \in \mathbb{R}^+$, then
    (a) $\mathcal{M}(e^{-i\theta t}) = (i\theta)^{-s}\Gamma(s)$, where $0 < \mathrm{Re}(s) < 1$.
    (b) $\mathcal{M}(\cos(\theta t)) = \theta^{-s}\Gamma(s)\cos(\pi s/2)$, where $0 < \mathrm{Re}(s) < 1$.
    (c) $\mathcal{M}(\sin(\theta t)) = \theta^{-s}\Gamma(s)\sin(\pi s/2)$, where $-1 < \mathrm{Re}(s) < 1$.

3. If $\zeta(s)$, $\mathrm{Re}(s) > 1$, is the well-known Riemann's zeta function.
    (a) $\mathcal{M}\left(\left(e^t - 1\right)^{-1}\right) = \Gamma(s)\zeta(s)$.
    (b) $\mathcal{M}\left(\left(e^t + 1\right)^{-1}\right) = \Gamma(s)\zeta(s)\left(1 - 2^{1-s}\right)$.                                 □

### 6.11.4  Wigner-Ville Transform

Wigner-Ville distribution is a second order or bilinear transform that performs the mapping of time-domain functions into time-frequency space. It is an alternative to the short term Fourier transform for nonstationary and transient signal (function) analysis. A nonstationary signal is a time varying signal in a statistical sense. This transform is named after E. P. Wigner and J. Ville.

**Definition 6.53.** *The Wigner-Ville transform (distribution) of $f(t), t \in \mathbb{R}$ is specified by*

$$\widetilde{W}_f(\tau, \omega) = \int_{-\infty}^{\infty} f\left(\tau + \frac{t}{2}\right)\overline{f\left(\tau - \frac{t}{2}\right)}e^{-i\omega t}dt; \quad \tau, \omega \in \mathbb{R} \tag{6.75}$$

□

#### Properties of Wigner-Ville Transform

Some useful results about this transform are listed below. Let $f(t) \leftrightarrow F(\omega), t, \omega \in \mathbb{R}$. In these observations $\tau, t, t_1, t_2, \omega, \omega_1, \omega_2 \in \mathbb{R}$.

1. $\widetilde{W}_f(\tau,\omega) = \frac{1}{2\pi} \int_{-\infty}^{\infty} F\left(\omega + \frac{\xi}{2}\right) \overline{F\left(\omega - \frac{\xi}{2}\right)} e^{i\tau\xi} d\xi$

2. $f(t_1) \overline{f(t_2)} = \frac{1}{2\pi} \int_{-\infty}^{\infty} \widetilde{W}_f\left(\frac{t_1+t_2}{2},\omega\right) e^{i\omega(t_1-t_2)} d\omega$

3. $f(t)\overline{f(0)} = \frac{1}{2\pi} \int_{-\infty}^{\infty} \widetilde{W}_f\left(\frac{t}{2},\omega\right) e^{i\omega t} d\omega$

4. $F(\omega_1)\overline{F(\omega_2)} = \int_{-\infty}^{\infty} \widetilde{W}_f\left(\tau,\frac{\omega_1+\omega_2}{2}\right) e^{-i(\omega_1-\omega_2)\tau} d\tau$

5. $F(\omega)\overline{F(0)} = \int_{-\infty}^{\infty} \widetilde{W}_f\left(\tau,\frac{\omega}{2}\right) e^{-i\omega\tau} d\tau$

6. $|f(t)|^2 = \frac{1}{2\pi} \int_{-\infty}^{\infty} \widetilde{W}_f(t,\omega)\, d\omega$

7. $|F(\omega)|^2 = \int_{-\infty}^{\infty} \widetilde{W}_f(\tau,\omega)\, d\tau$

8. $\int_{-\infty}^{\infty} |f(t)|^2\, dt = \frac{1}{2\pi} \int_{-\infty}^{\infty} |F(\omega)|^2\, d\omega = \frac{1}{2\pi} \int_{-\infty}^{\infty} \int_{-\infty}^{\infty} \widetilde{W}_f(\tau,\omega)\, d\omega d\tau$     □

**Examples 6.15.** In the following examples $t, \omega \in \mathbb{R}$.

1. If $f(t) = \delta(t - t_0)$, then $\widetilde{W}_f(\tau,\omega) = \delta(\tau - t_0)$.

2. If $F(\omega) = \delta(\omega - \omega_0)$, then $\widetilde{W}_f(\tau,\omega) = \delta(\omega - \omega_0)/(2\pi)$.

3. If $f(t) = \frac{1}{\sqrt{2\pi}\alpha} e^{-t^2/(2\alpha^2)}$, $\alpha \in \mathbb{R}^+$, then $\widetilde{W}_f(\tau,\omega) = \frac{1}{\sqrt{\pi}\alpha} e^{-\alpha^2\omega^2 - \tau^2/\alpha^2}$.     □

We shall use the Wigner-Ville transform to study and characterize Internet traffic.

### 6.11.5 Wavelet Transform

Wavelet transform is a technique for local analysis of nonstationary signals. This transform is an alternative but not a replacement of the Fourier transform. The building blocks in wavelet analysis are derived by translation and dilation of a mother function. It uses wavelets (short waves) instead of long waves. These wavelets are localized functions. Instead of oscillating forever, as in the case of the basis functions used in Fourier analysis (trigonometric functions), wavelets eventually drop to zero. Wavelet transforms can be either continuous or discrete. We initially study continuous wavelet transform, and then its discrete counterpart.

We only provide a partial history of important developments in wavelet transforms. Alfréd Haar described an orthonormal basis in 1910 in his doctoral dissertation. His work was a precursor to wavelet transform techniques. Jean Morlet introduced the concept of wavelet in 1981. Jean Morlet, and Alex Grossmann introduced the phrase *wavelet* in 1984. Actually they used an equivalent French word *ondelette*, meaning "small wave." Yves Meyer discovered orthogonal wavelets in 1985. Stéphane Mallat and Yves Meyer discovered the concept of multiresolution analysis in 1988. Ingrid Daubechies discovered compactly supported orthogonal wavelets in 1988. Finally, Mallat introduced fast wavelet transform in 1989.

The wavelet transform is a mapping of a function defined in time domain, into a function which has a *time-scale* representation. That is, the wavelet transformation is a two-dimensional representation of a one-dimensional function. In the following definition of wavelet transform, $L^2(\mathbb{R})$ is the space of square-integrable functions. Let $f(t), t \in \mathbb{R}$ be the signal that has to be transformed, where $f(\cdot) \in L^2(\mathbb{R})$. That is, $\int_{-\infty}^{\infty} |f(t)|^2\, dt < \infty$. The wavelet transform of the function $f(\cdot)$ is defined below. It is computed by shifting and scaling of the mother wavelet function $\psi(t), t \in \mathbb{R}$, where $\psi(\cdot) \in L^2(\mathbb{R})$.

**Definitions 6.54.** *Continuous wavelet transform.*

1. *The signal to be transformed is:* $f(t) \in \mathbb{R}$, $t \in \mathbb{R}$, *and* $f(\cdot) \in L^2(\mathbb{R})$.

2. *The function $\psi : \mathbb{R} \to \mathbb{C}$, where $\psi(\cdot) \in L^2(\mathbb{R})$, is called the mother wavelet or the prototype function.*

   (a) *$\Psi(\cdot)$ is the Fourier transform of $\psi(\cdot)$, that is $\psi(t) \leftrightarrow \Psi(\omega)$.*

   (b) *The function $\Psi(\cdot)$ should also satisfy the following condition*

$$C_\psi = \int_{-\infty}^{\infty} \frac{|\Psi(\omega)|^2}{|\omega|} d\omega < \infty \tag{6.76a}$$

   *This relationship is also called the admissibility condition. It is required for recovering $f(t)$ from the wavelet transform.*

3. *Let $a, b \in \mathbb{R}$, and $a \neq 0$. Define*

$$\psi_{a,b}(t) = \frac{1}{\sqrt{|a|}} \psi\left(\frac{t-b}{a}\right) \tag{6.76b}$$

*The continuous wavelet transform of the function $f(\cdot)$ is*

$$W_f(\psi, a, b) = \int_{-\infty}^{\infty} f(t) \overline{\psi_{a,b}(t)} dt \tag{6.76c}$$

□

The admissibility condition implies that $\Psi(0) = 0$, that is

$$\int_{-\infty}^{\infty} \psi(t) \, dt = 0$$

The variables $a$ and $b$, are the *scale* and *translation* parameters respectively. Generally $a$ is positive. Therefore for large values of $a$, the function $\psi_{a,b}(\cdot)$ becomes a stretched version (long-time duration) of $\psi(\cdot)$. In this case, $\psi_{a,b}(\cdot)$ is a low-frequency function. However, for small values of $a$, the function $\psi_{a,b}(\cdot)$ becomes a contracted version (short-time duration) of $\psi(\cdot)$. In this case, $\psi_{a,b}(\cdot)$ is a high-frequency function. The parameter $b$ simply shifts the mother wavelet. In order to preserve smoothness, the mother wavelet is also required to have zero values for the first few moments. This requirement is termed the *regularity condition*.

**Observation 6.32**. The inversion formula of the wavelet transform is

$$f(t) = \frac{1}{C_\psi} \int_{-\infty}^{\infty} \int_{-\infty}^{\infty} W_f(\psi, a, b) \, \psi_{a,b}(t) \frac{da\,db}{a^2} \tag{6.77}$$

□

See the problem section for a proof of the validity of this inversion formula.

**Examples of Wavelets**

Some commonly used wavelets are discussed below. In these examples $t \in \mathbb{R}$.

*Haar wavelet*: The Haar wavelet, named after Alfred Haar (1885-1933) is defined as

$$\psi(t) = \begin{cases} 1, & 0 \le t < 1/2 \\ -1, & 1/2 \le t < 1 \\ 0, & \text{else where} \end{cases}$$

Note that this wavelet satisfies the admissibility condition $\int_{-\infty}^{\infty} \psi(t)\, dt = 0$. Also its Fourier transform is given by

$$\Psi(\omega) = 2ie^{-i\omega/2} \frac{\left(1 - \cos \frac{\omega}{2}\right)}{\omega}$$

*Morlet wavelet*: Morlet wavelet is a complex exponential and Gaussian-windowed. This function is not a wavelet as per the definition.

$$\psi(t) = \frac{1}{\sqrt{2\pi}} e^{-i\omega_0 t} e^{-t^2/2}$$

Its Fourier transform is given by

$$\Psi(\omega) = e^{-(\omega + \omega_0)^2/2}$$

The function $\psi(\cdot)$ does not satisfy the admissibility condition $\int_{-\infty}^{\infty} \psi(t)\, dt = 0$. However, $\omega_0$ can be chosen such that $\Psi(0)$ is very close to zero. Consider the real part of $\psi(t)$, which is $\mathrm{Re}(\psi(t)) = \psi_c(t)$. It is given by

$$\psi_c(t) = \frac{e^{-t^2/2}}{\sqrt{2\pi}} \cos \omega_0 t$$

If $\psi_c(t) \leftrightarrow \Psi_c(\omega)$, then

$$\Psi_c(\omega) = \left[ e^{-(\omega+\omega_0)^2/2} + e^{-(\omega-\omega_0)^2/2} \right]/2$$

Note that $\Psi_c(0) = e^{-\omega_0^2/2} \neq 0$. However the value of $\omega_0$ can be chosen large enough such that $\Psi_c(0) \simeq 0$. In this case the $\psi(\cdot)$ is said to be "approximately analytic."

*Mexican-hat wavelet*: The Mexican-hat wavelet is defined as

$$\psi(t) = \left(1 - t^2\right) e^{-t^2/2}$$

This function is related to the second derivative of the Gaussian function $g(t) = e^{-t^2/2}$. That is

$$\psi(t) = -\frac{d^2}{dt^2} g(t)$$

Also

$$\Psi(\omega) = \sqrt{2\pi}\omega^2 e^{-\omega^2/2}$$

This wavelet satisfies the admissibility condition $\Psi(0) = 0$. All the derivatives of $\psi(t)$ and $\Psi(\omega)$ exist. Furthermore, this function has superb localization in both time and frequency domains. It is widely used in image processing. The Mexican-hat wavelet was originally introduced by the physicist Dennis Gabor (1900-1979).

### Discrete Wavelet Transform

The discrete wavelet transform is also useful in analysing time series.

**Definitions 6.55.** *Let $f(t), t \in \mathbb{R}$ where $f(\cdot) \in L^2(\mathbb{R})$. Also let $a_0 \in \mathbb{R}^+$, and $b_0 \in \mathbb{R}$.*

1. *Discrete wavelet transform of the function $f(\cdot)$, is*

$$d\left(m,n\right) = \int_{-\infty}^{\infty} f\left(t\right)\overline{\psi_{m,n}\left(t\right)}dt, \quad \forall\, m,n \in \mathbb{Z} \tag{6.78a}$$

$$\psi_{m,n}\left(t\right) = a_0^{\frac{m}{2}}\,\psi\left(a_0^m t - nb_0\right), \quad \forall\, m,n \in \mathbb{Z} \tag{6.78b}$$

*where $\psi\left(\cdot\right)$ is the mother wavelet. The values $d\left(m,n\right)$ are called the wavelet coefficients, and the $\psi_{m,n}\left(\cdot\right)$'s are called the wavelets.*
2. *If the wavelets form an orthonormal basis of $L^2\left(\mathbb{R}\right)$, then*

$$f(t) = \sum_{m,n \in \mathbb{Z}} d\left(m,n\right)\psi_{m,n}\left(t\right) \tag{6.79a}$$

*Note that the wavelets $\left\{\psi_{m,n}\left(t\right) \mid m,n \in \mathbb{Z}\right\}$ form an orthonormal basis if*

$$\int_{-\infty}^{\infty} \psi_{m,n}\left(t\right)\overline{\psi_{m',n'}\left(t\right)}dt = \delta_{mm'}\delta_{nn'}, \quad \forall\, m,n \in \mathbb{Z} \tag{6.79b}$$

$$\delta_{jk} = \begin{cases} 1, & j = k \\ 0, & j \neq k \end{cases}, \quad \forall\, j,k \in \mathbb{Z} \tag{6.79c}$$

$\delta_{jk}$ *is called the Kronecker's delta function.* □

Observe that the discrete wavelet transform is the transform of a continuous time function, but the scale and translation parameters are discretized. Also note that some authors have elected to define $\psi_{m,n}\left(t\right)$ as $a_0^{-\frac{m}{2}}\,\psi\left(a_0^{-m}t - nb_0\right)$, for all $m,n \in \mathbb{Z}$.

### 6.11.6 Hadamard Transform

The Hadamard transform is alternately referred to as Walsh-Hadamard transform, named after the American mathematician Joseph L. Walsh (1895-1973), and the French mathematician Jacques S. Hadamard (1865-1963).

**Definitions 6.56.** *The Hadamard transform is defined in terms of the Hadamard matrix.*

1. *The $2 \times 2$ Hadamard matrix $W_2$ is*

$$W_2 = \frac{1}{\sqrt{2}}\begin{bmatrix} 1 & 1 \\ 1 & -1 \end{bmatrix} \tag{6.80a}$$

2. *Hadamard matrix and transform. Let $x$ be a real vector of length $2^n = N$, where $n \in \mathbb{P}$. The Hadamard transform of this vector $x$ is a unitary map defined by the matrix $W_N$, where*

$$W_N = W_2^{\otimes n} = \underbrace{W_2 \otimes W_2 \otimes \ldots \otimes W_2}_{n\ \text{times}} \tag{6.80b}$$

*and $y = W_N x$. Note that $\otimes$ is the Kronecker product operator. The matrix $W_N$ is also called the Hadamard matrix.* □

See the chapter on matrices and determinants for the definition of the Kronecker product operation.

**Example 6.16.** We have

$$W_4 = W_2 \otimes W_2 = \frac{1}{2} \begin{bmatrix} 1 & 1 & 1 & 1 \\ 1 & -1 & 1 & -1 \\ 1 & 1 & -1 & -1 \\ 1 & -1 & -1 & 1 \end{bmatrix}$$

Similarly $W_8 = W_2 \otimes W_2 \otimes W_2 = W_4 \otimes W_2$ can be generated $\qquad\square$

**Lemma 6.4.** *Let* $y = W_N x, N = 2^n$ *where*

$$x = \begin{bmatrix} x_0 & x_1 & \cdots & x_{N-1} \end{bmatrix}^T, \quad and \quad y - \begin{bmatrix} y_0 & y_1 & \cdots & y_{N-1} \end{bmatrix}^T \qquad (6.81a)$$

*and* $W_N$ *is the Hadamard matrix. Assume that* $0 \leq i, j \leq (N-1)$, *and let the binary expansion of* $i$ *be* $i_{n-1} i_{n-2} \ldots i_0$, *where* $i_k \in \{0, 1\}$ *for* $0 \leq k \leq (n-1)$. *Similarly the binary expansion of* $j$ *is* $j_{n-1} j_{n-2} \ldots j_0$, *where* $j_k \in \{0, 1\}$ *for* $0 \leq k \leq (n-1)$. *Denote the dot product of the vectors* $(i_{n-1}, i_{n-2}, \ldots, i_0)$, *and* $(j_{n-1}, j_{n-2}, \ldots, j_0)$ *by* $i \cdot j = \sum_{k=0}^{n-1} i_k j_k$. *Then*

$$y_j - \frac{1}{\sqrt{N}} \sum_{i=0}^{N-1} (-1)^{i \cdot j} x_i, \quad 0 \leq j \leq (N-1) \qquad (6.81b)$$

*Proof.* The proof by induction is left to the reader. $\qquad\square$

In the computation of the Hadamard transformation, the exponent $i \cdot j = \sum_{k=0}^{n-1} i_k j_k$ could be replaced by $\bigoplus_{k=0}^{n-1} i_k j_k$, where the summation operator $\bigoplus$ uses modulo 2 arithmetic.

Some of the areas in which the Hadamard transform finds application are: data compression, coding theory, and experimental design.

### 6.11.7  Discrete Fourier Transform

The discrete Fourier transform (DFT) is an important tool in the study of signals. It essentially approximates a periodic sequence of discrete set of points by finite sums of weighted trigonometric (sine and cosine) functions. We next study the computation of the Fourier integral via DFT. This also provides us with the genesis of the DFT.

#### Computation of Fourier Integral via DFT or the Genesis of DFT

Define $\omega_0 = 2\pi/T_0$ and $\Omega_0 = 2\pi/\tau$, where $T_0$ and $\tau$ are positive real numbers. Also let $T_0 = N\tau$, then $\Omega_0 = N\omega_0$. Let $f(\cdot)$ be an arbitrary function such that $f(t) \leftrightarrow F(\omega), t, \omega \in \mathbb{R}$. Next define

$$f_s(t) \triangleq \sum_{n \in \mathbb{Z}} f(t + nT_0), \quad t \in \mathbb{R}$$

$$F_s(\omega) \triangleq \sum_{n \in \mathbb{Z}} F(\omega + n\Omega_0), \quad \omega \in \mathbb{R}$$

Use of Poisson's theorem yields

$$f_s(m\tau) = \frac{1}{T_0} \sum_{j \in \mathbb{Z}} F(j\omega_0) e^{ijm\omega_0\tau}, \quad \forall\, m \in \mathbb{Z}$$

Substituting $j = (Nk + n)$ in the above equation, and defining $\omega_N = e^{i2\pi/N}$ yields

$$f_s(m\tau) = \frac{1}{T_0} \sum_{n=0}^{N-1} \omega_N^{mn} \sum_{k \in \mathbb{Z}} F(n\omega_0 + k\Omega_0)$$

$$= \frac{1}{T_0} \sum_{n=0}^{N-1} \omega_N^{mn} F_s(n\omega_0)$$

Let $\mathbb{Z}_N = \{0, 1, 2, \ldots, N-1\}$. Substituting $m = 0, 1, \ldots, (N-1)$ in the above equation yields $N$ number of equations. From these $N$ equations $F_s(\omega_0 n)$'s can be calculated as a function of $f_s(m\tau)$'s, where $m, n \in \mathbb{Z}_N$. Note that it is not always possible to recover $F(\omega_0 n)$'s from $F_s(\omega_0 n)$'s. If for some $\omega_c \in \mathbb{R}^+$, $F(\omega) = 0$ for $|\omega| > \omega_c$, and $\Omega_0 > 2\omega_c$; then $F(\omega) = F_s(\omega)$ for $|\omega| < \omega_c$. If these conditions are not met, then we have the so-called aliasing error defined by $(F(\omega_0 n) - F_s(\omega_0 n))$.

The computation of $F_s(\omega_0 n)$'s can be performed formally within the framework of discrete Fourier transform and its inverse. The discrete Fourier transform is defined below. It is followed by a discussion of computationally efficient techniques to compute it.

**Definition 6.57.** *Let* $N \in \mathbb{P}$, $\pi = 3.1415926535897\ldots$, $\omega_N = e^{2\pi i/N}$, *and* $i = \sqrt{-1}$. *The discrete Fourier transform of the sequence of complex numbers*

$$\{y(0), y(1), \ldots, y(N-1)\} \tag{6.82a}$$

*is a sequence of complex numbers*

$$\{Y(0), Y(1), \ldots, Y(N-1)\} \tag{6.82b}$$

*where*

$$Y(m) = \frac{1}{\sqrt{N}} \sum_{n=0}^{(N-1)} y(n)\, \omega_N^{mn}, \quad \forall\, m \in \mathbb{Z}_N \tag{6.82c}$$

$\square$

In the above definition, the arguments of $y(\cdot)$ and $Y(\cdot)$ are computed modulo $N$.

**Observation 6.33.** The inverse of DFT is

$$y(n) = \frac{1}{\sqrt{N}} \sum_{m=0}^{(N-1)} Y(m)\, \omega_N^{-mn}, \quad \forall\, n \in \mathbb{Z}_N \tag{6.83}$$

$\square$

Sometimes, the abbreviations $\Im_N[y(n)] \triangleq Y(m)$, and $\Im_N^{-1}[Y(m)] \triangleq y(n)$ are also used. Some elementary properties of DFT are summarized below.

**Properties of DFT**

Let $N \in \mathbb{P}, \alpha_1, \alpha_2 \in \mathbb{C}$, and

$$\Im_N [y(n)] = Y(m), \quad \Im_N [y_1(n)] = Y_1(m), \quad \Im_N [y_2(n)] = Y_2(m)$$

1. Periodicity: $Y(m) = Y(m+N)$
2. Linearity: $\Im_N [\alpha_1 y_1(n) + \alpha_2 y_2(n)] = \alpha_1 Y_1(m) + \alpha_2 Y_2(m)$
3. Time reversal: $\Im_N [y(-n)] = Y(-m)$
4. Conjugate function: $\Im_N \left[\overline{y(n)}\right] = \overline{Y(-m)}$
5. Symmetry or duality: $\Im_N [Y(n)] = y(-m)$
6. Time shift: $\Im_N [y(n-n_0)] = \omega_N^{mn_0} Y(m), \quad \forall\, n_0 \in \mathbb{Z}$
7. Frequency shift: $\Im_N \left[\omega_N^{nk} y(n)\right] = Y(m+k), \quad \forall\, k \in \mathbb{Z}$
8. Circular convolution: Let

$$\{x(0), x(1), \ldots, x(N-1)\} \quad \text{and} \quad \{y(0), y(1), \ldots, y(N-1)\}$$

be two periodic complex sequences of period $N$ each. The circular convolution of these two sequences is a periodic sequence of period $N$. Let this convolved sequence be

$$\{w(0), w(1), \ldots, w(N-1)\},$$

where

$$w(n) = \sum_{k=0}^{(N-1)} x(k)\, y(n-k), \quad \forall\, n \in \mathbb{Z}_N$$

In the above equation, $(n-k)$ is computed modulo $N$. Therefore this convolution is circular. It can be shown that if $\Im_N [x(n)] - X(m), \; \Im_N [y(n)] = Y(m)$, and $\Im_N [w(n)] = W(m)$ then

$$W(m) = \sqrt{N} X(m)\, Y(m), \quad \forall\, m \in \mathbb{Z}_N$$

Similarly discrete Fourier transform of the sequence $x(n)\, y(n), n \in \mathbb{Z}_N$ is the sequence

$$\frac{1}{\sqrt{N}} \sum_{k=0}^{(N-1)} X(k)\, Y(m-k), \quad m \in \mathbb{Z}_N$$

9. Parseval's relationships:

$$\sum_{k=0}^{(N-1)} x(k)\, \overline{y(k)} = \sum_{k=0}^{(N-1)} \overline{y(k)} \frac{1}{\sqrt{N}} \sum_{j=0}^{(N-1)} X(j)\, \omega_N^{-jk}$$

$$= \frac{1}{\sqrt{N}} \sum_{j=0}^{(N-1)} X(j) \sum_{k=0}^{(N-1)} \overline{y(k)} \omega_N^{-jk}$$

Thus

$$\sum_{k=0}^{(N-1)} x(k)\, \overline{y(k)} = \sum_{j=0}^{(N-1)} X(j)\, \overline{Y(j)}$$

Therefore

$$\sum_{k=0}^{(N-1)} |x(k)|^2 = \sum_{j=0}^{(N-1)} |X(j)|^2$$

$\square$

## Computation of the DFT

A direct computation of the DFT of the complex sequence

$$\{y(0), y(1), \ldots, y(N-1)\} \triangleq \{y(n) \mid y(n) \in \mathbb{C},\ n \in \mathbb{Z}_N\}$$

requires up to $N^2$ complex multiplication and addition operations. Therefore the computational complexity of a direct computation of DFT of a size-$N$ sequence is $\Theta(N^2)$ operations.

Computationally efficient algorithms exist to compute DFT. Such algorithms are called fast Fourier transforms (FFT). It is assumed in these algorithms that it is more expensive to perform multiplication, than either an addition or subtraction operation. Two computationally efficient algorithms to compute DFT are discussed below. These are:

(a) A FFT algorithm originally due to the celebrated German mathematician J. C. F. Gauss, and later rediscovered independently by James W. Cooley (1926- ) and John W. Tukey (1915-2000). Cooley and Tukey devised a computerized algorithm to implement the discrete Fourier transform.

(b) A prime factor FFT algorithm, which uses the number-theoretic Chinese remainder theorem.

These FFT algorithms are generally regarded as some of the most important algorithms developed in the last century.

## Cooley-Tukey FFT Algorithm

The Cooley-Tukey FFT algorithm achieves reduction in the number of computations by using the principle of divide and conquer. The genesis of the FFT algorithm is first given. Let $N = 2D$, and split the sequence

$$\{y(n) \mid y(n) \in \mathbb{C}, n = 0, 1, \ldots, (N-1)\}$$

into two sequences:

$$\{p(n) \mid p(n) = y(2n), n = 0, 1, \ldots, (D-1)\}$$
$$\{q(n) \mid q(n) = y(2n+1), n = 0, 1, \ldots, (D-1)\}$$

These are the sequences with even and odd indices respectively. If we let $\Im_D[p(n)] = P(m)$, $\Im_D[q(n)] = Q(m)$,

$$\omega_N^{2km} = \omega_D^{km}$$
$$\omega_N^{(2k+1)m} = \omega_D^{km}\omega_N^m$$

It follows that

$$Y(m) = P(m) + \omega_N^m Q(m), \quad 0 \le m \le (N-1)$$

It should be noted that in the computation of $Y(m)$'s, $P(m)$ and $Q(m)$ are each periodic in $m$ with period $D$. Also

$$P(m+D) = P(m), \quad \forall\, m \in \mathbb{Z}_D$$
$$Q(m+D) = Q(m), \quad \forall\, m \in \mathbb{Z}_D$$
$$\omega_N^{D+m} = -\omega_N^m, \quad \forall\, m \in \mathbb{Z}_D$$

The transform coefficients $Y(m)$ for $0 \leq m \leq (N-1)$ can be computed as follows.

$$Y(m) = P(m) + \omega_N^m Q(m), \quad \forall\, m \in \mathbb{Z}_D$$
$$Y(m+D) = P(m) - \omega_N^m Q(m), \quad \forall\, m \in \mathbb{Z}_D$$

Computation of $P(m)$ and $Q(m)$, $\forall\, m \in \mathbb{Z}_D$, requires $(D-1)^2$ multiplications each. Therefore computation of $Y(m)$'s after this splitting requires $2(D-1)^2 + (D-1)$ multiplication operations, while a direct computation requires $(2D-1)^2$ such operations. Therefore there is a reduction in the multiplicative complexity, approximately by a factor of two. Let the complexity of computing DFT of size $N$ be $\mathcal{C}(N)$. Therefore, if such splitting operations are used, then

$$\mathcal{C}(N) \sim 2\mathcal{C}\left(\frac{N}{2}\right) + \frac{N}{2}$$

and $\mathcal{C}(2) = 1$.

Let $N = 2^K$, and successively use the splitting operation to compute $P(m)$'s and $Q(m)$'s, and so on. Then it can be shown that $\mathcal{C}(N) \sim NK/2$. Thus the computational complexity of the Cooley-Tukey FFT algorithm is $\Theta(N \log N)$.

### Coprime-Factorization FFT Algorithm

Another fast algorithm to compute DFT is obtained by factorizing $N$ into its coprime factors. This algorithm is based upon the Chinese remainder theorem, which is discussed in the chapter on number theory. Let

$$N = \prod_{k=1}^{K} N_k, \quad \text{where } N_k \in \mathbb{P},\ 1 \leq k \leq K$$

and $\gcd(N_k, N_j) = 1$, $k \neq j$, $1 \leq k, j \leq K$. That is, the factors $N_k$'s are relatively prime in pairs. Define

$$P_k = \frac{N}{N_k}, \quad 1 \leq k \leq K$$

Also let $Q_1, Q_2, \ldots, Q_K \in \mathbb{P}$ such that

$$(P_k Q_k) \equiv 1 \pmod{N_k}, \quad 1 \leq k \leq K$$

Let $n \equiv n_k \pmod{N_k}$, $1 \leq k \leq K$, where $n \in \mathbb{P}$, then $n \pmod{N}$ is mapped into $(n_1, n_2, \ldots n_K)$. This is called Map-1 mapping. That is

$$n \pmod{N} \rightarrow (n_1, n_2, \ldots n_K), \quad 0 \leq n \leq (N-1)$$
$$n \equiv \sum_{k=1}^{K} n_k P_k Q_k \pmod{N}$$

Let $\eta_k \equiv n_k Q_k \pmod{N_k}$, $1 \leq k \leq K$, then $n \pmod{N}$ is mapped into $(\eta_1, \eta_2, \ldots \eta_K)$. This is called Map-2 mapping. That is

$$n \ (\mathrm{mod}\, N) \rightarrow (\eta_1, \eta_2, \ldots \eta_K), \quad 0 \leq n \leq (N-1)$$

$$n \equiv \sum_{k=1}^{K} \eta_k P_k \ (\mathrm{mod}\, N)$$

Note that in the definition of the DFT, the data and frequency elements are indexed by $n$ and $m$ respectively, where $0 \leq m, n \leq (N-1)$. The maps for the frequency indexing variable are as follows. If $m \equiv m_k \ (\mathrm{mod}\, N_k)$, $1 \leq k \leq K$, then $m \ (\mathrm{mod}\, N)$ is mapped into $(m_1, m_2, \ldots m_K)$. This is Map-1 mapping. Thus

$$m \ (\mathrm{mod}\, N) \rightarrow (m_1, m_2, \ldots m_K), \quad 0 \leq m \leq (N-1)$$

$$m \equiv \sum_{k=1}^{K} m_k P_k Q_k \ (\mathrm{mod}\, N)$$

Let $\mu_k \equiv m_k Q_k \ (\mathrm{mod}\, N_k)$, $1 \leq k \leq K$, then $m \ (\mathrm{mod}\, N)$ is mapped into $(\mu_1, \mu_2, \ldots \mu_K)$. This is Map-2 mapping. Thus

$$m \ (\mathrm{mod}\, N) \rightarrow (\mu_1, \mu_2, \ldots \mu_K), \quad 0 \leq m \leq (N-1)$$

$$m \equiv \sum_{k=1}^{K} \mu_k P_k \ (\mathrm{mod}\, N)$$

These maps would result in four different implementations of DFT computation. These are:

(a)  Indexing variables $m$ and $n$ are both mapped as per Map-1.
(b)  Indexing variable $m$ is mapped as Map-1, while indexing variable $n$ is mapped as Map-2.
(c)  Indexing variable $m$ is mapped as Map-2, while indexing variable $n$ is mapped as Map-1.
(d)  Indexing variables $m$ and $n$ are both mapped as per Map-2.

All these implementations are conceptually similar. We demonstrate the coprime-factorization algorithm for a fast implementation of DFT via scheme number (b). In this scheme the indexing variable $m$ is mapped as per Map-1, while indexing variable $n$ is mapped as per Map-2. Then for $0 \leq m, n \leq (N-1)$

$$(mn) \ (\mathrm{mod}\, N) \equiv \left\{ \sum_{k=1}^{K} m_k \eta_k P_k^2 Q_k \right\} (\mathrm{mod}\, N) \equiv \left\{ \sum_{k=1}^{K} m_k \eta_k P_k \right\} (\mathrm{mod}\, N)$$

$$\equiv \left\{ \sum_{k=1}^{K} \frac{N m_k \eta_k}{N_k} \right\} (\mathrm{mod}\, N)$$

Therefore

$$\omega_N^{mn} = \prod_{k=1}^{K} \omega_{N_k}^{m_k \eta_k}$$

And for $m_k = 0, 1, \ldots, (N_k - 1)$, $1 \leq k \leq K$,

$$Y(m_1, m_2, \ldots m_K) = \frac{1}{\sqrt{N_K}} \sum_{\eta_K=0}^{(N_K-1)} \cdots$$

$$\cdots \left\{ \frac{1}{\sqrt{N_2}} \sum_{\eta_2=0}^{(N_2-1)} \left[ \frac{1}{\sqrt{N_1}} \sum_{\eta_1=0}^{(N_1-1)} y(\eta_1, \eta_2, \ldots \eta_K) \omega_{N_1}^{m_1 \eta_1} \right] \omega_{N_2}^{m_2 \eta_2} \right\} \cdots \omega_{N_K}^{m_K \eta_K}$$

Notice that this scheme converts DFT in a single dimension into a multidimensional DFT. Furthermore, the success of this implementation depends upon efficient implementation of DFT's of size $N_k$'s. This can be made true by having optimized DFT's for small values of $N_k$'s. Also if $N_k$'s are composite numbers, then using Cooley-Tukey type implementation of DFT of size $N_k$ gives further improvement in its computational efficiency.

In addition to the Cooley-Tukey and coprime-factorization FFT algorithms, there are other computationally efficient discrete Fourier transform algorithms. Nevertheless, the salient features of the Cooley-Tukey and coprime-factorization FFT algorithms permeate these other algorithms.

Furthermore, there are useful families of discrete transforms which are related to the discrete Fourier transform. A prominent example is the discrete cosine transform. This transform is used extensively in signal processing. Several fast versions of this algorithm are related to the FFT algorithms.

## 6.12 Faà di Bruno's Formula

Faà di Bruno's formula (theorem) specifies the $n$th derivative of a composite function explicitly. This formula was discovered by Faà di Bruno (1825-1888). Let $f(\cdot)$ and $g(\cdot)$ be functions of $t$ with a sufficient number of derivatives in an interval $(a, b)$. The Faà di Bruno's formula computes the $n$th derivative of $f(g(t))$. In establishing this derivative of a composite function, the Taylor series expansion of these functions is assumed to exist.

**Theorem 6.16.** *(Faà di Bruno's formula) Let $f(t)$ and $g(t)$, $t \in \mathbb{R}$ be functions with a sufficient number of derivatives. Denote the $m$th derivative of these functions with respect to $t$ by $f^{(m)}(t)$ and $g^{(m)}(t)$ respectively. Then for $n \in \mathbb{P}$*

$$
\frac{d^n}{dt^n} f(g(t)) = \sum_{\substack{\kappa=1 \\ \kappa_1,\kappa_2,\ldots,\kappa_n \geq 0 \\ \kappa_1+\kappa_2+\ldots+\kappa_n=\kappa \\ \kappa_1+2\kappa_2+\ldots+n\kappa_n=n}}^{n} \frac{n!}{\kappa_1!\kappa_2!\ldots\kappa_n!} f^{(\kappa)}(g(t)) \prod_{j=1}^{n} \left\{ \frac{g^{(j)}(t)}{j!} \right\}^{\kappa_j}
$$

(6.84)

*The above expansion requires constrained summation over integer partitions of $\kappa$.*

*Proof.* Taylor series expansion of $g(t)$ and $f(t)$ is

$$
g(t) = \sum_{m \in \mathbb{N}} g^{(m)}(t_0) \frac{(t-t_0)^m}{m!}
$$

$$
f(t) = \sum_{m \in \mathbb{N}} f^{(m)}(t_0) \frac{(t-t_0)^m}{m!}
$$

Let $g(t_0) \triangleq g_0, f(g_0) \triangleq f_0$, and

$$
a_m \triangleq \frac{g^{(m)}(t_0)}{m!}, \qquad \forall\, m \in \mathbb{P}
$$

$$
b_m \triangleq f^{(m)}(g_0), \qquad \forall\, m \in \mathbb{P}
$$

Then

$$g(t) = g_0 + \sum_{m \in \mathbb{P}} a_m (t - t_0)^m$$

$$f(g(t)) = f_0 + \sum_{\kappa \in \mathbb{P}} \frac{b_\kappa}{\kappa!} (g(t) - g_0)^\kappa$$

Consequently

$$f(g(t)) = f_0 + \sum_{\kappa \in \mathbb{P}} \frac{b_\kappa}{\kappa!} \left\{ \sum_{m \in \mathbb{P}} a_m (t - t_0)^m \right\}^\kappa$$

Note that for $\kappa \in \mathbb{P}$

$$\left\{ \sum_{m \in \mathbb{P}} a_m (t - t_0)^m \right\}^\kappa$$

$$= \sum_{\substack{\kappa_1, \kappa_2, \kappa_3, \dots \geq 0 \\ \kappa_1 + \kappa_2 + \kappa_3 + \dots = \kappa}} \frac{\kappa!}{\kappa_1! \kappa_2! \kappa_3! \dots} a_1^{\kappa_1} a_2^{\kappa_2} a_3^{\kappa_3} \dots (t - t_0)^{\kappa_1 + 2\kappa_2 + 3\kappa_3 + \dots}$$

Let $\kappa_1 + 2\kappa_2 + 3\kappa_3 + \dots + j\kappa_j = n$, where the largest value of $j = n$. Substituting the above expansion in the expression for $f(g(t))$ yields

$$f(g(t)) = f(g(t_0)) + \sum_{n \in \mathbb{P}} \frac{(t - t_0)^n}{n!} \sum_{\substack{\kappa = 1 \\ \kappa_1, \kappa_2, \dots, \kappa_n \geq 0 \\ \kappa_1 + \kappa_2 + \dots + \kappa_n = \kappa \\ \kappa_1 + 2\kappa_2 + \dots + n\kappa_n = n}}^{n} b_\kappa \frac{n!}{\kappa_1! \kappa_2! \dots \kappa_n!} \prod_{j=1}^{n} a_j^{\kappa_j}$$

The result follows.                                                                                             $\square$

**Examples 6.17.** Some elementary cases of the Faà di Bruno's formula are explicitly stated.

1. $n = 1$:
$$\frac{d}{dt} f(g(t)) = f^{(1)}(g(t)) g^{(1)}(t)$$

2. $n = 2$:
$$\frac{d^2}{dt^2} f(g(t)) = f^{(2)}(g(t)) \left\{ g^{(1)}(t) \right\}^2 + f^{(1)}(g(t)) g^{(2)}(t)$$

3. $n = 3$:
$$\frac{d^3}{dt^3} f(g(t)) = f^{(3)}(g(t)) \left\{ g^{(1)}(t) \right\}^3 + 3 f^{(2)}(g(t)) g^{(2)}(t) g^{(1)}(t)$$
$$+ f^{(1)}(g(t)) g^{(3)}(t)$$

4. $n = 4$:
$$\frac{d^4}{dt^4} f(g(t)) = f^{(4)}(g(t)) \left\{ g^{(1)}(t) \right\}^4 + 6 f^{(3)}(g(t)) g^{(2)}(t) \left\{ g^{(1)}(t) \right\}^2$$
$$+ 4 f^{(2)}(g(t)) g^{(3)}(t) g^{(1)}(t) + 3 f^{(2)}(g(t)) \left\{ g^{(2)}(t) \right\}^2$$
$$+ f^{(1)}(g(t)) g^{(4)}(t)$$

$\square$

Definitions of special mathematical functions are listed in the next section. These are also accompanied by their useful properties.

## 6.13 Special Mathematical Functions

Some useful and special mathematical functions are defined in this section. These are: gamma and incomplete gamma function, beta function, Riemann's zeta function, polylogarithm function, Bessel function, exponential integral, and the error function.

### 6.13.1 Gamma and Incomplete Gamma Functions

The gamma function is a generalization of the factorial of an integer for nonintegral values. This generalization opens up a vast new garden of applications. The gamma function $\Gamma(a)$, $a \in \mathbb{C}$ is defined as

$$\Gamma(a) = \int_0^\infty t^{a-1}e^{-t}dt, \qquad \mathrm{Re}\,(a) > 0 \tag{6.85a}$$

Note the recursion $\Gamma(a+1) = a\Gamma(a)$. It can also be shown that $\Gamma(1/2) = \sqrt{\pi}$. The gamma function reduces to the factorial function for integer values of its argument. That is, $\Gamma(n+1) = n!$, $\forall n \in \mathbb{N}$. An alternate expression for $\Gamma(a)$ due to Euler is

$$\Gamma(a) = \frac{1}{a} \prod_{n=1}^\infty \left\{ \left(1 + \frac{1}{n}\right)^a \left(1 + \frac{a}{n}\right)^{-1} \right\}, \qquad a \neq 0, -1, -2, \ldots \tag{6.85b}$$

The important point to note is that $\Gamma(a)$ is analytic for all finite $a \in \mathbb{C}$ except at points $a = 0, -1, -2, \ldots$. The above result can also be stated as

$$\Gamma(a) = \lim_{n\to\infty} \frac{n!\,n^a}{a\,(a+1)\ldots(a+n)}, \qquad a \neq 0, -1, -2, \ldots \tag{6.85c}$$

The above result also yields $\Gamma(1) = 1$. Using the above convenient form of $\Gamma(a)$, the recursive relation $\Gamma(a+1) = a\Gamma(a)$ can be reestablished. The equivalence between the integral form of $\Gamma(a)$ and the above expression is established in the problem section. Another useful relationship is

$$\Gamma(a)\,\Gamma(1-a) = \frac{\pi}{\sin \pi a}, \qquad a \in \mathbb{C}\backslash\mathbb{Z}$$

This result is proved in the problem section. There are two types of incomplete gamma functions. These are the lower incomplete gamma function $\gamma(\cdot,\cdot)$, and the upper incomplete gamma function $\Gamma(\cdot,\cdot)$. These functions are defined as

$$\gamma(a,x) = \int_0^x e^{-t}t^{a-1}dt; \qquad \mathrm{Re}\,(a) > 0, \quad |\arg(x)| < \pi \tag{6.85d}$$

$$\Gamma(a,x) = \int_x^\infty e^{-t}t^{a-1}dt, \qquad |\arg(x)| < \pi \tag{6.85e}$$

Thus

$$\Gamma(a) = \gamma(a,x) + \Gamma(a,x)$$

The normalized (or regularized) lower incomplete gamma function $P(\cdot,\cdot)$ is defined as

$$P(a, x) = \frac{\gamma(a, x)}{\Gamma(a)}; \quad \mathrm{Re}\,(a) > 0, \quad |\arg(x)| < \pi \tag{6.85f}$$

The following recursive relations are generally useful.

$$P(a+1, x) = P(a, x) - \frac{x^a e^{-x}}{\Gamma(a+1)}$$

$$\gamma(a+1, x) = a\gamma(a, x) - x^a e^{-x}$$

$$\Gamma(a+1, x) = a\Gamma(a, x) + x^a e^{-x}$$

Also

$$P(n, x) = 1 - e^{-x} \sum_{j=0}^{n-1} \frac{x^j}{j!}, \quad n \in \mathbb{P}$$

Another useful representation of $\gamma(a, x)$ is the series

$$\gamma(a, x) = x^a \sum_{n \in \mathbb{N}} \frac{(-x)^n}{(a+n)\, n!}; \quad |x| < \infty, \quad a \neq 0, -1, -2, \ldots$$

The above array of functions and different series representations are amenable to convenient numerical implementation via robust recursive techniques.

### 6.13.2  Beta Function

Beta function is defined as

$$B(x, y) = \int_0^1 t^{x-1}(1-t)^{y-1}\, dt; \quad \mathrm{Re}\,(x) > 0, \ \mathrm{Re}\,(y) > 0 \tag{6.86}$$

The beta function is symmetric in $x$ and $y$, that is, $B(x, y) = B(y, x)$. Also

$$B(x, y) = 2 \int_0^{\pi/2} (\sin t)^{2x-1} (\cos t)^{2y-1}\, dt$$

$$= \frac{\Gamma(x)\,\Gamma(y)}{\Gamma(x+y)}$$

The later result is established in the problem section.

### 6.13.3  Riemann's Zeta Function

One of the most celebrated functions in number theory is the Riemann's zeta function $\zeta(s)$, where $s$ is a complex number such that its real part is greater than 1.

$$\zeta(s) = \sum_{n \in \mathbb{P}} \frac{1}{n^s}, \quad s \in \mathbb{C}, \ \mathrm{Re}\,(s) > 1 \tag{6.87}$$

It is well-known that $\zeta(1) \to \infty$, and

$$\zeta(2) = \frac{\pi^2}{6}$$

$$\zeta(4) = \frac{\pi^4}{90}$$

See the problem section for a proof of these results via Fourier series expansion.

### 6.13.4  Polylogarithm Function

The $\alpha$-th polylogarithm of $z \in \mathbb{C}$ is defined as

$$Li_\alpha (z) = \sum_{n \in \mathbb{P}} \frac{z^n}{n^\alpha}, \qquad |z| < 1 \tag{6.88}$$

Note that this function is a generalization of the Riemann's zeta function. Some useful relationships are:

$$\frac{d}{dz} Li_\alpha (z) = \sum_{n \in \mathbb{P}} \frac{z^{n-1}}{n^{\alpha-1}}$$

$$Li_s (1) = \zeta (s), \qquad \mathrm{Re}\,(s) > 1$$

$$Li_{-1} (x) = \frac{x}{(1 - x)^2}, \qquad |x| < 1$$

$$Li_0 (x) = \frac{x}{(1 - x)}, \qquad |x| < 1$$

$$Li_1 (x) = -\ln (1 - x), \qquad |x| < 1 \ \text{ and } \ x = -1$$

### 6.13.5  Bessel Function

Bessel function of the first kind and order one, $J_1 (\cdot)$ is defined as:

$$J_1 (z) = \frac{z}{2} \sum_{k \in \mathbb{N}} \frac{(-1)^k}{k! \, (k + 1)!} \left(\frac{z}{2}\right)^{2k}, \quad z \in \mathbb{C} \tag{6.89a}$$

Also

$$J_1 (x) = \frac{x}{\pi} \int_{-1}^{+1} e^{ixt} \sqrt{1 - t^2} \, dt, \qquad x \in \mathbb{R} \tag{6.89b}$$

It is relatively straightforward to establish an equivalence between the above two expressions for $J_1 (x)$, where $x \in \mathbb{R}$.

### 6.13.6  Exponential Integral

An exponential integral $E_1 (z)$, $z \in \mathbb{C}$, and $|\arg (z)| < \pi$ is defined as

$$E_1(z) = \int_{z}^{\infty} \frac{e^{-t}}{t} dt \tag{6.90a}$$

The following two representations of $E_1 (z)$ can be used in the computations.

$$E_1 (z) = -\gamma - \ln z - \sum_{j \in \mathbb{P}} \frac{(-z)^j}{jj!} \tag{6.90b}$$

$$E_1 (z) = \cfrac{e^{-z}}{z + \cfrac{1}{1 + \cfrac{1}{z + \cfrac{2}{1 + \cfrac{2}{z + \ldots}}}}} \tag{6.90c}$$

where $\gamma = 0.5772156649\ldots$ is the Euler's constant. If $z$ is a real number, the series representation of $E_1(z)$ is useful when the value of $z$ is small ($0 < z \leq 1$). The continued fraction representation of $E_1(z)$ can be used to compute $E_1(z)$ for larger values of $z$ (that is for $z > 1$).

It is possible to have a more general definition of the exponential integral.

$$E_n(z) = \int_1^\infty \frac{e^{-zt}}{t^n} dt; \quad n \in \mathbb{N}, \quad \mathrm{Re}\,(z) > 0 \qquad (6.91a)$$

It can be recursively computed via the following relationships.

$$E_0(z) = \frac{e^{-z}}{z} \qquad (6.91b)$$

$$E_{n+1}(z) = \frac{1}{n}\left\{e^{-z} - zE_n(z)\right\}, \quad \forall\, n \in \mathbb{P} \qquad (6.91c)$$

where $|\arg(z)| < \pi$. Also note that

$$E_n(0) = \frac{1}{(n-1)}, \quad \forall\, n \in \mathbb{P} \setminus \{1\}$$

It can also be shown that

$$\frac{d}{dz} E_n(z) = -E_{n-1}(z), \quad \forall\, n \in \mathbb{P}$$

### 6.13.7 Error Function

Error function is defined as

$$\mathrm{erf}\,(z) = \frac{2}{\sqrt{\pi}} \int_0^z e^{-t^2} dt, \quad z \in \mathbb{C} \qquad (6.92)$$

This function occurs in several branches of mathematics, including probability theory, and theory of errors. A series representation of this function is

$$\mathrm{erf}\,(z) = \frac{2}{\sqrt{\pi}} \sum_{n \in \mathbb{N}} \frac{(-1)^n z^{2n+1}}{n!\,(2n+1)}$$

This series converges for all values of $z \in \mathbb{C}$. Therefore $\mathrm{erf}\,(z)$ for $z \in \mathbb{C}$ is an analytic function. Also observe that

$$\mathrm{erf}\,(-z) = -\,\mathrm{erf}\,(z), \quad \mathrm{erf}\,(\bar{z}) = \overline{\mathrm{erf}\,(z)}, \quad \mathrm{erf}\,(0) = 0, \quad \text{and} \quad \mathrm{erf}\,(\infty) = 1$$

## Reference Notes

Real variable concepts are from Spiegel (1969). Basic ideas about limits, continuity, and derivatives can be found in any standard book on analysis. The subsection on partial derivatives closely follows the lucid exposition given in Spiegel (1963). The section on differential equations is based upon Kreyszig (2011), and Bronshtein, and Semendyayev (1985). The section on complex variables largely follows Spiegel (1964), and Conway (1978). Elegant and concise introduction to the saddle point technique of integration is given in Papoulis (1962), and Spiegel (1964). Basics of Moebius transformations can be found in several books, including Conway (1978), Henle (1997), Anderson (2005), and Pressley (2010).

Properties of polynomials are from the textbooks by Mostowski, and Stark (1964), Walker (1950), and Cox, Little, and O'Shea, (2007). Origin of the word "algorithm" has been stated in the book by none other than the eminent mathematician Villani (2016). The definition of asymptotics follows Horowitz, and Sahni (1978), Sedgewick, and Flajolet (1996), Cormen, Leiserson, Rivest, and Stein (2009), and Janson, Luczak, and Rucinski (2000). The results on inequalities related to the binomial coefficients are from Palmer (1985). The subsection on vector algebra follows Spiegel (1959), and do Cormo (1976).

The discussion on Fourier series is from Papoulis (1962), Hsu (1984), Pinkus, and Zafrany (1997), and Vretblad (2003). A scholarly and elegant textbook on the subject of Fourier analysis is the work of Körner (1988). Laplace transforms have been discussed in several books including Kleinrock (1975), Hsu (1995), and Poularkis (1999). The book by Poularkis (1999) has discussion of Fourier transform, single-sided and bilateral Laplace transforms, Mellin transform, Wigner-Ville transform, wavelet transform, and several other transforms.

The definition and formulae about special mathematical functions are from Abramowitz, and Stegun (1965), Gradshteyn, and Ryzhik (1980), and Lebedev (1972). An alternate expression for the $n$th derivative of a composite function can be found in Gradshteyn, and Ryzhik (1980).

## Problems

1. Establish the Cauchy-Riemann equations for an analytic function $f(z) = u(x, y) + iv(x, y)$, where $z$ is in region $\mathcal{R}$.

    Hint: See Spiegel (1964). For $w = f(z)$ to be analytic in a region $\mathcal{R}$, the limit

$$\lim_{\Delta z \to 0} \frac{f(z + \Delta z) - f(z)}{\Delta z}$$

   must exist independent of the way in which $\Delta z$ approaches zero. Let $\Delta z = (\Delta x + i\Delta y)$. In the first approach, let $\Delta y = 0$, and $\Delta x \to 0$, and in the second approach let $\Delta x = 0$, and $\Delta y \to 0$. Assume that the partial derivatives of $u(\cdot, \cdot)$ and $v(\cdot, \cdot)$ with respect to both $x$ and $y$ exist.

   *Case* 1: Let $\Delta y = 0$, and $\Delta x \to 0$. This yields

$$\frac{\partial w}{\partial x} = \frac{\partial u}{\partial x} + i\frac{\partial v}{\partial x}$$

   *Case* 2: Let $\Delta x = 0$, and $\Delta y \to 0$. This yields

$$\frac{\partial w}{i\partial y} = \frac{1}{i}\frac{\partial u}{\partial y} + \frac{\partial v}{\partial y}$$

   For $f(z)$ to be analytic, the expressions on the right-hand side of the above two equations should be identical. This is required, because $f'(z)$ exists irrespective of the manner in which $\Delta z$ approaches zero. Equating these two expressions yields the Cauchy-Riemann equations.

2. Prove that if $z = x + iy$, and $w = f(z) = u(x,y) + iv(x,y)$ is analytic in a region $\mathcal{R}$, then

$$f'(z) \triangleq \frac{dw}{dz} = \frac{\partial u}{\partial x} + i\frac{\partial v}{\partial x} = \frac{\partial v}{\partial y} - i\frac{\partial u}{\partial y}$$

Hint: See Spiegel (1964). Let $\Delta z = (\Delta x + i\Delta y)$, and $\Delta w = (\Delta u + i\Delta v)$. Therefore

$$\Delta w = (\Delta u + i\Delta v) = \left\{\frac{\partial u}{\partial x}\Delta x + \frac{\partial u}{\partial y}\Delta y\right\} + i\left\{\frac{\partial v}{\partial x}\Delta x + \frac{\partial v}{\partial y}\Delta y\right\}$$

Substituting Cauchy-Riemann relationships in the above equation yields

$$\Delta w = \left\{\frac{\partial u}{\partial x} + i\frac{\partial v}{\partial x}\right\}(\Delta x + i\Delta y)$$

Taking limits as $\Delta z \to 0$ yields the first result, and the second result can be similarly derived.

3. Consider the mapping $f : \Omega \to \mathbb{C}$, where $\Omega \subseteq \mathbb{C}$. Also let $f(z) = w$ for $z \in \Omega$. Assume that $f(z)$ is analytic at $z_0 \in \Omega$.

   (a) Let $f'(z_0) \neq 0$, and a curve $C$ in the $z$-plane passes through $z_0$. Show that the tangent to the curve at $z_0$, is rotated by an angle $\arg(f'(z_0))$ in the $w$-plane.

   (b) Let two curves $C_1$ and $C_2$ pass through the point $z_0 \in \Omega$ in the $z$-plane. Show that the angle between the two curves $C_1$ and $C_2$ is preserved in magnitude and sense at $z_0$ under the mapping $f(z) = w$. This implies that, if $f(z)$ is analytic at $z_0$, and $f'(z_0) \neq 0$, the mapping $f(\cdot)$ is conformal.

   Hint: See Spiegel (1964).

   (a) Let $t$ be the parameter which describes the curves. This means that for the path $z = z(t) = (x(t) + iy(t))$ in the $xy$-plane, there is a path $w = w(t) = (u(t) + iv(t))$ in the $uv$-plane. Then, assuming that $f(z)$ is analytic at $z_0$ we have

$$\frac{dw}{dt}\bigg|_{w=w_0} = f'(z_0)\frac{dz}{dt}\bigg|_{z=z_0}$$

   Let the arguments of $\frac{dw}{dt}\big|_{w=w_0}$, $\frac{dz}{dt}\big|_{z=z_0}$, and $f'(z_0)$ be $\alpha_0, \beta_0$, and $\arg(f'(z_0))$ respectively. Then

$$\alpha_0 = \beta_0 + \arg(f'(z_0))$$

   Observe that, if $f'(z_0) = 0$, then $\arg(f'(z_0))$ is indeterminate.

   (b) From part (a) of the problem, each curve is rotated through an angle $\arg(f'(z_0))$. This implies that the angle is preserved in magnitude and sense at $z_0$ under the mapping $f(\cdot)$.

4. Prove Green-Ostrogradski theorem, if the boundary $C$ of a region $\mathcal{R}$ is a simple closed curve. Hint: See Spiegel (1964). The boundary $C$ of a region $\mathcal{R}$ is a simple closed curve. Assume that any straight line parallel to the $x$-axis cuts it in at most two points. The same is true for any straight line parallel to the $y$-axis. For this type of region $\mathcal{R}$ and boundary $C$, it can be shown that

$$\oint_C P(x,y)\,dx = -\iint_{\mathcal{R}}\frac{\partial P}{\partial y}dxdy, \quad \text{and} \quad \oint_C Q(x,y)\,dy = \iint_{\mathcal{R}}\frac{\partial Q}{\partial x}dxdy$$

The theorem follows by adding these two results. Proof of a more general version of the Green-Ostrogradski theorem can be found in the reference.

5. Prove Cauchy-Goursat integral theorem, with the additional restriction that $f'(z)$ is continuous at all points inside and on a simple closed curve $C$.

Hint: See Spiegel (1964). As per the hypothesis of the theorem $f(z) = \{u(x,y) + iv(x,y)\}$ is analytic and has a continuous derivative. This implies

$$f'(z) = \frac{\partial u}{\partial x} + i\frac{\partial v}{\partial x} = \frac{\partial v}{\partial y} - i\frac{\partial u}{\partial y}$$

Therefore the partial derivatives of $u(\cdot,\cdot)$ and $v(\cdot,\cdot)$ with respect to both $x$ and $y$ are continuous on and inside the contour $C$. Finally, using Green-Ostrogradski theorem and Cauchy-Riemann conditions yields

$$\oint_C f(z)\,dz = \oint_C \{u(x,y) + iv(x,y)\}(dx + idy)$$
$$= \oint_C \{u(x,y)\,dx - v(x,y)\,dy\} + i\oint_C \{v(x,y)\,dx + u(x,y)\,dy\}$$
$$= \iint_{\mathcal{R}} \left\{-\frac{\partial v}{\partial x} - \frac{\partial u}{\partial y}\right\}dxdy + i\iint_{\mathcal{R}} \left\{\frac{\partial u}{\partial x} - \frac{\partial v}{\partial y}\right\}dxdy = 0$$

6. Use Laplace's method of saddle point to derive the following results.

   (a) $n! \sim \sqrt{2\pi}n^{n+1/2}e^{-n}$ for large values of $n$. This is the well-known Stirling's approximation of $n!$.

   (b) Assume that a real-valued function $f(\cdot)$ has a single maximum value in the interval $(a,b)$. Let $f'(x)$ denote the first derivative of $f(x)$ with respect to $x$. Prove that

$$\lim_{n\to\infty} \left[\int_a^b |f(x)|^n\,dx\right]^{1/n} = c$$

   where $c = \max_{x\in(a,b)} |f(x)| = |f(x_0)|$, $f'(x_0) = 0$, and $x_0 \in (a,b)$.

   Hint: See Papoulis (1962).

   (a) The stated result was published by James Stirling (1692-1770), but was known earlier to De Moivre. From the definition of the gamma function

$$n! = \Gamma(n+1) = \int_0^\infty t^n e^{-t}dt$$

   Substitute $t = nz$ in the above integral. This yields

$$n! = n^{n+1} \int_0^\infty e^{n(\ln z - z)}dz$$

   The result follows immediately by using Laplace's method.

   (b) Without any loss of generality, assume that $f(x) \geq 0$. Therefore

$$\int_a^b \{f(x)\}^n\,dx = \int_a^b e^{n\ln f(x)}dx$$

   Let $h(x) = \ln f(x)$, and denote the second derivative of $h(x)$ with respect to $x$ by $h''(x)$. As $n \to \infty$ the above integral is approximated by

$$e^{n \ln f(x_0)} \left( \frac{-2\pi}{nh''(x_0)} \right)^{1/2} = \{f(x_0)\}^n \frac{A}{\sqrt{n}}$$

where $x_0 \in (a, b)$, $f'(x_0) = 0$, and the constant $A = (-2\pi/h''(x_0))^{1/2}$ is independent of $n$. Therefore as $n \to \infty$

$$\left[ \int_a^b \{f(x)\}^n \, dx \right]^{1/n} \sim f(x_0) \left( \frac{A}{\sqrt{n}} \right)^{1/n}$$

It can be observed that $\lim_{n \to \infty} (A/\sqrt{n})^{1/n} = 1$. The final result follows.

7. Establish the equations of a straight line and a circle in the complex $z$-plane. These are:
   (a) Straight line:
   $$\bar{b}z + b\bar{z} + c = 0, \quad \text{where } c \in \mathbb{R}, \ b \in \mathbb{C} \backslash \{0\}$$

   (b) Circle:
   $$az\bar{z} + \bar{b}z + b\bar{z} + c = 0, \quad \text{where } a, c \in \mathbb{R}, \ a \neq 0, \ b \in \mathbb{C}, \ |b|^2 > ac$$

   Hint: See Pressley (2010).
   (a) The equation of a straight line in the $xy$-plane is
   $$px + qy + r = 0, \quad p, q, r \in \mathbb{R}$$

   where $p$ and $q$ are not simultaneously equal to zero. Use of $z = (x + iy)$, $b = (p + iq)$, and $c = 2r$ results in the stated equation.
   (b) The equation of a circle in the $xy$-plane is
   $$(x + p)^2 + (y + q)^2 = r^2, \quad p, q \in \mathbb{R}, \text{ and } r > 0$$

   Use of $z = (x + iy)$, $b = a(p + iq)$, and $c = a\{(p^2 + q^2) - r^2\}$ results in the stated equation.

8. Show that the Moebius transformation is bijective.
   Hint: See Flanigan (1983). The result follows from the existence of the inverse. For a given $w_0$ it is possible to obtain a unique $z_0$, such that $w_0 = f(z_0)$ as $z_0 = f^{-1}(w_0)$.
   Next consider the value of $f(\infty)$. If $f(z) = (az + b)$, then $f(\infty) = \infty$. If $f(z) = 1/(cz + d)$, then $f(\infty) = 0$. It can be shown that in general, if $ac \neq 0$, then $f(\infty) = a/c$.

9. Show that the Moebius transformation
   $$f(z) = \frac{az + b}{cz + d}$$

   is a composition of translations, dilations, and the inversion.
   Hint: See Conway (1978), and Henle (1997). Consider two cases.
   (a) Suppose $c = 0$. Therefore $f(z) = (a/d)z + (b/d)$. If $f_1(z) = (a/d)z$, and $f_2(z) = z + (b/d)$, then $f(z) = f_2 \circ f_1(z)$.
   (b) Suppose $c \neq 0$. Therefore
   $$f(z) = \frac{a}{c} - \frac{(ad - bc)}{c^2} \left\{ \frac{1}{z + \dfrac{d}{c}} \right\}$$

   Next let $f_1(z) = (z + d/c)$, $f_2(z) = 1/z$, $f_3(z) = -(ad - bc)z/c^2$, and $f_4(z) = (z + a/c)$. Then $f(z) = f_4 \circ f_3 \circ f_2 \circ f_1(z)$.

10. Prove that the Moebius transformation is conformal, except at $z = -d/c, \infty$.

    Hint: See Flanigan (1983). The Moebius transformation $f(z)$ is obtained by a sequence of translation, dilation, and inversion. These transformations are conformal, except at the pole (zero of the denominator) of the Moebius transformation. As the sequence of these transformations are conformal, $f(z)$ is conformal.

    This conclusion can also be obtained alternately. Computation of the derivative of $f(z)$ with respect to $z$ yields

    $$f'(z) = \frac{ad - bc}{(cz + d)^2}$$

    Note that $f'(z)$ is finite and nonzero except at $z = -d/c, \infty$. This is a requirement for conformality.

11. Show that a Moebius transformation maps a "circle" in $\mathbb{C}^*$ to a "circle" in $\mathbb{C}^*$. Here a "circle" means either a straight line or a circle.

    Hint: See Flanigan (1983), and Spiegel (1964). In Moebius geometry straight lines are circles of very large (infinite) radius. Therefore, in a Moebius transformation, it is possible for a circle to be transformed into either a circle, or a straight line. It is also possible for a straight line to be transformed into either a straight line or a circle. The equation of a circle is of the form

    $$az\bar{z} + bz + \bar{b}\bar{z} + c = 0, \quad \text{where } a, c \in \mathbb{R}, \ a \neq 0, \ b \in \mathbb{C}, \ |b|^2 > ac$$

    If $a = 0$, we have a straight line.

    Consider the inversion transformation $w = 1/z$ or $z = 1/w$. The above equation becomes $cw\bar{w} + \bar{b}w + b\bar{w} + a = 0$, which is the equation of a circle in the $w$-plane.

    Consider the dilation transformation $w = pz$ or $z = w/p$, where $p \neq 0$. The above equation becomes $aw\bar{w} + (b\bar{p})w + (\bar{b}p)\bar{w} + cp\bar{p} = 0$, which is also the equation of a circle.

    It can similarly be shown that under the translation transformation, circles are transformed into circles.

    Finally, it has been shown in an earlier problem that the Moebius transformation is a composition of translations, dilations, and the inversion. Thus the stated result follows.

12. Let $z_1, z_2$, and $z_3$ be three distinct points in $\mathbb{C}^*$. Also let $w_1, w_2$, and $w_3$ be three distinct points in $\mathbb{C}^*$. Prove that there exists a unique Moebius transformation $f(\cdot)$ such that $f(z_k) = w_k$ for $k = 1, 2$, and $3$. This result is often called the fundamental theorem of Moebius transformation.

    Hint: See Conway (1978), Flanigan (1983), and Henle (1997). Uniqueness is first established, and then existence.

    Three distinct points in a plane uniquely determine a circle. Let the three points $z_1, z_2$, and $z_3$ lie on the circle $C_z$. Similarly, let $w_1, w_2$, and $w_3$ lie on the circle $C_w$. Therefore the mapping $f(z_k) = w_k$ for $k = 1, 2$, and $3$ exists, and is unique.

    An alternate way to prove uniqueness is as follows. Let the Moebius transformations $f(\cdot)$ and $f_1(\cdot)$ have the given property. Then $f^{-1}(f_1(z_k)) = z_k$ for $k = 1, 2$, and $3$. That is, the composition $f^{-1} \circ f_1(\cdot)$ has three fixed points. Therefore, it must be the identity transformation. This implies $f_1(\cdot) = f(\cdot)$.

    Existence is established as follows.

    *Step* 1: Initially assume that there is a Moebius transformation $f_1(\cdot)$ that maps the distinct points $z_1, z_2$, and $z_3$ to $1, 0, \infty$ respectively.

    *Step* 2: Using the assumption in Step 1, we can construct a Moebius transformation $f_2(\cdot)$ that maps the distinct points $w_1, w_2$, and $w_3$ to $1, 0, \infty$ respectively.

*Step* 3: From Steps 1 and 2, we have the composition transformation $f\left(\cdot\right) = f_2^{-1}\left(f_1\left(\cdot\right)\right)$ that satisfies $f\left(z_k\right) = w_k$ for $k = 1, 2$, and 3.

*Step* 4: We need to justify the assumption made in Step 1. Let $z_1, z_2, z_3 \in \mathbb{C}^*$, and define $f_1 : \mathbb{C}^* \to \mathbb{C}^*$ as

$$f_1\left(z\right) = \frac{\left(z_1 - z_3\right)\left(z - z_2\right)}{\left(z_1 - z_2\right)\left(z - z_3\right)}, \quad \text{if } z_1, z_2, z_3 \in \mathbb{C}$$

$$f_1\left(z\right) = \frac{\left(z - z_2\right)}{\left(z - z_3\right)}, \quad \text{if } z_1 = \infty$$

$$f_1\left(z\right) = \frac{\left(z_1 - z_3\right)}{\left(z - z_3\right)}, \quad \text{if } z_2 = \infty$$

$$f_1\left(z\right) = \frac{\left(z - z_2\right)}{\left(z_1 - z_2\right)}, \quad \text{if } z_3 = \infty$$

In the above definition, $f_1\left(z_1\right) = 1$, $f_1\left(z_2\right) = 0$, and $f_1\left(z_3\right) = \infty$.

13. Let $z_1, z_2$, and $z_3$ be three distinct points in $\mathbb{C}^*$. Also let $w_1, w_2$, and $w_3$ be three distinct points in $\mathbb{C}^*$. Show that the Moebius transformation $w = f\left(z\right)$ such that $f\left(z_k\right) = w_k$ for $k = 1, 2$, and 3 is determined by

$$\frac{\left(w_1 - w_3\right)\left(w - w_2\right)}{\left(w_1 - w_2\right)\left(w - w_3\right)} = \frac{\left(z_1 - z_3\right)\left(z - z_2\right)}{\left(z_1 - z_2\right)\left(z - z_3\right)}$$

Hint: See Spiegel (1964). Let

$$w = f\left(z\right) = \frac{az + b}{cz + d}$$

Then

$$w - w_k = \frac{az + b}{cz + d} - \frac{az_k + b}{cz_k + d} = \frac{\left(ad - bc\right)\left(z - z_k\right)}{\left(cz + d\right)\left(cz_k + d\right)}, \quad \text{for } k = 1, 2, 3$$

Use of the last equation and simplification of $\left(w_1 - w_3\right)\left(w - w_2\right) / \left\{\left(w_1 - w_2\right)\left(w - w_3\right)\right\}$ leads to the stated result.

14. Prove that the Moebius transformation $f\left(\cdot\right)$ preserves cross ratio. That is,

$$\left(z_0, z_1, z_2, z_3\right) = \left(f\left(z_0\right), f\left(z_1\right), f\left(z_2\right), f\left(z_3\right)\right)$$

Hint: See Henle (1997). This problem is a restatement of the last problem. Nevertheless, an alternate proof is provided. Let

$$g\left(z\right) = \left(z, z_1, z_2, z_3\right) = \frac{\left(z - z_2\right)\left(z_1 - z_3\right)}{\left(z - z_3\right)\left(z_1 - z_2\right)}$$

Therefore $g\left(z_1\right) = 1$, $g\left(z_2\right) = 0$, and $g\left(z_3\right) = \infty$. Thus, the transformation $g \circ f^{-1}\left(\cdot\right)$ maps: $f\left(z_1\right)$ to 1, maps $f\left(z_2\right)$ to 0, and maps $f\left(z_3\right)$ to $\infty$.

However, the transformation $f\left(z_1\right)$ to 1, $f\left(z_2\right)$ to 0, and $f\left(z_3\right)$ to $\infty$ is also given by the cross ratio $\left(z, f\left(z_1\right), f\left(z_2\right), f\left(z_3\right)\right)$. Furthermore, this transformation is unique by the fundamental theorem of Moebius transformation. This implies

$$g\left(f^{-1}\left(z\right)\right) = \left(z, f\left(z_1\right), f\left(z_2\right), f\left(z_3\right)\right)$$

Consequently,

$$\left(z_0, z_1, z_2, z_3\right) = g\left(z_0\right) = \left(f\left(z_0\right), f\left(z_1\right), f\left(z_2\right), f\left(z_3\right)\right)$$

15. Prove that the cross ratio $(z_0, z_1, z_2, z_3)$ is real if and only if the points $z_0, z_1, z_2, z_3$ lie on a Euclidean circle or a straight line.

    Hint: See Henle (1997). Let the Moebius transformation

    $$f(z) = \frac{az + b}{cz + d} = (z, z_1, z_2, z_3) = \frac{(z - z_2)(z_1 - z_3)}{(z - z_3)(z_1 - z_2)}$$

    The cross ratio is real if

    $$\frac{az + b}{cz + d} = \frac{a\bar{z} + \bar{b}}{\bar{c}\bar{z} + \bar{d}}$$

    Simplification yields

    $$(a\bar{c} - c\bar{a})\, z\bar{z} + (a\bar{d} - c\bar{b})\, z + (b\bar{c} - d\bar{a})\, \bar{z} + (b\bar{d} - d\bar{b}) = 0$$

    The above equation describes a Euclidean circle, if $(a\bar{c} - c\bar{a}) \neq 0$. However, if $(a\bar{c} - c\bar{a}) = 0$, then it is the equation of a straight line. The result follows.

16. Establish the following results about conjugate-Moebius transformations.

    (a) The composition of Moebius and conjugate-Moebius transformations in either order yields a conjugate-Moebius transformation. Further, the composition of two conjugate-Moebius transformations is a Moebius transformation.

    (b) Conjugate-Moebius transformations map "circles" to "circles" and are also conformal.

    Hint: See Pressley (2010).

    (a) Let $g(\cdot)$ and $h(\cdot)$ be Moebius transformations, and $c_{conj}(\cdot)$ be the complex conjugation transformation. Therefore $g \circ c_{conj}(\cdot)$ and $h \circ c_{conj}(\cdot)$ are conjugate-Moebius transformations.

    Note that $c_{conj} \circ g \circ c_{conj}(\cdot)$ is a Moebius transformation, and $c_{conj} \circ c_{conj}(\cdot)$ is the identity map.

    (i) Consider the composition of Moebius and conjugate-Moebius transformations

    $$g \circ (h \circ c_{conj}(\cdot)) = (g \circ h) \circ c_{conj}(\cdot)$$

    The right-hand side of the above equation is a conjugate-Moebius transformation, as $g \circ h(\cdot)$ is a Moebius transformation.

    (ii) Consider the composition of conjugate-Moebius and Moebius transformations

    $$(h \circ c_{conj}) \circ g(\cdot) = h \circ (c_{conj} \circ g \circ c_{conj}) \circ c_{conj}(\cdot)$$

    The right-hand side of the above equation is a conjugate-Moebius transformation, as $c_{conj} \circ g \circ c_{conj}(\cdot)$ and $h \circ (c_{conj} \circ g \circ c_{conj})(\cdot)$ are each Moebius transformations.

    (iii) Consider the composition of two conjugate-Moebius transformations

    $$(g \circ c_{conj}) \circ (h \circ c_{conj})(\cdot) = g \circ (c_{conj} \circ h \circ c_{conj})(\cdot)$$

    The right-hand side of the above equation is a Moebius transformation.

    (b) This result follows from the following facts:

    (i) Moebius transformation maps "circles" to "circles" and are also conformal.

    (ii) The conjugate map $c_{conj}(\cdot)$ is conformal, a reflection in the real axis, and maps lines to lines, and circles to circles.

17. Establish the following results about unitary Moebius transformations.

    (a) Composition of two unitary Moebius transformations is also unitary.

    (b) Inverse of unitary Moebius transformation is also unitary.

    Hint: See Pressley (2010).

    (a) Consider the unitary Moebius transformations

    $$f(z) = \frac{az+b}{-\overline{b}z+\overline{a}}, \quad \left(|a|^2 + |b|^2\right) \neq 0,$$

    $$g(z) = \frac{cz+d}{-\overline{d}z+\overline{c}}, \quad \left(|c|^2 + |d|^2\right) \neq 0$$

    Their composition

    $$f \circ g(z) = f(g(z)) = \frac{ag(z)+b}{-\overline{b}g(z)+\overline{a}} = \frac{\left(ac-b\overline{d}\right)z + (ad+b\overline{c})}{-\left(\overline{a}\overline{d}+\overline{b}c\right)z + \left(\overline{a}\overline{c}-\overline{b}d\right)}$$

    Also

    $$\left|ac-b\overline{d}\right|^2 + |ad+b\overline{c}|^2 = \left(|a|^2+|b|^2\right)\left(|c|^2+|d|^2\right) \neq 0$$

    (b) The inverse of $f(z)$ is

    $$f^{-1}(z) = \frac{\overline{a}z - b}{\overline{b}z + a}$$

    which is clearly unitary.

18. Prove that a polynomial $f(x, y)$ in two indeterminates $x$ and $y$ is homogeneous if and only if

    $$f(tx, ty) = t^h f(x, y), \text{ for any } t \text{ other than } 0 \text{ and } 1$$

    where $h$ is the degree of the polynomial.

19. Prove that the polynomials $f(x)$ and $g(x)$ of degrees $m$ and $n$ respectively, have a nonconstant common factor if and only if there exist nonzero polynomials $f_1(x)$ and $g_1(x)$ of degrees $(m-1)$ and $(n-1)$ respectively, such that $f(x)g_1(x) = f_1(x)g(x)$.

    Hint: See Walker (1962). Let the polynomials $f(x)$ and $g(x)$ have a nonconstant common factor $h(x)$. This implies $f(x) = h(x)f_1(x), g(x) = h(x)g_1(x)$, and $f(x)g_1(x) = f_1(x)g(x)$.

    Conversely, assume that $f(x)g_1(x) = f_1(x)g(x)$, and factorize $f(x)$ into its irreducible factors. The nonconstant factors of $f(x)$ will be among the nonconstant factors of $f_1(x)g(x)$. All these nonconstant factors cannot appear among the factors of $f_1(x)$ as $\deg f_1(x) < \deg f(x)$. Therefore one of these factors of $f(x)$ is indeed equal to a nonconstant factor of $g(x)$. That is, $f(x)$ and $g(x)$ have a nonconstant common factor.

20. Establish the quasi-homogeneity property of the resultant of two polynomials.

    Hint: See Gelfand, Kapranov, and Zelevinsky (1994).

21. Let $f(x) = (x-a)$, and $g(x) = \sum_{j=0}^{n} b_j x^j$, where $\deg g(x) = n \geq 1$. The coefficients of these polynomials are complex numbers. Prove that $R(f, g) = g(a)$.

22. Let $f(x), g(x) \in \mathbb{C}[x]$, where the degree of these polynomials are $m$ and $n$ respectively. Assume that these polynomials are nonconstant. Prove that there exist polynomials $s(x), t(x) \in \mathbb{C}[x]$ such that

    $$\{s(x)f(x) + t(x)g(x)\} = R(f, g)$$

Hint: See Mostowski, and Stark (1964). Multiply the first column of the determinant expression for $R(f, g)$ by $x^{n+m-1}$ and add it to the last column. Next multiply the second column of this determinant by $x^{n+m-2}$ and add it to the last column, and so on. Note that the penultimate column is multiplied by simply $x$ and added to the last column. The elements of the last column are $f(x)x^{n-1}, f(x)x^{n-2}, \ldots, f(x)x, f(x), g(x)x^{m-1}, g(x)x^{m-2}, \ldots, g(x)x,$ and $g(x)$. Expand the determinant by elements of the last column. This yields the required result.

Note that if the polynomials $f(x)$ and $g(x)$ have a common factor, then $R(f, g) = 0$. However, if there is no common factor, the stated result can be expressed as

$$\{\widetilde{s}(x)f(x) + \widetilde{t}(x)g(x)\} = 1$$

23. Let $f(x) \in \mathbb{C}[x]$, $f(x) = \sum_{i=0}^{m} a_i x^i$, $\deg f(x) = m \geq 2$, and $f'(x)$ be the derivative of $f(x)$ with respect to $x$. Prove that $R(f, f') = (-1)^{m(m-1)/2} a_m \Delta(f)$.
Hint: Let $f(x) = a_m \prod_{i=1}^{m} (x - \alpha_i)$, then

$$f'(x) = a_m \sum_{j=1}^{m} \prod_{i=1, i \neq j}^{m} (x - \alpha_i)$$

Therefore $f'(\alpha_j) = a_m \prod_{i=1, i \neq j}^{m} (\alpha_j - \alpha_i)$. Consequently

$$R(f, f') = a_m^{m-1} \prod_{j=1}^{m} f'(\alpha_j) = a_m^{2m-1} \prod_{\substack{1 \leq i, j \leq m \\ i \neq j}} (\alpha_j - \alpha_i)$$

$$= (-1)^{m(m-1)/2} a_m \Delta(f)$$

24. Let $f(x, y; t) = \sum_{i=0}^{m} a_{m-i}(x, y)t^i$ and $g(x, y; t) = \sum_{j=0}^{n} b_{n-j}(x, y)t^j$ be polynomials with real coefficients, where $m, n \in \mathbb{P}$. Also $a_i(x, y)$ is either equal to zero or a homogeneous polynomial of degree $i$ in variables $x$ and $y$, where $0 \leq i \leq m$. Similarly, $b_j(x, y)$ is either equal to zero or a homogeneous polynomial of degree $j$ in variables $x$ and $y$, where $0 \leq j \leq n$. Let $a_r(x, y)$ and $b_s(x, y)$ be the first nonvanishing coefficients of the polynomials $f(x, y; t)$ and $g(x, y; t)$ respectively, where $0 \leq r \leq m$ and $0 \leq s \leq n$. Prove that the resultant $R(f, g)(x, y)$ with respect to $t$ is a homogeneous polynomial in $x$ and $y$ of degree $(mn - rs)$.
Hint: See Mostowski, and Stark (1964). Observe that

$$f(x, y; t) = \sum_{i=0}^{m-r} a_{m-i}(x, y)t^i, \quad \text{and} \quad g(x, y; t) = \sum_{j=0}^{n-s} b_{n-j}(x, y)t^j$$

Next consider the roots of the polynomial $f(x, y; t)$ in $t$. Let these be $\alpha_i, 1 \leq i \leq (m - r)$. Similarly let the roots of the polynomial $g(x, y; t)$ in $t$, be $\beta_j, 1 \leq j \leq (n - s)$. The resultant $R(f, g)(x, y)$ of the two polynomials in $t$ is

$$R(f, g)(x, y) = \{a_r(x, y)\}^{n-s} \{b_s(x, y)\}^{m-r} \prod_{i=1}^{m-r} \prod_{j=1}^{n-s} (\alpha_i - \beta_j)$$

Let $c$ be a nonzero number. Denote $f(cx, cy; t)$ and $g(cx, cy; t)$, by $\widehat{f}(x, y; t)$ and $\widehat{g}(x, y; t)$ respectively. Because of the homogeneity property of the $a_i(x, y)$'s and $b_j(x, y)$'s we have

$$c^{-m} \widehat{f}(x, y; t) = \sum_{i=0}^{m-r} a_{m-i}(x, y)(t/c)^i, \quad \text{and} \quad c^{-n} \widehat{g}(x, y; t) = \sum_{j=0}^{n-s} b_{n-j}(x, y)(t/c)^j$$

Therefore the roots of the polynomial $\widehat{f}(x, y; t)$ in $t$ are $c\alpha_i, 1 \le i \le (m - r)$. Similarly, the roots of the polynomial $\widehat{g}(x, y; t)$ in $t$ are $c\beta_j, 1 \le j \le (n - s)$. The resultant of the two polynomials $\widehat{f}(x, y; t)$ and $\widehat{g}(x, y; t)$ in $t$ is

$$R\left(\widehat{f}, \widehat{g}\right)(x, y)$$

$$= \{c^r a_r(x, y)\}^{n-s} \{c^s b_s(x, y)\}^{m-r} c^{(m-r)(n-s)} \prod_{i=1}^{m-r} \prod_{j=1}^{n-s} (\alpha_i - \beta_j)$$

$$= c^{mn-rs} R(f, g)(x, y)$$

Thus $R(f, g)(x, y)$ is a homogeneous polynomial in $x$ and $y$, and it is of degree $(mn - rs)$.

25. Establish Stirling's approximation of the factorial function.

$$n! = \sqrt{2\pi n} \left(\frac{n}{e}\right)^n \exp\left\{\frac{\theta}{12n}\right\}$$

where $0 < \theta < 1$ and $\theta$ depends on $n$. Therefore

$$n! = \sqrt{2\pi n} \left(\frac{n}{e}\right)^n \{1 + o(1)\}$$

Thus $n! \sim \sqrt{2\pi} n^{n+1/2} e^{-n}$. Note that this approximation gives $1! \approx 0.92214$ and $2! \approx 1.919$. Hint: See Palmer (1985). Obtain upper and lower bounds for $\ln n!$ by approximating the integral $\int_1^n \ln x \, dx$ with circumscribed and inscribed trapezoids. This yields

$$n! = d\sqrt{n} \left(\frac{n}{e}\right)^n \{1 + o(1)\}$$

for some constant $d > 0$. Use of the expression for $\int_0^{\pi/2} \sin^n x \, dx$ yields:

$$\frac{\pi}{2} = \lim_{n \to \infty} \frac{2^{4n} (n!)^4}{\{(2n)!\}^2 (2n + 1)}$$

This formula is generally attributed to John Wallis (1616-1703). The constant $d$ can be determined by substituting the approximation for $n!$ in the above equation.

26. Prove that for any $n \in \mathbb{P}$, and $1 \le k \le n$.

$$\binom{n}{k} \le \frac{n^k}{k!} \le \left(\frac{en}{k}\right)^k$$

Hint: See Palmer (1985).

$$\binom{n}{k} = \frac{(n)_k}{k!} = \frac{n^k}{k!} \frac{(n)_k}{n^k} \le \frac{n^k}{k!} \le \left(\frac{en}{k}\right)^k$$

27. Prove that for any $n \in \mathbb{P}$, and $1 \leq k \leq n$.

$$\frac{(n)_k}{n^k} \leq \exp\left\{-(k-1)\sum_{m \in \mathbb{P}} \frac{1}{m(m+1)}\left(\frac{k-1}{n}\right)^m\right\}$$

$$\frac{(n)_k}{n^k} \sim \exp\left\{-\frac{k^2}{2n} - \frac{k^3}{6n^2}\right\}, \quad k = o\left(n^{3/4}\right)$$

$$\frac{(n)_k}{n^k} = 1 + o(1), \quad k = o\left(n^{1/2}\right)$$

$$\frac{(n)_k}{n^k} = O(1)\exp\left\{-\frac{k^2}{2n} - \frac{k^3}{6n^2}\right\}, \quad \text{any } k$$

Hint: See Palmer (1985). The first inequality is established as follows. Note that

$$\frac{(n)_k}{n^k} = \left(1 - \frac{1}{n}\right)\left(1 - \frac{2}{n}\right)\cdots\left(1 - \frac{k-1}{n}\right)$$

Observe that $\ln(1-x) = -\sum_{m \in \mathbb{P}} x^m/m$. Thus

$$\frac{(n)_k}{n^k} = \exp\left\{-\sum_{i=1}^{k-1}\sum_{m \in \mathbb{P}} \frac{(i/n)^m}{m}\right\}$$

By using the trapezoidal rule of integration, it can be shown that

$$\sum_{i=1}^{k-1} i^m \geq \frac{(k-1)^{m+1}}{(m+1)}$$

Thus

$$\sum_{i=1}^{k-1}\sum_{m \in \mathbb{P}} \frac{(i/n)^m}{m} = \sum_{m \in \mathbb{P}} \frac{1}{mn^m}\sum_{i=1}^{k-1} i^m \geq \sum_{m \in \mathbb{P}} \frac{(k-1)^{m+1}}{m(m+1)n^m}$$

The second inequality follows by noting that $\sum_{i=1}^{k-1} i^m = O\left(k^{m+1}\right)$.

28. Let $a \in \mathbb{R}_0^+$, $b \in (1, \infty)$, $c \in \mathbb{R}^+$, $d \in (1, \infty)$; and $n \in \mathbb{P}$ tends towards infinity. Show that

$$O(a) \subseteq O(\log_b n) \subseteq O(n^c) \subseteq O(d^n) \subseteq O(n!) \subseteq O(n^n)$$

Hint: See Rich (2008).

29. Let $a, b, c, d, e, f \in \mathbb{R}^3$. Show that

$$[a\ b\ c]\,[d\ e\ f] = \det A$$

where

$$A = \begin{bmatrix} a \circ d & b \circ d & c \circ d \\ a \circ e & b \circ e & c \circ e \\ a \circ f & b \circ f & c \circ f \end{bmatrix}$$

Hint: See Lipschutz (1969).

30. Let $\mathcal{V}_X = (V_X, \mathcal{F}_X, +, \times)$ be a vector space over a field $\mathcal{F}_X = (F_X, +, \times)$, and $(V_X, d_X)$ be a metric space. Similarly, let $\mathcal{V}_Y = (V_Y, \mathcal{F}_Y, +, \times)$ be a vector space over a field $\mathcal{F}_Y = (F_Y, +, \times)$, and $(V_Y, d_Y)$ be a metric space. A mapping $T : V_X \to V_Y$ is continuous. Let $A \subseteq V_X$ be a compact set, then prove that its image $T(A) \subseteq V_Y$ is also compact.

    Hint: See Milne (1980). Let $y_k = T(x_k)$ be a sequence in $T(A) \subseteq V_Y$. It is hypothesized that $A \subseteq V_X$ is a compact set, then there is a subsequence $x_{k,p}$ of $x_k$ which converges to a limit $x \in A$. As $T(\cdot)$ is a continuous mapping, the subsequence $y_{k,p} = T(x_{k,p})$ of the sequence $y_k$ converges to $y = T(x) \in T(A)$. Hence $T(A)$ is compact.

31. Let $u, v \in \mathbb{R}^n$. Prove the famous Minkowski's inequality.

$$\|u + v\|_2 \leq \|u\|_2 + \|v\|_2$$

    Hint: This inequality is named after Hermann Minkowski (1864-1909). To get insight into this problem, first establish the above result in the two-dimensional space $\mathbb{R}^2$.

32. Establish the Bunyakovsky-Cauchy-Schwartz inequality, $|\langle u, v \rangle| \leq \|u\| \cdot \|v\|$, where $u, v \in V$.

    Hint: Let $(\mathcal{V}, \langle \cdot, \cdot \rangle)$ be an inner product vector space. If $v = 0$, then the inequality is valid since $\langle u, 0 \rangle = 0$. Let $v \neq 0$. For all $\alpha \in \mathbb{C}$

$$0 \leq \|u - \alpha v\|^2 = \langle u - \alpha v, u - \alpha v \rangle$$
$$= \langle u, u \rangle - \overline{\alpha} \langle u, v \rangle - \alpha \left[ \langle v, u \rangle - \overline{\alpha} \langle v, v \rangle \right]$$

    The expression inside the $[\cdots]$ brackets is zero if $\overline{\alpha} = \langle v, u \rangle / \langle v, v \rangle$. Consequently

$$0 \leq \langle u, u \rangle - \frac{\langle v, u \rangle}{\langle v, v \rangle} \langle u, v \rangle = \|u\|^2 - \frac{|\langle u, v \rangle|^2}{\|v\|^2}$$

    where the equality $\langle v, u \rangle = \overline{\langle u, v \rangle}$ is used. The result follows. In the above derivation, equality holds if $u = \alpha v$, which shows a linear dependence.

33. Let $(\mathcal{V}, \langle \cdot, \cdot \rangle)$ be an inner product vector space. Prove that the inner product $\langle \cdot, \cdot \rangle$ defined on the vector space $\mathcal{V}$ induces a norm $\|\cdot\|$ on $\mathcal{V}$. This norm is given by the relationship $\|u\| = \sqrt{\langle u, u \rangle}$ for all $u \in V$.

    Hint: Essentially, it needs to be established that the induced norm $\|\cdot\|$ satisfies the norm-defining properties, $[N1]$ through $N[3]$. The property $[N1]$ follows from $[I1]$. Furthermore, $\|au\|^2 = \langle au, au \rangle = a \langle u, au \rangle = a\overline{a} \langle u, u \rangle = |a|^2 \langle u, u \rangle$. After taking the square root of both sides, $[N2]$ is obtained. The property $[N3]$ follows directly from the Bunyakovsky-Cauchy-Schwartz inequality. As

$$\|u + v\|^2 = \langle u + v, u + v \rangle = \langle u, u \rangle + \langle u, v \rangle + \langle v, u \rangle + \langle v, v \rangle$$
$$= \|u\|^2 + 2\operatorname{Re}(\langle u, v \rangle) + \|v\|^2$$
$$\leq \|u\|^2 + 2|\langle u, v \rangle| + \|v\|^2$$
$$\leq \|u\|^2 + 2\|u\|\|v\| + \|v\|^2 = (\|u\| + \|v\|)^2$$

    Thus $\|u + v\| \leq \|u\| + \|v\|$. This is the triangle inequality.

34. Let $(\mathcal{V}, \langle \cdot, \cdot \rangle)$ be an inner product space. Also let $u, v \in V$. Then the *polar forms* of $\langle u, v \rangle$ are:

    (a) $\langle u, v \rangle = \frac{1}{4} \|u + v\|^2 - \frac{1}{4} \|u - v\|^2$ over field $\mathbb{R}$.

    (b) $\langle u, v \rangle = \frac{1}{4} \|u + v\|^2 - \frac{1}{4} \|u - v\|^2 + \frac{i}{4} \|u + iv\|^2 - \frac{i}{4} \|u - iv\|^2$ over field $\mathbb{C}$.

    Prove these statements.

Hint: See Bachman, and Narici (2000).

(a) $\|u + v\|^2 = \langle u, u \rangle + \langle u, v \rangle + \langle v, u \rangle + \langle v, v \rangle$, and $\|u - v\|^2 = \langle u, u \rangle - \langle u, v \rangle - \langle v, u \rangle + \langle v, v \rangle$. Subtracting the second equation from the first, and noting that the field is $\mathbb{R}$ yields the result.

(b) Expand the right-hand side as in part (a) of the problem.

35. Prove the generalized Pythogorean theorem.
36. Establish Bessel's inequality.
37. Prove the observations on projections.

(a) The mapping $E$ is linear.
(b) $E^2 = E$.

Hint: See Lipschutz (1968).

(a) The sum $v = \sum_{i=1}^{n} u_i$ where $v \in V$ and $u_i \in U_i$ is well-defined. Also let $v' = \sum_{i=1}^{n} u_i'$ where $v' \in V$ and $u_i' \in U_i$. Then $(v + v') = \sum_{i=1}^{n} (u_i + u_i')$ and $kv = \sum_{i=1}^{n} ku_i$ where $(v + v')$, $kv \in V$ and $(u_i + u_i')$, $ku_i \in U_i$. Thus $E(v + v') = u_j + u_j' = E(v) + E(v')$ and $E(kv) = ku_j = kE(v)$. This implies that the mapping $E$ is linear.

(b) Note that $E(v) = u_j$. Also $E^2(v) = E(E(v)) = E(u_j) = u_j = E(v)$.

38. A working definition of the Dirac's delta function $\delta(\cdot)$ is provided in this problem. Prove that

$$\delta(x) = \lim_{\epsilon \to 0} \frac{\epsilon}{\pi(x^2 + \epsilon^2)}$$

Hint: Prove that

$$\int_{-\infty}^{\infty} \delta(x)\,dx = 1$$

39. Prove that

$$\frac{1}{1^2} + \frac{1}{2^2} + \frac{1}{3^2} + \cdots = \frac{\pi^2}{6}$$

$$\frac{1}{1^4} + \frac{1}{2^4} + \frac{1}{3^4} + \cdots = \frac{\pi^4}{90}$$

Hint: See Hsu (1984). Let $f(\cdot)$ be a periodic function with period $2\pi$, where $f(t) = t^2$, for $-\pi \le t \le \pi$. This function can be expanded using the trigonometric Fourier series. The coefficients of this series are

$$a_0 = \frac{2\pi^2}{3}$$

$$a_n = (-1)^n \frac{4}{n^2}, \quad b_n = 0; \quad \forall\, n \in \mathbb{P}$$

Therefore

$$f(t) = \frac{\pi^2}{3} + 4 \sum_{n \in \mathbb{P}} (-1)^n \frac{\cos nt}{n^2}$$

Substituting $t = \pi$ yields the first result. The second result is obtained by using the Parseval's relationship.

40. Establish the following Fourier transform pairs.

(a)

$$\int_{-\infty}^{t} f(\tau)\,d\tau \leftrightarrow \frac{F(\omega)}{i\omega} + \pi F(0)\,\delta(\omega)$$

(b) Let $\delta_{T_0}(t) = \sum_{n \in \mathbb{Z}} \delta(t - nT_0)$ and $\delta_{w_0}(\omega) = \sum_{n \in \mathbb{Z}} \delta(\omega - n\omega_0)$, where $\omega_0 = 2\pi/T_0$, then $\delta_{T_0}(t) \leftrightarrow \omega_0 \delta_{w_0}(\omega)$.

Hints: See Hsu (1984).

(a) $\int_{-\infty}^{t} f(\tau) d\tau = f(t) * u(t)$, where $*$ is the convolution operator, and $u(\cdot)$ is the unit step function. Then

$$\int_{-\infty}^{t} f(\tau) d\tau \leftrightarrow F(\omega) U(\omega)$$

where

$$U(\omega) = \pi \delta(\omega) + \frac{1}{i\omega}$$

The result follows.

(b) The Fourier series expansion of $\delta_{T_0}(t) = \frac{1}{T_0} \sum_{n \in \mathbb{Z}} e^{in\omega_0 t}$. Therefore

$$\delta_{T_0}(t) \leftrightarrow \frac{1}{T_0} \sum_{n \in \mathbb{Z}} 2\pi \delta(\omega - n\omega_0) = \omega_0 \delta_{w_0}(\omega)$$

41. Establish the following Fourier transform pairs. In all these functions $\alpha, t \in \mathbb{R}$.

(a) Let $\alpha > 0$ then

$$\frac{1}{\sqrt{2\pi}\alpha} e^{-\frac{t^2}{2\alpha^2}} \leftrightarrow e^{-\frac{\alpha^2 \omega^2}{2}}$$

(b) Let $\alpha > 0$ then

$$e^{-\alpha t} u(t) \leftrightarrow (\alpha + i\omega)^{-1}$$

(c) Let $\alpha > 0$ then

$$e^{-\alpha|t|} \leftrightarrow 2\alpha (\alpha^2 + \omega^2)^{-1}$$

(d) Let $\alpha > 0$ then

$$\frac{1}{(\alpha^2 + t^2)} \leftrightarrow \frac{\pi}{\alpha} e^{-\alpha|\omega|}$$

(e) Let

$$g(t) = \begin{cases} \dfrac{t^{n-1}}{(n-1)!}, & t \geq 0 \\ 0, & \text{otherwise} \end{cases} \quad , \quad \forall\, n \in \mathbb{P}$$

then

$$g(t) \leftrightarrow \frac{1}{(i\omega)^n}$$

(f) Let $f(t), t \in \mathbb{R}$ be the probability density function of a continuously distributed random variable. A useful technique for computing the $n$th moment of the random variable is via the use of the Fourier transform. If

$$m_n(t) \triangleq \int_{-\infty}^{\infty} t^n f(t) dt$$

show that

$$m_n(t) = i^n \frac{d^n}{d\omega^n} F(\omega) \Big|_{\omega=0}, \quad \forall\, n \in \mathbb{N}$$

where $f(t) \leftrightarrow F(\omega)$.

42. Let $T_0$ and $\tau$ be positive real numbers. Also let $\omega_0 = 2\pi/T_0$, $\Omega_0 = 2\pi/\tau$, $t \in \mathbb{R}$ and $f(\cdot)$ be an arbitrary function such that $f(t) \leftrightarrow F(\omega)$. Define

$$f_s(t) = \sum_{n \in \mathbb{Z}} f(t + nT_0), \quad t \in \mathbb{R}, \quad \text{and} \quad F_s(\omega) = \sum_{n \in \mathbb{Z}} F(\omega + n\Omega_0), \quad \omega \in \mathbb{R}$$

Note that $f_s(t)$ and $F_s(\omega)$ are not a Fourier transform pair. Establish Poisson's summation formulae.

(a) $f_s(t) = \frac{1}{T_0} \sum_{n \in \mathbb{Z}} F(n\omega_0) e^{in\omega_0 t}$

(b) $F_s(\omega) = \tau \sum_{n \in \mathbb{Z}} f(n\tau) e^{-in\omega\tau}$

Hint: See Hsu (1984).

(a) It is evident that

$$f_s(t) = f(t) * \delta_{T_0}(t), \quad \delta_{T_0}(t) = \sum_{n \in \mathbb{Z}} \delta(t - nT_0)$$

It is known that

$$\delta_{T_0}(t) \leftrightarrow \omega_0 \delta_{\omega_0}(\omega), \quad \delta_{\omega_0}(\omega) = \sum_{n \in \mathbb{Z}} \delta(\omega - n\omega_0)$$

Thus

$$\Im(f_s(t)) = F(\omega) \omega_0 \delta_{\omega_0}(\omega)$$
$$= \omega_0 \sum_{n \in \mathbb{Z}} F(n\omega_0) \delta(\omega - n\omega_0)$$

Since $e^{in\omega_0 t} \leftrightarrow 2\pi\delta(\omega - n\omega_0)$, $\forall \, n \in \mathbb{Z}$, the result follows by taking the inverse Fourier transform of both sides.

(b) Observe that $F_s(\omega) = F(\omega) * \delta_{\Omega_0}(\omega)$, where

$$\delta_{\Omega_0}(\omega) = \sum_{n \in \mathbb{Z}} \delta(\omega - n\Omega_0)$$

Since $\delta_\tau(t) \leftrightarrow \Omega_0 \delta_{\Omega_0}(\omega)$, taking the Fourier inverse of $F_s(\omega)$ yields

$$\Im^{-1}(F_s(\omega)) = \frac{2\pi}{\Omega_0} f(t) \delta_\tau(t)$$
$$= \tau \sum_{n \in \mathbb{Z}} f(n\tau) \delta(t - n\tau)$$

Since $\delta(t - n\tau) \leftrightarrow e^{-in\omega\tau}$, $\forall \, n \in \mathbb{Z}$, the result follows by taking the Fourier transform of both sides.

43. If $\zeta(s)$, $\text{Re}(s) > 1$, is the well-known Riemann's zeta function, prove the following results

(a) $\mathcal{M}\left((e^t - 1)^{-1}\right) = \Gamma(s)\zeta(s)$.

(b) $\mathcal{M}\left(2(e^{2t} - 1)^{-1}\right) = 2^{1-s}\Gamma(s)\zeta(s)$.

(c) $\mathcal{M}\left((e^t + 1)^{-1}\right) = \Gamma(s)\zeta(s)(1 - 2^{1-s})$.

Hint: See Debnath, and Bhatta (2007).

(a) Note that $(e^t - 1)^{-1} = \sum_{n \in \mathbb{P}} e^{-nt}$. Therefore

$$\mathcal{M}\left((e^t - 1)^{-1}\right) = \int_0^\infty t^{s-1} \sum_{n \in \mathbb{P}} e^{-nt} dt = \sum_{n \in \mathbb{P}} \int_0^\infty t^{s-1} e^{-nt} dt$$

$$= \sum_{n \in \mathbb{P}} \frac{\Gamma(s)}{n^s} = \Gamma(s)\zeta(s)$$

(b) Note that $2\left(e^{2t} - 1\right)^{-1} = 2\sum_{n \in \mathbb{P}} e^{-2nt}$. Therefore

$$\mathcal{M}\left(2\left(e^{2t} - 1\right)^{-1}\right) = 2\int_0^\infty t^{s-1} \sum_{n \in \mathbb{P}} e^{-2nt} dt = 2\sum_{n \in \mathbb{P}} \int_0^\infty t^{s-1} e^{-2nt} dt$$

$$= 2\sum_{n \in \mathbb{P}} \frac{\Gamma(s)}{(2n)^s} = 2^{1-s}\Gamma(s)\zeta(s)$$

(c) The result follows from parts (a) and (b), and by noting that

$$\frac{1}{e^t - 1} - \frac{1}{e^t + 1} = \frac{2}{e^{2t} - 1}$$

44. Let $f(t) \leftrightarrow F(\omega)$, $t, \omega \in \mathbb{R}$. Prove the following result about Wigner-Ville transform.

$$\widetilde{W}_f(\tau, \omega) = \frac{1}{2\pi} \int_{-\infty}^\infty F\left(\omega + \frac{\xi}{2}\right) \overline{F\left(\omega - \frac{\xi}{2}\right)} e^{i\tau\xi} d\xi, \qquad \tau, \omega \in \mathbb{R}$$

Hint:

$$\widetilde{W}_f(\tau, \omega)$$

$$= \int_{-\infty}^\infty f\left(\tau + \frac{t}{2}\right) \overline{f\left(\tau - \frac{t}{2}\right)} e^{-i\omega t} dt$$

$$= \frac{1}{(2\pi)^2} \int_{-\infty}^\infty \int_{-\infty}^\infty \int_{-\infty}^\infty F(\omega_1) \overline{F(\omega_2)} e^{i(\omega_1 - \omega_2)\tau + i(\omega_1 + \omega_2 - 2\omega)\frac{t}{2}} dt\, d\omega_1\, d\omega_2$$

Note that

$$\frac{1}{2\pi} \int_{-\infty}^\infty e^{it\left(\frac{\omega_1 + \omega_2}{2} - \omega\right)} dt = \delta\left(\frac{\omega_1 + \omega_2}{2} - \omega\right)$$

Therefore

$$\widetilde{W}_f(\tau, \omega)$$

$$= \frac{1}{2\pi} \int_{-\infty}^\infty \int_{-\infty}^\infty F(\omega_1) \overline{F(\omega_2)} e^{i(\omega_1 - \omega_2)\tau} \delta\left(\frac{\omega_1 + \omega_2}{2} - \omega\right) d\omega_1\, d\omega_2$$

In the above equation substitute $(\omega_1 - \omega_2) = \xi$ and $\left(\frac{\omega_1 + \omega_2}{2} - \omega\right) = y$. Thus

$$\widetilde{W}_f(\tau, \omega)$$

$$= \frac{1}{2\pi} \int_{-\infty}^\infty \int_{-\infty}^\infty F\left(\frac{2y + \xi + 2\omega}{2}\right) \overline{F\left(\frac{2y - \xi + 2\omega}{2}\right)} e^{i\tau\xi} \delta(y)\, dy\, d\xi$$

The result follows immediately.

45. Establish the inversion formula of the continuous wavelet transform.

Hint: Let $f(t) \leftrightarrow F(\omega)$, $\psi(t) \leftrightarrow \Psi(\omega)$, and $\psi_{a,b}(t) \leftrightarrow \Psi_{a,b}(\omega)$. Then

$$\Psi_{a,b}(\omega) = \sqrt{|a|}\Psi(a\omega)e^{-ib\omega},$$

$$\psi_{a,b}(t) = \frac{\sqrt{|a|}}{2\pi}\int_{-\infty}^{\infty}\Psi(av)e^{iv(t-b)}dv.$$

Also using a generalized Parseval's type of relationship yields

$$W_f(\psi,a,b) = \frac{\sqrt{|a|}}{2\pi}\int_{-\infty}^{\infty}F(\omega)\overline{\Psi(a\omega)}e^{ib\omega}d\omega$$

Thus

$$\frac{1}{C_\psi}\int_{-\infty}^{\infty}\int_{-\infty}^{\infty}W_f(\psi,a,b)\,\psi_{a,b}(t)\frac{dadb}{a^2}$$

$$= \frac{1}{4\pi^2 C_\psi}\int_{-\infty}^{\infty}\int_{-\infty}^{\infty}\int_{-\infty}^{\infty}\int_{-\infty}^{\infty}F(\omega)\overline{\Psi(a\omega)}\Psi(av)e^{ivt}e^{ib(\omega-v)}d\omega dv\frac{dadb}{|a|}$$

$$= \frac{1}{2\pi C_\psi}\int_{-\infty}^{\infty}\int_{-\infty}^{\infty}\int_{-\infty}^{\infty}F(\omega)\overline{\Psi(a\omega)}\Psi(av)e^{ivt}\delta(\omega-v)\,d\omega dv\frac{da}{|a|}$$

$$= \frac{1}{2\pi C_\psi}\int_{-\infty}^{\infty}\int_{-\infty}^{\infty}F(v)|\Psi(av)|^2 e^{ivt}dv\frac{da}{|a|}$$

$$= \frac{1}{2\pi C_\psi}\int_{-\infty}^{\infty}\frac{|\Psi(x)|^2}{|x|}dx\int_{-\infty}^{\infty}F(v)e^{ivt}dv = f(t)$$

46. Develop a motivation for the following expression for gamma function

$$\Gamma(a) = \lim_{n\to\infty}\frac{n!n^a}{a(a+1)\ldots(a+n)}, \quad a \neq 0, -1, -2, \ldots$$

Hint: Let $a$ and $n$ be integers. Write $(a+n)!$ the following two different ways.

$$(a+n)! = (a+n)(a+n-1)\ldots(a+1)a!$$
$$(a+n)! = (a+n)(a+n-1)\ldots(n+1)n!$$

Therefore for $a \neq -1, -2, \ldots, -n$

$$a! = \frac{n!(n+1)(n+2)\ldots(n+a)}{(a+1)(a+2)\ldots(a+n)}$$

That is, for $a \neq 0, -1, -2, \ldots, -n$

$$(a-1)! = \frac{n!n^a}{a(a+1)\ldots(a+n)} \times \frac{(n+1)(n+2)\ldots(n+a)}{n^a}$$

Therefore for $a \neq 0, -1, -2, \ldots$

$$\Gamma(a) = (a-1)! = \lim_{n\to\infty}\frac{n!n^a}{a(a+1)\ldots(a+n)}$$

47. Prove that

$$\Gamma\left(a\right) = \int_0^\infty t^{a-1} e^{-t} dt = \lim_{n \to \infty} \frac{n!n^a}{a\left(a+1\right)\ldots\left(a+n\right)}, \quad a \neq 0, -1, -2, \ldots$$

Hint: See Wilf (1978). It is known that

$$e^{-t} = \lim_{n \to \infty} \left(1 - \frac{t}{n}\right)^n$$

Define

$$A\left(a,n\right) = \int_0^n t^{a-1} \left(1 - \frac{t}{n}\right)^n dt, \quad n \in \mathbb{P}$$

where $\Gamma\left(a\right) = \lim_{n \to \infty} A\left(a,n\right)$. It can be shown that

$$A\left(a,1\right) = \frac{1}{a\left(a+1\right)}$$

Using integration by parts yields

$$A\left(a,n\right) = \frac{1}{a}\left\{\frac{n}{n-1}\right\}^{a+1} A\left(a+1,n-1\right), \quad n \geq 2$$

Successive application of the above recursion and use of the expression for $A\left(a+n-1,1\right)$ yields

$$A\left(a,n\right) = \frac{n!n^a}{a\left(a+1\right)\ldots\left(a+n\right)}$$

48. If $a \in \mathbb{C}\backslash\mathbb{Z}$, prove that

$$\Gamma\left(a\right)\Gamma\left(1-a\right) = \frac{\pi}{\sin \pi a}$$

Hint: See Wilf (1978). As $a \notin \mathbb{Z}$

$$\Gamma\left(a\right)\Gamma\left(-a\right)$$
$$= -\frac{1}{a^2} \prod_{n=1}^\infty \left\{\left(1+\frac{1}{n}\right)^a \left(1+\frac{a}{n}\right)^{-1}\right\}\left\{\left(1+\frac{1}{n}\right)^{-a} \left(1-\frac{a}{n}\right)^{-1}\right\}$$
$$= -\frac{1}{a^2} \prod_{n=1}^\infty \left(1-\frac{a^2}{n^2}\right)^{-1} = -\frac{1}{a^2}\left\{\frac{\sin \pi a}{\pi a}\right\}^{-1} = -\frac{\pi}{a \sin \pi a}$$

The result is obtained by observing that $\Gamma\left(1-a\right) = -a\Gamma\left(-a\right)$.

49. Prove that the beta function $B\left(x,y\right) = \Gamma\left(x\right)\Gamma\left(y\right)/\Gamma\left(x+y\right)$, for $\mathrm{Re}\left(x\right) > 0$, and $\mathrm{Re}\left(y\right) > 0$.

Hint: We write

$$\Gamma\left(x\right)\Gamma\left(y\right) = \int_0^\infty u^{x-1} e^{-u} du \int_0^\infty v^{y-1} e^{-v} dv$$
$$= \int_0^\infty \int_0^\infty u^{x-1} v^{y-1} e^{-u-v} du dv$$

Substitute $u = wt$, and $v = w\left(1-t\right)$ in the above expression. This yields

$$\Gamma\left(x\right)\Gamma\left(y\right) = \int_0^\infty w^{x+y-1} e^{-w} dw \int_0^1 t^{x-1}\left(1-t\right)^{y-1} dt$$

The final result follows.

# References

1. Abramowitz, M., and Stegun, I. A., 1965. *Handbook of Mathematical Functions*, Dover Publications, Inc., New York, New York.
2. Anderson, J. W., 2005. *Hyperbolic Geometry*, Second Edition, Springer-Verlag, Berlin, Germany.
3. Arora, S., and Barak, B., 2009. *Computational Complexity: A Modern Approach*, Cambridge University Press, Cambridge, Great Britain.
4. Atallah, M. J., and Blanton, M., Editors, 2010. *Algorithms and Theory of Computation Handbook: General Concepts and Techniques*, Second Edition, Chapman and Hall/CRC Press, New York, New York.
5. Atallah, M. J., and Blanton, M., Editors, 2010. *Algorithms and Theory of Computation Handbook: Special Topics and Techniques,* Second Edition, Chapman and Hall/CRC Press, New York, New York.
6. Bachman, G., and Narici, L., 2000. *Functional Analysis*, Dover Publications, Inc., New York, New York.
7. Barbeau, E. J., 1989. *Polynomials*, Springer-Verlag, Berlin, Germany.
8. Briggs, W. L., and Henson, V. E., 1995. *The DFT: An Owner's Manual for the Discrete Fourier Transform*, Society of Industrial and Applied Mathematics, Philadelphia, Pennsylvania.
9. Bronshtein, I. N., and Semendyayev, K. A., 1985. *Handbook of Mathematics*, Van Nostrand Reinhold Company, New York, New York.
10. Conway, J. B., 1978. *Functions of One Complex Variable*, Second Edition, Springer-Verlag, Berlin, Germany.
11. Cormen, T. H., Leiserson, C. E., Rivest, R. L., and Stein, C., 2009. *Introduction to Algorithms*, Third Edition, The MIT Press, Cambridge, Massachusetts.
12. Cox, D., Little J., and O'Shea D., 2007. *Ideals, Varieties, and Algorithms,* Third Edition, Springer-Verlag, Berlin, Germany.
13. do Carmo, M. P., 1976. *Differential Geometry of Curves and Surfaces*, Prentice-Hall, Englewood Cliffs, New Jersey.
14. Daubechies, I., 1992. *Ten Lectures on Wavelets*, Society for Industrial and Applied Mathematics, Philadelphia.
15. Debnath, L., and Bhatta, D., 2007. *Integral Transforms and Their Applications*, Second Edition, Chapman and Hall/CRC Press, New York, New York.
16. Flanigan, F. J., 1983. *Complex Variables: Harmonic and Analytic Functions*, Dover Publications, Inc., New York, New York.
17. Garrity, T. A., 2002. *All the Mathematics You Missed*, Cambridge University Press, Cambridge, Great Britain.
18. Gelfand, I. M., Kapranov, M. M., and Zelevinsky, 1994. *Determinants, Resultants, and Multidimensional Determinants*, Birkhauser, Boston, Massachusetts.
19. Gradshteyn, I. S., and Ryzhik, I. M., A., 1980. *Tables of Integrals, Series, and Products*, Academic Press, New York, New York.
20. Henle, M., 1997. *Modern Geometries: The Analytical Approach*, Prentice-Hall Inc., Simon & Schuster/A Viacom Company, Upper Saddle River, New Jersey.
21. Hight, D. W., 1977. *A Concept of Limits,* Dover Publications, Inc., New York, New York.
22. Hoffman, K., 1975. *Analysis in Euclidean Space,* Dover Publications, Inc., New York, New York.
23. Horowitz, E., and Sahni, S., 1978. *Fundamentals of Computer Algorithms*, Computer Science Press, Maryland.

24. Hsu, H. P., 1984. *Applied Fourier Analysis,* Harcourt Brace College Publishers, New York, New York.

25. Hsu, H. P., 1995. *Signals and Systems,* Schaum's Outline Series, McGraw-Hill Book Company, New York, New York.

26. Ivanov, O. A., 1999. *Easy as* $\pi$. An Introduction to Higher Mathematics, Springer-Verlag, Berlin, Germany.

27. Janson, S., Łuczak, T., and Ruciński, A., 2000. *Random Graphs*, John Wiley & Sons, Inc., New York, New York.

28. Kantorovich, L. V., and Akilov, G. P., 1982. *Functional Analysis,* Second Edition, Pergamon Press, New York, New York.

29. Kleinrock, L., 1975. *Queueing Systems, Volume I*, John Wiley & Sons, Inc., New York, New York.

30. Kolmogorov, A. N., and Fomin, S. V., 1970. *Introductory Real Analysis,* Dover Publications, Inc., New York, New York.

31. Korn, G. A., and Korn, T. M., 1968. *Mathematical Handbook for Scientists and Engineers*, Second Edition, McGraw-Hill Book Company, New York, New York.

32. Körner, T. W., 1988. *Fourier Analysis,* Cambridge University Press, Cambridge, Great Britain.

33. Kreyszig, E., 2011. *Advanced Engineering Mathematics*, Tenth Edition, John Wiley & Sons, Inc., New York, New York.

34. Lebedev, N. N., 1972. *Special Functions and their Applications,* Dover Publications, Inc., New York, New York.

35. Levinson, N., and Redheffer, R. M., 1970. *Complex Variables,* Holden-Day, San Francisco, California.

36. Lipschutz, S., 1968. *Linear Algebra*, Schaum's Outline Series, McGraw-Hill Book Company, New York, New York.

37. Lipschutz, M. M., 1969. *Differential Geometry*, Schaum's Outline Series, McGraw-Hill Book Company, New York, New York.

38. Lipschutz, S., 1998. *Set Theory and Related Topics,* Schaum's Outline Series, McGraw-Hill Book Company, New York, New York.

39. Loan, C. V., 1992. *Computational Frameworks for the Fast Fourier Transform*, Society of Industrial and Applied Mathematics, Philadelphia, Pennsylvania.

40. Mallat, S., 1998. *A Wavelet Tour of Signal Processing,* Academic Press, New York, New York.

41. Meyer, Y., 1993. *Wavelets Algorithms and Applications*, Society of Industrial and Applied Mathematics, Philadelphia, Pennsylvania.

42. Milne, R. D., 1980. *Applied Functional Analysis*: *An Introductory Treatment*, Pitman Advanced Publishing Program, London, Great Britain.

43. Moore, R. E., 1985. *Computational Functional Analysis*, John Wiley & Sons, Inc., New York, New York.

44. Mostowski, A., and Stark, M., 1964. *Introduction to Higher Algebra*, Pergamon Press, New York, New York.

45. Palmer, E. M., 1985. *Graphical Evolution*, John Wiley & Sons, Inc., New York, New York.

46. Papoulis, A., 1962. *The Fourier Integral and its Applications*, McGraw-Hill Book Company, New York.

47. Papoulis, A., 1977. *Signal Analysis*, McGraw-Hill Book Company, New York, New York.

48. Pinkus, A., and Zafrany, S. 1997. *Fourier Series and Integral Transforms*, Cambridge University Press, Cambridge, Great Britain.

49. Poularkis, A. D., 1999. *The Handbook of Formulas and Tables for Signal Processing,* CRC Press: New York, New York.

50. Pressley, A., 2010. *Elementary Differential Geometry,* Second Edition, Springer-Verlag, Berlin, Germany.

51. Rich, E., 2008. *Automata, Computability, and Computing*: *Theory and Applications*, Pearson Prentice-Hall, Upper Saddle River, New Jersey.

52. Rosen, K. H., Editor-in-Chief, 2018. *Handbook of Discrete and Combinatorial Mathematics,* Second Edition, CRC Press: Boca Raton, Florida.

53. Seaborn, J. B., 1991. *Hypergeometric Functions and Their Applications*, Springer-Verlag, Berlin, Germany.

54. Sedgewick, R., and Flajolet, P., 1996. *An Introduction to the Analysis of Algorithms*, Addison-Wesley Publishing Company, New York, New York.

55. Serpedin, E., Chen, T., and Rajan, D., Editors, 2012. *Mathematical Foundations for Signal Processing, Communications, and Networking,* CRC Press: New York, New York.

56. Shilov, G. E., 1996. *Elementary Real and Complex Analysis,* Dover Publications, Inc., New York, New York.

57. Sirovich, L., 1988. *Introduction to Applied Mathematics,* Springer-Verlag, Berlin, Germany.

58. Spiegel, M. R., 1959. *Vector Analysis*, Schaum's Outline Series, McGraw-Hill Book Company, New York, New York.

59. Spiegel, M. R., 1963. *Advanced Calculus*, Schaum's Outline Series, McGraw-Hill Book Company, New York, New York.

60. Spiegel, M. R., 1964. *Complex Variables*, Schaum's Outline Series, McGraw-Hill Book Company, New York, New York.

61. Spiegel, M. R., 1969. *Real Variables*, Schaum's Outline Series, McGraw-Hill Book Company, New York, New York.

62. Tolstov, G. P., 1962. *Fourier Series,* Dover Publications, Inc., New York, New York.

63. Toth, G., 2002. *Finite Moebius Groups, Minimal Immersions of Spheres, and Moduli*, Springer-Verlag, Berlin, Germany.

64. Villani, C., 2016. *Birth of a Theorem*: *A Mathematical Adventure*, Farrar, Straus, and Giroux, New York, New York.

65. Vretblad, A., 2003. *Fourier Analysis and Its Applications,* Springer-Verlag, Berlin, Germany.

66. Walker, R. J., 1962. *Algebraic Curves*, Dover Publications, Inc., New York, New York.

67. Wilf, H. S., 1978. *Mathematics for the Physical Sciences,* Dover Publications, Inc., New York, New York.

# Optimization, Stability, and Chaos Theory

$$B(0) = 0$$

$$(B(t) - B(s)) \sim \mathcal{N}\left(0, \sigma^2(t - s)\right),$$

$$s \leq t, \ s, t \in \mathbb{R}$$

Process has independent increments.
$\{B(t), t \in \mathbb{R}\}$ is a Wiener process

**Norbert Wiener.** Wiener was born on 26 November, 1894 in Columbia, Missouri, USA. He completed his high school at the age of eleven, and graduated with a B.A. from Tufts College in 1909 at the age of fourteen. Subsequently, he entered Harvard University to study zoology. He later switched his field of studies to mathcmatical logic. Wiener received his Ph.D. at the age of eighteen. He next went to Cambridge, England, and Göttingen, Germany, for further studies. Wiener joined the Massachusetts Institute of Technology in 1915, and remained there until his death.

Wiener made seminal contributions to the study of Brownian motion, general stochastic processes, Fourier theory, generalized harmonic analysis, Dirichlet's problem, Tauberian theorems, communication theory, and cybernetics. He was responsible for coining the word *cybernetics*. He is the author of: *The Fourier Integral and Certain of its Applications* (1933), *Cybernetics or Control and Communication in the Animal and the Machine* (1948), *Extrapolation, Interpolation, and Smoothing of Stationary Time Series with Engineering Applications* (1949), *Nonlinear Problems in Random Theory* (1958), and *God and Golem, Inc.: A Comment on Certain Points where Cybernetics Impinges on Religion* (1964). His textbook on cybernetics is regarded as one of the most influential books of the twentieth century. This book contributed to the popularization of communication theory terms such as feedback, information, control, input, output, stability, prediction, and filtering. Wiener died on 18 March, 1964 in Stockholm, Sweden.

## 7.1 Introduction

Any complex system demands that its resources be optimally utilized. The resources of the system have to be optimized for it to work efficiently. Therefore a careful and relevant mathematical model of the system is necessary. The phenomenon of Internet is also like any other complex system. Mathematical prerequisites to build and analyse optimization models of the Internet are outlined in this chapter. The following topics are discussed in this chapter: basics of optimization theory, certain useful inequalities, elements of linear programming, optimization techniques, calculus of variations, stability theory, and chaos theory.

Optimization theory by itself is a profound discipline. However, in this chapter we initially study theory of linear programming, and the work of the mathematician Joseph Louis Lagrange (1736-1813) on optimization theory. Elements of calculus of variations are also outlined.

Stability theory plays a prominent role in engineering. Specifically, stability problems arise in the study of dynamical systems, of which the Internet is both a prime and a modern example. The stability of dynamical systems like the Internet is studied from two different perspectives. These are the stability of a system from the input-output point of view, and the equilibrium point of view. These two different notions of stability are used to study and characterize congestion control of the Internet in a different chapter.

The input-output stability theory that is outlined in this chapter is based upon the work of the engineer H. Nyquist (1889-1976). The equilibrium stability theory that is outlined is based upon the work of the Russian mathematician and engineer A. M. Lyapunov (1857-1918). Several aspects of the Internet dynamics are nonlinear. In some nonlinear dynamical systems, a very small change in initial conditions produces very large divergence in the observed results. Such phenomena can be modeled by using chaos theory. This theory also appears to be a useful paradigm for modeling and understanding this complex federation of networks called the Internet.

## 7.2  Basics of Optimization Theory

Some of the building blocks of optimization theory are the concepts of convex and concave functions, and convex sets. These are also central to addressing nonlinear dynamical problems.

### 7.2.1  Convex and Concave Functions

Proofs of some of the inequalities in probability theory and information theory depend upon the use of the Jensen's inequality. This inequality is established in this subsection.

A function $f(\cdot)$ is *concave*, if for any two points $A$ and $B$ of the curve $y = f(x)$, the chord $AB$ lies below the curve. A function $f(\cdot)$ is *convex*, if $-f(\cdot)$ is concave. That is, a $f(\cdot)$ is convex, if $A$ and $B$ are any two points on the curve $y = f(x)$, then the chord $AB$ lies above the curve. Figure 7.1 shows concave and convex functions.

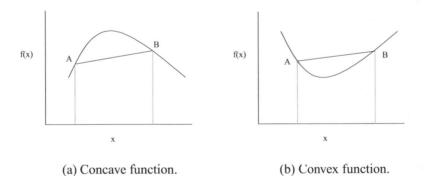

(a) Concave function.                    (b) Convex function.

Figure 7.1. Concave and convex functions.

**Definition 7.1.** *A real-valued continuous function $f(\cdot)$ is concave on an interval $I$ of the real line if*

$$f\left(\frac{x+y}{2}\right) \geq \frac{f(x) + f(y)}{2}, \quad \forall\, x, y \in I \tag{7.1a}$$

*The function $f(\cdot)$ is strictly concave on the interval $I$, if*

$$f\left(\frac{x+y}{2}\right) > \frac{f(x) + f(y)}{2}, \quad \forall\, x, y \in I, \ x \neq y \tag{7.1b}$$

□

There are several other equivalent definitions of a concave function. The above definition is called the *midpoint concavity* definition. Convex and strictly-convex continuous functions on an interval on the real line can be similarly defined by reversing the inequalities in the above definition. *Concave* and *convex* functions are sometimes referred to as *concave downwards,* and *concave upwards* respectively.

Some examples of concave functions are $\ln x$ for $x > 0$ and $\sqrt{x}$ for $x \geq 0$. Examples of convex functions are $x^2, |x|$, and $e^x$ for $x \in \mathbb{R}$. See Figure 7.2. Note that the *linear* function $(mx + b)$ is both concave and convex.

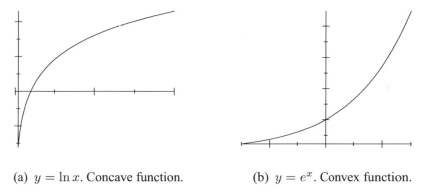

(a) $y = \ln x$. Concave function.          (b) $y = e^x$. Convex function.

Figure 7.2. Examples of convex and concave functions.

The following theorem states the celebrated Jensen's inequality.

**Theorem 7.1.** *Jensen's inequality. Let $f\,(\cdot)$ be a concave function on an open interval $I$ of the real line. Also let*

$$\sum_{j=1}^{m} \lambda_j = 1, \quad \lambda_j \geq 0, \ 1 \leq j \leq m \qquad\qquad (7.2a)$$

*and $x_1, x_2, \ldots, x_m \in I$, then*

$$f\left(\sum_{j=1}^{m} \lambda_j x_j\right) \geq \sum_{j=1}^{m} \lambda_j f\,(x_j) \qquad\qquad (7.2b)$$

*If $f\,(\cdot)$ is strictly concave, and $\lambda_j > 0,$ for $1 \leq j \leq m$, then equality holds if and only if $x_1 = x_2 = \ldots = x_m$.*

*Proof.* The result can be proved by mathematical induction. Details are left to the reader.   $\square$

The above result also has a probabilistic interpretation. For this version of the inequality refer to the chapter on probability theory. Note that if a function $f\,(\cdot)$ is convex over an interval $(a, b)$ for all values of $x_1, x_2 \in (a, b)$ and $0 \leq \lambda \leq 1$, then

$$f\,(\lambda x_1 + (1 - \lambda) x_2) \leq \lambda f\,(x_1) + (1 - \lambda)\,f\,(x_2) \qquad\qquad (7.3)$$

This function is strictly convex on the open interval $(a, b)$, if inequality holds in the above result $\forall$ $x_1, x_2 \in (a, b)$, with $x_1 \neq x_2$, and $\forall\,\lambda \in (0, 1)$.

**Theorem 7.2.** *A function $f(\cdot)$ has nonnegative (positive) second derivative for all values of $x$ in its domain, then the function is convex (strictly convex).*

*Proof.* Denote the first and second derivatives of $f(x)$ with respect to $x$, by $f'(x)$ and $f''(x)$ respectively. Use of Taylor series expansion of the function around $\alpha$ yields

$$f(x) = f(\alpha) + f'(\alpha)(x - \alpha) + \frac{f''(\xi)}{2}(x - \alpha)^2$$

where $\xi \in (\alpha, x)$. As per the hypothesis $f''(\xi) \geq 0$. Therefore the third term on the right-hand side, $f''(\xi)(x - \alpha)^2/2$ is always nonnegative for all values of $x$. Let $\alpha = \lambda x_1 + (1 - \lambda)x_2$ and $x = x_1$, then

$$f(x_1) \geq f(\alpha) + f'(\alpha)(1 - \lambda)(x_1 - x_2)$$

Similarly, letting $x = x_2$ yields

$$f(x_2) \geq f(\alpha) + f'(\alpha)\lambda(x_2 - x_1)$$

Multiplying the above two equations by $\lambda$ and $(1 - \lambda)$ respectively, and adding results in

$$f(\lambda x_1 + (1 - \lambda)x_2) \leq \lambda f(x_1) + (1 - \lambda)f(x_2)$$

Steps to prove strict convexity are similar.  □

A similar theorem can be stated about concave functions.

**Example 7.1.** Consider the function $\ln x$ where $x > 0$. The inequality $\ln x \leq (x - 1)$ is next established.

The first and second derivatives of $\ln x$ with respect to $x$ are $x^{-1}$ and $-x^{-2}$ respectively. Therefore, its second derivative is negative for all values of $x$ in its domain. Consequently this function is strictly concave. Using an analog of the last theorem yields

$$\ln x \leq \ln \alpha + \frac{(x - \alpha)}{\alpha}, \quad \alpha \in \mathbb{R}^+$$

where equality holds for $x = \alpha$. Letting $\alpha = 1$ results in $\ln x \leq (x - 1)$.  □

**Examples 7.2.** Some examples of convex functions.

1. Linear function: $f(x) = ax + b$ for $x \in \mathbb{R}$, and any $a, b \in \mathbb{R}$.
2. Exponential function: $e^{ax}$ for $x \in \mathbb{R}$, and any $a \in \mathbb{R}$.
3. Powers of absolute values: $|x|^a$ for $x \in \mathbb{R}$, and any $a \geq 1$.
4. Powers: $x^a$ for $x \in \mathbb{R}^+$, and any $a \geq 1$ or $a < 0$.
5. Even powers: $x^{2p}$ for $x \in \mathbb{R}$, and any $p \in \mathbb{P}$.
6. Powers: $-x^a$ for $x \in \mathbb{R}_0^+$, and $a \in [0, 1]$.
7. Negative logarithm: $-\ln x$ for $x \in \mathbb{R}^+$.
8. Negative entropy: $x \ln x$ for $x \in \mathbb{R}^+$.  □

**Examples 7.3.** Some examples of concave functions.

1. Linear function: $f(x) = ax + b$ for $x \in \mathbb{R}$, and any $a, b \in \mathbb{R}$.
2. Powers: $x^a$ for $x \in \mathbb{R}_0^+$, and $a \in [0, 1]$.

3. Logarithm: $\ln x$ for $x \in \mathbb{R}^+$.                                          □

**Examples 7.4**. Some examples of convex functions on $\mathbb{R}^n$.

1. Linear function: $f(x) = a^T x + b$ for $x \in \mathbb{R}^n$, and any $a \in \mathbb{R}^n, b \in \mathbb{R}$.
2. Norms:

    (a) $l_p$ norm: Let $x \in \mathbb{R}^n$, then $\|x\|_p = \left( \sum_{j=1}^n |x_j|^p \right)^{1/p}$, $\quad 1 \le p < \infty$.

    (b) $l_\infty$ norm: Let $x \in \mathbb{R}^n$, then $\|x\|_\infty = \max_{1 \le j \le n} \{|x_j|\}$.          □

### 7.2.2 Convex Sets

The concept of convexity and concavity of a function can also be easily extended to a function of several variables. However, these are more easily defined on *convex sets*. We study convex sets, because a fair amount of optimization theory is based upon the use of convex sets.

**Definitions 7.2.**

1. *Let $x, y \in \mathbb{R}^n$. The line segment joining these two points is the set of points*

$$\{w \mid w = \lambda x + (1 - \lambda) y, \ 0 \le \lambda \le 1\} \qquad (7.4a)$$

   *A set $C \subseteq \mathbb{R}^n$ is convex if and only if $\forall \, x, y \in C$, the line segment joining $x$ and $y$ also lies in $C$.*
2. *A convex combination of the points in the set*

$$\left\{ x^{(j)} \in \mathbb{R}^n \mid 1 \le j \le m \right\} \qquad (7.4b)$$

   *is the point $\sum_{j=1}^m \lambda_j x^{(j)} \in \mathbb{R}^n$, where $\lambda_1, \lambda_2, \ldots, \lambda_m$ are nonnegative numbers such that $\sum_{j=1}^m \lambda_j = 1$.*          □

The empty set and sets containing only a single point are also considered to be convex. The next result demonstrates, that a convex combination of two or more vectors in a convex set also belongs to the convex set.

**Lemma 7.1.** *Let $C$ be a convex set in $\mathbb{R}^n$, and $x^{(1)}, x^{(2)}, \ldots, x^{(m)} \in C$. If there exist nonnegative numbers $\lambda_1, \lambda_2, \ldots, \lambda_m$ such that $\sum_{j=1}^m \lambda_j = 1$, then the convex combination $\sum_{j=1}^m \lambda_j x^{(j)} \in C$.*
*Proof.* The proof, via mathematical induction, is left to the reader.          □

**Lemma 7.2.** *Intersection of finite number of convex sets is a convex set.*
*Proof.* The proof is left to the reader.          □

**Definition 7.3** *A convex hull of a set of points is the smallest convex set containing the given points. That is, the convex hull of a set $S$ is the intersection of all convex sets containing $S$ as a subset.*          □

Thus the convex hull is a convex set. If a set of points $T$ is finite, then the convex hull of $T$, is the set of all convex combinations of the points in $T$. Convex and concave functions of several variables can be defined on convex sets.

**Definitions 7.4.** *Let* $f(\cdot)$ *be a real-valued function* $f : C \to \mathbb{R}$, *where* $C \subseteq \mathbb{R}^n$ *is a convex set. Then*

1. *The function* $f(\cdot)$ *is convex on* $C$ *if*

$$f(\lambda x + (1 - \lambda) y) \leq \lambda f(x) + (1 - \lambda) f(y) \tag{7.5a}$$

   $\forall \, x, y \in C$ *and all* $\lambda \in [0, 1]$.
2. *The function* $f(\cdot)$ *is strictly convex on* $C$ *if*

$$f(\lambda x + (1 - \lambda) y) < \lambda f(x) + (1 - \lambda) f(y) \tag{7.5b}$$

   $\forall \, x, y \in C$ *with* $x \neq y$ *and all* $\lambda \in (0, 1)$.
3. *Concave and strictly concave functions are analogously defined, if the inequalities in the above definitions are reversed. Alternately, a function* $f(\cdot)$ *defined on a convex set is a concave or strictly concave function, if* $-f(\cdot)$ *is a convex or strictly convex function.*  □

The following results are used later in the chapter.

**Theorem 7.3.** *Let* $x \in \mathbb{R}^n$, $g : \mathbb{R}^n \to \mathbb{R}$, *and* $a \in \mathbb{R}$.

(a) *If* $g(\cdot)$ *is a convex function, then* $C_1 = \{x \mid g(x) \leq a\}$ *is a convex set.*
(b) *If* $h(\cdot)$ *is a concave function, then* $C_2 = \{x \mid h(x) \geq a\}$ *is a convex set.*

*Proof.* The proof is left to the reader.  □

**Lemma 7.3.** *Let* $x \in \mathbb{R}^n$, $g : \mathbb{R}^n \to \mathbb{R}$, *and* $a \in \mathbb{R}$. *If*

(a) $g_i(\cdot)$ *is a convex function, and* $C_i = \{x \mid g_i(x) \leq a_i\}, 1 \leq i \leq u$.
(b) $g_i(\cdot)$ *is a concave function, and* $C_i = \{x \mid g_i(x) \geq a_i\}, (u + 1) \leq i \leq v$.
(c) $g_i(\cdot)$ *is a linear function, and* $C_i = \{x \mid g_i(x) = a_i\}, (v + 1) \leq i \leq m$.

*Each* $C_i, 1 \leq i \leq m$ *is a convex set, and the intersection of all the convex sets* $C_i, 1 \leq i \leq m$ *is a convex set.*
*Proof.* The proof is left to the reader.  □

Note that the specified conditions to form a convex constraint set are not necessary, because it is possible to produce a convex set by the intersection of several nonconvex sets.

**Observations 7.1.**

1. Let $f(\cdot)$ and $g(\cdot)$ be convex functions defined on a convex set $S$, then their sum

$$h(x) = f(x) + g(x), \quad \text{for } x \in S$$

   is also convex.

2. A generalization of the above result is as follows. Let $f_i(\cdot)$, $1 \leq i \leq n$ be convex functions, each defined on a convex set $S$. Let $a_i \in \mathbb{R}^+$, for $1 \leq i \leq n$. Then $f(\cdot) = \sum_{i=1}^{n} a_i f_i(\cdot)$ is a convex function.

   Similarly, let $g_i(\cdot)$, $1 \leq i \leq n$ be concave functions, each defined on a convex set $S$. Let $a_i \in \mathbb{R}^+$, for $1 \leq i \leq n$. Then $g(\cdot) = \sum_{i=1}^{n} a_i g_i(\cdot)$ is a concave function.

3. Let $f : \mathbb{R}^n \to \mathbb{R}$, $A$ be a real-valued $n \times m$ matrix, and $b \in \mathbb{R}^n$. Also let $g : \mathbb{R}^m \to \mathbb{R}$, where $g(x) = f(Ax + b)$. If $f(\cdot)$ is a convex function, then $g(\cdot)$ is also a convex function. However, the domain of function $g(\cdot)$ is different than the domain of $f(\cdot)$. Actually, the domain of $g(\cdot)$ is $\{x \mid (Ax + b) \in \text{domain of } f(\cdot)\}$. Also note that, if $f(\cdot)$ is a concave function, then so is $g(\cdot)$. $\qquad\square$

Some of these observations are proved in the problem section. The details developed in this section serve as a foundation for optimization problems in the rest of the chapter.

## 7.3 Inequalities

Jensen's inequality has been stated in an earlier section. The Bunyakovsky-Cauchy-Schwartz inequality, and the Minkowski's inequality for the Euclidean norm have also been discussed in the chapter on applied analysis. Some more useful inequalities are summarized in this section. The arithmetic mean-geometric mean-harmonic mean inequality, Young's inequality, Hölder's inequality, Minkowski's inequality for a $p$-norm, and log sum inequality are established in this section.

### Arithmetic Mean-Geometric Mean-Harmonic Mean Inequality

Let $a_1, a_2, a_3, \ldots, a_n$ be $n$ positive real numbers. The arithmetic, geometric, and harmonic means of these numbers are $A_n, G_n$, and $H_n$ respectively, where

$$A_n = \frac{1}{n} \sum_{i=1}^{n} a_i, \tag{7.6a}$$

$$G_n = \left\{ \prod_{i=1}^{n} a_i \right\}^{1/n} \tag{7.6b}$$

$$\frac{1}{H_n} = \frac{1}{n} \sum_{i=1}^{n} \frac{1}{a_i} \tag{7.6c}$$

Then $G_n \leq A_n$. Equality occurs if and only if the $a_i$'s are all equal. If the arithmetic mean-geometric mean inequality is applied to the reciprocals

$$a_1^{-1}, a_2^{-1}, \ldots, a_n^{-1}$$

we obtain $H_n \leq G_n$. In this case also, equality occurs if and only if the $a_i$'s are all equal. In summary, we have

$$H_n \leq G_n \leq A_n \tag{7.6d}$$

**Young's Inequality**

Young's inequality is named after William Henry Young (1863-1942). Let $a, b \in \mathbb{R}_0^+$; and $p, q > 1$, where $\left( p^{-1} + q^{-1} \right) = 1$. Then

$$ab \leq \frac{a^p}{p} + \frac{b^q}{q} \tag{7.7}$$

Equality occurs if and only if $b = a^{p-1}$.

**Hölder's Inequality**

Hölder's inequality is named after Otto Ludwig Hölder (1859-1937). Let $a_i, b_i \in \mathbb{R}^+$ for $1 \leq i \leq n$; and $p, q > 1$, where $\left( p^{-1} + q^{-1} \right) = 1$. Then

$$\sum_{i=1}^{n} a_i b_i \leq \left\{ \sum_{i=1}^{n} a_i^p \right\}^{1/p} \left\{ \sum_{i=1}^{n} b_i^q \right\}^{1/q} \tag{7.8a}$$

Let

$$\alpha = \left\{ \sum_{i=1}^{n} a_i^p \right\}^{1/p}, \quad \text{and} \quad \beta = \left\{ \sum_{i=1}^{n} b_i^q \right\}^{1/q} \tag{7.8b}$$

Equality occurs, if and only if

$$b_i = \beta \left\{ \frac{a_i}{\alpha} \right\}^{p-1}, \quad i = 1, 2, \ldots n \tag{7.8c}$$

**Observations 7.2.** Let $p, q > 1$, where $\left( p^{-1} + q^{-1} \right) = 1$.

1. For $a_i, b_i \in \mathbb{R}$ for $1 \leq i \leq n$.
   (a) We have

   $$\sum_{i=1}^{n} |a_i b_i| \leq \left\{ \sum_{i=1}^{n} |a_i|^p \right\}^{1/p} \left\{ \sum_{i=1}^{n} |b_i|^q \right\}^{1/q} \tag{7.9a}$$

   (b) It is possible to obtain the Bunyakovsky-Cauchy-Schwartz inequality from the above result. Substitute $p = q = 2$ in the above inequality. It yields

   $$\sum_{i=1}^{n} |a_i b_i| \leq \left\{ \sum_{i=1}^{n} a_i^2 \right\}^{1/2} \left\{ \sum_{i=1}^{n} b_i^2 \right\}^{1/2}$$

   Note that

   $$\left| \sum_{i=1}^{n} a_i b_i \right| \leq \sum_{i=1}^{n} |a_i b_i|$$

   Combining the last two inequalities results in the Bunyakovsky-Cauchy-Schwartz inequality. It is

   $$\left| \sum_{i=1}^{n} a_i b_i \right| \leq \left\{ \sum_{i=1}^{n} a_i^2 \right\}^{1/2} \left\{ \sum_{i=1}^{n} b_i^2 \right\}^{1/2} \tag{7.9b}$$

(c) Let $n \to \infty$. We have

$$\sum_{i=1}^{\infty} |a_i b_i| \leq \left\{ \sum_{i=1}^{\infty} |a_i|^p \right\}^{1/p} \left\{ \sum_{i=1}^{\infty} |b_i|^q \right\}^{1/q} \tag{7.9c}$$

The above inequality is true provided the two summations on the right hand side converge.

2. An integral version of this result is

$$\int_a^b |f(x) g(x)| \, dx \leq \left\{ \int_a^b |f(x)|^p \, dx \right\}^{1/p} \left\{ \int_a^b |g(x)|^q \, dx \right\}^{1/q} \tag{7.9d}$$

The above inequality is true provided the two integrals on the right hand side exist.   □

### Minkowski's Inequality

Minkowski's inequality is named after Hermann Minkowski (1864-1909). Let $a_i, b_i \in \mathbb{R}$ for $1 \leq i \leq n$. For $p \geq 1$

$$\left\{ \sum_{i-1}^{n} |a_i + b_i|^p \right\}^{1/p} \leq \left\{ \sum_{i=1}^{n} |a_i|^p \right\}^{1/p} + \left\{ \sum_{i=1}^{n} |b_i|^p \right\}^{1/p} \tag{7.10}$$

If $n = p = 1$, we obtain the triangle inequality. If $p = 2$ we obtain the triangle inequality for vectors. Equality occurs if and only if $b_i = k a_i$, for $1 \leq i \leq n$, and $k$ is some real constant. The above inequality can also be extended to infinite sums (provided the series converge), and also to integrals provided they exist. The direction of inequality is reversed for $0 < p < 1$.

The $p$-norm of a vector $x = (x_1, x_2, \ldots, x_n) \in \mathbb{R}^n$, for $1 \leq p < \infty$ is defined as

$$\|x\|_p = \left\{ \sum_{i=1}^{n} |x_i|^p \right\}^{1/p}$$

Let $a, b \in \mathbb{R}^n$. Minkowski's inequality is an assertion of the fact that the vector $p$-norm satisfies the triangle inequality. That is

$$\|a + b\|_p \leq \|a\|_p + \|b\|_p$$

### Log Sum Inequality

Let $a_1, a_2, \ldots, a_n$ and $b_1, b_2, \ldots, b_n$ be nonnegative numbers. Then

$$\sum_{i=1}^{n} a_i \log \frac{a_i}{b_i} \geq \left\{ \sum_{i=1}^{n} a_i \right\} \log \frac{\sum_{i=1}^{n} a_i}{\sum_{i=1}^{n} b_i} \tag{7.11}$$

where equality occurs if and only if $a_i/b_i = k$ (a constant) for each $i$, where $1 \leq i \leq n$.

It is assumed that $0 \log 0 = 0$, $0 \log (0/0) = 0$, and $a \log (a/0) = \infty$ for $a > 0$. These assumptions follow from continuity.

## 7.4 Elements of Linear Programming

Some results from the theory of linear programming (LP) are summarized in this section. A linear programming problem optimizes an objective function which is linear in decision variables. This optimization is performed subject to certain linear equality/inequality constraints. The topics discussed in this section are: hyperplanes, Farkas' alternative, and relationship between the primal and dual of a linear programming problem.

### 7.4.1 Hyperplanes

A hyperplane is simply a plane in a higher-dimensional vector space. We represent a vector as a row matrix in this subsection.

**Definitions 7.5.** *Hyperplane in a vector space, and normal vector to a hyperplane are defined. Let $z = (z_1, z_2, \ldots, z_t)$ be a vector in $\mathbb{R}^t$, where $t \in \mathbb{P} \backslash \{1\}$.*

1. *Let $w = (w_1, w_2, \ldots, w_t)$ be a vector, where $w_i \in \mathbb{R}, 1 \leq i \leq t$, not all of which are zeros, and also let $b \in \mathbb{R}$. The set of vectors $\widehat{Z} = \left\{ z \mid wz^T = \sum_{i=1}^{t} w_i z_i = b \right\}$ is called a hyperplane in $\mathbb{R}^t$. This is a $t$-dimensional plane.*
2. *A nonzero vector $\eta = (\eta_1, \eta_2, \ldots, \eta_t) \in \mathbb{R}^t$ is called normal to the hyperplane $\widehat{Z}$, if $\sum_{i=1}^{t} \eta_i z_i = 0, \forall z \in \widehat{Z}.$*   □

**Observations 7.3.** Let $wz^T = b$ be a hyperplane $\widehat{Z}$ in $\mathbb{R}^t$, where $z \in \widehat{Z}$, and $w \neq 0$. Also let $\|\cdot\|$ be the Euclidean norm defined on the vector space $\mathbb{R}^t$.

1. A straight line in $\mathbb{R}^2$ is a hyperplane in it. Similarly, a plane in $\mathbb{R}^3$ is a hyperplane.
2. The hyperplane $\widehat{Z}$ is a subset of $\mathbb{R}^t$.
3. The vector $w$ is normal to the hyperplane $\widehat{Z}$.
4. Note that the hyperplanes $wz^T = b_1$ and $wz^T = b_2$, where $b_1, b_2 \in \mathbb{R}$, are parallel to each other.
5. Let $q \in \mathbb{R}^t$ be a point which does not lie in the hyperplane $\widehat{Z}$. The shortest distance between the point $q$ and the hyperplane $\widehat{Z}$ is $\left| wq^T - b \right| / \|w\|$. Thus, the perpendicular distance from the origin to the hyperplane is $|b| / \|w\|$.   □

The separating hyperplane theorem is useful in the study of linear programming. This theorem is due to Hermann Minkowski (1864-1909). We first give the definition of a separating hyperplane.

**Definition 7.6.** *Let $z = (z_1, z_2, \ldots, z_t)$ be a vector in $\mathbb{R}^t$, where $t \in \mathbb{P} \backslash \{1\}$. The set of points $\widehat{Z} = \left\{ z \mid wz^T = \sum_{i=1}^{t} w_i z_i = b \right\}$ is a hyperplane in $\mathbb{R}^t$, where $b \in \mathbb{R}, w = (w_1, w_2, \ldots, w_t)$ is a constant vector, $w_i \in \mathbb{R}, 1 \leq i \leq t$, not all of which are zeros.*
*Let $X$ and $Y$ be subsets of $\mathbb{R}^t$, where $X \cap Y$ is not necessarily empty.*

(a) *The hyperplane $\widehat{Z}$ separates $X$ from $Y$, if $wx^T \leq b, \forall x \in X$, and $wy^T \geq b, \forall y \in Y$.*
(b) *The hyperplane $\widehat{Z}$ strictly separates $X$ from $Y$, where $X \cap Y = \varnothing$; if $wx^T < b, \forall x \in X$, and $wy^T > b, \forall y \in Y$.*   □

If the two sets $X$ and $Y$ are disjoint, then there may or may not be a separating hyperplane. However, if the two sets $X$ and $Y$ are both disjoint and convex, then there exists a separating hyperplane.

**Theorem 7.4.** *Separating hyperplane theorem. Let $X \subset \mathbb{R}^t$ be a nonempty closed, and convex set. Also let $h \in \mathbb{R}^t \backslash X$. Then there exists a hyperplane which strictly separates $X$ from $h$.*
    *Proof.* See the problem section.                                                                          □

### 7.4.2 Farkas' Alternative

We next study Farkas' theorem. This theorem is often referred to as the Farkas' alternative. It is useful in the study of linear programming problems. This result is named after the Hungarian mathematician Gyula Farkas, who published it in the year 1902. Its proof requires the use of separating hyperplane theorem.

**Theorem 7.5.** *Farkas' alternative. Let $A$ be an $m \times n$ matrix of real numbers, $b$ is a column vector of real numbers of size $m$; and $m, n \in \mathbb{P}$. Then either:*

(i) *There exists a column vector $x$ of real numbers of size $n$ such that $Ax = b$ and $x \geq 0$; or* (*exclusive*)
(ii) *There exists a column vector $y$ of real numbers of size $m$ such that $y^T A \geq 0$ and $y^T b < 0$.*

   *Proof.* See the problem section.                                                                           □

In the statement of the Farkas' theorem, "or" is exclusive. This means that either alternative (i) is true or alternative (ii) is true, but both alternatives are not simultaneously true. Alternative (i) is called the *obverse,* and alternative (ii) the *reverse* of the Farkas' theorem.

**Corollary 7.1.** *Farkas' corollary. Let $A$ be an $m \times n$ matrix of real numbers, $b$ is a column vector of real numbers of size $m$; and $m, n \in \mathbb{P}$. Then either:*

(i) *There exists a column vector $x$ of real numbers of size $n$ such that $Ax \leq b$ and $x \geq 0$; or* (*exclusive*)
(ii) *There exists a column vector $y$ of real numbers of size $m$ such that $y^T A \geq 0$, $y^T b < 0$, and $y \geq 0$.*

   *Proof.* See the problem section.                                                                           □

Alternative (i) is called the *obverse,* and alternative (ii) the *reverse* of the Farkas' corollary.

### 7.4.3 Primal and Dual Problems

Linear programming is a popular and useful optimization technique. An example of LP problem is:

$$\text{Maximize } f(x) = c^T x$$
$$\text{Subject to} \ : \ Ax = b, \quad \text{and} \quad x \geq 0$$

where $A$ is an $m \times n$ matrix of real numbers, $b$ and $c$ are column vectors of real numbers of size $m \in \mathbb{P}$, and $n \in \mathbb{P}$ respectively. The goal of this problem is to determine the column vector $x$ of real numbers of size $n$, so that $f(x) = c^T x$ is maximized; subject to the $m$ equality *constraints* specified by $Ax = b$, and nonnegativity constraints $x \geq 0$. We introduce terminology used in discussing such linear programming problems.

- *Decision variables*: Let

$$x = \begin{bmatrix} x_1 & x_2 & \ldots & x_n \end{bmatrix}^T$$

  The elements of the vector $x$ are called decision variables. These take real values.
- *Objective function*: The real-valued function $f(\cdot)$ is called the objective function. The LP problem optimizes this function.
- *Feasibility and infeasibility*: The set of possible values of $x$ that satisfy the constraints is called the *feasible set of points* or the *constraint region* of the optimization problem. It is possible for the feasible region to be empty. In this case the LP problem is said to be *infeasible*.
- *Optimal solution*: If the maximization (minimization) LP problem has a feasible region, then the feasible point which maximizes (minimizes) the objective function is called the optimal solution. Optimal solution of a LP problem is not necessarily unique.
- *Unbounded LP problem*: The LP problem is said to be *unbounded*, if there exist feasible points so that the maximizing (minimizing) objective function diverges to $+\infty$ ($-\infty$).

The word "linear" in "linear programming" comes from the fact that the objective function $f(x)$ is linear in the decision variables $x_j$'s; and also the constraints specified by $Ax = b$ are *linear equations*. A popular and practical method to solve LP problems is via the *simplex* method. Several well-known textbooks on linear programming describe this method. Associated with each LP problem, there exists a *dual linear programming problem* which has both theoretical and practical implications. Maximization and minimization problems which are related are termed dual problems. The duality theory is sometimes used in solving the LP problems efficiently. The basic idea of duality is illustrated in the following example.

**Example 7.5.** Let $f(x, y)$ be a continuous function of $x$ and $y$, where $x \in X \subseteq \mathbb{R}^n$, and $y \in Y \subseteq \mathbb{R}^m$. Define

$$\min_{x \in X} f(x, y) \triangleq \tilde{a}(y), \quad \text{and} \quad \max_{y \in Y} f(x, y) \triangleq \tilde{A}(x)$$

Thus for any $x \in X$ and $y \in Y$,

$$\tilde{a}(y) \leq f(x, y) \leq \tilde{A}(x)$$

It follows from the above relationships that

$$\max_{y \in Y} \tilde{a}(y) \leq \tilde{A}(x), \quad \forall\, x \in X$$

$$\tilde{a}(y) \leq \min_{x \in X} \tilde{A}(x), \quad \forall\, y \in Y$$

Therefore

$$\max_{y \in Y} \min_{x \in X} f(x, y) \leq \min_{x \in X} \max_{y \in Y} f(x, y)$$

The following two problems are called dual problems.

(a) $\max_{y \in Y} \{\min_{x \in X} f(x, y)\}$, that is $\max_{y \in Y} \tilde{a}(y)$
(b) $\min_{x \in X} \{\max_{y \in Y} f(x, y)\}$, that is $\min_{x \in X} \tilde{A}(x)$

If $y^*, x^*; \widetilde{a}\,(y^*)$, and $\widetilde{A}\,(x^*)$ are the optimum values of the variables and functions respectively, then $\widetilde{a}\,(y^*) \leq \widetilde{A}\,(x^*)$. $\qquad\qquad$ □

Consider the following canonical LP problem. It is called the primal problem $\mathcal{P}_{LP}$. Associated with it is a dual problem $\mathcal{D}_{LP}$. These are:

*Primal problem $\mathcal{P}_{LP}$:*
*Objective*: max  $c^T x$
*Subject to*: $Ax \leq b$, and $x \geq 0$
where $A$ is an $m \times n$ matrix of real numbers, $b$ and $c$ are column vectors of real numbers of size $m \in \mathbb{P}$, and $n \in \mathbb{P}$ respectively; and $x$ is a variable column vector of size $n$. Let $z = \max \ c^T x$.

*Dual problem $\mathcal{D}_{LP}$:*
*Objective*: min $b^T y$
*Subject to*: $A^T y \geq c$, and $y \geq 0$
where $y$ is a variable column vector of size $m$. Let $w = \min \ b^T y$.

Observe that the primal problem has $m$ inequality constraints, and $n$ decision variables. However, the dual problem has $n$ inequality constraints, and $m$ decision variables. Also note that:

(a)  The primal problem maximizes its objective function, and the dual problem minimizes its corresponding objective function.
(b)  In the primal problem, $i$th constraint is $\leq$; and in the dual problem, $i$th variable is $\geq 0$; where $i \in [1, m]$.
(c)  In the primal problem, $j$th variable is $\geq 0$; and in the dual problem, $j$th constraint is $\geq$; where $j \subset [1, n]$.

Some relevant observations about the primal and dual formulations of a LP problem are listed below.

**Observations 7.4.** Consider the above canonical primal and dual LP problems.

1. The dual of a dual is the primal.
2. *Weak duality property*: If $x$ is a feasible solution to the primal problem, and $y$ is a feasible solution to the dual problem, then $c^T x \leq b^T y$. That is, $z \leq w$.
3. *Optimality property*: If $x^*$ is a feasible solution to the primal problem, and $y^*$ is a feasible solution to the dual problem such that $c^T x^* = b^T y^*$, then $x^*$ and $y^*$ are optimal solutions to the primal and dual problems respectively. $\qquad\qquad$ □

These observations are established in the problem section. The primal and dual of the LP problem has following possibilities.

*Primal problem $\mathcal{P}_{LP}$*: Objective function is maximized:
*Case P1*: Solution is infeasible $(z = -\infty)$.
*Case P2*: Solution is feasible and unbounded $(z = +\infty)$.
*Case P3*: Solution is feasible and bounded (finite) $(z = \text{finite real number})$.

*Dual problem $\mathcal{D}_{LP}$*: Objective function is minimized:

*Case D1*: Solution is infeasible ($w = +\infty$).
*Case D2*: Solution is feasible and unbounded ($w = -\infty$).
*Case D3*: Solution is feasible and bounded (finite) ($w =$ finite real number).

Following four cases can occur:

1.  Both the primal and dual problems have feasible optimal solutions. That is, $z = w$.
2.  Neither the primal nor the dual problems have feasible solutions. That is, Cases P1 and D1 occur simultaneously.
3.  The primal problem has a feasible and unbounded solution. That is, $z = +\infty$. From weak duality, we have $z \leq w$. This implies that dual objective function cannot be minimized. That is, dual has no feasible solution. Thus Case P2 implies Case D1.
4.  The dual problem has a feasible and unbounded solution. That is, $w = -\infty$. From weak duality, we have $z \leq w$. This implies that primal objective function cannot be maximized. That is, primal has no feasible solution. Thus Case D2 implies Case P1.                                   □

The following result is called the strong duality theorem.

**Theorem 7.6.** *Strong duality theorem. Let $A$ be an $m \times n$ matrix of real numbers, $b$ and $c$ are column vectors of real numbers of size $m \in \mathbb{P}$, and $n \in \mathbb{P}$ respectively. Also let $x$ and $y$ be column vectors of size $n$ and $m$ respectively. The elements of the vectors $x$ and $y$ take real values. The primal LP problem is*:

$$z = \text{Maximize } c^T x$$
$$\text{Subject to:} \quad Ax \leq b, \quad \text{and} \quad x \geq 0$$

*and its dual is*:

$$w = \text{Minimize } b^T y$$
$$\text{Subject to:} \quad A^T y \geq c, \quad \text{and} \quad y \geq 0$$

*If either the primal or dual has a feasible solution, then $z = w$.*
    *Proof.* See the problem section.                                                □

| Primal | *maximize* | Dual | *minimize* |
|---|---|---|---|
| constraints | $v^T x = b_i$<br>$v^T x \leq b_i$<br>$v^T x \geq b_i$ | variables | $y_i$ unrestricted<br>$y_i \geq 0$<br>$y_i \leq 0$ |
| variables | $x_j \geq 0$<br>$x_j \leq 0$<br>$x_j$ unrestricted | constraints | $u^T y \geq c_j$<br>$u^T y \leq c_j$<br>$u^T y = c_j$ |

Table 7.1 Guideline for specifying the dual linear programming problem from its primal.

The initially stated canonical primal problem is a specific example of a LP problem. There can be other types of LP problems, where the linear objective function can be either a maximization or a minimization. Further, the linear constraint sets can be inequalities in either direction, or simply be equalities. Also, the sign of the decision variables, can be either restricted or unrestricted.

The Table 7.1 provides a guideline to specify the dual problem from its primal. In this table, $u$ and $v$ are real-valued column vectors of size $m \in \mathbb{P}$ and $n \in \mathbb{P}$ respectively, $b_i, c_j$ are real numbers, and $x_j$ and $y_i$ are components of the real-valued column vectors $x$ and $y$ respectively.

---

## 7.5 Optimization Techniques

Optimization techniques are developed to determine extreme (maxima and minima) points of unconstrained and constrained functions. This goal is achieved by using only elementary results of differential calculus. The following topics are discussed in this section: Taylor's series of several variables, minimization and maximization of a function of a single variable, and minimization and maximization of a function of a several variables. Constrained optimization problem with equality constraints, constrained optimization problem with inequality constraints, and nonlinear optimization via duality theory are also discussed. This branch of analysis which deals with techniques and algorithms to solve optimization problems is generally called mathematical programming.

### 7.5.1  Taylor's Series of Several Variables

A function of several variables can be conveniently described, if the variables are represented as vector elements. In this subsection, a Taylor's series expansion of a function of several variables is discussed. Analogous to a function of a single variable, it is possible to have a Taylor's series representation of a function of several variables. These results are useful in characterizing the maxima or minima of several variables.

**Definitions 7.7.** *Some useful definitions.*

1. *The $n$-dimensional vector space of real elements is $\mathbb{R}^n$. If $x, y \in \mathbb{R}^n$, the distance between $x$ and $y$ is*

$$d(x,y) = \|x - y\| = \left\{ \sum_{i=1}^{n} (x_i - y_i)^2 \right\}^{1/2} \tag{7.12}$$

   *where $x = (x_1, x_2, \ldots, x_n), y = (y_1, y_2, \ldots, y_n),$ and $\|\cdot\|$ is called the Euclidean norm.*
2. *Let $x \in \mathbb{R}^n, r \in \mathbb{R}^+,$ and $B(x,r)$ be a set of points in $\mathbb{R}^n$, where*

$$B(x,r) = \{y \in \mathbb{R}^n \mid \|x - y\| < r\} \tag{7.13}$$

   *The set of points in $B(x,r)$ form a ball of radius $r$ centered at $x$.*
3. *A point $x \in \Omega \subseteq \mathbb{R}^n$ is an interior point of $\Omega$ if $\exists$ a real number $r > 0$ such that the ball $B(x,r)$ is contained within $\Omega$.*
4. *Let $\Omega \subseteq \mathbb{R}^n$, and $f(\cdot)$ be a real-valued function; where $f : \Omega \to \mathbb{R}, x = (x_1, x_2, \ldots, x_n),$ and $x \in \Omega$. The gradient of $f(x)$, denoted by $\nabla f(x)$ is a vector of partial derivatives*

$$\nabla f\left(x\right) = \left(\frac{\partial f\left(x\right)}{\partial x_1}, \frac{\partial f\left(x\right)}{\partial x_2}, \cdots, \frac{\partial f\left(x\right)}{\partial x_n}\right) \triangleq \frac{\partial f\left(x\right)}{\partial x} \tag{7.14}$$

*provided the partial derivatives exist.*                                                                                    $\square$

In the above definition, gradient is represented as a row vector. A Taylor's series for a function of several variables is next defined. This series expansion is described in terms of the Hessian matrix of $f\left(\cdot\right)$ at $x \in \mathbb{R}^n$. This matrix is denoted by $H\left(x\right)$ in honor of the German mathematician, Ludwig Otto Hesse (1811-1874).

**Definition 7.8.** *Let $f\left(\cdot\right)$ be a real-valued function $f : \Omega \to \mathbb{R}$, where $\Omega \subseteq \mathbb{R}^n$, and $x = \left(x_1, x_2, \ldots, x_n\right)$. Assume that the second partial derivatives of $f\left(\cdot\right)$, that is $\partial^2 f\left(x\right)/\partial x_i \partial x_j$ exist for all $1 \leq i, j \leq n$. The Hessian matrix $H\left(x\right)$ is defined as*

$$H\left(x\right) = \begin{bmatrix} \partial^2 f\left(x\right)/\partial x_1^2 & \partial^2 f\left(x\right)/\partial x_1 \partial x_2 & \cdots & \partial^2 f\left(x\right)/\partial x_1 \partial x_n \\ \partial^2 f\left(x\right)/\partial x_2 \partial x_1 & \partial^2 f\left(x\right)/\partial x_2^2 & \cdots & \partial^2 f\left(x\right)/\partial x_2 \partial x_n \\ \vdots & \vdots & \ddots & \vdots \\ \partial^2 f\left(x\right)/\partial x_n \partial x_1 & \partial^2 f\left(x\right)/\partial x_n \partial x_2 & \cdots & \partial^2 f\left(x\right)/\partial x_n^2 \end{bmatrix}$$

$$\tag{7.15}$$

$\square$

Since $\partial^2 f\left(x\right)/\partial x_i \partial x_j = \partial^2 f\left(x\right)/\partial x_j \partial x_i$, for $1 \leq i, j \leq n$, it follows that $H\left(x\right)$ is a symmetric matrix.

**Theorem 7.7.** *Taylor's series for a function of several variables. Let $f\left(\cdot\right)$ be a real-valued function $f : \Omega \to \mathbb{R}$, where $\Omega \subseteq \mathbb{R}^n$, and $x = \left(x_1, x_2, \ldots, x_n\right)$. Assume that $f\left(\cdot\right)$ has continuous partial derivatives through second order. Also if $x, \left(x + h\right) \in \Omega$, where $h = \left(h_1, h_2, \ldots, h_n\right)$, then*

$$f\left(x + h\right) = f\left(x\right) + \nabla f\left(x\right) h^T + \frac{1}{2} h H\left(\theta x + \left(1 - \theta\right)\left(x + h\right)\right) h^T \tag{7.16}$$

*for some $\theta \in [0, 1]$.*

*Proof.* The proof is left to the reader.                                                                                    $\square$

The above theorem is useful in establishing key observations about global minimizers and maximizers in several variables.

### 7.5.2  Minimization and Maximization of a Function

We investigate the minimization and maximization of a function of a single variable, and also of multiple variables. This will set up the foundation for analysing optimization problems with constraints.

#### Minimization and Maximization of a Function of a Single Variable

Optimization theory for a function of a single variable is described in this subsection. This will provide sufficient intuition to extend it to a function of several variables. Conditions for determining extreme points of a function are explored.

**Definitions 7.9.** *Let $f(\cdot)$ be a real-valued function $f : I \to \mathbb{R}$, where the interval $I \subseteq \mathbb{R}$ may be finite or infinite, open or closed, or half-open. The first derivative of $f(x)$ with respect to $x$ is denoted by $f'(x)$. Let $x^* \in I$.*

1. *The global minimum of $f(\cdot)$ occurs at a point $x^*$, on the interval $I$, if $f(x^*) \leq f(x)$ for all values of $x \in I$.*
2. *The strict global minimum of $f(\cdot)$ occurs at a point $x^*$, on the interval $I$, if $f(x^*) < f(x)$ for all values of $x \in I \backslash \{x^*\}$.*
3. *The local minimum of $f(\cdot)$ occurs at a point $x^*$, if $\exists$ a real number $\delta > 0$, such that $f(x^*) \leq f(x)$ for all values of $x \in (x^* - \delta, x^* + \delta) \subseteq I$.*
4. *The strict local minimum of $f(\cdot)$ occurs at a point $x^*$, if $\exists$ a real number $\delta > 0$, such that $f(x^*) < f(x)$ for all values of $x \in (x^* - \delta, x^* + \delta) \subseteq I \backslash \{x^*\}$.*
5. *The point $x^*$ is a critical or stationary point of $f(x)$ if $f'(x^*)$ exists and is equal to zero.*
6. *A point $(b, f(b))$ on the curve is called an inflection point, if it changes from convex to concave, or vice-versa at $x = b$.* □

The *global maximum, strict global maximum, local maximum,* and *strict local maximum* of $f(x)$ can be similarly defined. Therefore local stationarity yields either local minima or maxima. Also note that maximization of $f(\cdot)$ is the same as minimizing $-f(\cdot)$.

Let $f''(x)$ denote the second derivative of $f(x)$ with respect to $x$. The point of inflection on a twice differentiable curve, is a point $x$ where $f''(x)$ is positive on one side and negative on the other. Therefore, at such points, $f''(x) = 0$. The following observations summarize some useful facts about minimization and maximization of a function of a single variable.

**Observations 7.5.** Minimization and maximization of a function of a single variable.

1. Let $f(\cdot)$ be a differentiable function defined on an interval $I$. If $f(\cdot)$ has either a local maximum or minimum at $x^*$, then $f'(x^*) = 0$ or $x^*$ is an end-point of the interval $I$.
2. Let $f(\cdot), f'(\cdot)$, and $f''(\cdot)$ be all continuous on an interval $I$, and $x^* \in I$ be a critical point of $f(\cdot)$.
   (a) If $f''(x) \geq 0, \forall x \in I$, then a global minimum of $f(\cdot)$ on $I$ occurs at point $x^*$.
   (b) If $f''(x) > 0, \forall x \in I \backslash \{x^*\}$, then a strict global minimum of $f(\cdot)$ on $I$ occurs at point $x^*$.
   (c) If $f''(x^*) > 0$ then a strict local minimum of $f(\cdot)$ on $I$ occurs at point $x^*$.
3. Corresponding statements about maximization can be made by reversing the direction of inequalities in the last observation. □

Therefore $f''(x^*) > 0$ is a sufficient condition for $x^*$ to produce a minimum. Similarly, $f''(x^*) < 0$ is a sufficient condition for $x^*$ to produce a maximum. However, if $f''(x^*) = 0$, higher order derivatives must be investigated.

**Theorem 7.8.** *If at a critical point $x^*$, the first $(n-1)$ derivatives vanish but the $n$th derivative $f^{(n)}(x^*) \neq 0$, then the function $f(\cdot)$ has at $x = x^*$:*

(a) *An extreme (maximum or minimum) point if $n$ is even. The extreme point is a maximum if $f^{(n)}(x^*) < 0$ and a minimum if $f^{(n)}(x^*) > 0$.*
(b) *An inflection point if $n$ is odd.*

*Proof.* The proof is left to the reader.                                                                    □

**Examples 7.6.** Some illustrative examples.

1. Consider the function $f(x) = x^4$ defined over $\mathbb{R}$. Observe that $f'(x) = 4x^3$. Its critical point is $x^* = 0$. Also

$$f'(0) = f''(0) = f'''(0) = 0$$

   But $f^{(4)}(0) = 24 > 0$, therefore $x^* = 0$ is an extreme point, and it is a minimum.
2. Consider the function $f(x) = x^3$ defined over $\mathbb{R}$. Observe that $f'(x) = 3x^2$. Its critical point is $x^* = 0$. Also

$$f'(0) = f''(0) = 0$$

   But $f'''(0) = 6$, therefore $x^* = 0$ is an inflection point.                                           □

**Observations 7.6.** Let $f(x)$ be a twice differentiable curve in an interval $I \subseteq \mathbb{R}$.

1. $f''(x) \geq 0,\ \forall\, x \in I \Rightarrow f(x)$ is convex in the interval $I$.
2. $f''(x) \leq 0,\ \forall\, x \in I \Rightarrow f(x)$ is concave in the interval $I$.
   These results were established in an earlier section. However, we repeat them here for completeness.                                                                                           □

### Minimization and Maximization of a Function of Several Variables

The computation of minimum or maximum of a function of several variables is conceptually a simple extension of its one-dimensional counterpart.

**Definitions 7.10.** *Let $f(\cdot)$ be a real-valued function $f : \Omega \to \mathbb{R}$, where $\Omega \subseteq \mathbb{R}^n$, $x = (x_1, x_2, \ldots, x_n)$, and $x^* \in \Omega$.*

1. *The global minimum of $f(\cdot)$ occurs at a point $x^*$, on $\Omega$, if $f(x^*) \leq f(x)$ for all $x \in \Omega$.*
2. *The strict global minimum of $f(\cdot)$ occurs at a point $x^*$, on $\Omega$, if $f(x^*) < f(x)$ for all $x \in \Omega \backslash \{x^*\}$.*
3. *The local minimum of $f(\cdot)$ occurs at a point $x^*$, if $\exists$ a real number $\delta > 0$, such that $f(x^*) \leq f(x)$ for all $x \in B(x^*, \delta) \subseteq \Omega$.*
4. *The strict local minimum of $f(\cdot)$ occurs at a point $x^*$, if $\exists$ a real number $\delta > 0$, such that $f(x^*) < f(x)$ for all $x \in B(x^*, \delta) \subseteq \Omega \backslash \{x^*\}$.*
5. *The point $x^*$ is a critical or stationary point of $f(\cdot)$ if $\nabla f(x^*)$ exists and is equal to zero.*
6. *Let the function $f(\cdot)$ have continuous second partial derivatives on the set $\Omega$. If the point $x^*$ is an interior point of $\Omega$, and also a critical point, and the Hessian matrix $H(x^*)$ is indefinite, then $x^*$ is called a saddle point of $f(\cdot)$.*                                                □

The *global maximum, strict global maximum, local maximum,* and *strict local maximum* of $f(\cdot)$ can be similarly defined. The following observations summarize some useful facts about minimization and maximization of a function of several variables.

**Observations 7.7.** Let $f(\cdot)$ be a real-valued function $f : \Omega \to \mathbb{R}$, where $\Omega \subseteq \mathbb{R}^n$, and $x \in \Omega$.

1. Let $\nabla f(x)$ exist $\forall\, x \in \Omega$. If $x^*$ is an interior point of $\Omega$, and a local minima or maxima occurs at this point, then $\nabla f(x^*) = 0$.

2. Let $f(\cdot)$ possess continuous first and second partial derivative in $\mathbb{R}^n$, $H(x)$ be its Hessian matrix, and $x^* \in \Omega$ be a critical point of $f(\cdot)$.

   (a) The global minimum of $f(\cdot)$ occurs at a point $x^*$ on $\Omega$, if

   $$(x - x^*) H(y) (x - x^*)^T \geq 0; \quad \forall\, x \in \Omega, \quad \forall\, y \in [x^*, x]$$

   (b) The strict global minimum of $f(\cdot)$ occurs at a point $x^*$ on $\Omega$, if

   $$(x - x^*) H(y) (x - x^*)^T > 0; \quad \forall\, x \in \Omega \backslash \{x^*\}, \quad \forall\, y \in [x^*, x]$$

3. Corresponding statements about maximization can be made by reversing the direction of inequalities in the last observation. $\qquad\qquad\square$

The above observations can be immediately transformed into statements about the definiteness of the Hessian matrix $H(x)$.

**Theorem 7.9.** *Let $f(\cdot)$ be a real-valued function $f : \Omega \to \mathbb{R}$, where $\Omega \subseteq \mathbb{R}^n$, and $x \in \Omega$. The function $f(\cdot)$ possesses continuous first and second partial derivatives, and $H(x)$ is its Hessian matrix. Also $x^*$ is its critical point.*

(a) *The global minimum of $f(\cdot)$ occurs at a point $x^*$ on $\Omega$, if $H(x)$ is positive semidefinite.*
(b) *The strict global minimum of $f(\cdot)$ occurs at a point $x^*$ on $\Omega$, if $H(x)$ is positive definite.*
(c) *The global maximum of $f(\cdot)$ occurs at a point $x^*$ on $\Omega$, if $H(x)$ is negative semidefinite.*
(d) *The strict global maximum of $f(\cdot)$ occurs at a point $x^*$ on $\Omega$, if $H(x)$ is negative definite.* $\quad\square$

**Theorem 7.10.** *Let $f(\cdot)$ be a real-valued convex function $f : \Omega \to \mathbb{R}$, where $\Omega$ is a convex subset of $\mathbb{R}^n$. Then, any local minimizer $x^* \in \Omega$, of $f(\cdot)$ is also its global minimizer. Also, any local minimizer of a strictly convex function $f(\cdot)$ defined on a convex set $\Omega \subseteq \mathbb{R}^n$ is the unique and strict global minimizer of $f(\cdot)$.*

*Proof.* See the problem section. $\qquad\qquad\square$

The result about global maximization of a concave function is similar.

### 7.5.3  Nonlinear Constrained Optimization

Basics of nonlinear constrained optimization are examined in this subsection. More specifically, certain *nonlinear programming problems* are characterized. A nonlinear programming problem is a problem in which a function $f(\cdot)$ is maximized or minimized subject to a set of constraints, in which either the function $f(\cdot)$ or at least a single constraint is a nonlinear function. Recall that if $f(\cdot)$ is a linear function of a real variable $x$, then it is of the form $f(x) = ax + b$, where $a$ and $b$ are any real numbers. A function $f(\cdot)$ which is not of this form is called a *nonlinear* function of $x$. This definition of a nonlinear function can also be extended to $x \in \mathbb{R}^n$. A general nonlinear programming problem is stated as follows.

$$\textit{Maximize or Minimize } f(x), \quad x \in \mathbb{R}^n$$
$$\textit{Subject to}\,:\, g_i(x) \leq a_i, \quad 1 \leq i \leq u$$
$$g_i(x) \geq a_i, \quad (u+1) \leq i \leq v$$
$$g_i(x) = a_i, \quad (v+1) \leq i \leq m$$

where $a_i, g_i(x) \in \mathbb{R}$, $x \in \mathbb{R}^n$ and $1 \leq i \leq m$. The goal in this nonlinear programming problem, is to maximize or minimize a function $f(x) \in \mathbb{R}$, $x \in \mathbb{R}^n$ subject to the given set of constraints. In this problem $f(x)$ is called the *objective* or *cost*, and $g_i(x) \leq a_i$, $1 \leq i \leq u$; $g_i(x) \geq a_i$, $(u+1) \leq i \leq v$; and $g_i(x) = a_i$, $(v+1) \leq i \leq m$ are called the *constraints*. Furthermore, this problem is nonlinear, because the objective function or at least one of the functions appearing among the constraint sets is nonlinear. Nonlinear optimization techniques with equality and inequality constraints are studied in the next two subsections.

### 7.5.4  Optimization Problem with Equality Constraints

We investigate optimization problem with equality constraints in this subsection. Initially a simple example is considered in which the celebrated Lagrangian technique is introduced. This is followed by a more general example.

#### A Simple Example

We find the stationary points of the function $f(x_1, x_2) \in \mathbb{R}$, where $(x_1, x_2) \in \mathbb{R}^2$, subject to the constraint $g(x_1, x_2) = 0$. Note that $f(\cdot, \cdot)$ is the objective or cost function. Further assume that the functions $f(\cdot, \cdot)$ and $g(\cdot, \cdot)$ are smooth.

In this example it might be possible to eliminate one of the two variables by using the equality $g(x_1, x_2) = 0$, and then determine the stationary points of $f(\cdot, \cdot)$. If the function $g(\cdot, \cdot)$ is nonlinear, then this approach may become complex. Further, this approach might result in a restricted range of permissible values of $(x_1, x_2)$. In order to overcome this problem, we use the so-called method of constrained variation. Let the stationary point be $(x_1^*, x_2^*)$. At this point

$$df = \frac{\partial f}{\partial x_1} dx_1 + \frac{\partial f}{\partial x_2} dx_2 = 0$$

Furthermore

$$g(x_1^* + dx_1, x_2^* + dx_2) = 0, \quad \text{and} \quad g(x_1^*, x_2^*) = 0$$

Use of Taylor's series expansion of $g(x_1^* + dx_1, x_2^* + dx_2)$ yields

$$dg - \frac{\partial g}{\partial x_1} dx_1 + \frac{\partial g}{\partial x_2} dx_2 = 0$$

If $\partial g/\partial x_2 \neq 0$, the above equation yields

$$dx_2 = -\frac{\partial g/\partial x_1}{\partial g/\partial x_2} dx_1$$

Substitution of this result in the expression for $df = 0$ leads to

$$\frac{\partial f}{\partial x_1} \frac{\partial g}{\partial x_2} - \frac{\partial f}{\partial x_2} \frac{\partial g}{\partial x_1} = 0$$

evaluated at $(x_1^*, x_2^*)$. This result is a necessary condition for the existence of a local stationary point of $f(\cdot, \cdot)$. These algebraic details can be simplified by introducing a quantity $\lambda$, called the Lagrange multiplier, as

$$\lambda \triangleq -\frac{\partial f/\partial x_2}{\partial g/\partial x_2} \quad \text{evaluated at} \quad (x_1^*, x_2^*)$$

Substitution of this expression for $\lambda$ in the necessary condition for the existence of a local stationary point at $(x_1^*, x_2^*)$ yields

$$\frac{\partial f}{\partial x_1} + \lambda \frac{\partial g}{\partial x_1} = 0$$

The explicit expression for $\lambda$ yields

$$\frac{\partial f}{\partial x_2} + \lambda \frac{\partial g}{\partial x_2} = 0$$

Further, the constraint equation $g(x_1, x_2) = 0$ also has to be satisfied at $(x_1^*, x_2^*)$. The above results can be derived compactly by introducing the idea of *Lagrange function* and *multipliers*.

A Lagrange function or *Lagrangian* $\mathcal{L}(\cdot, \cdot, \cdot)$ for this example is defined as $\mathcal{L} : \mathbb{R}^2 \times \mathbb{R} \to \mathbb{R}$, where

$$\mathcal{L}(x_1, x_2, \lambda) = f(x_1, x_2) + \lambda g(x_1, x_2)$$

In the Lagrangian $\mathcal{L}(\cdot, \cdot, \cdot)$, $(x_1, x_2) \in \mathbb{R}^2$; and $\lambda \in \mathbb{R}$ is called the *Lagrange multiplier*. The stationary point $(x_1^*, x_2^*)$ is determined from the set of equations

$$\frac{\mathcal{L}(x_1, x_2, \lambda)}{\partial x_1} = 0, \quad \frac{\mathcal{L}(x_1, x_2, \lambda)}{\partial x_2} = 0, \quad \text{and} \quad \frac{\mathcal{L}(x_1, x_2, \lambda)}{\partial \lambda} = 0$$

These yield, as before

$$\frac{\partial f}{\partial x_1} + \lambda \frac{\partial g}{\partial x_1} = 0, \quad \frac{\partial f}{\partial x_2} + \lambda \frac{\partial g}{\partial x_2} = 0, \quad \text{and} \quad g(x_1, x_2) = 0$$

The important point to note is that by the introduction of Lagrange multiplier, constraint is removed in the modified objective function of the problem. This artifice was discovered by the mathematician Joseph Louis Lagrange (1736-1813) in the year 1761.

### A General Example

The problem is to find the stationary points of the function $f(x)$, $x \in \mathbb{R}^n$, subject to the equality constraints $g_j(x) = 0$ for $1 \le j \le m$. Note that $m < n$. That is, the number of constraints is less than the number of unknowns. Further assume that the objective and constraint functions are smooth. Also let $x = (x_1, x_2, \ldots, x_n)$, and the stationary point be $x^* = (x_1^*, x_2^*, \ldots, x_n^*)$. At the stationary point, as in the last example, we should have

$$df = 0, \quad \text{and} \quad dg_j = 0, \quad 1 \le j \le m$$

Therefore

$$df = \sum_{i=1}^{n} \frac{\partial f}{\partial x_i} dx_i = 0$$

$$dg_j = \sum_{i=1}^{n} \frac{\partial g_j}{\partial x_i} dx_i = 0, \quad 1 \le j \le m$$

In the above set of equations, the unknowns are $dx_i$ for $1 \le i \le n$. From $dg_j = 0$, $1 \le j \le m$ set of equations; $m$ out of $n$ number of $dx_i$'s are selected and expressed in terms of the remaining

$(n - m)$ number of $dx_i$'s. These are then substituted in the equation $df = 0$. Subsequently, equating to zero the coefficient of the $(n - m)$ number of $dx_i$'s yields the necessary condition for the determination of the stationary points of $f(\cdot)$. Note that the total number of $m$ possible selections of $dx_i$'s out of $n$ is $\binom{n}{m}$. This binomial coefficient can possibly be a very large number. Therefore this procedure is not convenient.

To partially overcome this difficulty, the artifice of Lagrange multipliers is used. The stationary points are found by first setting up an auxiliary function $\mathcal{L}(\cdot, \cdot)$. This auxiliary function, called the Lagrangian, is defined as $\mathcal{L} : \mathbb{R}^n \times \mathbb{R}^m \to \mathbb{R}$, where

$$\mathcal{L}(x, \lambda) = f(x) + \sum_{j=1}^{m} \lambda_j g_j(x)$$

and $x \in \mathbb{R}^n$, $\lambda \in \mathbb{R}^m$, $\lambda = (\lambda_1, \lambda_2, \ldots, \lambda_m)$. The scalars $\lambda_i$'s are the Lagrange multipliers. The stationary points are determined from the following set of $(m + n)$ necessary equations

$$\frac{\partial \mathcal{L}(x, \lambda)}{\partial x_i} = 0, \quad 1 \le i \le n, \quad \text{and} \quad \frac{\partial \mathcal{L}(x, \lambda)}{\partial \lambda_j} = 0, \quad 1 \le j \le m$$

Thus the introduction of the Lagrangian multipliers enables $x_i$ for $1 \le i \le n$ to be treated as though these were independent within the context of a modified objective function $\mathcal{L}(\cdot, \cdot)$. Let $x^* = (x_1^*, x_2^*, \ldots, x_n^*)$, and $\lambda^* = (\lambda_1^*, \lambda_2^*, \ldots, \lambda_m^*)$ satisfy the above $(m + n)$ set of equations. In order to determine $x^*$ and $\lambda^*$ we assumed that $m$ out of $n$ $dx_i$'s can be expressed in terms of the remaining $(n - m)$ differentials. This implies that the rank of the following $m \times n$ matrix

$$\begin{bmatrix} \partial g_1(x)/\partial x_1 & \partial g_1(x)/\partial x_2 & \cdots & \partial g_1(x)/\partial x_n \\ \partial g_2(x)/\partial x_1 & \partial g_2(x)/\partial x_2 & \cdots & \partial g_2(x)/\partial x_n \\ \vdots & \vdots & \ddots & \vdots \\ \partial g_m(x)/\partial x_1 & \partial g_m(x)/\partial x_2 & \cdots & \partial g_m(x)/\partial x_n \end{bmatrix}$$

is equal to $m$ at $x^*$. The above informal discussion yields the following theorem.

**Theorem 7.11.** *Necessary conditions for nonlinear optimization problem with equality constraints.*

$$\text{Maximize or minimize } f(x) \tag{7.17a}$$

$$\text{Subject to: } g_j(x) = a_j, \quad 1 \le j \le m \tag{7.17b}$$

*where $x = (x_1, x_2, \ldots, x_n) \in \mathbb{R}^n$, the objective $f(x) \in \mathbb{R}$; and $g_j(x), a_j \in \mathbb{R}$, for $1 \le j \le m$. Also, $m < n$. That is, the number of constraints is less than the number of variables. The objective function $f(\cdot)$, and the constraint functions $g_j(\cdot)$'s are continuously differentiable.*

*Let $x^* \in \mathbb{R}$ be a local stationary point, and the gradient vectors $\nabla g_j(x^*)$'s for $1 \le j \le m$ be linearly independent. Also let $\lambda = (\lambda_1, \lambda_2, \ldots, \lambda_m) \in \mathbb{R}^m$, where $\lambda_j$'s are the Lagrange multipliers, and*

$$\mathcal{L}(x, \lambda) = f(x) + \sum_{j=1}^{m} \lambda_j (g_j(x) - a_j) \tag{7.17c}$$

*The vector $x^*$ is a stationary point if there exists a vector $\lambda^* \in \mathbb{R}^m$ which satisfies the corresponding Lagrangian equations. These are:*

$$\nabla f\left(x^*\right) + \sum_{j=1}^{m} \lambda_j^* \nabla g_j\left(x^*\right) = 0 \tag{7.17d}$$

$$g_j\left(x^*\right) = a_j, \quad 1 \le j \le m \tag{7.17e}$$

$\square$

In general, the Lagrangian equations can be nonlinear and have more than one solution. The requirement that the vectors $\nabla g_j\left(x^*\right)$, for $1 \le j \le m$ be linearly independent is called a *constraint qualification*. This is required for the existence of $\left(x^*, \lambda^*\right)$. Each solution $\left(x^*, \lambda^*\right)$ is called a *Lagrange point*. Observe that the Lagrange multipliers $\lambda_j^*$'s can have either sign.

**Corollary 7.2.** *In the last theorem, let the equality constraint functions form a convex set, and the stationary point* $x^*$ *satisfy the Lagrangian equations.*

(a) *If the function* $f\left(\cdot\right)$ *is concave, then the solution* $x^*$ *is a global maximum.*
(b) *If the function* $f\left(\cdot\right)$ *is convex, then the solution* $x^*$ *is a global minimum.*   $\square$

If the first and second derivatives of the objective function $f\left(\cdot\right)$ and the constraint functions $g_j\left(\cdot\right)$'s are continuous, then sufficient condition for the existence of a relative maxima (minima) of $f\left(x\right)$ at $x^*$ is obtained by examining the quadratic $Q\left(x, \lambda\right)$, where

$$Q\left(x, \lambda\right) = \sum_{i=1}^{n}\sum_{k=1}^{n} \frac{\partial^2 \mathcal{L}\left(x, \lambda\right)}{\partial x_i \partial x_k} dx_i dx_k$$

The function $f\left(\cdot\right)$ has a maxima (minima) at $x = x^*$ if the quadratic $Q\left(x, \lambda\right)$ is negative (positive) definite for all values of $dx_i$'s.

### 7.5.5 Optimization Problem with Inequality Constraints

This subsection develops the celebrated Karush-Kuhn-Tucker (KKT) conditions for finding stationary points of a nonlinear optimization problem subject to certain inequality constraints. These conditions were first determined by W. Karush in 1939; and independently by H. W. Kuhn, and A. W. Tucker in 1951. We initially introduce the idea of KKT conditions informally via a multidimensional optimization problem with inequality constraints.

*Minimize* $f\left(x\right)$

*Subject to:* $g_j\left(x\right) \le a_j, \quad 1 \le j \le m$

where $x = \left(x_1, x_2, \ldots, x_n\right) \in \mathbb{R}^n$, the objective $f\left(x\right) \in \mathbb{R}$; and $g_j\left(x\right), a_j \in \mathbb{R}$, for $1 \le j \le m$. The objective function $f\left(\cdot\right)$, and the constraint functions $g_j\left(\cdot\right)$'s are continuously differentiable. The inequalities in this problem are converted to equalities by adding nonnegative slack variables $s_j^2$ to the left-hand-side of the inequality. That is

$$g_j\left(x\right) + s_j^2 = a_j; \quad s_j \in \mathbb{R}, \, 1 \le j \le m$$

A function $\mathcal{L}\left(\cdot, \cdot, \cdot\right)$ called Lagrangian is introduced, where

$$\mathcal{L}\left(x, \lambda, s\right) = f\left(x\right) + \sum_{j=1}^{m} \lambda_j \left(g_j\left(x\right) + s_j^2 - a_j\right)$$

and $\lambda, s \in \mathbb{R}^m$, $\lambda = (\lambda_1, \lambda_2, \ldots, \lambda_m)$, $s = (s_1, s_2, \ldots, s_m)$. The stationary points of the Lagrangian function $\mathcal{L}(\cdot, \cdot, \cdot)$ can be determined from

$$\frac{\partial \mathcal{L}(x, \lambda, s)}{\partial x_i} = \frac{\partial f(x)}{\partial x_i} + \sum_{j=1}^{m} \lambda_j \frac{\partial g_j(x)}{\partial x_i} = 0, \quad 1 \le i \le n$$

$$\frac{\partial \mathcal{L}(x, \lambda, s)}{\partial \lambda_j} = \left(g_j(x) + s_j^2 - a_j\right) = 0, \quad 1 \le j \le m$$

$$\frac{\partial \mathcal{L}(x, \lambda, s)}{\partial s_j} = 2\lambda_j s_j = 0, \quad 1 \le j \le m$$

These are $(2m + n)$ equations in the $(2m + n)$ unknowns $x$, $\lambda$, and $s$. Let the optimized $x$, $\lambda$ and $s$ vectors be $x^*$, $\lambda^*$ and $s^*$ respectively.

Note that the relationship $\lambda_j s_j = 0$ implies $\lambda_j \left(g_j(x) - a_j\right) = 0$, for $1 \le j \le m$. Also the equation $\left(g_j(x) + s_j^2 - a_j\right) = 0$ ensures that $g_j(x) \le a_j$, for $1 \le j \le m$. Consider the set of equations $\lambda_j s_j = 0$, for $1 \le j \le m$. For a given index $j$, two possibilities can occur, either $\lambda_j = 0$, or $s_j = 0$.

- If $s_j = 0$, then the $j$th inequality constraint is *active*. That is, $\left(g_j(x) - a_j\right) = 0$.
- If $\lambda_j = 0$, then the $j$th inequality constraint is *inactive*. Consequently it can be ignored.

Let $J_1$ and $J_2$ be the sets of indices with active and inactive inequality constraints respectively. Therefore $J_1 \cup J_2 = \{1, 2, \ldots, m\}$, and

$$-\frac{\partial f(x)}{\partial x_i} = \sum_{j \in J_1} \lambda_j \frac{\partial g_j(x)}{\partial x_i}, \quad 1 \le i \le n$$

The above set of equations can be written compactly in terms of the gradients. Thus

$$-\nabla f(x) = \sum_{j \in J_1} \lambda_j \nabla g_j(x)$$

It is next demonstrated that in this minimization problem, $\lambda_j$ for all $j \in J_1$ is positive.

A point $x \in \mathbb{R}^n$ is feasible, if it satisfies the constraints. A feasible region is a set of contiguous points which satisfy the constraints. A row vector $y \in \mathbb{R}^n$ of unit length is called a feasible direction from a feasible point $x \in \mathbb{R}^n$, if at least a small increment can be taken along $y$ so that it is still in the feasible region. Post-multiplying the above equation by $y^T$ yields

$$-\nabla f(x) y^T = \sum_{j \in J_1} \lambda_j \nabla g_j(x) y^T$$

As the vector $y$ has a feasible direction, we should have $\nabla g_j(x) y^T < 0$ for each $j \in J_1$ at the optimum point. If $\lambda_j > 0$ for each $j \in J_1$ then $\nabla f(x) y^T > 0$. It is a property of the gradients that $f(x)$ increases most rapidly in the direction of the gradient vector $\nabla f(x)$. Therefore, $\nabla f(x) y^T$ is the component of the increment of $f(\cdot)$ at point $x$ in the direction of $y$. Thus, if $\nabla f(x) y^T > 0$, the value of the function $f(\cdot)$ increases in the direction of $y$. Therefore, if $\lambda_j > 0$ for each $j \in J_1$, then it is not possible to find any direction, in the feasible region near the optimal point $x^*$, along which the value of the function can be decreased further. Thus in this minimization problem $\lambda_j \ge 0$, for $1 \le j \le m$.

In order to determine $x^*$, $\lambda^*$, and $s^*$, it is required that $f(\cdot)$, and $g_j(\cdot)$'s are continuously differentiable. In addition, the condition of *constraint qualification* has to be met at $x^*$. This condition means that the gradients of the active inequality constraints at $x^*$ are linearly independent. This informal discussion is summarized in the following theorem.

**Theorem 7.12.** *Necessary conditions for nonlinear optimization problem with inequality constraints.*

$$\text{Minimize } f(x) \tag{7.18a}$$

$$\text{Subject to: } g_j(x) \le a_j, \quad 1 \le j \le m \tag{7.18b}$$

*where $x = (x_1, x_2, \ldots, x_n) \in \mathbb{R}^n$, the objective $f(x) \in \mathbb{R}$; and $g_j(x), a_j \in \mathbb{R}$, for $1 \le j \le m$. The objective function $f(\cdot)$, and the constraint functions $g_j(\cdot)$'s are continuously differentiable.*

*Let $x^* \in \mathbb{R}^n$ be a locally minimizing point point, and $J_1$ be the set of indices of active inequality constraints at $x^*$. The gradient vectors $\nabla g_j(x^*)$'s for $j \in J_1$ are assumed to be linearly independent. Also let $\lambda = (\lambda_1, \lambda_2, \ldots, \lambda_m) \in \mathbb{R}^m$, where the $\lambda_j$'s are the Lagrange multipliers. For the existence of a minima at $x^*$, there must exist $\lambda^* \in \mathbb{R}^m$ which should satisfy the following necessary relationships.*

$$\frac{\partial f(x^*)}{\partial x_i} + \sum_{j=1}^{m} \lambda_j^* \frac{\partial g_j(x^*)}{\partial x_i} = 0, \quad 1 \le i \le n \tag{7.18c}$$

$$g_j(x^*) \le a_j, \quad 1 \le j \le m \tag{7.18d}$$

$$\lambda_j^* \ge 0, \quad 1 \le j \le m \tag{7.18e}$$

$$\lambda_j^* (g_j(x^*) - a_j) = 0, \quad 1 \le j \le m \tag{7.18f}$$

*These relationships are called the Karush-Kuhn-Tucker (KKT) conditions.*                                                                 □

**Corollary 7.3.** *In the last theorem, let each inequality constraint function be convex, and the stationary point $x^*$ satisfy the KKT conditions. If the function $f(\cdot)$ is convex, then the solution $x^*$ is a global minimum.*                                                                 □

If the goal of the nonlinear constrained optimization problem is maximization of the objective function $f(\cdot)$, then the corresponding Karush-Kuhn-Tucker conditions can be readily inferred from the following observations.

**Observations 7.8.** The objective function $f(\cdot)$ specified in the last theorem is maximized (instead of minimized). The inequality constraints and other specifications remain unchanged.

1. Let $x^*$ be a local maximizing point. The following Karush-Kuhn-Tucker conditions are satisfied at $(x^*, \lambda^*)$

$$-\frac{\partial f(x^*)}{\partial x_i} + \sum_{j=1}^{m} \lambda_j^* \frac{\partial g_j(x^*)}{\partial x_i} = 0, \quad 1 \le i \le n$$

$$g_j(x^*) \le a_j, \quad 1 \le j \le m$$

$$\lambda_j^* \ge 0, \quad 1 \le j \le m$$

$$\lambda_j^* (g_j(x^*) - a_j) = 0, \quad 1 \le j \le m$$

2. Let each inequality constraint function be convex, and the stationary point $x^*$ satisfy the KKT conditions. If the function $f(\cdot)$ is concave, then the solution $x^*$ is a global maxima.                                                    □

### Constrained Qualification

The Karush-Kuhn-Tucker conditions specify the necessary conditions for optimality. However, in order to determine $x^*, \lambda^*$, and $s^*$, the condition of *constraint qualification* has to be met at $x^*$. Consider an optimization problem with only inequality constraints. This condition of constraint qualification means that the gradients of the active inequality constraints at $x^*$ are linearly independent. This requirement is also called *linear independence constraint qualification* (LICQ). If a maximization or minimization problem has both equality and inequality constraints, then LICQ implies that the gradients of the equality and active inequality constraints at $x^*$ are linearly independent.

Thus $x^*$ does not exist, if the constraint qualification is not satisfied at $x^*$. Note that it is not possible to directly verify the constraint qualification without a knowledge of $x^*$. However, the constraint qualification is always satisfied in the following cases.

- Both equality and inequality constraint functions are linear functions of their arguments.
- In the optimization problem, let inequality constraint functions $g_j(\cdot)$ for $1 \leq j \leq m$ be convex, and the equality constraint functions $h_k(\cdot)$ for $1 \leq k \leq p$ be linear in their arguments. Further, there exists at least a single feasible point $\widetilde{x}$ such that $g_j(\widetilde{x}) < a_j$ for $1 \leq j \leq m$, and $h_k(\widetilde{x}) = b_k$ for $1 \leq k \leq p$. That is, the constraint set has a nonempty interior.

Linear independence constraint qualification (LICQ) is formally defined.

**Definitions 7.11**. *Consider the optimization problem*:

$$\text{Maximize or minimize } f(x) \tag{7.19a}$$

$$\text{Subject to:} \quad g_j(x) \leq a_j, \quad j \in \mathcal{I}_{inequality} \tag{7.19b}$$

$$g_k(x) = b_k, \quad k \in \mathcal{I}_{equality} \tag{7.19c}$$

*where $x = (x_1, x_2, \ldots, x_n) \in \mathbb{R}^n$, the objective $f(x) \in \mathbb{R}$; $g_j(x), a_j \in \mathbb{R}, j \in \mathcal{I}_{inequality}$; and $g_k(x), b_k \in \mathbb{R}, k \in \mathcal{I}_{equality}$.*

1. *A point $x \in \mathbb{R}^n$ is feasible, if it satisfies all the constraints.*
2. *The active set $\mathcal{A}(x)$ at any feasible point $x$ is the set of all indices from the equality constraints, which is $\mathcal{I}_{equality}$; and the set of indices from the inequality constraints $j$ for which $g_j(x) = a_j$. Thus*

$$\mathcal{A}(x) = \mathcal{I}_{equality} \cup \{j \in \mathcal{I}_{inequality} \mid g_j(x) = a_j\} \tag{7.19d}$$

3. *Let $x$ be a feasible point. The inequality constraint $j \in \mathcal{I}_{inequality}$ is said to be active if $g_j(x) = a_j$; and inactive if $g_j(x) < a_j$.*
4. *Let $x$ be a feasible point, and $\mathcal{A}(x)$ be an active set. Linear independence constraint qualification (LICQ) is said to occur at point $x$ if members of the set of constraint gradients $\{\nabla g_j(x) \mid j \in \mathcal{A}(x)\}$ are linearly independent.*                                                    □

**Optimization with Inequality Constraints and Nonnegative Variables**

At the beginning of this subsection, we considered minimization of a function with inequality constraints. In this nonlinear constrained optimization problem, we next impose the additional condition that the $x_i$'s are nonnegative. That is, $x_i \geq 0$, for $1 \leq i \leq n$. The optimization problem is thus:

$$Minimize\ f(x)$$

$$Subject\ to:\ g_j(x) \leq a_j,\quad 1 \leq j \leq m$$

where $x = (x_1, x_2, \ldots, x_n) \in \mathbb{R}^n$, the objective $f(x) \in \mathbb{R}$; $g_j(x), a_j \in \mathbb{R}$, for $1 \leq j \leq m$; and $x_i \geq 0$, for $1 \leq i \leq n$. The objective function $f(\cdot)$, and the constraint functions $g_j(\cdot)$'s are continuously differentiable.

The nonnegative constraints can be converted into an additional $n$ number of inequality constraints as $-x_i \leq 0$, for $1 \leq i \leq n$. This requires the introduction of $n$ number of additional Lagrange multipliers, and $n$ number of additional slack variables. Let these additional Lagrange multipliers and nonnegative slack variables be $\mu_i$ and $t_i^2$ respectively, for $1 \leq i \leq n$.

Let $\left(g_j(x) + s_j^2\right) = a_j$, where $s_j \in \mathbb{R}$, and $1 \leq j \leq m$. Also let $\left(-x_i + t_i^2\right) = 0$, where $t_i \in \mathbb{R}$, and $1 \leq i \leq n$. The Lagrangian function $\mathcal{L}(\cdot, \cdot, \cdot, \cdot, \cdot)$ is

$$\mathcal{L}(x, \lambda, s, \mu, t) = f(x) + \sum_{j=1}^{m} \lambda_j \left(g_j(x) + s_j^2 - a_j\right) + \sum_{i=1}^{n} \mu_i \left(-x_i + t_i^2\right)$$

and $\lambda, s \in \mathbb{R}^m$, $\lambda = (\lambda_1, \lambda_2, \ldots, \lambda_m)$, $s = (s_1, s_2, \ldots, s_m)$; $\mu, t \in \mathbb{R}^n$, $\mu = (\mu_1, \mu_2, \ldots, \mu_n)$, $t = (t_1, t_2, \ldots, t_n)$. The stationary points of the Lagrangian function $\mathcal{L}(\cdot, \cdot, \cdot, \cdot, \cdot)$ can be determined from

$$\frac{\partial \mathcal{L}(x, \lambda, s, \mu, t)}{\partial x_i} = \frac{\partial f(x)}{\partial x_i} + \sum_{j=1}^{m} \lambda_j \frac{\partial g_j(x)}{\partial x_i} - \mu_i = 0,\quad 1 \leq i \leq n$$

$$\frac{\partial \mathcal{L}(x, \lambda, s, \mu)}{\partial \lambda_j} = \left(g_j(x) + s_j^2 - a_j\right) = 0,\quad 1 \leq j \leq m$$

$$\frac{\partial \mathcal{L}(x, \lambda, s, \mu)}{\partial s_j} = 2\lambda_j s_j = 0,\quad 1 \leq j \leq m$$

$$\frac{\partial \mathcal{L}(x, \lambda, s, \mu)}{\partial \mu_i} = \left(-x_i + t_i^2\right) = 0,\quad 1 \leq i \leq n$$

$$\frac{\partial \mathcal{L}(x, \lambda, s, \mu)}{\partial t_i} = 2\mu_i t_i = 0,\quad 1 \leq i \leq n$$

These are $(2m + 3n)$ equations in $(2m + 3n)$ unknowns $x, \lambda, s, \mu$, and $t$. As in earlier discussion, $\lambda_j \geq 0$, for $1 \leq j \leq m$; and $\mu_i \geq 0$, for $1 \leq i \leq n$. As $\mu_i \geq 0$, for $1 \leq i \leq n$, we should have

$$\frac{\partial f(x)}{\partial x_i} + \sum_{j=1}^{m} \lambda_j \frac{\partial g_j(x)}{\partial x_i} \geq 0,\quad 1 \leq i \leq n$$

The equations $\left(g_j(x) + s_j^2 - a_j\right) = 0$, and $\lambda_j s_j = 0$ imply $\lambda_j \left(g_j(x) - a_j\right) = 0$, for $1 \leq j \leq m$. Similarly, the equations $\left(-x_i + t_i^2\right) = 0$, and $\mu_i t_i = 0$ imply $\mu_i x_i = 0$, for $1 \leq i \leq n$. This in turn implies

$$\left\{ \frac{\partial f(x)}{\partial x_i} + \sum_{j=1}^{m} \lambda_j \frac{\partial g_j(x)}{\partial x_i} \right\} x_i = 0, \quad 1 \le i \le n$$

This discussion is condensed in the following theorem.

**Theorem 7.13**. *Necessary conditions for nonlinear optimization problem with inequality constraints, and nonnegative variables.*

$$Minimize \ f(x) \tag{7.20a}$$

$$Subject\ to:\ g_j(x) \le a_j, \quad 1 \le j \le m \tag{7.20b}$$

*where* $x = (x_1, x_2, \ldots, x_n) \in \mathbb{R}^n$, *the objective* $f(x) \in \mathbb{R}$; *and* $g_j(x), a_j \in \mathbb{R}$, *for* $1 \le j \le m$; *and* $x_i \ge 0$, *for* $1 \le i \le n$. *The objective function* $f(\cdot)$, *and the constraint functions* $g_j(\cdot)$'s *are continuously differentiable.*

*Let* $x^* \in \mathbb{R}^n$ *be a locally minimizing point, and the gradients of all active inequality constraints be linearly independent at* $x^*$. *Also let* $\lambda = (\lambda_1, \lambda_2, \ldots, \lambda_m) \in \mathbb{R}^m$, *where* $\lambda_j$'s *are the Lagrange multipliers. For the existence of a minima at* $x^*$, *there must exist* $\lambda^* \in \mathbb{R}^m$ *which should satisfy the following necessary relationships.*

$$\frac{\partial f(x^*)}{\partial x_i} + \sum_{j=1}^{m} \lambda_j^* \frac{\partial g_j(x^*)}{\partial x_i} \ge 0, \quad 1 \le i \le n \tag{7.20c}$$

$$\left\{ \frac{\partial f(x^*)}{\partial x_i} + \sum_{j=1}^{m} \lambda_j^* \frac{\partial g_j(x^*)}{\partial x_i} \right\} x_i^* = 0, \quad 1 \le i \le n \tag{7.20d}$$

$$g_j(x^*) \le a_j, \quad 1 \le j \le m \tag{7.20e}$$

$$\lambda_j^* \ge 0, \quad 1 \le j \le m \tag{7.20f}$$

$$\lambda_j^* (g_j(x^*) - a_j) = 0, \quad 1 \le j \le m \tag{7.20g}$$

$$x_i^* \ge 0, \quad 1 \le i \le n \tag{7.20h}$$

*These relationships are called the Karush-Kuhn-Tucker (KKT) conditions.*           □

If in the last theorem: the stationary point $x^*$ satisfies the KKT conditions, the function $f(\cdot)$ is convex, and the constraint set is convex; then the solution $x^*$ is a global minimum. A result similar to that stated in the above theorem can be obtained if the objective function has to be maximized.

**Observations 7.9**. The objective function $f(\cdot)$ specified in the last theorem is maximized (instead of minimized). The inequality constraints and other specifications remain unchanged.

1. Let $x^*$ be a local stationary point. The following Karush-Kuhn-Tucker conditions are satisfied at $(x^*, \lambda^*)$

$$-\frac{\partial f(x^*)}{\partial x_i} + \sum_{j=1}^{m} \lambda_j^* \frac{\partial g_j(x^*)}{\partial x_i} \ge 0, \quad 1 \le i \le n$$

$$\left\{ -\frac{\partial f(x^*)}{\partial x_i} + \sum_{j=1}^{m} \lambda_j^* \frac{\partial g_j(x^*)}{\partial x_i} \right\} x_i^* = 0, \quad 1 \le i \le n$$

$$g_j\left(x^*\right) \leq a_j, \quad 1 \leq j \leq m$$
$$\lambda_j^* \geq 0, \quad 1 \leq j \leq m$$
$$\lambda_j^*\left(g_j\left(x^*\right) - a_j\right) = 0, \quad 1 \leq j \leq m$$
$$x_i^* \geq 0, \quad 1 \leq i \leq n$$

2. Let the stationary point $x^*$ satisfy the KKT conditions. If the function $f\left(\cdot\right)$ is concave, and the constraint set is convex; then the solution $x^*$ is a global maxima.    $\square$

**Example 7.7.** In this example

$$\textit{Minimize } f\left(x_1, x_2\right) = \left(2x_1^2 + 3x_2^2\right)$$
$$\textit{Subject to:} \quad 4x_1 + 3x_2 \leq -6$$

Here the function $f\left(\cdot, \cdot\right)$ and the constraint is convex. The Lagrangian is

$$\mathcal{L}\left(x_1, x_2, \lambda\right) = \left(2x_1^2 + 3x_2^2\right) + \lambda\left(4x_1 + 3x_2 + 6\right)$$

The partial derivatives of $\mathcal{L}\left(x_1, x_2, \lambda\right)$ with respect to $x_1, x_2,$ and $\lambda$ are

$$\frac{\partial \mathcal{L}\left(x_1, x_2, \lambda\right)}{\partial x_1} = 4x_1 + 4\lambda, \quad \frac{\partial \mathcal{L}\left(x_1, x_2, \lambda\right)}{\partial x_2} = 6x_2 + 3\lambda, \quad \frac{\partial \mathcal{L}\left(x_1, x_2, \lambda\right)}{\partial \lambda} = \left(4x_1 + 3x_2 + 6\right)$$

Equating the above partial derivatives to zero, and solving for $x_1, x_2,$ and $\lambda$ yields

$$x_1^* = -\frac{12}{11}, \quad x_2^* = -\frac{6}{11}, \text{ and } \lambda^* = \frac{12}{11}$$
$$f\left(x_1^*, x_2^*\right) = \mathcal{L}\left(x_1^*, x_2^*, \lambda^*\right) = \frac{36}{11}$$

$\square$

### 7.5.6 Nonlinear Optimization via Duality Theory

The concept of duality plays a very important theoretical and practical role in optimization theory. We have encountered duality theory earlier in the chapter when we studied linear programming. Its role in nonlinear constrained optimization problems is analogous to that in linear programming problems. A nonlinear constrained optimization problem, called the *primal*, can be converted into its *dual* form. This dual optimization problem can sometimes be computationally more efficient than the original problem. Further, the dual problem can also provide theoretical insight into the corresponding primal problem. Two types of dual problems are discussed in this subsection. These are the Lagrange-dual and Wolfe-dual problems. We initially outline Lagrange-duality theory, and subsequently Wolfe-duality theory.

#### Lagrange-Duality

We discuss classical Lagrangian duality. Consider the following nonlinear programming problem.

*Problem* $\mathcal{P}_L$: Consider the functions $f(\cdot)$ and $g_j(\cdot)$, where $f(x), g_j(x) \in \mathbb{R}, x \in C \subseteq \mathbb{R}^n$, and $1 \leq j \leq m$. Assume that these functions are continuous on $C$ and that the set $C$ is compact.

$$\underset{x \in C}{\textit{Minimize }} f(x) \tag{7.21a}$$

$$\textit{Subject to: } g_j(x) \leq 0, \quad 1 \leq j \leq m \tag{7.21b}$$

$\qquad\qquad\qquad\qquad\qquad\qquad\qquad\qquad\qquad\qquad\qquad\qquad\qquad\qquad\quad$ □

This problem is addressed by introducing the Lagrangian $\mathcal{L}(x, \lambda)$, where

$$\mathcal{L}(x, \lambda) = f(x) + \sum_{j=1}^{m} \lambda_j g_j(x)$$

and $\lambda \geq 0$. The problem $\mathcal{P}_L$ can be solved if a saddle point $(x^*, \lambda^*)$ of the Lagrangian function can be found. In this problem, minimization is performed with respect to $x$ and maximization with respect to $\lambda$. The notion of a saddle point within the context of problem $\mathcal{P}_L$ is made precise via the following definition.

**Definition 7.12.** *Let $x \in C$ and $\lambda \geq 0$. The vector $(x^*, \lambda^*)$ is a saddle point of the Lagrangian function $\mathcal{L}(\cdot, \cdot)$ if*

$$\mathcal{L}(x^*, \lambda^*) \leq \mathcal{L}(x, \lambda^*), \quad \forall\, x \in C$$
$$\mathcal{L}(x^*, \lambda) \leq \mathcal{L}(x^*, \lambda^*), \quad \forall\, \lambda \geq 0$$

*That is, $\mathcal{L}(x, \lambda)$ has a minimum with respect to $x$, but a maximum with respect to $\lambda$ at the point $(x^*, \lambda^*)$.* □

Note that if $x$ and $\lambda$ are scalar quantities, the shape of the function $\mathcal{L}(\cdot, \cdot)$ resembles a saddle. Hence the name saddle point. Based upon the above definition of saddle point, following observations can be made.

**Observations 7.10.** Saddle point related observations.

1. Let $x^* \in C$ and $\lambda^* \geq 0$. The vector $(x^*, \lambda^*)$ is a saddle point of the Lagrangian function $\mathcal{L}(\cdot, \cdot)$ iff:
   (a) $\mathcal{L}(x^*, \lambda^*) = \min_{x \in C} \mathcal{L}(x, \lambda^*)$
   (b) $g_j(x^*) \leq 0, 1 \leq j \leq m$
   (c) $\lambda_j^* g_j(x^*) = 0, 1 \leq j \leq m$
2. If $(x^*, \lambda^*)$ is a saddle point of the Lagrangian function $\mathcal{L}(\cdot, \cdot)$, then $x^*$ is a global optimum of the problem $\mathcal{P}_L$. □

Proofs of the above observations are provided in the problem section. Equipped with a precise notion of a saddle point of $\mathcal{L}(\cdot, \cdot)$, we are now ready to discuss details of Lagrangian duality. Define a function $w(\lambda)$ for $\lambda \geq 0$ by

$$w(\lambda) \triangleq \min_{x \in C} \mathcal{L}(x, \lambda)$$

That is, $w(\lambda) = \mathcal{L}(x^*, \lambda)$. Then if a saddle point exists for the problem $\mathcal{P}_L$, it can also be found by solving the following problem $\mathcal{D}_L$.

*Problem $\mathcal{D}_L$*: Lagrange-dual problem of the primal $\mathcal{P}_L$ is

$$\underset{\lambda}{Maximize}\ w(\lambda) \tag{7.22}$$

Subject to $\lambda \geq 0$.                                                                                    □

The problem $\mathcal{D}_L$ is called the *Lagrange-dual problem* of $\mathcal{P}_L$. The original problem $\mathcal{P}_L$ is called the *primal problem,* and $w(\cdot)$ is called the *dual function*. Observe that the Lagrange-dual problem $\mathcal{D}_L$ and the dual function $w(\cdot)$ can also be defined if there is no saddle point of the primal problem $\mathcal{P}_L$.

**Observations 7.11.** Observations related to Lagrange-duality.

1. For all $\lambda \geq 0$

$$w(\lambda) \leq w(\lambda^*) \leq f(x^*)$$

That is, $w(\lambda)$ is a lower bound of the absolute optimum value $f(x^*)$ of the primal problem $\mathcal{P}_L$. Also $w(\lambda^*)$ is the optimal value of the Lagrange-dual problem. This result is called the *weak duality theorem.*

2. The dual function $w(\cdot)$ is a concave function of $\lambda$. This observation does not require any assumption about convexity of functions $f(\cdot)$, and $g_j(\cdot)$, for $1 \leq j \leq m$. It also does not require any assumption about the convexity of the set $\mathcal{C}$.

3. The following result is called the *duality theorem.*

   (a) If the problem $\mathcal{P}_L$ has a saddle point $(x^*, \lambda^*)$, then $w(\lambda^*) = f(x^*)$. That is, the optimal value of the primal problem $\mathcal{P}_L$ is equal to the optimal value of the Lagrange-dual problem $\mathcal{D}_L$.

   (b) Conversely, if there is a solution $x^*$ of problem $\mathcal{P}_L$ and $\lambda^* \geq 0$ such that $w(\lambda^*) = f(x^*)$, then the primal problem $\mathcal{P}_L$ has a saddle point $(x^*, \lambda^*)$.                                    □

The correctness of these observations is established in the problem section.

**Example 7.8.** In this example

$$Minimize\ f(x_1, x_2) = \left(2x_1^2 + 3x_2^2\right)$$
$$Subject\ to: \qquad 4x_1 + 3x_2 \leq -6$$

This problem was addressed in the last example. However, the use of duality theory is demonstrated in this example. The Lagrangian is

$$\mathcal{L}(x_1, x_2, \lambda) = \left(2x_1^2 + 3x_2^2\right) + \lambda(4x_1 + 3x_2 + 6)$$

The partial derivatives of $\mathcal{L}(x_1, x_2, \lambda)$ with respect to $x_1$, and $x_2$ are

$$\frac{\partial \mathcal{L}(x_1, x_2, \lambda)}{\partial x_1} = 4x_1 + 4\lambda$$
$$\frac{\partial \mathcal{L}(x_1, x_2, \lambda)}{\partial x_2} = 6x_2 + 3\lambda$$

Equating the above partial derivatives to zero, and solving for $x_1$ and $x_2$, in terms of $\lambda$ yields

$$x_1 = -\lambda, \text{ and } x_2 = -\frac{\lambda}{2}$$

Therefore

$$w(\lambda) = -\frac{11}{4}\lambda^2 + 6\lambda$$

Note that $w(\cdot)$ is a concave function of $\lambda$. The maximum of $w(\cdot)$ is obtained if

$$\frac{\partial w(\lambda)}{\partial \lambda} = -\frac{11}{2}\lambda + 6 = 0$$

This yields $\lambda^* = 12/11$. Consequently

$$x_1^* = -\frac{12}{11}, \text{ and } x_2^* = -\frac{6}{11}$$

$$f(x_1^*, x_2^*) = \mathcal{L}(x_1^*, x_2^*, \lambda^*) = w(\lambda^*) = \frac{36}{11}$$

As expected, this result is identical to the last example.                          □

### Wolfe-Duality

Wolfe-duality is named after P. Wolfe. Consider the following nonlinear constrained optimization problem.

*Problem* $\mathcal{P}_W$: Consider the functions $f(\cdot)$ and $g_j(\cdot)$, where $f(x), g_j(x) \in \mathbb{R}, x \in \mathcal{C} = \mathbb{R}^n$, and $1 \leq j \leq m$. Also, $x = (x_1, x_2, \ldots, x_n)$. Assume that all these functions are convex on $\mathcal{C}$, and have continuous first partial derivatives with respect to the $x_i$'s.

$$\min_x f(x) \tag{7.23a}$$

$$\text{Subject to: } g_j(x) \leq 0, \quad 1 \leq j \leq m \tag{7.23b}$$

□

The above problem is called a *convex optimization problem*, because the objective and the constraint functions are all convex. Wolfe-duality is applicable to *only* such problems. For nonconvex problems, Lagrange-duality should be used. The Lagrangian $\mathcal{L}(x, \lambda)$ associated with this problem is

$$\mathcal{L}(x, \lambda) = f(x) + \sum_{j=1}^{m} \lambda_j g_j(x)$$

where $\lambda \geq 0$. Recall that the corresponding Lagrange-dual problem $\mathcal{D}_L$ is

$$\max_{\lambda \geq 0} \min_x \mathcal{L}(x, \lambda)$$

For any fixed $\lambda \geq 0$ the

$$\min_x \mathcal{L}(x, \lambda)$$

problem is convex. Therefore, minimum is attained if and only if the gradient of $\mathcal{L}(x, \lambda)$ with respect to $x$ is equal to zero. That is,

$$\frac{\partial \mathcal{L}(x, \lambda)}{\partial x_i} = 0, \quad 1 \leq i \leq n$$

Therefore the Wolfe-dual $\mathcal{D}_W$ of the problem $\mathcal{P}_W$ is

$$\max_{x, \lambda} \mathcal{L}(x, \lambda)$$

subject to $\nabla f(x) + \sum_{j=1}^{m} \lambda_j \nabla g_j(x) = 0$, and $\lambda \geq 0$. Note that the objective function $\mathcal{L}(\cdot, \cdot)$ of the Wolfe-dual is not necessarily concave. This problem is next stated formally.

*Problem $\mathcal{D}_W$*: Wolfe-dual problem of the primal $\mathcal{P}_W$ is

$$\max_{x, \lambda} \mathcal{L}(x, \lambda) \tag{7.24a}$$

$$\textit{Subject to: } \nabla f(x) + \sum_{j=1}^{m} \lambda_j \nabla g_j(x) = 0, \quad \text{and} \quad \lambda \geq 0 \tag{7.24b}$$

$\square$

The problem $\mathcal{D}_W$ is called the *Wolfe-dual problem* of $\mathcal{P}_W$. The original problem $\mathcal{P}_W$ is called the *primal problem*. Note that the equation $\nabla f(x) + \sum_{j=1}^{m} \lambda_j \nabla g_j(x) = 0$ is generally nonlinear. Further, $\mathcal{D}_W$ is not a convex optimization problem in general.

## 7.6 Calculus of Variations

Calculus of variations is a branch of mathematics which studies techniques used in optimizing (maximization or minimization) *functionals*. A real-valued function $f$ whose domain itself is a function (or a group of functions) is known as a functional. Functionals are generally stated as definite integrals whose integrands are stated in terms of functions and their derivatives. The goal is to determine an *extremal function* which optimizes the integral.

We next state a canonical problem in calculus of variations, in which the goal is to find a curve $y = Y(x)$, with end points at $x = x_1$ and $x = x_2$ such that the integral

$$\mathcal{I} = \int_{x_1}^{x_2} F(x, y, y') \, dx, \quad \text{where } y' = dy/dx, \; x_1 < x_2$$

is either a maximum or minimum. The corresponding value of the integral is called an *extreme value* or *extremum*. It is shown in the problem section that the necessary condition for the existence of extremum of the integral $\mathcal{I}$ is

$$\frac{d}{dx}\left(\frac{\partial F}{\partial y'}\right) - \frac{\partial F}{\partial y} = 0$$

The above equation is called the Euler-Lagrange equation. It does not tells us if this relationship is also sufficient to specify whether the integral $\mathcal{I}$ achieves either a minima or a maxima. This formulation can also be extended to optimize the integral

$$I = \int_{x_1}^{x_2} F\left(x, y_1, y_1', y_2, y_2', \ldots, y_n, y_n'\right) dx$$

where $y_1, y_2, \ldots, y_n$ are each functions of $x$; and $y_1', y_2', \ldots, y_n'$ are the corresponding derivatives with respect to $x$. Again, the goal is to determine extremal functions which optimizes the integral. The necessary conditions for optimizing this integral are the set of Euler-Lagrange equations.

$$\frac{d}{dx}\left(\frac{\partial F}{\partial y_i'}\right) - \frac{\partial F}{\partial y_i} = 0, \quad 1 \le i \le n$$

**Examples 7.9.** Some illustrative examples.

1. We determine a smooth curve on an $xy$-plane between two points, so that the distance between the two points is an extremum. Assume that the curve lies above the $x$-axis.
   Let $y = f(x)$ be a smooth curve in the $xy$-plane that passes through two endpoints $(x_1, y_1)$ and $(x_2, y_2)$ such that $x_1 < x_2$. The length of the curve between these two endpoints is

$$I(y(x)) = \int_{x_1}^{x_2} \left\{1 + y'^2\right\}^{1/2} dx$$

where $y'$ is the first derivative of $y(x)$ with respect to $x$. The goal is to find $y = f(x)$ which determines an extremum of $I(y(x))$. For this to be true, the Euler-Lagrange equation

$$\frac{d}{dx}\left(\frac{\partial F}{\partial y'}\right) - \frac{\partial F}{\partial y} = 0$$

has to be satisfied, where

$$F = \left\{1 + y'^2\right\}^{1/2}$$

We have

$$\frac{\partial F}{\partial y} = 0$$

$$\frac{\partial F}{\partial y'} = \frac{y'}{\left\{1 + y'^2\right\}^{1/2}}$$

and

$$\frac{d}{dx}\left\{\frac{y'}{\left\{1 + y'^2\right\}^{1/2}}\right\} = 0$$

This implies

$$\frac{y'}{\left\{1 + y'^2\right\}^{1/2}} = k$$

where $k$ is a constant. Thus

$$y' = \left\{\frac{k^2}{1 - k^2}\right\}^{1/2} = m$$

where $m$ is a constant. Therefore $y = (mx + c)$, where $c$ is a constant. This is the equation of a straight line.
   This curve only provides an extremum of the distance between the two points. It does not yet indicate whether the distance is either a minimum or a maximum.

2. We determine a smooth curve on an $xy$-plane between two points, so that the surface area generated by this curve by revolving it about the $x$-axis is an extremum.

Let $y = f(x)$ be a smooth curve in the $xy$-plane that passes through two endpoints $(x_1, y_1)$ and $(x_2, y_2)$ such that $x_1 < x_2$. The surface area generated by revolving this curve around the $x$-axis is

$$S(y(x)) = \int_{x_1}^{x_2} 2\pi y \, ds$$

where $ds = \left(dx^2 + dy^2\right)^{1/2}$. Therefore

$$S(y(x)) = \int_{x_1}^{x_2} 2\pi y \left\{1 + y'^2\right\}^{1/2} dx$$

where $y'$ is the first derivative of $y(x)$ with respect to $x$. The goal is to find $y = f(x)$ which minimizes $S(y(x))$. As $2\pi$ is a constant, the initial goal is to find an extremum of the functional

$$\mathcal{I}(y(x)) = \int_{x_1}^{x_2} y \left\{1 + y'^2\right\}^{1/2} dx$$

For this to be true, the Euler-Lagrange equation

$$\frac{d}{dx}\left(\frac{\partial F}{\partial y'}\right) - \frac{\partial F}{\partial y} = 0$$

has to be satisfied, where $F = y \left\{1 + y'^2\right\}^{1/2}$, and

$$\frac{\partial F}{\partial y} = \left\{1 + y'^2\right\}^{1/2}$$

$$\frac{\partial F}{\partial y'} = \frac{yy'}{\left\{1 + y'^2\right\}^{1/2}}$$

Therefore we have

$$\frac{d}{dx}\left\{\frac{yy'}{\left\{1 + y'^2\right\}^{1/2}}\right\} - \left\{1 + y'^2\right\}^{1/2} = 0$$

This implies

$$yy'' = 1 + y'^2$$

where $y''$ is the second derivative of $y$ with respect to $x$. The above equation is equivalent to

$$\frac{d}{dx}\left\{\frac{y}{\left\{1 + y'^2\right\}^{1/2}}\right\} = 0$$

Therefore

$$y = c_1 \left\{1 + y'^2\right\}^{1/2}$$

where $c_1$ is a constant. Substitute $y' = \sinh t$ in the above equation. This yields $y = c_1 \cosh t$. Therefore, we have $dy = c_1 \sinh t \, dt$ and $dy = \sinh t \, dx$. This in turn implies $dx = c_1 dt$. That is, $x = (c_1 t + c_2)$, where $c_2$ is a constant. Finally, we have

$$y = c_1 \cosh\left(\frac{x - c_2}{c_1}\right)$$

where the constants $c_1$ and $c_2$ are determined from the end-points $(x_1, y_1)$ and $(x_2, y_2)$, and $c_1 \neq 0$. This curve only provides an extremum for the surface area. It does not yet indicate whether the surface area is either a minimum or a maximum. $\qquad\square$

## 7.7 Stability Theory

Recall that the Internet is a huge feedback system. Feedback is used to control congestion in the Internet. Therefore, it is important to study stability theory meticulously. Concepts and notation which are used in describing a dynamical system are first described. This is followed by a discussion of stability theory.

### 7.7.1 Notions Used in Describing a System

In this subsection, concepts such as: relaxed system, linearity, causality, and time-invariance are introduced. A mathematical description of a dynamical system can be given via state variables and transfer-function matrices. These are subsequently used in the study of stability of such systems.

#### Description of Dynamical Systems

Before the stability of a system is discussed, the system has to be described carefully. A system is an interconnection of physical components that execute a specific function. It can sometimes be treated as a black box. In this case, its terminal properties are specified by the *external* or *input-output* description of the system.

In order for the system to be useful, it has to produce a useful output. This is accomplished via input signals. The desired output signal of the system can be achieved by varying its input signal in time. The input signal to the system can be electrical current or voltages, or mechanical forces. The input signals are also called *excitations*. The output signals of the system are called *responses*.

The internal and external properties of a continuous-time system can sometimes be described by a set of differential equations. Therefore the differential equations constitute the *internal* or *state-variable description* of the system.

**Definitions 7.13.** *Single and multivariable systems.*

1. *A single-variable system, has a single input terminal and a single output terminal.*
2. *A multivariable system can have*:
   (a) *A single input terminal and multiple output terminals.*
   (b) *Multiple input terminals and a single output terminal.*
   (c) *Multiple input and output terminals.*
3. *A single-variable system is initially at rest, if an input $u(t) \in \mathbb{R}$, where $t \in \mathbb{R}$ applied at $t \to -\infty$, produces an output that depends uniquely upon $u(t)$.*                                      □

Concepts such as relaxedness of a system, linearity, causality, and time-invariance are next examined. A system that is initially at rest at time $-\infty$ is an initially relaxed system, or simply a relaxed system. Denote the input signal in a single-variable system by $u(\cdot)$ or $u$. Let the corresponding output signal be $y(\cdot)$ or $y$. Therefore for a relaxed system, the output $y$ can be written as

$$y = Hu$$

where $H$ is either a matrix or differential or integral or stochastic operator. The operator $H$ and the input $u$ uniquely specify the output $y$. In the rest of this subsection, linear systems are studied. Such systems are generally amenable to useful analysis.

**Definition 7.14.** *A relaxed system is linear if and only if for any inputs* $u_1$ *and* $u_2$

$$H\left(\alpha_1 u_1 + \alpha_2 u_2\right) = \alpha_1 H u_1 + \alpha_2 H u_2, \quad \alpha_1, \alpha_2 \in \mathbb{R} \tag{7.25}$$

*otherwise it is said to be nonlinear.*                                                                        □

Sometimes, the above condition is expressed equivalently as:

(a) Additivity property: $H\left(u_1 + u_2\right) = H u_1 + H u_2$.
(b) Homogeneity property: $H\left(\alpha u\right) = \alpha H u$, for any $\alpha \in \mathbb{R}$.

A relaxed system possessing the above two properties is said to satisfy the *principle of superposition*. Next consider single-variable, relaxed, and linear systems. Such systems are best described via their response to Dirac's delta function $\delta\left(t\right), t \in \mathbb{R}$. The Dirac's delta function is sometimes called the impulse function in engineering literature. If the input to this system is an impulse function $\delta\left(t - \tau\right), \tau \in \mathbb{R}$, then its output is defined as

$$H\delta\left(t - \tau\right) \triangleq g\left(\cdot, \tau\right)$$

Observe that $g\left(\cdot, \cdot\right)$ is a real or complex-valued function of two variables. The first variable is the time at which the output is observed, and the second variable is the time at which the $\delta$-function is applied. Let the input to the system be $u\left(\cdot\right)$, then using the principle of superposition, the output of the system is given by $y\left(t\right)$, where

$$y\left(t\right) = \int_{-\infty}^{\infty} \left\{H\delta\left(t - \tau\right)\right\} u\left(\tau\right) d\tau = \int_{-\infty}^{\infty} g\left(t, \tau\right) u\left(\tau\right) d\tau$$

Therefore, if $g\left(\cdot, \tau\right)$ is known for all values of $\tau \in \mathbb{R}$, then for any input $u\left(\cdot\right)$ the output of the system $y\left(t\right)$ can be determined. The above integral is called the *superposition integral*, and $g\left(\cdot, \tau\right)$ is called the *impulse-response* of the system. These concepts can be easily extended to multivariable systems. Consider a relaxed and linear system with $p$ input and $q$ output terminals. Let the input and output column vectors of size $p$ and $q$ be $U\left(t\right)$ and $Y\left(t\right)$ respectively. The elements of these vectors are real numbers. Then

$$Y\left(t\right) = \int_{-\infty}^{\infty} G\left(t, \tau\right) U\left(\tau\right) d\tau, \quad t \in \mathbb{R}$$

where $G\left(t, \tau\right)$ is a real $q \times p$ matrix. It is called the *impulse-response matrix*. If $G\left(t, \tau\right) = \left[g_{ij}\left(t, \tau\right)\right]$, then $g_{ij}\left(t, \tau\right)$ is the impulse response of the system at time $t$ at the $i$th output terminal due to an impulse function applied at time $\tau$ to the $j$th input terminal.

The concept of *causality* is next developed. Causality implies that the output of a system at time $t$ does not depend upon inputs applied after time $t$. Equivalently, the output of a system at time $t$ depends only on the inputs applied in the interval $(-\infty, t]$. It is assumed that all physical systems are causal. Thus for a multivariable system

$$G\left(t, \tau\right) = 0, \quad \forall \tau \in \mathbb{R} \text{ and } \forall t < \tau$$

Thus

$$Y(t) = \int_{-\infty}^{t} G(t, \tau) U(\tau) \, d\tau, \quad t \in \mathbb{R}$$

Using this concept of causality, a *dynamic system* can be defined properly. A dynamic system is one in which the current output depends upon the current and the past inputs. In contrast, a *static system* is one in which the current output strictly depends only on the current inputs. The definition of a relaxed system is next generalized.

**Definition 7.15.** *A linear system is relaxed at time $t_0 \in \mathbb{R}$ if and only if the output $Y(t), t \in [t_0, \infty)$ is completely and uniquely determined by the input $U(t), t \in [t_0, \infty)$.*

*Alternately, the linear system is relaxed at time $t_0 \in \mathbb{R}$, if in such system $U(t) = 0, \forall t \in [t_0, \infty) \Rightarrow Y(t) = 0, \forall t \in [t_0, \infty)$.*                    □

Thus, if a system is causal, linear and relaxed at time $t_0$

$$Y(t) = \int_{t_0}^{t} G(t, \tau) U(\tau) \, d\tau, \quad t \geq t_0$$

Sometimes the characteristics of a system do not change with time. Such systems are called *time-invariant* or *autonomous*. If the system is not autonomous, then it is called *time varying* or *nonautonomous*. Time invariance is described in terms of a shifting operator $Q_\alpha$.

**Definitions 7.16.** *Shifting operator.*

1. *If the shifting operator $Q_\alpha$ is applied to a function $u(t), t \in \mathbb{R}$, then $Q_\alpha u(t) = u(t - \alpha)$, for $\alpha \in \mathbb{R}$.*
2. *A relaxed system is time-invariant if and only if $HQ_\alpha U = Q_\alpha HU$, for any input vector $U(\cdot)$, and any $\alpha \in \mathbb{R}$, otherwise it is time-varying.*                    □

For a linear, time-invariant, causal, and relaxed system

$$G(t, \tau) = G(t - \tau, 0)$$

With a little misuse of notation, $G(t, \tau)$ is denoted by $G(t - \tau)$ for a time-invariant system. Therefore the output of a linear, time-invariant, causal, and relaxed system is

$$Y(t) = \int_{t_0}^{t} G(t - \tau) U(\tau) \, d\tau, \quad t \geq t_0$$

If $t_0 = 0$, then for $t \geq 0$

$$Y(t) = \int_{0}^{t} G(t - \tau) U(\tau) \, d\tau$$

$$= \int_{0}^{t} G(\tau) U(t - \tau) \, d\tau$$

The integrals in the above equations are called the convolution integrals. Integrals of this type can be conveniently studied via single-sided Laplace transforms.

### Transfer-Function, Poles and Zeros

It is convenient to study relaxed, linear, causal, and time-invariant systems via Laplace transforms. Consider a single-variable linear, time-invariant, and causal system. In addition, assume that it is relaxed at time $t = 0$. Let the input and output functions be $u(\cdot)$ and $y(\cdot)$ respectively. Also denote the impulse response function of the system by $g(\cdot)$, then

$$y(t) = \int_0^t g(t - \tau) u(\tau) \, d\tau, \quad t \geq 0$$

If the single-sided Laplace transforms of $u(t), g(t)$, and $y(t)$, where $t \geq 0$ are $\widehat{u}(s), \widehat{g}(s)$, and $\widehat{y}(s)$ for $s \in \mathbb{C}$ respectively, then $\widehat{y}(s) = \widehat{g}(s) \widehat{u}(s)$.

Note that $\widehat{g}(s), s \in \mathbb{C}$ is called the *transfer-function*. Similarly, if the system is multivariable, linear, causal, time-invariant, and relaxed at time $t = 0$, then

$$Y(t) = \int_0^t G(t - \tau) U(\tau) \, d\tau$$

Define $\widehat{U}(s), \widehat{G}(s)$, and $\widehat{Y}(s)$ to be matrices which are obtained by taking the single-sided Laplace transforms of individual elements of the matrices $U(t), G(t)$, and $Y(t)$ respectively. In analogy with a single-variable system, $\widehat{G}(s), s \in \mathbb{C}$ is called the *transfer-function matrix*. The above matrix-convolutional equation yields $\widehat{Y}(s) = \widehat{G}(s) \widehat{U}(s)$, for $s \in \mathbb{C}$.

It is convenient to study these functions via their poles and zeros. Poles and zeros of a rational function $\widehat{g}(s), s \in \mathbb{C}$ are next defined. The function $\widehat{g}(\cdot)$ is said to be rational, if it can be expressed as a ratio of two polynomials in $s$. It is assumed that this function is irreducible. That is, there is no common factor (except possibly a constant) between numerator and denominator of $\widehat{g}(s)$.

**Definitions 7.17.** *Let $\widehat{g}(s)$ be an irreducible rational function in $s \in \mathbb{C}$.*

1. *A number $\beta \in \mathbb{C}$ is a pole of $\widehat{g}(s)$ if and only if $|\widehat{g}(\beta)| \to \infty$.*
2. *A number $\beta \in \mathbb{C}$ is a zero of $\widehat{g}(s)$ if and only if $\widehat{g}(\beta) = 0$.*                           ☐

### State-Variable Description of a System

A system can be described in terms of its states and dynamical equations.

**Definition 7.18.** *The state of a system at time $t_0$ is the information about the system, which with the real input vector $U(t), t \in [t_0, \infty)$ uniquely determines the output vector $Y(t), t \in [t_0, \infty)$.*                           ☐

It is assumed in this chapter, that the state of a system at time $t$ is described by a real column vector $X(t)$ of finite size. The vector $X(t)$ is called the *state vector*, and its components are called the *state variables*. Let $dX(t)/dt \triangleq \dot{X}(t)$ be a column vector whose elements are the first derivatives (with respect to time) of the elements of the vector $X(t)$.

**Definition 7.19.** *The set of equations which describe the unique relationships between the input and output state vectors are called dynamical equations. The state vector $X(t_0)$ and the input vector $U(t), t \in [t_0, \infty)$ are specified. The dynamical equations are of the form:*

(a) *State equation*: $\dot{X}(t) = H(X(t), U(t), t)$.
(b) *Output equation*: $Y(t) = G(X(t), U(t), t)$.                                    ☐

If the system is linear, then the dynamical equations are of the form

$$\dot{X}(t) = A(t)X(t) + B(t)U(t)$$
$$Y(t) = C(t)X(t) + D(t)U(t)$$

In the above set of equations, the column vectors $X(t)$ and $U(t)$ are of size $n$ and $p$ respectively. The output column vector $Y(t)$ is of size $q$. A sufficient condition for the above set of equations to have a unique solution is that all the elements of the matrix $A(t)$ are continuous functions of $t$ defined over the real line $\mathbb{R}$. In addition, assume for convenience that the elements of $B(t), C(t)$, and $D(t)$ are also continuous for all $t \in \mathbb{R}$. The matrices $A(t), B(t), C(t)$, and $D(t)$ are $n \times n$, $n \times p$, $q \times n$, and $q \times p$ respectively.

If the matrices $A(t), B(t), C(t)$, and $D(t)$ are independent of time, the system is both linear and time-invariant. Denote these matrices by $A, B, C$, and $D$ for convenience. The corresponding dynamical equations are

$$\dot{X}(t) = AX(t) + BU(t)$$
$$Y(t) = CX(t) + DU(t)$$

Next assume that the system is relaxed at time $t_0 = 0$. Let $s \in \mathbb{C}$, and $\widehat{X}(s), \widehat{U}(s)$, and $\widehat{Y}(s)$ be column vectors, whose elements are the single-sided Laplace transforms of the elements of the column vectors $X(t), U(t)$, and $Y(t)$ respectively. Then

$$s\widehat{X}(s) - X(0) = A\widehat{X}(s) + B\widehat{U}(s)$$
$$\widehat{Y}(s) = C\widehat{X}(s) + D\widehat{U}(s)$$

Simplification yields

$$\widehat{X}(s) = (sI_n - A)^{-1}X(0) + (sI_n - A)^{-1}B\widehat{U}(s)$$
$$\widehat{Y}(s) = C(sI_n - A)^{-1}X(0) + \left\{C(sI_n - A)^{-1}B + D\right\}\widehat{U}(s)$$

where $I_n$ is the identity matrix of size $n$. In the above set of equations, if $X(0)$ and $\widehat{U}(s)$ are known, then $\widehat{X}(s)$ and $\widehat{Y}(s)$ can be determined. Define

$$\widehat{G}(s) \triangleq \left\{C(sI_n - A)^{-1}B + D\right\}$$

The matrix $\widehat{G}(s)$ is called the *transfer-function matrix* of the system. Observe that the transfer-function matrix $\widehat{G}(s)$ can be expressed as

$$\widehat{G}(s) = \frac{P(s)}{\det(sI_n - A)}$$

where $P(s)$ is a matrix polynomial in $s$. That is, $P(s)$ is a polynomial in $s$ whose coefficients are matrices. The roots of the $n$th degree polynomial in the denominator are the eigenvalues of the matrix $A$. The linear, time-invariant, causal, and relaxed (at time $t = 0$) system is stable if and only if none of the $n$ eigenvalues of $A$ is located in the right-half of the $s$-plane, including the imaginary axis. That is, the eigenvalues are located in the left-half of the $s$-plane, excluding the imaginary axis. Under this condition, the state vector $X(t)$ tends to zero for large values of time $t$, for any finite initial value $X(0)$. This type of interpretation of stability is sometimes referred to as *asymptotic stability in the large*. It is possible in principle to give several definitions of stability. These and similar concepts are discussed at length in the next subsection.

### 7.7.2  Stability Concepts

Stability of a dynamical system can be studied from several different points of view. These are:

(a) Input-output stability. The system is considered as a black box, and its stability is studied.
(b) Total stability of a system. The scope of this stability is not only the input-output stability, but also the stability of the states of the system.
(c) Stability of *feedback systems*. In a feedback system, there is a *closed sequence* of cause-and-effect relationships among system variables. Feedback has a significant effect upon the stability of a system.
(d) Stability of *equilibrium points*. A point in the state space of a dynamical system is an equilibrium point if the system starts at this point, and remains at it for all future time, if no input is applied. The stability of equilibrium points is studied via techniques introduced by A. M. Lyapunov.

#### Input-Output Stability

The stability of a system can be specified via its input and output signals.

**Definition 7.20.** *A relaxed system with external input, is bounded-input bounded-output (BIBO) stable if and only if for any bounded input, the output is bounded.*                               □

Some observations about BIBO stability of single-variable system are stated below without proof.

**Observations 7.12.** Assume that the input to the system $u(t)$ is bounded for all values of $t$.

1. A relaxed single-variable system is specified by

$$y(t) = \int_{-\infty}^{t} g(t, \tau) u(\tau) \, d\tau$$

This system is BIBO stable if and only if

$$\int_{-\infty}^{t} |g(t, \tau)| \, d\tau < \infty, \quad \forall \, t \in \mathbb{R}$$

2. A relaxed, causal, and time-invariant single-variable system is specified by

$$y(t) = \int_{0}^{t} g(t - \tau) u(\tau) \, d\tau$$

This system is BIBO stable if and only if

$$\int_0^\infty |g(\tau)|\, d\tau < \infty$$

3. Assume that a relaxed, causal, and time-invariant single-variable system is specified by

$$\widehat{y}(s) = \widehat{g}(s)\,\widehat{u}(s)$$

where $\widehat{g}(s)$ is an irreducible rational function of $s \in \mathbb{C}$. This system is BIBO stable if and only if all the poles of $\widehat{g}(s)$ have negative real parts. That is, the poles lie in the open left-half $s$-plane excluding the imaginary axis. Note that the stability of this system is independent of the zeros of $\widehat{g}(s)$. □

Some observations about BIBO stability of multivariable system are also stated below without proof.

**Observations 7.13.** A multivariable system is described by $U(t)$ and $Y(t)$ which are input and output column vectors of size $p$ and $q$ respectively. Also $G(t,\tau) = [g_{ij}(t,\tau)]$ is a $q \times p$ impulse-response matrix.

1. A relaxed, and causal multivariable system is specified by

$$Y(t) = \int_{-\infty}^t G(t,\tau)\, U(\tau)\, d\tau$$

Assume that the input to the system $u_j(t)$ where $1 \le j \le p$, is bounded for all values of $t$. This system is BIBO stable if and only if

$$\int_{-\infty}^t |g_{ij}(t,\tau)|\, d\tau < \infty, \quad \forall\, t \in \mathbb{R}$$

for all values of $i$ and $j$, where $1 \le i \le q$ and $1 \le j \le p$.

2. A relaxed, causal, and time-invariant multivariable system is specified by

$$Y(t) = \int_0^t G(t-\tau)\, U(\tau)\, d\tau$$

This system is BIBO stable if and only if

$$\int_0^\infty |g_{ij}(\tau)|\, d\tau < \infty$$

for all values of $i$ and $j$, where $1 \le i \le q$ and $1 \le j \le p$.

3. Assume that a relaxed, causal, and time-invariant multivariable system is specified by

$$\widehat{Y}(s) = \widehat{G}(s)\,\widehat{U}(s)$$

where

$$\widehat{G}(s) = [\widehat{g}_{ij}(s)]$$

and each $\widehat{g}_{ij}(s)$ is an irreducible rational function of $s \in \mathbb{C}$. This system is BIBO stable if and only if all the poles of $\widehat{g}_{ij}(s)$ have negative real parts (excluding the imaginary axis) for all values of $i$ and $j$, where $1 \le i \le q$ and $1 \le j \le p$. □

### Total Stability

If the system is physical, then it is important to study all aspects of its stability. This includes not only the input-output stability of the dynamical system, but also the stability of its internal states.

**Definition 7.21.** *A linear dynamical system is totally stable, or T-stable, if and only if for any initial state and for any bounded input, the output and all the state variables are bounded.*   □

It is immediate from the definition of total stability that its requirements are much more stringent than the BIBO stability. For example, a system which is BIBO stable might have states which might increase with time indefinitely. Consequently a physical system should be designed to be T-stable. The following observations are stated without proof.

**Observations 7.14.** We make a note of some important facts about T-stability of a linear dynamical system.

1. Consider an $n$-dimensional linear time-varying dynamical system specified by

$$\dot{X}(t) = A(t)X(t) + B(t)U(t), \quad X(t_0) \in \mathbb{R}^n$$
$$Y(t) = C(t)X(t)$$

where the entries in the matrices $A(t), B(t)$, and $C(t)$ are continuous functions of $t \in \mathbb{R}$. This system is T-stable if the matrices $B(t)$ and $C(t)$ are bounded for all values of $t \in \mathbb{R}$, and the response of the system

$$\dot{X}(t) = A(t)X(t), \quad X(t_0) = 0$$

is asymptotically stable in the sense of Lyapunov. Lyapunov stability is discussed later in this subsection. The conditions stated in this observation are sufficient but not necessary for this system to be T-stable.

2. Consider an $n$-dimensional linear time-invariant dynamical system specified by

$$\dot{X}(t) = AX(t) + BU(t), \quad X(t_0) \in \mathbb{R}^n$$
$$Y(t) = CX(t)$$

where the entries in matrices $A, B$, and $C$ are constant real numbers. This system is T-stable if all the eigenvalues of the matrix $A$ have negative real parts (excluding the imaginary axis).   □

### Stability of Feedback Systems

Sometimes, it is useful to examine a system as a conglomeration of several smaller subsystems. Such systems are called composite systems. Larger systems are generally formed via three types of connections. These are parallel, tandem, and feedback connections. Of immediate interest to us is the system formed via feedback. In its simplest form, a feedback system is obtained if the output of the system is fed back to the input of the system via other systems, thus forming a closed path between its input and output.

It is assumed in this subsection that the systems are represented by their transfer-functions or transfer-matrix functions. See the Figure 7.3 for an example of a single-variable feedback system.

In this system $\widehat{u}(s)$ and $\widehat{y}(s)$, $s \in \mathbb{C}$ are the single-sided Laplace transforms of the input and output signals respectively. The transfer-function of the subsystem in the forward path is $\widehat{g}(s)$, $s \in \mathbb{C}$. The transfer-function of the subsystem in the feedback path is $\widehat{h}(s)$, $s \in \mathbb{C}$. The symbol $\oplus$ performs either addition or subtraction operation on the input and the fed back signal.

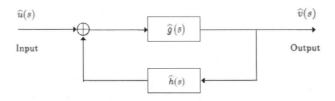

Figure 7.3. Single-variable feedback system.

The feedback is said to be *positive*, if the output signal that is fed back adds to the input signal. Similarly, the feedback is said to be *negative*, if the output signal that is fed back subtracts from the input signal.

Thus the net input to the subsystem in the $s$-domain with transfer-function $\widehat{g}(s)$ is

$$\left\{ \widehat{u}(s) \pm \widehat{y}(s)\,\widehat{h}(s) \right\}$$

The feedback is said to be positive (negative), if the positive (negative) sign is used. Therefore

$$\widehat{y}(s) = \left\{ \widehat{u}(s) \pm \widehat{y}(s)\,\widehat{h}(s) \right\} \widehat{g}(s)$$

Simplification yields $\widehat{g}_{cl}(s)$, $s \in \mathbb{C}$, which is the transfer-function of the complete feedback system. Thus

$$\widehat{g}_{cl}(s) = \frac{\widehat{y}(s)}{\widehat{u}(s)} = \frac{\widehat{g}(s)}{1 \mp \widehat{g}(s)\,\widehat{h}(s)}$$

The system shown in the Figure 7.3 is also said to be a *closed-loop* system, because there is a closed loop between the input and output signals. In contrast, an *open-loop* system has no loop. Actually the phrase open-loop is an oxymoron. The transfer-functions $\widehat{g}(s)\,\widehat{h}(s)$ and $\widehat{g}_{cl}(s)$ are called the *open-loop transfer-function* and *closed-loop transfer-function* respectively. Some text-books define $\widehat{g}(s)$ to be the open-loop transfer-function. This example, with a single input and a single output, can be easily extended to multivariable systems.

Study of feedback systems is important in its own right. This branch of knowledge has recently found use in the study and control of congestion in the Internet. Some of the basic reasons for using feedback are as follows:

(a) It is easier to control and adjust the transient response of the system.
(b) Decrease in the sensitivity of the system to variations in the parameters specified by the system $\widehat{g}(s)$.
(c) The system becomes less immune to external disturbances (noise signals). Noise signals in a system are usually the unaccountable and undesired signals.

In the year 1932, H. Nyquist, an engineer discovered a graphical technique to characterize the stability of linear and time-invariant feedback systems. Consider a negative feedback system, with the closed-loop transfer-function specified by

$$\widehat{g}_{cl}(s) = \frac{\widehat{g}(s)}{1 + \widehat{g}(s)\widehat{h}(s)} \tag{7.26}$$

This system becomes unstable if $\widehat{f}(s) = 0$, where

$$\widehat{f}(s) \triangleq \left\{ 1 + \widehat{g}(s)\widehat{h}(s) \right\} \tag{7.27}$$

That is, the closed-loop system becomes unstable at the zeros of $\widehat{f}(s)$. A qualitative discussion of Nyquist's criteria of stability of a closed-loop system is next given.

For $s \in \mathbb{C}$, consider the mapping of the function $\widehat{f}(s)$ from the $s$-plane to the $\widehat{f}(s)$-plane. Recall that in the $s$-plane, $\text{Re}(s)$ and $\text{Im}(s)$ are plotted on the horizontal and vertical axis respectively. Similarly, in the $\widehat{f}(s)$-plane, $\text{Re}\left(\widehat{f}(s)\right)$ and $\text{Im}\left(\widehat{f}(s)\right)$ are plotted on the horizontal and vertical axis respectively. Draw a closed contour $C_s$ in the $s$-plane. For simplicity assume that this contour does not pass through either the zeros or poles of $\widehat{f}(s)$. Also assume that this contour does not cross itself, and that it arbitrarily encircles only some zeros or poles of $\widehat{f}(s)$. Let $Z$ and $P$ be the number of zeros and poles of $\widehat{f}(s)$ (counting the multiplicities) that are encircled by the contour $C_s$ in the $s$-plane.

For each point $s$ on the contour $C_s$ evaluate $\widehat{f}(s)$ and plot it on the $\widehat{f}(s)$-plane. All these points in the $\widehat{f}(s)$-plane lie on a closed contour $C_{\widehat{f}}$. Using techniques from complex variables, it can be shown that the contour $C_{\widehat{f}}$ encircles the origin in the $\widehat{f}(s)$-plane $N = (Z - P)$ times. Furthermore, these encirclements are in the same direction (orientation) as $C_s$. Note that, if the value of $N$ is negative, then $-N$ number of encirclements of the origin in the $\widehat{f}(s)$-plane are in the opposite direction of the contour $C_s$.

Generally, $\widehat{f}(s)$ is not given in factored form. Therefore, it is not convenient to determine its zeros and poles. Typically, $\widehat{g}(s)\widehat{h}(s)$ is given in factored form. Therefore the above statements can be restated. Define $\widehat{v}(s)$ to be equal to $\widehat{g}(s)\widehat{h}(s)$. For each point $s$ on the contour $C_s$ evaluate $\widehat{v}(s)$ and plot it on the $\widehat{v}(s)$-plane. All these points in the $\widehat{v}(s)$-plane lie on a closed contour $C_{\widehat{v}}$. It can be shown that the net number of encirclements $N$, of the point $(-1, 0)$ in the $\widehat{v}(s)$-plane is equal to $(Z - P)$. In addition, these encirclements are in the same direction as $C_s$.

Without any loss of generality assume that the contour $C_s$ is drawn in a clockwise direction. In addition, select the contour $C_s$ such that it encloses the entire right-hand $s$-plane. This special contour $\Gamma_s$, is called the *Nyquist contour*. This contour passes through the imaginary axis from $-i\infty$ to $+i\infty$, and it is completed by a semicircular path of radius $r$, where $r$ approaches $\infty$. In addition, this contour should bypass (if any) the poles of $\widehat{f}(s)$ on the imaginary axis.

Therefore, if the closed-loop system is stable, the number of zeros of $\widehat{f}(s)$ inside and on the Nyquist contour should be equal to zero. These are the zeros in the closed right-half $s$-plane. Consequently for the system to be stable, the value of $Z$ should be equal to zero. Also note that the poles of $\widehat{f}(s)$ and $\widehat{v}(s)$ are identical. Consequently $N$, the number of encirclements of the point $(-1, 0)$ in the $\widehat{v}(s)$-plane should be equal to $-P$ for stability. This conclusion is formally stated as a theorem.

**Theorem 7.14.** (*H. Nyquist's theorem*) *Consider a linear, time-invariant, and closed-loop feedback system. Assume that its close-loop transfer-function exists, and is specified by*

$$\widehat{g}_{cl}(s) = \frac{\widehat{g}(s)}{\widehat{f}(s)}, \quad \widehat{f}(s) = \{1 + \widehat{v}(s)\} \qquad (7.28)$$

*Let $\Gamma_s$ be its Nyquist-contour drawn in the clockwise direction. Let the number of poles of $\widehat{v}(s)$ in the open right-half s-plane be equal to $P \in \mathbb{N}$.*

*The feedback system is T-stable if and only if the contour $C_{\widehat{v}}$ in the $\widehat{v}(s)$-plane does not pass through the point $(-1, 0)$, and:*

(a) *If $P = 0$, the contour $C_{\widehat{v}}$ in the $\widehat{v}(s)$-plane does not encircle the point $(-1, 0)$.*
(b) *If $P \in \mathbb{P}$, the contour $C_{\widehat{v}}$ in the $\widehat{v}(s)$-plane encircles the point $(-1, 0)$ exactly $P$ number of times in the counter-clockwise direction.*

  *Proof.* See the problem section.                                        □

The above result is also called *Nyquist's stability criteria*. The Nyquist-diagram, which shows the Nyquist's contour also provides additional information besides stability. It also provides information about the *degree of stability* of the feedback system. These are quantified by *gain* and *phase margins*. These measures which quantify relative stability of the system, are specified in terms of open-loop transfer function. More precisely, gain and phase margins are measures of proximity of the Nyquist's plot to the point $(-1, 0)$ in the $\widehat{v}(s)$-plane, where $s$ takes values on the Nyquist contour.

Thus a system is less stable, the closer it is to the point $(-1, 0)$ in the $\widehat{v}(s)$-plane. Gain margin is a factor by which $|\widehat{v}(i\omega)|$ would have to be increased to reach a unit value, when the phase of $\widehat{v}(i\omega)$ is equal to $\pm\pi$. Phase margin is described as the angle through which the $\widehat{v}(i\omega)$ plot would have to change to move to the point $(-1, 0)$, when $|\widehat{v}(i\omega)| = 1$. Therefore, if the phase margin is large, the system is more stable. Note that the phase margin is zero degrees (or radians) if the $\widehat{v}(i\omega)$ plot passes through $(-1, 0)$.

**Definitions 7.22.** *Let $\widehat{v}(i\omega) = |\widehat{v}(i\omega)| e^{i\varphi(\omega)}$*

1. *Gain margin $M_{gain}$ is the relative gain needed to drive the system into instability.*

$$M_{gain} = \frac{1}{|\widehat{v}(i\omega_\pi)|} \qquad (7.29a)$$

  *where $\widehat{v}(i\omega_\pi) = |\widehat{v}(i\omega_\pi)| e^{i\varphi(\omega_\pi)}$ and $\varphi(\omega_\pi) = \pm\pi$. Also $\omega_\pi$ is sometimes called the phase crossover frequency.*
2. *Phase margin $M_{phase}$ is the additional phase through which the $\widehat{v}(i\omega)$ locus must be rotated, such that the unity magnitude point $|\widehat{v}(i\omega_g)| = 1$ passes through the point $(-1, 0)$ in the $\widehat{v}(s)$-plane.*

$$M_{phase} = \pi - |\varphi(\omega_g)| \qquad (7.29b)$$

  *where $\widehat{v}(i\omega_g) = |\widehat{v}(i\omega_g)| e^{i\varphi(\omega_g)}$, $|\widehat{v}(i\omega_g)| = 1$, and $\varphi(\omega_g) \in [-\pi, \pi]$. Also $\omega_g$ is sometimes called the unity-gain crossover frequency.* □

In summary, gain margin is a measure of how much $|\widehat{v}(i\omega_\pi)|$ can be increased before instability sets in. In several cases increase in gain results in instability. In such cases the system is stable if

$$|\widehat{v}(i\omega_\pi)| < 1$$

Phase margin measures the proximity of $|\varphi(\omega_g)|$ to $\pi$. Some textbooks give similar but alternate definitions of gain and phase margins.

### Lyapunov Stability

This subsection is concerned with the stability of equilibrium points of a system. The concepts developed in this subsection were first stated by Lyapunov in 1892. In several scenarios, it is not necessary to obtain a complete solution of equations describing a system. It is simply required to know whether the system will function properly without any burn up, or crash, or oscillations. In some cases, it might be sufficient to know whether there exists any input which could cause any type of improper system response. Lyapunov's results give sufficient conditions for stability, and asymptotic stability of systems. However, these do not provide necessary conditions.

Note that the system discussed in the rest of this subsection can be either linear or nonlinear. Consider a system whose state equation is $\dot{X}(t) = H(X(t), U(t), t)$. If the input $U(t)$ is *known and fixed*, then $\dot{X}(t)$ can be expressed as

$$\dot{X}(t) = F(X(t), t)$$

The above equation also corresponds to a system with zero external input. The system with zero external input is generally called an unforced system.

**Definition 7.23.** *The state equation* $\dot{X}(t) = F(X(t), t)$ *describes an $n$-dimensional system. A vector $\chi \in \mathbb{R}^n$ is said to be an equilibrium point or state at time $t_0 \in \mathbb{R}^+$ if*

$$F(\chi, t) = 0, \quad \forall\, t \geq t_0 \tag{7.30}$$

*That is, $X(t) = \chi$ is a constant $\forall\, t \geq t_0$.*                                                  □

From the above definition, it follows that if the system is autonomous (time-invariant), then $\chi \in \mathbb{R}^n$ is an equilibrium point of the system at some time if and only if it is an equilibrium point of the system at all subsequent times. The terms *singular point* and *stationary point* are also used in the literature instead of equilibrium point (or state).

**Examples 7.10.** Some illustrative examples.

1. Let a single-variable and time-invariant system be specified by $\dot{x} = F(x)$. The equilibrium points of this system are the real zeros of the equation $F(x) = 0$.
2. For linear systems $0 \in \mathbb{R}^n$ is always an equilibrium state.                        □

Consider an autonomous system specified by the differential equation

$$\dot{X}(t) = F(X(t)), \quad X(0) = X_0$$

It is assumed that $F(X(t))$ satisfies proper continuity conditions such that this differential equation has a unique solution. The path followed by this system is called a *trajectory*. An equilibrium point is stable if all trajectories starting at nearby points stay close to it, otherwise it is unstable. Furthermore, the equilibrium point is asymptotically stable if all trajectories starting at nearby points stay nearby, and also approach the equilibrium point as time tends to infinity. These concepts are made precise in the following definitions.

**Definitions 7.24.** *Let an autonomous system (time-invariant) be specified by*

$$\dot{X}(t) = F(X(t)), \ \ and \ X(0) = X_0 \tag{7.31a}$$

*where its equilibrium point is $X(t) = X_e$. Let $\|\cdot\|$ denote the Euclidean norm.*

1. *The equilibrium state $X_e$ is stable at $t_0$ iff for any given $\epsilon > 0$, there exists a $\delta(\epsilon, t_0) > 0$ such that*

$$\|X(t_0) - X_e\| < \delta(\epsilon, t_0) \Rightarrow \|X(t) - X_e\| < \epsilon, \ \ \forall\, t \geq t_0 \tag{7.31b}$$

    *The stability is uniform if $\delta(\cdot, \cdot)$ is independent of $t_0$.*
    *The point $X_e$ is unstable if it is not stable.*
2. *The equilibrium point $X_e$ is asymptotically stable if it is stable and there exists $\eta > 0$ such that*

$$\|X(t_0) - X_e\| < \eta \Rightarrow \lim_{t \to \infty} \|X(t) - X_e\| = 0 \tag{7.31c}$$

    *That is, every solution $X(t)$ which starts sufficiently close to the equilibrium point $X_e$ converges to $X_e$ as $t \to \infty$.*
3. *The equilibrium point $X_e$ is globally asymptotically stable or asymptotically stable in the large if it is asymptotically stable for any finite initial condition, that is $\eta$ is any finite number.* □

A dynamical system can sometimes be described by either linear or nonlinear equations. As stated earlier, Lyapunov's approach to the study of stability of equilibrium states of such systems does not require that the solution of these equations be known. The basic idea behind Lyapunov's method is as follows. Consider a physical system in which its state is specified by a state $X(t)$. Let the system be in an equilibrium state at time $t_0$, and its energy be specified by $E(X(t))$. Furthermore, let the rate of change of energy $dE(X(t))/dt$ be negative for all possible values of the state $X(t)$ except at the equilibrium state $X(t_0)$. If this system is disturbed from its equilibrium state, then the system loses energy till it reaches the equilibrium state $X(t_0)$. Thus if this system is disturbed from its equilibrium, it *loses* energy in the process and returns to it equilibrium state. Lyapunov generalized this notion by introducing the idea of positive definite functions in his doctoral dissertation to study the stability properties of an equilibrium point.

**Definitions 7.25.** *Let $Y \in \mathbb{R}^n$ be a vector.*

1. *Also let $V(Y)$ be a real-valued function.*
    (a) *The function $V(\cdot)$ is positive definite if $V(Y) > 0$ for all $Y \neq 0$ and $V(0) = 0$.*
    (b) *The function $V(\cdot)$ is positive semidefinite if $V(Y) \geq 0$ for all $Y$, and $V(Y) = 0$ for some $Y \neq 0$.*
    (c) *The function $V(\cdot)$ is negative definite if $-V(Y)$ is positive definite.*
    (d) *The function $V(\cdot)$ is negative semidefinite if $-V(Y)$ is positive semidefinite.*
2. *Let $V(Y(t), t)$ be a real-valued time-varying function, such that $V(0, t) = 0$. The function $V(\cdot, \cdot)$ is said to be positive definite if there exists a positive definite function $W(Y(t))$ which is independent of $t$, such that $V(Y(t), t) \geq W(Y(t))$ for all $t$ and $Y(t)$.*
3. *Let $X(t) = (x_1(t), x_2(t), \ldots, x_n(t)) \in \mathbb{R}^n$ be a vector, where $t$ is time. Also $V(X(t), t)$ is a real-valued function.*

(a) *The function* $V(\cdot,\cdot)$, *is continuous, and its partial derivatives with respect to its arguments are also continuous.*
(b) *Furthermore,* $V(\cdot,\cdot)$ *is positive definite in a region* $\Re$ *containing the origin.*
(c) *The time derivative of* $V(X(t),t)$ *is* $\dot{V}(X(t),t)$.

$$\dot{V}(X(t),t) = \frac{\partial V(X(t),t)}{\partial t} + \sum_{i=1}^{n} \frac{\partial V(X(t),t)}{\partial x_i} \frac{dx_i}{dt} \tag{7.32}$$

*and* $\dot{V}(X(t),t)$ *is negative semidefinite.*
(d) *The time derivative of* $X(t)$ *is* $\dot{X}(t)$, *and a system is described by* $\dot{X}(t) = F(X(t),t)$. *This system has an equilibrium state at the origin.*

*Then* $V(X(t),t)$ *is called a Lyapunov function of the system. The surface* $V(X(t),t) = c$, *for some* $c \in \mathbb{R}^+$ *is called a Lyapunov surface or a level surface.*  □

The celebrated Lyapunov's stability theorem is stated below.

**Theorem 7.15.** (*A. M. Lyapunov's theorem, L1*) *A dynamical system is described by* $\dot{X}(t) = F(X(t),t)$. *It has an equilibrium state at the origin.*

(a) *If a Lyapunov function* $V(X(t),t)$ *exists in some sufficiently small neighborhood* $\Re$ *of the origin, then the equilibrium state is stable.*
(b) *If a Lyapunov function* $V(X(t),t)$ *exists, and* $\dot{V}(X(t),t)$ *is negative definite in some sufficiently small neighborhood* $\Re$ *of the origin (but excluding the origin), then the equilibrium state is asymptotically stable.*

*Proof.* See the problem section.                                                                □

Observe that this theorem provides a sufficient condition for the stability and asymptotic characteristics of the equilibrium point of a system. Furthermore, Lyapunov's theorem can be used without solving the differential equation. Therefore, if one can discover a Lyapunov function, then its stability can be characterized. However, if Lyapunov function has not been found, then it does not imply that the system is either stable or unstable. Technique for constructing Lyapunov functions for non-linear systems is generally hard. However construction of Lyapunov's function for time-invariant linear systems is always possible. The next theorem is about Lyapunov functions for time-invariant linear systems.

**Theorem 7.16.** (*A. M. Lyapunov's theorem, L2*) *Consider a time-invariant continuous-time linear system described by* $\dot{X}(t) = AX(t)$. *The equilibrium point at the origin of this system is asymptotically stable if and only if for any symmetric positive definite matrix* $Q$, *there exists a symmetric positive definite matrix* $M$ *which satisfies*

$$Q = -(MA + A^T M) \tag{7.33}$$

*The above relationship is called the Lyapunov's equation.*
*Proof.* See the problem section.                                                                □

## 7.8 Chaos Theory

Chaos means disorder. It is also a term used to describe ostensibly-complex and erratic behavior of simple and well-behaved dynamical systems. Chaos theory can be used in describing systems whose behavior lies between systems which are completely regular and systems which are completely random. Furthermore, chaos theory is a paradigm, which along with stochastic processes is useful in describing complex dynamical systems.

A dynamical system is described by differential equations, its physical parameters, and initial set of values. These differential equations can be either linear or nonlinear. A chaotic dynamical system is in addition nonlinear, deterministic, sensitive to initial conditions, and has a sustained level of irregularity in its behavior. These statements and concepts are further elaborated and restated below more precisely.

(a)  A system which is not linear is said to be *nonlinear*. The time evolution equations of a nonlinear system are nonlinear. Consequently, such systems are described by nonlinear functions.
(b)  *Determinism* implies that a system's future is completely described by its past behavior.
(c)  *Sensitivity to initial conditions*. This qualification means that infinitesimal changes in the initial state of the system result in drastically different behavior in its final state.
(d)  *Irregular behavior*. Such systems exhibit hidden order in what deceptively appears to be a random process. That is, there is *order in disorder*.
(e)  Long-term prediction due to sensitivity to initial conditions is generally impossible. Furthermore, initial conditions are known only to a finite degree of precision.

It can safely be said that the French mathematician Henri Poincaré (1854-1912) is responsible for the birth of the science of chaos. In November 1890, Poincaré won a first prize in an international contest to celebrate the sixtieth birthday of King Oscar II of Sweden and Norway, by submitting an essay on the three-body problem. The three-body problem is the problem of determining the dynamical motion of three bodies, given their initial coordinates, masses, and velocities. Poincaré introduced novel techniques to study such dynamical systems via qualitative and geometric features, and not just simply analytical formulae.

This subject of chaos theory almost lay dormant until the year 1963, when meteorologist Edward Lorenz analysed relatively harmless looking differential equations which he used in predicting weather. He discovered that the solution of the differential equations that he was studying could not be solved by any finite number of frequencies. Furthermore, these solutions were extremely sensitive to initial conditions. Sensitivity to initial conditions is also called the *butterfly effect*, after Lorenz. See notes at the end of this chapter regarding the choice of this phrase. This effect essentially implies that long-term prediction about the behavior of such systems (like weather) is nearly impossible.

The reader is forewarned that the interface between deterministic chaos and probabilistic random systems may not always be clear. It is quite possible that seemingly random systems could be described by hitherto undiscovered deterministic rules.

In the rest of this section, some preliminaries and definitions are first outlined. This is followed by a description of some measures to characterize chaotic maps. Finally, some examples which are useful in describing packet-source dynamics are given.

### 7.8.1  Preliminaries

Some notation and mathematical preliminaries are developed in order to define chaos precisely.

**Definitions 7.26.** *Let $f : \mathbb{R} \to \mathbb{R}$ and $g : \mathbb{R} \to \mathbb{R}$ be two functions, which are in $C^{\infty}$. Recall that a function $f\,(\cdot)$ is in $C^{\infty}$ if all its derivatives exist and are continuous on the specified interval.*

1. *The composition of functions $f\,(\cdot)$ and $g\,(\cdot)$ is denoted by $f \circ g\,(x) = f\,(g\,(x))$.*
2. *The $n$-fold composition of $f\,(\cdot)$ with itself is denoted by*

$$f^{n}\,(x) = \underbrace{f \circ f \circ \cdots \circ f}_{n \text{ times}}\,(x) \tag{7.34}$$

   *Note that $f^{0}\,(x) \triangleq x$, that is the zeroth composition of $f\,(\cdot)$ is the identity map.*
3. *A point $x \in \mathbb{R}$ is called a fixed point for $f\,(\cdot)$, if $f\,(x) = x$.*
4. *The set of points $x_0, x_1, x_2, \ldots, x_n, \ldots$ is called the orbit or trajectory of $x_0$ under iteration of $f\,(\cdot)$ where $x_1 = f\,(x_0)\,, x_2 = f\,(x_1)\,, \ldots, x_n = f\,(x_{n-1})\,, \ldots$.*
5. *A point $x \in \mathbb{R}$ is a periodic point of period $n \in \mathbb{P}$ if $f^{n}\,(x) = x$. The least positive $n$ for which $f^{n}\,(x) = x$ is called the prime period (or simply period) of $x$. Periodic orbits of period $n$ are called $n$-cycles.* □

**Examples 7.11.** Let $x \in \mathbb{R}$.

1. Let $f\,(x) = e^{x}$, then

$$f^{2}\,(x) = e^{e^{x}}$$
$$f^{3}\,(x) = e^{e^{e^{x}}}$$

   That is, the exponential function $e^{x}$ is iterated.
2. Let $f\,(x) = \sin x$, then

$$f^{2}\,(x) = \sin\,(\sin x)$$
$$f^{3}\,(x) = \sin\,(\sin\,(\sin x))$$

   It can be shown that $f^{n}\,(x_0) \to 0$ for large values of $n$ and any $x_0 \in \mathbb{R}$. This result can be easily checked on a hand-held calculator. The argument $x$ can be either in degrees or radians.
3. Let $f\,(x) = \cos x$, then $f^{n}\,(x_0) \to 0.99984\ldots$ for large values of $n$, and any $x_0 \in \mathbb{R}$, where $x_0$ is in degrees. If $x_0$ is in radians, then $f^{n}\,(x_0) \to 0.73908\ldots$, as $n \to \infty$. These results can again be checked easily on a hand-held calculator.
4. Let $f\,(x) = x^{3} + 3$. The orbit of the seed $0$ is the sequence

$$x_0 = 0,\ x_1 = 3,\ x_2 = 30,\ x_3 = 27003, \ldots,\ x_n = \text{ big}, x_{n+1} = \text{ bigger}, \ldots$$

   It can be observed, that this orbit tends to $\infty$ as $n \to \infty$.
5. The fixed points of the function $f\,(x) = x^{2m+1}$ are $x = 0, \pm 1$, where $m \in \mathbb{P}$.
6. Consider a function $f\,(x) = -x^{2m+1}$, where $m \in \mathbb{P}$. The fixed point of this function is at $0$. Its period is $2$ at $x = \pm 1$. This is true because $f\,(1) = -1$, and $f\,(-1) = 1$. Consequently $f^{2}\,(\pm 1) = \pm 1$. □

Chaos is next formally defined. It is based upon the concept of *topologically transitive* maps.

**Definition 7.27.** *A map* $f : I_{ch} \rightarrow I_{ch}$ *is said to be topologically transitive if for any two nonempty open sets* $U_{ch}, V_{ch} \subset I_{ch}$ *there exists* $n \in \mathbb{P}$ *such that* $f^n (U_{ch}) \cap V_{ch} \neq \varnothing$. $\qquad\square$

Therefore the map $f (\cdot)$ defined on $I_{ch}$ is transitive if: given any two nonempty subintervals $U_{ch}$ and $V_{ch}$ in $I_{ch}$, there is a point $x_0 \in U_{ch}$ and an $n \in \mathbb{P}$ such that $f^n (x_0) \in V_{ch}$. This definition of topological transitivity implies that this map has points which eventually move under successive iterations form one arbitrarily small neighborhood to another. Thus the map cannot be split into two disjoint open sets which are invariant under the map. Therefore, the transitivity property is an irreducibility condition, where the set $I_{ch}$ cannot be decomposed into subsets that are disjoint under repeated action of the mapping. Some mathematicians call this property *mixing*. Another property of chaotic dynamical systems is the *sensitivity to initial conditions* (SIC).

**Definition 7.28.** *A map* $f : I_{ch} \rightarrow I_{ch}$ *is said to have sensitive dependence on initial conditions, if there exists a sensitivity constant* $\beta > 0$ *such that, for any* $x \in I_{ch}$ *and any neighborhood* $N_{ch}$ *of* $x$, *there exists* $y \in N_{ch}$ *and* $n \in \mathbb{P}$ *such that* $|f^n (x) - f^n (y)| > \beta$. $\qquad\square$

This definition implies that a map possess the SIC property, if there exist points arbitrarily close to $x$ which subsequently separate form $x$ by at least $\beta$ after iterations of $f (\cdot)$. Consequently, if the mapping $f (\cdot)$ represents a dynamical system, then two trajectories which begin with almost identical initial conditions will eventually evolve along distinctly different trajectories. The reader should note that the SIC property requires the existence of only a single point in every neighborhood of $x$ so that the trajectories diverge under iteration. Furthermore, not all $y \in N_{ch}$ need exhibit this behavior.

After the above preliminary notions about chaos, we are ready to define it. Chaos can be defined in several different ways. The following definition is due to the mathematician Robert L. Devaney. It uses the concept of dense set. A subset $U_{ch} \subset X_{ch}$ is said to be dense in $X_{ch}$ if there exist points in $U_{ch}$ which are arbitrarily close to any point in $X_{ch}$.

**Definition 7.29.** *Let* $\varrho$ *be an interval, and* $f : \varrho \rightarrow \varrho$ *is a chaotic map on* $\varrho$ *if:*

(a) *The map* $f (\cdot)$ *has sensitive dependence on initial conditions.*
(b) *The map* $f (\cdot)$ *is topologically transitive.*
(c) *Periodic points of* $f (\cdot)$ *are dense in* $\varrho$. $\qquad\square$

Therefore a chaotic map has three characteristics: unpredictability, indecomposability, and a semblance of regularity. Unpredictability of the system is a consequence of the SIC property. Indecomposability implies that the system cannot be decomposed into two subsystems which have no interaction amongst each other. This characteristic is a consequence of the topologically transitive property. Finally, in the midst of apparent randomness is an element of regularity due to the presence of dense periodic points.

It has recently been established that, for maps defined on an interval, the first condition in the above definition follows from the other two. Generally speaking, a popular definition of chaos is simply SIC, which is the first condition of the above definition.

**Example 7.12.** Let $\varrho = [0, 1)$. A discontinuous map $D_{ch} : \varrho \rightarrow \varrho$ is defined as

$$D_{ch}\left(x\right) = \begin{cases} 2x, & \text{if } 0 \le x < 1/2 \\ \left(2x - 1\right), & \text{if } 1/2 \le x < 1 \end{cases}$$

This function is also called a doubling map. Note that this function can be expressed as $D_{ch}\left(x\right) \equiv 2x \pmod 1$. Therefore $D_{ch}^n\left(x\right) \equiv 2^n x \pmod 1$. This result implies that the plot of $D_{ch}^n\left(x\right)$ in the interval $\varrho$ consists of $2^n$ straight lines, each with a slope of $2^n$. It is next demonstrated that this map is chaotic on the interval $\varrho$. Define an interval

$$\varrho_k = [k/2^n, (k+1)/2^n), \quad \text{for } k = 0, 1, 2, \ldots, (2^n - 1)$$

The function $D_{ch}^n\left(\cdot\right)$ maps $x \in \varrho_k$ onto the interval $\varrho$. Thus the straight line corresponding to the function $D_{ch}^n\left(\cdot\right)$ in this interval $\varrho_k$ intersects the line $y = x$ at some point, which results in a periodic point. These periodic points are dense in $\varrho$, because these intervals are each of length $1/2^n$. The property of transitivity follows immediately, because for any open interval $I_{ch}$, it is always possible to find an interval of the type $\varrho_k$ inside $I_{ch}$ for sufficiently large values of $n$. Consequently the function $D_{ch}^n\left(\cdot\right)$ maps $I_{ch}$ onto all values in $\varrho$. This discussion also establishes sensitivity, provided the sensitivity constant is selected as $\beta = 1/2$.                                                                          □

A stronger form of SIC characterization is called expansiveness.

**Definition 7.30.** *Let $\varrho$ be an interval of real numbers, and $f : \varrho \to \varrho$ be a chaotic map on $\varrho$. This mapping is expansive if there exists $\lambda > 0$ such that, for any $x, y \in \varrho, x \ne y$, there exists $n \in \mathbb{P}$ such that $|f^n\left(x\right) - f^n\left(y\right)| > \lambda$.*                                                                          □

This characterization is different from SIC in that *all* nearby points of the trajectory eventually differ by at least $\lambda$. The following observation is used in the next subsection.

**Observation 7.15.** Let $\varrho = [0, 1]$, and $f : \varrho \to \varrho$ satisfy the following conditions:

(a) Let $a_0, a_1, \ldots, a_r \in \varrho$, where $0 = a_0 < a_1 < \cdots < a_r = 1$.
(b) Define a function $f_i\left(\cdot\right)$ in $C^2$ such that it is a restriction of $f\left(\cdot\right)$ to the interval $(a_{i-1}, a_i)$ for $1 \le i \le r$.
(c) Also $f\left((a_{i-1}, a_i)\right) = (0, 1)$, for $1 \le i \le r$. That is, the function $f_i\left(\cdot\right)$ is mapped onto $(0, 1)$.
(d) There exists $\lambda > 1$ such that $|f'\left(x\right)| \ge \lambda$ for $x \ne a_i, i = 0, 1, 2 \ldots, r$; where $f'\left(x\right)$ is the derivative of $f\left(x\right)$ with respect to $x$.

Then the map $f\left(\cdot\right)$ is chaotic and expansive.                                                                          □

See the problem section for a justification of the above observation.

## 7.8.2  Characterization of Chaotic Dynamics

Chaotic dynamics is characterized in this subsection. This is done via:

- Lyapunov's exponents
- Attractors
- Invariant density and Frobenius-Perron operators

**Lyapunov Exponent**

Chaotic dynamical systems are sensitive to initial conditions. This has been quantified by the sensitivity constant in the last subsection. There is another useful measure, called Lyapunov's exponent which also quantifies it. Recall that a dynamical system can be quantified by: the time-evolution equations, the initial conditions, and the values of the parameters which describe the system. Once these three ingredients are specified, it is hoped that the future behavior of the system can be computed for all time. Initial conditions are generally specified with only finite accuracy. For a chaotic dynamical system, these uncertainties in the initial conditions are magnified at an exponential rate. Lyapunov's exponent quantifies this phenomenon.

**Definition 7.31.** *Let $\varrho$ be a set. Consider a chaotic map defined by $f : \varrho \to \varrho$, and two orbits with nearly identical initial conditions $x_0$ and $(x_0 + \varepsilon)$, where $\varepsilon \to 0$. Then after $n \in \mathbb{P}$ iterations, the separation between the orbits is*

$$|f^n(x_0 + \varepsilon) - f^n(x_0)| = \varepsilon e^{n\widetilde{\lambda}(n, x_0)} \tag{7.35}$$

*The parameter $\widetilde{\lambda}(n, x_0)$ describes exponential divergence of the orbit. As $n \to \infty$, $\widetilde{\lambda}(n, x_0)$ tends to $\lambda(x_0)$. The parameter $\lambda(x_0)$ is called Lyapunov's exponent.* □

For a chaotic map, the parameter $\lambda(x_0)$ should be positive for "almost all" values of $x_0$. Let $f'(x)$ denote the derivative of $f(x)$ with respect to $x$. As $\varepsilon \to 0$, and for any given initial point $x_0$, the above definition yields

$$\widetilde{\lambda}(n, x_0) = \frac{1}{n} \ln |f^{n\prime}(x_0)|$$

$$= \frac{1}{n} \ln |f'(x_{n-1}) f'(x_{n-2}) \cdots f'(x_1) f'(x_0)|$$

$$= \frac{1}{n} \sum_{i=0}^{n-1} \ln |f'(x_i)|$$

Consequently, an alternate definition of Lyapunov's exponent for any given initial point $x_0$, can be given by

$$\lambda(x_0) = \lim_{n \to \infty} \frac{1}{n} \sum_{i=0}^{n-1} \ln |f'(x_i)|$$

provided the limit exists. If any of the derivatives in the above expression is zero, then $\lambda(x_0) \to -\infty$.

**Attractors**

In a dynamical system, attractor is the set of points to which orbits approach as the number of iterations tend towards infinity. It is possible for a chaotic system to have more than a single attractor for specified parameter values. An attractor is a special type of fixed point of a chaotic system. Recall that fixed point of a chaotic system is any value of $x$ which satisfies the relationship $f(x) = x$. Fixed points can also be classified as being either stable or unstable.

**Definitions 7.32.** *Stable, unstable, attractor, repeller, and neutral fixed points; and basin of attraction.*

1. *Let $f : \varrho \to \varrho$ be a chaotic map on an interval $\varrho$. Let $\widetilde{x}$ be a fixed point of $f$.*

   (a) *The point $\widetilde{x}$ is a stable fixed point, if there is a neighborhood $N_{ch} \subseteq \varrho$ of $\widetilde{x}$ such that if $x_0 \in N_{ch}$, then $f^n(x_0) \in N_{ch}$ for all values of $n$, and in addition $f^n(x_0) \to \widetilde{x}$ as $n \to \infty$.*

   (b) *The point $\widetilde{x}$ is an unstable fixed point, if all orbits beginning at any point in $N_{ch} \backslash \{\widetilde{x}\}$ leave the neighborhood $N_{ch}$ under iterations.*

2. *A stable fixed point is called an attractor (or sink), and an unstable fixed point a repeller (or source).*

3. *A fixed point is said to be neutral or indifferent, if it is neither attracting nor repelling.*

4. *The basin of attraction is the set of all initial states $x_0$ whose orbits converge to a given attractor.*     □

**Observations 7.16.** Let $\widetilde{x}$ be a fixed point of a function $f(x)$ defined on an interval $\varrho$. Denote the first derivative of $f(x)$ with respect to $x$ by $f'(x)$. Then

1. $\widetilde{x}$ is an attractor if $|f'(\widetilde{x})| < 1$.
2. $\widetilde{x}$ is a repeller if $|f'(\widetilde{x})| > 1$.
3. Nothing can be said about the type of $\widetilde{x}$ if $|f'(\widetilde{x})| = 1$.     □

See the problem section for proofs of these observations.

**Examples 7.13.** We consider several illustrative examples.

1. Consider the function $f(x) = x/3$. This function has $x = 0$ as its fixed point. Select a point $x_0$ near 0, but not equal to 0. It can be observed that its orbit tends to 0 as the number of iterations tend towards infinity. Therefore 0 is an attracting fixed point of $f(\cdot)$.
   Also note that $f'(x) = 1/3 < 1$. This fact also implies that 0 is an attracting fixed point of $f(\cdot)$.

2. Consider the function $g(x) = 3x$. This function has $x = 0$ as its fixed point. Select a point $x_0$ near 0, but not equal to 0. It can be observed that its orbits move away form 0 as the number of iterations tend towards infinity. Therefore 0 is a repelling fixed point of $g(\cdot)$.
   Also note that $g'(x) = 3 > 1$. This fact also implies that 0 is a repelling fixed point of $f(\cdot)$.

3. Let $h(x) = x^m$, where $m \geq 2$. The fixed point of these functions is $x = 0$ and the interval $(-1, 1)$ is its basin of attraction.

4. Let $f(x) = 2.5x(1 - x)$, where $x \in [0, 1]$. Its maximum value occurs at $x = 0.5$, and its fixed points are 0 and 0.6.

   (a) At the fixed point $x = 0$, $f'(0) = 2.5 > 1$. Therefore this fixed point is unstable, and its Lyapunov exponent is $\ln|f'(0)| = \ln|2.5| > 0$.

   (b) At the fixed point $x = 0.6$, $f'(0.6) = -0.5 < 0$. Therefore this fixed point is stable, and its basin of attraction is the interval $(0, 1)$. Its Lyapunov exponent is $\ln|f'(0.6)| = \ln|-0.5| < 0$.     □

**Invariant Density and Frobenius-Perron Operator**

A quantity of basic interest in a chaotic map is its *invariant density*. Consider the evolution of the trajectory of a chaotic map. As the map evolves, the frequency with which the trajectory visits a given neighborhood, $(x, x + dx)$ over an observation interval of $n$ iterations can be determined as follows.

Let $f : \varrho \to \varrho$ be a chaotic map defined on an interval $\varrho$, where $\varrho = [0, 1]$. Also let $x_0$ be the initial point of the orbit and $f^n(x_0)$ is the $n$th point in the orbit. The invariant density, $\rho(x)$ of the map is defined as

$$\rho(x) = \lim_{n \to \infty} \frac{1}{n} \sum_{i=0}^{n-1} \delta(x - f^i(x_0))$$

where $\delta(\cdot)$ is the Dirac's delta function. If $\rho(x)$ is independent of $x_0$ then the map is called ergodic.

Let the points on the trajectory of the ergodic map be $x_0, x_1, \ldots, x_n, \ldots$, and $\rho_n(x)$ be the density after $n$ iterations. Next observe that the effect of a single iteration of the map is to move a point $x_n$ to $f(x_n)$. That is, a density contribution of $\delta(x - x_n)$ is transformed to $\delta(x - f(x_n))$. Thus

$$\rho_{n+1}(x) = \int_0^1 \delta(x - f(z)) \rho_n(z)\, dz$$

If the density $\rho_n(x)$ is independent of $n$, it is called the invariant density of the map $f(x)$, and it is denoted by $\rho(x)$. This density function describes the density of the iterates of $x_n$ in the interval $\varrho$ as $n \to \infty$. Thus

$$\rho(x) = \int_0^1 \delta(x - f(z)) \rho(z)\, dz \tag{7.36}$$

The above operator is referred to as the Frobenius-Perron operator, and the corresponding equation, the Frobenius-Perron equation. It can have several solutions. Let $\widetilde{x}$ be a fixed point of the map $f(\cdot)$, then $\delta(x - \widetilde{x})$ is a solution of this equation.

### 7.8.3  Examples of Chaotic Maps

Some examples of chaotic maps are introduced below. These will help us to model packet generation processes. Two types of maps are considered. These are piecewise-linear and nonlinear maps.

#### Piecewise-Linear Maps

Chaotic piecewise-linear maps have special properties. These maps provide a stepping stone to study maps which model packet generation processes in the Internet. The following observation about piecewise-linear maps is useful in the study of the so-called Bernoulli map. Its validity is established in the problem section.

**Observation 7.17.** Let $\varrho = [0, 1]$, and $f : \varrho \to \varrho$ satisfy the following conditions:

(a) Let $a_0, a_1, \ldots, a_r \in \varrho$, where $0 = a_0 < a_1 < \cdots < a_r = 1$.
(b) Define a function

$$f_i(x) = c_i x + d_i,$$

where $c_i = (a_i - a_{i-1})^{-1}, d_i = -c_i a_{i-1}$, and $x \in (a_{i-1}, a_i)$. The function $f_i(\cdot)$ is the restriction of $f(\cdot)$ to the interval $(a_{i-1}, a_i)$, where $1 \leq i \leq r$.
(c) $f((a_{i-1}, a_i)) = (0, 1)$, for $1 \leq i \leq r$.

Then the invariant density is uniform, that is $\rho(x) = 1$ for $x \in \varrho$.                    □

An example of a piecewise linear map is the Bernoulli map.

### Bernoulli Maps

A Bernoulli map is a chaotic piecewise-linear map with two segments. See Figure 7.4. It is a generalization of the doubling map. This map is also called a Bernoulli-shift in the literature.

**Definition 7.33.** *A Bernoulli map is*

$$x_{n+1} = \begin{cases} x_n/d, & if \quad 0 \le x_n < d \\ (x_n - d)/(1 - d), & if \quad d \le x_n \le 1 \end{cases} \tag{7.37}$$

*where* $0 < d < 1$. $\qquad\qquad\qquad\qquad\qquad\qquad\qquad\qquad\qquad\qquad\qquad\qquad\qquad\qquad$ □

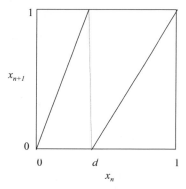

Figure 7.4. Bernoulli map.

The Bernoulli map is parameterized by a single parameter $d$. These maps can be conveniently generalized into nonlinear maps. The nonlinear maps are also called intermittency maps. These later maps have more than a single parameter. The rich tapestry of nonlinear maps is next examined.

### Nonlinear Maps

An example of a nonlinear map is the single intermittency map. It is called an intermittency map because, it has been used to model a phenomenon called intermittency in turbulence. Infrequent behaviors which dominate a system can be described by such intermittency maps. The single intermittency map is a generalization of the Bernoulli map. This map has two segments, of which one is linear, and the other nonlinear. See Figure 7.5 (a).

**Definition 7.34.** *A single intermittency map is*

$$x_{n+1} = \begin{cases} \epsilon + x_n + cx_n^m, & if \quad 0 \le x_n < d \\ (x_n - d)/(1 - d), & if \quad d \le x_n \le 1 \end{cases} \tag{7.38a}$$

*where*

$$c = \frac{1 - \epsilon - d}{d^m}, \quad 0 \le \epsilon \ll d, \ 0 < d < 1, \ and \ m > 1 \tag{7.38b}$$

$\qquad\qquad\qquad\qquad\qquad\qquad\qquad\qquad\qquad\qquad\qquad\qquad\qquad\qquad\qquad\qquad\qquad\qquad$ □

If $\epsilon = 0$ and $m = 1$, the single intermittency map is a Bernoulli map. A map with two nonlinear segments is called a double intermittency map. See Figure 7.5 (b).

**Definition 7.35.** *A double intermittency map is*

$$x_{n+1} = \begin{cases} \epsilon_1 + x_n + c_1 x_n^{m_1}, & if \quad 0 \leq x_n < d \\ x_n - \epsilon_2 - c_2 \left(1 - x_n\right)^{m_2}, & if \quad d \leq x_n \leq 1 \end{cases} \tag{7.39a}$$

$$c_1 = \frac{1 - \epsilon_1 - d}{d^{m_1}}, \quad and \quad c_2 = \frac{d - \epsilon_2}{\left(1 - d\right)^{m_2}} \tag{7.39b}$$

*where* $0 \leq \epsilon_1, \epsilon_2 \ll d,\ 0 < d < 1,\ and\ m_1, m_2 > 1.$ □

If $\epsilon_1 = \epsilon$, $\epsilon_2 = 0$, $m_1 = m$, and $m_2 = 1$, the double intermittency map reduces to a single intermittency map. These nonlinear maps and their extensions are further studied in the chapter on Internet traffic.

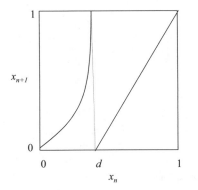

(a) Single intermittency map.

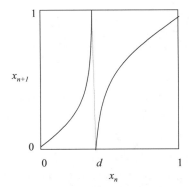

(b) Double intermittency map.

Figure 7.5. Nonlinear maps.

# Reference Notes

The goal of this chapter was to provide the mathematical underpinnings of linear programming techniques, optimization of nonlinear functions, and stability and chaos theory. A readable introduction to inequalities can be found in Iosevich (2007), Alsina, and Nelsen (2009), and Cloud, Drachman, and Lebedev (2009). The textbook by Franklin (1980) provides an erudite description of results on linear programming problems. The discussion on classical optimization techniques follows Cooper, and Steinberg (1970), Greig (1980), Foulds (1981), Nocedal, and Wright (2006), Peressini, Sullivan, and Uhl (1988), and Rao (2009). The subsection on Lagrangian duality is based upon Minoux (1986). The original work of H. W. Kuhn, and A. W. Tucker can be found in Kuhn, and Tucker (1951). The section on calculus of variations is based upon the textbook by Spiegel (1967).

Lucid accounts of stability theory can be found in the textbooks by Chen (1970), and Sarachik (1997). Erudite discussions about Lyapunov's stability theory can be found in Vidyasagar (1978), Levine (1996), and Khalil (2002). A readable and rigorous description of feedback systems is given in Rowland (1986), and Dorf (1992).

An accessible account of chaos theory can be found in the textbooks by Schuster (1995), Devaney (1989), Sandefur (1990), Hilborn (1994), Davies (1999), and Hirsch, Smale, and Devaney (2003). It has been demonstrated by Banks, Brooks, Cairns, Davis, and Stacey (1992) that in the Devaney's definition of chaos, the SIC condition follows from the other two conditions.

The phrase "butterfly effect," is attributed to the meteorologist E. N. Lorenz. A title of one of his talks was: "Predictability: Does the Flap of a Butterfly's Wings in Brazil set off a Tornado in Texas?" This title is in reference to the sensitivity of chaotic dynamical systems to initial conditions. The examples of linear and nonlinear chaotic maps are from Samuel (1999).

## Problems

1. Prove Jensen's inequality.
2. For $x > 0$ establish the inequality $\ln x \leq (x - 1)$.

   Hint: This result was established in an example. An alternate solution is given below. Consider the function $f(x) = \ln x - (x - 1)$. Its first derivative with respect to $x$ is equal to $(1/x - 1)$ and its second derivative is equal to $-x^{-2}$. Furthermore, the first derivative is equal to $0$ at $x = 1$, and the second derivative is equal to $-1$ at $x = 1$. Therefore, $f(x)$ has a maximum value at $x = 1$, which is equal to $0$. Thus $f(x) \leq 0$ for all positive values of $x$.
3. Let $f(\cdot)$ and $g(\cdot)$ be convex functions defined on a convex set $S$, then their sum

$$h(x) = f(x) + g(x), \quad \text{for } x \in S$$

is also convex. Establish this result.

Hint: For any $\lambda \in [0, 1]$; $x_1, x_2 \in S$; and by convexity of $f$ and $g$ we have

$$f(\lambda x_1 + (1 - \lambda) x_2) \leq \lambda f(x_1) + (1 - \lambda) f(x_2)$$
$$g(\lambda x_1 + (1 - \lambda) x_2) \leq \lambda g(x_1) + (1 - \lambda) g(x_2)$$

Adding the above two inequalities leads to

$$f(\lambda x_1 + (1 - \lambda) x_2) + g(\lambda x_1 + (1 - \lambda) x_2)$$
$$\leq \lambda f(x_1) + (1 - \lambda) f(x_2) + \lambda g(x_1) + (1 - \lambda) g(x_2)$$
$$= \lambda \{f(x_1) + g(x_1)\} + (1 - \lambda) \{f(x_2) + g(x_2)\}$$

That is

$$f(\lambda x_1 + (1 - \lambda) x_2) + g(\lambda x_1 + (1 - \lambda) x_2)$$
$$\leq \lambda \{f(x_1) + g(x_1)\} + (1 - \lambda) \{f(x_2) + g(x_2)\}$$

Thus

$$h(\lambda x_1 + (1 - \lambda) x_2) \leq \lambda h(x_1) + (1 - \lambda) h(x_2)$$

The result follows.

4. Let $f : \mathbb{R}^n \to \mathbb{R}$, $A$ be a real-valued $n \times m$ matrix, and $b \in \mathbb{R}^n$. Also let $g : \mathbb{R}^m \to \mathbb{R}$, where $g(x) = f(Ax + b)$. Then, prove that if $f(\cdot)$ is a convex function, then $g(\cdot)$ is also a convex function.

Hint: For $x_1, x_2 \in \mathbb{R}^m$ and $\lambda \in [0, 1]$ we have

$$
\begin{aligned}
g(\lambda x_1 + (1 - \lambda) x_2) &= f(A(\lambda x_1 + (1 - \lambda) x_2) + b) \\
&= f(\lambda(Ax_1 + b) + (1 - \lambda)(Ax_2 + b)) \\
&\leq \lambda f(Ax_1 + b) + (1 - \lambda) f(Ax_2 + b) \\
&= \lambda g(x_1) + (1 - \lambda) g(x_2)
\end{aligned}
$$

This implies that the function $g(\cdot)$ is convex.

5. Characterization of functions at end-points. Prove the following assertions.

(a) Let $f(x)$ be a convex function for $x \in [a, b]$. Then $f(\cdot)$ attains a maximum in the interval $[a, b]$. The maximum value is either $f(a)$ or $f(b)$.

(b) Let $f(x)$ be a concave function for $x \in [a, b]$. Then $f(\cdot)$ attains a minimum in the interval $[a, b]$. The minimum value is either $f(a)$ or $f(b)$.

6. Establish the arithmetic mean-geometric mean inequality. Let $a_i \in \mathbb{R}^+$, where $1 \leq i \leq n$. The arithmetic and geometric means of these numbers are $A_n$ and $G_n$ respectively, where

$$
A_n = \frac{1}{n} \sum_{i=1}^{n} a_i, \quad \text{and} \quad G_n = \left\{ \prod_{i=1}^{n} a_i \right\}^{1/n}
$$

Prove that $G_n \leq A_n$. Equality occurs if and only if the $a_i$'s are all equal.

Hint: See Iosevich (2007). The proof uses Jensen's inequality applied to convex functions. Observe that $e^x$ is a convex function for $x \in \mathbb{R}$. Therefore

$$
\sum_{i=1}^{n} \lambda_i = 1, \quad \lambda_i \geq 0, \ \ 1 \leq i \leq n
$$

$$
\exp\left( \sum_{i=1}^{n} \lambda_i x_i \right) \leq \sum_{i=1}^{n} \lambda_i e^{x_i}
$$

and $x_1, x_2, \ldots, x_n \in \mathbb{R}$. Substitute $\lambda_i = 1/n$, and $x_i = \ln a_i$ for $1 \leq i \leq n$ in the above result.

7. Let $a$ and $b$ be two positive numbers. Also, let their arithmetic and geometric means be $A_2$ and $G_2$ respectively. The harmonic mean $H_2$ of these two numbers is the reciprocal of the arithmetic mean of the reciprocal of these two numbers. Thus

$$
A_2 = \frac{a + b}{2}, \quad G_2 = \sqrt{ab}, \quad \text{and} \quad H_2 = \frac{2}{a^{-1} + b^{-1}}
$$

Prove that

$$
\min(a, b) \leq H_2 \leq G_2 \leq A_2 \leq \sqrt{\frac{a^2 + b^2}{2}} \leq \max(a, b)
$$

Hint: See Cloud, Drachman, and Lebedev (2009).

8. Establish Young's inequality. Let $a, b \in \mathbb{R}_0^+$; and

$$
(p^{-1} + q^{-1}) = 1, \quad \text{where} \ \ p, q > 1
$$

Then

$$ab \leq \frac{a^p}{p} + \frac{b^q}{q}$$

Equality occurs if and only if $b = a^{p-1}$.

Hint: The result is evident, if either $a$ and/or $b$ is equal to zero. Assume that $a, b \in \mathbb{R}^+$. In this case, the proof uses Jensen's inequality applied to convex functions. Observe that $e^x$ is a convex function for $x \in \mathbb{R}$. Therefore

$$\exp\left\{\frac{x_1}{p} + \frac{x_2}{q}\right\} \leq \left\{\frac{e^{x_1}}{p} + \frac{e^{x_2}}{q}\right\}$$

where $x_1 = p \ln a$, and $x_2 = q \ln b$.

Equality occurs if and only if $a^p = b^q$, which implies $b = a^{p-1}$.

9. Establish Hölder's inequality. Let $a_i, b_i \in \mathbb{R}^+$ for $1 \leq i \leq n$; and $p, q > 1$, where $\left(p^{-1} + q^{-1}\right) = 1$. Then

$$\sum_{i=1}^{n} a_i b_i \leq \left\{\sum_{i=1}^{n} a_i^p\right\}^{1/p} \left\{\sum_{i=1}^{n} b_i^q\right\}^{1/q}$$

Hint: See Tolsted (1964). Let

$$\alpha = \left\{\sum_{i=1}^{n} a_i^p\right\}^{1/p}$$

$$\beta = \left\{\sum_{i=1}^{n} b_i^q\right\}^{1/q}$$

and

$$a_i' = \frac{a_i}{\alpha}, \quad \text{and} \quad b_i' = \frac{b_i}{\beta}, \quad \text{for } 1 \leq i \leq n$$

We have to prove that $\sum_{i=1}^{n} a_i' b_i' \leq 1$. Using Young's inequality, we have

$$a_i' b_i' \leq \frac{a_i'^p}{p} + \frac{b_i'^q}{q}, \quad \text{for } 1 \leq i \leq n$$

Summing these inequalities yields

$$\sum_{i=1}^{n} a_i' b_i' \leq \frac{1}{p} \sum_{i=1}^{n} a_i'^p + \frac{1}{q} \sum_{i=1}^{n} b_i'^q$$

$$= \frac{1}{p} + \frac{1}{q} = 1$$

In the last step, we used the result $\sum_{i=1}^{n} a_i'^p = \sum_{i=1}^{n} b_i'^q = 1$.
Equality occurs if and only if for each $i = 1, 2, \ldots n$

$$a_i'^p = b_i'^q \Leftrightarrow b_i' = a_i'^{p-1} \Leftrightarrow \frac{b_i}{\beta} = \left\{\frac{a_i}{\alpha}\right\}^{p-1}$$

That is, equality occurs if and only if for each $i = 1, 2, \ldots n$

$$b_i = \beta \left\{\frac{a_i}{\alpha}\right\}^{p-1}$$

10. Establish Minkowski's inequality. Let $a_i, b_i \in \mathbb{R}$ for $1 \leq i \leq n$. For $p \geq 1$

$$\left\{ \sum_{i=1}^{n} |a_i + b_i|^p \right\}^{1/p} \leq \left\{ \sum_{i=1}^{n} |a_i|^p \right\}^{1/p} + \left\{ \sum_{i=1}^{n} |b_i|^p \right\}^{1/p}$$

Hint: See Cloud, Drachman, and Lebedev (2009). If $p = 1$, the result is triangle inequality. Next consider the case where $p > 1$. The result is proved via the use of Hölder's inequality. Let $q = p/(p-1)$. We have

$$\sum_{i=1}^{n} |a_i + b_i|^p \leq \sum_{i=1}^{n} |a_i| \, |a_i + b_i|^{p-1} + \sum_{i=1}^{n} |b_i| \, |a_i + b_i|^{p-1}$$

$$\leq \left\{ \sum_{i=1}^{n} |a_i|^p \right\}^{1/p} \left\{ \sum_{i=1}^{n} |a_i + b_i|^{(p-1)q} \right\}^{1/q}$$

$$+ \left\{ \sum_{i=1}^{n} |b_i|^p \right\}^{1/p} \left\{ \sum_{i=1}^{n} |a_i + b_i|^{(p-1)q} \right\}^{1/q}$$

$$= \left[ \left\{ \sum_{i=1}^{n} |a_i|^p \right\}^{1/p} + \left\{ \sum_{i=1}^{n} |b_i|^p \right\}^{1/p} \right] \left\{ \sum_{i=1}^{n} |a_i + b_i|^p \right\}^{1/q}$$

Observe that the result is true if $\sum_{i=1}^{n} |a_i + b_i|^p = 0$. Next assume that $\sum_{i=1}^{n} |a_i + b_i|^p \neq 0$. Then divide both sides of the above inequality by $\{\sum_{i=1}^{n} |a_i + b_i|^p\}^{1/q}$ to obtain the stated result.

11. Let $a_1, a_2, \ldots, a_n$ and $b_1, b_2, \ldots, b_n$ be nonnegative numbers. Then

$$\sum_{i=1}^{n} a_i \log \frac{a_i}{b_i} \geq \left\{ \sum_{i=1}^{n} a_i \right\} \log \frac{\sum_{i=1}^{n} a_i}{\sum_{i=1}^{n} b_i}$$

where equality occurs if and only if $a_i/b_i = k$ (a constant) for each $i$, where $1 \leq i \leq n$. Prove this log sum inequality.

Hint: See Cover and Thomas (1991). Without any loss of generality, assume that $a_i > 0$ and $b_i > 0$. Observe that the function $f(t) = t \log t$ is strictly convex for $t > 0$. This is true because $f''(t) = (1/t) \log e > 0$ for all positive values of $t$. Use of Jensen's inequality leads to

$$\sum_{i=1}^{n} \beta_i f(t_i) \geq f\left( \sum_{i=1}^{n} \beta_i t_i \right)$$

where $\beta_i \in \mathbb{R}_0^+$, for $1 \leq i \leq n$; and $\sum_{i=1}^{n} \beta_i = 1$. Substitution of

$$t_i = \frac{a_i}{b_i}, \quad \text{and} \quad \beta_i = \frac{b_i}{\sum_{i=1}^{n} b_i}, \quad \text{for } 1 \leq i \leq n$$

in the Jensen's inequality yields the stated result.

12. Let $wz^T = b$ be a hyperplane $\widehat{Z}$ in $\mathbb{R}^t$, $z \in \widehat{Z}$, $w \neq 0$, and $\|\cdot\|$ be the Euclidean norm defined on the vector space $\mathbb{R}^t$. A vector is represented as a row matrix. Prove the following results.

   (a) The vector $w$ is normal to the hyperplane $\widehat{Z}$.

(b) Let $q \in \mathbb{R}^t$ be a point which does not lie in the hyperplane $\widehat{Z}$. The shortest distance between the point $q$ and the hyperplane $\widehat{Z}$ is $\left| wq^T - b \right| / \|w\|$.

Hint:

(a) Let $\eta$ be a vector normal to the hyperplane $\widehat{Z}$, and $p$ be a point in $\widehat{Z}$. Therefore, the vector $(z - p)$ lies in the hyperplane $\widehat{Z}$. Thus $\eta (z - p)^T = 0$. If $\eta$ is selected such that $\eta p^T = b$, we obtain $\eta z^T = b$. As $wz^T = b$, the result follows.

(b) Let $u$ be a vector in the hyperplane $\widehat{Z}$. This implies $wu^T = b$. Also, the shortest distance between the point $q$ and the hyperplane $\widehat{Z}$ is $\left| w (q - u)^T \right| / \|w\| = \left| wq^T - b \right| / \|w\|$.

13. Prove the separating hyperplane theorem.

    Hint: We represent a vector as a row matrix. Let $X \subset \mathbb{R}^t$ be a nonempty closed, and convex set. Also let $h \in \mathbb{R}^t \backslash X$. Then as per the theorem, there exists a hyperplane which strictly separates $X$ from $h$. We prove that there is a hyperplane $xa^T = b$ for $b \in \mathbb{R}$, and $a \neq 0$ a constant vector; where $xa^T \leq b, \forall x \in X$, and $ha^T > b$. This theorem is proved in two steps.

    *Step* 1: In this step, it is established that there exists $\overline{x} \in X$ with minimum distance from $h \in \mathbb{R}^t \backslash X$; and also $(x - \overline{x})(h - \overline{x})^T \leq 0, \forall x \in X$.

    As $X$ is closed and compact, there exists a point $\overline{x} \in X$ which has a minimum distance from $h$. Thus

    $$\|h - \overline{x}\|^2 \leq \|h - x\|^2, \quad \forall x \in X$$

    Further, as $X$ is a convex set, for any $x \in X$, we have $\{(1 - \epsilon)\overline{x} + \epsilon x\} \in X, \forall \epsilon \in (0, 1)$. Therefore

    $$\|h - \overline{x}\|^2 \leq \|h - \overline{x} - \epsilon (x - \overline{x})\|^2$$
    $$= \|h - \overline{x}\|^2 + \epsilon^2 \|x - \overline{x}\|^2 - 2\epsilon (x - \overline{x})(h - \overline{x})^T$$

    That is, $(x - \overline{x})(h - \overline{x})^T \leq 0.5\epsilon \|x - \overline{x}\|^2$. Letting $\epsilon \to 0_+$ yields

    $$(x - \overline{x})(h - \overline{x})^T \leq 0, \quad \forall x \in X$$

    *Step* 2: Select $a \triangleq (h - \overline{x}) \neq 0$. Step 1 yields $(x - \overline{x}) a^T \leq 0$. As $\overline{x}a^T = b$, we have $xa^T \leq b$, $\forall x \in X$.

    Also $ha^T = (a + \overline{x}) a^T = \|a\|^2 + b > b$. That is, $ha^T > b$.

14. Prove Farkas' theorem (alternative).

    Hint: See Franklin (1980). We first show that the alternatives (i) and (ii) cannot be true simultaneously. Observe that $y^T A \geq 0$, and $x \geq 0$ imply $y^T Ax \geq 0$. This in turn implies $y^T b \geq 0$. This implies that alternative (ii) is false.

    It is next shown that of the two alternatives, one of them must be true. Suppose that alternative (i) is false. We show that alternative (ii) is true.

    If alternative (i) is false, then the point $b \notin S$, where $S$ is the closed convex set generated by the columns of the matrix $A$. That is, $S = \{Ax \mid x \geq 0\}$. As per the separating hyperplane theorem, there is a plane that *strictly* separates the set $S$ from the point $b$. Let the equation of the hyperplane in $z$ be $(a^T z + \beta) = 0$; where the constant column vector $a \neq 0$, and column vector $z$ are each of size $n \in \mathbb{P}$, and $\beta \in \mathbb{R}$. Therefore

    $$(a^T z + \beta) > 0, \quad z \in S$$
    $$(a^T z + \beta) < 0, \quad z = b$$

    Let $x \geq 0$ be any fixed column vector of size $n$, and $\lambda > 0$. For $z = A(\lambda x) \in S$, we obtain

$$a^T A \left( \lambda x \right) + \beta > 0, \quad x \geq 0, \text{ and } \lambda > 0$$

Divide the above inequality by $\lambda$, and let $\lambda \to +\infty$. This yields

$$a^T A x \geq 0, \quad x \geq 0$$

As the above result is true for any $x \geq 0$, we have

$$a^T A \geq 0$$

As the origin $0 \in S$, substituting $z = 0$ in the inequality $\left( a^T z + \beta \right) > 0$ shows that $\beta > 0$. At $z = b$, we have $\left( a^T z + \beta \right) < 0$. That is, $\left( a^T b + \beta \right) < 0$. Therefore

$$a^T b < -\beta$$

As $\beta > 0$, we have $a^T b < 0$. Substituting $a = y$ in the results $a^T A \geq 0$, and $a^T b < 0$ yields

$$y^T A \geq 0, \quad \text{and} \quad y^T b < 0$$

which is alternative (ii) of the Farkas' theorem.

15. Prove Farkas' corollary.

Hint: See Franklin (1980). Express $Ax \leq b$ as $(Ax + u) = b$. The alternative (i) of the corollary can be restated as

$$\begin{bmatrix} A & I \end{bmatrix} \begin{bmatrix} x \\ u \end{bmatrix} = b, \quad \text{has a solution} \quad \begin{bmatrix} x \\ u \end{bmatrix} \geq 0$$

where $I$ is an identity matrix of size $m$. As per Farkas' theorem, the above statement is true or (exclusive) $y^T \begin{bmatrix} A & I \end{bmatrix} \geq 0$, $y^T b < 0$. As $y^T \begin{bmatrix} A & I \end{bmatrix} = \begin{bmatrix} y^T A & y^T \end{bmatrix}$, the above statement is alternative (ii) of the corollary.

16. Prove the observations about the linear programming primal and dual problems.

Hint: See Cooper, and Steinberg (1970), and Foulds (1981).

(a) We prove that the dual of a dual is the primal. The canonical dual $\mathcal{D}_{LP}$ can be rewritten as:

Objective: $\max -b^T y$, subject to: $-A^T y \leq -c$, and $y \geq 0$.

The dual of this is:

Objective: $\min -c^T q$, subject to: $-\left( A^T \right)^T q \geq -b$, and $q \geq 0$.

This simplifies to:

Objective: $\max c^T q$, subject to: $Aq \leq b$, and $q \geq 0$. This is the original primal problem $\mathcal{P}_{LP}$.

(b) Weak duality property. We have

$$c^T x \leq \left( A^T y \right)^T x = y^T A x \leq y^T b = b^T y$$

Therefore $c^T x \leq b^T y$.

(c) Optimality property. From part (b) of the problem, for any feasible $x$ we have

$$c^T x \leq b^T y^*$$

However $c^T x^* = b^T y^*$. Therefore

$$c^T x \leq c^T x^*, \quad \text{for any feasible } x$$

Therefore $x^*$ is optimal for the primal problem $\mathcal{P}_{LP}$. It can be similarly shown that $y^*$ is optimal for the dual problem $\mathcal{D}_{LP}$.

17. Establish the strong duality theorem.

    Hint: The proof of the strong duality theorem uses the Farkas' corollary, and the weak duality property. Existence of the optimal solution is equivalent to the following inequalities.

    $$Ax \leq b, \quad x \geq 0$$
    $$-A^T y \leq -c, \quad y \geq 0$$
    $$c^T x \geq b^T y$$

    The inequality $c^T x \geq b^T y$ is listed because, the optimal solution $c^T x = b^T y$ follows from it and the weak duality result $c^T x \leq b^T y$. These inequalities can be compactly stated in matrix notation as

    $$\widetilde{A}\widetilde{x} \leq \widetilde{b}, \quad \widetilde{x} \geq 0$$

    where

    $$\widetilde{A} = \begin{bmatrix} A & 0 \\ 0 & -A^T \\ -c^T & b^T \end{bmatrix}, \quad \widetilde{b} = \begin{bmatrix} b \\ -c \\ 0 \end{bmatrix}, \quad \widetilde{x} = \begin{bmatrix} x \\ y \end{bmatrix}$$

    Assume that these inequalities are unsolvable. Then, as per the Farkas' corollary, there exists a column vector $\widetilde{y} \geq 0$ of real numbers of size $(m + n + 1)$ such that

    $$\widetilde{y}^T \widetilde{A} \geq 0, \quad \text{and} \quad \widetilde{y}^T \widetilde{b} < 0$$

    This is the reverse of the Farkas' corollary. Let

    $$\widetilde{y} = \begin{bmatrix} u \\ v \\ r \end{bmatrix}$$

    where $u \geq 0$, $v \geq 0$ are column vectors of real numbers of size $m$ and $n$ respectively; and $r \in \mathbb{R}_0^+$. Therefore $\widetilde{y}^T \widetilde{b} < 0$ yields $v^T c > u^T b$. That is

    $$c^T v > b^T u$$

    Similarly $\widetilde{y}^T \widetilde{A} \geq 0$ yields

    $$u^T A \geq rc^T, \quad \text{and} \quad rb^T \geq v^T A^T$$

    Observe that $rb^T \geq v^T A^T$ and $u \geq 0$ imply

    $$rb^T u \geq v^T A^T u \geq rv^T c > ru^T b = rb^T u$$

    Therefore we have $rb^T u > rb^T u$. This is true if $r = 0$. Therefore

    $$Av \leq 0, \quad \text{and} \quad A^T u \geq 0$$

    Let $x_\mu = (x + \mu v)$, and $y_\mu = (y + \mu u)$ where $\mu > 0$. It can be checked that $x_\mu$ and $y_\mu$ are indeed feasible for the primal and dual problems respectively. For

    $$Ax_\mu = A(x + \mu v) = Ax + \mu Av \leq b \Rightarrow Ax_\mu \leq b$$
    $$A^T y_\mu = A^T(y + \mu u) = A^T y + \mu A^T u \geq c \Rightarrow A^T y_\mu \geq c$$

From weak duality we have $c^T x_\mu \le b^T y_\mu$. Therefore

$$
\begin{aligned}
0 &\le b^T y_\mu - c^T x_\mu \\
&= b^T \left( y + \mu u \right) - c^T \left( x + \mu v \right) \\
&= \mu \left( b^T u - c^T v \right) + \left( b^T y - c^T x \right)
\end{aligned}
$$

That is

$$
\left( b^T u - c^T v \right) \ge \frac{1}{\mu} \left( c^T x - b^T y \right)
$$

As $\mu \to \infty$, we obtain $\left( b^T u - c^T v \right) \ge 0$. This contradicts an earlier result $c^T v > b^T u$. Therefore the reverse of the Farkas' corollary is not true. That is, its obverse must be true.

18. Let $f \left( \cdot \right)$ be a real-valued convex function $f : \Omega \to \mathbb{R}$, where $\Omega$ is a convex subset of $\mathbb{R}^n$. Then, any local minimizer $x^* \in \Omega$, of $f \left( \cdot \right)$ is also its global minimizer. Also, any local minimizer of a strictly convex function $f \left( \cdot \right)$ defined on a convex set $\Omega \subseteq \mathbb{R}^n$ is the unique and strict global minimizer of $f \left( \cdot \right)$. Prove this result.

Hint: See Peressini, Sullivan, and Uhl (1988). Let $x^* \in \Omega$ be a point at which a local minima of the convex function $f \left( \cdot \right)$ occurs. Therefore, there exists $r \in \mathbb{R}^+$ such that $f \left( x \right) \ge f \left( x^* \right)$, when $x \in \Omega$ and $\| x - x^* \| < r$.

Next select $\lambda \in \left( 0, 1 \right)$ so that $\left( x^* + \lambda \left( y - x^* \right) \right) \in \Omega$ and

$$
\| \left( x^* + \lambda \left( y - x^* \right) \right) - x^* \| < r
$$

Therefore because of convexity of $f \left( \cdot \right)$ on $\Omega$ we have

$$
\begin{aligned}
f \left( x^* \right) &\le f \left( x^* + \lambda \left( y - x^* \right) \right) = f \left( \lambda y + \left( 1 - \lambda \right) x^* \right) \\
&\le \lambda f \left( y \right) + \left( 1 - \lambda \right) f \left( x^* \right)
\end{aligned}
$$

Observe that the last inequality is strict, if $y \ne x^*$ and $f \left( \cdot \right)$ is strictly convex on $\Omega$. Simplification of $f \left( x^* \right) \le \left\{ \lambda f \left( y \right) + \left( 1 - \lambda \right) f \left( x^* \right) \right\}$ yields $0 \le \left( f \left( y \right) - f \left( x^* \right) \right)$. This inequality is strict if $f \left( \cdot \right)$ is strictly convex on $\Omega$ and $y \ne x^*$.

19. Let $x = \left( x_1, x_2, \ldots, x_n \right) \in \mathbb{R}^n$. Consider the following minimization problem.

$$
\textit{Minimize } f \left( x \right)
$$

$$
\begin{aligned}
\textit{Subject to:} \quad & g_j \left( x \right) \le a_j, \quad 1 \le j \le m \\
& x_i \ge 0, \quad 1 \le i \le n
\end{aligned}
$$

where $f \left( x \right)$ and $g_j \left( x \right)$'s have continuous first partial derivatives.

Define $\lambda = \left( \lambda_1, \lambda_2, \ldots, \lambda_m \right)$, and let the minimum of $f \left( x \right)$ occur at $x^*$ and $\lambda^*$. Show that the minimum of $f \left( x \right)$ can be determined from

$$
\frac{\partial f \left( x \right)}{\partial x_i} + \sum_{j=1}^{m} \lambda_j \frac{\partial g_j \left( x \right)}{\partial x_i} = 0, \ \text{ if } x_i > 0
$$

$$
\frac{\partial f \left( x \right)}{\partial x_i} + \sum_{j=1}^{m} \lambda_j \frac{\partial g_j \left( x \right)}{\partial x_i} \ge 0, \ \text{ if } x_i = 0
$$

$x_i \ge 0$, for $1 \le i \le n$; and

$$g_j(x) \leq a_j, \ \lambda_j \geq 0, \ \lambda_j(g_j(x) - a_j) = 0, \ \text{for } 1 \leq j \leq m$$

Hint: See Bunday (1984). A slightly different outline from that in the main text is given. The inequality constraint is initially transformed into an equality constraint by the addition of a nonnegative *slack variable* $s_j^2$. Let

$$g_j(x) + s_j^2 = a_j, \quad 1 \leq j \leq m$$

Let $s = (s_1, s_2, \ldots, s_m)$. Next set up the Lagrangian $\mathcal{L}(x, \lambda, s)$

$$\mathcal{L}(x, \lambda, s) = f(x) + \sum_{j=1}^{m} \lambda_j \left( g_j(x) + s_j^2 - a_j \right)$$

The stationary point of the Lagrangian function $\mathcal{L}(\cdot, \cdot, \cdot)$ is determined, subject to certain constraints on $x$ and $\lambda$. Also by hypothesis, $x \geq 0$. Let $x^*$, $\lambda^*$, and $s^*$ be the values of $x$, $\lambda$, and $s$ respectively which minimize $\mathcal{L}(x, \lambda, s)$ under these constraints. The following conditions must occur at $x = x^*$ and $\lambda = \lambda^*$. For $1 \leq i \leq n$ :

(a) If $x_i^* > 0$, then $\partial \mathcal{L}(x, \lambda, s) / \partial x_i = 0$.

(b) If $x_i^* = 0$, then $\partial \mathcal{L}(x, \lambda, s) / \partial x_i \geq 0$. That is, the minimum of the Lagrangian $\mathcal{L}(x, \lambda, s)$ occurs at a boundary.

The above two cases can be expressed explicitly in terms of $f(x)$ and $g_j(x), 1 \leq j \leq m$, at $x = x^*$, for $1 \leq i \leq n$.

$$\frac{\partial \mathcal{L}(x, \lambda, s)}{\partial x_i} = \frac{\partial f(x)}{\partial x_i} + \sum_{j=1}^{m} \lambda_j \frac{\partial g_j(x)}{\partial x_i} = 0, \ \text{if } x_i > 0$$

$$\frac{\partial \mathcal{L}(x, \lambda, s)}{\partial x_i} = \frac{\partial f(x)}{\partial x_i} + \sum_{j=1}^{m} \lambda_j \frac{\partial g_j(x)}{\partial x_i} \geq 0, \ \text{if } x_i = 0$$

Also at $x = x^*$, for $1 \leq j \leq m$

$$\frac{\partial \mathcal{L}(x, \lambda, s)}{\partial \lambda_j} = g_j(x) + s_j^2 - a_j = 0$$

$$\frac{\partial \mathcal{L}(x, \lambda, s)}{\partial s_j} = 2\lambda_j s_j = 0$$

Observe that the equation $g_j(x) + s_j^2 - a_j = 0$ implies $g_j(x) \leq a_j$. Also multiplication of the last equation by $s_j/2$ yields $\lambda_j s_j^2 = 0$. This yields

$$\lambda_j(g_j(x) - a_j) = 0, \ \text{for } 1 \leq j \leq m$$

Next observe that $\partial \mathcal{L}(x, \lambda, s) / \partial a_j = -\lambda_j$, for $1 \leq j \leq m$. Therefore, if $a_j, 1 \leq j \leq m$ is increased, the constraint region is enlarged. This in turn cannot yield a higher value of $f(x^*)$, although $f(x)$ can be smaller. Thus $\mathcal{L}(x, \lambda, s)$ does not increase as expected if $\lambda_j \geq 0$. That is, $\lambda_j \geq 0$, for $1 \leq j \leq m$. The stated conditions for this problem follow.

20. Consider the nonlinear constrained optimization problem

$$\textit{Maximize} \ \ f(x)$$

where $x = (x_1, x_2, \ldots, x_n) \in \mathbb{R}^n$

$$\text{Subject to:} \quad g_j(x) \leq a_j, \quad 1 \leq j \leq u$$
$$g_j(x) \geq a_j, \quad (u+1) \leq j \leq v$$
$$x_i \geq 0, \quad 1 \leq i \leq n$$

The functions $f(\cdot)$, and $g_j(\cdot)$'s are not necessarily linear, but have continuous first partial derivatives. Prove that the necessary conditions for $x^* \in \mathbb{R}^n$ to be an optimal solution are: existence of vectors $\lambda^*$ and $\mu^*$ where

$$\lambda^* = (\lambda_1^*, \lambda_2^*, \ldots, \lambda_u^*), \quad \text{and} \quad \mu^* = \left(\mu_{u+1}^*, \mu_{u+2}^*, \ldots, \mu_v^*\right)$$
$$\lambda_j^* \geq 0, \quad 1 \leq j \leq u, \quad \text{and} \quad \mu_j^* \leq 0, \quad (u+1) \leq j \leq v$$

Also, at $x = x^*$, for $1 \leq i \leq n$

$$\frac{\partial f(x)}{\partial x_i} - \sum_{j=1}^{u} \lambda_j^* \frac{\partial g_j(x)}{\partial x_i} - \sum_{j=u+1}^{v} \mu_j^* \frac{\partial g_j(x)}{\partial x_i} = 0, \quad \text{if } x_i^* > 0$$

$$\frac{\partial f(x)}{\partial x_i} - \sum_{j=1}^{u} \lambda_j^* \frac{\partial g_j(x)}{\partial x_i} - \sum_{j=u+1}^{v} \mu_j^* \frac{\partial g_j(x)}{\partial x_i} \leq 0, \quad \text{if } x_i^* = 0$$

For $1 \leq j \leq u$

$$(a_j - g_j(x^*)) = 0, \quad \text{if } \lambda_j^* > 0$$
$$(a_j - g_j(x^*)) \geq 0, \quad \text{if } \lambda_j^* = 0$$

For $(u+1) \leq j \leq v$

$$(a_j - g_j(x^*)) = 0, \quad \text{if } \mu_j^* < 0$$
$$(a_j - g_j(x^*)) \leq 0, \quad \text{if } \mu_j^* = 0$$

Hint: See Cooper, and Steinberg (1970). Set up the Lagrangian $\mathcal{L}(x, \lambda, \mu)$, by initially defining

$$\lambda = (\lambda_1, \lambda_2, \ldots, \lambda_u), \quad \text{and} \quad \mu = \left(\mu_{u+1}, \mu_{u+2}, \ldots, \mu_v\right)$$

and

$$\mathcal{L}(x, \lambda, \mu) = f(x) + \sum_{j=1}^{u} \lambda_j(a_j - g_j(x)) + \sum_{j=u+1}^{v} \mu_j(a_j - g_j(x))$$

The Lagrangian function $\mathcal{L}(\cdot, \cdot, \cdot)$ is maximized, subject to certain constraints on $x$, $\lambda$, and $\mu$. Also by hypothesis, $x \geq 0$. Furthermore, in order to maximize $\mathcal{L}(x, \lambda, \mu)$ each term in the sum $\sum_{j=1}^{u} \lambda_j(a_j - g_j(x))$ should be nonnegative. This occurs provided

$$\lambda_j \geq 0, \quad 1 \leq j \leq u$$

Similarly, each term in the sum $\sum_{j=u+1}^{v} \mu_j(a_j - g_j(x))$ should be nonnegative. This occurs provided

$$\mu_j \leq 0, \quad u + 1 \leq j \leq v$$

Therefore, a necessary condition for the maximization of the Lagrangian function $\mathcal{L}(\cdot, \cdot, \cdot)$ is that $x \geq 0$, $\lambda_j \geq 0$, for $1 \leq j \leq u$, and $\mu_j \leq 0$, for $u + 1 \leq j \leq v$. In the absence of these constraints, the necessary conditions for the existence of maximum of the unconstrained function $\mathcal{L}\left(\cdot,\cdot,\cdot\right)$ are

$$\frac{\partial\mathcal{L}\left(x,\lambda,\mu\right)}{\partial x_i}=0,\quad 1\le i\le n$$

$$\frac{\partial\mathcal{L}\left(x,\lambda,\mu\right)}{\partial\lambda_j}=0,\quad 1\le j\le u$$

$$\frac{\partial\mathcal{L}\left(x,\lambda,\mu\right)}{\partial\mu_j}=0,\quad \left(u+1\right)\le j\le v$$

However, these conditions have to be modified if $\mathcal{L}\left(\cdot,\cdot,\cdot\right)$ is maximized under the constraints $x\ge 0$, $\lambda\ge 0$, and $\mu\le 0$. Let $x^*,\lambda^*$, and $\mu^*$ be the values of $x,\lambda$, and $\mu$ which maximize $\mathcal{L}\left(x,\lambda,\mu\right)$ under these constraints. The following conditions must occur at $x=x^*,\lambda=\lambda^*$, and $\mu=\mu^*$.

(a) For $1\le i\le n$:

   (i) If $x_i^*>0$, then $\partial\mathcal{L}\left(x,\lambda,\mu\right)/\partial x_i=0$.

   (ii) If $x_i^*=0$, then $\partial\mathcal{L}\left(x,\lambda,\mu\right)/\partial x_i\le 0$. The maximum of the Lagrangian function $\mathcal{L}\left(\cdot,\cdot,\cdot\right)$ occurs at a boundary.

(b) For $1\le j\le u$:

   (i) If $\lambda_j^*>0$, then $\partial\mathcal{L}\left(x,\lambda,\mu\right)/\partial\lambda_j=0$. That is, $\left(a_j-g_j\left(x^*\right)\right)=0$. The corresponding constraint is said to be active at the point $x^*$.

   (ii) If $\lambda_j^*=0$, the corresponding constraint has no effect upon the maximization of $\mathcal{L}\left(\cdot,\cdot,\cdot\right)$, and the inequality is said to be inactive at the point $x^*$. In this subcase, $\left(a_j-g_j\left(x^*\right)\right)>0$.

(c) For $\left(u+1\right)\le j\le v$:

   (i) If $\mu_j^*<0$, then $\partial\mathcal{L}\left(x,\lambda,\mu\right)/\partial\mu_j=0$. That is, $\left(a_j-g_j\left(x^*\right)\right)=0$. The corresponding constraint is said to be active at the point $x^*$.

   (ii) If $\mu_j^*=0$, the corresponding constraint has no effect upon the maximization of $\mathcal{L}\left(\cdot,\cdot,\cdot\right)$, and the inequality is said to be inactive at the point $x^*$. In this subcase, $\left(a_j-g_j\left(x^*\right)\right)\le 0$.

The above cases can be expressed compactly in terms of $f\left(x\right)$, and $g_j\left(x\right),1\le j\le v$. At $x=x^*$, for $1\le i\le n$

$$\frac{\partial\mathcal{L}\left(x,\lambda,\mu\right)}{\partial x_i}=\frac{\partial f\left(x\right)}{\partial x_i}-\sum_{j=1}^{u}\lambda_j^*\frac{\partial g_j\left(x\right)}{\partial x_i}-\sum_{j=u+1}^{v}\mu_j^*\frac{\partial g_j\left(x\right)}{\partial x_i}=0,\ \text{if}\ x_i^*>0$$

$$\frac{\partial\mathcal{L}\left(x,\lambda,\mu\right)}{\partial x_i}=\frac{\partial f\left(x\right)}{\partial x_i}-\sum_{j=1}^{u}\lambda_j^*\frac{\partial g_j\left(x\right)}{\partial x_i}-\sum_{j=u+1}^{v}\mu_j^*\frac{\partial g_j\left(x\right)}{\partial x_i}\le 0,\ \text{if}\ x_i^*=0$$

At $x=x^*$, for $1\le j\le u$

$$\frac{\partial\mathcal{L}\left(x,\lambda,\mu\right)}{\partial\lambda_j}=\left(a_j-g_j\left(x^*\right)\right)=0,\ \text{if}\ \lambda_j^*>0$$

$$\left(a_j-g_j\left(x^*\right)\right)\ge 0,\ \text{if}\ \lambda_j^*=0$$

At $x=x^*$, for $\left(u+1\right)\le j\le v$

$$\frac{\partial \mathcal{L}(x, \lambda, \mu)}{\partial \mu_j} = (a_j - g_j(x^*)) = 0, \quad \text{if } \mu_j^* < 0$$

$$(a_j - g_j(x^*)) \leq 0, \quad \text{if } \mu_j^* = 0$$

The above KKT conditions can be suitably modified if the objective function has to be minimized.

21. Note that the KKT conditions are necessary but not sufficient, for the optimum solution (global maximum or minimum) of the general nonlinear programming problem. However, the KKT conditions are sufficient for determining a globally optimal solution, if it is assumed that the constraint set is convex, and the objective function is concave (convex) for its maximization (minimization).

    In the last problem, let $g_j(\cdot)$ be a convex function for $1 \leq j \leq u$; and $g_j(\cdot)$ be a concave function for $(u + 1) \leq j \leq v$. If the objective function $f(\cdot)$ is a concave (convex) function, then prove that the KKT conditions are both necessary and sufficient for $f(x^*)$ to be the global maximum (minimum).

22. Consider the primal problem $\mathcal{P}_L$ specified in the chapter. Let $x^* \in \mathcal{C}$ and $\lambda^* \geq 0$. The vector $(x^*, \lambda^*)$ is a saddle point of the Lagrangian $\mathcal{L}(x, \lambda) \Leftrightarrow$

    (a) $\mathcal{L}(x^*, \lambda^*) = \min_{x \in \mathcal{C}} \mathcal{L}(x, \lambda^*)$
    (b) $g_j(x^*) \leq 0, 1 \leq j \leq m$
    (c) $\lambda_j^* g_j(x^*) = 0, 1 \leq j \leq m$

    Establish the above result.

    Hint: See Minoux (1986). The $\Rightarrow$ part of the proof.

    Since $(x^*, \lambda^*)$ is a saddle point, then (a) is true by its definition. Also $\lambda \geq 0$ and $\mathcal{L}(x^*, \lambda^*) \geq \mathcal{L}(x^*, \lambda)$. This yields

$$f(x^*) + \sum_{j=1}^{m} \lambda_j^* g_j(x^*) \geq f(x^*) + \sum_{j=1}^{m} \lambda_j g_j(x^*)$$

That is

$$\sum_{j=1}^{m} (\lambda_j - \lambda_j^*) g_j(x^*) \leq 0$$

If for any $j$, $g_j(x^*) > 0$, then we can select $\lambda_j > 0$ sufficiently large so that the above inequality does not hold. This implies that (b) is true.

The above inequality implies that if $\lambda = 0$ then $\sum_{j=1}^{m} \lambda_j^* g_j(x^*) \geq 0$.

However $\lambda_j \geq 0$ and $g_j(x^*) \leq 0 \Rightarrow \sum_{j=1}^{m} \lambda_j^* g_j(x^*) \leq 0$. Therefore $\sum_{j=1}^{m} \lambda_j^* g_j(x^*) = 0$. This implies $\lambda_j^* g_j(x^*) = 0$ for $1 \leq j \leq m$. This is result (c).

The $\Leftarrow$ part of the proof. In this part, it is hypothesized that the statements (a), (b), and (c) are true. Statement (a) implies

$$\mathcal{L}(x^*, \lambda^*) \leq \mathcal{L}(x, \lambda^*), \quad \forall x \in \mathcal{C}$$

Statement (c) implies $\mathcal{L}(x^*, \lambda^*) = f(x^*)$. Also since $\lambda \geq 0$

$$\mathcal{L}(x^*, \lambda) = f(x^*) + \sum_{j=1}^{m} \lambda_j g_j(x^*) \leq f(x^*) = \mathcal{L}(x^*, \lambda^*), \quad \forall \lambda \geq 0$$

Thus

$$\mathcal{L}\left(x^{*}, \lambda\right) \leq \mathcal{L}\left(x^{*}, \lambda^{*}\right) \leq \mathcal{L}\left(x, \lambda^{*}\right), \quad \forall\, x \in C \text{ and } \forall\, \lambda \geq 0$$

That is, $\left(x^{*}, \lambda^{*}\right)$ is a saddle point of the Lagrangian $\mathcal{L}\left(x, \lambda\right).$

23. Consider the primal problem $\mathcal{P}_{L}$ specified in the chapter. If $\left(x^{*}, \lambda^{*}\right)$ is a saddle point of the Lagrangian $\mathcal{L}\left(x, \lambda\right),$ then prove that $x^{*}$ is a global optimum of the problem $\mathcal{P}_{L}.$
Hint: See Minoux (1986). The result $\mathcal{L}\left(x^{*}, \lambda^{*}\right) = \min_{x \in C} \mathcal{L}\left(x, \lambda^{*}\right)$ implies

$$f\left(x^{*}\right) + \sum_{j=1}^{m} \lambda_{j}^{*} g_{j}\left(x^{*}\right) \leq f\left(x\right) + \sum_{j=1}^{m} \lambda_{j}^{*} g_{j}\left(x\right), \quad \forall\, x \in C$$

Note that $\lambda_{j}^{*} g_{j}\left(x^{*}\right) = 0,$ for $1 \leq j \leq m.$ Therefore

$$f\left(x^{*}\right) \leq f\left(x\right) + \sum_{j=1}^{m} \lambda_{j}^{*} g_{j}\left(x\right), \quad \forall\, x \in C$$

Since $\lambda^{*} \geq 0$ and $g_{j}\left(x\right) \leq 0,$ for $1 \leq j \leq m$ the above inequality yields $f\left(x^{*}\right) \leq f\left(x\right), \forall\, x \in C.$

24. Establish the weak Lagrangian duality theorem.
Hint: See Minoux (1986). Since $x \in C$, for any $\lambda \geq 0$

$$w\left(\lambda\right) \leq \mathcal{L}\left(x, \lambda\right) = f\left(x\right) + \sum_{j=1}^{m} \lambda_{j} g_{j}\left(x\right)$$

Also if $x$ is a solution of the primal problem $\mathcal{P}_{L}$, then $\lambda_{j} g_{j}\left(x\right) \leq 0,$ for $1 \leq j \leq m.$ Therefore $\sum_{j=1}^{m} \lambda_{j} g_{j}\left(x\right) \leq 0$ and

$$w\left(\lambda\right) \leq f\left(x\right)$$

If $x^{*}$ is indeed a global optima of the primal problem $\mathcal{P}_{L}$, then for all $\lambda \geq 0$

$$w\left(\lambda\right) \leq w\left(\lambda^{*}\right) \leq f\left(x^{*}\right)$$

25. Prove the concavity of the dual function.
Hint: See Minoux (1986). For any two vectors $\lambda^{1}$ and $\lambda^{2}$, and any $\beta \in [0, 1]$ let

$$\lambda = \left\{\beta \lambda^{1} + (1 - \beta) \lambda^{2}\right\}$$

Also $\exists\, x'$ such that

$$w\left(\lambda\right) = f\left(x'\right) + \sum_{j=1}^{m} \lambda_{j} g_{j}\left(x'\right)$$

Furthermore, the definition of $w\left(\lambda^{1}\right)$ and $w\left(\lambda^{2}\right)$ yields

$$w\left(\lambda^{1}\right) \leq f\left(x'\right) + \sum_{j=1}^{m} \lambda_{j}^{1} g_{j}\left(x'\right), \text{ and } w\left(\lambda^{2}\right) \leq f\left(x'\right) + \sum_{j=1}^{m} \lambda_{j}^{2} g_{j}\left(x'\right)$$

Therefore

$$\left\{\beta w\left(\lambda^{1}\right) + (1 - \beta) w\left(\lambda^{2}\right)\right\} \leq f\left(x'\right) + \sum_{j=1}^{m} \left\{\beta \lambda_{j}^{1} + (1 - \beta) \lambda_{j}^{2}\right\} g_{j}\left(x'\right)$$

$$= f\left(x'\right) + \sum_{j=1}^{m} \lambda_{j} g_{j}\left(x'\right) = w\left(\lambda\right)$$

Note that this proof did not require any assumption about convexity of functions $f(\cdot)$, and $g_j(\cdot)$, for $1 \leq j \leq m$. It also did not require any assumption about the convexity of the set $\mathcal{C}$.

26. Prove the Lagrangian duality theorem.

   Hint: See Minoux (1986).

   (a) Let $(x^*, \lambda^*)$ be a saddle point of the primal problem $\mathcal{P}_L$. Then

   $$\mathcal{L}(x^*, \lambda^*) = f(x^*) + \sum_{j=1}^{m} \lambda_j^* g_j(x^*) = f(x^*)$$
   $$= \min_{x \in \mathcal{C}} \mathcal{L}(x, \lambda^*) = w(\lambda^*)$$

   However use of the weak duality property for all $\lambda \geq 0$, implies $w(\lambda) \leq f(x^*)$. Consequently

   $$f(x^*) = w(\lambda^*) = \max_{\lambda \geq 0} w(\lambda)$$

   (b) Assume that there exists a solution $x^*$ of problem $\mathcal{P}_L$ and $\lambda^* \geq 0$ such that $w(\lambda^*) = f(x^*)$. Using the definition of $w(\lambda^*)$ yields

   $$w(\lambda^*) = f(x^*) \leq f(x^*) + \sum_{j=1}^{m} \lambda_j^* g_j(x^*)$$

   Therefore $\sum_{j=1}^{m} \lambda_j^* g_j(x^*) \geq 0$. Also since $\lambda_j^* \geq 0$ and $g_j(x^*) \leq 0$ for all values of $j$, yields $\lambda_j^* g_j(x^*) \leq 0$, consequently $\sum_{j=1}^{m} \lambda_j^* g_j(x^*) \leq 0$, and therefore $\sum_{j=1}^{m} \lambda_j^* g_j(x^*) = 0$. Furthermore, $\sum_{j=1}^{m} \lambda_j^* g_j(x^*)$ is the sum of all nonpositive terms. Therefore $\lambda_j^* g_j(x^*) = 0$ for all values of $j$. Finally, since $\mathcal{L}(x^*, \lambda^*) = \min_{x \in \mathcal{C}} \mathcal{L}(x, \lambda^*)$, $g_j(x^*) \leq 0$, $1 \leq j \leq m$, and $\lambda_j^* g_j(x^*) = 0$, $1 \leq j \leq m$, it follows that $(x^*, \lambda^*)$ is a saddle point.

27. Consider a quadratic optimization problem. Let $A$ be an $m \times n$ matrix, $Q$ is an $n \times n$ positive semidefinite symmetric matrix, $b$ is a column vector of size $m$; and $c$ and $x$ are column vectors of size $n$. All elements of these matrices belong to the real field $\mathbb{R}$. The primal quadratic optimization problem is

   $$\min \left\{ \frac{1}{2} x^T Q x + c^T x \right\}$$

   *Subject to*: $Ax \geq b$, and $x \geq 0$

   The objective function and the constraint functions have continuous first derivatives. This is indeed a convex optimization problem. Determine Wolfe-dual of the stated primal problem.

   Hint: See Griva, Nash, and Sofer (2009). Let the Lagrange multiplier vectors corresponding to the inequalities $Ax \geq b$, and $x \geq 0$ be $\lambda = (\lambda_1, \lambda_2, \ldots, \lambda_m)$, and $\mu = (\mu_1, \mu_2, \ldots, \mu_n)$ respectively. Note that $\lambda$ and $\mu$ are row vectors. The Wolfe-dual is

   $$\max_{x, \lambda, \mu} \left\{ \frac{1}{2} x^T Q x + c^T x + \lambda(b - Ax) + \mu(-x) \right\}$$

   *Subject to*: $\left\{ Qx + c - A^T \lambda^T - \mu^T \right\} = 0; \quad \lambda \geq 0, \mu \geq 0$

   Substituting $c = \left( A^T \lambda^T + \mu^T - Qx \right)$ in the objective simplifies the Wolfe-dual to be

$$\max_{x,\lambda}\left\{\lambda b - \frac{1}{2}x^T Q x\right\}$$

*Subject to:* $c = \left(A^T\lambda^T + \mu^T - Qx\right);\quad \lambda \geq 0,\ \mu \geq 0$

Thus the Wolfe-dual is also a quadratic optimization problem.

28. Show that the necessary condition for the integral

$$\mathcal{I} = \int_{x_1}^{x_2} F\left(x, y, y'\right) dx$$

to have an extremum (maximum or minimum) is

$$\frac{d}{dx}\left(\frac{\partial F}{\partial y'}\right) - \frac{\partial F}{\partial y} = 0$$

Th above equation is called the Euler-Lagrange equation.

Hint: See Spiegel (1967), and Weinstock (1974). Let the function $y$ which determines the extremum of $\mathcal{I}$ be

$$y = Y\left(x\right),\quad x_1 \leq x \leq x_2$$

Let the perturbed value of $Y\left(x\right)$ be $y\left(x\right) = \left(Y\left(x\right) + \epsilon\eta\left(x\right)\right)$, where $x \in [x_1, x_2]$, $\epsilon$ is independent of $x$, and $\eta\left(x_1\right) = \eta\left(x_2\right) = 0$. The perturbed value of the integral is

$$\mathcal{I}\left(\epsilon\right) = \int_{x_1}^{x_2} F\left(x, Y + \epsilon\eta, Y' + \epsilon\eta'\right) dx$$

where $Y'$ and $\eta'$ are derivatives of $Y$ and $\eta$ with respect to $x$ respectively. The above integral achieves its extremum at $\epsilon = 0$. We next take derivative of the above integral with respect to $\epsilon$, let $\epsilon \to 0$, and set the derivative equal to $0$. Therefore

$$\left.\frac{d\mathcal{I}\left(\epsilon\right)}{d\epsilon}\right|_{\epsilon=0} = \int_{x_1}^{x_2}\left(\frac{\partial F}{\partial y}\eta + \frac{\partial F}{\partial y'}\eta'\right) dx = 0$$

where we used the fact $y\left(x\right) = Y\left(x\right)$ at $\epsilon = 0$. Using integration by parts, the right-hand side expression can be written as

$$\int_{x_1}^{x_2}\frac{\partial F}{\partial y}\eta dx + \left.\frac{\partial F}{\partial y'}\eta\right|_{x_1}^{x_2} - \int_{x_1}^{x_2}\eta\frac{d}{dx}\left(\frac{\partial F}{\partial y'}\right) dx = \int_{x_1}^{x_2}\eta\left\{\frac{\partial F}{\partial y} - \frac{d}{dx}\left(\frac{\partial F}{\partial y'}\right)\right\} dx = 0$$

where we used the result $\eta\left(x_1\right) = \eta\left(x_2\right) = 0$. As $\eta$ is arbitrary, we obtain the stated result.

29. Prove Nyquist's stability criteria rigorously.

Hint: See Spiegel (1964).

30. Prove Lyapunov's L1 theorem.

Hint: See Sarachik (1997), and Khalil (2002).

Part (a): Lyapunov's function exists in region $\Re$. Then

$$V\left(X\left(t\right), t\right) = V\left(X\left(t_0\right), t_0\right) + \int_{t_0}^{t}\dot{V}\left(X, \tau\right) d\tau$$

Define $V_0 \triangleq V\left(X\left(t_0\right), t_0\right)$. If $X\left(t\right) \in \Re$, then by hypothesis $\dot{V}\left(X\left(t\right), t\right) \leq 0$, therefore $V_0 \triangleq V\left(X\left(t_0\right), t_0\right) \geq V\left(X\left(t\right), t\right) > W\left(X\left(t\right)\right)$.

In order to establish stability, it must be proved that for any $\epsilon > 0$, and for all values of $t \geq t_0$, $\|X(t_0)\| < \delta \Rightarrow \|X(t)\| < \epsilon$.

For any $\epsilon > 0$, let $\alpha = \min_{\|X\|=\epsilon} W(X(t))$. This minimum exists, because $W(X(t))$ is a continuous function on the compact set $\|X\| = \epsilon$. Since $W(X(t))$ is positive definite in the region $\Re$, $\alpha > 0$. Therefore $W(X(t)) < \alpha \Rightarrow \|X(t)\| < \epsilon$.

At $t = t_0$ a $\delta > 0$ can be found such that $\|X(t_0)\| < \delta$ makes $V(X(t_0), t_0) < \alpha$. Note that this is always possible since $V(0, t) = 0$ (by hypothesis) and $V(X(t), t)$ is continuous in $X(t)$. Therefore if $\epsilon$ is selected such that the set $\|X(t)\| < \epsilon$ lies entirely in the region $\Re$ then

$$\|X(t_0)\| < \delta \Rightarrow \alpha > V_0 \geq V(X(t), t) > W(X(t))$$
$$\Rightarrow \|X(t)\| < \epsilon, \quad \forall t \geq t_0$$

This shows that the equilibrium point 0 is stable. This concludes proof of part (a) of the theorem.

Part (b): It has to be demonstrated that $\lim_{t\to\infty} \|X(t)\| \to 0$, for the system to be asymptotically stable. This is proved by contradiction.

As $V(X(t), t)$ is a monotonically decreasing function, it follows that $V(X(t), t) \geq \zeta$, $\forall t \geq t_0$, where $\lim_{t\to\infty} V(X(t), t) = \zeta$. Note that $\zeta > 0 \Rightarrow \|X(t)\| > \eta(\zeta) > 0, \forall t \geq t_0$. However $\dot{V}(X(t), t)$ is negative definite in this $\Re$, therefore $\dot{V}(X(t), t) \leq -\beta$ (for some positive value $\beta$), $\forall t \geq t_0$. This leads to

$$V(X(t), t) = V(X(t_0), t_0) + \int_{t_0}^{t} \dot{V}(X, \tau) \, d\tau \leq V_0 - \beta(t - t_0)$$

This result implies that $V(X(t), t)$ can have a negative value for sufficiently large values of time. However, this is not possible. Therefore $\lim_{t\to\infty} \|X(t)\| \to 0$.

31. The dynamical equations of a system are

$$\dot{x}_1 = (x_2 - x_3) - bx_1(x_1^2 + x_2^2 + x_3^2)$$
$$\dot{x}_2 = (x_3 - x_1) - bx_2(x_1^2 + x_2^2 + x_3^2)$$
$$\dot{x}_3 = (x_1 - x_2) - bx_3(x_1^2 + x_2^2 + x_3^2)$$

where $dx_i/dt \triangleq \dot{x}_i$, for $1 \leq i \leq 3$, and $b \in \mathbb{R}^+$. Prove that the system is asymptotically stable. Hint: See Sarachik (1997). Check if $V(X(t), t) = (x_1^2 + x_2^2 + x_3^2)$ is a candidate for Lyapunov function. Then

$$\dot{V}(X(t), t) = \frac{\partial V(X(t), t)}{\partial x_1} \dot{x}_1 + \frac{\partial V(X(t), t)}{\partial x_2} \dot{x}_2 + \frac{\partial V(X(t), t)}{\partial x_3} \dot{x}_3$$
$$= -2b(x_1^2 + x_2^2 + x_3^2)^2$$

The function $\dot{V}(X(t), t)$ is negative definite. Therefore the system is asymptotically stable.

32. Prove Lyapunov's L2 theorem.

Hint: See Sarachik (1997). Consider the linear system $\dot{X}(t) = AX(t)$. Select a Lyapunov function

$$V(X(t)) = X^T(t) M X(t)$$

where $M$ is a symmetric positive definite matrix. Taking derivative of $V(X(t))$ with respect to time yields

$$\dot{V}(X(t)) = X^T(t)(MA + A^T M) X(t)$$

Define $Q = -\left(MA + A^T M\right)$, then $\dot{V}\left(X\left(t\right)\right) = -X^T\left(t\right)QX\left(t\right)$. Observe that the matrix $Q$ is symmetric. Therefore, if for any symmetric positive definite matrix $Q$, there exists a symmetric positive definite matrix $M$ which satisfies the relationship $Q = -\left(MA + A^T M\right)$, then by theorem $L1$, the linear system is asymptotically stable.

Next assume that the linear system is asymptotically stable, then for any symmetric positive definite matrix $Q$ select

$$V\left(X\left(t\right)\right) = \int_t^\infty X^T\left(\tau\right)QX\left(\tau\right)d\tau > 0$$

Note that a solution of the equation $\dot{X}\left(t\right) = AX\left(t\right)$ is

$$X\left(\tau\right) = e^{A(\tau-t)}X\left(t\right), \quad e^{At} \triangleq \sum_{j=0}^\infty \frac{\left(At\right)^j}{j!}$$

Therefore substituting this value in the above equation yields

$$V\left(X\left(t\right)\right) = X^T\left(t\right)\int_0^\infty \left(e^{A\tau}\right)^T Qe^{A\tau}d\tau X\left(t\right)$$

Define

$$M \triangleq \int_0^\infty \left(e^{A\tau}\right)^T Qe^{A\tau}d\tau$$

where $M$ is symmetric and finite, as exponentially decreasing time functions are being integrated. Therefore

$$V\left(X\left(t\right)\right) = X^T\left(t\right)MX\left(t\right) > 0 \quad \text{for } X\left(t\right) \neq 0$$

As demonstrated earlier $\dot{V}\left(X\left(t\right)\right) = \left\{X^T\left(t\right)\left(MA + A^T M\right)X\left(t\right)\right\}$.

However taking the derivative of $V\left(X\left(t\right)\right) = \int_t^\infty X^T\left(\tau\right)QX\left(\tau\right)d\tau$ with respect to $t$ yields $\dot{V}\left(X\left(t\right)\right) = -X^T\left(t\right)QX\left(t\right)$.

In summary, it has been demonstrated that if the time-invariant linear system is asymptotically stable, the matrix $M$ which satisfies $Q = -\left(MA + A^T M\right)$ for any positive definite matrix $Q$, is symmetric and positive definite.

33. Let $f : \mathbb{R} \to \mathbb{R}$ and $g : \mathbb{R} \to \mathbb{R}$ be two functions. Let $f'\left(x\right)$ be the derivative of $f\left(x\right)$ with respect to $x$. Also $x_1 = f\left(x_0\right); x_2 = f\left(x_1\right) = f^2\left(x_0\right); x_3 = f\left(x_2\right) = f^3\left(x_0\right); \ldots; x_n = f\left(x_{n-1}\right) = f^n\left(x_0\right); \ldots$. Prove that:

(a) $df\left(g\left(x\right)\right)/dx = f'\left(g\left(x\right)\right)g'\left(x\right)$

(b) $df^n\left(x\right)/dx = f'\left(f^{n-1}\left(x\right)\right) \cdot f'\left(f^{n-2}\left(x\right)\right) \cdots \cdot f'\left(x\right)$

(c) $\left.\frac{df^n\left(x\right)}{dx}\right|_{x=x_0} = f'\left(x_{n-1}\right)f'\left(x_{n-2}\right)\cdots f'\left(x_1\right)f'\left(x_0\right)$

Hint: See Devaney (1989).

34. Let $\varrho = \left[0, 1\right]$, and $f : \varrho \to \varrho$ satisfy the following conditions:

(a) Let $a_0, a_1, \ldots, a_r \in \varrho$, where $0 = a_0 < a_1 < \cdots < a_r = 1$.

(b) Define a function $f_i\left(\cdot\right)$ in $C^2$ such that it is a restriction of $f\left(\cdot\right)$ to the interval $\left(a_{i-1}, a_i\right)$ for $1 \leq i \leq r$.

(c) Also $f\left(\left(a_{i-1}, a_i\right)\right) = \left(0, 1\right)$, for $1 \leq i \leq r$. That is, the function $f_i\left(\cdot\right)$ is mapped onto $\left(0, 1\right)$.

(d)  There exists $\lambda > 1$ such that $|f'(x)| \geq \lambda$ for $x \neq a_i$, $i = 0, 1, 2 \ldots, r$; where $f'(x)$ is the derivative of $f(x)$ with respect to $x$.

Prove that the map $f(\cdot)$ is chaotic and expansive.

Hint: See Erramilli, Singh, and Pruthi (1995).

*Step* 1: Let $x \in \varrho$ and $N_{ch}$ be a neighborhood of $x$. Also, let $y \in N_{ch}$, and $y \neq x$, then use of the mean value theorem yields

$$|f^n(x) - f^n(y)| = f^{n\prime}(\xi)|x - y|, \quad \xi \in (x, y)$$

Also

$$f^{n\prime}(\xi) = \prod_{j=0}^{(n-1)} f'\left(f^j(\xi)\right) \geq \lambda^n$$

Thus

$$|f^n(x) - f^n(y)| \geq \lambda^n |x - y|$$

Consequently the mapping $f(\cdot)$ satisfies the SIC property and it is expansive. That is, the SIC property is satisfied for all $(x, y)$.

*Step* 2: The topological transitive property is next established. Note from the above inequality, that for several iterations of the map on an interval $(x, y)$, there exists an integer $n \in \mathbb{P}$ such that $\lambda^n |x - y| \geq 1$. Therefore $n \geq -\{\ln |x - y|\} / \ln \lambda$ such that the interval $(x, y)$ is mapped onto $(0, 1)$ by $f^n(\cdot)$. Actually every interval $(x, y)$ of $(0, 1)$ is mapped onto every other interval within $n$ successive iterations. Thus the topological transitive property of the chaotic map is satisfied.

*Step* 3: Using the result in Step 2, there exists an $n \in \mathbb{P}$ such that $f^n(B_{ch}) = (0, 1)$, where $B_{ch} \subset (x, y)$. As a specific instance, there exists $z \in B_{ch}$ such that $f^n(z) = z \in (0, 1)$. Thus $z \in (x, y)$ is a fixed point of the map $f^n(\cdot)$ and a periodic point of period $n$ of the mapping $f(\cdot)$. Consequently the periodic points are dense.

35.  Let $\widetilde{x}$ be a fixed point of a function $f(x)$ defined on an interval $\varrho$. Denote the first derivative of $f(x)$ with respect to $x$ by $f'(x)$. Then

(a)  $\widetilde{x}$ is an attractor if $|f'(\widetilde{x})| < 1$.

(b)  $\widetilde{x}$ is a repeller if $|f'(\widetilde{x})| > 1$.

(c)  Nothing can be said about the type of $\widetilde{x}$ if $|f'(\widetilde{x})| = 1$.

Establish the above results.

Hint: See Gulick (1992), and Hirsch, Smale, and Devaney (2003).

(a)  Let $|f'(\widetilde{x})| = \alpha < 1$. Select $\beta$ such that $\alpha < \beta < 1$. Since $f'(\cdot)$ is continuous, find $\delta > 0$ such that $|f'(x)| < \beta$ for all $x \in I_{ch} \triangleq [\widetilde{x} - \delta, \widetilde{x} + \delta]$. For any $x \in I_{ch}$ and some $\xi$ between $x$ and $\widetilde{x}$, use of the mean value theorem yields,

$$\frac{f(x) - \widetilde{x}}{x - \widetilde{x}} = \frac{f(x) - f(\widetilde{x})}{x - \widetilde{x}} = f'(\xi)$$

Thus

$$|f(x) - \widetilde{x}| < \beta |x - \widetilde{x}|$$

This inequality implies that $f(x)$ is closer to $\widetilde{x}$ than $x$. Therefore $f(x) \in I_{ch}$. Application of this result again yields

$$\left|f^2(x) - \widetilde{x}\right| < \beta |f(x) - \widetilde{x}| < \beta^2 |x - \widetilde{x}|$$

Similarly $|f^n(x) - \widetilde{x}| < \beta^n |x - \widetilde{x}|$ for $n \in \mathbb{P}$. That is, $f^n(x) \to \widetilde{x}$ for $x \in I_{ch}$, as $0 < \beta < 1$.

(b) The proof of this observation is similar to that of part (a) of the problem.

(c) Consider the functions $p(x) = (x + x^5)$, $q(x) = (x - x^5)$, and $r(x) = x + x^2$. Each of these functions has a fixed point at $x = 0$ and $p'(0) = q'(0) = r'(0) = 1$. It can be graphically demonstrated that $p(\cdot)$ has a repeller at $x = 0$, and $q(\cdot)$ has an attractor at $x = 0$. It can also be shown that for the function $r(\cdot)$, $0$ is attracting from one side and repelling from the other side of $x = 0$.

36. Let $\varrho = [0, 1]$, and $f : \varrho \to \varrho$ satisfy the following conditions:

(a) Let $a_0, a_1, \ldots, a_r \in \varrho$, where $0 = a_0 < a_1 < \cdots < a_r = 1$.

(b) Define a function $f_i(x) = c_i x + d_i$, where $c_i = (a_i - a_{i-1})^{-1}$, $d_i = -c_i a_{i-1}$, and $x \in (a_{i-1}, a_i)$. The function $f_i(\cdot)$ is the restriction of $f(\cdot)$ to the interval $(a_{i-1}, a_i)$, where $1 \le i \le r$.

(c) $f((a_{i-1}, a_i)) = (0, 1)$, for $1 \le i \le r$.

Prove that the invariant density is uniform, that is $\rho(x) = 1$ for $x \in \varrho$.

Hint: See Erramilli, Singh, and Pruthi (1995). It has been shown in an earlier problem that this map is expansive.

$$\rho(x) = \sum_{i=1}^{r} \int_{a_{i-1}}^{a_i} \delta(x - f_i(z)) \rho(z) \, dz = \sum_{i=1}^{r} \frac{\rho(f_i^{-1}(x))}{|f_i'(f_i^{-1}(x))|}$$

Observe that $f_i'(f_i^{-1}(x)) = c_i$. Since $\sum_{i=1}^{r}(a_i - a_{i-1}) = 1$, the above equation is satisfied if $\rho(x) = 1$.

---

# References

1. Alsina, C., and Nelsen, R. B., 2009. *When Less is More*: *Visualizing Basic Inequalities*, The Dolciani Mathematical Expositions, No. 36, The Mathematical Association of America, Washington DC.

2. Avriel, M. 1976. *Nonlinear Programming*: *Analysis and Methods*, Prentice-Hall, Englewood Cliffs, New Jersey.

3. Banks, J., Brooks, J., Cairns, G., Davis, G., and Stacey, P., 1992. "On Devaney's Definition of Chaos," The American Mathematical Monthly, Vol. 99, No. 4, pp. 332-334.

4. Bellman, R., 1961. *Adaptive Control Processes*: *A Guided Tour*, Princeton University Press, Princeton, New Jersey.

5. Bellman, R., and Kalaba, R., Editors, 1964. *Selected Papers on Mathematical Trends in Control Theory*, Dover Publications, Inc., New York, New York.

6. Beveridge, G. S. G., and Schechter, R. S., 1970. *Optimization*: *Theory and Practice*, McGraw-Hill Book Company, New York, New York.

7. Bunday, B. D., 1984. *Basic Optimisation Methods*, Edward Arnold, London, United Kingdom.

8. Chen, C. T., 1970. *Introduction to Linear Systems Theory*, Holt, Rinehart and Winston, New York, New York.

9. Cloud, M. J., Drachman, B. C., and Lebedev, L. P., 2009. *Inequalities with Application to Engineering*, Second Edition, Springer-Verlag, Berlin, Germany.

10. Cooper, L., and Steinberg, D., 1970. *Introduction to Methods of Optimization*, W. B. Saunders Company, Philadelphia, Pennsylvania.

11. Cover, T. M., and Thomas, J. A., 1991. *Elements of Information Theory*, John Wiley & Sons, Inc., New York, New York.

12. Davies, B., 1999. *Exploring Chaos, Theory and Experiment,* Perseus Books, Reading, Massachusetts.

13. Devaney, R. L., 1989. *An Introduction to Chaotic Dynamical Systems,* Second Edition, Addison-Wesley Publishing Company, New York, New York.

14. Ditto, W., and Munakata, T., 1995. "Principles and Applications of Chaotic Systems," Comm. of the ACM, Vol. 38, No. 11, pp. 96-102.

15. Dorf, R. C., 1992. *Modern Control Systems*, Sixth Edition, Addison-Wesley Publishing Company, New York, New York.

16. Ecker, J. G., and Kupferschmid, M., 1988. *Introduction to Operations Research*, John Wiley & Sons, Inc., New York, New York.

17. Erramilli, A., Singh, R. P., and Pruthi, P., 1995. "An Application of Deterministic Chaotic Maps to Model Packet Traffic," Queueing Systems, Vol. 20, Nos. 1-2, pp. 171-206.

18. Faigle, U., Kern, W., and Still, G., 2002. *Algorithmic Principles of Mathematical Programming*, Kluwer Academic Publishers, Norwell, Massachusetts.

19. Forst, W., and Hoffmann, D, 2010. *Optimization - Theory and Practice,* Springer-Verlag, Berlin, Germany.

20. Foulds, L. R., 1981. *Optimization Techniques, An Introduction*, Springer-Verlag, Berlin, Germany.

21. Franklin, J., 1980. *Methods of Mathematical Economics*: *Linear and Nonlinear Programming, Fixed-Point Theorems,* Springer-Verlag, Berlin, Germany.

22. Grieg, D. M., 1980. *Optimisation,* Longman, New York, New York.

23. Griva, I., Nash, S. G., and Sofer, A., 2009. *Linear and Nonlinear Optimization,* Second Edition, Society for Industrial and Applied Mathematics, Philadelphia, Pennsylvania.

24. Gulick, D., 1992. *Encounters with Chaos*, McGraw-Hill Book Company, New York, New York.

25. Hespanha, J. P., 2009. *Linear System Theory*, Princeton University Press, Princeton, New Jersey.

26. Hilborn, R. C., 1994. *Chaos and Nonlinear Dynamics*: *An Introduction to Scientists and Engineers,* Oxford University Press, Oxford, Great Britain.

27. Hirsch, M. W., Smale, S., and Devaney, R. L., 2003. *Differential Equations, Dynamical Systems, and an Introduction to Chaos*, Second Edition, Academic Press, New York, New York.

28. Iosevich, A., 2007. *A View from the Top*: *Analysis, Combinatorics and Number Theory*, American Mathematical Society, Providence, Rhode Island.

29. Kandiller, L., 2007. *Principles of Mathematics in Operations Research*, Springer-Verlag, Berlin, Germany.

30. Khalil, H. K., 2002. *Nonlinear Systems,* Third Edition, Prentice-Hall, Upper Saddle River, New Jersey.

31. Koo, D., 1977. *Elements of Optimization, With Application in Economics and Business*, Springer-Verlag, Berlin, Germany.

32. Kreyszig, E., 2011. *Advanced Engineering Mathematics,* Tenth Edition, John Wiley & Sons, Inc., New York, New York.

33. Kuhn, H. W., and Tucker, A. W., 1951. "Nonlinear Programming," in Proceedings of the Second Berkeley Symposium on Mathematical Statistics and Probability. Editor: J. Neyman, University of California Press, Berkeley and Los Angeles, pp. 481-492.

34. Levine, W. S., Editor-in-Chief, 1996. *The Control Handbook*, CRC Press: New York, New York.

35. McCauley, J. L., 1993. *Chaos, Dynamics, and Fractals, An Algorithmic Approach to Deterministic Chaos,* Cambridge University Press, Cambridge, Great Britain.

36. Minoux, M., 1986. *Mathematical Programming*, translated by S. Vajda, John Wiley & Sons, Inc., New York, New York.

37. Nocedal, J., and Wright, S. J., 2006. *Numerical Optimization*, Second Edition, Springer-Verlag, Berlin, Germany.

38. Peressini, A. L., Sullivan, F. E., and Uhl Jr., J. J., 1988. *The Mathematics of Nonlinear Programming*, Springer-Verlag, Berlin, Germany.

39. Rao, S. S., 2009. *Engineering Optimization: Theory and Practice*, Fourth Edition, John Wiley & Sons, Inc., New York, New York.

40. Resende, G. C. M., and Pardalos, P. M., Editors, 2006. *Handbook of Optimization in Telecommunications*, Springer-Verlag, Berlin, Germany.

41. Rowland, J. R., 1986. *Linear Control Systems: Modeling, Analysis, and Design*, John Wiley & Sons, Inc., New York, New York.

42. Samuel, L. G., 1999. "The Application of Nonlinear Dynamics to Teletraffic Modeling," Ph. D. dissertation, University of London, England.

43. Sandefur, J. T., 1990. *Discrete Dynamical Systems, Theory and Applications,* Oxford University Press, Oxford, Great Britain.

44. Sarachik, P. E., 1997. *Principles of Linear Systems,* Cambridge University Press, Cambridge, Great Britain.

45. Schuster, H. G., 1995. *Deterministic Chaos: An Introduction*, Third Edition, John Wiley & Sons, Inc., New York, New York.

46. Sontag, E. D., 1990. *Mathematical Control Theory*, Springer-Verlag, Berlin, Germany.

47. Spiegel, M. R., 1964. *Complex Variables*, Schaum's Outline Series, McGraw-Hill Book Company, New York, New York.

48. Spiegel, M. R., 1967. *Theoretical Mechanics*, Schaum's Outline Series, McGraw-Hill Book Company, New York, New York.

49. Tolsted, E., 1964. "An Elementary Derivation of the Cauchy, Hölder, and Minkowski Inequalities from Young's Inequality," Mathematics Magazine, Vol. 37, No. 1, pp. 2-12.

50. Vidyasagar, M., 1978. *Nonlinear Systems Analysis*, Prentice-Hall, Englewood Cliffs, New Jersey.

51. Weinstock, R., 1974. *Calculus of Variations, with Applications to Physics and Engineering*, Dover Publications, Inc., New York, New York.

52. Young, W. H., 1912. "On classes of summable functions and their Fourier series," Proc. Royal Soc., ($\Lambda$), Vol. 87, Issue 594, pp. 225-229.

53. Zadeh, L. A., and Desoer, C. A., 1963. *Linear System Theory: The State Space Approach*, McGraw-Hill Book Company, New York, New York.

# Probability Theory

$$\frac{\partial}{\partial t} p(x_0, x; t) = -m \frac{\partial}{\partial x} p(x_0, x; t) + \frac{1}{2} \sigma^2 \frac{\partial^2}{\partial x^2} p(x_0, x; t)$$

**Kolmogorov's forward diffusion equation**

**Andrey Nikolaevich Kolmogorov**.
Kolmogorov was born on 25 April, 1903 in Tambov, Tambov province, Russia. He was raised by his mother's sister Vera Yakovlena. Kolmogorov entered Moscow State University in 1920 and graduated from it in 1925. He completed his doctorate in 1929, and was appointed professor at the same university in 1931.

He made long lasting contributions to the foundations of probability theory, set-theoretic topology, theory of Markov random processes, approximation theory, the theory of turbulent flows, motion of planets, functional analysis, the foundations of geometry, and the history and general methodology of mathematics. He also worked on the theory of dynamical systems, with application to Hamiltonian dynamics. Kolmogorov made a major contribution to Hilbert's sixth problem. Furthermore, he completely solved Hilbert's thirteenth problem in 1957. Later in his life, he took special interest in the education of gifted children. For his work, Kolmogorov was awarded several Russian and international prizes. He was also elected to several prestigious academies. Kolmogorov died on 20 October, 1987 in Moscow, Russia.

## 8.1 Introduction

A quick tour of probabilistic tools and techniques to study nondeterministic events is provided in this chapter. Initially axioms of probability theory are stated. The concept of random variable is next introduced. This is followed by a description of average measures such as expectation of a random variable. Typical second order measures, and the concept of independent random variables are also introduced. Common tools for studying distributions are $z$-transforms and moment generating functions. These ideas are further clarified via examples of discrete and continuous random variables. The well-known multivariate Gaussian distribution is also defined. Some well-known results like: Bienaymé-Chebyshev inequality, Chernoff bounds, Jensen's inequality, weak and strong law of large numbers, Gaussian and Lévy's central limit theorems, and stable distributions are also described. Distribution of range of a sequence of random variables is also studied. This is particularly useful in modeling Internet traffic. Elements of the theory of large deviation are outlined in the last section.

It was the French philosopher and mathematician Rene Descartes (1596-1650) who said, "It is a truth very certain that, when it is not in our power to determine what is true, we ought to follow what is most probable." It is in this spirit that the subject of probability theory and its applications are studied in this work.

Probability theory is the study of random phenomena. Such phenomena are empirical in that the outcomes of experiments are not necessarily identical (deterministic), however the outcomes of the experiments follow a certain regularity in a statistical sense. This intuitive notion is made more precise in probability theory. It should be noted that probability theory and statistics are interrelated. Statistics analyzes data from the real world. On the other hand, probability theory is an axiomatic branch of mathematics. Basics of applied analysis and optimization theory are required to study this chapter.

## 8.2  Axioms of Probability Theory

Probability theory can be developed on the basis of a set of axioms, which were first promulgated by A. N. Kolmogorov in the year 1933. In these axioms, an *experiment* is a mental or physical activity which produces a *measurable* outcome.

**Axioms of Probability Theory.** *Probability is defined as the triple* $(\Omega, \mathcal{F}, P)$, *where*:

(a) $\Omega$ *is the sample space. It is the set of all possible mutually exclusive outcomes of a specified experiment. Each such possible outcome $\omega$, is called a sample point.*

(b) $\mathcal{F}$ *is a family of events.* $\mathcal{F} = \{A, B, C, \ldots\}$, *where each event is a set of sample points* $\{\omega \mid \omega \in \Omega\}$. *Thus an event is a subset of $\Omega$. All subsets of $\Omega$ are not necessarily events in the set $\mathcal{F}$. The events in the set $\mathcal{F}$ observe the following rules.*

   (i) $\Omega \in \mathcal{F}$.
   (ii) *If $A \in \mathcal{F}$, then $A^c \in \mathcal{F}$.*
   (iii) *If $A_i \in \mathcal{F}, \forall\, i \in \mathbb{P}$; then $\bigcup_{i \in \mathbb{P}} A_i \in \mathcal{F}$.*
   *Such collection of events is called an algebra.*

(c) $P$ *is a real-valued mapping (function) defined on $\mathcal{F}$, where $P(A)$ is the probability of the event A. It is also called the probability measure. The function $P(\cdot)$ also has to satisfy the following axioms.*

$[Axiom\ 1]$ *For any event A,* $P(A) \geq 0$.
$[Axiom\ 2]$ $P(\Omega) = 1$.
$[Axiom\ 3]$ *If $A \cap B = \varnothing$, that is A and B are mutually exclusive events, then*

$$P(A \cup B) = P(A) + P(B) \tag{8.1a}$$

$[Axiom\ 3']$ *Let $A_1, A_2, \ldots$ be a sequence of events, such that $A_j \cap A_k = \varnothing, j \neq k$, where $j, k \in \mathbb{P}$, then*

$$P(A_1 \cup A_2 \cup \ldots) = P(A_1) + P(A_2) + \ldots \tag{8.1b}$$

$\square$

Observe that the Axiom $3'$ does not follow from Axiom 3. However Axiom $3'$ is superfluous if the sample space $\Omega$ is finite.

**Observations 8.1.** Let $A$, and $B$ be any events. Then

1. Let $\mathcal{F}$ be the family of events.
   (a) $\varnothing \in \mathcal{F}$, where $\varnothing$ is called the null event.
   (b) If $A, B \in \mathcal{F}$, then $A \cup B \in \mathcal{F}$, and $A \cap B \in \mathcal{F}$.
2. $P(\varnothing) = 0$.
3. $P(A^c) = (1 - P(A))$, where $A^c$ is the complement of the event $A$.
4. $P(A) \leq 1$.

5. $P(A) \leq P(B)$, if $A \subseteq B$.

6. An extension of the simple form of inclusion-exclusion principle used in combinatorics.

$$P(A \cup B) = P(A) + P(B) - P(A \cap B)$$

7. Given any $N$ events $A_1, A_2, \ldots A_N$,

$$
\begin{aligned}
&P(A_1 \cup A_2 \cup \ldots \cup A_N) \\
&= \sum_{1 \leq j \leq N} P(A_j) - \sum_{1 \leq j < k \leq N} P(A_j \cap A_k) + \sum_{1 \leq j < k < l \leq N} P(A_j \cap A_k \cap A_l) \\
&\quad - \sum_{1 \leq j < k < l < m \leq N} P(A_j \cap A_k \cap A_l \cap A_m) + \ldots \\
&\quad + (-1)^{N+1} P(A_1 \cap A_2 \cap \ldots \cap A_N)
\end{aligned}
\tag{8.2}
$$

The above equation uses the generalized form of the inclusion-exclusion principle.

8. The last observation yields the following set of Bonferroni's inequalities. These inequalities are named after the probabilist Carlo Emilio Bonferroni (1892-1960).

$$P(A_1 \cup A_2 \cup \ldots \cup A_N) \leq \sum_{1 \leq j \leq N} P(A_j) \tag{8.3a}$$

$$P(A_1 \cup A_2 \cup \ldots \cup A_N) \geq \sum_{1 \leq j \leq N} P(A_j) - \sum_{1 \leq j < k \leq N} P(A_j \cap A_k) \tag{8.3b}$$

$$
\begin{aligned}
P(A_1 \cup A_2 \cup \ldots \cup A_N) &\leq \sum_{1 \leq j \leq N} P(A_j) - \sum_{1 \leq j < k \leq N} P(A_j \cap A_k) \\
&\quad + \sum_{1 \leq j < k < l \leq N} P(A_j \cap A_k \cap A_l)
\end{aligned}
\tag{8.3c}
$$

$$\cdots \geq \cdots$$
$$\cdots \leq \cdots$$

$\square$

The following definitions are related to conditional probability and independence of two events. Conditional probability is the probability that an event occurs, given that another event has occurred. Independent events, as the name says, are events which do not affect the outcome of one another.

**Definitions 8.1.** *Conditional probability and independence.*

1. *The conditional probability of an event $A$, given event $B$ has occurred, is given by*

$$P(A \mid B) = \frac{P(A \cap B)}{P(B)}, \quad P(B) \neq 0 \tag{8.4a}$$

2. *Events $A$ and $B$ are said to be independent of each other if and only if*

$$P(A \cap B) = P(A) P(B) \tag{8.4b}$$

*If the above relationship does not hold, then the events $A$ and $B$ are said to be dependent.*  $\square$

If the events $A$ and $B$ are independent of each other then

$$P(A \mid B) = P(A), \text{ and } P(B \mid A) = P(B)$$

The concept of independence can be extended to more than two events. For example, the three events $A, B$, and $C$ are independent of each other, if:

$$P(A \cap B) = P(A) P(B)$$
$$P(B \cap C) = P(B) P(C)$$
$$P(C \cap A) = P(C) P(A)$$
$$P(A \cap B \cap C) = P(A) P(B) P(C)$$

In the next section random variables are discussed.

## 8.3 Random Variables

A random variable generally takes real values. However, it can sometimes take logical, complex, vector, tree, set, and other types of values. A random variable, distribution function, probability mass function, and probability density function are defined. A real-valued random variable is either discrete or continuous.

**Definitions 8.2.** *Basic definitions of random variable and related functions.*

1. *A random variable is a function $X$ that maps a sample point $\omega \in \Omega$ into the real line. That is, $X(\omega) \in \mathbb{R}$. The random variable is often simply denoted as $X$.*
2. *The distribution function $F_X(\cdot)$ of the random variable $X$ is defined for any $x \in \mathbb{R}$ as $F_X(x) = P(X \leq x)$. It is also sometimes referred to as the cumulative distribution function. The complementary cumulative distribution function $F_X^c(\cdot)$ of the random variable $X$, is specified by $F_X^c(x) = P(X > x) = (1 - F_X(x))$.*
3. *A random variable $X$ is discrete, if its set of possible values is countable. If the random variable $X$ takes on values $x_j, j = 1, 2, 3, \ldots$, then the probabilities $P(X = x_j) \triangleq p_X(x_j), j = 1, 2, 3, \ldots$, are called the probability mass function (or distribution) of the random variable $X$. The corresponding cumulative distribution function is said to be discrete.*
4. *A random variable $X$ is continuous, if its image set $X(\Omega)$ is a continuum of numbers. It is assumed that there exists a piecewise-continuous function $f_X(\cdot)$ that maps real numbers into real numbers such that*

$$P(a < X \leq b) = \int_a^b f_X(x)\, dx, \ \ \forall\, a < b \tag{8.5}$$

*The function $f_X(\cdot)$ is called the probability density function. The corresponding cumulative distribution function is said to be continuous.*                                        □

**Observations 8.2.** Some basic observations.

1. The distribution function $F_X(\cdot)$ of the random variable $X$, is a monotonically nondecreasing function. That is, if $x < y$ then $F_X(x) \leq F_X(y)$. Also $0 \leq F_X(x) \leq 1$. Furthermore, $\lim_{x \to -\infty} F_X(x) = 0$, and $\lim_{x \to \infty} F_X(x) = 1$. In addition, for $h > 0$, $F_X(x) = \lim_{h \to 0} F_X(x + h) = F_X(x_+)$.

2. Let $X$ be a discrete random variable, which takes on values $x_j, j = 1, 2, 3, \ldots$. The probabilities $P(X = x_j) = p_X(x_j), j = 1, 2, 3, \ldots$ satisfy

$$p_X(x_j) \geq 0, \quad \forall\, j \in \mathbb{P}; \quad \text{and} \quad \sum_{j \in \mathbb{P}} p_X(x_j) = 1 \tag{8.6a}$$

$$F_X(x) = \sum_{x_j \leq x} p_X(x_j) \tag{8.6b}$$

3. Let $X$ be a continuous random variable, and its probability density function be $f_X(x), x \in \mathbb{R}$. The probability density function satisfies the following relationships.

$$F_X(x) = \int_{-\infty}^{x} f_X(t)\, dt, \quad f_X(x) = \frac{d}{dx} F_X(x) \tag{8.7a}$$

$$\int_{\mathbb{R}} f_X(x)\, dx = 1 \tag{8.7b}$$

It is assumed that the derivative exists. It follows from the monotonicity of $F_X(\cdot)$, that $f_X(x) \geq 0$ for each $x \in \mathbb{R}$. $\qquad\square$

### Jointly Distributed Random Variables

Jointly distributed random variables are first defined for two random variables. This is then extended to $N$ random variables.

**Definitions 8.3.** *Let $X$ and $Y$ be jointly distributed random variables which take real values.*

1. *Joint distributions.*
   (a) *The joint cumulative distribution function of the two random variables $X$ and $Y$ is $F_{X,Y}(\cdot, \cdot)$, where*
   $$F_{X,Y}(x, y) = P(X \leq x, Y \leq y) \tag{8.8a}$$
   (b) *If $X$ and $Y$ are two discrete random variables, then the joint probability mass function of the two random variables $X$ and $Y$ is $p_{X,Y}(\cdot, \cdot)$, where*
   $$p_{X,Y}(x, y) = P(X = x, Y = y) \tag{8.8b}$$
   (c) *Let the two random variables $X$ and $Y$ be continuous. The random variables $X$ and $Y$ are jointly continuous if there exists a function $f_{X,Y}(\cdot, \cdot)$ such that*
   $$P\left(X \in \widetilde{A}, Y \in \widetilde{B}\right) = \int_{\widetilde{B}} \int_{\widetilde{A}} f_{X,Y}(x, y)\, dx dy \tag{8.8c}$$
   *where $\widetilde{A}$ and $\widetilde{B}$ are any subsets of real numbers. The function $f_{X,Y}(\cdot, \cdot)$ is called the joint probability density function.*

2. *Marginal distributions.*
   (a) *As $y$ tends to $\infty$, $F_{X,Y}(x, y)$ tends to $F_X(x)$. Similarly as $x$ tends to $\infty$, $F_{X,Y}(x, y)$ tends to $F_Y(y)$. $F_X(\cdot)$ and $F_Y(\cdot)$ are called marginal cumulative distribution functions of $X$ and $Y$ respectively.*

(b) *Let $X$ and $Y$ be both discrete random variables with joint probability mass function $p_{X,Y}(\cdot,\cdot)$. Then*

$$p_X(x) = \sum_y p_{X,Y}(x,y) \tag{8.9a}$$

$$p_Y(y) = \sum_x p_{X,Y}(x,y) \tag{8.9b}$$

*where $p_X(\cdot)$ and $p_Y(\cdot)$ are called the marginal probability mass functions of $X$ and $Y$ respectively.*

(c) *Let $X$ and $Y$ be both continuous random variables with joint probability density function $f_{X,Y}(\cdot,\cdot)$. Then*

$$f_X(x) = \int_{-\infty}^{\infty} f_{X,Y}(x,y)\,dy \tag{8.9c}$$

$$f_Y(y) = \int_{-\infty}^{\infty} f_{X,Y}(x,y)\,dx \tag{8.9d}$$

*where $f_X(\cdot)$ and $f_Y(\cdot)$ are called the marginal probability density functions of $X$ and $Y$ respectively.*

3. *Conditional distributions.*

(a) *Let $X$ and $Y$ be both discrete random variables with joint probability mass function $p_{X,Y}(\cdot,\cdot)$. Conditional probability mass function of $X$, given $Y = y$ and $p_Y(y) > 0$, is defined by*

$$p_{X|Y}(x \mid y) = P(X = x \mid Y = y), \quad \forall\, x \tag{8.10a}$$

(b) *Let $X$ and $Y$ be both continuous random variables with joint probability density function $f_{X,Y}(\cdot,\cdot)$. Conditional probability density function of $X$, given $Y = y$ and $f_Y(y) > 0$ is defined by*

$$f_{X|Y}(x \mid y) = \frac{f_{X,Y}(x,y)}{f_Y(y)}, \quad \forall\, x \tag{8.10b}$$

4. *Let $X_1, X_2, \ldots, X_N$ be $N \in \mathbb{P} \setminus \{1\}$ jointly distributed random variables. Then $F(\cdot,\cdot,\ldots,\cdot)$ is their joint cumulative distribution function, where*

$$F(x_1, x_2, \ldots, x_N) = P(X_1 \leq x_1, X_2 \leq x_2, \ldots, X_N \leq x_N) \tag{8.11}$$

*Joint probability mass function (for discrete random variables) and joint probability density function (for continuous random variables) for $N$ random variables can be similarly defined.*

□

## 8.4 Average Measures

Expectation of a random variable, expectation of a function of a random variable, and common second order expectations are defined and discussed in this section.

### 8.4.1  Expectation

The expectation of a discrete and continuous random variable is defined below.

**Definition 8.4.** *The expectation or mean or average value of a random variable $X$ is denoted by $\mathcal{E}(X)$. It is*

$$\mathcal{E}(X) = \int_{-\infty}^{\infty} x \, dF_X(x) \tag{8.12a}$$

*Specifically:*

(a) *If $X$ is a discrete random variable.*

$$\mathcal{E}(X) = \sum_{x:p_X(x)>0} x p_X(x) \tag{8.12b}$$

*provided the summation exists.*

(b) *If $X$ is a continuous random variable.*

$$\mathcal{E}(X) = \int_{-\infty}^{\infty} x f_X(x) \, dx \tag{8.12c}$$

*provided the integral exists.* □

Let $g(\cdot)$ be a function of a random variable $X$. The expectation of $g(X)$ is determined as follows.

(a) If $X$ is a discrete random variable: $\mathcal{E}(g(X)) = \sum_{x:p_X(x)>0} g(x) p_X(x)$.
(b) If $X$ is a continuous random variable: $\mathcal{E}(g(X)) = \int_{-\infty}^{\infty} g(x) \, dF_X(x)$.

Also if $X_1, X_2, \ldots X_N$ are $N$ jointly distributed random variables, and $b_1, b_2, \ldots b_N \subset \mathbb{R}$, then

$$\mathcal{E}(\sum_{j=1}^{N} b_j X_j) - \sum_{j=1}^{N} b_j \mathcal{E}(X_j) \tag{8.13}$$

The mean of a random variable $X$ is also called its first moment. Higher moments of the random variable are similarly defined.

**Definitions 8.5.** *Let $X$ be a random variable, and $r \in \mathbb{P}$. The $r$th moment of $X$ is $\mu_r = \mathcal{E}(X^r)$, and the $r$th central moment of $X$ is $\mathcal{E}((X - \mu_1)^r)$. The parameter $r$ is called the order of the moment.* □

The indicator function of an event is another useful concept. It is a useful artifice to define the probability of an event.

**Definition 8.6.** *$I_A$ is called an indicator function of an event $A$ if*

$$I_A = \begin{cases} 1, & \text{if } A \text{ occurs} \\ 0, & \text{if } A^c \text{ occurs} \end{cases} \tag{8.14}$$

*where $A^c$ is the complement of the event $A$.* □

Thus if $I_A$ is the indicator function of an event $A$, then $\mathcal{E}(I_A) = P(A)$. That is, the expectation of the indicator function of an event is its probability.

### 8.4.2  Common Second Order Expectations

The common second order expectations of a single random variable are variance, standard deviation, and squared coefficient of variation. Similarly, the common second order expectations of two jointly distributed random variables are covariance, and correlation coefficient.

**Definitions 8.7.** *Common second order expectations.*

1. *The variance $Var(X)$ of a random variable $X$ is*

$$Var(X) = \mathcal{E}\left((X - \mathcal{E}(X))^2\right) \tag{8.15a}$$

*That is, $Var(X) = \mathcal{E}\left(X^2\right) - (\mathcal{E}(X))^2 = \left(\mu_2 - \mu_1^2\right)$.*

2. *The standard deviation $\sigma_X$ of a random variable $X$, is $\sigma_X = \sqrt{Var(X)}$.*

3. *The squared coefficient of variation $C_X^2$, of a random variable $X$ where $\mathcal{E}(X) \neq 0$ is*

$$C_X^2 = \frac{Var(X)}{\{\mathcal{E}(X)\}^2} \tag{8.15b}$$

4. *The covariance $Cov(X,Y)$ of two jointly distributed random variables $X$ and $Y$ is*

$$Cov(X,Y) = \mathcal{E}\left((X - \mathcal{E}(X))(Y - \mathcal{E}(Y))\right) \tag{8.15c}$$

*That is*

$$Cov(X,Y) = \mathcal{E}(XY) - \mathcal{E}(X)\mathcal{E}(Y) \tag{8.15d}$$

5. *Let $\sigma_X$ and $\sigma_Y$ be the standard deviation of the jointly distributed random variables $X$ and $Y$ respectively, where $\sigma_X \neq 0$ and $\sigma_Y \neq 0$. The correlation coefficient $Cor(X,Y)$ of these random variables is*

$$Cor(X,Y) = \frac{Cov(X,Y)}{\sigma_X \sigma_Y} \tag{8.15e}$$

*If $Cor(X,Y) = 0$, then the random variables $X$ and $Y$ are uncorrelated.*                □

---

## 8.5  Independent Random Variables

A precise definition of stochastic independence is as follows.

**Definition 8.8.** *Random variables $X$ and $Y$ are stochastically independent (or simply independent) random variables if for all values of $x$ and $y$*

$$F_{X,Y}(x,y) = F_X(x)F_Y(y) \tag{8.16}$$

*where $F_{X,Y}(\cdot,\cdot)$ is the joint cumulative distribution function of the random variables $X$ and $Y$. Also $F_X(\cdot)$ and $F_Y(\cdot)$ are the marginal cumulative distribution functions of the random variables $X$ and $Y$ respectively.*                □

**Observations 8.3.** Let $X$ and $Y$ be independent random variables.

1. We have

$$\mathcal{E}(XY) = \mathcal{E}(X)\mathcal{E}(Y)$$

Note that the reverse is not true. That is, $\mathcal{E}(XY) = \mathcal{E}(X)\mathcal{E}(Y)$ does not imply the independence of random variables $X$ and $Y$.

2. Also

$$Var(X + Y) = Var(X) + Var(Y)$$

3. Further

$$Cov(X, Y) = Cor(X, Y) = 0$$

4. Given any $N$ mutually independent random variables $X_1, X_2, \ldots X_N$, and $b_1, b_2, \ldots, b_N \in \mathbb{R}$

$$\mathcal{E}(\prod_{j=1}^{N} X_j) = \prod_{j=1}^{N} \mathcal{E}(X_j)$$

$$Var(\sum_{j=1}^{N} b_j X_j) = \sum_{j=1}^{N} b_j^2 Var(X_j)$$

$\square$

## 8.6 Transforms and Moment Generating Functions

Transform techniques and moment generating functions sometimes make it easier to derive properties of random variables. The $z$-transform and moment generating function are discussed below.

### 8.6.1 $z$-Transform

The $z$-transform technique is a specially useful vehicle for studying discrete random variables. It is generally helpful in determining the mean value, second or higher moments of the random variable.

**Definition 8.9.** *Let $X$ be a discrete random variable with probability mass function $p_X(x)$, for $x \in \mathbb{N}$. The $z$-transform of the random variable $X$ is given by*

$$G_X(z) = \sum_{x \in \mathbb{N}} p_X(x) z^x \tag{8.17}$$

*where $z \in \mathbb{C}$, and $|z| \leq 1$.* $\square$

**Observations 8.4.** Some useful observations about $z$-transforms.

1. $G_X(1) = 1$.
2. The probabilities $p_X(x)$'s can be obtained from the $z$-transform as follows.

$$p_X(0) = G_X(0), \quad \text{and} \quad p_X(x) = \frac{1}{x!} \frac{d^x}{dz^x} G_X(z) \bigg|_{z=0}, \quad \forall x \in \mathbb{P}$$

3. The first and second moments are

$$\mathcal{E}(X) = \frac{d}{dz}G_X(z)\bigg|_{z=1}, \quad \text{and} \quad \mathcal{E}(X^2) = \frac{d^2}{dz^2}G_X(z)\bigg|_{z=1} + \mathcal{E}(X)$$

Higher moments can be similarly expressed.

4. Let $X_1, X_2, \ldots, X_N$ be a sequence of $N$ mutually independent random variables. Their $z$-transforms are $G_{X_i}(z)$ for $1 \leq i \leq N$. Also define $X = \sum_{i=1}^{N} X_i$. If its $z$-transform is $G_X(z)$. Then

$$G_X(z) = \prod_{i=1}^{N} G_{X_i}(z)$$

$\square$

### 8.6.2  Moment Generating Function

The moment generating function of a random variable is another convenient technique to determine its moments.

**Definition 8.10.** *Let $X$ be a random variable, and its $r$th moment be $\mu_r$, where $r \in \mathbb{P}$. The moment generating function of $X$ is given by $\mathcal{M}_X(t) = \mathcal{E}\left(e^{tX}\right)$.*

(a) *If $X$ is discrete random variable, then $\mathcal{M}_X(t) = \sum_x e^{tx}p_X(x)$.*
(b) *If $X$ is continuous random variable, then $\mathcal{M}_X(t) = \int_{-\infty}^{\infty} e^{tx}f_X(x)\,dx$.*

*It is assumed that $\mathcal{M}_X(t)$ exists for all $t \in (-h, h)$, for some $h > 0$.*  $\square$

From these definitions it follows that

$$\mu_r = \frac{d^r}{dt^r}\mathcal{M}_X(t)\bigg|_{t=0}, \qquad \forall\, r \in \mathbb{P}$$

Consider two random variables $X$ and $Y$. Assume that the moment generating functions of these two random variables exist, and are $\mathcal{M}_X(\cdot)$ and $\mathcal{M}_Y(\cdot)$ respectively. If $\mathcal{M}_X(t)$ and $\mathcal{M}_Y(t)$ are equal for all $t \in (-h, h)$ for some $h > 0$, then the cumulative distribution functions of the two random variables $X$ and $Y$ are identical. This result can be established via transform theory. Thus, if the moment generating function exists, then this moment generating function uniquely determines the corresponding distribution function.

The moments of a random variable are determined from its probability mass function, or probability density function. Note that a sequence of moments $\mu_1, \mu_2, \mu_3, \ldots$ does not always uniquely determine a distribution function. Only in some cases the corresponding probability mass function, or probability density function can also be determined uniquely if all the moments are known. However, if the moment generating function of a random variable did exist, then the corresponding distribution function can be determined uniquely.

The $z$-transform of a discrete random variable can be obtained by substituting $z$ for $e^t$ in the definition of the moment generating function. It can therefore be concluded that while the $z$-transform technique is applicable to discrete random variables, the moment generating function can be computed for both the discrete and continuous random variables. The above definitions are next extended to define the moment generating function of jointly distributed random variables.

**Definition 8.11.** *Let $X_1, X_2, \ldots, X_N$ be $N$ jointly distributed random variables. The joint moment generating function of $X_1, X_2, \ldots, X_N$ is*

$$\mathcal{M}_{X_1, X_2, \ldots, X_N}(t_1, t_2, \ldots, t_N) = \mathcal{E}\left(\exp\left\{\sum_{j=1}^{N} t_j X_j\right\}\right) \tag{8.18}$$

*It is assumed that the above expectation exists for all values of $t_1, t_2, \ldots, t_N$ such that $t_j \in (-h, h)$, for some $h > 0$, and $j = 1, 2, \ldots, N$.* □

**Observation 8.5.** Let $X_1, X_2, \ldots, X_N$ be a sequence of $N$ mutually independent random variables. The moment generating function of $X_i$ is $\mathcal{M}_{X_i}(\cdot)$ for $1 \le i \le N$. Also define $X = \sum_{i=1}^{N} X_i$. If its moment generating function is $\mathcal{M}_X(\cdot)$. Then

$$\mathcal{M}_X(t) = \prod_{i=1}^{N} \mathcal{M}_{X_i}(t)$$

□

The *characteristic function* $\varphi_X(\cdot)$ of a random variable $X$ is defined by replacing $t$ by $iu$, in the definition of moment generating function, where $i = \sqrt{-1}$ and $u \in \mathbb{R}$. An advantage of characteristic function over moment generating function is that it always exists, since the trigonometric values $\cos(ux)$ and $\sin(ux)$ are bounded for any $u, x \in \mathbb{R}$. Properties of characteristic functions are discussed at length, later in the chapter.

If a continuously distributed random variable takes nonnegative values, then it is sometimes convenient to use one-sided Laplace transform instead of the characteristic function. Let $f_X(x), x \in \mathbb{R}_0^+$ be the probability density function of a random variable $X$. Also let $s \in \mathbb{C}$. The single-sided Laplace transform of $f_X(\cdot)$, is

$$\widehat{f}_X(s) = \mathcal{L}[f_X(x)] = \int_0^\infty f_X(x) e^{-sx} dx$$

Note that $\mathcal{L}[f_X(x)] = \mathcal{E}\left(e^{-sX}\right)$. The $r$th moment of $X$ is

$$\mu_r = (-1)^r \frac{d^r}{ds^r} \widehat{f}_X(s)\bigg|_{s=0}, \qquad \forall\, r \in \mathbb{P}$$

Some well-known and useful probability distributions are discussed in the next section.

## 8.7  Examples of Some Distributions

Examples of discrete and continuous distributions are outlined in this section. Multivariate Gaussian distribution is also described.

### 8.7.1  Discrete Distributions

Properties of discrete distributions such as the Bernoulli distribution, binomial distribution, geometric distribution, Poisson distribution, Zipf distribution, and multinomial distribution are listed below.

*Bernoulli distribution*: Let $X$ be a random variable with Bernoulli distribution. Its parameter is $p$, where $0 \leq p \leq 1$. The probability mass function of $X$ is given by

$$p_X(x) = \begin{cases} q, & x = 0 \\ p, & x = 1 \end{cases} \tag{8.19}$$

where $q = (1 - p)$. Also $\mathcal{E}(X) = p$, $Var(X) = pq$, $G_X(z) = (q + pz)$, and $\mathcal{M}_X(t) = (q + pe^t)$.

*Binomial distribution*: Let $X$ be a random variable with binomial distribution. Its parameters are $p$ and $n$, where $0 \leq p \leq 1$, and $n \in \mathbb{P}$. The probability mass function of $X$ is given by

$$p_X(x) = \begin{cases} \binom{n}{x} p^x q^{n-x}, & x = 0, 1, 2, \ldots, n \\ 0, & \text{otherwise} \end{cases} \tag{8.20}$$

where $q = (1 - p)$. Also $\mathcal{E}(X) = np$, $Var(X) = npq$, $G_X(z) = (q + pz)^n$, and $\mathcal{M}_X(t) = (q + pe^t)^n$. Note that $n = 1$, yields a Bernoulli distribution.

*Geometric distribution*: Assume that $X$ is a random variable with geometric distribution. Its parameter is $p$, where $0 < p \leq 1$. The probability mass function of $X$ is given by

$$p_X(x) = \begin{cases} pq^x, & \forall \, x \in \mathbb{N} \\ 0, & \text{otherwise} \end{cases} \tag{8.21}$$

where $q = (1 - p)$. Also $\mathcal{E}(X) = q/p$, $Var(X) = q/p^2$, $G_X(z) = p/(1 - qz)$, and $\mathcal{M}_X(t) = p/(1 - qe^t)$. Also

$$P(X \geq m + n \mid X \geq m) = P(X \geq n), \quad \forall \, m, n \in \mathbb{N}$$

The above result is called *memoryless property* of the geometric distribution. Geometric distribution is the only discrete distribution that possesses the memoryless property.

*Poisson distribution*: Assume that $X$ is a random variable with Poissonian distribution. Its parameter is $\lambda \in \mathbb{R}^+$. The probability mass function of $X$ is given by

$$p_X(x) = \begin{cases} e^{-\lambda} \dfrac{\lambda^x}{x!}, & \forall \, x \in \mathbb{N} \\ 0, & \text{otherwise} \end{cases} \tag{8.22}$$

Also $\mathcal{E}(X) = \lambda$, $Var(X) = \lambda$, $G_X(z) = e^{\lambda(z-1)}$, and $\mathcal{M}_X(t) = e^{\lambda(e^t - 1)}$.

*Zipf distribution*: Let $X$ be a random variable with Zipf distribution. Its parameters are $\gamma \in \mathbb{R}^+$ and $n \in \mathbb{P}$. The probability mass function of $X$ is given by

$$p_X(x) = \kappa x^{-\gamma}, \quad x \in \{1, 2, \ldots, n\}, \quad \kappa^{-1} = \sum_{x=1}^{n} x^{-\gamma} \tag{8.23}$$

The parameter $\gamma$ is called the exponent of the distribution. For large values of $n$

$$\kappa^{-1} \simeq \begin{cases} \ln(n), & \gamma = 1 \\ \left(1 - n^{-(\gamma-1)}\right)/(\gamma - 1), & \gamma \in \mathbb{R}^+ \setminus \{1\} \end{cases}$$

This distribution is an example of power-law distribution, because of its form. The moments of this random variable are generally obtained numerically. The parameter value $\gamma = 1$ yields the classical Zipf distribution. For other values of the parameter $\gamma$, this distribution is called the generalized Zipf distribution. This distribution is named after the American linguist and philologist, George Kingsley Zipf (1902-1950). It has wide ranging applications in Internet engineering.

*Multinomial distribution*: An example of a discrete joint distribution is the multinomial distribution. Consider an experiment, in which there are $r$ possibilities. These possibilities occur with probabilities $p_1, p_2, \ldots, p_r$ such that $\sum_{i=1}^{r} p_i = 1$. Let $n$ such experiments be performed independent of each other. Also, let $X_i$ denote the number of times (out of $n$) the $i$th outcome occurs, where $1 \leq i \leq r$. Then

$$P(X_1 = x_1, X_2 = x_2, \ldots, X_r = x_r) = \frac{n!}{\prod_{i=1}^{r} x_i!} \prod_{i=1}^{r} p_i^{x_i} \tag{8.24a}$$

where $x_i \in \mathbb{N}$, for $1 \leq i \leq r$, and $\sum_{i=1}^{r} x_i = n$. Using multinomial theorem it can be shown that

$$\sum_{\substack{x_i \in \mathbb{N}, 1 \leq i \leq r, \\ (x_1 + x_2 + \ldots + x_r) = n}} P(X_1 = x_1, X_2 = x_2, \ldots, X_r = x_r) = 1 \tag{8.24b}$$

Let

$$\widetilde{n} = \binom{n + r - 1}{r - 1} \tag{8.24c}$$

then the multinomial distribution has $\widetilde{n}$ terms. Use of moment generating function of the jointly distributed random variables yields $\mathcal{E}(X_i) = np_i$, and $Var(X_i) = np_i(1 - p_i)$ for $1 \leq i \leq r$. The $Cov(X_i, X_j) = -np_i p_j$, for $i \neq j, 1 \leq i, j \leq r$.

### 8.7.2 Continuous Distributions

Properties of continuous distributions like: uniform distribution, exponential distribution, gamma distribution, normal distribution, Cauchy distribution, Pareto distribution, lognormal distribution, Weibull distribution, and Wigner distribution are listed below.

*Uniform distribution*: A random variable $X$ has a uniform distribution, if the probability density function of $X$ is given by

$$f_X(x) = \begin{cases} \dfrac{1}{(b-a)}, & x \in [a, b] \\ 0, & \text{otherwise} \end{cases} \tag{8.25}$$

Its parameter space is $a, b \in \mathbb{R}$, where $a < b$. Also $\mathcal{E}(X) = (a+b)/2$, $Var(X) = (b-a)^2/12$. Also

$$F_X(x) = \begin{cases} 0, & x < a \\ \dfrac{(x-a)}{(b-a)}, & x \in [a,b] \\ 1, & x > b \end{cases}$$

$$\mathcal{M}_X(t) = \frac{(e^{bt} - e^{at})}{(b-a)\,t}$$

*Exponential distribution*: A random variable $X$ has an exponential distribution, if the probability density function of $X$ is given by

$$f_X(x) = \begin{cases} 0, & x \in (-\infty, 0) \\ \lambda e^{-\lambda x}, & x \in [0, \infty) \end{cases} \tag{8.26}$$

Its parameter is $\lambda \in \mathbb{R}^+$. Also $\mathcal{E}(X) = 1/\lambda$, $Var(X) = 1/\lambda^2$. And

$$F_X(x) = \begin{cases} 0, & x \in (-\infty, 0) \\ \left(1 - e^{-\lambda x}\right), & x \in [0, \infty) \end{cases}$$

$$\mathcal{M}_X(t) = \frac{\lambda}{(\lambda - t)}, \quad t < \lambda$$

If $X$ is an exponentially distributed random variable, with parameter $\lambda$, then

$$P\left(X \geq t + u \mid X \geq u\right) = P\left(X \geq t\right), \quad t, u \in [0, \infty)$$

The above result is called the *memoryless property* of the exponential distribution. Exponential distribution is the only continuous distribution that possesses the memoryless property.

*Gamma distribution*: The probability density function of a random variable $X$ with gamma distribution is given by

$$f_X(x) = \begin{cases} 0, & x \in (-\infty, 0) \\ \dfrac{r\lambda\,(r\lambda x)^{r-1}\,e^{-r\lambda x}}{\Gamma(r)}, & x \in [0, \infty) \end{cases} \tag{8.27}$$

where $\Gamma(\cdot)$ is the gamma function. The parameters of this distribution are $\lambda, r \in \mathbb{R}^+$. If $r = 1$, then $X$ is an exponentially distributed random variable. The average value is $\mathcal{E}(X) = \lambda^{-1}$, $Var(X) = \left(r\lambda^2\right)^{-1}$. Also

$$\mathcal{M}_X(t) = \left\{\frac{\lambda r}{\lambda r - t}\right\}^r, \quad t < \lambda r$$

If $r$ is a positive integer greater than 1, then the distribution is called an Erlang distribution. It can be shown that the sum of $r$ independent exponentially distributed random variables each with parameter $r\lambda$, has an Erlang distribution. Note that some authors give a slightly modified version of the above definition of gamma distribution. The mean of the random variable $X$ in the above definition was adjusted such that it was independent of $r$.

*Normal distribution*: A random variable $X$ has a normal (or Gaussian) distribution, if the probability density function of $X$ is given by

$$f_X(x) = \frac{1}{\sqrt{2\pi}\sigma} \exp\left\{-\frac{1}{2}\left(\frac{x - \mu}{\sigma}\right)^2\right\}, \quad x \in \mathbb{R} \tag{8.28}$$

where $\mu \in \mathbb{R}$ and $\sigma \in \mathbb{R}^+$ are its parameters. Also $\mathcal{E}(X) = \mu$, and $Var(X) = \sigma^2$. Its moment generating function is given by

$$\mathcal{M}_X(t) = \exp\left(\mu t + \frac{\sigma^2 t^2}{2}\right)$$

Use of this moment generating function yields

$$\mathcal{E}\{(X - \mu)^r\} = \begin{cases} 0, & r \text{ is an odd integer} \\ \dfrac{r!}{(r/2)!} \dfrac{\sigma^r}{2^{r/2}}, & r \text{ is an even integer} \end{cases}$$

A normally distributed random variable $X$ with mean $\mu$ and variance $\sigma^2$ is generally denoted by

$$X \sim \mathcal{N}\left(\mu, \sigma^2\right)$$

If a normal random variable has $\mu = 0$, then it is called a centered normal random variable. If in addition $\sigma = 1$, then it is called a *standard normal random variable*. Its probability density function $\phi(\cdot)$, and cumulative distribution function $\Phi(\cdot)$ are given by

$$\phi(x) = \frac{1}{\sqrt{2\pi}} e^{-\frac{x^2}{2}}, \qquad x \in \mathbb{R}$$

$$\Phi(x) = \frac{1}{\sqrt{2\pi}} \int_{-\infty}^{x} e^{-\frac{y^2}{2}} dy, \quad x \subset \mathbb{R}$$

Also note that $\Phi(x) = (1 - \Phi(-x))$. The function $\Phi(\cdot)$ is generally evaluated numerically. The cumulative distribution function $\Phi(\cdot)$ can also be expressed in terms of the error function. Error function is defined as

$$\text{erf}(z) = \frac{2}{\sqrt{\pi}} \int_0^z e^{-t^2} dt, \quad z \in \mathbb{C}$$

Therefore

$$\Phi(x) = \frac{1}{2}\left\{1 + \text{erf}\left(\frac{x}{\sqrt{2}}\right)\right\}, \quad x \in \mathbb{R}$$

*Cauchy distribution*: A random variable $X$ has a Cauchy distribution, if the probability density function of $X$ is given by

$$f_X(x) = \frac{1}{\pi\beta\left[1 + \left(\dfrac{x - \alpha}{\beta}\right)^2\right]}, \quad x \in \mathbb{R} \tag{8.29}$$

where $\alpha \in \mathbb{R}$, and $\beta \in \mathbb{R}^+$ are its parameters. This distribution is symmetrical about the parameter $\alpha$. However, the mean, variance, and higher moments of this random variable do not exist. Its cumulative distribution function $F_X(\cdot)$ is

$$F_X(x) = \frac{1}{2} + \frac{1}{\pi} \tan^{-1}\left(\frac{x - \alpha}{\beta}\right), \quad x \in \mathbb{R}$$

The moment generating function of this random variable also does not exist. Nevertheless its characteristic function can be obtained by using an appropriate contour-integral technique that equals to $\exp(i\alpha t - \beta|t|)$.

*Pareto distribution*: A random variable $X$ has a Pareto distribution, if the probability density function of $X$ is given by

$$f_X(x) = \begin{cases} 0, & x \le x_0 \\ \dfrac{\alpha}{x_0}\left(\dfrac{x_0}{x}\right)^{\alpha+1}, & x > x_0 \end{cases} \tag{8.30}$$

where $x_0, \alpha \in \mathbb{R}^+$ are its parameters. Note that $x_0$ is called the location parameter, and $\alpha$ is called the shape parameter or tail-index. Also

$$\mathcal{E}(X) = \frac{\alpha x_0}{(\alpha - 1)}, \quad \alpha > 1$$

$$Var(X) = \frac{\alpha x_0^2}{(\alpha - 2)} - \left(\frac{\alpha x_0}{\alpha - 1}\right)^2, \quad \alpha > 2$$

$$F_X(x) = \begin{cases} 0, & x \le x_0 \\ 1 - \left(\dfrac{x_0}{x}\right)^{\alpha}, & x > x_0 \end{cases}$$

The moment generating function of this random variable does not exist. Pareto density function decays much more slowly than the exponential distribution. Because of its form, this distribution is an example of power-law distribution. This distribution has recently found application in modeling Internet traffic.

*Lognormal distribution*: A random variable $X$ has a lognormal distribution, if $X$ is a positive-valued random variable, and $Y = \ln X$ has a normal distribution. If $Y \sim \mathcal{N}(\mu, \sigma^2)$, then the probability density function of random variable $X$ is given by

$$f_X(x) = \begin{cases} 0, & x \le 0 \\ \dfrac{1}{x\sqrt{2\pi}\sigma} \exp\left\{-\dfrac{1}{2\sigma^2}(\ln x - \mu)^2\right\}, & x > 0 \end{cases} \tag{8.31}$$

where $\mu \in \mathbb{R}$ and $\sigma \in \mathbb{R}^+$ are its parameters. Conversely, if $X$ has a lognormal distribution, then $\mathcal{E}(\ln X) = \mu$, and $Var(\ln X) = \sigma^2$. Also

$$\mathcal{E}(X^r) = \exp\left(r\mu + \frac{1}{2}r^2\sigma^2\right), \quad r \in \mathbb{P}$$

This is an example of heavy-tailed distribution.

*Weibull distribution*: A random variable $X$ has a Weibull distribution, if the probability density function of $X$ is given by

$$f_X(x) = \begin{cases} 0, & x \le 0 \\ \lambda b x^{b-1} e^{-\lambda x^b}, & x > 0 \end{cases} \tag{8.32}$$

where $\lambda, b \in \mathbb{R}^+$ are its parameters. If $b = 1$, the Weibull probability density function reduces to the exponential probability density function with parameter $\lambda$. Also

$$\mathcal{E}(X^r) = \lambda^{-r/b}\Gamma\left(1 + \frac{r}{b}\right), \quad r \in \mathbb{P}$$

$$F_X(x) = \begin{cases} 0, & x \le 0 \\ 1 - \exp\left\{-\lambda x^b\right\}, & x > 0 \end{cases}$$

where $\Gamma\left(\cdot\right)$ is the gamma function. The random variable $X$ can be obtained via the transformation $X = Y^{1/b}$, where the random variable $Y$ has an exponential distribution with parameter $\lambda \in \mathbb{R}^+$. This is an example of heavy-tailed distribution.

*Wigner distribution*: A random variable $X$ has a Wigner distribution, if the probability density function of $X$ is given by

$$f_X\left(x\right) = \begin{cases} \dfrac{2}{\pi}\sqrt{1 - x^2}, & |x| \leq 1 \\[2mm] 0, & |x| \geq 1 \end{cases} \tag{8.33}$$

Also $\mathcal{E}\left(X\right) = 0, Var(X) = 1/4$. Its $r$th moment $\mu_r, r \in \mathbb{P}$ is

$$\mu_r = \mathcal{E}\left(X^r\right) = \begin{cases} 0, & r \text{ is an odd number} \\[2mm] \dfrac{r!}{(r/2)!\,(r/2 + 1)!}\dfrac{1}{2^r}, & r \text{ is an even number} \end{cases}$$

Thus

$$\mu_0 = 1,\ \mu_2 = \frac{1}{4},\ \mu_4 = \frac{1}{8},\ \mu_6 = \frac{5}{64},\ \mu_8 = \frac{7}{128},\ \mu_{10} = \frac{21}{512}, \ldots$$

There also exists a relationship between these moments and the Bessel function $J_1\left(\cdot\right)$ of the first kind of order one.

$$\sum_{r=0}^{\infty} \frac{\mu_r}{r!}\left(it\right)^r = \frac{2}{t}J_1\left(t\right)$$

where $i = \sqrt{-1}$. This distribution is named after the physicist Eugene P. Wigner (1902-1995). It is useful in the study of random matrices and spectral properties of random networks.

### 8.7.3  Multivariate Gaussian Distribution

Definition and properties of multivariate Gaussian (or normal) distribution are given in this subsection. This in turn is used in defining a Gaussian process. Gaussian process is discussed in the chapter on stochastic processes.

**Definition 8.12.** *Let* $Y_1, Y_2, \ldots, Y_n$ *be a set of* $n$ *independent standard normal random variables. Let*

$$X_i = \eta_i + \sum_{j=1}^{n} a_{ij}Y_j, \quad 1 \leq i \leq m \tag{8.34a}$$

*where* $a_{ij} \in \mathbb{R}, 1 \leq i \leq m, 1 \leq j \leq n$ *and* $\eta_i \in \mathbb{R}, 1 \leq i \leq m$ *are constants. The random variables* $X_1, X_2, \ldots, X_m$ *are said to have a multivariate normal distribution. Its joint probability density function exists provided its covariance matrix* $\Xi$ *has a positive determinant, where*

$$\Xi = \left[\xi_{ij}\right], \quad \xi_{ij} = Cov\left(X_i, X_j\right), \quad 1 \leq i, j \leq m \tag{8.34b}$$

$\square$

The above definition is valid because the sum of independent normal random variables is also a normal random variable. Therefore, each $X_i$ is a normal random variable.

**Observations 8.6.** Some basic observations about multivariate Gaussian distribution.

1. $\mathcal{E}(X_i) = \eta_i$ and $Var(X_i) = \xi_{ii} = \sum_{j=1}^{n} a_{ij}^2$, for $1 \le i \le m$.
2. The covariance $\xi_{ij} = Cov(X_i, X_j)$ is given by

$$\xi_{ij} = \sum_{k=1}^{n} a_{ik} a_{jk}, \quad \text{for } 1 \le i, j \le m$$

3. $\xi_{ij} = \xi_{ji}$, for $1 \le i, j \le m$. That is, the covariance matrix $\Xi$ is symmetric.
4. The covariance matrix $\Xi$ is positive definite. Therefore, its diagonal elements are all positive.
5. Let

$$X = \begin{bmatrix} X_1 \ X_2 \ \cdots \ X_m \end{bmatrix}^T$$
$$x = \begin{bmatrix} x_1 \ x_2 \ \cdots \ x_m \end{bmatrix}^T \in \mathbb{R}^m$$
$$\eta = \begin{bmatrix} \eta_1 \ \eta_2 \ \cdots \ \eta_m \end{bmatrix}^T \in \mathbb{R}^m$$

The joint probability density function of the random variables $X_1, X_2, \ldots, X_m$ is

$$f_X(x) = \frac{1}{(2\pi)^{m/2} (\det \Xi)^{1/2}} \exp\left\{ -\frac{1}{2} (x - \eta)^T \Xi^{-1} (x - \eta) \right\}$$

The elements of the random vector $X$ are said to have a multivariate Gaussian distribution. This is denoted compactly as $X \sim \mathcal{N}(\eta, \Xi)$.

6. Let $t = \begin{bmatrix} t_1 \ t_2 \ \cdots \ t_m \end{bmatrix}^T$. The joint moment generating function of the random variables $X_1, X_2, \ldots, X_m$ is

$$\mathcal{M}_X(t) = \exp\left\{ \eta^T t + \frac{1}{2} t^T \Xi t \right\}$$

$\square$

The concept of independent and uncorrelated random vectors is similar to that of independent and uncorrelated random variables. Also, the definitions of conditional distributions, and probability density functions are similar to those of random variables. Therefore, these properties of random vectors are not defined formally.

**Observations 8.7.** Some basic observations.

1. Let $X$ and $Y$ be random vectors of the same size, where $X \sim \mathcal{N}(\eta, \Xi)$, and $Y \sim \mathcal{N}(\mu, \Psi)$. Also, let the elements of the vector $X$ be uncorrelated with the elements of the vector $Y$. If $Z = (X + Y)$, then $Z \sim \mathcal{N}((\eta + \mu), (\Xi + \Psi))$.
2. Let $X$ be a random vector of size $n \in \mathbb{P}$, where $X \sim \mathcal{N}(\eta, \Xi)$. Also, let $A$ be a real-valued $r \times n$ matrix of rank $r \le n$; and $c$ be a real-valued column matrix of size $r$. If $Y = AX + c$, then $Y \sim \mathcal{N}(\mu, \Psi)$, where $\mu = A\eta + c$, and $\Psi = A\Xi A^T$.
   This result implies that under linear transformation, Gaussian distributions are preserved.
3. Let $X$ and $Y$ be jointly distributed random vectors (not necessarily with multivariate Gaussian distributions).
   (a) If the random vectors $X$ and $Y$ are independent of each other, then $X$ and $Y$ are uncorrelated.
   (b) If the random vectors $X$ and $Y$ are uncorrelated and each has a multivariate Gaussian distribution, then $X$ and $Y$ are independent.    $\square$

## 8.8  Some Well-Known Results

Some elementary and well-known results in probability theory like Markov's and Bienaymé-Chebyshev inequalities, first and second moment methods, Chernoff bounds, Jensen's inequality, weak and strong law of large numbers, and Gaussian central limit theorem are stated and proved.

### 8.8.1  Well-Known Inequalities

Well-known inequalities named after Markov, Bienaymé-Chebyshev, Chernoff, and Jensen are discussed in this subsection. A generalized Bienaymé-Chebyshev inequality is first established.

**Theorem 8.1.** *Let $X$ be a continuous random variable, and $h\left(\cdot\right)$ be a nonnegative function defined on the real line. Then*

$$P\left(h\left(X\right) \geq k\right) \leq \frac{\mathcal{E}\left(h\left(X\right)\right)}{k}, \quad \forall\, k > 0 \tag{8.35}$$

*Proof.* Let the probability density function of the random variable $X$ be $f_X\left(\cdot\right)$, then

$$\mathcal{E}\left(h\left(X\right)\right) = \int_{-\infty}^{\infty} h\left(x\right) f_X\left(x\right) dx$$

$$= \int_{\{x \mid h(x) \geq k\}} h\left(x\right) f_X\left(x\right) dx + \int_{\{x \mid h(x) < k\}} h\left(x\right) f_X\left(x\right) dx$$

$$\geq \int_{\{x \mid h(x) \geq k\}} h\left(x\right) f_X\left(x\right) dx \geq \int_{\{x \mid h(x) \geq k\}} k f_X\left(x\right) dx$$

$$= kP\left(h\left(X\right) \geq k\right)$$

The result follows. $\qquad\square$

This result is also valid, if the random variable $X$ is discrete. Direct consequences of this result are the well-known Markov's and Bienaymé-Chebyshev inequalities.

**Corollary 8.1.** (*Markov's inequality*) *Let $X$ be a random variable which takes only nonnegative values, then for any value $a > 0$*

$$P\left(X \geq a\right) \leq \frac{\mathcal{E}\left(X\right)}{a} \tag{8.36}$$

*Proof.* The result is a direct application of the above theorem. $\qquad\square$

**Corollary 8.2.** (*Bienaymé-Chebyshev inequality*) *Let $X$ be a random variable with mean $\mathcal{E}\left(X\right) = \mu$, and finite variance $\sigma_X^2$, then for any $a > 0$*

$$P\left(\left|X - \mu\right| \geq a\right) \leq \frac{\sigma_X^2}{a^2} \tag{8.37}$$

*Proof.* Observe that $P\left(|X - \mu| \geq a\right) = P\left((X - \mu)^2 \geq a^2\right)$. The result is immediate, if $h(X) = (X - \mu)^2$ and $k = a^2$ are substituted in the above theorem. $\qquad\square$

A convenient form of the above inequality is obtained by substituting $a = k\sigma_X$. This yields

$$P\left(\mu - k\sigma_X < X < \mu + k\sigma_X\right) \geq 1 - \frac{1}{k^2}$$

Basic results related to *first and second moment methods* are next established. These are essentially applications of Markov's and Bienaymé-Chebyshev inequalities.

**Corollaries 8.3.** *Let $X$ be a discrete random variable with mean and variance equal to $\mu$ and $\sigma_X^2$ respectively.*

1. *First moment method: If $X$ is nonnegative and integer-valued, then*:
   (a) $P(X > 0) \leq \mu$. *This is a useful result provided $\mu < 1$.*
   (b) $\mu \to 0$ *implies* $P(X = 0) \to 1$.
2. *Second moment method: Let $\mu \neq 0$, then*

$$P(X = 0) \leq \frac{\sigma_X^2}{\mu^2} \qquad\qquad (8.38)$$

3. *If $\mathcal{E}\left(X^2\right) \to \mu^2$, that is $\sigma_X \to 0$, then $P(X = 0) \to 0$.*

*Proofs.*

1. Observe that
$$P(X > 0) = \sum_{x \in \mathbb{P}} P(X = x) \leq \sum_{x \in \mathbb{P}} x P(X = x) = \mu$$

   Part (b) follows easily.
2. For $a > 0$ Bienaymé-Chebyshev inequality yields

$$P\left(|X - \mu| \geq a\right) \leq \frac{\sigma_X^2}{a^2}$$

   Substitution of $a = |\mu|$ in the above inequality results in

$$P\left(|X - \mu| \geq |\mu|\right) \leq \frac{\sigma_X^2}{\mu^2}$$

   If $|X - \mu| \geq |\mu|$ then $X = 0$ satisfies the inequality. That is

$$P(X = 0) \leq P\left(|X - \mu| \geq |\mu|\right)$$

   The result follows.
3. Use the last part. $\qquad\square$

The second moment method is used in proving some results in the classical random graph theory. If the moment generating function $\mathcal{M}_X(\cdot)$ of a random variable $X$ is known, then sometimes a more useful bound on probabilities is possible.

**Theorem 8.2.** (*Chernoff's bound*) *Let $X$ be a random variable, with moment generating function $\mathcal{M}_X(\cdot)$. Then*

$$P(X \geq x) \leq e^{-xt} \mathcal{M}_X(t), \quad \forall\, t \geq 0 \qquad (8.39a)$$

$$P(X \leq x) \leq e^{-xt} \mathcal{M}_X(t), \quad \forall\, t \leq 0 \qquad (8.39b)$$

*Proof.* See the problem section. $\qquad\qquad\qquad\qquad\qquad\qquad\qquad\qquad\qquad\qquad$ $\square$

**Example 8.1.** Let $X$ be a standard normal random variable. Its moment generating function $\mathcal{M}_X(t) = e^{t^2/2}$. Therefore the Chernoff's bound on $P(X \geq x)$ is

$$P(X \geq x) \leq e^{-xt} e^{t^2/2}, \quad \forall\, t \geq 0$$

where $e^{-xt} e^{t^2/2}$ is minimized if $\left(xt - t^2/2\right)$ is maximized. Maximum value of $\left(xt - t^2/2\right)$ is $x^2/2$. It occurs at $t = x$. Therefore $P(X \geq x) \leq e^{-x^2/2}$. It can be similarly shown that $P(X \leq x) \leq e^{-x^2/2}$. $\qquad\qquad$ $\square$

**Theorem 8.3.** (*Jensen's inequality*) *Let $X$ be a random variable, and $g(\cdot)$ be a convex function, then $\mathcal{E}(g(X)) \geq g(\mathcal{E}(X))$, provided the expectations exist.*

*Proof.* See the problem section. $\qquad\qquad\qquad\qquad\qquad\qquad\qquad\qquad\qquad\qquad$ $\square$

Two special cases of Jensen's inequality are: $\mathcal{E}(|X|) \geq |\mathcal{E}(X)|$, and $\mathcal{E}(X^2) \geq \{\mathcal{E}(X)\}^2$.

### 8.8.2  Law of Large Numbers

There are two versions of law of large numbers. These are:

- The weak law of large numbers
- The strong law of large numbers.

#### Weak Law of Large Numbers

The weak law of large numbers is next established. This law implies that, if an experiment is repeated a large number of times, the average of the results differs only slightly from the expected value of each experiment. The law assumes that the mean and the variance of the experimental results are finite.

**Theorem 8.4.** *Let $X_1, X_2, \ldots, X_n$ be a sequence of $n$ independent random variables, each with mean and variance equal to $\mu < \infty$ and $\sigma^2 < \infty$ respectively. Define $\overline{X}_n = \frac{1}{n} \sum_{j=1}^{n} X_j$, then for any positive number $\delta$*

$$\lim_{n \to \infty} P\left(\left|\overline{X}_n - \mu\right| < \delta\right) \to 1 \qquad (8.40)$$

*Proof.* Observe that $\mathcal{E}(\overline{X}_n) = \mu$ and $Var(\overline{X}_n) = \sigma^2/n$. Use of Bienaymé-Chebyshev inequality, for any $\delta > 0$ yields

$$P\left(\left|\overline{X}_n - \mu\right| \geq \delta\right) \leq \frac{\sigma^2}{n\delta^2}$$

For a given value of $\sigma$ and $\delta$, the ratio $\sigma^2/\left(n\delta^2\right)$ tends to zero as $n \to \infty$. Therefore

$$\lim_{n \to \infty} P\left(\left|\overline{X}_n - \mu\right| \geq \delta\right) \to 0$$

That is, $\lim_{n \to \infty} P\left(\left|\overline{X}_n - \mu\right| < \delta\right) \to 1.$ □

A sharper version of the weak law of larger numbers is stated in the problem section.

**Strong Law of Large Numbers**

The strong law of large numbers is perhaps one of the most intuitive results in probability theory. Consider the sequence $X_1, X_2, \ldots, X_n$ of $n$ independent, and identically distributed random variables, each with finite mean $\mu$. Define $\overline{X}_n = \frac{1}{n} \sum_{j=1}^{n} X_j$. The strong law of large numbers says that as $n$ tends to infinity, $\overline{X}_n$ tends to $\mu$ simultaneously for all $\overline{X}_n$'s except for only finitely many exceptions.

**Theorem 8.5.** *Let $X_1, X_2, \ldots, X_n$ be a sequence of $n$ independent, and identically distributed random variables, each with mean equal to $\mu < \infty$. Define $\overline{X}_n = \frac{1}{n} \sum_{j=1}^{n} X_j$, then*

$$P\left(\lim_{n \to \infty} \overline{X}_n = \mu\right) = 1 \tag{8.41}$$

*Proof.* See the textbook by Chung (1974a). □

The proof of the above theorem was first given by A. N. Kolmogorov. A proof of more restrictive strong law of large numbers is given in the problem section.

### 8.8.3  Gaussian Central Limit Theorem

Gaussian central limit theorem is perhaps one of the most important theorems in probability theory. The phrase *central limit theorem* was introduced by George Pólya (1887-1985) to emphasize the fact that such results were central to the work of probability theorists since the eighteenth century. This theorem provides an approximate distribution of the sample-mean of a sequence of random variables. The sample-mean of a sequence of $n$ random variables, $X_1, X_2, \ldots, X_n$ is given by $\overline{X}_n = \frac{1}{n} \sum_{j=1}^{n} X_j.$

**Theorem 8.6.** *Let $X_1, X_2, \ldots, X_n$ be a sequence of $n$ independent, and identically distributed random variables, each with finite mean and finite variance equal to $\mu$ and $\sigma^2$ respectively. Define $\overline{X}_n = \frac{1}{n} \sum_{j=1}^{n} X_j$, then $\mathcal{E}\left(\overline{X}_n\right) = \mu$ and $Var\left(\overline{X}_n\right) = \sigma^2/n$. Let*

$$Y_n = \frac{\left(\overline{X}_n - \mathcal{E}\left(\overline{X}_n\right)\right)}{\sqrt{Var\left(\overline{X}_n\right)}}$$

*then*

$$Y_n = \frac{\left(\overline{X}_n - \mu\right)}{\sigma/\sqrt{n}} \tag{8.42}$$

*and the cumulative distribution function $F_{Y_n}\left(\cdot\right)$ of the random variable $Y_n$ converges to the cumulative distribution function $\Phi\left(\cdot\right)$ of the standard normal random variable, as $n$ tends to infinity. That is, the distribution of this random variable approaches the standard normal distribution.*

*Proof.* Note that $Y_n$ can be written as $\left(\sum_{j=1}^{n} W_j\right)/\sqrt{n}$, where $W_j = (X_j - \mu)/\sigma$. The random variables $W_j, 1 \leq j \leq n$ are independent, and identically distributed, each with zero mean and unit variance. Denote the moment generating function of the random variable $Y_n$ by $\mathcal{M}_{Y_n}(\cdot)$, and those of $W_j$'s by $\mathcal{M}_W(\cdot)$, for $1 \leq j \leq n$. Then

$$\mathcal{M}_{Y_n}(t) = \left[\mathcal{M}_W\left(\frac{t}{\sqrt{n}}\right)\right]^n$$

Define $\mathcal{E}\left((X_j - \mu)^r\right) = \zeta_r$, for $0 \leq r < \infty$ and $1 \leq j \leq n$. Then

$$\mathcal{M}_W(t) - \sum_{r=0}^{\infty} \frac{\zeta_r}{r!}\left(\frac{t}{\sigma}\right)^r$$

Therefore

$$\mathcal{M}_W\left(\frac{t}{\sqrt{n}}\right) = \sum_{r=0}^{\infty} \frac{\zeta_r}{r!}\left(\frac{t}{\sigma\sqrt{n}}\right)^r$$

As $\zeta_1 = 0$, and $\zeta_2 = \sigma^2$

$$\mathcal{M}_W\left(\frac{t}{\sqrt{n}}\right) = 1 + \frac{1}{n}\left(\frac{1}{2}t^2 + \frac{1}{3!\sqrt{n}}\frac{\zeta_3}{\sigma^3}t^3 + \frac{1}{4!n}\frac{\zeta_4}{\sigma^4}t^4 + \cdots\right)$$

Therefore for large values of $n$,

$$\mathcal{M}_W\left(\frac{t}{\sqrt{n}}\right) \simeq \left(1 + \frac{t^2}{2n}\right)$$

which implies

$$\mathcal{M}_{Y_n}(t) \simeq \left(1 + \frac{t^2}{2n}\right)^n \simeq e^{t^2/2}$$

However $e^{t^2/2}$ is the moment generating function of a normally distributed random variable with zero mean and unit variance. It is well-known that, if two random variables have the same moment generating function, then their probability distribution functions are equal. Therefore it follows that the random variable $Y_n$ has a normal distribution as $n$ approaches infinity.    □

**Corollary 8.4.** *Let $X_1, X_2, \ldots, X_n$ be a sequence of $n$ independent, and identically distributed random variables, each with mean and variance equal to $\mu$ and $\sigma^2$ respectively. Define*

$$\overline{X}_n = \frac{1}{n}\sum_{j=1}^{n} X_j \tag{8.43a}$$

*If $a < b$, then*

$$P\left[a < \frac{(\overline{X}_n - \mu)}{\sigma/\sqrt{n}} < b\right] \simeq \Phi(b) - \Phi(a) \tag{8.43b}$$

*where $\Phi(\cdot)$ is the cumulative distribution function of the standard normal random variable.*    □

**Examples 8.2.** These examples illustrate the application of Gaussian central limit theorem to specific distributions.

1. Let $X$ be a random variable with binomial distribution. Let its parameters be $p$ and $n$. If $q = (1 - p)$ then $\mathcal{E}(X) = np$, $Var(X) = npq$. For $a < b$ the binomial distribution is approximated by

$$P\left[a < \frac{(X - np)}{\sqrt{npq}} < b\right] \simeq \Phi(b) - \Phi(a)$$

   as $n$ approaches infinity. Alternately, if $c = \left(np + a\sqrt{npq}\right)$, and $d = \left(np + b\sqrt{npq}\right)$ then

$$P[c < X < d] \simeq \Phi\left(\frac{d - np}{\sqrt{npq}}\right) - \Phi\left(\frac{c - np}{\sqrt{npq}}\right)$$

   as $n$ approaches infinity. This statement is often called the De Moivre-Laplace limit theorem.

2. Let $X$ be a random variable with Poisson distribution, with parameter $\lambda$. It is known that $\mathcal{E}(X) = Var(X) = \lambda$. For $a < b$ the Poisson distribution is approximated by

$$P\left[a < \frac{(X - \lambda)}{\sqrt{\lambda}} < b\right] \simeq \Phi(b) - \Phi(a)$$

   as $\lambda$ approaches infinity.                                                                            $\square$

Observe that the weak law of large numbers follows simply from the Gaussian central limit theorem. Furthermore, the Gaussian central limit theorem and the strong law of large numbers do not imply each other. A generalized central limit theorem is discussed in the next section.

---

## 8.9 Generalized Central Limit Theorem

In order to study highly varying data, an understanding of generalized central limit theorem is required. The Gaussian central limit theorem was studied earlier in the chapter. This theorem asserts that suitably normalized sums of independent random variables with finite variance have a normal distribution in the limit. Relaxation of the condition of finite variance in the Gaussian central limit theorem leads to a more generalized version of the theorem.

The generalized central limit theorem is based upon a knowledge of characteristic functions, infinitely divisible distributions, and stable distributions. The generalized central limit theorem is based upon the work of A. N. Kolmogorov, Paul Pierre Lévy (1886-1971), and Alexander Yakovlevich Khinchin (1894-1959). The generalized central limit theorem is also sometimes called Lévy's central limit theorem.

### 8.9.1 Characteristic Function

Characteristic function of a random variable is a convenient mathematical artifice to study its properties. It is especially useful in studying sums of independent random variables.

**Definition 8.13.** *The characteristic function of a random variable $X$ is defined as $\varphi_X(u) = \mathcal{E}\left(e^{iuX}\right)$, where $u \in \mathbb{R}$, and $i = \sqrt{-1}$. Sometimes $\varphi_X(\cdot)$ is simply denoted by $\varphi(\cdot)$.*                    $\square$

**Observations 8.8.** Some useful facts about characteristic functions are listed. Denote the characteristic function of the random variables $X$ and $Y$ by $\varphi_X(\cdot)$, and $\varphi_Y(\cdot)$ respectively.

1. Characteristic function $\varphi(\cdot)$ exists for any random variable, because $\left|e^{iux}\right|$ is bounded and continuous for all $u, x \in \mathbb{R}$. Actually, there is a one-to-one relationship between a distribution function and the corresponding characteristic function.

2. $\varphi_X(0) = 1$.

3. $\left|\varphi_X(u)\right| \leq 1, \ \forall\, u \in \mathbb{R}$.

4. $\varphi_X(u) = \overline{\varphi_X(-u)}, \ \forall\, u \in \mathbb{R}$. This is termed the Hermitian property of the characteristic function.

5. A characteristic function is uniformly continuous.

6. If the characteristic functions of two random variables are identical, then the corresponding distribution functions are also identical.

7. Assume that the first and second derivatives of $\varphi_X(u)$ with respect to $u$ exist. Denote these by $\varphi'_X(u)$ and $\varphi''_X(u)$ respectively. Then

$$\mathcal{E}(X) = -i\varphi'_X(0), \quad \text{and} \quad \mathcal{E}(X^2) = -\varphi''_X(0)$$

8. The characteristic function of $Y = (aX + b)$ is $\varphi_Y(u) = e^{ibu}\varphi_X(au), \ \forall\, u \in \mathbb{R}$.

9. The characteristic function of $-X$ is the complex conjugate $\overline{\varphi}_X(u), \ \forall\, u \in \mathbb{R}$.

10. The characteristic function of a random variable is real-valued, if and only if its probability density function is symmetric about $x = 0$.

11. If $X$ and $Y$ are independent random variables and $Z = (X + Y)$, then

$$\varphi_Z(u) = \varphi_X(u)\,\varphi_Y(u), \quad \forall\, u \in \mathbb{R}$$

$\square$

## 8.9.2  Infinitely Divisible Distributions

Infinitely divisible distributions are studied briefly in this subsection. These are useful in studying stable distributions. Stable distributions have heavy-tails.

**Definition 8.14.** *The distribution and characteristic functions of a random variable $X$ are denoted by $F(\cdot)$, and $\varphi(\cdot)$ respectively.*

(a) *The random variable $X$ is infinitely divisible, if for each $n \in \mathbb{P}\backslash\{1\}$,*

$$X = (X_{n1} + X_{n2} + \cdots + X_{nn}) \tag{8.44}$$

*is the sum of $n$ independent, and identically distributed random variables.*

(b) *The distribution function $F(\cdot)$ is infinitely divisible, if for each $n \in \mathbb{P}\backslash\{1\}$, there exists a distribution function $F_n(\cdot)$, whose $n$-fold convolution is equal to $F(\cdot)$.*

(c) *The characteristic function $\varphi(\cdot)$ is infinitely divisible, if for each $n \in \mathbb{P}\backslash\{1\}$, there exists a characteristic function $\varphi_n(\cdot)$, such that $\varphi(u) = \{\varphi_n(u)\}^n$.* $\square$

The above definitions are essentially equivalent. It can be shown that the normal, Poisson, gamma, and Cauchy distributions are all infinitely divisible. Furthermore, there does not exist a nontrivial infinitely divisible distribution which is concentrated on a bounded interval.

**Observations 8.9.** Some useful facts about infinitely divisible distributions are listed.

1. If the random variable $X$ is infinitely divisible, then so is $-X$.

2. If $X$ and $Y$ are independent, and infinitely divisible random variables, then $(X + Y)$ is also infinitely divisible.

3. Let $Y, X_1, X_2, \ldots$ be independent random variables. The random variable $Y$ has a Poisson distribution with parameter $\lambda \in \mathbb{R}^+$. The random variables $X_1, X_2, \ldots$ are identically distributed with characteristic function $\varphi(\cdot)$. Then $Z = \sum_{i=1}^{Y} X_i$ is an infinitely divisible random variable with characteristic function

$$\varphi_Z(u) = \exp\{\lambda(\varphi(u) - 1)\}, \quad \text{where } u \in \mathbb{R}$$

The random variable $Z$ is said to have compound Poisson distribution.

4. If $\varphi(\cdot)$ is an infinitely divisible characteristic function, then $\varphi(u) \neq 0, \forall\, u \in \mathbb{R}$. That is, an infinitely divisible characteristic function is never equal to zero.

5. If $\varphi(\cdot)$ is an infinitely divisible characteristic function, and if for each $n \in \mathbb{P}$, there exists a characteristic function $\varphi_n(\cdot)$, such that $\varphi(u) = \{\varphi_n(u)\}^n$, then
   (a) $\lim_{n \to \infty} n\{\varphi_n(u) - 1\} = \ln \varphi(u)$
   (b) $\lim_{n \to \infty} \varphi_n(u) = 1$                                                                 $\square$

Some of the above observations are proved in the problem section. Canonical representations of an infinitely divisible characteristic function due to Lévy and Khinchin, Lévy, and Kolmogorov are next obtained. The following canonical representation of an infinitely divisible characteristic function is generally attributed to Lévy and Khinchin.

**Theorem 8.7.** *A characteristic function $\varphi(u)$, $u \in \mathbb{R}$ is infinitely divisible if and only if*

$$\ln \varphi(u) = i\gamma u + \int_{-\infty}^{\infty} L(x, u)\, dG(x) \tag{8.45a}$$

*where $\gamma \in \mathbb{R}$, $G(\cdot)$ is a bounded distribution function defined on the real line $\mathbb{R}$, and*

$$L(x, u) = \begin{cases} \left(e^{iux} - 1 - \dfrac{iux}{1 + x^2}\right)\dfrac{1 + x^2}{x^2}, & x \neq 0 \\[2mm] -\dfrac{u^2}{2}, & x = 0 \end{cases} \tag{8.45b}$$

*The real number $\gamma$ is called the centering constant, and $(\gamma, G)$ is called the Lévy-Khinchin pair. Furthermore, this representation is unique.*                                                                 $\square$

The real constant $\gamma$ is called the centering constant for the following reason. If the characteristic function of a random variable $X$ has a Lévy-Khinchin representation $(\gamma, G)$, then the Lévy-Khinchin representation of the characteristic function of $(X + c)$ is $(\gamma + c, G)$, where $c \in \mathbb{R}$. An alternative canonical representation of an infinitely divisible characteristic function is provided by a theorem due to Lévy. This is accomplished by introducing a function $M(\cdot)$ defined on $\mathbb{R} \setminus \{0\}$, and a constant $\sigma \in \mathbb{R}_0^+$. The functions $G(\cdot)$ and $M(\cdot)$, and $\sigma$ are related as

$$M(x) = \begin{cases} \displaystyle\int_{-\infty}^{x} \frac{1 + y^2}{y^2}\, dG(y), & x < 0 \\[3mm] -\displaystyle\int_{x}^{\infty} \frac{1 + y^2}{y^2}\, dG(y), & x > 0 \end{cases} \tag{8.46a}$$

$$\sigma^2 = G(0_+) - G(0_-) \tag{8.46b}$$

**Theorem 8.8**. *A characteristic function* $\varphi(u)$, *$u \in \mathbb{R}$ is infinitely divisible if and only if*

$$\ln \varphi(u) = i\gamma u - \frac{1}{2}\sigma^2 u^2 + \int_{\mathbb{R}\setminus\{0\}} H(x, u)\, dM(x) \tag{8.47a}$$

*where $\gamma \in \mathbb{R}$, $\sigma \in \mathbb{R}_0^+$,*

$$H(x, u) = \left(e^{iux} - 1 - \frac{iux}{1 + x^2}\right), \quad x \in \mathbb{R}\setminus\{0\} \tag{8.47b}$$

*and $M(x)$ is defined on $x \in \mathbb{R}\setminus\{0\}$. Also $M(x)$ is nondecreasing and continuous on the right on $(-\infty, 0)$, and also on $(0, \infty)$. Further,*

$$\lim_{x \to -\infty} M(x) = \lim_{x \to \infty} M(x) = 0 \tag{8.47c}$$

*and*

$$\int_{(-1,0)\cup(0,1)} x^2 dM(x) < \infty \tag{8.47d}$$

*The function $M(\cdot)$, is called the Lévy spectral function, and $(\gamma, \sigma^2, M)$ is called the Lévy triple. Furthermore, this representation is unique.* □

A canonical representation of an infinitely divisible characteristic function of a random variable with finite variance is due to Kolmogorov. This is accomplished by introducing a function $K(\cdot)$ defined on $\mathbb{R}$, and a constant $\beta \in \mathbb{R}$. The functions $G(\cdot)$ and $K(\cdot)$; and $\gamma$ and $\beta$ are related as

$$K(x) = \int_{-\infty}^{x} \left(1 + y^2\right) dG(y), \quad x \in \mathbb{R} \tag{8.48a}$$

$$\beta = \gamma + \int_{-\infty}^{\infty} y\, dG(y) \tag{8.48b}$$

**Theorem 8.9**. *A characteristic function $\varphi(u)$, $u \in \mathbb{R}$ of a random variable with finite variance is infinitely divisible if and only if*

$$\ln \varphi(u) = i\beta u + \int_{-\infty}^{\infty} J(x, u)\, dK(x) \tag{8.49a}$$

*where $\beta \in \mathbb{R}$,*

$$J(x, u) = \begin{cases} \left(e^{iux} - 1 - iux\right) \dfrac{1}{x^2}, & x \neq 0 \\[2mm] -\dfrac{u^2}{2}, & x = 0 \end{cases} \tag{8.49b}$$

*$K(x)$, $x \in \mathbb{R}$, is a nondecreasing bounded function,*

$$\lim_{x \to -\infty} K(x) = 0 \tag{8.49c}$$

*and $(\beta, K)$ is called the Kolmogorov pair. Furthermore, this representation is unique.* □

The plausibility of the above three theorems is provided in the problem section. Geometric, Poisson, gamma, normal, and Cauchy distributions are all infinitely divisible. Bernoulli, binomial, and uniform distributions are not infinitely divisible.

**Examples 8.3.** Lévy-Khinchin, Lévy, and Kolmogorov representations of characteristic functions of normal, Poisson, and gamma distributions.

1. Consider a normally distributed random variable $\mathcal{N}\left(m, \alpha^2\right)$. Its characteristic function is $\exp\left(imu - \alpha^2 u^2/2\right), u \in \mathbb{R}$. This characteristic function has a:
   (a) Lévy-Khinchin representation $(\gamma, G)$, where $\gamma = m$, $G\left(x\right) = 0$ if $x < 0$, and $G\left(x\right) = \alpha^2$ if $x \geq 0$.
   (b) Lévy representation $\left(\gamma, \sigma^2, M\right)$, where $\gamma = m$, $\sigma^2 = \alpha^2$, $M\left(x\right) = 0$ if $x \in \mathbb{R} \backslash \{0\}$.
   (c) Kolmogorov representation $(\beta, K)$, where $\beta = m$, $K\left(x\right) = 0$ if $x < 0$, and $K\left(x\right) = \alpha^2$ if $x \geq 0$.

2. A random variable has a Poisson distribution with parameter $\lambda \in \mathbb{R}^+$. Its characteristic function is $\exp\left\{\lambda\left(e^{iu} - 1\right)\right\}, u \in \mathbb{R}$. This characteristic function has a:
   (a) Lévy-Khinchin representation $(\gamma, G)$, where $\gamma = \lambda/2$, $G\left(x\right) = 0$ if $x < 1$, and $G\left(x\right) = \lambda/2$ if $x \geq 1$.
   (b) Lévy representation $\left(\gamma, \sigma^2, M\right)$, where $\gamma = \lambda/2$, $\sigma^2 = 0$, $M\left(x\right) = 0$ if $x < 0$, $M\left(x\right) = -\lambda$ if $0 < x < 1$, and $M\left(x\right) = 0$ if $x \geq 1$.
   (c) Kolmogorov representation $(\beta, K)$, where $\beta = \lambda$, $K\left(x\right) = 0$ if $x < 1$, and $K\left(x\right) = \lambda$ if $x \geq 1$.

3. A gamma distribution is described by parameters $\lambda, r \in \mathbb{R}^+$. Assume that $\lambda = r^{-1}$. Therefore its characteristic function is $\varphi\left(u\right) = \left(1 - iu\right)^{-r}, u \in \mathbb{R}$. This characteristic function has a:
   (a) Lévy-Khinchin representation $(\gamma, G)$, where

   $$\gamma = r \int_0^\infty \frac{e^{-y}}{1 + y^2} dy$$

   and

   $$G\left(x\right) = \begin{cases} 0, & x < 0 \\ r \displaystyle\int_0^x \frac{ye^{-y}}{1 + y^2} dy, & x > 0 \end{cases}$$

   (b) Lévy representation: $\left(\gamma, \sigma^2, M\right)$, where

   $$\gamma = r \int_0^\infty \frac{e^{-y}}{1 + y^2} dy, \quad \sigma = 0,$$

   and

   $$M\left(x\right) = \begin{cases} 0, & x < 0 \\ -r \displaystyle\int_x^\infty \frac{e^{-y}}{y} dy, & x > 0 \end{cases}$$

   (c) Kolmogorov representation $(\beta, K)$, where $\beta = r$, and

   $$K\left(x\right) = \begin{cases} 0, & x < 0 \\ r \displaystyle\int_0^x ye^{-y} dy, & x > 0 \end{cases}$$

   $\square$

### 8.9.3  Stable Distributions

Stable distributions are a subclass of the family of infinitely divisible distribution functions. These distributions are characterized by skewness, and heavy-tails. Stable distributions are important in studying certain limit theorems. These distributions were first studied in the 1930's by Lévy and Khinchin.

Let the random variables $X_1, X_2$, and $X$ be identically distributed normal random variables. In addition, the random variables $X_1$, and $X_2$ are independent of each other. For any positive real numbers $b_1$ and $b_2$, it is well-known that

$$b_1 X_1 + b_2 X_2 \stackrel{d}{=} bX + c$$

for some $b \in \mathbb{R}^+$ and $c \in \mathbb{R}$. The symbol $\stackrel{d}{=}$ means equality in distribution. This fact motivates the following definition of a stable random variable.

**Definition 8.15.** *A random variable $X$ is stable, if for any $b_1, b_2 \in \mathbb{R}^+$, there exist $b \in \mathbb{R}^+$ and $c \in \mathbb{R}$ such that*

$$b_1 X_1 + b_2 X_2 \stackrel{d}{=} bX + c \tag{8.50}$$

*where $X_1$, and $X_2$ are independent random variables with the same distribution as $X$. The corresponding distribution and characteristic functions are also said to be stable.*

(a) *The random variable $X$ is strictly stable, if the above relationship is true with $c = 0$.*
(b) *The random variable $X$ is stable in the broad sense, if the above relationship is true with $c \neq 0$.*
(c) *The random variable $X$ is symmetric stable, if $X$ and $-X$ have the same distribution.* □

The word stable is used to reflect the fact that the shape of the distribution is unchanged under the above type of weighted summation of random variables. Note that a symmetric stable random variable is also strictly stable.

**Observation 8.10.** Let the characteristic function of a stable random variable $X$ be $\varphi(u)$, $u \in \mathbb{R}$. Then for $b_1, b_2, b \in \mathbb{R}^+$, and $c \in \mathbb{R}$ we have $\varphi(b_1 u) \varphi(b_2 u) = e^{icu} \varphi(bu)$. □

**Example 8.4.** Let $X \sim \mathcal{N}(\mu, \sigma^2)$ be a normal random variable. If $X_1$, and $X_2$ are independent random variables with $X_1 \sim \mathcal{N}(\mu, \sigma^2)$, $X_2 \sim \mathcal{N}(\mu, \sigma^2)$, and $(b_1 X_1 + b_2 X_2) \stackrel{d}{=} (bX + c)$.

$$b^2 = (b_1^2 + b_2^2), \quad \text{and} \quad c = \mu(b_1 + b_2 - b)$$

Thus random variable $X$ has a stable distribution. □

**Definition 8.16.** *A random variable $X$ which is concentrated at one point is said to have a degenerate distribution.* □

A degenerate distribution is always stable. It is always assumed henceforth in this section that the random variable $X$ has a nondegenerate distribution.

**Observations 8.11.** Some properties of a stable random variable are listed, and include possible alternate definitions of a stable random variable.

1. A stable characteristic function is infinitely divisible.
2. A nondegenerate random variable $X$ has a stable distribution, if and only if for all $n \in \mathbb{P} \setminus \{1\}$, there exist constants $b \in \mathbb{R}^+$ and $c \in \mathbb{R}$ such that

$$X_1 + X_2 + \cdots X_n \stackrel{d}{=} bX + c \tag{8.51}$$

where $X_1, X_2, \ldots, X_n$, are independent random variables with the same distribution as $X$. This observation is sometimes taken as the definition of a stable random variable.

3. An equivalent definition of a stable random variable $X$ can also be given in terms of *domain of attraction*. A random variable $X$ has a stable distribution if and only if it has a domain of attraction. That is, if there is a sequence of independent, and identically distributed random variables $X_1, X_2, \ldots$ and sequences of positive real numbers $\{b_n\}$ and real numbers $\{a_n\}$ such that

$$\frac{X_1 + X_2 + \cdots + X_n}{b_n} + a_n \stackrel{d}{\to} X, \quad \text{as} \quad n \to \infty \tag{8.52}$$

where $\stackrel{d}{\to}$ denotes convergence in distribution.

   (a) It can be shown that $b_n = n^{1/\alpha} h(n)$, where $\alpha \in (0, 2]$. The parameter $\alpha$ is called the index of the stable random variable. In addition, the function $h(\cdot)$ varies slowly at infinity. That is, $\lim_{x \to \infty} h(ux)/h(x) = 1$ for all $u > 0$. An example of a function which varies slowly at infinity is $h(\cdot) = \ln(\cdot)$.

   (b) Note that, if the $X_i$'s are independent and identically distributed random variables with finite variance, then $X$ is normally distributed, and this observation is the Gaussian central limit theorem. In this case, the value of the parameter $\alpha = 2$, and $b_n = n^{1/2}$.

   (c) The stated result is a generalized central limit theorem. It is often attributed to Lévy.   □

Lévy's generalized central limit theorem is stated explicitly in a slightly modified form.

**Theorem 8.10.** *Let $X_1, X_2, \ldots, X_n$ be a sequence of independent, and identically distributed random variables, $\{b_n\}$ is a sequence of positive real numbers, and $\{a_n\}$ is a sequence of real numbers. Let*

$$S_n = \frac{(X_1 + X_2 + \cdots + X_n)}{b_n} + a_n \tag{8.53}$$

*Then, as $n \to \infty$, $X$ is the limit in distribution of $S_n$ if and only if $X$ is stable.*   □

A canonical representation of stable laws is established in the next two theorems.

**Theorem 8.11.** *The characteristic function of a stable random variable has a canonical representation*

$$\ln \varphi(u) = i\gamma u - \frac{1}{2}\sigma^2 u^2 + \int_{\mathbb{R} \setminus \{0\}} H(x, u) \, dM(x), \quad u \in \mathbb{R} \tag{8.54a}$$

*where $\gamma \in \mathbb{R}$, $\sigma \in \mathbb{R}_0^+$,*

$$H(x, u) = \left\{ e^{iux} - 1 - \frac{iux}{1 + x^2} \right\}, \quad x \in \mathbb{R} \setminus \{0\} \tag{8.54b}$$

*with either*

$$\sigma^2 \neq 0, \quad and \quad M(x) = 0 \quad for \quad x \in \mathbb{R}\backslash\{0\} \tag{8.54c}$$

*or*

$$\sigma^2 = 0,$$
$$M(x) = \xi_1 |x|^{-\alpha} \quad for \quad x < 0, \quad M(x) = -\xi_2 x^{-\alpha} \quad for \quad x > 0,$$
$$\alpha \in (0,2), \quad \xi_1 \geq 0, \quad \xi_2 \geq 0, \quad (\xi_1 + \xi_2) > 0 \tag{8.54d}$$

*Conversely, any characteristic function with the above representation is stable.*
   *Proof.* See the problem section.                                                      □

The integral in the above theorem can be evaluated explicitly.

**Theorem 8.12.** *The characteristic function of a stable random variable $X$ has the simplified canonical representation if and only if*

$$\ln \varphi(u) = i\varrho u - \xi |u|^{\alpha} \{1 - i\beta sign(u)\,\omega(u,\alpha)\}, \quad u \in \mathbb{R} \tag{8.55a}$$

*where constants $0 < \alpha \leq 2, |\beta| \leq 1, \varrho \in \mathbb{R}, \xi \in \mathbb{R}_0^+$,*

$$\omega(u,\alpha) = \begin{cases} \tan\left(\dfrac{1}{2}\pi\alpha\right), & \alpha \neq 1 \\ -2\pi^{-1}\ln|u|, & \alpha = 1 \end{cases} \tag{8.55b}$$

$$sign(u) = \begin{cases} 1, & if \quad u > 0 \\ 0, & if \quad u = 0 \\ -1, & if \quad u < 0 \end{cases} \tag{8.55c}$$

   *Proof.* See the problem section.                                                      □

Note that $\omega(u,2) = 0$, corresponds to normal distribution. In this case the value of $\beta$ can be any number in the interval $[-1,1]$. Some authors specify the constant $\xi$ as equal to $\sigma^\alpha$, where $\sigma \geq 0$.
   The random variable $X$ is called $\alpha$-stable random variable, and it is specified by four parameters $\alpha, \sigma, \beta$, and $\varrho$. The stable random variable $X$ is specified formally as $S_\alpha(\sigma, \beta, \varrho)$ where:

(a) $\alpha \in (0,2]$ is the exponent or index.
(b) $\sigma \in \mathbb{R}_0^+$ is the scale parameter.
(c) $\beta \in [-1,1]$ is the skewness parameter.
(d) $\varrho \in \mathbb{R}$ is the location or shift parameter.

**Observations 8.12.** Let $X \sim S_\alpha(\sigma, \beta, \varrho)$, be a stable random variable.

1. Let $a \in \mathbb{R}$, and $X \sim S_\alpha(\sigma, \beta, \varrho)$, then $(X + a) \sim S_\alpha(\sigma, \beta, \varrho + a)$.
2. Let $X_i \sim S_\alpha(\sigma_i, \beta_i, \varrho_i)$ for $i = 1, 2$, and $X_1$ and $X_2$ are independent random variables. Then

$$(X_1 + X_2) \sim S_\alpha(\sigma, \beta, \varrho) \tag{8.56a}$$

with

$$\sigma = (\sigma_1^\alpha + \sigma_2^\alpha)^{1/\alpha} \tag{8.56b}$$

$$\beta = \frac{\beta_1 \sigma_1^\alpha + \beta_2 \sigma_2^\alpha}{\sigma_1^\alpha + \sigma_2^\alpha} \tag{8.56c}$$

$$\varrho = (\varrho_1 + \varrho_2) \tag{8.56d}$$

This result can be obtained by using the logarithmic representation of the characteristic functions of the two random variables $X_1$ and $X_2$.

3. The tail probabilities for $\alpha \in (0, 2)$ have a power-law relationship.

$$\lim_{x \to \infty} x^\alpha P(X > x) = \kappa_p \tag{8.57a}$$

$$\lim_{x \to \infty} x^\alpha P(X < -x) = \kappa_n \tag{8.57b}$$

where $\kappa_p$ and $\kappa_n$ are positive constants.

4. The moments of stable random variables can be either finite or infinite.

   (a) If $X \sim S_\alpha(\sigma, 0, \varrho)$, and $\alpha \in (1, 2]$, then $\mathcal{E}(X) = \varrho$.
   (b) Let $X \sim S_\alpha(\sigma, \beta, \varrho)$, and $\alpha \in (0, 2)$.

   (i) $\mathcal{E}\left(|X|^\delta\right) < \infty$, if $0 \le \delta < \alpha$.

   (ii) $\mathcal{E}\left(|X|^\delta\right) = \infty$, if $\delta \ge \alpha$.

   Therefore if $\alpha \in (0, 1]$, then $\alpha$-stable random variables have no first or higher order moments. However, if $\alpha \in (1, 2)$, then $\alpha$-stable random variables have first moment, and all fractional order moments of order $\delta$, which are less than $\alpha$. In this case the first moment $\mathcal{E}(X) = \varrho$.

   More specifically, all non-Gaussian stable random variables have infinite variance.

   (c) $\mathcal{E}\left(|X|^\delta\right) < \infty$, for all $\delta \ge 0$, if $\alpha = 2$. That is, if $\alpha = 2$, all moments exist.    □

Probability density functions of $\alpha$-stable random variables rarely exist in closed form. Some exceptions are Gaussian, Cauchy, and Lévy random variables.

**Examples 8.5.**

1. Gaussian distribution: $S_2(\sigma, 0, \varrho) = \mathcal{N}(\varrho, 2\sigma^2)$.
2. Cauchy distribution: $S_1(\sigma, 0, \mu)$. The probability density function is:

$$f_X(x) = \frac{1}{\pi\sigma \left[1 + \left(\dfrac{x - \mu}{\sigma}\right)^2\right]}, \quad x \in \mathbb{R}$$

3. Lévy distribution: $S_{1/2}(\sigma, 1, \mu)$. The probability density function is:

$$f_X(x) = \left(\frac{\sigma}{2\pi}\right)^{1/2} \frac{1}{(x - \mu)^{3/2}} \exp\left\{-\frac{\sigma}{2(x - \mu)}\right\}, \quad x > \mu$$

where $f_X(x) = 0$ for all $x \le \mu$, and $\sigma > 0$. As the index parameter $\alpha = 1/2$, the mean and variance of a Lévy distributed random variable do not exist. Its characteristic function $\varphi(u)$, $u \in \mathbb{R}$ is

$$\ln \varphi \left( u \right) = i\mu u - \left| \sigma u \right|^{1/2} \left\{ 1 - isign \left( u \right) \right\}, \quad u \in \mathbb{R}$$

The Lévy probability density function is specified by the pair $\left( \mu, \sigma \right)$. If $Y \sim \mathcal{N} \left( \mu, c^2 \right)$, then the random variable

$$\left( Y - \mu \right)^{-2}$$

has a Lévy probability density function specified by the pair $\left( 0, c^{-2} \right)$.                  □

The next section describes the properties of range of random variables. Asymptotic properties of the range are also explored.

## 8.10  Range Distribution

Let $X_1, X_2, \ldots, X_n$ be $n$ independent and identically distributed continuous random variables. Let their probability density and cumulative distribution functions be $f \left( \cdot \right)$ and $F \left( \cdot \right)$ respectively. Define

$$U = \min \left\{ X_1, X_2, \ldots, X_n \right\}$$
$$V = \max \left\{ X_1, X_2, \ldots, X_n \right\}$$
$$R = \left( V - U \right)$$

The random variable $R$ is called the range of the sequence of random variables $X_1, X_2, \ldots, X_n$. This random variable can be considered to be the range of values over which darts fall on a dart board. The darts are assumed to fall on a straight line.

Let $F_U \left( \cdot \right)$ and $F_V \left( \cdot \right)$ be the cumulative distributive functions of the random variables $U$ and $V$ respectively. Also let $f_U \left( \cdot \right)$ and $f_V \left( \cdot \right)$ be the probability density functions of the random variables $U$ and $V$ respectively. Then it can be shown that

$$F_U \left( x \right) = 1 - \left\{ 1 - F \left( x \right) \right\}^n$$
$$f_U \left( x \right) = n \left\{ 1 - F \left( x \right) \right\}^{n-1} f \left( x \right)$$
$$F_V \left( x \right) = \left\{ F \left( x \right) \right\}^n$$
$$f_V \left( x \right) = n \left\{ F \left( x \right) \right\}^{n-1} f \left( x \right)$$

### 8.10.1  Joint Distribution of $U$ and $V$

The joint distribution function of the random variables $U$ and $V$ is determined in this subsection. Let their joint cumulative distribution and probability density functions be $F_{U,V} \left( \cdot, \cdot \right)$ and $f_{U,V} \left( \cdot, \cdot \right)$ respectively. These functions are determined as follows. We consider two cases for the values of $F_{U,V} \left( u, v \right)$ and $f_{U,V} \left( u, v \right)$. These are: $u \geq v$ and $u < v$.

*Case* 1:  If $u \geq v$ then $F_{U,V} \left( u, v \right)$ is the probability that $X_i \leq v$, simultaneously for $1 \leq i \leq n$. This implies that $F_{U,V} \left( u, v \right) = \left\{ F \left( v \right) \right\}^n$.
*Case* 2:  If $u < v$ then $F_{U,V} \left( u, v \right)$ is the probability that:
   (a) Simultaneously $X_i \leq v$, for $1 \leq i \leq n$.
   (b) But not simultaneously $u < X_i \leq v$, for $1 \leq i \leq n$.
   Thus $F_{U,V} \left( u, v \right) = \left\{ F \left( v \right) \right\}^n - \left\{ F \left( v \right) - F \left( u \right) \right\}^n$.

Combining the two cases yields

$$
F_{U,V}(u,v) = \begin{cases} \{F(v)\}^n - \{F(v) - F(u)\}^n, & u < v \\ \{F(v)\}^n, & u \geq v \end{cases}
$$

Consequently

$$
f_{U,V}(u,v) = \begin{cases} n(n-1)\{F(v) - F(u)\}^{n-2} f(u) f(v), & u < v \\ 0, & u \geq v \end{cases}
$$

### 8.10.2  Distribution of Range

Some properties of the random variable $R$ are computed. Let the cumulative distribution and probability density functions of this random variable be $F_R(\cdot)$ and $f_R(\cdot)$ respectively. Then

$$
F_R(r) = P(V - U \leq r) = \int\int_A f_{U,V}(u,v)\,dudv
$$

where $A$ is the region on the $uv$-plane such that $0 \leq (v - u) \leq r$. It follows that

$$
F_R(r) = \begin{cases} \int_{-\infty}^{\infty} \int_u^{u+r} f_{U,V}(u,v)\,dvdu, & r > 0 \\ 0, & r \leq 0 \end{cases}
$$

The probability density function $f_R(\cdot)$ is given by

$$
f_R(r) = \begin{cases} \int_{-\infty}^{\infty} f_{U,V}(u, u+r)\,du, & r > 0 \\ 0, & r \leq 0 \end{cases}
$$

which yields

$$
f_R(r) = \begin{cases} n(n-1)\int_{-\infty}^{\infty}\{F(u+r) - F(u)\}^{n-2} f(u) f(u+r)\,du, & r > 0 \\ 0, & r \leq 0 \end{cases}
$$

Integrating both sides with respect to $r$ results in

$$
F_R(r) = \begin{cases} n\int_{-\infty}^{\infty}\{F(u+r) - F(u)\}^{n-1} f(u)\,du, & r > 0 \\ 0, & r \leq 0 \end{cases}
$$

It is shown in the problem section that the expected value of the random variable $R$ is

$$
\mathcal{E}(R) = \mathcal{E}(V) - \mathcal{E}(U) = \int_{-\infty}^{\infty} [1 - \{F(x)\}^n - \{1 - F(x)\}^n]\,dx
$$

### 8.10.3  A Property of the Average Value of Range

An upper bound for $\mathcal{E}(R)$ is obtained, when the probability density function $f(\cdot)$ is symmetric about its mean. For simplicity assume that $\mathcal{E}(X_i) = 0$ and $Var(X_i) = \sigma^2$. If $f(x)$ is symmetric then $f(-x) = f(x)$, and $F(-x) = (1 - F(x))$. It can also be proved under this assumption that $\mathcal{E}(U) = -\mathcal{E}(V)$. This in turn yields $\mathcal{E}(R) = 2\mathcal{E}(V)$. Therefore it remains to obtain an upper bound of $\mathcal{E}(V)$. This is done via the use of the so-called Bunyakovsky-Cauchy-Schwartz inequality, which states

$$\left(\int_a^b g(x) h(x) dx\right)^2 \le \int_a^b \{g(x)\}^2 dx \int_a^b \{h(x)\}^2 dx$$

where the equality occurs when $g(x)$ is a constant multiple of $h(x)$. Thus

$$\mathcal{E}(V) = n \int_{-\infty}^{\infty} x \{F(x)\}^{n-1} f(x) dx$$

$$= n \int_{-\infty}^{\infty} x \{F(x)\}^{n-1} dF(x)$$

$$\{\mathcal{E}(V)\}^2 \le n^2 \int_{-\infty}^{\infty} \{F(x)\}^{2n-2} dF(x) \int_{-\infty}^{\infty} x^2 dF(x)$$

Simplifying

$$\{\mathcal{E}(V)\}^2 \le n^2 \sigma^2 \int_0^1 y^{2n-2} dy$$

$$= \frac{n^2 \sigma^2}{(2n-1)}$$

$$\mathcal{E}(V) \le \frac{n\sigma}{(2n-1)^{\frac{1}{2}}}$$

Therefore

$$\frac{\mathcal{E}(R)}{\sigma} \le \frac{2n}{(2n-1)^{\frac{1}{2}}}$$

That is, for large values of $n$, $\mathcal{E}(R)$ is bounded by a constant times $\sigma\sqrt{n}$. This discussion is summarized in the following lemma.

**Lemma 8.1.** *Let $X_1, X_2, \ldots, X_n$ be $n$ independent and identically distributed random variables, each with mean 0, standard deviation $\sigma$, and symmetric probability density function. Define*

$$V = \max\{X_1, X_2, \ldots, X_n\}, \quad U = \min\{X_1, X_2, \ldots, X_n\}, \quad and \quad R = (V - U) \qquad (8.58)$$

*Then for large values of $n$, $\mathcal{E}(R)$ is bounded by a number which is proportional to $\sigma\sqrt{n}$.*   $\square$

Next consider a sequence of $n$ independent and identically distributed random variables, each with mean zero. Let $X$ be a representative of these random variables. The above result suggests that, in general for large values of $n$, the expected value of the range of these random variables is bounded by a number which is proportional to $n^\varrho$, where $\varrho$ depends upon the probability distribution of the random variable $X$.

## 8.11  Large Deviation Theory

Large deviation theory is the study of rare (extreme) events. It can be said that this theory is used for studying tails of probability density (or mass) functions. Theory of large deviation applies to especial types of rare events. These events are unlikely, rare, and occur together. A winning lottery is not a rare event, because it is simply a single event, and it cannot be further broken down into atomic subevents.

Recall that as per the Gaussian central limit theorem, the sum of a sequence of an infinite number of independent and identically distributed random variables has a normal probability distribution. However, if this sum has only a finite yet large number of random variables, then its distribution is approximately normal. In this later case, normal probability distribution is a good approximation within a standard deviation of its mean. For values of this sum which are several deviations away from the mean, normal probability distribution is not a good approximation. The values of sums which are several standard deviations away from its mean occur rarely. These are termed *large deviations,* and large deviation theory focuses on such rare events.

The above discussion is next made precise as follows. Let $\widetilde{X}_n$ be the sum of $n$ independent, and identically distributed random variables, each with finite mean $\mu$ and finite standard deviation $\sigma$. As per the law of large numbers, the mean of $\widetilde{X}_n$ is approximately $n\mu$. In addition, the Gaussian central limit theorem asserts that the deviation of $\widetilde{X}_n$ from $n\mu$ is typically of the order $\sqrt{n}\sigma$. It is also possible for these deviations to be much larger than $\sqrt{n}\sigma$, say $n^\alpha\sigma$, where $\alpha > 1/2$, and the probability

$$P\left(\left|\widetilde{X}_n - n\mu\right| > n^\alpha\sigma\right), \quad \alpha > 1/2$$

tends to zero for large values of $n$. Large deviation theory is concerned with the study of such probabilities. The probability of such large fluctuations generally decays exponentially in $n$. That is, the probability of the rare event to be determined is of the form $P\,(\text{rare event}) \sim e^{-n\cdot\text{const}}$.

### 8.11.1  A Prelude via Saddle Point Technique

Significant insight into the large deviation theory can be gained by using saddle point approximation technique. Let $X_1, X_2, \ldots, X_n$ be a sequence of $n$ independent and identically distributed random variables, where $\widetilde{X}_n = \sum_{j=1}^n X_j$. Denote a representative of these $n$ random variables by $X$ and its mean by $\mu$. Saddle point technique is used to approximate the probability $P\left(\widetilde{X}_n \geq na\right)$, where the constant $a > \mu$. We begin by defining the *cumulant* of a random variable.

**Definition 8.17.** *Let the moment generating function of the random variable $X$ be $\mathcal{M}_X\,(t) = \mathcal{E}\left(e^{tX}\right)$. The cumulant or log moment generating function of the random variable $X$ is*

$$\Upsilon\,(t) = \ln\mathcal{M}_X\,(t), \quad t \in \mathbb{R} \tag{8.59}$$

$\square$

The characteristic function of the random variable $X$ has been defined as $\mathcal{M}_X\,(it)$ where $i = \sqrt{-1}$. As $X_1, X_2, \ldots, X_n$ is a sequence of $n$ independent and identically distributed random variables, it can readily be inferred that the characteristic function of $\widetilde{X}_n$ is $\mathcal{M}_{\widetilde{X}_n}\,(it) = \left\{\mathcal{M}_X\,(it)\right\}^n$. Assume that the random variables $X_1, X_2, \ldots, X_n$ are continuous, and denote the probability density function of $\widetilde{X}_n$ by $f_{\widetilde{X}_n}\,(x)$, $x \in \mathbb{R}$. Use of Fourier transform theory yields

$$f_{\widetilde{X}_n}\,(x) = \frac{1}{2\pi}\int_{-\infty}^{\infty}\left\{\mathcal{M}_X\,(it)\right\}^n e^{-itx}dt$$

Therefore

$$f_{\widetilde{X}_n}(na) = \frac{1}{2\pi} \int_{-\infty}^{\infty} e^{n(\Upsilon(it)-iat)}\,dt$$

The above integral can be approximated via saddle point technique for large values of $n$. Define $g(u) = (\Upsilon(u) - au)$, where $u = it$. Let the first and second derivatives of $\Upsilon(u)$ with respect to $u$ be $\Upsilon'(u)$ and $\Upsilon''(u)$ respectively. Similarly, let the first and second derivatives of $g(u)$ with respect to $u$ be $g'(u)$ and $g''(u)$ respectively. As $n \to \infty$, use of the saddle point approximation yields

$$f_{\widetilde{X}_n}(na) \sim \frac{e^{ng(u_c)}}{\{2\pi n\,|g''(u_c)|\}^{1/2}}$$
$$g'(u_c) = (\Upsilon'(u_c) - a) = 0$$

Thus

$$f_{\widetilde{X}_n}(y) \sim \frac{e^{(n\Upsilon(u_c)-u_c y)}}{\{2\pi n\,|g''(u_c)|\}^{1/2}}$$

The above result is a good approximation of the tail of the probability density function. Therefore for $a > \mu$

$$P\left(\widetilde{X}_n \geq na\right) = \int_{na}^{\infty} f_{\widetilde{X}_n}(y)\,dy \sim \frac{e^{-n(au_c-\Upsilon(u_c))}}{\{2\pi n\,|\Upsilon''(u_c)|\}^{1/2}\,u_c}$$

Define $\ell(a) = (au_c - \Upsilon(u_c))$, and $\sigma_c^2 = \Upsilon''(u_c)$. Then

$$P\left(\widetilde{X}_n \geq na\right) \sim \frac{1}{\sqrt{2\pi n}\,\sigma_c u_c}\exp\left(-n\ell(a)\right)$$

Thus, as $n \to \infty$, the probability $P\left(\widetilde{X}_n \geq na\right)$ decays exponentially. Therefore, if the nonexponential term is ignored, $\ln P\left(\widetilde{X}_n \geq na\right)$ turns out to be a linear function of $n$. This result is summarized in the following observation.

**Observation 8.13.** Let $X_1, X_2, \ldots, X_n$ be a sequence of $n$ independent and identically distributed random variables. Also let a representative of these random variables be $X$, where $\mathcal{E}(X) = \mu$ is its mean, $\mathcal{M}_X(t) < \infty, \forall\, t \in \mathbb{R}$ is its moment generating function, $\Upsilon(t) = \ln \mathcal{M}_X(t)$ is its cumulant. Let the first and second derivatives of $\Upsilon(u)$ with respect to $u$ be $\Upsilon'(u)$ and $\Upsilon''(u)$ respectively. Also let $\widetilde{X}_n = \sum_{j=1}^{n} X_j$. Then $\forall\, a > \mu$, and as $n \to \infty$

$$P\left(\widetilde{X}_n \geq na\right) \sim \frac{1}{\sqrt{2\pi n}\,\sigma_c u_c}\exp\left(-n\ell(a)\right) \tag{8.60}$$

where $u_c$ satisfies $(\Upsilon'(u_c) - a) = 0$, $\ell(a) = (au_c - \Upsilon(u_c))$, and $\sigma_c^2 = \Upsilon''(u_c)$.    □

In summary, saddle point approximations can be used to successfully study the tail of distribution of a sum of random variables.

### 8.11.2 Cramér and Bahadur-Rao Theorems

The approximations obtained via saddle point technique are next rederived within a probabilistic framework. The well-known Cramér's theorem, rate functions, and Bahadur-Rao theorem are next introduced. These results are used in studying probabilities of large fluctuations. Cramér's theorem gives exponential rate of decay of the probability $P\left(\widetilde{X}_n \geq na\right)$. This is obtained via the Chernoff bound.

$$P\left(\widetilde{X}_n \geq na\right) \leq e^{-nat}\left\{\mathcal{M}_X\left(t\right)\right\}^n = e^{-n(at-\Upsilon(t))}, \;\; t \geq 0$$

Thus

$$\ln P\left(\widetilde{X}_n \geq na\right) \leq \inf_{t \geq 0}\left\{-n\left(at-\Upsilon\left(t\right)\right)\right\} = -n\sup_{t \geq 0}\left(at-\Upsilon\left(t\right)\right)$$

This result is summarized in the following lemma.

**Lemma 8.2.** *Let $X_1, X_2, \ldots, X_n$ be a sequence of $n$ independent and identically distributed random variables. Also let a representative of these random variables be $X$, and its cumulant be $\Upsilon\left(t\right)$. If $\widetilde{X}_n = \sum_{j=1}^n X_j$, then*

$$\ln P\left(\widetilde{X}_n \geq na\right) \leq -n\sup_{t \geq 0}\left(at-\Upsilon\left(t\right)\right) \tag{8.61}$$

$\square$

The above result motivates the definition of the *large deviation rate function*.

**Definition 8.18.** *Let $X$ be a random variable, $\mathcal{M}_X\left(\cdot\right)$ be its moment generating function, and $\Upsilon\left(t\right) = \ln \mathcal{M}_X\left(t\right)$ be its cumulant. The large deviation rate function $\ell\left(\cdot\right)$ of the random variable $X$ is*

$$\ell\left(x\right) = \sup_{t \in \mathbb{R}}\left(xt - \Upsilon\left(t\right)\right) \tag{8.62}$$

*The function $\ell\left(\cdot\right)$ takes values in the set of extended real numbers $\mathbb{R}^* = \mathbb{R} \cup \{+\infty\}$.* $\square$

The function $\ell\left(\cdot\right)$ is also called the *Fenchel-Legendre transform* of the cumulant function $\Upsilon\left(\cdot\right)$. Some properties of the large deviation rate function are listed below.

**Observations 8.14.** Let $\ell\left(\cdot\right)$ be the large deviation rate function of a random variable $X$ with finite mean $\mu$, and finite standard deviation $\sigma$.

1. $\ell\left(x\right) \geq 0$.
2. $\ell\left(x\right)$ is convex.
3. $\ell\left(x\right)$ has its minimum value at $x = \mu$. Also $\ell\left(\mu\right) = 0$.
4. Let $\ell''\left(x\right)$ be the second derivative of $\ell\left(x\right)$ with respect to $x$. Then $\ell''\left(\mu\right) = 1/\sigma^2$.
5. $\ell\left(x\right) = \sup_{t \geq 0}\left(xt - \Upsilon\left(t\right)\right)$ for $x \geq \mu$, and $\ell\left(x\right) = \sup_{t \leq 0}\left(xt - \Upsilon\left(t\right)\right)$ for $x \leq \mu$.
6. For $x \leq \mu$, $\ell\left(x\right)$ is nonincreasing, and for $x \geq \mu$, $\ell\left(x\right)$ is nondecreasing. $\square$

See the problem section for justification of these observations.

**Observations 8.15.** The large deviation rate functions of some discrete and continuous random variables are listed below.

1. Bernoulli random variable with parameter $p \in (0, 1)$.

$$\ell\left(x\right) = x\ln\left(\frac{x}{p}\right) + (1-x)\ln\left(\frac{1-x}{1-p}\right), \;\; x \in (0, 1)$$

Also $\ell(0) = -\ln(1-p)$, $\ell(1) = -\ln p$, and infinite for $x \in \mathbb{R} \backslash [0,1]$. Note that $\ell(p) = 0$.

2. Poisson random variable with parameter $\lambda > 0$.

$$\ell(x) = \lambda - x + x \ln\left(\frac{x}{\lambda}\right), \quad x > 0$$

and $\ell(x)$ is infinite for $x < 0$. Note that $\ell(0) = \lambda$, and $\ell(\lambda) = 0$.

3. Normal random variable with parameters $\mu \in \mathbb{R}$ and $\sigma^2 > 0$.

$$\ell(x) = \frac{1}{2}\left(\frac{x-\mu}{\sigma}\right)^2, \quad x \in \mathbb{R}$$

Note that $\ell(\mu) = 0$.

4. Exponentially distributed random variable with parameter $\lambda > 0$.

$$\ell(x) = \lambda x - 1 - \ln(\lambda x), \quad x > 0$$

and infinite elsewhere. Note that $\ell(\lambda^{-1}) = 0$.                                   $\square$

The next theorem is due to H. Cramér. It is the first significant result in the theory of large deviations.

**Theorem 8.13.** *Let $X_1, X_2, \ldots, X_n$ be a sequence of $n$ independent and identically distributed random variables. Also let a representative of these random variables be $X$, where $\mathcal{E}(X) - \mu$ is its mean, $\mathcal{M}_X(t) < \infty$, $\forall\, t \in \mathbb{R}$ is its moment generating function, $\Upsilon(t) = \ln \mathcal{M}_X(t)$ is its cumulant, and $\ell(\cdot)$ is its large deviation rate function. Let $\widetilde{X}_n = \sum_{j=1}^n X_j$. Then $\forall\, a > \mu$*

$$\lim_{n\to\infty} \frac{1}{n} \ln P\left(\widetilde{X}_n \geq na\right) = -\ell(a) \tag{8.63}$$

*Proof.* See the problem section.                                                            $\square$

This theorem states that for $a > \mu$ and large values of $n$, $P\left(\widetilde{X}_n \geq na\right)$ is approximated reasonably well by $\exp\left(-n\ell(a)\right)$. Alternately, for large values of $n$, we have $P\left(\widetilde{X}_n \geq na\right) = \exp\left(-n\ell(a) + o(n)\right)$, $\forall\, a > \mu$. Recall that $o(n)$ is a term, which when divided by $n$ goes to zero. A sharper result than that of Cramér's is the following theorem due to R. R. Bahadur and R. Ranga Rao.

**Theorem 8.14.** *Let $X_1, X_2, \ldots, X_n$ be a sequence of $n$ independent and identically distributed random variables. Also let a representative of these random variables be $X$, where $\mathcal{E}(X) = \mu$ is its mean, $\mathcal{M}_X(t) < \infty$, $\forall\, t \in \mathbb{R}$ is its moment generating function, $\Upsilon(t) = \ln \mathcal{M}_X(t)$ is its cumulant, and $\ell(\cdot)$ is its large deviation rate function. Let $\widetilde{X}_n = \sum_{j=1}^n X_j$. Then $\forall\, a > \mu$*

$$\lim_{n\to\infty} P\left(\widetilde{X}_n \geq na\right) = \frac{1}{\sqrt{2\pi n}\sigma_c t_c} \exp\left(-n\ell(a)\right)(1 + o(1)) \tag{8.64}$$

*where $t_c$ is the value of $t$ at which the supremum in the rate function $\ell(\cdot)$ occurs. That is, $a = \Upsilon'(t_c)$, $\ell(a) = (at_c - \Upsilon(t_c))$, and $\sigma_c^2 = \Upsilon''(t_c)$. Note that $\Upsilon'(\cdot)$ and $\Upsilon''(\cdot)$ are the first and second derivatives of $\Upsilon(\cdot)$ respectively.*

*Proof.* See the problem section.                                                            $\square$

Not surprisingly, the above result was also obtained by the saddle point approximation. Observe that the Bahadur-Rao result provides a nonexponential correction to Cramér's result. For large values of $n$, only the exponential term is dominant. However the presence of the nonexponential term is useful when $n$ is not quite large.

## Reference Notes

This chapter on probability theory is based upon the books by Grimmett, and Stirzaker (2001), Grinstead, and Snell (1997), Lipschutz (1965), Mood, Graybill, and Boes (1974), Papoulis (1965), Parzen (1960), Ross (1970, 2009, and 2010), Spiegel, Schiller, and Srinivasan (2000), Stirzaker (2003), and Sveshnikov (1968). A relatively advanced textbook has been written by Durrett (2005). This list is incomplete without a mention of the two-volume treatise by Feller (1968, and 1971).

The credit for Chernoff bounds is generally assigned to H. Chernoff (1952). However this bound was discovered earlier by H. Cramér (1938). The exposition on generalized central limit theorem is based upon the textbooks by Moran (1968), Lukacs (1970), Ibragimov, and Linnik (1971), Burrill (1972), Hall (1981), and Nikias, and Shao (1995). The reader can also refer to the remarkable textbooks by Breiman (1992), Chung (1974a,b), Gnedenko, and Kolmogorov (1968). Stable distributions are discussed with remarkable clarity in Samorodnitsky, and Taqqu (1994). A readable introduction to the theory of large deviation can be found in the textbooks by Bucklew (1990), Shwartz, and Weiss (1995), and Hollander (2000). H. Cramér's theorem on large deviation first appeared in 1938. The statement of this theorem is from Hollander (2000). R. R. Bahadur, and R. Ranga Rao's result on large deviation was published in the year 1960.

## Problems

1. A point is selected at random inside a circle. Prove that the probability that the point is closer to the center of the circle than to its circumference is equal to $0.25$.

2. A die is a cube with indentations on each of the six faces. One face has a single indentation. The second, third, fourth, fifth, and sixth face, each have $2, 3, 4, 5$, and $6$ indentations respectively. That is, there are different number of indentations on each face. A die is generally used for gambling. Find the probability that the sum of the faces of two independently thrown dice totals 7. Find the expected value of the sum.

   Hint: Let the sum of the two faces be $i$, where $2 \leq i \leq 12$. Denote the probability of the sum of the two faces by $p(i)$. Then

   $$p(7) = 6/36 = 1/6, \quad \text{and} \quad \sum_{i=2}^{12} ip(i) = 7$$

3. What is the smallest number of persons that can be assembled in a room so that there is a probability greater than $1/2$ of a duplicate birthday? Assume that there are 365 days in a year. Also assume that all birthdays are equally likely to occur. That is, each date has a probability of $1/365$. The occurrences of birthdays are assumed to be independent.

Hint: See Parzen (1960). This is the celebrated *birthday problem*. Define $Q(k) =$ probability that $k$ persons have no duplicate birthday. Then

$Q(1) = (365/365)$, since first person can be chosen in 365 equally likely ways.

$Q(2) = (365/365) \cdot (364/365)$, since second person can be chosen in 364 equally likely ways, so that no duplication occurs.

$Q(3) = (365/365) \cdot (364/365) \cdot (363/365)$, the third person should not have his or her birthday on either of the first two dates.

Similarly

$$Q(k) = \left(\frac{365}{365}\right)\left(\frac{364}{365}\right) \cdots \left(\frac{365 - k + 1}{365}\right)$$

For $1 \le k \le 365$, define

$$P(k) = (1 - Q(k)) = \text{the probability that there is birthday duplication.}$$

Some values of $P(k)$'s are listed in Table 8.1.

Thus $k = 23$ is the least number of people in a room that gives better than a fifty-fifty chance of getting some birthdays on same days!

| $k$ | $P(k)$ |
|---|---|
| 5 | 0.027 |
| 10 | 0.117 |
| 20 | 0.411 |
| 22 | 0.476 |
| 23 | 0.507 |
| 24 | 0.538 |
| 30 | 0.706 |
| 40 | 0.891 |
| 60 | 0.994 |

Table 8.1. Values of $k$ and the corresponding probability $P(k)$.

4. A student types $m$ letters, and seals them in different envelopes. He or she types different addresses randomly on each of them. Find the probability that at least one of the envelopes has the correct address typed on it.

Hint: See Sveshnikov (1968). Let $B_k$ be the event that the $k$th envelope has the correct address, where $k = 1, 2, \ldots, m$. The probability $p = P\left(\bigcup_{k=1}^{m} B_k\right)$ has to be determined. For $1 \le i, j, k \le m$

$$P(B_i) = \frac{1}{m} = \frac{(m-1)!}{m!}$$
$$P(B_i B_j) = P(B_i) P(B_j \mid B_i)$$
$$= \frac{1}{m} \cdot \frac{1}{(m-1)}$$
$$= \frac{(m-2)!}{m!}, \quad i \ne j$$

$$P\left(B_i B_j B_k\right) = \frac{(m-3)!}{m!}, \quad i \neq j \neq k$$

The formula for the probability of $m$ events yields

$$p = P\left(\bigcup_{k=1}^{m} B_k\right)$$

$$= \binom{m}{1}\frac{(m-1)!}{m!} - \binom{m}{2}\frac{(m-2)!}{m!} + \binom{m}{3}\frac{(m-3)!}{m!} - \ldots + (-1)^{m-1}\frac{1}{m!}$$

$$= 1 - \frac{1}{2!} + \frac{1}{3!} - \ldots + (-1)^{m-1}\frac{1}{m!}$$

For large values of $m$, $p \approx \left(1 - e^{-1}\right) \approx 0.6321$.

5. If $X$ and $Y$ are two jointly distributed random variables, then show that

$$|Cov(X,Y)| \leq \sigma_X \sigma_Y, \quad \text{and} \quad -1 \leq Cor(X,Y) \leq 1$$

6. Let $p \to 0, n \to \infty$, and $np < \infty$. Prove that the binomial distribution with parameters $n$ and $p$ can be approximated by the Poisson distribution.

   Hint: See Parzen (1960). If $p \to 0$, $n \to \infty$, and $np \to \lambda$, where $\lambda \in \mathbb{R}^+$, then

   $$\binom{n}{x}p^x\left(1-p\right)^{n-x} \to \frac{\lambda^x}{x!}e^{-\lambda}, \quad \forall\, x \in \mathbb{N}$$

7. Let $X$ and $Y$ be independent random variables with Gaussian distribution. Prove that the random variable $Z = (aX + bY)$ has a Gaussian distribution, where $a, b \in \mathbb{R}$.

8. Show that the probability density function $\phi\left(\cdot\right)$, and the cumulative distribution function $\Phi\left(\cdot\right)$ of the standard normal random variable satisfy

   $$\left(\frac{1}{x} - \frac{1}{x^3}\right) < \frac{(1-\Phi(x))}{\phi(x)} < \frac{1}{x}, \quad x \in \mathbb{R}^+$$

   $$\left(\frac{1}{x} - \frac{1}{x^3}\right) < \frac{(1-\Phi(x))}{\phi(x)} < \left(\frac{1}{x} - \frac{1}{x^3} + \frac{3}{x^5}\right), \quad x \in \mathbb{R}^+$$

   Hint: See Grimmett, and Stirzaker (2001). Let $\phi'\left(x\right)$ be the first derivative of $\phi\left(x\right)$ with respect to $x$. Using the relationship $\phi'\left(x\right) = -x\phi\left(x\right)$, and integration by parts successively, establish that

   $$1 - \Phi\left(x\right) = \int_x^\infty \phi\left(t\right)dt$$

   $$= \frac{\phi\left(x\right)}{x} - \int_x^\infty \frac{\phi\left(t\right)}{t^2}dt$$

   $$= \frac{\phi\left(x\right)}{x} - \frac{\phi\left(x\right)}{x^3} + 3\int_x^\infty \frac{\phi\left(t\right)}{t^4}dt$$

   $$= \frac{\phi\left(x\right)}{x} - \frac{\phi\left(x\right)}{x^3} + \frac{3\phi\left(x\right)}{x^5} - 15\int_x^\infty \frac{\phi\left(t\right)}{t^6}dt$$

Proceeding in this manner repeatedly, a series of sharper results can be further obtained. Thus, the tail probability of standard normal random variable is

$$P\left(X > x\right) \sim \frac{\exp\left(-x^2/2\right)}{x\sqrt{2\pi}}, \quad \text{as} \quad x \to \infty$$

9. The cumulative distribution function of the standard normal variable $\Phi(\cdot)$ can be approximated as

$$\Phi(x) - 0.5 \simeq \begin{cases} x(4.4-x)/10, & 0 \le x \le 2.2 \\ 0.49, & 2.2 < x < 2.6 \\ 0.5, & 2.6 \le x \end{cases}$$

where the error in approximation is at most 0.0052. Check the accuracy of this numerical claim. Hint: See Shah (1985).

10. $X$ is a continuously distributed random variable with probability density function $f_X(x)$, $x \in \mathbb{R}$. Let $Y = |X|$, and $f_Y(\cdot)$, be its probability density function. Prove that

$$f_Y(y) = \begin{cases} f_X(y) + f_X(-y), & y > 0 \\ 0, & y < 0 \end{cases}$$

$$\mathcal{E}(Y) = \int_0^\infty y f_X(y)\,dy + \int_0^\infty y f_X(-y)\,dy$$

Hint: See Parzen (1960). $P(Y \le y) = P(|X| \le y) = P(-y \le X \le y) = \int_{-y}^y f_X(x)\,dx$. The result follows, by differentiating both sides with respect to $y$.

11. Let $X$ be a continuous random variable, with cumulative distribution function $F_X(x)$, $x \in \mathbb{R}$. If $Y = X^2$, find $F_Y(\cdot)$ and the density function $f_Y(\cdot)$ if it exists. Hint: See Parzen (1960). Note that $F_Y(y) = 0$ for $y < 0$. If $y \ge 0$ then

$$\begin{aligned} F_Y(y) &= P(X^2 \le y) = P(-\sqrt{y} \le X \le \sqrt{y}) \\ &= F_X(\sqrt{y}) - F_X(-\sqrt{y}) + p_X(-\sqrt{y}) \end{aligned}$$

If the probability density function of the random variable $X$ is $f_X(\cdot)$, then

$$f_Y(y) = \begin{cases} 0, & y < 0 \\ \dfrac{1}{2\sqrt{y}}\{f_X(\sqrt{y}) + f_X(-\sqrt{y})\}, & y > 0 \end{cases}$$

12. Jack and Jill decide to meet at a restaurant at a specified time for dinner. They agree to arrive at the restaurant with a delay ranging between zero and one hour. Also all pairs of delays are equally likely. The first person to arrive agrees to wait for $60\alpha$ minutes, where $\alpha \in [0, 1]$, and leave if the other person did not arrive during this interval. Determine the probability that Jack and Jill have a dinner date.

Hint: Let $X$ and $Y$ be the arrival times of Jack and Jill, where $0 \le X, Y \le 1$. The required probability is

$$P(|X - Y| \le \alpha) = 1 - (1-\alpha)^2$$

13. Buffon's needle problem. An infinite table has parallel lines drawn on it. These lines are a distance $2\delta$ apart. A needle of length $2\lambda$ is randomly thrown on the table. Find the probability that the needle intersects one of the lines by assuming that $\lambda \le \delta$.

Hint: See Ross (2009). The position of the needle is specified by the distance $Y$ from the middle point of the needle to the nearest line, and the acute angle $\theta$ between the needle and the parallel line. Denote the middle point of the needle by $A$, the point of intersection of the needle and the line by $B$, and the foot of the perpendicular to the line from $A$ by $C$. Therefore $BCA$ is a right triangle, with hypotenuse $AB$. In this triangle, angle $ABC$ is equal to $\theta$, the length of the side $AC$ is equal to $Y$, and the length of the hypotenuse $AB$ is equal to $Y/\sin\theta$. See Figure 8.1.

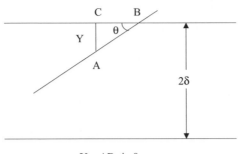

$$Y = AB \sin \theta$$

Figure 8.1. Buffon's needle.

The needle intersects the line if the length of the hypotenuse $AB$ is less than $\lambda$. That is, the needle intersects a line if $Y < \lambda \sin \theta$. Note that $0 \le Y \le \delta$ and $0 \le \theta \le \pi/2$. It is assumed that the random variables $Y$ and $\theta$ are independent, and uniformly distributed random variables. Thus

$$P\left(Y < \lambda \sin \theta\right) = \frac{2}{\pi \delta} \int_0^{\pi/2} \int_0^{\lambda \sin \theta} dx d\theta = \frac{2\lambda}{\pi \delta}$$

14. Let $Y_1, Y_2, \ldots, Y_n$ be independent outcomes of a trial process, each with identical Cauchy distribution. Let $A_n$ be the average of the $Y_i$'s, that is, $A_n = (1/n) \sum_{i=1}^n Y_i$. Prove that $A_n$ also has a Cauchy distribution.

Hint: Prove that the characteristic functions of the random variables $Y_i$ and $A_n$ are identical.

15. Let $X$ be a nonnegative and integer valued random variable. Prove that

$$\mathcal{E}\left(X\right) = \sum_{n=1}^\infty P\left(X \ge n\right)$$

Hint: See Ross (2009).

$$\mathcal{E}\left(X\right) = \sum_{j=1}^\infty j P\left(X = j\right) = \sum_{j=1}^\infty \sum_{n=1}^j P\left(X = j\right)$$

$$= \sum_{n=1}^\infty \sum_{j=n}^\infty P\left(X = j\right) = \sum_{n=1}^\infty P\left(X \ge n\right)$$

16. Let $W$ be a continuous random variable, which is defined over the real line $\mathbb{R}$. Prove that

$$\mathcal{E}\left(W\right) = \int_0^\infty P\left(W > w\right) dw - \int_0^\infty P\left(W < -w\right) dw$$

Hint: See Ross (2009). Let the probability density function of $W$ be $f_W\left(\cdot\right)$.

$$\int_0^\infty P\left(W > w\right) dw = \int_0^\infty \int_w^\infty f_W\left(x\right) dx dw$$

$$= \int_0^\infty \left(\int_0^x dw\right) f_W\left(x\right) dx = \int_0^\infty x f_W\left(x\right) dx$$

Similarly

$$\int_0^\infty P\left(W < -w\right) dw = \int_0^\infty \int_{-\infty}^{-w} f_W\left(x\right) dx dw$$

$$= \int_{-\infty}^0 \left(\int_0^{-x} dw\right) f_W\left(x\right) dx = -\int_{-\infty}^0 x f_W\left(x\right) dx$$

The result follows. It is also possible to state a discrete analog of this result.

17. Prove the inclusion and exclusion formula. Show that for any $N$ events $A_1, A_2, \ldots, A_N$,

$$P\left(A_1 \cup A_2 \cup \ldots \cup A_N\right)$$
$$= \sum_{1 \le j \le N} P\left(A_j\right) - \sum_{1 \le j < k \le N} P\left(A_j \cap A_k\right) + \sum_{1 \le j < k < l \le N} P\left(A_j \cap A_k \cap A_l\right)$$
$$- \sum_{1 \le j < k < l < m \le N} P\left(A_j \cap A_k \cap A_l \cap A_m\right)$$
$$+ \ldots + \left(-1\right)^{N+1} P\left(A_1 \cap A_2 \cap \ldots \cap A_N\right)$$

Hint: See Durrett (2005). Let $A = A_1 \cup A_2 \cup \ldots \cup A_N$, and $I_{A_i}$ be the indicator function of event $A_i$ for $1 \le i \le N$. Also let the indicator function of the event $A$ be $I_A$. Thus

$$I_A = 1 - \prod_{i=1}^N \left(1 - I_{A_i}\right)$$

The result follows by expanding the right-hand side of the above equation, and applying the expectation operator on both sides.

18. Let $Y$ be a normally distributed random variable with $\mathcal{E}\left(Y\right) = \mu$ and $Var\left(Y\right) = \sigma^2$. Define a positive random variable $X = e^Y$. Therefore $\ln X = Y$ has a normal distribution. The random variable $X$ is said to have a *lognormal* distribution. Prove that:

(a) The probability density function $f_X\left(\cdot\right)$, of the random variable $X$ is

$$f_X\left(x\right) = \frac{1}{x\sqrt{2\pi}\sigma} \exp\left\{-\frac{1}{2\sigma^2}\left(\ln x - \mu\right)^2\right\}, \quad x \in \mathbb{R}^+$$

where $\mu \in \mathbb{R}$, and $\sigma \in \mathbb{R}^+$.

(b) $\mathcal{E}\left(X\right) = \exp\left(\mu + \sigma^2/2\right)$.

(c) $Var\left(X\right) = \left\{\exp\left(2\mu + 2\sigma^2\right) - \exp\left(2\mu + \sigma^2\right)\right\}$.

(d) $\mathcal{E}\left(X^r\right) = \exp\left(r\mu + r^2\sigma^2/2\right), \forall r \in \mathbb{N}$.

Hint: These results can be found in any standard textbook on probability theory like Mood, Graybill, and Boes (1974).

19. Let $X$ be a random variable with lognormal probability density function. It is

$$f_X\left(x\right) = \frac{1}{x\sqrt{2\pi}} \exp\left\{-\frac{1}{2}\left(\ln x\right)^2\right\}, \quad x \in \mathbb{R}^+$$

Consider another random variable $Z$. For a fixed value of $a \in \left[-1, 1\right]$, its probability density function is

$$f_Z\left(z\right) = f_X\left(z\right)\left\{1 + a\sin\left(2\pi \ln z\right)\right\}, \quad z \in \mathbb{R}^+$$

Prove that $\mathcal{E}\left(X^r\right) = \mathcal{E}\left(Z^r\right) = \exp\left(r^2/2\right), \forall r \in \mathbb{N}$.

This problem demonstrates that the two random variables with different probability density functions can possess identical moments.

Hint: See Durrett (2005). The value of $a$ is restricted to the interval $[-1, 1]$ so that $f_Z(z) \geq 0$. We need to show that

$$I(r) \triangleq \int_0^\infty z^r f_X(z) \sin(2\pi \ln z)\, dz = 0, \quad \forall\, r \in \mathbb{N}$$

In the integral $I(r)$, make a change of variable $\ln z = u$. This results in

$$I(r) = \frac{\exp(r^2/2)}{\sqrt{2\pi}} \int_{-\infty}^\infty \exp\left\{ -\frac{1}{2}(u - r)^2 \right\} \sin(2\pi u)\, du$$

Again make a change of variable $(u - r) = v$. Thus

$$I(r) = \frac{\exp(r^2/2)}{\sqrt{2\pi}} \int_{-\infty}^\infty \exp\left\{ -\frac{1}{2}v^2 \right\} \sin\{2\pi(v + r)\}\, dv$$

As $r$ is a nonnegative integer, we have

$$I(r) = \frac{\exp(r^2/2)}{\sqrt{2\pi}} \int_{-\infty}^\infty \exp\left\{ -\frac{1}{2}v^2 \right\} \sin(2\pi v)\, dv$$

Recall that $f(x), x \in \mathbb{R}$ is an odd function of $x$, if $f(-x) = -f(x)$ for all values of $x \in \mathbb{R}$. The integrand is an odd function of $v$, therefore $I(r) = 0$.

20. Let $X$ be a random variable, and $a > 0$. Prove the following relative of Bienaymé-Chebyshev inequality.

$$P(|X - \mathcal{E}(X)| \geq a) \leq \frac{1}{a}\mathcal{E}(|X - \mathcal{E}(X)|)$$

21. Prove the theorem on Chernoff bounds.

Hint: See Ross (2009). The moment generating function of the random variable $X$ is defined as $\mathcal{M}_X(t) = \mathcal{E}(e^{tX})$. For $t > 0$,

$$P(X \geq x) = P(e^{tX} \geq e^{tx}) \leq e^{-xt}\mathcal{E}(e^{tX})$$

The last step follows via generalized Bienaymé-Chebyshev inequality. Similarly, for $t < 0$

$$P(X \leq x) = P(e^{tX} \geq e^{tx}) \leq e^{-xt}\mathcal{E}(e^{tX})$$

The above results are also true at $t = 0$.

22. Let $X$ be a random variable, and $g(\cdot)$ be a convex function. Prove that $\mathcal{E}(g(X)) \geq g(\mathcal{E}(X))$, provided the expectations exist. This relationship is also called Jensen's inequality.

Hint: See Ross (2009). Let $\beta = \mathcal{E}(X)$. Expand $g(x)$ in a Taylor's series about $\beta$. Thus

$$g(x) = g(\beta) + g'(\beta)(x - \beta) + \frac{g''(\xi)(x - \beta)^2}{2}$$

where $g'(\cdot)$, and $g''(\cdot)$ are the first and second derivatives of $g(\cdot)$ respectively. Furthermore, $\xi$ is some value that lies between $x$ and $\beta$. Since $g(\cdot)$ is a convex function, $g''(\xi) \geq 0$. Therefore

$$g(x) \geq g(\beta) + g'(\beta)(x - \beta)$$

Thus
$$g\left(X\right) \geq g\left(\beta\right) + g'\left(\beta\right)\left(X - \beta\right)$$

That is
$$\mathcal{E}\left(g\left(X\right)\right) \geq g\left(\beta\right) + g'\left(\beta\right)\left(\mathcal{E}\left(X\right) - \beta\right) = g\left(\beta\right) = g\left(\mathcal{E}\left(X\right)\right)$$

23. Expectations can also be computed by conditioning. Let $X$ and $Y$ be two random variables.
    (a) Prove that $\mathcal{E}\left(X\right) = \mathcal{E}\left(\mathcal{E}\left(X \mid Y\right)\right)$.
    (b) Define $Var\left(X \mid Y\right) = \mathcal{E}\left(\{X - \mathcal{E}\left(X \mid Y\right)\}^2 \mid Y\right)$, then prove that

$$Var\left(X\right) = \mathcal{E}\left(Var\left(X \mid Y\right)\right) + Var\left(\mathcal{E}\left(X \mid Y\right)\right)$$

Hint: See Ross (2009).

(a) Assume that $X$ and $Y$ are discrete random variables.

$$\mathcal{E}\left(\mathcal{E}\left(X \mid Y\right)\right) = \sum_y \mathcal{E}\left(X \mid Y = y\right) P\left(Y = y\right)$$

$$= \sum_y \sum_x x P\left(X = x \mid Y = y\right) P\left(Y = y\right)$$

$$= \sum_x x \sum_y P\left(X = x, Y = y\right) = \sum_x x P\left(X = x\right)$$

Similar result can also be obtained, if $X$ and $Y$ are continuous random variables.

(b) Note that $Var\left(X \mid Y\right) = \mathcal{E}\left(X^2 \mid Y\right) - \{\mathcal{E}\left(X \mid Y\right)\}^2$. Therefore

$$\mathcal{E}\left(Var\left(X \mid Y\right)\right) = \mathcal{E}\left(\mathcal{E}\left(X^2 \mid Y\right)\right) - \mathcal{E}\left(\{\mathcal{E}\left(X \mid Y\right)\}^2\right)$$

$$= \mathcal{E}\left(X^2\right) - \mathcal{E}\left(\{\mathcal{E}\left(X \mid Y\right)\}^2\right)$$

Also
$$Var\left(\mathcal{E}\left(X \mid Y\right)\right) = \mathcal{E}\left(\{\mathcal{E}\left(X \mid Y\right)\}^2\right) - \{\mathcal{E}\left(X\right)\}^2$$

Adding the last two equations yields the required result.

24. Let $X_1, X_2, \ldots, X_N$ be a sequence of independent and identically distributed random variables, each with mean $\mu_X$ and variance $\sigma_X^2$. Also $N$ is an integer valued random variable with mean $\mu_N$ and variance $\sigma_N^2$. Define $S_N = \sum_{i=1}^{N} X_i$. Prove that

$$E\left(S_N\right) = \mu_X \mu_N$$
$$Var\left(S_N\right) = \mu_N \sigma_X^2 + \mu_X^2 \sigma_N^2$$

Hint: See Mood, Graybill, and Boes (1974). Use the results of the last problem.

25. Let $X_1, X_2, \ldots, X_n$ be a sequence of $n$ independent random variables, each with mean $\mu < \infty$, and $\frac{1}{n}\sum_{j=1}^{n} Var\left(X_j\right) < \infty$. In addition, these random variables are pair-wise uncorrelated. Define $\overline{X}_n = \frac{1}{n}\sum_{j=1}^{n} X_j$, then prove that for any positive number $\delta$

$$\lim_{n \to \infty} P\left(\left|\overline{X}_n - \mu\right| < \delta\right) \to 1$$

This result is a refined statement of the weak law of large numbers.

26. Prove the strong law of large numbers by assuming that the fourth moment of the random variables $X_j$ is finite.

    Hint: See Ross (2009). Define $Y_j = (X_j - \mu)$, for $1 \le j \le n$, and $S_n = \sum_{j=1}^{n} Y_j$. Therefore $\mathcal{E}\left(Y_j^4\right) \triangleq K < \infty$. Expansion of $S_n^4$ consists of terms of the form

    $$Y_i^4, \quad Y_i^3 Y_j, \quad Y_i^2 Y_j^2, \quad Y_i^2 Y_j Y_k, \quad \text{and} \quad Y_i Y_j Y_k Y_l$$

    where $i \ne j \ne k \ne l$, and $1 \le i, j, k, l \le n$. As $\mathcal{E}(Y_j) = 0$, and the random variables $Y_j$'s are mutually independent random variables, we have

    $$\mathcal{E}\left(Y_i^3 Y_j\right) = \mathcal{E}\left(Y_i^3\right)\mathcal{E}\left(Y_j\right) = 0$$
    $$\mathcal{E}\left(Y_i^2 Y_j Y_k\right) = \mathcal{E}\left(Y_i^2\right)\mathcal{E}\left(Y_j\right)\mathcal{E}\left(Y_k\right) = 0$$
    $$\mathcal{E}\left(Y_i Y_j Y_k Y_l\right) = \mathcal{E}\left(Y_i\right)\mathcal{E}\left(Y_j\right)\mathcal{E}\left(Y_k\right)\mathcal{E}\left(Y_l\right) = 0$$

    For a given pair of values of $i$ and $j$, we have $\binom{4}{2} = 6$ terms in the expansion that equal $Y_i^2 Y_j^2$. Thus

    $$\mathcal{E}\left(S_n^4\right) = n\mathcal{E}\left(Y_j^4\right) + 6\binom{n}{2}\mathcal{E}\left(Y_i^2 Y_j^2\right)$$
    $$= nK + 6\binom{n}{2}\mathcal{E}\left(Y_i^2\right)\mathcal{E}\left(Y_j^2\right)$$

    Observe that
    $$0 \le Var\left(Y_j^2\right) = \mathcal{E}\left(Y_j^4\right) - \left\{\mathcal{E}\left(Y_j^2\right)\right\}^2$$

    That is
    $$\left\{\mathcal{E}\left(Y_j^2\right)\right\}^2 \le \mathcal{E}\left(Y_j^4\right) = K$$

    Thus
    $$\mathcal{E}\left(S_n^4\right) \le nK + 3n(n-1)K$$

    This in turn implies that
    $$\mathcal{E}\left\{\frac{S_n^4}{n^4}\right\} \le \frac{K}{n^3} + \frac{3K}{n^2}$$

    Therefore
    $$\mathcal{E}\left\{\sum_{n \in \mathbb{P}} \frac{S_n^4}{n^4}\right\} = \sum_{n \in \mathbb{P}} \mathcal{E}\left\{\frac{S_n^4}{n^4}\right\} < \infty$$

    The last result implies that with unit probability $\sum_{n \in \mathbb{P}} S_n^4/n^4 < \infty$. This in turn implies that the $n$th term of the series goes to $0$ in the limit. Therefore with unit probability, we have

    $$\lim_{n \to \infty} \frac{S_n^4}{n^4} \to 0 \Rightarrow \lim_{n \to \infty} \frac{S_n}{n} \to 0$$

    The result follows.

27. Prove that a characteristic function is uniformly continuous.

    Hint: See Burrill (1972). Let $\varphi(u)$, $u \in \mathbb{R}$ be the characteristic function of a random variable. Let the corresponding distribution function be $F(\cdot)$. Observe that

    $$|\varphi(u+h) - \varphi(u)| = \left|\int_{-\infty}^{\infty} e^{iux}\left(e^{ihx} - 1\right)dF(x)\right| \le \int_{-\infty}^{\infty}\left|e^{ihx} - 1\right|dF(x)$$

    for all $u, h \in \mathbb{R}$. Note that $\left|e^{ihx} - 1\right| \le 2$, and it tends to $0$ for each $x$ as $h \to 0$. Therefore $\lim_{h \to 0} \int_{-\infty}^{\infty}\left|e^{ihx} - 1\right|dF(x) = 0$.

28. If $\varphi(\cdot)$ is an infinitely divisible characteristic function, then prove that $\varphi(u) \neq 0$, $\forall u \in \mathbb{R}$.
Hint: See Burrill (1972), and Moran (1968). It is given that $\varphi(\cdot)$ is an infinitely divisible characteristic function, and therefore for each $n \in \mathbb{P}$, there exists a characteristic function $\varphi_n(\cdot)$, such that $\varphi(u) = \{\varphi_n(u)\}^n$. As $|\varphi(u)| \leq 1$, $\forall u \in \mathbb{R}$, we have

$$\lim_{n \to \infty} |\varphi_n(u)|^2 = \lim_{n \to \infty} |\varphi(u)|^{2/n} = h(u)$$

where $h(u) = 1$, if $\varphi(u) \neq 0$; and $h(u) = 0$, if $\varphi(u) = 0$. It is known that $\varphi(\cdot)$ is continuous, and $\varphi(0) = 1$. This implies that there exists a neighborhood of $0$ in which $\varphi(u) \neq 0$, and $h(u) = 1$. Thus $h(\cdot)$ is continuous at the origin, and it is also a characteristic function as it is the limit of the sequence $\{|\varphi_n(u)|^2\}$. As $h(\cdot)$ is a characteristic function, it is continuous. Furthermore, its only possible values are $0$ and $1$. Therefore $h(u)$ must have the value $1$, $\forall u \in \mathbb{R}$.

29. If $\varphi(\cdot)$ is an infinitely divisible characteristic function, and if for each $n \in \mathbb{P}$, there exists a characteristic function $\varphi_n(\cdot)$, such that $\varphi(u) = \{\varphi_n(u)\}^n$, then prove that:
(a) $\lim_{n \to \infty} n\{\varphi_n(u) - 1\} = \ln \varphi(u)$
(b) $\lim_{n \to \infty} \varphi_n(u) = 1$
Hint: See Burrill (1972).
(a) As $\varphi(u) \neq 0$, $\forall u \in \mathbb{R}$, we have

$$\varphi_n(u) = \{\varphi(u)\}^{1/n} = \exp\left\{\frac{1}{n} \ln \varphi(u)\right\}$$

As $\ln \varphi(u)$ is continuous, and $u$ is in a bounded interval, we therefore have $|\ln \varphi(u)| \leq d$ for some constant $d$. Thus

$$|n\{\varphi_n(u) - 1\} - \ln \varphi(\cdot)| = \left| n \sum_{j-2}^{\infty} \frac{1}{j!} \left\{\frac{1}{n} \ln \varphi(u)\right\}^j \right|$$

$$\leq \sum_{j=2}^{\infty} \frac{1}{j! n^{j-1}} |\ln \varphi(u)|^j$$

$$\leq d \sum_{j=2}^{\infty} \frac{1}{(j-1)!} \left\{\frac{d}{n}\right\}^{j-1} = d\left\{e^{d/n} - 1\right\}$$

The result follows as $\lim_{n \to \infty} \{e^{d/n} - 1\} = 0$.
(b) This result follows from part (a) of the problem, as

$$\lim_{n \to \infty} \{\varphi_n(u) - 1\} = \lim_{n \to \infty} \frac{1}{n} \{n(\varphi_n(u) - 1)\} = 0$$

30. Provide a justification of canonical representation of an infinitely divisible characteristic function of a random variable due to:
(a) Lévy and Khinchin.
(b) Lévy.
(c) Kolmogorov.
Hint: A complete derivation of the three related results (theorems) can be found in any of the books by: Breiman (1992), Burrill (1972), Chung (1974a), Gnedenko, and Kolmogorov (1968), and Moran (1968). We only provide plausibility of the stated results.

(a)  See Burrill (1972). Let the infinitely divisible characteristic function of the random variable be $\varphi\left(\cdot\right)$. Also, let $\varphi_n\left(u\right) = \varphi^{1/n}\left(u\right)$, and the distribution function which corresponds to $\varphi_n\left(\cdot\right)$ be $F_n\left(\cdot\right)$. From the last problem, we have

$$\ln\varphi\left(u\right) = \lim_{n\to\infty} n \int_{-\infty}^{\infty} \left\{e^{iux} - 1\right\} dF_n\left(x\right)$$

It would indeed be possible to obtain an integral representation of $\ln\varphi\left(u\right)$, if it were possible to interchange the limit and integral in the above expression. This is not possible, because as $nF_n\left(\infty\right) = n$, and the expression $nF_n\left(\infty\right) \to \infty$ as $n \to \infty$. To overcome this difficulty, a "damping factor" is introduced via a function $G_n\left(\cdot\right)$, where

$$G_n\left(x\right) = n \int_{-\infty}^{x} \frac{y^2}{1 + y^2} dF_n\left(y\right)$$

Use of the above two expressions yields

$$\ln\varphi\left(u\right) = \lim_{n\to\infty} \int_{-\infty}^{\infty} \left\{e^{iux} - 1\right\} \frac{1 + x^2}{x^2} dG_n\left(x\right)$$

Several properties of the function $G_n\left(\cdot\right)$ are next examined. Assume that the random variable with characteristic function $\varphi\left(\cdot\right)$ has mean $m$, variance $\sigma^2$, and distribution function $F\left(\cdot\right)$. As $\varphi_n\left(u\right) = \varphi^{1/n}\left(u\right)$, the random variable with characteristic function $\varphi_n\left(\cdot\right)$ has mean $m_n = m/n$, and variance $\sigma_n^2 = \sigma^2/n$. This implies that the probability density (mass) function with distribution function $F_n\left(\cdot\right)$ is concentrated near $0$, as $n$ tends towards infinity. Furthermore $x^2/\left(1 + x^2\right)$ is bounded. Thus for large $n$, the contribution to $G_n\left(\infty\right)$ for large $|x|$ can be ignored. Therefore

$$G_n\left(\infty\right) = n \int_{-\infty}^{\infty} \frac{x^2}{1 + x^2} dF_n\left(x\right)$$

$$\simeq n \int_{-\infty}^{\infty} x^2 dF_n\left(x\right) = n\left(\sigma_n^2 + m_n^2\right) = \sigma^2 + \frac{m^2}{n}$$

Thus

$$G_n\left(\infty\right) \simeq \sigma^2$$

In addition, Bienaymé-Chebyshev inequality implies

$$\int_{|x|\geq T} dG_n\left(x\right) = n \int_{|x|\geq T} \frac{x^2}{1 + x^2} dF_n\left(x\right) \simeq n \int_{|x|\geq T} dF_n\left(x\right)$$

$$\simeq n \int_{|x - m_n|\geq T} dF_n\left(x\right) \leq n\frac{\sigma_n^2}{T^2} = \frac{\sigma^2}{T^2}$$

Therefore $\lim_{T\to\infty} \int_{|x|\geq T} dG_n\left(x\right) = 0$. Using this result, and the approximation $G_n\left(\infty\right) \simeq \sigma^2$, it can be shown that

$$\lim_{n\to\infty} \int_{-\infty}^{\infty} g\left(x\right) dG_n\left(x\right) = \int_{-\infty}^{\infty} g\left(x\right) dG\left(x\right)$$

for any continuous bounded function $g\left(\cdot\right)$. Note that the integrand in the expression for $\ln\varphi\left(u\right)$ is not defined properly at $x = 0$. It actually behaves as $iu/x$ in the neighborhood of $x$. Therefore if $iu/x$ is removed from the integrand of $\ln \varphi(u)$ a bounded and continuous function for all values $x \in \mathbb{R}$ is obtained. Thus

$$\ln \varphi(u) = \int_{-\infty}^{\infty} \left\{ e^{iux} - 1 - \frac{iux}{1+x^2} \right\} \frac{1+x^2}{x^2} dG(x)$$

$$+ \lim_{n \to \infty} \int_{-\infty}^{\infty} \frac{iu}{x} dG_n(x)$$

Note that

$$\int_{-\infty}^{\infty} \frac{iu}{x} dG_n(x) = n \int_{-\infty}^{\infty} \frac{iux}{1+x^2} dF_n(x) \simeq niu \int_{-\infty}^{\infty} x dF_n(x) = imu$$

Therefore,

$$\lim_{n \to \infty} \int_{-\infty}^{\infty} \frac{iu}{x} dG_n(x) \simeq imu$$

The approximation for the above limit is not very accurate. It can actually be shown to be equal to $i\gamma u$, where $\gamma$ is some centering constant, which is not necessarily the mean $m$. The stated Lévy-Khinchin representation of the characteristic function $\varphi(\cdot)$ of a random variable is finally obtained.

(b) See Burrill (1972), and Moran (1968). The Lévy representation of the characteristic function $\varphi(\cdot)$ of a random variable can be obtained from its Lévy-Khinchin representation. Define a function $M(\cdot)$ on $x \in \mathbb{R} \setminus \{0\}$ as

$$M(x) = \begin{cases} \int_{-\infty}^{x} \frac{1+y^2}{y^2} dG(y), & x < 0 \\ -\int_{x}^{\infty} \frac{1+y^2}{y^2} dG(y), & x > 0 \end{cases}$$

Also let $\{G(0_+) - G(0_-)\} = \sigma^2$. The Lévy representation of the characteristic function $\varphi(\cdot)$ of a random variable follows.

(c) See Moran (1968). As the second moment is finite, $\varphi(u)$ and $\ln \varphi(u)$ can be differentiated twice. Therefore

$$\left| \lim_{h \to 0} h^{-1} \left\{ \ln \varphi(h/2) - \ln \varphi(-h/2) \right\} \right| < \infty$$

$$\left| \lim_{h \to 0} h^{-2} \left\{ \ln \varphi(h) - 2 \ln \varphi(0) + \ln \varphi(-h) \right\} \right| < \infty$$

Substituting the Lévy-Khinchin representation of the characteristic function $\varphi(\cdot)$ in the above inequalities yields

$$\int_{-\infty}^{\infty} (1+y^2) dG(y) < \infty, \quad \text{and} \quad \left| \int_{-\infty}^{\infty} y dG(y) \right| < \infty$$

Substitution of

$$K(x) = \int_{-\infty}^{x} (1+y^2) dG(y), \quad x \in \mathbb{R}, \quad \text{and} \quad \beta = \gamma + \int_{-\infty}^{\infty} y dG(y)$$

in the Lévy-Khinchin representation yields the Kolmogorov representation of the characteristic function $\varphi(\cdot)$.

31. Prove that a stable characteristic function is always infinitely divisible.

    Hint: See Moran (1968), and Lukacs (1970). If $X$ is a stable random variable, with a characteristic function $\varphi(u)$, $u \in \mathbb{R}$, then

    $$X_1 + X_2 + \cdots + X_n \stackrel{d}{=} bX + c, \quad n \in \mathbb{P} \setminus \{1\}$$

    where $b \in \mathbb{R}^+$ and $c \in \mathbb{R}$; and $X_1, X_2, \ldots, X_n$, are independent random variables with the same distribution as $X$. This implies

    $$\{\varphi(u)\}^n = \exp(icu)\,\varphi(bu)$$

    That is,

    $$\{\varphi(bu)\}^{1/n} = \exp\left(-icn^{-1}u\right)\varphi(u)$$

    This implies

    $$\{\varphi(u)\}^{1/n} = \exp\left(-icb^{-1}n^{-1}u\right)\varphi\left(b^{-1}u\right)$$

    The above equation demonstrates that the distribution is infinitely divisible.

32. Establish the canonical representation of a characteristic function of a stable random variable.

    Hint: See Moran (1968), Lukacs (1970), and Ibragimov, and Linnik (1971). The stated result is obtained in several steps.

    *Step* 0: Let $X$ be a stable random variable with characteristic function $\varphi(u)$, $u \in \mathbb{R}$. As the random variable $X$ is stable, for $b_1, b_2, b \in \mathbb{R}^+$, and $c \in \mathbb{R}$ we have

    $$\varphi(b_1 u)\,\varphi(b_2 u) = e^{icu}\varphi(bu)$$

    As a stable random variable is infinitely divisible, substitution of the above result in its Lévy representation yields

    $$\sigma^2 b^2 = \sigma^2\left(b_1^2 + b_2^2\right)$$
    $$M\left(xb^{-1}\right) = M\left(xb_1^{-1}\right) + M\left(xb_2^{-1}\right), \quad x \in \mathbb{R} \setminus \{0\}$$

    *Step* 1: Assume that $x < 0$. As $b^{-1} > 0$ it can be shown that for any finite set of positive real numbers $\alpha_i$ for $1 \le i \le n$, there exists a positive real number $\alpha_0$ such that

    $$M(\alpha_0 x) = \sum_{i=1}^{n} M(\alpha_i x), \quad \forall\, x > 0$$

    Note that $\alpha_0$ is a function of $\alpha_i$, for $1 \le i \le n$. Letting $\alpha_i = 1$ for $1 \le i \le n$ yields

    $$M(\alpha_0 x) = nM(x)$$

    That is

    $$n^{-1}M(x) = M(x/\alpha_0(n))$$

    Let $r = m/n$ be a positive rational number, where $m, n$ are positive integers. Therefore for every positive rational number $r$ there is a positive real number $D(r)$ such that

    $$rM(x) = M(D(r)x), \quad \text{where } x < 0$$

Note that $D(r)$ is defined for $r > 0$. It is next demonstrated that for $M(x)$ not identically equal to zero, $D(r)$ is nonincreasing for rational values of $r$. Let $r_1$ and $r_2$ be two rational numbers such that $r_1 < r_2$. As $M(x) \geq 0$, we have $r_1 M(x) \leq r_2 M(x)$. This implies $M(D(r_1)x) \leq M(D(r_2)x)$. As $M(x)$ is nondecreasing, and $x < 0$ we have $D(r_1) \geq D(r_2)$, and also $D(r)$ is strictly decreasing, if $M(x)$ is not identically equal to zero. In the rest of the discussion, assume that $M(x)$ is not identically equal to zero. It can also be shown that for any $\lambda > 0$, $D(\lambda_-)$ and $D(\lambda_+)$ must exist, where the limits are taken through rational values. Therefore, it follows that $D(\lambda)$ is continuous and strictly decreasing. As $M(x)$ is nondecreasing, and $\lim_{x \to -\infty} M(x) = 0$, we have

$$\lim_{x \to 0} D(x) \to \infty, \quad D(1) = 1, \quad \text{and} \quad \lim_{x \to \infty} D(x) = 0$$

The strictly decreasing function $D(\lambda) = z$ has an inverse function $B(z) = \lambda$. The function $B(\cdot)$ is defined for $z \geq 0$ and is single valued, and nonnegative. Therefore, we can write

$$\lambda M(x) = M(D(\lambda)x)$$

in terms of $B(z)$, and note that for every real $z > 0$ there corresponds a $B(z) > 0$ such that

$$B(z)M(x) = M(zx)$$

Let $M_1(x)$ and $M_2(x)$ be two solutions of the above equation, and suppose that $M_2(x) \neq 0$. Define

$$R(x) = \frac{M_1(x)}{M_2(x)}$$

Then

$$R(zx) = \frac{M_1(zx)}{M_2(zx)} = \frac{B(z)M_1(x)}{B(z)M_2(x)} = \frac{M_1(x)}{M_2(x)} = R(x)$$

As $z$ is arbitrary, the above result implies that $R(x)$ is a constant independent of $x$. Furthermore $M_1(x) = |x|^{-\alpha_1}$ is a solution, and $B(z) = |z|^{-\alpha_1}$. Consequently, a general solution of $B(z)M(x) = M(zx)$ is

$$M(x) = \xi_1 |x|^{-\alpha_1}$$

Observe that $\lim_{x \to -\infty} M(x) = 0$ implies $\alpha_1 > 0$, and as $M(x)$ is nondecreasing, we have $\xi_1 \geq 0$. Lévy's representation of the characteristic function requires that the integral $\int_{-1}^{0} x^2 dM(x)$ be finite. This implies that $\alpha_1 < 2$. In summary

$$M(x) = \xi_1 |x|^{-\alpha_1}, \quad \xi_1 \geq 0, \quad 0 < \alpha_1 < 2, \quad \text{for } x < 0$$

This solution also includes the possibility that $M(x)$ is identically equal to 0, because, it is possible that $\xi_1 = 0$. Substituting this result in the relationship $M(xb^{-1}) = M(xb_1^{-1}) + M(xb_2^{-1})$ yields

$$\xi_1(b^{\alpha_1} - b_1^{\alpha_1} - b_2^{\alpha_1}) = 0$$

*Step* 2: Assume that $x > 0$. The function $M(x)$ for $x > 0$ can be determined as in Step 1. Thus

$$M(x) = -\xi_2 x^{-\alpha_2}, \quad \xi_2 \geq 0, \quad 0 < \alpha_2 < 2, \quad \text{for } x > 0$$

and

$$\xi_2(b^{\alpha_2} - b_1^{\alpha_2} - b_2^{\alpha_2}) = 0$$

*Step* 3: In Step 0, the relationship $\sigma^2 b^2 = \sigma^2 \left(b_1^2 + b_2^2\right)$ was derived. Consider two cases.

*Case* (*a*): If $\sigma^2 \neq 0$, we have $\left(b^2 - b_1^2 - b_2^2\right) = 0$.

As $0 < \alpha_1 < 2$ and $\xi_1 \left(b^{\alpha_1} - b_1^{\alpha_1} - b_2^{\alpha_1}\right) = 0$ we infer that $\xi_1 = 0$. That is, $M(x) = 0$ for all $x < 0$.

Similarly, as $0 < \alpha_2 < 2$ and $\xi_2 \left(b^{\alpha_2} - b_1^{\alpha_2} - b_2^{\alpha_2}\right) = 0$ we infer that $\xi_2 = 0$. That is, $M(x) = 0$ for all $x > 0$.

*Case* (*b*): Consider scenarios under which $\sigma^2 = 0$. Note in this case that $\xi_1 > 0$ implies that $M(x)$ is not identically zero for $x < 0$. Similarly, $\xi_2 > 0$ implies that $M(x)$ is not identically zero for $x > 0$.

If $0 < \alpha_1 < 2$, we have $\xi_1 \left(b^{\alpha_1} - b_1^{\alpha_1} - b_2^{\alpha_1}\right) = 0$ and $\sigma^2 \left(b^2 - b_1^2 - b_2^2\right) = 0$. Letting $b_1 = b_2 = 1$, we have $\xi_1 \left(b^{\alpha_1} - 2\right) = 0$ and $\sigma^2 \left(b^2 - 2\right) = 0$. If $\xi_1 > 0$, then $\left(b^{\alpha_1} - 2\right) = 0$. This implies $\left(b^2 - 2\right) \neq 0$, which in turn implies $\sigma^2 = 0$.

If $0 < \alpha_2 < 2$, we have $\xi_2 \left(b^{\alpha_2} - b_1^{\alpha_2} - b_2^{\alpha_2}\right) = 0$ and $\sigma^2 \left(b^2 - b_1^2 - b_2^2\right) = 0$. Letting $b_1 = b_2 = 1$, we have $\xi_2 \left(b^{\alpha_2} - 2\right) = 0$ and $\sigma^2 \left(b^2 - 2\right) = 0$. If $\xi_2 > 0$, then $\left(b^{\alpha_2} - 2\right) = 0$. This implies $\left(b^2 - 2\right) \neq 0$, which in turn implies $\sigma^2 = 0$.

Let $\xi_1 > 0$, $\xi_2 > 0$, and $b_1 = b_2 = 1$. Then $\xi_1 \left(b^{\alpha_1} - b_1^{\alpha_1} - b_2^{\alpha_1}\right) = 0$ implies $b^{\alpha_1} = 2$. Similarly $\xi_2 \left(b^{\alpha_2} - b_1^{\alpha_2} - b_2^{\alpha_2}\right) = 0$ implies $b^{\alpha_2} = 2$. Thus $b^{\alpha_1} = b^{\alpha_2}$, which implies $\alpha_1 = \alpha_2$. If $\xi_1 = \xi_2 = 0$, we have $\ln \varphi(u) = i\gamma u$. This condition corresponds to the characteristic function of a degenerate random variable. Therefore $\left(\xi_1 + \xi_2\right) > 0$.

The converse of the theorem can be verified in a straightforward manner

33. Obtain an explicit canonical representation of a characteristic function of a stable random variable.

Hint: See Moran (1968), Lukacs (1970), and Ibragimov, and Linnik (1971). The stated result is obtained for three cases: $0 < \alpha < 1$, $1 < \alpha < 2$, and $\alpha = 1$.

*Case* 1: $0 < \alpha < 1$. Note that the integrals

$$\int_{-\infty}^{0} \frac{x}{1+x^2} \frac{dx}{|x|^{\alpha+1}}, \quad \text{and} \quad \int_{0}^{\infty} \frac{x}{1+x^2} \frac{dx}{x^{\alpha+1}}$$

are finite. Therefore for some $\gamma'$ we have

$$\ln \varphi(u) = i\gamma' u + \alpha\xi_1 \int_{-\infty}^{0} \left(e^{iux} - 1\right) \frac{dx}{|x|^{\alpha+1}} + \alpha\xi_2 \int_{0}^{\infty} \left(e^{iux} - 1\right) \frac{dx}{x^{\alpha+1}}$$

Consider $u > 0$, then

$$\ln \varphi(u) = i\gamma' u + \alpha u^{\alpha} \left\{ \xi_1 \int_{0}^{\infty} \left(e^{-ix} - 1\right) \frac{dx}{x^{\alpha+1}} + \xi_2 \int_{0}^{\infty} \left(e^{ix} - 1\right) \frac{dx}{x^{\alpha+1}} \right\}$$

Observe that the function $\left(e^{iz} - 1\right)/z^{\alpha+1}$, $z \in \mathbb{C}$ is analytic in a region bounded by the contour $\Gamma_c$. This contour consists of the segment $[r_c, R_c]$ of the real axis, where $r_c < R_c$; the circular arc of a circle with radius $R_c$ and center as the origin: $z = R_c e^{i\phi}$, where $0 \leq \phi \leq \pi/2$; the segment $[iR_c, ir_c]$ of the imaginary axis; and the circular arc of a circle with radius $r_c$ and center as the origin: $z = r_c e^{i\phi}$, where $\pi/2 \geq \phi \geq 0$. Use of Cauchy-Goursat theorem yields

$$\oint_{C} \frac{\left(e^{iz} - 1\right)}{z^{\alpha+1}} dz = 0$$

Note that the integrals over the circular arcs tend to zero as $r_c \to 0$, and $R_c \to \infty$. Therefore

$$\int_0^\infty \left(e^{ix} - 1\right) \frac{dx}{x^{\alpha+1}} = e^{-i\pi\alpha/2} L_1\left(\alpha\right)$$

where

$$L_1\left(\alpha\right) = \int_0^\infty \left(e^{-x} - 1\right) \frac{dx}{x^{\alpha+1}} = -\frac{\Gamma\left(1-\alpha\right)}{\alpha} < 0$$

Similarly

$$\int_0^\infty \left(e^{-ix} - 1\right) \frac{dx}{x^{\alpha+1}} = e^{i\pi\alpha/2} L_1\left(\alpha\right)$$

Therefore for $u > 0$, we have

$$\ln \varphi\left(u\right) = i\gamma' u + \alpha L_1\left(\alpha\right)\left(\xi_1 + \xi_2\right)\cos\frac{\pi\alpha}{2} u^\alpha \left\{1 + i\frac{\left(\xi_1 - \xi_2\right)}{\left(\xi_1 + \xi_2\right)} \tan\frac{\pi\alpha}{2}\right\}$$

Substituting

$$\varrho = \gamma', \quad \xi = -\alpha L_1\left(\alpha\right)\left(\xi_1 + \xi_2\right)\cos\frac{\pi\alpha}{2}$$

$$\beta = \frac{\left(\xi_2 - \xi_1\right)}{\left(\xi_1 + \xi_2\right)}$$

we obtain for $u > 0$

$$\ln \varphi\left(u\right) = i\varrho u - \xi u^\alpha \left\{1 - i\beta \tan\frac{\pi\alpha}{2}\right\}$$

For $u < 0$, we use the Hermitian property of the characteristic function. Thus,

$$\ln \varphi\left(u\right) = \ln \overline{\varphi\left(-u\right)} = i\varrho u - \xi \left|u\right|^\alpha \left\{1 + i\beta \tan\frac{\pi\alpha}{2}\right\}$$

Therefore, for every $u \in \mathbb{R}$

$$\ln \varphi\left(u\right) = i\varrho u - \xi \left|u\right|^\alpha \left\{1 - i\beta sign\left(u\right) \tan\frac{\pi\alpha}{2}\right\}$$

where constants $\left|\beta\right| \leq 1$, $\varrho \in \mathbb{R}$, and $\xi \in \mathbb{R}_0^+$.

*Case* 2: $1 < \alpha < 2$. Note that the integrals

$$\int_{-\infty}^0 \frac{x^3}{1+x^2} \frac{dx}{\left|x\right|^{\alpha+1}}, \quad \text{and} \quad \int_0^\infty \frac{x^3}{1+x^2} \frac{dx}{x^{\alpha+1}}$$

are finite. Therefore for some $\gamma''$ we have

$$\ln \varphi\left(u\right) = i\gamma'' u + \alpha\xi_1 \int_{-\infty}^0 \left(e^{iux} - 1 - iux\right) \frac{dx}{\left|x\right|^{\alpha+1}}$$

$$+ \alpha\xi_2 \int_0^\infty \left(e^{iux} - 1 - iux\right) \frac{dx}{x^{\alpha+1}}$$

Consider $u > 0$, then

$$\ln \varphi\left(u\right) = i\gamma'' u + \alpha u^\alpha \left\{\xi_1 \int_0^\infty \left(e^{-ix} - 1 + ix\right) \frac{dx}{x^{\alpha+1}}\right.$$

$$\left. + \xi_2 \int_0^\infty \left(e^{ix} - 1 - ix\right) \frac{dx}{x^{\alpha+1}}\right\}$$

Observe that the function $\left(e^{iz} - 1 - iz\right)/z^{\alpha+1}$, $z \in \mathbb{C}$ is analytic in a region bounded by the contour $\Gamma_c$ described in Case 1. Therefore

$$\int_0^\infty \left(e^{ix} - 1 - ix\right) \frac{dx}{x^{\alpha+1}} = e^{-i\pi\alpha/2} L_2\left(\alpha\right)$$

where

$$L_2\left(\alpha\right) = \int_0^\infty \left(e^{-x} - 1 + x\right) \frac{dx}{x^{\alpha+1}} = \frac{\Gamma\left(2 - \alpha\right)}{\alpha\left(\alpha - 1\right)} > 0$$

Similarly

$$\int_0^\infty \left(e^{-ix} - 1 + ix\right) \frac{dx}{x^{\alpha+1}} = e^{i\pi\alpha/2} L_2\left(\alpha\right)$$

Therefore for $u > 0$, we have

$$\ln\varphi\left(u\right) = i\gamma''u + \alpha L_2\left(\alpha\right)\left(\xi_1 + \xi_2\right)\cos\frac{\pi\alpha}{2}u^\alpha\left\{1 + i\frac{\left(\xi_1 - \xi_2\right)}{\left(\xi_1 + \xi_2\right)}\tan\frac{\pi\alpha}{2}\right\}$$

For $u < 0$, we use the Hermitian property of the characteristic function as in Case 1. Substituting

$$\varrho = \gamma'', \quad \xi = -\alpha L_2\left(\alpha\right)\left(\xi_1 + \xi_2\right)\cos\frac{\pi\alpha}{2}$$

$$\beta = \frac{\left(\xi_2 - \xi_1\right)}{\left(\xi_1 + \xi_2\right)}$$

we obtain for every $u \in \mathbb{R}$

$$\ln\varphi\left(u\right) = i\varrho u - \xi\left|u\right|^\alpha\left\{1 - i\beta sign\left(u\right)\tan\frac{\pi\alpha}{2}\right\}$$

where constants $\left|\beta\right| \leq 1, \varrho \in \mathbb{R}$, and $\xi \in \mathbb{R}_0^+$.

*Case* 3: $\alpha = 1$. In this case note that

$$\int_0^\infty \frac{1 - \cos x}{x^2} dx = \frac{\pi}{2}$$

This result is obtained as follows.

$$\int_0^\infty \frac{1 - \cos x}{x^2} dx = \int_0^\infty \frac{\sin x}{x} dx = \int_{-\infty}^\infty sinc(x)\, dx = \frac{\pi}{2}$$

For $u > 0$, we evaluate the integral $I$, where

$$I = \int_0^\infty \left\{e^{iux} - 1 - \frac{iux}{1 + x^2}\right\} \frac{dx}{x^2}$$

$$= \int_0^\infty \frac{\cos ux - 1}{x^2} dx + i\int_0^\infty \left\{\sin ux - \frac{ux}{1 + x^2}\right\} \frac{dx}{x^2}$$

$$= -\frac{\pi}{2}u + i\lim_{\epsilon \to 0}\left\{\int_\epsilon^\infty \frac{\sin ux}{x^2} dx - u\int_\epsilon^\infty \frac{dx}{x\left(1 + x^2\right)}\right\}$$

$$= -\frac{\pi}{2}u + iu\lim_{\epsilon \to 0}\left\{\int_{\epsilon u}^\infty \frac{\sin x}{x^2} dx - \int_\epsilon^\infty \frac{dx}{x\left(1 + x^2\right)}\right\}$$

$$= -\frac{\pi}{2}u + iu\lim_{\epsilon \to 0}\left\{-\int_\epsilon^{\epsilon u} \frac{\sin x}{x^2} dx + \int_\epsilon^\infty \frac{\sin x}{x^2} dx - \int_\epsilon^\infty \frac{dx}{x\left(1 + x^2\right)}\right\}$$

Note that

$$\widetilde{D} = \int_0^\infty \frac{\sin x}{x^2}\,dx - \int_0^\infty \frac{dx}{x\left(1+x^2\right)}$$

$$= \lim_{\epsilon \to 0} \left\{ \int_\epsilon^\infty \frac{\sin x}{x^2}\,dx - \int_\epsilon^\infty \frac{dx}{x\left(1+x^2\right)} \right\} < \infty$$

and

$$\lim_{\epsilon \to 0} \int_\epsilon^{\epsilon u} \frac{\sin x}{x^2}\,dx = \ln u$$

Thus

$$I = -\frac{\pi}{2}u + iu\left\{ -\ln u + \widetilde{D} \right\}$$

Also note that the integral $J$ is

$$J = \int_{-\infty}^0 \left\{ e^{iux} - 1 - \frac{iux}{1+x^2} \right\} \frac{dx}{x^2}$$

$$= \int_0^\infty \left\{ e^{-iux} - 1 + \frac{iux}{1+x^2} \right\} \frac{dx}{x^2}$$

Thus $J$ is equal to the complex conjugate of $I$. Therefore

$$J = -\frac{\pi}{2}u - iu\left\{ -\ln u + \widetilde{D} \right\}$$

Therefore for $u > 0$

$$\ln \varphi\left(u\right) = i\gamma u + \xi_1 \left\{ -\frac{\pi}{2}u + iu\ln u - iu\widetilde{D} \right\} + \xi_2 \left\{ -\frac{\pi}{2}u - iu\ln u + iu\widetilde{D} \right\}$$

$$= i\gamma'''u - \left(\xi_1 + \xi_2\right)\frac{\pi}{2}u + \left(\xi_1 - \xi_2\right)iu\ln u$$

where $\gamma''' = \left\{ \gamma + \widetilde{D}\left(\xi_2 - \xi_1\right) \right\}$. Using the Hermitian property of the characteristic function, and substituting

$$\varrho = \gamma''', \quad \xi = \frac{\pi}{2}\left(\xi_1 + \xi_2\right)$$

$$\beta = \frac{\left(\xi_2 - \xi_1\right)}{\left(\xi_1 + \xi_2\right)}$$

we obtain for every $u \in \mathbb{R}$

$$\ln \varphi\left(u\right) = i\varrho u - \xi\left|u\right|\left\{ 1 + i\beta sign\left(u\right)\frac{2}{\pi}\ln\left|u\right| \right\}$$

where constants $\left|\beta\right| \leq 1$, $\varrho \in \mathbb{R}$, and $\xi \in \mathbb{R}_0^+$.

34. Consider a symmetric probability density function $f\left(x\right)$, $x \in \mathbb{R}$. Let the corresponding characteristic function be $\varphi\left(u\right)$, $u \in \mathbb{R}$. Also $f\left(x\right) \sim \vartheta\left|x\right|^{-(\alpha+1)}$, $\vartheta \in \mathbb{R}^+$ for $\left|x\right| \to \infty$. Show that $\varphi\left(u\right) \sim 1 - b\left|u\right|^\alpha$, for $\left|u\right| \to 0$, where $b \in \mathbb{R}^+$ is a function of $\vartheta$ and $\alpha$.
Hint: For $h$ a large positive real number,

$$\varphi\left(u\right) = \vartheta \int_{-\infty}^{-h} e^{iux}\left|x\right|^{-(\alpha+1)}\,dx + \int_{-h}^h e^{iux}f\left(x\right)\,dx + \vartheta \int_h^\infty e^{iux}x^{-(\alpha+1)}\,dx$$

Consider three cases: $0 < \alpha < 1$, $1 < \alpha < 2$, and $\alpha = 1$.

*Case* 1: $0 < \alpha < 1$. For small positive values of $u$

$$\varphi(u) \sim \vartheta \int_{-\infty}^{-h} |x|^{-(\alpha+1)}\,dx + \int_{-h}^{h} e^{iux} f(x)\,dx + \vartheta \int_{h}^{\infty} x^{-(\alpha+1)}dx$$

$$+\vartheta \int_{-\infty}^{-h} \left\{ e^{iux} - 1 \right\} |x|^{-(\alpha+1)}\,dx + \vartheta \int_{h}^{\infty} \left\{ e^{iux} - 1 \right\} x^{-(\alpha+1)}dx$$

$$\sim 1 + \vartheta \int_{h}^{\infty} \left\{ e^{-iux} - 1 \right\} x^{-(\alpha+1)}dx + \vartheta \int_{h}^{\infty} \left\{ e^{iux} - 1 \right\} x^{-(\alpha+1)}dx$$

As in the last problem, and using its notation, for small positive values of $u$ we have

$$\int_{h}^{\infty} \left\{ e^{iux} - 1 \right\} x^{-(\alpha+1)}dx$$

$$= u^{\alpha} \int_{uh}^{\infty} \left\{ e^{ix} - 1 \right\} x^{-(\alpha+1)}dx \sim u^{\alpha} e^{-i\pi\alpha/2} L_1(\alpha)$$

where $L_1(\cdot)$ is defined in the last problem. It takes negative values. Therefore using the Hermitian property of a characteristic function, we have

$$\varphi(u) \sim 1 + 2\vartheta L_1(\alpha) \cos(\pi\alpha/2) |u|^{\alpha}, \quad \text{as } |u| \to 0$$

*Case* 2: $1 < \alpha < 2$. For small positive values of $u$

$$\varphi(u) \sim 1 + \vartheta \int_{-\infty}^{-h} \left\{ e^{iux} - 1 - iux \right\} |x|^{-(\alpha+1)}\,dx$$

$$+\vartheta \int_{h}^{\infty} \left\{ e^{iux} - 1 - iux \right\} x^{-(\alpha+1)}dx$$

This simplifies to

$$\varphi(u) \sim 1 + \vartheta \int_{h}^{\infty} \left\{ e^{-iux} - 1 + iux \right\} x^{-(\alpha+1)}dx$$

$$+\vartheta \int_{h}^{\infty} \left\{ e^{iux} - 1 - iux \right\} x^{-(\alpha+1)}dx$$

As in the last problem, and using its notation, for small positive values of $u$ we have

$$\varphi(u) \sim 1 + 2\vartheta L_2(\alpha) \cos(\pi\alpha/2) |u|^{\alpha}, \quad \text{as } |u| \to 0$$

where $L_2(\cdot)$ is defined in the last problem. It takes positive values.

*Case* 3: $\alpha = 1$. As in the last problem, it can be shown that

$$\varphi(u) \sim 1 - \pi\vartheta |u|, \quad \text{as } |u| \to 0$$

The corresponding values of the constant $b$ can be determined from the above results.

35. There is a sequence of independent and identically distributed random variables $X_1, X_2, \ldots,$ and sequences of positive real numbers $\{b_n\}$ and real numbers $\{a_n\}$ such that

$$\frac{X_1 + X_2 + \cdots + X_n}{b_n} + a_n$$

converges to a stable random variable with index $\alpha$. Show that $b_n \simeq n^{1/\alpha}$, where $\alpha \in (0,2]$. Hint: See Ibragimov, and Linnik (1971). Let the characteristic function of the random variable $X_i$ be $\psi(\cdot)$. Then for $\alpha \in (0,2]$, $\xi \in \mathbb{R}_0^+$, and $\forall u \in \mathbb{R}$ we have

$$\left| \psi\left( \frac{u}{b_n} \right) \right|^n = \exp\left\{ -\xi |u|^\alpha \right\} (1 + o(1))$$

That is

$$|\psi(v)| = \exp\left\{ -\frac{\xi}{n} |vb_n|^\alpha \right\} (1 + o(1))$$

For large $n$, we should have $b_n \simeq n^{1/\alpha}$.

36. Let $X_1, X_2, \ldots, X_n$ be a sequence of $n$ independent and identically distributed random variables. Denote a representative of these random variables by $X$. The probability density function of the random variable $X$ is denoted by $f(x)$, for $x \in \mathbb{R}$. The probability density function $f(\cdot)$ is symmetric about the $y$-axis. Define a random variable $Z_n$ as

$$Z_n = \frac{1}{n^{1/\alpha}} \sum_{i=1}^n X_i$$

where $\alpha \in (0,2)$. As $n \to \infty$, the random variable $Z_n$ tends to a stable random variable $Z$. Show that the tail of the probability density function $f(\cdot)$ is given by $f(x) \sim \vartheta |x|^{-(\alpha+1)}$, $\vartheta \subset \mathbb{R}^+$ as $|x| \to \infty$.

Hint: An informal explanation of this result is provided. Denote the characteristic function of the random variable $Z$ by $\varphi_Z(\cdot)$. The probability density function of $Z$ is symmetric about the $y$-axis. Therefore

$$\varphi_Z(u) \to e^{-b|u|^\alpha}, \quad b \in \mathbb{R}^+, \quad u \in \mathbb{R}$$

The function $e^{-b|u|^\alpha}$ can be written as

$$e^{-b|u|^\alpha} = \lim_{n \to \infty} \left\{ 1 - b\frac{|u|^\alpha}{n} \right\}^n$$

$$= \lim_{n \to \infty} \left\{ 1 - b\left| \frac{u}{n^{1/\alpha}} \right|^\alpha \right\}^n$$

Therefore, from the definition of the random variable $Z_n$, we observe that as $n \to \infty$ the characteristic function of $X/n^{1/\alpha}$ is $\left\{ 1 - b|u/n^{1/\alpha}|^\alpha \right\}$ for $|u| = o(n^{1/\alpha})$. That is, the characteristic function of the random variable $X$ is $\varphi_X(u) \sim \left\{ 1 - b|u|^\alpha \right\}$ for small values of $|u|$. It has been shown in an earlier problem that for $\alpha \in (0,2)$, this characteristic function corresponds to a probability density function $f(x) \sim \vartheta |x|^{-(\alpha+1)}$, $\vartheta \in \mathbb{R}^+$, where $|x| \to \infty$, and $\vartheta$ is some function of $b$ and $\alpha$.

37. Let $X$ be a $\alpha$-stable random variable, where $X \sim S_\alpha(\sigma, \beta, \varrho)$. Establish the following results.

(a) $\mathcal{E}\left( |X|^\delta \right) < \infty$, if $0 \le \delta < \alpha$, and $\alpha \in (0,2)$.

(b) $\mathcal{E}\left( |X|^\delta \right) = \infty$, if $\delta \ge \alpha$, and $\alpha \in (0,2)$.

(c) $\mathcal{E}\left( |X|^\delta \right) < \infty$, for all $\delta \ge 0$, if $\alpha = 2$.

Hint: See Nikias, and Shao (1995).

(a) Let $0 \leq \delta < \alpha$, and $\alpha \in (0, 2)$. As $|X|$ is a nonnegative random variable,

$$
\mathcal{E}\left(|X|^{\delta}\right) = \int_{0}^{\infty} P\left(|X|^{\delta} > t\right) dt = \int_{0}^{\infty} P\left(|X| > t^{1/\delta}\right) dt
$$
$$
= \int_{0}^{\infty} \delta y^{\delta-1} P\left(|X| > y\right) dy
$$

Note that

$$
y^{\delta-1} P\left(|X| > y\right) = O\left(y^{\delta-1}\right), \qquad \text{as } y \to 0
$$
$$
y^{\delta-1} P\left(|X| > y\right) = O\left(y^{\delta-\alpha-1}\right), \quad \text{as } y \to \infty
$$

Therefore $\mathcal{E}\left(|X|^{\delta}\right) < \infty$ if $0 \leq \delta < \alpha$, and $\alpha \in (0, 2)$.

(b) Let $\delta \geq \alpha$, and $\alpha \in (0, 2)$. Observe that for some $w > 0$

$$
\mathcal{E}\left(|X|^{\delta}\right) \geq \int_{w}^{\infty} \delta y^{\delta-\alpha-1} dy
$$

The right-hand side of the above inequality tends to infinity if $\delta \geq \alpha$. The result follows.

(c) The $\alpha$-stable random variable for $\alpha = 2$ is the normal random variable, as $\omega\left(u, 2\right) = 0$.

38. Let $X_1, X_2, \ldots, X_n$ be a sequence of independent and continuously distributed random variables. If

$$
U = \min\left\{X_1, X_2, \ldots, X_n\right\}, \text{ and } V = \max\left\{X_1, X_2, \ldots, X_n\right\}
$$

find the cumulative distribution and probability density functions of the random variables $U$ and $V$.

Hint: See Mood, Graybill, and Boes (1974). Let the cumulative distribution and probability density functions of the random variable $X_i$, be $F_{X_i}\left(\cdot\right)$ and $f_{X_i}\left(\cdot\right)$ respectively, for $1 \leq i \leq n$. Also, let the cumulative distribution and probability density functions of the random variable $U$, be $F_U\left(\cdot\right)$ and $f_U\left(\cdot\right)$ respectively. We have

$$
1 - F_U\left(u\right) = P\left(U > u\right) = P\left(\min\left\{X_1, X_2, \ldots, X_n\right\} > u\right)
$$
$$
= P\left(X_1 > u, X_2 > u, \ldots, X_n > u\right)
$$
$$
= P\left(X_1 > u\right) P\left(X_2 > u\right) \ldots P\left(X_n > u\right)
$$

The last step follows from the independence hypothesis of the sequence of the random variables $X_1, X_2, \ldots, X_n$. This leads to

$$
1 - F_U\left(u\right) = \prod_{i=1}^{n}\left\{1 - F_{X_i}\left(u\right)\right\}, \quad u \in \mathbb{R}
$$

Thus

$$
F_U\left(u\right) = 1 - \prod_{i=1}^{n}\left\{1 - F_{X_i}\left(u\right)\right\}, \quad u \in \mathbb{R}
$$

The corresponding probability density function is obtained by differentiating both sides of the above expression with respect to $u$. Further assume that the random variables $X_i$'s are distributed as random variable $X$. Let the cumulative distribution and probability density functions of the random variable $X$, be $F\left(\cdot\right)$ and $f\left(\cdot\right)$ respectively. In this case

$$F_U(u) = 1 - \{1 - F(u)\}^n, \text{ and } f_U(u) = n\{1 - F(u)\}^{n-1} f(u), \quad u \in \mathbb{R}$$

Similarly, let the cumulative distribution and probability density functions of the random variable $V$, be $F_V(\cdot)$ and $f_V(\cdot)$ respectively. We have

$$
\begin{aligned}
F_V(v) &= P(V \le v) \\
&= P(\max\{X_1, X_2, \ldots, X_n\} \le v) \\
&= P(X_1 \le v, X_2 \le v, \ldots, X_n \le v) \\
&= P(X_1 \le v) P(X_2 \le v) \ldots P(X_n \le v)
\end{aligned}
$$

The last step follows from the independence hypothesis of the sequence of random variables $X_1, X_2, \ldots, X_n$. This leads to

$$F_V(v) = \prod_{i=1}^{n} F_{X_i}(v), \quad v \subset \mathbb{R}$$

The corresponding probability density function is obtained by differentiating both sides of the above expression with respect to $v$. Further assume that the random variables $X_i$'s are distributed as random variable $X$. Let the cumulative distribution and probability density functions of the random variable $X$, be $F(\cdot)$ and $f(\cdot)$ respectively. In this case

$$F_V(v) = \{F(v)\}^n, \text{ and } f_V(v) = n\{F(v)\}^{n-1} f(v), \quad v \in \mathbb{R}$$

39. Let $X_1, X_2, \ldots, X_n$ be $n$ independent and identically distributed continuous random variables, with mean 0, and standard deviation $\sigma$. The cumulative distribution and probability density functions of $X_i$ are $F(\cdot)$ and $f(\cdot)$ respectively. The density function $f(\cdot)$ is symmetric about its mean. Define $V = \max\{X_1, X_2, \ldots, X_n\}$, and $U = \min\{X_1, X_2, \ldots, X_n\}$. Prove that $\mathcal{E}(U) = -\mathcal{E}(V)$.
Hint: The probability density function $f(\cdot)$ is symmetric about its mean. Therefore $f(x) = f(-x)$, and $F(-x) = 1 - F(x)$. Then

$$
\begin{aligned}
\mathcal{E}(V) &= n \int_{-\infty}^{\infty} x \{F(x)\}^{n-1} f(x)\, dx \\
\mathcal{E}(U) &= n \int_{-\infty}^{\infty} x \{1 - F(x)\}^{n-1} f(x)\, dx \\
&= n \int_{-\infty}^{\infty} x \{F(-x)\}^{n-1} f(-x)\, dx \\
&= -n \int_{-\infty}^{\infty} y \{F(y)\}^{n-1} f(y)\, dy = -\mathcal{E}(V)
\end{aligned}
$$

40. Let $X_1, X_2, \ldots, X_n$ be $n$ independent and identically distributed continuous random variables. Let the probability density and cumulative distribution function of the random variable $X_i$ be $f(\cdot)$ and $F(\cdot)$ respectively. Define

$$
\begin{aligned}
U &= \min\{X_1, X_2, \ldots, X_n\} \\
V &= \max\{X_1, X_2, \ldots, X_n\} \\
R &= (V - U)
\end{aligned}
$$

Prove that the expected value of $R$ is

$$\mathcal{E}(R) = \int_{-\infty}^{\infty} [1 - \{F(x)\}^n - \{1 - F(x)\}^n]\, dx$$

Hint: Let $f_U(\cdot)$ and $f_V(\cdot)$ be the probability density functions of the random variables $U$ and $V$ respectively.

$$\mathcal{E}(R) = \mathcal{E}(V) - \mathcal{E}(U)$$
$$= \int_{-\infty}^{\infty} x\,(f_V(x) - f_U(x))\, dx$$
$$= n \int_{-\infty}^{\infty} x \left[\{F(x)\}^{n-1} - \{1 - F(x)\}^{n-1}\right] f(x)\, dx$$

We first evaluate

$$\mathcal{I}(N) = n \int_{-N}^{N} x \left[\{F(x)\}^{n-1} - \{1 - F(x)\}^{n-1}\right] f(x)\, dx$$

Note that if $N \to \infty$, then $\mathcal{I}(N) \to \mathcal{E}(R)$. Integration by parts yields

$$\mathcal{I}(N) = \int_{-N}^{N} x\, d\,[\{F(x)\}^n + \{1 - F(x)\}^n]$$

$$= (x\,[\{F(x)\}^n + \{1 - F(x)\}^n])\,|_{-N}^{N} - \int_{-N}^{N} [\{F(x)\}^n + \{1 - F(x)\}^n]\, dx$$

As $x \to \pm\infty$, we have $[\{F(x)\}^n + \{1 - F(x)\}^n] \to 1$. Therefore if $N \to \infty$, the result follows.

41.  Establish the list of observations about the large deviation rate function $\ell(\cdot)$, of the random variable $X$. The random variable $X$ has finite mean $\mu$, and finite standard deviation $\sigma$.

(a) $\ell(x) \geq 0$.
(b) $\ell(x)$ is convex.
(c) $\ell(x)$ has its minimum value at $x = \mu$. Also $\ell(\mu) = 0$.
(d) Let $\ell''(x)$ be the second derivative of $\ell(x)$ with respect to $x$. Then $\ell''(\mu) = 1/\sigma^2$.
(e) $\ell(x) = \sup_{t \geq 0}(xt - \Upsilon(t))$ for $x \geq \mu$, and $\ell(x) = \sup_{t \leq 0}(xt - \Upsilon(t))$ for $x \leq \mu$.
(f) For $x \leq \mu$, $\ell(x)$ is nonincreasing, and for $x \geq \mu$, $\ell(x)$ is nondecreasing.

Hint: See Bucklew (1990), and Hollander (2000).

(a) This follows from the observation that at $t = 0$, $xt - \Upsilon(t) = 0$.
(b) The large deviation rate function $\ell(\cdot)$ is a supremum of a collection of convex functions. Hence $\ell(\cdot)$ is a convex function. Alternately, let $\lambda \in [0, 1]$; then

$$\ell(\lambda x_1 + (1 - \lambda) x_2)$$
$$= \sup_t \{t(\lambda x_1 + (1 - \lambda) x_2) - \Upsilon(t)\}$$
$$= \sup_t \{t\lambda x_1 - \lambda \Upsilon(t) + t(1 - \lambda) x_2 - (1 - \lambda)\Upsilon(t)\}$$
$$\leq \sup_t \{\lambda(t x_1 - \Upsilon(t))\} + \sup_t \{(1 - \lambda)(t x_2 - \Upsilon(t))\}$$
$$= \lambda \ell(x_1) + (1 - \lambda)\ell(x_2)$$

(c) Observe that $\mathcal{M}_X(0) = 1$. Therefore $\Upsilon(0) = 0$, and

$$\ell(x) \geq \{0x - \Upsilon(0)\} = 0$$

As $e^{tx}$ is convex in $x$ for any $t \in \mathbb{R}$, use of Jensen's inequality yields $\mathcal{M}_X(t) \geq e^{\mu t}$ for all $t$. Consequently $\Upsilon(t) \geq \mu t$, that is $\{\mu t - \Upsilon(t)\} \leq 0$ for all $t$. Thus $\ell(\mu) = 0$ and $\ell(x) \geq \ell(\mu)$ for all values of $x$.

(d) Let $m(t) = (xt - \Upsilon(t))$. Denote the first and second derivatives of $\ell(x)$ with respect to $x$ by $\ell'(x)$ and $\ell''(x)$ respectively. Similarly denote the first derivative of $m(t)$, $\mathcal{M}_X(t)$, and $\Upsilon(t)$ with respect to $t$ by $m'(t)$, $\mathcal{M}'_X(t)$, and $\Upsilon'(t)$ respectively. Denote the second derivative of $\mathcal{M}_X(t)$, and $\Upsilon(t)$ with respect to $t$ by $\mathcal{M}''_X(t)$, and $\Upsilon''(t)$ respectively. The third derivative of $\Upsilon(t)$ with respect to $t$ is $\Upsilon'''(t)$.
Therefore $\ell(x) = \sup_{t \in \mathbb{R}} m(t)$ at $m'(t) = 0$. That is

$$x = \Upsilon'(t) = \frac{\mathcal{M}'_X(t)}{\mathcal{M}_X(t)}$$

Thus $\ell(x) = (t\Upsilon'(t) - \Upsilon(t))$, and

$$\ell'(x) = \frac{d\ell(x)}{dx} = \frac{d\ell(x)}{dt}\frac{dt}{dx} = t\Upsilon''(t)\frac{dt}{dx}$$

$$\ell''(x) = \frac{d\ell'(x)}{dx}$$

$$= \Upsilon''(t)\left(\frac{dt}{dx}\right)^2 + t\Upsilon'''(t)\left(\frac{dt}{dx}\right)^2 + t\Upsilon''(t)\frac{d^2t}{dx^2}$$

Observe that at $t = 0$, $\Upsilon'(0) = \mu$. Thus if $x = \mu$, then $t = 0$. It can be shown that $dt/dx = 1/\Upsilon''(t)$, $\Upsilon''(0) = \sigma^2$, and therefore $dt/dx|_{t=0} = 1/\sigma^2$. Finally

$$\ell''(\mu) = \Upsilon''(t)\left(\frac{dt}{dx}\right)^2\bigg|_{t=0} = \frac{1}{\sigma^2}$$

(e) Recall that $e^x$ is an increasing function of $x$. Therefore $e^{tx}$ is convex in $x$ for any $t \in \mathbb{R}$. Consequently by Jensen's inequality $\mathcal{E}(e^{tX}) \geq e^{t\mu}$, where equality holds for $t = 0$. Thus $\Upsilon(t) \geq t\mu$, and therefore $(tx - \Upsilon(t)) \leq t(x - \mu)$.
Consider two subcases.
In the first subcase, let $x > \mu$. If $t \geq 0$, the above inequality yields $tx - \Upsilon(t) \leq 0$. Equality occurs at $t = 0$. Thus $(tx - \Upsilon(t))$ is nonpositive and achieves a maximum value of $0$ at $t = 0$.
In the second subcase, let $x = \mu$. Then the above inequality yields $tx - \Upsilon(t) \leq 0$ for any $t \in \mathbb{R}$.
Combining the two subcases yields $\ell(x) = \sup_{t \geq 0}(xt - \Upsilon(t))$ for $x \geq \mu$.
The result $\ell(x) = \sup_{t \leq 0}(xt - \Upsilon(t))$ for $x \leq \mu$ is proved similarly.

(f) Follows from parts (b) and (c).

42. Determine the large deviation rate function for the following random variables.
    (a) Bernoulli random variable.
    (b) Poisson random variable.
    (c) Normally distributed random variable.

(d) Exponentially distributed random variable.

43. Prove Cramér's theorem.

   Hint: See Bucklew (1990), Shwartz, and Weiss (1995), and Walrand, and Varaiya (2000).

   In order to demonstrate the existence of the limit stated in the theorem, upper and lower bounds are established.

   Upper bound: We prove that for all $n \geq 1$, and $\forall\, a > \mu$

   $$P\left(\tilde{X}_n \geq na\right) \leq \exp\left(-n\ell\left(a\right)\right)$$

   This result follows immediately from the lemma and observation about $\ell\left(a\right)$.

   Lower bound: We prove that for every $\epsilon > 0$ there exists an integer $n_0$ such that, for all $n > n_0$

   $$P\left(\tilde{X}_n \geq na\right) \geq \exp\left(-n\left(\ell\left(a\right) + \epsilon\right)\right)$$

   Let $t_c$ be the value of $t$ for which the supremum in the rate function $\ell\left(\cdot\right)$ occurs. That is

   $$\ell\left(a\right) = at_c - \Upsilon\left(t_c\right)$$

   Also $a = \Upsilon'\left(t_c\right)$, where $\Upsilon'\left(t\right)$ is the first derivative of $\Upsilon\left(t\right)$ with respect to $t$. In the next step we transform the random variable $X$ to $Y$. Let $f_X\left(\cdot\right)$ and $f_Y\left(\cdot\right)$ be the probability density functions of the random variables $X$ and $Y$ respectively. Define

   $$f_Y\left(y\right) = \frac{e^{t_c y} f_X\left(y\right)}{\mathcal{M}_X\left(t_c\right)}$$

   Note that the expression $\mathcal{M}_X\left(t_c\right)$ is in the denominator to normalize the probability density function of the random variable $Y$. This distribution is also called the *twisted distribution*. The mean of the random variable $Y$ is

   $$
   \begin{aligned}
   \mathcal{E}\left(Y\right) &= \int_{-\infty}^{\infty} y f_Y\left(y\right) dy \\
   &= \int_{-\infty}^{\infty} y \frac{e^{t_c y} f_X\left(y\right)}{\mathcal{M}_X\left(t_c\right)} dy \\
   &= \frac{1}{\mathcal{M}_X\left(t_c\right)} \frac{d}{dt} \int_{-\infty}^{\infty} e^{ty} f_X\left(y\right) dy \Big|_{t=t_c} \\
   &= \frac{\mathcal{M}_X'\left(t_c\right)}{\mathcal{M}_X\left(t_c\right)} = \Upsilon'\left(t\right)\big|_{t=t_c} = a
   \end{aligned}
   $$

   where $\mathcal{M}_X'\left(t\right)$ is the first derivative of $\mathcal{M}_X\left(t\right)$ with respect to $t$. Thus $\mathcal{E}\left(Y\right) = a$. A lower bound on $P\left(\tilde{X}_n \geq na\right)$ is determined as follows.

   $$P\left(\tilde{X}_n \geq na\right) = \underset{x_1 + x_2 + \ldots + x_n \geq na}{\int \int \cdots \int} \prod_{i=1}^{n} f_X\left(x_i\right) \prod_{i=1}^{n} dx_i$$

   Thus

   $$P\left(\tilde{X}_n \geq na\right) = \left\{\mathcal{M}_X\left(t_c\right)\right\}^n \underset{x_1 + x_2 + \ldots + x_n \geq na}{\int \int \cdots \int} e^{-t_c \sum_{i=1}^{n} x_i} \prod_{i=1}^{n} f_Y\left(x_i\right) \prod_{i=1}^{n} dx_i$$

Let $\epsilon' > 0$ and use it to restrict the range of the above multiple integrals. Therefore

$$P\left(\widetilde{X}_n \geq na\right)$$

$$\geq \{\mathcal{M}_X(t_c)\}^n \int\int\cdots\int_{n(a+\epsilon')\geq x_1+x_2+\ldots+x_n\geq na} e^{-t_c \sum_{i=1}^n x_i} \prod_{i=1}^n f_Y(x_i) \prod_{i=1}^n dx_i$$

$$\geq \{\mathcal{M}_X(t_c)\}^n e^{-nt_c(a+\epsilon')} \int\int\cdots\int_{n(a+\epsilon')\geq x_1+x_2+\ldots+x_n\geq na} \prod_{i=1}^n f_Y(x_i) \prod_{i=1}^n dx_i$$

As the random variables $Y_i$'s have mean $a$ and finite variance, as $n \to \infty$ the Gaussian central limit theorem implies

$$\int\int\cdots\int_{n(a+\epsilon')\geq x_1+x_2+\ldots+x_n\geq na} \prod_{i=1}^n f_Y(x_i) \prod_{i=1}^n dx_i$$

$$= P\left(n(a+\epsilon') \geq Y_1 + Y_2 + \ldots + Y_n \geq na\right)$$

$$= P\left(\sqrt{n}\epsilon' \geq \{(Y_1 + Y_2 + \ldots + Y_n) - na\}/\sqrt{n} \geq 0\right) \to \frac{1}{2}$$

Select $n_0$ such that the probability exceeds $1/4$ whenever $n \geq n_0$. Note that $n_0$ is dependent upon $\epsilon'$. Therefore for $n \geq n_0$

$$P\left(\widetilde{X}_n > na\right) \geq \frac{1}{4}\exp\left(-n\left(\ell(a) + \epsilon' t_c\right)\right)$$

As $t_c > 0$, select $\epsilon'$ such that $\frac{1}{4}\exp\left(-n\epsilon' t_c\right) \geq \exp\left(-n\epsilon t_c\right)$ whenever $n \geq n_0$, where $\epsilon > 0$. We finally obtain

$$P\left(\widetilde{X}_n \geq na\right) \geq \exp\left(-n\left(\ell(a) + \epsilon\right)\right)$$

This result is indeed the lower bound for the probability $P\left(\widetilde{X}_n \geq na\right)$. Combination of the lower and upper bound yields the final result.

44. Prove Bahadur-Rao theorem.

Hint: See Bucklew (1990), Shwartz, and Weiss (1995), and Walrand, and Varaiya (2000).
Let $t_c$ be the value of $t$ at which the supremum in the rate function $\ell(\cdot)$ occurs. That is, $\ell(a) = (at_c - \Upsilon(t_c))$. Also $a = \Upsilon'(t_c)$, where $\Upsilon'(t)$ is the first derivative of $\Upsilon(t)$ with respect to $t$. In the next step we transform the random variable $X$ to $Y$. Let $f_X(\cdot)$ and $f_Y(\cdot)$ be the probability density functions of the random variables $X$ and $Y$ respectively. Define

$$f_Y(y) = \frac{e^{t_c y} f_X(y)}{\mathcal{M}_X(t_c)}$$

Note that the expression $\mathcal{M}_X(t_c)$ is in the denominator to normalize the probability density function of the random variable $Y$. The mean of the random variable $Y$ can be calculated as in the last problem. It is $\mathcal{E}(Y) = a$. Similarly, the second moment of the random variable $Y$ is

$$\mathcal{E}(Y^2) = \int_{-\infty}^{\infty} y^2 f_Y(y)\,dy = \int_{-\infty}^{\infty} y^2 \frac{e^{t_c y} f_X(y)}{\mathcal{M}_X(t_c)}\,dy$$

$$= \frac{1}{\mathcal{M}_X(t_c)} \frac{d^2}{dt^2} \int_{-\infty}^{\infty} e^{ty} f_X(y)\,dy\bigg|_{t=t_c}$$

$$= \frac{\mathcal{M}_X''(t_c)}{\mathcal{M}_X(t_c)} = \left\{\Upsilon''(t) + (\Upsilon'(t))^2\right\}\bigg|_{t=t_c}$$

where $\mathcal{M}''_X(\cdot)$ and $\Upsilon''(\cdot)$ are the second derivatives of $\mathcal{M}_X(\cdot)$ and $\Upsilon(\cdot)$ respectively. Thus the variance of $Y$ is given by $\sigma_c^2 = \Upsilon''(t_c)$. Also define a random variable $W_j = (Y_j - a)$. Then $\mathcal{E}(W_j) = 0$ and $Var(W_j) = \sigma_c^2$. Define

$$\widetilde{Y}_n = \sum_{j=1}^n Y_j, \text{ and } \widetilde{W}_n = \sum_{j=1}^n W_j$$

Therefore $\widetilde{W}_n = \left(\widetilde{Y}_n - na\right)$. Thus

$$\mathcal{E}\left(\widetilde{Y}_n\right) = na, \ Var\left(\widetilde{Y}_n\right) = n\sigma_c^2, \ \mathcal{E}\left(\widetilde{W}_n\right) = 0, \text{ and } Var\left(\widetilde{W}_n\right) = n\sigma_c^2$$

Recall that we have defined the indicator function of an event $A$ by $I_A$, such that the expectation of this function is equal to its probability. The probability $P\left(\widetilde{X}_n \geq na\right)$ is determined as follows.

$$\begin{aligned}
P\left(\widetilde{X}_n \geq na\right) &= \mathcal{E}\left(I_{\widetilde{X}_n \geq na}\right) \\
&= \{\mathcal{M}_X(t_c)\}^n \mathcal{E}\left(e^{-t_c \widetilde{Y}_n} I_{\widetilde{Y}_n \geq na}\right) \\
&= e^{-n(at_c - \Upsilon(t_c))} \mathcal{E}\left(e^{-t_c \widetilde{W}_n} I_{\widetilde{W}_n \geq 0}\right) \\
&= e^{-n\ell(a)} \mathcal{E}\left(e^{-t_c \widetilde{W}_n} I_{\widetilde{W}_n \geq 0}\right)
\end{aligned}$$

Next define a random variable $V = \widetilde{W}_n / (\sigma_c \sqrt{n})$. Therefore the mean and variance of this random variable are $\mathcal{E}(V) = 0$ and $Var(V) = 1$ respectively. If $n \to \infty$, then the distribution of the random variable $V$ has a standard normal distribution as per the Gaussian central limit theorem. Therefore

$$\begin{aligned}
P\left(\widetilde{X}_n \geq na\right) &= e^{-n\ell(a)} \mathcal{E}\left(e^{-t_c \sigma_c \sqrt{n} V} I_{V \geq 0}\right) \\
&= e^{-n\ell(a)} \frac{1}{\sqrt{2\pi}} \int_0^\infty e^{-t_c \sigma_c \sqrt{n} v} e^{-v^2/2} dv \\
&= \frac{1}{\sqrt{2\pi n} \sigma_c t_c} \exp\left(-n\ell(a)\right)(1 + o(1))
\end{aligned}$$

where we used the approximation

$$\int_b^\infty e^{-z^2/2} dz \sim \frac{1}{b} e^{-b^2/2}, \text{ for } b \gg 1$$

## References

1. Allen, A., 1990. *Probability, Statistics, and Queueing Theory*, Second Edition, Academic Press, New York, New York.

2. Alon, N., and Spencer, J. H., 2000. *The Probabilistic Method*, Second Edition, John Wiley & Sons, Inc., New York, New York.

3. Bahadur, R. R., and Rao, R. R., 1960. "On Deviations of the Sample Mean," Ann. Math. Statis., Vol. 31, No. 4, pp. 1015-1027.

4. Breiman, L., 1992. *Probability*, Society of Industrial and Applied Mathematics, Philadelphia, Pennsylvania.

5. Bucklew, J. A., 1990. *Large Deviation Techniques in Decision, Simulation, and Estimation,* John Wiley & Sons, Inc., New York, New York.

6. Burrill, C. W., 1972. *Measure, Integration, and Probability*, McGraw-Hill Book Company, New York, New York.

7. Chernoff, H., 1952. "A Measure of Asymptotic Efficiency for Tests of a Hypothesis Based on the Sum of Observations," Ann. Math. Statis., Vol. 23, No. 4, pp. 493-507.

8. Chung, K. L., 1974a. *A Course in Probability Theory*, Second Edition, Academic Press, New York, New York.

9. Chung, K. L., 1974b. *Elementary Probability Theory with Stochastic Processes*, Springer-Verlag, Berlin, Germany.

10. Cramér, H., 1938. "Sur un Nouveaux Theorème-Limite de la Théorie des Probabilités," Actualités Scientifiques et Industrielles 736, Colloque Consacré à la Théorie des Probabilités, Hermann, Paris, October 1937, Vol. 3, pp. 5-23.

11. Debnath, L., and Bhatta, D., 2007. *Integral Transforms and Their Applications*, Second Edition, Chapman and Hall/CRC Press, New York, New York.

12. Durrett, R., 2005. *Probability Theory and Examples*, Third Edition, Thomas Learning - Brooks/Cole, Belmont, California.

13. Feller, W., 1968. *An Introduction to Probability Theory and Its Applications*, *Vol. I*, Third Edition, John Wiley & Sons, Inc., New York, New York.

14. Feller, W., 1971. *An Introduction to Probability Theory and Its Applications*, *Vol. II*, Second Edition, John Wiley & Sons, Inc., New York, New York.

15. Gnedenko, B. V., and Kolmogorov, A. N., 1968. *Limit Distributions for Sums of Independent Random Variables,* Addison-Wesley Publishing Company, New York, New York.

16. Gnedenko, B. V., 1978. *The Theory of Probability*, Mir Publishers, Moscow.

17. Grimmett, G. R., and Stirzaker, D. R., 2001. *Probability and Random Processes,* Third Edition, Oxford University Press, Oxford, Great Britain.

18. Grinstead, C. M., and Snell, J. L., 1997. *Introduction to Probability*, Second Edition, American Mathematical Society, Providence, Rhode Island.

19. Gubner, J. A., 2006. *Probability and Random Processes for Electrical and Computer Engineers*, Cambridge University Press, Cambridge, Great Britain.

20. Hall, P., 1981. "A Comedy of Errors: The Canonical Form for a Stable Characteristic Function," Bulletin London Mathematical Society, Vol. 31, Issue 1, pp. 23-27.

21. Hamming, R. W., 1991. *The Art of Probability for Scientists and Engineers*, Addison-Wesley Publishing Company, New York, New York.

22. Hollander, F. D., 2000. *Large Deviations*, American Mathematical Society, Providence, Rhode Island.

23. Ibragimov, I. A., and Linnik, Y. V., 1971. *Independent and Stationary Sequences of Random Variables,* Wolters-Noordhoff Publishing, Groningen, The Netherlands.

24. Itô, K., 1986. *Introduction to Probability Theory*, Cambridge University Press, Cambridge, Great Britain.

25. Kendall, M. G., and Stuart, A., 1963. *The Advanced Theory of Statistics, Volume 1*, Hafner Publishing Company, New York, New York.

26. Leon-Garcia, A., 1994. *Probability and Random Processes for Electrical Engineering*, Second Edition, Addison-Wesley Publishing Company, New York, New York.

27. Lindgren, B. W., 1968. *Statistical Theory*, Second Edition, Macmillan Company, New York, New York.

28. Lipschutz, S., 1965. *Probability*, Schaum's Outline Series, McGraw-Hill Book Company, New York, New York.

29. Loève, M., 1977. *Probability Theory I,* Fourth Edition, Springer-Verlag, Berlin, Germany.

30. Lukacs, E., 1970, *Characteristic Functions*, Second Edition, Charles Griffin and Company Limited, London.

31. Mood, A. M., Graybill, F. A., and Boes, D. C., 1974. *Introduction to the Theory of Statistics,* Third Edition, McGraw-Hill Book Company, New York, New York.

32. Moran, P. A. P., 1968. *An Introduction to Probability Theory,* Oxford University Press, Oxford, Great Britain.

33. Nikias, C. L., and Shao, M., 1995. *Signal Processing with Alpha-Stable Distributions, and Applications,* John Wiley & Sons, Inc., New York, New York.

34. Papoulis, A., 1965. *Probability, Random Variables, and Stochastic Processes*, McGraw-Hill Book Company, New York, New York.

35. Parzen, E., 1960. *Modern Probability Theory and Its Applications,* John Wiley & Sons, Inc., New York, New York.

36. Ross, S. M., 1970. *Applied Probability Models with Optimization Applications,* Holden-Day, Inc., San Francisco, California.

37. Ross, S. M., 2009. *A First Course in Probability*, Eighth Edition, Macmillan Publishing Company, New York, New York.

38. Ross, S. M., 2010. *Introduction to Probability Models,* Tenth Edition, Academic Press, New York, New York.

39. Sachkov, V. N., 1997. *Probabilistic Methods in Combinatorial Analysis*, Cambridge University Press, Cambridge, Great Britain.

40. Samorodnitsky, G. and Taqqu, M. S. 1994. *Stable Non-Gaussian Random Processes*, Chapman and Hall, New York, New York.

41. Sanov, I., 1957. "On the Probability of Large Deviations of Random Variables," (in Russian), Mat. Sb., Vol. 42, pp. 11-44. (English Translation in: Selected Translations in Mathematical Statistics and Probability: I, pp. 213-244, 1961).

42. Shah, A. K., 1985. "A Simpler Approximation for Areas Under the Statistical Normal Curve," The American Statistician, Vol. 39, Issue 1, p. 80.

43. Shwartz, A., and Weiss, A., 1995. *Large Deviations for Performance Analysis,* Chapman and Hall/CRC Press, New York, New York.

44. Spiegel, M. R., Schiller, J., and Srinivasan, R. A., 2000. *Probability and Statistics*, Second Edition, Schaum's Outline Series, McGraw-Hill Book Company, New York, New York.

45. Stirzaker, D. R., 2003. *Elementary Probability,* Second Edition, Cambridge University Press, Cambridge, Great Britain.

46. Sveshnikov, A. A., 1968. *Problems in Probability Theory, Mathematical Statistics, and Theory of Random Functions*, Dover Publications, Inc., New York, New York.

47. Walrand, J., and Varaiya, P., 2000. *High-Performance Communication Networks*, Morgan Kaufmann Publishers, San Francisco, California.

# Stochastic Processes

$$P\left(X\left(t\right) \leq x \mid X\left(t_1\right) = x_1, X\left(t_2\right) = x_2, \ldots, X\left(t_n\right) = x_n\right)$$

$$= P\left(X\left(t\right) \leq x \mid X\left(t_n\right) = x_n\right), \quad \text{for each state } x$$

whenever $n \in \mathbb{P}, t_1 < t_2 < \ldots < t_n < t,$
and $t_i \in T$ for $1 \leq i \leq n.$
$\{X\left(t\right), t \in T\}$ is a Markov process

**Andrei Andreyevich Markov.**
Markov was born on 14 June, 1856 in Ryazan, Russia. He studied under Pafnuty L. Chebyshev (1821-1894), at the St. Petersburg University. He established the validity of central limit theorem under general conditions. His fame rests upon the study of Markov chains which are aptly named after him.

This work essentially initiated a systematic study of stochastic processes. Subsequently, A. N. Kolmogorov and N. Wiener provided a rigorous foundation for the general theory of Markov processes. Markov passed away on 20 July, 1922 in Petrograd (now St. Petersburg), Russia.

## 9.1 Introduction

Sequence of random variables which are indexed by some parameter, say time, are called stochastic processes. Systems whose properties vary in a random manner can best be described in terms of stochastic processes. Behavior of such systems is sometimes termed *dynamic indeterminism*. Terminology and basic definitions, elements of measure theory, and examples of stochastic processes are provided. Point processes and renewal processes are also discussed in this chapter. The theory of Markov processes is also developed. Some examples of stochastic processes like: Poisson process, shot noise process, Gaussian process, Gaussian white noise process, and Brownian motion process are also discussed. The topics discussed in the section on renewal theory are: ordinary renewal process, modified renewal process, alternate renewal process, backward and forward recurrence-times of a renewal process, equilibrium renewal process, and equilibrium alternate renewal process. The section on Markov processes deals with discrete-time Markov chains, continuous-time Markov chains, and continuous-time Markov processes.

The theory of stochastic processes depends upon probability theory and certain elements of applied analysis. Probability theory and applied analysis are covered in different chapters. Stochastic processes have found application in several diverse areas, such as: information theory, cryptography, traffic characterization, stochastic description of networks, and queueing theory. We initially introduce the basic terminology and definitions used in the rest of the chapter.

## 9.2 Terminology and Definitions

The terminology used in specifying stochastic processes is first given. This is followed by some types of stochastic processes. In the rest of this chapter $P\left(\cdot\right)$ indicates a probability function defined on the family of events. Also recall from the chapter on probability theory that $\mathcal{E}\left(\cdot\right)$, $Var\left(\cdot\right)$, and $Cov\left(\cdot,\cdot\right)$ are the expectation, variance, and covariance operators respectively.

**Definitions 9.1.** *Basic definitions.*

1.  *A stochastic or random process is a family of random variables. Let $T$ be the index set of the stochastic process, and $\Omega$ be its sample space (or state space). Then*

$$\{X(t,\omega), t \in T, \omega \in \Omega\}$$

    *is a stochastic process, where the variable $t$ is typically time.*
    *Thus a stochastic process is a function of two variables. For a specific value of time $t$, it is a random variable. For a specific value of random outcome $\omega \in \Omega$, it is a function of time. This function of time is called a trajectory or sample path of the stochastic process.*
    *Usually the dependence of $X\left(\cdot,\cdot\right)$ on $\omega$ is dropped. Therefore the stochastic process is generally denoted by $\{X(t), t \in T\}$.*

2.  *The stochastic process is a discrete-time stochastic process, if the set $T$ is countable. The stochastic process is a continuous-time stochastic process, if the set $T$ is an interval (either closed or open) on the real line.*

3.  *An index set $T$ is linear, if for any $t, h \in T$, then $(t + h) \in T$. Examples of linear index set are $\mathbb{Z}, \mathbb{N}, \mathbb{P}$ and $\mathbb{R}$.*

4.  *The index set $T$, of a stochastic process is linear, and $t_i \in T$, for $0 \leq i \leq n$. Denote a random vector $\mathcal{X}\left(t_0, t_1, t_2, \ldots, t_n\right)$ as*

$$\mathcal{X}\left(t_0, t_1, t_2, \ldots, t_n\right) = \left(X(t_0), X(t_1), \ldots, X(t_n)\right) \tag{9.1}$$

    *A stochastic process is a stationary process, if for any $n \in \mathbb{N}$ and $\tau \in T$, and for all values of $t_0, t_1, t_2, \ldots, t_n \in T$, the vectors*

$$\mathcal{X}\left(t_0 + \tau, t_1 + \tau, t_2 + \tau, \ldots, t_n + \tau\right)$$

    *and $\mathcal{X}\left(t_0, t_1, t_2, \ldots, t_n\right)$ have identical distribution. In other words, the stochastic process is stationary if the distribution of the random vector $\mathcal{X}\left(t_0, t_1, t_2, \ldots, t_n\right)$ is invariant with respect to translations in time for all values of its arguments.*

5.  *A stochastic process is second-order stationary if: its index set is linear, the first moment $\mathcal{E}\left(X(t)\right)$ is independent of time $t$, and the second moment $\mathcal{E}\left(X(t)X(t+\tau)\right)$ depends only on $\tau$ and not on $t$.*
    *This stationarity is also referred to as wide-sense stationarity.*

6.  *For a wide-sense stationary process, let:*
    $\mathcal{E}\left(X(t)\right) \triangleq \mu$, *and* $\sigma^2 \triangleq \mathcal{E}\left(\left(X\left(t\right) - \mu\right)^2\right)$.

    (a) *The autocovariance $\gamma\left(\tau\right)$ of this process is*

$$\gamma\left(\tau\right) = \mathcal{E}\left(\left(X\left(t + \tau\right) - \mu\right)\left(X\left(t\right) - \mu\right)\right) \tag{9.2a}$$

    *Observe that $\gamma\left(0\right) = \sigma^2$. The autocovariance $\gamma\left(\tau\right)$ measures the covariance between elements of the stochastic process separated by an interval of $\tau$ time units.*

    (b) *The autocorrelation function $\zeta\left(\cdot\right)$ of this process is*

$$\zeta\left(\tau\right) = \frac{\gamma\left(\tau\right)}{\gamma\left(0\right)} \tag{9.2b}$$

(c) *The Fourier transform of the autocovariance function of this stochastic process is called its spectral density function.*

7. *For every set of time instants $t_i \in T, 0 \leq i \leq n$, where $t_0 < t_1 < t_2 < \ldots < t_n$, let $D(t_{i-1}, t_i) = (X(t_i) - X(t_{i-1})), 1 \leq i \leq n$. If the random variables $D(t_{i-1}, t_i), 1 \leq i \leq n$ are independent of each other, then the stochastic process is said to have independent increments.*

8. *The stochastic process $\{X(t), t \in T\}$ has stationary increments, if the random variables $(X(u+a) - X(v+a))$ and $(X(u) - X(v))$ are identically distributed for all values of $u, v \in T$, and time shift $a \in T$.*

   *If in addition, the increments $(X(u+a) - X(v+a))$ and $(X(u) - X(v))$ are stochastically independent of each other, then the process has stationary independent increments.*

9. *Let $\{A(t), t \in T_A\}$ and $\{B(t), t \in T_B\}$ be two stochastic processes defined on the same sample space. These two processes are independent of each other, if the two events*

$$\{A(t_{A0}) \leq a_0, A(t_{A1}) \leq a_1, \ldots, A(t_{Am}) \leq a_m\} \tag{9.3a}$$

$$\{B(t_{B0}) \leq b_0, B(t_{B1}) \leq b_1, \ldots, B(t_{Bn}) \leq b_n\} \tag{9.3b}$$

*are independent of each other $\forall\, t_{Ai} \in T_A, \forall\, a_i \in \mathbb{R}$, where $0 \leq i \leq m$; and $\forall\, t_{Bj} \in T_B, \forall\, b_j \in \mathbb{R}$, where $0 \leq j \leq n$; and $\forall\, m, n \in \mathbb{N}$.* □

Notice in the above definitions that all stationary processes are stationary in the second-order, but the converse is not true in general. Furthermore, the concept of independence of random processes is a generalization of the concept of independence of random variables.

---

## 9.3 Measure Theory

Elements of $\sigma$-algebra (sigma-algebra), measurable sets, measurable space, Borel set, and measure are outlined in this section. Basics of measurable functions are also discussed. These concepts are useful in description of point processes.

### 9.3.1 Sigma-Algebra, Measurable Sets, and Measurable Space

We define $\sigma$-algebra (sigma-algebra), measurable sets, and measurable space. Some examples are also provided.

**Definition 9.2.** *$\sigma$-algebra, measurable sets, and measurable space. Let $S$ be a nonempty set, $\mathcal{P}(S)$ be its power set, and $\widetilde{S}$ be a nonempty subset of $\mathcal{P}(S)$. A collection $\widetilde{S}$ of subsets of $S$ is a $\sigma$-algebra (or $\sigma$-field) if:*

(a) $S \in \widetilde{S}$.

(b) $A \in \widetilde{S} \Rightarrow S \backslash A \in \widetilde{S}$.

(c) If $A_1, A_2, A_3, \ldots \in \widetilde{S}$, is a countable sequence of sets in $\widetilde{S}$; we have $\bigcup_{n \in \mathbb{P}} A_n \in \widetilde{S}$.

   *Elements of the set $\widetilde{S}$ are called measurable sets.*

   *The two-tuple $\left(S, \widetilde{S}\right)$ is called a measurable space.* □

**Examples 9.1.** Let $S$ be a nonempty set, and $\mathcal{P}(S)$ be its power set.

1. $\{\varnothing, S\}$ is a $\sigma$-algebra. This is the smallest $\sigma$-algebra on $S$.
2. $\mathcal{P}(S)$ is a $\sigma$-algebra. This is the largest $\sigma$-algebra on $S$.
3. Let $A \subset S$, then $\{\varnothing, S, A, S \backslash A\}$ is a $\sigma$-algebra. $\qquad\square$

**Observations 9.1.** Let $\left(S, \widetilde{S}\right)$ be a measurable space.

1. $\varnothing \in \widetilde{S}$.
2. $A, B \in \widetilde{S}$, then $A \backslash B \in \widetilde{S}$.
3. The algebra $\widetilde{S}$, is closed under both finite and countable number of intersections. It is also closed under finite unions. $\qquad\square$

### 9.3.2 Borel Sets

Borel $\sigma$-algebra is named after Émile Borel (1871-1956).

**Definition 9.3.** *Borel $\sigma$-algebra. Let $(S, d)$ be a compact metric space, with the metric $d$ defined on the set of elements $S$. Also let $\mathcal{B}(S)$ be the $\sigma$-algebra generated by the family of all open sets in $S$. The Borel $\sigma$-algebra is $\mathcal{B}(S)$.*
*Members of the set $\mathcal{B}(S)$ are called Borel sets, and $\mathcal{B}(S)$ is called the $\sigma$-algebra of Borel sets.* $\qquad\square$

In the above definition, if $S$ is equal to $\mathbb{R}^m$, the metric is Euclidean. That is, for $\eta, \xi \in \mathbb{R}^m$, the metric is $\|\eta - \xi\|$.

**Observations 9.2.** Basic observations about Borel sets.

1. All closed sets are Borel sets.
2. If $C_1, C_2, C_3, \dots \subset S$ are closed sets, then $\bigcup_{n \in \mathbb{P}} C_n$ is a Borel set.
3. If $O_1, O_2, O_3, \dots \subset S$ are open sets, then $\bigcap_{n \in \mathbb{P}} O_n$ is a Borel set.
4. The Borel $\sigma$-algebra $\mathcal{B}(S)$, is generated by the family of all open sets in $S$, and it is the smallest such $\sigma$-algebra. The phrase "smallest" means that if $\mathcal{O}$ is another $\sigma$-algebra that contains all open sets of $S$, then $\mathcal{B}(S) \subset \mathcal{O}$.
5. The two-tuple $(S, \mathcal{B}(S))$ is a measurable space. $\qquad\square$

### 9.3.3 Measure

A measure on a set is a technique to assign a real number (including infinity) to each suitable subset of that set. Thus, measure is a generalization of the concept of length, area, and volume.

**Definition 9.4.** *Measure. Let $\left(S, \widetilde{S}\right)$ be a measurable space. The mapping (function) $\mu : \widetilde{S} \to \mathbb{R}_0^+ \cup \{\infty\}$ is a (positive) measure if:*

(a) $\mu(\varnothing) = 0$.

(b) *The function $\mu(\cdot)$ is countably additive. Let $\{A_n \in \widetilde{S} \mid n \in \mathbb{P}\}$ be a collection of pairwise disjoint sets. That is, $A_m \cap A_n = \varnothing$ for $m \neq n$, and $m, n \in \mathbb{P}$. We have*

$$\mu\left(\bigcup_{n \in \mathbb{P}} A_n\right) = \sum_{n \in \mathbb{P}} \mu(A_n) \tag{9.4}$$

*The triple $\left(S, \widetilde{S}, \mu\right)$ is called a measure space.*

*If $\mu(S) < \infty$, then $\mu(\cdot)$ is called a finite measure.*

*If $\mu(S) = 1$, then $\mu(\cdot)$ is called a probability measure, and $\left(S, \widetilde{S}, \mu\right)$ is called a probability space.* $\square$

Observe in the above definition that, the condition (a) means that an empty set has zero "length," and condition (b) implies that the "length" of a disjoint union of subsets is simply the sum of the "lengths" of individual sets.

**Observations 9.3.** Basic properties of measures. Let $\left(S, \widetilde{S}, \mu\right)$ be a measure space.

1. The sets $A_1, A_2, \ldots, A_n \in \widetilde{S}$ are pairwise disjoint, then

$$\mu(A_1 \cup A_2 \cup \ldots \cup A_n) = \mu(A_1) + \mu(A_2) + \cdots + \mu(A_n)$$

2. If $A, B \in \widetilde{S}$, and $A \subset B$, then $\mu(A) \leq \mu(B)$.
3. If $A, B \in \widetilde{S}$, $A \subset B$, and $\mu(B) < \infty$, then $\mu(B \backslash A) = \mu(B) - \mu(A)$.
4. If $A_1, A_2, A_3, \ldots \in \widetilde{S}$, then

$$\mu\left(\bigcup_{n \in \mathbb{P}} A_n\right) \leq \sum_{n \in \mathbb{P}} \mu(A_n)$$

5. If $A_1, A_2, A_3, \ldots \in \widetilde{S}$, and $\mu(A_n) = 0$ for each $n \in \mathbb{P}$, then $\mu\left(\bigcup_{n \in \mathbb{P}} A_n\right) = 0$. $\square$

### 9.3.4 Measurable Function

The notion of a measurable function is developed in this subsection. Measurable functions preserve structure between measurable spaces.

**Definition 9.5.** *Let $\left(S, \widetilde{S}\right)$ and $\left(T, \widetilde{T}\right)$ be measurable spaces. A function $f : S \to T$ is measurable, if $f^{-1}(M) \in \widetilde{S}$ for every $M \in \widetilde{T}$.* $\square$

The idea of measurable function implicitly depends upon the $\sigma$-algebras $\widetilde{S}$ and $\widetilde{T}$. In order to emphasize this relationship, the measurable functions are often specified as $f : \left(S, \widetilde{S}\right) \to \left(T, \widetilde{T}\right)$.

All continuous functions defined on $\mathbb{R}$ are measurable functions. However, the converse is not true. That is, all measurable functions on $\mathbb{R}$ are not necessarily continuous.

**Observations 9.4.** Some basic observations.

1. Let $\left(S, \widetilde{S}\right)$ and $\left(T, \widetilde{T}\right)$ be measurable spaces. A function $f : S \to T$ is measurable. Then $f^{-1}\left(\widetilde{T}\right) \subset \widetilde{S}$ is a sub-$\sigma$-field of $\widetilde{S}$. It is called the $\sigma$-field generated by the function $f(\cdot)$, and is usually denoted by $\sigma(f)$.

2. Let $S$ be equal to the sample space $\Omega$, $\widetilde{S}$ be equal to the family of events, $T$ be equal to the real line $\mathbb{R}$, and $\widetilde{S}$ also be equal to the Borel $\sigma$-field; then the measurable function $f\left(\cdot\right)$ is a *random variable*.

This interpretation of a random variable coincides with the random variable $X\left(\cdot\right)$ defined on the probability space $\left(\Omega, \mathcal{F}, P\right)$; where $\Omega$ is the sample space, $\mathcal{F}$ is the family of events, and $P$ is the probability function.

Recall that a random variable is a convenient technique to represent the sometimes qualitative elements of the set $\Omega$.                                                                               $\square$

**Examples 9.2.** Some examples of measurable functions. Let $\left(S, \widetilde{S}\right)$ be a measurable space.

1. The *indicator function* $I_A : S \to \mathbb{R}$. For $x \in S$, it is defined as

$$I_A\left(x\right) = \begin{cases} 1, & x \in A \\ 0, & x \in A^c \end{cases} \tag{9.5}$$

where $A \in \widetilde{S}$, and $T = \{0, 1\}$. The indicator function is a measurable function.

2. A *simple function* $f : S \to \mathbb{C}$ is defined as

$$f\left(x\right) = \sum_{n=1}^{N} \alpha_n I_{A_n}\left(x\right), \quad x \in S \tag{9.6}$$

where $\alpha_n \in \mathbb{C}$, and $A_n \in \widetilde{S}$ for $1 \leq n \leq N$ are disjoint measurable sets. Observe that the representation of $f\left(x\right)$ is not unique. A simple function is a measurable function.

It can be shown that the sum, difference, and product of two simple functions are still simple functions. Further, multiplication of a simple function by a constant results in a simple function.                                                                               $\square$

## 9.4  Examples of Stochastic Processes

Some stochastic processes are briefly described in this section. These are: Poisson process, shot noise process, Gaussian process, Brownian motion process, and Gaussian white noise process.

### 9.4.1  Poisson Process

A Poisson process $\left\{N(t), t \in \mathbb{R}_0^+\right\}$ is a *counting random process*. In this process, $N(t)$ counts the total number of events which have occurred up to time $t$.

**Definition 9.6.** *The random process* $\left\{N(t), t \in \mathbb{R}_0^+\right\}$ *is a homogeneous (or stationary) Poisson process, with rate parameter (intensity)* $\lambda > 0$, *if*:

(a) $N(0) = 0$.
(b) *The process has independent increments.*

(c) *The distribution of the number of events in any interval $(t, t + h]$ of length $t$, has a Poisson distribution with parameter $\lambda t$. Thus for any $h \in \mathbb{R}_0^+$, and $t \in \mathbb{R}^+$*

$$P\left(N(h+t) - N(h) = n\right) = e^{-\lambda t} \frac{(\lambda t)^n}{n!}, \quad \forall\, n \in \mathbb{N} \tag{9.7}$$

$\square$

There are several other equivalent definitions of the Poisson process.

**Observations 9.5.** Some important observations about Poisson process.

1. The Poisson process has stationary increments. It follows from part (c) of the above definition.
2. $\mathcal{E}\left(N(t)\right) = \lambda t$, and $Var(N(t)) = \lambda t$.
3. Sum of two independent Poisson processes is a Poisson process. If the parameters of the two Poisson processes are $\lambda_1$ and $\lambda_2$, then the Poisson parameter of their sum is $(\lambda_1 + \lambda_2)$.
4. Conversely, independent splitting of a single Poissonian stream results in several independent Poisson process streams.
5. The time interval between two successive events of a Poisson process is called *interarrival time*. Interarrival times have an exponential distribution with parameter $\lambda$. Furthermore, these interarrival times are stochastically independent of each other. Let the occurrence of the $n$th event be at time $S_n$,

$$S_n = \sum_{i=1}^{n} X_i, \quad n \in \mathbb{P}$$

where the $X_i$'s are independent and identically distributed as an exponentially distributed random variable $X$ with parameter $\lambda$. The random variable $S_n$ has an Erlang distribution.
6. A useful generalization of the Poisson process occurs, if $\lambda$ is made a function of time $t$. The resulting process is called a *nonhomogeneous or inhomogeneous Poisson process*. In this process, if $\Lambda(t) = \int_0^t \lambda(x)\, dx$, we have

$$P\left(N(t) = n\right) = e^{-\Lambda(t)} \frac{\{\Lambda(t)\}^n}{n!}, \quad \forall\, n \in \mathbb{N}$$

In this description of the nonhomogeneous Poisson process, $\lambda(\cdot)$ is called the stochastic intensity function of the process. Thus $\lambda(t)\, h$ represents the approximate probability that an event occurs in the time interval $(t, t + h)$. Also note that $\mathcal{E}(N(t)) = \Lambda(t)$. It is also possible to give an alternate definition of the nonhomogeneous Poisson process. $\square$

The Poisson process is used extensively in queueing theory and stochastic geometry because of its mathematical amenability.

**Poisson Processes and Uniform Distribution**

Relationship between Poisson processes and uniform distribution is explored. This is useful in studying shot noise process.

**Observation 9.6.** Let $\left\{N(t), t \in \mathbb{R}_0^+\right\}$ be a homogeneous Poisson process with parameter $\lambda$. Assume that only a single event has occurred by epoch $t > 0$. Let $Y$ be the arrival instant of the single event in the interval $[0, t]$. Show that $Y$ is uniformly distributed over the interval $[0, t]$. That is

$$P\left(Y \leq y\right) = P\left(N\left(y\right) = 1 \mid N\left(t\right) = 1\right) = \frac{y}{t}, \quad y \in [0, t] \tag{9.8}$$

$\square$

The above observation is next generalized by using the concept of order statistics.

**Definition 9.7.** *Order statistics. Let $Z_1, Z_2, \ldots, Z_n$ be $n$ random variables. Then*

$$\left(Z_{(1)}, Z_{(2)}, \ldots, Z_{(n)}\right) \triangleq \mathcal{Z} \tag{9.9}$$

*are the order statistics corresponding to $Z_1, Z_2, \ldots, Z_n$, where $Z_{(i)}$ is the ith smallest value among $Z_1, Z_2, \ldots, Z_n$, where $1 \leq i \leq n$.* $\square$

The next observation is about sequence of random variables with uniform distributions.

**Observation 9.7.** Let $Z_1, Z_2, \ldots, Z_n$ be independent and identically distributed continuous random variables. Each of these is distributed as random variable $Z$. which has a probability density function $f_Z\left(\cdot\right)$. Let the order statistics of these random variables be $\left(Z_{(1)}, Z_{(2)}, \ldots, Z_{(n)}\right) \triangleq \mathcal{Z}$.

(a) The joint probability density function of the order statistics $\mathcal{Z}$ is given by

$$f\left(z_1, z_2, \ldots, z_n\right) = n! \prod_{i=1}^{n} f_Z\left(z_i\right), \qquad z_1 < z_2 < \cdots < z_n \tag{9.10a}$$

(b) Let the random variables $Z_i$'s for $i = 1, 2, \ldots, n$ be uniformly distributed over the interval $(0, t)$ then the joint probability density function of the order statistics $\mathcal{Z}$ is given by

$$f\left(z_1, z_2, \ldots, z_n\right) = \frac{n!}{t^n}, \ 0 < z_1 < z_2 < \cdots < z_n < t \tag{9.10b}$$

These result are proved in the problem section. $\square$

The following result has application in shot noise processes, which are described below.

**Theorem 9.1.** *Let the occurrence times of the $n$ events in a homogeneous Poisson process of rate $\lambda$ be $S_1, S_2, \ldots, S_n$, and $N\left(t\right) = n$. Then the occurrence times of these $n$ events have the same distribution as the order statistics of $n$ independent random variables which are uniformly distributed over the interval $(0, t)$.*
*Proof.* See the problem section. $\square$

It is possible to extend the definition of homogeneous Poisson process to all points on the real line $\mathbb{R}$.

**Definition 9.8.** *Poisson process on real line $\mathbb{R}$.*
*The random process $\{N(t), t \in \mathbb{R}\}$ is a homogeneous Poisson process, with rate parameter (intensity) $\lambda > 0$, if:*

(a) *The process has independent increments.*

(b) *The distribution of the number of events in any real interval* $(a, b]$ *of length* $(b - a)$, *has a Poisson distribution with parameter* $\lambda (b - a)$. *That is, for* $a < b$

$$P\left(N(b) - N(a) = n\right) = e^{-\lambda(b-a)} \frac{\left(\lambda\left(b - a\right)\right)^n}{n!}, \quad \forall\, n \in \mathbb{N} \tag{9.11}$$

$\square$

The above definition leads to the definition of Poisson point processes which are described later in the chapter.

### 9.4.2 Shot Noise Process

An application of the Poisson process is the shot noise. The shot noise process was introduced by Norman R. Campbell in the study of thermionic noise in vacuum tubes. This process was initially developed to model fluctuations in flow of electrical current in certain electronic devices. Models similar to this find application in physics and communication engineering. Electrical current is made up of flows of electrons. Electrons arrive at the terminal of a device (anode) according to a homogeneous Poisson process $\left\{N(t), t \in \mathbb{R}_0^+\right\}$ of rate $\lambda$. The arriving electrons produces a current of intensity $w\left(x\right)$, at $x$ time units after arriving at the terminal of the device. Thus the intensity of current at time $t \in \mathbb{R}_0^+$ is the shot noise

$$I\left(t\right) = \begin{cases} 0, & N\left(t\right) = 0 \\ \sum_{i=1}^{N(t)} w\left(t - S_i\right), & N\left(t\right) \geq 1 \end{cases}$$

where $S_1, S_2, \ldots$ are the successive arrival times of the electrons. A more general definition of shot noise is next given.

**Definition 9.9.** *A stochastic process* $\left\{X(t), t \in \mathbb{R}_0^+\right\}$ *is a shot noise or filtered Poisson process if*

$$X\left(t\right) = \begin{cases} 0, & N\left(t\right) = 0 \\ \sum_{m=1}^{N(t)} w\left(t - S_m, Y_m\right), & N\left(t\right) \geq 1 \end{cases} \tag{9.12}$$

*where:*

(a) $\left\{N(t), t \in \mathbb{R}_0^+\right\}$ *is a homogeneous Poisson process if the rate (or intensity) parameter* $\lambda > 0$ *is a real constant.*
   *It is a nonhomogeneous Poisson process, if* $\lambda\left(\cdot\right)$ *is a function of time, and* $\lambda\left(t\right) > 0$ *for all values of* $t$.
   *The arrival instants are* $S_m \in \mathbb{R}^+$, *where* $m = 1, 2, 3, \ldots$.
(b) $Y_1, Y_2, Y_3, \ldots$ *is a sequence of independent and identically distributed random variables. Each of the* $Y_m$'s *is distributed as random variable* $Y$. *This sequence is also independent of the Poisson process* $\left\{N(t), t \in \mathbb{R}_0^+\right\}$.
(c) *The response function* $w\left(t - \tau_m, y\right) = 0$ *for* $t < \tau_m$. *A possible interpretation of it is as follows. The value* $w\left(t - \tau_m, y\right)$ *is the value of the function at time* $t$, *which originated at time* $\tau_m$ *with an amplitude* $Y_m = y$.
   *Further,* $X\left(t\right)$ *is the sum of all signals* $w\left(\cdot, \cdot\right)$'s *which originated in the time interval* $(0, t]$. $\square$

The first and second order statistics of a stochastic process can be evaluated by first computing its Laplace functional.

**Definition 9.10.** *Consider a stochastic process* $\{X(t), t \in \mathbb{R}_0^+\}$. *Its Laplace functional is* $\widehat{f}_{X(t)}(s) = \mathcal{E}(\exp(-sX(t)))$, *where* $s \in \mathbb{C}$. □

The above definition yields

$$\mathcal{E}(X(t)) = -\left. \frac{d}{ds} \widehat{f}_{X(t)}(s) \right|_{s=0}$$

$$\mathcal{E}\left((X(t))^2\right) = \left. \frac{d^2}{ds^2} \widehat{f}_{X(t)}(s) \right|_{s=0}$$

The following theorem specifies the Laplace functional of the shot noise process.

**Theorem 9.2.** *Consider a shot noise process* $\{X(t), t \in \mathbb{R}_0^+\}$ *in which arrivals occur with a rate parameter* $\lambda > 0$. *That is, the arrival process is a homogeneous Poisson process. The Laplace functional of the shot noise process is*

$$\widehat{f}_{X(t)}(s) = \mathcal{E}(\exp(-sX(t))) = \exp\left\{\lambda \int_0^t \varphi(s, \tau) \, d\tau\right\} \tag{9.13a}$$

*where*

$$\varphi(s, \tau) = \mathcal{E}(\exp(-sw(\tau, Y))) - 1, \quad \tau > 0, \ s \in \mathbb{C} \tag{9.13b}$$

*and the expectation* $\mathcal{E}(\cdot)$ *in the above equation is with respect to the random variable* $Y$.
*Proof.* See the problem section. □

**Corollary 9.1.** *The arrival process of the shot noise process* $\{X(t), t \in \mathbb{R}_0^+\}$ *is homogeneous Poisson. Its mean and variance are*:

$$\mathcal{E}(X(t)) = \lambda \int_0^t \mathcal{E}(w(\tau, Y)) \, d\tau \tag{9.14a}$$

$$Var(X(t)) = \lambda \int_0^t \mathcal{E}\left((w(\tau, Y))^2\right) d\tau \tag{9.14b}$$

*and the expectation* $\mathcal{E}(\cdot)$ *in the above equations is with respect to the random variable* $Y$. □

**Corollary 9.2.** *Consider the shot noise process* $\{X(t), t \in \mathbb{R}_0^+\}$. *Its limiting distribution is obtained by letting* $t \to \infty$. *Thus*

$$\lim_{t \to \infty} \widehat{f}_{X(t)}(s) = \lim_{t \to \infty} \mathcal{E}(\exp(-sX(t))) = \exp\left\{\lambda \int_0^\infty \varphi(s, \tau) \, d\tau\right\} \tag{9.15a}$$

*where*

$$\varphi(s, \tau) = \mathcal{E}(\exp(-sw(\tau, Y))) - 1, \quad \tau > 0, \ s \in \mathbb{C} \tag{9.15b}$$

*The expectation* $\mathcal{E}(\cdot)$ *in the above equation is with respect to the random variable* $Y$. □

**Example 9.3.** Consider a shot noise process $\left\{X(t), t \in \mathbb{R}_0^+\right\}$, where

$$w\left(t - S_m, Y_m\right) = Y_m \exp\left(-\alpha\left(t - S_m\right)\right), \quad \alpha > 0, \quad \text{and } m = 1, 2, 3, \ldots$$

and the $Y_m$'s are independent and exponentially distributed with parameter $\theta$. Further, these $Y_m$'s are independent of the Poisson process. Let $U$ be a uniformly distributed random variable over the interval 0 to $t$. Therefore

$$\mathcal{E}\left(\exp\left(-sw\left(t - U, Y\right)\right)\right) = \frac{1}{t}\int_0^t \mathcal{E}\left(\exp\left(-sY\exp\left(-\alpha\tau\right)\right)\right) d\tau$$

In the above equation, the expectation operator $\mathcal{E}(\cdot)$ on the left hand side is with respect to random variables $U$ and $Y$, and that on the right hand side is with respect to the random variable $Y$. As

$$\mathcal{E}\left(\exp\left(-sY\right)\right) = \frac{\theta}{\theta + s}$$

we obtain

$$\mathcal{E}\left(\exp\left(-sw\left(t - U, Y\right)\right)\right) = \frac{1}{t}\int_0^t \mathcal{E}\left(\exp\left(-sY\exp\left(-\alpha\tau\right)\right)\right) d\tau$$

$$= \frac{1}{t}\int_0^t \frac{\theta}{\left(\theta + s\exp\left(-\alpha\tau\right)\right)} d\tau$$

Observe that

$$\mathcal{E}\left(\exp\left(-sw\left(\tau, Y\right)\right)\right) = \frac{\theta}{\left(\theta + s\exp\left(-\alpha\tau\right)\right)}$$

Thus

$$\varphi\left(s, \tau\right) = \mathcal{E}\left(\exp\left(-sw\left(\tau, Y\right)\right)\right) - 1 = \frac{\theta}{\left(\theta + s\exp\left(-\alpha\tau\right)\right)} - 1$$

Consequently

$$\lim_{t \to \infty} \mathcal{E}\left(\exp\left(-sX(t)\right)\right) = \exp\left[\lambda\int_0^\infty \left\{\frac{\theta}{\left(\theta + s\exp\left(-\alpha\tau\right)\right)} - 1\right\} d\tau\right]$$

$$= \exp\left[\frac{\lambda}{\alpha}\ln\left\{\frac{\theta}{\theta + s}\right\}\right]$$

$$= \left\{\frac{\theta}{\theta + s}\right\}^{\lambda/\alpha}$$

Recall that $\left\{\theta/(\theta + s)\right\}^{\lambda/\alpha}$ is the one-sided Laplace transform of the probability density function of a gamma distributed random variable. Therefore the probability density function of $X(t)$ as $t \to \infty$ is

$$f\left(x\right) = \frac{\theta\left(\theta x\right)^{\lambda/\alpha - 1} e^{-\theta x}}{\Gamma\left(\lambda/\alpha\right)}, \quad x \in [0, \infty)$$

where $\Gamma(\cdot)$ is the gamma function.                                                                        □

The next theorem assumes that the arrival instants follow a nonhomogeneous Poisson process.

**Theorem 9.3.** *Consider a shot noise process* $\{X(t), t \in \mathbb{R}_0^+\}$ *in which arrivals occur according to a nonhomogeneous Poisson process, with intensity function* $\lambda(\cdot)$. *The Laplace functional of the shot noise process is*

$$\widehat{f}_{X(t)}(s) = \mathcal{E}(\exp(-sX(t))) = \exp\left\{\int_0^t \lambda(\tau)\,\varphi(s,\tau)\,d\tau\right\} \tag{9.16a}$$

*where*

$$\varphi(s,\tau) = \mathcal{E}(\exp(-sw(\tau,Y))) - 1, \quad \tau > 0, \ s \in \mathbb{C} \tag{9.16b}$$

*and the expectation* $\mathcal{E}(\cdot)$ *in the above equation is with respect to the random variable* $Y$.

*Proof.* See the problem section. $\qquad\qquad\qquad\qquad\qquad\qquad\qquad\qquad\qquad\qquad\square$

**Corollary 9.3.** *The arrival process of the shot noise process* $\{X(t), t \in \mathbb{R}_0^+\}$ *is nonhomogeneous Poisson. Its mean and variance are*:

$$\mathcal{E}(X(t)) = \int_0^t \lambda(\tau)\,\mathcal{E}(w(\tau,Y))\,d\tau \tag{9.17a}$$

$$Var(X(t)) = \int_0^t \lambda(\tau)\,\mathcal{E}\left((w(\tau,Y))^2\right)d\tau \tag{9.17b}$$

*and the expectation* $\mathcal{E}(\cdot)$ *in the above equations is with respect to the random variable* $Y$. *These results are often called Campbell's theorem.* $\qquad\qquad\qquad\qquad\qquad\qquad\square$

### 9.4.3  Gaussian Process

Several physical processes are modeled by a Gaussian (or normal) process.

**Definition 9.11.** $\{X(t), t \in T\}$ *is a Gaussian random process, if and only if* $\forall\, n \in \mathbb{P}$, *given any set of time instants* $t_1, t_2, \ldots t_n \in T$, *the sequence of random variables* $X(t_1), X(t_2), \ldots, X(t_n)$ *has a multivariate Gaussian distribution.* $\qquad\qquad\qquad\qquad\qquad\qquad\qquad\qquad\square$

An alternate definition of a Gaussian process is as follows. The process $\{X(t), t \in T\}$ is a Gaussian random process, if every finite linear combination of the random variables $X(t), t \in T$ has a normal distribution. This process has some simple and useful properties. For example, if a Gaussian process is second-order stationary, then it is also stationary.

### 9.4.4  Brownian Motion Process

*Brownian motion process* (BMP) describes the random movement of microscopic particles in a liquid or gas. These processes are so named after the botanist Robert Brown (1773-1858). This process is also called the *Wiener process*, or *Wiener-Levy process*. Mathematicians Norbert Wiener (1894-1964) and Paul Pierre Levy (1886-1971) are generally credited with developing a mathematical theory of this process.

**Definition 9.12.** *The random process* $\{B(t), t \in \mathbb{R}\}$ *is a Wiener (or Brownian motion or Wiener-Levy) process if*:

(a) $B(0) = 0$.

(b) *The process has independent increments.*

(c) $(B(t) - B(s))$ *has normal distribution with mean* $0,$ *and*

$$Var(B(t) - B(s)) = \sigma^2(t - s), \quad s \leq t \tag{9.18}$$

(d) *The process has continuous sample paths with probability* $1.$

*If* $\sigma = 1,$ *the process is called standard Brownian motion process.*                    □

Based upon this definition, the following observations are made about the BMP.

**Observations 9.8.** Some basic observations about BMP.

1. $\mathcal{E}(B(t)) = 0$ and $Var(B(t)) = \sigma^2 |t|, \forall t \in \mathbb{R}$. Therefore $B(t)$ has a normal distribution $\forall$ $t \in \mathbb{R}$. It is denoted by $\mathcal{N}(0, \sigma^2 |t|)$.

2. Let $c(s, t) \triangleq Cov(B(s), B(t))$ be the covariance of the process. Then

$$c(s, t) = \begin{cases} \sigma^2 \min(|s|, |t|), & \text{if } st > 0 \\ 0, & \text{if } st \leq 0 \end{cases}$$

3. If $a \geq 0$, $s \geq a$ and $t \geq a$, then $\mathcal{E}((B(s) - B(a))(B(t) - B(a))) = \sigma^2 \min(s - a, t - a)$.

4. If $t_1 \leq t_2 \leq t_3 \leq t_4$, then $Cov\{(B(t_2) - B(t_1)), (B(t_4) - B(t_3))\} = 0$.

5. The Wiener process is a Gaussian process.
   That is, if $t_1 \leq t_2 \leq \ldots \leq t_n$, and $b_1, b_2, \ldots, b_n$ are real constants, then the random variable $\sum_{i=1}^{n} b_i B(t_i)$ has a Gaussian distribution.

6. Let $t_1 \leq t_2 \leq \ldots \leq t_n$, then

$$(B(t_2) - B(t_1)), (B(t_3) - B(t_2)), \ldots, (B(t_n) - B(t_{n-1}))$$

   are independent of each other.

7. Define

$$\vartheta = \int_a^b f(t) \, dB(t)$$

   where $f(\cdot)$ is a deterministic function. This integral is called the Wiener integral. Then

$$\mathcal{E}(\vartheta) = 0, \quad \text{and} \quad Var(\vartheta) = \sigma^2 \int_a^b \{f(t)\}^2 \, dt$$

□

The justification of some of these observations is provided in the problem section. A generalization of the Brownian motion process is called a fractional Brownian motion process. This later process is studied in the chapter on Internet traffic.

### 9.4.5 Gaussian White Noise Process

A Gaussian white noise process is only a mathematical abstraction. However, it is a convenient description of several physical processes.

**Definition 9.13.** *The random process* $\{n(t), t \in \mathbb{R}\}$ *is a Gaussian white noise process if*:

(a) *It is a stationary Gaussian process.*
(b) $\mathcal{E}\left(n(t)\right) = 0.$
(c) *Its covariance function is* $w\left(\cdot\right),$ *where* $w\left(t\right) = K\delta\left(t\right), K > 0$ *and its Fourier transform* $W\left(\omega\right) = K$ *for all values of* $\omega \in \mathbb{R}.$

*This definition is not mathematically plausible because of the presence of Dirac's delta function* $\delta\left(\cdot\right).$ *It is modified as follows. Let the covariance function of a Gaussian process be* $w\left(t\right), t \in \mathbb{R}$ *such that*

$$w\left(t\right) = K\frac{\alpha}{2}e^{-\alpha|t|}, \quad \alpha > 0, \ t \in \mathbb{R} \tag{9.19a}$$

*Let the Fourier transform of* $w\left(t\right)$ *be* $W\left(\omega\right).$ *It is also called the spectral density of the process.*

$$W\left(\omega\right) = \frac{K}{\left\{1 + \left(\dfrac{\omega}{\alpha}\right)^2\right\}}, \quad \omega \in \mathbb{R} \tag{9.19b}$$

*If* $\alpha$ *is a very large number, then the spectral density function effectively has a constant value* $K,$ *and the covariance function is equal to Dirac's delta function times* $K.$                    □

The relationship $w\left(t\right) = K\delta\left(t\right), t \in \mathbb{R}$ implies that any two samples of white Gaussian noise are stochastically independent of each other.

### 9.4.6  Brownian Motion and Gaussian White Noise

Sometimes, it is convenient to think of BMP as an integral of Gaussian white noise process. For simplicity in discussion, let $\left\{B(t), t \in \mathbb{R}_0^+\right\}$ be a BMP with $Var\left(B\left(1\right)\right) = \sigma^2,$ and $g\left(\cdot\right)$ be a function having continuous derivative in the interval $[a, b].$ The stochastic integral $\int_a^b g\left(t\right) dB\left(t\right)$ is defined as:

$$\int_a^b g\left(t\right) dB\left(t\right) = \lim_{\substack{n\to\infty \\ \Delta t \to 0}} \sum_{j=1}^n g\left(t_{j-1}\right)\left\{B\left(t_j\right) - B\left(t_{j-1}\right)\right\}$$

where $a = t_0, b = t_n, \Delta t = \left(b - a\right)/n,$ and $t_j = t_0 + j\Delta t$ for $1 \le j \le n.$ Using the definition of BMP, it can be shown that

$$\mathcal{E}\left[\int_a^b g\left(t\right) dB\left(t\right)\right] = 0, \quad \text{and} \quad Var\left[\int_a^b g\left(t\right) dB\left(t\right)\right] = \sigma^2 \int_a^b \left\{g\left(t\right)\right\}^2 dt$$

The above equations imply that the process $\left\{dB(t), t \in \mathbb{R}_0^+\right\}$ is an operator that transforms the function $g\left(\cdot\right)$ into $\int_a^b g\left(t\right) dB\left(t\right).$ This transformation is often called the Gaussian white noise transformation, and the process $\left\{dB(t), t \in \mathbb{R}_0^+\right\}$ is called Gaussian white noise. Thus the derivative of the BMP can be "viewed" as a Gaussian white noise process, even though the derivative of a BMP does not exist. Alternately, it can be said that the integral of a Gaussian white noise process is indeed a BMP. This integral also has a physical interpretation. That is, if a time varying function $g\left(\cdot\right)$ travels through a Gaussian white noise medium during the interval $[a, b],$ the output at time $b$ is equal to $\int_a^b g\left(t\right) dB\left(t\right).$

As Gaussian white noise is physically nonexistent, an approximation to it can be called *colored noise*. It is defined as

$$X(t) = \frac{B(t+h) - B(t)}{h}, \quad t \in \mathbb{R}_0^+$$

where $h > 0$ is some fixed constant. Denote its mean, variance, and covariance by $\mu_X(t)$, $\sigma_X^2(t)$, and $c_X(s,t) \triangleq Cov(X(s), X(t))$, $s,t \in \mathbb{R}_0^+$ respectively. Assume that $0 \leq s \leq t$, and define $(t-s) = u \in \mathbb{R}_0^+$. Thus $\mu_X(t) = 0$ and

$$c_X(s,t) = \frac{\sigma^2}{h^2}\{h - \min(h,u)\}$$

Consider the following cases.

(a) If $u = 0$, that is $s = t$, then $\sigma_X^2(t) = \sigma^2/h$. Therefore the variation in the colored noise process becomes larger as $h$ becomes smaller.
(b) If $u \geq h$, then $c_X(s,t) = 0$. That is, $X(s)$ and $X(t)$ are independent of each other.
(c) If $u < h$, then $c_X(s,t) = \sigma^2(h-u)/h^2$.

Thus a colored noise process can reasonably approximate a Gaussian white noise process for small values of $h$.

The next section is concerned with the study of renewal theory. It is the study of sums of independent nonnegative random variables. As we shall see, renewal theory is a generalization of Poisson process.

## 9.5  Point Process

Point processes are stochastic-theoretic models of points distributed in some space in some random order. The term "process" generally implies some type of evolution of a phenomenon over time. The term is used to indicate evolution over time, or description of some time-independent phenomenon.

A point pattern in space $S$ is a collection of points distributed in some random order. A point process (PP), can also be viewed as a counting process that describes a randomly distributed set of points in some space $S$. The space $S$ is typically the Euclidean space $\mathbb{R}^m$, where $m \in \mathbb{P}$. If $m$ is equal to one, the space is a real line; and if $m$ is equal to two, then the space is a plane.

A well-known and useful example of a point process is the Poisson point process (PPP). Some terminology related to point processes is as follows:

- A PP can either be *simple* or not. A PP is said to be simple if no two points in the PP are at the same location in the space $S$.
- A PP can either be *stationary* or not. A PP is stationary, if the properties of the PP are invariant under translation.
- A PP can either be *isotropic* or not. Isotropy implies that the properties of the PP are invariant under rotation.
- A PP can either be Poissonian or not. A PPP often provides a convenient mathematical description of several physical phenomena.

A point process can be described in two equivalent ways. These are the random set, and the random measure formalisms. The random measure formulation turns out to be mathematically more convenient.

In the *random set formalism*, a point process is considered as a countable random set $\Phi = \{X_1, X_2, X_3, \ldots\}$, where the random variables $X_n$, $n \in \mathbb{P}$ take values in the set $S$. Value of an $X_n$ specifies the location of a point in space $S$. A realization of the point process $\Phi$ is

$$\phi = \{x_1, x_2, x_3, \ldots\} \subset S$$

Let $\widetilde{S}$ denote the Borel $\sigma$-algebra of the set $S$. A point process can also be specified by counting the number of points which lie in sets $B \in \widetilde{S}$. The number of points in $B$ is denoted by $N(B)$. Thus $N(B)$ is a random variable which takes values in the set of nonnegative integers $\mathbb{N}$. The point process can be described as a collection of random variables $N(B)$ which are indexed by $B \in \widetilde{S}$. This is the *random measure formulation* for describing the point process, and $N(\cdot)$ is a counting measure.

We shall assume that the point processes are *simple* and *locally finite* (or *boundedly finite*). A point process is simple if

$$N(\{x\}) \leq 1, \quad \forall\, x \in S$$

Each subset $B \in \widetilde{S}$ is bounded. That is, the number of points in such set is finite. A set is bounded, if it is contained in a compact set. A point process is said to be locally finite, if and only if

$$|B| < \infty \Rightarrow N(B) < \infty, \quad \forall\, B \in \widetilde{S}$$

There exists a one-to-one mapping between $\Phi$ and $N(\cdot)$. Informally, we have

$$N(B) = |\phi \cap B|, \quad \text{and} \quad \phi = \{x \in S \mid N(x) = 1\}$$

A useful artifice to describe a point process is the so-called Dirac measure.

**Definition 9.14.** *Dirac measure. Let $\left(S, \widetilde{S}\right)$ be a measurable space. The Dirac measure of a point $x \in S$ is*

$$\zeta_x(B) = \begin{cases} 1, & x \in B \\ 0, & x \notin B \end{cases} \tag{9.20}$$

*where $B \in \widetilde{S}$.*                                                                                   $\square$

The astute reader will notice the similarity between the definitions of an indicator function, and the Dirac measure. That is, $\zeta_x(B) = I_B(x)$, where $x \in S$, and $B \in \widetilde{S}$.

**Example 9.4.** Let

$$\Phi = \{X_1, X_2, X_3, \ldots\}$$

be a set of vector-valued random variables, where the random variables $X_n$, $n \in \mathbb{P}$, take values in the set $S$. The family of Borel sets of the space $S$ is $\widetilde{S}$. Then

$$N(B) = \sum_{n \in \mathbb{P}} \zeta_{X_n}(B), \quad B \in \widetilde{S}$$

is equal to the random number of points in the set $B$. Assume that the number of points in the set $B$ is bounded. Note that $N(B)$ is a random variable, and takes values in the set $\mathbb{N}$. Thus $N(\cdot)$ is a *counting measure.*                                                                                   $\square$

Observe that the counting measure $N$ takes values in the set $M_p$ of all locally finite counting measures $\nu$ on $\left(S, \widetilde{\mathcal{S}}\right)$. That is, $\nu\left(B\right) < \infty$, for each bounded set $B$. Also, each measure $\nu \in M_p$ has the form

$$\nu\left(B\right) = \sum_{n \in \mathbb{P}} \zeta_{x_n}\left(B\right), \quad B \in \widetilde{\mathcal{S}}$$

where $\{x_1, x_2, \ldots, x_k\}$ is its associated set of points, and $0 \leq k \leq \infty$.

The $\sigma$-field associated with the set $M_p$ is denoted by $\mathcal{M}_p$. It is generated by the sets

$$\{\nu \in M_p \mid \nu\left(B\right) = n\}$$

for $B \in \widetilde{\mathcal{S}}$, and $n \in \mathbb{N}$.

**Definition 9.15.** *Point process (PP). A point process is a countable random set*

$$\Phi = \{X_1, X_2, X_3, \ldots\} \tag{9.21a}$$

*where the random variables $X_n$, $n \in \mathbb{P}$, take values in the set $S$. The $X_n$'s denote the location of points in the space $S$. A realization of the point process $\Phi$ is $\phi = \{x_1, x_2, x_3, \ldots\} \subset S$. Let $\widetilde{\mathcal{S}}$ be the Borel $\sigma$-algebra of the set $S$.*

*Typically the set $S$ is equal to the Euclidean space $\mathbb{R}^m$, where $m \in \mathbb{P}$. The number of points in any bounded set $B \in \widetilde{\mathcal{S}}$ is specified by a locally finite and simple counting measure $N\left(\cdot\right)$. This counting measure is nonnegative, and integer-valued, where*

$$N\left(B\right) = \sum_{n \subset \mathbb{P}} \zeta_{X_n}\left(B\right), \quad B \in \widetilde{\mathcal{S}} \tag{9.21b}$$

*Further, $N$ takes values in the set $M_p$ of all counting measures $\nu$ on $\left(S, \widetilde{\mathcal{S}}\right)$ that are locally finite $\left(\nu\left(B\right) < \infty$, for bounded sets $B\right)$ and simple. Each measure $\nu \in M_p$ has the form*

$$\nu\left(B\right) = \sum_{n \in \mathbb{P}} \zeta_{x_n}\left(B\right), \quad B \in \widetilde{\mathcal{S}} \tag{9.21c}$$

*where $\{x_1, x_2, \ldots, x_k\}$ is its associated set of points, and $0 \leq k \leq \infty$.*

(a) *The point process on space $S$ is a measurable map $N$ from a probability space $(\Omega, \mathcal{F}, P)$ to the space $(M_p, \mathcal{M}_p)$. Note that $\mathcal{F}$ is the $\sigma$-field associated with the sample space $\Omega$, and $P\left(\cdot\right)$ is the probability function.*

(b) *The probability distribution of $N$ is determined by it finite-dimensional multivariate mass functions*

$$P\left(N\left(B_1\right) = n_1, N\left(B_2\right) = n_2, \ldots, N\left(B_k\right) = n_k\right) \tag{9.21d}$$

*where $B_i \in \widetilde{\mathcal{S}}$, and $n_i \in \mathbb{N}$, for $1 \leq i \leq k$.*

(c) *The mean of the random variable $N\left(B\right)$ is $\mathcal{E}\left(N\left(B\right)\right) = \Lambda\left(B\right)$. Also, $\Lambda\left(\cdot\right)$ is called the intensity measure of the point process.*

*Often $\Lambda\left(B\right) = \int_B \lambda\left(x\right) dx$, where $\lambda\left(\cdot\right)$ is called the rate or intensity function of the point process.*

*The point process is simply referred to as either $\Phi$ or $N$.*                                    $\square$

In describing a point process, it is often sufficient to simply define the probabilities on sets $B_i$ that generate $\widetilde{S}$.

**Notation.** Following the standard convention, and when appropriate, we denote the differential of a function, say $\Lambda(x)$, as $\Lambda(dx)$ and not $d\Lambda(x)$. $\qquad\square$

**Observation 9.9.** Let $N(B) = \sum_{n \in \mathbb{P}} \zeta_{X_n}(B)$, $B \in \widetilde{S}$ be the counting measure of a point process $\Phi$. The mean of the random variable $N(B)$ is $\mathcal{E}(N(B)) = \Lambda(B)$. For a function $f : S \to \mathbb{R}$, assume that $\int_S |f(x)| \Lambda(dx) < \infty$. Then

$$\mathcal{E}\left(\sum_{n \in \mathbb{P}} f(X_n)\right) = \int_S f(x) \Lambda(dx)$$

$\qquad\square$

See the problem section for a justification of this result.

### 9.5.1 Poisson Point Process

Poisson process was defined earlier in the chapter on the real line $\mathbb{R}$. The concept of occurrence of events at points on the real line $\mathbb{R}$ and counting the number of points in an interval $(a, b] \subset \mathbb{R}$, does not directly extend itself into Poisson processes in higher dimensions. A generalization of the Poisson process to dimension $m = 2, 3, 4, \ldots$ is called the Poisson point process. It finds application in the study of stochastic geometry. The Poisson process studied earlier can be considered to be a single dimension PPP.

**Definition 9.16.** *Poisson point process* (*PPP*). *A Poisson point process $\Phi$ is a countable set of random variables $X_i$'s, which take values in the set $S$. That is, $\Phi = \{X_n \mid X_n, n \in \mathbb{P}\}$. The set $S$ is typically equal to the Euclidean space $\mathbb{R}^m$, where $m \in \mathbb{P}$. Let $\widetilde{S}$ be the Borel $\sigma$-algebra of the set $S$. Also*

(a) *For every set $B \in \widetilde{S}$, the number of points $N(B)$, in it is*

$$N(B) = \sum_{n \in \mathbb{P}} \zeta_{X_n}(B) \tag{9.22a}$$

*The random variable $N(B)$ has a Poisson distribution with mean equal to $\Lambda(B) \in \mathbb{R}^+$. That is*

$$P(N(B) = k) = \exp(-\Lambda(B)) \frac{(\Lambda(B))^k}{k!}, \quad k \in \mathbb{N} \tag{9.22b}$$

*Also, $\Lambda(\cdot)$ is called the intensity measure of the point process, and it is locally finite.*

(b) *If $B_1, B_2, \ldots, B_m$ are disjoint sets in $\widetilde{S}$, then $N(B_1), N(B_2), \ldots, N(B_m)$ are mutually independent random variables.* $\qquad\square$

A PPP can either be homogeneous or nonhomogeneous. In a homogeneous PPP, the mean of the random variable $N(B)$ equals to $\Lambda(B) = \lambda |B|$, where $\lambda \in \mathbb{R}^+$, and $|B|$ denotes the hypervolume of the set $B$. Note that the homogeneous PPP is stationary, simple, and isotropic. In a nonhomogeneous PPP, $\Lambda(B) = \int_B \lambda(u)\, du$, and $\lambda(\cdot)$ is a positive function. That is, $\lambda(u) \in \mathbb{R}^+$ for all $u \in B$. The function $\lambda(\cdot)$ is also called the intensity or rate function of the PPP.

**Observation 9.10.** A Poisson point process $\Phi$, with counting measure $N\left(\cdot\right)$, and mean $\Lambda\left(B\right) \in \mathbb{R}^+$, $B \in \widetilde{S}$ is described by its finite-dimensional distribution as

$$P\left(N\left(B_1\right) = n_1, N\left(B_2\right) = n_2, \ldots, N\left(B_k\right) = n_k\right)$$

$$= \prod_{i=1}^{k}\left\{\exp\left(-\Lambda\left(B_i\right)\right)\frac{\left(\Lambda\left(B_i\right)\right)^{n_i}}{n_i!}\right\}$$

where $B_1, B_2, \ldots, B_k \in \widetilde{S}$ are mutually disjoint compact sets, and $n_i \in \mathbb{N}$, for $1 \leq i \leq k$. $\qquad\square$

**Observations 9.11.** Some elementary facts about PPPs.

1. Superposition of two independent homogeneous PPPs with intensities $\lambda_1$ and $\lambda_2$ results in a homogeneous PPP with intensity $\left(\lambda_1 + \lambda_2\right)$.
2. The *thinning* of homogeneous PPP. Select a point of the process with probability $p \in \left(0, 1\right)$ independently of other selections, and discard it with probability $\left(1 - p\right)$. This procedure results in two independent homogeneous PPPs of intensities $p\lambda$ and $\left(1 - p\right)\lambda$. $\qquad\square$

**Slivnyak's Theorem**

A remarkable result about homogeneous PPP's is the Slivnyak's theorem. The theorem states that a homogeneous PPP behavior is stochastically same whether it is observed from a randomly chosen point in the process, or from an independent location in space. This result is true only for homogeneous PPP's, and no other point process. The phrase "randomly chosen" point deserves special attention. This phrase simply means that every point in the process has equal chance of being selected.

We provide a motivation and qualitative explanation of this result via the next two examples. The following terminology is used in these examples. Denote the origin in space $\mathbb{R}^m$, $m \in \mathbb{P}$; by $o$. That is, $\left(0, 0, \ldots, 0\right) \triangleq o$. Let $b\left(x, r\right)$ be a hypersphere (or ball), centered at the point $x \in \mathbb{R}^m$, and of radius $r \in \mathbb{R}^+$. The volume of the hypersphere $b\left(o, r\right)$ is $\left|b\left(o, r\right)\right| = c_m r^m$, where $c_m = \pi^{m/2}/\Gamma\left(m/2 + 1\right)$, and $\Gamma\left(\cdot\right)$ is the gamma function. A hypersphere of radius $r \in \mathbb{R}^+$ is also sometimes denoted as $\mathcal{H}_r$.

**Example 9.5.** The distribution function of distance from $o$ to the nearest point of the homogeneous PPP $\Phi$ is computed. Denote this function by $H_s\left(r\right)$, where $r \geq 0$. We have

$$H_s\left(r\right) = 1 - P\left(N\left(b\left(o, r\right)\right) = 0\right)$$
$$= 1 - \exp\left(-\lambda\left|b\left(o, r\right)\right|\right)$$
$$= 1 - \exp\left(-\lambda c_m r^m\right)$$

where $\left|b\left(o, r\right)\right| = c_m r^m$. This distribution function is also called *spherical contact distribution*. $\qquad\square$

**Example 9.6.** The nearest-neighbor distance distribution function $D\left(\cdot\right)$ of the homogeneous PPP $\Phi$ is computed. It specifies the probability distribution of the distance of a typical point $x \in \Phi$ to the nearest point in $\Phi \backslash \left\{x\right\}$ (which is the nearest neighbor of $x$ in $\Phi$).

More specifically, let $0 \leq \varepsilon \leq r$. Assume that there is a point of homogeneous PPP $\Phi$ in $b\left(o, \varepsilon\right)$. We compute the probability that the distance from a point in the small sphere $b\left(o, \varepsilon\right)$ to its nearest neighbor in $\Phi$ is smaller than $r$, given that there is actually a point belonging to $\Phi$ exists in $b\,(o, \varepsilon)$. We denote this probability by $D_{\varepsilon}\,(r)$, where $r \geq 0$.

$$D_{\varepsilon}\,(r) = 1 - P\,(N\,(b\,(o, r) \setminus b\,(o, \varepsilon)) = 0 \mid N\,(b\,(o, \varepsilon)) = 1)$$

Using the definition of conditional probability, and the independence property of the PPP, we obtain

$$D_{\varepsilon}\,(r) = 1 - \frac{P\,(N\,(b\,(o, r) \setminus b\,(o, \varepsilon)) = 0)\, P\,(N\,(b\,(o, \varepsilon)) = 1)}{P\,(N\,(b\,(o, \varepsilon)) = 1)}$$

$$= 1 - P\,(N\,(b\,(o, r) \setminus b\,(o, \varepsilon)) = 0)$$

$$= 1 - \exp\,(-\lambda\,(|b\,(o, r)| - |b\,(o, \varepsilon)|))$$

$$= 1 - \exp\,(-\lambda c_m\,(r^m - \varepsilon^m))$$

The nearest-neighbor distance distribution function can be defined as $D\,(r) = \lim_{\varepsilon \to 0} D_{\varepsilon}\,(r)$. Thus

$$D\,(r) = 1 - \exp\,(-\lambda c_m r^m), \quad r \geq 0$$

$\square$

Note in the above two examples that $D\,(r) = H_s\,(r)$, where $r \geq 0$. As per Slivnyak's theorem, if a point $x$ is in an homogeneous PPP $\Phi$, then the probability law of $\Phi - \{x\}$ conditioned on the fact that $x \in \Phi$ is the same as the probability law of simply $\Phi$. As an example, the average number of points within a distance of $r$ from $x \in \mathbb{R}^m$ is:

$$\mathcal{E}\,(N\,(b\,(x, r) \setminus \{x\}) \mid x \in \Phi) = \mathcal{E}\,(N\,(b\,(x, r) \setminus \{x\}))$$

$$= \Lambda\,(b\,(x, r))$$

As per this theorem, for any homogeneous PPP $\Phi$, and for any function $f\,(\cdot)$ defined on the space $S$, we have: $P\,(f\,(\Phi) \mid x \in \Phi) = P\,(f\,(\Phi \cup \{x\}))$.

### 9.5.2  Laplace Functional

Characteristic functions, Laplace transforms, and moment generating functions are used with probability distribution functions. An analogous tool exists for working with point processes. This is the Laplace functional.

We denote a realization of the point process $\Phi$ as $\phi = \{x_1, x_2, x_3, \ldots\} \subset S$, where the space $S$ is typically $\mathbb{R}^m$. The corresponding counting measure is

$$N\,(B) = \sum_{n \in \mathbb{P}} \zeta_{x_n}\,(B), \quad B \in \tilde{S}$$

Let $f : S \to \mathbb{R}$, and

$$N\,(f) = \int_S f\,(x)\, N\,(dx)$$

It can be shown that

$$\int_S f\,(x)\, N\,(dx) = \sum_{n \in \mathbb{P}} f\,(X_n)$$

where it is assumed that the sum is finite. An informal justification of this result is provided in the problem section. We are ready to define Laplace functional of a point process $\Phi$.

**Definition 9.17.** *The function $f : S \to \mathbb{R}_0^+$ is continuous and has compact support. The Laplace functional of the point process $\Phi$ with counting measure $N(\cdot)$ is*

$$\mathcal{L}_N(f) = \mathcal{E}\left(\exp\left(-N(f)\right)\right) \tag{9.23}$$

*where $N(f) = \int_S f(x) N(dx)$.* □

In the above definition, the function $f$ is a "variable" of this expectation. This is similar to the parameter $s \in \mathbb{C}$ in the Laplace transform of the probability density function of a continuously distributed random variable $Y$, which is $\mathcal{E}\left(\exp\left(-sY\right)\right)$. Also the function $f$ has a compact support. That is, its support is a compact set.

**Observation 9.12.** The Laplace functional of the point process $\Phi$ is

$$\mathcal{L}_N(f) = \mathcal{E}\left(\exp\left(-\sum_{n \in \mathbb{P}} f(X_n)\right)\right), \qquad f : S \to \mathbb{R}_0^+$$

□

It is possible to show that the Laplace functional of a point process uniquely determines its distribution. Therefore, if

$$f(x) = \sum_{i=1}^{k} s_i I_{B_i}(x)$$

where $B_1, B_2, \ldots, B_k \in \widetilde{S}$ are disjoint, and $s_1, s_2, \ldots, s_k \in \mathbb{R}_0^+$; we have

$$N(f) = \sum_{i=1}^{k} s_i N(B_i)$$

Therefore

$$\mathcal{L}_N(f) = \mathcal{E}\left(\exp\left(-\sum_{i=1}^{k} s_i N(B_i)\right)\right)$$

The above result is justified in the problem section. Observe that the Laplace functional is a function of the vector $(s_1, s_2, \ldots, s_k)$. It is the joint Laplace transform of the random vector $(N(B_1), N(B_2), \ldots, N(B_k))$, whose distribution can be derived via this transform. Moreover, if $(B_1, B_2, \ldots, B_k)$ is specified over all bounded subsets of the space $S$, it is possible to obtain finite-dimensional distribution of the point process. The Laplace functional of the Poisson point process is next determined.

**Observation 9.13.** The Laplace functional of a nonhomogeneous Poisson point process with intensity measure $\Lambda(\cdot)$, and $f : S \to \mathbb{R}_0^+$ is

$$\mathcal{L}_N(f) = \exp\left\{-\int_S \left(1 - e^{-f(x)}\right) \Lambda(dx)\right\} \tag{9.24a}$$

The integral in the above expression is finite if and only if

$$\int_S \min\left(f\left(x\right),1\right) \Lambda\left(dx\right) < \infty \qquad (9.24\mathrm{b})$$

$\square$

See the problem section for a justification of this result. The mean and variance of $N\left(f\right)$ can also be conveniently determined via the following theorem.

**Theorem 9.4.** *Let* $f : S \to \mathbb{R}_0^+$ *be a continuous function with compact support, and* $\Phi = \{X_n \mid X_n, n \in \mathbb{P}\}$ *be a nonhomogeneous Poisson point process with intensity measure* $\Lambda\left(\cdot\right)$, *counting measure* $N\left(\cdot\right)$, *and* $N\left(f\right) = \int_S f\left(x\right) N\left(dx\right)$. *Then*:

(a) $\mathcal{E}\left(\sum_{n \in \mathbb{P}} f\left(X_n\right)\right) = \int_S f\left(x\right) \Lambda\left(dx\right)$, *provided the integral exists.*
(b) $Var\left(\sum_{n \in \mathbb{P}} f\left(X_n\right)\right) = \int_S \left(f\left(x\right)\right)^2 \Lambda\left(dx\right)$, *provided the integral exists.*

*Proof.* See the problem section.                                                               $\square$

### 9.5.3 Marked Point Process

It is often useful to assign a mark to each point in a point process. Consider a point process $\Phi = \{X_n \mid n \in \mathbb{P}\}$, where the random variables $X_n, n \in \mathbb{P}$ take values in the set $S$. Let a realization of the point process $\Phi$ be $\phi = \{x_1, x_2, x_3, \ldots\} \subset S$. A possible example of space $S$ is $\mathbb{R}^m$. Let $\widetilde{S}$ be the Borel $\sigma$-algebra of the set $S$. For every closed and bounded set $B \in \widetilde{S}$, the number of points $N\left(B\right)$ in it is

$$N\left(B\right) = \sum_{n \in \mathbb{P}} \zeta_{X_n}\left(B\right), \quad B \in \widetilde{S}$$

Assume that $\Phi$ is a PPP, and let the random variables $N\left(B\right)$ have a Poisson distribution with locally finite intensity function $\lambda\left(\cdot\right)$.

In a marked Poisson point process, each point $X_n \in \Phi$ is assigned a mark $M_n$, which takes values in the set $\mathbb{M}$. The set $\mathbb{M}$ is called the *space of marks*, or simply *mark space*. The random set

$$\Phi_M = \{(X_n, M_n) \mid X_n \in \Phi, n \in \mathbb{P}\}$$

is called the marked point process. Thus $\Phi_M$ is a subset of the product space $S \times \mathbb{M}$. Its corresponding unmarked process $\Phi$ is called the *ground process*.

It is possible for the set $\Phi_M$ to have a richer structure, if the $M_n$'s are random variables which take values in the set $\mathbb{M}$. The marked point process $\Phi_M$ is said to be *independently marked*, if $M_n$ is independent of $M_k$ for all $k \neq n$, where $k, n \in \mathbb{P}$; however $M_n$ may depend upon $X_n$. It is also possible for all $M_n$'s to be identically distributed as the random variable $M$. In this case, the process $\Phi_M$ is said to be *independent and identically marked*.

Let $\widetilde{\mathcal{M}}$ be the Borel $\sigma$-algebra of the set $\mathbb{M}$.

Assume that a point $x \in S$ has a value in a set $G \in \widetilde{\mathcal{M}}$ as per a probability kernel $p\left(x, G\right)$. This kernel is a mapping $p : S \times \widetilde{\mathcal{M}} \to [0, 1]$ such that $p\left(\cdot, G\right)$ is a measurable function on $S$, and $p\left(x, \cdot\right)$ is a probability measure on $\mathbb{M}$. The goal is to understand such marked point process defined on $S \times \mathbb{M}$.

It is indeed proper to ask whether the marked process $\Phi_M$ can be modeled as simply an ordinary point process defined over the product space $S \times \mathbb{M}$. The utility of the formulation of the marked point process might be demonstrated in the following example. Consider a marked point process

$\Phi_M$, in which each $x_n \in \phi$ is translated into $(x_n + x)$, where $x \in S$; however the mark of each point $x_n$ remains unchanged. The new marked point process will simply be

$$\Phi_{M_x} = \Phi_M + x$$

Similarly, it is possible to obtain a modified point process by rotating each point $x_n \in \phi$, without altering its mark.

**Definition 9.18.** *Marked point process. Let* $\Phi = \{X_n \mid n \in \mathbb{P}\}$, *be a Poisson point process, where the random variables* $X_n, n \in \mathbb{P}$, *take values in the set* $S$. *Also let* $\widetilde{S}$ *be the family of Borel sets of space* $S$. *That is, in every closed and bounded set* $B \in \widetilde{S}$, *the number of points* $N(B)$ *in it is*

$$N(B) = \sum_{n \in \mathbb{P}} \zeta_{X_n}(B) \tag{9.25a}$$

*where the random variable* $N(B)$ *has a Poisson distribution with locally finite intensity function* $\lambda(\cdot)$; *and* $\zeta_{X_n}(\cdot)$ *is a Dirac measure.*

*Also let* $\mathbb{M}$ *be the space of marks. The set*

$$\Phi_M = \{(X_n, M_n) \mid X_n \in \Phi, n \in \mathbb{P}\} \tag{9.25b}$$

*is a marked point process defined on* $S \times \mathbb{M}$. *The random variable* $M_n$ *takes values in the set* $\mathbb{M}$, *and*

$$P(M_n \in G \mid N) = p(X_n, G), \quad G \in \widetilde{\mathcal{M}}, \ n \leq N(S) \tag{9.25c}$$

*where* $p(x, G)$ *is a probability kernel from* $S$ *to* $\mathbb{M}$, *and* $\widetilde{\mathcal{M}}$ *is the family of Borel sets of* $\mathbb{M}$. *Also,* $|A \times \mathbb{M}|$ *is bounded for all* $A \subset \Phi$. *The counting measure of the point process* $\Phi_M$ *is*

$$M'(B \times G) = \sum_{n \in \mathbb{P}} \zeta_{X_n, M_n}(B, G), \quad B \in \widetilde{S}, \ G \in \widetilde{\mathcal{M}} \tag{9.25d}$$

*Typically,* $S$ *is equal to the Euclidean space* $\mathbb{R}^m$. $\qquad\square$

The next theorem demonstrates that $\Phi_M$ is a PPP defined over the product space $S \times \mathbb{M}$.

**Theorem 9.5.** *The marked point process* $\Phi_M$ *is a PPP defined over the product space* $S \times \mathbb{M}$, *where* $S$ *is equal to the Euclidean space* $\mathbb{R}^m$. *The intensity measure* $\Lambda_M$ *of this process is defined by*

$$\Lambda_M(B \times G) = \int_B p(x, G) \Lambda(dx), \quad B \in \widetilde{S}, G \in \widetilde{\mathcal{M}} \tag{9.26a}$$

*Consequently, the point process of marks*

$$\Phi' = \{M_n \mid M_n \in \mathbb{M}, n \in \mathbb{P}\} \tag{9.26b}$$

*with counting measure*

$$N'(G) = \sum_{n \in \mathbb{P}} \zeta_{M_n}(G), \quad G \in \widetilde{\mathcal{M}} \tag{9.26c}$$

*is a Poisson process on* $\mathbb{M}$ *with intensity measure* $\Lambda'(G) = \int_S p(x, G) \Lambda(dx), G \in \widetilde{\mathcal{M}}$, *assuming it is locally finite.*

*Proof.* See the problem section. $\qquad\square$

### 9.5.4 Transformation of Poisson Point Processes

It is possible for a PPP to get transformed into another PPP under suitable mappings. This phenomenon is next examined. The two types of transformations that we study are mapping and displacement.

#### Mapping

It is possible for a certain PPP defined on some underlying space and with some mean value, to get transformed into another PPP possibly of different dimension and a different mean value via some mapping. The mapping theorem specifies this type of transformation.

**Theorem 9.6.** *Mapping theorem. Let $\Phi$ be a PPP defined on space $\mathbb{R}^m$ with intensity function $\lambda(\cdot)$, and intensity measure $\Lambda(\cdot)$. Also let $f : \mathbb{R}^m \to \mathbb{R}^d$ be a measurable function such that $\Lambda\left(f^{-1}\{y\}\right) = 0$; for all $y \in \mathbb{R}^d$, and $f(\cdot)$ does not map a (non-singleton) compact set to a singleton. Then the process*

$$\widetilde{\Phi} = f(\Phi) \triangleq \bigcup_{X \in \Phi} \{f(X)\} \tag{9.27a}$$

*is also a PPP with intensity measure*

$$\widetilde{\Lambda}\left(\widetilde{B}\right) = \Lambda\left(f^{-1}\left(\widetilde{B}\right)\right) = \int_{f^{-1}(\widetilde{B})} \lambda(x)\,dx, \quad \text{for all compact sets } \widetilde{B} \subset \mathbb{R}^d \tag{9.27b}$$

*Proof.* See the problem section.                                                                                      $\square$

#### Displacement

It is possible for points of a point process to move either in a deterministic or random manner. If the underlying process is a PPP, and the points in it move or are displaced in some random order, then it is possible for the resulting point locations to also be a PPP. Let the initial point process be $\Phi$, and the displaced point process be $\widetilde{\Phi}$. Then

$$\widetilde{\Phi} = \{Y_n \mid Y_n = X_n + V_{X_n}, n \in \mathbb{P}\}$$

where the random variable $V_{X_n}$ is possibly dependent upon $X_n$. However the $V_{X_n}$'s are independent of each other. Let some realization of the point processes $\Phi$ and $\widetilde{\Phi}$ be $\phi$ and $\widetilde{\phi}$ respectively. Then $\widetilde{\phi} = \{y \mid y = x + V_x, x \in \phi\}$, where the random variable $V_x$ possibly depends upon $x$, and the $V_x$'s are independent of each other. If $\Phi$ is a PPP, then $\widetilde{\Phi}$ is also a PPP as per the displacement theorem.

**Theorem 9.7.** *Displacement theorem. Let $\Phi$ be a PPP defined on $\mathbb{R}^m$ with intensity function $\lambda(\cdot)$. Further, all points are independently displaced, and the distribution of the displaced location of a point $x$ in the point process $\Phi$ has a probability density function $\rho(x, \cdot)$. Then the displaced points form a PPP $\widetilde{\Phi}$ with intensity function*

$$\widetilde{\lambda}(y) = \int_{\mathbb{R}^m} \lambda(x)\,\rho(x, y)\,dx \tag{9.28}$$

*More specifically, if $\lambda(x)$ is equal to a constant $\lambda \in \mathbb{R}^+$, and $\rho(\cdot, \cdot)$ depends only on $(y - x)$, then $\widetilde{\lambda}(y)$ is equal to the constant $\lambda$.*

*Proof.* See the problem section. □

Observe that $\rho(x, \cdot)$ is the probability density function of the new location $(x + V_x)$. The density function $\rho(\cdot, \cdot)$ is also called the *displacement kernel*.

## 9.6 Renewal Theory

Recall that the Poisson process is an independent-increment counting process, where the interarrival times are independent and identically distributed exponential random variables. If the interarrival times are generalized to an arbitrary distribution, a counting process called a *renewal process* is obtained.

Consider the *counting process* $\{N(t), t \in \mathbb{R}_0^+\}$. It is a stochastic process, where $N(t)$ represents the total count of events that have occurred from time $0$ till time $t$. Observe that $N(t)$ takes integer values. Furthermore, if $u < v$, then $N(u) \leq N(v)$. Consequently the total number of events in the time interval $(u, v]$ is equal to $(N(v) - N(u))$. A counting process is an independent-increment process, if the number of events in disjoint time intervals occur independent of each other.

There are several types of renewal processes. Some of these are: ordinary renewal process, modified renewal process, alternate renewal process, equilibrium renewal process, and equilibrium alternate renewal process.

### 9.6.1 Ordinary Renewal Process

Definition of ordinary renewal process is given below.

**Definition 9.19.** *Let $\{N(t), t \in \mathbb{R}_0^+\}$ be a counting process, where $N(t)$ represents the total count of events that have occurred from time $0$ till time $t$. Denote the time interval between the $(n-1)$th and the $n$th event by $X_n$ where $n \in \mathbb{P}$. If the sequence of nonnegative random variables $\{X_n, n \in \mathbb{P}\}$ are independent and identically distributed, then $\{N(t), t \in \mathbb{R}_0^+\}$ is a renewal process.*

*Thus, $N(t)$ is the number of renewals till time $t$. This renewal process is called the ordinary renewal process.* □

Other types of renewal processes are defined subsequently. Define $X$ to be a generic random variable of the sequence $\{X_n, n \in \mathbb{P}\}$. It is a continuously distributed random variable. Also define the probability density function, the cumulative distribution function, and the complementary cumulative distribution function of the random variable $X$ to be $f(t), F(t),$ and $F^c(t)$ respectively where $t \in \mathbb{R}_0^+$. For simplicity, assume that $P(X > 0) > 0$, and $P(X < \infty) = 1$. Let $S_n$ be the time of the $n$th renewal. Then

$$S_0 = 0, \text{ and } S_n = \sum_{j=1}^{n} X_j, \ n \geq 1 \tag{9.29}$$

Let the probability density function of the random variable $S_n$ be $f_n(t)$, and the corresponding cumulative distribution function be $F_n(t)$ for $t \in \mathbb{R}_0^+$.

Also define $M_o(t) = \mathcal{E}(N(t))$ for $t \in \mathbb{R}_0^+$. It is called the *mean value* or the *renewal function*. The *renewal density* of the ordinary renewal process is $m_o(t) = dM_o(t)/dt$. Backward and forward recurrence times of the ordinary renewal process are next defined.

*Backward recurrence-time or the age of the renewal process*: Let $t$ be a fixed time-point. The backward recurrence time $U(t)$ is the time, measured backwards from time $t$ to the last renewal at or before $t$. However, if there have been zero number of renewals up to time $t$, then $U(t)$ is defined to be equal to $t$. Thus this value is the age of the renewal process. That is, $U(t) = (t - S_{N(t)})$.

*Forward recurrence-time or the excess of renewal process*: Let $t$ be a fixed time-point. The forward recurrence time $V(t)$ is the time, measured forwards from time $t$ to the next renewal to occur after $t$. It is also sometimes called the *residual life-time* or the excess of the renewal process. It is given by $V(t) = (S_{N(t)+1} - t)$.

The above definitions yield the following observations. Some of these observations are proved in the problem section.

**Observations 9.14.** Some basic observations about renewal processes.

1. $N(t) = \max\{n \mid S_n \leq t\}$
2. $S_n \leq t$ if and only if $N(t) \geq n$
3. $P(N(t) \leq n) = P(S_{n+1} > t)$
4. $P(N(t) = n) = F_n(t) - F_{n+1}(t)$. This relationship is useful in evaluating the moments of the random variable $N(t)$.
5. $M_o(t) = \mathcal{E}(N(t)) = \sum_{n=1}^{\infty} F_n(t)$
6. $M_o(t) < \infty, \ \forall\, t \in \mathbb{R}_0^+$
7. $m_o(t) = \sum_{n=1}^{\infty} f_n(t)$
8. $\lim_{t \to \infty} N(t)/t \to 1/\mathcal{E}(X)$
9. $\lim_{t \to \infty} S_{N(t)}/N(t) \to \mathcal{E}(X)$ or $\lim_{n \to \infty} S_n/n \to \mathcal{E}(X)$                          □

It is convenient to study some of these concepts in the Laplace-transform domain. Define the Laplace-transform pairs as follows.

$$f(t) \Longleftrightarrow \widehat{f}(s), \quad F(t) \Longleftrightarrow \widehat{F}(s)$$
$$f_n(t) \Longleftrightarrow \widehat{f}_n(s), \quad F_n(t) \Longleftrightarrow \widehat{F}_n(s)$$
$$m_o(t) \Longleftrightarrow \widehat{m}_o(s), \quad M_o(t) \Longleftrightarrow \widehat{M}_o(s)$$

where $s \in \mathbb{C}$. Following observations can be made from these definitions.

**Observations 9.15.** Some basic observations.

1. $\widehat{f}(s) = s\widehat{F}(s), \widehat{f}_n(s) = s\widehat{F}_n(s)$, and $\widehat{m}_o(s) = s\widehat{M}_o(s)$.
2. Since convolution in time domain is equivalent to product in $s$-domain, it can be inferred that

$$\widehat{f}_n(s) = \left(\widehat{f}(s)\right)^n$$

3. As

$$\widehat{m}_o(s) = \sum_{n=1}^{\infty} \left(\widehat{f}(s)\right)^n$$

we have

$$\widehat{m}_o(s) = \frac{\widehat{f}(s)}{\left(1 - \widehat{f}(s)\right)} \tag{9.30a}$$

$$\widehat{M}_o(s) = \frac{\widehat{f}(s)}{s\left(1 - \widehat{f}(s)\right)} \tag{9.30b}$$

4. The well-known *ordinary renewal equation* in time domain is now derived. The above equations lead to

$$\widehat{M}_o(s) = \frac{\widehat{f}(s)}{s} + \widehat{M}_o(s)\,\widehat{f}(s) \tag{9.31a}$$

$$M_o(t) = F(t) + \int_0^t M_o(t - u)\, f(u)\, du \tag{9.31b}$$

$$m_o(t) = f(t) + \int_0^t m_o(t - u)\, f(u)\, du \tag{9.31c}$$

5. $\lim_{t \to \infty} m_o(t) = 1/\mathcal{E}(X)$                                                                                    □

### 9.6.2  Modified Renewal Process

In the modified renewal process, the first interarrival time has a different distribution from the other interarrival times. This scenario might occur when the observation of a renewal process starts at some time $t > 0$, after the renewal process has already started. Consequently, the distribution of the first interarrival time might be different than those of the subsequent interarrival times. This process is sometimes called the *delayed renewal counting process*.

Let $\{N(t), t \in \mathbb{R}_0^+\}$ be a counting process. Also let the interarrival time of the events be given by the sequence $\{X_n, n \in \mathbb{P}\}$. The random variables $X_1, X_2, X_3, \ldots$ are continuously distributed, and independent of each other. Furthermore, the random variables $X_2, X_3, \ldots$ are identically distributed. Let $X$ be a generic nonnegative random variable of the sequence $\{X_n, n \geq 2\}$. Denote the probability density function and the cumulative distribution function of the random variable $X$ by $f(t)$ and $F(t)$ respectively where $t \in \mathbb{R}_0^+$. The probability density function of the random variable $X_1$ is $f_1(t), t \in \mathbb{R}_0^+$. The Laplace transform of $f(t)$ and $f_1(t)$ are $\widehat{f}(s)$ and $\widehat{f}_1(s)$ respectively, where $s \in \mathbb{C}$.

Also define the modified renewal function by $M_m(t) = \mathcal{E}(N(t)), t \in \mathbb{R}_0^+$. Its renewal density is $m_m(t) = dM_m(t)/dt$. The corresponding Laplace transform of these functions are $\widehat{M}_m(s)$ and $\widehat{m}_m(s)$ respectively. It can be shown as in the case of ordinary renewal process, that

$$\widehat{m}_m(s) = \frac{\widehat{f}_1(s)}{\left(1 - \widehat{f}(s)\right)} \tag{9.32a}$$

$$\widehat{M}_m(s) = \frac{\widehat{f}_1(s)}{s\left(1 - \widehat{f}(s)\right)} \tag{9.32b}$$

where $s \in \mathbb{C}$. The modified renewal equations can be derived from the above results.

### 9.6.3  Alternating Renewal Process

Alternating renewal process has two states, "on" and "off." Assume *initially* that the process is in "on" state. It remains in this state for a time $A_1$; it then goes into an "off" state and remains in it for a time $B_1$; it then goes into "on" state for a time $A_2$; then again into "off" state for a time $B_2$. This cycle of "on" and "off" states continues ad infinitum. See Figure 9.1.

$$\vdash -A_1- \dashv -- B_1 - \dashv -- A_2- \dashv -- B_2 - \dashv -- A_3- \dashv -- B_3 - \dashv \cdots\cdots$$

$$\quad On \qquad\qquad Off \qquad\qquad On \qquad\qquad Off \qquad\qquad On \qquad\qquad Off$$

Figure 9.1 Alternating renewal process.

Assume that the sequence of continuous random variables $\{A_n, n \in \mathbb{P}\}$ are independent and identically distributed nonnegative random variables. Let their probability density function, cumulative distribution function, and complementary cumulative distribution function be $f_1(t), F_1(t),$ and $F_1^c(t)$ respectively, where $t \in \mathbb{R}_0^+$.

Also, assume that the sequence of continuous random variables $\{B_n, n \in \mathbb{P}\}$ are independent, and identically distributed nonnegative random variables. Let their probability density function, cumulative distribution function, and complementary cumulative distribution function be $f_2(t), F_2(t),$ and $F_2^c(t)$ respectively, where $t \in \mathbb{R}_0^+$.

Further assume that the sequences, $A_n$'s and $B_n$'s are independent of each other. Let the Laplace transforms of $f_1(t)$ and $f_2(t)$ be $\widehat{f}_1(s)$ and $\widehat{f}_2(s)$ respectively, where $s \in \mathbb{C}$.

Let $C_n = (A_n + B_n)$ for each $n \in \mathbb{P}$. The time $C_n$ corresponds to the length of a cycle of "on" and "off" times. At the end of each such cycle, there is a transition from "off" to "on" state. Denote the generic random variables corresponding to the random variables $A_n, B_n,$ and $C_n$ by $A, B,$ and $C$ respectively. The Laplace transform of the probability density function of the random variable $C = (A + B)$ is $\widehat{f}_1(s)\widehat{f}_2(s)$. Also let $\mathcal{E}(A) = \mu_1,$ and $\mathcal{E}(B) = \mu_2$.

Define $N_1(t)$ to be the number of renewals of the "on" state process. This is a modified renewal process with the initial distribution that of the random variable $A$, and the subsequent distributions equal to that of the random variable $C$.

Define $N_2(t)$ to be the number of renewals of the "off" state process. This is an ordinary renewal process generated by a sequence of random variables $\{(A_n + B_n), n \in \mathbb{P}\}$. The distribution of the interarrival times of the process is that of the random variable $C$.

Let $M_j(t) = \mathcal{E}(N_j(t))$ and $m_j(t) = dM_j(t)/dt,$ for $j = 1, 2$. The Laplace transform of $M_j(t)$ is $\widehat{M}_j(s), s \in \mathbb{C}$ for $j = 1, 2$. Use of renewal theory yields

$$\widehat{M}_1(s) = \frac{\widehat{f}_1(s)}{s\left\{1 - \widehat{f}_1(s)\widehat{f}_2(s)\right\}} \tag{9.33a}$$

$$\widehat{M}_2(s) = \frac{\widehat{f}_1(s)\widehat{f}_2(s)}{s\left\{1 - \widehat{f}_1(s)\widehat{f}_2(s)\right\}} \tag{9.33b}$$

The corresponding Laplace transforms of the renewal densities are

$$\widehat{m}_j(s) = s\widehat{M}_j(s), \quad \text{for } j = 1, 2$$

Let $\pi_1(t)$ and $\pi_2(t)$ be the probability that the system is in "on" and "off" state respectively at time $t$.

**Theorem 9.8.** *For an alternating renewal process starting in "on" state at time $t = 0$*

$$\pi_1(t) = (M_2(t) - M_1(t) + 1) \tag{9.34a}$$
$$\pi_2(t) = (M_1(t) - M_2(t)) \tag{9.34b}$$

*and*

$$\lim_{t \to \infty} \pi_1(t) = \frac{\mu_1}{(\mu_1 + \mu_2)} \tag{9.35a}$$

$$\lim_{t \to \infty} \pi_2(t) = \frac{\mu_2}{(\mu_1 + \mu_2)} \tag{9.35b}$$

*Proof.* The $r$th renewal epoch of the renewal process $\{N_2(t), t \in \mathbb{R}_0^+\}$ is given by $S_r = \sum_{j=1}^{r} C_j$. The probability density function of $S_r$ is obtained by the $r$-fold convolution of the density function of the random variable $C$. Denote it by $g_r(t)$. Let its Laplace transform be $\hat{g}_r(s)$, which yields $\hat{g}_r(s) = \left\{ \hat{f}_1(s) \hat{f}_2(s) \right\}^r, r \in \mathbb{P}$.

The probability $\pi_1(t)$ that the system is in "on" state at time $t$ is next determined. The event that the system is in "on" state at time $t$ is equivalent to the following two mutually exclusive events.

(a) The length of the initial "on" state in which the system started at $t = 0$ is greater than $t$. The probability of this event is $P(A > t) = F_1^c(t)$.
(b) For some $x < t$, in the interval $(x, x + dx)$, a renewal epoch $S_r$ takes place and the process remains in "on" state for the time $(t - x)$. Note that the index $r$ takes values $1, 2, 3, \ldots$. The probability of this event is

$$\int_0^t \left\{ \sum_{r=1}^{\infty} P(x \le S_r < x + dx) \right\} P(A > t - x)$$

$$= \int_0^t \sum_{r=1}^{\infty} g_r(x) F_1^c(t - x) \, dx = \int_0^t m_2(x) F_1^c(t - x) \, dx$$

Combining the two probabilities results in

$$\pi_1(t) = F_1^c(t) + \int_0^t m_2(x) F_1^c(t - x) \, dx$$

If the Laplace transform of $\pi_j(t)$ is $\hat{\pi}_j(s)$, for $j = 1, 2$ then

$$\hat{\pi}_1(s) = \left( \frac{1 - \hat{f}_1(s)}{s} \right) (1 + \hat{m}_2(s))$$

$$= \frac{\left( 1 - \hat{f}_1(s) \right)}{s \left\{ 1 - \hat{f}_1(s) \hat{f}_2(s) \right\}}$$

It follows via algebraic manipulation that

$$\hat{\pi}_1(s) = \left( \widehat{M}_2(s) - \widehat{M}_1(s) + 1/s \right)$$

Inverting this equation yields

$$\pi_1(t) = (M_2(t) - M_1(t) + 1)$$

As $\pi_1(t) + \pi_2(t) = 1$, we have

$$\pi_2(t) = (M_1(t) - M_2(t)), \quad \text{and} \quad \widehat{\pi}_2(s) = \left(\widehat{M_1}(s) - \widehat{M_2}(s)\right)$$

Evaluating the limit as $t$ approaches infinity results in

$$\lim_{t\to\infty} \pi_1(t) = \lim_{s\to 0} s\widehat{\pi}_1(s) = \frac{\mu_1}{(\mu_1 + \mu_2)}$$

$$\lim_{t\to\infty} \pi_2(t) = \lim_{s\to 0} s\widehat{\pi}_2(s) = \frac{\mu_2}{(\mu_1 + \mu_2)}$$

$\square$

### 9.6.4 Backward and Forward Recurrence-Times

The distribution functions of the backward and forward recurrence-times of an ordinary renewal process as time $t$ approaches infinity are determined by using the results of the last subsection. Therefore, we initially summarize the results of the last subsection. Consider an alternating renewal process. In this process, each renewal interval $C_n$, consists of two successive intervals $A_n$ and $B_n$ respectively, such that $C_n = (A_n + B_n)$ for each $n \in \mathbb{P}$. During the interval $A_n$, the process is considered to be in an "on" state, while during the interval $B_n$, the process is in an "off" state. The tuples $(A_n, B_n)$ for each $n \in \mathbb{P}$ are independent and identically distributed. Also the sequences of random variables $\{A_n, n \in \mathbb{P}\}$ and $\{B_n, n \in \mathbb{P}\}$ are stochastically independent of each other. Thus each renewal cycle consists of "on" and "off" intervals. Let the three corresponding generic random variables be $A, B$, and $C$ respectively. Thus $\mathcal{E}(A) = \mathcal{E}(A_n) = \mu_1, \mathcal{E}(B) = \mathcal{E}(B_n) = \mu_2$, and $\mathcal{E}(C) = \mathcal{E}(C_n) = (\mu_1 + \mu_2)$ for all $n \in \mathbb{P}$. Also denote the proportion of time the process is in "on" state by $P_{on}$, and the proportional of time the process is in "off" state by $P_{off}$. Then $P_{on} = \mathcal{E}(A)/\mathcal{E}(C), P_{off} = \mathcal{E}(B)/\mathcal{E}(C)$, and $(P_{on} + P_{off}) = 1$.

Recall that the age of the renewal process is equal to $U(t)$. Its cumulative distribution $P(U(t) \leq y)$, as $t \to \infty$ is determined as follows. In this model, a cycle corresponds to a renewal interval. Let the system be in "on" state at time $t$ if the age at $t$ is less than or equal to $y$, and it is in "off" state if the age at $t$ is greater than $y$. That is, the system is in "on" state during the initial $y$ time units of the cycle, and "off" the remaining time. Recall that the random variable $X$ denotes a renewal interval. Also denote the complementary cumulative distribution function of the random variable $X$, by $F^c(\cdot)$. The limiting distribution of $U(t)$ is

$$\lim_{t\to\infty} P(U(t) \leq y) = \frac{\mathcal{E}(\min(X, y))}{\mathcal{E}(X)} = \frac{\int_0^\infty P(\min(X, y) > x)\, dx}{\mathcal{E}(X)}$$

$$= \frac{\int_0^y P(X > x)\, dx}{\mathcal{E}(X)} = \frac{\int_0^y F^c(x)\, dx}{\mathcal{E}(X)}$$

Therefore, as $t \to \infty$, $P(U(t) \leq y) = \int_0^y F^c(x)\, dx/\mathcal{E}(X)$. It can also be shown that as $t \to \infty$, $\mathcal{E}(U(t)) = \mathcal{E}(X^2)/(2\mathcal{E}(X))$.

It can be similarly demonstrated that the cumulative distribution function of the forward recurrence time $P(V(t) \leq y)$ as $t \to \infty$ is equal to $\int_0^y F^c(x)\, dx / \mathcal{E}(X)$. This is identical to the corresponding cumulative distribution function of the backward recurrence time. Therefore as $t \to \infty$, $\mathcal{E}(V(t)) = \mathcal{E}(X^2)/(2\mathcal{E}(X))$. Summarizing

$$\lim_{t \to \infty} P(U(t) \leq y) = \lim_{t \to \infty} P(V(t) \leq y) = \frac{\int_0^y F^c(x)\, dx}{\mathcal{E}(X)} \qquad (9.36a)$$

$$\lim_{t \to \infty} \mathcal{E}(U(t)) = \lim_{t \to \infty} \mathcal{E}(V(t)) = \frac{\mathcal{E}(X^2)}{2\mathcal{E}(X)} \qquad (9.36b)$$

Also observe that, as $t \to \infty$ the probability density function of the random variables $U(t)$ and $V(t)$ is $F^c(x)/\mathcal{E}(X)$, $x \in \mathbb{R}_0^+$.

### 9.6.5  Equilibrium Renewal Process

The equilibrium renewal process involves a system that has been running a long time before it is first observed.

It is a special case of the modified renewal process, where the probability density function $f_1(x) = F^c(x)/\mathcal{E}(X)$, $x \in \mathbb{R}_0^+$. Note that this is the probability density function of the age of the ordinary renewal process, as time $t$ approaches infinity. This result is used in the next subsection.

### 9.6.6  Equilibrium Alternating Renewal Process

The equilibrium alternating renewal process is considered in this subsection. In these systems, observation of the renewal process begins much later than the beginning of the process. Following three cases can be conveniently studied.

(a) The time origin $t = 0$ begins a long time after the process has started. The process is in "on" state at time $t = 0$. Therefore, the distribution of the random variable $A_1$ is $F_1^c(x)/\mu_1$, $x \in \mathbb{R}_0^+$. All other distributions of interarrival times $A_n$'s for $n \geq 2$ and $B_n$'s for $n \geq 1$ remain unchanged.

(b) This case is like the previous case, except that the process is in "off" state at time $t = 0$.

(c) The state of the process when it started is not known.

The equilibrium probability of the state of the renewal process to be in "on" state at $t = 0$, and at time $t > 0$ is next determined. Define this probability to be $\pi_{11}(t)$. Also recall that $N_2(t)$ is the number of renewals of the "off" state process. Let $m_{12}(t)$ be the renewal density of the $N_2(t)$ process, given that at equilibrium, the process was in "on" state at $t = 0$. Thus

$$\pi_{11}(t) = \int_t^\infty \frac{F_1^c(x)}{\mu_1}\, dx + \int_0^t m_{12}(x) F_1^c(t - x)\, dx$$

Let the Laplace transform of $\pi_{11}(t)$ and $m_{12}(t)$ be $\widehat{\pi}_{11}(s)$ and $\widehat{m}_{12}(s)$ respectively. Therefore

$$\widehat{m}_{12}(s) = \frac{\widehat{f}_2(s)\left(1 - \widehat{f}_1(s)\right)}{\mu_1 s \left\{1 - \widehat{f}_1(s)\, \widehat{f}_2(s)\right\}}$$

and

$$\widehat{\pi}_{11}(s) = \frac{\left\{\mu_1 s - 1 + \widehat{f}_1(s)\right\}}{\mu_1 s^2} + \widehat{m}_{12}(s) \frac{\left(1 - \widehat{f}_1(s)\right)}{s}$$

Finally

$$\widehat{\pi}_{11}(s) = \frac{1}{s} - \frac{\left(1 - \widehat{f}_1(s)\right)\left(1 - \widehat{f}_2(s)\right)}{\mu_1 s^2 \left\{1 - \widehat{f}_1(s)\,\widehat{f}_2(s)\right\}} \tag{9.37}$$

It can be shown that

$$\lim_{t\to\infty} \pi_{11}(t) = \lim_{s\to 0} s\widehat{\pi}_{11}(s) = \frac{\mu_1}{(\mu_1 + \mu_2)}$$

**Example 9.7.** The parameters discussed in this section on renewal theory are evaluated for the Poisson process. In this process, the random variable $X$ is exponentially distributed with parameter $\lambda$. Therefore $\mathcal{E}(X) = \lambda^{-1}$, and $\mathcal{E}(X^2) = 2\lambda^{-2}$. Also

$$f(t) = \lambda e^{-\lambda t}$$
$$F(t) = \left(1 - e^{-\lambda t}\right), \quad t \geq 0$$
$$P(N(t) = n) = \frac{(\lambda t)^n}{n!} e^{-\lambda t}, \quad \forall n \in \mathbb{N}$$
$$m_o(t) = \lambda$$
$$M_o(t) = \lambda t, \quad t \geq 0$$

and

$$\widehat{f}(s) = \frac{\lambda}{(\lambda + s)}$$
$$\widehat{F}(s) = \frac{\lambda}{s(\lambda + s)}$$
$$\widehat{f}_n(s) = \frac{\lambda^n}{(\lambda + s)^n}$$
$$\widehat{F}_n(s) = \frac{\lambda^n}{s(\lambda + s)^n}$$
$$\widehat{m}_o(s) = \frac{\lambda}{s}$$
$$\widehat{M}_o(s) = \frac{\lambda}{s^2}$$

As $t \to \infty$, the backward and forward recurrence times have an exponential distribution, each with parameter $\lambda$. It can also be shown that this counting process is an equilibrium renewal process.                                                                                                         □

## 9.7  Markov Processes

A *Markov process* (MP) is a stochastic process $\{X(t), t \in T\}$, with the special property that given $X(t)$, the probability of $X(s+t)$ where $s > 0$, is independent of the values of $X(w), w < t$.

That is, the future of the stochastic process depends only on the present and not upon its past history. Such processes are named after the probabilist Andrei Andreyevich Markov, who first studied such processes systematically. A formal definition of a Markov process is given below.

**Definition 9.20.** *A stochastic process* $\{X(t), t \in T\}$ *is a Markov process if for each state* $x$,

$$P\left(X\left(t\right) \le x \mid X\left(t_1\right) = x_1, X\left(t_2\right) = x_2, \ldots, X\left(t_n\right) = x_n\right)$$
$$- P\left(X\left(t\right) \le x \mid X\left(t_n\right) = x_n\right) \tag{9.38}$$

*whenever* $n \in \mathbb{P}$, $t_1 < t_2 < \ldots < t_n < t$, *and* $t_i \in T$ *for* $1 \le i \le n$. ☐

The index set $T$ is generally assumed to be time. Time (parameter values in the set $T$) can be either discrete or continuous. Also, the state-space can be either discrete or continuous. A Markov process is called a *Markov chain* (MC) if its state-space has discrete values. If the state-space has continuous values, then the process is simply called a *Markov process*. This classification of Markov processes is conveniently summarized below.

1. *Discrete-time process*:
   (a) *Discrete state-space*: A discrete-time discrete state-space Markov process is called a *discrete-time Markov chain*.
   (b) *Continuous state-space*: A discrete-time continuous state-space Markov process is called a *discrete-time Markov process*.
2. *Continuous-time process*:
   (a) *Discrete state-space*: A continuous-time discrete state-space Markov process is called a *continuous-time Markov chain*.
   (b) *Continuous state-space*: A continuous-time continuous state-space Markov process is called a *continuous-time Markov process*.

### 9.7.1 Discrete-Time Markov Chains

A Markov chain has discrete state-space. For convenience label the states of a MC from the set $\{0, 1, 2, \ldots\}$. This set of states is the set of integers $\mathbb{N}$. If the MC is defined at *discrete times* (DT), assume that the *state transitions* (change of state) occur at times $t_n$, for all $n \in \mathbb{N}$. Let $X\left(t_n\right) \triangleq X_n$, then this *discrete-time Markov chain* (DTMC) is denoted by $\{X_n, n \in \mathbb{N}\}$. Recall that all possible values of the random variable $X_n, n \in \mathbb{N}$ is called its state-space. Also if $X_n = k$, then the process is said to be in state $k$ at time $n$. A formal definition of a DTMC is given below.

**Definition 9.21.** *The stochastic process* $\{X_n, n \in \mathbb{N}\}$ *is a discrete-time Markov chain, if each random variable* $X_n$ *takes values in a finite or countable number of values. Assume that the state-space of these random variables is the set* $\mathbb{N}$. *These random variables satisfy the following Markov property. That is*

$$P\left(X_{n+1} = x_{n+1} \mid X_n = x_n, X_{n-1} = x_{n-1}, \ldots, X_0 = x_0\right)$$
$$= P\left(X_{n+1} = x_{n+1} \mid X_n = x_n\right), \quad \forall\, n \in \mathbb{N},\ \forall\, x_i \in \mathbb{N},\ 0 \le i \le (n+1) \tag{9.39}$$

*The conditional probability* $P\left(X_{n+1} = x_{n+1} \mid X_n = x_n\right), \forall\, n \in \mathbb{N}$ *is called the transition probability of the DTMC.* ☐

Thus in a DTMC, we have a random sequence in which dependency is by only a single unit backward in time. The state-space in the above definition of DTMC could have been any other countable state-space, instead of $\mathbb{N}$.

A framework for specifying transition probabilities of a DTMC is developed below. Criteria for classification of the states of a DTMC are also outlined. These are followed by equations to determine stationary probabilities of an ergodic DTMC. Finally, these ideas are clarified via some well-known examples.

**Transition Probabilities of a DTMC**

The transition probabilities of a DTMC essentially define the structure of a DTMC. The specification of these probabilities in turn culminates in the so-called Chapman-Kolmogorov equations of the DTMC.

**Definitions 9.22.** *Let $\{X_n, n \in \mathbb{N}\}$ be a DTMC with state-space $\mathbb{N}$.*

1. *The single-step transition probabilities of a DTMC are*

$$P\left(X_{n+1} = j \mid X_n = i\right), \quad \forall\, n, i, j \in \mathbb{N} \tag{9.40a}$$

*If the transition probabilities are independent of $n$, then denote*

$$P\left(X_{n+1} = j \mid X_n = i\right) = p_{ij}, \quad \forall\, n, i, j \in \mathbb{N} \tag{9.40b}$$

*This Markov chain is said to have homogeneous transition probabilities. Alternately the Markov chain is said to be a DTMC homogeneous in time.*

2. *The transition probabilities of a time-homogeneous DTMC can be represented by the transition probability matrix $\mathcal{P}$. This is a square matrix. If the number of states is infinite, then it is a matrix of infinite size. However if the number of states is finite then the matrix $\mathcal{P}$ is of finite size. Let $\mathcal{P} = [p_{ij}]$, where*

$$p_{ij} \geq 0, \quad \forall\, i, j \in \mathbb{N} \tag{9.41a}$$

$$\sum_{j \in \mathbb{N}} p_{ij} = 1, \quad \forall\, i \in \mathbb{N} \tag{9.41b}$$

$$\mathcal{P} = \begin{bmatrix} p_{00} & p_{01} & p_{02} & \cdots \\ p_{10} & p_{11} & p_{12} & \cdots \\ \vdots & \vdots & \vdots & \ddots \\ p_{n0} & p_{n1} & p_{n2} & \cdots \\ \vdots & \vdots & \vdots & \ddots \end{bmatrix} \tag{9.41c}$$

3. *The $n$-step transition probabilities are*

$$P\left(X_{m+n} = j \mid X_m = i\right), \quad \forall\, m, n, i, j \in \mathbb{N} \tag{9.42a}$$

*It is the probability that the process goes from state $i$ to state $j$ in $n$ steps. If the transition probabilities are independent of $m$, then denote*

$$P\left(X_{m+n} = j \mid X_m = i\right) = p_{ij}^{(n)}, \quad \forall\, m, n, i, j \in \mathbb{N} \tag{9.42b}$$

where $p_{ij}^{(1)} = p_{ij}$. Also $p_{ij}^{(0)} = \delta_{ij}$, where $\delta_{ij} = 1$ if $i = j$, and $0$ otherwise. Denote the $n$-step transition matrix by

$$\mathcal{P}^{(n)} = \left[p_{ij}^{(n)}\right] \tag{9.42c}$$

The transition matrix $\mathcal{P}^{(1)} = \mathcal{P}$, and $\mathcal{P}^{(0)} = I$ is the identity matrix.

4. Denote the probability that at time $t_n$ the system is in state $j \in \mathbb{N}$ by $P\left(X_n = j\right) = \xi_j^{(n)}$, where $n \in \mathbb{N}$. The corresponding probability vector of the state of the system at time $t_n$ is

$$\xi^{(n)} = \left[\xi_0^{(n)} \; \xi_1^{(n)} \; \xi_2^{(n)} \; \cdots\right], \quad \sum_{j \in \mathbb{N}} \xi_j^{(n)} = 1, \; \text{ and } \; n \in \mathbb{N} \tag{9.43}$$

$\xi^{(0)}$ is the initial probability vector. $\qquad \square$

The matrix $\mathcal{P}$ is a *stochastic matrix*, because the sum of the elements of each row is equal to unity.

**Theorem 9.9.** *For all* $n, m \in \mathbb{N}$

$$p_{ij}^{(n+m)} = \sum_{k \in \mathbb{N}} p_{ik}^{(n)} p_{kj}^{(m)} \tag{9.44a}$$

$$\mathcal{P}^{(m+n)} = \mathcal{P}^{(m)} \mathcal{P}^{(n)} \tag{9.44b}$$

*This equation is called the Chapman-Kolmogorov equation.*
*Proof.* The proof is left to the reader. $\qquad \square$

This theorem yields a method for computing $\mathcal{P}^{(n)}$. It is equal to the $n$th power of the matrix $\mathcal{P}$, that is $\mathcal{P}^{(n)} = \mathcal{P}^n$. The following observation is immediate from the above definitions and theorem.

**Observation 9.16.** For all $n \in \mathbb{N}$

$$\xi^{(n+1)} = \xi^{(n)} \mathcal{P}$$

$$\xi^{(n)} = \xi^{(0)} \mathcal{P}^{(n)} = \xi^{(0)} \mathcal{P}^n$$

$\qquad \square$

### Classification of States of a DTMC

A Markov chain is specified by its transition matrix. There are different measures to study and classify states of Markov chains. Classification of states also helps in identifying different types of Markov chains. Classification of the states of a homogeneous DTMC is next discussed. As mentioned earlier, we assume that the set of states is $\mathbb{N}$.

**Definitions 9.23.** *Basic definitions.*

1. *A state* $j \in \mathbb{N}$ *is accessible from state* $i \in \mathbb{N}$ *if* $p_{ij}^{(n)} > 0$ *for some integer* $n \in \mathbb{P}$. *This is denoted by* $i \to j$.

2. *States $i, j \in \mathbb{N}$ communicate with each other if they are mutually accessible. That is, $i \rightarrow j$ and $j \rightarrow i$. This is denoted by $i \leftrightarrow j$.*                                                    □

**Observation 9.17.** Communication between states.

(a) For any $i \in \mathbb{N}$, $i \leftrightarrow i$.
(b) For any $i, j \in \mathbb{N}$, if $i \leftrightarrow j$ then $j \leftrightarrow i$.
(c) Let $i, j, k \in \mathbb{N}$, where $i \leftrightarrow j$ and $j \leftrightarrow k$, then $i \leftrightarrow k$.                        □

The states which communicate with each other form a *class*. Actually communication is an equivalence relationship. Recall that an equivalence relationship is reflexive, symmetric, and transitive. Therefore the states which communicate with each other are said to be in the same *communicating (or equivalence) class*.

**Definition 9.24.** *If all the states of a DTMC communicate with each other (that is, every state can be reached from every other state) in a finite number of steps, then the Markov chain is irreducible, otherwise it is reducible.*                                                    □

A state $i \in \mathbb{N}$ is either periodic or aperiodic. These are next defined.

**Definitions 9.25.** *The periodicity or aperiodicity of a state $i \in \mathbb{N}$.*

1. *The period of a state $i \in \mathbb{N}$ is the greatest common divisor of all integers $n \in \mathbb{P}$, such that $p_{ii}^{(n)} > 0$. That is, state $i \in \mathbb{N}$ has a period $d$ if $p_{ii}^{(n)} = 0$ except when $n = d, 2d, 3d, \ldots$ and $d$ is the greatest such integer.*
2. *If $p_{ii}^{(n)} = 0$ for all integers $n \in \mathbb{P}$, then the period of state $i$ is equal to $0$.*
3. *If the period of a state $i \in \mathbb{N}$ is greater than one, then the state is periodic.*
4. *If the period of a state $i \in \mathbb{N}$ is equal to one, then the state is aperiodic.*                    □

If $p_{ii} > 0$, then the state $i$ clearly has a period. A state $i \in \mathbb{N}$ can also be classified as either recurrent (persistent) or transient. Furthermore, a recurrent state is classified as either positive-recurrent or null-recurrent.

**Definitions 9.26.** *Recurrent and transient states, and types of recurrent states.*

1. *For any state $i \in \mathbb{N}$, denote $f_i$ as the probability that, starting in state $i$, the Markov process renters state $i$.*
   (a) *The state $i$ is recurrent if $f_i = 1$.*
   (b) *The state $i$ is transient if $f_i < 1$.*
2. *A recurrent state $i \in \mathbb{N}$, is:*
   (a) *Positive-recurrent if the average number of steps until the state $i$ is returned to, is finite.*
   (b) *Null-recurrent if the average number of steps until the state $i$ is returned to, is infinite.*    □

**Observations 9.18.** Some basic observations.

1. Observations about communicating classes of a Markov chain:

(a) All the states in a communicating class have the same period. Therefore periodicity is a class property.

(b) All states in a communicating class are all simultaneously positive-recurrent or null-recurrent or transient.

2. Recurrence criteria of a state $i \in \mathbb{N}$.

   (a) The state $i$ is recurrent if and only if $\sum_{n=1}^{\infty} p_{ii}^{(n)} = \infty$.

   (b) The state $i$ is transient if and only if $\sum_{n=1}^{\infty} p_{ii}^{(n)} < \infty$.

3. Observations about finite DTMC:

   (a) An irreducible DTMC, which has finite number of states is positive-recurrent.

   (b) Null-recurrent states are not found in a finite-state discrete-time Markov chain.

   (c) The period of a state in a finite irreducible Markov chain is at most equal to the total number of states.

4. Consider an irreducible Markov chain, in which the number of states is either finite or denumerable. In this Markov chain, either all states are positive-recurrent, or all states are null-recurrent, or all states are transient. That is, exactly one of the three classifications holds true. $\quad\square$

The proofs of the above observations can be found in any standard textbook on stochastic processes.

**Ergodic DTMC**

We introduce the concept of ergodicity, and stationary probabilities of DTMC.

**Definitions 9.27.** *Let $\{X_n, n \in \mathbb{N}\}$ be a homogeneous DTMC with state-space $\mathbb{N}$.*

1. *A state $i \in \mathbb{N}$ is ergodic, if it is positive-recurrent, and aperiodic. An irreducible DTMC is said to be ergodic, if all of its states are ergodic.*

2. *The stationary probability distribution of a DTMC, if it exists, is specified by the row vector*

$$\xi = \begin{bmatrix} \xi_0 & \xi_1 & \xi_2 & \cdots \end{bmatrix} \tag{9.45}$$

*It satisfies the equations $\xi = \xi\mathcal{P}$, and $\sum_{j \in \mathbb{N}} \xi_j = 1$, where $\xi_j \geq 0$ for each $j \in \mathbb{N}$.* $\quad\square$

**Observations 9.19.** Consider an ergodic DTMC

1. The DTMC possesses a unique stationary distribution, where

$$\xi_j = \lim_{n \to \infty} p_{ij}^{(n)}, \quad \forall\, i, j \in \mathbb{N}$$

   Also $\xi_j > 0, \forall\, j \in \mathbb{N}$. It also satisfies $\sum_{j \in \mathbb{N}} \xi_j = 1$, and $\xi = \xi\mathcal{P}$.

2. $\lim_{n \to \infty} \mathcal{P}^n \to E\xi$, and $E = \begin{bmatrix} 1 & 1 & 1 & \cdots \end{bmatrix}^T$. That is, as $n$ tends towards infinity, the matrix $\mathcal{P}^n$ has nearly identical rows.

3. In the long run, the fraction of the total visits to state $j$ is given by $\xi_j$, irrespective of the initial state. Furthermore, the average number of transitions between two consecutive visits to state $j$ is $\xi_j^{-1}$.

4. Observe that the stationary probability distribution is called the stationary distribution because, if $\xi_j^{(0)} = \xi_j$ for all values of $j \in \mathbb{N}$, then for each $j \in \mathbb{N}$, $\xi_j^{(n)} = \xi_j$ for all $n \in \mathbb{N}$. That is, the probabilities $\xi_j^{(n)}$'s do not change over time, but are stationary. $\quad\square$

Some of the above theoretical results are illustrated via examples. A gambler's ruin problem, a random walk model, and the well-known Ehrenfest urn model are discussed.

**Examples 9.8.** Some illustrative examples.

1. *Gambler's ruin problem*: A gambler has a probability $p > 0$ of winning one unit at each play, and probability $q = (1 - p)$ of losing one unit. Assuming that successive plays of the game are independent, the probability that starting with $i$ units, the gambler accumulates $N$ units, before reaching 0 units is computed.

   Let $X_n$ be the player's fortune at time $n$. The process $\{X_n, n = 0, 1, 2, \ldots\}$ is a DTMC with state-space equal to $\{0, 1, 2, \ldots, N\}$. This state-space can be classified into three classes $\{0\}, \{1, 2, \ldots, N - 1\}$, and $\{N\}$. The first class corresponds to losing, and the third class corresponds to winning. The first and third classes are reachable from the second class of states, but their return is not possible. Consequently, the first and third classes are recurrent, and the states in the second class are transient. The states in the second class each have a period 2. Therefore, after a finite amount of time the gambler reaches either state 0 or state $N$. That is, the gambler cannot play the game indefinitely. Also the states 0 and $N$ are called *absorbing states*. These states are called absorbing, because if the process enters such states, it never leaves them.

   The transition matrix of the DTMC is a $(N + 1) \times (N + 1)$ matrix $\mathcal{P} = [p_{ij}]$, where $0 \leq i, j \leq N$.

$$p_{00} = p_{NN} = 1$$
$$p_{i,i+1} = p, \quad p_{i,i-1} = q, \quad 1 \leq i \leq (N - 1)$$

All other $p_{ij}$'s are zeros. Let $P_i$ be the probability that starting with $i$ units, the gambler's fortune reaches the goal of $N$ units, for $0 \leq i \leq N$. As per this definition, $P_0 = 0$, and $P_N = 1$. Also

$$P_i = pP_{i+1} + qP_{i-1}, \quad 1 \leq i \leq (N - 1)$$

Simplification of this equation yields

$$P_{i+1} - P_i = \frac{q}{p}(P_i - P_{i-1}), \quad 1 \leq i \leq (N - 1)$$

Using the boundary conditions and this difference equation results in

$$P_i = \begin{cases} \dfrac{1 - \left(\dfrac{q}{p}\right)^i}{1 - \left(\dfrac{q}{p}\right)^N}, & p \neq \dfrac{1}{2} \\[3ex] \dfrac{i}{N}, & p = \dfrac{1}{2} \end{cases}$$

If $N \to \infty$

$$P_i \to \begin{cases} 1 - \left(\dfrac{q}{p}\right)^i, & p > \dfrac{1}{2} \\[2ex] 0, & p \leq \dfrac{1}{2} \end{cases}$$

Therefore, if $p > 0.5$, the gambler's fortune increases significantly. However, if $p \leq 0.5$, the gambler loses all the money with high probability.

2. *Random walk model*: A MC whose state-space is equal to $\mathbb{Z}$ is said to be a random walk, if the transition probabilities are given by

$$p_{i,i+1} = p, \quad p_{i,i-1} = q, \quad \forall\, i \in \mathbb{Z}$$

where $0 < p < 1$ and $q = (1 - p)$. All other transition probabilities are each equal to zero. This is a model for an individual walking randomly along a straight line. This individual walks one step to the right with probability $p$, and one step to the left with probability $q$.

Note that each state is reachable from the other. Thus either all states are recurrent or all states are transient. Therefore let us determine if $\sum_{n=1}^{\infty} p_{00}^{(n)}$ exists or does not exist. It is clear that

$$p_{00}^{(2n+1)} = 0, \quad \forall\, n \in \mathbb{P}$$

$$p_{00}^{(2n)} = \binom{2n}{n} p^n q^n, \quad \forall\, n \in \mathbb{P}$$

The expression for $p_{00}^{(2n)}$ is obtained as follows. After $2n$ steps, $n$ of these have to be in either direction. Use of Stirling's approximation

$$n! \sim \sqrt{2\pi}\, n^{n+1/2} e^{-n}$$

leads to

$$p_{00}^{(2n)} \sim \frac{(4pq)^n}{\sqrt{\pi n}}$$

Recall that $a_n \sim b_n$ when $\lim_{n \to \infty} a_n/b_n = 1$. Thus, $\sum_n a_n < \infty$ iff $\sum_n b_n < \infty$.

That is, the summation $\sum_{n=1}^{\infty} p_{00}^{(n)}$ exists iff $\sum_{n=1}^{\infty} (4pq)^n / \sqrt{\pi n}$ exists. Consequently, if $p = q = 1/2$, the summation does not exist. Thus the Markov chain is recurrent if $p = 1/2$, and transient otherwise.

3. *The Ehrenfest Urn Model*: This model is named after the physicists Paul (1880-1933) and Tatyana (1876-1964) Ehrenfest. They used it to describe diffusion of molecules through a membrane. Let $M$ molecules be distributed among two urns A and B. Therefore, if there are $i$ molecules in urn A, then there are $(M - i)$ molecules in urn B. At each time instant, one of the $M$ molecules is selected at random and is then removed from its urn and placed in the other one. Let $\{X_n, n \in \mathbb{P}\}$ be a DTMC describing the number of molecules in urn A. The state-space of the DTMC is equal to $\{0, 1, 2, \ldots, M\}$. The transition probabilities are given by

$$p_{00} = 0 = p_{MM}$$

$$p_{01} = 1 = p_{M,M-1}$$

$$p_{i,i+1} = \frac{M - i}{M}, \quad p_{i,i-1} = 1 - p_{i,i+1}, \quad 1 \leq i \leq (M - 1)$$

all other $p_{ij}$'s are each equal to zero.

Let $\xi_i$, for $0 \leq i \leq M$ be the steady state distribution of the molecules in urn A. Then it can be shown that

$$\xi_i = \binom{M}{i} \left(\frac{1}{2}\right)^M, \quad 0 \leq i \leq M$$

The distribution of molecules in urn B is the same. This follows by the symmetry in the model.

□

### 9.7.2  Continuous-Time Markov Chains

The continuous-time Markov chains (CTMC) are continuous-time Markov processes with discrete state-space. The transition probabilities of a CTMC are initially defined in this subsection.

The corresponding Chapman-Kolmogorov equations are also derived. This is followed by definition of the transition density matrix, and classification of the states of a CTMC. The backward and forward Chapman-Kolmogorov equations are next stated. Subsequently, an alternate definition of CTMC is given, and the limiting probabilities of a CTMC are also examined. Finally, some well-known examples like birth-and-death and Poisson processes are outlined. A definition of a CTMC is given below.

**Definition 9.28.** *A continuous-time stochastic process $\left\{ X\left(t\right), t \in \mathbb{R}_0^+ \right\}$, with the set of integers $\mathbb{N}$ as the state-space, is a continuous-time Markov chain if for all values of $s, t \in \mathbb{R}_0^+; i, j, x\left(u\right) \in \mathbb{N}$, and $0 \leq u < s$*

$$P\left(X\left(t+s\right) = j \mid X\left(s\right) = i, X\left(u\right) = x\left(u\right), 0 \leq u < s\right)$$
$$= P\left(X\left(t+s\right) = j \mid X\left(s\right) = i\right) \tag{9.46}$$

*That is, the CTMC is a stochastic process with the special condition that the conditional distribution of $X\left(t+s\right)$ given the present state $X\left(s\right)$, and the past states $X\left(u\right), 0 \leq u < s$, depends only on the present state and is independent of the past states.*

*The conditional probability $P\left(X\left(t+s\right) = j \mid X\left(s\right) = i\right)$ is called the transition probability of the CTMC.*                                                                                                              □

The properties of transition probabilities of a CTMC are examined below.

**Transition Probabilities of a CTMC**

The basic structure of a CTMC is provided in terms of its transition probabilities.

**Definitions 9.29.** *Let $\left\{ X\left(t\right), t \in \mathbb{R}_0^+ \right\}$ be a CTMC with state-space equal to the set of integers $\mathbb{N}$.*

1. *If the conditional probability $P\left(X\left(t+s\right) = j \mid X\left(s\right) = i\right)$ is independent of all values of $s$, then the CTMC is said to have homogeneous transition probabilities. These homogeneous transition probabilities are expressed as*

$$p_{ij}\left(t\right) = P\left(X\left(t+s\right) = j \mid X\left(s\right) = i\right) \geq 0, \quad \forall\, i, j \in \mathbb{N} \tag{9.47a}$$

$$\sum_{j \in \mathbb{N}} p_{ij}\left(t\right) = 1, \quad \forall\, i \in \mathbb{N} \tag{9.47b}$$

$$\lim_{t \to 0} p_{ij}\left(t\right) = \begin{cases} 1, & i = j \\ 0, & i \neq j \end{cases}, \quad \forall\, i, j \in \mathbb{N} \tag{9.47c}$$

*Denote the matrix of transition probabilities by*

$$\mathcal{P}\left(t\right) = \left[p_{ij}\left(t\right)\right] \tag{9.48a}$$
$$\mathcal{P}\left(0\right) = I \tag{9.48b}$$

*where $I$ is the identity matrix.*

2. *Denote the probability that the system is in state $j \in \mathbb{N}$ at time t by $\xi_j (t) = P(X(t) = j)$. The corresponding probability vector of the state of the system at time $t \in \mathbb{R}_0^+$ is*

$$\xi(t) = \left[ \xi_0(t) \, \xi_1(t) \, \xi_2(t) \cdots \right], \quad where \quad \sum_{j \in \mathbb{N}} \xi_j(t) = 1 \tag{9.49}$$

*$\xi(0)$ is the initial probability vector.*                                                                              □

**Theorem 9.10.** *For all values of $s, t \in \mathbb{R}_0^+$*

$$p_{ij}(t+s) = \sum_{k \in \mathbb{N}} p_{ik}(t) \, p_{kj}(s), \quad \forall \, i, j \in \mathbb{N} \tag{9.50a}$$

*In matrix notation, this relationship is expressed as*

$$\mathcal{P}(t+s) = \mathcal{P}(t) \, \mathcal{P}(s) \tag{9.50b}$$

*This is the Chapman-Kolmogorov equation of a homogeneous CTMC.*
    *Proof.* See the problem section.                                                                                      □

**Lemma 9.1.** *For a homogeneous CTMC*

$$\xi(t) - \xi(0) \, \mathcal{P}(t), \quad t \in \mathbb{R}_0^+ \tag{9.51}$$

*Proof.* See the problem section.                                                                                          □

**Transition Density Matrix**

Single-step transition probabilities were used in describing a homogeneous DTMC. A similar role is played by transition intensities in describing a homogeneous CTMC.

**Definitions 9.30.** *Definitions related to transition density matrix.*

1. *The transition intensity is defined as*

$$q_{ij} = \lim_{h \to 0} \frac{p_{ij}(h) - p_{ij}(0)}{h} = \lim_{h \to 0} \frac{p_{ij}(h)}{h}, \quad i \neq j \tag{9.52a}$$

$$q_{ii} = \lim_{h \to 0} \frac{p_{ii}(h) - p_{ii}(0)}{h} = \lim_{h \to 0} \frac{p_{ii}(h) - 1}{h} \tag{9.52b}$$

*for all values of $i, j \in \mathbb{N}$. Let $-q_{ii} \triangleq q_i$.*
2. *Let $\mathcal{Q} = [q_{ij}]$, then in matrix notation*

$$\mathcal{Q} = \lim_{h \to 0} \frac{\mathcal{P}(h) - I}{h} \tag{9.52c}$$

*The matrix $\mathcal{Q}$ is called the transition density matrix, or infinitesimal generator, or simply the rate matrix of the CTMC.*                                                                                                        □

**Observations 9.20.** Some basic observations about CTMC.

1. The transition intensity $q_{ij}$, for $i \neq j$ is always finite, and nonnegative. However, $q_i$ always exists and is finite when the state-space is finite. If the state-space is denumerably infinite, then $q_i$ can be infinite. The value $q_i$ is always nonnegative.

2. The diagonal elements of the matrix $Q$ are nonpositive, and the off-diagonal elements are nonnegative. Furthermore, the sum of each row is equal to zero. That is, for any $i \in \mathbb{N}$

$$\sum_{\substack{j \in \mathbb{N} \\ j \neq i}} q_{ij} = q_i \tag{9.53}$$

$\square$

### Classification of States of a CTMC

Similar to the classification of the states of a homogeneous DTMC, the states of a homogeneous CTMC can be classified. Recall that the set of states of a CTMC is $\mathbb{N}$. Following terminology and definitions are introduced before we study classification of the states of a CTMC.

**Definitions 9.31.** *These definitions and notation are about accessibility of states.*

1. *A state $j \in \mathbb{N}$ is accessible or reachable from state $i \in \mathbb{N}$ if $p_{ij}(t) > 0$ for some $t > 0$. This is denoted by $i \rightarrow j$.*
2. *States $i, j \in \mathbb{N}$ communicate if they are mutually accessible. This is denoted by $i \leftrightarrow j$.*
3. *If all the states of a CTMC communicate with each other, (that is, every state can be reached from every other state) then the Markov chain is said to be irreducible.* $\square$

A state $i \in \mathbb{N}$ is either *recurrent* (*persistent*) or *transient*. Furthermore, a recurrent state is either *positive-recurrent* or *null-recurrent*. A state $i$ is said to be positive-recurrent if its mean recurrence time is finite. However, it is said to be null-recurrent if its mean recurrence time is infinite. This type of classification of a state can be stated precisely in terms of the $p_{ii}(t)$'s.

**Definition 9.32.** *Consider a state $i \in \mathbb{N}$. The state $i$ is transient iff $\int_0^\infty p_{ii}(t) \, dt < \infty$. Similarly, a state $i$ is positive-recurrent (or nonnull-persistent) if $\lim_{t \to \infty} p_{ii}(t) > 0$. And a state $i$ is null-recurrent (or null-persistent) if $\lim_{t \to \infty} p_{ii}(t) = 0$.* $\square$

The Chapman-Kolmogorov equations of the homogeneous CTMC can be elegantly combined with its rate matrix to obtain the backward and forward Chapman-Kolmogorov equations.

### Backward and Forward Chapman-Kolmogorov Equations

In the next theorem, the backward and forward Chapman-Kolmogorov equations which describe the CTMC are determined. These equations are defined in terms of the derivatives of a matrix and a row vector. The derivative of a matrix (row vector) is simply equal to a matrix (row vector) which is obtained by taking the derivative of each element of the matrix (row vector). For brevity, a prime ($'$) is used to denote the derivative with respect to time. The Chapman-Kolmogorov equations relate $\mathcal{P}'(t)$, $\mathcal{P}(t)$, and the rate matrix $Q$. The matrix $\mathcal{P}'(t)$ is the derivative of matrix $\mathcal{P}(t)$ with respect to time. The derivative of $\mathcal{P}(t)$ and the row vector $\xi(t)$, with respect to time are defined below.

**Definitions 9.33.** *The derivatives of the transition probability matrix $\mathcal{P}(t)$, and the row vector $\xi(t)$ are:*

1. *Let $\mathcal{P}(t) = [p_{ij}(t)]$, and*

$$p'_{ij}(t) = \frac{dp_{ij}(t)}{dt} \tag{9.54a}$$

*The matrix $\mathcal{P}'(t) = [p'_{ij}(t)]$. More compactly*

$$\mathcal{P}'(t) = \frac{d\mathcal{P}(t)}{dt} = \lim_{h \to 0} \frac{\mathcal{P}(t+h) - \mathcal{P}(t)}{h} \tag{9.54b}$$

2. *Also*

$$\xi'(t) = \frac{d\xi(t)}{dt}; \quad and \;\; \xi'_j(t) = \frac{d\xi_j(t)}{dt}, \;\; \forall\, j \in \mathbb{N} \tag{9.55a}$$

$$\xi'(t) = \begin{bmatrix} \xi'_0(t) \;\, \xi'_1(t) \;\, \xi'_2(t) \;\cdots \end{bmatrix} \tag{9.55b}$$

□

**Theorem 9.11.** *The backward Chapman-Kolmogorov is*

$$\mathcal{P}'(t) = \mathcal{Q}\mathcal{P}(t), \quad t \in \mathbb{R}_0^+ \tag{9.56a}$$

*Similarly the forward Chapman-Kolmogorov equation is*

$$\mathcal{P}'(t) = \mathcal{P}(t)\mathcal{Q}, \quad t \in \mathbb{R}_0^+ \tag{9.56b}$$

*Proof.* See the problem section.                                                            □

**Corollary 9.4.** *The above theorem yields*

$$\xi'(t) = \xi(t)\mathcal{Q}, \quad t \in \mathbb{R}_0^+ \tag{9.57}$$

*Proof.* See the problem section.                                                            □

**Another Definition of CTMC**

The following alternate definition of homogeneous CTMC uses the memoryless property of exponentially distributed random variable.

**Definition 9.34.** *A stochastic process with state-space $\mathbb{N}$, is a homogeneous CTMC if:*

(a) *When it enters state $i \in \mathbb{N}$, the time it spends in this state before making a transition to a different state $j \in \mathbb{N}$ is exponentially distributed with parameter $\nu_i \in \mathbb{R}^+$. This time is also called
the sojourn time of the state. The parameter $\nu_i$ depends only on $i$, and not on $j$. Furthermore,
the sojourn times in different states are independent random variables.*

(b) *The process leaves state $i$, and enters state $j$ (not equal to $i$) with probability $p_{ij}$, then*

$$p_{ii} = 0, \;\; \forall\, i \in \mathbb{N} \tag{9.58a}$$

$$\sum_{j \in \mathbb{N}} p_{ij} = 1, \;\; \forall\, i \in \mathbb{N} \tag{9.58b}$$

□

Therefore the CTMC is a stochastic process such that its transition from one state to another is as in a DTMC. Furthermore, these sojourn times in different states are independent random variables with exponential distribution.

**Observations 9.21.** Some basic observations.

1. $q_{ij} = \nu_i p_{ij}$, for any $i, j \in \mathbb{N}$ and $i \neq j$. Use of the definition of $q_{ij}$ yields

$$p_{ij}(h) = h\nu_i p_{ij} + o(h), \quad i \neq j$$

2. $q_i = \nu_i$, for all values of $i \in \mathbb{N}$. Use of the definition of $q_i$ yields

$$1 - p_{ii}(h) = h\nu_i \sum_{j \in \mathbb{N}} p_{ij} + o(h)$$
$$= h\nu_i + o(h)$$

3. If the state-space of the CTMC is $\mathcal{S} = \{0, 1, 2, \ldots, m\}$, then

$$Q = \begin{bmatrix} -\nu_0 & \nu_0 p_{01} & \nu_0 p_{02} & \cdots & \nu_0 p_{0m} \\ \nu_1 p_{10} & -\nu_1 & \nu_1 p_{12} & \cdots & \nu_1 p_{1m} \\ \nu_2 p_{20} & \nu_2 p_{21} & -\nu_2 & \cdots & \nu_2 p_{2m} \\ \vdots & \vdots & \vdots & \ddots & \vdots \\ \nu_m p_{m0} & \nu_m p_{m1} & \nu_m p_{m2} & \cdots & -\nu_m \end{bmatrix}$$

□

**Limiting Probabilities of a CTMC**

Limiting probabilities of a CTMC are defined in analogy with those of a DTMC.

**Definitions 9.35.** *Let $\{X(t), t \in \mathbb{R}_0^+\}$ be a homogeneous CTMC with state-space $\mathbb{N}$.*

1. *The stationary or limiting or equilibrium or long-run probability distribution of a CTMC, if it exists as $t \to \infty$, is specified by the row vector*

$$\xi = \begin{bmatrix} \xi_0 & \xi_1 & \xi_2 & \cdots \end{bmatrix}$$

   *where $\lim_{t \to \infty} \xi_j(t) \to \xi_j, \forall j \in \mathbb{N}$, and $\sum_{j \in \mathbb{N}} \xi_j = 1$.*
2. *If the stationary probability distribution of a CTMC exists, then the corresponding MC is said to be ergodic.*    □

If the limiting probabilities $\xi_j, \forall j \in \mathbb{N}$ exist, then these probabilities can be interpreted as the long-run proportion of time the process is in state $j$. Also, if the limiting probabilities of a CTMC exist, then it can be shown that

$$\lim_{t \to \infty} p_{ij}(t) = \xi_j, \quad \forall i, j \in \mathbb{N}, \quad \text{and} \quad \sum_{j \in \mathbb{N}} \xi_j = 1$$

Thus $\xi_j$ is the limiting value of the state transition probability $p_{ij}(t)$ as $t \to \infty$, and it is independent of the initial state $i \in \mathbb{N}$.

Sufficient conditions for the existence of limiting probabilities of a CTMC are established as follows. Consider a homogeneous CTMC $\{X(t), t \in \mathbb{R}_0^+\}$ with state-space $\mathbb{N}$. In addition, assume that all states of the CTMC communicate with each other (irreducibility). Then all states belong to the same class. The following three cases have to be examined.

(a)  If the states are all transient, then $\xi_j = \lim_{t\to\infty} p_{ij}(t) = 0$, $\forall\, i, j \in \mathbb{N}$.

(b)  If the states are all null-recurrent, then $\xi_j = \lim_{t\to\infty} p_{ij}(t) = 0$, $\forall\, i, j \in \mathbb{N}$.

(c)  If the states are all positive-recurrent, then $\xi_j = \lim_{t\to\infty} p_{ij}(t)$, $\forall\, i, j \in \mathbb{N}$, and $\sum_{j\in\mathbb{N}} \xi_j = 1$.
This is the condition for ergodicity of the CTMC.

In an ergodic homogeneous CTMC, $p_{ij}(\cdot)$ is a bounded function between 0 and 1, its derivative with respect to time, $p'_{ij}(t)$ converges as $t \to \infty$, $\forall\, i, j \in \mathbb{N}$. Moreover $\lim_{t\to\infty} p'_{ij}(t) = 0$, $\forall\, i, j \in \mathbb{N}$. Similarly $\lim_{t\to\infty} \xi'_j(t) = 0$, $\forall\, j \in \mathbb{N}$.

**Theorem 9.12.** *If the CTMC is ergodic*

$$\xi Q = 0, \quad and \quad \xi E = 1 \tag{9.59}$$

*where* $0$ *is a row vector of all zeros, and* $E = \begin{bmatrix} 1\ 1\ 1 \cdots \end{bmatrix}^T$.

*Proof.* The result follows from the relationship $\xi'(t) = \xi(t)\, Q$. $\qquad\qquad\square$

The equations $\xi Q = 0$, and $\sum_{j\in\mathbb{N}} \xi_j = 1$, can be used to determine the limiting probabilities. These equations have an interesting interpretation when written explicitly. That is

$$\sum_{i\in\mathbb{N}} \xi_i q_{ij} = 0, \quad \forall\, j \in \mathbb{N}$$

Thus

$$\sum_{\substack{i\in\mathbb{N} \\ i\neq j}} \xi_i q_{ij} = \xi_j q_j, \quad \forall\, j \in \mathbb{N}$$

Also when the process is in state $i$, it enters state $j$ at a rate $q_{ij}$. Since the proportion of time the process is in state $i$ is equal to $\xi_i$, the net rate at which the process leaves state $i$ and enters state $j$ is equal to $\xi_i q_{ij}$. Consequently, the net rate at which the process *enters* state $j$ is equal to $\sum_{\substack{i\in\mathbb{N} \\ i\neq j}} \xi_i q_{ij}$.

When the process is in state $j$, it leaves this state at rate $q_j$. Since the proportion of time the process is in state $j$ is equal to $\xi_j$, the net rate at which the process *leaves* state $j$ is equal to $\xi_j q_j$. Thus the above equation represents a balance of in-flow and out-flow rates. Therefore these relations are sometimes referred to as *balance equations*.

An important example of a CTMC is the birth-and-death process. This process is next studied in detail. These processes are used extensively in modeling queueing scenarios. Application of such processes to queues is studied in the chapter on queueing theory. A special case of a birth-and-death process is the Poisson process. It is simply a pure birth process.

**Examples 9.9.** The above concepts are further clarified via several examples.

1.  *Birth-and-death process*: The birth-and-death process is an important subclass of CTMC. Its a homogeneous CTMC $\{ X(t), t \in \mathbb{R}_0^+ \}$ with state-space $\mathbb{N}$. The elements of the rate matrix are defined as follows.

$$q_{i,i+1} = \lambda_i, \quad \forall\, i \in \mathbb{N}$$
$$q_{i,i-1} = \mu_i, \quad \forall\, i \in \mathbb{P}$$
$$q_{ij} = 0; \quad j \neq (i \pm 1),\, j \neq i, \quad \forall\, i, j \in \mathbb{N}$$
$$q_i = (\lambda_i + \mu_i), \quad \forall\, i \in \mathbb{N}, \quad \mu_0 = 0$$

Note that the parameters $\lambda_i$'s and $\mu_i$'s are all nonnegative. Use of the forward Chapman-Kolmogorov equations yields the following transition probabilities equations

$$p'_{ij}(t) = \lambda_{j-1}p_{i,j-1}(t) - (\lambda_j + \mu_j)p_{ij}(t) + \mu_{j+1}p_{i,j+1}(t),$$
$$\forall\, i \in \mathbb{N},\ \forall\, j \in \mathbb{P}$$
$$p'_{i0}(t) = -\lambda_0 p_{i0}(t) + \mu_1 p_{i1}(t), \quad \forall\, i \in \mathbb{N}$$

The initial conditions are $p_{ii}(0) = 1$, $p_{ij}(0) = 0$, $i \neq j$, $\forall\, i,j \in \mathbb{N}$. The state probabilities $\xi_j(t)$, $\forall\, j \in \mathbb{N}$ are obtained from the forward equations

$$\xi'_j(t) = \lambda_{j-1}\xi_{j-1}(t) - (\lambda_j + \mu_j)\xi_j(t) + \mu_{j+1}\xi_{j+1}(t), \quad \forall\, j \in \mathbb{P}$$
$$\xi'_0(t) = -\lambda_0\xi_0(t) + \mu_1\xi_1(t)$$

Assume that the process starts in state $i \in \mathbb{N}$. The initial conditions are $\xi_i(0) = 1$, and $\xi_j(0) = 0$, $i \neq j$, $\forall\, j \in \mathbb{N}$.
If the $\lambda_i$'s and $\mu_i$'s are positive, the CTMC is irreducible. In addition, it can also be shown that this CTMC is positive-recurrent. Therefore the stationary probabilities $\xi = \begin{bmatrix} \xi_0 & \xi_1 & \xi_2 & \cdots \end{bmatrix}$ exist as $t \to \infty$. These are obtained from the equations

$$0 = \lambda_{j-1}\xi_{j-1} - (\lambda_j + \mu_j)\xi_j + \mu_{j+1}\xi_{j+1}, \quad \forall\, j \in \mathbb{P}$$
$$0 = -\lambda_0\xi_0 + \mu_1\xi_1$$

Write the above equations as

$$\lambda_{j-1}\xi_{j-1} + \mu_{j+1}\xi_{j+1} = (\lambda_j + \mu_j)\xi_j, \quad \forall\, j \in \mathbb{P}$$
$$\mu_1\xi_1 = \lambda_0\xi_0$$

Observe the so-called *rate equality principle* in these equations. According to this principle, the rate at which a process enters a state $j \in \mathbb{N}$ is equal to the rate at which the process leaves the state $j$. Briefly stated *rate-in* equals *rate-out* for any state $j \in \mathbb{N}$. The left-hand sides of the above equations represent the rate-in and the right-hand sides represent the rate-out. That is, in state $j \in \mathbb{P}$:

(a) Rate-in is equal to: $\{\lambda_{j-1}\xi_{j-1} + \mu_{j+1}\xi_{j+1}\}$.
(b) Rate-out is equal to: $(\lambda_j + \mu_j)\xi_j$.

Similarly $\mu_1\xi_1$ is the rate-in, and $\lambda_0\xi_0$ is the rate-out in state 0. Next define

$$\alpha_0 = 1$$
$$\alpha_j = \frac{\lambda_0\lambda_1\dots\lambda_{j-1}}{\mu_1\mu_2\dots\mu_j}, \quad \forall\, j \in \mathbb{P}$$

If $\sum_{j\in\mathbb{N}}\alpha_j < \infty$, then

$$\xi_j = \frac{\alpha_j}{\sum_{j\in\mathbb{N}}\alpha_j}, \quad \forall\, j \in \mathbb{N}$$

It can be shown that $\sum_{j\in\mathbb{N}}\alpha_j < \infty$ is a sufficient condition for the CTMC to be ergodic. It is also possible to obtain the following qualitative interpretation of the birth-and-death process.

(a) If it is given that the state of the process $X(t) = i \in \mathbb{N}$, then the current probability distribution of the *remaining* time until the next *birth* is exponential with parameter $\lambda_i$.

(b) If it is given that the state of the process $X(t) = i \in \mathbb{P}$, then the current probability distribution of the *remaining* time until the next *death* is exponential with parameter $\mu_i$.

(c) Either a *single* birth or a *single* death can occur at a time.

2. *Poisson process*: The homogeneous Poisson process $\{X(t), t \in \mathbb{R}_0^+\}$ is a special case of a birth-and-death process. Its state-space is $\mathbb{N}$, and $\mu_i = 0$ for all values of $i \in \mathbb{N}$. Assume that $\lambda_i = \lambda$ for all values of $i \in \mathbb{N}$. In this case the process is known as a homogeneous Poisson process. The state probabilities $\xi_j(t), \forall\, j \in \mathbb{N}$ can be obtained from

$$\xi_j'(t) = \lambda \xi_{j-1}(t) - \lambda \xi_j(t), \quad \forall\, j \in \mathbb{P}$$
$$\xi_0'(t) = -\lambda \xi_0(t)$$

Assume that the process starts in state $i \in \mathbb{N}$. The initial conditions are $\xi_i(0) = 1$, and $\xi_j(0) = 0, i \neq j, \forall\, j \in \mathbb{N}$. Define

$$\xi(z, t) = \sum_{j \in \mathbb{N}} \xi_j(t) z^j$$

then $\xi(z, 0) = z^i$. Also

$$\frac{\partial}{\partial t} \xi(z, t) = \sum_{j \in \mathbb{N}} \frac{\partial}{\partial t}\{\xi_j(t)\} z^j = \sum_{j \in \mathbb{N}} \xi_j'(t) z^j$$

Use of the above equations yields

$$\frac{\partial}{\partial t} \xi(z, t) = \lambda(z - 1)\xi(z, t)$$

Thus

$$\xi(z, t) = z^i e^{\lambda(z-1)t}$$

The coefficient of $z^j$ in this equation is $\xi_j(t)$. Therefore

$$\xi_j(t) = \begin{cases} e^{-\lambda t}\dfrac{(\lambda t)^{j-i}}{(j-i)!}, & j = i, i+1, \ldots \\ 0, & j = 0, 1, \ldots, i-1 \end{cases}$$

In this derivation, observe how the Poisson process fits within the general framework of a CTMC. This can be compared with the axiomatic definition of the Poisson process given in an earlier section. Using a slightly different notation, the Poisson process was defined as a counting process. Thus a single count of the Poisson process corresponds to a single birth.

3. *Mean population size*: We determine the mean population size $M(t), t \in \mathbb{R}_0^+$ in the birth-death process. It is

$$M(t) = \mathcal{E}(X(t))$$
$$= \sum_{j \in \mathbb{P}} j\xi_j(t)$$

Assume that $\lambda_i = i\lambda$ for all values of $i \in \mathbb{N}$, and $\mu_i = i\mu$ for all values of $i \in \mathbb{P}$. Thus

$$\sum_{j \in \mathbb{P}} j\xi_j'(t) = \lambda \sum_{j \in \mathbb{P}} j(j-1)\xi_{j-1}(t) - (\lambda + \mu)\sum_{j \in \mathbb{P}} j^2 \xi_j(t) + \mu \sum_{j \in \mathbb{P}} j(j+1)\xi_{j+1}(t)$$

Let the derivative of $M(t)$ with respect to time be $M'(t)$. After simplification, the above equation yields

$$M'(t) = (\lambda - \mu) M(t)$$

If $M(0) = a$, we obtain

$$M(t) = ae^{(\lambda - \mu)t}, \quad \text{for } t \in \mathbb{R}_0^+$$

$\square$

### 9.7.3  Continuous-Time Markov Processes

The *continuous-time Markov process* (CTMP) which is continuous in both time parameter and the state-space is studied in this subsection. A special type of continuous-time MP has already been discussed. It was the Brownian motion process or the Wiener-Levy process. A general definition of a Markov process $\{X(t), t \in T\}$ was given at the beginning of this section. It holds true for a CTMP if it is assumed that the set $T$ takes continuous values, and the random variables $X(t)$'s are continuous. Analogous to Markov chains, the concept of transition probabilities can also be developed for such processes.

**Definitions 9.36.** *Let $\{X(t), t \in \mathbb{R}_0^+\}$ be a continuous-time MP.*

1. *Its cumulative transition probability is*

$$F(x_0, s; x, t) = P(X(t) \leq x \mid X(s) = x_0), \quad s < t \tag{9.60a}$$

2. *Its transition probability density $p(\cdot)$ is*

$$p(x_0, s; x, t)\, dx = P(x \leq X(t) < x + dx \mid X(s) = x_0), \quad s < t \tag{9.60b}$$

$\square$

In a homogeneous continuous-time MP the transition probability depends only on the length of the interval $(t - s)$.

**Definition 9.37.** *The transition probability density $p(\cdot)$ of a homogeneous continuous-time MP for any $t_0$ is denoted by*

$$p(x_0, x; t)\, dx = P(x \leq X(t + t_0) < x + dx \mid X(t_0) = x_0) \tag{9.61}$$

$\square$

The corresponding Chapman-Kolmogorov equation in terms of the transition probabilities is

$$p(x_0, s; x, t) = \int p(x_0, s; u, v)\, p(u, v; x, t)\, du \tag{9.62}$$

The continuous-time Markov processes can best be studied via their underlying partial differential equations. These partial differential equations are developed by first studying an example of these processes. In this example, the Brownian motion process discussed earlier is restricted to only positive values of time, and generalized to include nonzero values for the mean of the increment process. These are called *diffusion processes*.

### Diffusion Process

A diffusion process can be used to describe the motion of molecules in a gas. We initially define a diffusion process. This is followed by the development of the diffusion equations.

**Definition 9.38.** *The random process $\left\{X(t), t \in \mathbb{R}_0^+\right\}$ is a diffusion process if*:

(a) *The process has independent increments, that is for every pair of nonoverlapping intervals of time $(s, t)$ and $(u, v)$, such that $s < t \leq u < v$, the increments $(X(t) - X(s))$ and $(X(v) - X(u))$ are independent random variables.*

(b) *For $s < t$, the random variable $(X(t) - X(s))$ has normal distribution such that:*

$$\mathcal{E}\left(X(t) - X(s)\right) = m(t - s) \qquad (9.63a)$$
$$Var\left(X(t) - X(s)\right) = \sigma^2(t - s) \qquad (9.63b)$$

*The variables $m$ and $\sigma^2$ are called the infinitesimal mean and infinitesimal variance of the process respectively. Furthermore, these variables are constants. The variables $m$ and $\sigma^2$ are sometimes called the drift and diffusion coefficients respectively.* □

From this definition it can be inferred that the diffusion process is indeed a continuous-time MP. This is implied by the first property of the diffusion process. The requirement of independent increments is actually more restrictive than the Markovian property. Furthermore, this process is also Gaussian. Consequently

$$P\left(X(t) \leq x \mid X(s) = x_0\right) = P\left(X(t) - X(s) \leq x - x_0\right) = \Phi(\alpha)$$

$$\alpha = \frac{\left\{x - x_0 - m(t - s)\right\}}{\sigma(t - s)^{1/2}}$$

where $\Phi(\cdot)$ is the cumulative distribution function of a standard normal random variable. Note that

$$m = \lim_{\Delta t \to 0} \frac{\mathcal{E}\left(X(t + \Delta t) - X(t)\right)}{\Delta t}$$
$$\sigma^2 = \lim_{\Delta t \to 0} \frac{Var\left(X(t + \Delta t) - X(t)\right)}{\Delta t}$$

**Example 9.10.** If $X(0) = x_0$, then $\{X(t) - X(0)\}$ is normally distributed with mean $mt$ and variance $\sigma^2 t$, where $t \in \mathbb{R}^+$, then

$$p(x_0, x; t) = \frac{1}{\sigma\sqrt{2\pi t}} \exp\left\{-\frac{(x - x_0 - mt)^2}{2\sigma^2 t}\right\} \qquad (9.64)$$

□

Differential equations describing a diffusion process are next developed. Assume that the process started from state $x_0$ at time $t = 0$. Consider a time interval $(t - \Delta t, t)$ of length $\Delta t$. During this time interval, the process changes state by $\Delta x$ with probability $p'$, or changes state by $-\Delta x$ with probability $q' = (1 - p')$. The mean and the variance of the random variable $(X(t + \Delta t) - X(t))$ are then given by

$$\mathcal{E}(X(t + \Delta t) - X(t)) = (p' - q')\Delta x$$
$$Var(X(t + \Delta t) - X(t)) = 4p'q'(\Delta x)^2$$

As $\Delta t \to 0$ and $\Delta x \to 0$, the above equations yield

$$\frac{(p' - q')\Delta x}{\Delta t} \to m, \quad \text{and} \quad \frac{4p'q'(\Delta x)^2}{\Delta t} \to \sigma^2$$

This is possible if $\Delta x \to \sigma(\Delta t)^{1/2}$, and

$$p' = \frac{1}{2}\left\{1 + \frac{m(\Delta t)^{1/2}}{\sigma}\right\}, \quad \text{and} \quad q' = \frac{1}{2}\left\{1 - \frac{m(\Delta t)^{1/2}}{\sigma}\right\}$$

In the following development, it is assumed that $\Delta x = o\left((\Delta t)^{1/2}\right)$. Use of the law of total probability yields

$$p(x_0, x; t)\Delta x = p'p(x_0, x - \Delta x; t - \Delta t)\Delta x + q'p(x_0, x + \Delta x; t - \Delta t)\Delta x$$

Further assume that it is possible to expand $p(x_0, x; t)$ in a Taylor's series. That is

$$p(x_0, x \pm \Delta x; t - \Delta t) = p(x_0, x; t) - \Delta t\frac{\partial p(x_0, x; t)}{\partial t} \pm \Delta x\frac{\partial p(x_0, x; t)}{\partial x}$$
$$+ \frac{1}{2}(\pm \Delta x)^2\frac{\partial^2 p(x_0, x; t)}{\partial x^2} + o(\Delta t)$$

Use of the above two equations yields

$$-\Delta t\frac{\partial p}{\partial t} - (p' - q')\Delta x\frac{\partial p}{\partial x} + \frac{1}{2}(\Delta x)^2\frac{\partial^2 p}{\partial x^2} + o(\Delta t) = 0$$

Dividing both sides of the above equation by $\Delta t$ and letting $\Delta t \to 0$ and $\Delta x \to 0$ results in

$$\frac{\partial}{\partial t}p(x_0, x; t) = -m\frac{\partial}{\partial x}p(x_0, x; t) + \frac{1}{2}\sigma^2\frac{\partial^2}{\partial x^2}p(x_0, x; t) \tag{9.65a}$$

This equation is called the *forward diffusion equation* of the process. The *backward diffusion equation* of the process can be similarly obtained.

$$\frac{\partial}{\partial t}p(x_0, x; t) = m\frac{\partial}{\partial x_0}p(x_0, x; t) + \frac{1}{2}\sigma^2\frac{\partial^2}{\partial x_0^2}p(x_0, x; t) \tag{9.65b}$$

These equations are called forward and backward diffusion equations, because these have partial derivatives with respect to $x$ and $x_0$ respectively (besides $t$). These are partial differential equations of the first order in $t$. In addition, the forward and backward diffusion equations are of the second order in $x$ and $x_0$ respectively.

**Observations 9.22.** Some basic observations.

1. A function which satisfies both the forward and backward diffusion equations is the transition probability density

$$p\left(x_0, x; t\right) = \frac{1}{\sigma\sqrt{2\pi t}}\, \exp\left\{-\frac{\left(x - x_0 - mt\right)^2}{2\sigma^2 t}\right\}$$

2. The distribution of $X\left(t\right)$, where $X\left(0\right) = x_0$ as $t \to \infty$ is exponential with parameter $-2m/\sigma^2$, where $m < 0$.

Refer to the problem section for a justification of this statement.                    □

### Kolmogorov Equations

Let $\left\{X(t), t \in \mathbb{R}_0^+\right\}$ be a CTMP. Assume that for any $\delta > 0$,

$$P\left(|X(t) - X(s)| > \delta \mid X(t) = x\right) = o\left(t - s\right), \quad s < t$$

That is, if there is any small change in time, there is a small change in the corresponding state value. Furthermore, both mean and variance of this process are functions of $x$ and $t$. This is in contrast to the diffusion process, where these values are constants. That is

$$\lim_{\Delta t \to 0} \frac{\mathcal{E}\left(X\left(t + \Delta t\right) - X\left(t\right) \mid X\left(t\right) = x\right)}{\Delta t} = m\left(x, t\right)$$

$$\lim_{\Delta t \to 0} \frac{Var\left(X\left(t + \Delta t\right) - X\left(t\right) \mid X\left(t\right) = x\right)}{\Delta t} = \sigma^2\left(x, t\right)$$

Analogous to the forward and backward diffusion equations, the forward and backward *Kolmogorov's diffusion equations* satisfying the transition probability density $p\left(x_0, t_0; x, t\right)$ can be determined. The forward and backward Kolmogorov equations are

$$\frac{\partial}{\partial t}p\left(x_0, t_0; x, t\right) = -\frac{\partial}{\partial x}\left\{m\left(x, t\right)p\left(x_0, t_0; x, t\right)\right\}$$

$$+\frac{1}{2}\frac{\partial^2}{\partial x^2}\left\{\sigma^2\left(x, t\right)p\left(x_0, t_0; x, t\right)\right\} \qquad (9.66a)$$

$$\frac{\partial}{\partial t_0}p\left(x_0, t_0; x, t\right) = -m\left(x_0, t_0\right)\frac{\partial}{\partial x_0}p\left(x_0, t_0; x, t\right)$$

$$-\frac{1}{2}\sigma^2\left(x_0, t_0\right)\frac{\partial^2}{\partial x_0^2}p\left(x_0, t_0; x, t\right) \qquad (9.66b)$$

respectively. These partial differential equations exist provided the derivatives in these equations are continuous. The forward Kolmogorov equation is also called the *Fokker-Planck equation*.

---

## Reference Notes

The discussion on stochastic processes is based upon the books by Parzen (1962), Ross (1970, 1983, 2009, and 2010), Hoel, Port, and Stone (1972), and Sveshnikov (1968).

Elements of measure theory are covered in Oden, and Demkowicz (1996). A description of shot noise model is given in Parzen (1962), and Ross (1983). A scholarly introduction to point processes can be fond in Resnick (1992), Mikosch (2009), Serfozo (2009), and Pinsky and Karlin (2011). Kingman (1993) has written an exclusive monograph on Poisson point processes.

Readable accounts of renewal theory are given in Cox (1962), and Medhi (1994, and 2003). Proofs of observations about Markov chains can be found in Parzen (1962), Ross (2010), and Feller (1968).

We have not discussed stochastic partial differential equations or Ito's calculus in this chapter. A readable introduction to this fascinating subject is provided in the handbook by Gardiner (2004).

## Problems

1. Prove that the sum of two independent homogeneous Poisson processes is a Poisson process. Hint: See Medhi (1994). Note that the concept of independence of random processes is a generalization of the concept of independence of random variables. Let $N_i(t), i = 1, 2$ be two independent Poisson processes with parameters $\lambda_i > 0, i = 1, 2$. Let the sum of these two processes be $N(t) = N_1(t) + N_2(t)$, where $t \geq 0$. Then for $n \geq 0$

$$
P(N(t) = n)
$$
$$
= \sum_{j=0}^{n} P(N_1(t) = j) P(N_2(t) = n - j)
$$
$$
= \sum_{j=0}^{n} e^{-\lambda_1 t} \frac{(\lambda_1 t)^j}{j!} e^{-\lambda_2 t} \frac{(\lambda_2 t)^{n-j}}{(n-j)!} = e^{-(\lambda_1 + \lambda_2)t} \frac{\{(\lambda_1 + \lambda_2)t\}^n}{n!}
$$

Therefore $N(t), t \geq 0$ is a Poisson process with parameter $(\lambda_1 + \lambda_2)$.

2. Prove that the time interval between two successive events of a homogeneous Poisson process with parameter $\lambda$, has an exponential distribution with parameter $\lambda$. Furthermore, such successive different interarrival times are independent of each other.
   Hint: See Ross (2010). Let $t_1, t_2, t_3, \ldots$ be instants at which the event numbers $1, 2, 3, \ldots$ occur respectively. Define $X_1 = t_1$ and $X_i = t_i - t_{i-1}$ for $2, 3, 4, \ldots$. Therefore $\{X_1, X_2, X_3, \ldots\}$ is a sequence of successive interarrival times. It has to be proved that the random variables $X_i$'s, each have exponential distribution with parameter $\lambda$, and are independent of each other.
   Note that $P(X_1 > t) = P(N(t) = 0) = e^{-\lambda t}$. This implies that the random variable $X_1$ has an exponential distribution with parameter $\lambda$. Next observe that $P(X_2 > t) = \mathcal{E}(P\{X_2 > t \mid X_1\})$, however

$$
P(X_2 > t \mid X_1 = u) = P\{0 \text{ events in the interval } (u, u + t] \mid X_1 = u\}
$$

Use of the assumption of independent increments, and the stationary-increments property of the Poisson process yields

$$
P(X_2 > t \mid X_1 = u) = P\{0 \text{ events in the interval } (u, u + t]\} = e^{-\lambda t}
$$

Thus the interarrival time $X_2$ has an exponential distribution with parameter $\lambda$. In addition, $X_2$ is independent of $X_1$. It can be similarly proved that $X_3$ has an identical distribution, and it is independent of the time intervals $X_1$ and $X_2$. This argument can be repeated for successive interarrival time intervals.

3. Let $\{N(t), t \in \mathbb{R}_0^+\}$ be a homogeneous Poisson process with parameter $\lambda$. It is specified that only a single event has occurred by epoch $t > 0$. Let $Y$ be the arrival instant of the single event in the interval $[0, t]$. Show that $Y$ is uniformly distributed over the interval $[0, t]$.

   Hint: See Leon-Garcia (1994). Let $N(y), y \in [0, t]$ be the number of events up to time $y$. Also $(N(t) - N(y))$ is the increment in the time interval $[y, t]$. Then

$$
\begin{aligned}
P(Y \le y) &= P(N(y) = 1 \mid N(t) = 1) \\
&= \frac{P(N(y) = 1 \text{ and } N(t) = 1)}{P(N(t) = 1)} \\
&= \frac{P(N(y) = 1 \text{ and } (N(t) - N(y)) = 0)}{P(N(t) = 1)} \\
&= \frac{\lambda y e^{-\lambda y} e^{-\lambda(t-y)}}{\lambda t e^{-\lambda t}} = \frac{y}{t}
\end{aligned}
$$

This result implies that Poisson process models events which occur at "random" over $\mathbb{R}^+$. It can also be shown that if the number of events in the interval $[0, t]$ is $n$, then the individual arrival instants are distributed uniformly and independently in $[0, t]$.

4. Let $Z_1, Z_2, \ldots, Z_n$ be independent and identically distributed continuous random variables. Each of these is distributed as random variable $Z$. which has a probability density function $f_Z(\cdot)$. Let the order statistics of these random variables be $(Z_{(1)}, Z_{(2)}, \ldots, Z_{(n)}) \triangleq \mathcal{Z}$. Prove the following statements.

   (a) The joint probability density function of the order statistics $\mathcal{Z}$ is given by

$$
f(z_1, z_2, \ldots, z_n) = n! \prod_{i=1}^{n} f_Z(z_i), \qquad z_1 < z_2 < \cdots < z_n
$$

   (b) Let the random variables $Z_i$'s for $i = 1, 2, \ldots, n$ be uniformly distributed over the interval $(0, t)$ then the joint probability density function of the order statistics $\mathcal{Z}$ is given by

$$
f(z_1, z_2, \ldots, z_n) = \frac{n!}{t^n}, \quad 0 < z_1 < z_2 < \cdots < z_n < t
$$

   Hint: See Ross (1983), and Pinsky, and Karlin (2011).

   (a) Observe that $(Z_{(1)}, Z_{(2)}, \ldots, Z_{(n)}) = (z_1, z_2, \ldots, z_n)$ if $(Z_1, Z_2, \ldots, Z_n)$ is equal to any of the $n!$ permutations of $(z_1, z_2, \ldots, z_n)$. Further, the $Z_{(i)}$'s for $1 \le i \le n$ are independent of each other, as $Z_i$'s for $1 \le i \le n$ are independent of each other. Let $(z_{k_1}, z_{k_2}, \ldots, z_{k_n})$ be a permutation of $(z_1, z_2, \ldots, z_n)$. Then the probability density of $(Z_1, Z_2, \ldots, Z_n) = (z_{k_1}, z_{k_2}, \ldots, z_{k_n})$ is equal to $\prod_{i=1}^{n} f_Z(z_{k_i}) = \prod_{i=1}^{n} f_Z(z_i)$.

   (b) This result follows because $f_Z(z) = 1/t$ for $z \in (0, t)$.

5. Let the occurrence times of the $n$ events in a homogeneous Poisson process of rate $\lambda$ be $S_1, S_2, \ldots, S_n$, and $N(t) = n$. Then the occurrence times of these $n$ events have the same distribution as the order statistics of $n$ independent random variables which are uniformly distributed over the interval $(0, t)$. Prove this statement.

Hint: See Ross (1983), and Pinsky, and Karlin (2011). We compute the conditional joint probability density function of $S_1, S_2, \ldots, S_n$; given that $N(t) = n$. Let $0 < t_1 < t_2 < \cdots < t_n < t$. Also let $\Delta t_i$ be a small increment in time so that $(t_i + \Delta t_i) < t_{i+1}$, for $i = 1, 2, \ldots, n$, and $t_{n+1} = t$. We have

$$P\left(t_i \leq S_i \leq t_i + \Delta t_i,\ i = 1, 2, \ldots, n \mid N(t) = n\right)$$

$$= \frac{P\left(\begin{array}{c} \text{exactly one event in } [t_i, t_i + \Delta t_i],\ i = 1, 2, \ldots, n; \\ \text{and no other events in } [0, t] \end{array}\right)}{P(N(t) = n)}$$

$$= \frac{\left\{\displaystyle\prod_{i=1}^{n} \{\lambda \Delta t_i \exp(-\lambda \Delta t_i)\}\right\} \exp\{-\lambda(t - \Delta t_1 - \Delta t_2 - \cdots - \Delta t_n)\}}{\exp(-\lambda t)(\lambda t)^n / n!}$$

$$= \frac{n!}{t^n} \prod_{i=1}^{n} \Delta t_i$$

Therefore

$$\frac{P\left(t_i \leq S_i \leq t_i + \Delta t_i,\ i = 1, 2, \ldots, n \mid N(t) = n\right)}{\displaystyle\prod_{i=1}^{n} \Delta t_i} = \frac{n!}{t^n}$$

Letting $\Delta t_i$'s tend to zero, we obtain the conditional joint probability density function of $S_1, S_2, \ldots, S_n$; given that $N(t) = n$ as

$$f(t_1, t_2, \ldots, t_n) = \frac{n!}{t^n}, \quad 0 < t_1 < t_2 < \cdots < t_n < t$$

The stated result follows by using the result of the last problem.

6. Consider a shot noise process $\{X(t), t \in \mathbb{R}_0^+\}$ in which arrivals occur with a rate parameter $\lambda$. That is, the arrival process is a homogeneous Poisson process. The Laplace functional of the shot noise process is

$$\widehat{f}_{X(t)}(s) = \mathcal{E}(\exp(-sX(t))) = \exp\left\{\lambda \int_0^t \varphi(s, \tau)\, d\tau\right\}$$

where

$$\varphi(s, \tau) = \mathcal{E}(\exp(-sw(\tau, Y))) - 1, \quad \tau > 0,\ s \in \mathbb{C}$$

and the expectation $\mathcal{E}(\cdot)$ in the above equation is with respect to the random variable $Y$. Prove this result.

Hint: See Snyder, and Miller (1991). Let

$$P(N(t) - N(0) = n) = \frac{e^{-\lambda t}(\lambda t)^n}{n!} \triangleq \theta_n(t), \quad \text{for } n \in \mathbb{N}$$

We have

$$\widehat{f}_{X(t)}(s) = \mathcal{E}(\exp(-sX(t)))$$

$$= \sum_{n\in\mathbb{N}} \mathcal{E}\left(\exp\left(-sX(t)\right) \mid N\left(t\right) - N\left(0\right) = n\right) P\left(N\left(t\right) - N\left(0\right) = n\right)$$

$$= \sum_{n\in\mathbb{N}} \mathcal{E}\left(\exp\left(-sX(t)\right) \mid N\left(t\right) - N\left(0\right) = n\right) \theta_n\left(t\right)$$

We initially evaluate $\mathcal{E}\left(\exp\left(-sX(t)\right) \mid N\left(t\right) - N\left(0\right) = n\right)$. The conditional distribution of the times $0 \leq S_1 < S_2 < \cdots < S_n \leq t$ at which events have occurred, given that precisely $n$ events have occurred; is identical to the distribution of ordered and independent uniformly distributed random variables over the interval 0 to $t$. Therefore for $N\left(t\right) = 0$, that is $n = 0$, we have

$$\mathcal{E}\left(\exp\left(-sX(t)\right) \mid N\left(t\right) - N\left(0\right) = 0\right) = 1$$

and for $N\left(t\right) \geq 1$, that is $n \geq 1$, we have

$$\mathcal{E}\left(\exp\left(-sX(t)\right) \mid N\left(t\right) - N\left(0\right) = n\right)$$

$$= \mathcal{E}\left(\exp\left\{-s\sum_{m=1}^{n} w\left(t - S_m, Y_m\right)\right\} \mid N\left(t\right) - N\left(0\right) = n\right)$$

$$= \left\{\mathcal{E}\left(\exp\left(-sw\left(t - U, Y\right)\right)\right)\right\}^n$$

where $U$ is a uniformly distributed random variable over the interval 0 to $t$. As $(t - U)$ and $U$ have the same probability distribution we have

$$\mathcal{E}\left(\exp\left(-sw\left(t - U, Y\right)\right)\right) = \mathcal{E}\left(\exp\left(-sw\left(U, Y\right)\right)\right)$$

$$= \frac{1}{t}\int_0^t \mathcal{E}\left(\exp\left(-sw\left(\tau, Y\right)\right)\right) d\tau$$

For $n \geq 1$, this results in

$$\mathcal{E}\left(\exp\left(-sX(t)\right) \mid N\left(t\right) - N\left(0\right) = n\right)$$

$$= \left\{\frac{1}{t}\int_0^t \mathcal{E}\left(\exp\left(-sw\left(\tau, Y\right)\right)\right) d\tau\right\}^n$$

We have

$$\widehat{f}_{X(t)}\left(s\right) = \theta_0\left(t\right) + \sum_{n\in\mathbb{P}} \mathcal{E}\left(\exp\left(-sX(t)\right) \mid N\left(t\right) - N\left(0\right) = n\right) \theta_n\left(t\right)$$

$$= \sum_{n\in\mathbb{N}} e^{-\lambda t}\frac{(\lambda t)^n}{n!}\left\{\frac{1}{t}\int_0^t \mathcal{E}\left(\exp\left(-sw\left(\tau, Y\right)\right)\right) d\tau\right\}^n$$

$$= e^{-\lambda t}\sum_{n\in\mathbb{N}} \frac{1}{n!}\left\{\lambda\int_0^t \mathcal{E}\left(\exp\left(-sw\left(\tau, Y\right)\right)\right) d\tau\right\}^n$$

$$= \exp\left\{\lambda\int_0^t \left(\mathcal{E}\left(\exp\left(-sw\left(\tau, Y\right)\right)\right) - 1\right) d\tau\right\}$$

$$= \exp\left\{\lambda\int_0^t \varphi\left(s, \tau\right) d\tau\right\}$$

where

$$\varphi\left(s, \tau\right) = \mathcal{E}\left(\exp\left(-sw\left(\tau, Y\right)\right)\right) - 1, \quad \tau > 0$$

7. Consider a shot noise process $\{X(t), t \in \mathbb{R}_0^+\}$ in which arrivals occur according to a nonhomogeneous Poisson process, with intensity function $\lambda(\cdot)$. The Laplace functional of the shot noise process is

$$\widehat{f}_{X(t)}(s) = \mathcal{E}(\exp(-sX(t)))$$

$$= \exp\left\{\int_0^t \lambda(\tau)\,\varphi(s,\tau)\,d\tau\right\}$$

where

$$\varphi(s,\tau) = \mathcal{E}(\exp(-sw(\tau,Y))) - 1, \quad \tau > 0, \ s \in \mathbb{C}$$

and the expectation $\mathcal{E}(\cdot)$ in the above equation is with respect to the random variable $Y$. Prove this result.

Hint: See Snyder, and Miller (1991). Let

$$P(N(t) - N(0) = n) = \frac{e^{-\Lambda(t)}(\Lambda(t))^n}{n!} \triangleq \theta_n(t), \quad \text{for } n \in \mathbb{N}$$

We have

$$\widehat{f}_{X(t)}(s) = \mathcal{E}(\exp(-sX(t)))$$

$$= \sum_{n \in \mathbb{N}} \mathcal{E}(\exp(-sX(t)) \mid N(t) - N(0) = n)\, P(N(t) - N(0) = n)$$

$$= \sum_{n \in \mathbb{N}} \mathcal{E}(\exp(-sX(t)) \mid N(t) - N(0) = n)\, \theta_n(t)$$

$$= \theta_0(t) + \sum_{n \in \mathbb{P}} \mathcal{E}(\exp(-sX(t)) \mid N(t) - N(0) = n)\, \theta_n(t)$$

For $N(t) \geq 1$, that is $n \geq 1$, we have

$$\mathcal{E}(\exp(-sX(t)) \mid N(t) - N(0) = n)$$

$$= \mathcal{E}\left(\exp\left\{-s\sum_{m=1}^n w(t - S_m, Y_m)\right\} \Big| N(t) - N(0) = n\right)$$

The summation inside the expectation operator is unchanged, if a random reordering of the arrival times occurs. Assuming this reordering, the occurrence times $\{S_1, S_2, \ldots, S_n\}$ given $(N(t) - N(0)) = n$ are independent and identically distributed. Their common probability density function is

$$\frac{\lambda(\tau)}{\Lambda(t)}, \quad \text{where} \quad \Lambda(t) = \int_0^t \lambda(w)\,dw$$

As the random variables $Y_1, Y_2, \ldots, Y_n$ are independent and identically distributed, we have

$$\mathcal{E}(\exp(-sX(t)) \mid N(t) - N(0) = n)$$

$$= \left\{\frac{1}{\Lambda(t)}\int_0^t \lambda(\tau)\,\mathcal{E}(\exp(-sw(\tau,Y)))\,d\tau\right\}^n, \quad n \geq 1$$

where the expectation on the right hand side of the above equation is with respect to the random variable $Y$. Thus

$$\widehat{f}_{X(t)}(s) = \theta_0(t) + \sum_{n \in \mathbb{P}} \mathcal{E}\left(\exp\left(-sX(t)\right) \mid N(t) - N(0) = n\right)\theta_n(t)$$

$$= \sum_{n \in \mathbb{N}} e^{-\Lambda(t)} \frac{(\Lambda(t))^n}{n!} \left\{ \frac{1}{\Lambda(t)} \int_0^t \lambda(\tau)\,\mathcal{E}\left(\exp\left(-sw\left(\tau, Y\right)\right)\right) d\tau \right\}^n$$

$$= e^{-\Lambda(t)} \sum_{n \in \mathbb{N}} \frac{1}{n!} \left\{ \int_0^t \lambda(\tau)\,\mathcal{E}\left(\exp\left(-sw\left(\tau, Y\right)\right)\right) d\tau \right\}^n$$

$$= \exp\left\{ \int_0^t \lambda(\tau)\left(\mathcal{E}\left(\exp\left(-sw\left(\tau, Y\right)\right)\right) - 1\right) d\tau \right\}$$

$$= \exp\left\{ \int_0^t \lambda(\tau)\,\varphi(s, \tau)\, d\tau \right\}$$

where

$$\varphi(s, \tau) = \mathcal{E}\left(\exp\left(-sw\left(\tau, Y\right)\right)\right) - 1, \quad \tau > 0$$

8. $X$ is a random variable with Gaussian distribution, where $\mathcal{E}(X) = \mu$, and $Var(X) = \sigma^2$. Let $Y = |X|$, then prove that

$$\mathcal{E}(Y) = \mu\left\{2\Phi(\mu/\sigma) - 1\right\} + 2\sigma\phi(\mu/\sigma)$$

where $\phi(x) = e^{-x^2/2}/\sqrt{2\pi}$, $\Phi(x) = \int_{-\infty}^x \phi(t)\,dt$, and $x \in \mathbb{R}$.
Hint: Establish that $\mathcal{E}(Y) + \mu = 2\int_0^\infty y f_X(y)\,dy$, where $f_X(\cdot)$ is the probability density function of the random variable $X$.

9. Find an expression for the covariance function of the Brownian motion (Wiener) process.
Hint: See Hoel, Port, and Stone (1972).

10. $B(t), t \in \mathbb{R}$ is a Brownian motion process. Let $t_1 \leq t_2 \leq t_3 \leq t_4$. Show that

$$Cov\left\{(B(t_2) - B(t_1)), (B(t_4) - B(t_3))\right\} = 0$$

Hint:

$$Cov\left\{(B(t_2) - B(t_1)), (B(t_4) - B(t_3))\right\}$$
$$= \mathcal{E}\left\{(B(t_2) - B(t_1))(B(t_4) - B(t_3))\right\}$$
$$= \mathcal{E}(B(t_2) - B(t_1))\,\mathcal{E}(B(t_4) - B(t_3)) = 0$$

11. Let $\vartheta = \int_a^b f(t)\,dB(t)$, where $f(\cdot)$ is a deterministic function, and $B(t), t \in \mathbb{R}$ is a Brownian motion process. Show that

$$\mathcal{E}(\vartheta) = 0, \text{ and } Var(\vartheta) = \sigma^2 \int_a^b \left\{f(t)\right\}^2 dt$$

Hint: Approximate $\vartheta$ as $\sum_{i=1}^n a_i\left\{B(t_i) - B(t_{i-1})\right\}$. It is known that $\mathcal{E}(B(t)) = 0$, which implies $\mathcal{E}(\vartheta) = 0$. The variance of $\vartheta$ is

$$Var(\vartheta) = \sum_{i=1}^n \sum_{j=1}^n a_i a_j \mathcal{E}\left\{(B(t_i) - B(t_{i-1}))(B(t_j) - B(t_{j-1}))\right\}$$

Note that

$$\mathcal{E}\left\{(B\left(t_i\right) - B\left(t_{i-1}\right))(B\left(t_j\right) - B\left(t_{j-1}\right))\right\} = \begin{cases} \sigma^2 \left(t_i - t_{i-1}\right), & \text{if } i = j \\ 0, & \text{if } i \neq j \end{cases}$$

Thus

$$Var\left(\vartheta\right) = \sigma^2 \sum_{i=1}^{n} a_i^2 \left(t_i - t_{i-1}\right)$$

Therefore as $n \to \infty$, and $(t_i - t_{i-1}) \to 0$, $\forall\, i$, we obtain the desired result. See Hoel, Port, and Stone (1972) for a rigorous justification of the above steps.

12. Prove that the Wiener process is a Gaussian process.
    Hint: See Hoel, Port, and Stone (1972).

13. Let $f\left(\cdot\right)$ be a real-valued function defined on the space $S$. Also let $N\left(\cdot\right)$ be the counting measure of a point process $\Phi$. Define $N\left(f\right)$ as $\int_S f\left(x\right) N\left(dx\right)$. Show that $\int_S f\left(x\right) N\left(dx\right) = \sum_{n \in \mathbb{P}} f\left(X_n\right)$, where it is assumed that the sum is finite.
    Hint: Let $\delta\left(\cdot\right)$ be the Dirac's delta function. Informally, we have

$$\int_S f\left(x\right) N\left(dx\right) = \int_S f\left(x\right) \sum_{n \in \mathbb{P}} \delta\left(x - X_n\right) dx$$

$$= \sum_{n \in \mathbb{P}} \int_S f\left(x\right) \delta\left(x - X_n\right) dx = \sum_{n \in \mathbb{P}} f\left(X_n\right)$$

14. Let $N\left(B\right) = \sum_{n \in \mathbb{P}} \zeta_{X_n}\left(B\right)$, $B \in \widetilde{S}$ be the counting measure of a point process $\Phi$. The mean of the random variable $N\left(B\right)$ is $\mathcal{E}\left(N\left(B\right)\right) = \Lambda\left(B\right)$. For a function $f : S \to \mathbb{R}$, assume that $\int_S |f\left(x\right)| \Lambda\left(dx\right) < \infty$. Prove that

$$\mathcal{E}\left\{\sum_{n \in \mathbb{P}} f\left(X_n\right)\right\} = \int_S f\left(x\right) \Lambda\left(dx\right)$$

Hint: As in the last problem, we have

$$\mathcal{E}\left\{\sum_{n \in \mathbb{P}} f\left(X_n\right)\right\} = \mathcal{E}\left\{\int_S f\left(x\right) N\left(dx\right)\right\} - \int_S f\left(x\right) \Lambda\left(dx\right)$$

15. Show that, if $f\left(x\right) = \sum_{i=1}^{k} s_i I_{B_i}\left(x\right)$, then $N\left(f\right) = \sum_{i=1}^{k} s_i N\left(B_i\right)$, where $B_i \in \widetilde{S}$ for $1 \leq i \leq k$.
    Hint: Let $\delta\left(\cdot\right)$ be the Dirac's delta function. We have

$$N\left(f\right) = \int_S f\left(x\right) N\left(dx\right) = \int_S \sum_{i=1}^{k} s_i I_{B_i}\left(x\right) N\left(dx\right)$$

$$= \sum_{i=1}^{k} s_i \int_S I_{B_i}\left(x\right) N\left(dx\right)$$

The result follows, if it is shown that $\int_S I_{B_i}\left(x\right) N\left(dx\right) = N\left(B_i\right)$. Note that

$$\int_S I_{B_i}(x) N(dx) = \int_S I_{B_i}(x) \sum_{n \in \mathbb{P}} \delta(x - X_n) dx$$

$$= \sum_{n \in \mathbb{P}} \int_S I_{B_i}(x) \delta(x - X_n) dx$$

$$= \sum_{n \in \mathbb{P}} I_{B_i}(X_n) = N(B_i)$$

16. Show that the Laplace functional of a nonhomogeneous Poisson point process with intensity measure $\Lambda(\cdot)$, and $f : S \to \mathbb{R}_0^+$ is

$$\mathcal{L}_N(f) = \exp\left\{-\int_S \left(1 - e^{-f(x)}\right) \Lambda(dx)\right\}$$

The integral in the above expression is finite if and only if

$$\int_S \min\left(f(x), 1\right) \Lambda(dx) < \infty$$

Hint: See Resnick (1992), Mikosch (2009), and Serfozo (2009). We address this problem in two steps.

*Step* 1: In this step, expression for $\mathcal{L}_N(f)$ is established. Initially consider a simple function $f(x) = \sum_{i=1}^k \beta_i I_{B_i}(x)$, where $\beta_1, \beta_2, \ldots, \beta_k \in \mathbb{R}_0^+$, and $B_1, B_2, \ldots, B_k \in \widetilde{S}$ are disjoint. Then

$$N(f) = \int_S f(x) N(dx) = \sum_{i-1}^k \beta_i \int_S I_{B_i}(x) N(dx) = \sum_{i=1}^k \beta_i N(B_i)$$

The random variables $N(B_i)$'s are mutually independent of each other, and have Poissonian distribution. Use of the expression $\mathcal{E}(e^{tX}) = e^{\lambda(e^t - 1)}$ for the generating function of a Poisson distributed random variable $X$ with parameter $\lambda > 0$, results in

$$\mathcal{E}(\exp(-N(f))) = \prod_{i=1}^k \mathcal{E}(\exp(-\beta_i N(B_i)))$$

$$= \prod_{i=1}^k \exp\left\{-\left(1 - e^{-\beta_i}\right) \Lambda(B_i)\right\}$$

$$= \exp\left\{-\sum_{i=1}^k \left(1 - e^{-\beta_i}\right) \Lambda(B_i)\right\}$$

$$= \exp\left\{-\sum_{i=1}^k \int_{B_i} \left(1 - e^{-\beta_i}\right) I_{B_i}(x) \Lambda(dx)\right\}$$

$$= \exp\left\{-\int_S \left(1 - e^{-\sum_{i=1}^k \beta_i I_{B_i}(x)}\right) \Lambda(dx)\right\}$$

$$= \exp\left\{-\int_S \left(1 - e^{-f(x)}\right) \Lambda(dx)\right\}$$

If $f : S \to \mathbb{R}_0^+$ is any continuous function, then we approximate it as a simple function as follows. The function $f_n(\cdot)$ approximates $f(\cdot)$, if we let $n \to \infty$. The function $f_n(\cdot)$ is specified via $n2^n$ equally-spaced levels in the interval $[0, n)$. Let $a_i = (i-1)/2^n$, $A_i = [(i-1)/2^n, i/2^n)$, where $1 \le i \le n2^n$, and

$$f_n(x) = \sum_{i=1}^{n2^n} a_i I_{A_i}(f(x)) + n I_{[n,\infty)}(f(x))$$

Observe that the width of these levels $1/2^n \to 0$ as $n \to \infty$. The final result follows, as $e^{-f} \le 1$, and via monotonic convergence.

*Step* 2: In this step we establish that the integral in the Laplace functional is finite. The expression

$$\int_S \left(1 - e^{-f(x)}\right) \Lambda(dx)$$

is decomposed as

$$\left(\int_{\{x|f(x)>1\}} + \int_{\{x|f(x)\le1\}}\right) \left(1 - e^{-f(x)}\right) \Lambda(dx)$$
$$= I_1 + I_2$$

The integrand in $I_1$ takes values in the interval $(1 - e^{-1}, 1]$. Therefore $I_1$ is finite if and only if $\int_{\{x|f(x)>1\}} \Lambda(dx) < \infty$. Next, expand $\left(1 - e^{-f(x)}\right)$ in Taylor's series. It can be observed that $I_2$ is finite if and only if $\int_{\{x|f(x)\le1\}} f(x) \Lambda(dx) < \infty$. Therefore the integral in Laplace functional is finite if and only if $\int_S \min(f(x), 1) \Lambda(dx) < \infty$.

17. Let $f : S \to \mathbb{R}_0^+$ be a continuous function with compact support, and

$$\Phi = \{X_n \mid X_n, n \in \mathbb{P}\}$$

be a nonhomogeneous Poisson point process with intensity measure $\Lambda(\cdot)$, counting measure $N(\cdot)$, and $N(f) = \int_S f(x) N(dx)$. Prove the following results.

(a) $\mathcal{E}\left(\sum_{n\in\mathbb{P}} f(X_n)\right) = \int_S f(x) \Lambda(dx)$, provided the integral exists.

(b) $Var\left(\sum_{n\in\mathbb{P}} f(X_n)\right) = \int_S (f(x))^2 \Lambda(dx)$, provided the integral exists.

Hint: See Serfozo (2009). Let $\mathcal{E}(\exp(-\alpha N(f))) = \exp(-g(\alpha)) \triangleq w(\alpha)$, where $g(\alpha) = \int_S \left(1 - e^{-\alpha f(x)}\right) \Lambda(dx)$. Denote the first and second derivatives of $g(\alpha)$ with respect to $\alpha$, by $g'(\alpha)$ and $g''(\alpha)$ respectively. Similarly, denote the first and second derivatives of $w(\alpha)$ with respect to $\alpha$, by $w'(\alpha)$ and $w''(\alpha)$ respectively. It can be shown that

$$g(0) = 0, \ g'(0) = \int_S f(x) \Lambda(dx), \ \text{and } g''(0) = -\int_S (f(x))^2 \Lambda(dx)$$
$$w(0) = 1, \ w'(0) = -g'(0), \ \text{and } w''(0) = -g''(0) + (g'(0))^2$$

We next evaluate the first and second derivatives of $\mathcal{E}(\exp(-\alpha N(f)))$ with respect to $\alpha$, at $\alpha = 0$. These are

$$w'(0) = \frac{d}{d\alpha}\mathcal{E}(\exp(-\alpha N(f)))\bigg|_{\alpha=0} = -\mathcal{E}(N(f))$$
$$w''(0) = \frac{d^2}{d\alpha^2}\mathcal{E}(\exp(-\alpha N(f)))\bigg|_{\alpha=0} = \mathcal{E}\left((N(f))^2\right)$$

The results follow.

18. Prove that the marked point process $\Phi_M$ is a PPP defined over the product space $S \times \mathbb{M}$, where $S$ is equal to the Euclidean space $\mathbb{R}^m$. The intensity measure $\Lambda_M$ of the process is defined by

$$\Lambda_M (B \times G) = \int_B p(x, G) \Lambda(dx), \qquad B \in \widetilde{S}, G \in \widetilde{\mathcal{M}}$$

Consequently, the point process of marks $\Phi' = \{M_n \mid M_n \in \mathbb{M}, n \in \mathbb{P}\}$ with counting measure

$$N'(G) = \sum_{n \in \mathbb{P}} \zeta_{M_n}(G), \quad G \in \widetilde{\mathcal{M}}$$

is a Poisson process on $\mathbb{M}$ with intensity measure

$$\Lambda'(G) = \int_S p(x, G) \Lambda(dx), \quad G \in \widetilde{\mathcal{M}},$$

assuming it is locally finite.

Hint: See Serfozo (2009). It needs to be demonstrated that $\Phi_M$ is indeed a Poisson process on $S \times \mathbb{M}$ with the stated intensity measure. Let the counting measure of this process be $M'(B \times G)$. Consider the sum

$$M'(f) = \sum_{n \in \mathbb{P}} f(X_n, M_n)$$

where $f(\cdot, \cdot) \geq 0$ is any bounded measurable function. The Laplace functional of $\Phi_M$ is

$$\mathcal{E}\left(e^{-M'f}\right) = \mathcal{E}\left[\mathcal{E}\left\{\exp\left(-\sum_{n \in \mathbb{P}} f(X_n, M_n)\right) \mid N\right\}\right]$$

$$= \mathcal{E}\left[\prod_{n \in \mathbb{P}} \int_{\mathbb{M}} \exp\{-f(X_n, y)\} p(X_n, dy)\right]$$

$$= \mathcal{E}\left[\exp\left\{\sum_{n \in \mathbb{P}} \ln \int_{\mathbb{M}} e^{-f(X_n, y)} p(X_n, dy)\right\}\right]$$

Let

$$h(x) \triangleq -\ln \int_{\mathbb{M}} \exp(-f(x, y)) p(x, dy)$$

Note that $\int_{\mathbb{M}} \exp(-f(x, y)) p(x, dy) \leq 1$. Therefore $h \geq 0$. We have

$$\mathcal{E}\left(e^{-M'f}\right) = \mathcal{E}\left[\exp\left\{-\sum_{n \in \mathbb{P}} h(X_n)\right\}\right] = \mathcal{E}\left[\exp\left\{-\int_S h(x) N(dx)\right\}\right]$$

Use of Laplace functional of the PPP $\Phi$ with counting measure $N(\cdot)$ results in

$$\mathcal{E}\left(e^{-M'f}\right) = \exp\left\{-\int_S \left(1 - e^{-h(x)}\right) \Lambda(dx)\right\}$$

$$= \exp\left\{-\int_S \left(1 - \int_{\mathbb{M}} \exp(-f(x, y)) p(x, dy)\right) \Lambda(dx)\right\}$$

$$= \exp\left\{-\int_{S \times \mathbb{M}} (1 - \exp(-f(x, y))) p(x, dy) \Lambda(dx)\right\}$$

$$= \exp\left\{-\int_{S \times \mathbb{M}} (1 - \exp(-f(x, y))) \Lambda_M(dx, dy)\right\}$$

The last expression for $\mathcal{E}\left(e^{-M'f}\right)$ is a Laplace functional of a PPP. Therefore $\Phi_M$ is a PPP with intensity measure $\Lambda_M$. Note that $\Lambda_M\left(dx, dy\right) = p\left(x, dy\right)\Lambda\left(dx\right)$. Consequently

$$\Lambda_M\left(B \times G\right) = \int_B p\left(x, G\right)\Lambda\left(dx\right), \qquad B \in \widetilde{\mathcal{S}}, G \in \widetilde{\mathcal{M}}$$

Furthermore, $N'\left(G\right) = M'\left(B \times G\right)$ is the counting measure of the marked point process $\Phi'$. It has a Poissonian distribution with intensity measure $\Lambda'\left(G\right) = \Lambda_M\left(B \times G\right)$.

19. We establish the mapping theorem of PPP. Let $\Phi$ be a PPP defined on space $\mathbb{R}^m$ with intensity function $\lambda\left(\cdot\right)$, and intensity measure $\Lambda\left(\cdot\right)$. Also let $f : \mathbb{R}^m \to \mathbb{R}^d$ be a measurable function such that $\Lambda\left(f^{-1}\left\{y\right\}\right) = 0$; for all $y \in \mathbb{R}^d$, and $f\left(\cdot\right)$ does not map a (non-singleton) compact set to a singleton. Then the process

$$\widetilde{\Phi} = f\left(\Phi\right) \triangleq \bigcup_{X \in \Phi} \left\{f\left(X\right)\right\}$$

is also a PPP with intensity measure

$$\widetilde{\Lambda}\left(\widetilde{B}\right) = \Lambda\left(f^{-1}\left(\widetilde{B}\right)\right) = \int_{f^{-1}\left(\widetilde{B}\right)} \lambda\left(x\right) dx, \quad \text{for all compact sets } \widetilde{B} \subset \mathbb{R}^d$$

Hint: See Haenggi (2013). Let the counting measure of the process $\widetilde{\Phi}$ be $\widetilde{N}$. Observe that $\widetilde{N}\left(\widetilde{B}\right)$ has a Poisson distribution with mean $\Lambda\left(f^{-1}\left(\widetilde{B}\right)\right)$.
Further, if $\widetilde{B}_1, \widetilde{B}_2, \ldots, \widetilde{B}_n$ are pairwise disjoint, then their preimages

$$f^{-1}\left(\widetilde{B}_1\right), f^{-1}\left(\widetilde{B}_2\right), \ldots, f^{-1}\left(\widetilde{B}_n\right)$$

are also pairwise disjoint. Consequently the random variables

$$\widetilde{N}\left(\widetilde{B}_1\right) = N\left(f^{-1}\left(\widetilde{B}_1\right)\right),$$
$$\widetilde{N}\left(\widetilde{B}_2\right) = N\left(f^{-1}\left(\widetilde{B}_2\right)\right), \ldots,$$
$$\widetilde{N}\left(\widetilde{B}_n\right) = N\left(f^{-1}\left(\widetilde{B}_n\right)\right)$$

are independent.

20. We establish the displacement theorem of PPP. Let $\Phi$ be a PPP defined on $\mathbb{R}^m$ with intensity function $\lambda\left(\cdot\right)$. Further, all points are independently displaced, and the distribution of the displaced location of a point $x$ in the point process $\Phi$ has a probability density function $\rho\left(x, \cdot\right)$. Then the displaced points form a PPP $\widetilde{\Phi}$ with intensity function

$$\widetilde{\lambda}\left(y\right) = \int_{\mathbb{R}^m} \lambda\left(x\right)\rho\left(x, y\right) dx$$

More specifically, if $\lambda\left(x\right)$ is equal to a constant $\lambda \in \mathbb{R}^+$, and $\rho\left(\cdot, \cdot\right)$ depends only on $\left(y - x\right)$, then $\widetilde{\lambda}\left(y\right)$ is equal to the constant $\lambda$.

Hint: See Haenggi (2013). Let the displaced location of a point $x$ in the PPP $\Phi$ be $y_x$. As per the hypothesis of the theorem, the displacements of different points in $\Phi$ are independent of each other. Further, the probability density function of the new location is specified by $\rho\left(x, \cdot\right)$.

Define a marked process $\widetilde{\Phi}$ as follows. Let some realizations of the point processes $\Phi$ and $\widetilde{\Phi}$ be $\phi$ and $\widetilde{\phi}$ respectively. Then

$$\widetilde{\phi} = \{(x, y_x) \mid x \in \phi\}$$

Therefore, as per the marking theorem, $\widetilde{\Phi}$ is a PPP defined on $\mathbb{R}^{2m}$. Also the mapping theorem implies that the point process of only the markings is a PPP defined on $\mathbb{R}^m$. Therefore, the possibility that the point $y_x$ is located in the infinitesimal volume $dy$ at $y$ if its original location was at $x$ is simply $\lambda(x)\rho(x,y)\,dy$. Therefore, the intensity measure of the displaced process is

$$\widetilde{\Lambda}(G) = \int_G \int_{\mathbb{R}^m} \lambda(x)\rho(x,y)\,dx dy$$

The above result yields the expression for the intensity $\widetilde{\lambda}(y)$.

If $\lambda(x) = \lambda$, and $\rho(x,y)$ depends only on $(y - x)$, then

$$\widetilde{\lambda}(y) = \lambda \int_{\mathbb{R}^m} \rho(x,y)\,dx = \lambda \int_{\mathbb{R}^m} \rho(0, y - x)\,dx$$
$$= \lambda \int_{\mathbb{R}^m} \rho(0, x)\,dx = \lambda$$

The last step follows as $\rho(0, x)$ is a probability density function.

21. $\{N(t), t \in \mathbb{R}_0^+\}$ is a renewal process. Establish the following results.
   (a) $P(N(t) \le n) = P(S_{n+1} > t)$.
   (b) $P(N(t) = n) = F_n(t) - F_{n+1}(t)$.
   (c) $M_o(t) = \mathcal{E}(N(t)) = \sum_{n=1}^{\infty} F_n(t)$.
   (d) $\lim_{t \to \infty} m_o(t) = 1/\mathcal{E}(X)$.
   (e) If $\mathcal{E}(X) = \mu$ and $Var(X) = \sigma^2$, then

$$\lim_{l \to \infty} P\left(\frac{N(t) - \mathcal{E}(N(t))}{\sqrt{Var(N(t))}} \le x\right) = \Phi(x)$$

where

$$\frac{\mathcal{E}(N(t))}{t} \sim \frac{1}{\mu}$$
$$\frac{Var(N(t))}{t} \sim \frac{\sigma^2}{\mu^3}$$

This is the central limit theorem for renewal processes.

Hint:
(a) $S_n \le t \Leftrightarrow N(t) \ge n$. Therefore $P(S_n \le t) = P(N(t) \ge n)$. Also

$$P(N(t) \ge n) = 1 - P(N(t) < n) = 1 - P(N(t) \le n - 1)$$
$$P(S_n \le t) = 1 - P(S_n > t)$$

Thus $P(N(t) \le n - 1) = P(S_n > t)$. The result follows.
(b) Observe that

$$P(N(t) = n) = P(N(t) \ge n) - P(N(t) \ge n + 1)$$
$$= P(S_n \le t) - P(S_{n+1} \le t)$$

(c) $M_o(t) = \mathcal{E}(N(t)) = \sum_{n=1}^{\infty} P(N(t) \geq n) = \sum_{n=1}^{\infty} P(S_n \leq t) = \sum_{n=1}^{\infty} F_n(t)$.

(d) Apply the final value theorem of the Laplace transform to $\hat{m}_o(s)$ and use the l'Hospital's rule.

(e) See Medhi (1994). Apply the central limit theorem to $S_n = \sum_{j=1}^{n} X_j$, as $n \to \infty$. Therefore $S_n$ is asymptotically normal with mean $n\mu$ and variance $n\sigma^2$. Thus

$$P(S_n \leq t) = P\left\{\frac{S_n - n\mu}{\sigma\sqrt{n}} \leq \frac{t - n\mu}{\sigma\sqrt{n}}\right\}$$

Define $x = (t - n\mu)/(\sigma\sqrt{n})$. Thus

$$\Phi(x) = \lim_{n\to\infty} P\left\{\frac{S_n - n\mu}{\sigma\sqrt{n}} \leq x\right\}$$

and

$$t = n\mu + x\sigma\sqrt{n}$$

For large values of $n$, we have $t \sim n\mu$. Therefore $t \sim \left(n\mu + x\sigma\sqrt{t/\mu}\right)$. That is

$$n \sim \frac{t}{\mu} - x\sigma\sqrt{\frac{t}{\mu^3}}$$

We next use the relationship $P(S_n \leq t) = P(N(t) \geq n)$. Therefore as $n \to \infty$

$$\Phi(x) = \lim_{n\to\infty} P(N(t) \geq n)$$

$$= \lim_{n\to\infty} P\left\{N(t) \geq \frac{t}{\mu} - x\sigma\sqrt{\frac{t}{\mu^3}}\right\}$$

$$= \lim_{n\to\infty} P\left\{\frac{N(t) - t/\mu}{\sigma\sqrt{t/\mu^3}} \geq -x\right\}$$

As $x$ is continuous

$$\lim_{n\to\infty} P\left\{\frac{N(t) - t/\mu}{\sigma\sqrt{t/\mu^3}} \leq -x\right\} = 1 - \Phi(x)$$

That is

$$\lim_{n\to\infty} P\left\{\frac{N(t) - t/\mu}{\sigma\sqrt{t/\mu^3}} \leq x\right\} = 1 - \Phi(-x) = \Phi(x)$$

Therefore as $n \to \infty$, the number of renewals have a normal distribution with $\mathcal{E}(N(t)) = t/\mu$ and $Var(N(t)) = \sigma^2 t/\mu^3$.

22. Determine the stationary probability distribution of the Ehrenfest urn model.

23. In a random walk in one dimension, a walker walks along a straight line. The walker takes a single step per unit time, either to the right or left of its position, with probability $p$ and $(1-p)$ respectively. Assume that the walker initially starts walk at position 0 on the straight line. Find the probability that the walker returns to the initial position 0 after $t$ steps. For simplicity assume that $p = 1/2$.

Hint: See Newman (2005). This random walk is reminiscent of the gambler's ruin process. The required probability is called the *first return time* of the walk. It is the lifetime of the gambler's ruin process.

Let $f_t$ be the probability that the walker returns for the first time to the position 0 after $t$ steps. Also let $u_t$ be the probability that the walker is at position 0 as many times as it pleases, before returning to this initial position at time $t$. Observe that $f_t$ and $u_t$ are nonzero only for even values of $t$. Define $f_0 = 0$ and $u_0 = 1$. For any $n \in \mathbb{N}$

$$u_{2n} = \begin{cases} 1, & n = 0 \\ \sum_{m=1}^{n} f_{2m} u_{2n-2m}, & n \in \mathbb{P} \end{cases}$$

The above set of equations can be solved via the generating function approach. Define

$$\mathcal{U}(z) = \sum_{n \in \mathbb{N}} u_{2n} z^n, \quad \text{and} \quad \mathcal{F}(z) = \sum_{n \in \mathbb{P}} f_{2n} z^n$$

Using the above set of equations yields $\mathcal{U}(z) = (1 + \mathcal{F}(z) \mathcal{U}(z))$. From the random walk model, it is known that

$$u_{2n} = \frac{1}{2^{2n}} \binom{2n}{n}, \quad \forall n \in \mathbb{N}$$

This yields

$$\mathcal{U}(z) = \frac{1}{\sqrt{1-z}}, \quad \text{and} \quad \mathcal{F}(z) = 1 - \sqrt{1-z}$$

Therefore

$$f_{2n} = \frac{1}{(2n-1) 2^{2n}} \binom{2n}{n}, \quad \forall n \in \mathbb{P}$$

Stirling's approximation for $n!$ is $\sqrt{2\pi n} (n/e)^n$, as $n \to \infty$. Use of this result yields $f_{2n} \sim n^{-3/2}$, as $n \to \infty$. That is

$$f_t \sim t^{-3/2}$$

Therefore the probability distribution of return times of a one-dimensional random walk has a heavy-tail with tail-index $3/2$.

24. For all values of $s, t \geq 0$, establish the Chapman-Kolmogorov equation of the homogeneous CTMC.

$$p_{ij}(t+s) = \sum_{k \in \mathbb{N}} p_{ik}(t) p_{kj}(s), \quad \forall i, j \in \mathbb{N}$$

In matrix notation, the above result is

$$\mathcal{P}(t+s) = \mathcal{P}(t) \mathcal{P}(s)$$

Hint: See Medhi (1994).

$$\begin{aligned}
& p_{ij}(t+s) \\
&= P(X(t+s) = j \mid X(0) = i) \\
&= \sum_{k \in \mathbb{N}} P(X(t+s) = j, X(t) = k \mid X(0) = i) \\
&= \sum_{k \in \mathbb{N}} P(X(t+s) = j \mid X(t) = k, X(0) = i) P(X(t) = k \mid X(0) = i) \\
&= \sum_{k \in \mathbb{N}} p_{kj}(s) p_{ik}(t)
\end{aligned}$$

25. For a CTMC that is temporally homogeneous, prove that $\xi(t) = \xi(0)\mathcal{P}(t)$.
    Hint: See Medhi (2003). For each $j \in \mathbb{N}$

$$
\begin{aligned}
\xi_j(t) &= P(X(t) = j) \\
&= \sum_{i \in \mathbb{N}} P(X(t+u) = j \mid X(u) = i) P(X(u) = i) \\
&= \sum_{i \in \mathbb{N}} p_{ij}(t) P(X(0) = i) \\
&= \sum_{i \in \mathbb{N}} p_{ij}(t) \xi_i(0)
\end{aligned}
$$

The result follows.

26. Prove that the sum of each row of the transition density matrix $\mathcal{Q}$ of a homogeneous CTMC is equal to zero.
    Hint: See Medhi (2003). Note that for small values of $h$

$$
\begin{aligned}
p_{ij}(h) &= h q_{ij} + o(h), \quad i \neq j \\
p_{ii}(h) &= h q_{ii} + 1 + o(h)
\end{aligned}
$$

where $o(h)$ is a function of $h$ that tends to zero faster than $h$. That is

$$
\lim_{h \to 0} \frac{o(h)}{h} \to 0
$$

For any $i \in \mathbb{N}$, $\sum_{j \in \mathbb{N}} p_{ij}(h) = 1$. That is

$$
\sum_{\substack{j \in \mathbb{N} \\ j \neq i}} p_{ij}(h) + p_{ii}(h) = 1
$$

$$
\sum_{\substack{j \in \mathbb{N} \\ j \neq i}} q_{ij} + q_{ii} = 0
$$

Finally

$$
\sum_{\substack{j \in \mathbb{N} \\ j \neq i}} q_{ij} = q_i
$$

27. Prove the backward and forward Chapman-Kolmogorov equations of a CTMC. These are

$$
\mathcal{P}'(t) = \mathcal{Q}\mathcal{P}(t), \text{ and } \mathcal{P}'(t) = \mathcal{P}(t)\mathcal{Q}
$$

respectively.
    Hint: These equations can be established by using the definitions of $\mathcal{P}'(t)$, $\mathcal{Q}$, and the matrix form of the Chapman-Kolmogorov equation of the CTMC.

28. Prove that $\xi'(t) = \xi(t)\mathcal{Q}$.
    Hint: Use the forward Chapman-Kolmogorov equation and the relationship $\xi(t) = \xi(0)\mathcal{P}(t)$.

29. A homogeneous CTMC has $(m+1)$ number of states, such that its rate matrix $\mathcal{Q}$ has distinct eigenvalues $\lambda_i, 0 \leq i \leq m$. Let $\mathcal{Q} = RDR^{-1}$, where $D$ is a diagonal matrix with the eigenvalues on its diagonal, and the matrix $R$ is invertible. Also these matrices are $(m+1) \times (m+1)$

each. Assume that $\mathcal{P}(0) = I$, where $I$ is an identity matrix of size $(m+1)$. If $\Psi(t)$ is a diagonal matrix, with diagonal elements $e^{\lambda_i t}, 0 \leq i \leq m$, prove that

$$\mathcal{P}(t) = e^{\mathcal{Q}t} = I + \sum_{n \in \mathbb{P}} \frac{\mathcal{Q}^n t^n}{n!} = R\Psi(t)R^{-1}$$

Thus $\xi(t)$ can be computed from the relationship $\xi(t) = \xi(0)\mathcal{P}(t)$. This equation yields a transient solution of the process.

30. Derive the backward diffusion equation of the CTMP.

31. Verify that $p(x_0, x; t) = \exp\left\{ -(x - x_0 - mt)^2 / (2\sigma^2 t) \right\} / \left\{ \sigma\sqrt{2\pi t} \right\}$ satisfies both the forward and backward diffusion equations of the CTMP.

32. Consider a continuous-time MP $\left\{ X(t), t \in \mathbb{R}_0^+ \right\}$. Prove that the distribution of $X(t)$, where $X(0) = x_0$ as $t \to \infty$ is exponential with parameter $-2m/\sigma^2$, where $m < 0$.

    Hint: As $t \to \infty$, the first derivative of $p(x_0, x; t)$ with respect to $t$ vanishes. Therefore assume $p(x_0, x; t) = p(x)$ as $t \to \infty$, and use the forward diffusion equation.

---

# References

1. Bharucha-Reid, A. T., 1997. *Elements of the Theory of Markov Processes and their Applications*, Dover Publications, Inc., New York, New York.

2. Campbell, N. R., 1909. "The Study of Discontinuous Phenomenon," Proc. of the Camb. Phil. Soc., Vol. 15, pp. 117-136.

3. Campbell, N. R., 1910. "Discontinuities in Light Emission," Proc. of the Camb. Phil. Soc., Vol. 15, pp. 310-328.

4. Chung, K. L., 1974. *A Course in Probability Theory*, Second Edition, Academic Press, New York, New York.

5. Cox, D. R., 1962. *Renewal Theory*, Methuen & Co. Ltd., London, Great Britain.

6. Doob, J. L., 1953. *Stochastic Processes*, John Wiley & Sons, Inc., New York, New York.

7. Durrett, R., 2005. *Probability Theory and Examples*, Third Edition, Thomas Learning - Brooks/Cole, Belmont, California.

8. Feller, W., 1968. *An Introduction to Probability Theory and Its Applications, Vol. I*, Third Edition, John Wiley & Sons, Inc., New York, New York.

9. Feller, W., 1971. *An Introduction to Probability Theory and Its Applications, Vol. II*, Second Edition, John Wiley & Sons, Inc., New York, New York.

10. Gardiner, C. W., 2004. *Handbook of Stochastic Methods for Physics, Chemistry, and the Natural Sciences*, Third Edition, Springer-Verlag, Berlin, Germany.

11. Grimmett, G. R., and Stirzaker, D. R., 2001. *Probability and Random Processes*, Third Edition, Oxford University Press, Oxford, Great Britain.

12. Gubner, J. A., 2006. *Probability and Random Processes for Electrical and Computer Engineers*, Cambridge University Press, Cambridge, Great Britain.

13. Haenggi, M., 2013. *Stochastic Geometry for Wireless Networks*, Cambridge University Press, New York, New York.

14.  Hoel, P. G., Port, S. C., and Stone, C. J., 1972. *Introduction to Stochastic Processes*, Houghton Mifflin and Company, Palo Alto, California.

15.  Illian, J., Penttinen, A., Stoyan, H., and Stoyan, D., 2008. *Statistical Analysis and Modelling of Spatial Point Patterns*, John Wiley & Sons, Inc., New York, New York.

16.  Itô, K., 1986. *Introduction to Probability Theory*, Cambridge University Press, Cambridge, Great Britain.

17.  Kendall, M. G., and Stuart, A., 1963. *The Advanced Theory of Statistics, Volume 1,* Hafner Publishing Company, New York, New York.

18.  Kingman, J. F. C., 1993. *Poisson Processes*, Oxford University Press, Oxford, Great Britain.

19.  Leon-Garcia, A., 1994. *Probability and Random Processes for Electrical Engineering,* Second Edition, Addison-Wesley Publishing Company, New York, New York.

20.  Loève, M., 1977. *Probability Theory I,* Fourth Edition, Springer-Verlag, Berlin, Germany.

21.  Medhi, J., 1994. *Stochastic Processes,* Second Edition, John Wiley & Sons, Inc., New York, New York.

22.  Medhi, J., 2003. *Stochastic Models in Queueing Theory,* Second Edition, John Wiley & Sons, Inc., New York, New York.

23.  Mikosch, T., 1998. *Elementary Stochastic Calculus,* World Scientific, River Edge, New Jersey.

24.  Mikosch, T., 2009. *Non-Life Insurance Mathematics, An Introduction with the Poisson Process*, Springer-Verlag, Berlin, Germany.

25.  Newman, M. E. J., 2005. "Power Laws, Pareto Distributions and Zipf's Law," Contemporary Physics, Vol. 46, Issue 5, pp. 323-351.

26.  Oden, J. T., and Demkowicz, L. F., 1996. *Applied Functional Analysis*, Chapman and Hall/CRC Press, New York, New York.

27.  Papoulis, A., 1965. *Probability, Random Variables, and Stochastic Processes,* McGraw-Hill Book Company, New York, New York.

28.  Parzen, E., 1960. *Modern Probability Theory and Its Applications,* John Wiley & Sons, Inc., New York, New York.

29.  Parzen, E., 1962. *Stochastic Processes,* Holden-Day, San Francisco, California.

30.  Pinsky, M. A., and Karlin, S., 2011. *An Introduction to Stochastic Modeling*, Fourth Edition, Academic Press, Burlington, Massachusetts.

31.  Resnick, S. L., 1992. *Adventures in Stochastic Processes*, Birkhauser, Boston, Massachusetts.

32.  Ross, S. M., 1970. *Applied Probability Models with Optimization Applications*, Holden-Day, Inc., San Francisco, California.

33.  Ross, S. M., 1983. *Stochastic Processes*, John Wiley & Sons, Inc., New York, New York.

34.  Ross, S. M., 2009. *A First Course in Probability*, Eighth Edition, Macmillan Publishing Company, New York, New York.

35.  Ross, S. M., 2010. *Introduction to Probability Models,* Tenth Edition, Academic Press, New York, New York.

36.  Serfozo, R., 2009. *Basics of Applied Probability*, Springer-Verlag, Berlin, Germany.

37.  Snyder, D. L., and Miller, M. I., 1991. *Random Point Processes in Time and Space*, Springer-Verlag, Berlin, Germany.

38.  Stoyan, D., Kendall, W. S., and Mecke, J., 1995. *Stochastic Geometry and its Applications*, Second Edition, John Wiley & Sons, Inc., New York, New York.

39.  Sveshnikov, A. A., 1968. *Problems in Probability Theory, Mathematical Statistics, and Theory of Random Functions*, Dover Publications, Inc., New York, New York.

40.  Takács, L., 1960. *Stochastic Processes*, Methuen & Co. Ltd., London, Great Britain.

# Index